T0142192

Advances in Intelligent Systems and Computing

Volume 997

The series "Advances in Intelligent Systems and Computing" contains publications on theory, applications, and design methods of Intelligent Systems and Intelligent Computing. Virtually all disciplines such as engineering, natural sciences, computer and information science, ICT, economics, business, e-commerce, environment, healthcare, life science are covered. The list of topics spans all the areas of modern intelligent systems and computing such as: computational intelligence, soft computing including neural networks, fuzzy systems, evolutionary computing and the fusion of these paradigms, social intelligence, ambient intelligence, computational neuroscience, artificial life, virtual worlds and society, cognitive science and systems, Perception and Vision, DNA and immune based systems, self-organizing and adaptive systems, e-Learning and teaching, human-centered and human-centric computing, recommender systems, intelligent control, robotics and mechatronics including human-machine teaming, knowledge-based paradigms, learning paradigms, machine ethics, intelligent data analysis, knowledge management, intelligent agents, intelligent decision making and support, intelligent network security, trust management, interactive entertainment, Web intelligence and multimedia.

The publications within "Advances in Intelligent Systems and Computing" are primarily proceedings of important conferences, symposia and congresses. They cover significant recent developments in the field, both of a foundational and applicable character. An important characteristic feature of the series is the short publication time and world-wide distribution. This permits a rapid and broad dissemination of research results.

**** Indexing: The books of this series are submitted to ISI Proceedings, EI-Compendex, DBLP, SCOPUS, Google Scholar and Springerlink ****

More information about this series at http://www.springer.com/series/11156

Kohei Arai · Rahul Bhatia ·
Supriya Kapoor
Editors

Intelligent Computing

Proceedings of the 2019 Computing Conference, Volume 1

Editors
Kohei Arai
Faculty of Science and Engineering
Saga University
Saga, Japan

Rahul Bhatia
The Science and Information
SAI Organization
Bradford, West Yorkshire, UK

Supriya Kapoor
The Science and Information
SAI Organization
Bradford, West Yorkshire, UK

ISSN 2194-5357 ISSN 2194-5365 (electronic)
Advances in Intelligent Systems and Computing
ISBN 978-3-030-22870-5 ISBN 978-3-030-22871-2 (eBook)
https://doi.org/10.1007/978-3-030-22871-2

This Springer imprint is published by the registered company Springer Nature Switzerland AG
The registered company address is: Gewerbestrasse 11, 6330 Cham, Switzerland

Editor's Preface

On behalf of the Organizing Committee and Program Committee of the Computing Conference 2019, we would like to welcome you to the Computing Conference 2019 which will be held from July 16 to 17, 2019, in London, UK.

Despite the short history of computer science as a formal academic discipline, it has made a number of fundamental contributions to science and society—in fact, along with electronics, it is a founding science of the current epoch of human history called the Information Age and a driver of the Information Revolution. The goal of this conference is to give a platform to researchers with such fundamental contributions and to be a premier venue for industry practitioners to share new ideas and development experiences. It is one of the best respected conferences in the area of computer science.

Computing Conference 2019 began with an opening ceremony, and the conference program featured welcome speeches. It was a two-day conference, and each day started with keynote speeches from experts in the field. During the span of two days, a total of 18 paper presentation sessions and 4 poster presentation sessions were organized giving the opportunity to the authors to present their papers to an international audience. The conference attracted a total of 563 submissions from many academic pioneering researchers, scientists, industrial engineers, and students from all around the world. These submissions underwent a double-blind peer-review process. Of those 563 submissions, 170 submissions have been selected to be included in this proceedings. The published proceedings has been divided into two volumes covering a wide range of conference tracks, such as technology trends, computing, intelligent systems, machine vision, security, communication, electronics, and e-learning to name a few.

Deep appreciation goes to the keynote speakers for sharing their knowledge and expertise with us and to all the authors who have spent the time and effort to contribute significantly to this conference. We are also indebted to the Organizing Committee for their great efforts in ensuring the successful implementation of the conference. In particular, we would like to thank the Technical Committee for their constructive and enlightening reviews on the manuscripts in the limited timescale.

We hope that all the participants and the interested readers benefit scientifically from this book and find it stimulating in the process. We are pleased to present the proceedings of this conference as its published record.

Hope to see you in 2020, in our next Computing Conference, with the same amplitude, focus, and determination.

Kohei Arai

Contents

An Intelligent Highway Traffic Management System for Smart City . . . 1
Prasanta Mandal, Punyasha Chatterjee, and Arpita Debnath

Haptically-Enabled VR-Based Immersive Fire Fighting Training Simulator . . . 11
S. Nahavandi, L. Wei, J. Mullins, M. Fielding, S. Deshpande, M. Watson, S. Korany, D. Nahavandi, I. Hettiarachchi, Z. Najdovski, R. Jones, A. Mullins, and A. Carter

Fast Replanning Incremental Shortest Path Algorithm for Dynamic Transportation Networks . . . 22
Joanna Hartley and Wedad Alhoula

A Computational Scheme for Assessing Driving . . . 44
José A. Romero Navarrete and Frank Otremba

Secure Information Interaction Within a Group of Unmanned Aerial Vehicles Based on Economic Approach . . . 59
Iuliia Kim and Ilya Viksnin

Logic Gate Integrated Circuit Identification Through Augmented Reality and a Smartphone . . . 73
Carlos Aviles-Cruz, Juan Villegas-Cortez, Arturo Zuniga-Lopez, Ismael Osuna-Galan, Yolanda Perez-Pimentel, and Salomon Cordero-Sanchez

A Heuristic Intrusion Detection System for Internet-of-Things (IoT) . . . 86
Ayyaz-ul-Haq Qureshi, Hadi Larijani, Jawad Ahmad, and Nhamoinesu Mtetwa

A Framework to Evaluate Barrier Factors in IoT Projects . . . 99
Marcel Simonette, Rodrigo Filev Maia, and Edison Spina

Learning with the Augmented Reality EduPARK Game-Like App: Its Usability and Educational Value for Primary Education 113
Lúcia Pombo and Margarida M. Marques

An Expert Recommendation Model for Academic Talent Evaluation . . . 126
Feng Sun, Li Liu, and Jian Jin

A Comparative Study for QSDC Protocols with a Customized Best Solution Approach 140
Ola Hegazy

Fuzzy Sets and Game Theory in Green Supply Chain: An Optimization Model 149
Marwan Alakhras, Mousa Hussein, and Mourad Oussalah

A Technique to Reduce the Processing Time of Defect Detection in Glass Tubes 165
Gabriele Antonio De Vitis, Pierfrancesco Foglia, and Cosimo Antonio Prete

Direct *N-body* Code on Low-Power Embedded ARM GPUs 179
David Goz, Sara Bertocco, Luca Tornatore, and Giuliano Taffoni

OS Scheduling Algorithms for Improving the Performance of Multithreaded Workloads 194
Murthy Durbhakula

The Fouriest: High-Performance Micromagnetic Simulation of Spintronic Materials and Devices 209
I. Pershin, A. Knizhnik, V. Levchenko, A. Ivanov, and B. Potapkin

Consistency in Multi-device Environments: A Case Study 232
Luis Martín Sánchez-Adame, Sonia Mendoza, Amilcar Meneses Viveros, and José Rodríguez

OS Scheduling Algorithms for Memory Intensive Workloads in Multi-socket Multi-core Servers 243
Murthy Durbhakula

Towards Energy Efficient Servers' Utilization in Datacenters 254
Ahmed Osman, Assim Sagahyroon, Raafat Aburukba, and Fadi Aloul

An Evaluation of ICT Smart Systems to Reduce the Carbon Footprint . . . 263
Andreas Andressen, Lesley Earle, Ah-Lian Kor, and Colin Pattinson

Multisensory Real-Time Space Telerobotics 275
Marta Ferraz, Edmundo Ferreira, Emiel den Exter, Frank van der Hulst, Hannes Rovina, William Carey, Jessica Grenouilleau, and Thomas Krueger

Achieving a High Level of Open Market-Information Symmetry with Decentralised Insurance Marketplaces on Blockchains 299
Alex Norta, Risto Rossar, Mart Parve, and Liina Laas-Billson

Analysis of Argumentation Skills for Argumentation Training Support 319
Hayato Hirata, Shogo Okada, and Katsumi Nitta

Machine Autonomy: Definition, Approaches, Challenges and Research Gaps 335
Chinedu Pascal Ezenkwu and Andrew Starkey

Novel Recursive Technique for Finding the Optimal Solution of the Nurse Scheduling Problem 359
Samah Senbel

Override Control Based on NARX Model for Ecuador's Oil Pipeline System 376
Williams R. Villalba, Jose E. Naranjo, Carlos A. Garcia, and Marcelo V. Garcia

***MatBase* E-RD Cycles Associated Non-relational Constraints Discovery Assistance Algorithm** 390
Christian Mancas

Heuristic Search for Tetris: A Case Study 410
Giacomo Da Col and Erich C. Teppan

A Comparative Study of the Kinematic Response and Injury Metrics Associated with Adults and Children Impacted by an Auto Rickshaw . . . 424
A. J. Al-Graitti, G. A. Khalid, P. R. Berthelson, R. K. Prabhu, and M. D. Jones

A Discrete Cosine Transform Based Evolutionary Algorithm and Its Application for Symbolic Regression 444
Quanchao Liu and Yue Hu

Similarity Measurement of Handwriting by Alignment of Sequences . . . 463
Katalin Erdélyi and Bálint Molnár

Multi-channel Speaker Separation Using Speaker-Aware Beamformer . . . 474
Conggui Liu and Yinhua Liu

Revisiting Skip-Gram Negative Sampling Model with Rectification 485
Cun (Matthew) Mu, Guang Yang, and Yan (John) Zheng

LANA-I: An Arabic Conversational Intelligent Tutoring System for Children with ASD 498
Sumayh Aljameel, James O'Shea, Keeley Crockett, Annabel Latham, and Mohammad Kaleem

A Formal Model for Robot to Understand Common Concepts 517
Yuanxiu Liao, Jingli Wu, and Xudong Luo

Structure for Knowledge Acquisition, Use, Learning and Collaboration Inter Agents over Internet Infrastructure Domains 527
Juliao Braga, Joao Nuno Silva, Patricia Endo, and Nizam Omar

The Space Between Worlds: Liminality, Multidimensional Virtual Reality and Deep Immersion 548
Ralph Moseley

Predicting Endogenous Bank Health from FDIC Statistics on Depository Institutions Using Deep Learning 563
David Jungreis, Noah Capp, Meysam Golmohammadi, and Joseph Picone

A Novel Ensemble Approach for Feature Selection to Improve and Simplify the Sentimental Analysis 573
Muhammad Latif and Usman Qamar

High Resolution Sentiment Analysis by Ensemble Classification 593
Jordan J. Bird, Anikó Ekárt, Christopher D. Buckingham, and Diego R. Faria

Modelling Stable Alluvial River Profiles Using Back Propagation-Based Multilayer Neural Networks 607
Hossein Bonakdari, Azadeh Gholami, and Bahram Gharabaghi

Towards Adaptive Learning Systems Based on Fuzzy-Logic 625
Soukaina Ennouamani and Zouhir Mahani

Discrimination of Human Skin Burns Using Machine Learning 641
Aliyu Abubakar and Hassan Ugail

Sometimes You Want to Go Where Everybody Knows Your Name ... 648
Reuben Brasher, Justin Wagle, and Nat Roth

Fuzzy Region Connection Calculus and Its Application in Fuzzy Spatial Skyline Queries 659
Somayeh Davari and Nasser Ghadiri

Concerning Neural Networks Introduction in Possessory Risk Management Systems 678
Mikhail Vladimirovich Khachaturyan and Evgeniia Valeryevna Klicheva

Size and Alignment Independent Classification of the High-Order Spatial Modes of a Light Beam Using a Convolutional Neural Network 688
Aashima Singh, Giovanni Milione, Eric Cosatto, and Philip Ji

Automatic Induction of Neural Network Decision Tree Algorithms 697
Chapman Siu

Reinforcement Learning in *A Marketing Game* 705
Matthew G. Reyes

Pedestrian-Motorcycle Binary Classification Using Data Augmentation and Convolutional Neural Networks 725
Robert Kerwin C. Billones, Argel A. Bandala, Laurence A. Gan Lim, Edwin Sybingco, Alexis M. Fillone, and Elmer P. Dadios

Performance Analysis of Missing Values Imputation Methods Using Machine Learning Techniques 738
Omesaad Rado, Muna Al Fanah, and Ebtesam Taktek

Evolutionary Optimisation of Fully Connected Artificial Neural Network Topology 751
Jordan J. Bird, Anikó Ekárt, Christopher D. Buckingham, and Diego R. Faria

State-of-the-Art Convolutional Neural Networks for Smart Farms: A Review 763
Patrick Kinyua Gikunda and Nicolas Jouandeau

Neural Networks to Approximate Solutions of Ordinary Differential Equations 776
Georg Engel

Optimizing Deep Learning Model for Neural Network Topology 785
Sara K. Al-Ruzaiqi and Christian W. Dawson

Detecting Traces of Bullying in Twitter Posts Using Machine Learning 796
Caroline Jin, Harpreet Kaur, Amena Khatun, and Sitara Uppalapati

Credit Risk Analysis Applying Machine Learning Classification Models 804
Roy Melendez

Aligning Ground Truth Text with OCR Degraded Text 815
Jorge Ramón Fonseca Cacho and Kazem Taghva

Incremental Alignment of Metaphoric Language Model for Poetry Composition 834
Marilena Oita

A Trie Based Model for SMS Text Normalization 846
Niladri Chatterjee

Word Topic Prediction Model for Polysemous Words and Unknown Words Using a Topic Model 860
Keisuke Tanaka and Ayahiko Niimi

Improving Usability of Distributed Neural Network Training 867
Nathaniel Grabaskas

Towards a Better Model for Predicting Cancer Recurrence in Breast Cancer Patients 887
Nour A. AbouElNadar and Amani A. Saad

Inducing Clinical Course Variations in Multiple Sclerosis White Matter Networks 900
Giovanni Melissari, Aldo Marzullo, Claudio Stamile, Francesco Calimeri, Françoise Durand-Dubief, and Dominique Sappey-Marinier

The Efficacy of Various Machine Learning Models for Multi-class Classification of RNA-Seq Expression Data 918
Sterling Ramroach, Melford John, and Ajay Joshi

Performance Analysis of Feature Selection Methods for Classification of Healthcare Datasets 929
Omesaad Rado, Najat Ali, Habiba Muhammad Sani, Ahmad Idris, and Daniel Neagu

Towards Explainable AI: Design and Development for Explanation of Machine Learning Predictions for a Patient Readmittance Medical Application 939
Sofia Meacham, Georgia Isaac, Detlef Nauck, and Botond Virginas

A String Similarity Evaluation for Healthcare Ontologies Alignment to HL7 FHIR Resources 956
Athanasios Kiourtis, Argyro Mavrogiorgou, Sokratis Nifakos, and Dimosthenis Kyriazis

Detection of Distal Radius Fractures Trained by a Small Set of X-Ray Images and Faster R-CNN 971
Erez Yahalomi, Michael Chernofsky, and Michael Werman

Cloud-Based Skin Lesion Diagnosis System Using Convolutional Neural Networks 982
E. Akar, O. Marques, W. A. Andrews, and B. Furht

Color Correction for Stereoscopic Images Based on Gradient Preservation 1001
Pengyu Liu, Yuzhen Niu, Junhao Chen, and Yiqing Shi

Exact NMF on Single Images via Reordering of Pixel Entries Using Patches 1012
Richard M. Charles and James H. Curry

Significant Target Detection of Traffic Signs Based on Walsh-Hadamard Transform 1027
XiQuan Yang and Ying Sun

Fast Implementation of Face Detection Using LPB Classifier on GPGPUs 1036
Mohammad Rafi Ikbal, Mahmoud Fayez, Mohammed M. Fouad, and Iyad Katib

Optimized Grayscale Intervals Study of Leaf Image Segmentation 1048
Jianlun Wang, Shuangshuang Zhao, Rina Su, Hongxu Zheng, Can He, Chenglin Zhang, Wensheng Liu, Liangyu Jiang, and Yiyi Bu

The Software System for Solving the Problem of Recognition and Classification 1063
Askar Boranbayev, Seilkhan Boranbayev, Askar Nurbekov, and Roman Taberkhan

Vision Monitoring of Half Journal Bearings 1075
Iman Abulwaheed, Sangarappillai Sivaloganathan, and Khalifa Harib

Manual Tool and Semi-automated Graph Theory Method for Layer Segmentation in Optical Coherence Tomography 1090
Dean Sayers, Maged Salim Habib, and Bashir AL-Diri

Author Index 1111

An Intelligent Highway Traffic Management System for Smart City

Prasanta Mandal[1], Punyasha Chatterjee[1(✉)], and Arpita Debnath[2]

[1] School of Mobile Computing and Communication, Jadavpur University, Kolkata, India
prasanta.mtechdmc@gmail.com, punyasha.chatterjee@gmail.com

[2] The Calcutta Technical School, Kolkata, India
arpita.debnath03@gmail.com

Abstract. With the hasty expansion of urbanization and the overcrowding of cities, real-time traffic management system is considered to be very crucial for any smart city as it provides comfort, safety, security, and efficient time management to the people. However, while there is an increase of freight volume and the number of vehicles, length of the roads was not increasing adequately. It creates the problem of traffic congestion, prolonging traffic queues and unfortunate incidents like accidents etc leading to huge gridlocks. This issue can be addressed by incorporating smart and intelligent technologies in such a way that the overall traffic management system becomes capable of handling the traffic dynamically. In this paper, we have proposed a distributed approach to make a real time traffic analysis on the highways when an accident occurs. Based on it a smart decision can be taken by a vehicle to reduce the congestion on the lane. We have simulated our approach using the well-known network simulator ns-3.

Keywords: Smart city · Traffic congestion · Highway traffic · Vehicle-to-Vehicle communication · Road Side Units

1 Introduction

Smart City is a collection of intelligent control systems that makes it well organized by making a decision in more smart way to improve quality of living. In the past few years, tremendous growth in the Information and Communication Technology (ICT) field, is attracting the researchers to implement each and every component of smart city by making it more digitized. One of the key aspects for future Smart City is Intelligent Traffic Management System. With the hasty expansion of urbanization and the overcrowding of cities, real-time traffic management system is considered to be very crucial for any smart city as it provides travelers' safety and comfort. In many countries, the road traffic is considered as the most common mode of transportation due to its versatility and economical features. However, while there is an increase of freight volume

K. Arai et al. (Eds.): CompCom 2019, AISC 997, pp. 1–10, 2019.
https://doi.org/10.1007/978-3-030-22871-2_1

and the number of vehicles, length of the roads was not increasing adequately. It causes the problem of traffic congestion, prolonging traffic queues and unfortunate incidents like, accidents, etc. leading to huge gridlocks. The challenge is how to avoid these gridlocks without altering the existing infrastructure, rather than utilizing it upto its maximum capacity.

This issue can be addressed by incorporating smarter technologies in such a way that the overall traffic management system becomes capable of handling the traffic on real-time basis. The smart traffic management system can reduce congestion by considering the present situation of the traffic density and by providing the real-time information to the drivers about the most feasible route at any instant of time. By reducing the response time for the emergency services required due to accidents etc the automotive sector may also offer better assistance to the victims. With these requirements, there emerges the need of smart transportation system for providing comfort, safety, security, and efficient time management to the people.

In this paper, we have proposed a distributed approach to make a real time traffic analysis, based on which a smart decision can be taken by a vehicle when an accident takes place on the highways. The rest of this paper is organized as follows: Sect. 2 illustrates the literature review, Sect. 3 provides system architecture, Sect. 4 detailed the working principle, Sect. 5 analyses the results, and Sect. 6 contains conclusion and future works.

2 Related Work

In this section we discuss some existing works on smart traffic management systems.

In [1], Javaid et al. proposed a system in which they suggested a hybrid approach (combination of centralized and decentralized approach) to optimize traffic flow on roads. The system takes traffic density as input from camera/sensors, and manages traffic signals. For this purpose they used a centralized server to store the captured images and data. Hence, there is a chance of single point failure.

In [2], Swathi et al. proposed a smart traffic routing system where each vehicle chooses the best route with least congestion. Infrared sensors are used to collect data about traffic density by monitoring the reflected light from the vehicles. However, readings may change with the change in temperature and humidity. As a result proposed system may give an erroneous traffic density value depending upon the different weather situation.

In [3] Junping et al. introduced an intelligent transportation system which is able to accumulate a huge volume of data coming from numerous resources using video sensors. The main drawback of this system is the requirement of huge amount of memory for storing the videos.

In [4], Wiering et al. offered a learning algorithm based adaptive control system to control traffic light. A cumulative voting system is introduced here to determine the expected gain for setting the traffic light green signal by all the vehicles, waiting at intersection. By this technique, the drivers can take the lowest estimated waiting path. Though this method aids in dropping the waiting time at the intersections, it may be ineffective at the isolated intersections. Above all, if all the drivers opt for the same route, the optimal route may be overcrowded and thus, the system may become inefficient.

In [5], Calvert et al. improved the traffic management system by considering the uncertainty and stochastic behaviour of traffic flow. In [6] Kim et al. introduced a mobile edge cloud based traffic management.

In [7], Latif et al. modeled the intelligent traffic monitoring and guidance system using graph theory and formal methods. In heavy traffic situation, this model will be more effective as manual system does not work well in emergency situations. Further, the vehicle drivers are informed to choose alternative path to their destination in case of special timings. But if all the drivers choose the same path, the route may be congested.

In [8], Liu et al. proposed a service based Intelligent Transportation System Framework to manage heavy traffic, providing efficient road situation, systematic accident management and providing reliable information. In [9], Manikonda et al. proposed a RFID based intelligent traffic management system that creates a map of shortest time paths for the whole city.

3 System Architecture

In our proposed system we have considered highway traffic with two lanes as depicted in Fig. 1. In each lane, vehicles are moving in one-way and in opposite direction. Each lane is identified by unique identification number called *Lane Identifier* (LId) and it is predefined by the transportation management authority of the smart city. After some distance apart bypass roads are connected to the main highway. Road Side Units (RSUs) are installed in between the two lanes with some distance apart. Every RSU is equipped with two unidirectional antennae, such that each antenna transmits traffic information (like Lid, bypass road position, etc.) to the vehicles of a particular lane. When a vehicle comes under the transmission range of a RSU, it gets the traffic information of the lane on which it is running. In this way RSUs can provide the service on both lanes simultaneously with different antennae.

Each vehicle must be equipped with a high definition digital camera and it is responsible for capturing events in front of that. On Board Unit (OBU) is attached with each vehicle, which is responsible for real time decision making based on present information. When a front side camera of a vehicle captures a shot, OBU immediately performs some digital image processing techniques and finds out if there is any accident related event happened on its lane. GPS receiver is attached with every vehicle for finding its current location at any point of time. Each vehicle must be equipped with LED display, which shows the notification messages to the driver. Each vehicle is equipped with wireless transceiver that can send and receive messages.

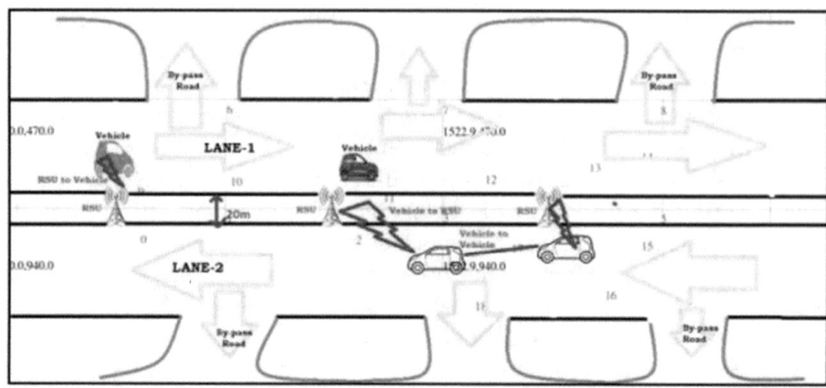

Fig. 1. Highway application scenario

4 Working Principle

When an accident takes place on any one of the lanes of the highway, the vehicle(s) which first detect(s) the event by camera sensor, immediately generate(s) a *warning message* and broadcast(s) it. These vehicles are called *source vehicles*. These messages may be directly received by vehicles and RSUs which are residing in one hop transmission zone of the *source vehicle(s)*.

The *warning messages* must be brief, restraining irrelevant information to reduce the message complexity as well as energy expenditure due to message communication. For our system, we have designed the warning message structure containing the following fields as depicted in Fig. 2. The description of the different fields of *warning message* are given below:

Source Vehicle Position (SPos)
Lane Id (LId)
Hop Limit (HLim)

Fig. 2. Warning message structure

Source Vehicle Position (SPos): It denotes the geographical position of the *source vehicle* taken from the GPS.

Lane ID (LId): It denotes the unique ID of the lane on which the accident has occurred.

Hop Limit (HLim): When a message is generated from a source vehicle, it sets this field by a fixed value to avoid the infinite broadcasting of a message. After arriving the packet in every vehicle, it simply reduces it by one and send the message; when this field becomes zero, broadcasting of that packet is off.

The actions taken by the vehicles and RSUs after receiving the *warning message* are given below:

A. Action Taken by Vehicles

If a warning message is received by a vehicle then it first checks the LId field to know whether the message contains information about its own lane or not. If LId is matched with its own lane's Id, it checks the SPos field, it compares its own position value with the message's SPos to identify whether the accident has happened in front of it or behind it. If its own position is ahead of *source vehicle*, it simply discards the message. If the source vehicle is ahead of the vehicle, it makes some decisions by following the steps as given below:

i. It calculates the distance from the SPos to itself. If this value is less than a predefined threshold, that means the vehicle is very closer to he accident spot it immediately displays a STOP signal on the LED display of the vehicle.

ii. If the calculated value is greater than the threshold value that means the vehicle is not very close to the accident spot, it searches the next bypass lane position which is periodically broadcasted by RSUs. If it does not exist, it again generates a STOP message on LED display, otherwise it takes that bypass road to avoid congestion. The vehicle will then broadcast the warning message within its 1-hop vehicles and RSUs. The warning message will be rebroadcasted by the vehicles which are within HLim distance from the source vehicle(s). The states are illustrated in the flow-chart in Fig. 3.

B. Action Taken by RSUs

If a warning message is received by a RSU, which is in front of the source vehicle, it discards the packets. Otherwise it simply directionally retransmits the warning message in the particular lane. At the same time, it starts broadcasting the nearby bypass lane information for the vehicles.

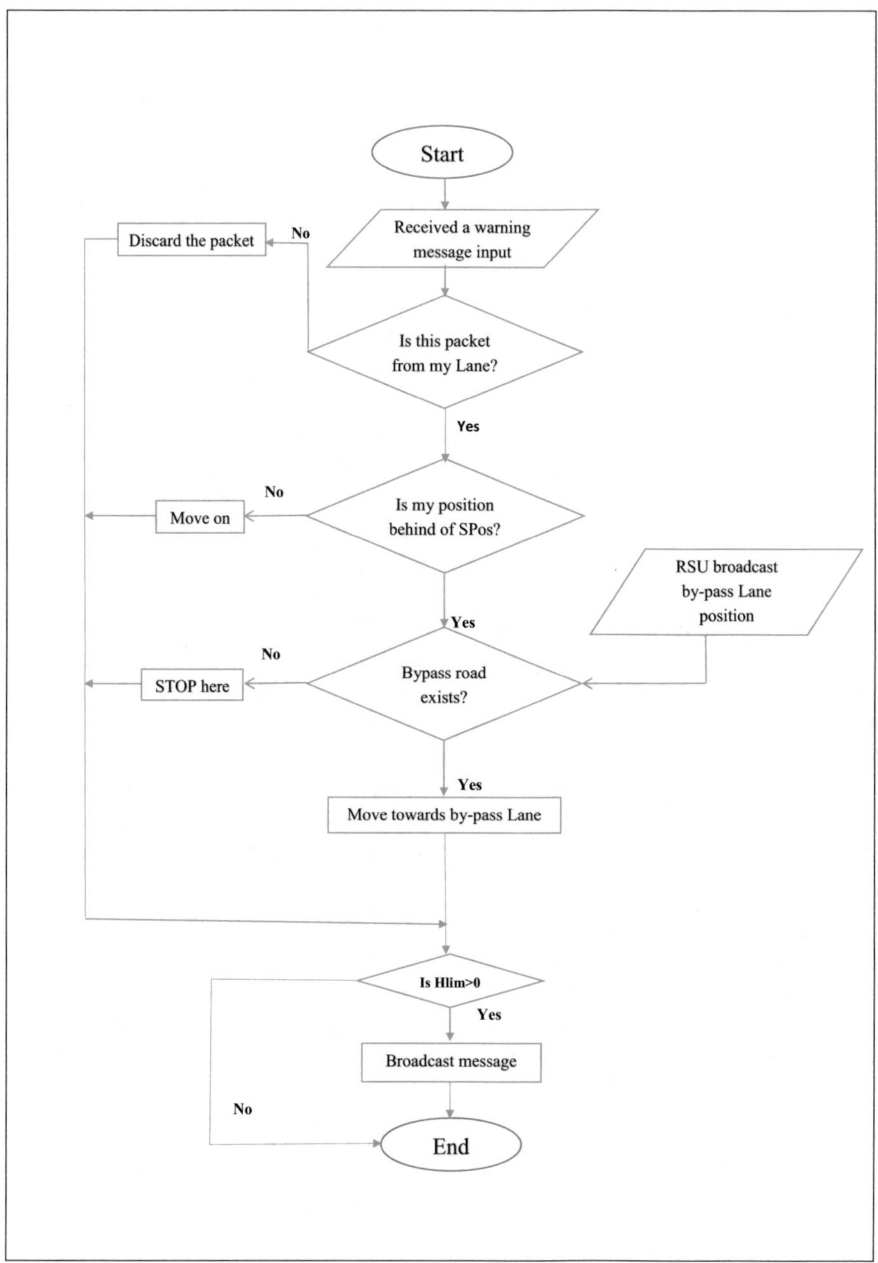

Fig. 3. Flow-chart for making decision by a vehicle

5 Results and Discussions

We have simulated our proposed algorithm for highway traffic management in discrete event based network simulator ns-3 in Linux. We are simulating with twelve node in which six are dynamic vehicles and another six are static RSU. The simulation parameters are shown in Table 1. The simulation snapshot is shown in Fig. 4. We have considered three performance metrics here:

Table 1. Simulation parameters

Parameter	Value
Packet size	1000 bytes
Vehicle transmission range	250 m
RSU transmission range	330 m
PhyMode	OFDM (DataRate-6 Mbps, BW-10 MHz)
WiFi standard	IEEE 802.11
Propagation delay model	Constant speed propagation delay
Propagation loss model	Range propagation loss model
Mobility model for vehicles	Constant velocity mobility model
Mobility model for RSU	Constant position mobility model
Wifi MAC type	AdhocWifiMac
Simulation time	60 s

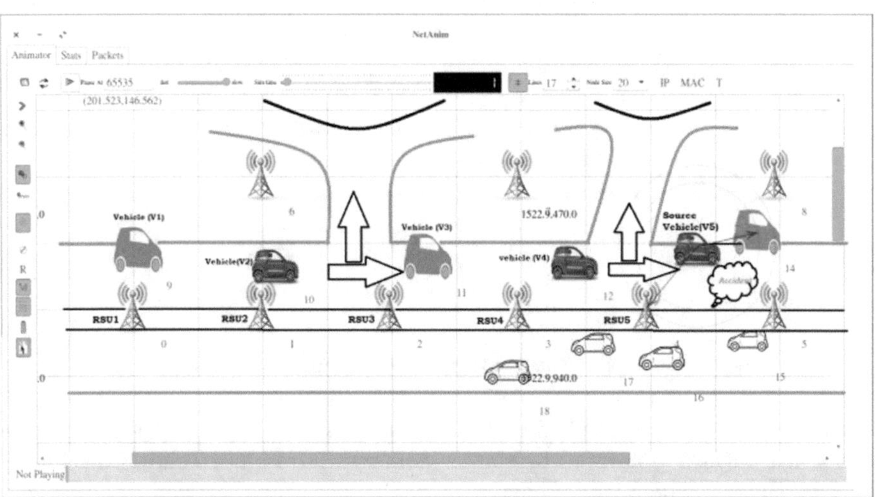

Fig. 4. Simulation snapshot of our proposed system

Packet Delivery Ratio (PDR): It is the ratio of total number of received packets by a vehicle to the total number of packets generated by source vehicle throughout the simulation time.

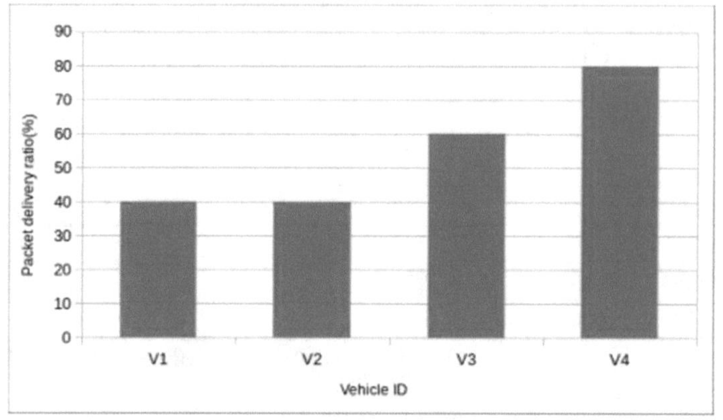

Fig. 5. Packet delivery ratio

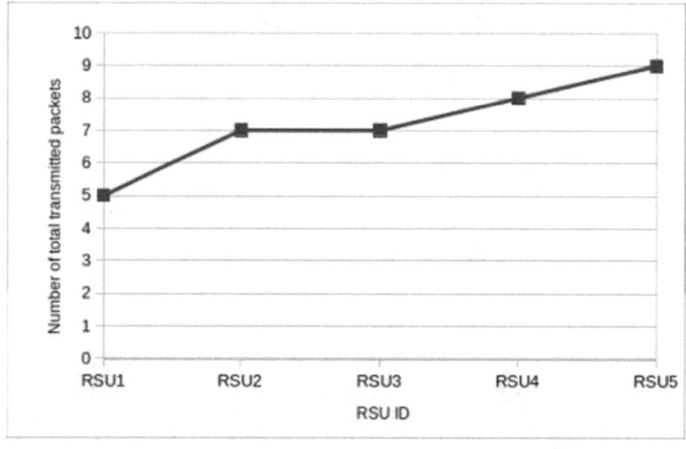

Fig. 6. Number of total transmitted warning packets by RSUs

PDR (%) = (total number of receive packets by a vehicle)/(total number of accident packets generated by source vehicle)*100. Figure 5 shows the Packet Delivery Ratio of the proposed system.

In Fig. 5, V4 is closest to the source vehicle and V1 is the farthest vehicle from the source vehicle in our simulation scenario. If the distance is less from the source node then the vehicle receives more number of packets from source node. With the increase in distance from source the total number of received packets decreases because of presence of some noise in the channel or sometimes inter-vehicular disconnections. That's why in the above graph, V4 have highest PDR% and V1 have lowest PDR%.

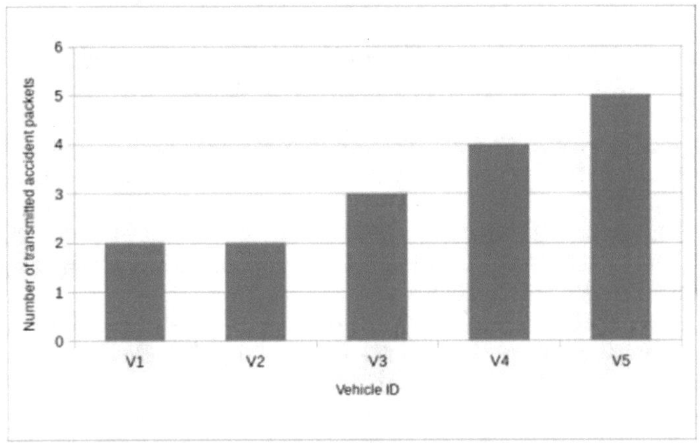

Fig. 7. Total number of transmitted accident packet by each vehicle

Number of Packets Transmitted: Figure 6, shows the total number of packets transmitted by RSUs. It justifies the fact that the source vehicle being the nearest to RSU5, receives the higher amount of transmitted packets from RSU5 with respect to the other RSUs in our simulation scenario.

Figure 7, shows the total number of packets transmitted by vehicles. Here it is evident that the total number of transmitted accident packets by vehicle V5 is higher than the others. The reason behind this type of behavior seems to be the distance. As the distance increases from the source vehicle to the other vehicles, the transmission rate decreases.

6 Conclusion and Future Work

The proposed work basically tried to reduce the congestion length in a lane, when accident occurs, by providing the relevant information to vehicles that are behind the accident region. Here we have considered the 2-lane highway traffic scenario.

In future we will extend it to congested urban road-traffic scenario. Also, we can consider Li-Fi (Light Fidelity) technology for V-2-V communication where communication between vehicles will be accomplished using light wave rather than radio wave because the radio wave is depleted too fast due to heavy use. Mainly here we are focusing on hop by hop communication between vehicles but in near future we will want to develop a novel position based routing protocol for our vehicular network that can create more automation in this implementation.

References

1. Javaid, S., Sufian, A., Pervaiz, S., Tanveer, M.: Smart traffic management system using Internet of Things. In: International Conference on Advanced Communication Technology, ICACT 2018, Chuncheon-si, Gangwon-do, Korea (South), pp. 393–398, February 2018
2. Swathi, K., Sivanagaraju, V., Manikanta, A.K.S., Kumar, D.: Traffic density control and accident indicator using WSN. Int. J. Mod. Trends Sci. Technol. **2**(4), 2455–3778 (2016)
3. Junping, Z., Feiue, W., Kunfeng, W., WeiHua, L., Xin, X., Cheng, C.: DataDriven intelligent transportation systems: survey. IEEE Trans. Intell. Transp. Syst. **12**(4), 1624–1639 (2011)
4. Wiering, M., Veenen, J., Vreeken, J., Koopman, A.: Intelligent traffic light control, pp. 1–30, Technical report, Department of Information and Computing Sciences, Universiteit Utrecht, 9 July 2004
5. Calvert, S.C., Taale, H., Snelder, M., Hoogendoorn, S.P.: Improving traffic management through consideration of uncertainty and stochastics in traffic flow. Case Stud. Transp. Policy **6**(1), 81–93 (2018)
6. Kim, S., Kim, D.-Y., Park, J.H.: Traffic management in the mobile edge cloud to improve the quality of experience of mobile video. Comput. Commun. **118**, 40–49 (2018)
7. Latif, S., Afzaal, H., Zafar, N.A.: Intelligent traffic monitoring and guidance system for smart city. In: 2018 International Conference on Computing, Mathematics and Engineering Technologies (iCoMET), Sukkur, pp. 1–6 (2018)
8. Liu, H.Y., Skjetne, E., Kobernus, M.: Mobile phone tracking: in support of modelling traffic-related air pollution contribution to individual exposure and its implications for public health impact assessment. Environ. Health **12**(1), 93 (2013)
9. Manikonda, P., Yerrapragada, A.K., Annasamudram, S.S.: Intelligent traffic management system. In: 2011 IEEE Conference on Sustainable Utilization and Development in Engineering and Technology, 20–21 October 2011, Malaysia, pp. 119–122 (2011)

Haptically-Enabled VR-Based Immersive Fire Fighting Training Simulator

S. Nahavandi(✉), L. Wei, J. Mullins, M. Fielding, S. Deshpande, M. Watson, S. Korany, D. Nahavandi, I. Hettiarachchi, Z. Najdovski, R. Jones, A. Mullins, and A. Carter

Institute for Intelligent Systems Research and Innovation, Deakin University, Geelong, VIC, Australia
saeid.nahavandi@deakin.edu.au

Abstract. Firefighting is a physically demanding task that requires extensive training. With the rising risks of global warming and its evident effects on spawning bush fire, there is an increasing need for recruiting new fire fighters. This imposes an unprecedented challenge of fast-tracking training procedures, especially in rural environments where most bush fires occur. Additionally, the current manual training procedures do not take into consideration the immersion factor, without which a novice fire fighter may be overwhelmed when facing a bush fire for the first time. This challenge has motivated us to harness the power of virtual reality and develop a portable firefighting training system. The developed firefighting training system, presented in this paper, is haptically enabled to allow the trainees to experience the jet reaction forces from the hose. The system also features realistic water dispersion and interaction with fire and smoke particles via accurate particle physics modelling.

Keywords: VR · Haptics · Firefighting

1 Introduction

The goal of firefighting is to save lives and property, however at the same time, it poses hazardous and extreme physical and mental challenges to fire fighters. It is of utmost importance to prepare them for not only the right skills and enough training to fight fire, but also for taking optimised actions, under extremely challenging conditions, to reduce injury and prevent property damage. Three drawbacks of traditional firefighting training have been identified as below [1, 2].

(1) Traditional firefighting simulations are mostly based on vision and sound, which provide only limited immersion and inadequate effectiveness. Although the latest research has attempted to integrate tracking systems for direct human-machine interaction, the single add-on solution provides arguably an increased level of immersion.
(2) Another drawback of traditional firefighting training is the lack of dynamics and interactivity. In reality, no two firefighting procedures are identical, and fire-fighters need to be trained to make reasonable decisions under intensely

K. Arai et al. (Eds.): CompCom 2019, AISC 997, pp. 11–21, 2019.
https://doi.org/10.1007/978-3-030-22871-2_2

challenging conditions. Therefore, the ability to generate dynamic training environments and consequences in response to trainee interactivity is critical.

(3) Last but not least, training simulations have always been criticised for their lack of accuracy and credibility. In many cases simulation has adversely affected the trainees on their performance.

To solve the first drawback, through analysis of the state of the art in firefighting simulation, we identified that an effective and immersive training system should include major human senses, (inputs received by human body), and interactions, (impact exerted by human body), directly involved in the firefighting activity, which work together to collaboratively contribute to the training realism. This includes vision, sound and specifically haptics, the sense of kinesthetic feedback. Haptics is a real-time bi-directional procedure that involves both sensing and interaction. It plays a critical role in the realism and immersion of the training. As most effort imparted during firefighting occurs on upper limbs, we chose to use a real fire hose augmented with haptics to provide kinesthetic feedback [3]. The fire hose is able to render adjustable magnitude of force based on actual water parameters sampled from real life scenarios. In addition, body sensing during firefighting also plays a critical role for the realistic rendering on the ambient environment. We propose to incorporate the HTC VIVE for visualisation, VIVE tracker for nozzle tracking, multiple interaction sensors will also be integrated to render body sensations and increase the immersion [4]. These include a rapid heat generation jacket and flash-hood to render the radiant heat from the ambient environment and conductive heat when handling objects. Also, effects under hazardous environments, (smoke, oxygen deficiency, elevated temperatures, and poisonous atmospheres), and extreme physical and psychological challenges will be accounted for.

To solve the second drawback, we developed a simulation eco-system to cover the planning, training and analysis stages of firefighting. Also, targeting not only the detailed skill training, but also the situation awareness and justification of the entire dynamic fire situation, and collectively working with multiple firefighters to achieve an overall mission success. Perhaps solving the third challenge is the most intriguing. In order to maintain the fidelity of the training scenarios, several environmental parameters, (e.g. oxygen levels, humidity, time, water), must be considered and modelled using particle physics simulations. In this paper, we propose a haptically enabled firefighting simulator to allow the firefighting trainees to experience jet reaction forces from the hose. The proposed simulator (Fig. 1), also features realistic water dispersion and interaction with fire and smoke particles via accurate particle physics modelling.

The rest of this paper is organised as follows. Section 2 describes the elements of firefighting training. Hardware and software configurations are described in Sects. 3 and 4, respectively. Experiments and results are discussed in Sect. 5. Finally, Sect. 6 derives conclusions and introduces to future work.

Fig. 1. Haptically-enabled firefighting VR training simulator.

2 Elements of Fire Fighting Training

There are three aspects that must be considered during a fire fighting session. First, a planning stage is invoked to develop a prioritised action plan to fight fire pockets depending on the nature, source and spread of the fire. Second, physical limitations should be considered in terms of materials used and the limits of the human body. Finally, post training analysis should be conducted to identify the skills that need improvement.

2.1 Planning

Although a key factor of firefighting is the actual fighting of fire with a hose, the planning stage before firefighters enter the danger zone is equally important. How to quickly analyse the fire situation, identify priority zones, deploy fire fighters towards missions, and make optimised decisions for action, can all contribute to risk reduction and potentially reduce casualties and property damage. Often referred to a Size Up or 360 review, one key part of the planning is the creation of realistic scenarios, including environment setup, (wind direction and speed, daytime/night, humidity, oxygen level), and object setup, (object materials such as paper/wood/metal/flammable liquid/Hazmat, which eventually determines the type of smoke and the type of fire being generated, such as electrical fire, flammable liquid, chemical fire, etc.). There are also specific targets and priorities of the targets defined in different scenarios: one goal may be for the retrieval of classified files, another may be for the rescue of delicate equipment. Plans can also be modified on the fly as fire situations change, such as when new origins of fire have been discovered, or when the fire situation has heavily been impacted such as structure collapse and explosion, or new personnel who are in danger have been identified. When a high priority task is unfeasible to proceed, the underlying firefighting plan can be dynamically overridden and trigger mission redeployment.

2.2 Training

During firefighting training, we chose a series of physical properties assignable to scenario objects which determine their behaviour when affected by fire and later by water. Examples of the developed objects physical properties include: fire point of origin, rate of spread, structural changes affected by fire, (melt, collapse, explode, liquid flow), type of smoke generated while burning, reactions to water or other agent, (simple physical reaction or chemical reaction). We also propose a series of parameters for the extinguishing agent, including: spray pattern, spray pressure, spray capacity, fluid viscosity, fluid drop size, fluid gravity, nozzle type, spray pattern and pressure drop, to model the detailed local impacts of the spray over fire and objects. These object-specific and spray-specific properties and parameters will be used to calculate how the fire will burn, what types of smoke, (solid particles), are generated and spread out, and how effective the fire will be controlled locally at object level, mid-level and indoor level and globally at scenario level. Through these developed property definitions and parameters, the developed system will be capable of handling collective firefighting training, where multiple firefighters work together to fight a single fire source, and their contributions are calculated separately (each water contribution on the overall fire situation) and aggregated against the same fire source, on the same affected objects.

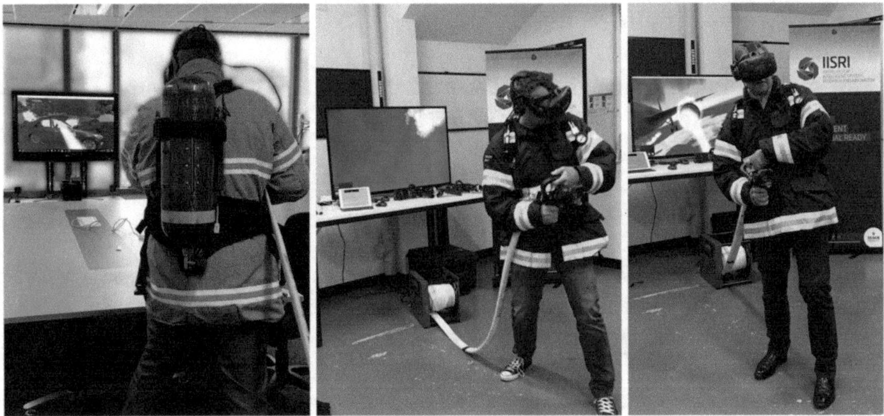

Fig. 2. Haptic VR firefighting system in action.

2.3 Post Training Analysis

Post training analysis also forms a critical part of the developed firefighting VR training system. As we plan to integrate a number of cutting-edge sensing and interaction technologies into the training procedure, the personalised data captured during training can be used for detailed analysis. To ensure the fidelity of the training simulation, physics engines are employed to ensure the laws of particle physics are properly governed in the virtual environment. This is achieved by addressing the following:

(1) Accurate physics-based formula on calculating detailed both visual rendering effects and event-driven scenario interactions. These include:
   - Environmental variables such as wind direction and speed, daytime/night, humidity, oxygen level.
   - Distinct fire and smoke appearance and impact results based on the physical properties defined on affected objects.
   - Time-dependent mathematical dynamic object visual and physical property modelling and impact rendering affected by fire and water.
   - Effect of water/agent on fire and the underlying objects based on physical and chemical activities.

(2) Detailed physics-based scenario dynamics and interactions driven by physics engine and haptic rendering engine, including movement, (objects falling and rolling), deformation (liquid flowing, objects melting and shattering) and total change of physical appearance (explosions and large collapse).

(3) Innovative implementations allowing for real-time per-particle collision detection among complex particle systems.

In order to address the identified elements of firefighting simulation based training, we designed the hardware and software framework described in the following sections.

3 Hardware Pipeline

Three major hardware components have been designed to enable intuitive control of firefighting and realistic force feedback [5–11]. Figure 2 shows the hardware setup of the proposed simulator.

3.1 Haptic Hose

The haptic hose is a device that simulates the dragging force, nozzle jet reaction force, handling and control difficulty that is experienced by the firefighter when using a hose. The haptic hose system is designed to use an actuated hose reel that provides haptic feedback to the firefighter trainee via the hand-held nozzle. A motor provides a variable torque to the reel through a low friction, low backlash motor system. Wound around the reel is a hose line that links the reel to the nozzle. Torque is applied to the hose line based on the gate action, flow rate and pattern settings on the nozzle. The torque simulates realistic jet reaction forces that would be imparted on the user when high-pressure water exits the nozzle in a real firefighting scenario.

The system provides enough torque to emulate high water volumes and pressures that would pull back the user. It actively complies with a user when they pull on the hose line. This allows a user to drag the hose from the reel when the gate on the nozzle is closed, approach and retreat from the elements in the scenario under force when the gate is open.

3.2 Branch or Nozzle

The branch/nozzle is a hand-held device that directs and shapes high-pressure water from a hose line. A typical nozzle has three main control functions:

- Water flow using a gate. The gate is mainly a valve that engages or disengages the water flow.
- Water flow rate using a rotary cuff. The rotary cuff determines how much water exits the nozzle.
- Water fog pattern using a rotary tip. The rotary end piece manipulates the water stream into a jet or wide fog pattern.

These functions are captured digitally to incorporate the real hardware into the virtual simulation environment.

3.3 CABA Mask

The Compressed Air Breathing Apparatus (CABA) mask or the Breathing Apparatus (BA) system is the device that is used to assist the firefighter with breathing in smoky conditions from compressed air cylinders. The designed mask unit comprises the following components:

- Mask.
- Off-the-shelf empty double cylinder unit modified.

- Computer retro-fit in to the double cylinder unit.
- Micro-controller with pressure sensor connected to the airline reaching the mask.
- Video transmitter with embedded audio.

Although during simulation training, there is no real compressed air being used, the CABA mask design simulates the firefighter realistic breathing allowing the instructor to hear and monitor the breathing behaviour of the firefighter trainee.

4 Software Pipeline

4.1 Haptic Rendering

To be able to render multiple sensor inputs captured during the training simulation along with haptic feedback, we designed a Hardware Abstraction Layer (HAL) to encapsulate hardware details from the software and implemented separate hardware managers to communicate with the software in a uniform way. This ensured the synchronisation issue of multiple sensor inputs at different update intervals and unified the different communication protocols required for different hardware components.

Fig. 3. Photo-realistic rendering of different firefighting scenarios -The last picture is an actual picture not a rendering.

4.2 Modelling Particle Dynamics

To be able to define highly dynamic and interactive training scenarios, the following guidelines have been setup:

- Accurate physics-based formula for the calculation of visual rendering effects and (Fig. 3) event-driven scenario.
- Detailed scenario dynamics driven by a physics and haptics engine, including object movement, deformation and total change of physical appearance.
- Innovative implementations allowing for real-time per-particle collision detection among complex particle systems.

At the training stage, events happen based on both time dependent mathematical models and each object's specific physical properties to calculate the gradual impact of fire on the object. Interactive planning and task rescheduling can be issued at training stage which overrides the initial plan, providing higher immersion and greater physical and psychological challenges towards the trainee. These physical properties, along with the real-time per-particle collision detection algorithm, seamlessly integrate with standard visual properties, and fundamentally enhanced the realism of both rendering and interactions among fire, smoke, object and water. Complex firefighting situations, (such as water and fire struggling against each other on objects, or fire being extinguished due to the temporary lack of oxygen and re-ignition over time), which are impractical to simulate in previous systems are made possible here.

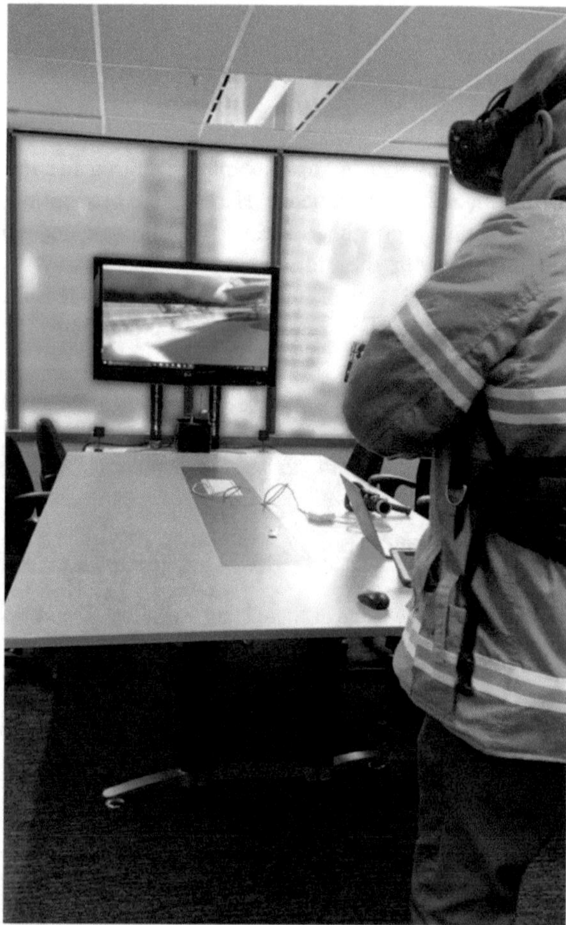

Fig. 4. Penciling task. Subject maintains a standing posture and puts out a fire wall.

Water, smoke and fire have been modelled and developed in isolation to examine and validate their realistic nature and interaction based on principles of physics. Particle systems have been adopted for real-time simulation per-particle interaction between different particles and scenario objects to enable rendering dynamic interaction and collision detection of various particles during the simulation.

4.3 Maintaining Fidelity

To be able to ensure accuracy and credibility in the training simulation, a very detailed fire progression algorithm has been developed. This algorithm is developed on top of the grid based spatial partitioning and provides detailed definition on not only the fire itself, but also fire spread in relation to the physical properties of ambient objects and oxygen supply. Firstly, the algorithm acquires user input on the initial fire status, including fire location, fire class, (i.e. class A solid fire, class B liquid fire, class C gas fire and class D metal fire), and combustible materials to initialise the primitive visual and thermal properties of the fire, such as smoke intensity, thermal progression and propagation, flame progression. Meanwhile, the environment will be deployed with inter-linked grids of fire nodes representing the specific combustible and thermal properties in those spatial locations [12–14]. Links between the nodes represent the heat transferred from the source of fire and each link can be individually defined and adjusted to represent environment dynamics. As the initial fire increases its thermal impact within the environment, heat is transferred along the linked nodes at different speed, increasing thermal impact to the adjacent nodes and eventually light those nodes up depending on the physical properties defined on those nodes, i.e. paper, electronics, metal, etc. In addition to this, environmental impacts have also been taken into consideration, such as oxygen level and wind direction and speed if the fire is in an open space. These in return also affect the head transfer function defined on each pair of linked nodes and collaboratively simulate realistic and accurate fire progression.

5 Experiments and Results

In order to evaluate the efficacy of the developed system we conducted a pilot study to assess the usability of the system. This experiments ethics approval was given by the Human Ethics Advisory Group (HEAG) of the Faculty of Science Engineering and Built Environment, Deakin University, Australia. The experiment was identified as low risk and was approved as complying with the National Statement on Ethical Conduct in Human Research (2007). Six voluntary participants, (4 males and 2 females), with an average age of 36 years, (SD 10 years)participated in the study. None of them had restricted movements inhibited by injury.

The participants were asked to use the firefighting nozzle, while connected to the haptic hose, to put out a virtual fire while maintaining a standing posture. This practice is commonly known as ‘Penciling’ in firefighting drills (Fig. 4). Before proceeding with the experiment, the penciling drill was demonstrated by an expert and the subjects were asked to rate how easy it is by filling out the 12-scale questionnaire below:

- How hard do you think using a haptic fire hose is?
- How do you rate your gaming experience?
- How do you rate your VR experience?

After the experiment, the subjects were asked to answer two more questions, detailed below:

- How hard did you find putting out virtual fire is?
- How much in control did you have putting out the virtual fire?

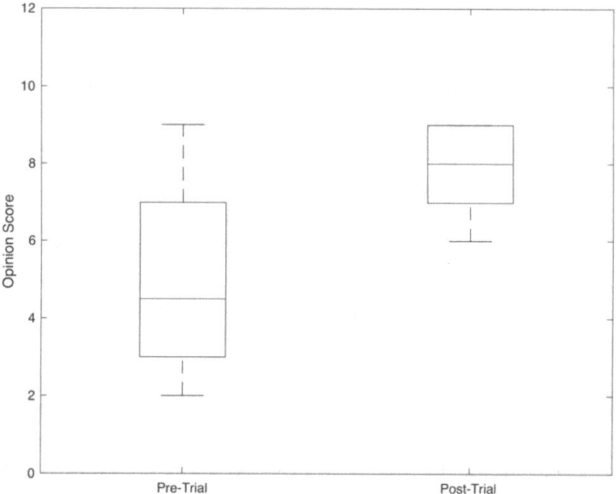

Fig. 5. Subjective scores pre- and post-trial. The results show a significant increase in the perceived complexity of performing the penciling drill after using the system ($p < 0.05$).

The results shown in Fig. 5 demonstrate the efficacy of the proposed system. Subjects recorded perceiving significant increase ($p < 0.05$) in complexity in performing the penciling drill in comparison to their perception of the complexity of the task after witnessing a subject matter expert performing it. There was no significant correlation between gaming and VR experience and the perceived complexity of the task.

6 Conclusion

In this paper we presented a haptically-enabled firefighting virtual training system. The developed system features a haptic feedback for simulating water (jet) reaction forces usually experienced during fighting a fire. The system also utilises particle physics simulation to ensure the fidelity of the interaction between water, fire and smoke particles. The system is designed to be expandable using a hardware abstraction layer (HAL) and is usable with most commercially-off-the-shelf VR headsets. The initial results show that the haptic hose does affect the perceived complexity of firefighting, regardless of the level of prior gaming or VR experience of the subjects. More experiments are needed to assess the efficacy of the developed system on training outcomes.

Acknowledgment. This research was fully supported by the Institute for Intelligent Systems Research and Innovation (IISRI) at Deakin University.

References

1. Yuan, D., Jin, X., Zhang, X.: Building a immersive environment for firefighting tactical training. In: Proceedings of 2012 9th IEEE International Conference on Networking, Sensing and Control, April 2012, pp. 307–309
2. Tate, D.L., Sibert, L., King, T.: Virtual environments for shipboard firefighting training. In: Proceedings of IEEE 1997 Annual International Symposium on Virtual Reality, March 1997, pp. 61–68
3. Lawson, W., Sullivan, K., Narber, C., Bekele, E., Hiatt, L.M.: Touch recognition and learning from demonstration (lfd) for collaborative human-robot firefighting teams. In: 2016 25th IEEE International Symposium on Robot and Human Interactive Communication (RO-MAN), August 2016, pp. 994–999
4. Vichitvejpaisal, P., Yamee, N., Marsertsri, P.: Firefighting simulation on virtual reality platform. In: 2016 13th International Joint Conference on Computer Science and Software Engineering (JCSSE), July 2016, pp. 1–5
5. Wei, L., Zhou, H., Nahavandi, S.: Haptically enabled simulation system for firearm shooting training. Virtual Real., June 2018. https://doi.org/10.1007/s10055-018-0349-0
6. Wei, L., Zhou, H., Nahavandi, S.: Haptic collision detection on disjoint objects with overlapping and inclusive bounding volumes. IEEE Trans. Haptics **11**(1), 73–84 (2018)
7. Wei, L., Zhou, H., Nahavandi, S., Wang, D.: Toward a future with human hands-like haptics: a universal framework for interfacing existing and future multipoint haptic devices. IEEE Syst. Man Cybern. Mag. **2**(1), 14–25 (2016)
8. Wei, L., Sourin, A.: Function-based approach to mixed haptic effects rendering. Vis. Comput. **27**(4), 321–332 (2011). https://doi.org/10.1007/s00371-011-0548-0
9. Wei, L., Najdovski, Z., Zhou, H., Deshpande, S., Nahavandi, S.: Extending support to customised multi-point haptic devices in chai3d. In: 2014 IEEE International Conference on Systems, Man, and Cybernetics (SMC), October 2014, pp. 1864–1867
10. Covaciu, F., Pisla, A., Carbone, G., Puskas, F., Vaida, C., Pisla, D.: Vr interface for cooperative robots applied in dynamic environments. In: 2018 IEEE International Conference on Automation, Quality and Testing, Robotics (AQTR), May 2018, pp. 1–6
11. Gommlich, F., Heumer, G., Vitzthum, A., Jung, B.: Simulation of standard control actuators in dynamic virtual environments. In: 2009 IEEE Virtual Reality Conference, March 2009, pp. 269–270
12. Ryge, A., Thomsen, L., Berthelsen, T., Hvass, J.S., Koreska, L., Vollmers, C., Nilsson, N.C., Nordahl, R., Serafin, S.: Effect on high versus low fidelity haptic feedback in a virtual reality baseball simulation. In: 2017 IEEE Virtual Reality (VR), March 2017, pp. 365–366
13. McMahan, R.P., Bowman, D.A., Zielinski, D.J., Brady, R.B.: Evaluating display fidelity and interaction fidelity in a virtual reality game. IEEE Trans. Vis. Comput. Graphics **18**(4), 626–633 (2012)
14. Lamata, P., Gomez, E.J., Bello, F., Kneebone, R.L., Aggarwal, R., Lamata, F.: Conceptual framework for laparoscopic vr simulators. IEEE Comput. Graph. Appl. **26**(6), 69–79 (2006)

Fast Replanning Incremental Shortest Path Algorithm for Dynamic Transportation Networks

Joanna Hartley(✉) and Wedad Alhoula(✉)

Department of Computing and Technology, Nottingham Trent University, Nottingham, UK
Joanna.Hartley@ntu.ac.uk,
wedad.alhoula2008@my.ntu.ac.uk

Abstract. Incremental search technique is used to solve fully dynamic shortest path problems. It uses heuristics to focus their search and reuse information from previous searches to find solutions to series of similar search tasks much faster than is possible by solving each search task from the scratch. This paper focused on improving the speed of computation time, a novel algorithm has been developed that combines the incremental search OLPA* algorithm and bi-directional heuristic search approach. The idea of using the bi-directional heuristic search is to reduce the search space and then reduce the number of node expansions. This novel algorithm has been called the Bidirectional Life-long A* algorithm [BiOLPA*]. The experimental results demonstrate that the BiOLPA* algorithm on road network is significantly faster than the LPA*, OLPA* and A* algorithms, not only in terms of number of expansion nodes but also in terms of computation time. Furthermore, this research provides some additional measurements to back our claims regarding the effect of blocked link location to the goal and the performance of our BiOLPA* algorithm approach.

Keywords: Lifelong planning A* · Dynamic shortest paths · OLPA* algorithm · Bi-directional search method · Road networks

1 Introduction

Traffic congestion is a serious problem that affects the mobility of people in society. People live in one area of the city and work in another. They also visit friends and family living in different parts of the country. Intelligent Transportation Systems (ITS) have worked towards improving the efficiency of the transportation networks using advanced processing and communication technology. The analysis and operation of these systems necessitates a variety of models and algorithms.

For example, the LPA* algorithm, in case of a blocked link that is close to a goal node and there is no alternative route available close to the blocked link, the second search (updated route) will take a long time to find a goal node because the LPA* algorithm will have lost its benefits from reusing the previous calculation. In this case, the A* algorithm that updated the route from scratch will be faster than the LPA* algorithm. This research shows that the BiOLPA* algorithm outperforms the existing LPA* and A* algorithms,

K. Arai et al. (Eds.): CompCom 2019, AISC 997, pp. 22–43, 2019.
https://doi.org/10.1007/978-3-030-22871-2_3

and always updates the route faster than all existing algorithms regardless of the location of the blocked link. The aim of this paper is to reduce the search space and speed up the computation time in dynamic networks when determining the optimal route and computing an alternative route (when network changes have occurred). The proposed approaches focus on both the search pace and shortest path queries. This paper is organized as follows. Next section presents literature reviews about the shortest path problem and reviews the A*, LPA* and OLPA* algorithms. Section 3 then describes and clarify the proposed algorithm (BiOLPA*). Section 4 discusses the experiments evaluation and the results of the experiments. Section 5 concludes the paper.

2 Related Work

The classic solution for both search space and shortest path queries is Dijkstra's algorithm [4]. Google Maps and most navigation applications (e.g., OpenTripPlanner) initially used Dijkstra's algorithm to find the most efficient route [6]. Dijkstra's algorithm works on a static network (where the edge weights on the network are static and deterministic). It works by examining the closest node to the start node. However, despite its simplicity, Dijkstra's algorithm is inefficient for large road networks. This algorithm has high time complexity and takes up a larger amount of storage space.

To achieve better performance, a variety of speeding up techniques have been proposed [2, 7, 12]. In particular in relation to search space, the first improvement of Dijkstra's algorithm was bidirectional search; starting the search from a start node and an additional search from a goal node, performed in a backwards direction, with the termination search occurring when both directions meet. The Bounded-hop Method reduces search space [1, 3]. In the field of the shortest path queries, the most efficient method is the Hierarchical method family, which recomputed a hierarchy of shortcuts and applies it to process the queries [5, 9]. All approaches focus on a single type of query, either search space or shortest path.

However, there are a number of transportation applications that use informed search algorithms, rather than one of the standard static shortest path algorithms. This is primarily due to the shortest paths needing to be rapidly identified either because an immediate response is required (e.g., in vehicle route guidance systems) or because the shortest path needs to be recomputed repeatedly (e.g., vehicle routing and scheduling). For this reason, a number of different heuristic shortest path algorithms have been investigated for reducing the execution time of the shortest path algorithms. For example, the A* algorithm [8] is widely used in artificial intelligence. Heuristic information (in the form of estimated distance to the destination) is used to focus the search towards the destination node. This results in finding the shortest path faster than the standard static search algorithms.

A. *A* Algorithm*

A* algorithm is normally used to solve an optimal route problem from a source location to a goal location. The evaluation function f is a sum of the two functions:

- The path cost function, which is the cost from the source node to the current node (denoted by $g(n)$).

- An admissible "heuristic estimate" of distance to the destination (denoted by $h(n)$). Each node n maintains f(n) where $f(n) = g(n) + h(n)$.

The function $f(n) = g(n) + h(n)$ is an estimate of this cost and is used by A* to decide which node should be expanded next.

The heuristic function h(n) has two important properties: it is admissible and consistent.

h(n) is an admissible heuristic if in place for every node n,

$h(n) \leq h*(n)$ Where h * (n) is the real cost to reach the destination state from, so, the heuristic function never overestimates the real cost. The other important property is that the heuristic function $h(n)$ is consistent if, for every node n, traveling through any successor n' of n will result in: (where c is the link-cost between node n and n') $h(n) \leq c(n, n') + h(n')$.

This can be viewed as a kind of triangle inequality as seen in Fig. 1. Each side of a triangle cannot be larger than the sum of the other two. Every consistent heuristic function is also admissible.

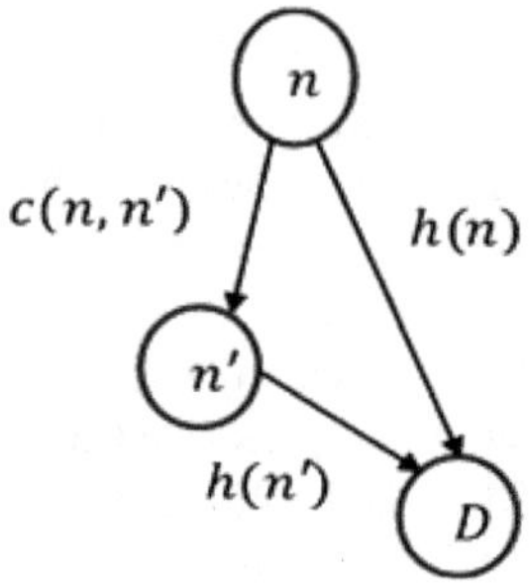

Fig. 1. Consistent heuristic function

By using a heuristic function, A* significantly reduces the number of nodes visited. Finding a solution without the loss of the solution optimality will depend on the type of heuristics used (i.e. an admissible will generate an optimal solution). In the worst-case scenario, the number of nodes visited will be exponential to the length of the optimal solution.

B. *Lifelong Planning A* (LPA*) Algorithm*

Among the algorithms suggested for the dynamic shortest path problem is the algorithm of Lifelong Planning A* algorithm (LPA*) [7]. This algorithm has been given this name because of its ability to reuse information from previous searches. The LPA* algorithm is a dynamic shortest path algorithm. It is used to adjust the shortest path to adapt to the dynamic transportation network. If arbitrary sequences of link insertions, deletions, and cost changes are allowed, then the dynamic shortest path problems are named "fully dynamic shortest path problems". LPA* is an incremental search technique that is used to solve a fully dynamic shortest path problems. It uses heuristics to focus its search and thus combines two different techniques to reduce its search efforts DynamicSWSF-FP algorithm and A* algorithm [7].

The LPA* algorithm is described as follows:

Assume S denotes the finite set of nodes of the graph and $succ(s) \subseteq S$ denotes the set of successors of node $s \in S$. Similarly, $pred(s) \subseteq S$ denotes the set of predecessors of node $s \in S$. In this case $0 < c(s, s') \leq \infty$ denotes the cost of moving from node s to node $s' \in succ(s)$ and $g(s)$ denotes the start distance of node s ∈ S, for, example, the cost of an optimal route from start node s to current node s'.

As A*, the heuristic approximates the goal distances of the nodes s. They need to be consistent, For example, satisfy $h(s_{goal}) = 0\, and\, h(s) < c(s, s') + h(s')$ for all nodes $s \in S$ and $s' \in succ(s)$ with s, s_{goal}.

There are three estimates used in the LPA* algorithm:

- g(s) Is the start distance of each node s, which directly corresponds to the g(value) of A* and can be reused in subsequent searches.
- h(s) Is the estimated distance to the goal node, which has the same meaning as the h-value in A* and is used to focus the search in the destination direction.
- rhs(value) Is one-step look-ahead value based on the g-value and thus is potentially better informed than the g(value).

To compute a node's rhs(value), firstly, we find the minimum g(value) between the nodes adjacent to the current node, and then take that value and add the movement cost of that node to it. They always satisfy the following relationship:

$$rhs(s) = \begin{cases} 0 & if\ s = s_{start} \\ min_{s' \in pred(s)}(\mathrm{g(s') + c(s', s)}) & otherwise \end{cases} \tag{1}$$

Node Expansion

- The node is locally consistent and it is removed from the queue if $g(s) = rhs(s)$
- The node is over-consistent if $g(s) > rhs(s)$
- The node is under-consistent if $g(s) < rhs(s)$

  So, if the node is under-consistent then $g(s) = \infty$ (Which makes the node either locally over consistent or locally consistent). If the node is then locally consistent, it's removed from the queue, else its key is updated.

The LPA* algorithm uses an open set which is a priority queue that always contains nodes that potentially need their g(value) to be updated to make them locally consistent. The LPA* algorithm always expands the node in the priority queue with the smallest key f (value). The key, k(s), of node s is a vector with two components:

$$k(s) = [k1(s); k2(s)]$$

$$Where\, k1(s) = min(g(s), rhs(s)) + h(s) \tag{2}$$

$$And\, k2(s) = min(g(s), rhs(s))$$

The termination conditions for the LPA* algorithm are that the LPA* algorithm expands nodes until the three following conditions are met

- The goal node is locally consistent.
- The key of the node set for expansion is no less than the key of the goal node.
- The priority queue is empty.

C. *Optimize Lifelong Planning A* (OLPA*) Algorithm*

The behavior of the OLPA* algorithm follows the same principles as the LPA* algorithm. An appropriate data structure is used to improve the efficiency of the dynamic algorithms implementations. A data structure is a methodical way of organizing and accessing data and an algorithm is a step-by-step procedure for performing some task in a limited amount of time. The OLPA* algorithm an Open set is implemented as a priority queue dictionary (pqdict) with Using a binary heap can significantly speed up the search strategy, particularly on a large road network instead of general priority queue. Using priority queue dictionary with the binary heap provides $O(1)$ to search and retrieve items with the highest priority regardless of the number of items in the queue. In addition, it supports inserting elements with priorities, and removing or updating a priority element, each operation takes $O(log\,n)$.

Furthermore, the OLPA* algorithm the set of predecessors of node has been implemented as a priority queue dictionary (pqdict) rather than open set. So, when the node is updated, the search find the predecessor with minimum value to be a parent for updated node, by searching for each a predecessor check until we find predecessor with the minimum key value. This takes $O(n)$. So the search can always find the predecessor with minimum key value at the front of the queue and this search can find the predecessor of node in $O(1)$ instead of finding it in $O(n)$.

The OLPA* algorithm outperforms the LPA* and the A* algorithm. This has been achieved by reducing route computation time in the first search and the second search (when some links blocked).

3 Proposed Algorithm

The proposed algorithm consists of two components:

(1) An improved OLPA* for computing dynamic shortest paths.
(2) A new search strategy. Each of the two components is described in detail below.

A. *Extend* OLPA* *to Bi-directional LPA* (BiOLPA*) Algorithm*

This paper presents a novel algorithm called the Bidirectional Lifelong A* algorithm [BiOLPA*]. The BiOLPA* algorithm searches forwards from the start node and backwards from the destination node using a novel search strategy. This proposed search strategy is called the autonomous strategy. It improves the strategy of node selection in the algorithm and increases the search speed by searching forwards and backwards simultaneously, searching alternatively such as via Poul's strategy [11], or based on the number of nodes in both priority queues. The side that has the fewest number of nodes in the priority queue is started and expanded first. This strategy was proposed by [10], named the cardinality comparison strategy. In this strategy, the algorithm will decide the direction (forward or backward) based on the number of nodes in both priority queues. The side that has the fewest number of nodes in the priority queue will start to expand first. This thesis proposes a novel strategy called an autonomous strategy. We chose this name because we do not implement exactly in the program which direction to start to search first, and the algorithm itself decides based on the heuristic values. The autonomous strategy chooses the most promising node that has the highest probability of being on the shortest path and has the smallest f (value) in relation to both priority queues.

$$f(n) = g(n) + h(n)$$

However, the BiOLPA* algorithm can adjust the shortest path to adapt to the dynamic transportation network and guarantee to find the shortest path faster than both the OLPA* and the bi-directional search methods individually because it combines their techniques. Consequently, in this paper, a novel algorithm has been developed that combines an incremental search algorithm with a bi-directional heuristic search approach (see Fig. 2).

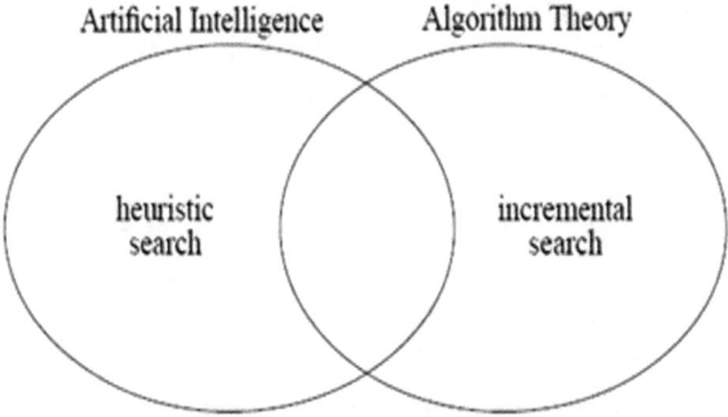

Fig. 2. Combination incremental search and heuristic search

- Heuristic search is how to search efficiently using heuristics to guide the search.
- The Incremental search is how to search efficiently by re-using previous calculation.

Two different ways of decreasing the search efforts for determining the shortest path have been investigated.

- Firstly, some of the edge costs are not affected by the changes, and thus do not need to be recomputed. Heuristic function knowledge, in the form of approximations of the goal distance, can be used to speed up the search. This is what the OLPA* algorithm does.
- Secondly, the heuristic searching strategy, using the heuristics from the start node to the goal node and the heuristics from the goal node to the start node, will reduce the search space and then speed up the search by half. This is what the bi-directional search method does.

(1) *Bi-directional Lifelong Planning A*: (BiOLPA*) the Variables*

Assume S indicates the finite set of nodes of the graph and $succ(s) \subseteq S$ indicates the set of successors of node s ∈ S. Similarly, $pred(s) \subseteq S$ indicates the set of predecessors of node s ∈ S. In this case, $0 < c(s, s') \leq \infty$ denotes the cost of moving from node s to node $s' \in succ(s)$ and $g(d, s)$ denotes the start distance of node $s \in S$. The search direction, d denotes the best selection of either the forward or backward direction, and is presented as (1) if the best search direction is forward and is presented as (−1) if the best search direction is backward. The significant difference between OLPA* and BiOLPA* that BiOLPA* uses d to indicate the right direction to search from. The start distances satisfy the following relationship:

$$g(d, s) = \begin{cases} 0 & if\ s = s_{start} \\ pop_{s' \in pred(s)}(g(s') + c(s', s)) & otherwise \end{cases} \tag{3}$$

As with the A* algorithm, the heuristic approximates the goal distances of the nodes s. They need to be consistent. For example, to satisfy $h(s_{goal}) = 0$ and $h(s) < c(s, s') + h(s')$ for all nodes $s \in S$ and $s' \in succ(s)$ with s, s_{goal} as in the OLPA* algorithm, there are three estimates in the BiOLPA* algorithm.

- $g(d, s)$ Is the start distance of each node s, which directly corresponds to the $g(value)$ of A* and can be reused in subsequent searches.
- $h(d, s)$ Is the estimated distance to the goal node, which has the same meaning as the h-value in A* and is used to focus the search in the destination direction.
- $rhs(d, s)$ Is the one-step look-ahead value based on the g-values and thus is potentially better informed than the g-values.

To compute a node's $rhs(value)$ the minimum $g(value)$ between the nodes adjacent to the current node is calculated and added to the movement cost of that node. This always satisfies the following relationship:

$$rhs(d, s) = \begin{cases} 0 & if\ s = s_{start} \\ pop_{s' \in pred(s)}(g(s') + c(s', s)) & otherwise \end{cases} \tag{4}$$

The important concept called local consistency needs to be defined. This concept is important because a local consistency check can be used to avoid node re-expansion. Additionally, the g-value of all nodes is equal to their start distances if all nodes are locally consistent. Whenever link costs are updated, the g(value) of the affected nodes will be changed. The nodes become locally inconsistent.

Node Expansion

- The node is locally consistent and it is removed from the queue if $g(d,s) = rhs(d,s)$
- The node is over-consistent if $g(d,s) > rhs(d,s)$

If the node is over-consistent, then it is assigned the new value of rhs (value) to be $g(value) = rhs(value)$. The $g(value)$ is changed to match the rhs(value), making the node locally consistent. The node is then removed from the queue.

- The node is under-consistent if $g(d,s) < rhs(d,s)$ If the node is under-consistent, then $g(d,s) = \infty$ (Which makes the node either locally over-consistent or locally consistent). If the node is locally consistent, its key is updated.

The proposed algorithm BiOLPA* contains a priority queue dictionary for each direction. This means that two priority queues will be used. Priority queues always contains nodes that potentially need their g(value) to be updated, in order to make them locally consistent.

The BiOLPA* algorithm always expands the promising node in both priority queues with the highest priority that means it has the smallest key f(value). The key, k (d, s), of node s is a vector with two components, while d (direction process) is the best selection of either the forward or backward direction by expanding on the most promising nodes from both searches.

$$k(d,s) = [k\_1(d,s); k\_2(d,s)]$$

$$\text{Where } k_1(d,s) = min(g(d,s), rhs(d,s)) + h(d,s) \tag{5}$$

$$\text{And } k_2(d,s) = min(g(d,s), rhs(d,s))$$

The Termination Conditions

The BiOLPA* algorithm expands nodes until the same node that was selected from both priority queues is locally consistent, or when the priority queues are empty. That way, there is no path from s_{start} to s_{goal}.

(2) *Bi-directional Lifelong Planning A*: (BiOLPA*) the Algorithm*

The BiOLPA* algorithm has the same overall structure as the OLPA* search, with some novel differences. For example, the BiOLPA* algorithm contains a priority queue dictionary for each direction.

Another novel difference is implementing both the sets of the predecessor's nodes and a set of successor nodes as a priority queue dictionary **(pqdict)** rather than an open set. Additionally, the OLPA* algorithm is designed to focus forwards on the goal state while the BiOLPA* search checks whether the frontiers of the two searches meet. Table 1 shows the pseudo-code of the BiOLPA* algorithm.

Note that d denotes the best selection of either the forward or backward direction, and is presented as (1) if the best search direction is a forward search and presented as (−1) if the best search direction is a backward search.

As seen in Table 1, the main function Main () first calls Initialise () to initialise the path-planning problem 21}. Initialise () sets the initial g-values of all nodes to infinity and sets their rhs-values according to Eq. (4) {03-04}. Accordingly, firstly s_{start} is the only locally inconsistent node and is inserted into the priority queue. Otherwise, it is an empty priority queue with a key calculated, as according to Eq. (5) {05}.

When the BiOLPA* algorithm starts the first search, it initialises the g(s) and rhs(s) of all nodes as infinity. The BiOLPA* algorithm then waits for changes in the edge costs {24}. If the weight of any link arbitrarily changes, it uses UpdateVertex () {27} to adapt to this change. It first checks the estimates (g, rhs) of the nodes around the link changed, which has the most potential to be affected by this change. It also checks their membership in the priority queue if they become locally consistent or inconsistent.

- The rhs-values of the nodes are updated.
- Nodes which have become locally consistent are removed from the queue.
- Nodes which remain locally inconsistent have their keys updated.

After that, the BiOLPA* algorithm continue to examine the nodes until the termination conditions are met.

Finally, a shortest path is recalculated {23} by calling ComputeShortestPath (), which repeatedly expands on the locally inconsistent nodes without visiting unnecessary nodes that are not affected by the changes in order of their priorities {14}.

As mentioned above, a locally inconsistent node means that the node is locally over-consistent if $g(d, s) > rhs(d, s)$. When ComputeShortestPath () expands a locally overconsistent node {16-17}, it is assigned the new value of $rhs(value)$ {16}, which makes the vertex locally consistent. Consequently, if the expanded node was locally over consistent, then the change of its $g(value)$ can affect the local consistency of its successors {17}. If ComputeShortestPath () expands a locally under consistent node {19-20}, then the $g(value)$ is assigned the infinity value {19}. This makes the node either locally consistent or overconsistent, and its successors can be affected {20}. To keep Invariants {1-3}, ComputeShortestPath () updates the $rhs(value)$ of these nodes, checks their local consistency, and inserts them to or deletes them from the priority queue accordingly {07-08-10}.

B. *A New Bi-directional Search Strategy*

A search strategy is defined by the order of the node expansions. The initial use of bi-directional algorithms combining forward and backward searches within a given search space was carried out by [11] for use in relation to pathfinding problems. These two search processes run alternatively. Author in [10] proposed a new hypothesis strategy called cardinality comparison strategy (monotonicity hypothesis).

The monotonicity hypothesis is a common strategy that chooses the direction with the smaller priority queue, rather than simply alternating the directions. The experimental results show that the proposed cardinality comparison strategy performs significantly better than the original bi-directional strategy. This research presents a novel hypothesis strategy to enhance the intelligence of node selection in the proposed

Table 1. Bidirectional OLPA* algorithm

```
U.TopKey() Returns the smallest priority of all vertices in priority queue U[d]. (If U[d] is
empty, then U[d].TopKey () returns [∞; ∞].)
U[d].Pop() Deletes the vertex with the smallest priority in priority queue U[d] and returns
the vertex.
U[d].Top() Returns the vertex with the smallest priority in priority queue U[d].
U[d].Insert(s, k) Inserts vertex s into priority queue U[d] with priority k.
U[d].Remove(s) Removes vertex s from priority queue U[d].
U[d].Update(s) Update the vertex with the new key value in priority queue U
Procedure CalculateKey(d, s)
return (min (g(d, s), rhs(d, s)) + h(d, s); min (g(d, s), rhs(d, s)));
Procedure Initialize ()
U[1] = ϕ; U[−1] = ϕ;
for all s ∈ S rhs(1, s) = g(1, s) = rhs(−1, s) = g(−1, s) = ∞;
rhs(1, s_start) = rhs(−1, s_goal) = 0;
U[1].Insert(s_start, [h(1, s_start); 0]; U[−1].Insert(s_goal, [h(−1, s_goal); 0];
Procedure UpdateVertex(d, u)
if(d = 1 AND u ≠ s_start)rhs(d, u) = pop_s'epred(u)(g(d, s') + c(s', u));
if(d = −1 AND u ≠ s_goal)rhs(d, u) = pop_s'esucc(u)(g(d, s') + c(u, s'));
if(g(d, u) ≠ rhs(d, u))U[d].Update(u, CalculateKey(d, u));
if(u ∈ U[d])
else
U.Update(u, CalculateKey(d, u));
else
U.pop(v, None)
Procedure ComputeShortestPath ()
while(rhs(1, (U[1] ∪ U[−1]).Top()) ≠ g(1, (U[1] ∪ U[−1]).Top())OR
      rhs(−1, (U[1] ∪ U[−1]).Top()) ≠ g(−1, (U[1] ∪ U[−1]).Top()))
ifU[1].TopKey() < U[−1].TopKey()d = 1 else d = −1
        u = U[d].Pop();
if(g(d, u) > rhs(d, u))
g(d, u) = rhs(d, u));
for all s ∈ (if d = 1succ(u) else pred(u))UpdateVertex(d, s);
else
g(d, u) = ∞;
for alls ∈ (ifd = 1succ(u)else pred(u)) ∪ (u)UpdateVertex(d, s);
Procedure Main ()
Initialize ();
forever
ComputerShortestPath ();
wait for changes in edge costs;
for all directed edges (u, v)with changed edge costs
update the cost c(u, v);
UpdateVertex(u);
```

algorithm. This strategy is called the autonomous strategy (it is named this because the algorithm decides which the best direction start to search based on the heuristic value). The difference between the proposed strategy and [10] strategy is that the selection between forward and backward direction is not based on the fewest number of nodes in both priority queues. Instead, it is based on the priority of the front node of both priority queues that has the highest probability of being on the shortest path with the smallest key f (value) In the BiOLPA* search, the following steps are repeatedly executed until the two searches meet somewhere.

Select a forward (Step 2) or a backward direction (Step 3), and then perform the actual node expansion.

- Forward search:
  The starting search from the start node moves towards the goal node.
- Backward search:
  The search from the goal node moves backwards from the goal node.

The autonomous strategy employs the minimal control necessary to guarantee the termination of the algorithm. Thus, the strategy determines the best node in the selection of either the forward or backward direction. For example, $h(f,s)$ if it forwards or $h(g,s)$, if it is in a backward direction. This research used a heuristic function with a control parameter $(d), h(d,s)$, where d indicates the best selection of either the forward or the backward directions (see Fig. 3).

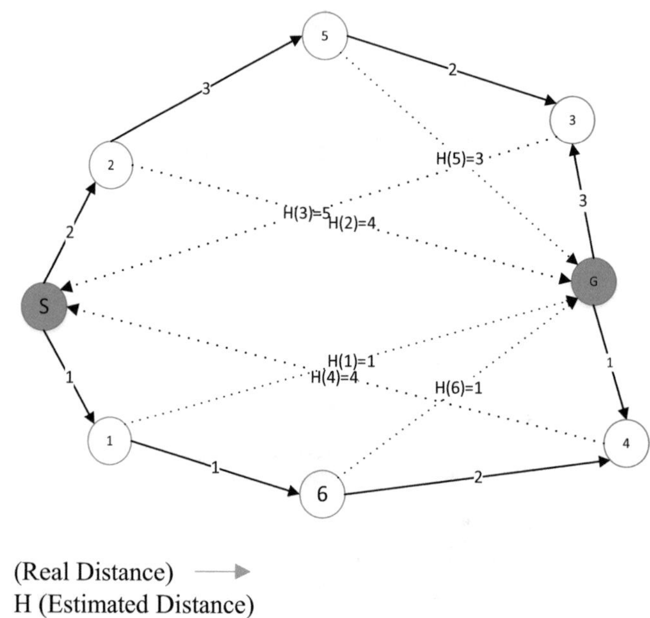

Fig. 3. Autonomous search strategy

The forward search node (1) is more promising than node (2) Because it has the smallest key $f(value) = g(1) + h(1) = 2$ than $f(value) = g(2) + h(2) = 7$. Similarly, in the backward direction, node (4) is more promising than node (3),

$$f(value) = g(3) + h(3) = 6$$
$$f(value) = g(4) + h(4) = 5.$$

The strategy then will compare node (1) (for forward direction) and node (4) (for backward direction) and choose the most promising node that has the highest probability of being on the shortest path with the smallest f (*value*). Based on Fig. 3, the search starts from node (1) so the forward direction search will start first. BiOLPA* always expands the promising node in both priority queues with the smallest key $f(value)$ regardless of the direction.

For example:

$$f(value) = g(5) + h(5) = 2 + 3 + 3 = 8$$
$$f(value) = g(6) + h(6) = 1 + 1 + 1 = 3$$

After node (1) the algorithm will search forward again because node (6) has the smallest $f(value)$. After node (6), the algorithm will start to search from the backward node (4) with the smallest $f(value)$. After this search, the two directions will meet in node (4) and this is the termination point of the algorithm.

The path is $S \rightarrow 1 \rightarrow 6 \rightarrow 4 \rightarrow G$.

4 Experimental Evaluation

To examine the efficiency of the novel approaches, the experiments are performed using real-world road networks of Nottingham city. Figure 4 shows Nottingham city network. To demonstrate the efficiency of our algorithms, we measure the performance of the BiOLPA*, OLPA*, LPA* and A* algorithm. To simulate real-time traffic conditions, this paper used an accident event that was simulated arbitrarily along the path as a cost of some node changes to adapt to the random changes in traffic conditions. To assess the run time performance, we measure the average response time over 1,000 for a randomly selected pair of start and goal nodes. All algorithms were implemented in Python. All

Fig. 4. Nottingham city network

experiments were executed on Windows XP (2.67 GHz) processors and 48 GB of RAM. To avoid biasing the experimental results in favor of the OLPA* and the BiOLPA* algorithms, this research uses the Euclidean distance as a heuristic between two nodes. For comparison purposes, the number of nodes expanded and computation time is taken as benchmarks to test the efficiency before and after an accident.

Experiment 1: The first experiment compares computational time (in seconds) versus the number of expansion node. Figure 5 shows the first search of both algorithms, It is clear to see that the BiOLPA* algorithm has the fewer number of expansion node than the LPA* algorithm.

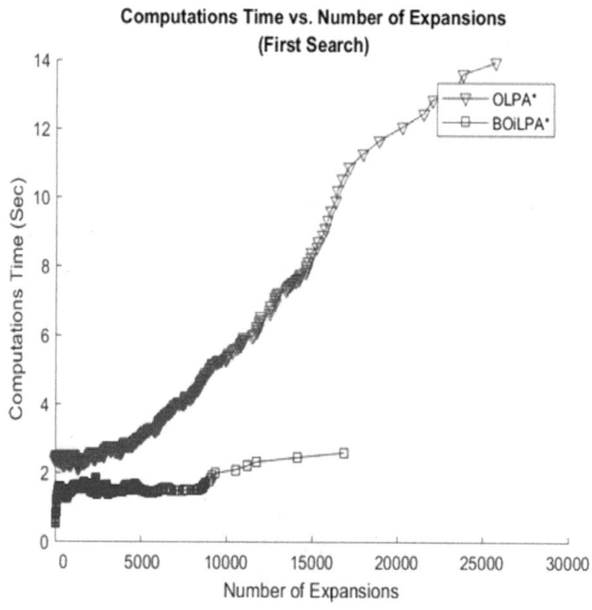

Fig. 5. First search - Computation time vs Node expansion

In Fig. 5, for the route of 30000 nodes, the BiOLPA* algorithm expanded around 16000 nodes with the time of 2 s. While OLPA* expanded more than 25000 nodes and spent 14 s. This is a very important observation because it means that an increase in the number of nodes in the route, results in BiOLPA* using less search space with less time than the other algorithm.

In the second search as in Fig. 6 shows that the BiOLPA* algorithm has a slight increase of computation time up to less than 1 s and reduce the number of expands nodes from 17000 nodes in the first search to only 5000 nodes in the second search. While the OLPA* algorithm in the second search, expands 10000 nodes in 5 s, So, BiOLPA* reduce the number of expansion nodes in the second search by a half.

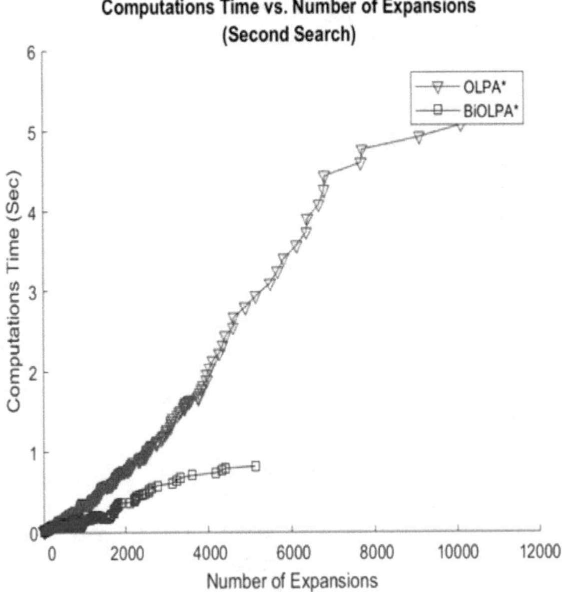

Fig. 6. Second search - Computation time vs Node expansion

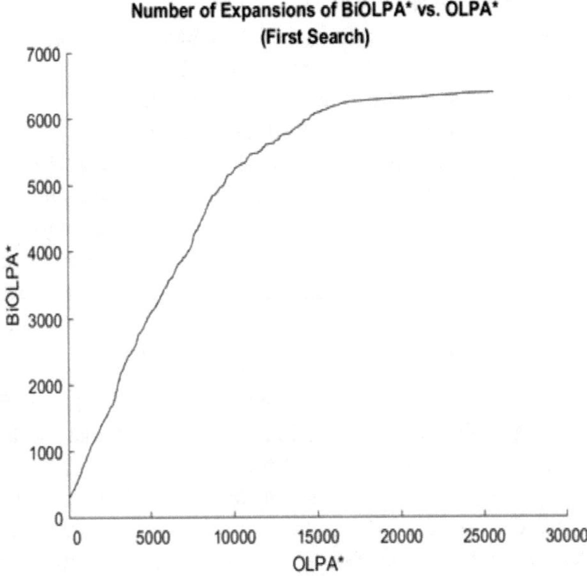

Fig. 7. First search - Node expansion of BiOLPA* vs OLPA*

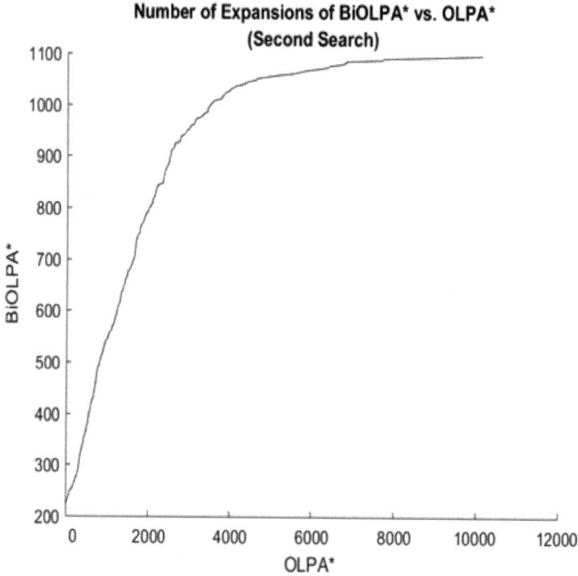

Fig. 8. Second search - Node expansion of BiOLPA* vs OLPA*

As a result that the BiOLPA* algorithm is able to show the advantage from using the BiOLPA* algorithm on road network that reduces the search space by reducing the number of node expansion of the search, which reduces the computation time of the algorithm. Figures 7 and 8 shows the comparison of node expansions between both algorithms in the first search and second search in order.

Figure 7 shows that the BiOLPA* algorithm expanded from 1000 up to 7000 nodes in the total node in the route 30000 nodes. But the OLPA* algorithm more nodes so when BiOLPA* algorithm expanded 7000 nodes the OLPA* expands 25000 nodes. In addition, the first search and the second search the BiOLPA* algorithm has a slight increase of the number of expansion for example in the first search the number of node expansion increases by 1000 nodes, while the OLPA* algorithm increases by 5000 nodes. Also in the second search Fig. 8, the BiOLPA* algorithm increases by 100 nodes while the OLPA* algorithm increases by 2000 nodes.

Experiment 2: In the second experiment, this research tests the performance of the BiOLPA* algorithm and the OLPA* algorithm when the blocked link is close to the goal node. The blocked link close to the goal means that the algorithm reuses almost all of the initial search in re-planning. So, the run time and number of node expansion by the BiOLPA*, OLPA* will be less than the A* algorithm. Moreover, if the accident happened near the alternative route, this also increases the advantage of the BiOLPA* algorithm and the OLPA* algorithm in our experiments. Finding an alternative route close to an accident is more significant than how close the accident is to the goal. For example, if the accident is just one step from the goal, but there is no nearby alternative, then re-planning will take a long time. This means that only a small portion of the previous search result can be reused by the OLPA* algorithm and it may lose its advantage (the advantage of the previous calculations).

Fig. 9. Updated blocked link

In this status, it has to search from scratch as the A* algorithm. Figure 9 shows that the alternative route is very close to the accident event. This gives a very good result in this experiment and the number expanding node is equal to zero in case of the BiOLPA* algorithm, OLPA* algorithms and the computation time is less than the A* algorithm. Zero expansion means that there is a very close alternative route.

In Fig. 9, initial shortest path depicted by a blue line. An accident event is depicted by a red colour. The updated shortest route depicted by a blue square line.

However, this research, normalize the location of the blocked link, that means it is better to divide the accident location by the route size rather than using the raw accident location, which recorded from running the program 1000 times. The raw location does not tell whether the accident is closer to the start or the goal, while the normalized location will be in the range of 0 to 1 and it can clearly tell. For example, assume that the

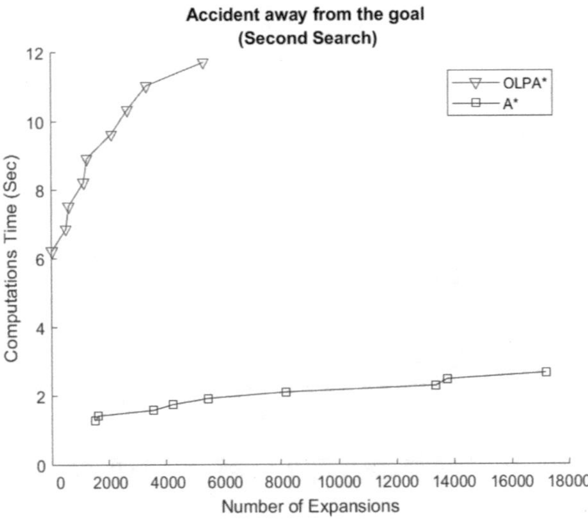

Fig. 10. Accident away from the goal (OLPA* vs A*)

route size is 5 and the accident is at 4. The normalized accident location is 4/5 = 0.8 > 0.5 closer to the goal. Now, assume that the route size is 10 and the accident is at 4 (also). The normalized accident location is 4/10 = 0.4 < 0.5 closer to the start.

Therefore, the accident in both examples was at 4. Figure 10 shows the experimental results when the accident event is far away from the goal. So, consider only the second search (updated route).

Figure 10, shows that only a small portion of the previous search result is reused by the OLPA* algorithm which caused the loss of its advantage then the re-planning will take a long time. In this status, the A* algorithm expands more nodes than the OLPA* algorithm. But the OLPA* computation time is more than the A* algorithm. In this case, searching from the scratch is faster than the update route from the blocked link. Because the OLPA* algorithm does not use the previous calculation then the updated route will take a long time to expands new nodes compared to the A* algorithm.

However, Figs. 11, 12 and 13 show that the BiOLPA* algorithm has fewer node expansion and still faster than both the OLPA*, LPA* and the A* algorithms. So, in case of the blocked link is away from the goal location the BiOLPA* algorithm does not lose it is the advantage as the OLPA* algorithm. The BiOLPA* algorithm expands less than 1000 nodes in less than 1 s. While the A* algorithm expands 16000 nodes in more than 2.5 s and the OLPA* and the LPA* algorithms expand the same node about 5000 nodes. But the OLPA* algorithm takes 12 s while the LPA* algorithm taks more than 20 s.

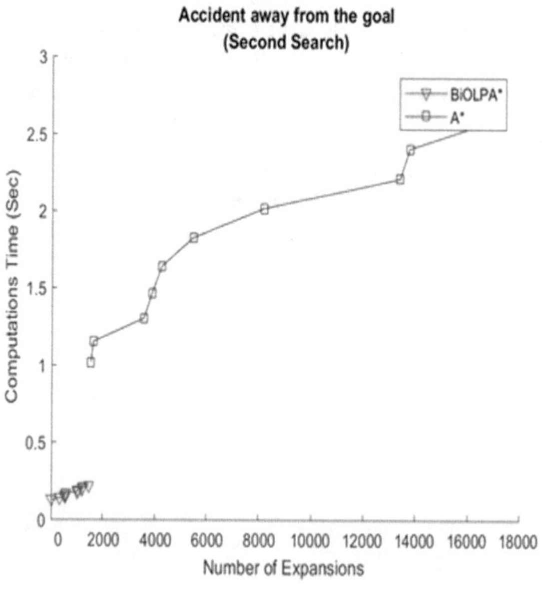

Fig. 11. Accident away from the goal (BiOLPA* vs A*)

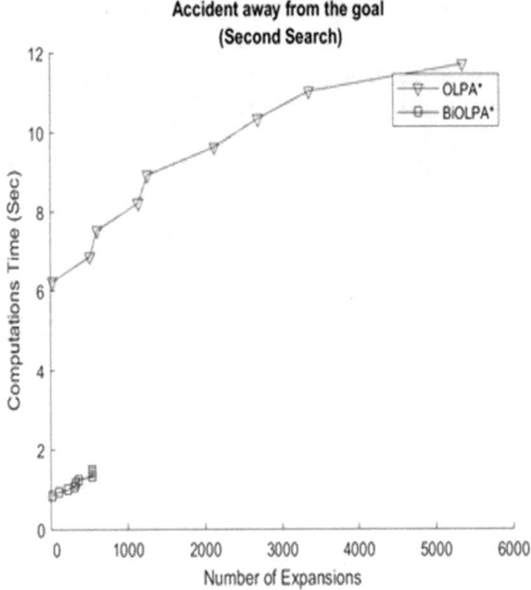

Fig. 12. Accident away from the goal (BiOLPA* vs OLPA*)

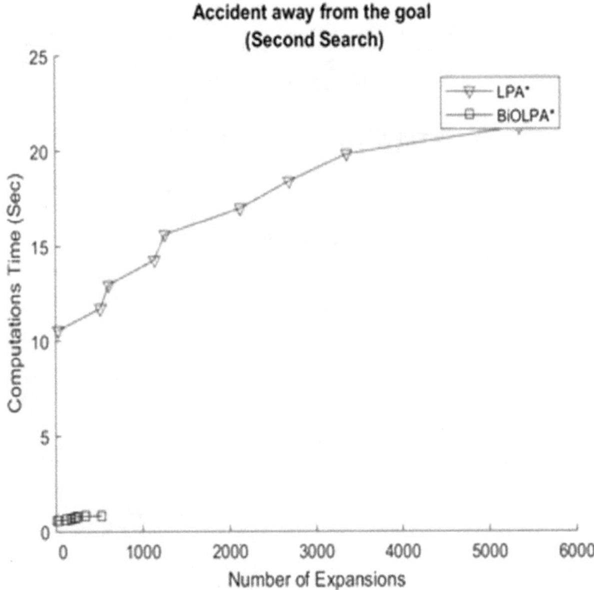

Fig. 13. Accident away from the goal (BiOLPA* vs LPA*)

This research also performed more experiments that evaluate the BiOLPA*, OLPA* and A*, algorithms in case that the blocked link is close to the goal node. Figures 14, 15, 16 and 17 show our experimental results. The closer the edge cost changes are to the goal node, the high the performance of the BiOLPA* algorithm in our experiments. This is an important insight since it proposes the use of BiOLPA* when the blocked link is close to the goal node.

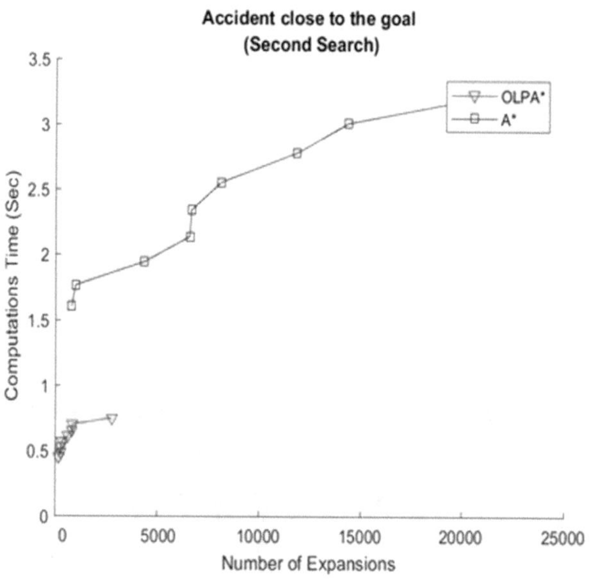

Fig. 14. Accident close to the goal (OLPA* vs A*)

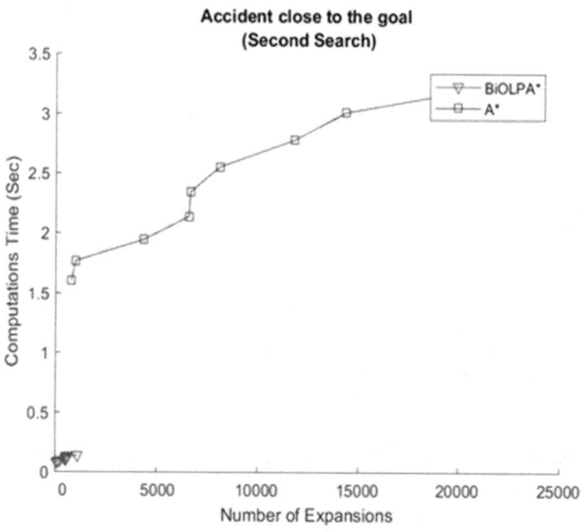

Fig. 15. Accident close to the goal (BiOLPA* vs A*)

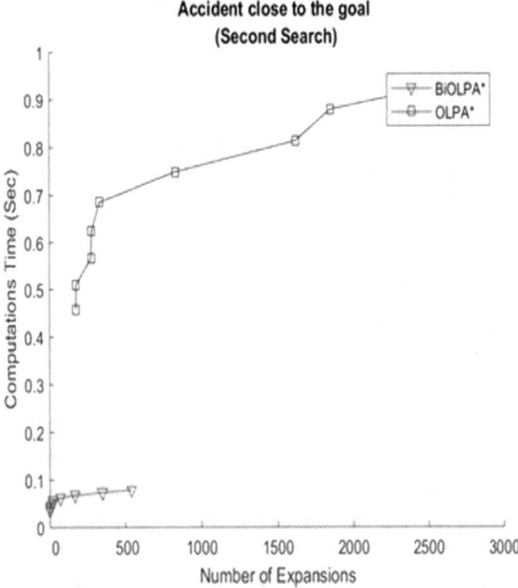

Fig. 16. Accident close to the goal (BiOLPA* vs OLPA*)

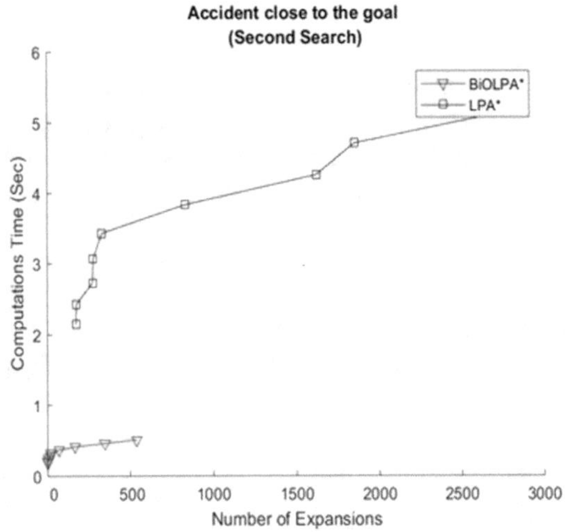

Fig. 17. Accident close to the goal (BiOLPA* vs LPA*)

Figure 14 illustrates that the OLPA* is faster than A* algorithm and expands less number of nodes than the A* algorithm.

The BiOLPA* algorithm achieves the biggest success in terms of the computation time and the number of nodes expands. Compared to Figs. 15, 16 and 17 the BiOLPA* algorithm are the fastest ones. For example, the A* takes 0.7 s to expand about 700 nodes, while the BiOLPA* spends approximately 0.001 s to expand less than 10 nodes. Also, the OLPA* and LPA* algorithms need to expand 2500 nodes in case that the blocked link near to the goal while the BiOLPA* algorithm expands 500 nodes in less than 0.01 s.

The experiment results indicated that regardless of the location of the accident, if it is close or away from the goal in all cases, the BiOLPA* algorithm is more efficient than the OLPA*, LPA* and A* algorithms. BiOLPA* is significantly faster than the OLPA*, LPA* and A* algorithms, in terms of the number of expansion nodes as well as in terms of computation time.

5 Conclusion

This paper presented the development of a novel approach that extends the existing Lifelong Planning A* algorithm (OLPA*) to solve the dynamic shortest-path problem. The experimental results show that the BiOLPA* algorithms are able to reduce the search space and then increase the speed of the computation time. To examine the efficiency of the novelties approaches, the experiments are performed using real-world road networks of Nottingham city. To simulate real-time traffic conditions, this research used an accident event that simulated arbitrarily along the path as a cost of some node change. The accident event is used to simulate the situation of a node submitting an en route for a new optimal route in order to adapt to the random change in traffic conditions.

The experimental result shows that if the accident event is close to the goal but there is no alternative route available. The second search will take a long time and the LPA* and OLPA* algorithms will lose their benefit from reusing previous calculation, and in this case, the A* algorithm will be faster than both the LPA* and the OLPA* algorithm. While in all cases, the BiOLPA* algorithm is more efficient than the OLPA*, LPA* and A* algorithms. The BiOLPA* algorithm is significantly faster than the OLPA*, LPA* and A* algorithms, in terms of a number of expansion nodes as well as in terms of computation time.

The experimental result shows that the BiOLPA* algorithm that combination of the OLPA* algorithm and the bidirectional search method is able to significantly reduce the search efforts.

References

1. Akiba, T., Iwata, Y., Kawarabayashi, K., Kawata, Y.: Fast shortest-path distance queries on road networks by pruned highway labelling. In: McGeoch, C.C., Meyer, U. (eds.) 2014 Proceedings of the 16th Workshop on Algorithm Engineering and Experiments (ALENEX), Portland, Oregon, 5 January 2014, pp. 147–154. Society for Industrial and Applied Mathematics, Philadelphia (2014)
2. Bast, H., Funke, S., Sanders, P., Schultes, D.: Fast routing in road networks with transit nodes. Science **316**(5824), 566 (2007)

3. Cohen, E., Halperin, E., Kaplan, H., Zwick, U.: Reachability and distance queries via 2-hop labels. In: Proceedings of the 13th ACM-SIAM Symposium on Discrete Algorithms, SODA 2002, 6–10 January 2002, pp. 937–946. Society for Industrial and Applied Mathematics, San Francisco (2002)
4. Dijkstra, E.: A note on two problems in connexion with graphs. Numer. Math. **1**(1), 269–271 (1959)
5. Geisberger, R., Sanders, P., Schultes, D., Delling, D.: Contraction hierarchies: faster and simpler hierarchical routing in road networks. In: Proceedings of the 7th International Workshop on Experimental Algorithms (WEA 2008), Provincetown, MA, 30 May–01 June 2008, pp. 319–333. Springer, Heidelberg (2008)
6. Klunder, G., Post, H.: The shortest path problem on large-scale real-road networks. Networks **48**(4), 182–194 (2006)
7. Koenig, S., Likhachev, M., Furcy, D.: Lifelong planning A*. Artif. Intell. **155**(1–2), 93–146 (2004)
8. Russell, S., Norvig, S.: Artificial Intelligence: A Modern Approach, 3rd edn. Prentice Hall, Englewood Cliffs (2009)
9. Sanders, P., Schultes, D.: Highway hierarchies hasten exact shortest path queries. In: ESA 2005 Proceedings of the 13th Annual European Conference on Algorithms, Palma de Mallorca, Spain, 03–06 October 2005, pp. 568–579. Springer, Heidelberg (2005)
10. Pohl, I.: Bi-directional search. In: Meltzer, B., Michie, D. (eds.) Machine Intelligence, vol. 6, pp. 127–140. Edinburgh University Press, Edinburgh (1971)
11. Pohl, I.: Bi-directional and heuristic search in path problems. Technical report 104, SLAC (Stanford Linear Accelerator Center), Stanford, California (1969)
12. Wu, L., Xiao, X., Deng, D., Cong, G., Zhu, A., Zhou, S.: Shortest path and distance queries on road networks: an experimental evaluation. Proc. VLDB Endow. **5**(5), 406–417 (2012)

A Computational Scheme for Assessing Driving

José A. Romero Navarrete(✉) and Frank Otremba

Federal Institute for Materials Research and Testing (BAM), Unter Den Eichen 44-46, 12203 Berlin, Germany
{jose-antonio.romero-navarrete, frank.otremba}@bam.de

Abstract. The driving style, characterized in terms of the acceleration dispersion has been recognized as a factor influencing the environmental un-friendliness of the driving. However, such acceleration variations could also represent different levels of crash propensity, as well as different levels of pavement damage potentials. Road crashes and road works affect the congestion levels, and the environmental impact of the driving. In this paper, different validated models are set together to create a computational platform to integrally assess the effect of the driving style on the environment, including the road safety and the pavement damage in the assessment, conforming an extended eco-driving. The results suggest that the greater dispersion of the driving acceleration induces a poor performance in the three indicators considered. A relative increase of 2.33 in the dispersion of the driving acceleration corresponds to the relative increases of 2.4, 2.05 and 1.52 in the safety, emissions and pavement damage performance indicators, respectively.

Keywords: Driving style · Pavement damage · Road safety · Performance measure · Emissions

1 Introduction

The concept of eco-driving has been centered on the level of pollutants emitted to the atmosphere, as a first-hand approach to mitigate the environmental impact of fossil fuel vehicles [1]. According to this approach, a diversity of studies has produced measures and recommendations to improve the driving behavior, and to reduce fuel consumption [2, 3]. The traveling speed and acceleration variation of the driving have been identified as fundamental factors to increase the emissions to the environment [4, 5]. However, the impact of a poor driving from the environmental perspective, has been identified also as a promotor for a poor performance from the infrastructure un-friendliness perspective, and has also been related to the level of acceleration's variation [6]. In this case, the acceleration variation was associated to variations of the tangential force at the wheel-pavement interface, that further create larger shearing stresses that combine with the normal stresses associated to the dynamic weight of the vehicles.

The parameters described above, concerning the environmental and infrastructure un-friendliness derived from the dispersion of the acceleration while driving, could also be associated to the safety of driving, as a fatigued driver could also drive at higher

K. Arai et al. (Eds.): CompCom 2019, AISC 997, pp. 44–58, 2019.
https://doi.org/10.1007/978-3-030-22871-2_4

levels of acceleration or deceleration. However, the road safety environmental effect to be considered in this paper, has to do with the consequences of the road crashes that usually involve affectations to the traffic, causing congestion and higher levels of pollution. On the other hand, an unfriendly driving could also be associated to infrastructure damage, whose reparation and rehabilitation imply affectations to the traffic and congestion, further increasing the levels of environmental pollution.

In this paper, three performance measures are proposed to assess the level of eco-driving, where the concept of eco-driving has been extended to the consideration of the effect of the driving style on both the road safety and the vehicle pavement damage potentials. The paper begins with a review of the literature, about the effects of the driving style on the environment, road safety and pavement damage. The different formulations are then described for the three performance measures proposed, related to road safety, environmental emissions and infrastructure damage. A study case is further presented in the section of results, concerning a segment of a driving history. The resultant computational scheme is further described.

2 Review of Literature

2.1 Congestion and Pollution

As in the case of road work, road crashes generate higher congestion levels to which a higher pollution is linked. Chapman [7] discusses the effect of congestion on emissions, outlying the economic consequences of that. Bigazzi et al. [8] report increments of 18% in pollution exposure of the road users, as a result of congestion. Gately et al. [9] report the potential effects of the elimination of congestion on the level of pollutants in the city of Boston, estimating that such elimination could reduce the level of pollutants per unit area, as a function of the type of pollutant, hour of the day, and location. For example, the elimination of congestion at 2 P M, would reduce the level of NOx from 48 mg/m^2 to 35 mg/m^2 (−27%), and the level of CO from 70 mg/m^2 to 50 mg/m^2 (−28%).

2.2 Road Safety and Pollution

Pasidis [10] reports that there would be a reduction in traffic speed as a result of a road crash; increasing the journey time by 27% during a crash event, averaging a 70 min traffic delay per km of affected road. The congestion derived from traffic crashes is reported by Pulugurtha [11], as a function of many features: the distance from the accident site; the type of accident; the elapsed time after the crash; and the traffic volume. According to their simulations, the travel time variation, for a road with a traffic volume of 100 vehicles per hour, can be up to 600% in case of a fatal crash. The effect of road crashes on environment has been recognized in the Second Phase of the Strategic Highway Research Program (SHRP 2) [12]. It is stated that crashes are the leading cause of nonrecurring congestion, arguing that collision prevention could reduce delay, fuel consumption and emissions.

According to Sager [13], there is a proved correlation between road accidents and pollution. While the author interprets the situation as the pollution inducing the accidents due to the harming of drivers' ability for a safe driving, the opposite can also be suggested. The author reports that an increase of 0.3 accidents per day, can lead to an increase of 1 $\mu g/m^3$ of the concentration of NO^2. The transient effect of a traffic crash on the quality of air, has been reported by Joo et al. [14], finding that the higher congestion derived from a road crash, would have caused an increase in about 6% the level of pollutants in near-freeway communities (NO_2 and $PM_{2.5}$).

2.3 Driving Style and Road Safety

An aggressive driving is characterized by a driver behavior that includes over-speeding and tailgating. According to the Foundation for Traffic Safety (AAA), half of the traffic fatalities derive from such kind of behavior [15]. In this respect, tailgating is a poor driving behavior that implies higher levels of deceleration and deceleration. In that sense, Berry et al. [16] associate an aggressive driving with higher levels of accelerations. Sagberg et al. [17] report that organizations have defined hard-braking events as indicators of unsafe driving, and Kim et al. [18] associate such levels of accelerations to rear-end crashes. Vaiana et al. [19] propose a two-dimensional "Driving Safe Diagram", where the longitudinal and lateral accelerations are registered, identifying areas in the diagram representing different driving styles. According to such scheme, greater longitudinal accelerations and decelerations, as well as greater lateral accelerations, imply an unsafe driving. According to the authors, an aggressive, un-safe driving is characterized by longitudinal decelerations and accelerations of 7 m/s^2 and 6 m/s^2, respectively, while a safe driving is characterized within a range from 2 m/s^2 for both, accelerating and decelerating.

2.4 Driving Style and Infrastructure Damage

A driving style can thus be described in terms of the level of vehicle accelerations and decelerations, involving traction and braking forces, respectively. The effect of higher traction and braking forces on the pavement has been considered when designing road intersections, implying higher specifications for such road segments [20]. With respect to surface shear forces, Kimura [21] describes the solution for these stresses' equations, explaining that the maximum shearing stress in the vicinity of the tire-pavement interface is significant. Consequently, an aggressive driving style, characterized by higher levels of acceleration, generates higher levels of pavement damage.

2.5 Driving Style and Fuel Consumption

The effect of the driving style on fuel consumption and pollution has been the subject of many efforts reported in the literature. According to Ma et al. [4], differences in driving style can induce variations in fuel consumption, from 10% to 20%. The fuel saving potentials of an optimized driving style is reported by Javanmardi et al. [3], with savings around 2.46% and 8.83% in the case of highway and urban infrastructures, respectively. Figure 1 illustrates the measured and optimized speed histories for an urban displacement. It can be observed in this figure, that the optimized driving involves the smoothing of the real measured speed profile, thus involving lower levels of accelerations.

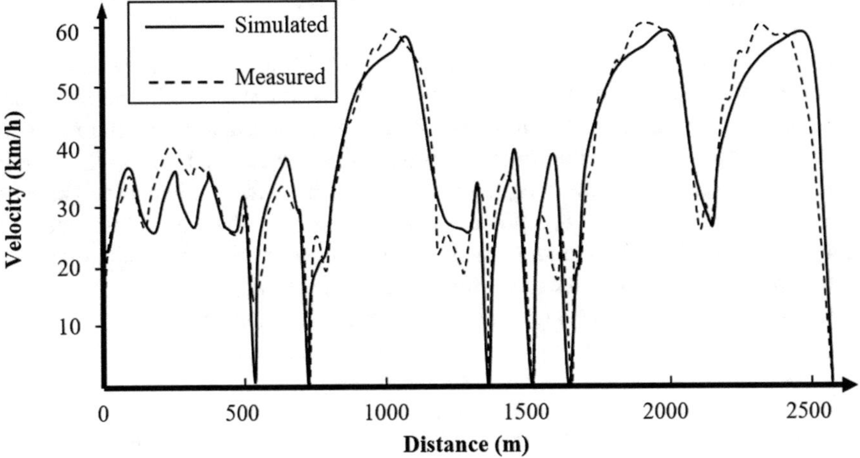

Fig. 1. Measured and simulated (optimized) driving style for an urban road. (with data from [3]).

2.6 Extended Concept for Eco-Driving

It has been illustrated in the previous section, that the driving style affects the environment in a direct and indirect ways. While the direct environmental impact of the driving style has to do with the level of emissions at the exhaust tube; indirectly, such driving style has been associated to road crashes and infrastructure damage that can induce higher levels of pollution, due to the resulting congestion when an accident occurs, and when an infrastructure must be repaired. Therefore, an extended concept for eco-driving, could consider not only the emissions at the exhaust tube, but also the level of safety when driving, as well as the level of unfriendliness against the infrastructure.

Exhaust tube emissions performance measure (ETI). Equation (1) describes the discrete fuel consumption ΔF during an interval of time equal to Δt, as a function of several experimental parameters and operating conditions, as reported in [22]:

$$\Delta F = \left(\alpha + \beta_1 R_T v + \left(\beta_2 M_v a_F^2 v/1000\right)_{a>0}\right)\Delta t \ \text{ for } \ R_T > 0 \tag{1}$$

$$\Delta F = \alpha \Delta t \ \text{ for } R_T \leq 0$$

where:
R_T: Total tractive force (kN) required to drive the vehicle,
M_v: Vehicle mass (kg),
v: Instantaneous speed (m/s)

a_F: Instantaneous acceleration rate (m/s^2), negative for deceleration,

α: Constant idle fuel rate (mL/s), as an estimate to maintain engine operation,

β_1: The efficiency parameter which relates fuel consumed to the energy provided by the engine (mL/kJ), and

β_2: The efficiency parameter which relates fuel consumed during positive acceleration (mL/(kJ m/s^2).

The key variable in this equation is the total tractive force, R_T. Such tractive force depends on the dynamic situation of the vehicle, that is, if it moves whether at constant speed, or accelerating or decelerating. When moving at constant speed, the necessary traction force must balance the rolling resistance force, F_r, in addition to the aerodynamic drag force F_d, and the longitudinal component of the vehicle's weight when travelling at positive slopes, F_w. When speeding - up or – down at known values of acceleration - deceleration, the total traction force will correspond to the inertial force, derived from the product of the mass times the acceleration.

At constant speed, the general equation for R_T, thus results as follows:

$$R_T = F_r + F_d + F_w \tag{2}$$

where the rolling resistance force F_r is given by the following expression [23]:

$$F_r = C_r(c_2V + c_3)(W)(1/1000) \tag{3}$$

where V is the vehicle speed (km/h); W is the instantaneous weight of the vehicle (kN); and C_r is the rolling resistance coefficient. The value of C_r depends on the pavement condition; and those for c_2 and c_3, on the type of tire.

F_d will depend on the aerodynamic coefficient C_D, the front area of the vehicle A_f (m^2), the density of the air (ρ, kg/m^3), and on the vehicle speed, V_v (m/s), as follows [24]:

$$F_v = C_D A_p \frac{\rho}{2} V_v^2 \tag{4}$$

Finally, the road slope-derived force, is as follows:

$$F_W = W \sin\lambda \tag{5}$$

where λ is the instantaneous road grade.

When accelerating/decelerating at a_F level, the tangential force F_L is described by the following equation:

$$F_L = M_v a_F \tag{6}$$

For accelerations, $R_T = F_L$. The emissions per kg of consumed fuel have been reported as 2.6 kg [2].

Pavement damage potentials – performance indicator. This formulation derives from what has been proposed by Romero et al. [6]. According to such formulation, the normal and shearing strain energies, which are transiently stored within a discrete pavement model, correlates with the pavement damage potentials of the vehicles. The pavement model considered for comparative evaluations is presented in Fig. 2. According to this approach, the vehicle tire is on a single pavement element at any given time.

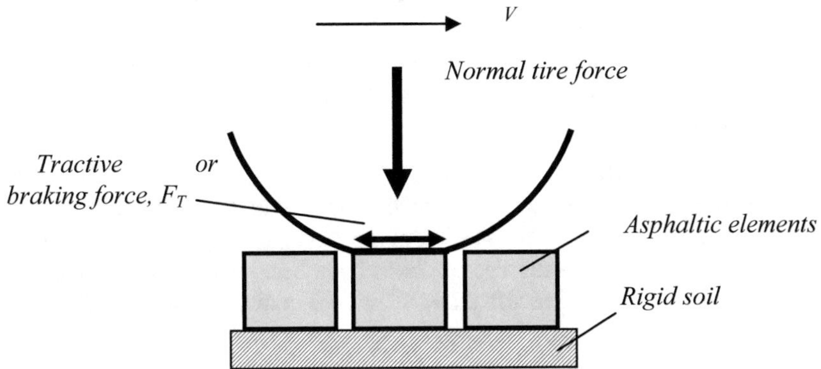

Fig. 2. Discrete pavement model to assess road damage potentials of vehicles.

The normal strain U_N and shearing strain U_T energies, transiently stored within the single elements, are given by the following equations [6]:

$$U_N = \frac{1}{2E_D}\left(\frac{W_t}{A_t}\right)^2 V_t \quad U_T = \frac{(1+\nu)}{E_D}\left(\frac{R_t}{A_t}\right)^2 V \tag{7}$$

where W_t is the instantaneous tire load on the pavement; V_t is the discrete pavement elements volume; E_D is the dynamic elastic module of the pavement elements, as a function of the loading time, pavement temperature (T), asphalt binder softening temperature T_s, and void content of the asphalt mix (VC). A_t is the contact patch area; and ν is the material's Poisson ration. The total strain energy stored within the pavement, U_T, is thus given by:

$$U_T = U_N + U_T \tag{8}$$

U_T is calculated in terms of the asphalt element height L_e, and A_t.

Road safety performance indicator. On the basis of the GPS data on which the performance is proposed to be assessed in this paper, three performance measures are put together to assess the safe driving performance of a transport.

Speed for current road segment. This is a critical variable for assessing the vehicle safety performance, as both its absolute value and its variations influence the road safety, in different circumstances. The relationship between the severity of the crashes and speed has been expressed in terms of the following equation [25]:

$$LO_2 = LO_1 \left(\frac{v_2}{v_1}\right)^2 \tag{9}$$

where LO_i refers to the number of injuries under an i speed circumstance, and v_i represents the speed at such different circumstances. The sensitivity of this equation to relatively small changes in speed is further considered [25], as the second-degree power in (9) can be adjusted between 1.1 to 4.1, as a function of the severity of the crash and of the type of road.

Accelerations/decelerations. Dispersion of the speed is a variable directly connected to the smoothness of the driving. While such variations of speed can reflect a driving style, they can also indicate the disposition of the driver to carry out the driving-related activities. In this respect, greater decelerations have been associated to driver fatigue [26]. The respective factor to assess the level of acceleration can thus be described by the following equation:

$$F_i = \gamma\, d_i + \phi\, a_i \tag{10}$$

where γ and φ represent sensitivity factors for the deceleration (d_i) and acceleration (a_i), respectively.

Hybrid speed – change of direction safety indicator. The travelling speed while negotiating a turn plays an important role for road safety. Consequently, both, the latitude- and the longitude- change rates should be combined with the speed, to create a performance measure that reflects the level of risk associated to the different turning maneuvers. Furthermore, any acceleration or deceleration during turning can also indicate an unsafe driving practice. The resulting equation in this case is the following:

$$R_i = v_i^2 \left(\frac{|\dot{Y}_i| + |\dot{X}_i|}{\left[1 + |\dot{Y}_i|\right]\left[1 + |\dot{X}_i|\right]}\right) \kappa |a_i| \tag{11}$$

where X_i and Y_i are the longitude and latitude coordinates of the driving at instant i. It should be noted that a_i represents the instantaneous acceleration and κ is a constant that depends on the type of vehicle carrying out the turning maneuver. For example, the most critical turning-braking situation would involve an articulated, partially loaded, road-tanker. In this context, the information about the type of vehicle, to define κ, would be available for setting the GPS analysis system. Combining the individual measures defined above, the following safety performance measure is obtained:

$$O_i = v_i^2 \left(\frac{\dot{Y}_i + \dot{X}_i}{\left[1 + \dot{Y}_i\right]\left[1 + \dot{X}_i\right]}\right) \kappa |a_i| + LO_1 \left(\frac{v_2}{v_1}\right)^2 + \gamma\, d_i + \phi\, a_i \tag{12}$$

3 Results and Discussion

The simulation of the different performance measures described above, involved several computer programming stages, from the reading of the raw GPS data, to the visualization of the different calculated performance measures. Two programs were written in *C* language, respectively for reading the GPS data and for calculating the performance measures, while the visualization of the outputs was prepared in Matlab. The input data for the calculation of the different performance measures includes the altitude, longitude, latitude coordinates, and the speed. Tables 1, 2 and 3 list the values for the different parameters involved in the three performance measures considered. Figure 3(a) illustrates the path followed by the vehicle in latitude – longitude coordinates, and part (b) of this figure, illustrates the altitude coordinate. The path corresponds to a trip from Berlin to Hanover, in Germany.

Table 1. Road safety parameters.

Parameter	Value	Parameter	Value	Parameter	Value
γ deceleration sensitivity	8	Injuries at speed 1 LO_1	13	κ vehicle stability factor	0.4
ϕ deceleration sensitivity	10.55	Speed 1, v_1	19 m/s		

Table 2. Pavement damage parameters.

Parameter	Value	Parameter	Value	Parameter	Value
Pavement temperature, T	30° C	Asphalt mix void content, VC	30%	Height of the pavement tiles, L_t	0.065 m
Pavement temperature, T	70° C	Area of the contact patch, A_t	0.093025 m^2	Normal tire force, W_t	100000 N

Table 3. Fuel consumption parameters.

Parameter	Value	Parameter	Value	Parameter	Value
Rolling resistance, C_r	2.25	c_2, c_3 for radial tires	0.0328, 4.575	Vehicle mass, M_v	10193 kg
β_1 efficiency parameter	0.08 mL/kJ	β_2 efficiency parameter	0.02 (mL/(kJm/s^2))	α idle fuel consumption	0.375 mL/s

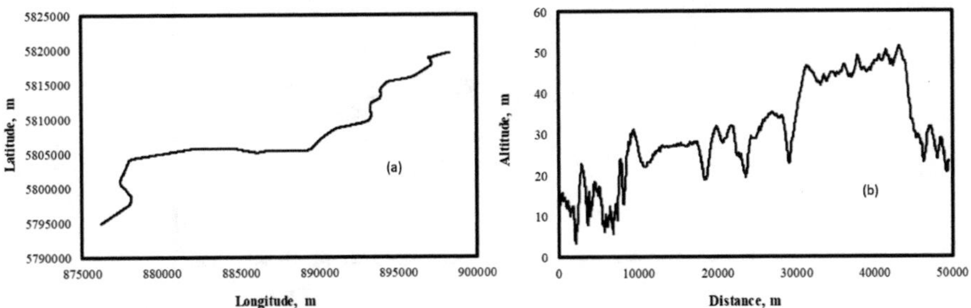

Fig. 3. Longitude – latitude coordinates of the travelling, and altitude profile.

The results are presented for four different speed profiles, corresponding to the nominal speed profile obtained from the GPS device, and three others, representing 85%, 115% and 130% of such measured speed profile (100%). Such variations in the speed profile thus implied different levels of accelerations. The effect of the different resulting accelerations, associated to the driver style, will be clarified when analyzing the global results. Out from the 50 km of the analyzed trip, only 2 km are taken into account to illustrate the effect of the speed and acceleration on the different performance measures. Figure 4 illustrates the speed profile for such road segment, while the corresponding accelerations are presented in Fig. 5. The combination of the different speeds and the resulting accelerations describes what is considered here as the driving style.

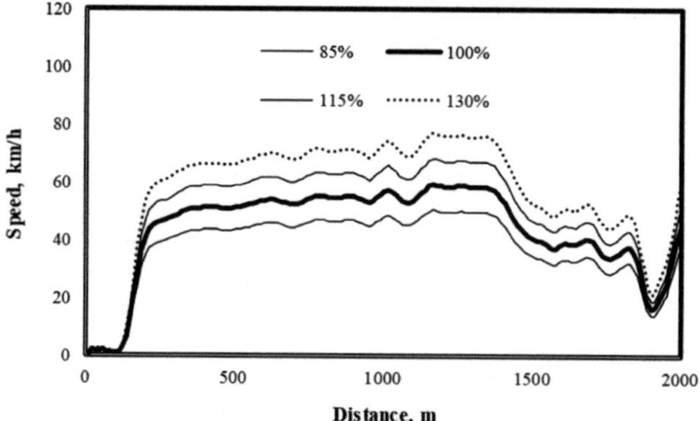

Fig. 4. Speed time history for 2 km length.

Pavement damage potentials. Figures 6, 7 and 8 illustrates the results for the normal, shear and total strain energies, respectively, for the road segment considered. The total strain energy is the quantity that in this paper is assumed as related to the pavement damage potentials of the driving. According to these results, the shear and normal strain energy do not correlate with each other. At the beginning of the trip, the normal stress energy decreases with the increase of the speed, and at the same time the shear stress increases as a result of the increasing speed, as speeding-up involves greater traction forces as a result of the increased rolling and air drag resistances.

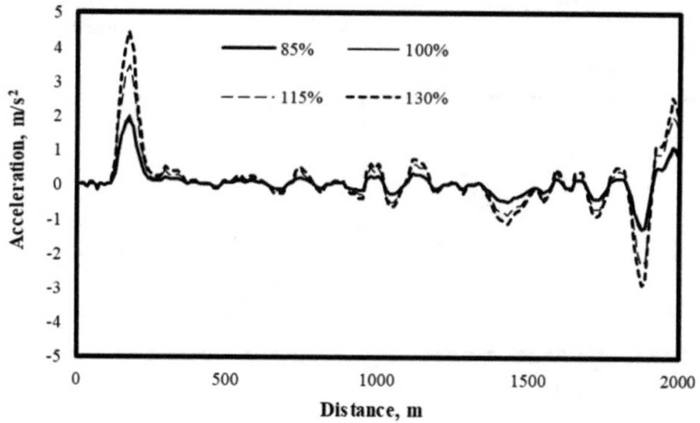

Fig. 5. Acceleration profiles.

It should be noted that the peak values of both energies, are of the same order of magnitude. Also, that the 85% speed is very close to the 100% speed profile.

CO^2 (CO2) emissions performance. Figure 9 illustrates the estimated CO2 emissions for the here different speeds considered. According to these results, the maximum fuel consumption correlates with the increase of the speed, and this performance is highly sensitive to the speed. As in the pavement damage performance measure, the 85% speed profile represents a very close behavior with the 100% speed profile.

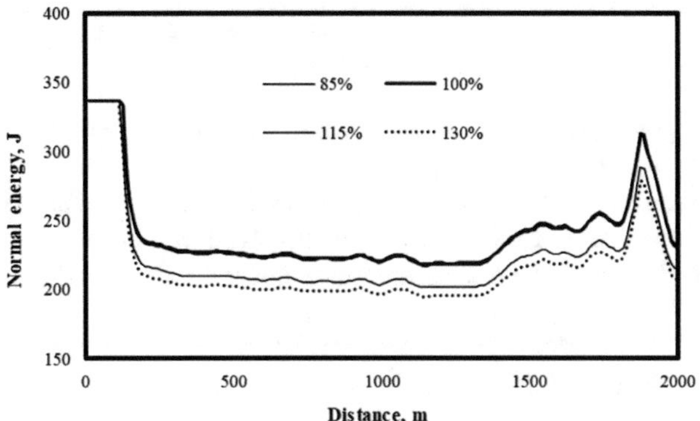

Fig. 6. Normal stress pavement stored energy.

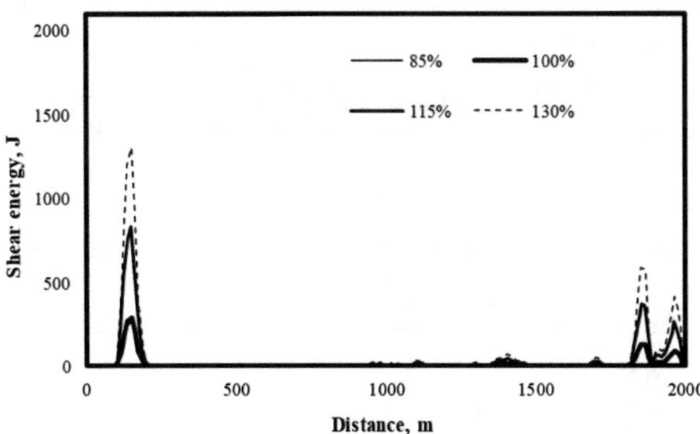

Fig. 7. Shearing stress pavement stored energy.

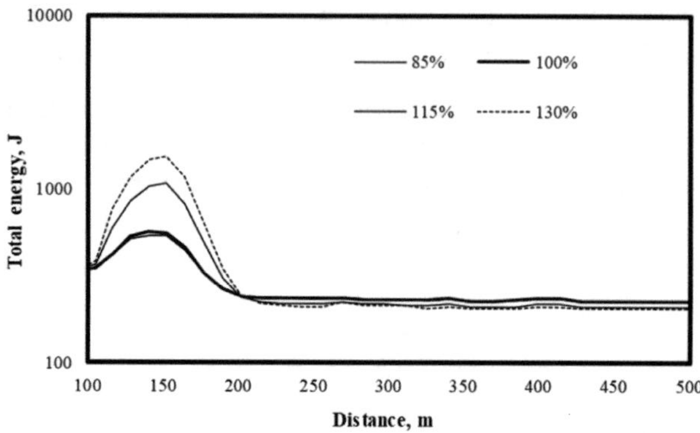

Fig. 8. Total (shear- plus normal- stress energies), indicating the road damage potentials.

Safety performance. Figure 10 illustrates the results for the road safety performance, indicating the propensity of the vehicle to be involved in a crash. According to these results, the speed increase represents the more unsafe driving condition.

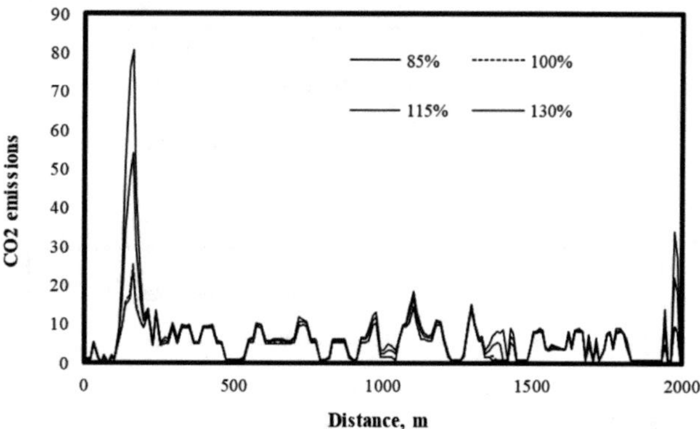

Fig. 9. CO2 emissions for three different levels of speed profile.

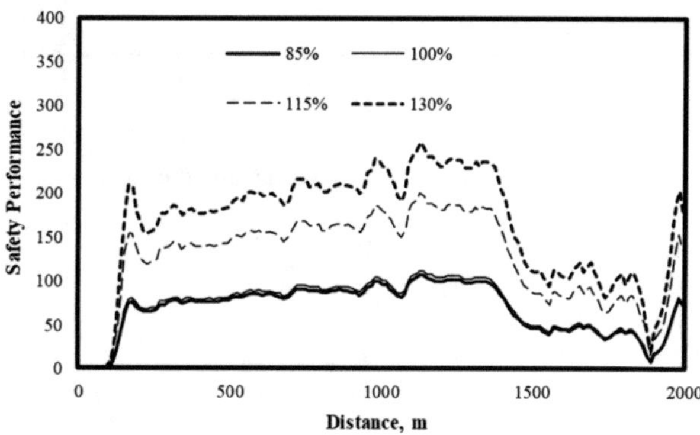

Fig. 10. Road safety performance as a function of the speed profile.

Aggregated values for the different performance's measures. Figure 11 illustrates the accumulated effect for the different performance measures proposed in this paper, normalized to the minimum values of such performance measures. According to these outputs, there are different trends. On the one hand, the shear- and normal- strain energy, exhibit an opposite effect when increasing the traveling speed, with the normal strain energy diminishing with an increase speed. However, the overall effect of this performance measure, in terms of the summation of energies, indicates that an increase in the speed corresponds to an increase in the pavement damaging potentials.

As far as the emissions performance is concerned, the results in this figure suggest a nonlinear effect. That is, an increase of 52% in the traveling speed (1.3/0.85), represents an increase of 100% in the emissions. This is associated to the increase in the traction force, resulting from greater rolling and drag resistances.

The safety performance of the driving in this figure, suggests that the unsafe condition increases also nonlinearly with speed. The standard deviation of the acceleration (part (f) of this figure), characterizing the driving style, also is included in this figure, revealing a harsher driving style with the increase of speed.

To identify the more influential factor for the different performance measures, the ratio of the maximum to minimum values of the different performance measures and factors, are presented in Fig. 12. These results indicate that the closer values of such different ratios are found in the case of dispersion of the acceleration, and the performance measures associated to the road safety performance and the emissions. That is, the acceleration dispersion increase is 2.33, while the increase in safety and CO2 emissions, are 2.4 and 2.05, respectively, while such ratio in the case of the speed, is only 1.52. The effect on the pavement damage is weakly correlated with the increase in acceleration dispersion, as the increase of pavement damage is only of 12%.

The driving style, measured in terms of the acceleration variation, is thus the most influential factor for the determination of the driving safety, infrastructure friendliness, and environmental effects.

Visualization. Figure 13 includes a screenshot of the outputs' visualization program. It describes the path of the vehicle along the trace of the road. In this case, the results correspond to a trip from Berlin to Hanover. The output variables include some of the input data for the program, that is, the height over sea level, and the speed.

It is considered that this type of output, exhibited to the driver in real time, or to the vehicle administrator as a postprocessing tool, could contribute to analyze the overall performance of the driver.

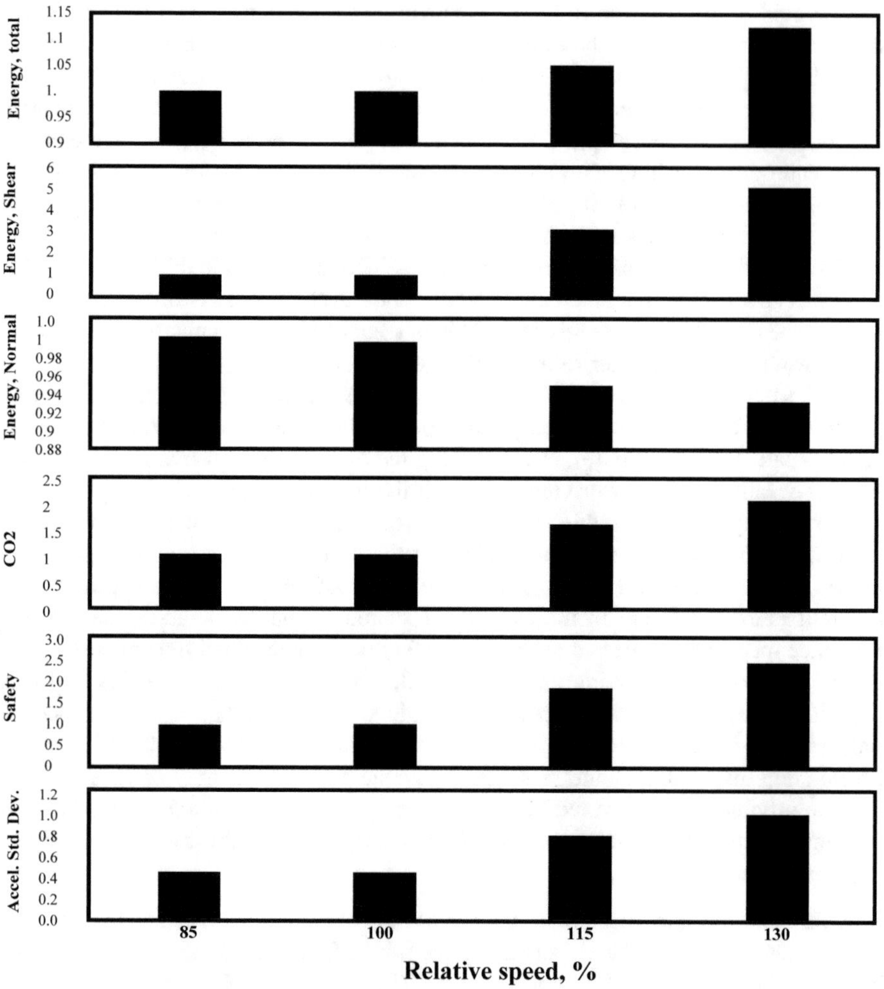

Fig. 11. Aggregation of the different performance measures.

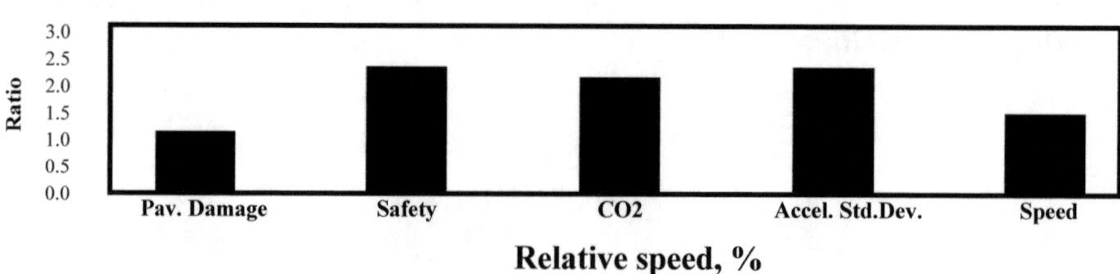

Fig. 12. Max/min values for the different performance measures.

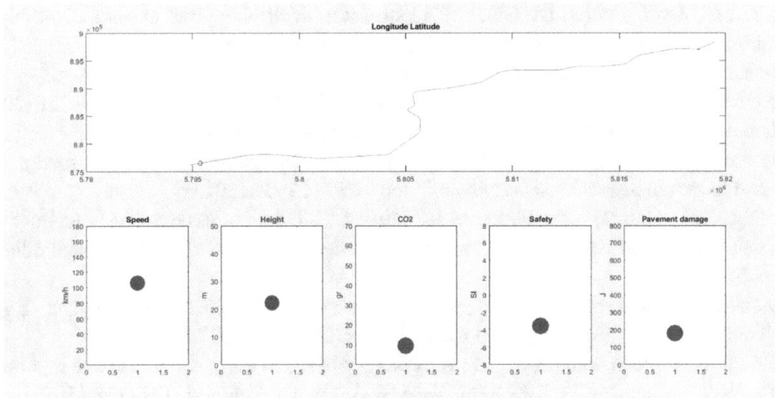

Fig. 13. Display of the visualization program.

4 Conclusions

The driving style has been recognized in the literature, as the most important factor contributing to different driving outputs, including pollution and safety. However, the influence of driving style on the pavement damage potentials had not been explicitly reported in the literature.

In this paper, several physical models have been put together to assess three driving outputs that have to do with the transport safety, pavement damage and emissions. The pavement damage was introduced, as the reparation of a damaged infrastructure implies road work that affects the transport efficiency and emissions. On the other hand, the road safety has been remarked as a prominent element that should be considered when considering the sustainability of the transportation, as road crashes involve traffic affectations that derive in congestion.

The results suggest that the driving style, measured in terms of the dispersion of the acceleration, negatively affects the three performance measures considered. The systematization of the analysis, based on the software proposed in this paper, could contribute to integrally improve the sustainability of the transportation.

References

1. Ho, S.-H., Wong, Y.-D., Chang, V.-W.: What can eco-driving do for sustainable road transport? Perspectives from a city (Singapore) eco-driving programme. Sustain. Cities Soc. **14**, 82–88 (2015)
2. Ayyildiz, K., Cavallaro, F., Nocera, S., Willenbrock, R.: Reducing fuel consumption and carbon emissions through eco-drive training. Transp. Res. Part F **46**, 96–110 (2017)
3. Javanmardi, S., Bideaux, E., Trégouët, J.F., Trigui, R., Tattegrain, H., Bourles, E.N.: Driving style modelling for eco-driving applications. IFAC Pap. Online **50–1**, 13866–13871 (2017)
4. Ma, H., Xie, H., Huang, D., Xiong, S.: Effects of driving style on the fuel consumption of city buses under different road conditions and vehicle masses. Transp. Res. Part D **41**, 205–216 (2015)

5. Li, G., Li, Sh-E, Cheng, B., Green, P.: Estimation of driving style in naturalistic highway traffic using maneuver transition probabilities. Transp. Res. Part C **74**, 113–125 (2017)
6. Romero, J.A., Lozano-Guzmán, A.A., Betanzo-Quezada, E., Obregón-Biosca, A.A.: A flexible pavement damage metric for a straight truck. Int. J. Heavy Veh. Syst. **20**(3), 209–221 (2013)
7. Chapman, L.: Transport and climate change: a review. J. Transp. Geogr. **15**, 354–367 (2007)
8. Bigazzi, A.Y., Figliozzi, M.A.: Marginal costs of freeway congestion with on-road pollution exposure externality. Transport research part A **57**, 12–24 (2013)
9. Gately, C.K., Hutzra, L.R., Peterson, S., Wing, I.S.: Urban emissions hotspots: quantifying vehicle congestion and air pollution using mobile phone GPS data. Environ. Pollut. **229**, 496–504 (2017)
10. Passidis, I.: Congestion by accident? A two-way relationship for highways in England. J. Transp. Geogr. (2017). https://doi.org/10.1016/j.jtrangeo.2017.10.006
11. Pulugurtha, S.A., Mahanthe, S.S.B.: Assessing spatial and temporal effects due to a crash on a freeway through traffic simulation. Case Stud. Transp. Policy **4**, 122–132 (2016)
12. Dingus, T.A., Hankey, J.M., et al.: Naturalistic driving study: technical coordination and quality control. SHRP 2 Report S2-S06-RW-1. The National Academies Press, Washington (2015)
13. Sager, L.: Estimating the effect of air pollution on road safety using atmospheric temperature inversions. Working paper No. 251, Grantham Research Institute on Climate Change and the Environment, October 2016
14. Joo, S., Oh, C., Lee, S., Lee, G.: Assessing the impact of traffic crashes on near freeway air quality. Transp. Res. Part D **57**, 64–73 (2017)
15. AAA: Aggressive driving, Foundation for Traffic Safety (2017). www.aaafoundation.org/aggressive-driving
16. Berry, I.M.: The effects of driving style and vehicle performance on the real-world fuel consumption of U.S. light-duty vehicles. Thesis, Massachusetts Institute of Technology (2007)
17. Sagberg, F., Piccinini, F.B., Engström, J.: A review of research on driving styles and road safety. Hum. Factors **57**(7), 1248–1275 (2015)
18. Kim, S., Song, T.-J., Rouphail, N.M., Aghdashi, S., Amaro, A., Gonçalves, G.: Exploring the association of rear-end crash propensity and micro-scale driver behavior. Saf. Sci. **89**, 45–54 (2016)
19. Vaiana, R., Luele, T., Astarita, V., Caruso, M.V., Tassitani, A., Zaffino, C., Giofré, V.P.: Driving behavior and traffic safety: an acceleration-based safety evaluation procedure for smartphones. Mod. Appl. Sci. **8**, 88–96 (2014)
20. Jones, W.: High performance intersections – a case study. Asphalt 02/11/11 (2011)
21. Kimura, T.: Studies on stress distribution in pavements subjected to surface shear forces. Proc. Jpn. Acad. Ser. B Phys. Biol. Sci. **90**(2), 47–55 (2014)
22. Akçelik, R., Besley, M.: Operating cost, fuel consumption, and emission models in aaSIDRA and aaMOTION. In: Proceedings, 25th Conference of Australian Institutes of Transport Research (CAITR2003), University of South Australia, Adelaide, Australia, 3–5 December 2003 (2008)
23. Rakha, H., Lucic, I., Demarchi, S., Van Aerde, M., Setti, J.: Vehicle kinematics model for predicting maximum truck acceleration levels. In: Transportation Research Board 2001 Annual Meeting, January, Washington (2001)
24. Shames, I.H.: Fluid Mechanics. Mc-Graw Hill, New York (1967)
25. SWOV: SWOV Fact sheet. The relation between speed and crashes. Institute for road safety research, Leidschendam, the Netherlands, April 2012
26. Blockey, P.N., Hartley, L.R.: Aberrant driving behaviour: errors and violations. J. Ergon. **38** (9), 1759–1771 (1995)

Secure Information Interaction Within a Group of Unmanned Aerial Vehicles Based on Economic Approach

Iuliia Kim(✉) and Ilya Viksnin

ITMO University, Lomonosov Street 9, 191002 Saint Petersburg, Russia
yulia1344@gmail.com

Abstract. Groups of unmanned aerial vehicles (UAVs) are considered in the context of territory monitoring. Preference is given to decentralized collective strategy of group management due to opportunity of providing consensual agent interaction with the help of common communication channel presence. For the correct group functioning there is a necessity in providing security of information transfer via communication channels. The emphasis is put on providing pragmatic information integrity, and to avoid violation occurrence in this category, method, based on credit theory was developed. The method implies regulation of information volume transferred by agents through establishment of fixed amount for information conventional units per time discrete (installment plan). In case of data retention by agent, subsequently its payment value is reduced, thus, its indebtedness is increased. At the end of installment plan period agent's level of trust and reputation is calculated. During introduction into the group of a new agent the credit is determined, according to which the new agent gets from other group members not full information but information reduced by the established interest rate. However, this agent must transmit data in accordance with predetermined installment plan conditions. At the end of credit period the decision whether the new agent is accepted or blocked is made. To assess the proposed method, the interaction in UAV group was modeled. Saboteur and correct agent were introduced into the group. Series of independent tests were conducted, in the 90,2% of them the saboteur was blocked. Calculated F-measure of saboteur recognition was equal to 0,94.

Keywords: Unmanned aerial vehicle · Credit theory · Pragmatic information integrity

1 Introduction

In the recent time multi-agent systems obtained a widespread recognition due to their resilience in comparison with centralized implementations. Such systems provide users with stronger fault tolerance, higher interaction speed and agent relative self-independence. There is a significant amount of ideas and already

K. Arai et al. (Eds.): CompCom 2019, AISC 997, pp. 59–72, 2019.
https://doi.org/10.1007/978-3-030-22871-2_5

existing projects dedicated to ways of multi-agent system implementation in the society life. In the work [1] the contribution of multi-agent approach providing intelligence in the distributed smart grids is discussed. The multi-agent prototype was proposed for production control in water fabs [2]. Relatively to recent projects, the demand side management strategy was offered for optimal energy distribution in smart houses [3]. In addition, self-organized multi-agent system consisted of various agent types and provided with possibility of feedback and coordination was suggested for smart factories [4]. In this article multi-agent systems are considered in the context of closed and open space monitoring by group of unmanned aerial vehicles (UAVs). The term "monitoring" is considered in this work as regular examination for nonstandard changes.

2 Related Work and Theoretical Background

2.1 Robotic Group Management

The majority of existing projects dedicated to management systems implied the presence of human factor, owing to which autonomous functioning in such groups was absent. Robotic system self-organization and agent group management is explained in the work [5]. The classification of group management strategies is presented in Fig. 1.

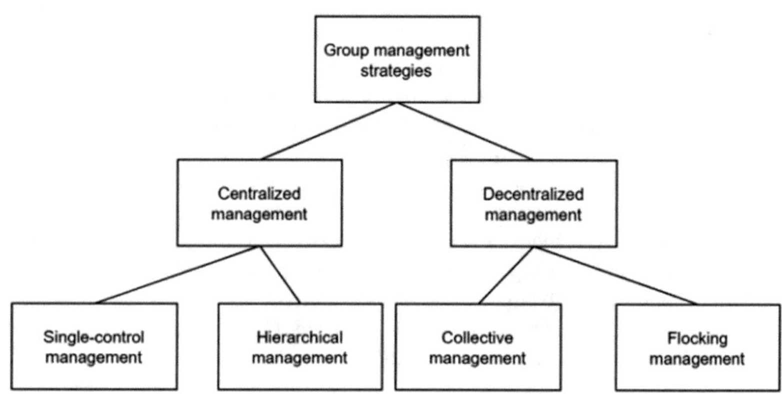

Fig. 1. Classification of group management strategies

Group management strategies can be divided into two types: centralized and decentralized. In its turn, centralized management can be single-control and hierarchical. In case of centralized single-control management, group has central control unit (commander), which plans and inspects actions of all the group members (agents). Centralized hierarchical management obeys the concept that one commander controls the determined group, whose members, in their turn, control their own subgroups.

The advantage of centralized management is its realization simplicity. However, implementation of centralized management is accompanied with a risk of system crashing due to weak fault tolerance: to destroy the functionality of all the system or its significant part it is enough to impact negatively only on the central unit.

Decentralized management strategy is divided into two types: collective and flocking. In case of collective decentralized management group, agents have a common channel for information exchange. In flocking groups members do not have communication channel and make decisions based on indirect information about environmental changes caused by actions of other agents.

The preference is given to collective management strategy, as presence of common channel provides communication among UAVs in order to find the optimal algorithm for achieving the stated group purposes. The structure of collective management consists in the following: in the group of n members each agent R_i $(i \in [1, n])$ possesses its own management system C_i, which is responsible for actions of this particular agent. These systems are united by common communication channel. Information about actions, chosen by C_i is transmitted to other C_j, $(j \in [1, n], i \neq j)$, and based on obtained data, the agents optimize their performance [5]. The strategy of collective decentralized management is depicted in Fig. 2.

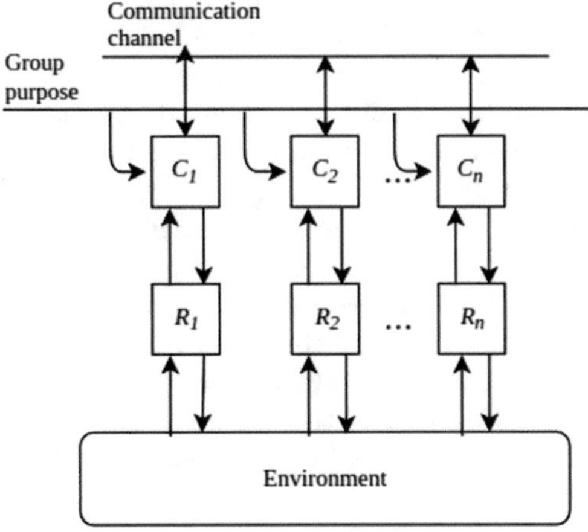

Fig. 2. Collective group management strategy

2.2 UAV Implementation

UAVs give a possibility to cover large territories in a short period of time by small group, and they have already been integrated in such processes as controlling sowing complexes, cartography [6], etc. Wide area of visibility, high speed of UAV make possible to quickly monitor, predict, prevent emergency situations and prospectively to avoid or reduce fatal consequences and significant material and human losses. In the work [7] use of UAV group for forest restoration was described. In such relevant circumstances it is vital to guarantee correct and effective group functioning. Primarily, it is necessary to provide security of information transmitted through the group common communication channel.

2.3 Information Security Provision

In this article the emphasis is made on providing information integrity of the transmitted data. The message model has three basic constituents: syntax (rules of encoding, decoding, interpretation), semantics (message content), pragmatics (message usefulness, knowledge). From these instances it is possible to conclude that syntactic integrity – property of the information presentation form, semantic integrity – property of message content characteristics, pragmatic integrity – property of information utility in the context of environment and own recipient state.

The mechanisms providing information security in multi-agent robotic systems are divided into two fundamental types: mechanisms providing hard security and soft security. Communication channel encryption with public key, use of mobile cryptography, agent authorization can be referenced to the first mechanism type. A demonstrative example of soft security mechanisms is trust and reputation model for objects [8]. However, one of the vulnerabilities of this model is situation, when saboteurs make up half or more of the group: in this case saboteurs can give each other high trust points discrediting other agents. The measurement of agent reputation during the whole interaction time was implemented as a solution method.

The enumerated methods are oriented, basically, to the semantic integrity preservation - content constituent [9,10]. It should be considered that in order not to undermine their trust level, saboteurs can transmit correct data but not in a full size. Due to such actions, the group does not possess enough information for task performing. In such case pragmatic integrity violation happens [11]. In this work the most attention is paid to pragmatic information integrity, which is category consisted in reliability and completeness of data that serves as basis for information message.

3 Research Task Statement

There is a group consisted of N UAVs: $R = \{r_1, r_2, ..., r_N\}$. Each agent possesses the aggregation of M properties: $S_i = \{r_i | s^i_j, j \in [1, M]\}$, $i \in [1, N]$. As an assumption, UAV homogeneity is considered: $\forall r_i$, $r_i \in R$, $i \neq j$, $i, j \in [1, N]$: $S_i = S_j$. It is supposed that hard security mechanisms are realized and perform correctly.

Agents interact with each other during the period T. The given group functions in the territory F, which consists of f equal sectors. The UAVs check these determined regions during equal intervals (discretes) of time. At the end of check process the agents exchange with the collected data. Information I_i, which is possessed by each ith UAV at the expiration of discrete $[t_{k-1}, t_k] \in T$, before the data exchange process, is divided into own and acquired. Own information is data about agent's own technical state at the current moment and technical state of the whole group for the previous time discrete. Acquired information is information collected by i^{th} agent about environment for the period $[t_{k-1}, t_k] \in T$. This relation is represented in (1)–(3):

$$I_{own_i} = I_{cts_{[t_{k-1}, t_k]i}} \cup \left(\cup_{j=1}^{N} I_{cts_{[t_{k-2}, t_{k-1}]j}} \right), \tag{1}$$

where I_{own_i} is i^{th} agent's own information, $I_{cts_{[t_{k-1}, t_k]i}}$ – i^{th} agent's information about its technical state for the current moment, $I_{cts_{[t_{k-2}, t_{k-1}]j}}$ – j^{th} agent's information about its technical state for the previous time discrete.

$$I_{acquired_i} = I_{es_{[t_{k-1}, t_k]i}}, \tag{2}$$

where $I_{acquired_i}$ is i^{th} agent's own information, $I_{es_{[t_{k-1}, t_k]i}}$ – i^{th} agent's data about current environment state.

$$I_i = I_{own_i} \cup I_{acquired_i} \tag{3}$$

After the group data exchange, each agent has in the acquired information composition data about the current technical state of other group members and information about environment state collected by the rest of the group. Data structure possessed by agents after the information exchange with area check results for the period $[t_{k-1}, t_k] \in T$ is illustrated in (4)–(6):

$$I_{own_i} = I_{cts_{[t_{k-1}, t_k]i}} \cup \left(\cup_{j=1}^{N} I_{cts_{[t_{k-2}, t_{k-1}]j}} \right) \tag{4}$$

$$I_{acquired_i} = I_{es_{[t_{k-1}, t_k]i}} \cup \left(\cup_{j=1}^{N, i \neq j} \left(I_{es_{[t_{k-1}, t_k]j}} \cup I_{cts_{[t_{k-1}, t_k]j}} \right) \right) \tag{5}$$

$$I_i = I_{own_i} \cup I_{acquired_i} \tag{6}$$

Thus, information obtained by joinder of data transmitted by all the agents, has to tend to the ability of full group and environment state reflection. This statement is expressed in (7) and (8):

$$I = \cup_{i \in [1,N]} I_i \tag{7}$$

$$I \to U, \tag{8}$$

where U is full information about group and environment for the current moment.

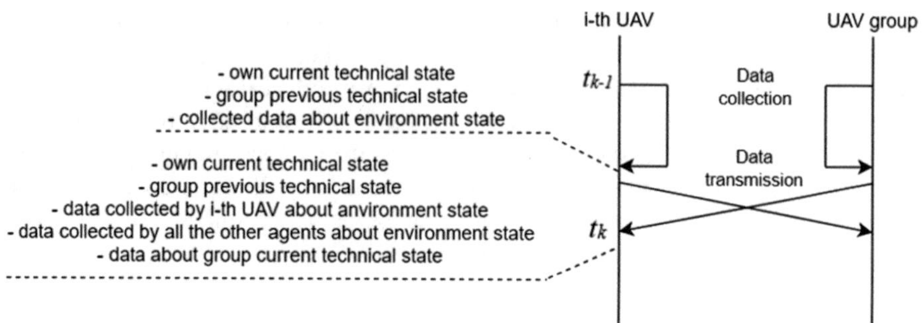

Fig. 3. Group information interaction scheme in the single time discrete relatively to the ith agent in assumption of destructive information influence absence

The scheme of UAV information possession during a single time discrete is represented in Fig. 3:

However, saboteurs pursuing the purpose of work sabotage can be present in the group. In order not to deteriorate its own reputation with spurious data transmission, the saboteur can retain information while getting information from other agents in full. In such way, the saboteur hinders from making a complete view about group and environment and can slow down work productivity of all the group. The saboteurs can retain information from other agents about:

1. their own technical state in case of insignificant malfunctions;
2. environment state;
3. both their own technical state and environment state.

Situations 1 and 2 at the end moment of the group data exchange for the discrete $[t_{k-1}, t_k]$ are represented by (9) and (10), respectively:

$$I_{cts_d} \in \widetilde{I}_{acquired_i} < I_{cts_i} \in I_{acquired_d}, \tag{9}$$

where I_{cts_d} is information about saboteur's current technical state, $d \in [1, N]$; I_{cts_i} – information about current technical state of the ith correct agent, $i \neq d$; $\widetilde{I}_{acquired_i}$ – acquired information of the ith correct agent in case of saboteurs' destructive information influence.

$$I_{es_{[t_{k-1},t_k]_d}} \in \widetilde{I}_{acquired_i} < I_{es_{[t_{k-1},t_k]_i}} \in I_{acquired_d}, \tag{10}$$

where $I_{es_{[t_{k-1},t_k]_d}}$ is information about current environment state transmitted by saboteur, $d \in [1, N]$; $I_{es_{[t_{k-1},t_k]_i}}$ – data about current environment state transmitted by i^{th} correct agent, $i \neq d$.

The situation 3 is resumptive for cases enumerated before, and it is illustrated by (11):

$$\widetilde{I}_{acquired_i} < I_{acquired_d}, i \neq d \tag{11}$$

Thus, saboteur impact on other agents' formation of $I_{acquired}$. This is reflected in (12):

$$\widetilde{I}_{acquired_i} < I_d \backslash I_{own_d}, i \neq d \tag{12}$$

Figure 4 depicts situations with information retain respectively to the correct UAV in a single time interval, where data pieces marked in red color are at risk of being reduced by saboteurs.

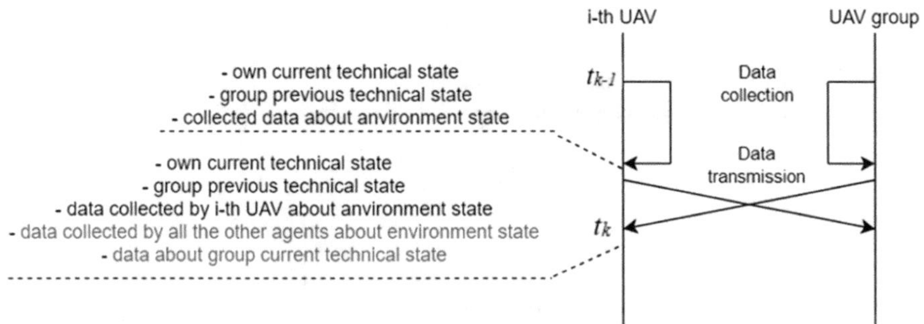

Fig. 4. Group information interaction scheme in the single time discrete relatively to the ith correct agent in assumption of destructive information influence presence

The research task consists in searching of effective methods allowing to minimize the risk of destructive information influence on the group from the saboteur side and in providing pragmatic integrity preservation of transmitted information: $\forall i \in [1, N]$ $\widetilde{I}_{acquired_i} = I_{acquired_d} + \epsilon, \epsilon \to 0, I \to U$.

4 Pragmatic Integrity Preservation of Transmitted Information, based on Economic Approach

4.1 Credit Theory Overview

To prevent information retention by agents it is proposed to integrate into the trust and reputation model a method based on credit theory. The term "credit" is understood as level of trust, which creditor expresses relatively to debtor [12, 13], and this term suitable for soft security mechanisms. Capital credit theory [14] affirms that crediting process contributes to the increase of population wealth level. In the context of information security, credit leads to the security raise of data transmitted within the group.

A debtor is given with a credit of D for the period PP (full credit term) with the annual interest rate p. At the end of the term debtor with considering of interest rate must pay the sum equal to θ. During the full credit term, after each time discrete U, debtor must make fixed annuity payment A, which consists of credit body and credit interest. Proportions of these two categories can be modified preserving fixed payment size. The calculation of annuity payment size is performed according to (13):

$$A = K \cdot D, \tag{13}$$

where K is annuity coefficient.

Annuity coefficient is represented in (14):

$$K = \frac{h \cdot (1+h)^U}{(1+h)^U - 1}, \tag{14}$$

where $h = p/12$ – monthly interest rate.

In this way, credit payout results can be expressed by (15):

$$\sum_{u=1}^{U} A_u = \theta \tag{15}$$

4.2 Credit Theory in the Context of Implementation in UAV Group

In the context of UAV group interaction it is suggested introducing terms:

1. installment plan;
2. credit.

1. Interaction time T is divided into U installment plan intervals of the determined length. For the installment plan interval $[t_{u-1}, t_u] \in T$ each agent has to transmit fixed information amount I. In its turn, interaction period $[t_{u-1}, t_u]$ is split into V equal time intervals, at the end of which all the agents must transmit information amount equal to $\frac{I}{V}$, as it is represented in (16):

$$\sum_{v=1}^{V} payment_v^u = I, U \in T, \tag{16}$$

where *payment* is information amount sent by agent in a single installment plan interval. In case if the UAV retains any information, it violates fixed payment, and a fine is imposed as a sanction, which reduces the value of this agent's future payments. Thus, the agent has indebtednesses. As installment plan period is over, debts of each agent are calculated, and agents with larger debt will have lower trust index and consequently - reputation index. Payment check is performed in the group by regularly randomly reelected responsible agent. Trust and reputation level of the agent in the group are represented with the dependencies (17) and (18), respectively:

$$TRUST_i = f\left(\sum_{v=1}^{V} payment_v^u\right)^+, U \in T \tag{17}$$

$$REPUTATION_i = f\left(\sum_{u=1}^{U}\sum_{v=1}^{V} payment_v^u\right)^{+}, U \in T \tag{18}$$

2. In case of integration of a new UAV at the moment t_0, it is necessary to perform its crediting relatively to transmitted information amount. Crediting process consists in the following steps:

- planning of interaction period T for the new UAV and the group;
- calculation of full credit term: $t < PP < T$ where t is durability of installment plan period;
- setting of initial size of the interest rate p;
- calculation of installment plan size *payment* for the period PP.

During the term PP the integrated UAV must transmit fixed information amount, which is declared in the group by the installment plan conditions. However, this new agent gets from other UAVs information about group and environment minus the given interest rate, i.e. during credit term for the time discrete $[t_{k-1}, t_k]$ after data exchange with other agents the jth integrated agent's acquired information is expressed by (19):

$$I_{acquired_j} = I_{es_{[t_{k-1},t_k]_j}} \cup (1-p) \cdot \left(\cup_{i=1}^{N,i\neq j}\left(I_{es_{[t_{k-1},t_k]_i}} \cup I_{cts_{[t_{k-1},t_k]_i}}\right)\right) \tag{19}$$

During interaction process the interest rate is reduced by a certain value, at the end of the full credit term the interest rate reaches zero. As the full credit term is finished, the checked UAV becomes a full member of the group, updates installment plan conditions, gets full information from other UAVs during monitoring and has a right to be elected as a responsible agent for payment checking.

The check of integrated agents is fulfilled by the enumerated procedures:

1. comparison of the stated credit size $payment_{must}$ and factual $payment_{got}$ at the end of PP;
2. comparison of information amount $payment_{must_{u_{[t_{v-1},t_v]}}}$, which have to be transmitted in a single time discrete within the framework of installment plan period, and factual $payment_{got_{u_{[t_{v-1},t_v]}}}$, $u \in [1, U]$, $[t_{v-1}, t_v] \in [t_{u-1}, t_u]$;
3. calculation of discrepancies of $payment_{must_{u_{[t_{v-1},t_v]}}}$ and $payment_{got_{u_{[t_{v-1},t_v]}}}$ at the end of PP.

The decision about blocking or accepting of the new UAV is made by the responsible agent. Temporal crediting process scheme is represented in Fig. 5.

The proposed method of pragmatic integrity preservation minimizes risk of destructive information influence on the group. It is not profitable for saboteurs to be integrated into the group with high interest rate, long credit term and short installment plan period: big amount of checks does not give a possibility to retain much information, and costs (energetic, temporal, informational) during

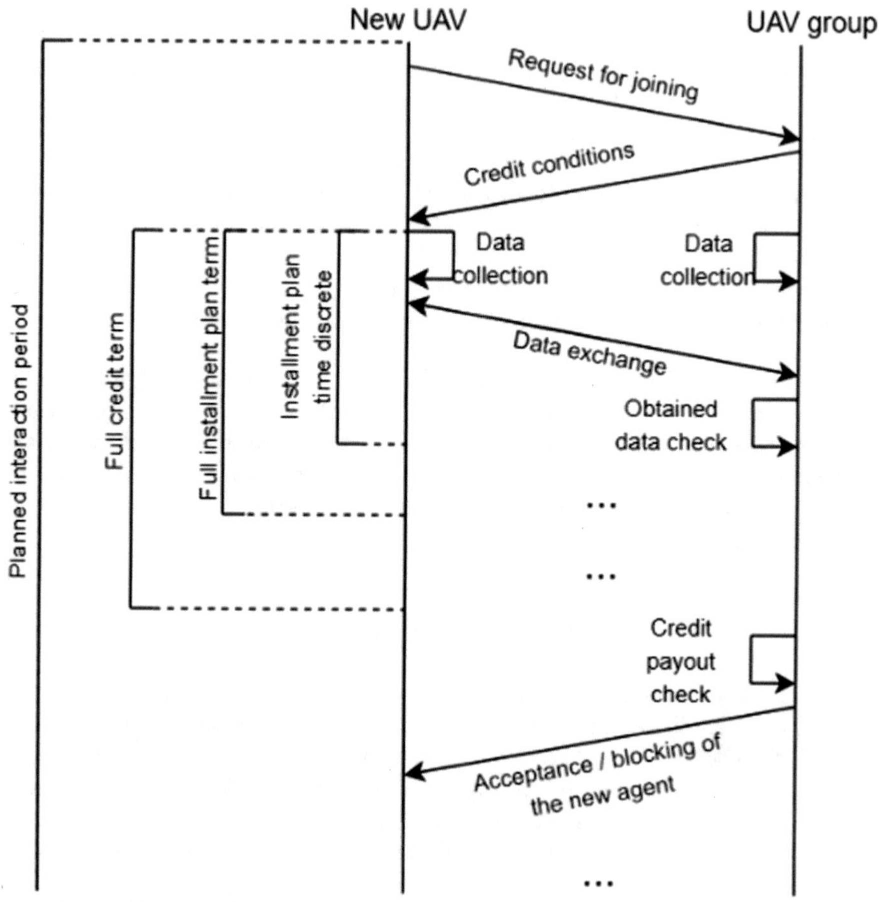

New UAV
UAV group
Request for joining
Credit conditions
Data collection
Data collection
Data exchange
Obtained data check
Credit payout check
Acceptance / blocking of the new agent
Planned interaction period
Full credit term
Full installment plan term
Installment plan time discrete

Fig. 5. Temporal crediting process scheme

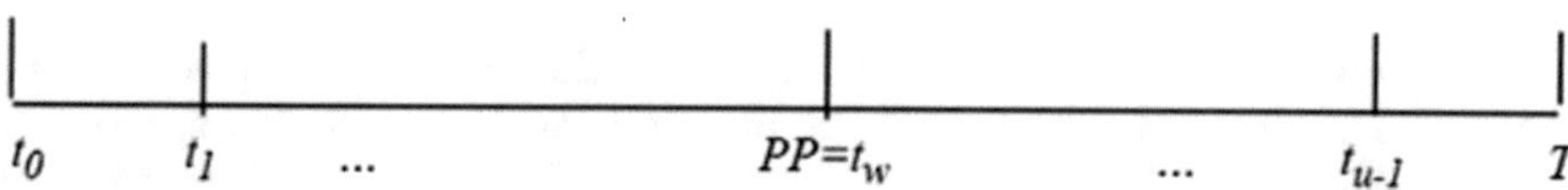

Fig. 6. Temporal interaction scale of group and integrated UAV

the credit exceed incomes, which can be acquired after credit repayment. The example of the described situation is depicted by Fig. 6 and (20):

$$I^{sent}_{[t_0,PP]} > I^{got}_{[PP,t_{w+1}]}, w \in [1, U-1], \tag{20}$$

where $I^{sent}_{[t_0,PP]}$ is information transmitted by the new UAV for the credit period, $I^{got}_{[PP,t_{w+1}]}$ – information received in the next installment plan discrete right after credit period is over.

5 Experiment

The proposed method of pragmatic information integrity preservation was tested on a specially programmed simulator for effectiveness in the context of risk of minimization of destructive information influence on pragmatic information constituent.

Initial parameters and work sequence of the experiment:

- group consisted of 10 UAVs;
- territory consisted of 20 equal sectors and monitored by the group;
- 1 conditional time unit, during which the group examines 1 sector;
- integration of 2 new UAVs: one of them is saboteur;
- full credit term is ten conditional time units;
- initial interest is 50%, at the end of each installment plan period the given interest rate is reduced by 5%;
- the marginal indebtedness size, with which a new agent can become a full group member is half from the stated credit sum;
- saboteurs depending on situations can either transmit the information in full or retain a part of it;
- correct UAVs can retain information due to technical malfunctions, malfunction emergence probability is considered equal to 10%.

Series of independent tests was conducted. Debt size distribution of saboteur and correct UAV is represented in Figs. 7 and 8, respectively:

Based on the obtained results precision, recall and F-measure were calculated according to (21)–(23), respectively:

$$precision = \frac{detected_{saboteur_{correct}}}{detected_{saboteur}} \approx 0,99, \tag{21}$$

where $detected_{saboteur_{correct}}$ is proportion of the total amount of tests, when saboteur was detected correctly, $detected_{saboteur}$ – general proportion of the total amount of tests, when saboteur presence was detected.

$$recall = \frac{detected_{saboteur_{correct}}}{presence_{saboteur}} \approx 0,90, \tag{22}$$

where $presence_{saboteur}$ is percentage of the total amount of tests, where saboteur is present.

$$F = 2 \cdot \frac{precision \cdot recall}{precision + recall} \approx 0,94 \tag{23}$$

The calculated statistical indicators reflect high ability of the developed method to disclose saboteurs during credit period by analyzing amount of transmitted information by the new UAV.

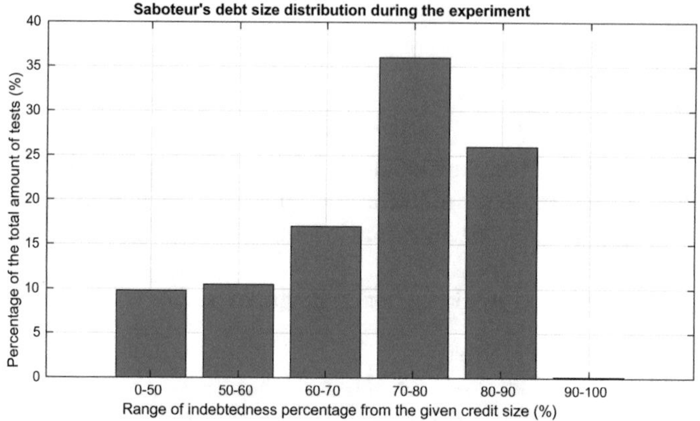

Fig. 7. Histogram of saboteur's indebtedness size during the experiment

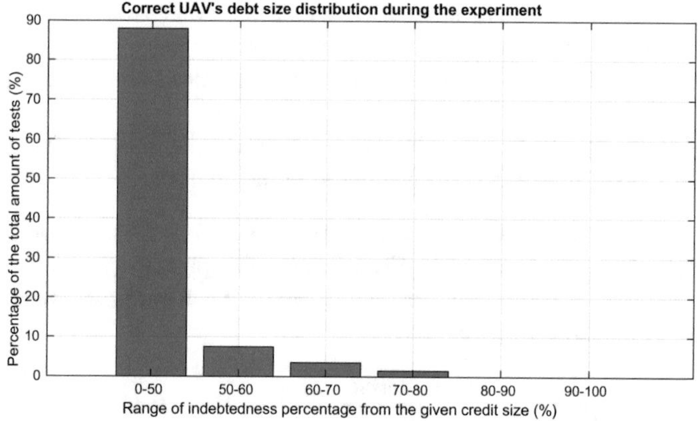

Fig. 8. Histogram of correct UAV's indebtedness size during the experiment

6 Conclusion

The stated research task was performed. By dint of the method, based on credit theory, it becomes possible to minimize destructive information influence of

UAVs-saboteurs on the group and provide preservation of pragmatic integrity in data transmitted through the common communication channel. In comparison with the existing model of trust and reputation, the proposed algorithm allows not only to assess message content but also message volume in the context of information security.

Information crediting makes unprofitable group work sabotage, as it significantly slows down the sabotage process and requires large costs from the saboteur side (time, energy, information). Thus, there is less probability that the group work will be rattened from the external way. Prospectively it is planned to implement and test the proposed method in a real UAV group, and consequently to develop UAV monitoring system providing security of transmitted data.

References

1. Pipattanasomporn, M., Feroze, H., Rahman, S.: Multi-agent systems in a distributed smart grid: design and implementation. In: 2009 IEEE/PES Power Systems Conference and Exposition, Seattle, WA, pp. 1–8 (2009). https://doi.org/10.1109/PSCE.2009.4840087
2. Mönch, L., Stehli, M., Zimmermann, J., Habenicht, I.: The FABMAS multi-agent-system prototype for production control of water fabs: design, implementation and performance assessment. Prod. Plann. Control **17**, 701–716 (2006). https://doi.org/10.1080/09537280600901269
3. Li, W., Logenthiran, T., Woo, W.L., Phan, V., Srinivasan, D.: Implementation of demand side management of a smart home using multi-agent system. In: 2016 IEEE Congress on Evolutionary Computation (CEC), Vancouver, BC, pp. 2028–2035 (2016). https://doi.org/10.1109/CEC.2016.7744037
4. Wang, S., Wan, J., Zhang, D., Li, D., Zhang, C.: Towards smart factory for industry 4.0: a self-organized multi-agent system with big data based feedback and coordination. In: Computer Networks, vol. 101, pp. 158–168 (2016). https://doi.org/10.1016/j.comnet.2015.12.017
5. Gaiduk A., Kapustyan S., Shapovalov I.: Self-organization in groups of intelligent robots. In: Robot Intelligence Technology and Applications 3. Advances in Intelligent Systems and Computing, vol. 345, pp. 171–181 (2015). https://doi.org/10.1007/978-3-319-16841-8_17
6. Zhuravlev, V., Zhuravlev, P.: Usage of unmanned aerial vehicles in general aviation: current situation and prospects. Civ. Aviat. High Technol. **226**, 156–164 (2016). https://doi.org/10.26467/2079-0619-2016-0-226-156-164
7. Reis, B.P., Martins, S.V., Filho, E.I.F., Sarcinelli, T.S., Gleriani, J.M., Leite, H.G., Halassy, M.: Forest restoration monitoring through digital processing of high resolution images. Ecol. Eng. **127**, 178–186 (2019). https://doi.org/10.1016/j.ecoleng.2018.11.022
8. Zikratov, I., Zikratova, T., Lebedev, I., Gurtov, A.: Trust and reputation model design for objects of multi-agent robotics systems with decentralized control. Sci. Tech. J. Inf. Technol. Mech. Opt. **91**, 30–38 (2014)
9. Jovanov, I., Pajic, M.: Sporadic data integrity for secure state estimation. In: 2017 IEEE 56th Annual Conference on Decision and Control (CDC), pp. 163–169 (2017). https://doi.org/10.1109/CDC.2017.8263660

10. Santra, P., Roy, A., Majumder, K.: Comparative analysis of cloud forensic techniques in IaaS. In: Advances in Computer and Computational Sciences. Advances in Intelligent Systems and Computing, vol. 554, pp. 207–215 (2017). https://doi.org/10.1007/978-981-10-3773-3_20
11. Komarov, I., Drannik, A., Yurieva, R.: Multiagent information security problem's simulation. World Sci. Discov. **52**, 61–71 (2014)
12. Lavrushin, O.: The theory of the credit basis and its use in modern economy. J. Econ. Regul. **8**, 6–15 (2017)
13. Al-Azzam, M.H., Mimouni, K., Ali, M.A.: The impact of socioeconomic factors and financial access on microfinance institutions. Int. J. Econ. Finance **4**, 61–71 (2012). https://doi.org/10.5539/ijef.v4n4p61
14. Hahn, L.A.: Economic theory of bank credit. J. Post Keynesian Econ. **37**, 309–335 (2015). https://doi.org/10.1093/acprof:oso/9780198723073.001.0001

Logic Gate Integrated Circuit Identification Through Augmented Reality and a Smartphone

Carlos Aviles-Cruz[1(✉)], Juan Villegas-Cortez[1(✉)], Arturo Zuniga-Lopez[1(✉)], Ismael Osuna-Galan[2(✉)], Yolanda Perez-Pimentel[2(✉)], and Salomon Cordero-Sanchez[3(✉)]

[1] Departamento de Electrónica, Universidad Autónoma Metropolitana, Unidad Azcapotzalco, Av. San Pablo 180 Col. Reynosa Tamaulipas, 02200 Ciudad de México, Mexico
{caviles,juanvc,azl}@azc.uam.mx

[2] Polytechnic University of Chiapas, Carretera Tuxtla Gutierrez - Portillo Zaragoza Km 21 + 500, Suchiapa, Chiapas, Mexico
{iosuna,ypimentel}@upchiapas.edu.mx

[3] Departamento de Química, Universidad Autónoma Metropolitana, Unidad Iztapalapa, San Rafael Atlixco 186, Vicentina, 09340 Ciudad de México, Mexico
scs@xanum.uam.mx
https://sites.google.com/site/cavilesc/

Abstract. The basic concepts of logic gates are now taught in high school and university. In laboratories, it is essential to understand the integrated circuits (ICs) that make digital systems and computers possible. This paper proposes an augmented reality (AR) system based on smart phones that automatically identifies 7 basic logic gate integrated circuits that are: AND gate, OR gate, NOT gate, NAND gate, NOR gate, XOR gate and XNOR gate. Starting from the IC image taken by the smartphone, it digitally generates a layer of virtual objects that are then mixed with the original image. Under the markerless paradigm, 2 virtual objects are placed over the IC: IC identification and pin information. Finally, the smartphone screen shows the image of the logic gate obtained. The entire evaluation of the AR system is presented with a technical efficiency of 100%.

Keywords: Augmented reality · Logic gates · Android · Application development · Electrical engineering

1 Introduction

Augmented Reality (AR) technology has occupied different areas of our lives, from games, movies, shows, entertainment, to education. AR has three common properties: the interactivity, the handling of 3D objects, and the combination

K. Arai et al. (Eds.): CompCom 2019, AISC 997, pp. 73–85, 2019.
https://doi.org/10.1007/978-3-030-22871-2_6

of the real and virtual world [2]. Due to the above properties, in this paper, it has been used for the identification of a logic gate integrated circuit (IC). Added to the profuse use of smartphones in high school and university students. The advantage of AR is the ability to add a layer of digitally generated information, facilitating in time and understanding the identification, connection and experimentation of logical circuits and Boolean logic. AR embedded in a mobile device, has the potential to attract the attention of students and great impact on the teaching-learning process [3]. In the literature, several AR studies are found focusing on improving elementary-middle school level education [5,6,13,17]. Beside, in higher education fields, the main related AR works are in: computer [15], control engineering [10], energy engineering [16], engineering [4] electronics engineering [8,14] and physics [9].

The most related work in the field of electronic circuits is [8] where authors applied interactive marker-based AR technique to implant the functions of each logic gates in the marker and, program these markers to interact with others just like in the real logic gates laboratory. Markers can be in any shape or have any pictures or symbol on it, which make user recognize them easily. Authors have created 9 markers that represent 7 basic Logic gates (AND, OR, NOT, NAND, NOR, XOR and XNOR). Authors have simulated a real digital circuit using these markers and the output is just the same as using the real logic gates.

The objective of this article is to make an auxiliary tool for the connection and experimentation of a typical electronics circuit in a laboratory, facilitating the identification of the IC, as well as providing the information of the pins and gates of an IC. It is expected that this tool will facilitate the cumbersome task of consulting IC manufacturers' specification sheets.

Proposal is developed on a smartphone running Android operating system. The main characteristic of this application is that belongs to markerless paradigm (in AR field), which means, it does not require a predefined mark, pattern, object, or badge placed in the real world that will then be identified and followed to properly orient and register virtual content into the space.

From a photo taken by a smartphone camera, through the horizontal and vertical profile projection, the license plate is located, then, a digital image processing and a pattern recognition tasks are developed in order to identify the type of logic gate of IC. Once IC type is identified, using AR technology, two layers of information are superimposed: (1) identification of the IC and (2) pin information (see Fig. 1). The proposed system automatically identifies 7 basic logic gate integrated circuits: AND gate, OR gate, NOT gate, NAND gate, NOR gate, XOR gate and XNOR gate.

The entire proposed project is composed of the following main stages: *Image capturing*, *Isolating digit/letter*, *Classifying digit/letter*, *Identifying IC* and *Generating virtual objects*, which are described in the system description section.

The rest of the paper is organized as follows. In Sect. 2, the system description is presented. The system evaluation is presented in Sect. 3. Finally, conclusions and future works are given in Sect. 4.

2 System Description

In order to facilitate the experimentation of logic gates through ICs, this tool is presented to assist in the rapid identification of ICs with the emerging AR technology. High school and college students may have the possibility of an application on a smartphone device that will save them time searching through the specification sheets of commercial ICs. The proposed system allows the automatic identification of ICs that enable basic logic operations.

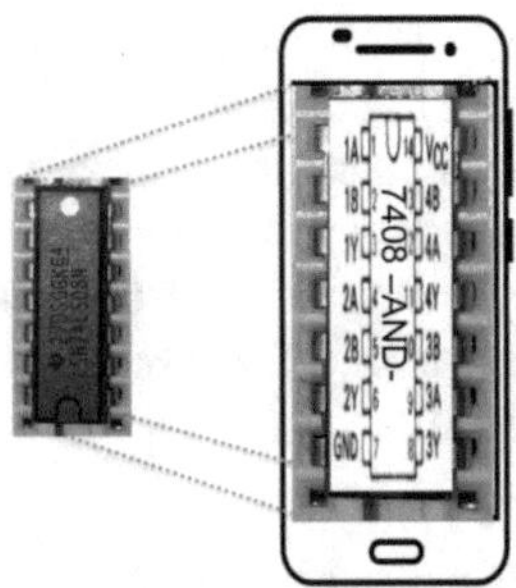

Fig. 1. Main idea of the proposed AR system.

The methodology begins with the taking of the IC's photograph, then, several stages of image processing are applied to allow the correct identification of the IC. With the identification of the IC the generation of virtual objects is carried out, then both information are mixed, the original image plus the virtual objects, and they are visualized in a smartphone. Virtual objects are IC identification and pins information. The main stages of this work are (see Fig. 2): *Image Capturing*, *Isolating digit/letter*, *Classifying digit/letter*, *Identifying IC* and *Generating virtual objects*, which are detailed as follows:

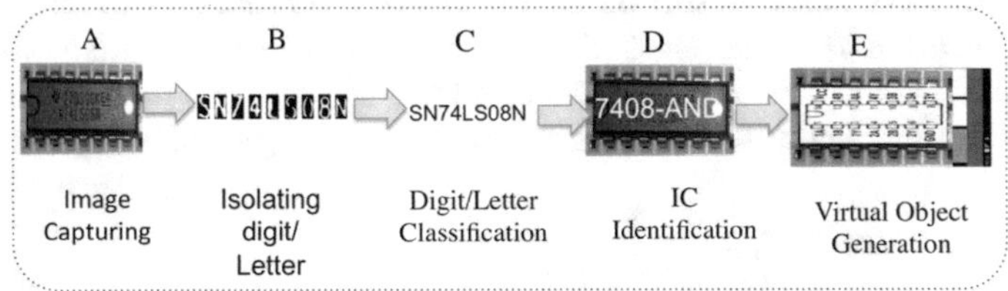

Fig. 2. Major stages of the augmented reality system.

2.1 Image Capturing

The application begins with the taking of a photography of the IC to be identified. Smartphones do not work images in the RGB format, they do it in the YcrCb color format. In this color space the image is represented by: luminance information (Y) and chrominance information (Cb and Cr) [1]. At all stages of image processing, only the luminance information Y is used, which allows the correct identification of the IC.

2.2 Isolating Letter/Digit

Once the coordinates where the IC is located have been obtained, the IC information zone is taken. Figure 3(a) shows an example of the IC 7408. In order to try to be as light-independent as possible, a histogram equalization is applied, as shown in Fig. 3(b). The next step is to try to remove as much noise as possible from the image, in this case, it appears that the noise is impulsive, therefore, a median statistical filter is used, with a 5×5 rectangular mask. Figure 3(c) shows the result of applying this filtering.

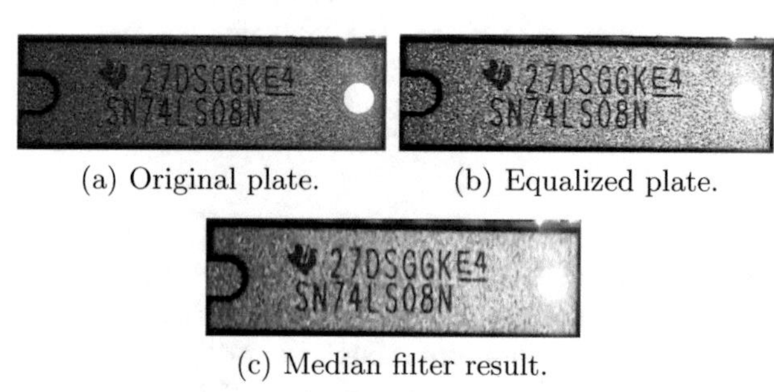

(a) Original plate. (b) Equalized plate.

(c) Median filter result.

Fig. 3. Example of an IC plate processing.

The next step is to obtain the area where only digits or letters are located, for which, an analysis of vertical (VPP) and horizontal (HPP) projection profiles is performed. In the case of the vertical projection profile, all the pixel values in each column are added and then divided by the total of values (see Eq. 1). For HPP, all the pixel values in each line are added and also, divided by the total of values (see Eq. 2). Figure 4 shows an example of such processing.

$$VPP(j) = \frac{\sum_{i=1}^{N} f(i,j)}{\sum_{j=1}^{M} \sum_{i=1}^{N} f(i,j)} \quad \forall\, j \in [1...M] \tag{1}$$

$$HPP(i) = \frac{\sum_{j=1}^{M} f(i,j)}{\sum_{j=1}^{M} \sum_{i=1}^{N} f(i,j)} \quad \forall\, i \in [1...N] \tag{2}$$

where i, j are de lines and columns, respectively; and $[N \times M]$ sub image trimmed size.

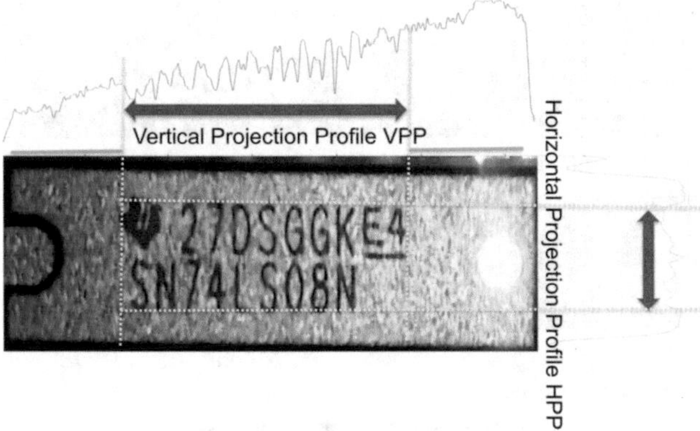

Fig. 4. Projection profiles analysis applied to locating digits and letters: vertical projection profile (VPP) and horizontal projection profile (HPP)

The analysis of vertical and horizontal projection profiles will allow trimming only the area of letters and digits. The applied logic analysis is explained as follows:

1. *On HPP:* Analysis starts from de left side of the curve and the key is to find the first minimum after the first maximum, finding the interval where the letters and digits are located. As an example of the HPP analysis, the right side of Fig. 4 shows the limits (red arrow).

2. *On VPP:* For this analysis, the key point is to find the maximum variation intensity (the maximum change in magnitude), going through the curve either from left to right or vice versa. Then, a maximum variation percentage (e.g. 20%) is defined. Having the defined percentage, a filtering (30 pixels wide, defined experimentally) is applied along the whole curve, with the following logic: leave the intervals intact above this threshold, remarking that the initial changes on both the left and right sides are discarded. The top of Fig. 4 depicts an example, where a red arrow indicates the interval.

In order to preserve the minimum zone where the digits and letters are located, the minimums and maximums values area taken from HPP and VPP analysis and then, the image is trimmed (see Fig. 5).

Fig. 5. Example of digit/letter trimming by vertical (VPP) and horizontal (HPP) projection profiles analysis.

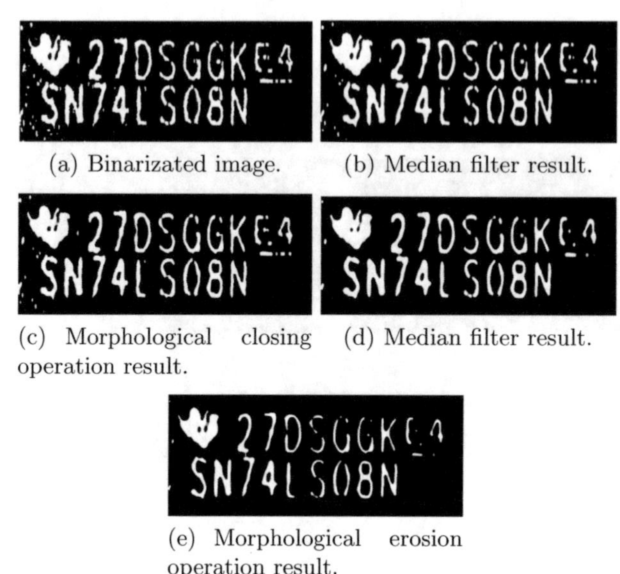

(a) Binarizated image. (b) Median filter result. (c) Morphological closing operation result. (d) Median filter result. (e) Morphological erosion operation result.

Fig. 6. Example of a binarization and cleaning up image.

The next task is to have a clean binary image to perform the segmentation of letters and digits. The binarization of image is done from the analysis of the histogram, finding a threshold that allows keeping the information of digits and letters as complete as possible. For the cleaning phase, the following chained tasks are carried out: a medium filter (with a 10×5 rectangular window), a morphological closing operation (with a 3×6 rectangular window), a medium filter (with a 12×8 rectangular window) and, finally, a morphological erosion operation (with a 5×2 rectangular window), see Fig. 6 as an example.

Finally, the last image-processing task is to take each of the two lines contained in the IC. Developing a horizontal profile projection analysis allows to find the overall minimum along the center of the curve, which must be the separation threshold between the two lines of information, as depicts in Fig. 7(a), see the dotted line in yellow color. Then, with this threshold value, the two lines of information are separated, as shown in Fig. 7(b) and (c), respectively.

At last, the remaining task is to separate the digits and letters from each line, for which, a VPP analysis was performed. Zero values in the VPP curve are indicating the limits between characters. Figure 8 depicts an example for each character limits. The isolated letters and digits are ready to be classified, an example can be seen in Fig. 9 for the IC 7408.

(a) Lines located by horizontal projection profile (HPP) analysis.

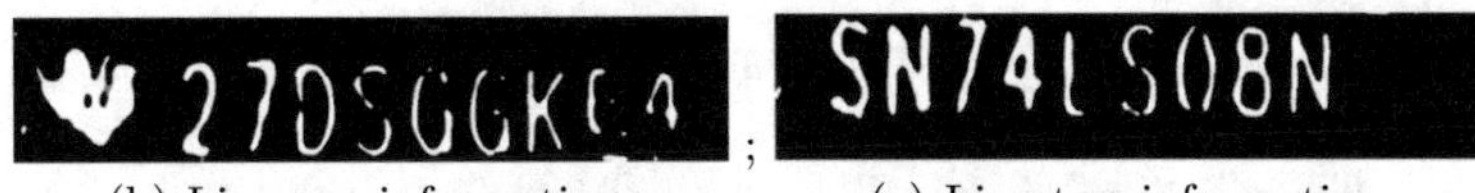

(b) Line one information. ; (c) Line two information.

Fig. 7. Example of lines separation information.

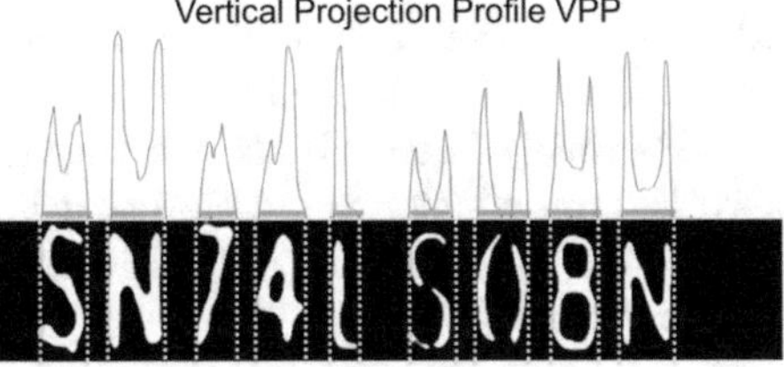

Fig. 8. Example of characters by vertical projection profile (VPP) analysis.

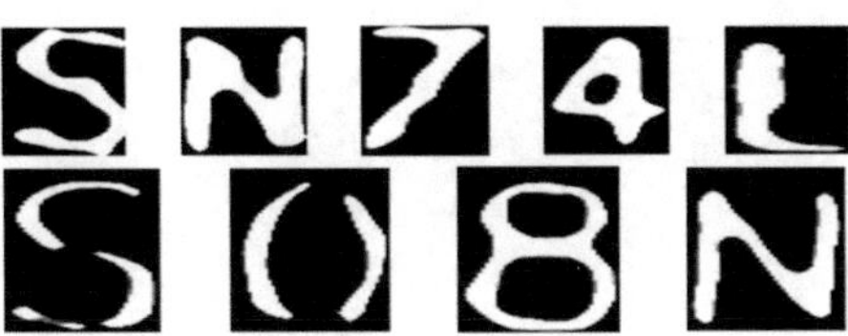

Fig. 9. Example of isolated characters for IC 7408.

2.3 Classifying Digit/Letter

A multilayer neural network (NN) [7,11] was used to develop the classification of letters and digits task. Two hidden layers neural network was implemented in order to get the best classification results (see Fig. 10). A preprocessing stage was applied to input images, all digits and letters were re-dimensioned to 256×256 pixels, thus, the input layer was set to $65,536$ neurons. Regarding the output layer, it is composed of 36 neurons (26 letters plus 10 digits giving a total of 36 outputs classification possibilities). The NN was trained using the back-propagation algorithm [12]. For the training stage, the algorithm was stopped until it was reached with zero classification error.

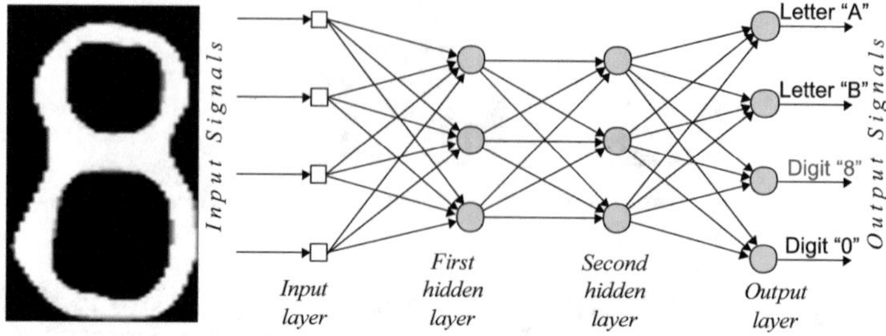

Fig. 10. Classification of letters and digits through a neural network of 2 hidden layers.

2.4 Identifying IC

Having the letters and digits isolated and identified, we move on to the identification of the sequence of the IC. Initially, all identified letters are discarded, for example: *"S"*, *"N"*, etc. Because of every IC starts with a number "74", then, the following two are worked on, which are the ones that indicate the gate type. Table 1 shows the correspondence between the IC ending and the type of logic gate, just for a quad 2-input type IC.

Table 1. Correspondence between ending number and logic gate ICs

32	$\Longrightarrow$	OR
08	$\Longrightarrow$	AND
02	$\Longrightarrow$	NOR
00	$\Longrightarrow$	NAND
04	$\Longrightarrow$	NOT
86	$\Longrightarrow$	XOR
266	$\Longrightarrow$	XNOR

2.5 Generating Virtual Objects

The virtual objects that will be placed over the originally taken photos are of two types, on the one hand, the identification of the IC and, on the other hand, the information of the pins. Virtual objects (generated digitally) can be seen in the figures: Fig. 11(a) and (b). All virtual objects are dynamically placed on the middle of the IC. The "IC identification" is a white text dynamically superimposed on the IC location. While, the "Pins IC information" is composed of a white rectangle with inputs and outputs pins information. Pin information is dynamically drawn on the middle of IC.

(a) Original IC picture.

(b) IC identification.

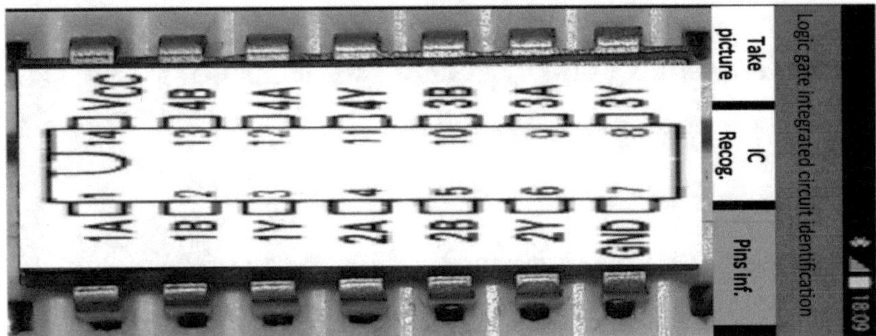

(c) IC Pin Information.

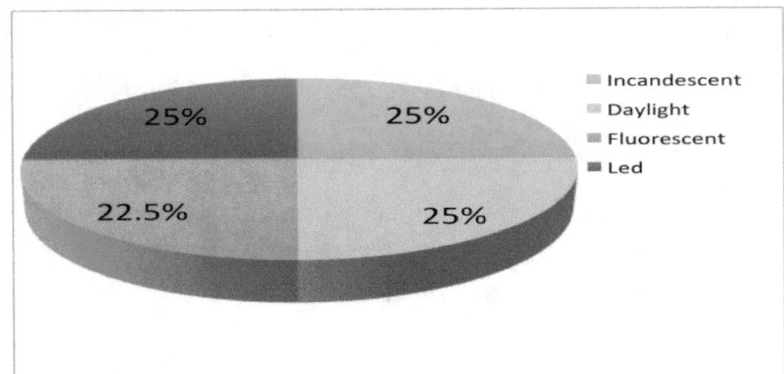

(d) Performance of the proposal against four different light conditions.

Fig. 11. Result of the methodology applied to IC 7400: NAND gate.

(a) Original IC picture.

(b) IC identification.

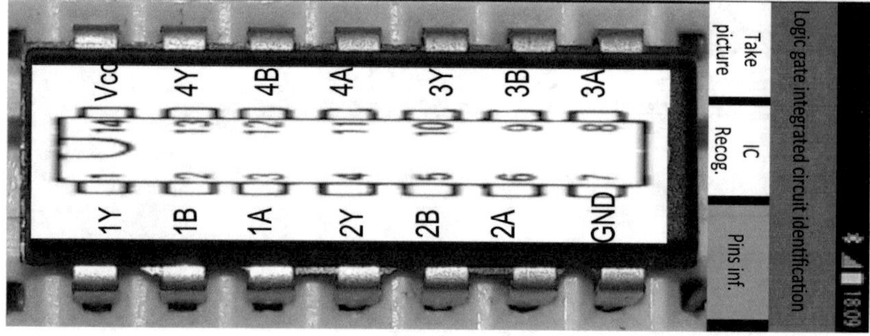

(c) IC Pin Information.

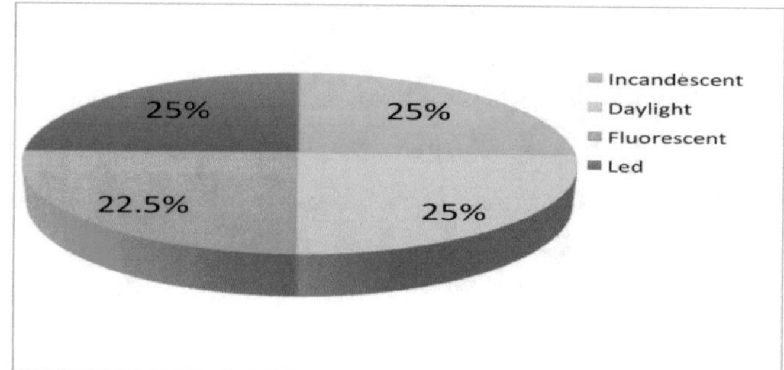

(d) Performance of the proposal against four different light conditions.

Fig. 12. Result of the methodology applied to IC 7402: NOR gate.

(a) Original IC picture.

(b) IC identification.

(c) IC Logic Information.

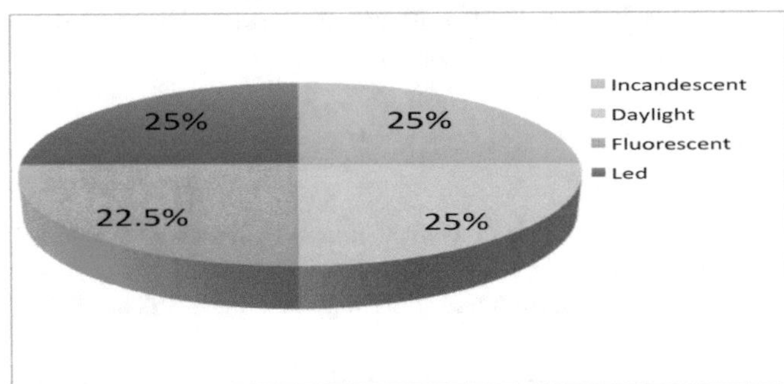

(d) Performance of the proposal against four different light conditions.

Fig. 13. Result of the methodology applied to IC 7408: AND gate.

3 Evaluation of the System

The proper functioning of the proposed AR mobile system to support the automatic identification of ICs was evaluated under different conditions of illumination: (a) daylight, (b) fluorescent light, (c) incandescent light and (d) LED light. The correct identification of the commercial logic gates IC was carried out under each of the 7 proposed logic gate ICs from the National Instruments Company. The smartphone used was a Samsung Galaxy 8, with which, 20 images were taken per IC under each of the lighting conditions. In order to avoid possible damage due to incorrect connections, it is suggested do not connect the circuit to the power supply.

As an example, only 3 logic gates IC of the 7 are presented: 7400-NAND gate IC, 7402-NOR gate IC and 7408-AND gate IC (see Figs. 11, 12 and 13); the results are very similar for the rest of the 7 logical gates. As shown in Figs. 11(e), 12(e) and 13(e), the system has an overall efficiency of 100% to identify the 7 logic gate ICs for the 4 lighting conditions. The system identifies ICs at 100% in all lighting conditions.

In Figs. 11(b), 12(b) and 13(b) the IC number plate identification can be seen. In Figs. 11(c), 12(c) and 13(c) pin information is shown.

4 Conclusion and Further Research

The proposed mobile AR system to identifier IC has worked perfectly at each of its stages. We hope that this application will be useful to the high school and college students in the rapid identification of ICs in laboratory practices.

From the technical point of view, the seven proposed logic gate ICs were correctly identified: AND, OR, XOR, NOT, NAND, NOR and XNOR by the proposal, having an overall efficiency of 100% to identify the 7 logic gate ICs under the four light conditions: daylight, incandescent light and LED light.

The markless method developed is capable of displaying 2 virtual objects: (1) Identification of IC; and (2) Information of pins. Finally, the AR remains a new tool for high school and undergraduate students for learning the theory and practice of logical gate ICs.

The future research includes the identification of other types of ICs such as operation amplifiers, oscillators, filters, etc.

References

1. Annadurai, S.: Fundamentals of Digital Image Processing, 4th edn. Pearson Education, London (2007)
2. Azuma, R.T.: A survey of augmented reality. Presence: Teleoper. Virtual Environ. **6**(4), 355–385 (1997)
3. Bal, E.: The future of augmented reality and an overview on the to researches: a study of content analysis. Qual. Quant. **52**(6), 2785–2793 (2018)

4. Bazarov, S.E., Kholodilin, I.Y., Nesterov, A.S., Sokhina, A.V.: Applying augmented reality in practical classes for engineering students. IOP Conf. Ser.: Earth Environ. Sci. **87**(3), 1–7 (2017)
5. Billinghurst, M., Duenser, A.: Augmented reality in the classroom. Computer **45**(7), 56–63 (2012)
6. Billinghurst, M., Kato, H., Poupyrev, I.: The magicbook - moving seamlessly between reality and virtuality. IEEE Comput. Graph. Appl. **21**(3), 6–8 (2001)
7. Bishop, C.M.: Neural Networks for Pattern Recognition, vol. 1, 1st edn. Clarendon Press, New York (1995)
8. Boonbrahm, P., Kaewrat, C., Boonbrahm, S.: Using augmented reality interactive system to support digital electronics learning. In: Learning and Collaboration Technologies. Technology in Education, LCT 2017. Lecture Notes in Computer Science, vol. 10296, no. 1, p. 20 (2017)
9. Daineko, Y., Dmitriyev, V., Ipalakova, M.: Using virtual laboratories in teaching natural sciences: an example of physics courses in university. Comput. Appl. Eng. Educ. **25**(1), 39–47 (2017)
10. Frank, J.A., Brill, A., Kapila, V.: Interactive mobile interface with augmented reality for learning digital control concepts. In: 2016 Indian Control Conference (ICC), pp. 85–92, January 2016
11. Gunawan, T., Mahamud, N., Kartiwi, M.: Development of offline handwritten signature authentication using artificial neural network, vol. 2018-March, pp. 1–4 (2018)
12. Li, Y., Fu, Y., Li, H., Zhang, S.: The improved training algorithm of back propagation neural network with self-adaptive learning rate. In: 2009 International Conference on Computational Intelligence and Natural Computing. vol. 1, pp. 73–76, June 2009
13. Rahmat, R.F., Akbar, F., Syahputra, M.F., Budiman, M.A., Hizriadi, A.: An interactive augmented reality implementation of hijaiyah alphabet for children education. J. Phys.: Conf. Ser. **978**(1), 012102 (2018)
14. Reyes-Aviles, F., Aviles-Cruz, C.: Handheld augmented reality system for resistive electric circuits understanding for undergraduate students. Comput. Appl. Eng. Educ. **26**(3), 602–616 (2018)
15. Teng, C.H., Chen, J.Y., Chen, Z.H.: Impact of augmented reality on programming language learning: efficiency and perception. J. Educ. Comput. Res. **56**(2), 254–271 (2018)
16. Tirado-Morueta, R., Sánchez-Herrera, R., Márquez-Sánchez, M., Mejías-Borrero, A., Andujar-Márquez, J.: Exploratory study of the acceptance of two individual practical classes with remote labs. Eur. J. Eng. Educ. **43**(2), 278–295 (2018)
17. Yilmaz, R.M.: Educational magic toys developed with augmented reality technology for early childhood education. Comput. Hum. Behav. **54**(1), 240–248 (2016)

A Heuristic Intrusion Detection System for Internet-of-Things (IoT)

Ayyaz-ul-Haq Qureshi(✉), Hadi Larijani, Jawad Ahmad, and Nhamoinesu Mtetwa

School of Computing, Engineering and Built Environment (SCEBE), Glasgow Caledonian University, Glasgow G4 0BA, UK
ayyaz.qureshi@gcu.ac.uk

Abstract. Today, digitally connected devices are involved in every aspect of life due to the advancements in Internet-of-Things (IoT) paradigm. Recently, it has been a driving force for a major technological revolution towards the development of advanced modern computer networks connecting physical objects around us. The emergence of IPv6 and installation of open access public networks is attracting cyber-criminals to compromise the user specific security information. This is why the security breaches in IoT devices are dominating the headlines lately. In this research we have developed a random neural network based heuristic intrusion detection system (RNN-IDS) for IoTs. Upon feature selection, the neurons are trained and further tested at different learning rates with NSL-KDD dataset. Two methods are adopted to analyse the proposed scheme where the accuracy of RNN-IDS increased from 85.5% to 95.25%. Results also suggest that upon comparison with other machine learning algorithms, the proposed intelligent intrusion detection has higher accuracy in recognition of anomalous traffic from normal patterns.

Keywords: Intrusion detection systems · NSL-KDD · Machine learning · Random Neural Networks · Cyber-Security · IoT security

1 Introduction

To provide intelligent services to the end users, Internet-of-Things (IoT) provides a platform where information networks are seamlessly integrated into the physical ones. To impart such ubiquitous services, data collected from participating sensor nodes must be fused and analysed. There are number of security threats in the way of successful implementation trust management in IoTs [1]. In order to achieve the effective defence against cyber-attacks, the Intrusion Detection Systems (IDS) certainly perform a crucial task. Today, the wide use of computer aided programs and networks provide the low cost solutions to the end user problems in a short interval of time which has made the digital world

K. Arai et al. (Eds.): CompCom 2019, AISC 997, pp. 86–98, 2019.
https://doi.org/10.1007/978-3-030-22871-2_7

an integrated part of the physical world. Extensive usage of internet connected smart devices means that, massive data is shared among them which gives rise to vulnerabilities [2]. Hence, the need to secure the end user information is now higher than ever. The extent of research work in the field of computer security has increased many folds over last few decades but rapid attacks on networks has made the mitigation of network attacks a challenging task.

A system which acts as a front line defence against network intrusion must satisfy the principles of information confidentiality, integrity, and availability commonly known as CIA architecture [3]. Hackers pose serious threats to existing networks after bypassing these three components in a deceptive manner. Such occurrences has made the availability of Intrusion Detection Systems (IDS) a vital task [4]. As discussed above, an efficient IDS has to perform a critical role in order to ensure safety towards user information. Based on the practicality, IDS are classified as mis-used based (MIDS) and anomaly based intrusion detection systems (AIDS) [5].

The mis-used based intrusion detection systems (MIDS) which are commonly known as signature based IDS, use the existing signatures to analyse the incoming network traffic. These are the universally known attack patterns which are collected based on the type of protocols and applications used hence constantly need updation [6]. On the contrary, anomaly based intrusion detection systems (AIDS) use the classification approach to detect the malicious activity happening in the network [3].

Although there are various challenges involved in the successful deployment of AIDS, but several of them are categorised as follows:

1. Complexity in establishment of profiles to distinguish between normal and sceptical traffic patterns.
2. Irrelevant feature selection and incomplete datasets result in high false positive rates.
3. Platform dependencies decline the performance in real-time detection of anomalies.
4. Ineffective design results in redeployment of IDS which results in significant performance degradation.

There are a lot of approaches that have been used to develop the intrusion detection systems. After the proposition of theory of deep learning [7], the era of machine learning has been revolutionised. In this paper we tend to use classification techniques to detect the anomaly from normal traffic patterns. In [5], the authors have reduced high number of false positive rates and false negative rates in intrusion detection systems by developing a hybridised approach to estimate the optimal performance. Dataset is pre-processed with Information Gain and Vote Algorithm to extract usable features. The dataset is then used to train several classifiers. The authors concluded that detection time is significantly reduced while the accuracy increased using the hybrid approach for data dimensionality reduction. J48 outperforms all other classifiers in detection of malicious patterns in both binary class and multi-class of NSL-KDD dataset.

In [8], the authors have proposed a novel attack detection mechanism to achieve high accuracy and low false positives rates. In order to transform correlated features in the dataset, principle component analysis (PCA) has been used. Long Short-Term Memory based Recurrent Neural Network (LSTM-RNN) based model is implemented on tensor flow and results are compared with KNN, SVM, GRNN and PNN classifiers. The results shows that higher accuracy is achieved with low false positive alarms and high true positive rates with the overall precision of 99.46%. But this technique is more dependent upon types of features selected via PCA before feeding them to train input layer neurons.

As mentioned before, most of the times, the intrusion detection systems produce high false alarms due to inefficient and incompetent datasets. In [9], a detailed analysis of KDDCUP'99 dataset is done. Several machine learning classifiers such as J48, SVM, NB Tree, MLP, RF and RF Tree are trained using the dataset. Based on the low accuracy achieved, the authors concluded that original KDDCUP'99 has many limitation due enormous number of redundant records and uneven distribution of data. Hence, a new benchmark dataset NSL-KDD is proposed which increased the accuracy of classifiers and reduce the training time subsequently.

In [10], an Artificial Neural Network (ANN) based IDS is developed and trained with NSL-KD dataset. The IDS is tested for both binary class and multi-class attack types of NSL-KDD dataset which include U2R, R2L, Probe and DoS. ANN-IDS is trained with quasi-Newton back propagation (BFGS) and Levenberg-Marquart (LM) algorithms. Feature selection is done and model is trained for both reduced and full features on different number of input and hidden layer neurons for all attack types. Results reveal that although the detection rate for U2R and R2L attacks is very low, the model has produced accuracy of 79.9% and 81.2% for multi-class and binary class respectively.

Since the real world data is huge in quantity and classified as 'Big Data', deep learning solutions are required to analyse the incoming traffic for malicious activities. In [11] the authors have used deep learning to develop a Recurrent Neural Network based intrusion detection system (Recurrent-NN). NSL-KDD dataset is used to train the proposed model. Network is trained with different number of hidden layer neurons and learning rates for both binary-class and multi-class of the dataset. The performance of proposed model is compared with SVM, MLP, RF, Bayesian and several other machine learning architectures. The results revealed that recurrent neural network based IDS has surpassed all other classifiers. Although high accuracy is achieved for different attack classes but excessive training time and vanishing gradient remained the key problems for this scheme.

It is evident from literature that, intrusion detection has been performed from various machine learning algorithms. In this paper we propose a heuristic intrusion detection system for IOT paradigm using Random Neural Network [12] (RNN-IDS). NSL-KDD dataset is used for training the feed-forward neural network. Attributes are selected and data is normalised before its trained and tested against malicious attacks on network. Although RNN is substantially used in the deployment of Heating, ventilation, and air conditioning (HVAC) [13,14], occupancy detection [15] and pattern recognition [16] etc. but there is a lot of potential to use its features in the implementation of scalable intrusion detection systems.

The main contributions of this research are:

- A novel heuristic intrusion detection system for IoTs using random neural networks (RNN-IDS) has been developed and implemented.
- Enhanced performance is achieved by comparing the reduced and complete features of the NSL-KDD dataset.
- Critical comparisons are conducted after training the system with randomised data using a fixed number of hidden layer neurons and different learning rates.
- Performance of the proposed RNN-IDS has been compared with Support vector machine, naive bayes, J48 (decision tree), multi layer perception and various other ML methods.

Paper organization is outlined as follows: Sect. 1 provides introduction to the intrusion detection and discussed about past research findings. Section 2 presents basic understanding of RNN and Gradient Decent Algorithm. Section 3 outlines the methodology adopted to implements IDS. Results are discussed in Sect. 4 while Sect. 5 concludes the paper.

2 Background

This section covers the essential knowledge related to Random Neural Networks (RNN) and Gradient Descent Algorithm (GDA).

2.1 Random Neural Network Model

In the intention to replicate how human learns the information, Artificial Neural Networks (ANN) [17] came into existence which revolutionised the area of machine learning. Gelenbe proposed a new class of ANN and named it as Random Neural Network (RNN) [12].

In RNN model, neuron are connected to each other in different layers and have excitation and inhibition states depending upon the signal potential it receives. In a network if neuron encounters a positive (+1) or negative (−1) signal, it goes into excited or inhibit state respectively. Sate of the neuron n_i at time t is shown as $S_i(t)$. Neuron n_i remains in idle state as long as value of $S_i(t) = 0$ and in order to get excitation signal, it changes to $S_i(t) > 0$ because $S_i(t)$ is always considered a non-negative integer.

Upon excitation, the neuron n_i transmits impulse signal towards other neuron n_j with rate of transmission h_i. The transmitted signal may reach neuron n_j with probability of $p^+(i,j)$ as a positive signal, or probability of $p^-(i,j)$ as a negative signal or it may also leave the network with probability of $k(i)$.

Where,

$$k(i) + \sum_{j=1}^{N} p^+(i,j) + p^-(i,j) = 1, \forall i, \tag{1}$$

Weights are updated on neuron n_i and n_j as:

$$w^+(i,j) = h_i p^+ + (i,j) \geq 0, \tag{2}$$

and

$$w^-(i,j) = h_i p^- + (i,j) \geq 0. \tag{3}$$

To predict the probability of signals in RNN, Poisson distribution is used. Hence, for the neuron n_i, Poisson rate $\Lambda(i)$ demonstrates the positive signal whereas negative signal is depicted by Poisson rate $\lambda(i)$
Mathematically,

$$\lambda^+(i) = \sum_{j=1}^{n} e(j)r(j)p^+(j,i) + \Lambda(i), \tag{4}$$

$$\lambda^-(i) = \sum_{j=1}^{n} e(j)r(j)p^-(j,i) + \lambda(i). \tag{5}$$

The output activation function $e(i)$ for neurons can be written as:

$$e(i) = \frac{\lambda^+(i)}{h(i) + \lambda^-(i)}, \tag{6}$$

where $h(i)$ is the transmission rate, which can be calculated by combining Eqs. 1, 2 and 3:

$$h(i) = (1 - k(i))^- 1 \sum_{j=1}^{N} [w^+(i,j) + w^-(i,j)], \tag{7}$$

In Eq. 7, since $h(i)$ is the gain of firing rate and the probabilities of positive and negative weights updated during training RNN model, hence it can also be written as:

$$h(i) = \sum_{j=1}^{N} [w^+(i,j) + w^-(i,j)]. \tag{8}$$

Interested reader can further understand the network operation in [12].

2.2 Gradient Descent Algorithm

Random Neural Network based Intrusion Detection System (RNN-IDS) proposed in this paper has been trained using Gradient Descent Algorithm (GD). It has been used to get the local minima of function so that overall mean square error can be reduced. Weights are updated and maximum training accuracy is achieved. This algorithm has been used by researchers for iterative optimization. Error function can be denoted as [18]:

$$E_p = \frac{1}{2}\sum_{i=1}^{n} \alpha_i (q_j^p - y_j^p)^2, \alpha_i \geq 0 \tag{9}$$

where $\alpha \in (0,1)$ shows the state of output neuron i, also q_j^p is an actual differential function and y_j^p is the predicted output value. From Eq. 9, after training the neurons a an b, weights are updated as $w^+(a,b)$ and $w^-(a,b)$, derived as [18]:

$$w_{a,b}^{+t} = w_{a,b}^{+(t-1)} - \eta \sum_{i=1}^{n} \alpha_i (q_j^p - y_j^p)[\frac{\partial q_i}{\partial w_{a,b}^+}]^{t-1}, \tag{10}$$

similarly:

$$w_{a,b}^{-t} = w_{a,b}^{-(t-1)} - \eta \sum_{i=1}^{n} \alpha_i (q_j^p - y_j^p)[\frac{\partial q_i}{\partial w_{a,b}^-}]^{t-1}. \tag{11}$$

3 Methodology

In this research, random neural network has been adopted to develop the intrusion detection system. Like any other neural network architecture, RNN is also inspired by human brain in where nodes which referred to as neurons are connected with each other. The model consists of input, hidden and output layers. To start learning, selected features becomes the input to input layer neurons. After initiating the calculation of suitable weights and biases, input layer pass this data to the hidden layer for further transformation. Learning at hidden layer is important because it has to play a vital role in prediction of the output from actual features upon testing. Hidden layer then forward the information to output layer so that a feasible output is mapped.

The proposed scheme (Fig. 1) is developed by following steps:

Data Set. In order to verify the effectiveness of proposed intrusion detection scheme, NSL-KDD dataset has been adopted. It is the refined version of KDD-CUP'99 dataset which contains various unnecessary features. Due to the deletion of redundant records the classification is reported to be less biased. Detection rates with adopted dataset is significantly higher due to less presence of duplicate records. There are 41 features containing different attributes. Corresponding labels are assigned to each feature which categorise them as normal or an attack. The 42nd feature contains information about various attack classes in dataset. These attacks are recognised as Denial-of-Service (DoS), User-to-root (U2R), Root-to-local (R2L) and Probe. Rest of the records are defined as normal patterns. In this research we are using *KDDTrain20* for training the classifier as it has good quantity of anomalous records. Further information about NSL-KDD can be found in Table 1.

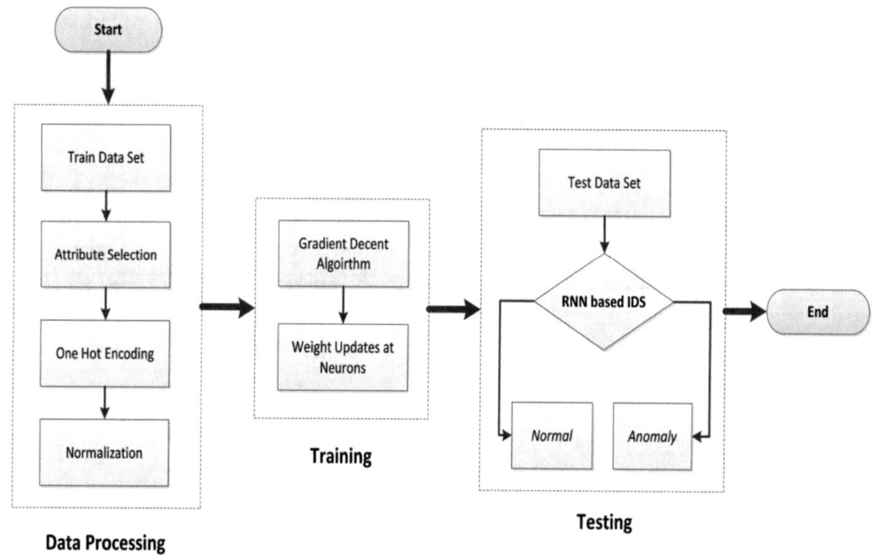

Fig. 1. Proposed RNN-IDS

Table 1. Description of dataset [6,19]

NSL-KDD data	Data instances	Normal traffic	Attack traffic (%)
Train20	25,192	13,499	46.6
Train+	125,973	67,343	46.5
Test+	22,544	9711	56.9
Test-	11,850	2152	81.8

Attribute Selection. There are total 41 input features and 1 output class feature in data space of NSL-KDD dataset. Some of the features have low sample to feature ratio which can result in high false positive rates during classification. Also, excluding the less important features would enable us to train RNN-IDS faster whilst reducing the training time. In [20], has summarised features such as *dst_host_rerror_rate*, *su_attempted*, *num_access_files*, *num_file_creations*, *num_outbound_cmds*, *dst_host_count*, *is_host_login* and a few others which either have zeroed values or less feature space. 29 features are extracted using different feature reduction techniques. We are training the proposed scheme with reduced features as well as complete feature of NSL-KDD.

Pre-processing with Encoding. A few features in NSL-KDD dataset such as Flag, Protocol Type and Service are not numeric and contain label values. Since the proposed IDS would be trained and tested with RNN, hence all the remaining features must be converted into numerics before training. To achieve this task, we have used one hot encoding and converted all nominal values to integer values based on their existence in dataset.

Normalization. Data Normalization is a technique used for the transformation of input data where its occurrence is highly divergent. The data is restructured before it is utilized, because without such pre-processing the classifier take more than normal time to train the proposed IDS. Min-Max Normalization technique has been utilized in this research so that input value can be mapped between [0 and 1] range effectively.

It can be denominated as:

$$v_i = \frac{u_i - \min(u)}{\max(u) - \min(u)}, \tag{12}$$

where $u = (u_1, \ldots, u_n)$ is the number of input values and $v_(i)$ is the output normalized data.

4 Experimental Results and Analysis

In this research, the proposed a heuristic intrusion detection system for IOT environment using Random Neural Network (RNN-IDS), has been trained and tested in controlled environment using MATLAB installed on Intel Core(i5) processor and 16 GB RAM. The algorithm used for training the network is Gradient Descent (GD).

The IDS would be considered accurate in classification of anomalous records from normal records if it has low false positive rates and it predicts the outcome with high precision [2]. The accuracy of IDS is interpreted by True Positives (TP) which is denominated as α, True Negatives (TN) as β, False Positives (FP) as γ while False Negative (FN) as δ, respectively. The actual intrusion patterns in NSL-KDD and predicted attacks by RNN-IDS could be represented in the form of confusion matrix as outline in Table 2. Total accuracy of proposed scheme can be calculated as:

Table 2. Confusion matrix of attack sequence

		RNN Attack Prediction		
		p	**n**	**aggregate**
Actual Intrusion	p′	True Positive	False Negative	P′
	n′	False Positive	True Negative	N′
	aggregate	P	N	

$$Accuracy\,(RNN\text{-}IDS) = \frac{\alpha + \beta}{\gamma + \delta + \alpha + \beta} \tag{13}$$

In order to completely estimate the performance, some other matrices are:

$$Precision = \frac{\alpha}{\alpha + \gamma} \tag{14}$$

$$Detection\,Rate = \frac{\alpha}{\alpha + \delta} \tag{15}$$

$$False\,Discovery\,Rate = \frac{\gamma}{\alpha + \gamma} \tag{16}$$

For the comparison of results, with same number of hidden layer neurons as used in [10], two methods have been are adopted for training and testing RNN-IDS, where system is trained with varient learning rates of 0.01, 0.1 and 0.4, respectively.

Method I. In first method, as per last contribution, the reduced features as reported in [20] are used. RNN-IDS has 29 input layer neurons which are connected by 21 hidden layer neurons. Since its a binary classification and we want to predict anomalies after testing the system with *KDDTest+* dataset, we have 1 output layer neuron which would predict the attacks from normal traffic based on information it receives from hidden layer.

Method II. In second method, the complete 41 features of NSL-KDD dataset are utilized. Here we have 41 input layer neurons connected to 21 hidden layer neurons while 1 output layer neuron is used to quantify network attack. The network is trained with given dataset and tested against *KDDTest+*.

To completely demonstrate the performance of proposed scheme, Table 3 highlights the statistics collected against different performance metrics. The increase in number of true positives with change in learning rate for both methods is significant due to the fact that RNN-IDS has correctly quantified the intrusions from normal patterns. Also, the decrease in false positives indicates that the performance of intrusion detection system is improving with change in learning rate. It is evident from the previous findings, any system with low false positives is considered to be accurate and method II of proposed scheme has reduce false discovery rate to 0.09%.

Also, RNN-IDS model proved its robustness in Method-II, where the precision to detect anomalies from normal traffic is increased from 93.6% to 99.02%. Analysis of collected results suggested that, the decrease in learning rate gradually increased the detection of network attacks up-to 95%. This happened due to the fact that even though network would converge slowly, but learning algorithm is not missing any local minima in calculation of weights and biases for the neurons.

After training the system and testing it against desired *Test+* dataset, the results as shown in Fig. 2 make clear indication that the performance of RNN-IDS is more accurate in Method-II for the prediction of unknown attacks. As it

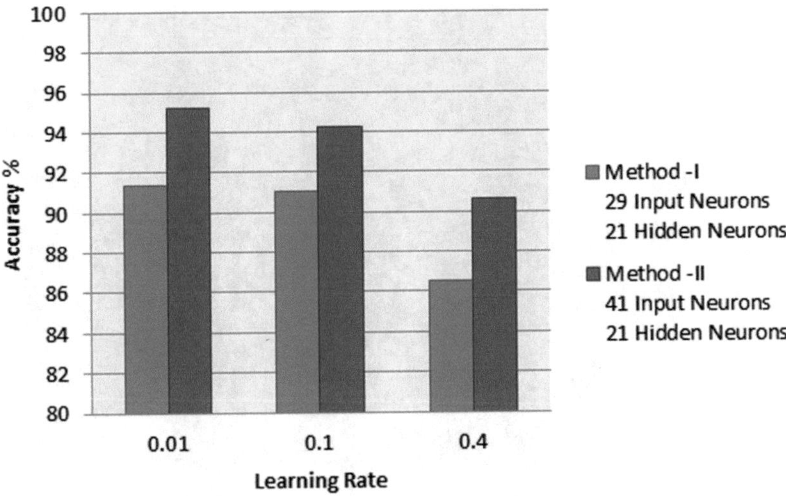

Fig. 2. Accuracy of proposed RNN-IDS model with different learning rates

Table 3. Results - RNN-IDS feature based comparison

Performance metrics	Learning rates					
	Method - I			Method - II		
	0.4	0.1	0.01	0.4	0.1	0.01
True Positive (TP)	91.72	93.36	94.24	93.96	95.68	95.58
True Negative (TN)	2.06	3.72	2.46	2.14	2.62	3.12
False Positive (FP)	6.24	2.94	3.32	3.92	1.72	0.92
False Negative (FN)	8.28	6.64	5.76	6.04	4.32	4.02
Detection Rate	91.7	93.3	94.2	93.96	95.6	95.90
Precision	93.6	96.5	96.6	96.0	98.2	99.02
False Discovery Rate	6.3	3.0	3.4	0.04	0.01	0.09
Mean Square Error	0.05	0.04	0.03	0.03	0.03	0.02
Accuracy	86.5%	91.0%	91.4%	90.6%	94.2%	95.2%

utilize full feature space and the total of 41 input neurons contributed to predict the anomaly.

In order to test the operation of proposed random neural network based IDS we have compared the results with different machine learning algorithms such as J48, Support Vector Machine (SVM), Naive Bayes (NB). Naive Bayes Tree, Multi Layer Perceptron (MLP), Random Forest (RF), Random Forest Tree, Recurrent Neural Network and Artificial Neural Network (ANN), respectively. Different performance matrices are used to estimate the overall efficiency of RNN-IDS such as detection rate, false negative rate, detection rate, precision,

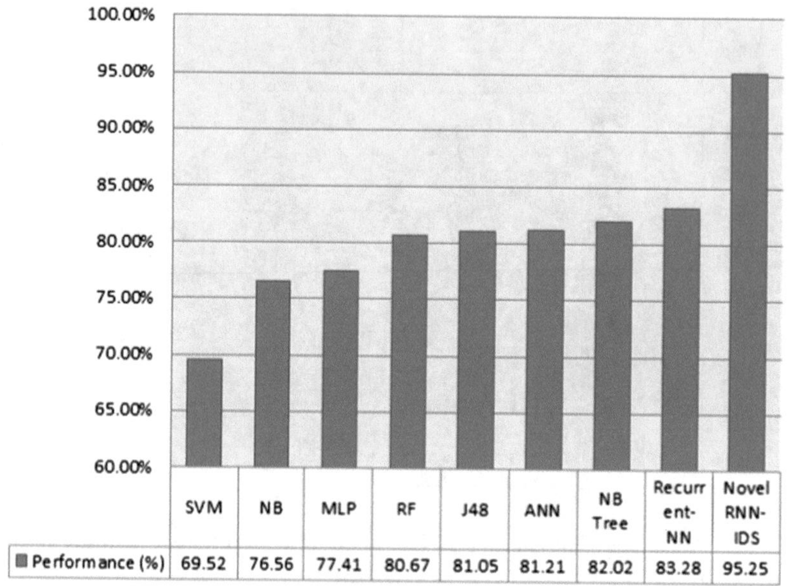

	SVM	NB	MLP	RF	J48	ANN	NB Tree	Recurr ent-NN	Novel RNN-IDS
Performance (%)	69.52	76.56	77.41	80.67	81.05	81.21	82.02	83.28	95.25

Fig. 3. Comparison of proposed RNN-IDS with several ML methods

false discovery rate and mean square error which would account towards the calculation of accuracy in detection of attacks. The results revealed that RNN-IDS has the highest accuracy of 95.2% for detecting novel attacks with next best of 83.2% in case of Recurrent Neural Networks.

Based on the results shown in Figs. 2, 3 and Table 3, the following facts can be inferred:

- Mean Square Error is decreased when learning rate is reduced. Although learning is slow but RNN-IDS has performed classification more accurately and false positives are reduced.

$$Mean\,Square\,Error = \frac{1}{n}\sum_{i=1}^{n}(\pi_{RNN} - \pi_a)^2 \tag{17}$$

  Where, π_{RNN} is the predicted intrusion based on trained RNN-IDS system while π_a is an actual intrusion.

- In comparison of reduced and complete features, the accuracy of RNN-IDS is increased from 86.5% to 95.25%, where it has classified intrusions in the network with high precision rate of 99.02%.
- The proposed RNN-IDS has performed many folds better than traditional machine learning algorithms such as J48, Support Vector Machine (SVM), Naive Bayes (NB). Naive Bayes Tree, Multi Layer Perceptron (MLP), Random Forest (RF), Random Forest Tree, Recurrent Neural Network and Arti-

ficial Neural Network (ANN), with higher accuracy and low false positive rates.

5 Conclusion

In this research we have proposed a novel intrusion detection system using the feed-forward nature of Random Neural Networks (RNN-IDS). Two methods were adopted to estimate the performance relating to different number of input, but identical hidden layer neurons. The proposed model was trained and further tested with NSL-KDD dataset. The comparison of empirical results trained with different learning rates revealed that RNN-IDS accuracy reached up to 95.2%. The performance is also compared with other machine learning algorithms such as J48, SVM, NB, NB Tree, MLP, RF, RF Tree, recurrent neural network and ANN, where proposed RNN-IDS scheme has surpassed all of them for the detection of anomalies in network.

References

1. Conti, M., Dehghantanha, A., Franke, K., Watson, S.: Internet of Things security and forensics: challenges and opportunities. Future Gener. Comput. Syst. **78**, 544–546 (2018). https://www.sciencedirect.com/science/article/pii/S0167739X17316667
2. Saeed, A., Ahmadinia, A., Javed, A., Larijani, H.: Intelligent intrusion detection in low-power IoTs. ACM Trans. Internet Technol. **16**(4), 1–25 (2016). http://dl.acm.org/citation.cfm?doid=3023158.2990499
3. Moustafa, N., Creech, G., Slay, J., Moustafa, N., Creech, G., Slay, J.: Big data analytics for intrusion detection system: statistical decision-making using finite Dirichlet mixture models (2017). https://www.unsw.adfa.edu.au/australian-centre-for-cyber-security/
4. Qureshi, A.U.H., Larijani, H., Ahmad, J., Mtetwa, N.: A novel random neural network based approach for intrusion detection systems. In: 2018 IEEE 10th International Computer Science and Electronic Engineering Conference (CEEC). IEEE, September 2018
5. Aljawarneh, S., Aldwairi, M., Yassein, M.B.: Anomaly-based intrusion detection system through feature selection analysis and building hybrid efficient model. J. Comput. Sci. **25**, 152–160 (2018). https://linkinghub.elsevier.com/retrieve/pii/S1877750316305099
6. Kwon, D., Kim, H., Kim, J., Suh, S.C., Kim, I., Kim, K.J.: A survey of deep learning-based network anomaly detection. Cluster Comput. 1–13 (2017). https://doi.org/10.1007/s10586-017-1117-8
7. LeCun, Y., Bengio, Y., Hinton, G.: Deep learning. Nature **521**(7553), 436–444 (2015). http://www.nature.com/articles/nature14539
8. Meng, F., Fu, Y., Lou, F., Chen, Z.: An effective network attack detection method based on kernel PCA and LSTM-RNN. In: 2017 International Conference on Computer Systems, Electronics and Control (ICCSEC), pp. 568–572. IEEE, December 2017. https://ieeexplore.ieee.org/document/8447022/

9. Tavallaee, M., Bagheri, E., Lu, W., Ghorbani, A.A.: A detailed analysis of the KDD CUP 99 data set. Technical report. http://nsl.cs.unb.ca/NSL-KDD/
10. Ingre, B., Yadav, A.: Performance analysis of NSL-KDD dataset using ANN. In: 2015 International Conference on Signal Processing and Communication Engineering Systems, pp. 92–96. IEEE, January 2015. http://ieeexplore.ieee.org/lpdocs/epic03/wrapper.htm?arnumber=7058223
11. Yin, C., Zhu, Y., Fei, J., He, X.: A deep learning approach for intrusion detection using recurrent neural networks. IEEE Access **5**, 21 954–21 961 (2017). http://ieeexplore.ieee.org/document/8066291/
12. Gelenbe, E.: Random neural networks with negative and positive signals and product form solution
13. Javed, A., Larijani, H., Ahmadinia, A., Emmanuel, R., Mannion, M., Gibson, D.: Design and implementation of a cloud enabled random neural network-based decentralized smart controller with intelligent sensor nodes for HVAC. IEEE Internet Things J. **4**(2), 393–403 (2017). http://ieeexplore.ieee.org/document/7740096/
14. Javed, A., Larijani, H., Ahmadinia, A., Gibson, D.: Smart random neural network controller for HVAC using cloud computing technology. IEEE Trans. Ind. Inform. **13**(1), 351–360 (2017). http://ieeexplore.ieee.org/document/7529229/
15. Ahmad, J., Larijani, H., Emmanuel, R., Mannion, M., Javed, A.: An intelligent real-time occupancy monitoring system using single overhead camera. In: Proceedings of SAI Intelligent Systems Conference, pp. 957–969. Springer (2018)
16. Simonyan, K., Zisserman, A.: Very deep convolutional networks for large-scale image recognition, September 2014. http://arxiv.org/abs/1409.1556
17. Khan, G.M.: Artificial Neural Network (ANNs), pp. 39–55. Springer, Cham (2018). https://doi.org/10.1007/978-3-319-67466-7_4
18. Ahmad, J., Larijani, H., Emmanuel, R., Mannion, M., Javed, A., Phillipson, M.: Energy demand prediction through novel random neural network predictor for large non-domestic buildings. In: 2017 Annual IEEE International Systems Conference (SysCon), pp. 1–6. IEEE, April 2017. http://ieeexplore.ieee.org/document/7934803/
19. NSL-KDD—Datasets—Research—Canadian Institute for Cybersecurity. http://www.unb.ca/cic/datasets/nsl.html. Accessed 03 May 2018
20. Bajaj, K., Arora, A.: Improving the intrusion detection using discriminative machine learning approach and improve the time complexity by data mining feature selection methods. Int. J. Comput. Appl. (975–8887) **76**(1) (2013). http://citeseerx.ist.psu.edu/viewdoc/download?doi=10.1.1.481.8435&rep=rep1&type=pdf

A Framework to Evaluate Barrier Factors in IoT Projects

Marcel Simonette[1(✉)], Rodrigo Filev Maia[2], and Edison Spina[1]

[1] Universidade de São Paulo, Cidade Universitária, São Paulo, Brazil
{marceljs, spina}@usp.br

[2] Centro Universitário FEI, SBC, Brazil
rfilev@fei.edu.br

Abstract. IoT services need to be accessed by their users anytime and anywhere, regardless of the access network technology or user device. These requirements bring several challenges to the design, development, and operation of IoT systems. To develop these systems, engineers must deal with multiple device technologies, engineering standards, human factors, business processes, and sustainability issues. The successful implementation of IoT systems requires more than engineers' technical skills; it is necessary to understand the management issues of the engineering effort. IoT systems have several dimensions, which demands a systemic management process to promote the system as a whole. We need to identify the relevant issues of the IoT project to properly execute it, dealing with the risks and barriers of implementation and operation. To identify the project feasibility barriers and the managerial approach to IoT projects, we use two frameworks: BRICS Mosaic model and NCTP framework. They are models that apply the experience of engineers and the analysis of application scenarios to identify barriers to IoT solutions and the managerial approach to steer the solution.

Keywords: Internet of Things · Systems engineering · Project management

1 Introduction

The processes of design, implementation and operation of IoT systems are complex. It is necessary to deal with multiple device technologies, engineering standards, business processes, human factors, and sustainability issues. As an answer to this complexity, we propose using two systems engineering disciplines, model-based systems engineering and systems engineering management.

Model-based systems engineering has concepts that allow engineers to develop models to understand problems, develop candidate solutions, and validate design decisions. Furthermore, systems models can be used to identify risks and obstacles to solution implementation [1, 2].

The successful implementation of IoT systems requires more than the technical skills of engineers. Managerial traits are necessary as well. Engineering projects have a set of basic management principles that is always present. However, the design and implementation of IoT systems push the boundaries of both systems engineering and

K. Arai et al. (Eds.): CompCom 2019, AISC 997, pp. 99–112, 2019.
https://doi.org/10.1007/978-3-030-22871-2_8

management. The intrinsic complexity of IoT systems requires the systems engineering management as the approach to deal with the risks and resources of IoT systems design and implementation, preventing accidents or systems failures that may occur not only due to engineering issues, but also as a result from the management of the design and implementation process [3–6].

The contributions of this article are the inclusion of a new plane, called sustainability, at the BRICS Mosaic model [7, 8]. Moreover, this work includes presenting how the BRICS Mosaic model and its Feasibility Barriers Factors (FBF) can be used to identify the barriers to the implementation and operation of IoT systems. Also, it provides an example of FBF used in conjunction with the Novel, Complexity, Technology, and Pace (NCTP) framework to identify the managerial traits necessary to steer the life cycle of an IoT system, from lust-to-dust. The case studied is IoT system for urban lighting infrastructure.

This article is organized as follows: we present an overview of the BRICS Mosaic model, a paradigm of model-based systems engineering, and an overview of the NCTP framework, a systems engineering management framework. Afterward, a case study is provided. Then, we conclude our article.

We chose to use a case study because this approach allows the characterization of real-life events, such as organizational and managerial processes, and requirements constraints. Furthermore, this approach enables a holistic view regarding the problem to be treated [9, 10].

2 BRICS Mosaic Model

BRICS Mosaic Model was developed using the concept plans of Next Generation Networks (NGN) [11], the ITU Y.2060 reference model [12], and the results from two projects of the 7th Framework Programme for Research and Technological Development (FP7): CSA for Global RFID-related Activities and Standardization (CASA-GRAS2) [13], and Internet of Things Architecture (IoT-A) [14]. BRICS Mosaic is a representative concept that organizes the IoT research areas into planes of a cylindrical mosaic, represented in Fig. 1. The planes are the reason for naming the model: BRICS, which is an acronym for: Building blocks of Research for the Internet-Connected objects.

2.1 The Mosaic Planes

Each one of the BRICS Mosaic planes is a representation of a specific solution view. They are considered "planes of functionality." The first plane, at the top of the cylinder of Fig. 1, represents the *Technological* view; the following planes represented in Fig. 1 are a representation of *Security, Business Process, Sustainability, Integrated Management and Control, Regulations,* and *Human Factors*. Each plane has the same set of dimensions that drive, influence and affect the development of the services provided by IoT systems, namely (Fig. 2): Environment, Context, Human Users, Devices, Applications, Access Networks, and Backbone. The initial version of the BRICS Mosaic model did not consider the sustainability of IoT projects [7, 8]. We include the

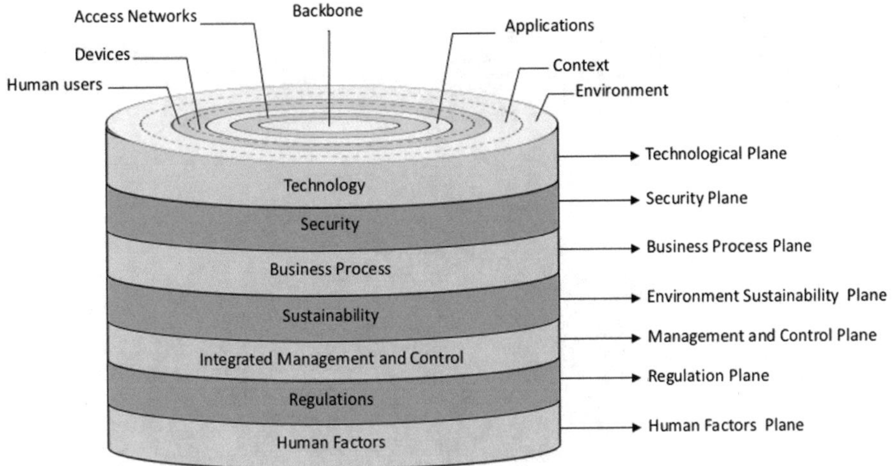

Fig. 1. BRICS Mosaic model.

Sustainability plane due to the relevance of the sustainability issues that IoT projects must consider; it is a relevant barrier factor to projects not treated in earlier versions of the model.

BRICS Mosaic must be considered as a whole. No single plane and no single dimension, or a single dimension of a plane, can produce a satisfactory model of an IoT system. The planes represented in Fig. 1 are not the unique solution view in an IoT system. They are the minimal set of views that drive, influence and affect IoT solutions. If any other view is identified in an IoT solution, it may be considered as another plane in Mosaic.

2.2 Feasibility Barrier Factors

A Feasibility Factor is a barrier, a restriction to be overcome. All the planes of functionality in the BRICS Mosaic model have the same set of Feasibility Factors, and each Factor in each plane is represented as the zone between two concentric circles in this plane. Furthermore, each zone represents a different medium in which information is carried over. The concentric circles in each plane represent the fact that data can transit from any point to any other point.

We use the *Sustainability* plane of the BRICS Mosaic model (the fourth plane represented in Fig. 2), to exemplify the planes zones. The peripheral zone of a plane is the physical environment itself; it can have Contexts or places. The following zone represents Human Users; this zone is separated from the Devices zone by a dotted circumference, which represents the separation between humans and their computation devices (mobile or not). The Applications zone represents the applications executed in those devices. The Access Network zone represents both wired and wireless networks, which use other networks to reach the Backbone of the internet, the centre zone, which represent the data routes among computer networks and internet routers.

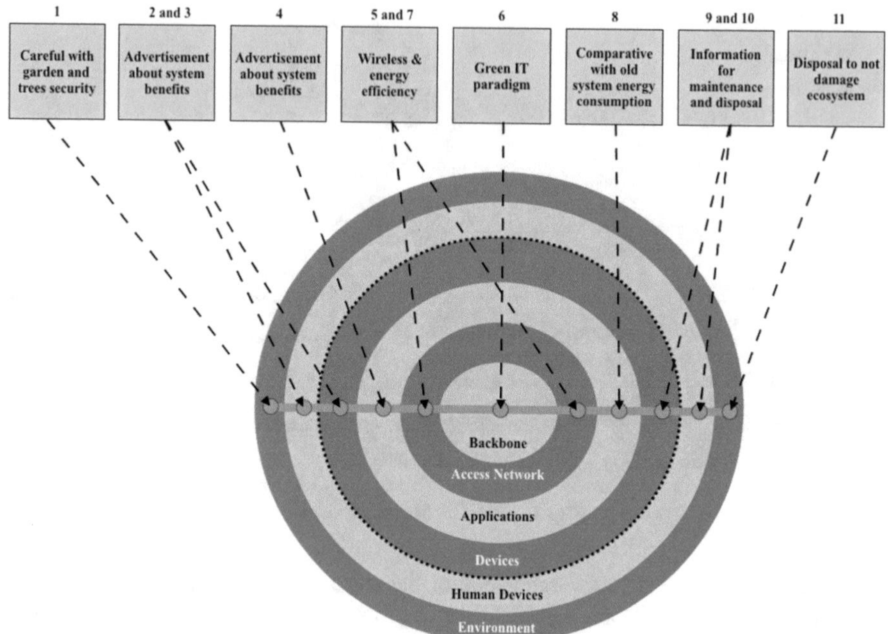

Fig. 2. The transversal path in the *Sustainability* plane of the BRICS MOSAIC model.

Although our walk through the *Sustainability* plane arrived at its centre, it does not mean the end of an IoT solution. IoT solutions use electronic devices that must be discarded properly. The data about the devices must go through the Access Network and Applications that control the maintenance of the devices, and, then, at the end of their lifecycle, they are disposed, returned to the environment. When we consider this scenario, we have a transversal path, represented by the green line in Fig. 2, which depicts the kinds of technologies, applications, users, and so forth that need to be involved in an IoT service. This transversal path has 11 components (zones) and each one is a Feasibility Barrier Factor (FBF). Note that some of these components may represent technological alternatives, while others may be requirements. For example, for component 5 in Fig. 2, the Access Network zone, technologies such as WiMAX, 3G, and LTE may be considered a solution to deploy the IoT service, or it may indicate that the IoT service requires these three technologies.

The analysis of each FBF is a qualitative analysis of the events of the services, which values the importance of the factor to the service execution. The possible values to an FBF are: Very Low (VL); Low (L); Moderate (M); High (H) and Very High (VH). The case study below presents how we perform this qualitative analysis.

3 NCTP Framework

Considering the intrinsic complexity of an IoT solution, there is not a single set of management practices that fits all IoT systems projects. Furthermore, the failures in engineering systems often occur due to organizational errors, which demands a link between project domain and management practices [15]. Saucer [6] proposes a framework that helps systems engineers to plan and to execute systems projects phases by using management practices including systems engineering principles. This framework, NCTP framework, uses four dimensions to classify a project: Novelty, Complexity, Technology, and Pace (NCTP), which are used to evaluate the environment and the tasks to identify the right management practice.

The dimensions and their subfactors of the NCTP framework are [6, 16]:

- *Novelty* is how new the system (product and, or service) is to the market:
  - *Derivative*: Extensions and improvements of existing systems.
  - *Platform*: New generation in existing systems families.
  - *Breakthrough*: Introduce a new concept, idea, or use.
- *Complexity* means how complex the system is:
  - *Assembly*: One unit composed of a collection of components, performing a single function.
  - *System*: A sophisticated collection of interactive elements and subsystems, jointly dedicated to a wide range of functions to meet a specific operational need.
  - *Array*: Collections of systems (system of systems) that function together to achieve a common purpose.
- *Technology* means an extension of a new technology used in the system:
  - *Low-tech*: Relies on existing and well-established technologies.
  - *Medium-tech*: Uses an existing or base technology and incorporates some new technology or new feature non-existent in previous systems.
  - *High-tech*: Employs new technologies; nevertheless, they already exist when the system life cycle starts.
  - *Super-high-tech*: Uses new technologies that do not exist in the initiation of system life cycle. The system mission is clear while the solution is not.
- *Pace* is related to project urgency and available time:
  - *Regular*: The time demanded by project execution is not critical to immediate organizational success.
  - *Fast-competitive*: Project conceived to address market opportunities, form new business lines, or create a strategic positioning. Missing a deadline may not be fatal; however, it could impact profits and competitive positioning.
  - *Time-critical*: Project completion is time critical with a window of opportunity.
  - *Critical-Blitz*: Urgent developments to solve crisis or emergencies.

NCTP framework allows classifying a system project by using characteristics that make the project endeavour unique in how it is managed. Figure 3 presents the four dimensions of the framework on a graph; the lines connecting the subfactors of NCTP form a diamond that represents a qualitative representation of the risk level.

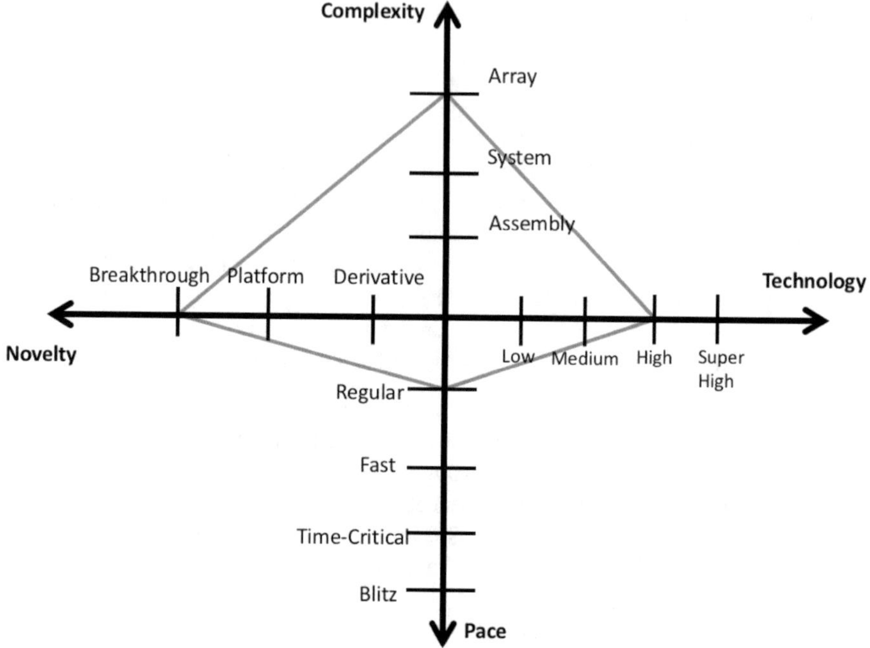

Fig. 3. Example of NCTP framework graph. The diamond in the graph gives a qualitative representation of the level of project risk

4 Case Study: IoT Lightning System

We present a case study to show how the BRICS Mosaic model and NCTP can help to identify the difficulties present in the deployment of an IoT service, which allows identifying the actions to be undertaken to enable the IoT system deployment and maintenance. It is a case study of an IoT system for urban lighting infrastructure.

Urban public lighting infrastructure is one of the systems composing the urban public infrastructure. The requirements for this urban lighting system are:

- The use of white led lamps.
- Regard public security at night.
- It must have minimum interference with gardens and trees.
- The light distribution should be appropriate for urban security cameras.
- Enhancing buildings architecture must be appropriate.
- It must be energy efficient.
- Remote management stations must monitor the lighting equipment.
- All the equipment should be programmed to operate with different levels of illumination, according to the place and daytime.

Our experience in systems engineering allows us to consider these requirements, classifying the project of this case study by using the four dimensions of the NCTP framework:

- *Novelty - Breakthrough*: Introduces a new concept, or a new idea, or new use.
- *Complexity - Array*: Large systems, widely dispersed collections of systems (system of systems) that function together to achieve a common purpose.
- *Technology - High-tech*: Represents situations in which most of the technologies employed are new; nevertheless, they already exist when the system life cycle starts.
- *Pace - Regular*: Efforts where time is not critical to immediate organizational success.

The above classification defines the project characteristics, which allows identifying the management practices that we may adopt. In a graph, Fig. 3 presents the four dimensions in which we classified the project, and, by connecting the four NCTP classifications with a straight line, we have a diamond. This diamond gives a qualitative representation of the risks to be considered in project management [6]. We use this qualitative information to reflect on the set of management practices to steer the project, as IoT projects may fail because systems engineer management may assume that IoT projects are "just another project." This assumption may happen if we disregard the complexity of IoT, which demands the identification of the appropriate management practices for each of the zones represented in the BRICS Mosaic planes.

To identify the difficulties in the urban public lighting infrastructure IoT systems project, and to identify if the preferred set of management practices to manage the project, we walk through the line represented in Fig. 2, from one side of the cylindrical model to the other, from FBF number one to FBF number eleven. The analysis of each FBF in any of the planes is a qualitative analysis of the events that occur in the plane; this analysis gives quantitative values to the importance of the FBF to the service execution. The BRICS Mosaic model defines five values to an FBF: Very low (VL); Low (L); Moderate (M); High (H) and Very High (VH).

We need to analyse each FBF, from number one to number eleven, in all the BRICS Mosaic planes. However, as we are giving a short example, we performed the analysis for only two planes: *Business Process* plane and *Environment Sustainability* plane. The analysis of each FBF is performed by a short story elaborated by using the requirements and the experience of the engineers performing the analysis.

4.1 FBF1: Environment - Context

The environment of the urban public lighting system infrastructure is a campus, in which there are several buildings surrounded by gardens. People transit through the several streets and avenues from 6:00 am to 11:00 pm on business days.

For the *Business Process* plane, the environment has to be continuously monitored to provide administration of the asset to avoid damages to buildings and equipment in the public areas. Moreover, this monitoring creates a comfortable and safe place for the people who transit the buildings and the streets of the campus. These users do not have direct interaction with the lighting control system; they only enjoy the services provided. *FBF evaluation*: M.

For the *Environment Sustainability* plane, the lighting system cannot cause damage to the species of plants and trees in garden areas. A balance between the security provided by illumination and the physiology and metabolism of plants in the garden areas. *FBF evaluation*: H.

4.2 FBF2 and 3: Human and Users-Devices

Although the people who transit the buildings and the streets of the campus may be considered "systems users", they do not have any action in it. Only the public administration staff and technical staff of the campus will interact with the systems by managing the lighting service operations.

For the *Business Process Plane*, the lighting system uses fibber optics and 3G/4G network to exchange data throughout the equipment. The users cannot have access to the system through any devices, such as mobile phones. Lighting systems must have mechanisms to control access to the equipment as well as auditing mechanisms. *FBF evaluation:* M.

For the *Environment Sustainability Plane*, the benefits of the system should be highlighted to users, to encourage them to preserve the system against depredation. *FBF evaluation:* VL.

4.3 FBF4: Application

People transiting buildings and the streets of the campus do not have any application to interact with the system directly. The technical staff must have equipment for remotely monitoring all the lightning system devices and the mobile devices, such as laptops and tablets to verify each equipment locally, such as microcontroller systems installed in the street lights.

For the *Business Process Plane*, no interaction with users, only the public administration staff and the technical staff must have access to the equipment. Advertisements of the new system are important as an informative action to the public. *FBF evaluation*: VL.

For the *Environment Sustainability Plane*, no interaction with users. Advertisements of the system environment sustainability features are important as informative action to the public. *FBF evaluation*: VL.

4.4 FBF5 and 7: Access Network

Every equipment exchanges data with a centralized control system by fibber optic and 3G/4G network provided by a private telecom operator. This network is responsible for the data exchange about device monitoring, billing, automation and control.

For the *Business Process* plane, the cost of implementation is highly important, and mobile communication is expensive, since it is uses 3G/4G communication. An appropriate business model has to be developed for services to become economically feasible. *FBF evaluation*: H.

For the *Environment Sustainability* plane, the wireless communication may require the minimum amount of energy for the microcontroller control mechanisms of the street lights. The messages to be transmitted by the equipment should be short to reduce the need for transmission. However, this must not compromise the robustness of the system. *FBF evaluation*: M.

4.5 FBF6: Backbone

The backbone will interconnect all servers and control equipment of the lighting system, besides connecting the computing system with maps (geo reference). The backbone must have a high level of availability and must have enough throughput to transmit data.

For the *Business Process* plane, the users must be defined with profiles based on business functions. A list of business requirements must be identified due to system security requirements. The cost of the system implementation is highly important, and the network equipment and management may become expensive. An appropriate business model has to be developed to meet business requirements with network features to make it an economically feasible project. *FBF evaluation*: H.

For the *Environment Sustainability* plane, the equipment should follow the green IT paradigm and not spend too much energy. The technical room should be properly refrigerated, and all shuck room environment should be designed to reduce the use of air conditioning while keeping all the equipment in the right operational temperature. *FBF evaluation*: H.

4.6 FBF8: Applications

The control application must be web-based and must have the following features:

- Monitor each equipment individually (lamps, luminaire).
- Remotely turn each luminaire on or off.
- Programming the automatic lighting system equipment activation (individually or groups of) based on GPS coordinates and time of day.
- Monitor the fails and alarms of each equipment.

For the *Business Process* plane, the system must be able to correlate data from sets of equipment to provide data for predictive maintenance, to prioritize critical repairs besides keeping several kinds of managerial reports to present information about the whole system, set equipment fails, and have complete logs of each piece of equipment to analyse its operational behaviour. The application must collect data from each equipment for the system to determine the consumption pattern of the entire system precisely. *FBF evaluation*: H.

For the *Environment Sustainability* plane, the system must have tools to provide information about the use of energy and the economy caused by the new system (compared to the previous lighting system). *FBF evaluation*: M.

4.7 FBF9 and 10: Human Users - Devices

There are two groups of users:

- Technicians and engineers responsible for operating the entire system and taking all the necessary actions to keep the system running.

- Managers who evaluate the system, make decisions regarding its benefits and decide its evolution.

For the *Business Process* plane, all the devices must work and keep their programmed tasks independent of the control system communication. Every field technician must have portable devices to check and to operate the system. All devices must monitor several parameters such as: lamp status, power supply status, operational status of all the components and total time of operation. *FBF evaluation*: M.

For the *Environment Sustainability* plane, all the devices must provide a set of data about the luminaire state, the cumulative operating hours, the power consumption and other relevant data to evaluate the energy efficiency of each equipment and possibly evaluate the maintenance interval to increase the equipment lifetime. All the devices must have an instruction set on how to dispose of it, taking into consideration how to replace the equipment and what the proper destination of the damaged or obsolete equipment is. Specially consider batteries or any other harmful piece or device. *FBF evaluation*: H.

4.8 FBF11: Environment - Context

Urban lighting systems are complex systems. They need to interact with other systems to be effective and must provide several interfaces to exchange data between systems. For instance, the security system (surveillance), telecommunication system and energy monitoring system must interact with the lighting system for its correct operation. Information from different systems must be integrated to be analysed by Big Data systems to provide a complete view about the systems on the campus.

For the *Business Process* plane, the cost of the system implementation is highly important. Expenses and revenues from different stakeholders have to be taken into account. It is hard to balance costs and guarantee an appropriate return on investment from this public service. *FBF evaluation*: H.

For the Environment Sustainability Plane, the disposal of all the equipment must be controlled so as not to damage the environment; also, the misbehaviour of one system cannot interfere with other systems. It is relevant to verify whether artificial light affects the physiology and metabolism of plants to keep the gardens of the campus healthy. It is also important to manage the infrastructure properly so as not to generate more waste than effectively needed, and to control the use of chemical products in the installation and maintenance for not damaging the soil on campus. *FBF evaluation*: H.

4.9 FBF Evaluation

The same set of FBF is presented in all BRICS Mosaic planes. In our reduced example, we elaborated short stories to enable a quick understanding of the feasibility factors of only two planes of functionality. In a full case, it is necessary to evaluate each FBF in all planes of functionality.

Following the evaluation process, after the qualitative FBF identification, it is necessary to compare the FBFs, to identify the set of factors that affect the system. We perform it by using a simple translation of the FBF evaluation to a numeric value, to transform the qualitative information into numerical values. It is worth noting that the values used in this transformation have been chosen to enable the following interpretation: if the FBF has been evaluated as VH in all seven planes, then the total evaluation is equal to seven, and this value corresponds to a 100% barrier, i.e., the FBF is a barrier, or an impediment to project execution. If the FBF has been evaluated as VL in all the seven planes, then the total evaluation is equal to 1.4, and this has been arbitrarily set to a 10% barrier. Intermediate cases are evaluated by linear interpolation. The relation between the FBF evolution and numerical value is presented in Table 1.

Table 1. FBF translation from qualitative value to numeric value.

FBF evaluation	FBF numerical value
VL	0.2
L	0.4
M	0.6
H	0.8
VH	1.0

We use the numerical values to analyse the distribution of each FBF for each Mosaic plane. This information is important because the representation of the FBF distribution in a graph allows systems engineers to perceive the planes of functionality that are critical to project development. Figure 4 presents the distribution of our example, and we can observe that FBF1: Environment – Context, FBF4: Application, FBF6: Backbone, FBF8: Applications, FBF9 and 10: Human Users – Devices, and FBF11: Environment – Context have a remarkable presence in the *Environment Sustainability* plane, indicating that the plane demands attention from the systems engineering management. If this plane does not receive the necessary attention, the technological development will be a waste of time and resources, as the system will not meet all its requirements.

4.10 Systems Engineering Management

According to our experience, the *Novelty*, *Complexity*, *Technology*, and *Pace* of the urban lighting system from our example is represented in Fig. 3. However, FBF analysis indicates that special attention with the *Environment Sustainability* plane of the BRICS Mosaic model is necessary, indicating that this project domain demands different management practices. To understand the dimensions that need attention of the systems engineering management, we use the NCTP framework. Considering the stories from the FBF analysis, the values of NCTP dimensions for the Environment Sustainability plane are:

- *Novelty = Breakthrough.*
- *Complexity = System.*
- *Technology = High-tech.*
- *Pace = Fast Competitive.*

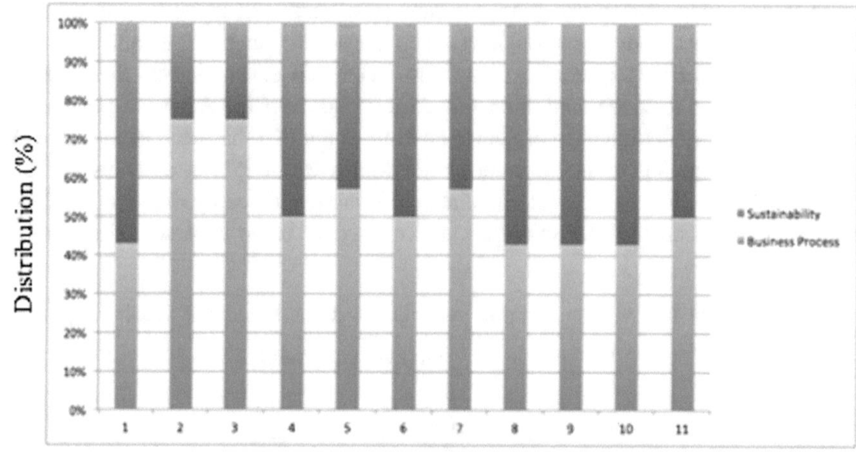

Fig. 4. Distribution of the eleven FBF in the BRICS Mosaic planes: *Business Process* and *Environment Sustainability*.

Figure 5 represents the NCTP framework; the dashed line represents the approach identified by using the FBF of BRICS Mosaic model. The solid line represents our initial approach to the IoT project development as a whole. The analyses in Fig. 5

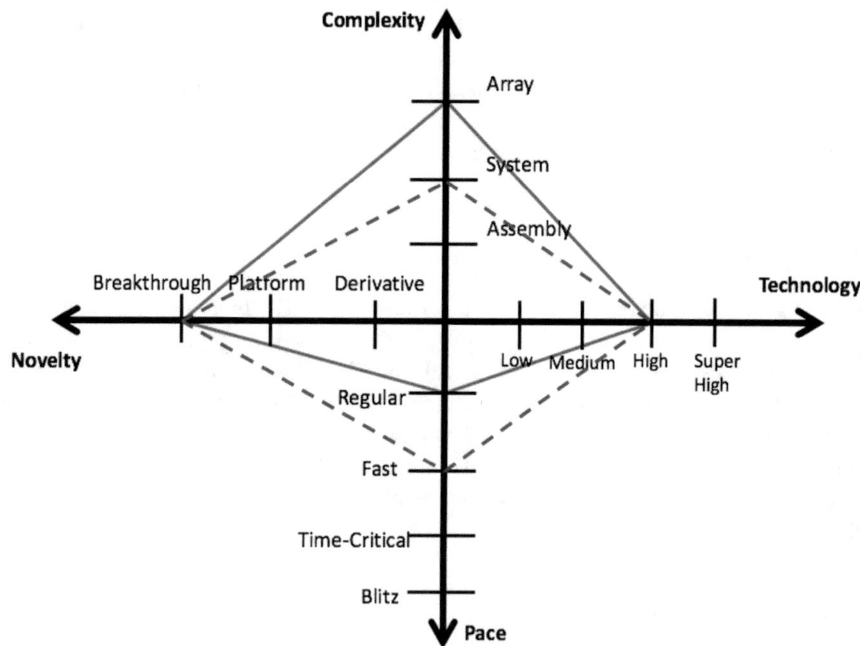

Fig. 5. NCTP framework for *Environment Sustainability* plane.

indicate that we need to consider the *Pace* of the project, as it is responsible for the area of the dashed diamond over the solid diamond area, which represents our initial insight about the project. We need management practices that deal with the features of *Fast-Competitive* value in the *Pace* dimension.

5 Conclusion

In this article, we have described the use of two systems engineering disciplines to deal with the complexity of IoT systems: model-based systems engineering and systems engineering management. We use a study case to present the use of the BRICS Mosaic model, and its FBF, to identify the planes of functionality in IoT systems that need attention in their implementation and operation. Moreover, we use the NCTP framework to identify the necessary managerial traits necessary to steer the life cycle of an IoT system.

5.1 Concluding Remarks

Our research has one objective: to reduce the ultimate failure point, i.e., managerial error, in systems projects. The management of IoT projects deals with several issues not always present in traditional project management. Despite the dimensions of the NCTP framework, it does not answer all the questions about mapping the risk factors in all the planes of IoT projects. The feasibility barriers factors of a BRICS Mosaic model provide insights about the planes of functionality and can be used to support the identification of better management practices for IoT projects.

While there is not a linear relationship between the diamond areas of the NCTP graph, it does represent a qualitative difference of risk. Combined with the BRICS Mosaic model, the difference between the diamond areas can support systems engineer to evaluate the gap between the preferred managerial approach and the actual approach.

Acknowledgments. This work was partly supported by the Society and Technology Study Centre (or, CEST – Centro de Estudos Sociedade e Tecnologia, in Portuguese) at the Universidade de São Paulo.

References

1. Baker, L., Clemente, P., Cohen, B., Permenter, L., Purves, B., Salmon, P.: Foundational concepts for model driven system design. INCOSE International Symposium **6**(1), 1179–1185 (1996). https://doi.org/10.1002/j.2334-5837.1996.tb02139.x
2. Piaszczyk, C.: Model based systems engineering with the department of defence architectural framework. Syst. Eng. **14**(3), 305–326 (2011). https://doi.org/10.1002/sys.20180
3. Sharon, À., Weck, O., Dori, D.: Project management vs. systems engineering management: A practitioners' view on integrating the project and product domains. Syst. Eng. **14**(4), 427–440 (2011). https://doi.org/10.1002/sys.20187

4. Sage, A.: Systems engineering: Fundamental limits and future prospects. Proc. IEEE **69**(2), 158–166 (1981). https://doi.org/10.1109/PROC.1981.11948
5. Shenhar, A.: Systems engineering management: a framework for the development of a multidisciplinary discipline. IEEE Trans. Syst. Man Cybern. **24**(2), 327–332 (1994). https://doi.org/10.1109/21.281431
6. Sauser, B.: Toward mission assurance: a framework for systems engineering management. Syst. Eng. **9**(3), 213–227 (2006). https://doi.org/10.1002/sys.20052
7. Simonette, M., Filev, R., Gabos, D., Amazonas, J., Spina, E.: BRICS mosaic model for IoT feasibility barriers. In: Vachtsevanos, G., Bulucea, C.A., Mastorakis, N.E., Natalianis, K. (eds.) Recent Researches in Electrical Engineering: Proceedings of the 13th International Conference on Circuits, Systems, Electronics, Control & Signal Processing (CSECS '14), Lisbon, Portugal, October 30–November 1, 2014. Atenas: WSEAS, 2014. pp. 180–188. http://www.wseas.us/e-library/conferences/2014/Lisbon/ELEL/ELEL-21.pdf. ISBN 978-9604743926. ISSN: 1790-5117
8. Simonette, M., Maia, R., Amazonas, J., Spina, E.: Toward IoT system project: BRICS Mosaic model and system engineering management. Int. J. Neural Netw. Adv. Appl. **2**, 1–11 (2015). http://www.naun.org/main/NAUN/neural/2015/a022016-089.pdf. ISSN: 2313-0563
9. Eisenhardt, K.: Building theories from case study research. Acad. Manag. Rev. **14**(4), 532–550 (1989)
10. Gillham, B.: Case Study Research Methods. Continuum Research Methods. Bloomsbury Academic, New York (2000)
11. ITU-T Y.2011, Next Generation Networks - Frameworks and functional architecture models, General principles and general reference model for Next Generation Networks, 10/2004
12. ITU-T Y.2060, Global Information Infrastructure, Internet Protocol Aspects and Next-Generation Networks, Next Generation Networks - Frameworks and functional architecture models. Overview of the Internet of Things, 06/2012
13. CASAGRAS2 - CSA for Global RFID-related Activities and Standardization. Project website: http://cordis.europa.eu/projects/rcn/85786_en.html
14. IoT-A – Internet of Things Architecture. Project website: http://cordis.europa.eu/projects/rcn/95713_en.html
15. Paté-Cornell, M.: Organizational aspects of engineering system safety: the case of offshore platforms. Science **250**(4985), 1210–1217 (1990)
16. Shenhar, A., Dvir, D.: How projects differ, and what to do about it. In: Morris, P.W.G., Pinto, J.K. (eds.) The Wiley Guide to Managing Projects. Wiley, Hoboken, NJ (2004). https://doi.org/10.1002/9780470172391.ch50

Learning with the Augmented Reality EduPARK Game-Like App: Its Usability and Educational Value for Primary Education

Lúcia Pombo(✉) and Margarida M. Marques

Research Centre on Didactics and Technology in the Education of Trainers (CIDTFF), Department of Education and Psychology, University of Aveiro, Campus Universitário de Santiago, 3810-193 Aveiro, Portugal
{lpombo,marg.marq}@ua.pt

Abstract. The EduPARK is a research and development project that intends to promote interdisciplinary mobile learning, supported by the development of an application to be used in an urban park, providing students' involvement, motivation and engagement to enhance authentic and contextualized learning. The EduPARK project follows a qualitative interpretative methodology fitting in a design-based research, a useful framework for developing technology-enhanced learning resources comprising various cycles of prototype refinement: a game-like app for mobile devices, with Augmented Reality (AR) contents that follows geocaching principles. After those refinement stages, the final version of the app was released for the public in the Google Play Store; and around 50 activities were organized to collect systematic data to better understand mobile learning in outdoor settings, not only in formal but also in non- and informal contexts. To date, EduPARK has involved about 800 students from primary to higher education, 200 teachers and 60 tourists; however, this paper reports a survey study focused particularly on data of participants attending the first four years of primary education. A total of 290 students, organized in 73 teams, completed the game and expressed their opinion about the usability and the educational value of the app in a questionnaire applied after the activity; and automatic app loggings were collected. Results show that the EduPARK app achieved a good usability and has educational value for primary education students. The present work proposes a data collection tool, inspired in Brooke's instrument, regarding the educational value of a game-based mobile AR resource.

Keywords: Mobile learning · Augmented reality · Outdoor activities · Game-based learning · Authentic learning

1 Introduction

Traditional classrooms are often described as teacher and textbook centered with students playing a passive role in learning. In traditional classrooms pupils are asked to turn off their mobile devices, which usually accompany them everywhere, before initiating the learning activities. This denotes a big gap between the use of mobile

K. Arai et al. (Eds.): CompCom 2019, AISC 997, pp. 113–125, 2019.
https://doi.org/10.1007/978-3-030-22871-2_9

devices inside and outside school that can lead to students' disengagement with learning activities, thus, impacting negatively their academic success [1]. Yet, at least in educational research related contexts, mobile devices can be easily incorporated in primary education classrooms and have learning potential when used in dynamic, collaborative and interdisciplinary activities in classroom environment [2]. These authors pointed gains on student attention, motivation and classroom climate, as well as an improvement in the development of students' key competencies.

The EduPARK project (http://edupark.web.ua.pt/) aims to promote interdisciplinary learning using mobile devices, augmented reality (AR) and games based on geocaching principles, in outdoor environments, particularly in an urban park, the Infante D. Pedro Park, in Aveiro - Portugal. For that purpose, the project developed a game-like app - the EduPARK app - an interactive application with four educational games, articulated with the national *curriculum*, for specific audiences: teachers, students from basic to higher education, and also by tourists/public in general in a lifelong learning perspective. In collaboration with Aveiro City Council, plant identification plaques were installed in the urban park with AR markers that trigger information in images, audios, videos, schemes, and interactive 3D plant leaves. That information was overlaid on top of a real-time camera feed of a feature within the park, augmenting the reality. Tiles, already located in the park, are also used as AR markers to augment information about historical and regional issues and virtual treasures can be found along the game both in plaques or in tiles [3]. The EduPARK app was developed through a design-based research with four development and refinement cycles that involved the app evaluation, *in loco*, by students and teachers [4]. The app is now freely available for Android devices, in the Google Play Store (http://edupark.web.ua.pt/app).

The EduPARK project organized about 50 sessions of park exploration with the EduPARK app, involving more than 800 students, 200 teachers and 60 visitors. In these sessions, players walked in the Park using mobile devices with the app installed, and followed the instructions of the project's mascot to experience the AR available in several interest points.

The main relevance of the project is its innovation in terms of outdoor learning strategies, in formal, informal and non-formal contexts, in an interdisciplinary way, combining it with technology. This will allow learning to move beyond traditional classroom environments to nature spaces that students can physically explore at the same time that they make connections with curricular contents. The project promotes articulation between research, teachers, professional practices, and initial and advanced training, constituting a very useful theoretical and practical framework, with impact not only in schools, but also in the community and in the tourism sector.

The purpose of the paper is to present a survey study that analyzes the usability and the educational value of the app focusing primary education data (6 to 10 years old pupils) with the overall goal of proposing methodologies of mobile learning in primary education schools. The rest of the paper contents: (i) addresses some keywords related to this approach, such as mobile learning, augmented reality, geocaching, outdoor activities, game-based learning, and authentic learning; (ii) briefly presents the EduPARK app, in what concerns its development process, its unique combination of innovative features, and game structure and mechanics, so the reader can have a concrete idea of this mobile learning approach; (iii) describes the methodology of the study; (iv) presents and discusses the main results; and (v) brings forward the core conclusions, limitations and the main contributions of this paper.

2 Theoretical Context

Mobile learning refers to a way of learning that comprises social and content interactions, supported by mobile devices such as smartphones or tablets, and, hence, it can occur across physical locations and educational contexts [5]. These devices are small and light enough to be easily carried to different places [6], support interactivity with others and with media content [7] and allow a high variety of contextual and situated learning activities, trough the proliferation of hardware and applications [8]. However, mobile learning may entrench digital divides regarding technology access, technological skills and learning competencies [8] and it requires an high preparation from teachers [6], who may not be tech-savvy. Nevertheless, it may be worthwhile to develop mobile teaching strategies, as the literature accumulated evidence that learners using a mobile device reached higher cognitive achievements than those not using these devices, particularly on kindergarten and elementary levels, and their affective learning outcomes also seem to be enhanced [6].

The dissemination of mobile devices allows the public to have access to Augmented Reality (AR) systems [9]. It is a technology that allows overlapping virtual elements (e.g., 3D models) with real objects of the physical world, in real-time, producing a new experience [10, 11]. AR content can be triggered by image recognition or by the user's location (from GPS or wireless network). In educational contexts, AR can make boring learning content more enjoyable, it can be used to provide immediate feedback and support autonomous learning and, thus, it has potential to increase learning performance [12, 13]. Among AR pitfalls are its usability and GPS related technical problems [12]. The GPS technical problems were also identified in the EduPARK during the early stages of app development. That was due of a lack of reliable GPS signal in outdoors with abundant treetops, so the project team decided to use image-based AR technology, with marker-based tracking [14] instead of GPS location-based markers.

The literature claims that mobile learning can also be combined with game-based learning (GBL) to increase the learner motivation, self-directedness, and social and inquiry skills [15]. GBL denotes the use of games to promote knowledge and skills acquisition, whilst providing players/learners with a sense of achievement [16]. Games can be designed to be powerful learning environments, particularly if they activate prior knowledge and offer instant feedback [17]. However, GBL requires careful design to balance the play and the learning outcomes [15] and, if the aimed learning and the game content are not integrated enough, the GBL approach using mobile devices is not likely to be effective [6]. The EduPARK project decided to use geocaching principles to ensure an engaging game play experience to the students. Finding hidden treasures promotes curiosity, which is known to be a powerful intrinsic motivator, as happened very successfully in the Pokémon Go [18].

The use of mobile devices liberates the learner from physical boundaries, beyond the typical classroom, and allows outdoor learning [7]. The literature claims that using mobile devices to learn in the outdoors and informal locations is more effective than using them in more formal places [6]. According to the literature, authentic learning experiences can be provided by mobile devices [7] and by AR technologies [19]. These technologies can situate the learner in a realistic physical and social context and scaffold learning processes [11]. Authentic learning builds on constructivist learning theories, particularly on situated learning, and includes features such as providing an authentic learning context and collaborative construction of knowledge [20], giving students an active role in learning as they experience and use information in ways that are grounded in reality, instead on memorizing facts in abstract situations.

3 The Game-Like App EduPARK

The creation of a user-friendly mobile game-like app with Augmented Reality (AR) contents on the park's biodiversity and other interdisciplinary information required a close iterative collaboration amongst the members of the project team, which are from different areas and professional contexts: Educational researchers and Science Teachers, Botanic specialists, Computer Science professors, and a Programmer. The Educational specialists identified curriculum related learning opportunities in the park and developed four educational guides for the app, targeting publics from basic to higher education and also for the general public. The Botanic experts identified and described a set of 70 native and exotic species in the Park and provided botanical contents for the AR triggers, giving priority to species with value to curriculum learning and to the promotion of conservation and environmental habits. The Computer Science professors developed the technological aspects of the AR app in an intense dialogue with the Educational experts, in order to define the relevant features to promote motivation and engagement with learning. The programmer technician developed the mobile application, for Android devices, using Unity 5, a popular cross-platform game engine [21]. As for the AR marker detection the Vuforia SDK for Unity was used, since Vuforia is currently the most widely adopted platform for AR technology [22].

The app development fits a design-based research methodology with four cycles of development involving user testing and evaluation. Several field tests were conducted, involving different convenience samples of potential users, from basic to higher education, including pre and in-service teacher training and park visitors in a lifelong learning perspective. The app was progressively refined according to the users' feedback in each cycle, concerning usability and educational value. Those refinements and final version of the EduPARK app are presented in detail in Pombo & Marques [23].

EduPARK app features integrates a set of aims/principles, as addressed by Pombo, Marques, Afonso, Dias & Madeira [24], including: (i) contextualized learning in the park, e.g., related with the botanical species or with the historical monuments in the park; (ii) interdisciplinary learning, e.g., using flowers to address issues about symmetry axis; (iii) recommendation to play in small teams, although it is perfectly playable individually as well, as generally research suggests that collaboration is more effective than competition for reaching achievements, as the app's educational challenges can promote collaborative discussion of ideas; (iv) the context is successful either in formal (school visits to the park), non-formal (e.g., in environmental education sessions of app use promoted by the local City Council) or informal learning contexts (e.g., a family visiting the park explores it through one of its games/educational guides); (v) user-friendly, so users, even children in primary education, can use it without support of the EduPARK team.

To develop the app it was important to carefully analyze the National Curriculum to identify multidisciplinary issues (e.g., integrating Biology and History or Biology and Math) that might be explored in the selected park. The identified issues were used to create four interdisciplinary educational guides, or quiz games, for the app. From these, three quizzes are intended for different levels of the Portuguese Education System: (i) the 1st Cycle (for 6 to 9 years-old students); (ii) the 2nd (for 10 and 11 years-old students) and 3rd Cycles (for 12 to 14 years-old students) of Basic Education; (iii) the Secondary (for 15 to 18 years-old students) and Higher Education; and (iv) one quiz is intended for any ordinary citizen visiting the city park, the Infante D. Pedro Park.

The Park is a large green area, with exotic and native botanic species, avifauna, a lake and several historical points of interest. It comprises educational value, in what concerns conservation and sustainable attitudes, since the ability to understand ecosystems is boosted by experiences in real environments influencing communities' attitudes about nature [3]. From an analysis of the park's patrimony, several teaching and learning opportunities have emerged and were explored in the educational guides. Additionally, the park spaces, which have physical barriers around most of their perimeters, are well defined safe environments for children to have mobile AR gaming experiences. This is an example of a truly authentic context for situated learning, where the location is essential for the learning [18]. The project has a mascot that is being used in the app to guide the players and give them immediate formative feedback after answering; e.g., when an incorrect answer is given, the mascot explains the right answer. The inspiration for the EduPARK mascot was the park's informal name: 'Monkey's park'. That name's origin is linked to a female monkey that lived in the park, for several decades. The mascot is also available as a physical plush doll, and it rapidly becomes a friendly figure for a 6 to 9 years old child (Fig. 1).

The project planned to promote school visits and other activities in the park for app test and exploration, so it acquired a set of smartphones and tablets to lend to the project's participants.

The game is organized in four stages; each stage corresponds to an area of the park, identified in the app's map (left picture in Fig. 1). In each stage there are a set of questions. Each question has 4 response options. The aim of the game is to gather as many points, bananas and virtual caches/treasures as possible. Points can be gained by answering correctly the questions that are contextualized with certain locations of the park. For that, pupils are invited to visualize AR resources underlying some questions, and try to find out, or to remember, for example, what kind of triangles exist in the Tea House chimney, which is the rock of the bandstand base or what are the reproductive structures of certain botanical species of the park. They are also challenged to find virtual treasures with bananas (Fig. 1) that can be used for extra support with the questions or to gain extra points at the end of the game.

Fig. 1. EduPARK activities with an app's screen (left picture), children playing in group (right top picture) and a virtual treasure displayed on the mobile screen (right bottom picture).

During the game, players can use different tools (Fig. 1 left): (i) AR camera, a tool to recognize the AR markers and to trigger the AR contents; (ii) compass, a tool that supports orientation in the park based on the display of the direction of magnetic north; (iii) bananas, a tool that shows the number of captured bananas and allows to switch bananas for extra help in the quiz questions; and (iv) map, a tool that also supports orientation in the park. In the EduPARK activities, each team of players was accompanied by an adult familiarized with the app and game, for safety reasons.

The EduPARK app game has awakened, systematically, interest and enthusiasm from users, who learn in a fun way while having a healthy walk in the park.

As the park does not have free internet coverage and not all Portuguese mobile device owners have a strong internet coverage service, the team decided to develop the app for offline use. The only requirement for having a mobile device with a fully functional EduPARK app, in an offline mode, is to download the app and its educational guides before going to the park to use it.

In the EduPARK project, the 32 plaques installed in the park have the same layout; however, the information in each one varies accordingly with the botanic specimen: the scientific and common names, its family (in biological classification), its origin and the AR marker, with the project's mascot [14]. The AR content associated with each plaque includes resources about the identified species (texts, photos, videos, 3D models) (Fig. 2).

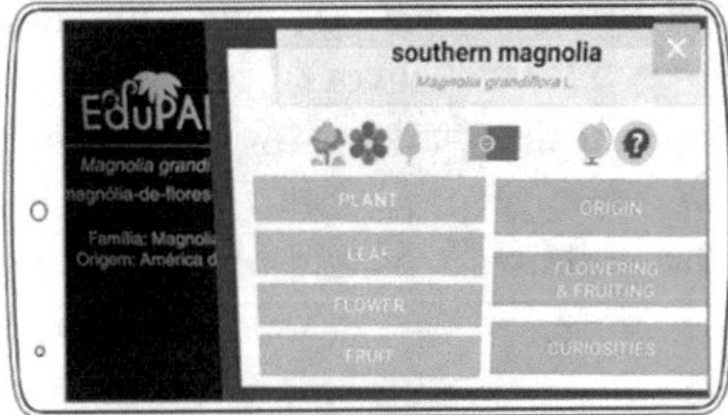

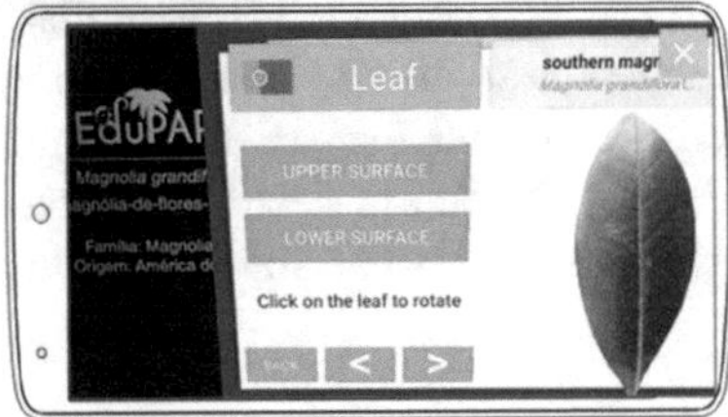

Fig. 2. The EduPARK app triggering AR content through image recognition (top picture); example of botanic AR content, the leaf can be digitally rotated to show its upper and lower surface (bottom pictures).

4 Methodology

The project EduPARK aims to study how playing a game, supported by an interactive mobile AR app, the EduPARK app, in a specific outdoor context may promote learning and motivation for learning, among other affective gains. At this stage the app is already developed and released for the public in the Google Play Store, therefore, the present paper gives continuity to previous works, such as [14, 24], by reporting a survey study focused on the analysis of the usability and educational value of the EduPARK app for children attending primary education.

Empirical data was gathered throughout the first year of app use in 15 outreach activities organized by the project involving schools and other educational entities with primary education children. Students were organized in teams of two to five members to collaboratively play with the EduPARK app in the Infante D. Pedro Park (Aveiro, Portugal). A total of 290 primary education children, both in formal and non-formal educational contexts, played with the app and participated in this study. Data collection included a questionnaire survey applied immediately after the app use and, also, automatic app logging mechanisms regarding the score attained, number of correct and incorrect answers, etc. Data were collected anonymously and research ethics principles were respected.

The questionnaire comprises four sections, with closed questions: 1 (strongly disagree) to 5 (strongly agree) Likert scale, multi-choice and item selection. Section one was about the perceived usability of the app and it had 10 items based on the 'European Portuguese Validation of the System Usability Scale (SUS)' [25], which is a scale translated for Portuguese from Brooke's [26] original instrument. Individual SUS scores were computed according to Brooke [26], with values varying from 0 to 100. In the present study, scale results were interpreted according to Sauro [27], who reviewed

500 studies and considered 68 to be an average SUS score, and to Bangor, Kortum and Miller [28], who empirically defined a qualitative classification for SUS. The second section gathered data about students' opinion on the educational value of the app. Hence, it included 12 items, carefully defined and revised by two educational researchers, on: (a) the app's learning value; (b) its contribution for intrinsic motivation; (c) engagement; (d) authentic learning; (e) lifelong learning; and (f) conservation and sustainability habits. These items are presented in Table 1. An Educational Value Scale (EVS) was computed in a similar way of the SUS. Section three includes one Likert scale question with one item about the respondents' level of interest regarding the activity of playing an educational game in the park using the EduPARK app. Finally, the fourth section collected basic demographic data (such as age and gender), as well as some information about the children's profile as mobile devices users.

Table 1. Educational Value Scale (EVS) items

EVS item
1. This app helps you learn more about topics we study at school
2. This app shows information in a confusing way
3. I feel motivated to learn when I use this app
4. I do not feel like using this app to learn
5. Even in the difficult quiz-questions, I try to find the right answers
6. Sometimes I respond randomly (without thinking)
7. This app shows real-world information that helps you learn
8. I will quickly forget what I have learnt from this app
9. Park visitors can learn from this app
10. This app promotes learning only in a school context
11. This app makes me feel like talking to others about nature protection
12. This app does not help to realize that it is important to protect nature

The app logging mechanisms were automatically and anonymously collected and include the following information from finished games: (i) number of questions answered correctly and incorrectly; (ii) number of hunted geocaching treasures; (iii) number of bananas collected in the treasures; (iv) final score, including the points gained through the collected bananas; and (v) time of gameplay.

Data from the app loggings and other data from the questionnaires, with the exception of the scales computing, were analyzed through descriptive statistics. These sets of data were triangulated in order to provide a comprehensive knowledge of the usability and educational value of the EduPARK app. This analysis will be presented in the next section.

5 Results and Discussion

This section starts with a brief characterization of this study's participants, followed by an analysis of the EduPARK app usability for primary education children and, finally, the analysis of its educational value.

Data collection allowed gathering information about the profile of primary education children who experienced the EduPARK app under an activity organized by the project during the first year of public implementation. The population age varies from 6 to 10 years old, being 48% of girls and 52% of boys. They all were attending grade one to four (7.6% in grade 1; 26.2% in grade 2; 42.9% in grade 3; and 23.3% in grade 4). The majority of students (74.4%) have Android mobile devices, such as smartphones or tablets. Most students claim they use mobile devices to learn either frequently (26.2%) or sometimes (49.7%); the others (24.1%) did not use them for that purpose at all. According to these results, most of these children are already quite familiar with mobile technologies and are used to employ them for learning. These results seem to support the literature, regarding the proliferation of mobile devices [29], especially in what concerns young population.

5.1 The EduPARK App's Usability

Figure 3 summarizes children's opinion regarding the usability of the EduPARK app. Their perception was positive, as, e.g., 223 students agreed or strongly agreed with the statement 'This app was easy to use' and 172 disagreed or strongly disagreed with the statement 'This app should not be so difficult to use.' SUS score values ranged from 30 to 100, with an average of 76.6, which is an higher value than the average SUS value (68) computed by Sauro (2011). Moreover, according to the classification of Bangor et al. (2009), the EduPARK app achieved a good usability for primary education students.

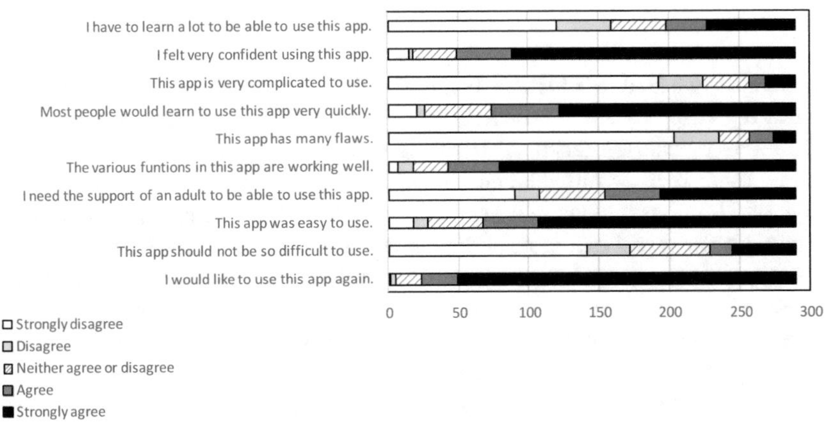

Fig. 3. Primary education students' opinion regarding the usability of the EduPARK app

5.2 The EduPARK App's Educational Value

Almost all the students (95.2%) considered the activity of playing with the EduPARK app in the park for learn very interesting.

Figure 4 summarizes children's opinion regarding the educational value of the EduPARK app. Their perception was positive, as, e.g., 270 students agreed or strongly agreed with the statement 'This app shows real-world information that helps you learn.' and 257 disagreed or strongly disagreed with the statement 'I do not feel like using this app to learn.' EVS score values ranged from 37.5 to 100, with an average of 84.7, which seems to be a high value, although more studies are needed to sustain that claim. Nevertheless, the results seem to reveal that the EduPARK app has educational value for primary education students.

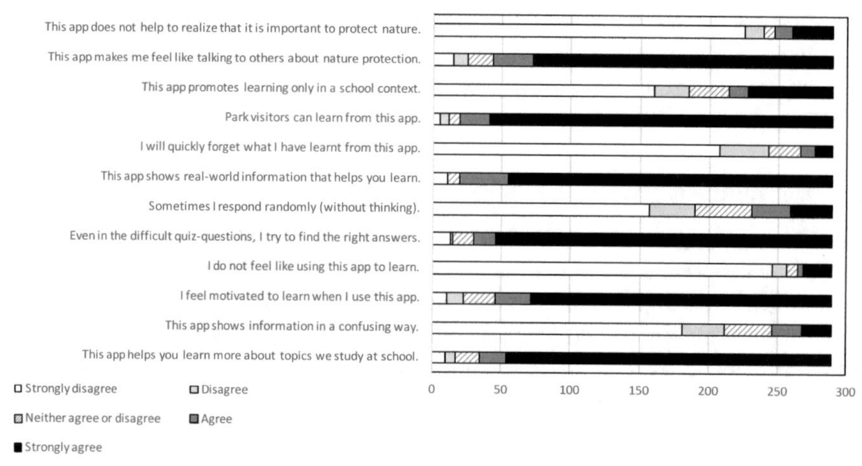

Fig. 4. Primary education students' opinion regarding the educational value of the EduPARK app

Students organized in 73 teams, completed the primary education game during the activities organised by the EduPARK, so the data collected by the app was automatically and anonymously generated from these completed games. These data are presented in Table 2. In 27 quiz-like questions, the young children answered correctly an average of 21.5 questions, with a maximum of 27 correct answers and a minimum of 11. These results reveal that most of the gamers were able to select the right answer of the quiz questions, either because they already knew the answer or because of the adequacy of the app's supporting information. By discovering a maximum of four treasures, players can earn a maximum of 20 bananas (five per treasure), according to how quickly they found them. Data reveals that most players found all the treasures and earned an average of 14.6 bananas, so the treasure hunt tasks were quite accessible to most players. The number of earned bananas, of correctly answered questions and of AR contents accessed are all considered to compute the game final score; data revealed that final scores reached an average of 237.8, maximum of 299 and minimum of 134.

Thus, the results about the children's performance in the game support their views regarding the high educational value of the EduPARK app.

Additionally, the time in game varied from nearly 53 min to 2 h and 8 min, with an average of 1 h and 24 min. This factor did not have impact on the players' scores. The gameplay time showed that students were able to stay commited with the game for long periods having a positive performance in the proposed tasks in the app.

Table 2. Automatic app loggings regarding 73 teams who completed the primary education game

	Number of correct answers (out of 27)	Number of incorrect answers (out of 27)	Number of treasures (out of 4)	Bananas (out of 20)	Final score	Game play time
Average	21.5	5.5	3.8	14.6	237.8	01:24:40
Maximum	27	16	4	19	299	02:08:47
Minimum	11	0	2	3	134	00:52:56

6 Conclusion

The EduPARK project developed an innovative interactive mobile AR game, freely available at Google Play Store, to promote authentic interdisciplinary learning in a specific urban park. This work reports on the first year of implementation of the EduPARK app with 290 primary education students in outdoor activities organized by the project. The focus is the usability and educational value of the app for this school level.

The results point to a good usability [28] and high educational value of the EduPARK app for primary education children, indicating that it is easy to use and promotes authentic learning in this target-public. Hence, resources that combine this set of innovative features, such as being mobile, explorable in the outdoors (namely in urban parks), with contextualized contents, supporting game-based geocaching activities and with AR contents, can be easy to use and may promote learning, both at a cognitive level and at an affective one (such as increased motivation for learning). These features [2, 6, 7, 12, 13, 16, 18] can be successfully integrated in methodologies to teach, in an authentic way, interdisciplinary and contextualized issues to primary education students in informal settings of their communities, such as urban parks.

Limitations of this study are related to the young age of participants. Although using a data collection tool adapted for the age range of the first four years of primary education, the level of excitement related with the timing and set where the questionnaires were applied (just after playing the game and in the park) may have hindered their concentration during the questionnaire filling. Moreover, possible reading difficulties of this target population may also have biased the results. Nevertheless, to reduce the impact of this factor, children were supported by the adults, who accompanied each group, to assure they understood the questions and answer options, whenever needed. Teams' constitution may also have influenced the results, as each student's level of participation in the game decreases as the number of team members increases. This factor may have impact on how the activity is perceived by the app players. However, this variation in teams' constitution could not be addressed, particularly with activities with a higher number of participants, as it was related with the human resources available to accompany each group of children in each session.

Additionally, the present work presents a new data collection tool, the Educational Value Scale (EVS), regarding the educational value of a game-based mobile resource that was inspired in the well-established SUS instrument [26]. The EVS can be useful both for educators and educational researchers for easily evaluate and compare different innovative resources and make decisions regarding which resources employ in their practice.

Future work under the EduPARK project will involve the study of the usability and educational value of the EduPARK app for other school levels, in order to analyze the potential of the proposed methodology in several formal and non-formal contexts. Finally, more studies of EVS, involving the rating of other educational software/ systems are needed, to both improve and consolidate this data collection toll.

Acknowledgment. This work is financed by FEDER - Fundo Europeu de Desenvolvimento Regional funds through the COMPETE 2020 - Operational Programme for Competitiveness and Internationalisation (POCI), and by Portuguese funds through FCT - Fundaçãopara a Ciência e a Tecnologia within the framework of the project POCI-01-0145-FEDER-016542.

References

1. Reyes, M.R., Brackett, M.A., Rivers, S.E., White, M., Salovey, P.: Classroom emotional climate, student engagement, and academic achievement. J. Educ. Psychol. **104**(3), 700–712 (2012)
2. Martí, M.C., Mon, F.M.E.: El uso de las tabletas y su impacto en el aprendizaje. Una investigación nacional en centros de Educación Primaria. Rev. Educ. (379), 170–191 (2018)
3. Pombo, L.: Learning with an app? It's a walk in the park. Prim. Sci. (153), 12–15 (2018)
4. Pombo, L., Marques, M.M.: The EduPARK game-like app with Augmented Reality for mobile learning in an urban park. In: 4.º Encontro sobre Jogos e Mobile Learning, pp. 393–407 (2018)
5. Crompton, H., Burke, D., Gregory, K.H.: The use of mobile learning in PK-12 education: a systematic review. Comput. Educ. **110**, 51–63 (2017)
6. Sung, Y.-T., Chang, K.-E., Liu, T.-C.: The effects of integrating mobile devices with teaching and learning on students' learning performance: a meta-analysis and research synthesis. Comput. Educ. **94**, 252–275 (2016)
7. Burden, K., Maher, D.: Mobile technologies and authentic learning in the primary school classroom. In: Burden, K., Leask, M., Younie, S. (eds.) Teaching With ICT in the Primary School, pp. 171–182. Routledge, London (2014)
8. Parsons, D.: The future of mobile learning and implications for education and training. In: Ally, M., Tsinakos, A. (eds.) Increasing Access Through Mobile Learning, pp. 217–229. Commonwealth of Learning and Athabasca University, Vancouver (2014)
9. Azuma, R.T.: The most important challenge facing augmented reality. Presence Teleoperators Virtual Environ. **25**(3), 234–238 (2016)
10. Azuma, R., Baillot, Y., Behringer, R., Feiner, S., Julier, S., MacIntyre, B.: Recent advances in augmented reality. IEEE Comput. Graph. Appl. **21**(6), 34–47 (2001)

11. Dunleavy, M., Dede, C.: Augmented reality teaching and learning. In: Spector, M., Merrill, M.D., Elen, J., Bishop, M.J. (eds.) The Handbook of Research for Educational Communications and Technology, 4th edn, pp. 735–745. Springer, New York (2014)
12. Akçayır, M., Akçayır, G.: Advantages and challenges associated with augmented reality for education: a systematic review of the literature. Educ. Res. Rev. **20**, 1–11 (2017)
13. Radu, I.: Why should my students use AR? A comparative review of the educational impacts of augmented-reality. In: 2012 IEEE International Symposium on Mixed and Augmented Reality (ISMAR), pp. 313–314 (2012)
14. Pombo, L., Marques, M.M.: Marker-based augmented reality application for mobile learning in an urban park - steps to make it real under the EduPARK project. In: XIX International Symposium on Computers in Education (SIIE) and VIII CIED Meeting/ III CIED International Meeting, pp. 174–178 (2017)
15. Giannakas, F., Kambourakis, G., Papasalouros, A., Gritzalis, S.: A critical review of 13 years of mobile game-based learning. Educ. Technol. Res. Dev. **66**(2), 341–384 (2017)
16. Qian, M., Clark, K.R.: Game-based learning and 21st century skills: a review of recent research. Comput. Human Behav. **63**, 50–58 (2016)
17. Ketelhut, D.J., Schifter, C.C.: Teachers and game-based learning: Improving understanding of how to increase efficacy of adoption. Comput. Educ. **56**(2), 539–546 (2011)
18. Laine, T.: Mobile educational augmented reality games: a systematic literature review and two case studies. Computers **7**(1), 19 (2018)
19. Mitchell, J.N., Sawyer, R.K.: Foundations of the learning sciences. In: Sawyer, R.K. (ed.) The Cambridge Handbook of the Learning Sciences, 2nd edn, p. 776. Cambridge University Press, New York (2014)
20. Herrington, A., Herrington, J.: What is an authentic learning environment? In: Authentic Learning Environments in Higher Education, pp. 1–14. IGI Global, Hershey (2006)
21. Unity Technologies: Unity (2017). https://unity3d.com/unity. Accessed 14 Feb 2017
22. PTC: PTC Releases Major Update to Vuforia – World's Leading Platform for Augmented Reality Development (2016). http://www.ptc.com/news/2016/ptc-releases-major-update-to-vuforia. Accessed 14 Feb 2017
23. Pombo, L., Marques, M.M.: The EduPARK mobile augmented reality game: learning value and usabiliy. In: 14th International Conference Mobile Learning 2018, pp. 23–30 (2018)
24. Pombo, L., Marques, M.M., Afonso, L., Dias, P., Madeira, J.: Evaluation of an augmented reality mobile gamelike application as an outdoor learning tool. Int. J. Mob. Blended Learn. (2019, forthcoming)
25. Martins, A.I., Rosa, A.F., Queirós, A., Silva, A., Rocha, N.P.: European portuguese validation of the system usability scale (SUS). Procedia Comput. Sci. **67**, 293–300 (2015)
26. Brooke, J.: SUS - a quick and dirty usability scale. In: Jordan, P.W., Thomas, B., Weerdmeester, B.A., McClelland, I.L. (eds.) Usability Evaluation in Industry, pp. 189–194. Taylor & Francis, London (1996)
27. Sauro, J.: MeasuringU: Measuring Usability with the System Usability Scale (SUS). MeasuringU (2011). http://measuringu.com/sus/. Accessed 10 Feb 2017
28. Bangor, A., Kortum, P., Miller, J.: Determining what individual SUS scores mean: adding an adjective rating scale. J. Usability Stud. **4**(3), 114–123 (2009)
29. Johnson, L., Becker, S.A., Estrada, V., Freeman, A.: NMC Horizon Report: 2014 K-12 Edition. The New Media Consortium, Austin (2014)

An Expert Recommendation Model for Academic Talent Evaluation

Feng Sun[1], Li Liu[1], and Jian Jin[2(✉)]

[1] South China University of Technology, Guangzhou, China
sunfeng@scut.edu.cn, seliushiya@mail.scut.edu.cn
[2] School of Government, Beijing Normal University, Beijing, China
jinjian.jay@bnu.edu.cn

Abstract. Talent recruitment is an essential step for institutes to improve organization structures of teachers and promote the development of teachers' capacity. To guarantee the reliability of talent recruitment, the department of human resources has to form a group of reviewers. The recommendation on a group of reviewers will definitely affect the effectiveness of talent recruitment. In this study, the importance of a topic is initially estimated by the frequency of topic occurrence within a specific research area, which helps to express candidates and reviewers' expertise. Next, in consideration of the topic importance in a candidate's capacity, an integer programming model is formulated to recommend a reviewer group which is comprised of a reviewer leader and several reviewers. Also, different practical constraints are reckoned in the optimization model, which includes the affinity between reviewers and candidates, reviewers' expertise and the burden assignment of each reviewer, etc. To evaluate the effectiveness of the proposed model, it is benchmarked with two baseline algorithms in terms of coverage, average number of reviewers, relevance between reviewers and topics, etc. Comparative experiment results were conducted and it shows that the proposed approach is capable to recommend reviewers effectively.

Keywords: Reviewer assignment · Topic importance · Integer optimization · Reviewer leader

1 Introduction

Talent recruitment is an important way for universities and institutes to build high-level faculty systems, and promote the benign flow and dynamic development of the talent team. In the recruitment process, the human resource department of universities usually requires candidates to submit their academic resumes with their several recent excellent papers. On the basis of preliminary qualification screening, the human resource department often needs to group the researchers to be evaluated according to the department, invites appropriate scholars as judges and establishes a group of experts to evaluate the academic level of the candidates. Therefore, how to select the appropriate review experts,

K. Arai et al. (Eds.): CompCom 2019, AISC 997, pp. 126–139, 2019.
https://doi.org/10.1007/978-3-030-22871-2_10

build a reviewer group, and make an objective evaluation of the reviewers' academic level, has became an important task for talent recruitment.

Regarding the issue of expert recommendation, most early related research regarded it as a specific form of information retrieval, and used the vector space model [9], language model [1,3] and other classic information models to analyze this problem. Since then, in view of thematic models based on rigorous mathematical reasoning such as Latent Dirichlet Analysis (LDA) and Author Topic Model (AT) [2,14], relevant researchers have began to use the topic model to recommend experts. However, less consideration is given to the actual requirements of expert recommendation in most of these studies. Hence, relevant researches have introduced experts, academic networks and social networks between experts and papers [7,21], extraction of features [8], conflicts of interest during the review process [5,18], etc., so that the recommendation model is applicable to reality.

In various methods and models, many studies utilize thematic analysis to analyze the topic distribution of expert expertise, and estimate the authority of experts and the relevance of experts and papers. However, in many analytical studies based on topic models, there is no distinction between the importance of different topics in particular subjects. Therefore, this study assumes the frequency of occurrence of the topic as its importance, and forms an expert recommendation optimization model that considers the "non-uniform topic". At the same time, the review of the article sometimes requires an expert as the leader to control the process and the quality of the review. Therefore, the review team leader requires a more rigorous screening process. Although some studies have proposed the concept of review leader recommendations [17], no specific recommendation method has been given. Therefore, this research will study how to efficiently complete expert group recommendations including a review team leader and multiple team members.

This study was based on the literature [11]. In the literature, a review expert model that incorporates the importance of the topic is proposed, without considering the recommendation of the "team leader". This study mainly includes two innovations. Firstly, for the different importance of the topic, this study makes the topic vector better express the characteristics of the article and optimize the effect recommended by the review experts. Secondly, this research proposes an integer optimization model for the recommendation of review leader and reviewers separately. A large number of experiments verify that our approach can better satisfy the practical demands.

2 Research Status

2.1 Expert Discovery and Expert Recommendation Based on Classic Information Retrieval Model

Correlation-based expert recommendations are designed to increase the semantic relevance of expert knowledge and content to be reviewed. Relevant research suggests that the recommended expert knowledge should match the research content of the paper. The general steps of the expert recommendation based on relevance are: (1) obtaining the texts of the articles published by the candidate review experts and the papers to be reviewed; (2)measuring their semantic relevance; and (3) recommending experts with high semantic relevance to the pending review papers as reviewers.

Hettich used TF-IDF algorithm to characterize the expert knowledge and research content of the papers to be reviewed, calculate the matching degree, and select the experts with the highest matching degree as the best review experts [9]. Yukawa used vector space models to decompose search documents and candidated expert documents into vectors, and then used cosines to measure similarity [20]. In the field of expert discovery, Cao pioneered a two-stage model framework [3]. They calculated the relevance of the search term to the document in the first stage, and calculated the relevance of the candidate expert to the search term in the second stage. Finally, two scores were combined to obtain the final expert ranking. Balog proposed an expert-centric language model (Model 1) and a document-centric model (Model 2) [1]. Model 1 uses all the description documents of the candidate experts to represent them, and sorts all the candidate experts by calculating the probability that each expert produces the search terms. Model 2 uses the relationship between the document and the candidate experts to calculate the probability of generating the search terms from the document. Finally, the document is used as the intermediate variable to obtain the relevance of the search term and the candidate experts to recommend relevant experts.

Many models based on classical information retrieval are mostly built on word-level matching or partial matching to achieve expert recommendation, while ignoring the phenomenon that different words may indicate the same topic. Moreover, the models based on classical information retrieval seldom consider the practical demands of the review such as the expert workload and the minimum number of review experts. Although the classic information retrieval models are easier for developers to understand and have high performance, these obvious shortcomings limit the wide application of these models in expert recommendation.

2.2 The Problem of Expert Recommendation and Expert Review Allocation

In fact, a document usually covers several topics. The topic model introduces the concept of implicit topics and can be understood as a highly abstract representation of semantics on a collection of corpora. Famous models on thematic analysis include LDA, AT model and so on. Therefore, many related studies have introduced these models and their variants into expert recommendations to analyze the research topics of experts.

Charlin used a combination of LDA models, linear regression and collaborative filtering to assign review experts [4]. Tang introduced conference information based on the AT model to form the Author Conference Topic (ACT) model [16]. The ACT model links the author's topic distribution to terms and meetings. This can be used to retrieve expert information and meeting information. Unlike the

ACT, the Citation Author Topic (CAT) model introduces citation relationships between papers to improve the consistency of terms in the topic and the relevance of the topic and the author [19]. In the AT model, each author only corresponds to a topic distribution. It remains a major constraint in mining experts' interest. So in the Author Persona Topic (APT) model [1] and the Latent Interest Topic (LIT) model [13], the author added more hidden layers to model the interests of experts easily and flexibly. Some researchers have conducted corresponding research on the topic evolution trend [6].

At the same time, the study also introduces expert academic networks, social networks and expert individual information, and models the factors such as expert authority, research background diversity and conflicts of interest, further improving the scientificity and accuracy. Based on the collaboration of academic data, citations and academic papers in ArnetMiner [16], Gollapalli constructed an Author Document Topic diagram to mine authoritative experts and retrieved to corresponding topic. Finally, the sort of candidate experts is obtained via "documents" [7]. Hirseh proposed the use of the H index to evaluate the achievements of experts [10]. Tayal considers the intensity of work, the conflict of interest between the author and the reviewer, the interest preferences of the reviewer and the keywords of the paper to be reviewed in the process of evaluating the expert assignment [18]. And the research considers four constraints to avoid conflicts of interest: whether they are relatives, colleagues, the coauthor of a paper or have teacher-student relationship.

3 Expert Recommendation Optimization Model

3.1 Problem Definition

Assume that there are n researchers to be reviewed in a certain discipline. In this study, the researchers submit a collection of representative papers to assess his research capabilities. At the same time, the expert database R contains m experts, namely, $R = \{r_i\}_{i=1}^{m}$. And we assume that the total number of review leader banks is e in the review library. We assume that an expert's published paper can represent his or her professional competence. Specifically, this study uses the collection d_i published by the expert r_i to indicate its academic level, and assumes that the papers published by experts constitute a set $D = \{d_i\}_{i=1}^{m}$. On this basis, this study assumes that the collection of expert papers contains h topics, which can constitute the vector $\tau = \{t_i\}_{i=1}^{h}$. Moreover, the correlation between topic j and each expert i is $C_{ij}(i \in (1,m), j \in (1,h))$, and the correlation between subject i and the researcher j to be evaluated $B_{ij}(i \in (1,n), j \in (1,h))$. B' represents the prominent subject relationship matrix of the researchers, and C' is the expert's prominent topic matrix. At the same time, the expert group recommendation system should also consider some restrictions in the real scene. For example, the amount of papers reviewed by each expert cannot exceed p, and the number of expert reviewers of each scientific researcher to be evaluated is greater than q. All symbols in this study are shown in Table 1.

Table 1. Notification definition

m	Total number of experts
n	Number of researchers to be reviewed
e	Total number of review team chiefs
h	Total Number of topics
p	Number of experts in each review team of researchers to be reviewed
q	The maximum number of researchers to be reviewed by each expert
$R = \{r_i\}_{i=1}^{m}$	Expert database R
$D = \{d_i\}_{i=1}^{m}$	Proceedings composed of papers published by experts
$\tau = \{t_i\}_{i=1}^{h}$	Theme vector composed of h themes
B	$n \times h$ matrix, B_{ij} indicates the probability that the subject j appears in the researcher i to be evaluated
C	$m \times h$ matrix, C_{ij} indicates the probability that topic j appears in expert i
m	Total number of experts
$F = \{f_t\}_{t=1}^{h}$	Topic t's Probability of occurrence in the representative of the researcher to be reviewed
S	$n \times h$ matrix, relevance of the representative of the researcher to be reviewed and the topic
B'	0-1 matrix of $n \times h$, B'_{ij} indicates that the subject j is a prominent subject of the researcher i to be evaluated
C'	0-1 matrix of $e \times h$, C'_{ij} indicates that topic j is the prominent topic of expert i
X	0-1 matrix of $m \times n$, indicates the distribution relationship between the review leader and the researcher to be reviewed
Y	0-1 matrix of $m \times n$, indicates the distribution relationship between the review team member and the researcher to be evaluated

3.2 Topic Extraction and Topic Importance Analysis

The primary issue of this study is that the recommended expert team leader and experts need to know the research of the researcher to be reviewed. This study assumes that the "topic" is the smallest unit that expresses the paper, and the paper can be described by the topic vector. In order to achieve the recommended research direction of the recommended experts, a primary issue is to analyze the representative work of the scientific researchers to be reviewed and the topics contained in the expert papers.

It is assumed that all articles of the expert r_i constitute the document d_i, and θ_{di} is the topic distribution of the document d_i. The papers published by m experts form the document set $D = \{d_1, d_2, ..., d_m\}$. This study first uses LDA to estimate the distribution of topics in the collection of documents. Specifically, according to the basic idea of LDA, the probability of occurrence of document set D can be described by Eq. (1).

$$\log p(D|\alpha,\beta) = \Sigma_{di \in D}\ \Sigma_{w \in di}\ c(w,d_i)\ \log(\Sigma_{z=1}^{h}\ p(w|z,\beta)\ p(z|d,\theta_{di})) \tag{1}$$

Among them, α and β are priori parameters of LDA. $c(w,d_i)$ is the number of words w. $p(w|z,\beta)$ is the probability that the topic z generates the word w, and $p(z|d,\theta_{di})$ is the probability that the document d_i contains the subject z. Then, this study uses Gibbs sampling algorithm for analyzing LDA models to estimate the distribution of topics in each document, represent each document as a topic vector, form h topics about the probability distribution C on document D, distribution B of the representative work of the researcher to be reviewed, and distribution F in a particular "discipline".

For a particular discipline, topics with a higher frequency of occurrence may be more popular or basic. For a particular document, however, if a topic appears lower frequently in a subject but appears at a higher frequency in the document, it can be assumed that the topic is more important than the document. Therefore, in order to ensure the review quality of the representative to be reviewed, it is necessary to arrange more professional review experts for the "minor topic". This method is described in detail in the literature [11]. According to this hypothesis, if two topics t_1 t_2 have the same probability of occurrence in the document i, the one whose probability of occurrence is relatively small within the scope of the subject, has stronger ability to identify the document's characteristics. This idea can be expressed by Eq. (2).

$$\begin{gathered} If \quad B_{it_1} = B_{it_2} \quad and \quad f_{t1} < f_{t2}, \\ then \quad P_{it_1} > P_{it_2} \end{gathered} \tag{2}$$

Among them, B and f are the topics and the probability of occurrence in document and subjects respectively. P_{it_1} and P_{it_2} are the correlations between the topics t_1 and t_2 considering the subject background and the article i. At the same time, according to the above assumptions, topics with higher occurrence frequency in the discipline should be given less weight, and vice versa. Therefore, according to the idea of Term Frequency-Inverse Document frequency (TFID), the correlation P_{ij} between document i and topics j can be defined as:

$$\forall i \in [1,n], \forall j \in [1,h] \quad P_{ij} = B_{ij} \times \log(\frac{1}{f_j}). \tag{3}$$

3.3 Review Team Leader Recommendation

In the actual review, it is necessary to select the "review leader" from several review experts for a researcher to be evaluated. The review team leader should be involved in all areas covered by the representative work of the researcher and have high authority. In addition, experts especially the team leader generally are required with higher qualifications. To this end, we selects the experts whose H index is located in the top 10% of the expert pool to form the "team leader" R', and assumes that the number of experts in the group leader library is e.

For a specific masterpiece of a scientific researcher to be evaluated, this study first uses a clustering algorithm to analyze the correlation s between different topics and the representative work. Then, we select the topic with the lowest s value as the "non-highlighting topic" in the representative work, and the rest of the topics as "highlighting topic". On this basis, this study establishes a 0-1 matrix of n researchers and n topics, and a 0-1 matrix of e candidate leaders and n topics. The element with a value of 0 indicates that the topic is not prominent in the masterpiece or expert knowledge structure, and vice verse. We assume that the expert i acts as the review leader of the researcher j only if the "important topics" of his masterpiece work are all included in the main research direction of the review leader. Expressed in mathematical notation:

$$\forall t \in [1, h], c'_{it} - b'_{jt} \geq 0. \tag{4}$$

In many academic activities, the number of scientific researchers to be reviewed is large and the number of experts is small. Therefore, it is necessary to set the maximum load p' for the review leader and establish an optimization plan to allocate the review leader reasonably. Supposing X is a 0-1 matrix representing distribution status of the group leader, $x_{ij} = 1$ indicates that the expert i is assigned as the review leader of the researcher j. Therefore, the optimal proposal of review team leader recommendation can be expressed as:

$$\begin{aligned} & \max \Sigma R'^{H} X \\ s.t.: \quad & C1: \forall i \in [1, e], \quad \forall j \in [1, n], \quad X_{ij}(c'_i - b'_j) \geq 0 \\ & C2: \forall j \in [1, n], \quad \Sigma_{i=1}^{\theta} x_{ij} = 1 \\ & C3: \forall i \in [1, e], \quad \Sigma_{j=1}^{n} x_{ij} \leq p' \\ & C4: \forall i \in [1, e], \quad \forall j \in [1, n], x_{ij} \in \{0, 1\}. \end{aligned} \tag{5}$$

R'^{H} is the H index of the experts in the review team leader. The input of the optimization problem includes the topics matrix representing the expert, the topics vector of the masterpiece of the scientific researchers to be evaluated, and the limited quantity of the review leader's review burden. And the output is the allocation matrix X of the review leader and researcher to be reviewed. This optimization scheme ensures that the sum of the H indices of the recommended review leader is maximized under the conditions that fully meet the constraints.

3.4 An Optimization Model for Solving Review Team Recommendation

Optimizing for review team members recommendation is independent of the recommendations of the review team leader. However, when recommending a reviewer for a researcher, his review leader should be removed from the expert database, and the review burden of other review leaders should also be considered, that is, the number of scientific researchers to be reviewed by each expert does not exceed q. The correlation between the experts and the important topics of the masterpiece of the researchers is strong. Experts should cover as many topics as possible for the representative of the researcher. The coverage of important topics by experts on the representative of the researcher should be higher than the sub-important topics. A detailed analysis can be found in the literature [11]. Finally, the recommended model for the review team members is:

$$\begin{aligned} & \max \Sigma C_{mh} S_{nh}^{T} Y_{mn}^{T} \\ s.t.:\ & C1: \forall j \in [1,n], \quad \Sigma_{i=1}^{m} y_{ij} = p \\ & C2: \forall i \in [1,m], \quad \Sigma_{j=1}^{n} y_{ij} \leq p \\ & C3: \forall i \in [1,m], \quad \forall j \in [1,n], y_{ij} \in \{0,1\}. \end{aligned} \tag{6}$$

4 Experiments

4.1 Dataset

This research takes the management discipline as the research object and analyzes the recommendation questions of the expert reviewer team. The experiment requires three parts of data. The first is the detailed information in the expert database, which includes the expert's personal information, the publication, the h index, the citation relationship, etc. The second is the collection of scientific researchers to be reviewed, including their personal information and masterpieces. The third is the subject information of each scientific researcher to be evaluated, including subject papers and subject journals. This study cannot obtain the true information of the expert database and the scientific researchers to be reviewed due to the limitations of the actual conditions. Therefore, the study will use the sampling below to analyze the recommended performance of the proposed model.

This experiment uses data described in the paper [11]. Specifically, the data comes from 5,226 scholars in the Wanfang database on the field of information management and information systems and their 75,880 articles. Scholar information includes name, unit, paper name, abstract, H index, published journals, and disciplines. The article information includes the title of the paper, the publication of the journal, and abstracts etc. According to the subject label of the Wanfang database, the research sample constitutes two datasets which are similar to the literature [11]. Among them, dataset A includes 100 articles as the masterpieces of the scholars and 200 potential experts as the expert database, while dataset B consists of 300 articles and 400 potential experts. In order to simplify the analysis, the experiment assumes that each researcher to be evaluated submits only one article as a masterpiece. However, this model still works when each researcher submits more papers. The only difference is that the model needs to combine these representative papers into one article as his masterpiece of the researcher first.

4.2 Inspection and Evaluation

For recommendation problems, manual evaluation may be a more effective method, and the manual evaluation model can be roughly divided into the following two categories. The first is to read the paper manually, find suitable experts, and then use it as a standard result to calculate the recall rate and accuracy of the recommended results [12]. The second is to ask relevant experts or users to evaluate the recommendation results according to the existing evaluation criteria [15].

The "optimal" results of manual selection and recommendation are usually limited by the knowledge structure and scope of the marker. Therefore, there is a lot of controversy as to whether the manually labeled data can be used as a standard result. Moreover, regarding the expert recommendation, recall rate and accuracy cannot measure the importance of different topics in the results. Therefore, this study uses several evaluation criteria proposed by the paper [11] to quantitatively compare the advantages and disadvantages of different methods.

Evaluation Indicator Definition

(1) Average coverage rate (avc): The topic that the expert is good at accounts for the proportion of all topics in the representative work of the researcher to be evaluated. This indicator measures the overall review ability of the review team recommended by the model for the researcher to be reviewed. avc is defined as follows:

$$avc = \frac{h_A}{h}. \tag{7}$$

h_A is the number of topics covered by the researchers' masterpieces, and h is the number of topics. If the "the researchers' masterpieces" of a masterpiece is included in the that of any of the review experts, the topic is considered to be covered by the recommended expert's knowledge.

(2) Per Capita Audit Rate of topics (avn): The average coverage frequency of each topic accounts for the proportion of the total number of judges in a representative work. It analyzes the effect of the review from the perspective of the topic. The ideal result is that the avn for important topics in the masterpiece is high, and even close to one. It means that all reviewers have the ability to review the researchers on important topics. avn is defined as follows:

$$avn = \frac{\Sigma n_{A'}}{n \times h \times p}. \tag{8}$$

$n_{A'}$ is the total coverage frequency of all topics in a representative work, n is the number of researchers to be reviewed, h is the number of topics, and p is the number of reviewers assigned to each researcher.

(3) Correlation degree considering the importance of the topic (cor_s): The Spearman correlation coefficient of the expert's professional level on each topic

and the importance of the corresponding topic in the researcher's representative. It can measure whether the more important topics in the researchers' masterpiece are prioritized.

$$cor_s = \frac{\Sigma r_s}{m \times n} \tag{9}$$

$$r_s = 1 - \frac{6\Sigma_{i=1}^{h} d_i^2}{h(h^2 - 1)} \tag{10}$$

r_s is the Spearman correlation coefficient of the topic distribution between the masterpiece of the researcher to be evaluated and the review expert, where d is the rank difference of their topic vectors. The higher the cor_s value is the more review experts can be assigned to more important topics.

Comparison Algorithm: This study selects a uniform topic model that does not consider the difference in topic importance and a non-uniform topic greedy algorithm that considers the difference in topic importance as comparison. The uniform topic model uses the integer programming algorithm to achieve the topic matching between the expert and the masterpieces of the researcher to be reviewed, without distinguishing the importance of topics.

The greedy recommendation model uses the basic principles of the greedy algorithm to complete the recommendation. Specific steps are as follows:

(1) Extract topics distribution B_{nt}^{greedy}, R_{mt}^{greedy} of experts and masterpieces of researchers to be evaluated;
(2) According to the non-uniform topic model, form topics distribution of experts and researchers' masterpieces $B_{nt}^{greedy'}$, $R_{mt}^{greedy'}$;
(3) Calculate the correlation degree between the $i(i \in [1, n])$ researcher and all m candidate experts;
(4) Sort all m experts according to the correlation degree. Select the top p experts with the highest score, and increase the corresponding expert's task by one;
(5) After the assignment of an expert, check whether his task is less than q. If it is equal to q, delete this expert;
(6) After n rounds of circulation, complete the distribution of the experts.

Results and Discussion: The results obtained by the three models for dataset A are shown in Figs. 1 and 2. In this set of experiments, the number of topics is set to 20. When $p = 5$ and $q = 5$, the highest allocation coverage and audit rate are obtained. The reason is that an increase in the number of review experts may result in more topics of each researcher being reviewed and the increasing review number of each topic. Therefore, better and higher avc and avn will be obtained. Figure 2 shows that when the number of researchers to be reviewed by each expert is increased, avc and avn are slightly reduced. The reason is that each expert may be assigned more material to be reviewed, which leads to more experts may participate in the review, which makes both indicators decline.

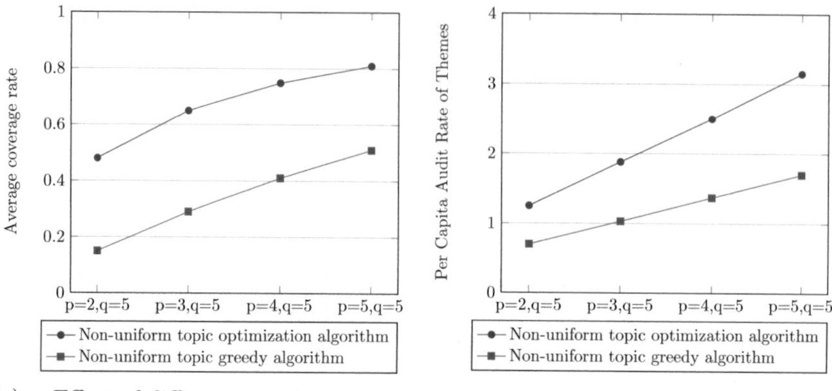

(a) Effect of different p and q on avc (b) Effect of different p and q on avn

Fig. 1. Effect of different value of p and q (q = 5)

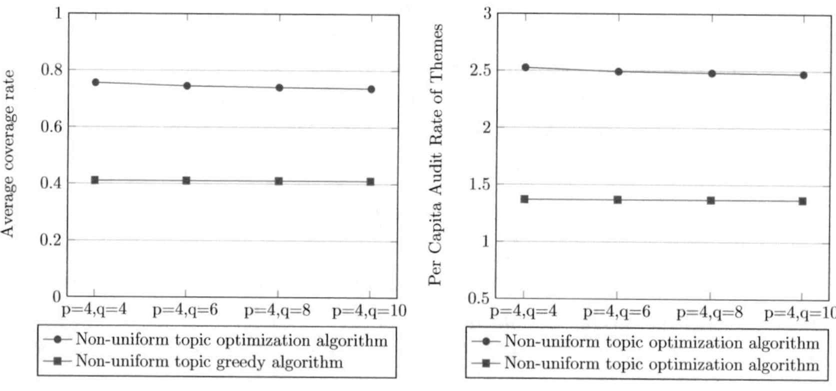

(a) Effect of different p and q on avc (b) Effect of different p and q on avn

Fig. 2. Effect of different value of p and q (p = 4)

Figure 3 shows the experimental results of the three models in the dataset A when $p = 4$, $q = 5$. It can be seen in Fig. 3 that two non-uniform topic models are obviously superior to the uniform one in terms of avc and avn. Meanwhile, Fig. 3 shows that in the dataset A, when the number of topics is 20, the obtained recommendation effect is better.

Figure 4 shows the difference in the relevance of the expert and article topics in the two datasets A and B for the three models under different p and q when the number of topics is 20. As can be seen from Fig. 4, the proposed model and the uniform topics model and the greedy recommendation model have a large gap in this indicator, while the greedy model does not perform well. The reason is that the greedy recommendation model selects the experts who are most suitable for reviewing the current researchers in each cycle, and the cycle order of the

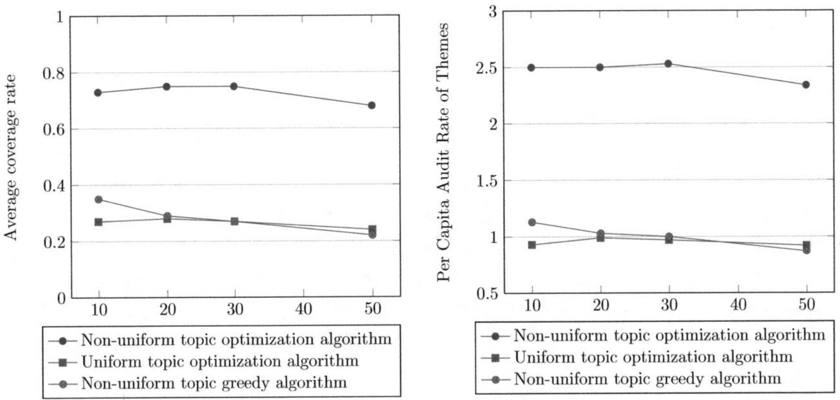

(a) Effect of the number of topics on *avc* (b) Effect of the number of topics on *avn*

Fig. 3. Effect of the number of topics

Fig. 4. Comparison of the proposed model, greedy recommendation models and uniform topic models.

researchers to be evaluated will have an impact on the recommendation results. This further illustrates the large instability of the greedy recommendation model, which makes it unsuitable for expert allocation problems.

5 Conclusion

This study analyzes the problem of recommending experts for the scientific research to be evaluated in the talent recruitment of colleges, analyzes the importance of different topics in different subjects, and proposes an expert recommendation model based on integer programming. The model considers the differences of different topic in specific subject, and integrates the topic relevance of the experts and the researchers, the minimum reviewer requirements of the researchers, the workload of the experts and other actual problems in the programming model. To test the validity of the model, this study uses two datasets to compare greedy recommendation models and models that did not consider the importance of the topic. Multiple sets of experiments found that the proposed model in this study was significantly better than the model without considering the importance of the topic and the greedy recommendation algorithm.

Based on current research, the expert recommendation model considering the importance of the topic has extensive application in many fields, such as recommending appropriate reviewers for journal meetings, recommending lawyers for specific cases, and recommending doctors for patients.

References

1. Balog, K., Azzopardi, L., de Rijke, M.: Formal models for expert finding in enterprise corpora. In: SIGIR (2006)
2. Blei, D.M., Ng, A.Y., Jordan, M.I.: Latent Dirichlet allocation. J. Mach. Learn. Res. **3**(Jan), 993–1022 (2003)
3. Cao, Y., Liu, J., Bao, S., Li, H.: Research on expert search at enterprise track of TREC 2005. In: TREC (2005)
4. Charlin, L., Zemel, R.S.: The Toronto paper matching system: an automated paper-reviewer assignment system. In: ICML (2013)
5. Chen, C.Y.: Conflict of interest detection in incomplete collaboration network via social interaction, p. 54. Master thesis. Computer Science and Information Engineering, National Taiwan University of Science and Technology, Taipei, Taiwan (2009)
6. Daud, A., Li, J., Zhou, L., Muhammad, F.: Temporal expert finding through generalized time topic modeling. Knowl.-Based Syst. **23**(6), 615–625 (2010)
7. Gollapalli, S.D., Mitra, P., Giles, C.L.: Ranking experts using author-document-topic graphs. In: JCDL, pp. 87–96 (2013)
8. Han, S., Jiang, J., Yue, Z., He, D.: Recommending program committee candidates for academic conferences. In: Proceedings of the Workshop on Computational Scientometrics: Theory & Applications, San Francisco, California, USA, pp. 1–6 (2013)
9. Hettich, S., Pazzani, M.J.: Mining for proposal reviewers: lessons learned at the national science foundation. In: KDD (2006)
10. Hirsch, J.E.: An index to quantify an individual's scientific research output. Proc. Natl. Acad. Sci. U.S.A. **102**(46), 16569–16572 (2005)
11. Jin, J., Yang, H.C., Li, N., Geng, Q.: A topic relevance aware model for reviewer recommendation. Digit. Libr. Forum **4**, 47–55 (2017)

12. Karimzadehgan, M., Zhai, C.X.: Integer linear programming for constrained multi-aspect committee review assignment. Inf. Process. Manag. **48**(4), 725–740 (2012)
13. Kawamae, N.: Latent interest-topic model: finding the causal relationships behind dyadic data. In: CIKM (2010)
14. Rosen-Zvi, M., Griffiths, T., Steyers, M., Smyth, P.: The author-topic model for authors and documents. In: UAI (2004)
15. Sun, Y.H., Ma, J., Fan, Z.P., et al.: A group decision support approach to evaluate experts for R&D project selection. IEEE Trans. Eng. Manag. **55**(1), 158–170 (2008)
16. Tang, J., Zhang, J., Yao, L., Li, J., Zhang, L., Su, Z.: ArnetMiner: extraction and mining of academic social networks. In: KDD (2008)
17. Tang, W., Tang, J., Tan, C.: Expertise matching via constraint-based optimization. In: IEEE/WIC/ACM International Conference on Web Intelligence and Intelligent Agent Technology (WI-IAT), vol. 1, pp. 34–41. IEEE (2010)
18. Tayal, D.K., Saxena, P.C., Sharma, A.: New method for solving reviewer assignment problem using type-2 fuzzy sets and fuzzy functions. Appl. Intell. **40**(1), 54–73 (2014)
19. Tu, Y., Johri, N., Roth, D., Hockenmaier, J.: Citation author topic model in expert search. In: ICCL (2010)
20. Yukawa, T., Kasahara, K., Kato, T., Kita, T.: An expert recommendation system using concept-based relevance discernment. In: Proceedings of the 13th International Conference on Tools with Artificial Intelligence. IEEE (2001)
21. Zhang, J., Tang, J., Li, J.: Expert finding in a social network. In: Advances in Databases: Concepts, Systems and Applications, pp. 1066–1069. Springer, Heidelberg (2007)

A Comparative Study for QSDC Protocols with a Customized Best Solution Approach

Ola Hegazy(✉)

Imam Abdulrahman Bin Faisal University, Dammam, KSA, Saudi Arabia
ohegazy@iau.edu.sa

Abstract. In the new era of quantum communications, the QSDC has granted a great acceptance, reassurance, inducement and comfort among the huge amount of transmission techniques. This is mainly because this technique is way better than all others; security wise and efficiency wise; which in turn because it relies on the nature of quantum elements that have been proven to be secure by nature - referring to the quantum mechanics laws. In our search we will introduce a specific technical comparison between our new protocol approach of QSDC that we proposed before, and some other well-known protocols that have been proposed in the last decades. Clarifying the advantages and disadvantages of each and the difference in the implementations of all them we will analysis the security represented in each of them. Finally, we will gather up all the issues related to this approach and come up with a recommended direction to follow in using this QSDC technique with a customize use of qubits and their transmission tools.

Keywords: Quantum Secure Direct Communication (QSDC) · One Time Pad (OTP)

1 Introduction

The idea of the One Time Pad (OTP) has quite exhausted classically or quantumly, and according to our research and study in the quantum field (cryptography and communication), we made sure that the discovery of the Quantum Secure Direct Communication (QSDC) technique was the best discovery ever in the field of digital communication; whereas this technique has great advantages over the ordinary transmission using the Key distribution techniques classically or quantumly.

The main advantages of this technique is that; it doesn't need for a prior key exchange operation among parities of the transmission; it doesn't have cipher text, and it grants high level of security through transmission and even on site; finally it could be the basic building block in some other cryptographic techniques [1] like: quantum signature [2]; quantum bidding [1]; quantum secret sharing [3]; and quantum dialogues [4] for example.

The first protocol using the QSDC has proposed in 2000 [5]. It was enriched by a block transmission technique. The data carriers in QSDC can be either entangled quantum structures, as in the "Efficient-protocol" [5] and the "Two-steps protocol" [6], or single particles transporters, like in Deng and Long protocol [7], Noor Novel

K. Arai et al. (Eds.): CompCom 2019, AISC 997, pp. 140–148, 2019.
https://doi.org/10.1007/978-3-030-22871-2_11

approach protocol [8], and our Super-dense coding protocol [9]. Also, Large dimensional quantum structure can also be operated as data carrier like in the high-dimensional QSDC protocol [10].

As long as the quantum communication area always have the researches applied in the quantitative approaches, so we chose in our research a comparative method that high lighted on the main benefits of our pre-posed protocol in applying the QSDC, which have great advances in security and efficiency than some other proposed protocols. This will be clarified by discussing the strengths/weakness of each protocol under investigation in a comparative way.

In this paper we will first introduce a short description of each of the sample protocols under investigation ([5–9]) that taken for the aim of the comparison, and then will analysis the advantages and disadvantages of each according to the ease of implementation, the security tools and overall efficiency.

Finally, we will deduce the reason of our recommendation to use our customizing protocol's method when applying the QSDC technique.

2 Protocols Descriptions

2.1 The Theoretically Efficient Protocol

As a first protocol assigning the idea of QSDC in 2000 [1], the Efficient protocol projects the QSDC technique by encoding the information to be sent in the quantum states of EPR pairs as one of the 4 Bell states:

$$\begin{aligned}
|\varphi 1\rangle &= \sqrt{1/2}(|00\rangle + |11\rangle) \\
|\varphi 2\rangle &= \sqrt{1/2}(|00\rangle - |11\rangle) \\
|\varphi 3\rangle &= \sqrt{1/2}(|10\rangle + |01\rangle) \\
|\varphi 4\rangle &= \sqrt{1/2}(|10\rangle + |01\rangle)
\end{aligned} \tag{1}$$

Beforehand, the two parties of the transmission "Alice and Bob" had agreed that the above quantum states $|\varphi 1\rangle$, $|\varphi 2\rangle$, $|\varphi 3\rangle$, $|\varphi 4\rangle$ are representing the classical data bits of 00, 01, 10, 11 respectively. In this protocol the sender Alice will construct a sequence of length N EPR pairs denoted by "($P_1(1)$, $P_1(2)$), ($P_2(1)$, $P_2(2)$), …, ($P_i(1)$, $P_i(2)$), …, ($P_N(1)$, $P_N(2)$)", with inherently inserting arbitrary EPR pairs for checking the security of the sequence transmission. Then she takes the 2nd particle of each pair a sequence of particles of these block states to transmit to Bob denote as: "$_1(2)$, $P_2(2)$, $P_3(2)$, …, $P_N(2)$". The rest of this block will constitute her other own sequence denoted as: "$P_1(1)$, $P_2(1)$, $P_3(1)$, …,$P_N(1)$".

On the reception of this sequence, Bob acknowledges Alice by this, whereas she starts to pick randomly some of the checking pairs on her sequence and makes measurements on them in either "0, 1" bases or "+, −" bases. Accordingly, the outcomes of this measurements should be either 0 or 1 respectively, where then she in turn acknowledges Bob with the places of these pairs in the sequence.

On the arrival of this acknowledge to Bob, he starts to make measurements on those chosen pairs by Alice, and also, using arbitrarily bases of $\{0, 1\}$ or $\{+, -\}$ for his measurements. Bob saves the remaining pairs of his sequence.

Bob reveals the bases and the results of his measurements in public, which based on it, Alice could evaluate the error-rate in the transmission and decide if there is an eavesdropping or not. This part of the procedure is known as the first eavesdropping test.

If Alice makes sure that there isn't eavesdropping, she sends Bob the rest of her EPR sequence: "$P_1(1)$, $P_2(1)$, $P_3(1)$,…,$P_i(1)$", of course after dropping the pairs that have been measured for the test. After Bob gets her sequence, he pairs up the remaining in both series to form the original order sequence and then applies Bell bases measurements on it and documents the outcomes.

Alice then requests Bob to reveal the measurements outcomes of the rest of the testing pairs in the sequence to evaluate the error-rate, if the rate they get is less thana specified threshold, then the outcomes of these measurements are the transferred message that she wants to exchange between both of them and by then the direct communication is done. This part of the procedure is known as the second eavesdropping test.

In this protocol the sending of photons sequence, or as it's denoted by the "block transmission", is critical to the safety of the direct transmission. After the first photons sequence is conducted safely, the second photons sequence will be allowed to move. In this way of transmission, the eavesdropper couldn't get the two photons in the same EPR at the same time, then consequently, this will prohibit her from getting the correct data.

2.2 The Two-Step Protocol

Although that the Two-step protocol is embracing the same idea of the Efficient QSDC protocol by using a block transmission to deliver the data from one side to another, but it is different in the way of its implementation [6]; as in this procedure the sender Alice produces an N order EPR pairs with the same status; $|\varphi 1\rangle_{AB} = \sqrt{1/2}(|0\rangle_A|0\rangle_B + |1\rangle_A|1\rangle_B)$. And similarly, to the previous protocol; Alice graps one photon from each EPR coupled pair (the same positioned photon) to constitute an EPR partner particles sequence that call it "S_A". The rest of the EPR coupled particles compose the other sequence, name it "S_B". Again, before start, it is approved by both sides that the quantum states $|\varphi 1\rangle$, $|\varphi 2\rangle$, $|\varphi 3\rangle$, $|\varphi 4\rangle$ are representing the classical data bits of 00, 01, 10, 11, respectively.

Now for assuring the safety of the communication, the transmission will be split in 2 phases:

Phase one; Alice transmits the series of particles S_B to Bob, then tests the safety of this communication with him by the following procedure:

a. Bob picks an arbitrary large number of photons that he received and measures them randomly by one of the two measuring bases, say Z and X that equal to "$|+\rangle = \sqrt{1/2}(|0\rangle \pm |1\rangle)$".

b. Bob announces Alice by the positions of the photons he has picked, and what was the measuring bases he chose, with the outcomes he got on this sample.
c. Accordingly, Alice takes the same measuring bases as Bob to make her measurements on the corresponding particles in her S_A sequence and compares the results. If both sender and receiver should get "0 or 1" when they evaluate the states of the same particles with the bases, then this assure that the transmission was safe. They call this the first check for eavesdropping.

(2) If the two authorized parties approved on the safety of sequence S_B broadcasting, then Alice encrypts her sequence S_A by the real information using the unitary operators U_i where ($i = 0, 1, 2, 3$). Right after she transmits it to Bob, whom extracts the private information straightly from this sequence using the measurements of:

$$\begin{aligned} U_0 &= I = |0\rangle\langle 0| + |1\rangle\langle 1|, \\ U_1 &= \sigma_z = |0\rangle\langle 0| - |1\rangle\langle 1|, \\ U_2 &= \sigma_x = |1\rangle\langle 0| + |0\rangle\langle 1|, \\ U_3 &= i\sigma_y = |0\rangle\langle 1| - |1\rangle\langle 0|. \end{aligned} \tag{2}$$

In the contrary of the first protocol, this one is sending all the information as one block in the second transmission after confirming the security of the first one. This attitude decreases the safety level of communication by giving Eve the chance of collecting the information if she can breach the second transmission only.

2.3 The Single-Photon Protocol

In 2004, Deng and Long had proposed their single-photon protocol, which was named by quantum onetime pad protocol [6]. In this protocol the two authorized parities, "Alice and Bob" have to share previously a series of single particles safely, and then the sender "Alice" encrypts her private information on her shared part of particles and sends them to the receiver "Bob". This is done by two phases procedure as follows:

(1) Phase one: "The secure doves sending phase"

The receiver Bob creates a sequence of polarized single particles, S. Then he transmits them to Alice. Every single particle of this sequence is erratically in one of the four states: $|0\rangle$, $|1\rangle$, $|+\rangle = \sqrt{1/2}(|0\rangle + |1\rangle)$ or $|-\rangle = \sqrt{1/2}(|0\rangle - |1\rangle)$.

Upon receival of this sequence, Alice with Bob going to check the validation of the transfer operation [7]. As in the previous protocol, Alice picks arbitrarily some sample particles from the arrived particles and applies haphazardly her measurements by using "Z or X" bases. Then she informs Bob with places and her measurements' outcomes of those sample particles. By comparing the results, if they match then the transmission was safe, if not then they assure that there is an eavesdropper. They denote this phase by the first test of eavesdropping. In case of the approval on the security of the transmission, they continue their communication up to the next phase; elsewise, they abort the communication.

(2) Phase two: "The message coding and doves returning phase"

In this phase, Alice encrypts her secret information on the single remaining particles of the sequence, that called it *S*'. In this encryption process she uses the unitary operators $U_0 = I = |0\rangle\langle 0| + |1\rangle\langle 1|$ or $U_3 = i\sigma y = |0\rangle\langle 1| - |1\rangle\langle 0|$ referring to classical data bits is 0 or 1, correspondingly.

After finishing the encryption of her sequence *S*', Alice sends them to Bob. Whereas the original sequence *S* was created by Bob, and due to the no change, that is happened by the two unitary operators U_o & U_3, then Bob can use the original measurement bases to measure received particle to reveal out the message sent by Alice.

The steps of checking the security in the first phase allows both sides of the transmission "Alice and Bob" to get an approximation if there is any intruder in the middle of the transmission or not.

Even if there is an intruder in the line, her eavesdropping in this case will never give her any good knowledge on the private secret because she doesn't recognize the initial status of the particles of the main sequence *S*, where the data of the private message is encrypted by the change of their states.

2.4 The Novel Approach of Secret Sharing Protocol

Depending on the two important phenomena of quantum entanglement swapping [10] with the non-local correlations created by the quantum teleportation, an absolutely secure {4 by 4} quantum message sharing technique for a classical secret transmission has been created by Ain in 2017 [8]. In his protocol he guarantees each of the confidentiality and authenticity, as it is said to be safe versus the internal and external eavesdropper attacks through exchanging and reconstructing the secrets for {4 by 4} threshold technique. The proposed technique is subdivided into four main stages: EPR pairs subdivision, Message propagation, Receivers validations and Message retrieval. These four stages contain a huge number of evaluations and quantum photons generation different EPR states.

Also, an essential condition for this protocol is to have 4 receivers for the only one sender Alice. In the first two stages the sender Alice should generate 8 EPR pairs and after transmission half of each to one for every receiver, she should generate another 4 new EPR pairs depending on the results of the first transmission on the receivers' sides. This other generation will be made by using not only the resulting classical bits of the receivers, but also using an OTP sequence that she should also prepare it for these operations, and then applying measurements on some of the created qubits to get classical bits outcomes to use part of them again in generating another state of EPR pair, and send the other part to some of the receivers.

In the last two stages, the receivers will operate on the results they got from the sender's qubits in the first and second stages, and then send these manipulated results back to the sender, who will apply their reverse operation to get the outcomes that verifying the identities of the receivers, so then at this point only when the sender verified the receivers identities, she will allow the recovery the secret message.

In this last stage every participant in the transmission (the 5 parities) should send the results of his possession after measurements he made to get the whole message revealed.

3 The Contributed Analysis

In the first protocol, the transmission of a block of photons is crucial to complete the main idea of the protocol. Although this guarantees the security, but it consumes a big amount of quantum generated particles to hide the message bits inside as haphazard checking pairs, and this is decreasing the efficiency of the transmission and wasting a lot of precious quantum particles and time. Also, every side has to save its sequence till the end after making sure that there is no eavesdropper to make the last recovery of the message's elements, that's lead to the need of quantum memory which is by the quantum mechanics laws is not feasible.

Note that; it is not clearly stated which state the generated photons will have, and if they are all the same state or not.

On the other hand, the second protocol is almost the same as the first, except that the latest one creates a block of photons N that all are in one state of the Bell states, and this feature in particular could expose the whole idea of security to be broken if the intruder Eve could discover this state (even by guessing)! Again, this protocol wastes the half of the sequence in checking and testing the eavesdropping in the first transmission, and also the ending operation there isn't a high level of security assurance as all the sequence qubits are in the same quantum state.

In the third protocol, it is important to have a pre-shared sequence that is leads us to dilemma of the pre-sharing like in the QKD techniques, which waste a lot of time and quantum resources that we try to avoid in the QSDC protocols. Also, the main big problem is the need of a quantum memory, as the sender has to store the quantum states in her possession, which is not yet applied as the storing of quantum data is contradicting with the main nature and advantage of the quantum media that's the no-cloning theorem. However, as it is mentioned by Long [1], this protocol is "not yet fully developed". Also, he provided that the storage of the qubits could be done by the light delay in a fiber optics line of the transmission by the delay period of time for testing the safety in transmitting sequence *S*, which needs a round-trip time of classical transmission. But if this could be approved, it will roughly give an estimate of the communication distance of this approach equals almost a quarter of the BB84 QKD approach with a similar number of bps.

At last, but not least, the fourth protocol of Ain, it is the one with greatest security as he mentioned; which is reached only by enormous amount of quantum particles and states generations in the sender side – that's the main backbone of this technique; and also by the aid of massive number of calculations and measurements in every one side of the transmission (the 5 sides for the 5 participants), and in addition to a huge number of transmissions back and forth between every participant and the sender side.

Also, throughout the 4 stages of the protocol that include several exchange of quantum states and/or classical bits results from one side to another, there maybe a need of saving some quantum information to be used again in generating a new quantum states as in the last phase, plus for sure the need of saving this OTP key of the sender side that used in the last phase, to be revealed by the end to recover the secret message, and of course this need of saving quantum data is not available yet as mentioned above.

Eventually, for all above reasons, this protocol could be theoretically the highest security, but it maybe not applicable and infeasible to established.

4 The Suggested Solution

As in our published work in 2009 [9], we have provided a new protocol for the QSDC with a new idea that's based on Bell's maximally entangled states and the super dense coding theorem. In our technique we tried to overcome most of the above problems by offering a comparative way to transmit the private message form one side to another (only one sender and one receiver), in a simple and efficient way of transmission that apply high level of security accompanied by large capacity of qubits in one way of communication.

In our procedure we suggested that by using the idea of the super dense coding theorem, the classical message bits will input to the operator ***U*** selector in Fig. 1. And then according to their values (one of the four possible values 00, 01, 10 or 11), one of the quantum four gates: *I, X, Y* and *Z* will be selected randomly. Where:

$$I = \begin{bmatrix} 1 & 0 \\ 0 & 1 \end{bmatrix}, X = \begin{bmatrix} 0 & 1 \\ 1 & 0 \end{bmatrix}, Y = \begin{bmatrix} 0 & -i \\ i & 0 \end{bmatrix}, Z = \begin{bmatrix} 1 & 0 \\ 0 & -1 \end{bmatrix} \tag{3}$$

With the output maximally entangled state from the Bell states generator; which will also be haphazardly generated to get one of Bell states based on the inputs $(|i_0\rangle, |i_1\rangle)$ of the quantum generator as follows:

$$|0\rangle, |0\rangle \rightarrow |\varphi^+\rangle = \frac{1}{\sqrt{2}}(|00\rangle + |11\rangle)$$

$$|1\rangle, |0\rangle \rightarrow |\varphi^-\rangle = \frac{1}{\sqrt{2}}(|00\rangle - |11\rangle)$$

$$|0\rangle, |1\rangle \rightarrow |\psi^+\rangle = \frac{1}{\sqrt{2}}(|01\rangle + |10\rangle)$$

$$|1\rangle, |1\rangle \rightarrow |\psi^-\rangle = \frac{1}{\sqrt{2}}(|01\rangle - |10\rangle)$$

Now inputting these two inputs to the *U* Pauli operator; one input of the classical bits of the M bits of the message, and the other input of the Bell state generated; we will drive it to perform one of the four unitary operations mentioned above of the *U* Pauli operators to encrypt the Bell state to create a new Bell state $|\psi^0$ that will carry the encrypted message to be transmitted to the other side. The new created $|\psi^0$ will be also another one of Bell states according to our analysis that we had presented before [9]. In this transmission the new encrypted quantum state will be subdivide into its two particles to transmit each one separately on spatially separated quantum channels as shown in Fig. 1.

When this transmitted state reaches the receiver side, Bob will apply the reverse operation of the encoder circuit to extract the M bits of the message from the quantum encrypted state. Of course, the encoding operations and decoding operations has been per-specified and approved by both transmission legitimated parities "Alice and Bob", and these operations are simply and application of sequence of quantum gate operations in series on the two qubits of each entangled state.

In this proposed algorithm, the transmission of the encoded message that is presented in the state $|\psi^0$ is very simple and secure, as the two spatially separated quantum channels grantees that Eve can not measure the two particles of the state simultaneously at the same moment to apply the decoding operation on them and extract the message, specially that she doesn't even know what is the decoder circuit or the encoder circuit, so accordingly no way for her to got original message by any means.

So, in here, we save the effort and time of retransmission in the above techniques and the rehearsal of the checking phases and stages just to grantee the safe communication or to validate the receiver identity. Also, it doesn't need any quantum memory to store any sequence or resend any pre-created or manipulated quantum state.

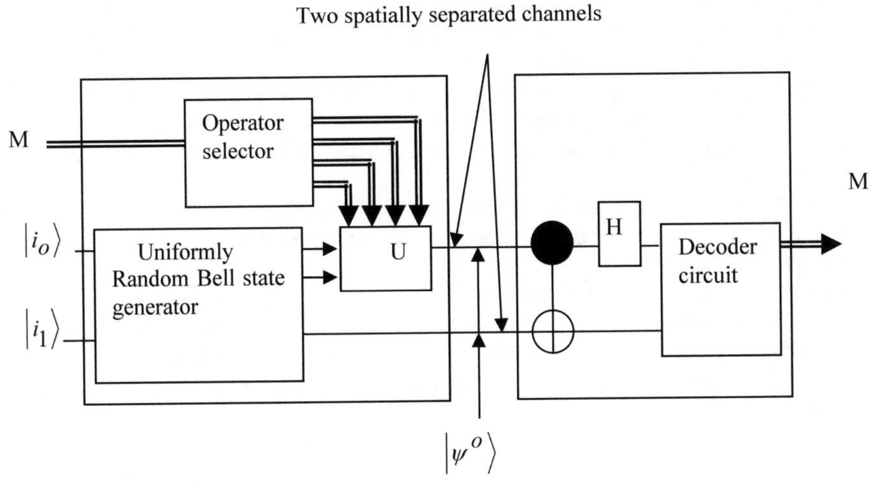

Fig. 1. The encoding and decoding circuits used in our proposed protocol.

5 Conclusion

In this paper we presented a comparative study among the most known protocols of the Quantum Secure Direct Communication QSDC technique, where we chose four different techniques that proposed different types of applying QSDC protocol, we explained briefly the theory of work of each of them, and we discussed their efficiency security wise and implementation wise, clarifying the feasibility and infeasibility of each. We showed that some of them are sacrificing of efforts and cost for the sack of promising security which was not realistically satisfied. We analysis and illustrate the technical reasons for that, and by the end we demonstrate our pre-offered algorithm that could overcome the faults and weakness in the implementation of those protocol with guarantees an ultimate security that raise from the well and precise implementation of the underlined protocol by making a good use of the quantum media and its precious features and phenomena, as the use of the quantum channels with designing them to be spatially separated is an essential requirement to reach the security target in more simpler way than other algorithms.

References

1. Long, G.: Quantum secure direct communication: principles, current status, perspectives. In: IEEE 85th Vehicular Technology Conference, IEEE *Xplore* (2017)
2. Yoon, C., Kang, M., Lim, J., Yank, H.: Quantum signature scheme based on quantum search algorithm. Pyhs. Scr. **90**, 15103–15108 (2015)
3. Zhang, Z.: Multiparty quantum secret sharing of secure direct communication. Phys. Lett. A **342**, 60–66 (2005)
4. Gao, G.: Two quantum dialogue protocols without information leakage. Opt. Commun. **283**, 2288–2293 (2010)
5. Long, G., Liu, X.: Theoretically efficient high-capacity quantum key distribution scheme. Phys. Rev. A **65**, 032302 (2002)
6. Deng, F., Long, G., Liu, X.: Two-step quantum direct communication protocol using the Einstein-Podolsky-Rosen pair block. Phys. Rev. A **68**, 042317 (2003)
7. Deng, F., Long, G.: Secure direct communication with a quantum one-time pad. Phys. Rev. A **69**, 052319 (2004)
8. Ain, N.: A novel approach for secure multi-party secret sharing scheme via quantum cryptography. In: International Conference on Communication, Computing and Digital Systems (C-CODE) (2017)
9. Hegazy, O., Bahaa, A., Dakroury, Y.: Quantum secure direct communication using entanglement and super dense coding. In: Proceedings of the International Conference on Security and Cryptography, SECRYPT, pp. 175–181, Milano, Italy (2009)
10. David, Mc., Quantum Computing Explained. IEEE Computer Society, Wiley, Canada (2008)

Fuzzy Sets and Game Theory in Green Supply Chain: An Optimization Model

Marwan Alakhras[1(✉)], Mousa Hussein[1], and Mourad Oussalah[2]

[1] EE, COE, United Arab Emirates University, Al Ain, United Arab Emirates
{alakhras,mihussein}@uaeu.ac.ae

[2] Department of Computer Science and Engineering, University of Oulu, Oulu, Finland
Mourad.Oussalah@oulu.fi

Abstract. An optimization model using fuzzy sets and game theory for three players is put forward in this wok. The decision model is influenced by customer demands in a green supply chain. The proposed model includes an empirical solution to enhance the confidence level of players to choose plausible green strategy. Initially, the strategies are formulated using the game theory as manufacturer, costumer and government, to be able to optimize the pay-off uncertainty conditions of demands, by combining computational fuzzy set with ability of sensitive analysis of related fuzzy parameters to enhance the calculations and problem solving, with presenting Nash equilibrium the problem solving part.

Keywords: Fuzzy sets · Game theory · Nash equilibrium · Green supply chain · Pay off function · Green strategy

1 Introduction

Due to the astonishing growth of green supply chain approaches, and sustainable development, this subject attracted many researchers. Traditionally, most of methods and models often were presented to conceptually justify green supply chain. However, the issue here is to be able to exactly find and plausible solutions, that motivate the main players in any supply chain to pick an accurate solution under uncertainty conditions. Any valid solution should definitely constitute of various mathematical models, which are sensitive to changes in the trading industries.

In our work, we considered merging of game theory and fuzzy sets to analyze and model green supply chain strategies, to make a more plausible fuzzy game model with deployment of players, decision-making parameters of pay-off functions. In which, major parameters, and players' pay-off functions are optimized in order to accurately analyze results for players, and achieve enhanced confidence in choosing a certain strategy. This model aims at the government and manufacturers abilities to analyze the customer inquiries, what strategy they usually pick considering the income and cost tradeoff to change status. Where an optimization model is put forward to choose the best cost/income parameters of players pay-off (pricing, subsidiaries…), based on customer demand. Due to the vagueness in the real world data, then a fuzzy inference system that depends on game theory looks promising, especially, in green supply chain,

K. Arai et al. (Eds.): CompCom 2019, AISC 997, pp. 149–164, 2019.
https://doi.org/10.1007/978-3-030-22871-2_12

this can build a better practical solution in order to optimize related functions and variables. And for more comprehensive analysis, a game model of three-players, namely (government, manufacturer and customer) has been considered. Hence, the final results of this work show that if the green strategies are considered by players, the game fuzzy model can provide more economic results in players' pay-off than the non-fuzzy game model.

The remaining part of this work is organized as follows: Sect. 2 lists some of the work done in the literature. In Sect. 3, the modelling of the problem is elaborated. Section 4 describes modelling of the problem with applying Fuzzy Set theory concepts with sensitive analysis. Section 5 details the simulation and application considered for the proposed fuzzy model with optimal solution. In Sect. 6, Numerical analysis with case study is provided. And finally, the conclusion is drawn.

2 Literature Review

Since the early 1980's the topics of supply chain management attracted many researchers [1–4], particularly to look at this problem from programming point of view. Where, concepts like material selection, production process, transportation, and definitely the interaction between various supply chain constitutes have always been investigated. Recently, due to reduction of resources, increase in environmental pollutions because of trade developments, and the growth of customers inquiries, green supply chain also affected by these changes, where, the challenge was to enhance environmental effects and increase economic revenue [5]. The definition of Green supply chain management was also considered in [6]. On one hand, it was claimed that; the enhancement of long-term economic revenue, in addition to enhancement of environmental impacts can be considered as win-win strategy [7, 8].

On another hand, Green supply chain management does not have solely impact on environment, but also it has a direct impact on manufacturers. One may consider the green product progress, increase in opportunities and re-innovation of products [9]. Moreover, using developed technologies in supply chain management and processes to decrease industrial pollution is another issue to consideration the impact on environment was discussed in [10]. For example, the some researchers [11] considered the carbon emission life cycle model as important tool help users in picking certain products, and likewise, can be considered as tool for manufacturers while Game theory can be considered as an important technique, especially, when conflicts of interest among players are rising. Essentially to assist decision makers to increase positive collaboration to satisfy common goals [12]. Practical game theory is often considered parallel in both streams the supply chain from one side and economic stability from the other side as in [13–16]. Also evolutional game models have been considered to establish rationale relations between various subsidies and governmental penalties, definitely this would affect environmental performance of companies [17]. In this particular work in 2007, the authors suggested imposing environmental performance and regulations by governments to have direct impact on subsidies through set of penalties. Later on in 2009 and 2011 other versions of game theory were proposed to assist in putting clear pricing policies on regulations on the environment performance

[18, 19]. Particularly, a symmetric bargaining model of the game theory was suggested. Another model in 2012 [20] was proposed based on dynamic evolutionary game theory. The proposed model studied potential coordination among various players, namely, retailers and manufacturers, through optimizing long term economic benefits, and environment impact, based on win-win green supply conditions. One particular case was considered in [21]. This case considered the relationship between optimizing the Carbon emission and economic benefits.

Those studies, essentially presented the use of game theory in supply chain, worth to mention that many of them presented the two-players model case, with simple techniques for the presented conditions. In our work, three players on the chain were simultaneously considered, costumer, manufacturer and government. This combination differentiates our work from the previous studies. Moreover, this distinction allows us to study the effectiveness of various parameters in the players pay-off function. In addition, the use of fuzzy set theory and fuzzy inference system allows us to generate more plausible model, especially with the use of sensitivity analysis which achieves optimal results under uncertainty conditions.

3 Modelling Principle

In order to be able to clarify the modelling of the problem, initially the description of main parameters have to be set to the common ground, this includes the main income and cost parameters for the Subsidiaries considered in this particular problem i.e., Government, Manufacturer, and Customer. For the income purposes, income acquired from some financial facilities, i.e., credits, loans…etc., can be considered as sources for income for the governments. Whereas, Tax reduction, Custom charges reduction, Loans, and any other special facilities can be considered as income for manufacturers. While, after sale services, special payment condition, discounts, etc. are considered as incomes for customers. The long term income variable like sustainable development benefits which has direct impact on players' decision-making especially for manufacturers and government.

Costs variable can vary, for instance: Environmental costs may include costs of miss-using of resources, environmental pollution, and the resultant human risks to human. Unemployment Cost variable, which is usually studied by governments due to change of technology used by certain manufacturer. Subsidiary cost variable, this can be the cost that a player considers as subside or decline of other members of the game. The variable cost of manufacturing technology, which can include training costs, maintaining costs, energy costs, etc. will be added as direct costs for the manufacturing. Losing credit costs, loss international or government credit in the event of non-implementation of international green industry related requirements has to be an added cost.

Now let N be the set of participating players and S the set of considered strategies that can be defined as:

$$S_i = \{S_1, S_2, \ldots, S_k\} \; i \in N = \{1, 2, 3\}$$

The variables of decision making strategies are described in the Table 1.

Table 1. Players considered strategies

Government (1)	Customers (2)	Manufacturer (3)
Passive	Inquiries increase	Maintain standard
Monitor	Passive	Update to tolerable standard
Sub system	Inquiries increase	Update to acceptable standard

Based on the above variables and relationships, then pay-off function can be modeled as follows:

$$\forall S_{xy},\; P_{xy} = \sum\nolimits_{a=1}^{k} (I_{xa})_{S_{xy}} - \sum\nolimits_{b=1}^{l} (C_{xb})_{S_{xy}} \tag{1}$$

$$x, y = 1, 2, 3$$

$$a = 1, 2, \ldots, k$$

$$b = 1, 2, \ldots, l$$

Where, S_{xy} is the strategy (x) that is chosen by player (y), P_{xy} is pay-off of player (y) if strategy (x) is selected, $(I_{xa})_{S_{xy}}$ is the income (a) of player (y) if strategy (x) is selected, $(C_{xb})_{S_{xy}}$ is cost (b) of player (y) if strategy (x) is selected, y is the number of players (1: government, 2: Customer, 3: Manufacturer), x is the number of strategies, a is the number of income, and b is the number of cost. Where details of income and cost variables for each player are described in Tables 2 and 3:

Table 2. The income variables of pay-off

Government (1)	Customers (2)	Manufacturer (3)
Penalties	Subsidiary	Sale
Int. Subsidiary	Sustainable development benefit	Subsidiary
Sustainable development benefit		Sustainable development benefit

Table 3. The cost variables of pay-off

Government (1)	Customers (2)	Manufacturer (3)
Monitor	Purchasing	Investment
Subsidiary	Environment	Production
Environment		Subsidiary
Losing credit		Overhead
Unemployment		Penalties

If the parameters; α and γ defined as, the penalty rate of manufacturer in relation to income due to selling of products, the parameter β is the rate of increase in price of products in case of technology change, γ is the manufacturer's percent of share from incentive budget of government, and ε is the customers' cost if increase of demand is seen in state of no change of technology. Then according to Eq. (2), Tables 2 and 3, the players pay-off functions could be formulated as:

$$P_1 = (I_{11} + I_{12} + I_{13}) - (C_{11} + C_{12} + C_{13} + C_{14} + C_{15})$$

Where, $I_{11} = \alpha(1+\beta)(1+\rho)I_{21}$, and $C_{12} = \gamma C_{12} + (1-\gamma)C_{12}$

$$\text{then: } P_1 = [\alpha(1+\beta)(1+\rho)I_{21} + I_{12} + I_{13}] - [(C_{11} + \gamma C_{12} + (1-\gamma)C_{12} + C_{13} + C_{14} + C_{15})] \tag{2}$$

$$P_2 = [\alpha(1+\beta)(1+\rho)I_{21} + I_{22} + I_{23}] - (C_{21} + \varphi(1+\rho)C_{22} + C_{23} + C_{24} + C_{25})$$

Where $I_{22} = \gamma C_{12}$, and $C_{25} = \alpha(1+\beta)(1+\rho)I_{21}$

$$P_2 = [(1+\beta)(1+\rho)I_{21} + \gamma C_{12} + I_{23}] - (C_{21} + \varphi(1+\rho)C_{22} + C_{23} + C_{24} + \alpha(1+\beta)(1+\rho)I_{21})$$

$$\text{then: } P_2 = [(1-\alpha)(1+\beta)(1+\rho)I_{21} + \gamma C_{12} + I_{23}] - (C_{21} + \varphi(1+\rho)C_{22} + C_{23} + C_{24}) \tag{3}$$

$P_3 = (I_{31} + I_{32}) - (C_{31} + C_{32})$, where $I_{31} = (1-\gamma)C_{12}$, and $C_{31} = (1+\beta)I_{21} + \beta\rho I_{21} + \varepsilon = (1+\beta+\beta\rho)I_{21} + \varepsilon$

$$\text{then: } P_3 = [(1-\gamma)C_{12} + I_{32}] - [(1+\beta+\beta\rho)I_{21} + \varepsilon + C_{32}] \tag{4}$$

Then the three player's matrix can be described as shown in Tables 4, 5, and 6.

Table 4. Matrix model for Customer's strategy = S_{31}

	Manufacturer strategy = $\boldsymbol{S}_{21}$	Manufacturer strategy = $\boldsymbol{S}_{22}$	Manufacturer strategy = $\boldsymbol{S}_{23}$
Government strategy = $\boldsymbol{S}_{11}$	$\sum_{i=1}^{3}(I_{1i})_{S_{11}}-\sum_{j=1}^{5}(C_{1j})_{S_{11}}$ $\sum_{i=1}^{3}(I_{2i})_{S_{21}}-\sum_{j=1}^{5}(C_{2j})_{S_{21}}$ $\sum_{i=1}^{2}(I_{3i})_{S_{31}}-\sum_{j=1}^{2}(C_{3j})_{S_{31}}$	$\sum_{i=1}^{3}(I_{1i})_{S_{11}}-\sum_{j=1}^{5}(C_{1j})_{S_{11}}$ $\sum_{i=1}^{3}(I_{2i})_{S_{22}}-\sum_{j=1}^{5}(C_{2j})_{S_{22}}$ $\sum_{i=1}^{2}(I_{3i})_{S_{31}}-\sum_{j=1}^{2}(C_{3j})_{S_{31}}$	$\sum_{i=1}^{3}(I_{1i})_{S_{11}}-\sum_{j=1}^{5}(C_{1j})_{S_{11}}$ $\sum_{i=1}^{3}(I_{2i})_{S_{23}}-\sum_{j=1}^{5}(C_{2j})_{S_{23}}$ $\sum_{i=1}^{2}(I_{3i})_{S_{31}}-\sum_{j=1}^{2}(C_{3j})_{S_{31}}$
Government strategy = $\boldsymbol{S}_{12}$	$\sum_{i=1}^{3}(I_{1i})_{S_{12}}-\sum_{j=1}^{5}(C_{1j})_{S_{12}}$ $\sum_{i=1}^{3}(I_{2i})_{S_{21}}-\sum_{j=1}^{5}(C_{2j})_{S_{21}}$ $\sum_{i=1}^{2}(I_{3i})_{S_{31}}-\sum_{j=1}^{2}(C_{3j})_{S_{31}}$	$\sum_{i=1}^{3}(I_{1i})_{S_{12}}-\sum_{j=1}^{5}(C_{1j})_{S_{12}}$ $\sum_{i=1}^{3}(I_{2i})_{S_{22}}-\sum_{j=1}^{5}(C_{2j})_{S_{22}}$ $\sum_{i=1}^{2}(I_{3i})_{S_{31}}-\sum_{j=1}^{2}(C_{3j})_{S_{31}}$	$\sum_{i=1}^{3}(I_{1i})_{S_{12}}-\sum_{j=1}^{5}(C_{1j})_{S_{12}}$ $\sum_{i=1}^{3}(I_{2i})_{S_{23}}-\sum_{j=1}^{5}(C_{2j})_{S_{23}}$ $\sum_{i=1}^{2}(I_{3i})_{S_{31}}-\sum_{j=1}^{2}(C_{3j})_{S_{31}}$
Government strategy = $\boldsymbol{S}_{13}$	$\sum_{i=1}^{3}(I_{1i})_{S_{13}}-\sum_{j=1}^{5}(C_{1j})_{S_{13}}$ $\sum_{i=1}^{3}(I_{2i})_{S_{21}}-\sum_{j=1}^{5}(C_{2j})_{S_{21}}$ $\sum_{i=1}^{2}(I_{3i})_{S_{31}}-\sum_{j=1}^{2}(C_{3j})_{S_{31}}$	$\sum_{i=1}^{3}(I_{1i})_{S_{13}}-\sum_{j=1}^{5}(C_{1j})_{S_{13}}$ $\sum_{i=1}^{3}(I_{2i})_{S_{22}}-\sum_{j=1}^{5}(C_{2j})_{S_{22}}$ $\sum_{i=1}^{2}(I_{3i})_{S_{31}}-\sum_{j=1}^{2}(C_{3j})_{S_{31}}$	$\sum_{i=1}^{3}(I_{1i})_{S_{13}}-\sum_{j=1}^{5}(C_{1j})_{S_{13}}$ $\sum_{i=1}^{3}(I_{2i})_{S_{23}}-\sum_{j=1}^{5}(C_{2j})_{S_{23}}$ $\sum_{i=1}^{2}(I_{3i})_{S_{31}}-\sum_{j=1}^{2}(C_{3j})_{S_{31}}$

Table 5. Matrix model for Customer's strategy = S_{32}

	Manufacturer strategy = $\boldsymbol{S}_{21}$	Manufacturer strategy = $\boldsymbol{S}_{22}$	Manufacturer strategy = $\boldsymbol{S}_{23}$
Government strategy = $\boldsymbol{S}_{11}$	$\sum_{i=1}^{3}(I_{1i})_{S_{11}}-\sum_{j=1}^{5}(C_{1j})_{S_{11}}$ $\sum_{i=1}^{3}(I_{2i})_{S_{21}}-\sum_{j=1}^{5}(C_{2j})_{S_{21}}$ $\sum_{i=1}^{2}(I_{3i})_{S_{32}}-\sum_{j=1}^{2}(C_{3j})_{S_{32}}$	$\sum_{i=1}^{3}(I_{1i})_{S_{11}}-\sum_{j=1}^{5}(C_{1j})_{S_{11}}$ $\sum_{i=1}^{3}(I_{2i})_{S_{22}}-\sum_{j=1}^{5}(C_{2j})_{S_{22}}$ $\sum_{i=1}^{2}(I_{3i})_{S_{32}}-\sum_{j=1}^{2}(C_{3j})_{S_{32}}$	$\sum_{i=1}^{3}(I_{1i})_{S_{11}}-\sum_{j=1}^{5}(C_{1j})_{S_{11}}$ $\sum_{i=1}^{3}(I_{2i})_{S_{23}}-\sum_{j=1}^{5}(C_{2j})_{S_{23}}$ $\sum_{i=1}^{2}(I_{3i})_{S_{32}}-\sum_{j=1}^{2}(C_{3j})_{S_{32}}$
Government strategy = $\boldsymbol{S}_{12}$	$\sum_{i=1}^{3}(I_{1i})_{S_{12}}-\sum_{j=1}^{5}(C_{1j})_{S_{12}}$ $\sum_{i=1}^{3}(I_{2i})_{S_{21}}-\sum_{j=1}^{5}(C_{2j})_{S_{21}}$ $\sum_{i=1}^{2}(I_{3i})_{S_{32}}-\sum_{j=1}^{2}(C_{3j})_{S_{32}}$	$\sum_{i=1}^{3}(I_{1i})_{S_{12}}-\sum_{j=1}^{5}(C_{1j})_{S_{12}}$ $\sum_{i=1}^{3}(I_{2i})_{S_{22}}-\sum_{j=1}^{5}(C_{2j})_{S_{22}}$ $\sum_{i=1}^{2}(I_{3i})_{S_{32}}-\sum_{j=1}^{2}(C_{3j})_{S_{32}}$	$\sum_{i=1}^{3}(I_{1i})_{S_{12}}-\sum_{j=1}^{5}(C_{1j})_{S_{12}}$ $\sum_{i=1}^{3}(I_{2i})_{S_{23}}-\sum_{j=1}^{5}(C_{2j})_{S_{23}}$ $\sum_{i=1}^{2}(I_{3i})_{S_{32}}-\sum_{j=1}^{2}(C_{3j})_{S_{32}}$
Government strategy = $\boldsymbol{S}_{13}$	$\sum_{i=1}^{3}(I_{1i})_{S_{13}}-\sum_{j=1}^{5}(C_{1j})_{S_{13}}$ $\sum_{i=1}^{3}(I_{2i})_{S_{21}}-\sum_{j=1}^{5}(C_{2j})_{S_{21}}$ $\sum_{i=1}^{2}(I_{3i})_{S_{32}}-\sum_{j=1}^{2}(C_{3j})_{S_{32}}$	$\sum_{i=1}^{3}(I_{1i})_{S_{13}}-\sum_{j=1}^{5}(C_{1j})_{S_{13}}$ $\sum_{i=1}^{3}(I_{2i})_{S_{22}}-\sum_{j=1}^{5}(C_{2j})_{S_{22}}$ $\sum_{i=1}^{2}(I_{3i})_{S_{32}}-\sum_{j=1}^{2}(C_{3j})_{S_{32}}$	$\sum_{i=1}^{3}(I_{1i})_{S_{13}}-\sum_{j=1}^{5}(C_{1j})_{S_{13}}$ $\sum_{i=1}^{3}(I_{2i})_{S_{23}}-\sum_{j=1}^{5}(C_{2j})_{S_{23}}$ $\sum_{i=1}^{2}(I_{3i})_{S_{32}}-\sum_{j=1}^{2}(C_{3j})_{S_{32}}$

Table 6. Matrix model for Customer's strategy = S_{33}

	Manufacturer strategy = $\boldsymbol{S}_{21}$	Manufacturer strategy = $\boldsymbol{S}_{22}$	Manufacturer strategy = $\boldsymbol{S}_{23}$
Government strategy = $\boldsymbol{S}_{11}$	$\sum_{i=1}^{3}(I_{1i})_{S_{11}}-\sum_{j=1}^{5}(C_{1j})_{S_{11}}$ $\sum_{i=1}^{3}(I_{2i})_{S_{21}}-\sum_{j=1}^{5}(C_{2j})_{S_{21}}$ $\sum_{i=1}^{2}(I_{3i})_{S_{33}}-\sum_{j=1}^{2}(C_{3j})_{S_{33}}$	$\sum_{i=1}^{3}(I_{1i})_{S_{11}}-\sum_{j=1}^{5}(C_{1j})_{S_{11}}$ $\sum_{i=1}^{3}(I_{2i})_{S_{22}}-\sum_{j=1}^{5}(C_{2j})_{S_{22}}$ $\sum_{i=1}^{2}(I_{3i})_{S_{33}}-\sum_{j=1}^{2}(C_{3j})_{S_{33}}$	$\sum_{i=1}^{3}(I_{1i})_{S_{11}}-\sum_{j=1}^{5}(C_{1j})_{S_{11}}$ $\sum_{i=1}^{3}(I_{2i})_{S_{23}}-\sum_{j=1}^{5}(C_{2j})_{S_{23}}$ $\sum_{i=1}^{2}(I_{3i})_{S_{33}}-\sum_{j=1}^{2}(C_{3j})_{S_{33}}$
Government strategy = $\boldsymbol{S}_{12}$	$\sum_{i=1}^{3}(I_{1i})_{S_{12}}-\sum_{j=1}^{5}(C_{1j})_{S_{12}}$ $\sum_{i=1}^{3}(I_{2i})_{S_{21}}-\sum_{j=1}^{5}(C_{2j})_{S_{21}}$ $\sum_{i=1}^{2}(I_{3i})_{S_{33}}-\sum_{j=1}^{2}(C_{3j})_{S_{33}}$	$\sum_{i=1}^{3}(I_{1i})_{S_{12}}-\sum_{j=1}^{5}(C_{1j})_{S_{12}}$ $\sum_{i=1}^{3}(I_{2i})_{S_{22}}-\sum_{j=1}^{5}(C_{2j})_{S_{22}}$ $\sum_{i=1}^{2}(I_{3i})_{S_{33}}-\sum_{j=1}^{2}(C_{3j})_{S_{33}}$	$\sum_{i=1}^{3}(I_{1i})_{S_{12}}-\sum_{j=1}^{5}(C_{1j})_{S_{12}}$ $\sum_{i=1}^{3}(I_{2i})_{S_{23}}-\sum_{j=1}^{5}(C_{2j})_{S_{23}}$ $\sum_{i=1}^{2}(I_{3i})_{S_{33}}-\sum_{j=1}^{2}(C_{3j})_{S_{33}}$
Government strategy = $\boldsymbol{S}_{13}$	$\sum_{i=1}^{3}(I_{1i})_{S_{13}}-\sum_{j=1}^{5}(C_{1j})_{S_{13}}$ $\sum_{i=1}^{3}(I_{2i})_{S_{21}}-\sum_{j=1}^{5}(C_{2j})_{S_{21}}$ $\sum_{i=1}^{2}(I_{3i})_{S_{33}}-\sum_{j=1}^{2}(C_{3j})_{S_{33}}$	$\sum_{i=1}^{3}(I_{1i})_{S_{13}}-\sum_{j=1}^{5}(C_{1j})_{S_{13}}$ $\sum_{i=1}^{3}(I_{2i})_{S_{22}}-\sum_{j=1}^{5}(C_{2j})_{S_{22}}$ $\sum_{i=1}^{2}(I_{3i})_{S_{33}}-\sum_{j=1}^{2}(C_{3j})_{S_{33}}$	$\sum_{i=1}^{3}(I_{1i})_{S_{13}}-\sum_{j=1}^{5}(C_{1j})_{S_{13}}$ $\sum_{i=1}^{3}(I_{2i})_{S_{23}}-\sum_{j=1}^{5}(C_{2j})_{S_{23}}$ $\sum_{i=1}^{2}(I_{3i})_{S_{33}}-\sum_{j=1}^{2}(C_{3j})_{S_{33}}$

4 Modelling with Fuzzy Sets and Sensitive Analysis

As described earlier, and due to nature of the problem, the decision making parameters and major uncertain variables can be interpolated to fuzzy sets. Then, we can claim that, the entire model and its achieved results are, somehow, more efficient and practical in order to achieve optimal results. The consideration of fuzzy set can increase of confidence level for players to make decision on changing the status. Among of the all existing variables, the ones of making decision by customers are said to be of higher membership degree. In reality, if the variables making decision by customer are major parameters in calculating pay-off function and can be regarded as certain states with membership degree of value equal to 1, then the exact optimal results will not be achieved.

In fact, any other value not necessarily zero -customers is not interested enough- which can be considered by the other subsidiaries of the game, where, their decisions will not be affected, even if the final achieved pay-off value is higher than what it was before. On the other hand, other variables, for instance, benefits resulting from sustainable development may not reflect the positive effect of the customers pay-off function. Accordingly, we can model these notices as:

$$if\ I_{31} + I_{32} - \partial C_{31} < B_1\ then\ S_{31}$$

$$if\ I_{31} + I_{32} - \partial C_{31} < B_1\ then\ S_{32}$$

$$if\ I_{31} + I_{32} - \partial C_{31} < B_1\ then\ S_{33}$$

Where B_1 and B_2 are attraction levels of sum of incomes resulting from facilities and long-term interests of sustainable development for decision-making and $B_2 > B_1$. For the matter of fact, we expect that all players (especially customers), have no accurate and detailed view to draw decision depending on the result of Pay-off function. Then we can assume that customers will draw their pay-off function qualitatively, rather than numerically.

Fuzzy set theory can be very applicable to this kind of analysis. Now if the customer pay-off variables are considered as the fuzzy sets, then customer behaviour will be linguistically predictable and thus reflects change in variable value. This principle motivates the other players, government and manufacturer to take the optimal value for their related variables in order to positively modify customer inquiries or at least keep it steady.

To achieve more practical game model, and to more towards the optimal solution, a model based on fuzzy set is constructed as follows, which is adapted from [22].

Step 1: Specify the variables to be represented as fuzzy sets:

Fuzzy sets:

$$C_{12}, C_{23}, \partial C_{31}, I_{32}, P_3 \tag{5}$$

Step 2: Specify fuzzy implications:

$$C_{12}, C_{23}, \partial C_{31}, I_{32} \rightarrow \partial P_3 \tag{6}$$

Step 3: Assign linguistic values to fuzzy sets.
Linguistic values set of each variable can be defined as:

$$X = \{x_1, x_2, \ldots, x_n\} \tag{7}$$

In this research, these variables can be defined by the following fuzzy sets:

$$\begin{aligned} &C_{12} = \{Unfavorable(UF),\ Favorite(F)\}\\ &C_{23} = \{Unfavorable(UF),\ Favorite(F)\}\\ &\partial C_{31} = \{Low(L),\ High(H)\}\\ &I_{32} = \{Low(L),\ High(H)\}\\ &P_3 = \{Negative(N),\ Zero(Z),\ Positive(P)\} \end{aligned} \tag{8}$$

For the above described sets: negative linguistic value in pay-off function represents no interest to change in demand, zero represents passive response in making decision and, positive represents willing to change demand.

Step 4: Calculation of membership degree for every linguistic value:

$$\begin{aligned} &\mu_{C_{12}(UF)} = \mu_1(x)\\ &\mu_{C_{12}(F)} = \mu_2(x)\\ &\mu_{C_{23}(UF)} = \mu_3(x)\\ &\mu_{C_{23}(F)} = \mu_4(x)\\ &\mu_{\partial C_{31}(L)} = \mu_5(x)\\ &\mu_{\partial C_{23}(H)} = \mu_6(x)\\ &\mu_{I_{32}(L)} = \mu_7(x)\\ &\mu_{I_{32}(H)} = \mu_8(x)\\ &\mu_{P_3(N)} = \mu_9(x)\\ &\mu_{P_3(Z)} = \mu_{10}(x)\\ &\mu_{P_3(P)} = \mu_{11}(x) \end{aligned} \tag{9}$$

Step 5: Implementing the fuzzy inference: Among the available fuzzy inference systems, Mamadani inference [22] is elected in this problem, where: the inputs: $C_{12}, C_{23}, \partial C_{31}$, and I_{32}, and the output is P_3, then Rule$_i$ can be defined as:

$$if\ C_{12i},\ C_{23i},\ \partial C_{31i},\ I_{32i}\ then\ P_{3i}$$

$$a_i = \min(\mu_i(C_{12}), \mu_i(C_{23}), \mu_i(\partial C_{31}), \mu_i I_{32})$$

$$\mu_i = \min\left(a_i, \mu_{P_{3i}}\right)$$

The overall system output can be calculated using the union operator as:

$$\mu = \cup \mu_i \tag{10}$$

Where μ is member function of fuzzy sets, a_i is the ith alpha cut of fuzzy set. Where, finally the Defuzzification process should take place. In our case the fuzzy result will be used in the next step.

Step 6: Specifying the logical relationships: group of relationships between the final result and the predicted customers' decisions. This relationship can be described as follows:

$$\begin{aligned} &If\, 0 \leq \theta \leq l \rightarrow S_{31} \\ &f\, l < \theta \leq m \rightarrow S_{32} \\ &If\, m < \theta \leq h \rightarrow S_{33} \\ &0 < l < m < h \end{aligned} \tag{11}$$

Where θ is the fuzzy result of the problem, l is the minimal qualitative pay-off value, possibly customers that would show of decrease in demand. m is the mean qualitative pay-off value, customers that show no interest in changing the demand. And h is the maximal qualitative pay-off value for customers that show satisfaction and possibly of increase in demand. This pattern analysis outperforms the sensitivity analysis of customers pay-off function which is described in the next part.

Step 7: Applying sensitivity analysis to customers' pay-off, and choosing optimal result:

To analyse pay-off function, we assume that a, b, c and, d are four input parameters with different values. In this case there will be $a \times b \times c \times d$. In a collaborative game and after achieving results of sensitivity analysis for customers' behaviour, the other game players undertake the optimal value of governmental incentive facilities parameters for customers, the optimal value of incentive facilities of manufacturer, and the optimal value of increase in products price due to change technology levels.

5 Simulation of the Proposed Fuzzy Model and Optimal Solution

To evaluate the proposed model, a Matlab program was constructed to compute the defuzzified output of this fuzzy model, the results as extracted to Microsoft excel sheet as shown in Fig. 1. Which contains the major parts of the system, first the input vector, Rule Base for the inference system which consists of set of 16 if-then rules as explained in Step 5, the membership value for each rule with respect to the input vector, the aggregated membership function value for the output, and finally computing the defuzzified result, which will be used by experts to make the decision. According to the example in Fig. 1, if $I_{32} = 0.5, \partial C_{31} = 1.5, C_{23} = 3$, and $C_{12} = 2$, then, the accumulated final result is 8%, which indicates the customers conceive to modify the corresponding pay-off. Nevertheless, the achieved 8% is not quantitative rather than qualitative measure that indicates enough information about the customers' ability for making-decision about certain product.

	C_{12}	C_{23}	∂C_{31}	I_{32}
Input Data:	2	3	1.5	0.5

						µ Result for inputs				
						C_{12}	C_{23}	∂C_{31}	I_{32}	MIN
R1:	U	U	H	L	N	0	0	1	1	0
R2:	U	U	H	H	N	0	0	1	0	0
R3:	U	U	L	L	Z	0	0	0	1	0
R4:	U	U	L	H	Z	0	0	0	0	0
R5:	U	F	H	L	N	0	1	1	1	0
R6:	U	F	H	H	Z	0	1	1	0	0
R7:	U	F	L	L	Z	0	1	0	1	0
R8:	U	F	L	H	Z	0	1	0	0	0
R9:	F	U	H	L	N	1	0	1	1	0.25
R10:	F	U	H	H	Z	1	0	1	0	0.25
R11:	F	U	L	L	Z	1	0	0	1	0.25
R12:	F	U	L	H	Z	1	0	0	0	0.25
R13:	F	F	H	L	Z	1	1	1	1	0.5
R14:	F	F	H	H	Z	1	1	1	0	0.25
R15:	F	F	L	L	P	1	1	0	1	0.33
R16:	F	F	L	H	P	1	1	0	0	0.25

	P_c Rate (%)																
	0	1	2	3	4	5	6	7	8	9	10	11	12	13	14	15	16
µ Result for Rule1	0	0	0	0	0	0	0	0	0	0	0	0	0	0	0	0	0
µ Result for Rule2	0	0	0	0	0	0	0	0	0	0	0	0	0	0	0	0	0
µ Result for Rule3	0	0	0	0	0	0	0	0	0	0	0	0	0	0	0	0	0
µ Result for Rule4	0	0	0	0	0	0	0	0	0	0	0	0	0	0	0	0	0
µ Result for Rule5	0	0	0	0	0	0	0	0	0	0	0	0	0	0	0	0	0
µ Result for Rule6	0	0	0	0	0	0	0	0	0	0	0	0	0	0	0	0	0
µ Result for Rule7	0	0	0	0	0	0	0	0	0	0	0	0	0	0	0	0	0
µ Result for Rule8	0	0	0	0	0	0	0	0	0	0	0	0	0	0	0	0	0
µ Result for Rule9	0.25	0.25	0.25	0.25	0.25	0.25	0.25	0.25	0.167	0	0	0	0	0	0	0	0
µ Result for Rule10	0	0	0	0	0.2	0.25	0.25	0.25	0.25	0.25	0.25	0.25	0	0	0	0	0
µ Result for Rule11	0	0	0	0	0.2	0.25	0.25	0.25	0.25	0.25	0.25	0.25	0	0	0	0	0
µ Result for Rule12	0	0	0	0	0.2	0.25	0.25	0.25	0.25	0.25	0.25	0.25	0	0	0	0	0
µ Result for Rule13	0	0	0	0	0.2	0.4	0.5	0.5	0.5	0.5	0.5	0.25	0	0	0	0	0
µ Result for Rule14	0	0	0	0	0.2	0.25	0.25	0.25	0.25	0.25	0.25	0.25	0	0	0	0	0
µ Result for Rule15	0	0	0	0	0	0	0	0	0.2	0.333	0.333	0.333	0.333	0.333	0.333	0	0
µ Result for Rule16	0	0	0	0	0	0	0	0	0	0.25	0.25	0	0	0.25	0.25	0.25	0.25
Membership Degree Function Result (max µi)	0.25	0.25	0.25	0.25	0.25	0.4	0.5	0.5	0.5	0.5	0.5	0.333	0.333	0.333	0.333	0.25	0.25
Final Result (Deffuzification)	8%																

Fig. 1. Program sample results

The optimal result of decision-making for all the players can be computed as:

$$\max_{s_i \in S_i} u_i (S_i, S_{-i}) \tag{12}$$

Games are seeking of the optimal pay-off function for player (i) in front of combination of strategies related to this particular player strategy (S_i) with other strategies of all players except player (S_{-i}).

Nash Equilibrium presented in [23, 25] is said to provide optimal result for those kind of problems, where Nash stated that "an n-tuples such that each player's mixed strategy maximizes his pay-off if the strategies of the others are held fixed. Thus each player's strategy is optimal against those of the others" [25]. In our case, there is a combination of strategies among all combinations in game model at least that players are not interested to change that in a logical condition. Since the game is a dynamic with the complete information, if the Nash equilibrium is iteratively earned more than once, then the optimal result would be achieved via complete equilibrium or backward Nash equilibrium [24]. This optimal result can be achieved by the analysis of the resultant Nash equilibrium. One of the techniques, which can be used to obtain this result, is by converting the original game model to a secondary game model. Finally is gaining final result by all results of the secondary game.

6 Numerical Analysis

Taking numerical example can ease the description of the process. Our example presumes that Alain Cement manufacturer in Alain city is planning to update the production technology in order to decrease the impact on the environment, and drop the flow gas of factory to the new standardized level. After 6 months of study by the research and development department of this manufacturer, the total principle to invest, possible benefits and income, machinery update cost, labor training cost, and other required variables are listed in the following table (Table 7):

Table 7. Variables and data assumption

$n = 10$ (years)	$i = 10\%$	$\alpha = 0.03$
$\beta_1 = 0.02$	$\beta_2 = 0.03$	$\rho_1 = -0.2$
$\rho_2 = 0.1$	$I_{12} = 0.45C_{12}$	$I_{13(A)} = 2m\$$
$I_{13(B)} = 5m\$$	$C_{11} = 1.25m\$$	$\gamma C_{12} = 0.40C_{21}$
$(1-\gamma)C_{12} = [0, 2.5]m\$$		$C_{13(A)} = 5m\$$
$C_{13(B)} = 2m\$$	$C_{14(A)} = 3m\$$	$C_{14(B)} = 1m\$$
$C_{15(A)} = 1m\$$	$C_{15(B)} = 2m\$$	$I_{21} : 65m\$$
$I_{23(A)} = 1m\$$	$I_{23(B)} = 3m\$$	$C_{21(A)} = 40m\$$
$C_{21(B)} = 60m\$$	$C_{22(A)} = 50m\$$	$C_{22(B)} = 48m\$$
$C_{22(C)} = 47m\$$	$C_{23(A)} = [0, 5]m\$$	$C_{24(A)} = 3m\$$
$C_{24(B)} = 1m\$$	$C_{24(C)} = 0.65m\$$	$\partial C_{22} = 0.5\rho * C_{22}$
$I_{32(A)} = 0.5m\$$	$I_{32(B)} = 1m\$$	$C_{32(A)} = 3m\$$
$C_{32(B)} = 1m\$$	$\varepsilon = 0.01m\$$	

Note, the definition of the variables in the above table follows the same pre-described definition. And, the following assumptions can be made to solve this problem:

The problem assumes two stages, first is classical model, and second is the fuzzy sets based model. In the classical model, the two players the government and the manufacturer enrolled in technology change, while customer tolerable level uses 50% of subsidiary total. Also, manufacturer assumes the highest value of production price.

Status of reduction of customer demand, product purchasing cost variable for the loyalty customers is calculated based on the new prices, on the other hand, for the lost customers is calculated according to product purchasing cost. While the Sub costs of the manufacturer and government is changing according to the consumers' demands.

Likewise, consumer's income from this update will change also according to the demand.

Assume that symbol A represents the intolerable level in sustainable development of the environment. Symbol B represents the tolerable level of this change, and symbol C represents acceptable level. The total update cost is converted to yearly constant cost as shown in the following formula:

$$A = P(A/P, i\%, n) \rightarrow A = P(A/P, 10\%, 10)$$

The second model is built based on fuzzy sets according to the pre-described relations (5–9), and then the final fuzzy membership degrees and their corresponding triangular membership functions are illustrated in Fig. 2. Using the Matlab program some results of (10) are shown in Fig. 3 and in Table 8.

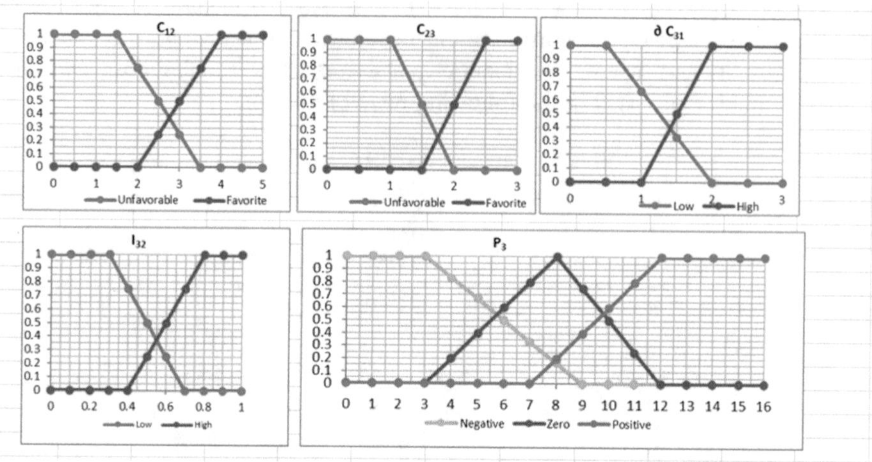

Fig. 2. The results membership functions

	C_{12}	C_{23}	∂C_{31}	I_{32}		μ Result for inputs C_{12}	C_{23}	∂C_{31}	I_{32}	MIN
Input Data:	3	5	1	0.5						
R1:	U	U	H	L	N	0	0	0	1	0
R2:	U	U	H	H	N	0	0	0	0	0
R3:	U	U	L	L	Z	0	0	1	1	0
R4:	U	U	L	H	Z	0	0	1	0	0
R5:	U	F	H	L	N	0	1	0	1	0
R6:	U	F	H	H	Z	0	1	0	0	0
R7:	U	F	L	L	Z	0	1	1	1	0
R8:	U	F	L	H	Z	0	1	1	0	0
R9:	F	U	H	L	N	1	0	0	1	0
R10:	F	U	H	H	Z	1	0	0	0	0
R11:	F	U	L	L	Z	1	0	1	1	0
R12:	F	U	L	H	Z	1	0	1	0	0
R13:	F	F	H	L	Z	1	1	0	1	0
R14:	F	F	H	H	Z	1	1	0	0	0
R15:	F	F	L	L	P	1	1	1	1	0.5
R16:	F	F	L	H	P	1	1	1	0	0.25

	P_c Rate (%)																
	0	1	2	3	4	5	6	7	8	9	10	11	12	13	14	15	16
μ Result for Rule1	0	0	0	0	0	0	0	0	0	0	0	0	0	0	0	0	0
μ Result for Rule2	0	0	0	0	0	0	0	0	0	0	0	0	0	0	0	0	0
μ Result for Rule3	0	0	0	0	0	0	0	0	0	0	0	0	0	0	0	0	0
μ Result for Rule4	0	0	0	0	0	0	0	0	0	0	0	0	0	0	0	0	0
μ Result for Rule5	0	0	0	0	0	0	0	0	0	0	0	0	0	0	0	0	0
μ Result for Rule6	0	0	0	0	0	0	0	0	0	0	0	0	0	0	0	0	0
μ Result for Rule7	0	0	0	0	0	0	0	0	0	0	0	0	0	0	0	0	0
μ Result for Rule8	0	0	0	0	0	0	0	0	0	0	0	0	0	0	0	0	0
μ Result for Rule9	0	0	0	0	0	0	0	0	0	0	0	0	0	0	0	0	0
μ Result for Rule10	0	0	0	0	0	0	0	0	0	0	0	0	0	0	0	0	0
μ Result for Rule11	0	0	0	0	0	0	0	0	0	0	0	0	0	0	0	0	0
μ Result for Rule12	0	0	0	0	0	0	0	0	0	0	0	0	0	0	0	0	0
μ Result for Rule13	0	0	0	0	0	0	0	0	0	0	0	0	0	0	0	0	0
μ Result for Rule14	0	0	0	0	0	0	0	0	0	0	0	0	0	0	0	0	0
μ Result for Rule15	0	0	0	0	0	0	0	0	0.2	0.4	0.5	0.5	0.5	0.5	0.5	0	0
μ Result for Rule16	0	0	0	0	0	0	0	0	0	0.25	0.25	0	0	0.25	0.25	0.25	0.25
Membership Degree Function Result (max μi)	0	0	0	0	0	0	0	0	0.2	0.4	0.5	0.5	0.5	0.5	0.5	0.25	0.25
Final Result (Deffuzification)	12%																

Fig. 3. Results of the fuzzy inference for the problem

Looking back at model (12), logical rules between fuzzy inference results and consumers decision-making are listed here:

$$if\ 0 \le \theta \le 5\% \to S_{31}$$

$$if\ 5\% < \theta < 10\% \to S_{32}$$

$$if\ 10 \le \theta \to S_{33}$$

Using the sensitivity analysis of consumers pay-offs for another data sets other than the ones used in Fig. 2, after implementing the aforementioned logical relations, then, optimal parameters of this change may look like:

$$C_{12} = 0, C_{23} = 0, \partial C_{31} = 1\%, I_{32} = 0.5, \partial P_3 = 8\%$$

This optimal solution considers that, governments are not going utilize incentive costs of facilities to update the technology in the acceptable and tolerable levels.

But if the governments are willing to implement the incentive costs, then the optimal solution for tolerable level, and acceptable level respectively can be re-modeled as follows:

$$C_{12} = 1.5,\ C_{23} = 0,\ \partial C_{31} = 1\%,\ I_{32} = 0.5,\ \partial P_3 = 8\%$$

$$C_{12} = 3,\ C_{23} = 3,\ \partial C_{31} = 1.5\%,\ I_{32} = 1,\ \partial P_3 = 10\%$$

Where, ∂P_3 shows increase in Customers' pay-off function than the previous level in that. Some of the results of consumers pay-offs are shown in Table 8:

Table 8. Consumers pay-offs

C_{12}	C_{23}	∂C_{31}	I_{32}	P_3
0.0	0.0	2.0	0.5	4.0%
1.0	4.0	1.5	0.5	5.0%
3.0	2.5	2.5	0.5	5.0%
3.0	3.0	2.5	0.5	6.0%
2.0	3.0	3.0	0.5	6.0%
3.0	2.5	1.5	0.5	7.0%
2.0	2.5	1.5	0.5	7.0%
0.0	1.0	0.0	0.5	8.0%
0.0	0.0	1.0	0.5	8.0%
1.5	0.0	1.0	0.5	8.0%
2.5	2.5	1.0	0.5	9.0%
3.0	3.0	1.5	1.0	10.0%
3.0	4.0	1.5	0.5	10.0%
3.0	3.0	1.0	0.5	11.0%
3.0	3.0	0.0	0.5	11.0%
3.0	5.0	1.0	0.5	12.0%

Then, the matrix of the game can be shown in Table 9.

Table 9. Fuzzy sets based model- Consumer Strategy = S_{31}

	S_{21}	S_{22}	S_{23}
S_{11}	−8, 4, −68	−3, 2.82, −66	2, 2.81, −64.7
S_{12}	−7.7, 2.4, −68	−3, 2.82, −66	2, 2.81, −64.7
S_{13}	−7.7, 2.4, −68	−2.7, 6.1, −65	0.1, 5.55, −59.4

After implementing the Nash Equilibrium and its Backward version based on the (2, 3, 4 and 12), and Tables 4, 5 and 6. Then Tables 10 and 11 show the optimal answers of each parameter.

Table 10. Fuzzy sets based model- Consumer Strategy = S_{32}

	S_{21}	S_{22}	S_{23}
S_{11}	−8, 12, −68	−3, 11.1, −66.2	2, 10.6, −64.7
S_{12}	−7.3, 10, −68	−3, 11.1, −66.2	2, 10.6, −64.7
S_{13}	−7.3, 10, −68	−3, 14.4, −66	−0.5, 14, −59

Table 11. Fuzzy sets based model- Consumer Strategy = S_{33}

	S_{21}	S_{22}	S_{23}
S_{11}	−8, 16, −68.01	−3, 15.3, −66.2	2, 15.4, −64.7
S_{12}	−7.1, 13.8, −68.01	−3, 15.3, −66.2	2, 15.4, −64.7
S_{13}	−7.1, 13.8, −68.1	−3, 18.6, −65	−0.8, 15.7, −57.1

Considering to above tables, the optimal answers of fuzzy set based is formulated as:

$$N(G)_1 = (S_{13}, S_{22}, S_{31}) = (-2.7, 6.1, -65)$$

$$N(G)_2 = (S_{13}, S_{22}, S_{33}) = (-3, 18.6, -65)$$

$SPE = (S_{13}, S_{22}, S_{33}) = (-3, 18.6, -65) \rightarrow$ optimal result based on fuzzy sets model

Carrying out our analysis separately, (for classical model), the optimal results can be:

$$N(G)_1 = (S_{12}, S_{21}, S_{31}) = (-7.7, 2.4, -68)$$

$$N(G)_2 = (S_{12}, S_{21}, S_{32}) = (-7.3, 10, -68)$$

$$N(G)_3 = (S_{13}, S_{22}, S_{32}) = (-3, 11.9, -62.8)$$

$SPE(G) = (S_{13}, S_{22}, S_{32}) = (-3, 11.9, -62.8) \rightarrow$ optimal result based on classical model

Considering both results the classical version and the fuzzy set based one, it is clear that in fuzzy set based version, the results of Nash equilibrium and its Backward version are different than the classical version. And, it can propose different strategies. In the optimal solution of the fuzzy set based version model, government adapts strategies of sub system and monitory process while, manufacturer can update its technology from intolerable to the tolerable level. Consumers demand will show positive attitude to this update status. Considering the classical based version, the optimal solution indicated that consumers choose passive strategy while both pay-offs of manufacturer and consumers are reduced.

7 Conclusion

The strategies among three players of a green supply chain have been solved and modelled by using the game theory. In order to achieve practical model, initially, the proposed model merged the consumers strategies and main variables of pay-offs for three players by using fuzzy logic relations. Later, using analytical models and relations sensitively analyse the consumers pay-offs by modifying the fuzzy variables. Moreover, the used problem solving method proposes Nash equilibrium and its backward version. Finally, a numerical analysis was done and problem modelling and solving was accomplished using fuzzy set based version and classical version, and their results were compared with each other. The achieved results showed that the fuzzy set based version optimizes pay-offs for the three players more than the classical version, which results to change strategies of the players and thus motivates to move forward into a green strategy.

References

1. Blanchard, D.: Supply Chain Management Best Practices, 2nd edn. Wiley, Hoboken (2010)
2. Harrison, A., van Hoek, R.I.: Logistics Management and Strategy: Competing Through the Supply Chain. Pearson/Financial Times Prentice Hall, Upper Saddle River (2011)
3. Hines, T.: Supply Chain Strategies: Customer Driven and Customer Focused. Elsevier Butterworth-Heinemann, Oxford (2004)
4. Oliver, R.K., Webber, M.D.: Supply-chain management: logistics catches up with strategy. In: The Roots of Logistics: A Reader of Classical Contributions to the History and Conceptual Foundations of the Science of Logistics, pp. 183–194. Springer, Berlin (2012)
5. Sheu, J.-B., Chou, Y.-H., Hu, C.-C.: An integrated logistics operational model for green-supply chain management. Transp. Res. Part E Logist. Transp. Rev. **41**(4), 287–313 (2005)
6. Basu, R.: Total Supply Chain Management. Routledge, Abingdon (2016)
7. Zhu, Q., Cote, R.P.: Integrating green supply chain management into an embryonic eco-industrial development: a case study of the Guitang Group. J. Clean. Prod. **12**(8–10), 1025–1035 (2004)
8. Zhu, Q., Sarkis, J., Lai, K.: Confirmation of a measurement model for green supply chain management practices implementation. Int. J. Prod. Econ. **111**(2), 261–273 (2008)
9. Wang, H.-F., Gupta, S.M.: Green Supply Chain Management: Product Life Cycle Approach. McGraw Hill, New York (2011)
10. Wang, H.-F.: Web-Based Green Products Life Cycle Management Systems: Reverse Supply Chain Utilization. Information Science Reference, Hershey (2009)
11. Zhao, R., Deutz, P., Neighbour, G., McGuire, M.: Carbon emissions intensity ratio: an indicator for an improved carbon labelling scheme. Environ. Res. Lett. **7**(1), 014014 (2012)
12. Cachon, G.P., Netessine, S.: Game Theory in Supply Chain Analysis, vol. 74. Springer, Boston, MA (2004)

13. Esmaeili, M., Aryanezhad, M.-B., Zeephongsekul, P.: A game theory approach in seller–buyer supply chain. Eur. J. Oper. Res. **195**(2), 442–448 (2009)
14. Li, S.X., Huang, Z., Zhu, J., Chau, P.Y.K.: Cooperative advertising, game theory and manufacturer–retailer supply chains. Omega **30**(5), 347–357 (2002)
15. Nagarajan, M., Sošić, G.: Game-theoretic analysis of cooperation among supply chain agents: review and extensions. Eur. J. Oper. Res. **187**(3), 719–745 (2008)
16. Yue, J., Austin, J., Wang, M.-C., Huang, Z.: Coordination of cooperative advertising in a two-level supply chain when manufacturer offers discount. Eur. J. Oper. Res. **168**(1), 65–85 (2006)
17. Zhu, Q., Dou, Y.: Evolutionary game model between governments and core enterprises in greening supply chains. Syst. Eng. - Theory Pract. **27**(12), 85–89 (2007)
18. Chen, Y.J., Sheu, J.-B.: Environmental-regulation pricing strategies for green supply chain management. Transp. Res. Part E Logist. Transp. Rev. **45**(5), 667–677 (2009)
19. Sheu, J.-B.: Bargaining framework for competitive green supply chains under governmental financial intervention. Transp. Res. Part E Logist. Transp. Rev. **47**(5), 573–592 (2011)
20. Barari, S., Agarwal, G., (Chris) Zhang, W.J., Mahanty, B., Tiwari, M.K.: A decision framework for the analysis of green supply chain contracts: An evolutionary game approach. Expert Syst. Appl. **39**(3), 2965–2976 (2012)
21. Nagurney, A., Yu, M.: Sustainable fashion supply chain management under oligopolistic competition and brand differentiation. Int. J. Prod. Econ. **135**(2), 532–540 (2012)
22. Kahraman, C.: Fuzzy Applications in Industrial Engineering. Springer-Verlag, Berlin (2006)
23. Hart, S., Mas-Colell, A.: Stochastic uncoupled dynamics and Nash equilibrium. Games Econ. Behav. **57**, 286–303 (2006)
24. Colell, A.M.: Bargaining Games. In: Cooperation: Game-Theoretic Approaches, pp. 69–90. Springer, Berlin, Heidelberg (1997)
25. Hart, S.: Nash Equilibrium and Dynamics (2008)

A Technique to Reduce the Processing Time of Defect Detection in Glass Tubes

Gabriele Antonio De Vitis(✉), Pierfrancesco Foglia, and Cosimo Antonio Prete

Dipartimento di Ingegneria dell'Informazione, Università di Pisa, Pisa, Italy
gabrieleantonio.devitis@ing.unipi.it, foglia@iet.unipi.it, antonio.prete@unipi.it

Abstract. The evolution of the glass production process requires high accuracy in defects detection and faster production lines. Both requirements result in a reduction in the processing time of defect detection in case of real-time inspection. In this paper, we present an algorithm for defect detection in glass tubes that allows such reduction. The main idea is based on the reduce the image areas to investigate by exploiting the features of images. In our experiment, we utilized two algorithms that have been successfully applied in the inspection of pharmaceutical glass tube: Canny algorithm and MAGDDA. The proposed solution, applied on both algorithms, doesn't compromise the quality of detection and allows us to achieve a performance gain of 66% in terms of processing time, and 3 times in term of throughput (frames per second), in comparison with standard implementations. An automatic procedure has been developed to estimate optimal parameters for the algorithm by considering the specific production process.

Keywords: Defect detection · Glass tube production · Real time inspection · Image processing · Inspection systems

1 Introduction

The evolution of the glass production process requires both high accuracy in defects detection and faster production lines.

The general schema of inspection systems for semi-finished glass production, based on machine vision [1–4], is constituted by the Image Acquisition Subsystem and the Host Computer (Fig. 1). The Image Acquisition Subsystem is devoted to the acquisition of the digitized images (frames); key components of such system are a LED-based illuminator, a line scan camera, and a frame grabber, which groups together single sequential lines captured by the camera into a single frame, transferring it to the Host Computer. The Host Computer implements defect detection and classification algorithms (in the Defect Detection and Classification subsystems). Discard decisions, sent [5] to a Cutting and Discarding Machine, are taken considering process parameters, some of which are settled via a high usable operator GUI [6]. Defect detection and classification algorithms usually are applied only to a part of the acquired image (Region Of Interest – ROI), as only a portion of the image represents the semi-finished glass.

K. Arai et al. (Eds.): CompCom 2019, AISC 997, pp. 165–178, 2019.
https://doi.org/10.1007/978-3-030-22871-2_13

The inspection system works in pipeline, and the Image Acquisition Subsystem feeds the pipeline at a rate which is determined by the sampling rate of the line scan camera divided by the number of lines in a frame. The Defect Detection and Classification module must work with the same rate, to avoid frame loss, i.e., the sampling rate of the line scan camera enforces an upper bound to the processing time of defect detection and classification algorithms. The current requirement of increasing the production speed involves the use of line scan cameras with increased sampling rate to keep constant or to improve the accuracy of defects detection. Consequently, the increase of production speed determines the need to reduce the processing time of defect detection and classification algorithms.

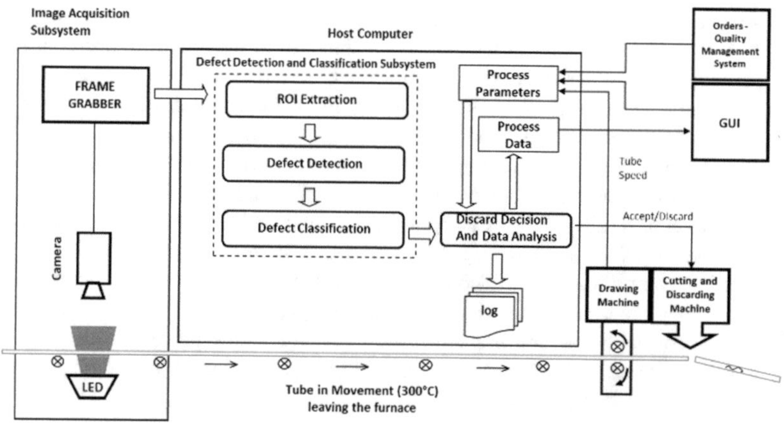

Fig. 1. The architecture of an inspection system for glass production.

In this paper, we propose an algorithm to reduce the processing time of defect detection by means of the size reduction of the images to investigate. We easy exclude the subareas that can be assumed to not include defects, by exploiting Detrend Standard Deviation of columns luminous intensity and double thresholding with hysteresis.

We consider the critical production of glass tubes, converted into pharmaceutical containers such as vials, syringes and carpules. In this case, to achieve a 360° inspection, 3 cameras and led illuminators are utilized in the system. Due to imperfections in the raw materials used in the furnace, the glass tube may have defects such as knot inclusions (blobs) or flexible fragments called lamellae, which can cause subsequent problems and pharmaceutical recalls [7–9]. The main classes of defects relevant for pharmaceutical glass production [1, 10] due to critical size features and their significant effects on the final quality of the tubes are:

(1) Air lines, due to the presence of air bubbles in the furnace which are pulled by the drawing machine; they appear as darker lines of long dimensions (Fig. 2) with a back illuminator, with the end parts thinner than the center one. This line, when it is too close to the tube surface, breaks and therefore is thinner and more difficult to detect.

(2) Knot inclusion (blobs), due to imperfections in the raw materials used in the furnace; they appear on the tube surface as circular lenses, while they appear on the captured image as dark patches, orthogonal to the frame (Fig. 3).

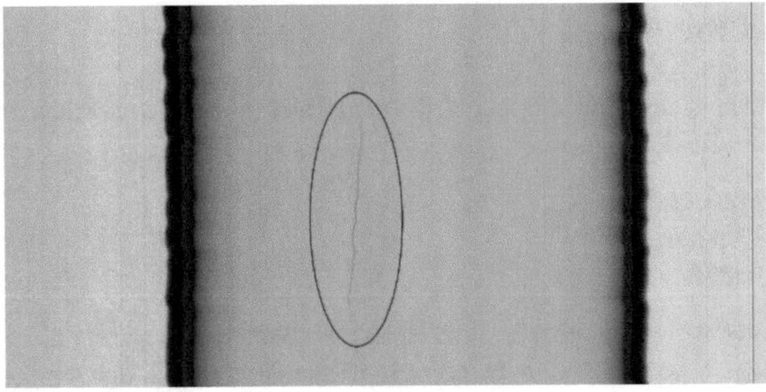

Fig. 2. Example of an air line defect in a glass tube. We must observe that the line is straight but appears as curved and irregular due to the oscillations and rotations of the tube.

Fig. 3. Example of a blob defect in a glass tube (highlighted by a red oval).

In our experiment, we utilized two algorithms that have been successfully applied in the inspection of pharmaceutical glass tube: the Canny algorithm [11] and the MAGDDA [12]. With Canny, the image is first smoothed with a Gaussian filter and then gradient magnitude is computed at each pixel; edges (marked pixels) are determined via non-maximum suppression, double threshold, and hysteresis. MAGDDA algorithm [12] works at row level and apply a moving average filter of an assigned windows size WS to the image.

The reduced area to be investigated has a direct impact on the processing time of the further stages of detection, and in the overall processing time. Results of the experimental evaluation show that the proposed solution achieves a 66% reduction in processing time during the detection/classification phase and improvement of 3x in term of frames per seconds, preserving the same accuracy in defect detection of standard solutions.

We characterize also the effects of the parameters of the algorithm on processing time and quality of detection and propose a procedure to determine upper bound to such parameters.

2 State of the Art

As usually done in inspection systems, the set of elaborations performed on each frame can mainly be divided into three stages [2, 13, 14]:

1. Image preprocessing
2. Defects detection
3. Defects classification

In the pre-processing phase, algorithms are used to prepare the image for the following stages, with the aim of reducing detection errors due to the acquisition process and/or speeding up the calculation by excluding the regions not to be investigated. The steps generally adopted concern noise reduction, contrast enhancement [15], elimination of unwanted regions and identification of the Region Of Interest (ROI).

State-of-the-art ROI extraction techniques consist in identifying the visible part of the glass inside the frame. In the case of the glass tube, ROI extraction consists in identifying the visible part of the tube inside the frame (called hereinafter internal part of the tube, Fig. 4), excluding also the edges of the tube. They appear dark as the light rays of the illuminator, having a critical angle of incidence on them according to Snell's law, are reflected on the glass tube and do not affect the camera sensor. An algorithm to extract the internal part of the tube has been proposed in [1], and it is based on detecting the minimums of columns luminous intensity and moving from them to detect not visible part of the tube.

In the defect detection stage, algorithms are used to determine image regions whose pixels may identify a defect. To extract these regions, segmentation techniques are adopted [13], typically based on thresholds [3] or on edge detection [11, 16].

The classification stage consists of algorithms that extract a series of characteristics of the segmented regions, eventually including them within predetermined classes of defects.

State of the art techniques for feature/defect detection and extraction are the edge detection techniques [16]. Edge detection aims to identify points in a digital image where the image brightness changes sharply compared to the rest. Among the various edge detection techniques, the algorithm proposed by Canny [11] (Canny algorithm) is considered the ideal one for images with noise [16]. The image is first smoothed with a Gaussian filter and then gradient magnitude is computed at each pixel of the smoothed image; edges are determined by applying non-maximum suppression, double threshold, and hysteresis. The algorithm has been usefully adopted in various applications domains (inspection of semiconductor wafer surface [17], detection of defects in satin glass [18], measuring icing shape on conductor [19], studies on bubble formation in co-fed gas-liquid flows [20]) and it has been also adopted for the detection of defects in glass tube inspection systems [1].

Other techniques for defect detection and extraction are based on thresholds, that can be global (fixed for the whole image) or local, i.e. they can be variable in different regions of the image [21].

As for global thresholds techniques, [3] presents an inspection method for float glass fabrication. The authors utilize a benchmark image to remove bright and dull stripes that are present in their glasses. Then, they utilize adaptive global thresholds based on the OTSU algorithm [22] to separate distortions from defects. The OTSU algorithm selects threshold values (one for each image) that maximize the inter-class variance of the image histogram [23]. It is useful for separating background from defects/foreground and produces satisfactory results when images present bimodal or multimodal histograms [13]. It has been successfully utilized in [13] to derive a configurable industrial vision system for surface inspection of transparent parts (in particular, it has been tested on headlamp lens) and again in [24] to detach defects from the background in a float glass defect classificatory and in [25] or glass inspection vision systems.

By considering the characteristics of the tube glass production, the use of single or multiple constant thresholds does not allow the detection of defects (Fig. 6b highlights that the luminous intensity of columns belonging to the defect is similar to the ones of columns not including a defect). Besides, techniques based on background subtraction or other template matching techniques [3, 26] cannot be utilized due to the tube vibration and the not perfect circular section of the tube ("sausage" shape).

As for local thresholds techniques, the Niblack's [21] binarization method is a local adaptive thresholding technique, based on varying threshold over the image by using local mean value and the standard deviation of gray level evaluated in a window centered in each pixel. This method can separate the object or text from the background effectively in the areas near to the object. Niblack method is one of the document segmentation methods and has shown good results in segmenting text from the background. Anyway, it can be applied also to images without text [27] and has been applied in a vision system for auto seeding and for observing the surface of the melt in the Ky method for the Sapphire Crystal Growth Process [28].

All these techniques extract a ROI from the acquired image. This extraction removes the part of the image that is not part of the inspection object (for example the portion of image that does not contain the tube, as in [1], or the image of the roller conveyor that is separated from the glass in [18]) but do not use information on defects to further reduce the ROI. Only in [25] is presented a technique that identifies not defective areas within the ROI (background) using a threshold on the local variance calculated in a window centered on a pixel. In the paper, statistics on areas without error are used to automatically calculate global thresholds to perform segmentation, more accurately than OTSU. In our work, we use the knowledge of the no defective areas to reduce the size of the ROI and improve the execution time. Unlike the previous work, in our case, it is not possible to use the standard deviation for the lighting effects. We besides use a double threshold mechanism with hysteresis to further limit the size of the ROI, and we perform columns elaboration as they have an almost constant distribution of luminous intensity, thus avoiding the introduction of further expensive filtering.

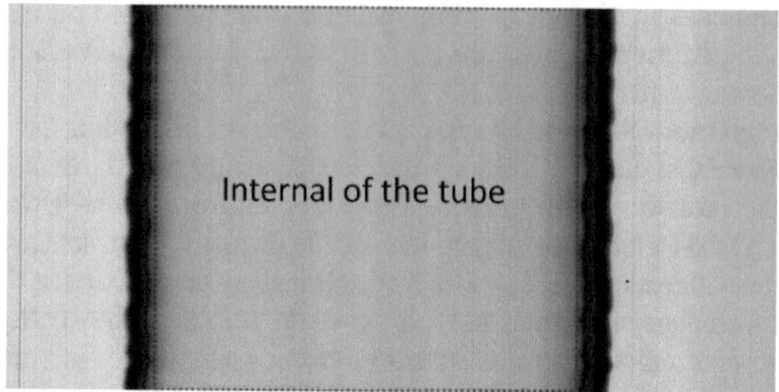

Fig. 4. Image taken by the line-scan camera. The array of CCD sensors in the line scan camera is orthogonal to the direction of the movement of the tube. The internal of the tube is also highlighted. It represents the portion of the image which is further analyzed for detecting defects. The frame is composed of 1000 × 2048 pixels.

3 Rational

State-of-the-art techniques for defect detection consider the ROI as the input image for their elaboration. Since not all the portions of the ROI include defects, the application of defect detection algorithm to these portions is useless and wastes processing time. Figure 5 shows the processing time of the 3 stages of detection when the Canny algorithm is applied to a defective image. Bar Canny Whole ROI represents the processing time when Canny is applied to the whole ROI, while Canny Defective ROI represents the processing time when Canny is applied only to the columns of the ROI that includes the defect (defective ROI). As can be seen, it is possible to reduce the processing time of about 60%. As our main requirement is to reduce the overall processing time, our idea is to remove, from the ROI, portions where it is possible to easily predict that no defects are present (reducing the size of the ROI).

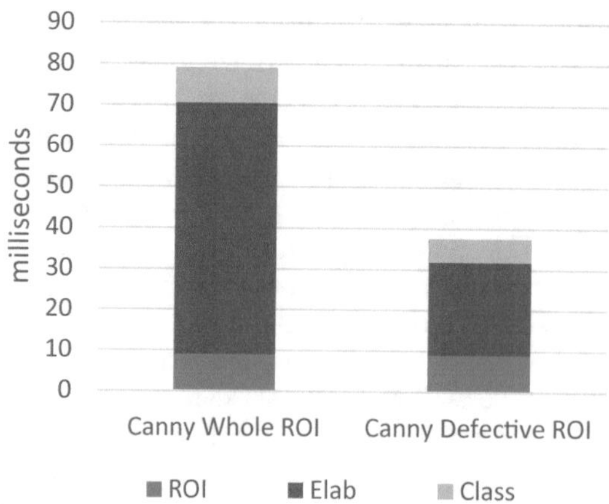

Fig. 5. Processing Time on an Intel i7-940 CPU (2.93 GHz) on a defective image. Canny Whole ROI represents the processing time when the Canny algorithm is applied to the whole ROI. Canny defective ROI represents the processing time when Canny is applied only to the columns of the ROI that include the defects.

To reduce the size of the ROI, we observe that the luminous intensity inside a column is almost constant except in pixels where there is noise or defects. So, the standard deviation of the column may be used as an indicator of defect presence on that column, as shown in Fig. 6b. Therefore, values of the standard deviation for column below a specific threshold indicate that the column can be excluded by the next processing. Anyway, the standard deviation of the columns in a glass object is influenced by the alignment of the illuminator with the tube. In case of not perfect alignment, usual in glass tube production, the standard deviation shows an increasing trend from one edge of the tube to the other, which can be approximated with a linear trend (Fig. 6d). In order not to have any influence from this factor, we consider the values of the standard deviation of the columns removing their linear trend (Detrended Standard Deviation - DSD).

Another relevant requirement concerns the ability to accurately detect the size of defects because it is a parameter for taking discard decisions. We have experimented that luminous intensity of the defects presents high values near the central area of the defects but tends to decrease away from it. Therefore, not all the columns including a defect present high standard deviation value. For these reasons, to detect these columns, we apply two hysteresis thresholds: t_L and t_H with $t_L < t_H$. Columns with DSD values less than t_L are not considered to belong to ROI, columns with values greater than t_H are considered to belong to ROI and columns with values between t_L and t_H are considered to belong to ROI only if they are adjacent to columns that belong to ROI (Fig. 7).

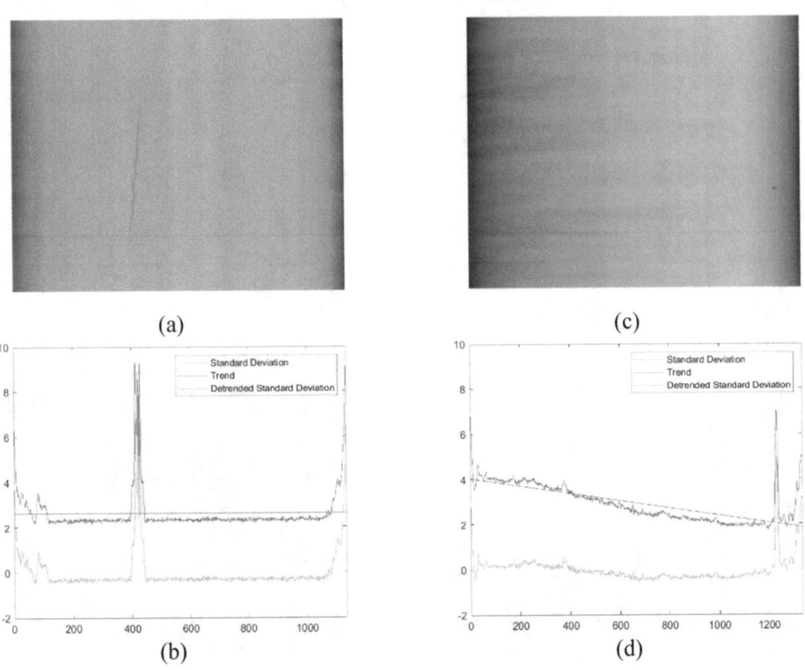

Fig. 6. (a) Internal part of the tube including an air line. (b) Standard Deviation, Linear Trend and DSD of columns of Fig. (a). (c) Internal part of the tube including a blob. (d) Standard Deviation, Linear Trend and DSD of columns of Fig. (c).

Using the DSD criterion, areas near the edge of the tube are often classified as ROI, as they have peaks of DSD greater than the peaks of the DSD of the columns where defects are placed. These values are caused by many effects as the vibration of the moving tube or the imperfect circular shape of the tube surface. In these areas, there may be defects that if included in the ROI will be recognized by the Defect Detection and Classification Subsystem. Two solutions are then possible. The first is to exclude the ROIs located near the edges. This solution is not destructive because areas near the edges for a camera appear in the central area for one of the other two cameras positioned at 120 degrees from it and the tube. The second solution is to include these areas in the ROI for the next stage of Defect Detection.

4 Algorithm and Choice of the Thresholds

The proposed algorithm (Detrended Standard Deviation ROI Reduction algorithm, DSDRR), starting from the image representing the internal part of the tube, calculates for each column the DSD. Then, the algorithm finds the columns whose DSD values are greater than the t_H threshold and promotes all these columns as belonging to the ROI. Next, for each of the columns belonging to the ROI, the algorithm finds the adjacent columns whose DSD values are greater than t_L and promotes also these columns as belonging to the ROI. Finally, if ROI of areas close to the edge of the tube must be excluded from the analysis, the algorithm removes from the ROIs the columns adjacent to the first and the last column.

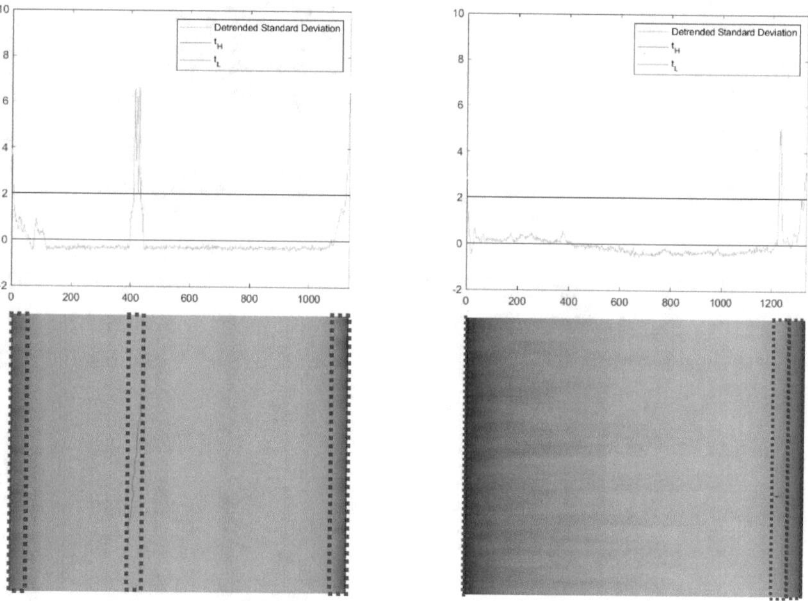

Fig. 7. Application of two thresholds with hysteresis to the DSD of columns of Fig. 6a and 6c. Above is shown the ROIs of images (in red boxes) in Fig. 6a and 6c as results by applying the proposed algorithm.

The choice of the threshold values determines algorithm performance in terms of detection. A bad choice may cause the exclusion of defects in the ROIs (causing a false negative in detection), or it can generate too large ROI with no reduction of the overall processing time.

The choice of threshold t_H must guarantee that at least one column of a defect belongs to the ROI, i.e. the DSD for that column must be higher than t_H. So, to ensure that all the defects are correctly included in the ROI, the threshold t_H must be lower than the maximum value of the DSD of the columns of each defect. Too high values of t_H can exclude from the ROI areas that include defects, and too low values of t_H, therefore, lead to extremely large ROI, that is, high processing times.

The threshold t_L must guarantee that all the columns of a defect belong to the ROI, otherwise portions of the defect are not detected, therefore, causing the incorrect measurement of the defect size. Too high value of t_L could exclude a portion of the shape of the defect from the ROI, and too low values of t_L could cause again the ROI to include the entire internal area of the tube. A possible solution is to set it to a value lower than the minimum value of the DSD of the columns of each defect. This ensures that, if a defect is detected using threshold t_H, its entire shape is always included in the ROI.

The thresholds t_L and t_H used by the algorithm are related also to properties of defects that depend on production-related parameters (such as glass tube size, diameter, thickness, opacity etc.). In an industrial application scenario, these parameters change with batches of production, so it is important to adapt the thresholds to the changes in production. We have developed a procedure that suggests upper bounds for the thresholds that must be adopted when the production parameters change.

The procedure consists of:

(1) collecting a number of frames from the acquisition system during the starting phase of a new batch of production,
(2) applying an algorithm for the detection and classification of defects to each frame,
(3) for each classified defect, calculating the maximum value and minimum value of the DSD of the columns of the defect and adding them respectively to the sets maxDSD and minDSD,
(4) the suggested upper bounds are: $t_L < \min(\text{minDSD})$ and $t_H < \min(\text{maxDSD})$.

Table 1. Host computer.

Hardware configuration	
Processor	Intel® Core™ i7-940 Processor (8 MB Cache, 2.93 GHz)
RAM	8 GB

Defect detection system			
Algorithm name	Without DSDRR	DSDRR with edges	DSDRR no edges
ROI extraction	Internal part	Internal part + DSDRR including edge with $t_L = 0$, $t_H = 2$	Internal part + DSDRR excluding edge with $t_L = 0$, $t_H = 2$
Defect Detection	Canny (35,80)	MAGDDA (ws = 145, k = 10.7)	
Implementation	OPENCV		

5 Results

The inspection system of Fig. 1 is working on production lines of a glass tube foundry [1]. The image processing stages have been implemented in a task activated by the frame grabber when a new frame is ready in main memory. Table 1 summarizes the main features of the Host Computer, the algorithms utilized for the various stages of defect detection and their configuration parameters. All the image processing algorithms have been implemented using the OpenCV library [29] and compiled with the Visual Studio compiler.

System performance has been evaluated using a dataset of 30 frames acquired during the production phase. All frames are composed of 1024 lines acquired by the line scan camera sensor (2 K pixels). These frames contain 6 air lines and 10 blobs; one of these frames contains 3 blobs and another one contains 2 air lines while 17 frames do not have any type of defects. The machine on which tests are executed has a configuration similar to the production one and is equipped with an Intel Core i7-940 CPU running at 2.93 GHz. As for processing time, we perform 1000 executions of the whole dataset of images [30], and we take the average execution time spent by each stage (ROI extraction/Defect detection/Defect classification) and the maximum total processing time.

In our experiment, we utilized two algorithms that have been successfully applied in the inspection of pharmaceutical glass tube: the Canny algorithm [11] and the MAGDDA [12]. With Canny, the image is first smoothed with a Gaussian filter and then gradient magnitude is computed at each pixel; edges (marked pixels) are determined via non-maximum suppression, double threshold, and hysteresis. MAGDDA algorithm [12] works at row level and apply a moving average filter of an assigned windows size WS to the ROI. Then applies a fixed threshold (k) to mark the pixel. As for the Defect Classification stage, we adopt an algorithm that groups adjacent marked pixels using connected-components labeling [31] and builds, for each group, the smallest rectangle that contains them. The rectangle features permit to individuate blobs and air lines.

In the tested image-set, the autotuning method suggests for t_H a value less than 2,0362 and for t_L a value less than 0,0126. We set $t_H = 2$ and $t_L = 0$. For Canny, we find that the optimal thresholds are 35, 80.

Table 2. Number of recognized defects/Defective Frames and their classification

	Expected value	Without DSDRR DSDRR with edge Canny	DSDRR no edge Canny	Without DSDRR DSDRR with edge MAGDDA	DSDRR no edge MAGDDA
	TP	TP/FP (FN)	TP/FP (FN)	TP/FP (FN)	TP/FP (FN)
Blobs	10	10/5 (0)	9/5 (1)	10/6 (0)	9/6 (1)
Air lines	6	6/0 (0)	6/0 (0)	6/0 (0)	6/0 (0)
Defective frames	13	13/2 (0)	12/2 (1)	13/2 (0)	12/2 (1)

Table 3. Air line length [pixels]

Frame	Without DSDRR DSDRR with edge DSDRR no edge Canny	Without DSDRR DSDRR with edge DSDRR no edge MAGDDA
1.1	81 / 81 / 81	72 / 72 / 72
1.2	473 / 473 / 473	456 / 456 / 456
2.1	709 / 709 / 709	758 / 758 / 758
2.1	337 / 337 / 337	402 / 402 / 402
2.6	502 / 502 / 502	509 / 509 / 509
3.9	450 / 450 / 450	513 / 513 / 513

With this setting, in terms of defect detection, the accuracy of each defect detection algorithm is the same with or without the application of the DSDRR algorithm when including also the edges of the tube (Tables 2 and 3 for air lines; the same results are achieved for blob area). When the DSDRR algorithm excludes the edges, there is a false negative due to a defect near the edge of the tube whose columns are excluded from the ROI. Anyway, the system is equipped with three cameras to guarantee a 360° inspection of the tube, and defects near edge of the tube are located by one of the other cameras near the center of the image.

The DSDRR algorithm is effective in reducing the ROI that is processed by the Defect Detection and Classification algorithms. DSDRR calculates 70 ROIs. 53 of these are in the proximity of the edge of tube (and 1 of these contains a defect), 14 contain defects (3 defects are in the same columns/ROI). Compared to the total area, the ROI is reduced on average by 88%, to a maximum of 96% and at least 69%. Excluding the parts near the edges of the tube, the ROI is reduced on average by 98% and at most by 100% (in frames without defects) and at least by 82%.

The reduced area of the ROI has a direct impact on the processing time of the further stages (Table 4) of detection and in the overall processing time (Fig. 8). As shown in Table 4, the average processing time of Canny without DSDRR is about 61 ms while the execution time of Canny with DSDRR with edges is about 20 ms and 10 ms with DSDRR no edges, with a decrease of 67% and 83%. For MAGDDA, the processing time without DSDRR is 8 ms while with DSDRR with edges is about 3 ms and 2 ms with DSDRR no edges. To further reduce the processing time, these results suggest excluding the investigation of the image edges. Indeed, the system is equipped with three cameras to guarantee a 360° inspection of the tube, and defects near an image edge for a camera are located by one of the other cameras near the center of the image. So, the inspection quality is guaranteed.

Similar improvements can be observed for the processing time of the Classification stage, as noisy pixels that do not belong to DSDRR ROI are not considered by the classification algorithm. The DSDRR algorithm increases the processing time for ROI extraction (from about 8 to 10 ms). Despite this, the total maximum processing time for Canny moves from 90 to 37/30 ms (with a 3x increase in throughput) and for MAGDDA from 34 to 19/17 ms (with a 2x increase in throughput).

Table 4. Average Processing Time of defect detection phases (ms)

		ROI	ELAB	CLASS
Canny	Without DSDRR	7,845	61,538	9,395
	DSDRR with edge	10,217	20,403	3,945
	DSDRR no edge	10,331	10,437	2,554
MAGDDA	Without DSDRR	7,695	8,333	8,114
	DSDRR with edge	10,221	2,817	3,924
	DSDRR no edge	10,331	1,442	2,446

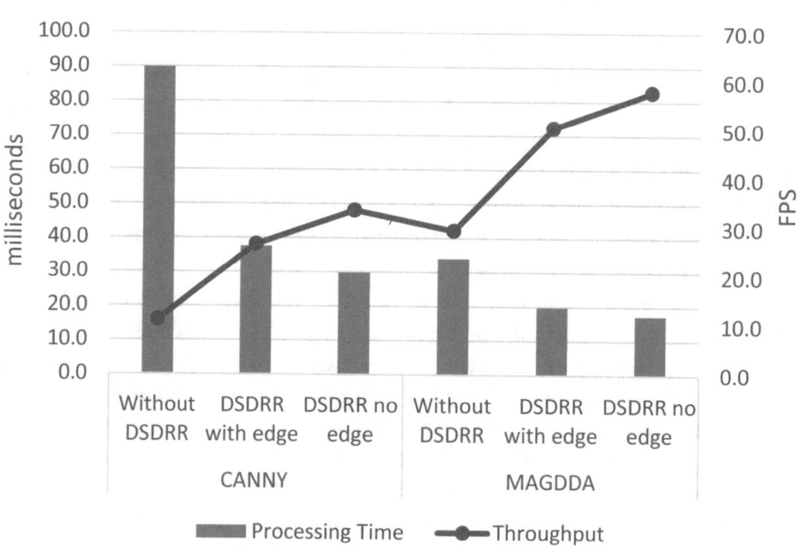

Fig. 8. Processing time and throughput.

6 Conclusion

A vision system can be exploited to inspect in real-time the glass tube quality during the production process. Improvements in such process and the need to increase the detection accuracy suggest the adoption of a solution that reduces the processing time of all the steps involved in defect detection and classification. A classical approach for dealing with inspection consists in extracting the whole internal part of the tube (ROI) and passing it to the defect detection and classification algorithm. By considering the features of defects and properties that are relevant in the images, we presented an algorithm that further reduces the ROI area by excluding columns that can be determined to not include defects.

We consider the production of glass tubes for pharmaceutical uses. Experimental results indicate that our proposal does not change the quality of detection of the system and significantly improves processing time of both defect detection and classification stages. The overall throughput (frames per second) is improved up to 3 times with respect to standard solutions.

As for future works, we plan to test our algorithm in other application domains and to investigate strategies to parallelize the algorithms by considering CMPs and GPU architectures and related memory systems [32].

Acknowledgments. This work has been partially supported by the Italian Ministry of Education and Research (MIUR) in the framework of the CrossLab project (Departments of Excellence – LAB Advanced Manufacturing and LAB Cloud Computing, Big data & Cybersecurity).

References

1. Foglia, P., Prete, C.A., Zanda, M.: An inspection system for pharmaceutical glass tubes. WSEAS Trans. Syst. **14**, Art. #12, 123–136 (2015)
2. Kumar, A.: Computer-vision-based fabric defect detection: a survey. IEEE Trans. Ind. Electron. **55**(1), 348–363 (2008)
3. Peng, X., et al.: An online defects inspection method for float glass fabrication based on machine vision. Inter. J. Adv. Man. Technol. **39**(11–12), 1180–1189 (2008)
4. Bradski, G., Kaehler, A.: Learning OpenCV: Computer vision with the OpenCV library. O'Reilly Media Inc, Sebastopol (2008)
5. Campanelli, S., Foglia, P., Prete, C.A.: An architecture to integrate IEC 61131-3 systems in an IEC 61499 distributed solution. Comput. Ind. **72**, 47–67 (2015)
6. Foglia, P., Prete, C.A., Zanda, M.: Relating GSR signals to traditional usability metrics: Case study with an anthropomorphic web assistant. In: Instrumentation and Measurement Technology Conference Proceedings, pp. 1814–1818. IMTC 2008. IEEE (May 2008)
7. Reynolds, G., Peskiest, D.: Glass delamination and breakage, new answers for a growing problem. BioProcess Int. **9**(11), 52–57 (2011)
8. Iacocca, R.G., Toltl, N., et al.: Factors affecting the chemical durability of glass used in the pharmaceutical industry. AAPS PharmSciTech **11**(3), 1340–1349 (2010)
9. Schaut, R.A., Weeks, W.P.: Historical review of glasses used for parenteral packaging. PDA J. Pharm. Sci. Technol. **71**(4), 279–296 (2017)
10. Berry, H.: Pharmaceutical aspects of glass and rubber. J. Pharm. Pharmacol. **5**(11), 1008–1023; Wiley (2011)
11. Canny, J.F.: A computational approach to edge detection. IEEE Trans. Pattern Anal. Mach. Intell. **8**(6), 679–698 (1986)
12. De Vitis, G.A., Foglia, P., Prete, C.A.: A Special Purpose Algorithm to Inspect Glass Tubes in the Production Phase. TR-DII-2018-01, University of Pisa, 2018
13. Martínez, S.S., et al.: An industrial vision system for surface quality inspection of transparent parts. Int. J. Adv. Manuf. Technol. **68**(5–8), 1123–1136 (2013)
14. Malamas, N., et al.: A survey on industrial vision systems, applications and tools. Image Vis. Comput. **21**(2), 171–188 (2003)
15. Li, D., Liang, L., Zhang, W.: Defect inspection and extraction of the mobile phone cover glass based on the principal components analysis. Int. J. Adv. Manuf. Technol. **73**(9–12), 1605–1614 (2014)
16. Kumar, M., Saxena, R.: Algorithm and technique on various edge detection: a survey. Signal Image Process. **4**(3), 65 (2013)

17. Shankar, N.G., Zhong, Z.W.: Defect detection on semiconductor wafer surfaces. Microelectron. Eng. **77**(3–4), 337–346 (2005)
18. Adamo, F., et al.: A low-cost inspection system for online defects assessment in satin glass. Measurement **42**(9), 1304–1311 (2009)
19. Huang, X., et al.: An online technology for measuring icing shape on conductor based on vision and force sensors. IEEE Trans. Instrum. Meas. **66**(12), 3180–3189 (2017)
20. de Beer, M.M., et al.: Bubble formation in co-fed gas–liquid flows in a rotor-stator spinning disc reactor. Int. J. Multiph. Flow **83**, 142–152 (2016)
21. Saxena, L.P.: Niblack's binarization method and its modifications to real-time applications: a review. Artif. Intell. Rev., 1–33 (2017)
22. Otsu, N.: A threshold selection using an iterative selection method. IEEE Trans. Syst. Man Cybern **9**, 62–66 (1979)
23. Wakaf, Z., Jalab, H.A.: Defect detection based on extreme edge of defective region histogram. J. King Saud Univ. – Comput. Inform. Sci. **30**(1), 33–40 (2018)
24. Huai-guang, L., et al.: A classification method of glass defect based on multiresolution and information fusion. Int. J. Adv. Manuf. Technol. **56**(9), 1079–1090 (2011)
25. Aminzadeh, M., Kurfess, T.: Automatic thresholding for defect detection by background histogram mode extents. J. Manuf. Syst. **37**(1), 83–92 (2015)
26. Kong, H., et al.: Accurate and efficient inspection of speckle and scratch defects on surfaces of planar products. IEEE Trans. Industr. Inf. **13**(4), 1855–1865 (2017)
27. Farid, S., Ahmed, F.: Application of Niblack's method on images. Emerging Technologies. ICET 2009. International Conference on. IEEE (2009)
28. Kim, C.M., Kim, S.R., Ahn, J.H.: Development of auto-seeding system using image processing technology in the sapphire crystal growth process via the kyropoulos method. Appl. Sci. **7**(4), 371 (2017)
29. https://opencv.org/
30. Abella, J., Padilla, M., et al.: Measurement-based worst-case execution time estimation using the coefficient of variation. ACM Trans. Des. Autom. Electron. Syst. **22**(4), Article 72 (June 2017)
31. Grana, C., Borghesani, D., Cucchiara, R.: Optimized block-based connected components labeling with decision trees. IEEE Trans. Image Process. **19**(6), 1596–1609 (2010)
32. Bartolini, S., Foglia, P., Prete, C.A.: Exploring the relationship between architectures and management policies in the design of NUCA-based chip multicore systems. Future Gener. Comput. Syst. **78**(2), 481–501 (2017)

Direct *N-body* Code on Low-Power Embedded ARM GPUs

David Goz(✉), Sara Bertocco(✉), Luca Tornatore(✉), and Giuliano Taffoni(✉)

INAF - OATs, via Tiepolo 11, Trieste, Italy
{david.goz,sara.bertocco,luca.tornatore,giuliano.taffoni}@inaf.it
http://www.oats.inaf.it

Abstract. This work arises on the environment of the ExaNeSt project aiming at design and development of an exascale ready supercomputer with low energy consumption profile but able to support the most demanding scientific and technical applications. The ExaNeSt compute unit consists of densely-packed low-power 64-bit ARM processors, embedded within Xilinx FPGA SoCs. SoC boards are heterogeneous architecture where computing power is supplied both by CPUs and GPUs, and are emerging as a possible low-power and low-cost alternative to clusters based on traditional CPUs. A state-of-the-art direct N-body code suitable for astrophysical simulations has been re-engineered in order to exploit SoC heterogeneous platforms based on ARM CPUs and embedded GPUs. Performance tests show that embedded GPUs can be effectively used to accelerate real-life scientific calculations, and that are promising also because of their energy efficiency, which is a crucial design in future exascale platforms.

Keywords: ExaNeSt · HPC · N-body solver · ARM SoC · GPU computing · Parallel algorithms · Heterogeneous architecture

1 Introduction

Nowadays ARM delivers technology to drive power-efficient System-on-Chip (hereafter SoC) solutions combining CPU and GPU into unified compute subsystem offering double-precision floating point arithmetic, and options for high performance I/O and memory interface. Those systems represent an excellent solution to build less expensive and more power-efficient computational clusters than standard High Performance Computing (HPC) facilities.

In the last years, some effort has been devoted to investigate the potential of SoCs for computationally intensive real-life scientific applications, comparing the performances with those obtained on a typical x86 HPC node (e.g. [15]). They conclude that considering SoCs for computationally intensive scientific applications seems very promising, and this technology might represent the next revolution in high performance community. However, scientific applications are usually ported on those platforms rather than to be re-engineered in order to

K. Arai et al. (Eds.): CompCom 2019, AISC 997, pp. 179–193, 2019.
https://doi.org/10.1007/978-3-030-22871-2_14

fully exploit the new hardware by solving the architecture-application performance gap, i.e. the gap between the capabilities of the hardware (HW) and the performance released by the HPC software (SW). This is crucial when designing the new generation of HPC supercomputer, the Exascale platform. The realization of an Exascale supercomputer requires significant advances in a variety of technologies, the aim of which is the energy efficiency. A number of projects has been financed in Europe to develop an exascale-class prototype system (e.g. ExaNeSt project[1], Montblanc project[2], and Mango project[3]).

The `ExaNest H2020` project [11] aims at the design and development of an exascale-class prototype system built upon power-efficient hardware able to execute ambitious real-world applications coming from a wide range of scientific and industrial domains, including also HPC for astrophysics [1]. Since the power-efficiency is the main concern, the ExaNeSt basic compute unit consists of low-energy-consumption ARM CPUs, embedded GPUs, FPGAs and low-latency high throughput interconnects [12]. An approach based on HW/SW co-design is crucial to design Exascale resources that can be effectively exploited by real scientific applications.

The work presented in this paper aims to study whether a direct N-body code, called `Hy-Nbody`, for real scientific production may benefit from embedded GPUs given that powerful high-end GPGPUs have already demonstrated to provide tremendous performance benefit. This is the first work to implement such algorithm on embedded GPU and to compare results with multi-core on SoC implementation. `Hy-Nbody` code is a re-engineered version of `HiGPUs` [6,7,20], a state-of-the-art direct N-body code, based on the Hermite 6th order time integrator, which has been widely used for scientific production in Astrophysics, i.e. for simulations of star clusters with up to ~8 million bodies [21,22], and of galaxy mergers [4]. Moreover, `HiGPUs` has been extensively tested on large supercomputer such as IBM iDataPlex DX360M3 Linux Infiniband Cluster provided by the Italian supercomputing consortium CINECA using up to 256 GPGPUs. All kernels of the `Hy-Nbody` code have been implemented using OpenCL language[4] in order to write efficient code for hybrid architecture. `Hy-Nbody` has been designed to fully exploit computational nodes in a cluster by means of an hybrid parallelization schema, MPI+OpenMP for host code and OpenCL for device code.

This paper is organized as follows. Section 2 describes direct N-body solvers used in Astrophysics. Section 3 presents the `Hy-Nbody` code and its optimizations in order to exploit ARM CPU and Mali GPU. Section 4 describes the computational test bed based on Rockchip Firefly RK3399 Soc boards based on Linux Operating system. Section 5 is devoted to present and discuss the results. Future development and scope of this work are presented in Sect. 6, and the conclusions in Sect. 7.

[1] http://www.exanest.eu.
[2] http://montblanc-project.eu.
[3] http://www.mango-project.eu.
[4] http://www.khronos.org/opencl/.

2 N-body Solvers Running on Hybrid Computing Platforms

N-body solvers provide the backbone for different scientific and engineering applications, such as astrophysics, nuclear physics, molecular dynamics, fluid mechanics and biology. In astrophysics, the N-body problem is the problem of predicting the individual motions of a group of celestial objects interacting with each other only gravitationally. The classical (i.e. non relativistic) direct N-body problem has analytic solution only with $N = 2$, so in general it must be simulated using numerical methods. The numerical solution of the direct N-body problem is still considered a challenge despite the significant advances in both hardware technologies and software development. The main drawback related to the direct N-body problem relies on the fact that the algorithm requires $O(N^2)$ computational cost. In practice many variations of the naive algorithm are used, for instance, implementing high order Hermite integration schemes [17] and block time-stepping. These variations can eliminate most of the standard parallelization method for N^2 algorithm, requiring huge effort to maximize performance.

There are some N-body codes designed to speed up the classical N-body problem for real scientific production in astrophysics using GPGPUs, for example, φ`GPU` code [2], φ`GRAPE` code [8], `NBODY6` code [16], `MYRIAD` code [13], and `HiGPUs` code [6,7,20]. In all of them, the GPGPU is fed by the host CPU with the gravity equation of data in the form of coordinates, velocities and masses of particles, and it handles calculating the forces for the data points.

None of the above has been optimized and ported on embedded GPUs, which is the scope tackled by this paper.

3 Code Implementation

The 6th order Hermite integrator for astrophysical N-body simulations consists of three stages (physical/mathematical aspects are described in [17]): a predictor step that predicts particle's positions and velocities; an evaluation step to evaluate new accelerations, their first order (*jerk*), second order (*snap*), and third order derivatives (*crackle*); a corrector step that corrects the predicted positions and velocities using the results of the previous steps.

In the following we describe `Hy-Nbody`, which has been conceived to fully exploit the compute capabilities of heterogeneous architecture:

1. **Device Exploitable:** The entire 6th order Hermite schema is implemented and optimized using OpenCL kernels, allowing to test the code on any OpenCL-compliant device (CPUs/GPUs/FPGAs).
2. **Parallelization Schema:** A one-to-one correspondence between MPI processes and computational nodes is established and each MPI process manages all the OpenCL-compliant devices of the same type available per node (device type is selected by the user). Inside of each shared-memory computational

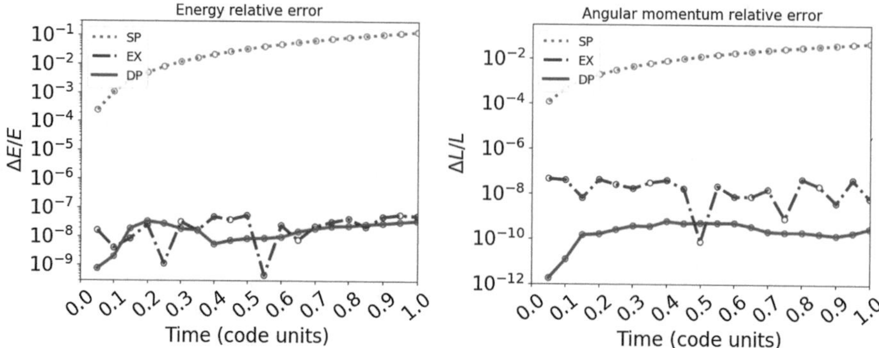

Fig. 1. Left panel: relative error $\Delta E/E$ for DP-arithmetic (continuous red line), EX-arithmetic (dot-dashed blue line) and SP-arithmetic (dotted green line) as a function of the integration time (in code unit). Right panel: relative error $\Delta L/L$ for DP-arithmetic (continuous red line), EX-arithmetic (dot-dashed blue line) and SP-arithmetic (dotted green line) as a function of the integration time (in code unit).

node parallelization is achieved by means of OpenMP environment. Hence, in `Hy-Nbody`, the host code is parallelized with hybrid MPI+OpenMP programming, while the device code is parallelized with OpenCL. The user is allowed to choose at compile time if the application uses MPI or OpenMP, or both, or neither.

3. **Decomposition Over Host and Device:** The Hermite integration is performed on the selected OpenCL-compliant device(s). The algorithm uses a share-time step scheme that integrates all particles. Thus, in a simulation with N-particles using n devices, during the evaluation stage each device deals with N/n particles and evaluates $N(N/n)$ accelerations and their derivatives, subsequently collected and reduced from the all set of computational nodes. In `Hy-Nbody` the evaluation of the time step, by means of the so called *generalized Aarseth criterion* [17], the total energy and the angular momentum of the system are performed on the device as well. The latest quantities are periodically evaluated during the simulation in order to check the accuracy of the integration schema.
   The implementation requires that particle data is communicated between the host and the device at each share-time step, which gives rise to synchronization points between host and device(s). Accelerations, *jerk*, *snap* and time step computed by the device(s) are retrieved by the host on every computational node, reduced and then sent back again to the device(s).
4. **Kernel Optimizations:** OpenCL kernels have access to distinct memory regions distinguished by access type and scope. *Local* memory provides read-and-write access to work-items within the same work-group (OpenCL terminology) and it is specifically designed to reduce the latency of data transactions. The *evaluation* kernel is the computationally most expensive part of the Hermite algorithm. This makes the calculation of the accelerations a good

candidate for exploiting the local memory of the device. When the *evaluation* kernel is issued to the device(s), each work-item goes through the following steps:

(i) Each work-item in the work-group caches one particle from *global* memory into the *local* memory. The total number of cached particles is therefore equal to the work-group size (selected by the user).

(ii) Partial acceleration, *jerk* and *snap* for each work-item are calculated and stored in registers using particles cached in local memory.

(iii) Steps (i) and (ii) are repeated until all particles handled by the device have been read (avoiding to sum up the self interaction).

(iv) Results are stored in global memory ready to be read by the host.

The previous schema implies $N(N/n)$ calculations performed by the device, requiring internal synchronization due to the fact that *local* memory is limited. However, the exploiting of local *memory* is generally accepted as the best method to reduce global memory latency in discrete GPUs.

5. **Kernel Vectorization:** Since the majority of OpenCL-compliant devices supports vector instruction set, all kernels of the application have been vectorized. Vectorizing code can effectively improve memory bandwidth because of regular memory access, better coalescing of these memory accesses and reducing the number of loads/stores (each load/store is larger).
6. **Precision:** High precision computations are necessary for many numerical and scientific applications. Indeed, the Hermite 6th order integration schema requires double precision (DP) arithmetic in the evaluation of inter-particles distance and acceleration in order to minimize the round-off error. Full IEEE-compliant DP-arithmetic is efficient in available CPUs and GPGPUs, but it is still extremely resource-eager and performance-poor in other accelerators like embedded GPUs or FPGAs. As an alternative, the extended-precision (EX) numeric type can represent a trade-off in porting `Hy-Nbody` on devices not specifically designed for scientific calculations, such as embedded GPUs or FPGAs. An EX-number provides approximately 48 bits of mantissa at single-precision exponent ranges. `Hy-Nbody` can be run using DP, EX or single precision (SP) arithmetic (user-defined at compile time). The EX-arithmetic is implemented as proposed by [23].

To test the effect of the arithmetic on the accumulation of the round-off error, the energy E and the angular momentum L of the N-body system during the simulation are compared with the values at the start of the simulation. Latest quantities must remain constant within an isolated system. The relative errors $\Delta E/E$ and $\Delta L/L$ are determined using the following equations:

$$\frac{\Delta E}{E} = \frac{|E_{start} - E(t)|}{E_{start}}, \quad \text{and} \quad \frac{\Delta L}{L} = \frac{|L_{start} - L(t)|}{L_{start}} \tag{1}$$

where E_{start}, L_{start} and $E(t)$, $L(t)$ are the energy and the angular momentum at the start and at a given time of the simulation, respectively. Figure 1 shows the relative errors of energy and angular momentum of the system as a function of time (in code unit). The simulation was carried out with 4096 particles.

As relevant result, adopting SP-arithmetic, round-off error accumulates during the simulation, while it is roughly constant using EX or DP-arithmetic. The test presented suggests that EX-arithmetic can be effectively adopted for N-body problem ensuring to keep control over the accumulation of the round-off error during the simulation. This approach requires only 32-bit compute capability to the computational device.

3.1 Tuning OpenCL Kernels for the Embedded ARM Mali GPU

OpenCL is a portable language but it not always performance portable, so existing OpenCL code is typically tuned for specific architecture. However, general purpose programming for embedded GPUs is still relatively new, and the associated runtime libraries are often immature.

ARM developer guide[5] says that for best performance on Mali-T864 (Mali Midgard family) the code should be vectorized to achieve the best performance. Regardless of the native width of the GPU's SIMD functional units, using wider vectors in the kernel may provide the GPU architecture more opportunity for exploiting data-level parallelism. Kernels in `Hy-Nbody` have already been vectorized, since also discrete GPUs show enhanced performances exploiting vectorization. On the Mali GPU, moreover, the global and local OpenCL address spaces are mapped to main host memory. This means that explicit data copies from global to local memory with associated barrier synchronizations are not necessary. Thus, using local memories as a cache can waste both performance and power on the Mali GPU. A specific ARM-GPU-optimized version of all kernels of `Hy-Nbody` has been implemented in which the local memory is not used.

4 Testbed Description

Waiting for the ExaNest prototype release, the `Hy-Nbody` code has been validated and tested on a deployed cluster based on heterogeneous ARM-hardware (CPUs + embedded GPUs). Each computational node is a Rockchip Firefly-RK3399 single board computer. It is a six core 64-bit High-Performance Platform, based on SoC with the ARM big.LITTLE architecture. ARM big.LITTLE technology features two sets of cores: a low performance energy-efficient cluster that is called "LITTLE" and power hungry high performance cluster that is called "big". Rockchip Firefly-RK3399 SoC is presented in Fig. 2. Each board contains (1) a cluster of four Cortex-A53 cores with 32 kB L1 cache and 512 L2 cache, and (2) a cluster of two Cortex-A72 high-performance cores with 32 kB L1 cache and 1M L2 cache. Each cluster operates at independent frequencies, ranging from 200 MHz up to 1.4 GHz for the LITTLE and up to 1.8 GHz for the big. The SoC contains 4 GB DDR3 - 1333 MHz RAM. The L2 caches are connected to the main memory via the 64-bit Cache Coherent Interconnect (CCI) 500

[5] http://infocenter.arm.com/help/topic/com.arm.doc.100614_0303_00_en/arm_mali_gpu_opencl_developer_guide_100614_0303_00_en.pdf.

that provides full cache coherency between big.LITTLE processor clusters and provides I/O coherency for the Mali-T864 GPU. The peculiarity of this board is that Mali-T864 is a OpenCL-compliant Quad-Core ARM Mali GPU.

The main characteristics of this cluster, named `INCAS`[6], are listed in Table 1. The cluster is managed by SLURM (Simple Linux Utility for Resource Management), a free and open-source job scheduler for Linux and Unix-like kernels, used by many of the world's supercomputers and computer clusters.

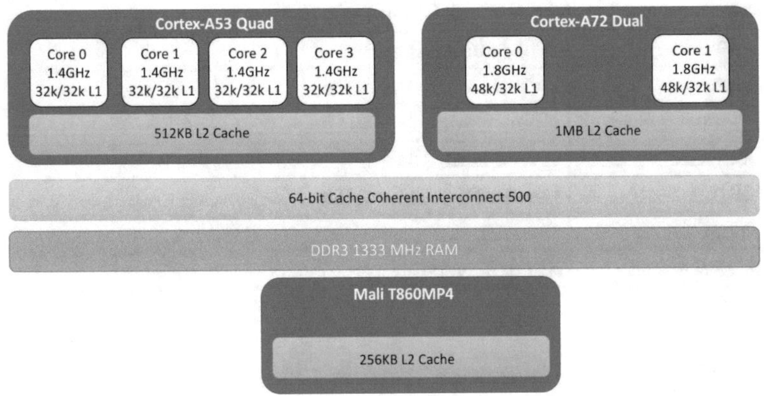

Fig. 2. Rockchip Firefly-RK3399 design.

Results presented in the following section have been carried out by means of `INCAS`, which is fully described in [3].

5 Performance Results

The 6th order Hermite integration schema implemented in `Hy-Nbody` relies on three different stages, described in Sect. 3. The *evaluation* stage is the most computationally demanding, considering that with N-bodies the algorithm requires $O(N^2)$ computational cost. The performance of the *evaluation* kernel is measured for both ARM CPU and GPU, testing how the running time (average of 10 runs of the kernel) changes as a function of the number of OpenMP threads in the CPU code, and of the work-group size in the GPU code. On the GPU side the impact of specific ARM-GPU-optimizations, as discussed in Sect. 3.1, are investigated. Performances have been measured for both DP and EX precision arithmetic.

[6] **IN**tensive **C**lustered **A**rm-**Soc**.

Table 1. The main characteristics of our cluster used to test the `Hy-Nbody` code.

Cluster name	`INCAS`
Nodes available	8
SoC	Rockchip RK3399 (28 nm HKMG Process)
CPU	Six-Core ARM 64-bit processor (Dual-Core Cortex-A72 and Quad-Core Cortex-A53)
GPU	ARM Mali-T864 MP4 Quad-Core GPU
Ram memory	4 GB Dual-Channel DDR3
Network	1000 Mbps Ethernet
Power	DC12V - 2A (per node)
Operating System	Ubuntu version 16.04
Compiler	gcc version 7.3.0
MPI	OpenMPI version 3.0.1
OpenCL	OpenCL 2.2
Job scheduler	SLURM version 17.11

5.1 ARM CPUs Performance Results

ARM big.LITTLE processors have three main software execution models: cluster migration (a single cluster is active at a time, and migration is triggered on a given workload threshold), CPU migration (pairing every big core with a LITTLE core), and heterogeneous multiprocessing mode (also known as Global Task Scheduling, which allows using all of the cores simultaneously).

The CPUs speedup, i.e. the ratio of the serial execution time to the parallel execution time utilizing multiple cores by means of OpenMP threads, is measured and studied. Kernel execution time on both ARM Cortex-A53x4 and Cortex-A72x2 CPUs have been obtained setting explicit CPU affinity and using the Linux system function `getrusage`, getting the total amount of time spent executing in user mode.

Figure 3 shows the speedup for both ARM Cortex-A53x4 and Cortex-A72x2 CPUs varying the number of OpenMP threads as a function of the number of particles. On the ARM Cortex-A53x4, for both DP-arithmetic and EX-arithmetic, some speedup is obtained only when the number of particles exceeds 4096 in number. As expected, the best performance is achieved with four OpenMP threads, where most likely there is one thread per available core. On the ARM Cortex-A72x2 one thread is always faster then multiple threads for DP-arithmetic and only a minor speedup is achieved with two threads adopting EX-arithmetic.

Figure 4 shows the ratio of the best running time achieved by the CPUs as a function of the number of particles for both arithmetic. ARM Cortex-A72x2 is faster than Cortex-A53x4 by approximately a factor of two.

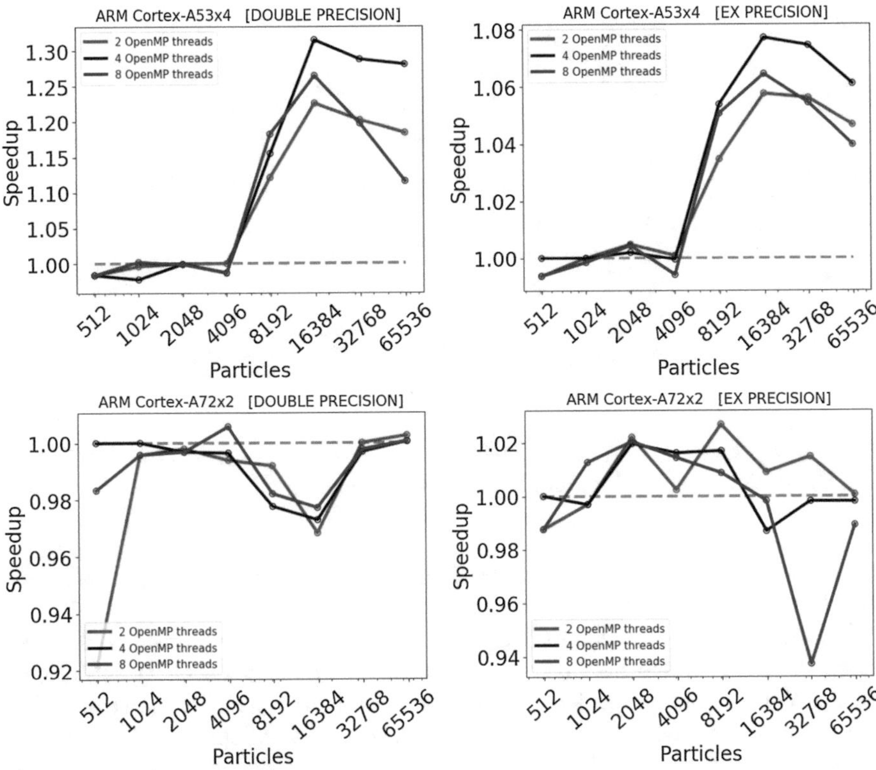

Fig. 3. Host speedup for DP-arithmetic (left panels) and EX-arithmetic (right panels) varying OpenMP threads as a function of the number of particles. Top panels for ARM Cortex-A53x4 CPU and bottom panels for ARM Cortex-A72x2 CPU.

5.2 ARM Embedded GPU Performance Results

The impact of the work-group size on the ARM Mali-T864 GPU performance is studied. Figure 5 shows the speedup achieved varying the OpenCL work-group size for GPGPU kernel code (top panels) and embedded-GPU-optimized kernel code (bottom panels) as a function of the number of particles. The speedup is normalized by the time to solution obtained with work-group size of four. Kernel execution times on the ARM-GPU have been obtained by means of OpenCL's built-in profiling functionality, which allows the host to collect runtime information. It is worth noting that work-group sizes of 128 and 256 cause a failure to execute the GPGPU kernel (top panels of Fig. 5) because of insufficient local memory on the GPU. Only ARM-GPU-optimized version of the kernel, which avoids the usage of local memory, can be run with those work-group sizes (the maximum possible work-group size on ARM Mali-T864 is 256). Despite ARM recommends for best performance using a work-group size that is between 4 and 64 inclusive, the results show that speedup is not driven by any specific work-group size, regardless the usage of local memory. These findings suggest to let

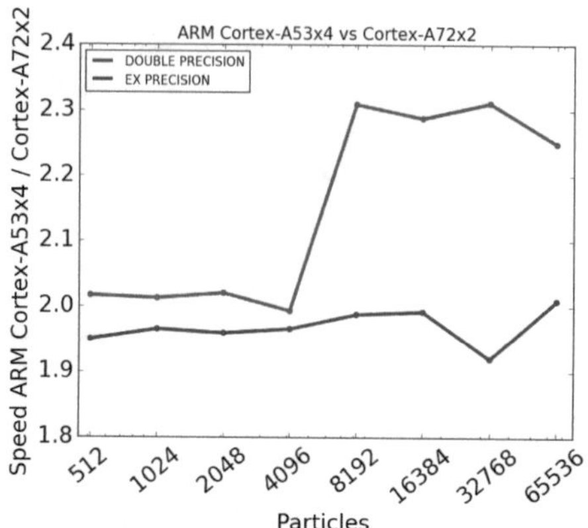

Fig. 4. Speed comparison between ARM Cortex-A53x4 and Cortex-A72x2 CPUs for both DP-arithmetic (red line) and EX-arithmetic (blue line).

the driver to pick the work-group size it thinks as best (the driver usually selects the work-group size as 64).

The impact of embedded-GPU-optimizations are also quantified. Figure 6 shows the ratio of the best time to solution achieved by GPGPU kernel code and ARM-GPU-optimized kernel code for both arithmetic. In the case of EX-arithmetic the speedup is approximately 10%, while adopting DP-arithmetic the speedup is nearing 5% increasing the number of particles. These findings reveal that adopting the same optimization strategies as those used for high-performance GPGPU computing might lead to worse performance on embedded GPUs. This is in agreement with what was found by [14], when they tested some non-graphic benchmarks on embedded GPUs.

5.3 ARM CPU-GPU Comparison

It is widely accepted that high-end GPGPUs can greatly speedup the solution of the direct N-body problem (see Sect. 2). However in this work we want also to evaluate the performance on low-power embedded GPUs for our kernel. Figure 7 shows the best running time on ARM Cortex-A72x2 as the ratio over the best execution time taken by the ARM-GPU-optimized implementation. The ARM-GPU-optimized implementation is as fast as the dual-core implementation on the ARM Cortex-A72x2 using DP-arithmetic, as long as the ARM-GPU is kept fed with enough particles, while is almost three times faster using EX-precision.

6 Future Development and Scope

HPC is currently facing, among others, the major technology challenge of the sustainable power consumption. Efficient hardware acceleration is the key to overcome this issue. However, for programmers it is not straightforward to take

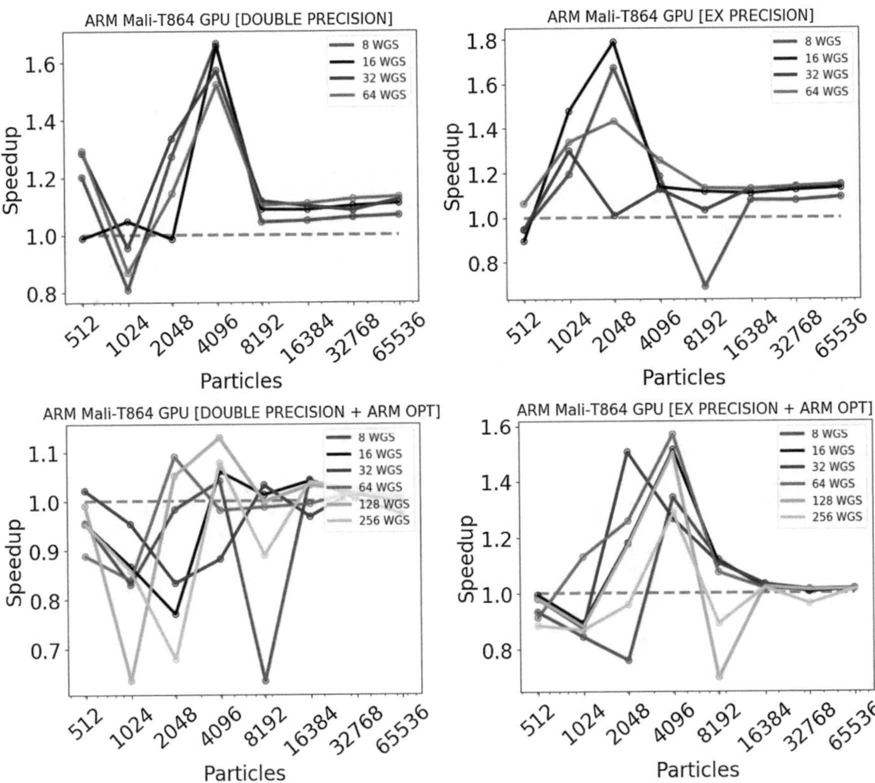

Fig. 5. GPU speedup for DP-arithmetic (left panels) and EX-arithmetic (right-panels) varying the OpenCL work-group size as a function of the number of particles. Speedup is normalized by the time to solution with work-group of size 4. Top panels for GPGPU kernel code and bottom panels for ARM-GPU-optimized kernel code.

the mapping decision of a given application to a multi-core CPU or accelerator while optimizing performance. For this reason, the next step of this research activity is also to quantitatively measure the power-efficiency of `Hy-Nbody`'s algorithms on ARM-SoC, possibly shedding some light on their suitability for exascale applications.

On the `ExaNeSt` prototype the HW acceleration is mainly issued by "unconventional"[7] FPGA devices, which in comparison to both CPUs and GPUs are more power-efficient (i.e. higher throughput per watt) for different class of applications as shown in the available literature (e.g. [5,19]). Unlike both CPUs and GPUs, FPGAs do not have any fixed architecture. On the contrary, they provide fine-grain grid of functional units, such as DSP and memory blocks, which can be interconnected to make any desired circuit. High-level synthesis allows

[7] Despite FPGAs have been invented in the 1980s, they only start recently to be used in HPC.

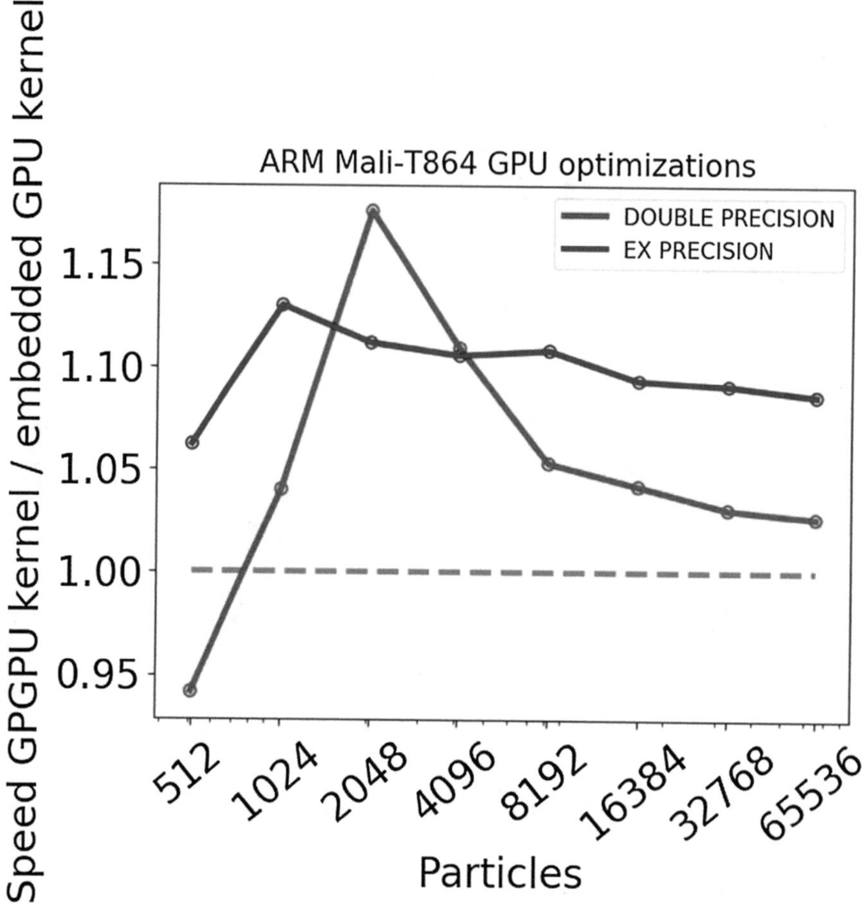

Fig. 6. Impact of ARM-GPU-optimizations on time to solution for Mali-T864 GPU as a function of the number of particles. Red line for DP-arithmetic and blue-line for EX-arithmetic.

the conversion of an algorithm description in high level languages, e.g. C/C++ or OpenCL, into a digital circuit. However, algorithms have to be modeled for FPGA implementation because the hardware features must be taken into account when attempting to optimize performance. In the case of a CPU or GPU, the programmer tries to achieve the best mapping of a kernel onto a fixed hardware architecture, while for FPGA the aim is to make optimized architecture for that kernel, balancing throughput and resource usage.

The future scope of this research activity is to port `Hy-Nbody` on FPGA exploiting the `ExaNeSt` prototype, which is based on Xilinx Zynq Ultrascale+ on SoC FPGA. A state-of-the-art direct N-body code, like `Hy-Nbody`, suitable for real-life astrophysical applications has never been ported on FPGA. The findings

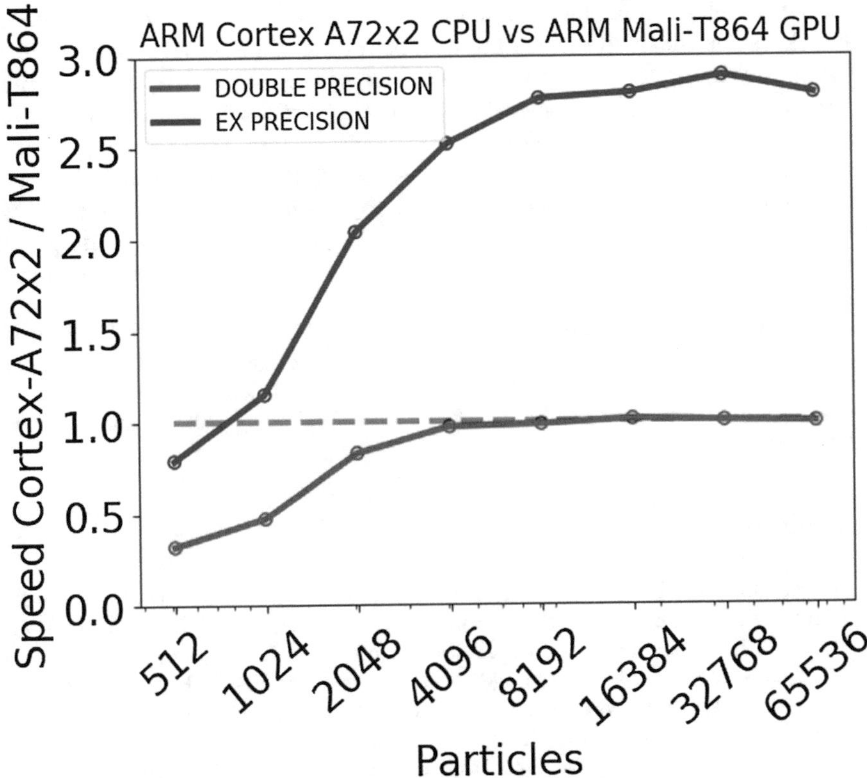

Fig. 7. Comparison of the time to solution between ARM Cortex-A72x2 CPU and Mali-T864 GPU for both DP-arithmetic (red line) and EX-arithmetic (blue line) as a function of the number of particles.

from this work on ARM SoC are fundamental in order to enhance our capabilities to obtain high performance energy-acceleration of kernels for scientific computing on FPGAs.

7 Conclusions

This research activity has shown that SoC boards can be successfully used to execute a state-of-the-art direct N-body code. The findings reveal that adopting the same optimization strategies as those employed for high-end GPGPUs might not be the best approach on embedded low-power GPUs, because of restricted hardware features. Secondly, embedded GPUs appear to be attractive from a performance perspective as soon as their double-precision compute capability increases. However, the emulated-double-precision approach can be a solution to supply enough power to execute scientific computation and benefit at maximum of SoC devices.

SoC technology will play a fundamental role on future Exascale heterogeneous platforms that will involve millions of specialized parallel compute units. Software developer for scientific applications will be forced to design power-efficient algorithms for heterogeneous systems with different devices, and likely with complex memory hierarchies.

Finally, commercial SoC devices such as the Firefly, are an excellent solution to build low cost testbeds to port codes and approach new heterogeneous ARM-based platforms, however they still lack of low latency network that is important when applications are communication bounded more than computing bounded.

Acknowledgments. This work was carried out within the ExaNeSt (FET-HPC) project (grant no. 671553) and the ASTERICS project (grant no. 653477), funded by the European Union's Horizon 2020 research and innovation programme.

This research has been made use of IPython [18], Scipy [10], Numpy [24] and MatPlotLib [9].

References

1. Ammendola, R., Biagioni, A., Cretaro, P., Frezza, O., Cicero, F.L., Lonardo, A., Martinelli, M., Paolucci, P.S., Pastorelli, E., Simula, F., Vicini, P., Taffoni, G., Pascual, J.A., Navaridas, J., Luján, M., Goodacree, J., Chrysos, N., Katevenis, M.: The next generation of Exascale-class systems: the ExaNeSt project. In: 2017 Euromicro Conference on Digital System Design (DSD), pp. 510–515, August 2017
2. Berczik, P., Nitadori, K., Zhong, S., Spurzem, R., Hamada, T., Wang, X., Berentzen, I., Veles, A., Ge, W.: High performance massively parallel direct N-body simulations on large GPU clusters. In: International conference on High Performance Computing, Kyiv, Ukraine, 8–10 October 2011, pp. 8–18, October 2011
3. Bertocco, S., Goz, D., Tornatore, L., Taffoni, G.: INCAS: INtensive Clustered ARM SoC - Cluster Deployment. INAF-OATs technical report, 222, August 2018
4. Bortolas, E., Gualandris, A., Dotti, M., Spera, M., Mapelli, M.: Brownian motion of massive black hole binaries and the final parsec problem. MNRAS **461**, 1023–1031 (2016)
5. Brodtkorb, A.R., Dyken, C., Hagen, T.R., Hjelmervik, J.M., Storaasli, O.O.: State-of-the-art in heterogeneous computing. Sci. Program. **18**(1), 1–33 (2010)
6. Capuzzo-Dolcetta, R., Spera, M.: A performance comparison of different graphics processing units running direct N-body simulations. Comput. Phys. Commun. **184**, 2528–2539 (2013)
7. Capuzzo-Dolcetta, R., Spera, M., Punzo, D.: A fully parallel, high precision, N-body code running on hybrid computing platforms. J. Comput. Phys. **236**, 580–593 (2013)
8. Harfst, S., Gualandris, A., Merritt, D., Spurzem, R., Portegies Zwart, S., Berczik, P.: Performance analysis of direct N-body algorithms on special-purpose supercomputers. New Astron. **12**, 357–377 (2007)
9. Hunter, J.: Matplotlib: a 2D graphics environment. Comput. Sci. Eng. **9**(3), 90–95 (2007)
10. Jones, E., Oliphant, T., Peterson, P., et al.: SciPy: open source scientific tools for Python (2001). Accessed 13 Sept 2015

11. Katevenis, M., Chrysos, N., Marazakis, M., et al.: The ExaNeSt project: interconnects, storage, and packaging for Exascale systems. In: 2016 Euromicro Conference on Digital System Design (DSD), pp. 60–67, August 2016
12. Katevenis, M., Ammendola, R., Biagioni, A., Cretaro, P., Frezza, O., Cicero, F.L., Lonardo, A., Martinelli, M., Paolucci, P.S., Pastorelli, E., Simula, F., Vicini, P., Taffoni, G., Pascual, J.A., Navaridas, J., LujÃn, M., Goodacre, J., Lietzow, B., Mouzakitis, A., Chrysos, N., Marazakis, M., Gorlani, P., Cozzini, S., Brandino, G.P., Koutsourakis, P., van Ruth, J., Zhang, Y., Kersten, M.: Next generation of Exascale-class systems: ExaNeSt project and the status of its interconnect and storage development. Microprocess. Microsyst. **61**, 58–71 (2018)
13. Konstantinidis, S., Kokkotas, K.D.: MYRIAD: a new N-body code for simulations of star clusters. Astron. Astrophys. **522**, A70 (2010)
14. Maghazeh, A., Bordoloi, U.D., Eles, P., Peng, Z.: General purpose computing on low-power embedded GPUs: has it come of age? In: 2013 International Conference on Embedded Computer Systems: Architectures, Modeling, and Simulation (SAMOS), pp. 1–10, July 2013
15. Morganti, L., Cesini, D., Ferraro, A.: Evaluating systems on chip through HPC bioinformatic and astrophysic applications, pp. 541–544, February 2016
16. Nitadori, K., Aarseth, S.J.: Accelerating NBODY6 with graphics processing units. MNRAS **424**, 545–552 (2012)
17. Nitadori, K., Makino, J.: Sixth- and eighth-order Hermite integrator for N-body simulations. New Astron. **13**, 498–507 (2008)
18. Perez, F., Granger, B.: IPython: a system for interactive scientific computing. Comput. Sci. Eng. **9**(3), 21–29 (2007)
19. Sirowy, S., Forin, A.: Where's the beef? Why FPGAs are so fast. Technical report, September 2008
20. Spera, M.: Using Graphics Processing Units to solve the classical N-body problem in physics and astrophysics. ArXiv e-prints, November 2014
21. Spera, M., Capuzzo-Dolcetta, R.: Rapid mass segregation in small stellar clusters. ArXiv e-prints, January 2015
22. Spera, M., Mapelli, M., Bressan, A.: The mass spectrum of compact remnants from the PARSEC stellar evolution tracks. MNRAS **451**, 4086–4103 (2015)
23. Thall, A.: Extended-precision floating-point numbers for GPU computation, p. 52, January 2006
24. van der Walt, S., Colbert, S., Varoquaux, G.: The NumPy array: a structure for efficient numerical computation. Comput. Sci. Eng. **13**(2), 22–30 (2011)

OS Scheduling Algorithms for Improving the Performance of Multithreaded Workloads

Murthy Durbhakula(✉)

Indian Institute of Technology Hyderabad, Hyderabad, India
cs15resch11013@iith.ac.in,
murthy.durbhakula@gmail.com

Abstract. Major chip manufacturers have all introduced multicore microprocessors. Multi-socket systems built from these processors are used for running various server applications. However to the best of our knowledge current commercial operating systems are not optimized for multi-threaded workloads running on such servers. Cache-to-cache transfers and remote memory accesses impact the performance of such workloads. This paper presents a unified approach to optimizing OS scheduling algorithms for both cache-to-cache transfers and remote DRAM accesses that also takes cache affinity into account. By observing the patterns of local and remote cache-to-cache transfers as well as local and remote DRAM accesses for every thread in each scheduling quantum and applying different algorithms, we come up with a new schedule of threads for the next quantum taking cache affinity into account. This new schedule cuts down both remote cache-to-cache transfers and remote DRAM accesses for the next scheduling quantum and improves overall performance. We present two algorithms of varying complexity for optimizing cache-to-cache transfers. One of these is a new algorithm which is relatively simpler and performs better when combined with algorithms that optimize remote DRAM accesses. For optimizing remote DRAM accesses we present two algorithms. Though both algorithms differ in algorithmic complexity they perform equally well for the workloads presented in this paper. We used three different synthetic workloads to evaluate these algorithms. We also performed sensitivity analysis with respect to varying remote cache-to-cache transfer latency and remote DRAM latency. We show that these algorithms can cut down overall latency by up to 16.79% depending on the algorithm used.

Keywords: Algorithms · Multiprocessor systems · Performance · OS scheduling

1 Introduction

Many commercial server applications today run on cache coherent NUMA (ccNUMA) based multi-socket multi-core servers. Performance of multithreaded applications running on such servers is impacted by both cache-to-cache transfers and remote DRAM accesses. One way to alleviate this problem is to rewrite the application. Alternatively the operating system scheduler can be optimized to reduce the impact of both cache-to-cache transfers and remote DRAM accesses. In this paper we present a

K. Arai et al. (Eds.): CompCom 2019, AISC 997, pp. 194–208, 2019.
https://doi.org/10.1007/978-3-030-22871-2_15

unified approach to optimizing OS scheduler for both cache-to-cache transfers and remote memory accesses that also takes cache affinity into account. In order to schedule a group of threads on a particular socket we take three factors into account: (i) Cache-to-cache transfers (ii) Remote DRAM accesses and (iii) Cache affinity. These factors do not always align with each other and hence forces us to make certain tradeoffs. In this paper we present various scheduling algorithms that differ from each other in how they make these tradeoffs. They also differ from each other in complexity and benefit.

Section 2 discusses the scheduling algorithms. Section 3 describes the methodology I used in evaluating the scheduling algorithms. Section 4 presents results. Section 5 describes related work and Sect. 6 presents conclusions.

2 Scheduling Algorithms

The general approach in all these algorithms is to first find optimal grouping of threads for reduced cache-to-cache transfers and then optimize that group for reduced remote DRAM accesses.

For cache-to-cache transfer latency optimization algorithms we assume the presence of performance counters that count cache-to-cache transfers between every pair of threads. For remote DRAM latency optimization algorithms we assume L DRAM access counters per thread that count the number of DRAM accesses to every node in a given scheduling quanta, where L is number of sockets in the system.

2.1 Algorithms for Optimizing Cache-to-Cache Transfers

Algorithm1

In this algorithm for every thread i, we find top three threads j, k, l with whom thread i has maximum cache-to-cache transfers. We then find a thread p which has highest cache-to-cache transfers among maximum pairs (i, j). Then we pick top three threads j, k, l with whom thread p has maximum cache-to-cache transfers. This forms our first group of four threads. We then remove p, j, k, l from the 16 threads. We repeat above algorithm with remaining 12 threads and get our second group of four threads. We then remove second group of four threads from 12 threads. We continue above process till we form all four groups of four threads.

We can see that in the above algorithm we have run the routine of finding maximum in N threads a constant number of times where the constant number is dependent on number of sockets. Hence the complexity of above algorithm is: O(NKL) where N is total number of threads, K is number of threads per socket, and L is total number of sockets. Below is high-level pseudo-code of this algorithm.

Input: Threads T0,...TN with cache-to-cache transfers count among themselves. Every pair of threads can have cache-to-cache transfers. Hence there are Nc2 cache-to-cache transfers. Each node has 4 cores and can run one thread per core. Current schedule S_current which has a grouping of N threads divided into L groups. Each group can run on one node. There are L nodes/sockets.

Output: A new schedule S_next with new grouping of threads for optimized cache-to-cache transfer latency

begin

1. For every thread i, find a thread j with which it has maximum cache-to-cache transfers. Let's call the maximum cache-to-cache transfers as Max-i. This can be done in 0(N) time.
2. Find a thread p which has highest maximum cache-to-cache transfer Max-p among all the N maximum cache-to-cache transfers; one per thread. This can also be done in 0(N) time.
3. Pick thread p and its pair j with which it has maximum cache-to-cache transfers. Now find other threads k, l with which thread p has second and third highest cache-to-cache transfers. These can also be done in 0(N) time. Threads (p,j,k,l) forms the first optimal group of threads in S_next. Remove this group from N threads and repeat steps 2 and 3 until all L groups are formed.

end

Overall complexity of the algorithm: Step 1 and 2 have 0(N) complexity. Step 3 has O(N(K-1)) where K is number of threads per core. Steps 2 and 3 are run L times where L is number of nodes. Hence overall complexity is 0(NKL).

Algorithm 1 for Optimized Cache-to-Cache Transfers

Algorithm2

In this algorithm we sort all pairs of cache-to-cache transfers. There are Nc2 such pairs. Then we find pair (i, j) which has highest cache-to-cache transfers. We then traverse the sorted list to find next highest pair (k, l). If neither k nor l matches (i, j) then (i, j, k, l) forms first group of four threads. If either k or l matches either of (i, j) then we do the following. Let us assume, without any loss of generality, that k is i and l does not match i or j. Then we find a thread p which has highest cache-to-cache transfers with thread l and does match i or j. Now (i, j, l, p) forms first group of four threads. Once we get first group of four threads we remove them from 16 threads. Then we run the above procedure for the remaining 12 threads to find second group of four threads and so on until we find all four groups of four threads.

The complexity of above algorithm is Nc2log(Nc2) because sorting of Nc2 pairs dominates the complexity of above algorithm. Below is high-level pseudo-code of this algorithm.

Input: Threads T0,...TN with cache-to-cache transfers count among themselves. Every pair of threads can have cache-to-cache transfers. Hence there are Nc2 cache-to-cache transfers. Each node has 4 cores and can run one thread per core. Current schedule S_current which has a grouping of N threads divided into L groups. Each group can run on one node. . There are L nodes/sockets.

Output: A new schedule S_next with new grouping of threads for optimized cache-to-cache transfer latency

begin

1) Sort all Nc2 pair of cache-to-cache transfers in descending order. This can be done in O(Nc2log(Nc2)) time.
2) Start from highest pair (i,j). Traverse the sorted pairs in descending order. As you traverse find next highest pair (k,l). If (k,l) does not overlap with (i,j) then (i,j,k,l) forms first optimal group of threads in S_next. Else if (k,l) overlaps then we find non-overlapping (l,p) as explained above. (i,j,l,p) forms first optimal grouping of four threads. Remove this group of threads from N threads and repeat Step 2 to find the next group of threads. This can be done in O(Nc2) time.

end

Overall complexity of the algorithm is: O(Nc2log(Nc2)).

Algorithm 2 for Optimized Cache-to-Cache transfers

2.2 Algorithms for Optimizing DRAM Accesses

Algorithm1

In this algorithm, we first sort all local and remote memory access counts for each group of threads along with node information in monotonically decreasing order. We start from top and assign the thread-group with highest access count to that node. Say thread-group G0 has highest access count of 10000 to node 2 then we assign G0 to node 2. Then we go to next element and assign the next thread-group G1 to its corresponding node. Once we assign a thread-group to one node we cannot assign it to another node. So in the above example once G0 is assigned to node 2, it cannot be assigned to any other node. The complexity of the algorithm is of order O((L*L)log (L*L)) where L is total number of thread groups as well as nodes. This is because the sorting algorithm dominates the complexity and we use merge-sort. There are specialized algorithms such as counting sort which could further reduce complexity; however their application is data dependent. Below is high-level pseudo-code of this algorithm.

Input: L groups of threads with DRAM accesses to L nodes. Each node has 4 cores and can run one thread per core. Current schedule S_current which has a mapping of L groups to L nodes. Each group has 4 threads.

Output: New schedule S_next with new mapping of groups to nodes.

begin

1. Each of L groups have L DRAM access counts; one per node. Sort L*L DRAM accesses in descending order. Complexity is O((L*L)log(L*L)).
2. Scan DRAM access counts. Start from highest DRAM access count and say it is coming from group L1 to node N2. Assign L1 to N2 in schedule S_next. Now L1 cannot be assigned to any other node.
3. Repeat step 2 until all groups are assigned a node in S_next. Complexity is O(L*L).

end

Overall complexity of the algorithm is: O((L*L)log(L*L)).

Algorithm 1 for Optimized DRAM Accesses

Algorithm 2

In this algorithm we first start off with node 0 and we sort all thread-group's accesses to that node in monotonically decreasing order. Then we pick first thread-group with highest accesses to node 0. Then we remove that thread-group from the list. We then sort thread accesses to node 1 out of 3 remaining thread-groups. We pick next top thread-group to node 1. We remove that thread-group from the list. We sort accesses to node 2 from the remaining two thread-groups. We pick next top thread-group to node 2. Remaining thread-group will go to node 3. The complexity of this algorithm is order O(Llog(L)) assuming sorting is done in parallel on all nodes, where L is total number of thread-groups. Even here it is sorting that dominates the complexity. Below is high-level pseudo-code of this algorithm.

Input: L groups of threads with DRAM accesses to L nodes. Each node has 4 cores and can run one thread per core. Current schedule S_current which has a mapping of L groups to L nodes. Each group has 4 threads.

Output: New schedule S_next with new mapping of groups to nodes.

begin

1. Sort all DRAM accesses from L groups of threads in descending order. If all nodes do it in parallel, Complexity is O(Llog(L)). Steps 2 and 3 have to be done serially by all nodes involved.
2. Start with node N0 and pick top group of threads and assign them to N0 in schedule S_next. Now this group of threads cannot be assigned to any other node.
3. Repeat step 2 for all nodes until all groups of threads are assigned a node in S_next. Complexity is O(L).

end

Overall complexity of the algorithm is: O(Llog(L)).

Algorithm 2 for Optimized DRAM Accesses

3 Methodology

We implemented each of these algorithms as a stand-alone C++ program and evaluated them by running synthesized data access patterns. These synthesized data access patterns vary both cache-to-cache transfers as well as local and remote dram access counts in a known pattern for each scheduling quantum. The data is then fed to each of the algorithms to see if they can track the pattern. The overall benefit for each algorithm depends on how well they can track the pattern. We chose synthetic workloads in order to clearly bring out benefits of each algorithm and its sensitivity to remote cache-to-cache transfer latency and remote DRAM latency. As part of future work I plan to study the impact of these algorithms on real workloads, similar to my earlier work [23]. In a real system we need performance counters to count number of cache-to-cache transfers between every pair of threads as well as performance counters to count local and remote DRAM accesses for every thread. This is the only hardware support we need for the scheduler optimizations.

Table 1 lists base configuration used for evaluation of different scheduling algorithms. On top of the base configuration we also perform sensitivity analysis of different algorithms with respect to varying remote cache-to-cache transfer latency from 100 to 175 to 250 cycles and varying remote DRAM latency from 250 to 375 to 500 cycles. Table 2 lists three synthetic cache-to-cache access patterns and Table 3 lists three synthetic DRAM access patterns used in this paper. A high level system diagram can be seen in Fig. 1. The circled portion represents one socket with 4 cores per socket.

Table 1. Configuration parameters

Parameter	Value
CPU Frequency	2 Ghz
Number of cores per socket	4
Number of sockets/nodes	4
Shared Cache	1 MB
Local DRAM Latency	125 cycles
Remote DRAM Latency	250 cycles
Cache-to-Cache Latency within socket	50 cycles
Cache-to-Cache Latency across sockets	100 cycles
Number of scheduling	16
Quanta	

4 Results

In this section we present results of various experiments done on algorithms described in Sect. 2 using synthetic workloads described in Sect. 3. This work can be extended to evaluate these algorithms on real workloads by running similar experiments.

4.1 Only Cache-to-Cache Transfers Optimization

First we run the three synthetic workloads Synth1_C_To_C-Synth1_DRAM, Synth2_C_To_C-Synth2_DRAM, and Synth3_C_To_C-Synth3_DRAM with only cache-to-cache transfer optimization algorithms. Those results are plotted in the graph below. The graph shows percentage improvement in overall latency.

As we can see Algorithm 2 consistently outperforms Algorithm 1. However, the extra improvement we get with Algorithm 2 is not always worth the high algorithmic complexity associated with Algorithm 2. We also observe that sometimes performance improves as we move from Synth1_C_To_C-Synth1DRAM to Synth2_C_To_C-Synth2 DRAM or from Synth2_C_To_C-Synth2DRAM to Synth3_C_To_C-Synth3DRAM and sometimes performance decreases. On further analysis we found that this is dependent on whether we get more or less benefit when there is a phase change in the workload. Sometimes due to phase change performance improves and sometimes performance decreases. It is workload and algorithm dependent. However, irrespective of workload, Algorithm 2 outperforms Algorithm 1 when only cache-to-cache transfer optimization is enabled.

4.2 Both Optimizations Together Without Taking Cache Affinity into Consideration

In this section we present results of running the synthetic workloads when both cache-to-cache transfer optimization and remote DRAM optimizations are on while not taking cache-affinity into account.

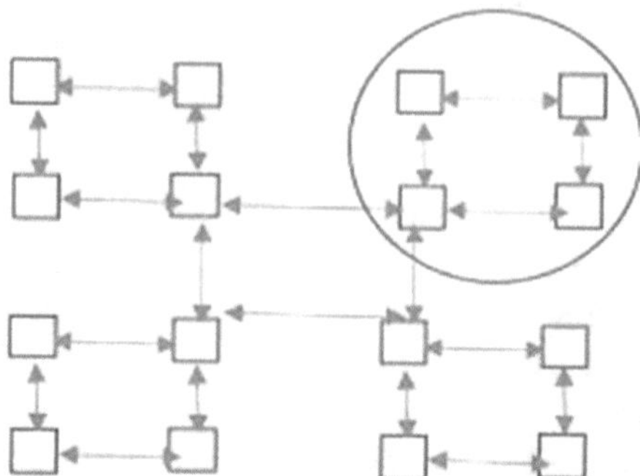

Fig. 1. High-level ccNUMA Multi-socket Multi-core System diagram with 4 sockets and 4 cores-per-socket

Table 2. Synthetic workloads for cache-to-cache transfers

Access pattern	Description
Synth1_C_To_C - Single phase access pattern	Same cache-to-cache access pattern for all quanta except first quanta. Single optimal grouping of threads
Synth2_C_To_C - Two phase access pattern	Two cache-to-cache access patterns equally distributed over all quanta. Optimal grouping of threads from 1st till 8th quanta. Second optimal grouping of threads from 9th till 16th quanta
Synth3_C_To_C - Four phase access pattern	Four cache-to-cache access patterns equally distributed over all quanta. Optimal grouping of threads from 1st till 4th quanta. Second optimal grouping of threads from 5th till 8th quanta. Third optimal grouping of threads from 9th till 12th quanta. Fourth optimal grouping of threads from 13th till 16th quanta

Table 3. Synthetic workloads for remote DRAM accesses

Access pattern	Description
Synth1_DRAM - Single phase access pattern	Same DRAM access pattern for all quanta except first quanta. Single optimal grouping of threads
Synth2_DRAM - Two phase access pattern	Two DRAM access patterns equally distributed over all quanta. Optimal grouping of threads from 1st till 8th quanta. Second optimal grouping of threads from 9th till 16th quanta
Synth3_DRAM - Four phase access pattern	Four cache-to-cache access patterns equally distributed over all quanta. Optimal grouping of threads from 1st till 4th quanta. Second optimal grouping of threads from 5th till 8th quanta. Third optimal grouping of threads from 9th till 12th quanta. Fourth optimal grouping of threads from 13th till 16th quanta

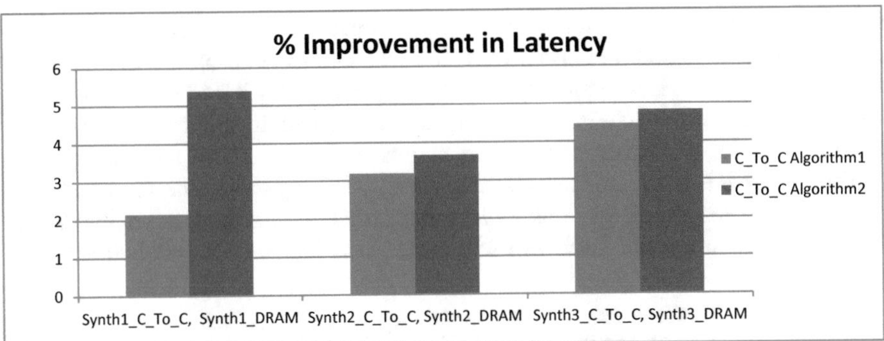

Fig. 2. Percentage improvement in overall latency with only cache-to-cache transfer optimization

Note: For rest of this paper, for remote DRAM optimization, we are only showing Algorithm 2 because we did not find any major difference in results between Algorithm 2 and Algorithm 1 and Algorithm 2 is of lower complexity than Algorithm1.

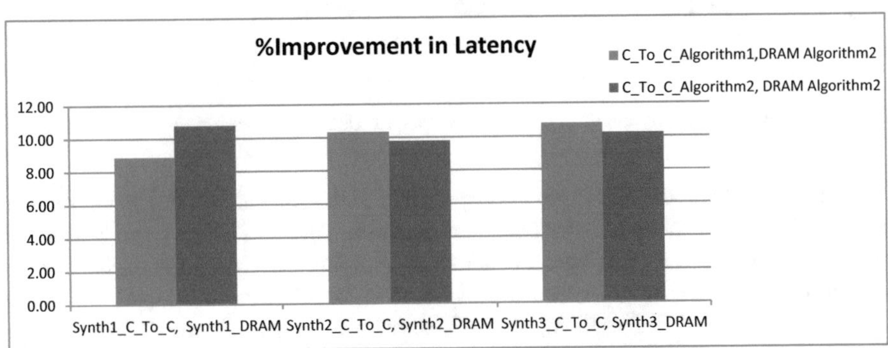

Fig. 3. Percentage improvement in overall latency with both optimizations without Cache Affinity

We can see from the above graph (Fig. 3) that with remote DRAM optimization enabled sometimes C_To_C_Algorithm1 outperforms C_To_C_Algorithm2. This is because remote DRAM latencies are typically higher than remote cache-to-cache latencies. However one key takeaway is that with remote DRAM optimization added percentage improvement in latency is much higher than percentage improvement in latency with only cache-to-cache transfer optimization on as shown in Fig. 2.

4.3 Both Optimizations Together with Cache Affinity

In this section we present results of running the synthetic workloads when both cache-to-cache transfer optimization and remote DRAM optimizations are enabled while taking cache-affinity into account. Whenever a thread shifts from its current socket to another socket we associate a penalty due to cache affinity. In each socket four threads share a 1 MB cache. Hence roughly each thread can use 256 KB of data. Of which we assume that in the next quanta when it gets scheduled it still can use 64 KB of worthwhile data. We experimented with various fractions of 256 KB data and concluded that 64 KB is a reasonable choice. Hence, assuming cache line size of 64 bytes, cache transfer penalty is same as transferring (1024) cache lines to another socket. Assuming remote transfer latency of 250 cycles, it is 250 * 1024 = 256000 cycles.

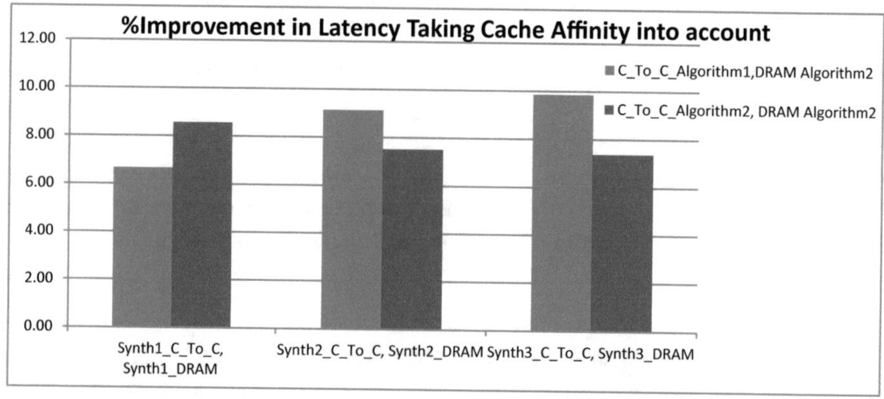

Fig. 4. Percentage improvement in overall latency with both optimizations with Cache Affinity

We can see from the above graph (Fig. 4) that with cache affinity the benefit due to combined cache-to-cache transfer latency and remote DRAM latency optimization has reduced. However it is still much better than applying only cache-to-cache transfer latency optimization. We can also observe that C_To_C Algorithm1 performance is very close to that of C_To_C Algorithm2, sometimes outperforming C_To_C Algorithm2. Hence given higher algorithmic complexity of C_To_C Algorithm2, C_To_C Algorithm1 is a better choice unless the number of threads N is small enough that it does not make much of a difference.

4.4 Sensitivity Analysis

In this section we show results of applying sensitivity analysis to combined algorithms with cache affinity while varying both remote cache-to-cache transfer latencies and remote DRAM latencies. We increased both remote cache-to-cache transfer latencies and remote DRAM latencies as follows:

Experiment 1: Remote Cache-to-Cache Latency: 175 cycles, Remote DRAM Latency: 375 cycles

Here we increased remote cache-to-cache transfer latency to 175 cycles keeping local cache-to-cache transfer latency unchanged. Further we have also increased remote DRAM latency to 375 cycles keeping local DRAM latency unchanged.

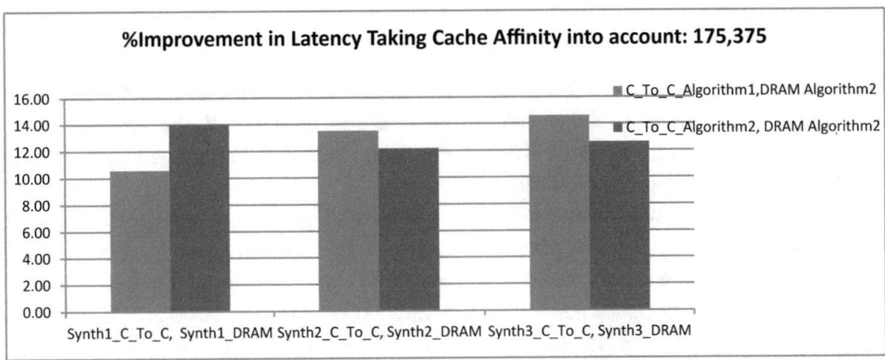

Fig. 5. Percentage improvement in overall latency with both optimizations and cache affinity - Sensitivity analysis: 175,375

We can see from the above graph (Fig. 5) that performance improves with increased remote cache-to-cache transfer latency and increased remote DRAM latency. This is to be expected since the algorithms are trying to reduce the number of remote cache-to-cache transfers and the number of remote DRAM accesses.

Experiment 2: Remote Cache-to-Cache Latency: 250 cycles, Remote DRAM Latency: 500 cycles
Here we increased remote cache-to-cache transfer latency to 250 cycles keeping local cache-to-cache transfer latency unchanged. Further, we have also increased remote DRAM latency to 500 cycles keeping local DRAM latency unchanged.

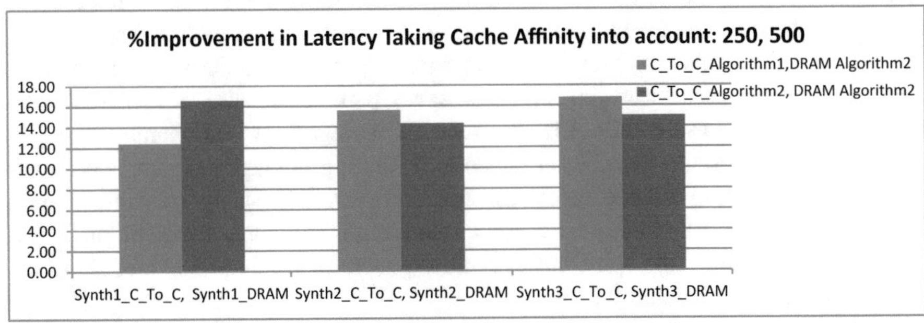

Fig. 6. Percentage improvement in overall latency with both optimizations and cache affinity: Sensitivity analysis: 250, 500

As expected we can see from the above graph (Fig. 6) that the performance improves even further compared to experiment 1. Again this is because the algorithms are trying to reduce the number of remote cache-to-cache transfers and the number of remote DRAM accesses.

5 Related Work

To the best of my knowledge this is the first study which presents a unified approach to optimizing both remote cache-to-cache transfers and remote DRAM accesses while taking cache affinity into account. There have been studies which tackled each of these problems separately which we discuss below.

Thread-clustering algorithms were examined by Thekkath and Eggers [1]. Their research dealt with finding the best way to group threads that share memory regions together onto the same processor so as to maximize cache sharing and reuse. However, the way they track sharing is address-based and can be prone to errors, particularly for migratory sharing patterns. If processes P0, P1, P2 are involved in migratory sharing, and if you select P0 and P1 to schedule on the same socket, and if the cache line actually migrates from P0 to P2 and then back to P1 then we would not have reduced inter-socket cache-to-cache transfers. In our work we track sharing using pair-wise counters. In our work, we assume that the number of threads matches the number of processors: in other words, the system is always load-balanced.

Our work most closely resembles the work done by Tam et al. [2]. Their paper also proposes using hardware performance counters for thread clustering in a shared-memory multiprocessor system. We use a different approach to collect inter-socket cache-to-cache transfers using our performance counters. We do not count remote cache stalls per address range (since it can be prone to the migratory sharing problem described above); instead, we count pair-wise remote cache transfers. This approach works for small and medium-range servers. For bigger machines (such as Cray Red Storm, Aprro International Atlas [3], etc.) we can use a coarse - grained approach such as combining close-by sockets as one "logical socket". This is so that, even if w e may not schedule highly interacting threads on the same socket we will schedule them on close-by sockets to reduce the number of hops. We assume that the home node for all cache-to-cache transfers for the selected cluster of threads falls in the same socket. One way to overcome this problem is to maintain a per-home-node-based counter for counting cache-to-cache transfers between every pair of threads, and then take the home node into consideration for the scheduling algorithm. Sridharan *et al.* [4] examined a technique to detect user space lock sharing among multi-threaded applications by annotating user-level synchronization libraries. Using this information, threads sharing the same highly contended lock are migrated onto the same processor. Our work adopts the same spirit but at a more general level that is applicable to any kind of memory region sharing.

Many researchers have investigated OS support for minimizing cache conflict and capacity problems of shared L2/L3 cache processors [5–7]. Our work, however, focuses on reducing the impact of communication misses that are inherent to applications. Our work is complementary to these past efforts.

Chandra et al. [8] have studied the impact of various OS scheduling policies on performance of both uniprocessor and multiprocessor workloads. Our work is similar to theirs in that we also study OS scheduling algorithms for improving performance of parallel workloads. However, our focus is more on memory intensive workloads for ccNUMA multi-socket multi-core servers. We assume that at-a-time only one multi-threaded application is running with each core assigned one thread for the sake of load-balancing. The algorithms we propose are completely different. Their work also evaluated benefits of page migration. However, page migration and replication are not always beneficial. For instance say thread T0 is scheduled on Node 1 with 1000 accesses to Node 1's DRAM and also accesses a page P2 on Node 2 DRAM, with intention-to-write, 500 times. At the same time thread T1 is scheduled on Node 3 with 2000 accesses to Node 3's DRAM and accesses same page P2 on Node 2 DRAM, with intention-to-write, 500 times. In such a situation it is not possible to decide to which node to migrate P2 to. In fact if there are such "hot" pages accessed by multiple threads it is actually better to schedule them on the same node where hot page is, rather than migrating the hot page.

Srikanthan et al. [19, 20] proposed a sharing-aware-mapper SAM that takes into account other factors such as latency tolerance and instruction level parallelism in addition to cache-to-cache transfers inherent in an application and then collocates threads accordingly. They evaluated their scheduler on a 2-socket 40-CPU and 4-socket 80-CPU machine and showed improvements of up to 61% over a default linux scheduler for mixed workloads. Similar to our approach they use hardware counters to enable optimized scheduling of threads. Unlike their approach the focus of this paper is on: (i) combining sharing-aware scheduler with (ii) remote DRAM access aware scheduler that also takes (iii) cache affinity into account. These three factors sometime act against each other and we have shown the unified approach to optimize scheduling of such threads.

Lepers et al. [21] studied the impact of asymmetric interconnect on thread and memory placement on NUMA systems. They found that different node interconnectivity can have as much as 2x performance impact under same distribution of thread and data across nodes. Based on this insight they implemented a dynamic thread and memory placement algorithm in Linux that delivers similar or better performance than the best static placement and up to 218% better performance than when the placement is chosen randomly. Unlike their approach we assume symmetric interconnect in this paper. Our focus is orthogonal to the focus of their paper. The focus of this paper is on reducing remote cache-to-cache transfers and remote DRAM accesses while taking cache affinity into account.

Kaseridis et al. [9] proposed a dynamic memory subsystem resource management scheme that considers both cache capacity and memory bandwidth contention in large multi-chip CMP systems. Their memory bandwidth contention algorithms monitor when a particular schedule of threads exceed the maximum bandwidth supported by a node and then try to schedule bandwidth demanding threads with those threads that need little memory bandwidth. Whereas our approach proactively tries to find the best pairing of threads for any scheduling quanta while keeping the overall bandwidth utilization within the maximum bandwidth limits. And the algorithms presented in this paper are different from their algorithms.

Ipek et al. [10] proposed using reinforcement learning based approach to tune DRAM scheduling policies to effectively utilize off-chip DRAM bandwidth. Their work differs from ours in two different ways. First, we are using OS scheduling algorithms to optimize DRAM bandwidth utilization. Second, we use parallel workloads to evaluate the scheduling policy where as they use multiprogramming workloads with no sharing.

Ahn et al. [11] studied the impact of DRAM organization on the performance of data parallel memory systems. In contrast to their work we focus on novel OS scheduling algorithms that will improve DRAM performance of parallel applications on general purpose multiprocessors. Zhu et al. [12] proposed using novel DRAM scheduling algorithms for SMT processors. In contrast to their work this paper proposes using new OS scheduling algorithms with minimal hardware support.

Tang et al. [14] studied the impact of co-locating threads of different multi-threaded applications in a data-center environment on overall performance. They proposed heuristics for co-locating different workloads. Their focus is mostly on data-center related workloads and the algorithms presented in this work are different from their heuristics.

More recently Kim and Huh [17] proposed OS scheduling algorithms for implementing fairness in multi-core systems. They proposed a few algorithms that improve fairness in heterogeneous multi-core systems while improving overall throughput. Their focus is on mainly improving fairness using OS scheduling policies while the focus of this paper is on reducing remote cache-to-cache transfers and remote DRAM accesses thereby improving overall performance. Further Kim and Huh use SPEC workloads for the evaluation of their algorithms whereas we focus on parallel workloads.

Sahoo and Dehury [18] proposed job scheduling algorithms for CPU intensive workloads in a health-care cloud environment. Their focus is mainly on health-care applications in a cloud-based environment whereas the focus of this paper is on improving performance of parallel workloads.

Harris et al. [22] proposed and implemented Callisto which is a run-time system for OpenMP and other task parallel programming models. It handles synchronization and load balancing on multiple cores. Their work is complementary to the work presented in this paper. We can combine load balancing with collocation of threads optimized for reducing both remote cache-to-cache transfers and remote DRAM accesses.

There are many other studies [13, 15, 16] which focused on tuning DRAM scheduling policies or memory access ordering for better overall performance. Our work is different from these in two ways. First we focus on OS scheduling algorithms to reduce the impact of remote cache-to-cache transfers and remote DRAM accesses. Second, these studies [13] focus on optimal DRAM utilization for co-located single threaded workloads where as our work focuses on improving performance of the multi-threaded parallel workloads.

6 Conclusion

Many commercial server applications today run on cache coherent NUMA (ccNUMA) based multi-socket multi-core servers. Performance of multithreaded applications running on such servers is impacted by both cache-to-cache transfers and remote DRAM accesses. Often times optimizing for both these issues requires us to make certain tradeoffs. In this paper we have presented a unified approach to optimizing OS scheduling algorithm for both cache-to-cache transfers and remote DRAM accesses that also takes cache affinity into account.

We presented two algorithms of varying complexity for optimizing cache-to-cache transfers. One of these is a new algorithm which is relatively simpler and performs better when combined with algorithms that optimize remote DRAM accesses. For optimizing remote DRAM accesses we presented two algorithms. Though both algorithms differ in algorithmic complexity we find that for our workloads they perform equally well. We used three different synthetic workloads for both cache-to-cache transfers and DRAM accesses to evaluate these algorithms. Further we also performed sensitivity analysis with respect to varying remote cache-to-cache transfer latency and remote DRAM latency. We showed that these algorithms can cut down overall latency by up to 16.79% depending on the algorithm used.

This work could be extended in several ways. One way is to use machine learning algorithms to learn any kind of phase behavior among prior scheduling quanta and incorporate that into scheduling decision for the next quanta. Another way is to monitor the benefit of these algorithms every scheduling quanta and turn off/on the optimization adaptively if benefit falls below a specified threshold. In general for any long running application with stable patterns, hardware could provide feedback to the OS, which could in turn use that information to adapt its policies to benefit application performance.

Acknowledgments. I would like to thank Prof. Alan Cox of Rice University for initially discussing with me the concept of optimizing OS scheduling algorithms for improving the performance of various workloads. I would also like to thank various reviewers of this work for their comments and feedback. Finally I would like to thank my parents, wife, and kids for supporting me morally during the course of this work.

References

1. Thekkath, R., Eggers, S.J.: Impact of sharing-based thread placement on multi-threaded architectures. In: International Symposium on Computer Architecture (1994)
2. Tam, D., Azimi, R., Stumm, M.: Thread clustering: sharing-aware scheduling on SMP-CMP-SMT multiprocessors. In: ACM SIGOPS Operating System Review, June 2007
3. www.top500.org
4. Sridharan, S., et al.: Thread migration to improve synchronization performance. In: Workshop on Operating System Interference in High Performance Applications (2006)
5. Nakajima, J., et al.: Enhancements for hyper-threading technology in the operating system – seeking the optimal micro-architectural scheduling. In: International Parallel and Distributed Processing Symposium (2005)
6. Snavely, A., et al.: Symbiotic job scheduling for a simultaneous multithreading processor. In: Conference on Architectural Support for Programming Languages and Operating Systems (ASPLOS) (2000)
7. El-Moursy, A., et al.: Compatible phase co-scheduling on a CMP of multi-threaded processors. In: International Parallel and Distributed Processing Symposium. 2006
8. Chandra, R., Devine, S., Verghise, B., Gupta, A., Rosenblum, M.: Scheduling and page migration for multiprocessor compute servers. In: Proceedings of ASPLOS (1994)

9. Kaseridis, D., Stuecheli, J., Chen, J., John, L.K.: A bandwidth-aware memory-subsystem resource management using non-invasive resource profilers for large CMP systems. In: Proceedings of Sixteenth International Symposium on High Performance Computer Architecture (2010)
10. İpek, E., Mutlu, O., Martínez, J.F., Caruana, R.: Self-optimizing memory controllers: A reinforcement learning approach, In: Proceedings of International Symposium on Computer Architecture, Beijing, China, June 2008
11. Ahn, J.H., Erez, M., Dally, W.J.: The design space of data - parallel memory systems. In: Proceedings of SC, 2006
12. Zhu, Z., Zhang, Z.: A performance comparison of DRAM memory system optimizations for SMT processors. In: Proceedings of HPCA-11 (2005)
13. Nauman, R., Lim, W.-T., Thottethodi, M.: Effective management of DRAM bandwidth in multicore processors. In: Proceedings of PACT-2007
14. Tang, L., Mars, J., Vachharajani, N., Hundt, R., Soffa, M.L.: The impact of memory subsystem resource sharing on datacenter applications. In: Proceedings of International Symposium on Computer Architecture (2011)
15. Hur, I., Lin, C.: Adaptive history-based memory schedulers. In: Proceedings of the International Symposium on Microarchitecture (2004)
16. Rixner, S., Dally, W.J., Kapasi, U., Mattson, P.R., Owens, J.D.: Memory access scheduling. In: Proceedings of International Symposium on Computer Architecture (2000)
17. Kim, C., Huh, J.: Fairness-oriented OS scheduling support for multicore system. In: Proceedings of 2016 International Conference on Supercomputing
18. Sahoo, P.K., Dehury, C.K.: Efficient data and CPU-intensive job scheduling algorithms for healthcare cloud. Elsevier Comput. Electr. Eng. **68**, 119–139 (2018)
19. Srikanthan, S., Dwarkadas, S., Shen, K.: Data sharing or resource contention: toward performance transparency on multicore systems. In: USENIX Annual Technical Conference (2015)
20. Srikanthan, S., Dwarkadas, S., Shen, K.: Coherency stalls or latency tolerance: informed CPU scheduling for socket and core sharing. In: USENIX Annual Technical Conference (2016)
21. Lepers, B., Quema, V., Fedorova, A.: Thread and memory placement on NUMA systems: asymmetry matters. In: USENIX Annual Technical Conference (2015)
22. Harris, T., Maas, M., Marathe, V.J.: Callisto: co-scheduling parallel runtime systems. In: 9th EuroSys Conference (2014)
23. Durbhakula, M.: Sharing aware OS scheduling algorithms for multi-socket multi-core servers. In: Proceedings of First International Forum on Next-Generation Multicore/Manycore Technologies (2008)

The Fouriest: High-Performance Micromagnetic Simulation of Spintronic Materials and Devices

I. Pershin[1,2,3(✉)], A. Knizhnik[3], V. Levchenko[2], A. Ivanov[2], and B. Potapkin[3]

[1] Moscow Institute of Physics and Technology, 141701 Moscow, Russia
pershin2010@gmail.com
[2] Keldysh Institute of Applied Mathematics, 125047 Moscow, Russia
[3] Kintech Lab Ltd., 123298 Moscow, Russia

Abstract. Micromagnetic modeling is a powerful tool for analysis of spintronic materials and devices. We have developed a new software named The Fouriest designed for micromagnetic modeling on Nvidia GPUs. Basically, the program solves the Landau-Lifshitz Equation on a 3-D grid, using Fast Fourier Transform for calculation demagnetization fields. The key advantage of the new code is that it can model not only a single magnetic system, but also an ensemble of ones, which is often required in spintronics. The performance of such calculations via our software is significantly higher than using other programs that do not support concurrent modeling of multiple systems. This performance gain is obtained by batching Fast Fourier Transforms of ensemble systems, giving a full utilization of all GPU parallelism levels. Systems in the ensemble being processed can differ from each other in their shape and physical parameters, and can even interact in various ways.

Keywords: The Fouriest · Landau–Lifshitz Equation · Micromagnetics · Demagnetization · Fast Fourier Transform · GPU · CUDA · cuFFT · Zero-padding · Spintronics · Write-Error-Rate

1 Introduction

Micromagnetic modeling is a powerful tool for analysis of spintronic materials and devices [19]. OOMMF (Object Oriented MicroMagnetic Framework) [7] is the standard implementation in this field. In recent years, a few programs have appeared that use a GPU for the modeling, since the performance gain is very significant. Open-source software MuMax3 [27] is the most common one, also there are FastMag [5], GpMagnet [18] and others.

However, these implementations are not efficient for analysis of sets of independent or coupled small magnetic systems, which often exist in real spintronic devices. Such tasks include the Write-Error-Rate calculation [15], MTJ-based magnetic sensor devices [12], the Nudged Elastic Band method [4], parametric computations and phase diagrams calculations. The problem with existing

K. Arai et al. (Eds.): CompCom 2019, AISC 997, pp. 209–231, 2019.
https://doi.org/10.1007/978-3-030-22871-2_16

implementations is that GPUs have a high parallelism level, and therefore special software support is needed to model these magnetic systems sets in order to completely utilize GPUs processing power. The solution to this problem is our program The Fouriest, which is a new implementation of 3D micromagnetic solver optimized for high performance calculation of ensemble of small magnetic systems. The Fouriest is GPU-accelerated and written in C++ and Python using CUDA Toolkit [20,21].

In this paper, firstly we will describe the problem statement and equations (Sect. 2), then the computation methods used (Sect. 3). Particular attention will be paid to the implementation (Sect. 4) and performance (Sect. 4.2). After that, we will give an overview of physical tasks The Fouriest was tested on (Sect. 5), and will finish with practical applications (Sect. 6).

2 Problem Statement and Equations

Let's consider a ferromagnetic sample of arbitrary shape. We will use the continuum micromagnetic approach ignoring the real atomic structure. This implies that there is a function of magnetization $\mathbf{m}(\mathbf{r}, t)$ on the sample, and $|\mathbf{m}(\mathbf{r}, t)| \equiv 1$. To calculate this function, we use a 3-D Cartesian grid superimposed on the sample, then the dynamic of a magnetization $\mathbf{m}_i$ of each grid cell satisfies the Landau-Lifshitz equation [16] with the modification of adding the stochastic term for the temperature to be taken into account [30].

$$\dot{\mathbf{m}}_i = -\gamma^* \left[\mathbf{m}_i \times \mathbf{H}_i\right] - \frac{\gamma^* \alpha}{|\mathbf{m}_i|} \left[\mathbf{m}_i \times \left[\mathbf{m}_i \times \mathbf{H}_i\right]\right] + 2\sqrt{\frac{\gamma^* \alpha k_B T}{M_s V_i}} \boldsymbol{\xi}(\mathbf{m}_i, t), \qquad \gamma^* = \frac{\gamma}{1+\alpha^2}, \tag{1}$$

where γ is a gyromagnetic ratio, α is a dissipation factor, $\mathbf{H}_i$ is an effective field, k_B is the Boltzmann constant, T is a temperature, V_i is the volume of the cell, M_s is a saturation magnetization, and $\boldsymbol{\xi}(\mathbf{m}_i, t)$ is a random process of unit intensity with the normal distribution preserving $|\mathbf{m}_i|$. In the case of equilibrium $\boldsymbol{\xi}(\mathbf{m}_i, t)$ provides the Boltzmann distribution. As for the length of $\mathbf{m}_i$, it doesn't change over time, and $|\mathbf{m}_i|$ has a value from 0 to 1 depending on the area of intersection of the ferromagnet and the grid cell, so $\mathbf{m}_i$ will be a unit length vector for all the internal cells.

The effective field $\mathbf{H}_i$ may include

- External magnetic field

$$\mathbf{H}_{ext} = -\frac{\partial E_{ext,i}}{\partial \mathbf{m}_i}$$

- Uniaxial anisotropy field

$$\mathbf{H}_{anis} = -\frac{\partial E_{anis,i}}{\partial \mathbf{m}_i} = \frac{2K}{M_s}(\mathbf{m}_i \cdot \mathbf{n}_K)\mathbf{n}_K,$$

where K is an anisotropy constant, $\mathbf{n}_K$ is a unit vector corresponding to the direction of $\mathbf{K}$.

- Exchange interaction between adjacent cells based on 6-neighbours scheme [6],

$$\mathbf{H}_{exch} = -\frac{\partial E_{exch,i}}{\partial \mathbf{m}_i} = 2\frac{A_{ex}}{M_s} \sum_{j:adjacent} \frac{\mathbf{m}_j - \mathbf{m}_i}{\Delta_i^2},$$

  where A_{ex} is an exchange interaction energy, Δ_i is a distance between i cell and j cell. If a neighbour is absent, the $\mathbf{m}_i$ value is used instead of the $\mathbf{m}_j$, implementing Neumann boundary conditions.
- Demagnetization field [26]

$$\mathbf{H}_{demag} = -\frac{\partial E_{demag,i}}{\partial \mathbf{m}_i} = M_s \sum_j \mathsf{N}_{ij}\mathbf{m}_j, \qquad N_{ij}^{lm} = \oint_{\partial V_j} \frac{(\mathbf{r}_i - \mathbf{r}')_l n_m}{|\mathbf{r}_i - \mathbf{r}'|^3} ds',$$

  where j traverses through all of the grid cells, and N_{ij} is demagnetization tensor (symmetric, 3-dimensional, rank-2). This will be detailed in the Sect. 3.3.
- Spin-transfer torque [15]

$$\mathbf{H}_{STT} = \frac{\hbar \eta_{\|} j}{2eM_s d}\left[\mathbf{m}_i \times \mathbf{m}_{ref}\right],$$

  where d is a thickness of the magnetic sample, $\hbar$ is the Planck constant, $\eta_{\|}$ is a parallel ratio of spin-polarization, j is a current density, e is the elementary charge, $\mathbf{m}_{ref}$ is a magnetization of a reference layer.

The physical values' units are based on CGS: $[\gamma] = (\text{Oe} \times \text{ns})^{-1}$, $[\mathbf{H}_i] = [\mathbf{H}_{ext}] = \text{Oe}$, $[k_B] = \text{erg/K}$, $[T] = \text{K}$, $[V_i] = \text{cm}^3$, $[M_s] = \text{emu/cm}^3$, $[K] = \text{erg/cm}^3$, $[A_{ex}] = \text{erg/cm}$, $[d] = [\Delta_i] = \text{cm}$, $[\hbar] = \text{erg} \times \text{s}$, $[j] = \text{MA/cm}^2$, $[e] = \text{MC}$.

3 Computational Methods

Now we will describe how exactly a micromagnetic sample of arbitrary shape can be split into magnetic moments (Sect. 3.1), then the numerical scheme for (1) will be specified (Sect. 3.2), and the calculation of the demagnetization field will be told about (Sect. 3.3).

3.1 Grid Superimposing

To specify the shape of a magnet, we use an approach called constructive solid geometry [24]. Firstly, the primitive are to be defined: cylinders, cuboids, spheres, etc. Each primitive A corresponds to the indicator function 1_A, defined as

$$1_A(\mathbf{r}) = \begin{cases} 1, & \mathbf{r} \in A \\ 0, & \mathbf{r} \notin A \end{cases} \tag{2}$$

Furthermore, some operators on these function are used in order to obtain complex figures: union, intersection, difference, displacement, scaling and rotation. For example, for intersection it will be $1_{A\cap B}(\mathbf{r}) = 1_A(\mathbf{r}) \times 1_B(\mathbf{r})$, and for displacement $1_{A+\mathbf{r}'}(\mathbf{r}) = 1_A(\mathbf{r} - \mathbf{r}')$. Here are some examples of shapes which can be set with this method (Fig. 1).

A figure is constructed at the initial stage, after which the grid is superimposed on it. For each grid cell it is checked whether it is inside the figure, directly using the constructed function $1_A(\mathbf{r})$. The most plain way in to assign $|\mathbf{m}_i|$ to 0 or 1 as a result of the checking. However, sometimes boundary cells need more accurate approach. We use $|\mathbf{m}_i| \in [0, 1]$, and the concrete value is determined by using a subgrid inside each cell of the grid. We found that $8 \times 8 \times 8$ subgrid is generally enough for the boundary cells to be handled properly. It should be noted that there are approaches that not only adjust $|\mathbf{m}_i|$, but correct $\mathbf{H}_{demag,i}$ too [8].

As for the choice of the grid step Δ, one can follow the rule $\Delta \ll l_{ex}$, where $l_{ex} = \sqrt{\frac{A_{ex}}{2\pi M_{sat}^2}}$ and is called the exchange length [2]. Such choice of the step should be enough at least to resolve domain walls.

Fig. 1. Constructive solid geometry: a dumbbell, a cylinder united with a cuboid, displacement and scaling of a cuboid

3.2 The Numerical Scheme for the Landau-Lifshitz Equation

The Eq. (1) can be written in the following divided form

$$\dot{\mathbf{m}}_i = \mathbf{f}_i(\mathbf{m}_i, \mathbf{H}_i(\mathbf{m}_1, \ldots, \mathbf{m}_N)) + \mathbf{g}_i(\mathbf{m}_i, t), \tag{3}$$

where $\mathbf{f}_i$ contains the precession and the damping, and $\mathbf{g}_i$ contains the stochastic term, and the effective field $\mathbf{H}_i$ depends on all the cells. Let's define two subproblems.

$$\dot{\mathbf{m}}_i = \mathbf{f}_i(\mathbf{m}_i, \mathbf{H}_i(\mathbf{m}_1, \ldots, \mathbf{m}_N)) \equiv -\gamma^* \left[\mathbf{m}_i \times \mathbf{H}_i\right] - \frac{\gamma^* \alpha}{|\mathbf{m}_i|} \left[\mathbf{m}_i \times \left[\mathbf{m}_i \times \mathbf{H}_i\right]\right], \quad (4)$$

$$\dot{\mathbf{m}}_i = \mathbf{g}_i(\mathbf{m}_i, t) \equiv 2\sqrt{\frac{\gamma^* \alpha k_B T}{M_s V_i}} \boldsymbol{\xi}(\mathbf{m}_i, t), \quad (5)$$

It turns out that we can solve these subproblems separately on each time step, using the fractional-step method [17] to approximate the solution of (3) in two stages

$$\widetilde{\mathbf{m}}_i^{n+1} = e^{\mathbf{f}_i \tau} \mathbf{m}_i^n, \qquad \mathbf{m}_i^{n+1} = e^{\mathbf{g}_i \tau} \widetilde{\mathbf{m}}_i^{n+1}. \quad (6)$$

In this context, τ is a time step, $e^{\mathbf{f}_i \tau}$ and $e^{\mathbf{g}_i \tau}$ are corresponding numerical schemes for the subproblems, $\mathbf{m}_i^n$ and $\mathbf{m}_i^{n+1}$ are the values of magnetic moments that differ by one time step τ, and $\widetilde{\mathbf{m}}_i^{n+1}$ is an intermediate value. Now we will describe how each of (4) and (5) equations is solved.

The Scheme for the Precession and Damping. Accounting of thermal fluctuations increases the requirements imposed on the approximation order of the scheme of $e^{\mathbf{f}_i \tau}$, because the $I_{scheme} \ll I_{stoch}$ is to be satisfied in addition to the $I_{scheme} \ll I_{dissip}$, and the first condition is stricter on actual problems, especially at low temperatures [30]. Here we denoted I_{stoch} as a thermal energy source, I_{dissip} as a dissipative energy loss, I_{scheme} as a energy source/loss caused by a numerical error of the scheme.

Therefore, the classical explicit Runge-Kutta method with a fourth-order approximation is used to calculate $\widetilde{\mathbf{m}}_i^{n+1}$ deterministically based on $\mathbf{m}_i^n$ (6)

$$\begin{aligned}
\mathbf{k}_i^1 &= \mathbf{f}_i(\mathbf{m}_i^n, \mathbf{H}_i(\mathbf{m}_1^n, \ldots, \mathbf{m}_N^n)) \\
\mathbf{k}_i^2 &= \mathbf{f}_i(\mathbf{m}_i^n + \frac{\tau}{2}\mathbf{k}_i^1, \mathbf{H}_i(\mathbf{m}_1^n + \frac{\tau}{2}\mathbf{k}_1^1, \ldots, \mathbf{m}_N^n + \frac{\tau}{2}\mathbf{k}_N^1)) \\
\mathbf{k}_i^3 &= \mathbf{f}_i(\mathbf{m}_i^n + \frac{\tau}{2}\mathbf{k}_i^2, \mathbf{H}_i(\mathbf{m}_1^n + \frac{\tau}{2}\mathbf{k}_1^2, \ldots, \mathbf{m}_N^n + \frac{\tau}{2}\mathbf{k}_N^2)) \\
\mathbf{k}_i^4 &= \mathbf{f}_i(\mathbf{m}_i^n + \tau\mathbf{k}_i^3, \mathbf{H}_i(\mathbf{m}_1^n + \tau\mathbf{k}_1^3, \ldots, \mathbf{m}_N^n + \tau\mathbf{k}_N^3)) \\
\widetilde{\mathbf{m}}_i^{n+1} &= \mathbf{m}_i^n + \frac{\tau}{6}\left(\mathbf{k}_i^1 + 2\mathbf{k}_i^2 + 2\mathbf{k}_i^3 + \mathbf{k}_i^4\right).
\end{aligned} \quad (7)$$

The total accumulated error of the method is $O(\tau^4)$. However, the application of the RK4 requires large computational costs, because the (4) problem turns out to be a stiff system of equations due to the presence of the $\mathbf{H}_{exch}$ in (1). If the space step Δ is sufficiently small, the $\mathbf{f}_i(\mathbf{m}_i, \mathbf{H}_i(\mathbf{m}_1, \ldots, \mathbf{m}_N))$ can be very large:

$$\max|\mathbf{H}_i| \geq \max|\mathbf{H}_{exch}| = \max\left|2\frac{A_{ex}}{M_s}\sum_{j:adjacent}\frac{\mathbf{m}_j - \mathbf{m}_i}{\Delta_i^2}\right| = \frac{A_{ex}}{M_s}\frac{12}{\Delta^2} \sim \frac{1}{\Delta^2} \quad (8)$$

It's well known that explicit Runge-Kutta methods has a small domain of absolute stability on such problems, making the suitable τ to be very small too.

Unfortunately, the implicit methods can not be used here because of the large number of cell in the grid and strong connectivity of them. Alternatively, explicit schemes with an adaptive time step can be used, Runge–Kutta–Fehlberg method for example [10]. We plan to implement one of them in the future development of The Fouriest.

The Scheme for the Temperature Source. The approximation of the temperature source (1, 5) may be done by two random rotations [14]. Let $\Xi(\mathbf{o}, \delta)$ be the rotation matrix by an angle of δ about a unit vector $\mathbf{o}$

$$\Xi(\mathbf{o},\delta) = \begin{pmatrix} \left(o_x^2 - o_y^2 - o_z^2\right)\sin^2\frac{\delta}{2} + \cos^2\frac{\delta}{2} & 2\sin\frac{\delta}{2}\left(o_x o_y \sin\frac{\delta}{2} - o_z\cos\frac{\delta}{2}\right) & 2\sin\frac{\delta}{2}\left(o_x o_z \sin\frac{\delta}{2} + o_y\cos\frac{\delta}{2}\right) \\ 2\sin\frac{\delta}{2}\left(o_x o_y \sin\frac{\delta}{2} + o_z\cos\frac{\delta}{2}\right) & \left(o_y^2 - o_x^2 - o_z^2\right)\sin^2\frac{\delta}{2} + \cos^2\frac{\delta}{2} & 2\sin\frac{\delta}{2}\left(o_y o_z \sin\frac{\delta}{2} - o_x\cos\frac{\delta}{2}\right) \\ 2\sin\frac{\delta}{2}\left(o_x o_z \sin\frac{\delta}{2} - o_y\cos\frac{\delta}{2}\right) & 2\sin\frac{\delta}{2}\left(o_y o_z \sin\frac{\delta}{2} + o_x\cos\frac{\delta}{2}\right) & \left(o_z^2 - o_x^2 - o_y^2\right)\sin^2\frac{\delta}{2} + \cos^2\frac{\delta}{2} \end{pmatrix}$$

Then in one time step τ a magnetic moment changes as:

$$\mathbf{m}_i^{n+1} = \Xi\left(\Xi(\widetilde{\mathbf{m}}_i^{n+1}, 2\pi\alpha_i^n)\mathbf{o}_i^n, 2\sqrt{\frac{\gamma^*\alpha k_B T\tau}{M_s V_i}}\beta_i^n\right)\widetilde{\mathbf{m}}_i^{n+1}, \tag{9}$$

where α_i^n is a random number uniformly distributed on $[0, 1)$, independent of the others, β_i^n is a random number normally distributed with mean of 0 and standard deviation of 1, independent of the others, and $\mathbf{o}_i^n$ is a somehow produced nonzero vector satisfying $\mathbf{o}_i^n \perp \widetilde{\mathbf{m}}_i^{n+1}$. There are many ways to obtain a non-zero vector perpendicular to a given one, we use the following:

$$\mathbf{o}(\mathbf{m}) = \begin{cases} (-m_y, m_x, 0), & m_x \neq 0 \text{ or } m_y \neq 0 \\ (1, 0, 0), & \text{otherwise} \end{cases}$$

The (9) formula can easily be understood using Fig. 2 and following explanations. So, we need to rotate the vector $\mathbf{m}$ of the magnetic moment. Since there are two degrees of freedom of $\mathbf{m}$ on its sphere, a rotation plane has to be equiprobably chosen first. We take the starting plane, which is characterized by its normal $\mathbf{o}$, and we take a random number α uniformly distributed in $[0, 1)$. Then we get a random plane by rotating the $\mathbf{o}$ about the $\mathbf{m}$, that is $\Xi(\mathbf{m}, 2\pi\alpha)\mathbf{o}$. There are three of possible resulted planes in the picture. Finally, we rotate $\mathbf{m}$ itself in this fixed plane just obtained, and the angle of this rotation is a centered normally distributed random value $\propto \beta$.

3.3 Demagnetization

The demagnetization field created in the $\mathbf{r}$ by the volume V by definition [26] is

$$\mathbf{H}_{demag}(\mathbf{r}) = -\int_V \frac{(\mathbf{r} - \mathbf{r}')\nabla \cdot \mathbf{M}(\mathbf{r}')}{|\mathbf{r} - \mathbf{r}'|^3} d\tau' + \oint_{\partial V} \frac{(\mathbf{r} - \mathbf{r}')\mathbf{M}(\mathbf{r}') \cdot \mathbf{n}(\mathbf{r}')}{|\mathbf{r} - \mathbf{r}'|^3} ds'$$

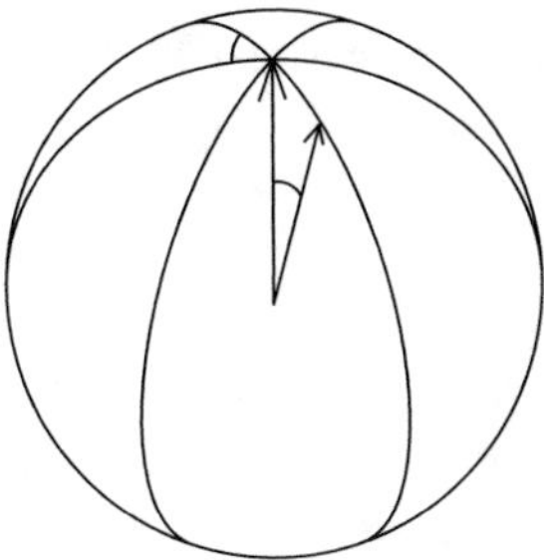

Fig. 2. The rotation of a magnetic moment due to the temperature

Let the ferromagnetic sample volume be $V = \sum_j V_j$, where V_i is a grid cell volume. Then $\mathbf{M}(\mathbf{r}') \equiv const = M_s\mathbf{m}_j$ within a j grid cell and $\nabla \cdot \mathbf{M}(\mathbf{r}') \equiv 0$, so we get

$$\mathbf{H}_{demag}(\mathbf{r}) = M_s \sum_j \oint_{\partial V_j} \frac{(\mathbf{r}-\mathbf{r}')\mathbf{m}_j \cdot \mathbf{n}(\mathbf{r}')}{|\mathbf{r}-\mathbf{r}'|^3} ds'$$

For each i grid cell we have to define what to assume for the $\mathbf{H}_{demag,i}$. Generally, there are two ways to do this. We use the approach in which $\mathbf{H}_{demag,i} = \mathbf{H}_{demag}(\mathbf{r}_i)$ [26], where $\mathbf{r}_i$ is the center of the cell. Alternatively, one can average $\mathbf{H}_{demag}(\mathbf{r})$ within the i cell [3], which is a bit more accurate, but leads to more cumbersome formulae. The former way is sufficient for most tasks, so we settled on it. Now can express the grid cells' demagnetization fields as

$$\mathbf{H}_{demag,i} = M_s \sum_j \mathsf{N}_{ij}\mathbf{m}_j \tag{10}$$

$$N_{ij}^{lm} = \oint_{\partial V_j} \frac{(\mathbf{r}_i - \mathbf{r}')_l n_m}{|\mathbf{r}_i - \mathbf{r}'|^3} ds' \tag{11}$$

These N_{ij} are called demagnetization tensors and can be expressed in closed form as analytical functions of i, j and grid step size. Note that N_{ij} don't change over time since they depend only on the grid and therefore can be calculated only once in advance.

So, the naive approach is to directly use (10) at each stage of each time step of numerical scheme in order to get $\mathbf{H}_{demag,i}$. Unfortunately, this is very expensive to compute. If the number of grid cells is R, then on each time step $\mathbf{H}_{demag}$ implemented naively takes $O(R^2)$ floating point operations, while other fields in the effective field $\mathbf{H}_i$ in (1) take only $O(R)$ operations since they are local.

However, there are some ways to compute $\mathbf{H}_{demag}$ much faster. One can use either the fast multipole method [26] or the fast Fourier transform with help of the convolution theorem [3]. We decided to use the latter, because it's rather easy to implement the algorithm using various FFT libraries, and FFT parallelism level is high, which is important when implementing on GPU. Using either of these two methods, computation complexity drops to $O(R \log R)$. This is still the most costly field to compute though.

FFT and Zero-Padding. We will simplify the case from 3-D to 1-D to illustrate the method of the convolution and zero-padding, since vectors and tensors become scalar values in 1-D. So, firstly let's pay attention to the fact that $N_{ij} = N_{i-j}$ due to translation symmetry (11), so in 1-D we can rewrite (10) as

$$H_{demag,i} = \sum_{j=0}^{R-1} N_{i-j} M_j, \quad i = 0 \ldots R-1 \tag{12}$$

Now M and N are sequences of length R and $2R-1$ accordingly. If two sequences a and b are given, then their convolution can be obtained. There are two kind of discrete convolutions: linear one and circular one.

Linear convolution:

$$a = \ldots, 0, 0, a_0, a_1, \ldots, a_{N-2}, a_{N-1}, 0, 0, \ldots$$

$$b = \ldots, 0, 0, b_0, b_1, \ldots, b_{M-2}, b_{M-1}, 0, 0, \ldots$$

$$s(n) = a * b = \sum_{m=0}^{n} a(m) b(n-m), \quad n = 0 \ldots N+M-2$$

Circular convolution:

$$a = \ldots, a_{N-2}, a_{N-1}, a_0, a_1, \ldots, a_{N-2}, a_{N-1}, a_0, a_1, \ldots$$

$$b = \ldots, b_{N-2}, b_{N-1}, b_0, b_1, \ldots, b_{N-2}, b_{N-1}, b_0, b_1, \ldots$$

$$s(n) = a * b = \sum_{m=0}^{N-1} a(m) b(n-m), \quad n = 0 \ldots N-1$$

Also, there is the convolution theorem, claiming that

$$a * b = iDFT(DFT(a) \times DFT(b)), \tag{13}$$

where DFT is the direct discrete Fourier transform, $iDFT$ is an inverse one, $(\times)$ is a pointwise multiplication, and $(*)$ is a circular convolution.

The key is to not calculate convolutions directly, but instead use the theorem (13), and one of the fast Fourier transform algorithms to compute DFT and $iDFT$. Unfortunately for us, $H_{demag,i}$ has a form of linear convolution (12), not a circular one, since $|M| = R \neq |N| = 2R-1$, so we can't directly apply the theorem. However, there is a method called zero-padding [3] of turning linear convolution to circular by adding right amount of zeros. What we call zero-padded sequences are $\tilde{M}$ and $\tilde{N}$, $|\tilde{M}| = |\tilde{N}| = 2R$, where additional elements are zeros.

$\tilde{M} = \dots$	0	M_0	$\dots$	M_{R-1}	0	0	$\dots$	0	M_0	$\dots$
$\tilde{N} = \dots$	N_{-1}	N_0	$\dots$	N_{R-1}	0	N_{-R+1}	$\dots$	N_{-1}	N_0	$\dots$

It's easy to prove that $H_{demag,i} = \sum_{j=0}^{R-1} N_{i-j} M_j =$ (first R elements of $\tilde{M} * \tilde{N}$), where $(*)$ is a circular convolution. Now we can apply the convolution theorem and finally obtain

$$H_{demag,i} = \text{first } R \text{ elements of } (iDFT(DFT(\tilde{M}) \times DFT(\tilde{N}))) \tag{14}$$

This is how the demagnetization is calculated. Since the fast Fourier transform requires $O(R \log R)$ operations, and $(\times)$ do only $O(R)$, total complexity of computing of H_{demag} is $O(R \log R)$ too. Returning to the 3-D case, DFTs become multi-dimensional [9], and each of three dimensions need to be zero-padded, but generally the method remains the same. It should be noted that in this case the padding increases the memory consumption of the grid by 8 times, though this is not a real trouble, because on modern GPUs tens of millions of grid cells can be processed, which is enough for most tasks.

4 Implementation

We have developed a software named The Fouriest to implement all the ideas described above. As for internal design, it consists of two parts. The program computational core (the back-end) is written in C++ using CUDA Toolkit [20], and has to be run on Nvidia GPU. Python language is used as the scripting language (the front-end), which gives an easy and flexible interface to the computational core. The bindings between C++ and Python are generated automatically via SWIG (Simplified Wrapper and Interface Generator).

Post-processing and visualization are important parts of micromagnetic modeling. In our program, magnetic moments, fields and energies can be dumped 'in raw' in various formats, both textual and binary. We support VTK (Visualization ToolKit) format, files in which can be visualized using Paraview [11] (e.g. Fig. 7) or others. Additionally, we have our own high-performance viewer called MagView with its binary format (e.g. Fig. 1). And the data can be dumped in a simple text form for Gnuplot [28] or Matplotlib [13] (e.g. Fig. 8). Post-processing of the data is also available, it includes a distribution function over an ensemble (e.g. Sect. 5.4), various statistics (e.g. Sect. 6.1), and more. New formats as well as new post-processing routines can be added via Python or C++ modules.

4.1 Script Examples

Here is the Python script for calculating μMAG Standard Problem 4 (Sect. 5.3). Firstly, the processing object is initialized with the necessary parameters, the other parameters are assumed to have default values. Then two dumping objects are created for saving the average magnetization and the total energy later on.

After that the relaxation function is started, and only the energy dumping object is passed to it, and since the desired saving time step is not specified, only the initial and final values of energy will be dumped on the disk. Then the elapsed time is reset, the external field is set, and the main stage of the numerical experiment begins, in which the magnetic system evolves for 1 ns, and every 10^{-3} ns the average magnetization and the energy are dumped on the disk.

```
proc = init(
    M_sat = 800.0,
    magnetization = float3(1, 0.1, 0),
    alpha = 0.02,
    A_ex = 1.3e-6,
    figure = Cuboid(500e-7, 125e-7, 3e-7),
    cells = int3(128, 32, 1),
    dir_name = 'autotest/problem4'
)

table_magn = table(proc, 'magnetization')
table_energy = table(proc, 'energy')

relax(proc, dumps = [table_energy])
proc.time = 0
proc.field_ext = float3(-246.0, 43.0, 0.0)
run(proc, time = 1.0, time_dump = 1e-3, dumps = [table_magn, table_energy])
```

Calculation of an ensemble of magnetic systems also can be done quite neatly. Here is an example of Write-Error-Rate estimating (Sect. 6.1) on 1000 systems. Again, the processing object is initialized, then the special dumping object is created for calculating and saving the value of WER. After that, the whole ensemble of magnetic systems evolves for 20 ns and every 10^{-2} ns the WER is computed and dumped on the disk.

```
proc = init(
    gyro = 2e-2,
    alpha = 0.01,
    M_sat = 1100,
    magnetization = float3(0, 0, 1),
    parallel = 0.4,
    temp = 300,
    A_ex = 2e-6,
    anis = float3(0, 0, 8e6),
    current_dens = float3(0, 0, -2.7),
    figure = Cylinder(40e-7, 1e-7),
    systems = 1000,
    d_time = 1e-4,
)

wer = WER(proc)
run(proc, time = 20.0, time_dump = 1e-2, dumps = [wer])
```

4.2 CUDA Batches and Performance

As we have already noted in Sect. 3.3, the calculation of the demagnetization is the most computationally expensive, and therefore must be done scrupulously. To implement (14), we use in-place 3D FFT from cuFFT library [21], which has low memory consumption and high performance. But, let's check this performance. If we have just calculated some *steps* of the Runge-Kutta-4 for a single magnetic system, some *cells* in size, and it took some *seconds*, then we can define the performance of the calculation as

$$perf = \frac{cells \times steps}{seconds}$$

Figure 3 shows this performance as a function from the grid size. Let's not pay attention to the 'fouriest, full load' line for a while, and look at the 'fouriest, single system' and 'mumax, single system' lines instead. The difference between these two lines will be discussed later, now it's important what do they have in common. The whole range can be divided into the region of small grids and the region of large grids. In the first one, the performance is rather small, and it grows with the growth of the number of cells. But eventually the saturation begins, the performance becomes constant, that means we reach the region of large systems.

So, can we somehow fix the first region? If we have only one magnetic system to be processed, the answer is no. The problem is that modern GPUs have a very high level of parallelism, and rather small FFT tasks just can't consume their capacities completely. But if there are many different data sets, on which the same FFT is to be done, then the answer is yes. This technology is called CUDA Batches. At a low level, we combine separate calculations into a large one, and immediately get a big increase of performance. And an ensemble of magnetic systems is exactly where CUDA Batches can be applied. It should be noted that the batching must be done in the code and cuFFT library calls, and the ensemble data must be placed in a single continuous piece of memory. So one can't just run several instances of a program 'in parallel' and hope that this will increase the total performance. This is the exact reason why we developed our own program The Fouriest that supports computations of an ensemble using CUDA Batches. Moreover, systems in the ensemble being processed can differ from each other in their shape and physical parameters, since this has no effect on the calculation of demagnetization. In addition, systems can even interact in various ways. The only limitation is that each magnetic system must have its own grid, and these grids must have the same dimensions and space steps.

Returning to the performance testing and Fig. 3, for an ensemble of magnetic systems we define the performance as

$$perf = \frac{systems \times cells \times steps}{seconds}$$

Let's calculate an ensemble of magnetic systems, the total number of cells in which is big enough to reach the values of the large systems region. It's clearly to see that 'fouriest, full load' performance remains almost constant across the entire range of grids, because now the power of the GPU is properly occupied. The smaller the grid size, the greater is the gain in performance as compared to single system computation or a computation without batching. For example, if the grid size is 1000 cells, then it's 10 times more effective to compute a large ensemble of magnetic systems than a single one, but we need to take a least 10 systems in the ensemble to get this gain. And when grids become large enough that even one of them can fully occupy the power of a GPU, the advantages of using CUDA Batches come to naught.

As for the difference between MuMax3 and The Fouriest in the single system mode, we believe that the matter is in a optimization mentioned in [27]. They do not use the 3D FFT from library, but do a set of 1D FFT instead, while removing redundant ones over zeros, which arise in the case of multidimensional zero-padding. This may be really effective for large grids, but on smaller ones, on the contrary, the performance decreases, apparently due to some internal optimization of 3D FFT in the library. Since the small grids are generally the main goal of our program, we consider the results (Fig. 3) to be quite good. It's interesting to note that the more new and powerful is a GPU, the bigger will be the region of small grids, which is in favor of our code. Also there is a tendency to miniaturize the spintronics technical process, which gives a more efficient modeling of an ensemble of such systems.

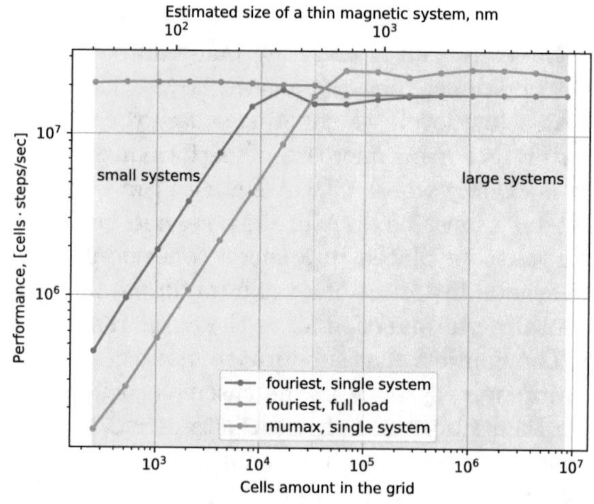

Fig. 3. Performance testing: GeForce GTX 1070; Cuda Toolkit v.8; The Fouriest v. from 2018-02-24; MuMax3 v. 3.10 (compiled from sources) in FixDt mode

5 Verification

Physical modeling programs verification is a very important task, since on real applications it's often not possible to check the correctness of the output data.

The Fouriest has been tested on a variety of tasks, including a few of μMAG standard problems [1] (Sects. 5.2 and 5.3). The most important tests will be described below.

5.1 Demagnetization Field on the Axis of Symmetry of a Cylinder

This test checks if the demagnetization field is correctly calculated at $t = 0$ for some uniform initial magnetizations. Consider a cylinder of radius R and thickness of 2δ placed in the center of a Cartesian grid.

Firstly, let's suppose that the cylinder has a uniform magnetization along its axis of symmetry, so $\mathbf{M} = (0, 0, M_z)$. The demagnetization field H_z on that axis can be expressed in closed form [29] as a function of z. Also we can specify initial magnetization as $\mathbf{M} = (M_x, 0, 0)$, and obtain H_x on the same axis then. These formulae will be

$$H_z(z) = -M_z\left(1 + \frac{1}{2}\frac{z-\delta}{\sqrt{R^2+(z-\delta)^2}} - \frac{1}{2}\frac{z+\delta}{\sqrt{R^2+(z+\delta)^2}}\right),$$

$$H_x(z) = -\frac{M_x}{4}\left(\frac{\delta-z}{\sqrt{R^2+(\delta-z)^2}} + \frac{\delta+z}{\sqrt{R^2+(\delta+z)^2}}\right)$$

Note that $H_z(z)$ and $H_x(z)$ are not the components of the same vector, since the initial conditions are different for these functions.

We have checked whether our computations are the same as the analytical results and they are (Fig. 4), if the grid is fine enough. Additionally, we have found out that the dependence of the error $|\Delta H| = \max\limits_z |\Delta H(z)| \sim d^2$, where d is a grid cell size (Fig. 5).

5.2 μMAG Standard Problem 2

This problem includes both demagnetization and exchange interactions. No other effects are assumed. Consider a cuboid of thickness t, width d and length L. We fix its proportions by setting $\frac{t}{d} = 0.1$, $\frac{L}{d} = 5$, and d is a only one parameter to be scaled (Fig. 6). If initial uniform magnetization is given, remanent magnetization can be calculated by doing a relaxation of the magnet for a sufficiently long time. Let's define the exchange length as

$$l_{ex} = \sqrt{\frac{A_{ex}}{2\pi M_{sat}^2}}$$

An interesting fact is that if the width is much greater than the exchange length, then the remanent magnetization of the magnet is not uniform, but has

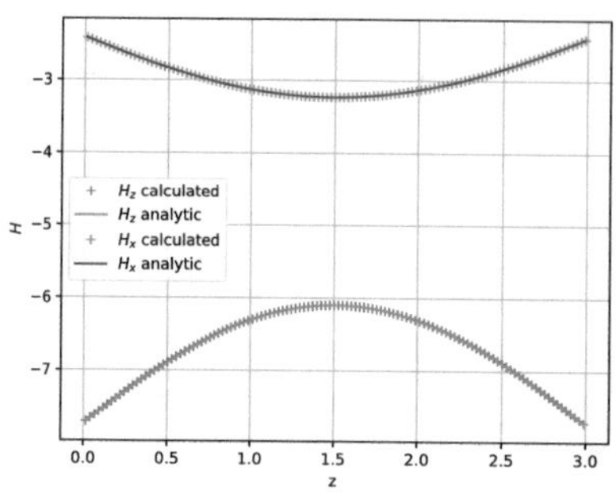

Fig. 4. $\mathbf{H}_{demag}$ on the cylinder axis

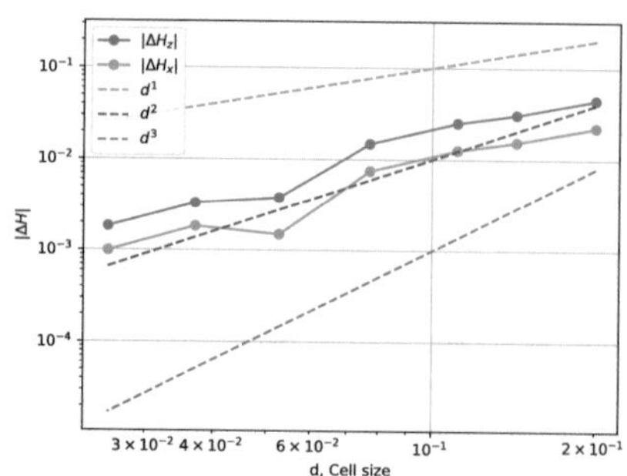

Fig. 5. Convergence of the $\mathbf{H}_{demag}$ when decreasing grid step

a form of so-called s-state (Fig. 7). So we run a set of relaxations where d ranged from l_{ex} to $30l_{ex}$, and plot a dependence of $\langle m_x \rangle$ and $\langle m_y \rangle$ on d/l_{ex} (Fig. 8). The coincidence with the MuMax3 results MuMax3 on the same problem can be considered good enough, although in our results the remanent $\langle m_y \rangle$ at the beginning of the d/l_{ex} range is not truly equal to zero, and in MuMax3 results it is. This may depend on the choice of the initial magnetization, and on the exact method of relaxation, since we don't know exactly how they did the experiment, we have only the output data of their remanent magnetization.

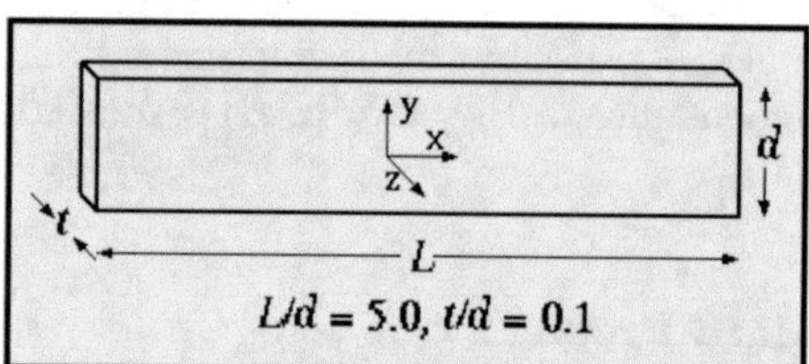

Fig. 6. Problem 2 geometry

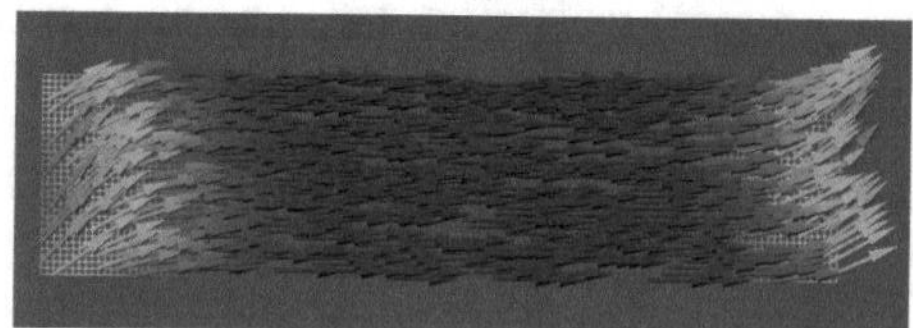

Fig. 7. S-state

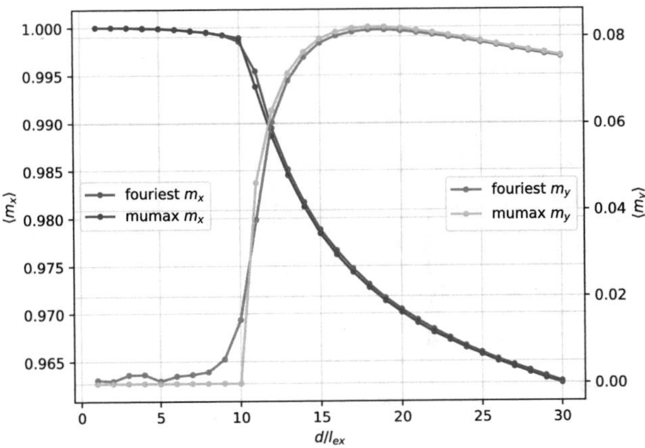

Fig. 8. Relaxation of a set of cuboids

5.3 μMAG Standard Problem 4

This test in a sense continues the previous one, but adds dynamic aspects to be tested. Now we have only one cuboid of exact geometry, it is $t = 3$ nm, $L = 500$ nm, $d = 125$ nm (Fig. 9). The simulation contains two stages. At the first one, the relaxation is going on, and the s-state (Fig. 7) is eventually obtained. At the second one, the external field $\mathbf{H}_{ext} = (-246.0, 43.0, 0.0)$ is applied, and the dynamics of the average magnetization is looked on for 1 ns. The material parameters are the following: $A_{ex} = 1.3 \times 10^{-6}$ erg/cm, $M_{sat} = 800.0$ emu/cm^3, $\alpha = 0.02$, $\gamma = 0.0176$ (Oe $\times$ ns)$^{-1}$. The results of magnetization evolution were compared with the MuMax3 ones (Fig. 10) and had found to be well matched.

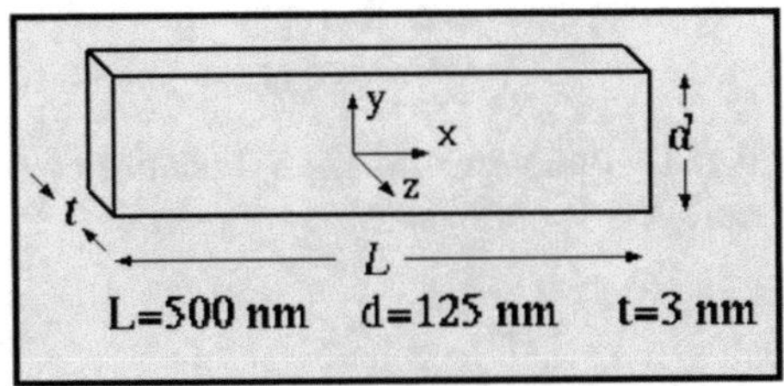

Fig. 9. Problem 4 geometry

5.4 Boltzmann Distribution of an Ensemble of Magnetic Moments

The goal of this test was checking correctness of the temperature source and its statistical properties. The main idea is to run a large set of identical magnetic systems for a time about the relaxation time ($t_{relax} \gg (\gamma^* \alpha |\mathbf{H}|_{max})^{-1}$), then there should be a Boltzmann distribution $\propto \exp\left(\frac{-E}{E_T}\right)$ for magnetic moments, where $|\mathbf{H}|_{max}$ is the absolute maximum of an effective field.

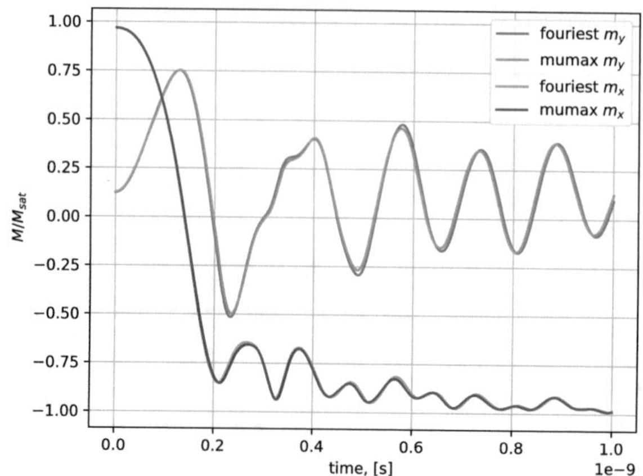

Fig. 10. The dynamics of a cuboid

Let there be a cylindrical magnet at a temperature of T. We assume that the magnet can be described by one grid cell, so there is the macrospin case. The magnet has a strong anisotropy in comparison with the temperature ($K \gg \frac{k_B T}{V}$), the axis of which coincides with the cylinder's axis and with the z-axis of a Cartesian coordinate system, $\mathbf{K} = (0, 0, K)$. Furthermore, there are no external influences ($\mathbf{H}_{STT} = \mathbf{H}_{ext} = 0$). Due to the fact that there is only one cell, there is no exchange interaction to be considered. In addition, the magnet has a demagnetization which can be described by the symmetric tensor N. However, it will be shown that in this case the demagnetization can be included in the effective anisotropy.

$$\mathsf{N} = \begin{pmatrix} N^{xx} & N^{xy} & N^{xz} \\ N^{xy} & N^{yy} & N^{yz} \\ N^{xz} & N^{yz} & N^{zz} \end{pmatrix}$$

Analyzing the integral (11), one can find that, because of a few symmetries, in our case $N^{xx} = N^{yy}$ and $N^{xy} = N^{xz} = N^{yz} = 0$. So,

$$\mathsf{N} = \begin{pmatrix} N^{xx} & 0 & 0 \\ 0 & N^{xx} & 0 \\ 0 & 0 & N^{zz} \end{pmatrix}$$

Actually, these N^{xx} and N^{zz} can be expressed by the closed expressions depending on the radius and the height of the cylinder [29], but they won't be needed.

In our case, the total energy is

$$E = \cancelto{0}{E_{ext}} + E_{anis} + \cancelto{0}{E_{exch}} + E_{demag} = -\frac{(\mathbf{m}\cdot\mathbf{K})^2}{M_s K} + \frac{M_s}{2}\mathbf{m}^T\mathbf{N}\mathbf{m}$$
$$= -\frac{K m_z^2}{M_s} + \frac{M_s}{2}\left(N^{xx} m_x^2 + N^{xx} m_y^2 + N^{zz} m_z^2\right).$$

So as far as $|\mathbf{m}| \equiv 1$,

$$E(\mathbf{m}) = -\frac{K m_z^2}{M_s} + \frac{M_s}{2}\left(N^{xx}(1-m_z^2) + N^{zz} m_z^2\right) = -\left(\frac{K}{M_s} + \frac{M_s}{2}\left(N^{xx} - N^{zz}\right)\right) m_z^2 + const$$

The effective field $\mathbf{H} = -\frac{\partial E}{\partial \mathbf{m}}$ doesn't change because of adding an arbitrary constant to the energy, so finally we get

$$\dot{\mathbf{m}} = -\gamma^* \left[\mathbf{m} \times \mathbf{H}\right] - \gamma^* \alpha \left[\mathbf{m} \times \left[\mathbf{m} \times \mathbf{H}\right]\right] + 2\sqrt{\frac{\gamma^* \alpha k_B T}{M_s V_i}}\boldsymbol{\xi}(\mathbf{m}_i, t), \qquad \mathbf{H} = \frac{2K_{eff}}{M_s}(\mathbf{m}\cdot\mathbf{n}_{K_{eff}})\mathbf{n}_{K_{eff}} \tag{15}$$

where K_{eff} is the effective anisotropy $K + \frac{M_s^2}{2}\left(N^{xx} - N^{zz}\right)$.

Now let's consider a large set of magnetic moments satisfying (15) with the initial value of $\mathbf{m} = (0,0,1)^T$. Then, at times of $t_{relax} \gg (\gamma^* \alpha |\mathbf{H}|_{max})^{-1}$, there should be the Boltzmann distribution for the amount of the moments dN fitting a solid angle of $d\Omega$.

$$dN(\Omega) \propto \exp\left(\frac{-E_{anis}}{E_T}\right) d\Omega = \exp\left(\frac{K_{eff} m_z^2}{M_s}\frac{M_s V}{k_B T}\right) d\Omega = \exp\left(\frac{K_{eff} V \cos^2\theta}{k_B T}\right) \sin\theta d\theta d\varphi$$

Paying attention to the symmetry with respect to a rotation by an azimuthal angle of φ ($\mathbf{H}$ depends only on a polar angle of $\theta = \arccos(m_z)$), distribution function turns out to be

$$f(\theta) = \frac{\exp\left(\frac{K_{eff} V \cos^2\theta}{k_B T}\right)\sin\theta}{\int_0^{\frac{\pi}{2}} \exp\left(\frac{K_{eff} V \cos^2\theta}{k_B T}\right)\sin\theta d\theta}$$

The interval $[0, \frac{\pi}{2}]$ is used as the limits due to the fact that the moments are localized in only one potential well near the initial value $\mathbf{m} = (0,0,1)^T$ because of the strong anisotropy ($K_{eff} \gg \frac{k_B T}{V}$). The second potential well near $\mathbf{m} = (0,0,-1)^T$ is unreachable.

The function $f(\theta)$ and the experimental distribution of the ensemble can be seen in Fig. 11 on a logarithmic scale. Also, there is a curve of the potential energy $E_{anis}(\theta) = -\frac{K_{eff}\cos^2\theta}{M_s}$ on a linear scale.

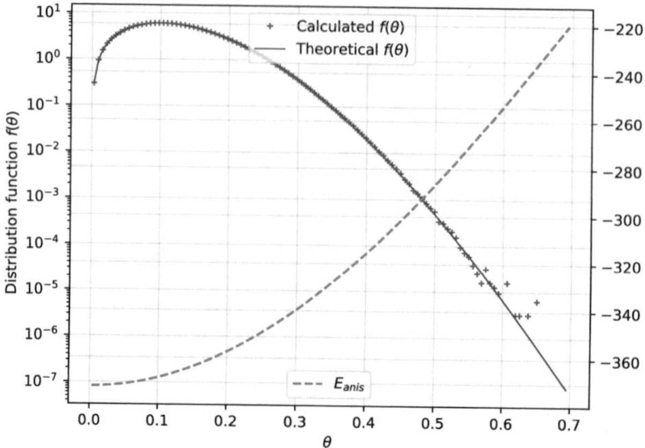

Fig. 11. The localization of the magnetic moments in the potential well under the strong anisotropy. The amount of the magnetic moments is 5×10^7 and $\frac{K_{eff}V}{k_B T} = 50.0$ and $\frac{K_{eff}}{M_s} = 372.0$ erg/emu

5.5 Deterministic Switching of a Magnetic Moment by Spin-Polarized Current

This test checks the dynamic correctness of the spin-polarized current. Consider a magnetic moment provided with a vertical anisotropy $\mathbf{n}_K = (0, 0, 1)$, exposed by a spin-polarized current with $\mathbf{m}_{ref} = (0, 0, -1)$. It's assumed that there are no other fields. In our case, the Landau-Lifshitz Equation (1) will simplify to

$$\dot{\mathbf{m}} = -\gamma^* [\mathbf{m} \times \mathbf{H}] - \gamma^* \alpha [\mathbf{m} \times [\mathbf{m} \times \mathbf{H}]], \qquad \mathbf{H} = \mathbf{H}_{STT} = \frac{\hbar \eta_{\|} j}{2eM_s d} [\mathbf{m} \times \mathbf{m}_{ref}],$$

We can transform it to the following form without excess vector products, using the fact that $\mathbf{m} \times \mathbf{m} \times \mathbf{m} \times \mathbf{m}_{ref} = -\mathbf{m} \times \mathbf{m}_{ref}$

$$\dot{\mathbf{m}} = -\gamma^* [\mathbf{m} \times (-\mathbf{a}_J)] - \gamma^* \alpha [\mathbf{m} \times [\mathbf{m} \times \mathbf{b}_J]], \qquad \mathbf{a}_J = \frac{\alpha \hbar \eta_{\|} j}{2eM_s d} \mathbf{m}_{ref}, \quad \mathbf{b}_J = \frac{\hbar \eta_{\|} j}{2e\alpha M_s d} \mathbf{m}_{ref} \tag{16}$$

Now we will pick a dimensionless unit system for our quantities and equations in order to simplify the subsequent calculations. First, we assume that the magnetic field unit is $\Delta_H = H_{anis}$. For the current unit, it's natural to choose that $\Delta j = \frac{j \Delta H}{|b_J|}$, because this vector $\mathbf{b}_J$ is responsible for the dissipation which we will be looking into. As for the unit of time, we will take inverse precession frequency in a unit field, so $\Delta_\tau = \frac{1}{\gamma^* \Delta_H}$.

The Eq. (16) splits into two independent ones in spherical coordinates for the azimuthal angle φ and polar angle θ. For the θ in our new units it will be

$$\dot{\theta} = \alpha \gamma^* \big(b_J - \Delta_H \cos\theta\big) \sin\theta.$$

Let the initial magnetization be deflected by an angle $\theta_0 \ll 1$ from the vertical axis. We will look for the time necessary for the magnetic moment to reach a value of $\frac{\pi}{2}$, which we call the switching time.

$$\tau = \int_{\theta_0}^{\pi/2} \frac{d\theta}{\alpha\gamma^*\big(b_J - \Delta_H \cos\theta\big)\sin\theta}$$

One can check that there is the following expression for the integral we need

$$\int \frac{A\,dx}{\sin x(a + b\cos x)} = \frac{A}{a^2 - b^2}\left\{a \ln\tan\frac{x}{2} + b \ln\frac{a + b\cos x}{\sin x}\right\}$$

So, now we finally obtain

$$\tau = \int_{\theta_0}^{\pi/2} \frac{d\theta}{\alpha\gamma^*(b_J - \Delta_H\cos\theta)\sin\theta} = \frac{1}{\alpha\gamma^*(b_J^2 - 4K^2)}\left\{-b_J \ln\tan\frac{\theta_0}{2} - \Delta_H \ln b_J + \Delta_H \ln\frac{b_J - \Delta_H\cos\theta_0}{\sin\theta_0}\right\} \approx$$

$$\frac{1}{\alpha\gamma^*(b_J^2 - \Delta_H^2)}\left\{-b_J \ln\frac{\theta_0}{2} + \Delta_H \ln\frac{b_J - \Delta_H}{b_J\theta_0}\right\} = \frac{\Delta_\tau}{\alpha(j^{*2} - 1)}\left\{-j^* \ln\frac{\theta_0}{2} + \ln\frac{j^* - 1}{j^*\theta_0}\right\}, \quad j^* = j/\Delta_j$$

Figure 12 shows the dependence of the inverse switching time on the electric current density. As one might expect, the stronger is the current, the faster is the switching. The parameters are the following: $\mathbf{H}_{anis} = 6000$ Oe, $M_s = 1000$ emu/cm^3, $d = 2$ nm, $R = 12.5$ nm, $\gamma = 0.02$, $\mathbf{m}_0 = (0.01, 0, 1)$, $\alpha = 0.005$ and $\eta_{\|} = 0.5$. The simulation results coincide with the analytical formula completely.

6 Applications

As already mentioned, there are a number of tasks where it can be useful to quickly process a lot of magnetic systems, including

- Write-Error-Rates in spin-torque transfer (STT) magnetic memory (MRAM) devices (Sect. 6.1) [15]
- Modeling of MTJ-based magnetic sensor devices [12]. Basically, this is an array of identical ferromagnets located far enough from each other. Therefore, a reasonable approximation may be not to calculate the entire array on one large grid, but use an individual grid for each magnet instead. In this approximation, the interaction between individual sensors is absent or done in a simplified manner like for multipoles, therefore these sensors can be effectively processed via The Fouriest.
- Parametric computations and phase diagrams. Often one needs to test a range of one or more system parameters, like size or shape, checking each configuration for suitability in some way. If the grids are small enough, then the use of our program definitely will speed up such calculations.
- Nudged Elastic Band for potential barriers modeling [4].

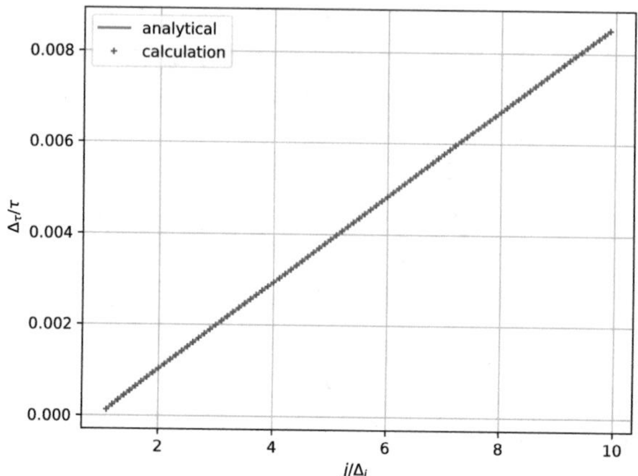

Fig. 12. Dependence of inverse switching time on the electric current density

6.1 Write-Error-Rate

Consider a flat magnetic cylinder with out-of-plane anisotropy which has a spin-polarized current applied. We can treat this as an elementary MRAM cell, where the uniform magnetization $\mathbf{m} = (0, 0, 1)$ is identified with '1', $\mathbf{m} = (0, 0, -1)$ is with '0' [15]. The electric current is what causes the MRAM cell to switch from '1' to '0' and vice versa (Fig. 13). Unfortunately, if we fix the time of current application, then there is a probability that the cell won't have enough time to switch, because the process is non-deterministic if $T > 0$. This probability of non-switching is called a Write-Error-Rate.

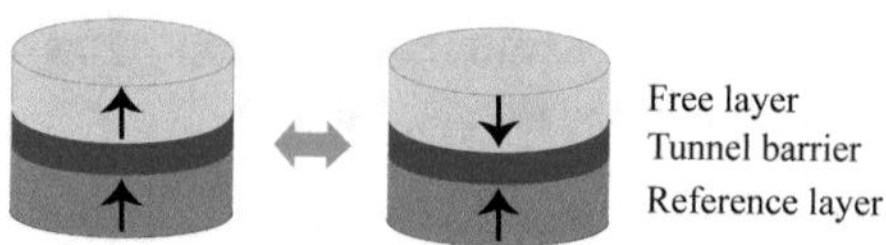

Fig. 13. Spin-torque transfer of MRAM

To calculate this value, we need to simulate a large ensemble of magnetic systems, then count the proportion of those that did not switch. The macrospin approach is widely used for that purpose, where the entire magnetic system corresponds to a single magnetic vector $\mathbf{m}$. However, the resulting value of WER is overestimated if compared to the one obtained by micromagnetic modeling [25]. This is a good opportunity for the application of The Fouriest, as it was designed for such calculations of an ensemble of small micromagnetic systems.

The parameters are selected as follows $\mathbf{K} = (0, 0, K), K = 8 \times 10^6$ erg/cm^3, $\alpha = 0.01$, $\gamma = 0.02$ (Oe $\times$ ns)$^{-1}$, $\mathbf{M}_{sat} = 1100$ emu/cm^3, $T = 300$ K, $A_{ex} =$

2×10^{-6} erg/cm, $\eta_{\|} = 0.4$. Let's introduce the critical current density j_0 via thermal stability factor Δ as

$$j_0 = \frac{4\alpha e \Delta k_B T}{\eta_{\|} \hbar} \frac{d}{V}, \qquad \Delta = \frac{\mathbf{K}_{eff} V}{k_B T}, \qquad K_{eff} = K_z - \frac{M_s^2}{2}(N_{zz} - N_{xx})$$

As in [25], we have taken two cylinders of different sizes, and for each of them plot the time dependence of Write-Error-Rate for two values of j normalized by the critical current density j_0 (Figs. 14 and 15). One can see that the larger is the size of a magnetic system, the greater is the difference between macrospin and micromagnetic approaches of WER calculation. Increasing of the current density reduces the WER, as one would expect.

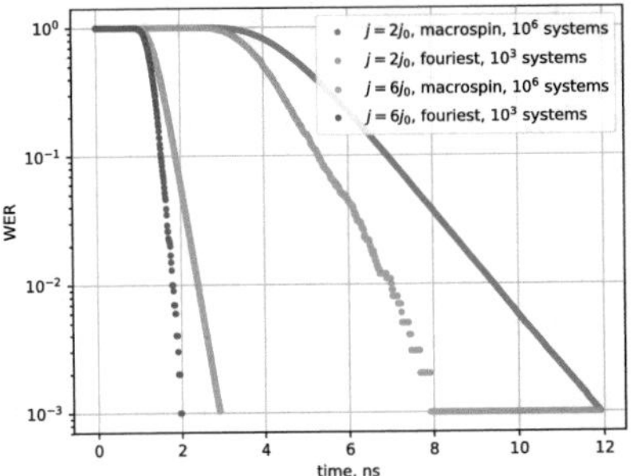

Fig. 14. Cylinder of $R = 20$ nm, $d = 1$ nm

Fig. 15. Cylinder of $R = 40$ nm, $d = 1$ nm

7 Conclusion

We have presented a new micromagnetic solver called The Fouriest. This software was developed for modeling of many micromagnetic tasks, including the calculation of an ensemble of magnetic system. The problem statement has been described, including the Landau-Lifshitz equation with the stochastic term, and various effective field terms. A numerical scheme for the equation has been chosen, as well as a method of specifying the magnet shape and a method of the demagnetization field calculation. The description of the software and scripts examples have been given, and the performance has been tested. It turned out that on small magnetic ensembles the total performance is significantly higher than the MuMax3 one on these tasks. The code verification has been done on various physical problems. The Write-Error-Rate for STT MRAM was calculated as the first application of the program.

As for the future work, there are many possible applications for The Fouriest in which it's needed to process a large number of independent or coupled magnetic systems. If we talk about the already mentioned Write-Error-Rate task, then there is still room for research. In Sect. 6.1 the WER is simply calculated as $\frac{\text{not switched}}{\text{all systems}}$. Even with the acceleration provided by our program, it still takes quite a lot of time to model a large enough ensemble for actual WER values. However, there are advanced techniques with which you can estimate the same WER values on much smaller ensembles [22,23] and thus significantly reduce the computational costs. It is also interesting to apply The Fouriest to problems where systems in the ensemble are interconnected. For example, there is a promising Nudge Elastic Band method [4] used to work with potential barriers. The systems are connected by artificial elastic forces, which make them form a certain trajectory in the configuration space.

We also want to highlight the limitation of the study. First of all, a task to be processed should contain multiple magnetic systems in order to obtain a significant performance boost compared to other programs. These systems should also be small enough (Fig. 3), although the proper grids depend on the GPU. The micromagnetic approach also imposes some restrictions on the possible tasks, compared to atomistic modeling. However, we are optimistic about the further development and practical applications of The Fouriest, and we are open to new suggestions.

References

1. MuMAG – Micromagnetic Modeling Activity Group. https://www.ctcms.nist.gov/~rdm/mumag.org.html. Accessed 23 July 2018
2. Abo, G.S., Hong, Y.K., Park, J., Lee, J., Lee, W., Choi, B.C.: Definition of magnetic exchange length. IEEE Trans. Mag. **49**(8), 4937–4939 (2013)
3. Bagnérés, A., Durbiano, S.: 3D computation of the demagnetizing field in a magnetic material of arbitrary shape. Comput. Phys. Commun. **130**(1–2), 54–74 (2000)
4. Bessarab, P.F., Uzdin, V.M., Jónsson, H.: Method for finding mechanism and activation energy of magnetic transitions, applied to skyrmion and antivortex annihilation. Comput. Phys. Commun. **196**, 335–347 (2015)
5. Chang, R., Li, S., Lubarda, M., Livshitz, B., Lomakin, V.: Fastmag: fast micromagnetic simulator for complex magnetic structures. J. Appl. Phys. **109**(7), 07D358 (2011)
6. Donahue, M.J.: A variational approach to exchange energy calculations in micromagnetics. J. Appl. Phys. **83**(11), 6491–6493 (1998)
7. Donahue, M.J.: OOMMF user's guide, version 1.0. Technical report (1999)
8. Donahue, M.J., McMichael, R.D.: Micromagnetics on curved geometries using rectangular cells: error correction and analysis. IEEE Trans. Magn. **43**(6), 2878–2880 (2007)
9. Dudgeon, D.E., Mersereau, R.M.: Multidimensional Digital Signal Processing Prentice-Hall Signal Processing Series. Prentice-Hall, Englewood Cliffs (1984)
10. Fehlberg, E.: Low-order classical Runge-Kutta formulas with stepsize control and their application to some heat transfer problems (1969)
11. Henderson, A., Ahrens, J., Law, C., et al.: The ParaView Guide, vol. 366. Kitware Clifton Park, Clifton Park (2004)
12. Huai, Y.: Spin-transfer torque MRAM (STT-MRAM): challenges and prospects. AAPPS Bull. **18**(6), 33–40 (2008)

13. Hunter, J.D.: Matplotlib: A 2D graphics environment. Comput. Sci. Eng. **9**(3), 90–95 (2007)
14. Ivanov, A.V.: Kinetic modeling of magnetic's dynamic. Matematicheskoe Modelirovanie **19**(10), 89–104 (2007)
15. Khvalkovskiy, A., Apalkov, D., Watts, S., Chepulskii, R., Beach, R., Ong, A., Tang, X., Driskill-Smith, A., Butler, W., Visscher, P., et al.: Basic principles of STT-MRAM cell operation in memory arrays. J. Phys. D: Appl. Phys. **46**(7), 074,001 (2013)
16. Landau, L.D., Lifshitz, E.: On the theory of the dispersion of magnetic permeability in ferromagnetic bodies. Phys. Z. Sowjet. **8**, 153 (1935). http://cds.cern.ch/record/437299
17. LeVeque, R.J.: Finite Volume Methods for Hyperbolic Problems, vol. 31. Cambridge University Press, Cambridge (2002)
18. Lopez-Diaz, L., Aurelio, D., Torres, L., Martinez, E., Hernandez-Lopez, M., Gomez, J., Alejos, O., Carpentieri, M., Finocchio, G., Consolo, G.: Micromagnetic simulations using graphics processing units. J. Phys. D: Appl. Phys. **45**(32), 323,001 (2012)
19. Miltat, J.E., Donahue, M.J.: Numerical micromagnetics: finite difference methods. Handb. Magn. Adv. Magn. Mater. **2**, 14–15 (2007)
20. NVIDIA Corporation: CUDA C Programming Guide, version 9.2.148 edn (2018)
21. NVIDIA Corporation: CUFFT LIBRARY USER'S GUIDE, version 9.2.148 edn (2018)
22. Pramanik, T., Roy, U., Jadaun, P., Register, L.F., Banerjee, S.K.: Write error rates of in-plane spin-transfer-torque random access memory calculated from rare-event enhanced micromagnetic simulations. J. Magn. Magn. Mater. **467**, 96–107 (2018)
23. Pramanik, T., et al.: Shape-engineered ferromagnets and micromagnetic simulation techniques for spin-transfer-torque random access memory. Ph.D. thesis (2018)
24. Requicha, A.A., Voelcker, H.B.: Constructive solid geometry (1977)
25. Roy, U., Kencke, D.L., Pramanik, T., Register, L.F., Banerjee, S.K.: Write error rate in spin-transfer-torque random access memory including micromagnetic effects. In: 2015 73rd Annual Device Research Conference (DRC), pp. 147–148. IEEE (2015)
26. Tan, X., Baras, J.S., Krishnaprasad, P.S.: Fast evaluation of demagnetizing field in three-dimensional micromagentics using multipole approximation. In: Smart Structures and Materials 2000: Mathematics and Control in Smart Structures, vol. 3984, pp. 195–202. International Society for Optics and Photonics (2000)
27. Vansteenkiste, A., Van de Wiele, B.: MUMAX: a new high-performance micromagnetic simulation tool. J. Magn. Magn. Mater. **323**(21), 2585–2591 (2011)
28. Williams, T., Kelley, C., Bröker, H., Campbell, J., Cunningham, R., Denholm, D., Elber, E., Fearick, R., Grammes, C., Hart, L.: Gnuplot 4.5: An interactive plotting program 2011 (2017). http://www.gnuplot.info
29. Wysin, G.M.: Demagnetization fields (2012)
30. Zipunova, E.V., Ivanov, A.V.: Selection of an optimal numerical scheme for simulation system of the landau-lifshitz equations considering temperature fluctuations. Matematicheskoe Modelirovanie **26**(2), 33–49 (2014)

Consistency in Multi-device Environments: A Case Study

Luis Martín Sánchez-Adame, Sonia Mendoza(✉), Amilcar Meneses Viveros, and José Rodríguez

Department of Computer Science, CINVESTAV-IPN, Av. IPN 2508, Col. San Pedro Zacatenco, Del. Gustavo A. Madero, 07360 Mexico City, Mexico
luismartin.sanchez@cinvestav.mx,
{smendoza,ameneses,rodriguez}@cs.cinvestav.mx

Abstract. The development of interactive environments has led to the creation of useful tools for controlling them. However, developers should be careful when considering the control level left to users, since the applications may become unusable. Whether the system or user decides the distribution of the available devices, Graphical User Interface (GUI) consistency must always be maintained. Consistency not only provides users with a stable framework in similar contexts but is an essential learning element and a lever to ensure the GUI efficient usage. Nevertheless, maintaining GUI consistency in an interactive environment is an open issue. Therefore, this paper proposes a set of consistency guidelines that serve as a tool for the construction of multi-device applications. As a case study, Spotify was evaluated by experts who identified consistency violations and assessed their severity.

Keywords: Design guidelines · GUI consistency · User eXperience · Interactive environments · Multi-device applications

1 Introduction

At present we are surrounded by devices, the most common are smartphones, tablets, and laptops. However, more and more devices are being added to create a truly interactive environment: sensors, cameras, microphones, smart watches, and screens are some examples that we can find in the most unexpected places. Today the number of devices connected to the Internet surpasses 7 billion [3]. The omnipresence of these devices, especially the mobile ones, is progressively changing the way people perceive, experience, and interact with products and each other [12]. This transition poses a significant challenge since we have to design User eXperiences (UX) according to each device.

Within this context, a topic of particular interest is when a single application can be executed on multiple devices. The ability to seamlessly connect multiple devices of varying screen size and capabilities has always been an integral part of the vision for distributed UX and Ubiquitous Computing [5]. That is, the

K. Arai et al. (Eds.): CompCom 2019, AISC 997, pp. 232–242, 2019.
https://doi.org/10.1007/978-3-030-22871-2_17

same application has to provide a similar UX regardless of the device or its environment. One way to preserve UX is to have a consistent application.

Consistency states that presentation and prompts should share as much as possible common features and refer to a common task, including using the same terminology across different inputs and outputs [26]. Several studies have shown that consistency is a crucial factor for multi-device experience, but they have also argued that it is a challenge for developers, since maintaining consistency of a multi-device system is an open problem [19,22,25,27]. Consistency is important because it reduces the learning curve and helps eliminate confusion, in addition to reducing production costs [8,20,31].

Microsoft is an excellent case to exemplify the importance of consistency. Windows 10 and Office (in its most recent versions) are two of the most important products of the company; It is notorious that both GUIs are a design statement since they follow the same layout. In both software products, we can see that their toolbars have a similar design, i.e., the grouping, positioning, and labelling of buttons and commands is identical. This is intended to allow users to focus on their productivity, without the need to learn a new tool panel for each software they use. For this reason, Microsoft developed a series of tools, including an API and design guidelines that are integrated into a framework called Ribbon [16], so that this design discourse propagates to all applications developed by third parties.

In this way, we can talk about three approaches to the design of applications which, although authors like Coutaz and Calvary [4] and Vanderdonckt [30] studied years ago, Levin [12] summarises them in her 3C framework:

- **Consistent design approach:** Each device acts as a solo player, creating the entire experience on its own.
- **Continuous design approach:** Multiple devices handle different pieces sequentially, advancing the user toward a common goal.
- **Complementary design approach:** Multiple devices play together as an ensemble to create the experience.

This framework presents a series of challenges, since it involves, among other things, the fragmentation of the GUI and business logic. Thus the task of the developers is to preserve a positive UX among all the devices.

By adding consistency elements to the design of multi-device environments, usability is improved, and the possibility of a scenario with negative UX is reduced [2,6]. The primary goal of our work is to propose a set of design guidelines that serve as a model in the creation of consistent applications. These guidelines are depicted through a case study: Spotify [29]. This application has been evaluated by five UX experts, who have identified a list of consistency violations and have assessed the severity of each one.

This paper is organised as follows. First, in Sect. 2, related work is studied. Section 3 describes the research methodology that we use. Then, in Sects. 4 and 5, we respectively define and implement our set of design guidelines. Finally, in Sect. 6, we discuss the achieved work and provide some ideas for future work.

2 Related Work

This section describes some of the investigations carried out in the field of multi-device UX. They serve to support the importance and necessity of our proposal.

After having interviewed 29 professionals in the area of interactive environments, Dong et al. [5] identified three key challenges that have prevented designers and developers from building usable multi-device systems: (1) the difficulty in designing interactions between devices, (2) the complexity of adapting GUIs to different platform standards, and (3) the lack of tools and methods for testing multi-device UX.

Marcus [14] was a pioneer in the description of good practices to develop GUIs. He claims that the *organisation*, *economisation*, and *communication* principles help GUI design. The highlights are his four elements of consistency: (1) *internal* (applying the same rules for all elements within the GUI), (2) *external* (following existing conventions), (3) *real-world* (following real-world experience), and (4) *no-consistency* (when to deviating from the norm).

Meskens et al. [15] presented a set of techniques to design and manage GUIs for multiple devices integrated into *Jelly*, a single multi-device GUI design environment. *Jelly* allows designers to copy widgets from one device design canvas to another, while preserving the consistency of their content across devices using linked editing.

O'Leary et al. [21] argue that designers of multi-device UX need tools to better address situated contexts of use, early in their design process through ideation and reflection. To address this need, they created and tested a reusable design kit that contains scenarios, cards, and a framework for understanding tradeoffs of multi-device innovations in realistic contexts of use.

Woodrow [32] defines and contextualises three critical concepts for usability in multi-device systems: (1) *composition* (distribution of functionality), (2) *consistency* (what elements should be consistent across which aspects), and (3) *continuity* (a clear indication of switching interactions). He makes a call for more active involvement by both the systems engineering and engineering management communities in advancing methods and approaches for interusability (interactions spanning multiple devices with different capabilities).

All these works show that the highly interactive environments formed by multi-device applications are a promising field with many issues to explore. However, they are also examples of a knowledge gap that we try to fill with our guidelines for consistency.

3 Research Methodology

The research methodology for the development of our design guidelines is based on the Design Science Research Methodology (DSRM) process model proposed by Peffers et al. [24] (see Fig. 1). We chose this methodology because its popularity in the state of the art, and it has proved to be useful in similar problems [10,11,23].

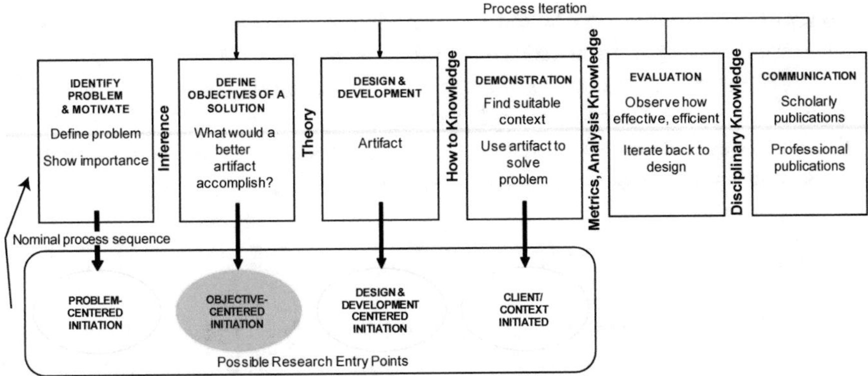

Fig. 1. We started from an *Objective-Centred Initiation* (coloured in orange) in the DSRM process model [24].

An *Objective-Centred Initiation* has been chosen as a research entry point because our goal is to improve the design of multi-device applications. As for *Identification & Motivation*, we have already described the importance of highly interactive environments, and the role of GUI consistency in that matter. The *Objective of a Solution*, the second step of the process, is to develop a set of design guidelines that helps developers to create consistent applications to improve UX. The third step *Design & Development* is the description of our guidelines for GUI consistency (see Sect. 4). *Demonstration* and *Evaluation* are described in our case study (see Sect. 5). This is the first iteration of the process. Subsequent iterations will begin in the *Design & Development* stage, in order to improve the proposed guidelines.

4 Consistency Guidelines

With the review of various works in the state of the art, and taking into account the challenges discovered and common characteristics of each one, we present our five design guidelines to maintain consistency in multi-device systems:

- **Honesty:** Interaction widgets have to do what they say and behave as expected. An honest GUI has the purpose of reinforcing the user's decision to use the system. When the widgets are confusing, misleading, or even suspicious, users' confidence will begin to wane.
- **Functional Cores:** These are indivisible sets of widgets. The elements that constitute a Functional Core form a semantic field, thus out of their field they lose meaning. The granularity level of interaction for a Functional Core depends on the utility of a particular set of widgets.
- **Multimodality:** Capability of multi-device systems to use different means of interaction whenever the execution context changes. In general, it is desirable that regardless of the input and output modalities, the user can achieve the same result.

- **Usability Limitations:** When multimodality scenarios exist, it is possible that situations of limited usability could be reached. When the interaction environment changes and its context is transformed, the environment can restrict the user's interaction with the system.
- **Traceability:** Denotes the situation in which users can observe and, in some cases, modify the evolution of the GUI over time.

5 Case Study: Spotify

In this section, we describe a case study to exemplify the proposed guidelines. We focus on Spotify [29], which was chosen as a case study because it is a well known commercial application, and many users around the world use it. We construct the analysis based on the works by Andrade et al. [1], Grice et al. [7], and Schmettow et al. [28].

Spotify is a cross-platform application for playing music via streaming. It allows users to play individual songs as well as playback by artist, album, or playlists created by other Spotify users. Data is streamed from both servers and a peer-to-peer network. There are clients for Mac OS and Windows along with several smartphone platforms and other devices, like video games consoles and Internet-connected speakers.

The GUI is similar to those found in other desktop music players. Spotify offers the possibility of listening to music on devices that users have linked to their account. This is achieved through a list of devices that can appear on the desktop or mobile clients. Thus, the user can play songs from one device and control the playback from a different one [9].

The evaluation has been worked out with the help of five UX experts. We chose the experts for their experience applying usability tests, and because they are familiar with the topics of our research. All the experts are university professors and have doctorates; two of them belong to our university. Their experience comes from both work in industry and from research centres. It should be noted that none is related to this work in addition to their participation in the evaluation.

Before starting the evaluation, we gathered and explained to the experts each of our design guidelines, their purpose, and discussed some examples so that everyone had a similar starting point. Each expert drafted a list of problems and violations of the guidelines that we propose. Once the evaluators have identified potential consistency problems, the individual lists have been consolidated into a single master list. The master list was then given back to the evaluators who independently have assessed the severity of each violation. The ratings from the individual evaluators are then averaged, and we present the results in Table 1. For the rating, we adapted the severity classification proposed by Zhang et al. [33]:

0 - Not a consistency problem at all.
1 - Cosmetic problem only. No need to be fixed unless extra time is available.

2 - Minor consistency problem. Fixing this, should be given a low priority.
3 - Major consistency problem. Important to fix, should be given a high priority.
4 - Consistency catastrophe. Imperative to fix this before the product can be released.

Table 1. Consistency problems and its rating in Spotify

Place	Problem	Guidelines[†]	Severity
Devices list	For a device to appear in the list, it has to be unlocked and with the client in the foreground	H, T	1.8
	Only the available devices are displayed, if there is none, the list cannot be displayed	H, T	1
	In the PC application, there is no settings option, just the icon next to the volume. The user has to click on it to find other devices on their network	H, T, M	2.4
	If a user wants to see a history of the devices with access to their account, to consult or withdraw the permission, they have to do it in the web version, there are no alternatives	H, M	1.6
	Sometimes, devices are not shown if they are not on the same WiFi network, in some cases they do. However, it is specified that all elements of the interactive space have to be on the same network	H, M, U	2.6
Login	To associate a device with an account, the user has to log in, so in devices such as televisions where the keyboard is not as intuitive as in a PC, this interaction can become cumbersome	F, U	1.6
Native clients	Not a problem *per se*; being all the GUIs native clients, many consistency problems are reduced; the development of each client carries a cost. Also, the user has a rather closed environment	M	0
Devices	For some devices (e.g., speakers) there is not a syncing process to manually add them. It either connects or does not	M, T, U	3.6
	Some devices need a dongle to be compatible with Spotify, thus the user must be sure that their system is powered up and turned to the correct input before listening to any music	M, T, U	3.4
	Devices with no GUI (e.g., speakers) will need a mobile device as a remote control	M, T	1

[†] Honesty (H), Functional Cores (F), Multimodality (M), Usability Limitations (U), Traceability (T).

A total of 10 problems were detected, and guidelines were violated 23 times. *Traceability* and *Multimodality* were the two most frequently violated guidelines, six and seven times, respectively (see Fig. 2). In contrast, the guidelines with less detected problems were *Usability Limitations* and *Functional Cores* with four and one violations respectively.

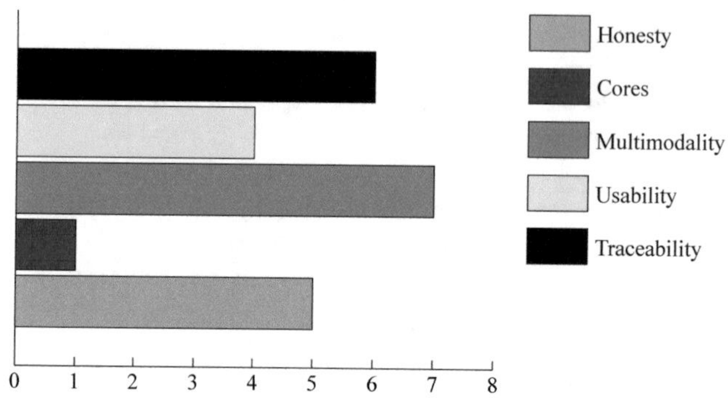

Fig. 2. Consistency violations in Spotify

Concerning the severity of the problems detected, we can see in Fig. 3 that severity level 2 - "Minor consistency problem" was the most frequent with 30%, closely followed by severity level 1 - "Cosmetic problem only" with 24% of occurrence. On the contrary, we can notice that the severest classification 4 - "Consistency catastrophe" just got 10%.

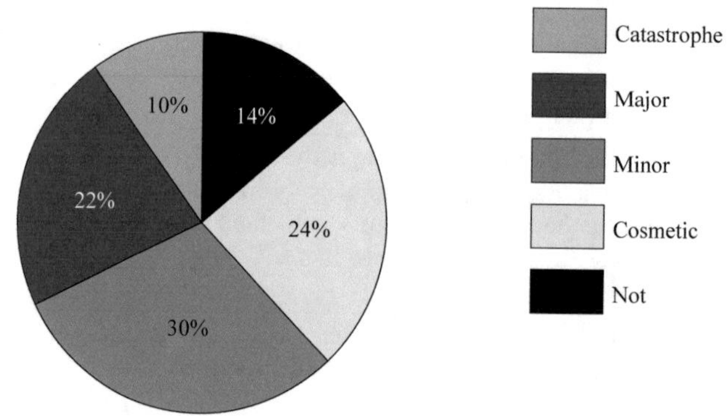

Fig. 3. Severity rating of consistency problems found in Spotify

In general, we can say that Spotify got positive evaluations because the most severe classifications were few. The evaluation was also fruitful, as several problems could be discussed, as well as scenarios that, if neglected, could cause conflicts in the future.

Something that we could emphasise is that the guideline of *Functional Cores* was only violated once. This remark could tell us that the widgets were well designed and that they fulfil their function, homogenously, through the devices. Contrarily, *Multimodality* was the guideline that more violations accumulated; this did not represent a surprise, because the more devices an application encompasses, the harder it will be to replicate the functionalities in each one.

The experts concurred that the design guidelines could be a useful tool to detect consistency problems. However, they also acknowledged that in order to be more effective, they have to be refined and detailed.

6 Conclusions

The main contribution of this paper is a series of consistency guidelines for the design of multi-device applications. This kind of applications represent the challenge of configuring the available resources and their role in the environment. When the users control the application, it allows them to explore their environment, identify the tasks and services compatible with itself, and combine independent resources in a significant manner, in order to perform tasks and interact with services. Consistency is the element that maintains the users in a stable base since it is the key to assist GUI distribution. Besides, it is an essential factor in maintaining a positive UX.

We chose this type of evaluation to be able to explore our proposal broadly. We knew that by working with experts in the area, we could get feedback on our work, benefit from their experience, and directly observe how other people use our design guidelines.

While it is true that an expert evaluation can give good results, it is not exempt from problems. For example, we recognise that we have few points of view since we only have the participation of five experts, which can lead to misleading results. However, we considered that this was the best way to carry out an exploratory study, since with a small group we can have more in-depth discussions and work for a longer time. In this sense, the results seem promising, because we realised how the experts interpreted the guidelines. In general, our expectations were fulfilled, but we also know that we have to refine and be more specific so that each guide is not too ambiguous.

It is also possible that expert evaluators can identify many consistency problems in multi-device applications without relying on our guidelines. However, using them provides evaluators with a structure that helps them take into account each major design dimension in turn, and to prevent them from becoming distracted by other design aspects, which could cause them to miss essential consistency problems.

As the nature of the guidelines is empirical, to prove their validity it is necessary to perform experiments, such as heuristic evaluations and UX tests with end-users.

Some ideas for future work include developing more consistency guidelines and validate them with prototypes and developers. To support our ideas, we plan to conduct a comparative study between our design guidelines and another set of guidelines in a similar domain. In this way, we could compare the number of problems found in each case. Also, we have the intention of enriching our work through revisions of GUI design patterns [13,17,18]. Finally, we expect that our consistency guidelines to continue evolving in the future as we gain more experience and insights from using them to evaluate other applications, such as those found in the Internet of Things domain.

Acknowledgment. We thank CONACyT (Consejo Nacional de Ciencia y Tecnología) for funding Luis Martín Sánchez Adame's doctoral fellowship. Scholarship number: 294598.

References

1. Andrade, F.O., Nascimento, L.N., Wood, G.A., Calil, S.J.: Applying heuristic evaluation on medical devices user manuals. In: Jaffray, D.A. (ed.) World Congress on Medical Physics and Biomedical Engineering, Toronto, Canada, 7–12 June 2015, pp. 1515–1518. Springer International Publishing, Cham (2015)
2. Anić, I.: The importance of visual consistency in UI design (2018). https://www.uxpassion.com/blog/the-importance-of-visual-consistency-in-ui-design/. Accessed Oct 2018
3. CiscoSystems: Cisco visual networking index: global mobile data traffic forecast update, 2016–2021, Technical report (2017). https://www.cisco.com/c/en/us/solutions/collateral/service-provider/visual-networking-index-vni/mobile-white-paper-c11-520862.html
4. Coutaz, J., Calvary, G.: HCI and software engineering: the case for user interface plasticity. In: Jacko, J.A. (ed.) The Human-Computer Interaction Handbook: Fundamentals, Evolving Technologies, and Emerging Applications-Human Factors and Ergonomics Series, chap. 56, pp. 1107–1118. CRC Press (2008)
5. Dong, T., Churchill, E.F., Nichols, J.: Understanding the challenges of designing and developing multi-device experiences. In: Proceedings of the 2016 ACM Conference on Designing Interactive Systems, DIS 2016, pp. 62–72. ACM, Brisbane (2016)
6. Gaffney, G.: Why consistency is critical (2018). https://www.sitepoint.com/why-consistency-is-critical/. Accessed Oct 2018
7. Grice, R.A., Bennett, A.G., Fernheimer, J.W., Geisler, C., Krull, R., Lutzky, R.A., Rolph, M.G., Search, P., Zappen, J.P.: Heuristics for broader assessment of effectiveness and usability in technology-mediated technical communication. Tech. Commun. **60**(1), 3–27 (2013)
8. Grosjean, J.C.: Design d'interface et critere ergonomique 9: Coherence (2018). http://www.qualitystreet.fr/2011/01/23/design-dinterface-et-critere-ergonomique-9-coherence/. Accessed Oct 2018
9. Kreitz, G., Niemela, F.: Spotify – large scale, low latency, P2P music-on-demand streaming. In: 2010 IEEE Tenth International Conference on Peer-to-Peer Computing (P2P), pp. 1–10 (2010). https://doi.org/10.1109/P2P.2010.5569963
10. Lamprecht, J., Siemon, D., Robra-Bissantz, S.: Cooperation isn't just about doing the same thing - using personality for a cooperation-recommender-system in online social networks. In: Yuizono, T., Ogata, H., Hoppe, U., Vassileva, J. (eds.) Collaboration and Technology, pp. 131–138. Springer, Cham (2016)

11. Laubis, K., Konstantinov, M., Simko, V., Gröschel, A., Weinhardt, C.: Enabling crowdsensing-based road condition monitoring service by intermediary. Electronic Markets (2018)
12. Levin, M.: Designing Multi-device Experiences: An Ecosystem Approach to Creating User Experiences Across Devices. O'Reilly, Newton (2014)
13. Luna, H., Mendoza, R., Vargas, M., Munoz, J., Alvarez, F.J., Rodriguez, L.C.: Using design patterns as usability heuristics for mobile groupware systems. IEEE Lat. Am. Trans. **13**(12), 4004–4010 (2015). https://doi.org/10.1109/TLA.2015.7404939
14. Marcus, A.: Principles of effective visual communication for graphical user interface design. In: Baecker, R.M., Grudin, J., Buxton, W.A., Greenberg, S. (eds.) Readings in Human-Computer Interaction, Interactive Technologies, pp. 425–441. Morgan Kaufmann (1995). https://doi.org/10.1016/B978-0-08-051574-8.50044-3. http://www.sciencedirect.com/science/article/pii/B9780080515748500443
15. Meskens, J., Luyten, K., Coninx, K.: Jelly: a multi-device design environment for managing consistency across devices. In: Proceedings of the International Conference on Advanced Visual Interfaces, AVI 2010, pp. 289–296. ACM, Roma (2010). https://doi.org/10.1145/1842993.1843044
16. Microsoft: Windows ribbon framework (2018). https://docs.microsoft.com/en-us/windows/desktop/windowsribbon/-uiplat-windowsribbon-entry. Accessed Oct 2018
17. Neil, T.: Mobile Design Pattern Gallery: UI Patterns for Smartphone Apps. O'Reilly, Newton (2014)
18. Nguyen, T.D., Vanderdonckt, J., Seffah, A.: Generative patterns for designing multiple user interfaces. In: Proceedings of the International Conference on Mobile Software Engineering and Systems, Texas, MOBILESoft 2016, pp. 151–159. ACM, Austin (2016). https://doi.org/10.1145/2897073.2897084
19. Nichols, J.: Automatically generating high-quality user interfaces for appliances (2006)
20. Nikolov, A.: Design principle: consistency (2017). https://uxdesign.cc/design-principle-consistency-6b0cf7e7339f. Accessed Oct 2018
21. O'Leary, K., Dong, T., Haines, J.K., Gilbert, M., Churchill, E.F., Nichols, J.: The moving context kit: designing for context shifts in multi-device experiences. In: Proceedings of the 2017 Conference on Designing Interactive Systems, DIS 2017, pp. 309–320. ACM, New York (2017). https://doi.org/10.1145/3064663.3064768
22. de Oliveira, R., da Rocha, H.V.: Consistency priorities for multi-device design. In: Baranauskas, C., Palanque, P., Abascal, J., Barbosa, S.D.J. (eds.) Human-Computer Interaction - INTERACT 2007, pp. 426–429. Springer, Heidelberg (2007)
23. Patrício, L., de Pinho, N.F., Teixeira, J.G., Fisk, R.P.: Service design for value networks: enabling value cocreation interactions in healthcare. Serv. Sci. **10**(1), 76–97 (2018)
24. Peffers, K., Tuunanen, T., Rothenberger, M., Chatterjee, S.: A design science research methodology for information systems research. J. Manag. Inf. Syst. **24**(3), 45–77 (2007)
25. Pyla, P.S., Tungare, M., Pérez-Quinones, M.: Multiple user interfaces: why consistency is not everything, and seamless task migration is key. In: Proceedings of the CHI 2006 Workshop on the Many Faces of Consistency in Cross-Platform Design (2006)

26. Reeves, L.M., Lai, J., Larson, J.A., Oviatt, S., Balaji, T.S., Buisine, S., Collings, P., Cohen, P., Kraal, B., Martin, J.C., McTear, M., Raman, T., Stanney, K.M., Su, H., Wang, Q.Y.: Guidelines for multimodal user interface design. Commun. ACM **47**(1), 57–59 (2004). https://doi.org/10.1145/962081.962106
27. Rowland, C., Goodman, E., Charlier, M., Light, A., Lui, A.: Designing Connected Products: UX for the Consumer Internet of Things. O'Reilly, Newton (2015)
28. Schmettow, M., Schnittker, R., Schraagen, J.M.: An extended protocol for usability validation of medical devices. J. Biomed. Inform. **69**(8), 99–114 (2017). https://doi.org/10.1016/j.jbi.2017.03.010
29. Spotify (2018). https://www.spotify.com/about-us/. Accessed Oct 2018
30. Vanderdonckt, J.: Distributed user interfaces: how to distribute user interface elements across users, platforms, and environments. AIPO, pp. 3–14 (2010)
31. Wong, E.: Principle of consistency and standards in user interface design (2018). https://www.interaction-design.org/literature/article/principle-of-consistency-and-standards-in-user-interface-design. Accessed Oct 2018
32. Woodrow, W.W.: Designing for interusability: methodological recommendations for the systems engineer gleaned through an exploration of the connected fitness technologies space. INSIGHT **19**(3), 75–77 (2016). https://doi.org/10.1002/inst.12115. https://onlinelibrary.wiley.com/doi/abs/10.1002/inst.12115. https://onlinelibrary.wiley.com/doi/pdf/10.1002/inst.12115
33. Zhang, J., Johnson, T.R., Patel, V.L., Paige, D.L., Kubose, T.: Using usability heuristics to evaluate patient safety of medical devices. J. Biomed. Inform. **36**(1), 23–30 (2003)

OS Scheduling Algorithms for Memory Intensive Workloads in Multi-socket Multi-core Servers

Murthy Durbhakula(✉)

Indian Institute of Technology Hyderabad, Hyderabad, India
cs15resch11013@iith.ac.in,
murthy.durbhakula@gmail.com

Abstract. Major chip manufacturers have all introduced multicore microprocessors. Multi-socket systems built from these processors are used for running various server applications. Depending on the application that is run on the system, remote memory accesses can impact overall performance. This paper presents a new operating system (OS) scheduling optimization to reduce the impact of such remote memory accesses. By observing the pattern of local and remote DRAM accesses for every thread in each scheduling quantum and applying different algorithms, we come up with a new schedule of threads for the next quantum. This new schedule potentially cuts down remote DRAM accesses for the next scheduling quantum and improves overall performance. We present three such new algorithms of varying complexity followed by an algorithm which is an adaptation of Hungarian algorithm. We used three different synthetic workloads to evaluate the algorithm. We also performed sensitivity analysis with respect to varying DRAM latency. We show that these algorithms can cut down DRAM access latency by up to 55% depending on the algorithm used. The benefit gained from the algorithms is dependent upon their complexity. In general higher the complexity higher is the benefit. Hungarian algorithm results in an optimal solution. We find that two out of four algorithms provide a good trade-off between performance and complexity for the workloads we studied.

Keywords: High performance computing · Algorithms · Multiprocessor systems · Performance · OS scheduling

1 Introduction

Many commercial server applications today run on cache coherent NUMA (ccNUMA) based multi-socket multi-core servers. Depending on the application, DRAM accesses can impact overall performance; particularly remote DRAM accesses. These are inherent to the application. One way to ameliorate this problem is to rewrite the application. Another way is to observe remote DRAM access patterns at run-time and adapt the OS scheduler to minimize the impact of these accesses. In this paper, we

K. Arai et al. (Eds.): CompCom 2019, AISC 997, pp. 243–253, 2019.
https://doi.org/10.1007/978-3-030-22871-2_18

present such an OS scheduling optimization which observes local and remote DRAM accesses for each scheduling quantum and then applies one of the four different algorithms to decide where to schedule each thread for the next scheduling quantum. The main idea is to schedule a thread with most remote accesses coming from, say Node N, to that node N. For example, in a four node system, if thread T0 is currently scheduled on Node N0 and we observe that there are only 10 DRAM accesses to node N0, 10 to Node N1, 10 to Node N2, and 1000 to Node N3, then we schedule thread T0 to node N3 in the next scheduling quantum. To the best of our knowledge, none of the current commercial operating systems optimize their scheduler for remote DRAM accesses. We show that some of our scheduling algorithms can cut down DRAM accesses by up to 55% depending on the workload and DRAM latencies. In this study we assume each socket/node has 4 cores and each core can run one thread of a parallel application at-a-time. The rest of the paper is organized as follows: Sect. 2 discusses the scheduling algorithms. Section 3 describes the methodology I used in evaluating the scheduling algorithms. Section 4 presents results. Section 5 describes related work and Sect. 6 presents conclusions.

2 Scheduling Algorithms

Various scheduling algorithms can help reduce remote memory accesses. We consider four different algorithms in this paper. Three of them are greedy algorithms and fourth one is based on the Hungarian algorithm. For all the algorithms we assume dedicated hardware performance counter support to count local and remote memory accesses.

For every thread there are as many counters as the number of nodes in the system. In each scheduling quantum for every read request to DRAM we keep track of which thread made the request and whether the access is local DRAM access or remote access and increment corresponding counter. At the end of the scheduling quantum OS reads all the performance counters and applies one of the four scheduling algorithms and makes scheduling decision for the next quantum. We describe these algorithms in the next sub-section. While describing the algorithms we assume a four-node system with each node capable of running four different threads on four cores. Hence the system can run a total of 16 threads.

2.1 Algorithm 1

In this algorithm, we first sort all local and remote memory access counts for each thread in monotonically decreasing order. We start from top and assign the thread with highest access count to that node. Say thread T0 has highest access count of 10000 to node 2 then we assign T0 to node 2. Then we go to next element and assign the next thread to its corresponding node. Once we assign a thread to a particular node we cannot assign it to another node. So in the above example once T0 is assigned to node

2, it cannot be assigned to any other node. Further once we reach a maximum of four threads assigned to a particular node then we no longer assign any more threads to that node and simply skip assigning more threads to that node. This is for load balancing reasons. The complexity of the algorithm is of order O(NLlog(NL)) where N is total number of threads and L is total number of nodes. This is because the sorting algorithm dominates the complexity and we use merge-sort. There are specialized algorithms such as counting sort which could further reduce complexity; however, their application is data dependent.

Input: Threads T0,…TN with DRAM accesses to nodes N0…NL. Each node has 4 cores and can run one thread per core. Current schedule S_current which has a mapping of N threads to L nodes.

Output: New schedule S_next with new mapping of threads to nodes.

begin

1. Each of N threads have L DRAM access counts; one per node. Sort N*L DRAM accesses in descending order. Complexity is O(NLlog(NL)).
2. Scan DRAM access counts. Start from highest DRAM access count and say it is coming from thread T1 to node N2. Assign T1 to N2 in schedule S_next. Now T1 cannot be assigned to any other node.
3. Repeat step 2 until all threads are assigned a node in S_next. Complexity is O(NL).

end

Overall complexity of the algorithm is O(NLlog(NL)).

Algorithm 1

2.2 Algorithm 2

In this algorithm we first start off with node 0 and we sort all thread accesses to that node in monotonically decreasing order. Then we pick first four threads with highest accesses to node 0. Then we remove those four threads from the list. We then sort thread accesses to node 1 out of 12 remaining threads. We pick top four threads to node 1. We remove these four threads from the list. We sort accesses to node 2 from the remaining 8 threads. We pick top 4 to node 2. Remaining 4 will go to node 3. The complexity of this algorithm is order O(Nlog(N)) assuming sorting is done in parallel on all nodes, where N is total number of threads. Even here it is sorting that dominates the complexity.

Input: Threads T0,…TN with DRAM accesses to nodes N0…NL. Each node has 4 cores and can run one thread per core. Current schedule S_current which has a mapping of N threads to L nodes.

Output: New schedule S_next with new mapping of threads to nodes.

begin

1. for(i=0; i<L;i=(i+1)) begin

   Sort all DRAM accesses from N threads in descending order

   end

   If all nodes do it in parallel. Complexity is O(Nlog(N)).

2. Start with node N0 and pick top 4 threads and assign them to N0 in schedule S_next. Now these threads cannot be assigned to any other node.

3. Repeat step 2 for all nodes until all threads are assigned a node in S_next. Complexity is O(N).

end

Overall complexity of the algorithm is O(Nlog(N)).

Algorithm 2

2.3 Algorithm 3

Unlike previous two algorithms here we form combinations of all four threads and their summation of DRAM accesses to each node. These algorithms are then sorted in monotonically decreasing order. We start by picking the topmost combination of four threads and assign it to corresponding node. We then remove all combinations where any of these threads occur from our list. Then the topmost combination from remaining elements is picked and scheduled on corresponding node. This process is repeated until all threads are scheduled. The complexity of this algorithm is clearly more than the previous two. It is O(L*N©4 log (L*N©4)) where N©4 is N combinatorial 4, N being total number of threads and L being total number of nodes. This is because of the combinations of four threads we form from N threads and for sorting of those combination.

Input: Threads T0,...TN with DRAM accesses to nodes N0...NL. Each node has 4 cores and can run one thread per core. Current schedule S_current which has a mapping of N threads to L nodes.

Output: New schedule S_next with new mapping of threads to nodes.

begin

Compute all combinations of group of 4 threads from N threads.
Complexity is O(N©4) where N©4 is N combinatorial 4.

For each combination of 4 threads compute total DRAM access count for every node. Complexity is O(L*N©4).

Sort the combinations obtained in step 2 in descending order.
Complexity is O(L*N©4log(L*N©4)).

Scan the sorted list created in step 3 starting from top. Assign the group of 4 threads with highest count to corresponding node. Once these threads are assigned they can no longer be assigned to any other node. Keep scanning and assigning threads to nodes until all threads are assigned to nodes in S_next. Complexity if O(L*N©4).

end

Overall complexity is o(L*N©4log(L*N©4))

Algorithm 3

2.4 Algorithm 4

Instead of combinatorial enumerations, we can model the problem as an assignment problem for which a more optimal polynomial time Hungarian algorithm can be applied. The complexity of optimized Hungarian algorithm is known to be O(N^3) which is better than that of Algorithm 3. For each of the N threads there are N places across L nodes where they can be assigned. The Hungarian algorithm finds optimal placement of threads for which overall DRAM access latency (both for local and remote accesses combined) is minimized. I am not showing pseudo-code here as it can be found in many places. Original Hungarian algorithm can be found here [10].

3 Methodology

We implemented each of these algorithms as a stand-alone C++ program and evaluated them by running synthesized data access patterns. These synthesized data access patterns vary the local/remote dram access counts in a known pattern for each scheduling quantum while the counts themselves are generated as a random number in the range 0 to 10000 accesses. The data is then fed to each of the algorithms to see if they can track the pattern. The overall benefit for each algorithm depends on how well they can track the pattern. We chose synthetic workloads in order to clearly bring out benefits of each algorithm and its sensitivity to remote DRAM latency. As part of future work we plan to study the impact of these algorithms on real workloads. Further in this study we do not take into consideration impact of other parameters such as cache-to-cache transfers, cache-affinity, etc. while coming with a new schedule. We purely focus on optimization for remote DRAM latency.

Table 1 lists base configuration used for evaluation of different scheduling algorithms. On top of the base configuration we also perform sensitivity analysis of different algorithms with respect to varying remote DRAM latency from 150 to 200 to 300 cycles. Table 2 lists three synthetic DRAM access patterns used in this paper. A high level system diagram can be seen in Fig. 1. The circled portion represents one socket with 4 cores per socket.

Table 1. Configuration parameters

Parameter	Value
CPU frequency	1 Ghz
Number of cores per socket	4
Number of sockets/nodes	4
Local DRAM latency	100 cycles
Remote DRAM latency	150 cycles
Number of scheduling quanta	16

Table 2. Synthetic workloads

Access pattern	Description
Constant access pattern– Synth1	Same access pattern except for the first quanta. Single optimum grouping for all 16 quanta except for the first one
2-phase access pattern –Synth 2	Two access patterns equally distributed over all quanta except for the first one. Optimal grouping 1 till 9th quanta following by optimal grouping 2 till 16th quanta
4-phase access pattern –Synth 3	Four access patterns equally distributed over all quanta except for the first one. Optimal grouping 1 till 5th quanta followed by optimal grouping 2 till 9th quanta followed by optimal grouping 3 till 13th quanta followed by optimal grouping 4 till 16th quanta

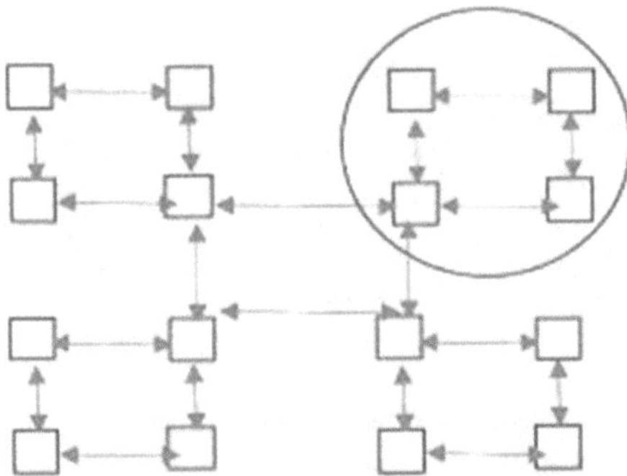

Fig. 1. High-level ccNUMA multi-socket multi-core system diagram with 4 sockets and 4 cores-per-socket

4 Results

All three workloads were run on four algorithms described in Sect. 2. Table 3 shows percentage benefit from each algorithm in terms of number of DRAM cycles saved for all sixteen threads running on 4 sockets for 16 scheduling quanta.

Table 3. Performance comparison of different scheduling algorithms (in % DRAM cycles saved). Remote DRAM latency of 150 cycles.

Workload	Algo 1	Algo 2	Algo 3	Algo 4
Synth1	13	16	25	25
Synth2	13	15	23	23
Synth3	9	11	19	19

As we can see from Table 3, for each algorithm there is a consistent decrease in benefit as we move from workload 1 to 2 to 3. This is expected because as we move from workload 1 to 3 there is more variance in DRAM access patterns. And since we use past behavior as an indication of future all the algorithms would have, in general, greatest benefit for workloads having less variance. Further we also see that Algorithm 3 and Algorithm 4 have highest benefit compared to other algorithms. This is because they are more sophisticated compared to other algorithms. Algorithm 4 gives as much benefit as Algorithm 3 with added benefit that it is less complex and hence more scalable. Depending on the usage of the system one needs to decide which Algorithm to use. Algorithm 2 and Algorithm 4 seems to provide a good trade-off between performance and complexity.

We also performed sensitivity analysis of benefit of these algorithms under varying remote DRAM latency. We varied the latency from 150 to 200 to 300 cycles. Tables 4 and 5 shows results for remote DRAM latency of 200 and 300 cycles, respectively.

Table 4. Performance comparison of different scheduling algorithms (in % DRAM cycles saved). Remote DRAM latency of 200 cycles.

Workload	Algo 1	Algo 2	Algo 3	Algo 4
Synth1	21	25	39	39.3
Synth2	21	24	36	36
Synth3	14	18	31	31

Table 5. Performance comparison of different scheduling algorithms (in % DRAM cycles saved). Remote DRAM latency of 300 cycles.

Workload	Algo 1	Algo 2	Algo 3	Algo 4
Synth1	30	35	55	55.2
Synth2	30	34	50	50
Synth3	20	25	43	43

As we can see from Tables 4 and 5, performance increases consistently in almost all cases as we increase remote DRAM latencies from 150 to 200 to 300 cycles. Similar to the base case we see that Algorithm 3 and Algorithm 4 have highest benefit compared to other algorithms. There is also a consistent decrease in benefit as we move from workload 1 to 2 to 3 due to reasons explained above. Algorithm 2 and Algorithm 4 again seems to provide a good trade-off between performance and complexity.

5 Related Work

Chandra et al. [1] have studied impact of various OS scheduling policies on performance of both uniprocessor and multiprocessor workloads. Our work is similar to theirs in that we also study OS scheduling algorithms for improving the performance of parallel workloads. However our focus is more on memory intensive workloads for ccNUMA multi-socket multi-core servers. We assume that at-a-time only one multi-threaded application is running with each core assigned one thread for the sake of load-balancing. The algorithms we propose are completely different. Their work also evaluated benefits of page migration. However page migration & replication are not always beneficial. For instance say thread T0 is scheduled on Node 1 with 1000 accesses to Node 1's DRAM and also accesses a page P2 on Node 2 DRAM, with intention-to-write, 500 times. At the same time thread T1 is scheduled on Node 3 with 2000 accesses to Node 3's DRAM and accesses same page P2 on Node 2 DRAM, with intention-to-write, 500 times. In such a situation it is not possible to decide to which node to migrate P2 to. In fact if there are such "hot" pages accessed by multiple threads it is actually better to schedule them on the same node where hot page is, rather than migrating the hot page.

Kaseridis et al. [2] proposed a dynamic memory subsystem resource management scheme that considers both cache capacity and memory bandwidth contention in large multi-chip CMP systems. Their memory bandwidth contention algorithms monitor when a particular schedule of threads exceed the maximum bandwidth supported by a node and then try to schedule bandwidth demanding threads with those threads that need little memory bandwidth. Whereas our approach proactively tries to find the best pairing of threads for any scheduling quanta while keeping the overall bandwidth utilization within the maximum bandwidth limits. And the algorithms presented in this paper are different from their algorithms.

Ipek et al. [3] proposed using reinforcement learning based approach to tune DRAM scheduling policies to effectively utilize off-chip DRAM bandwidth. Their work differs from ours in two different ways. First, we are using OS scheduling algorithms to optimize DRAM bandwidth utilization. Second, we use parallel workloads to evaluate the scheduling policy where as they use multiprogramming workloads with no sharing.

Ahn et al. [4] studied the impact of DRAM organization on the performance of data parallel memory systems. In contrast to their work we focus on novel OS scheduling algorithms that will improve DRAM performance of parallel applications on general purpose multiprocessors. Zhu et al. [5] proposed using novel DRAM scheduling algorithms for SMT processors. In contrast to their work this paper proposes using new OS scheduling algorithms with minimal hardware support.

Tang et al. [7] studied the impact of co-locating threads of different multi-threaded applications in a data-center environment on overall performance. They proposed heuristics for co-locating different workloads. Their focus is mostly on data-center related workloads and the algorithms presented in this work are different from their heuristics.

More recently Kim and Huh [11] proposed OS scheduling algorithms for implementing fairness in multi-core systems. They proposed a few algorithms that improve fairness in heterogeneous multi-core systems while improving overall throughput. Their focus is on mainly improving fairness using OS scheduling policies while the focus of this paper is on reducing remote DRAM accesses thereby improving overall performance. Further Kim and Huh use SPEC workloads for the evaluation of their algorithms whereas we focus on parallel workloads.

Sahoo and Dehury [12] proposed job scheduling algorithms for CPU intensive workloads in a health-care cloud environment. Their focus is mainly on health-care applications in a cloud-based environment whereas the focus of this paper is on improving performance of memory intensive parallel workloads.

Lepers et al. [13] studied the impact of asymmetric interconnect on thread and memory placement on NUMA systems. They found that different node interconnectivity can have as much as 2x performance impact under same distribution of thread and data across nodes. Based on this insight they implemented a dynamic thread and memory placement algorithm in Linux that delivers similar or better performance than the best static placement and up to 218% better performance than when the placement is chosen randomly. Unlike Lepers et al. [13] this paper assumes a symmetric interconnect. Further, the focus of this paper is on presenting and evaluating different dynamic OS scheduling algorithms, of varying complexity, that reduce the impact of remote DRAM accesses.

Harris et al. [14] proposed and implemented Callisto which is a run-time system for OpenMP and other task parallel programming models. It handles synchronization and load balancing on multiple cores. Their work is complementary to the work presented in this paper. We can combine load balancing with collocation of threads optimized for reducing remote DRAM accesses.

There are many other studies [6–9] which focused on tuning DRAM scheduling policies or memory access ordering for better overall performance. Our work is different from these in two ways. First we focus on OS scheduling algorithms to reduce the impact of remote DRAM accesses. Second, these studies [6] focus on optimal DRAM utilization for co-located single threaded workloads where as our work focuses on improving performance of a multi-threaded parallel workload.

6 Conclusion

Many commercial server applications today run on ccNUMA multi-socket multi-core based servers. These applications typically suffer from remote DRAM accesses that diminish their overall performance. This paper presented an operating system scheduling optimization to ameliorate the performance impact of remote DRAM accesses. By observing local and remote DRAM accesses for various threads and incorporating that into the OS scheduling decision, we come up with a new schedule for the next scheduling quantum. We presented three new scheduling algorithms followed by an adaption of an existing Hungarian algorithm. Depending on the scheduling algorithm used performance benefit varied across different synthetic workloads. We also performed sensitivity analysis of these algorithms by varying remote DRAM latency. We showed that some of the algorithms can cut down DRAM access latency by up to 55% depending on the workload used. The benefit gained from the algorithms is dependent on their complexity. Higher the complexity higher is the benefit. Hungarian algorithm and Algorithm 2 provide a good trade-off between performance and complexity for the workloads we studied.

This work could be extended in several ways. One way is to monitor the benefit resulting from the scheduling algorithms and if performance is reduced by applying these algorithms then the optimization can be turned off. The application behavior can still be monitored in the background to see if it enters a phase where it is beneficial to turn the scheduling optimization on again. Another way is to use machine learning algorithms to learn any kind of phase behavior among prior scheduling quanta and incorporate that into scheduling decision for the next quanta. In general for any long running application with stable patterns, hardware could provide feedback to the OS, which could in turn use that information to adapt its policies to benefit application performance.

Acknowledgments. I would like to thank Prof. Alan Cox of Rice University for initially discussing with me the concept of optimizing OS scheduling algorithms for improving the performance of various workloads. I would also like to thank various reviewers of this work for their comments and feedback. Finally, I would like to thank my wife and kids for supporting me morally during the course of this work.

References

1. Chandra, R., Devine, S., Verghise, B., Gupta, A., Rosenblum, M.: Scheduling and page migration for multiprocessor compute servers. In: Proceedings of ASPLOS (1994)
2. Kaseridis, D., Stuecheli, J., Chen, J., John, L.K.: A bandwidth-aware memory-subsystem resource management using non-invasive resource profilers for large CMP systems. In: Proceedings of Sixteenth International Symposium on High Performance Computer Architecture (2010)
3. İpek, E., Mutlu, O., Martínez, J.F., Caruana, R.: Self-optimizing memory controllers: a reinforcement learning approach. In: Proceedings of International Symposium on Computer Architecture, Beijing, China, June 2008
4. Ahn, J.H., Erez, M., Dally, W.J.: The design space of data - parallel memory systems. In: Proceedings of SC (2006)
5. Zhu, Z., Zhang, Z.: A performance comparison of DRAM memory system optimizations for SMT processors. In: Proceedings of HPCA-11 (2005)
6. Nauman, R., Lim, W.-T., Thottethodi, M.: Effective management of DRAM bandwidth in multicore processors. In: Proceedings of PACT-2007
7. Tang, L., Mars, J., Vachharajani, N., Hundt, R., Soffa, M.L.: The impact of memory subsystem resource sharing on datacenter applications. In: Proceedings of International Symposium on Computer Architecture (2011)
8. Hur, I., Lin, C.: Adaptive history-based memory schedulers. In: Proceedings of the International Symposium on Microarchitecture (2004)
9. Rixner, S., Dally, W.J., Kapasi, U., Mattson, P.R., Owens, J.D.: Memory access scheduling. In: Proceedings of International Symposium on Computer Architecture (2000)
10. Kuhn, H.W.: The Hungarian method for the assignment problem. Nav. Res. Logist. Q. **2**, 83–97 (1955)
11. Kim, C., Huh, J.: Fairness-oriented OS scheduling support for multicore system. In: Proceedings of 2016 International Conference on Supercomputing
12. Sahoo, P.K., Dehury, C.K.: Efficient data and CPU-intensive job scheduling algorithms for healthcare cloud. In: Elsevier Computers and Electrical Engineering, Vol. 68 (2018)
13. Lepers, B., Quema, V., Fedorova, A.: Thread and memory placement on NUMA systems: asymmetry matters. In: USENIX Annual Technical Conference (2015)
14. Harris, T., Maas, M., Marathe, V.J.: Callisto: co-scheduling parallel runtime systems. In: 9th EuroSys Conference (2014)

Towards Energy Efficient Servers' Utilization in Datacenters

Ahmed Osman(✉), Assim Sagahyroon, Raafat Aburukba, and Fadi Aloul

American University of Sharjah, Sharjah, UAE
b00067674@alumni.aus.edu

Abstract. Cloud computing datacenters contain hundreds of servers that host different kinds of services for a wide spectrum of customers. These datacenters have substantial energy demands for their operation, thus promoting the need for optimizing their power consumption and energy demands. Resources allocation and optimized scheduling of incoming tasks are at the heart of any successful power management technique used for datacenters. In this work we focus on the efficient utilization of servers in the datacenter to optimize power consumption. The goal is to develop task allocation techniques that contributes to the overall optimization of energy demands by optimizing the consumption of the datacenter servers. The allocation problem is modeled using Integer Linear Programming (ILP) techniques, where models are formulated with the objective of minimizing the total power consumed by the active and idle cores of the servers. Preliminary results show that an optimization driven servers' allocation strategy produces noticeable improvement in power consumption when compared to scheduling techniques such as round robin.

Keywords: Cloud computing · Energy optimization · Resource allocation · Scheduling · Integer linear programming (ILP)

1 Introduction

The shift to cloud computing necessitated the establishment and use of energy-hungry data centers to facilitate the execution of vast internet based services. Additionally, the expected growth of Internet of Things (IoT) applications, smart cities, and connected health devices, will also add to the expansion of datacenter demands. Therefore, energy consumption models and devising policies and strategies to optimize power consumption of data centers is of essence to their efficient and cost effective operation [1].

Servers are the power horse of data centers since they are responsible for processing all the load demands and the execution of the various tasks. Hence, they are the major contributor of power consumption when one considers the IT infrastructure of a datacenter. Exploring methods to model servers' power and its energy needs plays a fundamental role in designing energy-efficient datacenters.

Some of the published work approached the power estimation problem using measurement-based approaches. Lewis et al. [2] proposed to a system-wide energy consumption model for servers by making use of hardware performance counters and

K. Arai et al. (Eds.): CompCom 2019, AISC 997, pp. 254–262, 2019.
https://doi.org/10.1007/978-3-030-22871-2_19

experimental measurements. The model measures energy input to the system as a function of the work done for completing tasks being gauged and the residual thermal energy given off by the system as a result. Tudor et al. [3] who devised a model that represented server power as a function of energy used by CPU, memory and I/O devices, but practically speaking current server platforms do not allow for a detailed and precise measurement of each components power separately, and typically only the full system power can be measured. In [4] levels of resource utilization by each major component of a server are considered. Researchers developed two models to analyze power consumption for two types of access privileges. A fine-grained power model that requires access to utilization of all system components, on the other hand, CPU-based power model only needs utilization of the CPU in the system. Other approaches sought to develop models that would assist in estimating and optimizing power and energy demands using smart task allocation and scheduling techniques. In cloud computing, scheduling is the problem of allocating compute, network resources, or storage to tasks over a period of time, whether these resources are physical or virtual resources. In [5] a scheduling algorithm for online scheduling of servers of cloud computing providers was proposed. The algorithm follows three steps: first it predicts the workload assigned to the datacentre. The workload is anticipated from one of two resources: based on the cloud provider's information or using a new kind of multi-arm bandit workload prediction. Then, the algorithm modifies and optimizes the set of available servers to best fit the predicted workload. The final step is to schedule the jobs to the available processors using a modified version of Round Robin scheduling method. In [6] a green energy-efficient scheduling algorithm for cloud computing datacenters using DVFS technique was proposed. The algorithm aims to increase the utilization factor of the different servers in the datacenter to reduce its energy consumption. The authors assumed heterogeneous servers with each server having different processor, RAM, and hard disk. Each server has a defined minimum and maximum frequency and this frequency is reduced to minimize the energy while keeping in mind the requirements of the workload. VMs (virtual machines) are created from the servers and the jobs are assigned to these VMs using priority job scheduling. Xin Ye et al. [7] proposed a multi-objective VM placement model for cloud computing. The authors took into consideration the different requirements of both the cloud providers and users to build the model with the objectives of minimizing the energy consumption, maximizing load balance, maximizing resource utilization, and maximizing robustness. The proposed model is solved using an improved Energy-Efficient KnEA (Knee Point-Driven Evolutionary Algorithm) algorithm. In [8], a task scheduling algorithm was proposed with the objective of maximizing the profit by minimizing the make span and maximizing resources utilization. The cloud computing environment was represented by a set of heterogeneous clusters containing different types and number of servers. The requests submitted by clients were assigned to one cluster only. The proposed solution consisted of two steps: finding an initial solution that shows the fastest two clusters and assign jobs to them using the Min-Min algorithm, then within each cluster tasks were assigned to the individual servers. The authors assumed the tasks were independent and that the cluster can execute more than one job, but each client is assigned to a single cluster.

In this work, we attempt to optimize the energy consumption of a datacentre by modelling the problem as an ILP (Integer Linear Programming) problem with the objective being the minimization of its power consumption, using methods that would eventually lead to an optimum number of servers being made use of.

2 Developing the Model

Typically, in an IT infrastructure of a cloud computing datacenter, sources of power consumption include storage units, networking hardware and interfaces, and computing resources. The focus of the work presented in this paper is on the compute resources since they are responsible for about 26% of the total energy of the datacenter [9]. Therefore, we attempt to develop techniques that utilize the CPUs inside the servers efficiently. We seek to construct an additive server power model that represents the entire servers' power consumption as a summation of their individual subcomponents, namely, the cores of the CPUs.

In developing the ILP model, the following assumptions are made:

- A server can be in one of two states either ON or OFF.
- CPUs inside a server are assumed to have either two or four cores per CPU (i.e. dual or quad core).
- A core can be ON (thus executing a task), idle (on but not running any task), or completely switched OFF.
- Tasks are allocated and scheduled simultaneously upon their arrival (none of the tasks will be assigned at a later time).
- A server will be ON if at least a single core on one of its CPUs is executing a task and is assumed to be OFF otherwise.
- For any user task there is at least one core that meets the requirements needed to execute the task.
- At any point in time, a core can execute only one task at 100% utilization.
- The execution of a task on a certain core is assumed to consume a specific amount of power that we will call the *energy cost*.

The intended objective is to find the specific assignment of tasks to cores that gives the minimum energy value. The total power is that consumed by the active cores when a task is allocated to it, plus the power consumed by the idle cores. To construct an ILP-based model, variables and constraints are identified and derived. Table 1 presents the variables used in developing the model of the problem and the meaning of each variable. The model formulation starts with an objective function to minimize the energy consumption when assigning tasks to cores in a datacenter.

We observe that the energy consumption stems from the cost of active cores ($X_{i,j,k,n}$) and the cost of idle cores ($Y_{i,j,k}$), while the off cores are assumed to have zero energy consumption.

Table 1. Model variables and their definitions.

Variables	Description
S_i	Binary variable, equal to 1 if server i is on, and 0 otherwise
P_i	Binary variable, equal to 1 if CPU j in server i is active and 0 otherwise
$G_{i,j}$	Binary variable, equal to 1 if CPU j in server i- is idle and 0 otherwise
$C_{i,j,k}$	Binary variable, equal to 1 if core k in CPU j in server i is active, and 0 otherwise
$Q_{i,j,k}$	Binary variable, equal to 1 if core k in CPU j in server i is idle, and 0 otherwise
$T_{i,j,k,n}$	Binary variable, equal to 1 if task n is assigned to core k in CPU j in server i, and 0 otherwise
$X_{i,j,k,n}$	Binary variable, equal to 1 if task n is assigned to core k in CPU j in server i, and 0 otherwise
$Y_{i,j,k}$	The energy cost value of core k in CPU j in server i when it's idle

As represented by the function of Eq. (1) below:

$$Min\left(\sum\nolimits_{i=1}^{I}\sum\nolimits_{j=1}^{J}\sum\nolimits_{k=1}^{K}X_{i,j,k}C_{i,j,k}+\sum\nolimits_{i=1}^{I}\sum\nolimits_{j=1}^{J}\sum\nolimits_{k=1}^{K}Y_{i,j,k}Q_{i,j,k}\right) \quad (1)$$

The energy cost value typically depends on factors such as the size of the task and characteristics of the core executing the task. However, the function in Eq. (1) assumes all tasks have uniform or equivalent characteristics.

This objective function of Eq. (1) is subject to the following set of constraints:

- If a core k is active then the CPU j that contains the core k is considered active as well.

$$P_{i,j}=1 \qquad if\ \sum\nolimits_{k=1}^{K}C_{i,j,k}\geq 1 \quad \forall i,\forall j \quad (2)$$

- If a CPU j is active then at least one of its cores is active.

$$\sum\nolimits_{k=1}^{K}C_{i,j,k}\geq 1 \qquad if\ P_{i,j}=1 \quad \forall i,j\forall \quad (3)$$

- If a CPU is active, all the cores inside the CPU must be either active or idle but not switched off.

$$\prod\nolimits_{k=1}^{K}\left(C_{i,j,k}+Q_{i,j,k}\right)=1 \qquad if\ P_{i,j}=1 \quad \forall i,\forall j \quad (4)$$

- A core cannot be active and idle at the same time.

$$C_{i,j,k}+Q_{i,j,k}\leq 1 \qquad \forall i,\forall j,\forall k \quad (5)$$

- If a CPU is idle then all its cores must be idle.

$$\sum\nolimits_{k=1}^{K} Q_{i,j,k} = K \qquad if\ G_{i,j} = 1 \quad \forall i, \forall j \tag{6}$$

- If a server is off all the cores inside the server are off.

$$\sum\nolimits_{j=1}^{J} \sum\nolimits_{k=1}^{K} C_{i,j,k} + Q_{i,j,k} = 0 \qquad if\ S_i = 0 \quad \forall i \tag{7}$$

- If one or more of the CPUs in server i are active then server i must be active.

$$S_i = 1 \qquad if \sum\nolimits_{j=1}^{J} P_{i,j} \geq 1 \quad \forall i \tag{8}$$

- If a server is active then at least one of its CPUs is active.

$$\sum\nolimits_{j=1}^{J} P_{i,j} \geq 1 \qquad if\ S_i = 1 \quad \forall i \tag{9}$$

- If a server is active then all the CPUs in an active server are either active or idle but not switched off.

$$\prod\nolimits_{j=1}^{J} (P_{i,j} + G_{i,j}) = 1 \qquad if\ S_i = 1 \quad \forall i \tag{10}$$

- If a server i is off, all the CPUs inside the server are off.

$$\sum\nolimits_{j=1}^{J} P_{i,j} + G_{i,j} = 0 \qquad if\ S_i = 0 \quad \forall i \tag{11}$$

- A CPU cannot be active and idle at the same time.

$$P_{i,j} + G_{i,j} \leq 1 \quad \forall i, \forall j \tag{12}$$

- A task n must be assigned to one core only at any given time.

$$\sum\nolimits_{i=1}^{I} \sum\nolimits_{j=1}^{J} \sum\nolimits_{k=1}^{K} T_{i,j,k,n} = 1 \quad \forall n \tag{13}$$

- A core can execute a single task only at any given time or not execute any task.

$$\sum\nolimits_{n=1}^{N} T_{i,j,k,n} \triangleq \begin{cases} 1 & if\ C_{i,j,k} = 1 \\ 0 & if\ C_{i,j,k} = 0 \end{cases} \qquad \forall i, \forall j, \forall k \tag{14}$$

- The number of active cores is equal to the total number of tasks to be scheduled.

$$\sum\nolimits_{i=1}^{I} \sum\nolimits_{j=1}^{J} \sum\nolimits_{k=1}^{K} C_{i,j,k} = N \tag{15}$$

3 Validation and Testing of the Model

To validate the ILP formulation we use datacentre instances that are randomly generated and summarized in Table 2. These instances consist of different datacentre configurations where a certain number of tasks is to be scheduled to the available cores. All cores are assumed to be capable of executing any task. Each instance has different number of servers. Each server is assumed to contain 1 to 4 CPUs. Each CPU can be either *dual* or *quad* core. Initially, the *active* and *idle* energy costs of the cores were assigned randomly and are assumed to be the same for all cores on the same CPU.

The model and its usefulness is validated using the generic ILP solver CPLEX [10]. The solver computes the minimum energy consumption of the datacenter and the corresponding assignment of tasks to the cores that would achieve this power. Tests are conducted using an Intel (R) Core (TM) i5 machine @ 2.50 GHz – 2.70 GHz with 8 GB of RAM and Windows 10 Home 64-bit operating system.

Table 2. Tested instances.

Instance no.	No. of servers	Total no. of CPUS	Total no. of cores	No. of tasks
1	2	3	8	2
2	2	4	12	4
3	3	5	16	6
4	6	17	44	10
5	8	24	72	20
6	12	36	108	25

Table 3 shows the minimum energy consumption returned by the CPLEX solver and the average time taken. As presented in the Table, CPLEX found a minimum energy consumption in all the tested instances, and the time taken by CPLEX did not exceed 2.41 s for any instance.

Table 3. Results of model validation.

Instance no.	No. of tasks	Total no. of cores	Minimum energy	CPLEX CPU time (S)
1	2	8	9	0.15
2	4	12	14	0.18
3	6	16	39	0.20
4	10	44	46	0.37
5	20	72	172	1.21
6	25	108	262	2.41

To further validate the practicality of the model, results produced using this model are compared against results produced using two other scheduling techniques, namely, Round Robin, and Maximum Possible Value.

The round robin algorithm assigns the task to the first available core that can execute it. It does not consider minimizing the energy consumption while assigning tasks, and hence consumes minimal time in making these assignments. The maximum possible value scheduling technique represents the worst possible scenario that can happen if no scheduling algorithm is used at all. In this case, the tasks are assigned to the set of cores which would execute them with the maximum energy consumption possible. The maximum value is found by changing the minimization function of the model to a maximization function and then solving it using a solver.

Table 4 lists the energy consumption values found by CPLEX solver using the proposed model, and those returned by Round Robin, and Maximum Possible Value methods. Evidently, we see that the proposed model returned the minimum energy values in all of the tested instances.

Table 4. Comparison of the proposed model against other techniques.

Instance no.	Energy consumption			CPU time (seconds)
	Maximum possible value	Round robin	Proposed model	CPLEX
1	26	12	**9**	0.15
2	50	20	**14**	0.18
3	100	57	**39**	0.20
4	196	55	**46**	0.37
5	534	216	**172**	1.21
6	1483	422	**262**	2.41

Next, the objective function in Eq. (1) is modified by taking the size of the task into consideration when solving the problem. The cost $X_{i,j,k,n}$ of each task per core is assumed as a positive integer that is proportional to the task size. The following additional assumptions are applied:

- Task sizes are chosen randomly in the range of 500–10,000 instructions.
- The active energy cost for each CPU is chosen randomly in the range of 2 to 40.
- The idle energy cost for each CPU is assumed to be 50% of the active energy.
- All cores on the same CPU are identical, that is active cores and idle cores have the same energy needs.

The cost $X_{i,j,k,n}$ is calculated using Eq. (16), where $TaskSize_n$ *is* the size of task n in number of instructions and $U_{i,j,k}$ represents the active energy cost of the CPU.

$$X_{i,j,k,n} = \frac{TaskSize_n}{500} * U_{i,j,k} \tag{16}$$

The task size is divided by 500 just to scale the numbers down (500 is chosen because the smallest task size is assumed to be 500 instructions). The parameter $U_{i,j,k}$ is a randomly generated value that is assigned to each CPU in the datacenter. This value is equal in all the cores of the same CPU. The model is implemented and tested using the same testing instances in Table 2 above. The values of $X_{i,j,k,n}$ for each task per core were generated and assigned randomly for this preliminary test, again CPLEX is used to test the model. Table 5 shows the results of solving the model after introducing task sizes. CPLEX again found the minimum consumption but in instance 6 it took around 8000 s to reach the optimum Value.

The dramatic increase in solving time comes from the increase in the number of variables in the objective function as a result of adding the task size to the model and hence the huge search space.

4 Conclusions and Future Work

As discussed above the initial results are promising and our work will continue to incrementally enhance the base model of Eq. (1) to make it more comprehensive.

Table 5. Energy consumption and running times after adding task sizes to the objective function.

Instance no.	No. of tasks	Total no. of cores	Minimum energy	CPLEX CPU time (S)
1	2	8	9	0.15
2	4	12	14	0.18
3	6	16	39	0.20
4	10	44	46	0.37
5	20	72	172	1.21
6	25	108	262	2.41

Additional task and core characteristics such as: task's size, task's required frequency, and task's required memory that would guarantee the successful execution of the task will be included in future formulations of the model. Future work will also allow for a new batch of tasks to be scheduled at time t1 after the first batch is already scheduled at time t0. The new tasks will be assigned to the remaining free cores following the scheduling of the first batch. Furthermore, we will seek to identify methods to reduce the search space, for example, in instance 6 of Table 5, the total number of variables in the objective function has increased the search space, to work around this problem further modifications of the model need to be implemented and verified and perhaps pre-processing steps are warranted as well. Finally, and to approximate real life datacenters energy consumption scenarios, industry-based servers' specifications with typical power consumption values will be used in assessing the performance of the final model.

References

1. Beloglazova, A., Abawajy, J., Buyya, R.: Energy aware resource allocation heuristics for the efficient management of data centers for cloud computing. Futur. Gener. Comput. Syst. **8**(5), 755–768 (2012)
2. Lewis, A.W., Ghosh, S., Tzeng, N.F.: Run-time energy consumption estimation based on workload in server systems. HotPower **8**, 17–21 (2008)
3. Tudor, B., Teo, Y.: On understanding the energy consumption of arm-based multicore servers. SIGMETRICS Perform. Eval. Review **41**(1), 267–278 (2013)
4. Alan, I., Arslan, E., Kosar, T.: Energy-aware data transfer tuning. In: Proceedings of the 14th IEEE/ACM International Symposium on Cluster, Cloud and Grid Computing (CCGrid), pp. 626–634 (2014)
5. Dambreville, A., Tomasik, J., Cohen, J., Dufoulon, F.: Load prediction for energy-aware scheduling for cloud computing platforms. In: Proceedings of the IEEE 37th International Conference on Distributed Computing Systems (ICDCS), pp. 2604–2607 (2017)
6. Wu, C.M., Chang, R.S., Chan, H.Y.: A green energy-efficient scheduling algorithm using the DVFS technique for cloud datacenters. Futur. Gener. Comput. Syst. 37, 141–147 (2014)
7. Tian, H., Wu, J., Shen, H.: Efficient algorithms for VM placement in cloud data centers. In: Proceedings of the 18th International Conference on Parallel and Distributed Computing, Applications and Technologies (PDCAT), pp. 75–80 (2017)
8. Bey, K.B., Benhammadi, F., Sebbak, F., Mataoui, M.: New tasks scheduling strategy for resources allocation in cloud computing environment. In: Proceedings of the 6th International Conference on Modeling, Simulation, and Applied Optimization (ICMSAO), pp. 1–5 (2015)
9. Elnozahy, E.M., Kistler, M., Rajamony, R.: Energy-efficient server clusters. In: International Workshop on Power-Aware Computer Systems, pp. 179–197 (2002)
10. IBM: IBM ILOG CPLEX Optimizer. Internet. http://www-01.ibm.com/software/integration/optimization/cplex-optimizer/. Accessed 8 June 2018

An Evaluation of ICT Smart Systems to Reduce the Carbon Footprint

Andreas Andressen, Lesley Earle, Ah-Lian Kor(✉), and Colin Pattinson

School of Computing, Creative Technologies, and Engineering, Leeds Beckett University, Leeds LS4 2TL, UK
A.Roseschmidtandresen9742@student.leedsbeckett.ac.uk, {L.Earle,A.Kor,c.pattinson}@leedsbeckett.ac.uk

Abstract. This research aims to assess and evaluate the impact on sustainability in buildings through implementation of ICT Smart Systems. The setting for this research will be for a large global organisation's headquarters in Germany. The list of objectives is: to audit the ICT infrastructure used and to survey the existing smart systems implemented; to investigate the total energy expenditure and carbon footprint for ICT equipment during a yearly period; and to explore how to best transfer best green ICT practices to other buildings. Based on the findings in this paper, investing in energy-saving ICT equipment, or even a BMS, can be very cost beneficial to a company and reduce the carbon footprint of commercial buildings when implemented correctly.

Keywords: Lifecycle assessment methodology · Smart buildings · Carbon footprint · Building management system

1 Introduction

According to the Intergovernmental Panel on Climate Change, the world's buildings account for 32% of global final energy use [13] and nearly a third of global GHG emissions [1]. This number is expected to further increase as standards of living increase around the globe. [2] (PBL NEAA, 2016). A remarkably large part of this energy expenditure comes from Heating, Ventilation and Air-Conditioning (HVAC) systems, a staggering 60% of total consumption. Lighting accounts for roughly 11 to 20% of total building energy demand. The remainder is divided amongst a host of electrical appliances, such as phones, computers and other ICT equipment [3]. Considering these facts, it is crucial to explore developments within this sector to reduce the carbon footprint and energy usage. This study will focus on opportunities within the 'Smart Building' sector as there are many opportunities to be found here, particularly involving the usage of ICT. The aim of this research is to assess and analyse the impact on sustainability for ICT smart systems in buildings within a commercial setting for a large, global industrial manufacturing organisation's headquarters in Germany that has such systems implemented already. A set of objectives will support this aim, these are: (i) to audit the current ICT Infrastructure used by the organisation in their building;

K. Arai et al. (Eds.): CompCom 2019, AISC 997, pp. 263–274, 2019.
https://doi.org/10.1007/978-3-030-22871-2_20

(ii) to investigate the total energy usage from the ICT Infrastructure; (iii) to survey the existing smart system implementation and how it helps reduce the carbon footprint of ICT; (iv) to explore how to transfer best green ICT practices to other companies.

2 Literature Review

GHG emissions are growing rapidly as a result of the ICT sector, due to the more and more network connected devices being available everywhere, as well the increasing demand for web services. In a BAU scenario, these GHG emissions are expected to triple between 2007 and 2020. However, the ICT sector also contributes to reductions in GHG emissions and energy savings through innovative smart technologies. This is known as the 'enabling effect' of ICT [4].

ICT is viewed as a crucial part of transitioning to a more sustainable future. While the premise of using ICT smart systems to reduce the carbon footprint sounds very positive, the reality of the situation is that ICT has both significant positive and negative impacts on the environment as laid out in a study by the Organisation for Economic Cooperation and Development (OECD)[1] (see Table 1).

Table 1. ICT impacts on the environment [5]

Positive impacts	Negative impacts
Dematerialisation and online delivery	The production and distribution of ICT Equipment
Reduction in the need for travel	Energy consumption in use (directly and for cooling)
A host of modelling, monitoring and management applications	Short product lifecycle and e-waste
Greater energy efficiency in production, use and recycling	Potentially exploitative applications

Some of these impacts are more pronounced than others, but a big contributor to global emissions is the short product life-cycle sees in most ICT devices. The link between ICT and the broader sustainability goal is tenuous and not always well understood [ibid]. A real issue here is "incomplete substitution" of ICT, whereby ICT equipment adds to the overall problem rather than lessening it. OECD identifies this as a very significant issue. The Global e-Sustainability Initiative (GeSI) however, remains positive about the potential of ICT technologies. Through measurement of what is known as the 'mitigation potential' which is used to measure the avoided or reduced GHG emission of ICT, it is believed that the increased use of ICT technologies, such as those in smart building management, will cut overall emissions by 16.5% by 2020 [6]. The concept of 'smart buildings' have been around for at least two decades. As the

[1] http://www.oecd.org/site/stitff/45983022.pdf.

development of technology have progressed, so to have the definition and implementation of the concept of smart buildings. So what is a 'Smart building'? Arkin and [7] suggests that the 'smartness' of a building is not just the technologies utilised within the building (e.g. HVAC or lighting), but that it is also measured by how well these systems are integrated together. Author in [8] suggest that smart buildings are simply electronically enhanced buildings. According to a study laid out by CRC Construction Innovation, there are four primary elements that integrate in a smart building: ICT sensors, integrated information management system and performance models, actuators and a backbone or nervous system, which essentially connections the former elements together [9]. The impact from buildings is very significant. A survey conducted in 2012 by the U.S. Energy Information Administration on Commercial Buildings Energy Consumption shows how HVAC and lighting alone accounts for roughly 78% of a commercial buildings energy usage[2] (see Fig. 1).

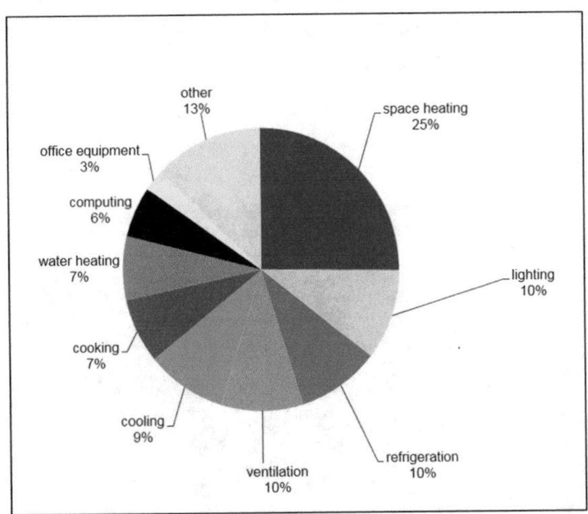

Fig. 1 Distribution of energy use in commercial buildings [10]

In another report: Climate Change: Implications for Buildings, it has been underlined that there may be a potential for up to 50–90% energy savings in both new and existing buildings, providing it is done right.[3] One of the key issues with a 'building-as-usual' strategy include is the rise of the global population, and with it, higher demand for energy and higher temperatures as a result of global warming which increases energy demand principally for HVAC systems [11]. The primary mitigation strategies comprise of: Energy efficient technology, such as building envelopes, energy-efficient appliances

[2] https://www.eia.gov/consumption/commercial/reports/2012/energyusage/.

[3] https://www.cisl.cam.ac.uk/business-action/low-carbon-transformation/ipcc-climate-science-business-briefings/pdfs/briefings/IPCC_AR5__Implications_for_Buildings__Briefing__WEB_EN.pdf.

and improved automation that can respond to changing conditions; Service demand reduction via personal and behavioural changes, which can be moderated through carbon pricing, carbon trading, property CO_2 taxing and absolute consumption limits for a building; System and infrastructure efficiency, which includes retrofitting older buildings and building new buildings with very low energy usage, passive building designs that can alleviate the need for HVAC systems entirely and prioritisation of energy performance and use-factions through building design [12], construction and commissioning; Carbon efficiency, or shifting fuel and energy supply infrastructure to buildings and evolution of the energy provision for those that lack access to modern energy carriers and equipment (estimated 2 billion people) [13]. As laid out in a report by the European Commission, it is crucial to reduce energy consumption in buildings when taking into account the targets set for 2020 by the European Council in 2007 in their SET Plan (European Strategic Energy Technology Plan). To achieve this, the one of the most important aims is the mobilisation of ICT to accelerate towards a sustainable and low-carbon economy, and increasing energy efficiency through ICT [14]. Going further, the European Commission has laid out a number of fundamental pillars when it comes to development of smart buildings (see Fig. 2).

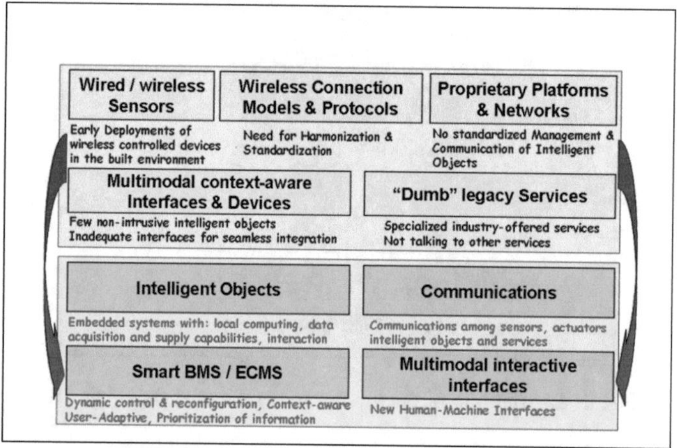

Fig. 2. Technologies for smart buildings [ibid]

There seems to be disagreement as to what sustainable and energy-efficient buildings are and little utilization of ICT-based informed decision-making in the process and use of sustainable and energy-efficient facilities. Some other problem areas that seem to exist surrounding smart buildings are the many standards regulations for such buildings, some of which are in conflict with others, the lack of standardization of environmental systems and their configurations and need for occupancy feedback to interact with behaviour modifications in smart systems [ibid]. Beyond this, there are several other barriers, such as lack of incentives for developments in smart building technologies, the buildings sector being slow to adapt new technology; a whole 15 year cycle for commercial buildings on average [ibid], and lack of skilled technicians and operators to handle BMS [15]. Possibly the most significant problem seems to be the confusion surrounding the enormous quantity of different unknown and often competing vendors that offer different kinds of equipment, components and other systems for smart buildings, most of which often do not work together. With this is mind, many organisations are delaying investments in ICT smart systems for their buildings [16]. It is crucial that organisations take time to assess the impact of ICT initiatives in buildings as a greater understanding of the benefits these offer will accelerate their normalisation and adoption.

3 Methodology

There are many different methodologies aiming to assess the balance between direct emissions and enabling effects of ICT. While no one true methodology exists that supports every case scenario, an appropriate methodology is one that is practical, user-friendly and enables the freedom of adapting the method of the methodology to the objectives of the organisation or organisation, while still yielding robust results [17]. As this research is aimed towards an organisation which is situated in a smart building with an extensive ICT network for energy efficiency, the following methodology has been chosen due to its adaptability and flexibility of use.

3.1 Lifecycle Assessment (LCA)

LCA is an approach to evaluating environmental impact that has been around since the 1980s. It is a methodology used to identify and evaluate environmental impacts, especially in relation to a product or material's life cycle. It is considered to be a valuable resource to provide insight and analysis of a product's energy, waste, emissions, etc. from a life-cycle perspective, and can provide further examination of any associated issues that might have an effect on the environment [18]. The methodology focuses on analysing and interpreting energy and/or physical requirements for the production and use of a product and any associated waste that may be released into the environment as an effect of the production or use of such a product. For this research, the ISO 14040 LCA is being adopted. LCA has four basic stages, these are: Goal and scope definition, inventory analysis, impact assessment and interpretation [19] (see Fig. 3). The case study is a global organisation based in Germany. For the purpose of this research, we will be focusing on their headquarters, which is a Leadership in Energy & Environmental Design (LEED) Platinum certified building.[4]

[4] http://leed.usgbc.org/leed.html.

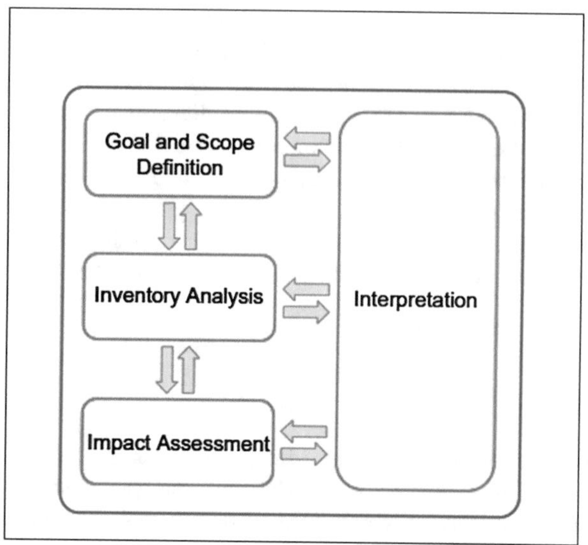

Fig. 3 The methodological framework for LCA [20]

For the purposes of this particular research, the four stages of LCA will be interpreted as such:

1. *Goal and Scope Definition*: Evaluate the environmental impact of the reported ICT equipment used in the organisation's headquarters.
2. *Inventory Analysis:* Provide an overview of the reported ICT inventory, including any ICT equipment that contribute to lower energy usage and by effect lower emissions.
3. *Impact Assessment*: Assess the current carbon footprint of the ICT equipment in the building using SusteIT tool[5].
4. *Interpretation*: Reporting the results, and making recommendations how to apply best green ICT practices to other organisations.

The baseline for this research will be data provided by the organisation. This will include a limited inventory list of ICT equipment in the building and an overview of the ICT smart systems that lower the energy expenditure of the building.

4 Results and Discussion

4.1 Inventory Audit

The organisation has provided an inventory list of the ICT equipment used in the building as well the sensors and systems that control usage and output of energy. The

[5] http://www.susteit.org.uk/files/.

office hours are 8–9 h per day, 5–6 days a week with an average number of working months of about 11 months. So as to better estimate energy usage and carbon footprint, it will be assumed a working day is 9 h, 6 days a week. This equates to 288 days/year where the office is open. Outside of office hours, all ICT equipment is switched off by the BMS. Certain equipment, such as routers, sensors and other control modules are left active throughout the year, equating to 8760 h/year. For company mobile phones, the power consumption has been averaged from that of charging a phone and the number of hours spent charging on average in a year.[6] Based on the data gathered and provided by the organisation, SusteIT will be used to calculate the energy usage, CO_2 emissions and energy cost of the ICT equipment.

The total kWh can be calculated into kWh/year by the formula, as is done in the SusteIT tool:

$$E(\mathrm{kWh}) = P(\mathrm{W}) \times \frac{t(\mathrm{hr})}{1000} \times 288(\mathrm{days})$$

Table 2 shows the ICT inventory list calculated in SusteIT. This includes estimated power consumption of single unit, both active and in power-saving/standby mode. This has been divided into four categories with their calculated total power consumption: PCs & Monitors (280,956 kWh/year), Networking (10,240 kWh/year), Phones (16,390 kWh/year), Printers (97,544 kWh/year) and Sensors & Control (267,500 kWh/year). Furthermore, the organisation has a large photovoltaic system consisting of 800 solar panels covering more than 13,900ft^2 of the building's roof. According to the client, this system provides roughly a third of the energy required to run the building, this has therefore been subtracted from the overall power usage and emissions of the ICT inventory.

Table 3 shows the data as it would be further inputted into SusteIT. To calculate the Energy Cost and CO_2 emissions per year, a country electricity price and country emissions number is required. This data has been retrieved from Eurostat, the statistical office of the European Union and from the European Environment Agency respectively.[7,8] As the office is based in Germany, all prices are displayed in Euro (€).

Figure 4 shows the distribution of kWh/y in percentage for the inventory. This shows that the biggest source of energy usage is from PCs & Monitors (34%), closely followed by Sensors & Controls. Networking, Phones and Printers do no contribute a large part to the energy usage. At the same time, the photovoltaic array that generates energy does indeed generate about a third of the energy used, thus subtracting from the overall usage.

[6] https://www.techadvisor.co.uk/how-to/gadget/how-much-does-it-cost-charge-phone-tablet-or-laptop-3632210/.

[7] http://ec.europa.eu/eurostat/statistics-explained/index.php/Electricity_price_statistics.

[8] https://www.eea.europa.eu/data-and-maps/indicators/overview-of-the-electricity-production-2/assessment.

Table 2 Inventory list

Device	Quantity	Brand - model	Watts (Active)	Watts (Power saving mode)	Hours (Active, idle) per year	Hours (Power saving mode) per year	kWh/y per unit	kWh/y per unit (Power saving)	Total kWh/y
Standard PCS	200	Fujitsu Esprimo	80	20	1728	864	138.2	17.28	31.096
Laptops	1,000	Fujitsu Lifebook	65	15	2016	576	131	8.64	1,39,640
LED monitors w. power saver functions	1,600	DELL	28	3	2016	576	56.4	1J	93,120
Conference room and other screens	180	Samsung LED TV Monitors	50	20	1440	1152	72	23	17,100
PCS & monitors sub-total									280,956
Wireless Access Points	320	ASUS BRT-AC828	8	1	4380	4380	27.6	4.4	10,240
Networking sub-total									10,240
PBX connected phones	650	Siemens OptiPoint	5	-	2592	-	13	-	8,450
Regular VOIP phones	200	Siemens Gigaset	2	-	2592	-	5.2		1,040
Business use mobile phones (charging)	1,000	Varies {Samsung or iPhone mainly)	8	-	864	-	6.9		6,900
Phones sub-total:									16,390
Small networked office printer/scanner	250	HP LaserJet Pro	470	3	576	2016	270.7	7.7	69,600
Networked office printer/scanner/fax machines	40	HP Laserjet Enterprise	620	6	1008	1584	625	19.71	27,944
Printers sub-total:									97,544
Building sensors for energy saving and optimization	30,000	varied	1	-	8760	-	8.8	-	264,000
RoomOptiControl modules	200	Siemens	2	-	8760	-	17.5	-	3,500
Sensors & control sub-total									267,500
Photovoltaic Array	800 Solar Panels	Siemens	-	-	-	-	-	-	221,967*
*Power generation sub-total:									
								Total	4,50,633

Table 3. Overview of energy use and CO_2 emissions for ICT equipment

Category	Energy use (kWh/year)	%	Energy cost (Euros/year)	CO_2 emissions (kg/year)
PCs & Monitors	280,956	34	42,143	119,406
Networking	4,351	1	1,536	4,352
Phones	6964	1	2,459	6,966
Printers	41,446	5	14,632	41,456
Sensors & Control	267,500	32	40,125	113,688
Power Generation	−221,967	−2742,143	−33,295	−94,336
Total	379,250	100	67,600	191,532

Germany electricity price (0.15 Euros per kWh)
Germany CO_2 emission (0.425 kg CO_2 per kWh)

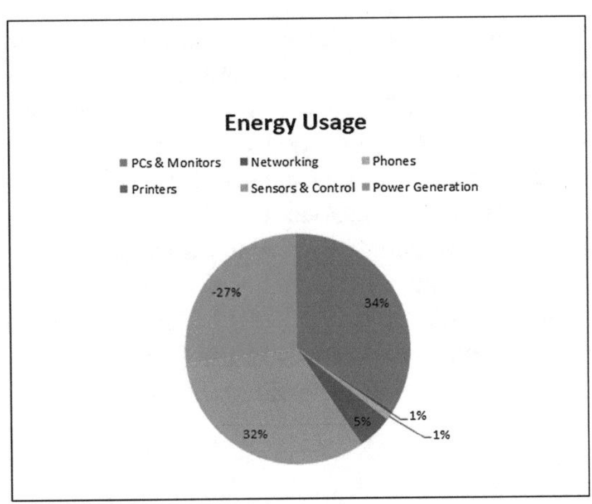

Fig. 4 Distribution of the energy usage for ICT equipment

4.2 Building Management System Desigo

The 30,000 sensors and control modules utilized in the building is connected to a Desigo[9] system from Siemens which is an advanced BMS that links up all subsystems in the building (light, HVAC, energy, video management, fire protection and access). The sensors in the building continually analyses and adjusts the subsystems for the BMS to effectively reduce the energy usage of the building. To save energy, the company has heavily invested in ICT equipment that use as little power as possible, or

[9] http://www.buildingtechnologies.siemens.com/bt/global/en/buildingautomation-hvac/building-automation/building-automation-and-control-system-europe-desigo/pages/desigo.aspx.

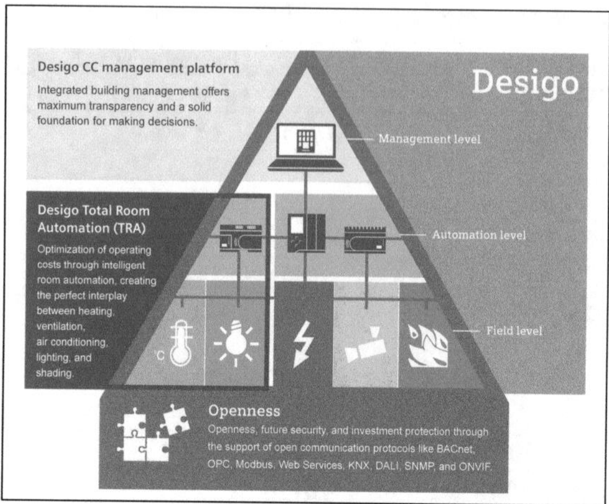

Fig. 5 DESIGO

equipment that has good power-saving modes to conserve energy when it is not being used. RoomOptiControl, a part of the Desigo system, has been installed in every room of the building (Fig. 5). These modules provide continuous feedback to users and will detect unnecessary energy consumption and alert users. At management level, the Desigo BMS is connected to Desigo CC; a management software platform. This allows personnel to continually monitor, respond to alarms and to make changes where needed, this software is used both on company smartphones as well as personal computers and is connected to the cloud, meaning operators do not have to be on site to monitor the BMS. Through this advanced BMS implementation and the use of their photovoltaic array, the company's headquarters use 90% less energy and have a 90% smaller carbon footprint compared to their old headquarters.

5 Conclusion

ICT Smart Systems such as BMS can be very beneficial in reducing energy usage, costs and emissions when implemented correctly. Buildings currently account for 32% of global energy use and 19% of GHG emissions from energy, and energy use of buildings could double or even triple by 2050 as standards of living and commerce increases [13]. Furthermore, many initiatives and policies are being put in place, such as the one the European Union in the Climate Action policy.[10] It is clear that new strategies for ICT smart systems must be considered to accelerate the adoption and normalization of such systems on a global scale. As the research has demonstrated, there are many benefits to a fully implemented BMS. Converging from field level into a centralized management system to constantly being able to monitor building metrics can be extremely cost effective. With these considerations in mind, why then is ICT

[10] https://ec.europa.eu/clima/policies/strategies_en.

smart systems not seen more commonly in buildings? Most likely, this is due to the many obstacles to implementing ICT smart systems into buildings, such as lack of incentives for smart buildings and the need for new business models where there are plenty of money for well-developed existing projects [21]. Another factor holding back the smart building revolution back seems to be a lack of advice and knowledge on the subject and the huge market full of equipment, systems, interfaces and components that often do not function together. A convergence of these systems is needed, for a simpler and cheaper overall solution. In turn companies should then be able to see more immediate benefits of introducing a ICT smart systems into a new or existing building.

Recommendations

- Switch to ICT equipment with power-saving functions, so when it is not being used, it will go to standby mode or similar. For portable equipment, not only will this conserve battery life, but tremendously decrease the overall energy usage as shown in Table 2.
- Switch off all non-essential equipment outside of office hours. Even with power-saving functions, the energy use for equipment is still significant for extended periods of time and can be costly.
- Install energy-saving kits for power-outlets that can be programmed to turn off when not in use.
- Invest in a BMS. Even a simple implementation could potentially be very cost-effective and help accelerate towards a more sustainable future.
- Invest in and install in-house energy generation such as a photovoltaic array, which are currently at an all-time low cost [22]. Combining this with an advanced, fully connected BMS could reduce energy use by up to 90% while reducing the carbon footprint.

References

1. Mardiana-Idayu, A., Riffat, S.B.: Building energy consumption and carbon dioxide emissions: threat to climate change (2015). https://www.omicsonline.org/open-access/building-energy-consumption-and-carbon-dioxide-emissions-threat-to-climate-change-2157-7617-S3-001.php?aid=53007. 17 Nov 2017
2. PBL Netherlands Environmental Assessment Agency (NEAA): Trends in global CO_2 emissions – 2016 report (2016). http://edgar.jrc.ec.europa.eu/news_docs/jrc-2016-trends-in-global-co2-emissions-2016-report-103425.pdf. 5 Jan 2018
3. EIA: International energy outlook (2017). https://www.eia.gov/outlooks/ieo/. 22 Nov 2017
4. GeSI: Evaluating the carbon-reducing impacts of ICT (2010). http://www.ecofys.com. 13 Nov 2017
5. Berkhout, F., Hertin, J.: OECD report: impacts of information and communication technologies on environmental sustainability: speculations and evidence. In: A Report to the OECD (2001). http://www.oecd.org/sti/inno/1897156.pdf. 24 Nov 2017
6. GeSI: SMART 2020: enabling the low carbon economy in the information age (2007). https://www.theclimategroup.org/sites/default/files/archive/files/Smart2020Report.pdf. 5 Jan 2018

7. Arkin, H., Paciuk, M.: Evaluating intelligent buildings according to level of service systems integration. Autom. Constr. **6**(5–6), 471–479 (September 1997). https://doi.org/10.1016/s0926-5805(97)00025-3
8. Kroner, W.M.: An intelligent and responsive architecture. Autom. Constr. **6**(5) (September 1997). https://doi.org/10.1016/s0926-5805(97)00017-4
9. CRC Construction Innovation: Smart buildings for healthy and sustainable workplaces: scoping study report (2003). https://eprints.qut.edu.au/26895/1/26895.pdf. 27 Dec 2017
10. U.S. Energy Information Administration: 2012 commercial buildings energy consumption survey: energy usage summary (2016). https://www.eia.gov/consumption/commercial/reports/2012/energyusage/. 7 Jan 2018
11. Buckman, A.H., Mayfield, M., Beck, S.: What is a smart building? (2014). http://eprints.whiterose.ac.uk/80714/7/Published%20PDF.pdf. 13 Dec 2017
12. Chalmers, P.: Climate change: implications for buildings (2014). https://www.cisl.cam.ac.uk/business-action/low-carbon-transformation/ipcc-climate-science-business-briefings/pdfs/briefings/IPCC_AR5__Implications_for_Buildings__Briefing__WEB_EN.pdf. 29 Nov 2017
13. European Commission: ICT for a low carbon economy – smart buildings (2009). http://ec.europa.eu/information_society/activities/sustainable_growth/docs/sb_publications/smartbuildings-ld.pdf. 2 Dec 2017
14. Ma, Z., Badi, A., Jørgensen, B.N.: Market opportunities and barriers for smart buildings. In: Proceedings of IEEE Green Energy and Systems Conference (IGSEC), 6–7 Nov, 2016, Long Beach, California, USA. https://doi.org/10.1109/igesc.2016.7790078
15. Weinschenk, C.: Smart buildings and the IoT are complex and challenging (2017). https://www.energymanagertoday.com/smart-buildings-iot-complex-challenging-0166257/. 9 Dec 2017
16. Öko-Institut e.V; Fraunhofer & Co KG; BIO Intelligence Service: Towards an overall measurement methodology of the carbon and energy footprints of the ICT sector (2013). https://publications.europa.eu/en/publication-detail/-/publication/9a79fd07-27af-4ad5-b39f-0fe11a49b9e5/language-en/format-PDF/source-60976487. 13 Nov 2017
17. ETSI: Environmental Engineering (EE): Life Cycle Assessment (LCA) of ICT equipment, networks and services; General methodology and common requirements (2011). https://www.etsi.org/deliver/etsi_ts/103100_103199/103199/01.01.01_60/ts_103199v010101p.pdf. 5 Dec 2017
18. ATIS Exploratory Group on Green: ATIS report on ICT LCA (2010). https://my.leedsbeckett.ac.uk/bbcswebdav/pid-1787676-dt-content-rid-5582400_1/courses/12689-1718/ATIS-LCA-Review%281%29.pdf. 6 Dec 2017
19. Lehtinen, H., Saarentaus, A., Rouhiainen, J., Pitts, M., Azapagic, A.: A review of LCA methods and tools and their suitability for SMEs (2011). http://www.biochem-project.eu/download/toolbox/sustainability/01/120321%20BIOCHEM%20LCA_review.pdf. 7 Dec 2017
20. Fawkes, S.: Common barriers to smart cities and energy efficiency (2015). https://www.2degreesnetwork.com/groups/2degrees-community/resources/common-barriers-smart-cities-and-energy-efficiency/. 8 Dec 2017
21. Wesoff, E.: Solar costs are hitting jaw-droppings lows in every region of the world (2017). https://www.greentechmedia.com/articles/read/solar-costs-are-hitting-jaw-dropping-lows-in-every-region-of-the-world. 10 Dec 2017
22. Mardiana-Idayu, A., Riffat, S.B.: Review on heat recovery technologies for building applications. Renew. Sustain. Energy Rev. **16**(2), 1241–1255 (2012)

Multisensory Real-Time Space Telerobotics

Earth-Ground Preliminary Results for the International Space Station Experiment

Marta Ferraz(✉), Edmundo Ferreira, Emiel den Exter, Frank van der Hulst, Hannes Rovina, William Carey, Jessica Grenouilleau, and Thomas Krueger

European Space Agency, ESTEC, Human-Robot Interaction Lab, Keplerlaan, 2201 AZ Noordwijk, The Netherlands
marta.ferraz@esa.int

Abstract. It is hypothesized that telerobotic operations, under altered gravity, can be improved by offering enhanced sources of sensory information to the astronaut. This paper summarizes preliminary results from Earth-ground testing carried out in preparation for an International Space Station (ISS) experiment with astronaut subjects (targeted for mid-to-late 2019). Cognitive and physical performance was compared in 10 operators remotely controlling a prototype rover on a lunar analogue, under different sensory conditions provided by a telerobotics user-interface - visual/visual plus auditory stimuli for a navigation task; visual plus auditory, visual plus somatosensory (force feedback), and visual plus auditory and somatosensory stimuli for a sample collection task. Three time-delay conditions were simulated: 0 s, 0.5 s and 1 s; corresponding to telerobotic operations on the lunar surface (via pressurized rovers or space habitats), from cis-lunar space and from the ISS, respectively. The audio feedback was delayed in all cases with 150 ms. Results indicate that auditory stimuli optimized operators' performance in navigation and sample collection tasks when proceeding visual and force feedback stimuli at 350 ms and 850 ms. Combined visual and auditory stimuli improved operators' performance in navigation tasks and 0 s time-delay conditions. At the same time combined visual and auditory feedback are lowering task completion times, mental and physical load when compared to visual stimuli only. Thus, this combination could improve telerobotic operations from pressurized rovers or space habitats. In contrast, not adapted force feedback was associated with great instability in operators' performance during a sample collection task with increased time-delays – also observed when auditory stimuli preceded visual and force feedback stimuli at 350 ms and 850 ms. Increased time-delays: 0.5 s and 1 s - were linked to increases in task completion times in operators.

Keywords: Space telerobotics · Human-Robot interaction · Multisensory · Human performance · Mental load · Physical load · Electroencephalography

K. Arai et al. (Eds.): CompCom 2019, AISC 997, pp. 275–298, 2019.
https://doi.org/10.1007/978-3-030-22871-2_21

1 Introduction

Recently, there has been a strong investment on the preparation of Crew Surface Telerobotics (CST) missions for space exploration - the crew remotely operates surface robots from a spacecraft or deep-space habitat [1]. These missions include Moon, Mars and Near-Earth Asteroid exploration [2–4]. CST missions include, for instance, the HERACLES project[1] [3] and the Exploration Mission-2 from NASA - expected to start in the first half of 2020 [5]. One of the central goals of these missions is to explore the geological features of the Moon's surface - the lunar far-side - via robotic systems: a crew inside the Orion Multi-Purpose Crew Vehicle and/or the Deep Space Gateway will teleoperate robotic rovers on the lunar surface.

A main objective of CST missions is to collect geological samples from the South Pole-Aitkin Basin, namely, the Schrödinger Basin, on the lunar far-side: an unexplored and geologically-rich region containing rocks from older events in lunar history [6].

Telerobotic tasks to be carried out in the Schrödinger basin, include, for example: analyzing surface features and sample collection; preparing landing sites; deploying retro-reflectors/beacons or markers on landing sites and key geological areas; deploying storage/habitation systems on the surface; and monitoring landing locations [3].

The latest envisioned scenario encompasses two phases for Moon exploration: phase A: preparation for direct human lunar exploration and phase B: human lunar surface exploration. During phase A (several years), CST missions will be performed from Earth (ground control - 90% of the time) and stationary orbit (10% of the time - approximately 10 days). In phase B (approximately 42 days), humans will operate robotic systems from pressurized rovers or space habitats [3].

The communication latency (round-trip time) between Near Rectilinear Halo Orbit (NRHO) and the lunar far-side is approximately 0.4 s (s) (including occasional signal interruptions) at most [*note, the NRHO will see the orbiter come closer to the surface; at its farthest point it will be at 70,000 km which yields the 0.4 s*]; when considering phase B, the latency between surface pressurized rovers or space habitats and robotic assets on the surface can be less than 0.4 s. These short communication latencies tolerate real-time (or on-line) remote control of robots on the lunar surface, allowing high-fidelity telepresence operations [7].

Research indicates that the maximum latency for real-time telepresence is approximately 0.5 s - "latencies within the cognitive window of the human reaction time" - beyond this point, perceptual functions are usually compromised [7, 10].

Research already indicates benefits that can be expected from robotic control under low-latency situation; e.g., "a factor of 2 improvement in speed operating from NRHO ($\sim$0.5 s) instead of Earth when testing multiple users on a lunar driving simulator using varying latency" [3]. These results seem to be particularly relevant, given the fact

[1] Human-Enhanced Robotic Architecture and Capability for Lunar Exploration and Science (HERACLES) project - a project to improve deep space exploration capabilities and generate groundbreaking opportunities for lunar exploration through human-robotic partnership, via cooperation of multiple space agencies.

that one of the main goals of lunar CST missions is achieving an as high as possible scientific return per capital investment for each mission.

Analyses and tests must be carried out in order to define the most efficient operating setup for real-time remote control of space robots. The central goal of the current research project is the development and analysis of real-time telerobotic control systems for exploration of planetary surfaces[2]. In this research project, CST feasibility experiments will be carried out in micro-g (LEO) - aboard the ISS. Since lunar geological exploration/surface preparation for direct human exploration are currently considered as priority missions for the international space community, it was chosen to run CST feasibility experiments to control rovers on Earth from the ISS (i.e. as an analogue to crew in cislunar orbit controlling robots on the lunar surface), including tasks such as navigation and geological sample collection (e.g., rocks), in lunar analogue sites on Earth.

Earth-ground experiments must be carried out prior to CST feasibility experiments (in LEO) in order to refine the telerobotic control systems to be sent to the ISS, as well as to accurately refine the experiments to be conducted in the ISS. This approach also makes it possible to compare operators' performance under Earth-ground and micro-g, as biological adaptation to altered gravity conditions may affect human performance.

This paper describes and compares cognitive and physical performance in 10 operators remotely controlling a prototype rover on a lunar analogue, under different sensory conditions provided by a telerobotics user-interface, visual/visual plus auditory stimuli for a navigation task; visual plus auditory, visual plus somatosensory (force feedback, FFB), and visual plus auditory and somatosensory stimuli for a sample collection task. Three time-delay conditions were simulated for the remote control of a prototype rover, 0, 0.5 s and 1 s, for both the navigation and sample collection tasks, corresponding to telerobotic operations on the lunar surface (e.g., via pressurized rovers or space habitats), from cis-lunar space and from the ISS, respectively. In this study, and due to implementation constraints, auditory stimuli preceded visual and FFB stimuli for increased time-delay conditions, differential of, 350 ms and −850 ms relative to visual and FFB input for 0.5 s and 1 s time-delay conditions, respectively.

2 Problem and Hypothesis

2.1 Problem: Real-Time Telerobotics: The Segmented Human Body

"The biggest challenge to developing telepresence is achieving that sense of "being there." Can telepresence be a true substitute for the real thing? Will we be able to couple our artificial devices naturally and comfortably to work together with the sensory mechanisms of human organisms?"

Marvin Minsky (1980)

[2] This research project is integrated in the METERON project: an European initiative to help prepare for future human-robot exploration missions to the moon, Mars and other celestial bodies.

Telepresence is a concept associated with interactive technology that "enables humans to feel present at remote locations even though not really there"; it allows for humans to better perceive and interact with remote environments via sensing and actuating interactive human interfaces [11, 12, p. 205]. According to [13, p. 11], "the human system interface has two important functions; first, it has to excite the operator senses so as to show the status of the executed task in the remote environment and second, it has to process the operator commands in order to properly control remote devices". Author in [14, p. 22] define *telepresence* as the "extent to which the human operator loses his or her awareness of being present at the site and instead feels present in the artificial environment".

Telepresence has been explored in the field of telerobotics in order to improve operator's efficacy and efficiency (including safety) during Earth-ground and space telerobotic operations. An example refers to real-time (or on-line) telerobotics via FFB control: a method to be applied when the operator experiences low communication latencies between the control station and the remote environment, i.e., in order of hundreds of ms. This approach targets facilitating tasks requiring control of complex robot motion in real-time (e.g., complex navigation tasks; fine motor control of a robot arm) [15].

FFB aims at inducing to the operator a feeling/impression of contact with the remote environment to increase operator's situational awareness [16, 17]. Real-time telerobotic operations via FFB typically proceed as follows: the operator receives visual FB from the remote location (via a GUI), at the same time he makes use of a physical interface, providing FFB (master, i.e., a force reflective hand controller) to control a desired trajectory on a slave robot (e.g., robot arm) in the remote location. The main goal of FFB is to provide back to the operator the forces that the operator imposes on the task while interacting with the remote environment. This is usually achieved via a force torque sensor between the gripper and the robot arm that reflects the forces back to the operator.

Present-day FFB control interfaces (e.g., force reflective hand controllers; exoskeletons) mainly stimulate the following sensory organs in the human body: proprioceptors and vestibular system (muscle spindles, joints and vestibular system; encode information relative to self- motion and body schema, posture and balance) [8].

The author in [15] showed that FFB, at the level of a human can be implemented with delays as high as 0.8 s - a micro-g study focusing on the suitability of FFB applied to a joystick controller (and not robot control per se) indicated a trend for no differences in human stiffness discrimination between Earth and micro-g - astronauts evaluated after a 3-month adaptation period to micro-g. Conversely, [7] stated that teleoperation, under Earth conditions, guided with haptic and visual FB, requires 100 ms latency or less in order to facilitate an operators' performance - haptic FB degrades performance when latencies of 1 s or longer are experienced (e.g., control of robot arms).

Author in [18, p. 8] indicated that tracking and control of movement impulses were deteriorated, in an astronaut, while manipulating a FFB joystick controller in micro-g, compared to Earth - manipulating the joystick in an upright position, "stabilizing his body with a handle for the left hand and module rails for the feet". The astronaut had to control a 2-DOF robot on Earth - manipulate a pointer (end effector) integrated on the robot, in order to accomplish tracking tasks. The astronaut was participating in his third space mission (a total of 410 days spent in space). The study occurred on the 45th flight-day.

Despite a long-lasting interest and only recent developments, it is still unknown at this time if FFB control per se (or any type of haptic feedback), represents an advantage for real-time teleoperation of space robots in altered gravity conditions.

Several Earth-ground studies indicate that FFB improves telerobotic operations compared to no FFB (visual stimulus only) [16, 17, p. 242] - "(…) yields a reduction in task completion time, energy consumption, error indices (…) and the user's cognitive workload". Interestingly, studies also indicate that FFB interfaces do not represent the full spectrum of natural haptic feedback from the environment, thus, compromising situational awareness, reason why FFB tends to have limited benefit on overall telerobotics performance (improvements in order of 50%) [16, 19].

According to [16], it takes significantly more time to perform a task with a teleoperation interface integrated with FFB, compared to when the task is performed manually - teleoperation also tends to be more prone to error. The author performed a meta-analysis of FFB studies, indicating that there is increased risk to damage the material in the remote environment, due to lack of precise force regulation; FFB mechanisms to prevent excessive forces tend to degrade operator performance.

Hence, FFB control per se, may not be sufficient to improve situational awareness in operators.

Studies in micro-g indicate impairments in sensorimotor functions, in astronauts, during short-term spaceflight, compared to Earth: fine manual control decrements (reduced speed and accuracy of movements); degraded postural control (including balance) and sensing of limb position. Sensorimotor impairments have been a primary cause of accidents in space [20, 21]. Astronauts show improvements in sensorimotor performance after 3 to 6 weeks of adaption to micro-g. However, there is no clear evidence that humans can perfectly adapt to sensorimotor perturbations in micro-g [21].

Impairments in sensorimotor functions have also been observed in astronauts exploring the Moon's surface - during the lunar Apollo missions. For example, "During rover operations, crewmembers often misperceived the angles of sloped terrain, and the bouncing from craters at times caused a feeling of nearly overturning while traveling cross-slope (…) this is not surprising given the evidence of tilt-translation disturbances following G-transitions, as incorrect perceptions of vehicle accelerations, tilted terrain, and uneven (bumpy) surfaces may cause inappropriate responsive actions" [22, p. 7].

The previous results are particularly concerning for telerobotic operations in real-time, which require accurate motor control from operators. Degraded sensorimotor functions, in altered gravity, may also change how operators perceive FFB information - decreased ability to discriminate counteracting forces and their relation to physical action; may be highly disturbing for short-term telerobotic space exploration missions (e.g., the ESA CST lunar missions are currently projected to last for 10 to 42 days).

A new approach to real-time space telerobotics is suggested - Multisensory Real-time Space Telerobotics - in order to improve perceptual functions in operators and surpass sensorimotor perturbations caused by altered gravity. It is argued that telerobotic operations, in altered gravity, can be improved by offering enhanced sources of sensory information to the operator - optimizing performance.

Multisensory integration regards the integration of multiple sources of sensory information in the nervous system, in order to generate coherent percepts from the environment and to facilitate goal-directed behavior within it. Combined sources of sensory information (e.g., visual, auditory, somatosensory) tend to facilitate perception and goal-directed behavior [23–25]. This will be the topic of the next sub-chapter.

2.2 Enhanced Sensory Experiences Improve Human Performance on Earth

Hominids evolved during millions of years actively exploring multiple sources of sensory information in the physical environment. Perception is made easier when combining information from multiple sensory modalities [26].

Humans react faster to trimodal stimuli (combined visual, auditory and somatosensory) compared to unimodal or bimodal stimuli (visual; visual plus auditory/somatosensory) - e.g., faster manual reaction times - observed when trimodal stimulus are presented simultaneously, or with small delays [27, 28].

Research in telerobotics has shown that bimodal experiences (combined visual and somatosensory or auditory stimuli) tend to optimize operators' performance, compared to unimodal experiences (visual stimulus only) (e.g., control of robot arms), improvements translated in decreases in task completion times/error rates, lower peak force, less variability and less damaging contact forces [29, 30].

Massimino and Sheridan [29] showed that replacing FFB with auditory information representing forces through sound (sensory substitution) provided a significant advantage in perception of contact forces in operators in peg-in-hole tasks. Auditory stimulus was also an advantage for telerobotic tasks with high temporal delays (3 s), reducing task completion times in operators even while experiencing obstructed views of the remote environment, where FFB was not useful due to increased operational errors (operators were unable to complete tasks due to high instability caused by FFB).

Studies comparing the effects of trimodal experiences in operators' performance are scarce. Research in telesurgery showed that trimodal stimulus (combined visual, auditory and somatosensory - FFB) improved situational awareness in operators [30].

The previous results suggest that enhanced sensory experiences tend to improve perception in humans (including reaction time), compared to restricted sensory experiences on Earth. No micro-g studies have evaluated this relation. In similar fashion, in this paper, it is reasoned that enhanced sensory experiences may improve perception in operators (including reaction time), compared to restricted sensory experiences, in altered gravity - e.g., combined visual, auditory and somatosensory stimuli. In turn, perceptual improvements may optimize space telerobotic operations.

2.3 Hypotheses

Enhanced Sensory Experiences May Improve Telerobotic Performance under both Earth-Ground and Altered Gravity Conditions
As previously mentioned, studies indicate impairments in sensorimotor functions (e.g., manual control) in astronauts during short-term spaceflight. Visual function seems also to be degraded in micro-g, namely during initial stages of spaceflight - e.g., degraded

visuo-motor performance, reduced contrast sensitivity, decrements in near-vision acuity (difficulties in evaluating distances, object's size and shape) [20].

The visual system has less temporal precision than the auditory and tactile systems [31]. Humans react faster to auditory compared to visual and somatosensory stimuli [8]. Auditory signals facilitate tactile perception [32].

Auditory function seems not to be altered in micro-g [20]. Hence, auditory stimulus may be a determinant factor for optimizing space telerobotic operations - the operator can rely on auditory information to better/faster perceive and act in the remote environment, when experiencing sensorimotor and visual degradation.

According to [33, p. 235], sensory feedback delays result in performance degradation (e.g., delays in visual FB significantly increase manipulation times - 1 to 3 s) - Earth. Humans react faster to trimodal stimuli - faster manual reaction time, which is critical for the control of remote robots, via physical interfaces. This response was observed when stimuli are presented simultaneously, or with small time-delays (a few hundreds of ms) - Earth. Hence, trimodal stimulus may improve reaction time in operators, even when experiencing temporal asynchronies between sources of sensory information. Faster reaction time to auditory stimuli may improve motor control of remote robots and decrease operational errors, in operators, when experiencing delays relative to visual and somatosensory information - the operator may faster perceive (interpret)/act in the remote environment (better connection of segmented time events for the prospective attainment of a goal, e.g., continuous motion of a robot arm).
Hence, it is hypothesized that:
(1) H_0: Enhanced sensory experiences (combined visual, auditory and somatosensory stimuli) increase operator's efficacy and efficiency (including safety), in real-time Earth/space telerobotic operations, compared to restricted sensory experiences (unimodal/bimodal stimulation - stimulating the visual and somatosensory or auditory senses), in different time-delay conditions - 0 s (telerobotic operations on the lunar surface via pressurized rovers or space habitats), 0.5 s (telerobotic operations from cis-lunar space) and 1 s (telerobotic operations from the ISS)".

Enhanced Sensory Experiences May Reduce Cognitive and Physical Fatigue in Operators under both Earth-Ground and Altered Gravity Conditions

Significant increases in fatigue denote cognitive and/or physical exhaustion, which is, in most cases, linked to cognitive and physical performance decrements - on both, Earth and micro-g [34–36].

Studies indicate that cognitive and physical fatigue is linked to sensorimotor performance degradation under both Earth and micro-g conditions - including increased risk of accidents [21, 37, 38]. Space telerobotic missions have been subject to incidents due to crew fatigue - several incidents have occurred during the control of the Canada arm (control errors and a near collision) [39]. The previous results are concerning for Earth/space telerobotic missions, which require optimal cognitive and physical performance from the operator in order to be successful.

Cognitive performance decrements are, in most cases, associated with decreases in alertness and increases in cognitive load (CL) [34, 40].

In this research it is argued that enhanced sensory experiences may optimize attention and memory functions in operators, on both, Earth and altered gravity - translated in increases in alertness and decreases in CL.

Alertness concerns the ability to respond to events in the environment and is characterized as a continuous function - maintaining focus during a task in the presence of distracting stimuli. Increased alertness tends to facilitate perception in humans [41]. Researchers have emphasized that alertness relies on bottom-up sensory input [34].

Hence, it is suggested that enhanced sensory experiences (e.g., combined visual, auditory and somatosensory stimuli) may increase alertness levels in operators, compared to restricted sensory experiences (e.g., visual; visual plus auditory or somatosensory), under both Earth and altered-gravity. In turn, increased alertness may improve situational awareness in operators by contributing to accurate mental representations (in working memory - WM) of the remote environment where teleoperation occurs - increased awareness about elements/event sequences (e.g., landscape; robot motion) in the remote site, thus, facilitating action-related functions.

WM capacity is evaluated by accessing CL levels [34, 40]. According to [40, p. 165], "Cognitive workload has been conceptualized as the allocation of mental resources or effort required to maintain adequate performance on one or more tasks". Enhanced sources of sensory information tend to reduce load in the WM system, because information load is spread over multiple sensory modalities - facilitating processing of information [25]. Exposure to enhanced sensory experiences can decrease CL in operators and thus increase their WM capacity - benefiting the retention/maintenance of information in the WM system and facilitating perceptual functions.

Hence, it is hypothesized that:
(2) H_0: "Enhanced sensory experiences (combined visual, auditory and somatosensory stimuli) reduce cognitive and physical fatigue in humans, in real-time Earth/space telerobotic operations, compared to restricted sensory experiences (unimodal/bimodal stimulation - stimulating the visual senses, visual and somatosensory or auditory senses), in different time-delay conditions - 0 s (telerobotic operations on the lunar surface via pressurized rovers or space habitats), 0.5 s (telerobotic operations from cis-lunar space) and 1 s (telerobotic operations from the ISS)". In turn, decreases in cognitive and physical fatigue may improve operators' performance - efficacy, efficiency and safety.

3 Methods

This is a descriptive, quantitative and cross-sectional pilot study - aiming at refining methods and procedures for a large-scale study and to observe if there are significant effects concerning the evaluated conditions. It describes and compares operators' cognitive and physical performance in telerobotic operations - navigation and sample collection tasks - under Earth-ground conditions, to different sensory conditions provided by a telerobotics user-interface in real-time:

- Navigation task - visual stimuli; combined visual and auditory stimuli;
- Sample collection task - combined visual and auditory stimuli; combined visual and FFB stimuli; and combined visual, auditory and (FFB) stimuli.

Three time-delay conditions were simulated for the remote control of a prototype rover - 0 s, 0.5 s and 1 s, for both the navigation and sample collection tasks - corresponding to telerobotic operations on the lunar surface (via pressurized rovers or space habitats), from cis-lunar space and from the ISS, respectively. In this study, auditory stimuli were presented first relative to the visual and FFB stimuli for increased time-delay conditions - differential of −350 ms and −850 ms relative to visual and FFB input for 0.5 s and 1 s time-delay. Due to implementation constraints in preliminary testing, it was not possible to simulate 0.5 s and 1 s time-delay for the auditory stimuli.

Specifically, the experiment evaluated cognitive and physical responses - CL and physical load (PL) - and performance - efficacy, efficiency and safety - in 10 test subjects remotely controlling a prototype rover on an artificial lunar analogue - at ESTEC, ESA, in The Netherlands.

3.1 Sample Selection

Ten participants (ages between - Min = 22 and Max = 67) - 6 males and 4 females - were selected from ESTEC, ESA, in The Netherlands: healthy adults with normal to corrected-to-normal vision, right-handed dominance, no story of neurological disorders and no prior experience with FFB. Two participants had to be replaced after the study adaptation phase due to implementation issues.

3.2 Environmental Conditions and Telerobotics Control Station

Subjects were evaluated inside a full-replica of the ISS Columbus module, sitting on an ergonomic chair (to remove linear acceleration); same environmental conditions: lighting, temperature, humidity, acoustic - to those of the ISS Columbus Module [20] (to further compare cognitive and physical performance under Earth and micro-g):

- Lighting: LED light tubes: 4500 K neutral light spectrum/210 cd;
- Temperature: 20° to 22° Celsius (C);
- Humidity conditions: 40% to 45% relative humidity;
- Acoustic conditions: 51.5 dBA background noise.

Each test subject used a telerobotics control station onboard the full-replica of the ISS Columbus module to command a prototype rover on an artificial lunar analogue (see Fig. 1); performing a navigation task and sample collection task by taking control of a robotic arm equipped with a gripper (located on the right-side of the rover), i.e., to collect rock samples from an artificial surface boulder (geological tasks).

The telerobotics control station for the current experiment, included (see Fig. 2):

- A laptop computer[3]: ZBook HP® 15 - offering *visual stimuli* - the operator was provided with real images of the remote environment (including rover and landscape) via a video stream - as seen from the prototype rover platform head;
- Ear plug monitors: Z3™ - connected to the laptop - offering *auditory stimuli* - the operator was provided with sounds from real world events via an audio stream;

[3] Similar to the laptop computer to be used in the ISS experiment.

- A hand FFB controller (Force Dimension Sigma.7®; bilateral control): *offering somatosensory (FFB) stimuli* - right-hand command interface to the prototype rover, allowing for elaborated control of the rover's robotic arm and gripper - i.e., to collect rock samples from an artificial surface boulder. The bilateral teleoperation system comprised a KUKA LWR 4+ robot arm that acted as a slave and was controlled via the Force Dimension Sigma.7 master device. The robot arm was operated in Cartesian impedance mode with 1000 N/m translational stiffness and 50 Nm/rad rotational stiffness. A Robotiq 2-Finger 85 gripper was mounted via an ATI Gamma force/torque sensor to the arm end-point to enable 6DOF force/torque sensing at the gripper. The implemented FFB technique to couple the master and the slave robot bilateral position-force controller. The references to FFB in this article refer mostly to this implementation:
- A custom-built 3DOF joystick (no FFB) - command interface to the rover, allowing for navigation control of the rover to the intended destination.

Fig. 1. Prototype rover on the artificial lunar analogue.

Fig. 2. Telerobotics control station.

3.3 Study Design

Data collection was conducted in two phases. The first phase concerned sample characterization. The following independent variables were analyzed: birth and chronological age, height (ft/in) and weight (lbs) and baseline neurophysiological - CL. Each test subject's baseline levels for CL were measured in a single 30-min baseline neurophysiological session (reaction time and accuracy tests running on a laptop computer) - to normalize the cognitive metrics for each test subject.

The second phase concerned adaptation stage and experimental tasks.

In the adaptation stage, each subject was familiarized with the telerobotics control station to be used in the experimental tasks. The principal investigator (PI) introduced the telerobotics control station and tasks to the test subject verbally - encouraging the test subjects to perform the tasks as fast as possible. Each test subject interacted with the control station to remotely control the rover on the artificial lunar analogue, performing:

- Simple and complex navigation tasks (60 min): Driving the prototype rover on a traverse with and without surface obstacles - 30 min for each traverse (within each traverse, 5 min for each sensory condition/time-delay condition);
- Simple sample collection task (63 min): Picking rocks from the surface of an artificial boulder with the robot gripper, followed by placing the rocks inside the rover sample container and bring back the rocks to the surface boulder (rather simple manipulative task - artificial boulder with surface inclinations) - 20 min for each sensory condition (within each sensory condition, 7 min for each time-delay condition).

There were two experimental tasks: navigation task and sample collection task. Each test subject was evaluated individually under the two tasks - attending to each experimental task one day a week - in the following sensory/time-delay conditions:

- Navigation task: Visual stimuli and combined visual and auditory stimuli, 0 s, 0.5 s and 1 s time-delay for the visual stimuli, auditory stimuli always presented at 150 ms;
- Sample collection task: Combined visual and auditory stimuli, combined visual and FFB stimuli and combined visual, auditory and FFB stimuli, 0 s, 0.5 s and 1 s time-delay for the visual and FFB stimuli, auditory stimuli always presented at 150 ms.

The sensory/time-delay conditions were counterbalanced in order to avoid conditioning effects on operator's cognitive and physical responses and performance. Evaluating sessions occurred between 10:00 and 11:30 PM, in summer - occasionally, some subjects had to be evaluated from 14:30 to 16:00 PM due to personal restrictions.

In the following paragraphs a detailed description of the experimental tasks is presented.

Navigation Task (accuracy): The PI instructed the test subject to perform a traverse with the rover - drive the rover from a landing site ("start line") following a yellow line (marker on the floor aligned with the rover center) and finishing the task at the "start line". The test subject was instructed to perform the traverse as fast as possible at the same time avoiding collisions with the yellow markers (yellow cones) (see Fig. 3).

Sample Collection Task: The PI instructed the test subject to collect a rock sample from the surface of an artificial boulder with the robot gripper, followed by delivering the rock sample to the sample container. The subject was instructed to perform the overall task as fast as possible at the same time avoiding collisions with the rover structure. Before starting the task, the location of the robot arm/gripper was similar for all subjects - gripper touching a marker ("x") on the surface of the sample container; the rock sample was placed inside a yellow circle on the artificial boulder (same position/location for all subjects). Test subjects delivered the rock sample to a similar location on the sample container ("x" marker); the position of the rover was similar for all the experimental trials (defined via markers on the floor) (see Fig. 4).

The following dependent variables were evaluated under the navigation and sample collection tasks for each sensory/time-delay condition:

- Neurophysiological: CL mean values (from 0 to 1) evaluated via the B-Alert X10® EEG headset system;

- Physical Load (PL): Vector Magnitude (VM) mean values (kCalories/min - energy expenditure) evaluated via the Actigraph wGTX3-BT® monitor;
- Performance:

- Operator's efficacy (sample collection task) - measured through attempt times to pick the rock samples (mean values);
- Operator's efficiency (navigation and sample collection task) - measured through task completion times for the overall experimental task (mean values);
- Operation safety (navigation task) - measured through frequency of collisions - how many times did the operator hit external obstacles with the rover (mean values);

- Behavioral: Collected individually after each task - open interview regarding the telerobotics control station/sensory conditions (subject was asked about the intuitiveness of the user-interface/if sensory conditions optimized task performance).

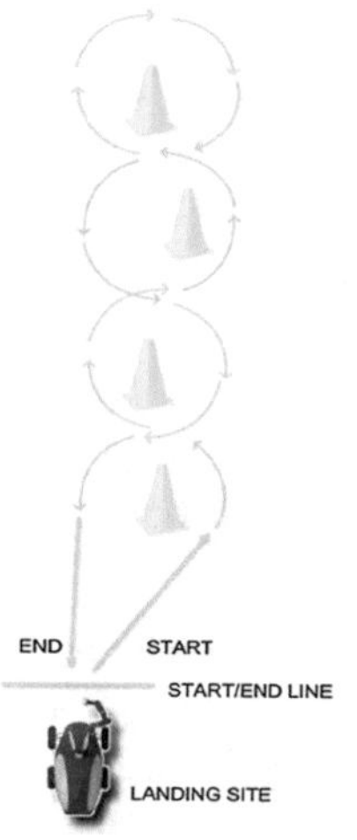

Fig. 3. Navigation Task.

Fig. 4. Sample collection task.

3.4 Measuring Instruments

The Neurophysiological variable CL (values from 0 to 1) - was evaluated via the B-Alert X10® wireless EEG headset system (validated medical device). The EEG acquired nine monopolar channels: Fz, F3, F4, Cz, C3, C4, POz, P3, P4 with fixed gain referenced to linked mastoids. Data acquisition at 256 samples/s. The EEG power spectral analysis quantified absolute and relative PSD for each 1 s epoch using Fast-Fourier transform (50% Kaiser window) for each of the 1 Hz bins from 1–40 Hz and computed the following frequency bands: Delta, Theta, Alpha and Beta. Data was transmitted to a portable computer using Bluetooth and continuously stored using the B-Alert software® at ~45 KB/Min/channel frequency [34, 40].

The PL variable VM[4] (kCalories/min - energy expenditure) was evaluated via the Actigraph wGTX3-BT® monitor - placed on the right wrist. Raw data was recorded at a frequency of 30 Hz - data processed in 1 s epochs and using a validated algorithm [42]. The Actigraph wGTX3-BT® monitor includes a solid state triaxial accelerometer that records motion data on three axes in real-time - vertical (Y), horizontal right–left (X) and horizontal front-back axis (Z). Acceleration from the three axes is converted into activity counts that increase linearly with the magnitude of the acceleration. The Actilife software 6.13.3® was used to analyze the VM data offline.

Performance variables: Efficacy, efficiency and safety evaluated via video recording, 4 GoPro The Hero® cameras placed on the right/left/front rover surfaces.

Behavioral variables: Open interviews were evaluated via audio recording.

4 Results

4.1 Navigation Task

Performance Variables

Results are shown in Figs. 5 and 6 (task completion times variable) and Tables 1 (navigation task) and 2 (sample collection task).

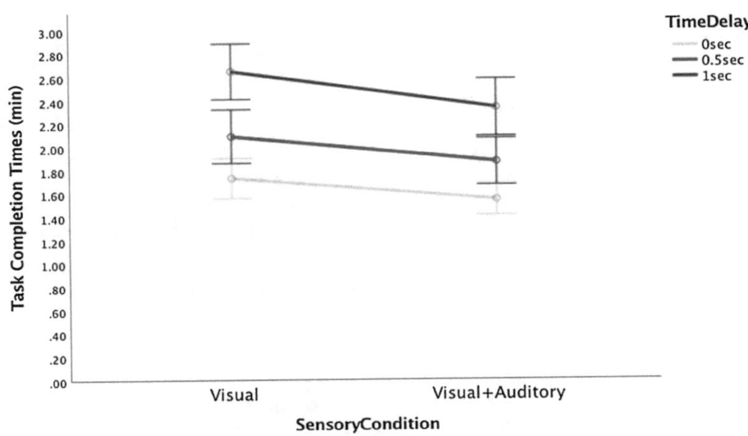

Fig. 5. Navigation task. Task completion times (min).

[4] Magnitude of acceleration represented via the square root of the sum of the squares of each axis of data (kCalories/min).

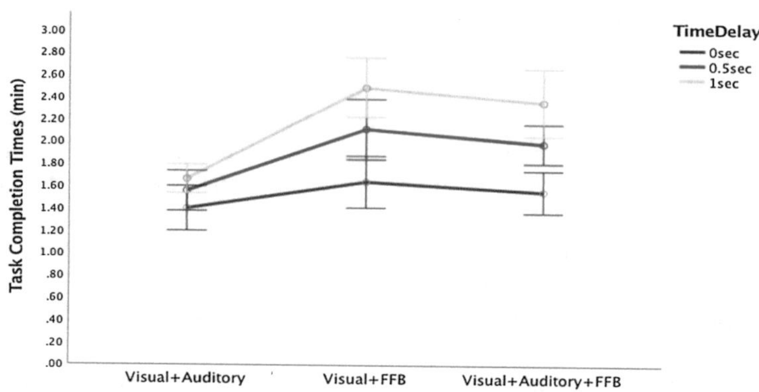

Fig. 6. Sample collection task. Task completion times (min)

Table 1. Navigation Task. Sensory conditions: Visual (V) and Visual Plus Auditory (V+A). Time-Delay: 0 s, 0.5 s and 1 s, auditory stimuli at 150 ms for each time-delay condition. Mean values (M) and Standard Deviation (SD) for Performance variables; Task Completion Times (minutes) and Frequency of Collisions; Neurophysiological variables Cognitive Load (metrics from 0 to 1); and Physical Load variables (Vector Magnitude).

Navigation task
Time-delay/Sensory conditions

	0 s delay		0.5 s delay		1 s delay	
	V	V+A	V	V+A	V	V+A
Dependent Variables	M/SD	M/SD	M/SD	M/SD	M/SD	M/SD
Dependent Variables	M/SD	M/SD	M/SD	M/SD	M/SD	M/SD
Task Completion Times (min)	1.73 ± 0.54	1.54 ± 0.42	2.09 ± 0.73	1.86 ± 0.63	2.64 ± 0.75	2.33 ± 0.77
Frequency of Collisions	0.3 ± 0.76	0.1 ± 0.31	0.1 ± 0.31	0 ± 0	1.4 ± 1.7	0.5 ± 0.5
Cognitive Load (0 to 1)	0.67 ± 0.10	0.65 ± 0.09	0.66 ± 0.08	0.65 ± 0.87	0.65 ± 0.07	0.62 ± 0.07
Physical Load (Vector Magnitude)	6.17 ± 10.5	2.15 ± 3.01	5.45 ± 5.08	2.06 ± 2.10	6.15 ± 6.01	3.36 ± 3.38

Comparison between the sensory/time-delay conditions - dependent variables - was conducted through the Shapiro-Wilk adherence to normality test (Lilliefors correction) followed by the Two-Way Repeated Measures ANOVA test - the latter allowed comparing the mean values at a confidence interval of 95% and 99% ($p < 0.05$ and $p < 0.01$). The Friedman test and the Non-Parametric Wilcoxon Signed-Rank test were used in the absence of normal distribution. The IBM SPSS software®, version 25, was used to perform statistical analysis.

Task Completion Times: Visual and Visual Plus Auditory - 0 s time-delay.

Higher mean values were observed for the visual condition, compared to the visual plus auditory condition, 0 s time-delay - $p < 0.05$.

Task Completion Times: Visual and Visual Plus Auditory - 0.5 s and 1 s time-delay.

Higher mean values were observed for the visual condition, compared to the visual plus auditory condition, 0.5 s and 1 s time-delay - $p < 0.05$; and $p < 0.01$, respectively.

Task Completion Times for Time-delay Conditions: 0 s, 0.5 s and 1 s (visual), 150 ms time-delay (auditory).

Significant differences were observed in the following time-delay conditions: visual 0 s and 0.5 s and 1 s time-delay ($p < 0.01$) - higher mean values for 1 s time-delay; visual plus auditory 0 s and 0.5 s and 1 s time-delay ($p < 0.01$) - higher mean values for 1 s; visual plus auditory 0 s and 0.5 s time-delay ($p < 0.05$) - higher mean values for 0.5 s.

Neurophysiological Variable

CL - Visual and Visual Plus Auditory - 0 s time-delay.

Higher mean values were observed for the visual condition, compared to the visual plus auditory condition, 0 s time-delay - $p < 0.05$.

CL - Visual and Visual Plus Auditory Conditions - 0.5 s and 1 s time-delay.

Higher mean values were observed for the visual condition, compared to the visual plus auditory condition, 1 s time-delay - $p < 0.01$.

Physical Load Variable

Vector Magnitude - Visual and Visual Plus Auditory - 0 s time-delay.

Higher mean values were observed for the visual condition, compared to the visual plus auditory condition, 0 s time-delay - $p < 0.05$.

Vector Magnitude - Visual and Visual Plus Auditory - 0.5 s and 1 s time-delay.

Higher mean values were observed for the visual condition, compared to the visual plus auditory condition, 0 s time-delay - $p < 0.01$; almost significant for the 1 s time-delay condition ($p = 0.074$).

Behavioral Variable

Nine ($n = 9$) test subjects stressed that auditory stimuli facilitated navigation tasks for all the time-delay conditions. One ($n = 1$) test subject emphasized that it was not clear if auditory stimuli facilitated or not navigation tasks for all of the time-delay conditions.

4.2 Sample Collection Task

Performance Variables

Task Completion Times - Visual Plus Auditory/Visual Plus FFB/Visual Plus Auditory Plus FFB - 0.5 s and 1 s time-delay (visual, FFB), 150 ms time-delay (auditory).

Significant differences were observed for 0.5 s time-delay, between the following conditions: visual plus auditory and visual plus FFB ($p < 0.01$) - higher mean values for the visual plus FFB condition; visual plus auditory and visual plus auditory plus FFB ($p < 0.01$) - higher mean values for the visual plus auditory plus FFB condition.

Significant differences were observed for 1 s time-delay, between the following conditions: visual plus auditory and visual plus FFB ($p < 0.01$) - higher mean values

Table 2. Sample Collection Task. Sensory conditions: Visual (V) and Visual Plus Auditory (V +A). Time-Delay Conditions: 0 s, 0.5 s and 1 s, auditory stimuli at 150 ms for each time-delay condition. Mean values (M) and Standard Deviation (SD) for Performance variables; Task Completion Times (min) and Attempts to Collect sample; Neurophysiological variables - Cognitive Load (metrics from 0 to 1); and Physical Load variables (Vector Magnitude).

Sample collection task Time-delay/Sensory conditions									
0 s delay				0.5 s delay			1 s delay		
	V+A	V+FFB	V+A+FFB	V+A	V+FFB	V+A+FFB	V+A	V+FFB	V+A+FFB
Dependent Variables	M/SD	M/SD	M/SD	M/SD	M/SD	M/SD	M/SD	M/SD	M/SD
Task Completion Times (min)	1.4 ± 0.63	1.64 ± 0.63	1.55 ± 0.59	1.55 ± 0.56	2.11 ± 0.85	1.98 ± 0.55	1.66 ± 0.40	2.49 ± 0.84	2.36 ± 0.96
Attempts to Collect Sample	1.06 ± 0.21	1.49 ± 0.73	1.28 ± 0.4	1.17 ± 0.37	1.86 ± 0.86	1.33 ± 0.61	1.17 ± 0.32	2.40 ± 1.07	1.9 ± 1.01
Cognitive Load (0 to 1)	0.64 ± 0.06	0.65 ± 0.82	0.64 ± 0.07	0.62 ± 0.07	0.66 ± 0.05	0.64 ± 0.07	0.63 ± 0.09	0.66 ± 0.08	0.64 ± 0.08
Physical Load (Vector Magnitude)	5.07 ± 2.98	5.31 ± 3.22	3.63 ± 2.1	3.96 ± 2.26	8.6 ± 0.46	5.3 ± 3.33	3.71 ± 3.45	7.18 ± 5.73	4.45 ± 3.18

for the visual plus FFB; visual plus auditory and visual plus auditory plus FFB ($p < 0.05$) - higher mean values for the visual plus auditory plus FFB.

Task Completion Times for Time-delay Conditions - 0 s, 0.5 s and 1 s (visual), 150 ms time-delay (auditory).

Significant differences were observed ($p < 0.01$) in the following conditions - visual plus FFB, 0 s and 1 s time-delay ($p < 0.01$) - higher values for 1 s; and visual plus auditory plus FFB, 0 s and 0.5 s and 1 s time-delay ($p < 0.05$) - higher values for 0.5 s and 1 s compared to 0 s.

Attempts to Collect Sample - Visual Plus Auditory, Visual Plus FFB and Visual Plus Auditory Plus FFB - 0.5 s and 1 s time-delay (visual, FFB), 150 ms time-delay (auditory).

Significant differences were observed for 0.5 s and 1 s time-delay, between the following conditions: visual plus auditory and visual plus FFB ($p < 0.05$) - higher mean values for visual plus FFB. Significant differences were observed for 1 s time-delay, between the following conditions: visual plus auditory and visual plus auditory plus FFB ($p < 0.05$) - higher mean values for visual plus auditory plus FFB.

Neurophysiological Variable

CL - Visual Plus Auditory, Visual Plus FFB and Visual Plus Auditory Plus FFB - 0.5 s and 1 s time-delay (visual, FFB), 150 ms time-delay (auditory).

Significant differences were observed for 0.5 s time-delay, between the following conditions: visual plus auditory and visual plus FFB ($p < 0.01$) - higher mean values for visual plus FFB; visual plus auditory and visual plus auditory plus FFB ($p < 0.05$) - higher mean values for visual plus auditory plus FFB.

Significant differences were observed for 1 s time-delay, between the following conditions: visual plus auditory and visual plus FFB ($p < 0.01$) - higher mean values for visual plus FFB; visual plus FFB and visual plus auditory plus FFB ($p < 0.05$) - higher mean values for visual plus FFB.

Physical Load Variable

Vector Magnitude - Visual Plus Auditory, Visual Plus FFB and Visual Plus Auditory Plus FFB - 0 s time-delay.

Significant differences were observed between the following conditions: visual plus auditory and visual plus FFB ($p < 0.05$) - higher mean values for visual plus FFB.

Vector Magnitude - Visual Plus Auditory, Visual Plus FFB and Visual Plus Auditory Plus FFB - 0.5 s and 1 s time-delay (visual, FFB), 150 ms time-delay (auditory)

Significant differences were observed for 0.5 s time-delay, between the following conditions: visual plus auditory and visual plus FFB ($p < 0.01$) - higher mean values for visual plus FFB; visual plus FFB and visual plus auditory plus FFB ($p < 0.05$) - higher mean values for visual plus FFB condition; visual plus auditory and visual plus auditory plus FFB ($p < 0.05$) - higher mean values for visual plus auditory plus FFB.

Significant differences were observed for 1 s time-delay, between the following conditions: visual plus auditory and visual plus FFB ($p < 0.05$) - higher mean values for visual plus FFB; visual plus FFB and visual plus auditory plus FFB ($p < 0.05$) - higher mean values the visual plus FFB.

Vector Magnitude - Time-delay Conditions - 0 s, 0.5 s and 1 s time-delay (visual, FFB), 150 ms time-delay (auditory).

Significant differences were observed between the following conditions: visual plus FFB 0 s and 0.5 s time-delay ($p < 0.05$) – higher mean values for 0.5 s.

Behavioral Variable

All test subjects revealed not to be aware of auditory stimuli during the sample collection task; that FFB created great instability during the sample collection task for increased time-delay conditions - 0.5 s and 1 s - translated in difficulties to mentally represent the robot arm position due to counteracting forces provided by FFB.

Table 2. Sample Collection Task. Sensory conditions: Visual (V) and Visual Plus Auditory (V+A). Time-Delay Conditions: 0 s, 0.5 s and 1 s, auditory stimuli at 150 ms for each time-delay condition. Mean values (M) and Standard Deviation (SD) for Performance variables; Task Completion Times (min) and Attempts to Collect sample; Neurophysiological variables - Cognitive Load (metrics from 0 to 1); and Physical Load variables (Vector Magnitude).

5 Discussion

5.1 Navigation Task

Test subjects took less time navigating the traverse with the rover in the visual plus auditory condition, compared to the visual condition, 0 s time-delay ($p < 0.05$) - indicating that visual plus auditory stimuli improved performance - time efficiency - in operators. CL values were lower in the visual plus auditory condition compared to the visual condition, 0 s time-delay ($p < 0.05$) - indicating that visual plus auditory stimuli reduced mental fatigue in operators. Neuroscience research also indicated that multimodal stimuli tend to reduce CL compared to unimodal stimuli [25]. PL (movement variability - energy expenditure in kCalories/min) was also lower in the visual plus auditory condition compared to the visual condition ($p < 0.05$) - indicating less movement variability in the subjects, hence higher motor efficiency (lower energy expenditure) in the visual plus auditory condition while driving the rover - using the 3DOF joystick. The former results - improved time efficiency and lower CL and PL in the visual plus auditory condition - are in accordance with the hypotheses 1 and 2 formulated in this study - enhanced sensory experiences increase efficiency (hypothesis 1) and reduce mental and physical effort (hypothesis 2) in operators, compared to restricted sensory experiences - in this case for a 0 s time-delay, corresponding to telerobotic operations from pressurized rovers or space habitats.

Higher task completion times, 0.5 s and 1 s time-delay, for the visual condition, compared to the visual plus auditory condition ($p < 0.05$; and $p < 0.01$, respectively; the effect increases for 1 s compared to 0.5 s) can be justified by auditory stimuli preceding visual stimuli with asynchronies in order of 350 ms and 850 ms. Auditory stimuli seemed to work as a predictor of future events in navigation tasks for both moderate and high time-delay asynchronies with visual stimuli - allowing for better accuracy in operators while controlling the rover, particularly, when there was an increase in temporal window asynchrony between the visual and visual plus auditory conditions - significant differences were higher for 1 s time-delay - $p < 0.01$ (CL values were also higher for the visual condition - $p < 0.01$). The previous result seems to suggest that the larger the temporal asynchrony the more auditory stimuli worked as a predictive factor of the rover behavior - allowing for perception to occur with relatively large time-delays - 850 ms. Auditory stimuli had a strong effect on visual processing to affect perception at relatively large time-delays, given the fact that the strength of multisensory integration tends to decrease when temporal asynchronies surpass approximately 100 ms [43].

Higher task completion times in both visual and visual plus auditory conditions, between 0/0.5 s and 1 s time-delay - higher mean values for 1 s ($p < 0.01$) - indicate that increases in time-delay in the order of 500 ms were linked to increases in task completion times independently of the sensory condition/asynchronies. Higher task completion times in the visual plus auditory condition, between 0 s and 0.5 s time-delay ($p < 0.05$) - higher mean values for 0.5 s, possibly indicate that auditory stimuli reduced completion times, within time-delay conditions, when asynchronies between auditory and visual stimuli were smaller - 350 ms. Previous research indicated that task completion times increase with increasing time-delays [7, 44].

There were no significant differences in frequency of collisions between visual plus auditory and the visual sensory conditions for all the time-delay conditions. Nonetheless, it was observed a trend for frequency collision values in the visual condition to be higher compared to the values observed in the visual plus auditory condition - 0 s time-delay. No significant differences in frequency of collisions may have occurred due to the low level of difficulty associated with the traverse - repeatedly driving the rover at a relatively long distance from the cones (from 1 to 2 m).

Higher PL in the visual compared to the visual plus auditory condition, 0.5 s time-delay ($p < 0.01$) (almost significant for 1 s - $p = 0.074$) - indicates that operators showed less movement variability - higher motor efficiency (lower energy expenditure) - in the visual plus auditory condition while driving the rover - using the 3DOF joystick. A lower time window asynchrony between visual and auditory input -

auditory input preceding visual input at 350 ms - seemed to have facilitated perception of the rover behavior - input to the joystick linked to quick confirmation of the rover behavior via auditory cues - hence, decreasing motor variability in the subjects while manipulating the joystick.

Most participants (n = 9) highlighted that auditory stimuli facilitated the navigation task compared to the visual condition - emphasizing feeling more aware/having better control of the rover in the visual plus auditory condition. This result seems to be linked to the fact that most events in the physical world have both an auditory and visual component.

5.2 Sample Collection Task

There were no significant differences between the visual plus auditory, visual plus FFB and visual plus auditory plus FFB conditions, 0 s time-delay, for task completion times, CL and attempts to pick up a sample - indicating that, for 0 s time-delay, there were no advantages for any of the sensory conditions for reducing task completion times and CL and increasing efficacy in rather simple manipulative tasks - collecting a rock from an artificial boulder with surface inclinations. Despite no significant differences, task completion times and attempts to pick up sample values associated with the visual plus FFB condition, showed a trend to be higher compared to the visual plus auditory and visual plus auditory plus FFB conditions. Furthermore, subjects stressed that they were not aware of auditory stimuli during the sample collection task for all the time-delay conditions. In effect, auditory input to the subjects materialized only after a failed attempt to pick up the rock (after the rock slipped - friction between the rock and the artificial boulder) or if the subjects dropped the rock sample - which worked as a weak sensory cue.

It was also observed that PL was higher in the visual plus FFB, compared to the visual plus auditory condition ($p < 0.05$), 0 s time-delay - not significantly different from the visual plus auditory plus FFB condition. The previous result suggests that FFB may increase physical load/auditory stimuli lowered PL, in operators, in rather simple manipulative tasks, 0 s time-delay, even when the subjects faced weak auditory cues.

It was observed significant differences for task completion times, 0.5 s and 1 s time-delay, between the following conditions - visual plus auditory and visual plus FFB ($p < 0.01$ - higher mean values for visual plus FFB) and visual plus auditory and visual plus auditory plus FFB ($p < 0.05$ - higher mean values for visual plus auditory plus FFB) - indicate that, (1) operators seemed to have relied mostly on auditory stimuli to predict the robot arm behavior in the visual plus auditory condition for both asynchronies of 350 ms and 850 ms, (2) FFB created significant instability in the task for both 0.5 s and 1 s time-delay. The fact that auditory stimuli did not improve task completion times in the visual plus auditory plus FFB condition compared to the visual plus auditory condition - even by preceding visual and FFB stimuli at 350 ms/850 ms - reveals that FFB created significant instability in the task, hence, increasing task completion times. This result is reinforced by no significant differences between the visual plus FFB condition and the visual plus auditory plus FFB condition, at both 0.5 s and 1 s time-delay. Furthermore, it was stressed by all subjects (and verified by video recording) that FFB created significant instability in the task for increased time-delay - 0.5 s and 1 s - translated in difficulties to mentally represent the robot arm position due to counteracting forces provided by FFB - creating significant oscillations of the robot arm, e.g., oscillations of the robot arm due to counteracting forces when the subject still thinks he's pressing a surface (overshooting). The previous result is also corroborated by the following outcomes:

- A trend for task completion times to be higher in the visual plus FFB compared to the visual plus auditory plus FFB condition, 0.5 s and 1 s time-delay - showing that auditory stimuli tended to reduce task completion times.
- A higher number of attempts to pick up the sample in the visual plus auditory plus FFB compared to the visual plus auditory condition, 1 s time-delay ($p < 0.05$) - reinforcing the conclusion that FFB created instability in the task given the fact that the delay between auditory and visual input was the same for both visual plus auditory and visual plus auditory plus FFB conditions - 850 ms.
- Reinforcement that auditory stimuli worked as a predictor of the robot arm behavior:
- Attempts to collect the sample were lower in the visual plus auditory compared to the visual plus FFB condition, 0.5 s and 1 s time-delay ($p < 0.05$) - reinforcing the conclusion that auditory stimuli worked as a predictor of the robot arm behaviour - reducing the number of attempts.
- Higher CL in the visual plus FFB compared to the visual plus auditory condition, 0.5 s and 1 s ($p < 0.01$) - reinforcing that auditory stimuli worked as a predictor of the robot arm behaviour - optimizing task completion times and thus lowering CL.
- Higher CL in the visual plus auditory plus FFB compared to the visual plus auditory condition, 0.5 s ($p < 0.05$) - reinforcing that FFB created instability in the task since auditory stimuli worked as a predictor of the robot arm behaviour - optimizing task completion times and thus lowering CL.
- Higher CL in the visual plus FFB compared to the visual plus auditory plus FFB condition, 1 s ($p < 0.05$) - reinforcing that auditory stimuli worked as a predictor of the robot arm behaviour - optimizing task completion times and thus lowering CL.
- Higher PL (movement variability - energy expenditure in kCalories/min) in the visual plus FFB compared to the visual plus auditory condition, 0.5/1 s ($p < 0.01$ and $p < 0.05$, respectively) - reinforcing that FFB created instability - oscillations of the robot arm - and auditory stimuli helped with the FFB instability as a predictive cue by reducing uncertainty in motor actions and thus PL.
- Higher PL in the visual plus FFB compared to the visual plus auditory plus FFB condition, 0.5 s/1 s ($p < 0.05$) - reinforcing that FFB created instability - oscillations of the robot arm - and that auditory stimuli helped with the FFB instability as a predictive cue by reducing uncertainty in motor actions and thus PL.

Higher task completion times, in the visual plus FFB condition, between 0 and 1 s time-delay ($p < 0.01$) - higher values for 1 s - and higher completion times, in visual plus auditory plus FFB, between 0/0.5 s and 1 s time-delay - ($p < 0.05$) - higher values for 0.5 s and 1 s - indicate that increases in time-delay in the order of 500 ms were linked to increases in task completion times independently of the sensory condition/asynchronies. Previous research indicates that task completion times increase with increasing time-delays [17, 38].

The fact that FFB implemented as position-force controller brings instability in telerobotic tasks with increased time-delays is well-known in telerobotic literature, a potential reason why experts in telerobotics have been suggesting new haptic techniques - e.g., cutaneous devices - in order to improve operators' performance in telerobotic tasks, while at the same time increasing transparency [29, 44, 45].

6 Conclusions and Future Work

Visual plus auditory stimuli improved performance and lowered mental and physical fatigue, in operators, in navigation tasks, compared to the visual stimuli only - 0 s time-delay, corresponding to telerobotic operations from pressurized rovers or space habitats. For 0.5 and 1 s time-delay it is not clear that visual plus auditory stimuli can benefit operators' performance given the fact that auditory stimuli preceded visual stimuli. Hence, it remains to be tested if enhanced sensory experiences - synchronized visual and auditory stimuli - improve operator's performance and lower mental and physical fatigue in increased time-delay conditions - telerobotic operations from cis-lunar space (0.5 s) and from the International Space Station (1 s), respectively.

Auditory stimuli had a strong effect on visual processing to affect perception - optimizing operators' performance in navigation and sample collection tasks - at relatively large time-delays - preceding visual stimuli at 350 ms and 850 ms.

Enhanced sensory experiences - visual plus auditory plus FFB stimuli - did not improve operators' performance, in sample collection tasks, compared to restricted sensory experiences - visual plus auditory or FFB - 0 s time-delay. Nonetheless, subjects stressed that they were not aware of auditory stimuli. In order to improve task performance in sample collection tasks - increase time efficiency and task efficacy and lower mental effort in operators - it is suggested an increase in the strength of the auditory cues via the sonification technique [29] - obtaining auditory stimuli in response to the rover data and interactions (e.g., representing forces through sound - sensory substitution).

Those sonifications may prove to be highly beneficial for operators' performance in sample collection tasks with increased time-delays, given the fact that FFB based on a position-force controller creates significant instability in operators' performance with increased time-delays. The anticipated advantage might be through replacing FFB (that causes difficulties in mental representation of the robot arm movements by representing forces or other robot events) with sound. In addition an alternative to the present FFB implementation - a position-force controller that was optimized for operation under 0 s time-delay conditions primarily – should be investigated. Possibilities are to use alternative control architectures and techniques that specifically deal with FFB under time-delayed conditions - e.g., increasing the stability margin by time-delay dependent controller tuning/damping, suppressing instability by use of a passivity observer, and decoupling by model mediation techniques. In future work, such techniques will be applied in order to diminish the intensity of counteracting forces and thus reduce the robot arm oscillations. In addition, complementing the current setup with force sensing on the gripper to enable both gripping feedback through FFB and sonifications to represent quality of grip, may allow the operator to evaluate and control gripping quality while handling the sample, improving the efficacy. It is then expected to observe improvements operators' performance in future studies. It may also be the case that visual plus auditory plus a new haptic technique - e.g., cutaneous feedback - improves operators' performance in telerobotic operations under increased time-delay.

Increased time-delays - 0.5 s and 1 s - are associated with increases in completion times in operators, in both navigation and sample collection tasks, compared to low time-delays - 0 s - expecting increases in completion times from telerobotic operations on the lunar surface (via pressurized rovers or space habitats) compared to telerobotic operations from cis-lunar space and telerobotic operations from the ISS.

In this study, auditory stimuli was constantly delayed with 150 ms in all configurations and thus preceded both visual and FFB stimuli due to implementation constraints. Nevertheless, results were obtained in order to guide future studies as well as to refine the telerobotic system to be sent to the ISS. In future work, the effect of relatively similar synchronies between visual, auditory and FFB stimuli - input to the operator at 0 s and 0.5 s and 1 s time-delay - will be tested; the network bandwidth is different for computational processing of visual, auditory and FFB data, in that auditory and FFB information usually precedes visual information.

Despite this pilot study's limitations - small sample size and a fixed delay of the auditory delay – methods seem to be adequate and significant results were observed. Nevertheless, future research is necessary to confirm the results.

Acknowledgments. This study was supported by METERON and Fundacao para a Ciencia e Tecnologia, Portugal. We thank to HRE-S and Dr. Julia Teles.

References

1. Fong, T., Zumbado, J.R., Currie, N., Mishkin, A., Akin, D.L.: Space telerobotics: unique challenges to human-robot collaboration in space. Rev. Hum. Factors Ergon. **9**, 6–56 (2013)
2. Bualat, M., Fong, T., Allan, M.B., Bouyssounouse, X., Cohen, T., Fluckiger, L., Gogna, R., Kobayashi, L., Lee, G., Lee, S., Provencher, C.: Surface telerobotics: development and testing of a crew controlled planetary rover system. In: AIAA SPACE 2013 Conference and Exposition, AIAA SPACE Forum, 1–10 (2013)
3. Grenouilleau, J., Carey, W.: Lunar exploration precursor (LEAP) mission description document. ESTEC - European Space Agency (2017)
4. Podnar, G., Dolan, J.M., Elfes, A.: Telesupervised robotic systems and the human exploration of Mars. J. Cosmol. **12**, 4058–4067 (2010)
5. NASA Spaceflight. NASA examines options and flight paths for SLS EM-2 mission. https://www.nasaspaceflight.com/2016/03/nasa-examines-options-flight-paths-sls-em-2/. Last accessed 7 Nov 2017
6. Ennis, M., Fagan, A., Pogue, J., Porter, S., Snape, J.: Feasibility assessment of all science concepts within south pole-Aitken basin. In: Kring, D. (eds.) A Global Lunar Landing Site Study to Provide the Scientific Context for Exploration of the Moon, pp. 477–562 (2012)
7. Lester, D., Thronson, H.: Human space exploration and human spaceflight: latency and the cognitive scale of the universe. Space Policy **27**(2), 89–93 (2011)

8. Giang, W., Santhakumaran, S., Masnavi, E., Glussich, D., Kline, J., Chui, F., Burns, C.M., Histon, J.M., Zelek, J.: Multimodal interfaces. literature review of ecological interface design, multimodal perception and attention, and intelligent adaptive multimodal interfaces. In: Defence R&D Canada Contract Report, 1-269 (2010)
9. Gutfreund, Y., King, A.: What is the role of vision in the development of the auditory space map? In: The Handbook of Multisensory Processing, pp. 574–587. The MIT Press, Cambridge (2012)
10. Lester, D., Hodges, K.V., Anderson, R.C.: Exploration telepresence: a strategy for optimizing scientific research at remote space destinations. Sci. Robot. **2**(7), 1–2 (2017)
11. Minsky, M.: Telepresence. In: Omni magazine, pp. 45–51 (1980)
12. Sheridan, T.B.: Teleoperation, telerobotics and telepresence: a progress report. Control. Eng. Pract. **3**(2), 205–214 (1995)
13. Aracil, R., Buss, M., Cobos, S., Ferre, M., Hirche, S., Kuschel, M., Peer, A.: The human role in telerobotics. Advances in telerobotics. In: STAR 31, pp. 11–24 (2007)
14. Durlach, N.I., Mavor, A.S.: Virtual Reality: Scientific and Technological Challenges. National Academy Press, Washington, DC (1995)
15. Schiele, A., Aiple, M., Kruger, T., van der Hulst, F., Kimmer, S., Sisek, J., Emiel, D.: Haptics-1: preliminary results from the first stiffness JND identification experiment in space. In: Bello, F., et al. (eds.) EuroHaptics 2016, I, pp. 13–22 (2016)
16. Nitsch, V., Farber, B.: A meta-analysis of the effects of haptic interfaces on task performance with teleoperation systems. IEEE Trans. Haptics **6**(4), 387–398 (2013)
17. Wildenbeest, J.G.: The impact of haptic feedback quality on the performance of teleoperated assembly tasks. IEEE Trans. Haptics **6**(2), 242–252 (2013)
18. Weber, B., Schätzle, S., Riecke, C., Brunner, B., Tarassenko, S., Artigas, J., Balachandran, R., Albu-Schäffer, A.: Weight and weightlessness effects on sensorimotor performance during manual tracking. In: Bello, F., et al. (eds.) EuroHpatics 2016, pp. 1–21 (2016)
19. Oosterhout, J.V., Wildenbeest, J.G., Boessenkool, H., Heemskerk, C.J., de Baar, M.R., van der Helm, F.C., Abbink, D.A.: Haptic shared control in tele-manipulation: effects of inaccuracies in guidance on task execution. IEEE Trans. Haptics **8**(2), 164–175 (2015)
20. Clément, G.: Fundamentals of Space Medicine. Springer Science and Business Media, New York (2011)
21. Manzey, D., Lorenz, T.B., Heuers, H., Sangals, J.: Impairments of manual tracking performance during spaceflight: more converging evidence from a 20-day space mission. Ergonomics **43**(5), 589–609 (2000)
22. Paloski, W.H., Oman, C.M., Bloomberg, J.J., Reschke, M.F., Wood, S.J., Harm, D.L., Peters, B.T., Mulavara, A.P., Locke, J.P., Stone, L.S.: Risk of sensory-motor performance failures affecting vehicle control during space missions: a review of the evidence. J. Gravit. Physiol. **15**, 1–29 (2008)
23. Schacter, D., Gilbert, D., Wegner, D.: Psychology. European Edition. Palgrave Macmillan, New York (2011)
24. Stein, B.: The New Handbook of Multisensory Processing. The MIT Press, Cambridge (2012)
25. Sweller, J., Ayres, P., Kalyuga, S.: Cognitive Load Theory. Springer Science, New York (2011)
26. Gutfreund, Y., King, A.: What is the role of vision in the development of the auditory space map? In: The Handbook of Multisensory Processing, pp. 574–587. The MIT Press, Cambridge (2012)
27. Hagmann, C.E., Russo, N.: Multisensory integration of redundant trisensory stimulation. Atten. Percept. Psychophys. **78**(8), 1–21 (2016)

28. Todd, J.W.: Reaction to multiple stimuli. In: Archives of Psychology, 25. Science Press, New York (1912)
29. Massimino, M.J., Sheridan, T.B.: Sensory substitution for force feedback in teleoperation. IFAC Proc. Vol. **25**(9), 109–114 (1992)
30. McMahan, W., Gewirtz, J., Standish, D., Martin, P., Kunkel, J.A., Lilavois, M., Wedmid, A., Lee, D.I., Kuchenbecker, K.: Tool contact acceleration feedback for telerobotic surgery. IEEE Trans. Haptics **4**(3), 210–220 (2011)
31. Wozny, D.R., Beierholm, U.R., Shams, L.: Human trimodal perception follows optimal statistical inference. J. Vis. **8**(3), 1–11 (2008)
32. Bresciani, J.P., Ernst, M.O., Drewing, K., Bouyer, G., Maury, V., Kheddar, A.: Feeling what you hear: auditory signals can modulate tactile tap perception. Exp. Brain Res. **162**, 172–180 (2005)
33. Smith, T.J., Smith, K.U.: The human factors of workstation telepresence. In: *Nasa, Third* Annual Workshop on Space Operations Automation and Robotics (SOAR 1989), pp. 235–250 (1990)
34. Berka, C., Levendowski, D.J., Lumicao, M.N., Yau, A., Davis, G., Zivkovic, V.T., Olmstead, R.E., Tremoulet, P.D., Craven, P.L.: EEG correlates of task engagement and mental workload in vigilance, learning, and memory tasks. Aviat., Space, Environ. Med. **78** (5), B231–B244 (2007)
35. Galvan, R.: Effects of fatigue on simulated space telerobotics performance: a preliminary study analysis. Masters dissertation, Department of Aeronautics and Astronautics. MIT (2012)
36. NASA Evidence Report. Risk of performance decrements and adverse health outcomes resulting from sleep loss, circadian desynchronization, and work overload. In: Human Research Program. Behavioral Health and Performance Element, pp. 1–83 (2016)
37. Fowler, B., Meehan, S., Singhal, A.: Perceptual-motor performance and associated kinematics in space. Hum. Factors **50**(6), 879–892 (2008)
38. Monk, T.H., Buysse, D.J., Billy, B.D., Kennedy, K.S., Willrich, L.M.: Sleep and circadian rhythms in four orbiting astronauts. J. Biol. Rhythm. **13**(3), 188–201 (1998)
39. Williamson, P.: MSS Lessons Learned, DX23/ISS Mechanical & Robotic Systems Training Lesson handout. Houston, TX, NASA JSC DX23 ISS Mechanical and Robotic Systems Training, 7 (2007)
40. Berka, C., Levendowski, D.J., Cvetinovic, M., Petrovic, M., Davis, G., Lumicao, M., Zivkovic, V.T., Popovic, M.V., Olmstead, R.: Real-time analysis of EEG indexes of alertness, cognition, and memory acquired with a wireless EEG headset. Int. J. Hum.-Comput. Interact. **17**, 151–170 (2004)
41. Posner, M.I.: Measuring Alertness. NY Acad. Sci. **1129**, 193–199 (2008)
42. Sasaki, J.E., John, D., Freedson, P.S.: Validation and comparison of ActiGraph activity monitors. J. Sci. Med. Sport **14**(5), 411–416 (2011)
43. Meredith, M.A., Nemitz, J.W., Stein, B.E.: Determinants of multisensory integration insuperior colliculus neurons. I. Temporal factors. J. Neurosci. **7**(10), 3215–3229 (1987)
44. Thomson, J.M., Ottensmeyer, M.P., Sheridan, T.B.: Human factors in telesurgery: effects of time-delay and asynchrony in video and control feedback with local manipulative assistance. Telemed. J. **5**(2), 129–130 (1999)
45. Onda, K., Osa, T., Sugita, N., Hashizume, M., Mitsuishi, M.: Asynchronous force and visual feedback in teleoperative laparoscopic surgical system. In: 2010 IEEE/RSJ International Conference on Intelligent Robots and Systems, pp. 844–849 (2010)

Achieving a High Level of Open Market-Information Symmetry with Decentralised Insurance Marketplaces on Blockchains

Alex Norta[1(✉)], Risto Rossar[2], Mart Parve[2], and Liina Laas-Billson[2]

[1] Department of Software Science, Tallinn University of Technology, Tallinn, Estonia
alex.norta.phd@ieee.org
[2] Black Foundation OÜ, Tallinn, Estonia
info@black.insure

Abstract. The legacy insurance industry is failing to cover risk with capital in a free-market way that is in the interest of the consumer. The status quo is highly discretionary with a small set of agents representing capital who decide about risk coverage in their own interest. Thus, customers and the insurance ecosystem face a wasteful context in the value chain of capital allocation. The advent of blockchain technology such as smart contracts, disruptively changes the status quo towards an improved disintermediated context with rapidly increased speed for bringing insurance products to market. Additionally, blockchain technology also allows for replacing a fraud riddled insurance ecosystem that suffers from qualitative highly bureaucratic decision making. Instead, blockchain technology eliminates inefficient layers of bureaucracy and establishes a novel quantitative organisational governance system based on transparent and traceable rules rooted in mathematics. Black Insurance fills this gap with a distributed blockchain-based insurance platform that overcomes currently existing information asymmetries to reinstate free-market forces for risk coverage with capital in the interest of customers.

Keywords: Insurance · Blockchain · Smart contract · Risk · Capital · Decentralisation · Disintermediation · Distribution

1 Introduction

The essence of the insurance industry is to allocate capital efficiently and effectively for risk coverage [5]. The challenge is to predict the occurrence of risk ahead of time for assured capital allocation [15]. It is not clear to which extent risk prediction is accurate yielding a misallocation of capital as a result. The cause of this mis-allocation is inherent inefficiency in the insurance industry that is offset via high transaction costs. In essence, the decision making for capital allocation does not follow free-market forces and is instead controlled by a

K. Arai et al. (Eds.): CompCom 2019, AISC 997, pp. 299–318, 2019.
https://doi.org/10.1007/978-3-030-22871-2_22

small set of colluding agents with high discretionary power. These agents decide over the capital allocation in their own interest and ignore the customers instead. The discretionary decision-making power is enabled by an overly bureaucratic organisation [31] of the insurance industry as it exists today. Thus, the additional consequences of excessive bureaucracy in the insurance industry is the total masking of free-market forces and a disregard of customer interests.

A more detailed check of the cost structure shows[1] that insurance companies have 20% of average costs for administrative expenses and net profit of gross written premium while for small-sized insurers, the average costs are considerable higher than for larger insurers. Overly bureaucratic decision making [2] is also responsible for the high cost structure. Important information travels through layers of bureaucracy to the key decision makers at the top of an insurance marketplace where information arrives late, distorted and filtered. The decisions are then inappropriate and arrive late on the operational level. With respect to the discretionary capital-allocation problem of a small set of agents, research shows that the situation will worsen [13,14]. The trends show more acquisitions and mergers between existing insurance companies will shrink the set of discretionary agents even more and consequently, free-market forces in the insurance industry predictably further decrease. Literature shows [10] that information asymmetry in the insurance industry is a given problem and discretionary agents exploit this status quo. Thus, we must infer based on the available evidence [9] that overly bureaucratic governance is the cause of this anti free market information asymmetry problem.

The status quo of the insurance industry shows bringing products to market is hindered by a lot of waste, bureaucracy and high transaction costs that are hindering free-market forces to unfold for a high degree of customer satisfaction. A gap exists in the insurance industry with respect to a platform for closing the currently existing information asymmetry that is the result of wasteful bureaucracy. This whitepaper fills the gap by exploring the question how to replace the current insurance ecosystem with a novel blockchain-based insurance platform? To establish a separation of concerns, we deduce the following subquestions: Who are the key stakeholders and their key interests in relationship to the insurance platform? What is the static architecture for the insurance platform? What are the blockchain-supported dynamic interaction protocols within the architecture?

The remainder of the paper is structured as follows. Section 2 introduces a running insurance case that shows the challenges of the status quo, followed by additional presuppositions. Section 3 presents the requirement sets and stakeholders of the BlackInsurance platform. Next, Sect. 4 derives the distributed system architecture from the requirement sets. Section 5 shows the dynamic governance lifecycle the BlackInsurance architecture enables together with the set of discovered blockchain transactions that enable trustless collaboration between the involved stakeholders. Section 6 evaluates the feasibility of a rapid Black-

[1] https://medium.com/BlackInsurance/costs-in-the-insurance-industry-today-b86e225eea0b.

Insurance platform deployment based on pre-existing blockchain technology. Section 7 concludes this whitepaper and presents future work. Finally, note that this paper is derived from a larger whitepaper [28] that is part of a recent startup fundraising event.

2 Motivating Example and Preliminaries

We present the status quo of the insurance-industry deficiencies in Sect. 2.1 with a motivating example. Next, Sect. 2.2 gives preliminaries that are relevant for the remainder of this whitepaper.

2.1 Motivating Example

Insurance companies control the industry with respect to the products that are sold, pricing and affiliated product terms. An insurance product is a set of conditions and pricing of premiums according to which insurers sell insurance contracts. For example, a motor vehicle insurance policy is backed by a corresponding insurance product that describes which accidents and losses are covered, how much premium any given insured has to pay and also the extent of deductibles. Note also that an insurance product is backed by reserve capital to cover risk that usually amounts to circa 40% of the charged capital. The capital purpose is to backstop losses, if the latter exceed received premiums. Consequently, the capital sees returns of the premiums exceed the losses. Capital combined with an insurance product forms the underwriting capacity, ie., the insurance carrier's commitment that a certain volume of an insurance product can be sold. Due to established insurance oligopolies that eliminate the need to change structurally, very little innovation occurs. Despite changing insurance-market needs, over-regulation and inefficient bureaucracy have made it near impossible to place insurance products in the market.

Still, insurance companies are approachable for new insurance products. As Fig. 1 depicts, there is a considerable delay, in average 6 to 18 months, before insurance companies decide whether to accommodate the broker with the desired insurance product, together with the underwriting capacity. The reasons for such a delay are manifold. First, authorities impose many laws and regulations on new insurance products. Secondly, insurance companies are plagued by outdated and costly, arcane bureaucracy that lacks adequate process automation with information technology. Insurance companies also enjoy government protection in the form of regulatory entry barriers and are thus, complacent as they lack an incentive to experiment with innovative products. Additionally, mostly excel sheets are in use that different departments maintain by themselves that over time comprise very diverging content and figures, rendering decision making very challenging.

The problem of diverging assessments per insurance company and department increases over time due to diverging technological standards of information technology in use. A considerable amount of manual bureaucratic labor

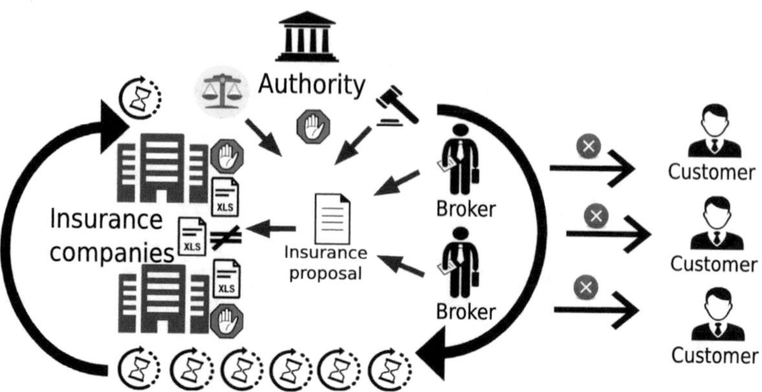

Fig. 1. Status quo of the insurance industry.

combined with diverging excel sheets yields more opportunities for errors. Consequently, entrepreneurs and insurance brokers fail in placing new insurance products on the markets for customers who are left unsatisfied. The oligopolistic insurance-industry structure is also a challenge for smaller- and retail investors who are not able to place their money into stable insurance portfolios, even though the interest is there. Thus, established insurance companies control the industry and retain the profits. Figure 1 also infers that brokers have relevant market know-how as they understand the customers, the industry and the market needs. Thus, brokers are able to design well-targeted insurance products that they cannot bring to market, also because of unavailable underwriting.

2.2 Preliminaries

We discuss next the employed methodology for this whitepaper. The development of the BlackInsurance platforms follows a model-driven design (MDD) methodology [4] that emphasizes the system-design process towards a rapid implementation deployment. A set of chained model types provide an increasingly deeper understanding of the BlackInsurance platform. The sequence of development commences with specifying the platform goals, followed by the static architecture-topology design and finally, we give the dynamic behaviour protocols that exist between architecture components and respective stakeholders. Thus, a goal-model [29] first describes the requirements of the system, a UML component [16] diagram outlines the static architecture topology and thirdly, with business process modelling notation BPMN [1] we specify the dynamic behaviour of the platform comprising key events stored on a blockchain.

Goal models as a notation taken from the agent-oriented modelling (AOM) method [29], capture the requirements of the BlackInsurance platform and comprise the simple notation in Fig. 2(a). Functional goals of a system are depicted as parallelograms, 'quality goals' are synonymously called non-functional requirements in software engineering and depicted as clouds, and agents with specified

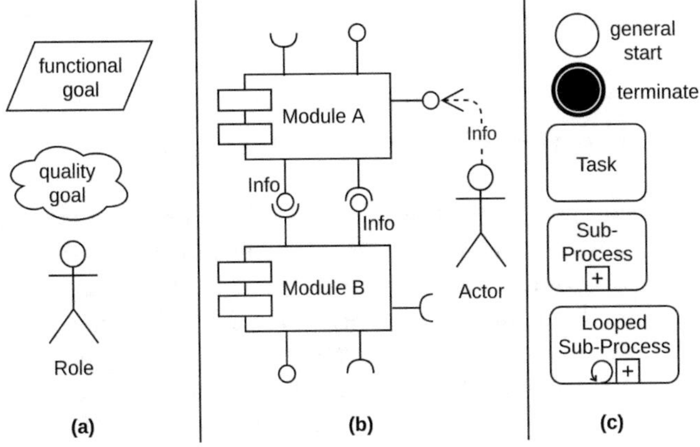

Fig. 2. Goal-model notation in (a), followed by component-diagram notation in (c) and finally, BPMN notation in (c).

roles may be human or artificial. The value proposition is the root of a hierarchically decomposed AOM goal model and denotes the overall systems goal. Quality goals and roles are attached to functional goals and inherited to lower-level functional goals.

Based on the goal model, we derive a UML component diagram [6] to specify the static structure of the BlackInsurance platform. Figure 2(b) depicts UML-notation elements with labeled, rectangle components. The latter, we further refine with sub-components and assign provided- and required interfaces. Figure 2(b) depicts a provided interfaces as a line with a circle and required interfaces are lines with a cup at the end. Actors interact with components of the BlackInsurance platform and are synonymous to roles in goal models.

The dynamic governance behaviour of the BlackInsurance platform we express with BPMN. While this notation has grown extensively and comprises a rich set of modelling elements, Fig. 2(c) depicts that small set of elements we consider for this whitepaper. A dynamic governance lifecycle commences with a general start element that is depicted as a white-filled circle, and terminates eventually for which we consider a black-filled circle. The governance lifecycles comprise atomic tasks that can not be further decomposed. On the other hand a rounded rectangle with a contained plus sign is a non-atomic sub-process that can be further decomposed with lower-level processes. Finally, sub-processes may also loop until a certain exit condition is met, which is denoted as a directed circle. The depicted elements in Fig. 2(c) are connected by directed arcs that we partially label with digital objects that are passed from one element to the next.

Pertaining to governance lifecycles we consider for the BlackInsurance platform, the foundation for this whitepaper is research for peer-to-peer (P2P) collaboration [25] in a service-oriented cloud computing context. The lifecycle is formalised in three separate publications, namely, the setup-phase in [23], the rollout- and enactment phase in [24] and finally, the rollback- and termination phase in [27].

3 Goals of the BlackInsurance Platform

The AOM method [29] is employed for defining the BlackInsurance-system requirements as it is a socio-technical approach to model dynamic and complex distributed systems comprising also human- and software agents. We provide the overall goal model in parts where Sect. 3.1 shows the value proposition for the BlackInsurance platform and Sect. 3.2 comprises the first refinement level for managing a syndicate. Note that due to page limitations in this paper, we provide two further refinement models in the extensive whitepaper [28] for managing a virtual insurance company and for matching insurance risk with syndicate capital that also show goals that require token-based monetisation.

3.1 Value Proposition

In the depiction of Fig. 3, the value proposition being the goal-model root, carries the label *friction-less risk-coverage capacity for insurance brokers.* The functional refinement goals on the first level are *manage a syndicate, manage a virtual insurance company, exchange tokens, viewing performance dashboard, manage black platform.* Note that all these first-level refinement goals are further detailed in subsequent goal models below.

The value proposition in Fig. 3 is the root for a set of goal models that we explain next. *Error handling* infers there exists system-support help for a user to recover from errors. *Error avoiding* is support to anticipate and prevent common errors that occur on the BlackInsurance platform. *Learnable* means that users are able to form syndicates, place insurance products, match both for underwriting, and so on. *Scalable* means the platform is capable of managing a large amount of simultaneously occurring processes. *Secure* as a quality goal decomposes into confidentiality, integrity, availability and authorised actions [3]. Confidentiality is the absence of unauthorised disclosure of information, integrity is the absence of improper system alterations and availability the readiness for correct service. Additionally, authorised actions assure that only a stakeholder with appropriate credentials is able to carry out an action using the BlackInsurance platform.

On the first refinement level of functions, Fig. 3 shows additional quality criteria the we explain below. Fast means that, e.g., the management of a syndicate is carried out without latency involved that is debilitating to platform usability. Performant means that there exists a high throughput of information artefacts, e.g., high-delivery number of tokens. Highly automated means the carrying out of specific functions should either not involve stakeholders at all, or only to a minimal degree. Private means the information related to a platform activity remains undisclosed. Transparent is an opposite goal in that a function's precondition, activity and postcondition are fully visible to the stakeholders. Modifiable means a system function is responsive to contextual changes. For example, if the data

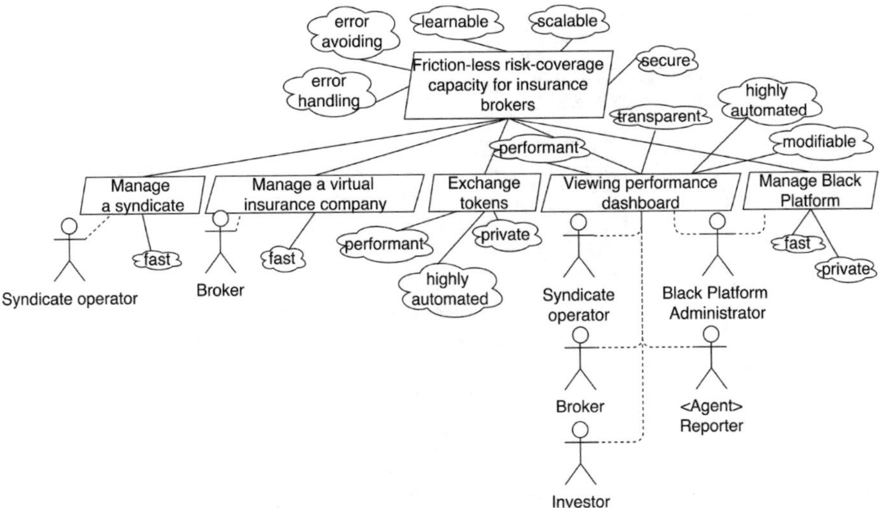

Fig. 3. The value proposition and first refinement level of the BlackInsurance goal model.

set for a specific insurance changes in a region then a viewing dashboard must be able to respond in the display.

The roles in Fig. 3 comprise a syndicate operator who manages a syndicate lifecycle. A broker manages a virtual insurance company and finally, most roles are associated with the function of viewing the performance dashboard, i.e., the syndicate operator, broker, investor black platform administrator and also the reporter agent. Note the latter is an artificial agent who facilitates a highly automated generation of dashboard views. Finally, the black platform administrator also is associated to the function of managing the black platform.

3.2 Syndicate-Management Goals

The refinement goal model in Fig. 4 shows the hierarchy below the functional goal for managing a syndicate. Note that the quality goals and roles in Fig. 3 are inherited down the refined functional-goal hierarchy.

The second functional-goal refinement in Fig. 4 comprises receiving ICO investment, allocating capacity to products, creating and editing syndicates with authenticated content. For the syndicate, new members should be invited and the syndicate itself must be managed according to a set of policies. Such policies must be issued and define the underwriting rules. Also claim rules must be defined. Furthermore, compensation needs to be payed to the claims fund that is recorded. Finally, it must be possible to also delete policies for syndicate management. Next, the progress in the syndicate lifecycle management must be observable, i.e., the performance is monitorable by checking information about policies and claims. A syndicate must also have profit-sharing rules defined for

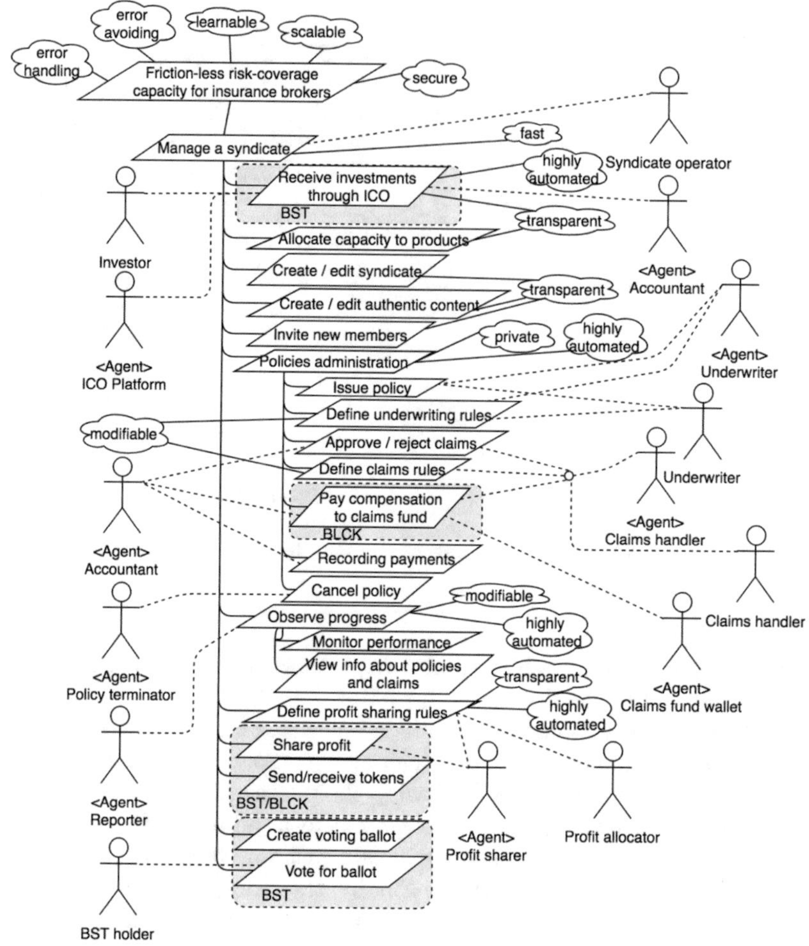

Fig. 4. First partial goal-refinements of the BlackInsurance platform.

sending and receiving tokens. Finally, voting ballots must be created in case of important decision making by a syndicate that requires voting.

Additional quality goals are part of the model in Fig. 4 that we explain in Sect. 3.1. Now these quality goals are assigned in a targeted way to specific functional goals in the hierarchy. There exist also additional roles in the refining goal model, being an investor, underwriter, claims handler, profit allocator and a black security token (BST) holder. Furthermore, to satisfy multiple assignments of the quality goal being highly automated, we introduce a set of software agents that support the human roles in interacting with the BlackInsurance platform. These software agents are an accountant, underwriter, claims agent, claims fund wallet, profit sharer, reporter of observed progress, accountant and ICO platform.

The goal model in Fig. 4 comprises grey areas that stretch over respective functional goals. These grey zones carry labels for the BST and/or the black utility token (BLCK). Thus, this way the goal models serve to start defining the structure of the token economics for the BlackInsurance platform.

4 BlackInsurance Architecture Model

The BlackInsurance architecture we derive from the goal models above. Note that due to page limitations for this paper, the extended whitepaper [28] explains the mapping heuristics from goal models to the architecture. Section 4.1 then shows the architecture as a component diagram together with an explanation.

4.1 Component Diagram

The architecture of the BlackInsurance platform in Fig. 5 shows the result of applying the heuristics that [28] explains. Since the information flow along providing- and receiving interfaces is an addition that goal models can not express, we focus on this explanation primarily and first describe the interaction of actors with respective components. Next, we explain the information exchanges between components too. For brevity, the information exchanges between embedded components is out of focus for this whitepaper.

Component-to-Actor Information Exchanges
The core component of Fig. 5 is the *syndicate manager* that comprises four tokenised sub-components being the investment manager, profit-share manager, policy administrator and ballot manager. There are eight actors who interface with the syndicate manager and we explain the information exchange starting with the investor that continues in a clockwise direction.

The investor in Fig. 5 checks what the total investments are into a syndicate itself and also into insurances a syndicate back with capital to cover insured risk. In return, investments are submitted to the investment manager if an investor chooses so. Furthermore, an investor can also take profit from the component labeled profit-share manager. Next, the syndicate operator submits information about insurance products to the capacity-allocator component and additionally, description information to the content manager. Such descriptions comprise details about the syndicate, the members, investments, adopted insurance products, and so on.

The interaction of BST token holders with the syndicate manager is limited to the ballot manager for casting votes, while syndicate operators receive performance reports from the progress observer and submit member-permission information to the member manager. The profit allocator submits profit-share rules via a receiving interface to the profit-sharing rules manager component. The policy-administrator component interfaces with two actors being the claims handler who submits claims rules and the underwriter who submits underwriting information. Finally, the syndicate operator also draws investment data and submits ICO terms to the investment manager. The syndicate manager also delivers profits to the profit-share manager.

The *virtual insurance company manager* comprises one tokenised sub-component labeled policy-and-claims manager and has one actor type exchange information. The broker submits description information to the insurance-product manager, profile facts to the company-profile manager and members-permission information to the embedded component for managing members of a virtual insurance company. Additionally, the broker also submits policy- and claims data while receiving capacity information from the component labeled capacity receiver.

The *black platform manager* has a ballot-manager component that is tokenised and interfaces with two actor types. The black platform administrator approves syndicates and insurance products via the corresponding respective components. Additionally, the BLCK token holder submits votes to the ballot-manager component.

The *tokens exchanger* depicted in Fig. 5 shows that all contained sub-components are tokenised and has only receiving interfaces for interacting actors. The broker submits purchase orders to the black-service payment tracker and bids to the exchange manager for both BST and BLCK tokens. Also the investor and syndicate operator submit bids to the exchange manager. The syndicate operator submits purchase-order information to the black-service payment tracker and finally, the black platform administrator submits fees to the black-service fee scheduler.

The last component remaining is the *performance dashboard manager* that does not show contained sub-components and interfaces with four respective actors. Both the syndicate operator and investor retrieve performance reports. The same holds for the black platform administrator and broker.

Inter-Component Information Exchange

As the most complex component of Fig. 5 is the *syndicate manager*, we commence with explaining the inter-component information exchange from this architecture location. First, the syndicate manager's wallet exchanges bidirectionally tokens with the token exchanger's embedded exchange manager. Next, the syndicate manager's wallet then also exchanges equally tokens bidirectionally with the wallets of the virtual insurance-company manager and the black platform manager.

Performance data the progress observer provides to the performance dashboard manager and continuing clockwise with the syndicate manager, the content manager provides syndicate descriptions to the syndicate approval manager that in turn provides voting details to the syndicate manager. The latter also receives information about syndicate members from the component labeled members manager.

The policy manager exchanges three types of information sets with the virtual insurance-company manager. First, the latter provides underwriting data to the policy manager that itself returns policy- and claims data. The policy

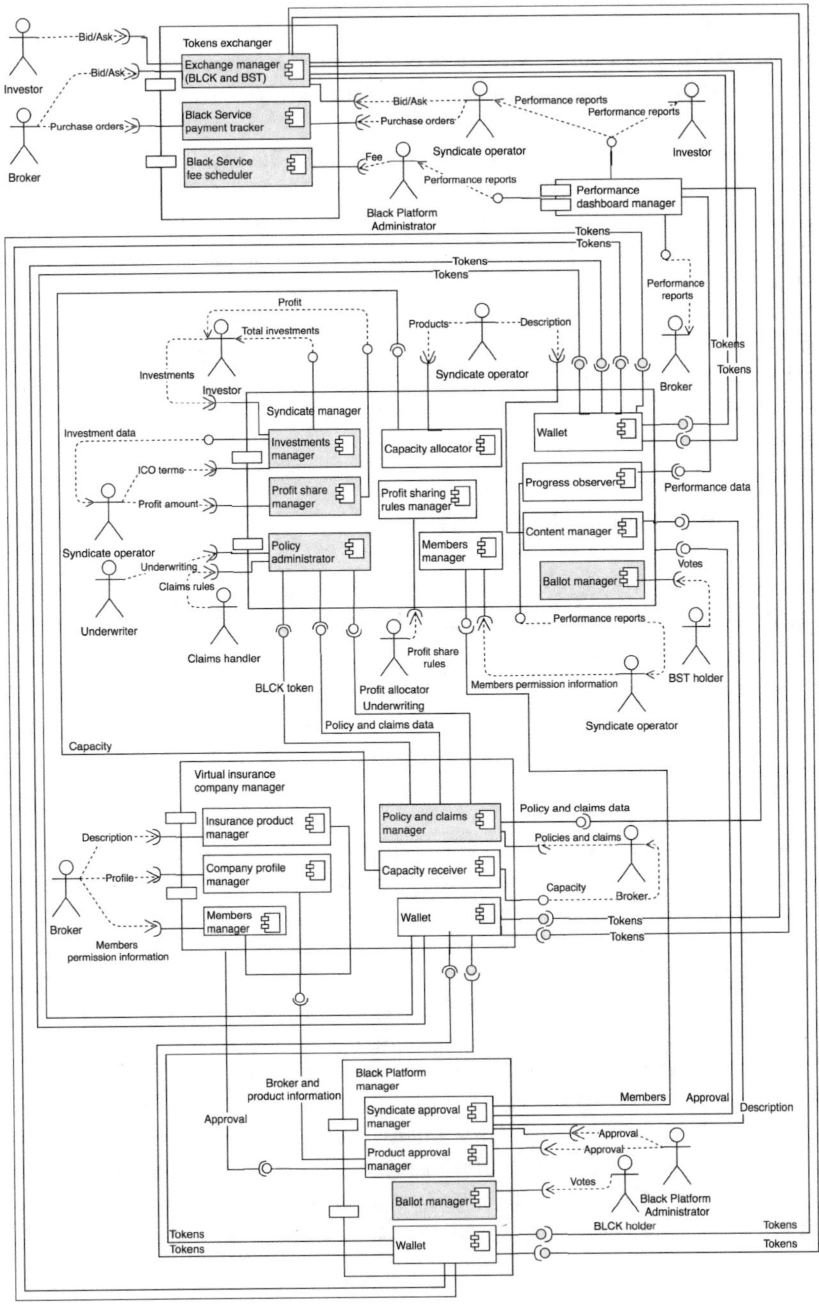

Fig. 5. The BlackInsurance platform system architecture.

administrator component also sends BLCK tokens to the capacity receiver of the virtual insurance-company manager. Finally, the syndicate manager requests capacity information from the capacity receiver of the virtual insurance company manager.

The *virtual insurance company manager* in Fig. 5, submits policy- and claims data to the component being the performance dashboard manager. The wallet bidirectionally sends and receives tokens to the token-exchanger component and the wallet of the black platform manager. Additionally, the company profile manager requests broker and product information from the component labeled product-approval manager. The latter also provides approval information to the virtual insurance company manager. The *token exchanger* component's contained exchange manager bidirectionally sends and receives BLCK- and BST tokens to the wallet of the black platform manager while the remaining information flow is already covered above.

5 Dynamic Blockchain Protocols

For describing the dynamic behaviour of the BlackInsurance system, we first present in Sect. 5.1 the on-chain transactions that are relevant for preventing, or limiting the need for litigation. Section 5.2 describes the governance lifecycles of the syndicates, insurance companies, -products and coverage matchings between insurance products and syndicates.

5.1 On-Chain Transaction Set

The columns in Table 1 list to the very left components that are also part of the architecture model in Fig. 5. Note, four components are listed where on-chain transactions occur that we provide in Table 1 and the component labeled performance dashboard manager is omitted. The next column labeled event shows numbers that identify the set of on-chain transactions. The explanation column describes briefly what the objective of a respective transaction is and finally, the column to the very right lists per transaction row the set of involved stakeholders that are part of Fig. 5, depicting the BlackInsurance architecture.

With respect to on-chain transactions in Table 1, the syndicate manager is the most blockchain-expensive component, followed by the virtual insurance-company manager and the black platform manager. The least blockchain expensive component in Table 1 is the token exchanger, while the performance dashboard manager does not perform any on-chain transactions at all.

5.2 Dynamic Governance

There are three aspects of interleaved, dynamic governance lifecycles that the architecture of Fig. 5 supports. These are first, the governance lifecycle of the syndicate and second, the creation of a virtual insurance company together with issued and managed insurance products. Note that due to page limitations in

Table 1. Set of on-chain transactions for litigation prevention, or -limitation.

component	event	explanation	stakeholder
Syndicate manager	1	Syndicate creation	Syndicate operator
	2	Syndicate investment commitment	Investor, Syndicate operator
	3	Syndicate investment fulfillment	Investor, Syndicate operator
	4	BST creation	Syndicate operator, Black admin
	5	BST issuance	Investor, Syndicate operator
	6	Offer capacity deal	Syndicate operator, Broker
	7	Reserve funds for a capacity deal	Syndicate operator, Broker
	8	Transfer BLCK between wallets	Any platform user
	9	Transfer BST between wallets	Investor, Syndicate operator, Black admin
	10	Create registered content (PoE)	Syndicate operator
	11	Create claim rules	Claims handler
	12	Create underwriting rules	Underwriter
	13	Create profit share rules	Profit allocator
	14	Allocate profit for BST	Investor, Syndicate operator, smart contract, oracle
	15	Add member to Syndicate team	Syndicate operator
	16	Remove member from Syndicate team	Syndicate operator
	17	Settle claim	Claims handler
	18	Underwrite policy	Underwriter
	19	Transfer compensation	Syndicate operator, Broker
	20	Create voting ballot	Syndicate operator
	21	Vote for ballot	Syndicate operator, Investor
Virtual insurance company manager	22	Create company profile	Broker
	23	Create a product	Broker, Black admin
	24	Add member to Company team	Broker
	25	Remove member from Company team	Broker
	26	Accept capacity deal	Syndicate operator, Broker
	27	Counteroffer capacity deal	Syndicate operator, Broker
	28	Create policy quota	Broker, Underwriter
	29	Issue policy	Broker, Underwriter
	30	Transfer premiums	Broker, Syndicate operator
	31	Submit claim	Broker, Claims handler
	32	Submit additional claim evidence	Broker, Claims handler
	33	Transfer BLCK between wallets	Any platform user
Tokens exchanger	34	Update service fee schedule	Black admin
	35	Pay service fee	Any platform user
	36	Submit bid/ask	Any platform user
	37	Accept bid/ask	Any platform user
	38	Settle trade	Black admin
	39	Move tokens through token gateway (atomic swap)	Any platform user, token gateway oracle
Black Platform manager	40	Approve syndicate	Black admin
	41	Approve product	Black admin
	42	Approve syndicate members	Black admin
	43	Approve virtual insurance company	Black admin
	44	Transfer BLCK between wallets	Any platform user
	45	Transfer BST between wallets	Investor, Syndicate operator, Black admin
	46	Mint BLCK tokens	Black admin
	47	Burn BLCK tokens	Black admin
	48	Create voting ballot	Black admin, Syndicate operator
	49	Vote for ballot	Black admin, Broker, Syndicate operator

this paper, the extended whitepaper [28] comprises the first two governance lifecycles. The third governance lifecycle in Fig. 6 describes the matching of insurance products with syndicates where capital of the latter covers the inherent risks in insurances issued by the former. The notation used for the three partial governance-lifecycle figures is explained in Fig. 2(c) of Sect. 2.2. The lifecycle models are derived from published governance research work [24–26] and given in this paper and the extended whitepaper [28] in a conceptual and abstracted format. Additionally, we use the event ID numbers from Table 1 in the governance lifecycles as inserted numbered, red circles, which is not native to pure BPMN notation. The three lifecycle models in the extended whitepaper [28] show that all on-chain transactions from Table 1 are used for governance.

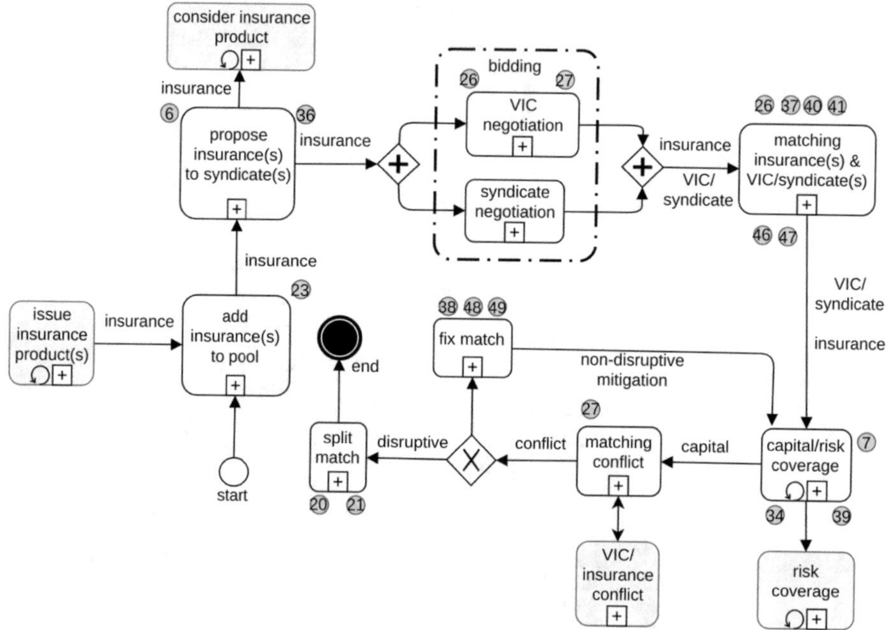

Fig. 6. Lifecycle protocol of insurance-product matching with syndicates.

The final lifecycle in Fig. 6 shows the matching of syndicates and the insurances of VICs that commences with adding insurances to a pool for consideration. The grey sub-process labeled issue insurance products is part of the virtual-insurance lifecycle [28]. The insurances are then proposed to the syndicates and passed on to the grey sub-process labeled consider insurance product by a syndicate of the syndicate lifecycle [28]. Next, a bidding process commences where syndicates and VICs engage in matching attempts that eventually succeed. Post-matching, capital provided by the syndicates covers risks of matched with insurances. Figure 6 depicts that this coverage affects the lifecycle of virtual-insurance governance, as the grey looping sub-process shows with the label risk coverage.

If a conflict occurs during the risk coverage with capital, the matching conflict sub-process exchanges facts with the grey sub-process of VIC/insurance conflict [28]. There are two possible outcomes to such conflicts, one being non-disruptive mitigation due to settling a respective conflict. On the other hand, a conflict results in a disruptive resolution that requires the splitting of a matched insurance with a syndicate.

The lifecycle in Fig. 6 shows additional on-chain transaction IDs that are part of Table 1. Transaction 23 is affiliated with adding insurances to the available pool and records on-chain the creation of a product. Next, for proposing insurances to syndicates, Transaction 6 records offered capacity deals of syndicates and Transaction 36 records on-chain the updated service-fee schedule. After that, during the bidding stage, Transactions 26 and 27 record the acceptance of capacity deals and also of corresponding counteroffers respectively. The subsequent lifecycle stage of concretely matching insurances with syndicates comprises the biggest affiliated set of on-chain transactions. Transaction 26 records first the accepted capacity deal again, followed by Transaction 37 for recording on-chain payed service fees. Transactions 40 and 41 then record atomic swaps for moving tokens through gateways and approving the syndicate respectively and finally, Transactions 46 and 47 record the transfer of BLCK tokens between wallets that respectively involve different sets of stakeholders. Next, the risk coverage of an insurance by capital involves Transaction 7 for recording the allocated reserve funds for a capacity deal and Transactions 43 and 39 for recording the adding of brokers and updating the service-fee schedule. A matching-conflict notification merely involves Transaction 27 for recording a counteroffer capacity deal. If the match can be fixed with a non-disruptive mitigation then Transaction 38 records on-chain submitted bids while Transactions 48 and 49 record the minting and burning of BLCK tokens respectively. Finally, when only a disruptive splitting of a match is the last option, Transactions 20 and 21 record on-chain the creation of a voting ballot and the set of concrete votes on that splitting decision respectively.

6 Rapid Deployment-Feasibility Evaluation

For the BlackInsurance platform architecture in Sect. 4, we evaluate in a paper-based way the availability of pre-existing and emerging technologies for a rapid system deployment. This technology stack also comprises blockchain products and -projects that allow to satisfy specific requirements the BlackInsurance platform must satisfy. In case no existing projects are available, research initiatives are discussed that promise to fill potential gaps in the technology stack. Note that the suggestions for rapid system deployment are tentative.

The remaining structure is as follows. Section 6.1 discusses smart-contract platforms and their verifiability. Next, Sect. 6.2 discusses options for data management on blockchains. Note that due to page limitations in this paper, the extended whitepaper [28] about the BlackInsurance platform comprises an complete discussion about the deployment options and related research, specifically about the important open research issues for identity authentication and smart-contract verification.

6.1 Smart-Contract Platforms

For the BlackInsurance platform, the Ethereum[2] smart-contract platform is a potential candidate. While Proof-of-Work (PoW) [34] is sensible for very critical and high-value transactions that require on-chain consensus establishment, a significant Ethereum disadvantage is that PoW creates performance and scalability challenges for the BlackInsurance platform, if employed for all transactions. As Table 1 shows, the BlackInsurance platform requires a considerable set of blockchain transactions to immutably store critical collaboration events that are litigation relevant.

With respect to a smart-contract platform that successfully validates transactions with Proof-of-Stake (PoS) [35], the equally non-permissioned Qtum[3] smart-contract platform uses simple payment verification (SPV) [22] and the unspent transaction output protocol (UTXP) [12] for supporting lite wallets that only manage transaction headers instead of the entire transaction body. Consequently, Qtum distributed applications (Dapp)[4] run on mobile devices that have low storage and processing capacity. On the other hand, Ethereum has not succeeded in a PoS-version delivery and the realisation date is unclear. While PoS is already fully functioning in the Qtum system, both Ethereum and Qtum offer Solidity as a smart-contract development language, which allows for easy adoption by experienced Solidity developers.

Hyperledger [32] is a permissioned alternative to Ethereum and Qtum for which Fabric [8] is a modular implementation for smart contracts that also offers pluggable extensions for additional functions. More recently, EOS [18] is a novel blockchain with the objective of solving the performance issues and lack of self-governance mechanism of other smart-contract systems. Developed as an operating system, EOS aims to allow for building large-scale Dapps comprising delegated PoS (DPoS) [33] for consensus finding where stakeholders elect delegates for generating and validating a block. While the ideas expressed in the EOS whitepaper [18] promise the delivery of the most advanced smart-contract and blockchain-based operating system, the introduction has been marred by many complications[5], worst of all, EOS reveals itself even as a system with unacceptable centralising tendencies[6]. An emerging alternative is IOTA[7] that uses directed acyclic graphs (DAG) instead of a blockchain to allow for highly scaling simultaneous transactions with fast confirmation times irrespectively of their size. Also Hashgraph[8] is an alternative DAG system comparable to IOTA.

[2] https://www.ethereum.org
[3] https://qtum.org/ens.
[4] https://qtumeco.io/dapps.
[5] https://www.bitrates.com/news/p/eos-mainnet-marred-by-mistakes-miscommunication-and-monopolies.
[6] https://www.bitrates.com/news/p/eos-mainnet-marred-by-mistakes-miscommunication-and-monopolies.
[7] https://iota.org/.
[8] https://hashgraph.com/.

Since the BlackInsurance platform with the architecture of Fig. 5 requires a considerable amount of smart contracts, it is important to check with tool support the soundness of these smart contracts before enactment and also eliminate concurrency conflicts and dependability issues [3]. We discover several pre-existing systems for evaluating contracts that are unfortunately still very unsatisfactory. The available beta-version of the Securify[9] online service claims to formally verify Solidity smart contracts for insecure code that comprises critical security issues. Securify also still lacks good documentation and does not allow to estimate how the online service checks formal properties. Based on an online Securify example, we discover the implemented heuristics that are available in the beta-version check for transaction recordings, recursive calls, insecure coding patterns, unexpected Ether flows and the use of untrusted inputs in security operations. Equally not mature are the Embark-framework[10] and Populus[11] for smart-contract development and deployment for satisfactory formal verification and evaluation.

A remedy for overcoming the currently existing lack of verification tools for Solidity is to consider a platform that uses a functional programming language for smart-contract development. For example, Cardano[12] uses functional Haskell [30] as a programming language for which verification tools are available [19]. Also Aeternity[13] uses Erlang [20] that also has been well researched [17] for verification. Even the use of matured programming languages such as Java for the NEO[14] platform pose the opportunity to consider pre-existing verification tools [11]. Such verifiability of smart contracts is specifically important for cases where legal compliance must be demonstrated. Also note that government-independent identity authentication with blockchain-technology support [36] is an essential issue for engaging in a digital and decentralised insurance marketplace with legal relevance.

6.2 Data Management with Blockchains

For blockchain-based mass data management of the BlackInsurance platform, the Inter Planetary File System[15] (IPFS) [7] is a suitable pre-existing system. IPFS is a peer-to-peer (P2P) and distributed hypermedia protocol where the content is addressable with blockchain use. IPFS is open source, highly performant and provides a decentralized block-storage model. Thus, IPFS comprises hyperlinks that address data sets in a block-storage model where data is distributed across several computers.

BigchainDB [21] is a suitable system for complex operations on large datasets and also for constructing profiles. The main features of BigchainDB are that

[9] https://securify.ch/.
[10] https://github.com/iurimatias/embark-framework.
[11] http://populus.readthedocs.io/en/latest/.
[12] https://www.cardano.org/en/home/.
[13] https://aeternity.com/.
[14] https://neo.org/.
[15] https://ipfs.io/.

it is a blockchain database with immutable decentralised control where it is possible to create and move digital assets. An advantage of BigchainDB for the BlackInsurance platform is the capability to integrate other decentralised blockchain systems such as IPFS, Ethereum, Qtum, etc.

7 Conclusion

This whitepaper presents the BlackInsurance platform that uses blockchain technology and smart contracts for the rapid deployment of syndicate-provided capital for the coverage of risks that insurance products cover. For that, we provide a running case about the traditional insurance-industry problems, followed by BlackInsurance requirement goal models that also arrange the relationship of stakeholders with the platform. We next derive the BlackInsurance architecture from the goal models where the data-exchange channels are specified between architecture components and stakeholders. Next, the dynamic system-protocol study commences with the set of component-specific on-chain transactions and three governance lifecycles for the establishment of syndicates and insurance-products issuing virtual insurance companies, and the matching between syndicates and respective insurance products. These governance lifecycles allow for the assignment of the on-chain transactions. We study also in detail the token structure and economics based on the requirements of the BlackInsurance platform and its dynamic behaviour protocols. Finally, we evaluate the availability of pre-existing technologies for the rapid implementation and deployment of the BlackInsurance platform focusing on important aspects such as smart-contract platforms, blockchain-based data management, solutions for identity authentication and finally, available research results realizing a high automation degree with the use of artificial agents and research that backs the lacking availability of tools for assuring the security of smart contracts.

We show a deployment of the BlackInsurance platform requires a technology stack that comprises diverse blockchain systems. Several smart-contract platforms are available that have respective advantages and disadvantages. Essential is to understand with which smart-contract platforms verified compliance is achievable versus others that offer for certain aspect of the BlackInsurance architecture suitable consensus mechanisms, mobile usability, and so on. We also consider it important to use separate blockchain-systems for data management and identity authentication. It is not credible to assume that only one blockchain system is able to cover all functional- and quality-goal aspects of the BlackInsurance platform. Finally, we stress that the topic of smart-contract verification for soundness and security is still an ongoing research issue as the currently available tools are not satisfactory.

References

1. Business Process Modeling Notation (BPMN), Version 2.0. Object Management Group (2011). http://www.bpmn.org/spec/BPMN/2.0/

2. Anbarci, N.: Insurance and behavioral economics: improving decisions in the most misunderstood industry. By Howard C. Kunreuther, Mark V. Pauly, and Stacey McMorrow. Cambridge and New York: Cambridge University Press, 2013. J. Econ. Literature, **53**, 124 (2015)
3. Avizienis, A., Laprie, J., Randell, B., Landwehr, C.: Basic concepts and taxonomy of dependable and secure computing. IEEE Trans. Dependable Secure Comput. **1**(1), 11–33 (2004)
4. Balasubramanian, K., Gokhale, A., Karsai, G., Sztipanovits, J., Neema, S.: Developing applications using model-driven design environments. Computer **39**(2), 33–40 (2006)
5. Baranoff, E., Sager, T.: The relations among asset risk, product risk, and capital in the life insurance industry. J. Banking Finan. **26**(6), 1181–1197 (2002)
6. Bell, D.: UML Basics: The Component Diagram. IBM Global Services (2004)
7. Benet, J.: IPFS-content addressed, versioned, P2P file system. arXiv preprint arXiv:1407.3561 (2014)
8. Cachin, C.: Architecture of the hyperledger blockchain fabric. In: Workshop on Distributed Cryptocurrencies and Consensus Ledgers (2016)
9. Cai, J., Liu, Y., Qian, Y., Yu, M.: Information asymmetry and corporate governance. Q. J. Finan. **5**(03), 1550014 (2015)
10. Chen, H., Lu, Y., Weiss, M.: Information risk and the cost of equity capital revisited: evidence from the US property-casualty insurance industry (2017)
11. Chimento, J., Ahrendt, W., Pace, G., Schneider, G.: StaRVOOrS: a tool for combined static and runtime verification of Java. In: Runtime Verification, pp. 297–305. Springer (2015)
12. Croman, K., Decker, C., Eyal, I., Gencer, A.E., Juels, A., Kosba, A., Miller, A., Saxena, P., Shi, E., Sirer, E.G., et al.: On scaling decentralized blockchains. In: International Conference on Financial Cryptography and Data Security, pp. 106–125. Springer (2016)
13. Cummins, J., Klumpes, P., Weiss, M.: Mergers and acquisitions in the global insurance industry: valuation effects. Geneva Pap. Risk Insur.-Issues Pract. **40**(3), 444–473 (2015)
14. Du, K., Sim, N.: Mergers, acquisitions, and bank efficiency: Cross-country evidence from emerging markets. Res. Int. Bus. Finan. **36**, 499–510 (2016)
15. Ellul, A., Jotikasthira, C., Lundblad, C., Wang, Y.: Mark-to-market accounting and systemic risk: evidence from the insurance industry. Econo. Policy **29**(78), 297–341 (2014)
16. Stevens, P., Pooley, R.: Using UML: Software Engineering with Objects and Components. Addison-Wesley Longman Publishing Co., Inc. (1999)
17. Fredlund, L., Gurov, D., Noll, T., Dam, M., Arts, T., Chugunov, G.: A verification tool for erlang. Int. J. Softw. Tools Technol. Transfer **4**(4), 405–420 (2003)
18. Grigg, I.: EOS – An Introduction. Whitepaper (2017). http://iang.org/papers/EOS_An_Introduction.pdf
19. Leucker, M., Noll, T., Stevens, P., Weber, M.: Functional programming languages for verification tools: a comparison of standard ml and haskell. Int. J. Softw. Tools Technol. Transfer **7**(2), 184–194 (2005)
20. Logan, M., Merritt, E., Carlsson, R.: Erlang and OTP in Action. Manning Publications Co. (2010)
21. McConaghy, T., Marques, R., Müller, A., De Jonghe, D., McConaghy, T., McMullen, G., Henderson, R., Bellemare, S., Granzotto, A.: BigchainDB: a scalable blockchain database. White Paper, BigChainDB (2016)

22. Nakamoto, S.: Bitcoin: a peer-to-peer electronic cash system (2008)
23. Norta, A.: Creation of smart-contracting collaborations for decentralized autonomous organizations. In: International Conference on Business Informatics Research, pp. 3–17. Springer (2015)
24. Norta, A.: Establishing distributed governance infrastructures for enacting cross-organization collaborations. In: International Conference on Service-Oriented Computing, pp. 24–35. Springer (2015)
25. Norta, A.: Designing a smart-contract application layer for transacting decentralized autonomous organizations. In: International Conference on Advances in Computing and Data Sciences, pp. 595–604. Springer (2016)
26. Norta, A.: Self-aware smart contracts with legal relevance. In: 2018 World Congress on Computational Intelligence (WCCI). IEEE (2018, forthcoming)
27. Norta, A., Othman, A., Taveter, K.: Conflict-resolution lifecycles for governed decentralized autonomous organization collaboration. In: Proceedings of the 2015 2nd International Conference on Electronic Governance and Open Society: Challenges in Eurasia, pp. 244–257. ACM (2015)
28. Norta, A., Risto, R., Parve, M., Laas-Billson, L.: A Decentralized Insurance Marketplace on Blockchains for Achieving a High Level of Open Market-Information Symmetry - Preprint Version 9b (2018). https://www.black.insure/scientific-whitepaper.pdf?pdf=scientific-whitepaper.pdf
29. Sterling, L., Taveter, K.: The Art of Agent-Oriented Modeling. MIT Press, Cambridge (2009)
30. Thompson, S.: Haskell: The Craft of Functional Programming, vol. 2. Addison-Wesley, Boston (2011)
31. Tran, T., Weiss, E., Ruhsam, C., Czepa, C., Tran, H., Zdun, U.: Enabling flexibility of business processes by compliance rules: a case study from the insurance industry (2015)
32. Vukolić, M.: Rethinking permissioned blockchains. In: Proceedings of the ACM Workshop on Blockchain, Cryptocurrencies and Contracts, BCC 2017, pp. 3–7. ACM, New York (2017)
33. Zheng, Z., Xie, S., Dai, H., Wang, H.: Blockchain challenges and opportunities: a survey. Work Pap. 2016 (2016)
34. Vukolić, M.: The quest for scalable blockchain fabric: proof-of-work vs. BFT replication. In: International Workshop on Open Problems in Network Security, pp. 112–125. Springer (2015)
35. Vasin, P.: Blackcoin's proof-of-stake protocol v2. **71** (2014). https://blackcoin.co/blackcoin-pos-protocol-v2-whitepaper.pdf
36. Leiding, B., Norta, A.: Mapping requirements specifications into a formalized blockchain-enabled authentication protocol for secured personal identity assurance. In: International Conference on Future Data and Security Engineering, pp. 181–196. Springer (2017)

Analysis of Argumentation Skills for Argumentation Training Support

Hayato Hirata[1(✉)], Shogo Okada[2], and Katsumi Nitta[3]

[1] Asahi University, Hozumi, Mizuho-City, Gifu 501-0296, Japan
hirahaya@alice.asahi-u.ac.jp
[2] Japan Advanced Institute of Science and Technology, Ishikawa 923-1292, Japan
[3] Tokyo Institute of Technology, Meguro-Ku, Tokyo 152-8550, Japan

Abstract. In law schools, to educate students in argumentation skills, argumentation training is often conducted. To reduce the burden of a supervisor, we developed an argumentation agent. By using this argumentation agent, students can join the training for themselves without help from a supervisor. However, after the training, evaluation of the quality of argumentation is necessary. In this paper, we propose a method to evaluate argumentation skills from argumentation records. At first, we attach annotation to the utterances in the record, and extract several feature pattern of annotations. Then, we ask a supervisor to make a score of argumentation skills for argumentation records. Finally, we analyze correlation among feature patterns and scores by multiple regression analysis, and find on what features the supervisors focused when they give a score.

Keywords: Argumentation skills · Competitive argumentation · Cooperative argumentation · Factor · Argumentation protocol

1 Introduction

In law schools, in order to educate students in argumentation skills, argumentation training sessions such as a moot court, a mock arbitration, a mock facilitation or a mock negotiation are often conducted. In these training sessions, two types of argumentation are included – a competitive (adversarial) argumentation and a cooperative argumentation. As the purpose of a competitive argumentation is to defeat the opponent logically, the logicalness of argumentation is important. On the contrary, as the purpose of the cooperative argumentation is to reach a consensus, not only the logicalness but keeping good relations with the opponent and selecting a proper strategy of compromise is also important. Therefore, we need various criteria to evaluate argumentation skills.

It is burden to a supervisor to conduct an argumentation training because it takes a lot of time to observe the argumentation and give advice to students. To reduce the burden of a supervisor, we have developed an argumentation agent using a humanoid robot. This agent is a speech interactive system with various argumentation strategies, and the student can argue with this agent by voice. By using this agent, a student can take part in the argumentation training without the help of a supervisor. However, there

K. Arai et al. (Eds.): CompCom 2019, AISC 997, pp. 319–334, 2019.
https://doi.org/10.1007/978-3-030-22871-2_23

is another job for the supervisor. After the argumentation, a supervisor has to evaluate argumentation skills and give a score. Evaluating argumentation skills is difficult problem for a supervisor because there are no definite common criteria. In the case of a debate contest, the criteria of scoring argumentation skills are simple because the debate is a typical competitive argumentation, and the logicalness, the persuasiveness and the manner of speech are mainly evaluated. However, in the case of the facilitation and negotiation which are typical cooperative argumentations, criteria of scoring are not clear. Therefore, different supervisors may give differing score to the same argumentation training. If we can define concrete criteria, it is possible to educate on argumentation skills effectively.

There are several avenues of research regarding evaluation argumentation skills. For example, Shibayama analyzed legal essays and found correlation between the score of essays and issue points included in the essays [1]. Lynch analyzed the legal sentences and found correlation between the score and features of logical graphs extracted from legal sentences [2]. Volha et al. developed a virtual debate coach which assesses the argumentation skills from not only utterances but nonverbal data [3]. However, these research approaches did not give so much attention to the evaluation of the cooperative argumentation, and they did not analyze the contents of argumentation topics and treatment of topics is superficial.

The objective of this paper is to propose a method of analyzing argumentation skill of both of competitive argumentation and cooperative argumentation, considering the background knowledge of topics of argumentation.

This paper is organized as follows. In Sect. 2, we introduce the overview of an argumentation agent. It is designed to argue with a student about a given problem, and make an argumentation record. In Sect. 3, we explain how an argumentation record is analyzed. To each utterance in an argumentation record, several annotations are attached by using an annotation editor. In Sect. 4, we consider criteria of argumentation skills by observing the score sheet of the Intercollegiate Negotiation Competition. In Sect. 5, we show a method to estimate scores of argumentation skills from a sequence of annotations of an argumentation record.

2 Argumentation Agent

2.1 Overview of Argumentation Agent

An argumentation agent is developed on the humanoid robot, Pepper, developed by SoftBank Co. (Fig. 1). *Pepper is the world's first social humanoid robot able to recognize faces and basic human emotions. Pepper was optimized for human interaction and is able to engage with people through conversation and his touch screen* [4].

The flow of argumentation agent is shown in Fig. 2 [5]. When a student conveys a message to an argumentation agent, by speech recognition module, the speech signal is converted to a text message. Then, by referring to the background knowledge and the argumentation protocol, it enumerates various candidates of responses, and based on the argumentation strategy, the best one is selected. We prepared several argumentation strategies such as aggressive strategy, compromising strategy, and so on.

While a student speaks, this agent observes features of behavior and voice signal of the student and recognizes whether his speech is enthusiastic or not. This information is also considered to select the best response.

Fig. 1. Communication with pepper

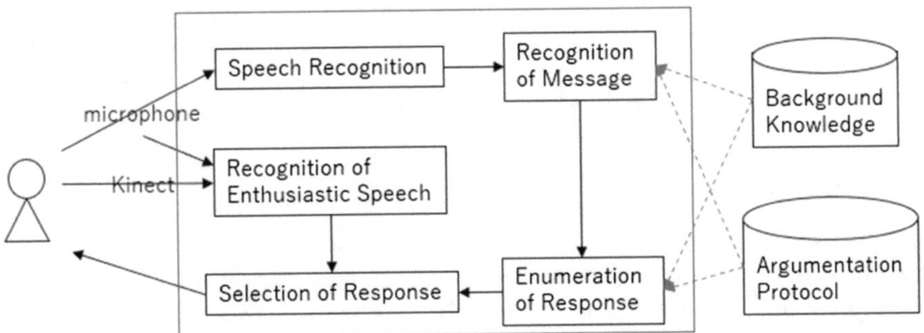

Fig. 2. Flow of an argumentation agent

Following is an example between a student (S) and Pepper (P). Response of Pepper may vary according to the student's voice and behavior.

S> If you leave garbage around your house, the living environment of neighbors starts to deteriorate. Please get rid of garbage as soon as possible.

P> I don't want to get rid of it because some of it is recyclable.

S> It is not recyclable because you have not separated recyclable items from others.

P> I will separate recyclable items when I need money.

S> I don't believe it because you have never separated anything up to now

2.2 Background Knowledge

Before the argumentation training, a supervisor has to decide the argumentation issue. In many cases, the issue is composed of several sub issues. To use the argumentation agent, the supervisor is required to input the background knowledge of the argumentation issue in the form of a factor hierarchy. A factor is an abstraction of a fact or a legal concept introduced by HYPO and CATO system [8, 9]. The main issue and sub issues are included in the factor hierarchy (Fig. 3). The argumentation agent refers to the factor hierarchy to enumerate candidates of response.

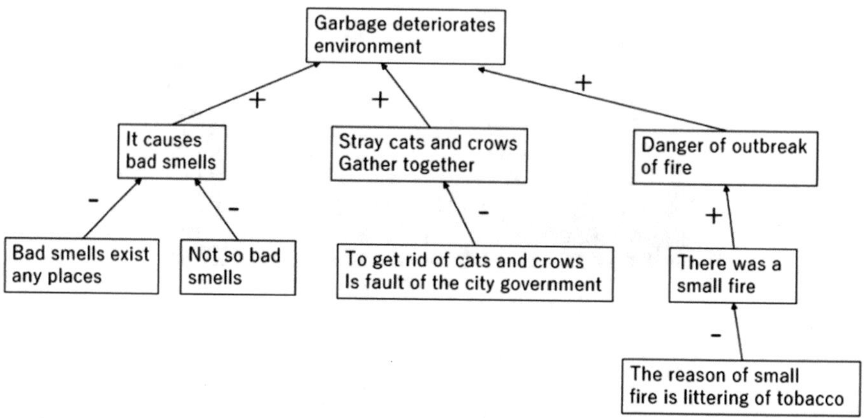

Fig. 3. An example of a factor hierarchy

2.3 Argumentation Protocol

To participate in the argumentation training, a student has to obey the argumentation rule. For example, if a participant makes a proposal, the other side may accept or reject the proposal immediately. In such a situation, the other side cannot claim other issues until he responds to the proposal. Such argumentation rules are given as an argumentation protocol of the Pleading Game [7]. This protocol is based on the pleadings procedure and Toulmin diagram, and it defines a condition and an effect for each speech act such as claim, concede, deny, defend and so on.

We modified the original protocol to fit the actual argumentation (Fig. 4). If both participants obey such argumentation rules, the argumentation will go smoothly.

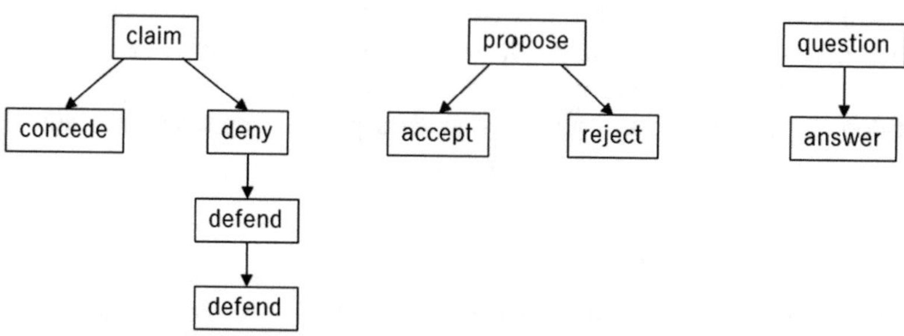

Fig. 4. A part of argumentation protocol

3 Analysis of Argumentation Skills

3.1 Overview of Argumentation Analysis Support System

In our argumentation support system, a student argues with the argumentation agent and its utterances are stored as argumentation records. The argumentation record is analyzed and its argumentation skills are evaluated. This analysis is composed of the learning phase (Fig. 5) and the evaluation phase (Fig. 6). In the learning phase, we gather sample argumentation records about the problem from students. The supervisor evaluates their argumentation skills by referring to the evaluation items. Then, annotations are attached to the argumentation records using an annotation support tool, CORTE, and from the sequence of annotations, several feature patterns are extracted. By correlation analysis between scores and feature patterns, we make the model of estimating score in the form of correlation coefficients (Fig. 7).

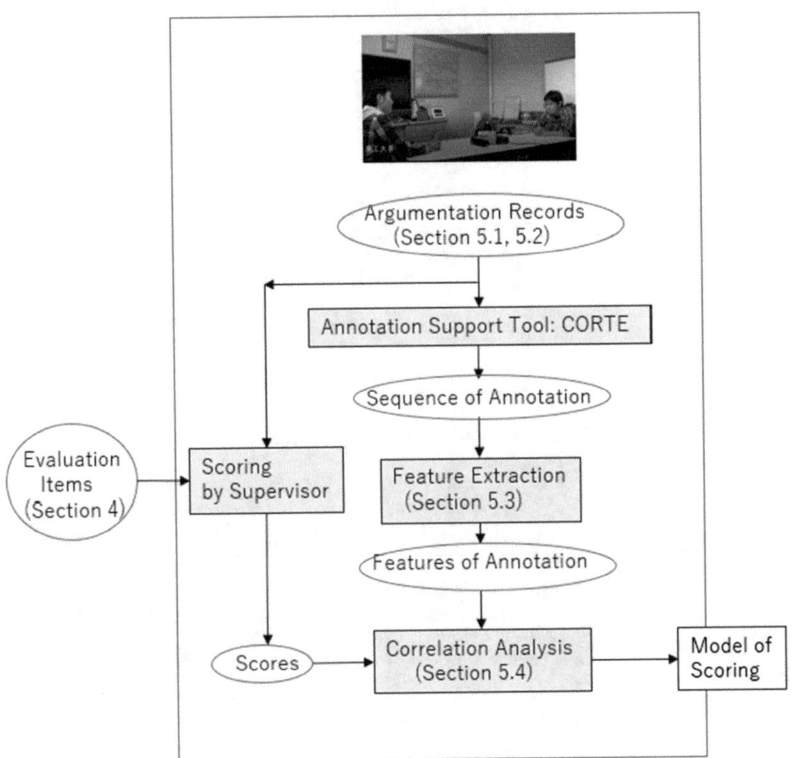

Fig. 5. Learning phase of argumentation analysis

In the evaluation phase, when a student argues with the argumentation agent, annotations are also attached to the argumentation record, and by using the model of scoring, our system estimates the score of the record.

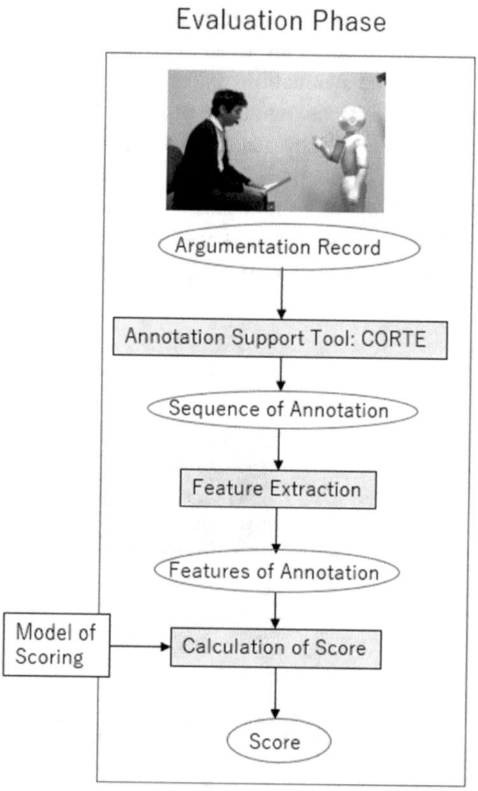

Fig. 6. Evaluation phase of argumentation analysis

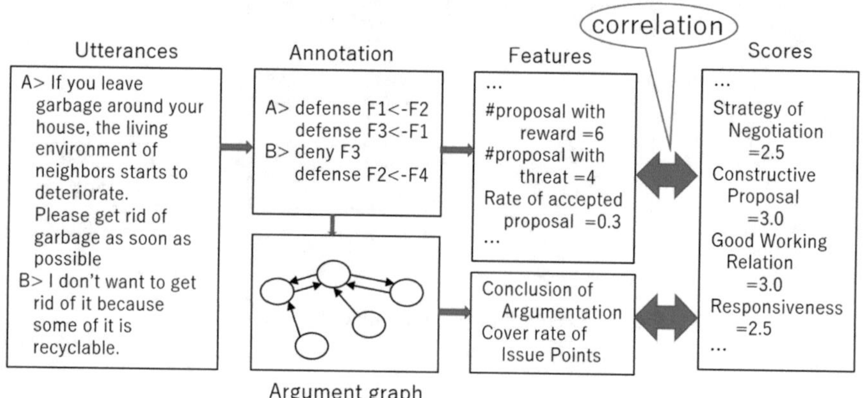

Fig. 7. Correlation analysis between features and scores

3.2 Annotation Support Module: CORTE

To analyze argumentation skills, we attach following labels to each utterance in the discussion record.

(i) Utterance ID and speaker ID
(ii) Speech acts
(iii) Argument Structure

Speech acts denote the role of the utterance such as claim, defend, concede, deny, close-ended-question, open-ended-question, answer, propose with reward, propose with threat and other. Close-ended-question is a form of question, which demands answer from multiple options. And open-ended-question demands an answer with free format appropriate to the question such as opinion, argument, agreement, denial, complement, close-ended-question, open-ended-question, answer, demand, proposal and other. One utterance may contain one or more of those speech acts. With speech act tags, this tool is able to handle features in discourse.

An Argument Structure consists of the *conclusion* part, the *data* part and the *warrant* part of the Toulmin model. If contents of these parts match the factors introduced in Subsect. 2.2, their factor IDs are also attached.

For example, let consider the following arguments between Mr. A (student) and Mr. B (Pepper), and let consider the following factors.

Arguments:

A20> "If you leave garbage around your house, the living environment of neighbors starts to deteriorate. Please get rid of garbage as soon as possible."

B21> "I don't want to get rid of it because some of it is recyclable."

Factors:

F1: Garbage deteriorates environment.
F2: Leave garbage.
F3: Get rid of garbage.
F4: Garbage is recyclable.

Both A20 and B21 are arguments, and following tags are attached in the form of XML documents.

```
<utterance id="20" speaker="A" speechAct="defend">
  <body>
    If you leave garbage around your house, the living environment of neighbors
    starts to deteriorate. Please get rid of garbage as soon as possible.
  </body>
  <element id="b123" factor="F1" role="conclusion" >
```

```
        the living environment of neighbors starts to deteriorate.
    </element>
    <element id=”b124” factor=”F2” role=”data” support=”b123”>
        you leave garbage around your house
   </element>
      <element id=”b125” factor=”F3” role=”conclusion”>
          Please get rid of garbage as soon as possible.
      </element>
      <element id=”b123” factor=”F1” role=”data” support=”b125”>
          the living environment of neighbors starts to deteriorate.
      </element>
  </utterance>

<utterance id=”21” speaker=”B” speechAct=”defend”>
  <body>
    I don’t want to get rid of it because some of it is recyclable
  </body>
  <element id=”b125” factor=”F2” role=”conclusion” attacks=”b123”>
      I don’t want to get rid of it
  </element>
  <element id=”b126” factor=”F4” role=”data” supports=”b125”>
      some of it is recyclable
  </element>
</utterance>
```

In order to help a supervisor to attach these tags to an argumentation record, we developed an annotation support editor, CORTE [10, 11]. Figure 8 shows an example of the screen shots of CORTE. In the left window, the utterance list is shown, and in the top window, the detail of a selected utterance is displayed. If a supervisor designates the data part and the conclusion part in the top window using a mouse, then its Toulmin diagram is displayed in the center window.

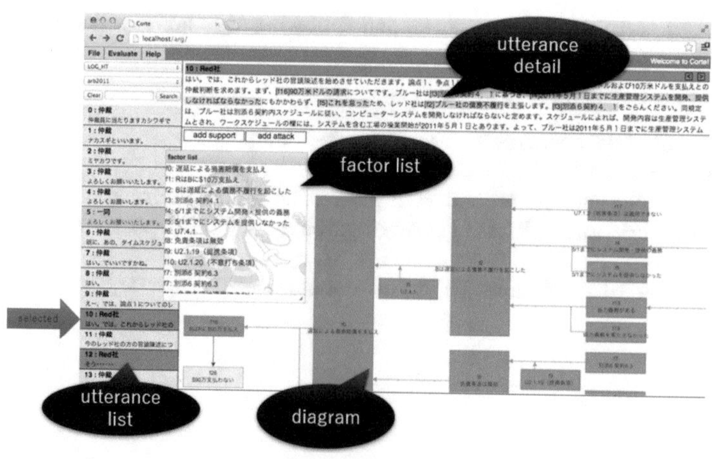

Fig. 8. Screen shot of CORTE

4 Criteria of Argumentation

The Intercollegiate Negotiation Competition (INC) is held every year in Japan, and about 50 teams of law school students take part in this competition [6]. In the competition, all of the two teams argue about the same mock business trouble between two companies in 25 rooms.

On the first day (Round A) of INC, they argue the arbitrating problem, and the second day (Round B), they negotiate for cooperation. Therefore, in Round A, a competitive argumentations are included, and in Round B, cooperative argumentations are included.

Marking score is based on 15 separate criteria for each of the arbitration round (Round A) and the negotiation round (Round B). Evaluation of each criterion will be made on a scale of 0 (minimum score) to 5 (highest score), in increments of 0.5 (except that there is no 0.5). This provides a total scale of 10 increments. Therefore, for each round, the total score given by each judge will be from 0 points to 75 points and the total score of all three judges will be from 0 points to 225 points.

The follwing lists are the criteria in the score sheets:

(a) Score Sheet of Round A: Arbitration

(1) Expression & Organization of the Briefs, (2) Persuasiveness of the Briefs,
(3) Opening Statement, (4) Sub Topic 1: Logical Presentation,
(5) Sub Topic 1: Persuasiveness, (6) Sub Topic 2: Logical Presentation,
(7) Sub Topic 2: Persuasiveness, (8) Legal Arguments,
(9) Facts, (10) Responsiveness to the Other Side,
(11) Responsiveness to the Arbitrators, (12) Closing Statement,
(13) Presentation and Speech, (14) Lawyerly Manner, (15) Teamwork

(b) Score Sheet of Round B: Negotiation

(1) Preliminary Memo, (2) Objective/Goal Setting, (3) Strategy for Negotiation,
(4) Constructive Proposal of Alternatives, (5) Effective Discussion,
(6) Responsiveness, (7) Communication/Mutual Understanding,
(8) Principled Negotiation, (9) Business Manner, (10) Teamwork/Role Assignments, (11) BATNA, (12) Good Working Relationship, (13) Outline of the Agreement,
(14) Negotiation Ethics, (15) Self-Evaluation

In Round A, as logicalness and persuasiveness are important views, *Persuasiveness of the Briefs* and *Legal Arguments* are included in the criteria. And also, *Persuasiveness* and *Logical Presentations* of two sub topics are included in the criteria. As these sub topics are important issues, if participants do not argue these topics properly, the score becomes low.

On the contrary, in Round B, keeping good relations is important to reach a consensus. Therefore, *Effective Discussion* and *Good Working Relationship* are included in the criteria. Furthermore, *Responsiveness* is included in both criteria.

5 Analysis of an Argumentation Record

Our approach to analyze the argumentation skills is shown in Figs. 5 and 6. At the learning stage, we gather a lot of argumentation records which argue the same topics. And lawyers give scores for given criteria of argumentation skills, then we attach annotation for each utterance in argumentation records. After that, we extract features (patterns) of a sequence of annotations, then finally we calculate correlation between features and scores. We will explain these steps using Garbage House Problem as an example.

5.1 Garbage House Problem

(a) Story

An old man, Mr. O, lives in Osaka City by himself. As he has no friends, Mr. O lost the motivation to live and so began to gather trash and garbage, and finally his house is called "Garbage House". Neighbors suffer from various inconveniences caused by the garbage such as bad smells, traffic obstruction, stray cats, small fires and so on. The staff member of the civil government, Mr. C, tried to persuade Mr. O to get rid of the garbage. But, Mr. O insisted that the items are not just garbage because he can sell or recycle some of them, so it is property. According to him, as he is old and his only income is pension, he needs property which he can sell anytime. Legally, it is possible for the city government to get rid of garbage forcefully, but if so, Mr. O will begin to gather garbage again. So, Mr. C has to persuade Mr. O to get rid of garbage voluntary until Mr. O understands the situation by showing some proposal to take care of him.

(b) Issue Point of Garbage House Problem

- Mr. O lives alone and he has no friends, so he may hate neighbors.
- Mr. O may not understand he has caused problems in the living environment.
- Mr. O need property to ensure his income in the future, or the city government will have to take care of him because of his age.
- Mr. O may sell some of the garbage, but he has not separated the useful item from garbage items.
- Removing garbage by force is authorized by law.
- If the city government removes the garbage forcibly, the government may claim compensation money.
- If Mr.O gets rid of garbage, the city goverment will support him.

(c) Factors for Garbage House Problem

To represent arguments of the garbage house problem, we listed up 72 factors among which 15 factors are basic level, 6 factors are proposal level and 51 factors are intermediate level knowledge.

5.2 Argumentation Records

For the "Garbage House" problem, we gathered 10 argumentation records each of which contains argumentation utterances between Mr. O and Mr. C. These argumentation experiments took about 20∼40 min, and 84∼435 utterances are included in them. The result of argumentation is shown in Table 1. About half of the negotiations succeeded in reaching an agreement.

For each argumentation record, we asked three lawyers, Mr. S1, Mr. S2 and Mr. S3, to give scores for following items.

Legality of three sub topics (competitive)

A1: Persuasiveness: Recognition of decreasing living environment

A2: Persuasiveness: Economical value of garbage

A3: Persuasiveness: Legality of forcible removal of garbage

Negotiation Skills (cooperative)

B1: Strategy for Negotiation

B2: Constructive Proposal of Alternatives

B3: Good Working Relationship

B4: Responsiveness

Table 1. Results of argumentation records

Record ID	Number of utterances	Result of negotiation
C1	84	Agreement. Get rid of the garbage voluntarily
C2	274	No agreement
C3	76	Agreement. Get rid of the garbage with help of government
C4	137	Agreement. Get rid of the garbage with help of government
C5	81	Part agreement. Get rid of oly the outside garbage
C6	258	No agreement
C7	278	Part agreement. Select recyclable garbage from others
C8	369	Agreement. Get rid of the garbage with help of government
C9	435	Part agreement. Get rid of oly the outside garbage
C10	197	No agreement

Table 2 shows a score of Mr. S1. The scores varied according to the personality characteristics of the three lawyers, S1, S2 and S3, because these are ambiguous criteria, and lawyers have to make score subjectively. For example, to evaluate (B4) *Good Working Relationship*, one lawyer may focus on the rate of acceptance of proposal, and another lawyer may focus on the number of proposal with reward. There is no answer on which all lawyers will agree. Therefore, it is not practical to obtain one score by merging these scores. Instead of unifying scores, analyzing these scores separately is more useful.

Table 2. Score of Mr. S1

Record ID	A1	A2	A3	B1	B2	B3	B4
C1	2.5	2.5	1.0	2.0	2.5	3.5	2.5
C2	3.0	3.5	3.0	2.0	1.0	2.5	3.0
C3	1.5	2.5	–	1.5	3.5	3.5	3.0
C4	3.0	2.5	2.5	3.0	3.0	3.5	3.0
C5	3.0	2.5	–	1.0	1.5	3.0	3.0
C6	3.0	2.5	–	1.5	1.0	1.5	3.0
C7	3.0	2.5	–	1.0	1.0	1.5	2.5
C8	3.5	2.5	3.5	3.5	3.0	2.5	3.0
C9	3.0	2.5	1.5	3.5	3.5	3.5	3.5
C10	3.5	3.5	3.5	2.0	2.5	2.5	3.0

5.3 Extraction of Features

From the sequence of annotations, we measured the following features:

(1) Main issues

(1.1) Recognition of Decreasing Living Environment

- Whether Following factors appeared or not.
  BadSmells, FearOfFire, SmallFireOccured, LitteringTobacco, Gathering Cats and Crows, Obstacles, DamageLandscape

(1.2) Value of Garbage

- Whether Following factors appeared or not.
  GarbageRecyclable, GarbageSalable, PropertyNecessarForAged, SeparateValuedThing.

(1.3) Legality of Forcible Removal of Garbage

- Whether Following factors appeared or not.
  LegalityCompulsoryExecution, FineCompulsoryExecution.

(2) Flow of issues

- Number of utterances which brought up the same issues again.
- Number of utterances which repeat the same message continuously.

(3) Flow of speech acts

- Number of Claimed utterances.
- Ratio of Conceded utterances to Claimed utterances.
- Ratio of Denied utterances to Claimed utterances.
- Ratio of utterances which obey the argumentation protocol.

(4) Proposals

- Number of Proposal utterances.
- Number of Proposal with Reward utterances.
- Number of Proposal with Threat utterances.
- Ratio of Accepted proposal utterances to Proposal utterances.

5.4 Correlation Analysis

We conducted Multiple Regression Analysis between features stated above and scores of lawyers, Mr. S1, Mr. S2 and Mr. S3.

(A1) Persuasiveness: Recognition of Decreasing Living Environment

Score of S1 =2.09+0.60[Fire]+0.81[Cat]

R^2=0.941

Score of S2 =2.25+0.81[Fire]

R^2=0.364

Score of S3 =1.14+1.98[Cat]+1,29[Obstacle]

R^2=0.655

Concerning the criterion of Recognition of Decreasing Living Environment, Mr. S1 and S2 focused on the subtopic of danger of fire and only Mr. S1 focused on cats.

And S3 focused on cats and obstacles. Though S1 and S2 focused on similar subtopic, value of R^2 is very different. Mr. S1's scoring is more consistent than Mr. S2.

(A2) Persuasiveness: Value of Garbage

Score of S1 =2.75+0.50[Memory]-0.75[Collection]+0.75[Security]

R^2=0.698

Score of S2 =2.50+0.59[Recyclable]+0.41[Security]–0.55[Separation] +0.68[Security feeling]

R^2=0.915

Score of S3 =1.15+2.75[Memory]+1.20[Recycle]+2.00[Security]

R^2=0.678

To make the score of Value of Garbage, Mr. S1 focused on whether it has sentimental value or not and that it is necessary for the aged person. Mr. S2 focused on whether the garbage is recyclable or not and that it is necessary for the aged person. Mr. S3 focused on whether it has sentimental value or not and it is recyclable or not and it is necessary for the aged person. Among them, Mr. S1 and Mr. S3 focused on the similar features, and only Mr. S2 focused on that the security feeling is necessary for the aged person. According to the value of R^2, Mr. S2's value judgement is more consistent.

(A3) Persuasiveness: Legality of Compulsory Execution

Score of S1 = Regression analysis failed

Score of S2 =2.88[PossibilityCompulsoryExecution] +1.75[FineCompulsoryExecution]

R^2=0.860

Score of S3 =0.750+2.55[Regulation]

R^2=0.526

Concerning the criteria of Legality of Compulsory Execution, we failed to analyze score of Mr. S1 because his scoring is not consistent for this criteria. Mr. S2 focused on PossibilityCompulsoryExecution and FineCompulsoryExecution. The pair of these two subtopics are important to persuade the owner of the garbage house. And R^2 value of Mr. S2 obtained high score because S2's judgement.is consistent.

(B1) Strategy of Negotiation

Score of S1 = −0.88+0.08[Protocol]
R^2=0.480

Score of S2 = −1.67+0.13[Protocol]
R^2=0.478

Score of S3 =2.55+0.31[Proposal with Reward]
R^2=0.763

The criteria of Strategy of Negotiation is very difficult to analyze because it concerns a lot of aspects and they depend on specific features of each case. Mr. S1 and S2 focused on Protocol which means to make the message according to the argumentation protocol is important. However, their R^2 value is not so high. Instead, Mr. S3 focused on Proposal with Reward, which means to make more proposals with reward is important. As R^2 value of Mr. S3 is high, his judgement is consistent and seems to be valid.

(B2) Constructive Proposal

Score of S1 =2.87−0.34[Repeated Issue]
R^2=0.467

Score of S2 =2.08+0.43[Proposal with Reward]
R^2=0.549

Score of S3 =2.84+0.30[Proposal with Reward]
R^2=0.738

The criterion of Constructive Proposal is also difficult to judge. Mr. S1 gave a high score if Repeated Issue is a few, but R^2 value is not so high. Mr. S2 and S3 focused on the number of proposals with reward. A proposal with reward is a typical example of constructive proposal, it is valid.

(B3) Good Working Relations

Score of S1 =2.51+0.04[Consent]+0.01[Repeated Continuously]
R^2=0.964

Score of S2 =3.44−0.33[Repeated]
R^2=0.617

Score of S3 =2.49+0.14[Proposal]−0.003[Utterances]
R^2=0.764

The criterion of Good Working Relations is also difficult to judge. Mr. S1 focused on the number of Consented claims, Mr. S2 focused on the number of Repeated issues and Mr. S3 focused on the number of proposals. All these features are important to estimate the good working relation. Among them, R^2 value of S1 is high, because S1's judge is consistent.

(B4) Responsiveness

$$\text{Score of S1} = 2.58 - 0.13[\text{Accept}] + 0.26[\text{Proposal with Reward}]$$
$$R^2 = 0.786$$
$$\text{Score of S2} = 4.322 - 0.12[\text{Repeated}] - 0.04[\text{Protocol}] + 0.06[\text{ProposalThreat}] - 0.02[\text{Consent}] + 0.002[\text{Denial}]$$
$$R^2 = 0.992$$
$$\text{Score of S3} = -0.21 + 0.09[\text{Protocol}]$$
$$R^2 = 0.340$$

The criterion of Responsiveness is an interesting one because different judges focus on different features. Mr. S1 focused on the number of Accepted Proposals and the number of proposals with reward. He thinks if the number of accepted proposals is more, the score decreases. Mr. S2 thought if the number of utterances which obey the argumentation protocol is more, the score decreases. This is one of the important aspects of Responsiveness.

5.5 Evaluation

By 10 fold cross validation, we tested the performance to estimate argumentation records whose score is more than 3.5 (shown as dark boxes in Table 2). The result is shown in Table 3.

For estimation of scores of Issue1, Issue2 and Issue3, the accuracy is not so high because several cases failed to estimate scores. This means our estimation method doesn't fit measuring persuasiveness. On the contrary, for the other four items which are related to cooperative argumentation, the accuracy is high enough. This shows that the proposed method is promising way to estimate the score of items for cooperative arguments.

Table 3. Performance of estimation of scores

	S1	S2	S3
Issue1: Persuasiveness Recognition of Decreasing Living Environment	100%	90%	–
Issue2: Persuasiveness Value of Garbage	–	100%	–
Issue3: Persuasiveness Legality of Compulsory Execution	–	–	67%
Strategy of Negotiation	80%	70%	90%
Constructive Proposal	60%	–	100%
Good Working Relation	90%	80%	100%
Responsiveness	90%	90%	–

6 Conclusion

In this paper, we proposed the method to estimate the argumentation skills from the argumentation record. First, we showed there are several criteria to evaluate skills from the score sheet of the Intercollegiate Negotiation Competition. Among them, we selected several criteria. Then, we showed a method which estimate scores of selected criteria based on a sequence of annotations. By an experiment of the Garbage House Problem, we showed it is possible to estimate the scores of cooperative argumentation records according to the personality of supervisors. By using an argumentation agent and the proposed scoring method, we expect the burden of argumentation training will be reduced.

As for future work, we will analyze the argumentation skills using not only utterance texts but non verbal data such as audio features, gesture and posture.

References

1. Shibayama, T., et al.: Technical report of law school admission test 2006 (in Japanese) (2006)
2. Linch, C., et al.: Comparing argument diagrams. In: JURIX 2012, pp. 81–90 (2012)
3. Petukhova, V., et al.: Virtual debate coach design: assessing multimodal argumentation performance. In: ICMI'17, pp. 41–50 (2017)
4. Pepper Web page. https://www.softbankrobotics.com/emea/en/pepper?q=emea/fr/pepper
5. Hamada, T., et al.: An argumentation agent using multimodal information. In: HAI Symposium 2017 (in Japanese) (2017)
6. Intercollegiate Negotiation Competition Homepage. http://www.negocom.jp/
7. Gordon, T.: Pleadings Game: An Artificial Intelligence Model of Procedural Justice. Kluwer Academic Publishers, Dordrecht (1995)
8. Ashley, K.D.: Modeling legal argument: reasoning with cases and hypothecals. The MIT Press, Cambridge (1990)
9. Aleven, V.: Teaching case-based argumentation through a model and examples. Ph.D. thesis (1997)
10. Kubosawa, S., Lu, Y., Okada, S., Nitta, K.: Argument analysis with factor annotation tool. In: Jurix 2012 (2012)
11. Kubosawa, S., et al.: A discussion training support system and its evaluation. In: ICAIL 2013 (2013)

Machine Autonomy: Definition, Approaches, Challenges and Research Gaps

Chinedu Pascal Ezenkwu(✉) and Andrew Starkey

School of Engineering, University of Aberdeen, Aberdeen, UK
{r01cpe17, a.starkey}@abdn.ac.uk

Abstract. The processes that constitute the designs and implementations of AI systems such as self-driving cars, factory robots and so on have been mostly hand-engineered in the sense that the designers aim at giving the robots adequate knowledge of its world. This approach is not always efficient especially when the agent's environment is unknown or too complex to be represented algorithmically. A truly autonomous agent can develop skills to enable it to succeed in such environments without giving it the ontological knowledge of the environment *a priori*. This paper seeks to review different notions of machine autonomy and presents a definition of autonomy and its attributes. The attributes of autonomy as presented in this paper are categorised into low-level and high-level attributes. The low-level attributes are the basic attributes that serve as the separating line between autonomous and other automated systems while the high-level attributes can serve as a taxonomic framework for ranking the degrees of autonomy of any system that has passed the low-level autonomy. The paper reviews some AI techniques as well as popular AI projects that focus on autonomous agent designs in order to identify the challenges of achieving a true autonomous system and suggest possible research directions.

Keywords: Autonomous agent · Machine autonomy · Automation · Robots · Artificial intelligence · Learning

1 Introduction

Autonomous machine intelligence has been a long time concern among roboticists and artificial intelligence researchers; however, there has not been a unifying definition of autonomy in artificial agents. Due to the severalties of its definitions, it is difficult to assess or rate the degree of autonomy of most disruptive and sophisticated systems such as self-driving cars, industrial robots or game playing agents. Importantly, the first step in building a truly autonomous system is the understanding of what it means for a system to be truly autonomous and what the attributes of a system that exhibits autonomy are. Having a coherent conceptual framework would help researchers know how much progress has been made in the syntheses of such systems. Some authors have presented autonomous system in a very broad sense that leaves no distinction between autonomy and automation. For example, Smithers et al. [1] viewed the autonomous system as the act of building robots. In the same vein, Franklin [2] referred to a remotely controlled mobile robot as an autonomous system. The most popular

K. Arai et al. (Eds.): CompCom 2019, AISC 997, pp. 335–358, 2019.
https://doi.org/10.1007/978-3-030-22871-2_24

view is the notion that autonomous robots are to interact with their environments without ongoing human intervention [3]. The use of the word 'ongoing' implies that even if the designer equips the robot *a priori* with all the prevailing knowledge it needs to operate in the environment it would still be labelled autonomous inasmuch as it is not assisted to perform its tasks after deployment.

According to US National Institute of Standards and Technology (NIST) in [4], a system is fully autonomous if it is capable of achieving its goal within a defined scope without human interventions while adapting to operational and environmental conditions. The definition of the autonomous system here is in respect to a well-defined scope. Similarly, Harbers et al. [5] classified robots, "which are able to perform well-constrained tasks such as surgery, driving on a highway, or vacuum cleaning" as being completely autonomous. It can be argued, however, that such systems are not completely autonomous. Autonomy is when these systems if introduced to an unknown scope or domain, given suitable sensors and actuators for that scope, without changing the algorithm in any way, learns how to handle the sensor signals, adapt their behaviours and act intelligibly. Bradshaw et al. [6] posits, "since there is no entity that could perform all possible tasks in all possible circumstances; full autonomy does not exist?" This is true even for the most intelligent entity - human. Rich et al. [7] defines artificial intelligence as "the study of how to make computers do things, which at the moment people do better". This places human-level cognition as the benchmark for building artificial intelligence. Although the computer has outperformed humans in some tasks such as games and some perception tasks, it turns out that our understanding of how humans achieve their high degree of autonomy is paramount in the subject of autonomous systems. Roe et al. [8] showed that the brain does not require different algorithms to perform different tasks. In their experiment, they rewired the brain of a ferret such that retinal inputs were routed to the auditory pathway. The result demonstrated that the auditory cortex learnt to process visual input after some time. This denotes that the brain is adaptive enough to respond to different forms of sensory data based on a single learning algorithm. In consonance to the preceding, there should be a single algorithm that could enable an autonomous agent to adapt to new situations given suitable sensors without having to teach it how to process and make sense from these sensors' data. To achieve this in any artificial system requires that the system is able to exhibit some attributes of autonomy. This paper seeks to present a definition of machine autonomy and its attributes; and, based on this definition, to provide a review of some underpinning methods and a number of popular researches that aim at autonomous agents. Finally, the paper highlights some research gaps and suggests possible research directions.

2 Attributes of Machine Autonomy

Leveraging on different viewpoints of different researchers on autonomy [9–11], this study considers the attributes of autonomy under two categories – the low-level and the high-level attributes. The low-level attributes are must-have for any autonomous system. They include perception, actuation, learning, context-awareness and decision-making. However, systems that possess only the aforementioned attributes are at the lowest level of autonomy. Conversely, if any of these attributes is missing in any system then such a system cannot be described as autonomous according to these definitions. In contrast, the high-level attributes of autonomy are more advanced attributes and they are the subject of numerous researches in autonomous systems in recent years. They include domain-independence, self-motivation, self-recovery and self-identification of goals. These have been summarised in Table 1 and described in Sects. 2.1 and 2.2.

Table 1. Attributes of machine autonomy

Low-level attributes	High-level attributes
Learning	Domain-independence
Context-awareness	Self-motivation
Actuation	Self-identification of goals
Perception	Self-recoverability
Decision-making	

2.1 Low-Level Attributes

Perception. For an agent to make the right decision in its environment it requires information from the environment. Perception is the problem of analysing and representing sensory inputs from dedicated purpose sensors. All autonomous agents must have a means of perceiving their worlds and analysing observations to extract or filter relevant features that would enable them to make sense of the environment. Human agents are naturally equipped with these abilities. However, to enable an artificial agent to exhibit human-level perception is indeed a challenging task. Machine perception has attracted huge research attention. In recent times, there are remarkable contributions in computer vision, natural language processing, feature extraction and dimensionality reduction techniques due to the emergence of some powerful methods in deep learning and machine learning in general.

Actuation. An agent requires a means of providing feedback to the world. Actuation is the ability of an agent to cause a change in both its environment state and/or its internal state. Actuators often "convert other sources of energy such as electric energy, hydraulic fluid or pneumatic pressure to mechanical motion" [12]. The motion is often in response to observations in the agent's environment. Every autonomous agent requires actuators to enable it to act suitably in its environment. Lomonova [12] presents a detailed review of the state-of-the-art actuation systems.

Learning. A learning agent has the ability to make sense from sensory inputs. The science of designing such algorithms is machine learning. Machine learning is an important aspect of computing due to its vast applications across domains. A learning agent can engage in supervised, unsupervised or reinforcement learning depending on the kind of environment it is to operate in and the nature of data that are available to it. A supervised learning agent requires labelled data points to derive a predictive model of its environment. This is different for unsupervised learning technique in which the agent tries to figure out internal structure within unlabelled dataset. The most applied of these three in interactional agent design is the reinforcement learning technique in which an agent makes decisions using unlabelled data points or observations depending on some reward function. Even though learning is one of the necessary attributes of machine autonomy not all learning agents are autonomous.

Context-awareness. Context is an alternative term for the state of an agent's environment. Context-awareness or situational-awareness is the ability of an agent to sense, interpret and adapt to the current context of its environment [13–15]. A context-aware agent has perception and learning abilities and can keep track of a dynamic environment. An autonomous agent should be context-aware i.e. it should have the ability to sense and interpret in real time the prevailing state of its environment and consequently, enabling learning to be contextually relevant.

Decision-making. An important attribute of any computing system is its ability to make decisions. Decision-making is the ability of an agent to map context or perceptual information to action. An intelligent agent should be able to select best actions for all situations. However, decision-making has been implemented in different ways. While some agents depend on hardcoded lookup tables to make their decisions, state-of-the-art techniques are concerned with giving agents the ability to learn optimal and robust decision-making policies. The approach of using hardcoded lookup tables is also referred to as symbolic artificial intelligence or GOFAI (Good Old-Fashioned Artificial Intelligence) [16] while the more recent approach is in non-symbolic artificial intelligence. However, it is difficult to achieve an autonomous GOFAI agent.

2.2 High-Level Attributes

Domain-independence. Domain-independent agents do not require the ontological knowledge of their environments at design time to succeed in the environment. This would enable the system to succeed even if the sensors' values change unpredictably [17]. The traditional approach of designing the agent whereby the engineer tries to figure out and account for all the possible problems the agent could encounter in the environment, either by hardcoding it or through a task-specific value system, has a huge limitation. The agent would certainly fail woefully if any kind of situation the engineer did not foresee surfaces [18]. A domain-independent algorithm would enable the robot to cope autonomously with any form of environment when given suitable sensors and actuators to sense observations from the environment and act based on the observations using the actuators.

Self-motivation. For an autonomous system to be able to handle an array of interesting tasks as they occur in the environment it should have some level of self-motivation. Typically, intelligent agents are given task-specific knowledge in the form of utility or reinforcement to drive their actions towards achieving a predefined goal. One of the drawbacks of this approach is that the designer must understand the environment and decide the best way to assign values to states and/or actions to enable the agent to act efficiently in the environment. The degree of autonomy of an explicit reward-driven agent would be low since the agent will not be able to adjust its behaviour to handle new interesting states in its environment unless the value system is modified to capture the emerging interests. Assuming a robot is designed to move from a point A to a point B in a navigational environment and the reward function is such that the robot earns a positive reward for avoiding an obstacle and a negative reward for bumping into it. Supposing an unforeseen situation occurs and an obstacle falls on the path to the robot's goal such that the only way this robot can make it to the goal is to displace the obstacle by bumping into it; the robot may not see this as the right thing to do since it is not captured in its reward function. Some researchers have approached this by implementing $\in$-greedy policies which enables the robot to perform some random action with small probability, $\in$. However, a more promising technique is to provide autonomous agents with some degree of self-motivation. Different terms have been used to refer to self-motivation in the literature. Some researchers have viewed self-motivation as intrinsic motivation [19–22], others as artificial curiosity [23–25] while Georgeon et al. [26] proposed interactional motivation. However, the key idea of self-motivation is to avoid explicit assignments of values to states or state-action pairs as often done in traditional reinforcement learning. A self-motivated agent is able to act intelligibly in an environment without being hardcoded or being given a task-specific value system. Such an agent uses curiosity to handle situations in the environment, while engaging in an open-ended or lifelong learning [27].

Self-recovery. An interesting application of autonomous agents is the use of robots in environments that are extremely hazardous to humans. The aim of deploying such robots instead of humans in these environments would be defeated if a human technician has to be physically present in the environment to troubleshoot and repair the robots upon failure or reprogram the robot to cope if the environment or the goal changes unpredictably. This necessitates the need for a self-recovery or self-programming mechanisms in autonomous agents. Self-recovery can be proactive, reactive or a fall-back mechanism. In proactive self-recovery, the agent is able to foresee possible causes of failure and devise a solution to abate it. Reactive self-recovery allows the agent to recover from failure after it has occurred. Chaput [18] proposed a fall-back mechanism by which an agent learns different hierarchies of knowledge such that if it encounters a completely strange situation that its higher-level knowledge cannot handle the agent falls back to its lower hierarchy of knowledge and starts building new knowledge that can handle the new situation from there. A self-recovery agent must have the ability to self-analyse itself based on the prevailing situation and build knowledge from the outcome of these analyses. The knowledge gathered in the course of the self-analyses may result in the autonomous reconfiguration of the agent's intelligence mechanism and eventually, an emergent behaviour to suit the situation. Such behaviours may not be explicitly traceable to the agent's program. This has also been referred to as self-programming [28–30].

Self-identification of Goal. Goal-oriented agents or robots are the commonest kinds of robots in the literature. Typically, these robots are given goal information either by hardcoding them or by using suitable value systems. Hardcoding goals in an agent program demands that the designer understands, and can formalise, the model of the environment with respect to the agent's actions. This approach will probably fail if this environment is not predictable or known to the designer. For example, it is difficult to hardcode in a self-driving car what constitutes safe driving for every scenario on the road. Furthermore, techniques like the model-free reinforcement learning have successfully applied value systems in numerous applications without the knowledge of transition dynamics of the environment. Nevertheless, it requires prior understanding of the environment and possible actions of the agent in order to be able to design a suitable reward function that would enable the agent to achieve specified goals. To avoid this challenge, there is a need for mechanisms that would enable the agent to self-identify goals and learn suitable skills to enable it to realise them. Simply put, an agent is able to self-identify goals in a given environment if it can develop suitable skills to enable it to achieve a goal that is not explicitly defined in the environment. Although efforts have been made in affordance learning [31–33] and intrinsically motivated learning research [20], self-identification of a goal is still a huge challenge in complex real world tasks.

3 AI Approaches to Autonomous Agents

While autonomy is not possible in symbolic or GOFAI agents, machine learning, evolutionary techniques and developmental artificial intelligence have been widely applied in numerous autonomous agent researches. Despite their promising characteristics, none of these techniques has been able to yield a truly autonomous outcome due to some limitations. These techniques as well as their strengths and weaknesses, in the context of machine autonomy, are discussed in this section. Machine learning is a broad term for techniques that make sense from data. They include supervised, unsupervised and reinforcement learning. Moreover, other techniques have emerged from the diverse ways of implementing the basic machine learning methods. These techniques include active learning and transfer learning. Nonetheless, instead of considering the broad family of machine learning as a topic, these techniques are reviewed individually since they are suitable for different kinds of problems and their strengths and weaknesses are equally different. Strengths and weaknesses of supervised learning are given in Table 2.

3.1 Supervised Learning

A supervised learning agent learns a function that maps input to output given some example of input-output pairs [34, 35]. In supervised learning, each data point in the training set is labelled. The learning algorithms analyse the training data and infers a function that can return the corresponding output for a given input. A supervised learning task is solved when the resulting function is able to generalise data points that are not in the training set, otherwise the function is said to overfit. Arrays of solutions have been devised to minimise overfitting during training. These include early stopping [34], cross-validation [36], regularisation [34, 37] and dropout technique [38]. Supervised learning algorithms are categorised as classification or regression techniques.

Classification is a form of supervised learning task in which the training samples belong to a finite set of classes. A classifier $f(x)$ is trained to predict the class y, belonging to a finite set of classes, to which an independent input feature vector x belongs [39]. Training the classifier requires a labelled training set, $(x^{(i)}, y^{(i)})_{i=1}^{m}$; where

Table 2. Strengths and weaknesses of supervised learning

Strengths	Weaknesses
Supervised learning is a suitable learning technique when a dataset of sensory inputs and the corresponding desirable outputs is available	Supervised learning algorithms are suitable only when there is a labelled dataset. In the absence of labelled samples, supervised learning cannot be used
Most supervised learning algorithms have better generalisation than other methods	Some supervised learning algorithms such as deep learning algorithms require large amounts of data and cannot generalise to simple problems [45]
Supervised learning algorithms such as ANN, SVM with kernel trick, decision tree, random forest and KNN yield non-linear models i.e. they fit into situations where a linear hyperplane is not able to represent the approximation function or the decision boundary for the training samples. This is a huge advantage because real world tasks are often non-linear	While the knowledge representations in some supervised learning algorithms such as decision trees and naïve Bayes algorithms are transparent, those of more sophisticated and powerful techniques such as ANN, SVM, random forest and deep learning are not interpretable and are considered black box in nature

m is the size of the dataset and $(x^{(i)}, y^{(i)})$ is the $i-th$ training example. The label $y^{(i)}$ for all $i \in \{1, 2, \ldots, m\}$ is discrete. Regular classification methods in the literature include k-Nearest Neighbour (KNN), logistic regression, Support Vector Machine (SVM), decision tree, random forest, naïve Bayes and Linear Discriminant Analysis (LDA).

In contrast, regression is the task of approximating a function $f(x)$ from an independent input feature vector x to a continuous output variable, y. Similar to a classifier, training the regressor, $f(x)$,requires a labelled training set, $(x^{(i)}, y^{(i)})_{i=1}^{m}$; where m is the size of the dataset and $(x^{(i)}, y^{(i)})$ is the ith training example. However, the label $y^{(i)}$ for all $i \in \{1, 2, \ldots, m\}$ is a real-value, such as an integer or floating-point value. Commonly used regression methods include linear regression, Linear Weighted Regression (LWR), Artificial Neural Networks (ANN), ridge regression and Support Vector Regression (SVR).

Some Applications of Supervised Learning in Autonomous Agent Design. Supervised learning has been largely applied in robotics and the development of other AI agents through imitation learning [40, 41]. Imitation learning techniques give an agent the ability to learn a policy using a training set obtained from an expert's demonstrations in a similar task [39]. Direct imitation learning is also known as behavioural cloning. Two major problem of direct imitation learning are the correspondence problem [42] and the difficulty for an imitation learning agent to generalise to situations that are not contained in the demonstration dataset. Correspondence problem occurs because of the difference in the architecture, skeleton or degrees of freedom between a human demonstrator and a robot. These two challenges have been better handled using indirect imitation learning or inverse reinforcement learning [42] whereby the agent learns the objectives behind the expert's action in form of a reward function and uses it along with its own experience to improve its behaviours.

The generalisation ability of some supervised learning algorithm such as the deep neural network has helped in improving perceptual tasks in high dimensional feature space. For example, deep learning has helped to improve reinforcement learning in video game playing agents due to its generalisation potential [43] and ability to scale to multi-dimensional feature space [44].

3.2 Unsupervised Learning

An unsupervised learning agent learns a pattern in an unlabelled dataset [34]. Unsupervised learning tasks are implemented as clustering techniques, such as k-means, self-organising maps (SOM), adaptive resonance theory (ART), hierarchical models or mixture models. A clustering algorithm is able to automatically figure out the internal structure of a set of inputs without any form of feedback.

Other approaches to unsupervised learning include the Hidden Markov Model (HMM) [46], blind source separation (BSS) [47] and association rule mining [48]. In HMM, the system is assumed to be a Markov process with hidden states while BSS techniques are feature extraction techniques for dimensionality reduction; examples include principal component analysis(PCA), independent component analysis (ICA), non-negative matrix factorization [47] and singular value decomposition. Association rule mining is popular for "discovering interesting relations among variables in a large database" [48]. Common algorithms used for association rule mining include FP-growth algorithm, a priori algorithm and Eclat algorithm [49]. Strengths and weaknesses of unsupervised learning are given in Table 3.

Table 3. Strengths and weaknesses of unsupervised learning

Strengths	Weaknesses
They are the most suitable approach to pattern recognition when there is no domain knowledge	It is difficult to decide the correct output or agent action given a set of unlabelled inputs or observations
They resemble learning in humans and animals [53] more than the supervised learning techniques	Evaluation of unsupervised learning algorithms is difficult unless there are some labelled samples so that the clustering can be interpreted
These techniques are often used as data quantisation and dimensionality reduction techniques [54–56] and they have been vastly applied in robotics [57–59]	Some unsupervised learning techniques require some hyperparameters to be chosen a priori before clustering data points. For example, k-means algorithm requires that the number of cluster centres or centroids, k, is chosen correctly

Some Applications of Unsupervised Learning in Autonomous Agent Design. Unsupervised learning algorithms like PCA, SOM, ICA and k-means algorithms are often applied in dimensionality reduction tasks in multimodal sensors tasks [50]. Furthermore, some attempts have been made in implementing the forward kinematics of a robot using unsupervised learning algorithms [51, 52]. Chaput in [51] implemented a self-recovery mechanism, using a hierarchy of SOMs, which enables a robot to fall back to a lower level of knowledge if its higher-level knowledge cannot handle a situation in the environment.

3.3 Reinforcement Learning

Reinforcement learning differs from the supervised and unsupervised learning in the sense that it is a kind of learning whereby an agent learns an optimal policy for sequential decision-making by interacting with its environment in a trial and error fashion [43, 60, 61]. A reinforcement learning agent receives "a state s_t, from a state space S, at time t and selects an action a_t from an action space A, following a policy $\pi(a_t|s_t)$, obtains a reward r_t, and transitions to next state s_{t+1}, in accordance with the model of the environment $T(s_{t+1,}r_t|s_t, a_t)$" [43, 60, 61]. The goal of the agent is to maximise expected long-term discounted rewards or the expectation of long-term return over a horizon (episodic or continual). The return is expressed as $R_t = \sum_{k=1}^{\infty} \gamma^k r_{t+k}$, where $\gamma \in (0,1]$ is the discount factor.

For the reinforcement-learning agent to choose the best action a in any state s it must have some function that predicts the measure of how good each state, or state-action pair, is. This function is called the value function. The state value function $v_\pi(s) = E[R|s]$ is the expected return following policy π starting from state s while the action-value function $Q_\pi(s, a) = E[R|s, a, \pi]$ is the expected return of taking action a in state s, then, following policy π. The goal of a reinforcement learning agent is to achieve an optimal policyπ^*. The optimal policy π^* chooses action from $Q^{\pi^*}(s, .)$ or $V^*(s)$ such that the value function is maximised at each state s. Reinforcement learning is the problem of arriving at the optimal policy for a particular task. Assuming the environment dynamics or model is known, two basic approaches to deriving optimal policy are value iteration and policy iteration. This reinforcement learning problem setting is referred to as model-based reinforcement learning. Methods such as the Monte Carlo methods, temporal difference TD and TD(λ) learnings are model-free reinforcement learning. They do not require the environment dynamics to arrive at the optimal policy for any given task. Model-free reinforcement learning is the basis for algorithms like Q-learning [62] and SARSA [60] algorithms which have been widely applied in the literature. To handle generalisation and memory issues reinforcement learning techniques are combined with function approximation such as the least square regression, artificial neural networks, deep learning and so on. This, especially function approximation with deep learning, has improved the applications of reinforcement learning across domains. In addition, methods that do not care about value function in realising the policy π has equally emerged. These methods are generally called policy gradient methods. An example of a policy gradient method is the REINFORCE algorithm. Likewise, the actor-critic method combines policy iteration, value iteration and policy gradient. This has been widely applied in practice. See [61] for detailed explanations of reinforcement techniques. Strengths and weaknesses of reinforcement learning are given in Table 4.

Table 4. Strengths and weaknesses of reinforcement learning

Strengths	Weaknesses
A reinforcement learning agent does not need labelled dataset like supervised learning methods	Adequate exploration of the state-action space is required for the agent to have enough experience and knowledge from the environment. This makes reinforcement learning impracticable for most real world tasks
A reinforcement learning agent is able to learn synchronously through interaction with the environment	Adequate exploration of the state-action space is required for the agent to have enough experience and knowledge from the environment. This makes reinforcement learning impracticable for most real world tasks
Model-free reinforcement learning methods do not need the transition dynamics of the environment in order to be effective	It is difficult to define the reward function for some real world domain [65]
	Sparse and delayed reward can be a problem in several cases

Some Applications of Reinforcement Learning in Autonomous Agent Design. Reinforcement learning is one of the most applied techniques in autonomous agent designs due to its ability to allow an agent to improve its behaviour through interactions with the environment. A popular application of reinforcement learning is in the game industry. For example, reinforcement learning was applied in the development of AlphaGo Low, which unlike AlphaGo, the first agent to beat a world Go champion, acquired a superhuman skill in difficult domains, starting *tabula rasa*. AlphaGo Low beat AlphaGo 100-0 in the game of Go [63]. Reinforcement learning has equally been applied in real life robots such as pancake flipping task, bipedal walking energy minimisation task and an archery-based aiming task [64]. Real world applications using *tabula rasa* reinforcement learning can be time consuming and difficult; as such most real world reinforcement learning applications rely on imitation learning and other techniques to hasten the speed of learning.

3.4 Active Learning

An active learning agent tries to learn a model using a dataset of labelled training samples. It estimates how confident the learned model is in predicting the output of a set of the training samples in the unlabelled dataset. The agent queries human experts for the correct outputs of the unlabelled data points which have a low confidence estimate as the case may be. The pool of the labelled dataset is updated with the new input-output pairs and learning continues. Several techniques exist in the literature for implementing active learning [66, 67]. Strengths and weaknesses of active learning are given in Table 5.

Table 5. Strengths and weaknesses of active learning

Strengths	Weaknesses
It fits into situations where there are few labelled data and some unlabelled data	It requires active participation of a human expert throughout the training and testing process
It is less expensive and less time consuming than supervised learning in situations where an experienced human annotator has to label all training and testing samples	The agent environment must be known, accessible, understandable and interpretable to the human expert

Some Applications of Active Learning in Autonomous Agent Design. Active learning is often applied in situations where there are limited annotated demonstration dataset for outright imitation learning. It has been employed to improve imitation learning in 3D navigation tasks [68] in MASH simulator [69]. In [70], active learning for outdoor obstacle detection in a mobile robot using a dataset with severely unbalanced class priors was demonstrated.

3.5 Transfer Learning

Transfer learning is a learning paradigm that allows the use of an already trained model as a starting point to learn a new task. Transfer learning is defined in [71] thus: "Given a source domain D_s and learning task T_s, a target domain D_T and learning task T_T, transfer learning aims to help improve the learning of the target predictive function $f_T(.)$ in D_T using the knowledge in D_s and T_s, where $D_s \neq D_T$, or $T_s \neq T_T$". In the case where $D_s \neq D_T$ holds, then at least one of the following is true: (i.) the domains are of different feature spaces; (ii.) probability distributions of the domains are different. In the case where $T_s \neq T_T$ holds, then at least one of the following is true: (i) the label spaces are different; (ii) predictive functions are different. The relevance of transfer learning is that samples are difficult and costly to acquire in most real world problems. Furthermore, it could be more efficient to build knowledge for a new task by tuning an existing knowledge [71]. For example, in [72] the knowledge for solving a simple version of a problem is transferred to a more complex one – transfer learning from 2D to 3D mountain car problem; transfer learning from a Mario game without enemies to a Mario game with enemies. More in-depth reviews of transfer learning techniques are provided in [71, 73, 74]. Strengths and weaknesses of transfer learning are given in Table 6.

Some Applications of Transfer Learning in Autonomous Agent Design. Transfer learning has been applied in [75] to significantly speed up and improve asymptotic performance of reinforcement learning in a physical robot. Kira [76] demonstrated that learning can be successfully sped up between two heterogeneous robots utilizing different sensors and representations by making them transfer support vector machine (SVM) classifiers among each other. Large datasets are often required in several applications of deep convolutional neural networks (CNN). In most cases, knowledge can be transferred from one domain to another by reusing pre-trained convolution layers. In [77], this technique is applied for synthetic aperture radar (SAR) target classification with limited labelled data.

Table 6. Strengths and weaknesses of transfer learning

Strengths	Weaknesses
Transfer learning may speed up learning in a new domain using experiences from a different but somehow similar domain	Transfer learning may not be an efficient approach when the type of data in the source domain and those from the target domain are too dissimilar
A transfer learning agent does not require a large amount of experiences in the target domain	There is no standard way of knowing the size of datasets both in the source and target domains for which transfer learning is suitable

3.6 Evolutionary Robotics

Evolutionary robotics is a methodology towards autonomous robot development that leverages evolutionary computation, a family of techniques that incorporates principles from biological population genetics to perform search, optimisation, and machine learning [78]. Popular evolutionary computation techniques are genetic algorithms [79], evolutionary strategy [80] and genetic programming [81]. The general idea of evolutionary robotics is to initialise a population of candidate controllers or policies for the robot. After each iteration, the controllers are modified according to a fitness function using some genetic operators such as mutation, crossover and selection of fitter candidates. Fitness function is a metric that reflects the desired performance for the task [78]. A detailed survey and analyses of fitness functions often applied in evolutionary robotics is presented in [82]. A candidate controller may represent a neural network, a collection of rules or a collection of parameter settings [78]. Strengths and weaknesses of evolutionary robotics are given in Table 7.

Table 7. Strengths and weaknesses of evolutionary robotics

Strengths	Weaknesses
Evolutionary techniques are adaptive and they are robust to changes in an agent environment	Evolutionary robotics techniques require fitness function, which is often difficult to craft for most tasks
Evolutionary robotics is a suitable approach towards autonomous coordinated and cooperative multi-agent systems [85, 86]	Fitness function is always task-specific
It applies to a wide variety of problems i.e. supervised, unsupervised and reinforcement learning problems	Training evolutionary algorithms can be computationally intensive depending on the task

Some Applications of Evolutionary Robotics in Autonomous Agent Design. Evolutionary robotic techniques have been widely applied in some current robotics research [83]. For example, deep neuroevolution, an algorithm that trains a deep neural network using genetic algorithms, has competed with some state-of-the-art algorithms for deep reinforcement learning such as deep Q-network (DQN), asynchronous advantage actor-critic (A3C) and Evolution strategies (ES) in challenging reinforcement learning tasks [84]. The paper reports that deep neuroevolution performed faster than the other three algorithms.

3.7 Developmental Artificial Intelligence

Developmental approach to autonomous systems seeks to replicate infant cognition in artificial systems. This term is often used interchangeably as developmental robotics, autonomous mental development or epigenetic robotics in the literature. In [27] developmental robotics is defined as "an interdisciplinary approach to the autonomous design of behavioural and cognitive capabilities in artificial agents that takes direct inspiration from the developmental principles and mechanisms observed in the natural cognitive system of children". The motivation of developmental artificial intelligence is in line with Turing's idea that "it is easier to build an artificial baby and train it to maturity than trying to build and simulate an adult mind" [87]. In addition to Turing's idea, several works in developmental psychology especially those by Piaget have contributed to the basis for research in developmental artificial intelligence [88, 89]. Fundamentally, the developmental artificial intelligence approach seeks to achieve an autonomously open-ended learning driven by intrinsic motivation or artificial curiosity in a way similar to how human infants learn. Given primitive sensors, motors, and a suitable learning algorithm without any prior knowledge of its environment, a developmental agent should be able to bootstrap to maturity out of its own curiosity leading to self-exploration in the environment. Extensive surveys of developmental artificial intelligence are provided in [27, 90–92]. Strengths and weaknesses of developmental artificial intelligence are given in Table 8.

Table 8. Strengths and weaknesses of developmental artificial intelligence

Strengths	Weaknesses
It makes use of computational models of curiosity, which could enable an agent to explore its environment intelligibly	It is difficult to train a completely self-motivated agent
Developmental agents learn in an open-ended manner and are able to adapt to changes in the environment	Developmental artificial intelligence employs techniques from other methods like reinforcement learning, evolutionary algorithms, etc. and may suffer the inherent weaknesses of these methods

Some Applications of Developmental Artificial Intelligence in Autonomous Agent Design. In [93], intelligent adaptive curiosity, an intrinsic motivation mechanism, was used to teach a real robot how to manipulate objects on a baby play mat in such a way that maximises its learning progress. Artificial curiosity has been used to improve reinforcement learning for motion planning on real world humanoids [23].

3.8 Summary

Overall, each of these methods has inherent weaknesses that have made them not suitable for realising full autonomy. However, researchers have adopted synergies of algorithms in an attempt towards developing a true autonomous agent. For example, deep neural networks is combined with reinforcement learning in deep reinforcement learning; deep learning has been combined with evolutionary technique in deep neuroevolution techniques; artificial curiosity as studied in developmental robotics has been integrated with reinforcement learning while active learning has been used with deep imitation learning. Although all of these current methods appear promising, they are still not able to realise full autonomy according to the definition given in this paper. This is largely due to over reliance of these algorithms on externally crafted motivation functions, a human expert experience or domain-specificity of the techniques. The following section presents detailed reviews of popular research papers that have implemented these techniques.

4 Review of Selected Papers

This section considers applications of the techniques or combination of techniques studied in Sect. 3. In order to keep the review as succinct as possible, the studies are summarised in Table 10. However, the works are evaluated only on the basis of the high-level attributes desired in the paper. Autonomous system technologies apply to different kinds of systems; nevertheless, the review in this paper considers only embodied and situated agents, either real world robots or simulated agents, which have demonstrated at least all the low-level attributes. The last column of Table 10 gives comments for cases to give further details. The papers are reviewed based on the high-level attributes presented in Sect. 2.2. Table 9 recaps the attributes and presents their abbreviations as used in Table 10.

5 Discussions

Table 10 summarises diverse techniques towards realisations of intelligent systems. However, these methods were implemented in different environments and/or for tasks other than true autonomous systems, as such they cannot be properly compared against each other. To be able to review these approaches for autonomy, the high-level attributes of autonomy, as defined in this paper, were used as the framework for the review. The works in the papers already satisfy all the low-level attributes and as such, using those attributes was not necessary for this review.

Each of these methods professes to be good in the task for which it is designed; however, none of them has demonstrated all the high-level attributes of autonomy. Based on our review, the four best performing methods where adopted for further discussions. These are the deep neuroevolution with novelty search (DNNS) [84]; deep reinforcement learning (DRL) [95]; Intrinsically motivated hierarchical reinforcement learning (IMHRL) [96]; and curiosity-driven reinforcement learning (CDRL) [98].

Table 9. Meanings of abbreviations used in Table 10

Abbreviation	Meaning	Definition
DI	Domain-independence	Domain-independence is the ability of an agent to cope autonomously with any environment if given suitable sensors and actuators for that environment without being reprogrammed by the designer
SM	Self-motivation	Self-motivation enables an agent to act intelligibly in an environment without being hardcoded or being given a task-specific value system. Such an agent uses curiosity to handle situations in the environment, while engaging in an open-ended or lifelong learning
SR	Self-recovery	Self-recovery is the ability of an agent to self-analyse and reconfigure itself so as to emerge behaviours that are suitable to the prevailing situations in a changing environment
SIG	Self-identification of goal	Self-identification of goal is an agent's ability to develop suitable skills to enable it achieve a goal that is not explicitly defined in the environment

DNNS, DRL and CDRL proved to generalise in more than one environment. However, these environments are closely related in terms of the kind of input signals that the agents receive from the environments and the values/scores the agents try to optimise. Further work is required to ensure that an agent generalises in a variety of significantly unrelated environments.

Although self-motivation was demonstrated by DNNS, IMHRL and CDRL, CDRL demonstrated how much progress an agent can make based on purely intrinsic motivation while other techniques explored the improvements in the agents' behaviours when intrinsic and extrinsic motivations are combined. In DNNS, extrinsic motivation is modelled as a fitness function as is typical with evolutionary techniques while IMHRL used external reward signal for the extrinsic motivation. Using only intrinsic motivation, the CDRL agent was able to make 30% progress in the level 1 of a super Mario game. This was impressive being the first attempt towards a purely curiosity-driven agent in a game environment. However, it requires an improvement.

According to the definition of self-recovery in this paper, none of the techniques demonstrated this attribute. Apart from IMHRL, which was tested in a simple playground environment, the other three were demonstrated in Open AI gym environments. These environments are less flexible to demonstrate self-recoverability of these techniques; as such, a more flexible environment is required for this evaluation.

Amongst the four agents only CDRL demonstrated self-identification of goal. The agent learnt how to navigate the vizdoom environment using only the intrinsic motivation. Similarly, the agent learnt how to make progress while killing the enemies in about 30% of level 1 of the super Mario game. However, there is still room for improvement in more complex and real world environments such as the real world self-driving car or industrial robots in unstructured environments.

Table 10. Summary of related literatures.

Citation	Aim	Method(s)	Contribution	Environment	Attributes demonstrated by the agent				Comment
					DI	SM	SR	SIG	
[68]	To improve the ability of imitation learning to generalise and learn from raw high dimensional data	Deep imitation learning + Active learning	An approach towards improving generalisation in imitation learning	Simulated environment	X	X	X	X	The agent was demonstrated in four different environments; however, it completely depends on expert demonstrations to learn how to succeed in each of these environments. This is in contrast to our definition of domain-independence. Other attributes are not demonstrated
[84]	To demonstrate that a combination of deep learning and genetic algorithms can compete with deep reinforcement learning algorithms in video games	Deep Neuroevolution + Novelty search	A combination of deep neural networks and genetic algorithms as an alternative for deep reinforcement learning. Moreover, the paper demonstrated how novelty search can improve and hasten learning	Simulated environment	√	√	X	X	The same algorithm was applied in different domains without an account of any change in the algorithm. Self-motivation was implemented as a novelty search, although there was no demonstration on how the agent would perform on the task using only its self-motivation without external reward
[94]	To implement autonomous mobile robot capable of	Reinforcement learning	Novel concepts called health and sub-health states were suggested	Real world environment	X	X	X	X	The robot's behaviours were entirely dependent on the reward function specified by the experimenter; so, there was no proof of self-motivation.

(continued)

Table 10. (*continued*)

Citation	Aim	Method(s)	Contribution	Environment	Attributes demonstrated by the agent				Comment
					DI	SM	SR	SIG	
	solving a spiral maze								Moreover, there was no attempt to demonstrate other attributes
[95]	The paper develops a novel strategy that teaches an agent to learn control policies that are suitable for succeeding in a range of Atari 2600 games	Deep reinforcement learning	A strategy for learning human-level control policies from high dimensional visual input.	Simulated environment	√	X	X	X	It was demonstrated that the agent learnt how to play seven Atari 2600 games with no adjustment of the architecture or learning algorithm. This meets our definition of domain-independence. However, other attributes were not demonstrated
[96]	To demonstrate that intrinsically motivated hierarchical reinforcement learning can enable an artificial agent to learn reusable skills that are needed for competent autonomy	Intrinsic motivation + Hierarchical reinforcement learning	The combination of intrinsic motivation with reinforcement learning	Simulated environment	X	√	X	X	The paper combines both intrinsic (self) and extrinsic (external) rewards or motivations. The intrinsic reward motivated the agent to learn complicated subtasks leading to the goal. The extrinsic motivation is activated after the agent attained the goal state. Furthermore, the paper did not show if the algorithm could adapt to a new environment, or a change in the original environment, without a manual alteration of the agent program or the reward system

(*continued*)

Table 10. *(continued)*

Citation	Aim	Method(s)	Contribution	Environment	Attributes demonstrated by the agent				Comment
					DI	SM	SR	SIG	
[97]	They develop as strategy that helps to train a robot to be able interact with an unknown environment	Evolutionary algorithms	A novel chromosome encoding for mobile robotics as well as the simulator for this task was achieved	Simulated and real world environments	X	X	X	X	A task-specific fitness function was crafted for the agent. The fitness function is not generic enough and would require to be adapted to any new environment if the agent must behave as desired by the designer
[98]	To implement a curiosity-based intrinsic reward that enables an agent to cope with high dimensional visual inputs	Intrinsic Curiosity model + reinforcement learning (A3C)	An intrinsic reward signal that "scales to high-dimensional continuous state spaces like images, bypasses the hard problem of predicting pixels and is unaffected by the unpredictable aspects of the environment that do not affect the agent" [98]	Simulated Environments	√	√	X	√	The work in this paper tried to show the possibility of achieving meaningful behaviour using only curiosity as reward. However, the agent progress towards the goal was not yet significant (the agent made a 30% progress in level 1 of the game) but there was a proof of the possibility of purely curiosity-driven learning agent. Moreover, an agent pre-trained at the level 1 of the super Mario bros using only intrinsic motivation as reward performs better in level 2 when fine-tuned than an agent that is trained from the scratch in level 2 using only curiosity

6 Conclusion and Research Directions

The paper has reviewed different notions of autonomy and has presented a schema for the classification of autonomy using multidimensional attributes of autonomy. The attributes of machine autonomy presented in the paper are classified into two major groups – low-level and high-level autonomy. The low-level autonomy constitutes the basic attributes that any autonomous system must have while the high-level attributes are bases for evaluating the system's degree of autonomy. It is challenging to think of how best to evaluate any machine's autonomy based on these attributes. Future work should consider creating evaluation metrics for these attributes.

Different AI approaches to autonomous agent design were reviewed and their key challenges highlighted. It is clear that none of the techniques is able to realise all the attributes of autonomy as defined in this paper. Further work is therefore required in developing a single algorithm that would enable an embodied *tabula rasa*, equipped with suitable sensors and actuators, to acquire adequate knowledge to succeed in different environments by adapting and reprogramming itself without any form of human guidance.

Furthermore, most researches in autonomous agents are often experimented in a simulated environment due to high sample complexities of the algorithms that underpin them. This has made it almost impracticable to replicate these algorithms in physical robots despite the ubiquity of high performance computers. Most successful works on physical robots often combine imitation or active learning with other techniques, which are still much dependent on human designers and cannot yield true autonomy. Another reason for this challenge is that robotic simulators for testing autonomy are excessively simple when compared to the real world situations; as such, often times, success in the simulated environment will not imply same in the real world. A way forward could be to develop simulators that present the same difficulty and complexity in training these agents as in the real world while giving the flexibility for researchers to test and evaluate the attributes of autonomy. These can serve as benchmarking tools for research in machine autonomy.

References

1. Smithers, T., Laboratory VUBAI: Taking Eliminative Materialism Seriously: A Methodology for Autonomous Systems Research. Artificial Intelligence Laboratory, Vrije Universiteit Brussel, Brussels (1992)
2. Franklin, S.: Artificial Minds. MIT Press, Cambridge, p. 449 (1995)
3. Nolfi, S., Floreano, D.: Evolutionary Robotics: The Biology, Intelligence, and Technology, p. 320, MIT Press, Cambridge (2000)
4. Froese, T., Virgo, N., Izquierdo, E.: Autonomy: a review and a reappraisal. In: Advances in Artificial Life, pp. 455–464. Springer, Heidelberg (2007)
5. Harbers, M., Peeters, M.M.M., Neerincx, M.A.: Perceived autonomy of robots: effects of appearance and context. In: Aldinhas Ferreira, M.I., Silva Sequeira, J., Tokhi, M.O., Kadar, E., Virk, G.S. (eds.) A World with Robots: International Conference on Robot Ethics: ICRE 2015, pp. 19–33. Springer, Cham (2017)

6. Bradshaw, J.M., Hoffman, R.R., Woods, D.D., Johnson, M.: The seven deadly myths of "autonomous systems". IEEE Intell. Syst. **28**, 54–61 (2013)
7. Rich, E., Knight, K.: Artificial Intelligence. McGraw-Hill Higher Education, p. 640 (1990)
8. Roe, A.W., Sur, M.: Visual projections routed to the auditory pathway in ferrets: receptive fields of visual neurons in primary auditory cortex. J. Neurosci. **12**(9), 3651–3664 (1992)
9. Scharre, P., Horowitz, M.: An introduction to autonomy in weapon systems. In: Ethical Autonomy—working paper (2015)
10. Lane, D.M.: Persistent autonomy artificial intelligence or biomimesis? 2012 IEEE/OES Autonomous Underwater Vehicles (AUV) pp. 1–8 (2012)
11. Pernar, S.: The Evolutionary Perspective-a Transhuman Philosophy (2015)
12. Lomonova, E.A.: Advanced actuation systems—state of the art: fundamental and applied research. In: International Conference on Electrical Machines and Systems pp. 13–24 (2010)
13. Hong, J., Suh, E., Kim, S.: Context-aware systems: a literature review and classification. Expert. Syst. Appl. **36**, 8509–8522 (2009)
14. Viterbo, J., Sacramento, V., Rocha, R., Baptista, G., Malcher, M., Endler, M.A.: Middleware architecture for context-aware and location-based mobile applications. In: 2008 32nd Annual IEEE Software Engineering Workshop pp. 52–61 (2008)
15. Gui, F., Zong, N., Adjouadi, M.: Artificial intelligence approach of context-awareness architecture for mobile computing. Sixth International Conference on Intelligent Systems Design and Applications **2**, 527–533 (2006)
16. Haugeland, J.: Artificial Intelligence: The Very Idea. Bradford, Cambridge, MA (1985)
17. Pierce, D., Kuipers, B.J.: Map learning with uninterpreted sensors and effectors. Artificial Intelligence **92**, 169–227 (1987)
18. Chaput, H.H.: The constructivist learning architecture: A model of cognitive development for robust autonomous robots. PhD. AI Laboratory, The University of Texas at Austin. Supervisors: Kuipers and Miikkulainen (2004). https://pdfs.semanticscholar.org/7bb4/18868b2c95443243f5f7a5b9a5a15d342570.pdf. Accessed 12 Aug 2018
19. Oudeyer, P.Y., Kaplan, F., Hafner, V.V.: Intrinsic motivation systems for autonomous mental development. IEEE Trans. Evol. Comput. **11**, 265–286 (2007)
20. Barto, A.G.: Intrinsic motivation and reinforcement learning. In: Baldassarre, G., Mirolli, M. (eds.) Intrinsically Motivated Learning in Natural and Artificial Systems. pp. 17–47. Springer, Heidelberg (2013)
21. Santucci, V.G., Baldassarre, G., Mirolli, M.: GRAIL: a goal-discovering robotic architecture for intrinsically-motivated learning. IEEE Trans. Cognit. Develop. Syst. **8**, 214–231 (2016)
22. Schmidhuber, J.: Formal theory of creativity, fun, and intrinsic motivation (1990–2010). IEEE Trans. Auton. Mental Develop. **2**, 230–247 (2010)
23. Frank, M., Leitner, J., Stollenga, M., Förster, A., Schmidhuber, J.: Curiosity driven reinforcement learning for motion planning on humanoids. Frontiers in Neurorobotics **7**, 25 (2013)
24. Ngo, H., Luciw, M., Forster, A., Schmidhuber, J.: Learning skills from play: artificial curiosity on a katana robot arm. In: The 2012 International Joint Conference on Neural Networks (IJCNN) pp. 1–8 (2012)
25. Gordon, G., Ahissar, E.: A curious emergence of reaching. Advances in Autonomous Robotics. In: TAROS 2012 Lecture Notes in Computer Science, Berlin, Heidelberg, 7429: 1–12 (2012)
26. Georgeon, O.L., Marshall, J.B., Gay, S.: Interactional motivation in artificial systems: between extrinsic and intrinsic motivation. In: 2012 IEEE International Conference on Development and Learning and Epigenetic Robotics (ICDL) pp. 1–2 (2012)
27. Cangelosi, A., Schlesinger, M.: Developmental Robotics: From Babies to Robots. The MIT Press, Cambridge, p. 408 (2014)

28. Froese, T., Ziemke, T.: Enactive artificial intelligence: investigating the systemic organization of life and mind. Artificial Intelligence **173**, 466–500 (2009)
29. Thórisson, K.R., Eric, N., Ricardo, S., Pei, W.: Editorial: approaches and assumptions of self-programming in achieving artificial general intelligence. J. Artif. Gen. Intell. **3**, 1 (2013)
30. Georgeon, O.L., Marshall, J.B.: Demonstrating sensemaking emergence in artificial agents: a method and an example. Int. J. Mach. Conscious. **05**, 131–144 (2013)
31. Stramandinoli, F., Tikhanoff, V., Pattacini, U., Nori, F.: A bayesian approach towards affordance learning in artificial agents. In: 2015 Joint IEEE International Conference on Development and Learning and Epigenetic Robotics (ICDL-EpiRob) pp. 298–299 (2015)
32. Wang, J.G., Mahendran, P.S., Teoh, E.K.: Deep affordance learning for single- and multiple-instance object detection. In: TENCON 2017 - 2017 IEEE Region 10 Conference pp. 321–326 (2017)
33. Glover, A.J., Wyeth, G.F.: Toward lifelong affordance learning using a distributed markov model. IEEE Trans. Cognit. Develop. Syst. **10**, 44–55 (2018)
34. Russell, S.J., Norvig, P.: Artificial Intelligence: A Modern Approach. Pearson Education, London, p. 1132 (2003)
35. Mohri, M., Rostamizadeh, A., Talwalkar, A.: Foundations of Machine Learning. The MIT Press, Cambridge, p. 480 (2012)
36. Ng, A.Y.: Preventing "overfitting" of cross-validation data. In: Proceedings of the Fourteenth International Conference on Machine Learning, pp. 245–253 (1997)
37. Bhlmann, P., Van De Geer, S.: Statistics for High-Dimensional Data: Methods, Theory and Applications. Springer, Berlin, Incorporated, p. 573 (2011)
38. Hinton, G.E., Srivastava, N., Krizhevsky, A., Sutskever, I., Salakhutdinov, R.: Improving neural networks by preventing co-adaptation of feature detectors. CoRR 2012; abs/1207.0580 (2012)
39. Hussein, A., Gaber, M.M., Elyan, E., Jayne, C.: Imitation learning: a survey of learning methods. ACM Comput. Surv. **50**, 21:1–21:35 (2017)
40. Tscherepanow, M., Hillebrand, M., Hegel, F., Wrede, B., Kummert, F.: Direct imitation of human facial expressions by a user-interface robot. In 2009 9th IEEE-RAS International Conference on Humanoid Robots, pp. 154–160. IEEE (2009)
41. Billard, A., Grollman, D.: Imitation Learning in Robots. In: Seel, N.M. (ed.) Encyclopedia of the Sciences of Learning, pp. 1494–1496. Springer US, Boston (2012)
42. Nehaniv, C.L., Dautenhahn, K.: Imitation in animals and artifacts. In: Dautenhahn, K., Nehaniv, C.L. (eds.) pp. 41–61. MIT Press, Cambridge (2002)
43. Szepesvari, C.: Algorithms for Reinforcement Learning. Morgan and Claypool Publishers, San Rafael (2010)
44. LeCun, Y., Bengio, Y., Hinton, G.: Deep learning. Nature **521**, 436 (2015)
45. Marcus, G.: Deep Learning: A Critical Appraisal, CoRR, abs/1801.00631 (2018)
46. Lanchantin, P., Pieczynski, W.: Unsupervised restoration of hidden nonstationary markov chains using evidential priors. IEEE Trans. Signal Process. **53**, 3091–3098 (2005)
47. Acharyya, R., Ham, F.M.: A new approach for blind separation of convolutive mixtures. In: 2007 International Joint Conference on Neural Networks pp. 2075–2080 (2007)
48. Ramezani, R., Saraee, M., Nematbakhsh, M.A.: MRAR: mining multi-relation association rules. J. Comp. Security **1**(2), 133–158 (2014)
49. Heaton, J.: Comparing dataset characteristics that favor the apriori, eclat or FP-growth frequent itemset mining algorithms. SoutheastCon **2016**, 1–7 (2016)
50. Fodor, I.K.: A survey of dimension reduction techniques. Center for Applied Scientific Computing, Lawrence Livermore National Laboratory **9**, 1–18 (2002)
51. Chaput, H.H.: The constructivist learning architecture: a model of cognitive development for robust autonomous robots (2004)

52. Zhong, C., Liu, S., Lu, Q., Zhang, B.: Continuous learning route map for robot navigation using a growing-on-demand self-organizing neural network. Int. J. Adv. Robot. Syst. **14**, 1729881417743612 (2017)
53. Karhunen, J., Raiko, T., Cho, K.: Unsupervised deep learning: a short review (2014). https://pdfs.semanticscholar.org/9a9a/9e32ca5cb15e00d0b5a1f2a2656905ba79df.pdf. Accessed 24 Nov 2018
54. Jolliffe, I.T., Cadima, J.: Principal component analysis: a review and recent developments. Philosophical Transactions, Series A, Mathematical, Physical, and Engineering Sciences **374** (2065), 20150202 (2016)
55. Pindah, W., Nordin, S., Seman, A., Mohamed Said, M.S.: Review of dimensionality reduction techniques using clustering algorithm in reconstruction of gene regulatory networks, pp. 172–176 (2015)
56. Xie, H., Li, J., Xue, H.: A survey of dimensionality reduction techniques based on random projection, CoRR, abs/1706.04371 (2017)
57. Ishii, K., Nishida, S., Ura, T.: A self-organizing map based navigation system for an underwater robot, robotics and automation. In: Proceedings of 2004 IEEE International Conference on ICRA 2004, vol. 5, pp. 4466–4471, April 2004
58. Kim, S., Park, F.C.: Fast robot motion generation using principal components: framework and algorithms. IEEE Trans. Ind. Elec. **55**(6), 2506–2516 (2008)
59. Finn, C., Tan, X.Y., Duan, Y., Darrell, T., Levine, S., Abbeel, P.: Learning visual feature spaces for robotic manipulation with deep spatial autoencoders, CoRR, abs/1509.06113 (2015)
60. Sutton, R.S., Barto, A.G.: Introduction to Reinforcement Learning. MIT Press, Cambridge, MA, USA (1998)
61. Sutton, R.S., Barto, A.G.: Introduction to Reinforcement Learning (2nd edition, in preparation). MIT Press, Cambridge (2017)
62. Watkins, C.J.C.H., Dayan, P.: Q-learning. Mach. Learn. **8**, 279–292 (1992)
63. Silver, D., Huang, A., Maddison, C.J., Guez, A., Sifre, L., van den Driessche, G., Schrittwieser, J., Antonoglou, I., Panneershelvam, V., Lanctot, M., Dieleman, S., Grewe, D., Nham, J., Kalchbrenner, N., Sutskever, I., Lillicrap, T., Leach, M., Kavukcuoglu, K., Graepel, T., Hassabis, D.: Mastering the game of go with deep neural networks and tree search. Nature **529**, 484–489 (2016)
64. Kormushev, P., Calinon, S., Caldwell, G.D.: Reinforcement learning in robotics: applications and real-world challenges. Robotics **2**(3), 122–148 (2013)
65. Abbeel, P., Ng, A.Y.: Apprenticeship learning via inverse reinforcement learning. In: Proceedings of the Twenty-First International Conference on Machine Learning, Banff, Alberta, Canada, p. 1 (2004)
66. Settles, B.: Active Learning Literature Survey. University of Wisconsin, Madison. (2009)
67. Olsson, F.: A literature survey of active machine learning in the context of natural language processing (2009)
68. Hussein, A., Elyan, E., Gaber, M.M., Jayne, C.: Deep imitation learning for 3D navigation tasks. Neural Comput. Appl. **29**, 389–404 (2018)
69. Mash-Simulator. https://github.com/idiap/mash-simulator. Accessed 10 July 2018
70. Dima, C., Hebert, M.: Active learning for outdoor obstacle detection (2005)
71. Pan, S.J., Yang, Q.: A survey on transfer learning. IEEE Trans. Knowled. Data Eng. **22**, 1345–1359 (2010)
72. Brys, T., Harutyunyan, A., Taylor, M.E., Nowé, A.: Ann.: Policy Transfer Using Reward Shaping, pp. 181–188 (2015)
73. Weiss, K., Khoshgoftaar, T.M, Wang, D.: A survey of transfer learning. J. Big Data **3**, 9 (2016)

74. Tsung, F., Zhang, K., Cheng, L., Song, Z.: Statistical transfer learning: a review and some extensions to statistical process control. Quality Engineering **30**, 115–128 (2018)
75. Barrett, S., Taylor, M.E., Stone, P.: Transfer learning for reinforcement learning on a physical robot, p. 1 (2010)
76. Kira, Z.: Inter-robot transfer learning for perceptual classification, pp. 13–20 (2010)
77. Huang, Z., Pan, Z., Lei, B.: Transfer learning with deep convolutional neural network for SAR target classification with limited labeled data. Remote Sensing **9**, 907 (2017)
78. Grefenstette, J.J.: Evolutionary algorithms in robotics†. In: Fifth International Symposium on Robotics and Manufacturing, ISRAM 94 (1994)
79. Holland, J.H.: Adaptation in Natural and Artificial Systems. MIT Press, Cambridge, MA, USA (1992)
80. Vent, W.: Rechenberg, ingo, evolutionsstrategie—optimierung technischer systeme nach prinzipien der biologischen evolution. 170 S. mit 36 abb. Frommann-Holzboog-Verlag. stuttgart 1973. broschiert. Feddes Repert 86, p. 337 (2008)
81. Luke, S., Hamahashi, S., Kitano, H.: Genetic programming, pp. 1098–1105 (1999)
82. Nelson, A.L., Barlow, G.J., Doitsidis, L.: Fitness functions in evolutionary robotics: a survey and analysis. Robot. Auton. Syst. **57**, 345–370 (2009)
83. Doncieux, S., Bredeche, N., Mouret, J., Eiben, A.E.G.: Evolutionary robotics: what, why, and where to. Frontiers in Robotics and AI **2**, 4 (2015)
84. Such, F.P., Madhavan, V., Conti, E., Lehman, J., Stanley, K.O., Clune, J.: Deep neuroevolution: genetic algorithms are a competitive alternative for training deep neural networks for reinforcement learning. CoRR, abs/1712.06567 (2017)
85. Duarte, M., Costa, V., Gomes, J.C., Rodrigues, T., Silva, F., Oliveira, S.M., Christensen, A. L.: Evolution of Collective Behaviors for a Real Swarm of Aquatic Surface Robots, CoRR, abs/1511.03154 (2015)
86. Schrum, J., Lehman, J., Risi, S.: Automatic evolution of multimodal behavior with multi-brain HyperNEAT, pp. 21–22. In: Proceedings of the 2016 on Genetic and Evolutionary Computation Conference Companion, Denver, Colorado, USA, (2016)
87. Turing, A.M.: Computers & Thought. In: Feigenbaum, E.A., Feldman, J. (eds.) MIT Press, Cambridge, pp. 11–35 (1995).
88. Piaget, J.: The origins of intelligence in children. International Universities Press, New York (1952)
89. Tsou, J.Y.: Genetic epistemology and piaget's philosophy of science: Piaget vs. kuhn on scientific progress. Theory & Psychology **16**, 203–224 (2006)
90. Asada, M., Hosoda, K., Kuniyoshi, Y., Ishiguro, H., Inui, T., Yoshikawa, Y., Ogino, M., Yoshida, C.: Cognitive developmental robotics: a survey. IEEE Trans. Auton. Mental Develop. **1**, 12–34 (2009)
91. Guerin, F.: Learning like a baby: A survey of artificial intelligence approaches. Knowl. Eng. Rev. **26**, 209–236 (2011)
92. Oudeyer, P.: Autonomous development and learning in artificial intelligence and robotics: scaling up deep learning to human-like learning. CoRR, abs/1712.01626 (2017)
93. Oudeyer, P., Kaplan, F., Hafner, V.V., Whyte, A.: The playground experiment: task-independent development of a curious robot. In: Proceedings of AAAI Spring Symposium on Developmental Robotics, pp. 42–47 (2005)
94. Zuo, B., Chen, J., Wang, L., Wang, Y.: A reinforcement learning based robotic navigation system. IEEE Int. Conf. Sys. Man Cyber. (SMC), pp. 3452–3457 (2014)
95. Mnih, V., Kavukcuoglu, K., Silver, D., Graves, A., Antonoglou, I., Wierstra, D., Riedmiller, M.A.: Playing atari with deep reinforcement learning. CoRR, abs/1312.5602 (2013)
96. Singh, S.P., Barto, A.G., Chentanez, N.: Intrinsically motivated reinforcement learning. In: Advances in Neural Information Processing Systems 17 (NIPS). MIT Press (2004)

97. Assis, LdS., Soares, AdS., Coelho, C.J., Van Baalen, J.: An evolutionary algorithm for autonomous robot navigation. Proc. Comp. Sci. **80**, 2261–2265 (2016)
98. Pathak, D., Agrawal, P., Efros, A.A., Darrell, T.: Curiosity-driven exploration by self-supervised prediction. CoRR, abs/1705.05363 (2017)

Novel Recursive Technique for Finding the Optimal Solution of the Nurse Scheduling Problem

Samah Senbel(✉)

Sacred Heart University, Fairfield, CT 06825, USA
senbels@sacredheart.edu

Abstract. Solving the Nurse Scheduling problem is a major research area in Operations Research. Due to it being an NP-Hard problem, most researchers develop a heuristic solution for it. The NSP has several constraints that need to be satisfied: several mandatory "hard" constraints that reflect hospital requirements, and several optional "soft" constraints that reflect the nurses' preferences. In this paper, we present a recursive solution to the problem that makes use of those constraints to shrink the search space and obtain results in a reasonable amount of time. We present two variations of the solution, a nurse-by-nurse method of building the optimal schedule, and a shift-by-shift approach. Both variations were implemented and tested with various scenarios and the shift-by-shift solution provided much better results. The solution can also be modified easily to provide fair long-term scheduling.

Keywords: Nurse scheduling problem · Optimal solution · Brute force solution · Scheduling

1 Introduction

The Nurse Scheduling Problem is an important part of hospitals' operations and has been addressed by mathematicians, computer scientists, and nursing professionals alike for years, due to its complex and sensitive nature. The goal is to generate a nurse roster table for a group of nurses over a certain scheduling period, while considering their needs and providing fairness to all. The roster table needs to satisfy two sets of constraints: Hard constraints are constraint that have to be satisfied and are usually related to hospitals' policies and regulations. Soft constraints are the needs and requirements of the nurses themselves. The degree of satisfaction of the soft constraints is the measure of the quality of the scheduling technique.

Nurses have a vital role to play in the healthcare system, and work-life balance is important for job satisfaction and improving performance. Any schedule should be respectful of their days off and shifts off requirement and their need for rest [1]. Scheduling is classically done by a nurse manager responsible for a certain group of nurses. Typically, a small group of 10–30 nurses over a period of 7–30 days usually.

There are several objectives and constraints to meet. The hospital constraints are usually mandatory and fixed over time, such as the required number of nurses per shift,

K. Arai et al. (Eds.): CompCom 2019, AISC 997, pp. 359–375, 2019.
https://doi.org/10.1007/978-3-030-22871-2_25

and the need to have non-consecutive shift assignments. Fairness is of the utmost importance as well, so nurses are to be assigned equal number of shifts if possible and their requirements are to be handled with equal importance. Thus, a computerized fair system would be better received by the entire team. It is also important to maintain fairness over multiple scheduling periods, so a nurse taking an unfavorable shift in a certain scheduling period should be compensated for that in a consecutive period.

Nurses' scheduling preferences are called "soft" constraints are not mandatory to fulfill. However, a good scheduling technique needs to satisfy as many of these preferences as possible. This is considered the measure of quality of the scheduling technique and the objective is to satisfy as many of them as possible while guaranteeing the hard constraints and maintaining fairness.

In this paper, we present a practical method to get the optimal possible schedule using two different techniques, and compare their performances. This paper is organized as follows: In Sect. 2, we provide a literature review of the different solutions to the NSP. In Sect. 3, we describe the NSP definition, constraints and data structures used. In Sect. 4, we explain our proposed solutions to the problem. Section 5 presents our implementation results, and Sect. 6 concludes our work.

2 Literature Review

Having been studied as a mathematical problem for over 40 years, the Nurse scheduling problem has an extensive literature. Due to its myriad of constraints and large search space, it is considered an NP-hard problem and solving the problem is far from easy. For this reason most attention has been paid to developing heuristic solutions to find a near-optimal solution in reasonable time.

During early research in the NSP, relatively simple mathematical models were proposed to optimize the assignment of nurses to jobs in order to perform various tasks [2]. Although these approaches are able to solve small sized problems, the computational time for large sized problems was usually prohibitive for most practical applications. Burke [3] presents an excellent survey of the different heuristic techniques developed over time to get near- optimal solutions.

Genetic and evolutionary techniques [4] are popular in solving the NSP. The basis of genetic algorithms consist of generating an initial set of schedules that satisfy the hard constraints, then doing n-point crossover, single-bit mutation and a rank-based selection to generate a new "generation" of schedules.

Tabu Search is another popular technique [5]. It is based on the premise that problem solving must incorporate adaptive memory of schedules according to the hard constraints, and responsive exploration to possible new schedules. Another interesting algorithm used is simulated annealing [6], which is a stochastic method that is able to conditionally accept worse schedules as a mean for reaching a more optimal global solution. The goal of this approach is to escape from possible sub-optimality and evade the local convergence.

Lots of other heuristic methods have been developed for dealing with the nurse scheduling problem: variable neighborhood search [7], iterated local search [8], particle swarm optimization [9], ant colony optimization [10] among others.

3 Nurse Scheduling Problem Definition

In this section, we define the system inputs, data structures, constraints and objective function used in our solution for the nurse scheduling problem.

3.1 System Inputs

- N: Number of nurses to schedule.
- D: The scheduling period is for D days, which is a repeated period.
- S: Each day has S equal-duration shifts to be assigned, typically two or three.
- C: Each shift needs C_j nurses to cover it. C_j could be the same for all shifts, or vary by shift. For clarity, we assume C is the same for all shifts in this paper,
- Pref: A table representing the scheduling preferences of all the nurses.

3.2 Data Structures

We use two main arrays for implementing our solution:

- The schedule table (Sch) is a table of size N rows (one per nurse) and S*D columns, one per shift for all D days. The value of each element is either 0 (not assigned this shift) or 1 (assigned to this shift). Equation 1 shows the matrix definition. This is the final output of the system.

$$\forall_{i=0,1,\ldots,N-1} \forall_{j=0,1,\ldots,(S*D-1)} \, \mathrm{Schi,j} \in \{0,1\} \tag{1}$$

- The preference table (Pref) is of similar dimensions to the scheduling table. It contains the preference for each shift by each nurse, represented as a penalty point for all assigned days. A value of 0 means that this shift is favorable and an increasing value signals an increasing amount of dislike for this shift. Any range of values can be used for this penalty. This table is supplied by the nurses themselves to represent their preferences for the scheduling period. This is the main input to the system.

3.3 Hard Constraints

We have four hard constraints to satisfy:

- Constraint 1: Guarantee hospital-required coverage for each shift. The number of nurses required per shift could be fixed for all shifts or it could vary between shifts.

$$\forall_{j=0,1,\ldots,(S*D-1)} C_j = ^{N-1} \sum_{i=0} \mathrm{Sch}_{i,j} \tag{2}$$

- Constraint 2: No consecutive shifts for any nurse. This is a natural constraint to guarantee a sufficient rest period. If there is a previously generated schedule, then its last shift has to be taken into consideration as well.

$$\forall_{i=0,1,\ldots,N-1}\forall_{j=0,1,\ldots,(S*D-2)}\ Sch_{i,j} + Sch_{i,j+1} \in \{0,1\} \tag{3}$$

- Constraint 3: One shift per day for all nurses. The sum of all shift assignments is 1 (working) or 0 (day off) per day for all nurses.

$$\forall_{i=0,1,\ldots,N-1}\forall_{d=0,1,..,D-1}^{S-1}\sum_{j=0} Sch_{i,j+d} * S \in \{0,1\} \tag{4}$$

- Constraint 4: Fairness: Approximately equal number of shifts per nurse. The shift of each nurse should be equal to the floor or the ceiling of the average load per nurse.

$$\text{AverageLoad} = (D * S * C) / N \tag{5}$$

$$\forall_{i=0,1,\ldots,N-1}^{N-1}\sum_{i=0} Sch_{i,j} \in \{\ \text{Floor}(\text{AverageLoad}),\ \text{Ceil}(\text{AverageLoad})\} \tag{6}$$

3.4 Soft Constraints

The soft constraints represent the nurses' personal preferences for days off, particular shifts off, and a preference for one or more of the different shifts. These constraints are not guaranteed to be satisfied. The Pref table contains all these preferences. A very high penalty (100 or more) is put for days and shifts off, and a lower penalty (1 or 2) for undesirable but acceptable shifts. The soft constraints, in order, are:

- Constraint 5: Required days off for all nurses. It is desirable that each nurse is allowed to choose the same number of days off as her peers per each scheduling period. This is done by assigning the maximum penalty for all shifts on that day.
- Constraint 6: Required shifts off for all nurses. There should a maximum number of shifts a nurse can request to be unassigned, once again to guarantee fairness. This is done by assigning a high penalty to that particular shift.
- Constraint 7: Shift preference applied for all nurses. This is input by the nurse for each individual shift. A preferred shift is assigned a value of zero and an un-preferred shift is given a low penalty.

3.5 Objective Function

Our objective is to minimize the maximum penalty given to any nurse, while fulfilling the four hard constraints and the three soft constraints. In case of two schedules with the same minimum maximum penalty, preference is given to the schedule with the lower total penalty. The NSP objective can be formulated as follows:

$$\textbf{Minimize}\ \text{Maximum\_Penalty} = \forall_{i=0,1,\ldots,N-1}\text{Maximum}\left(\sum_{j=0}^{S*D-1} Pref_{i,j} * Sch_{i,j}\right) \tag{7}$$

AND

$$\textbf{Minimize}\,\text{Total\_Penalty} = \sum_{i=0}^{N-1} \sum_{j=0}^{S*D-1} \text{Pref}_{i,j} * \text{Sch}_{i,j} \tag{8}$$

while satisfying Eqs. 1 thru 6.

4 Optimal Solution of the NSP

We propose a simple recursive solution to the nurse scheduling problem. We make use of the nurse preference table and the hard constraints to limit the search space of the problem rendering it feasible to find the optimal solution in reasonable time. The basic idea is to build up the scheduling table recursively row by row or column by column and to abandon any path that will not lead to a possible solution. In this section, we describe the two techniques we tested, the Top-to-Bottom Recursive solution and the Left-to-Right Recursive solution.

4.1 Top-to-Bottom Recursive Solution to the NSP

In this solution, we first generate a table of all the possible nurse schedules (rows) that can be in the schedule table. Each row has to satisfy three of the four hard constraints: No consecutive shifts, one shift per day and load fairness. In case the number of shifts per pay is two, then the one shift per day constraint is automatically satisfied by satisfying the no consecutive shifts constraint. Then, we recursively pick schedules from this table until we get a feasible solution. Algorithm 1 shows this process.

```
Algorithm 1: A Top-to-Bottom recursive solution to the NSP
main
{
   countL=0;                          // number of possible loads per nurse
   Loadmin = Math.floor(D*S*C/N);
   Loadmax = Math.ceil(D*S*C/N);
   generateLTable(0 new int[S*D] ,0, Loadmin);              //Algorithm 2
   if ( Loadmax != Loadmin)
          generateLTable(0, new int[S*D] ,0, Loadmax);
   generatePenTable();                                       //Algorithm 3
   MAX = maxInt;
   TOT = maxInt;
   generateSchedule(0, new int[N][S*D], new int[S*D], 0);  //Algorithm 4
}
```

We start by creating a table “Load” that has all the possible values a row can take. Figure 1 shows the Load table at D = 7, S = 2, C = 2, and N = 6 when Load_{max} is 5 and Load_{min} is 4. There are 582 possibilities in the “Load” table for these parameters.

The Load table is created by recursively generating all possible combinations of 0 and 1 in a possible row of length SxD. As the row is filled in left to right, we only progress if all three hard constraints are still achievable. If we get two 1s next to each

```
00 00 00 01 01 01 01
00 00 00 10 01 01 01
00 00 00 10 10 01 01
....
....
....
....
....
....

10 10 10 10 00 10 00
10 10 10 10 01 00 00
10 10 10 10 10 00 00
```

Fig. 1. The "Load" table at D = 7, S = 2, C = 2, and N = 6

other, that path is abandoned as it violated constraint 2. If we get two shifts in the same day when S = 3 or more, we abandon that path as it violates constraint 3. If the sum of 1's is bigger than the required load, or cannot be achieved within the row, we abandon that path as it violates constraint 4. Once a row is found to satisfy all three, it is added to the "Load" array and the count is incremented. Algorithm 2 summarizes this process.

Algorithm 2 : Pseudocode for generating the "Load" table

```
generateLTable(int index, int[] Possible, int sum, int L )
  {
   if( (index == size) && (sum==L )   )
      { // found a row that satisfies all 3 constraints, add it to "Load"
          for(int i = 0; i < S*D; i++)
               Load[countL][i] = Possible[i];
          countL++;
      }
    else { for(int i = 0; i < 2; i++)  {
             Possible[index] = i;
             if (i==1)
                 sum++;      // add 1 to load sum so far
             if (( index == 0)     // first shift, no constraints
                ||((((index>0)&&(Possible[index]==0||Possible[index-1]==0))
                                     // no consecutive shifts
                 && sum<=L)        // Sum <= required load
                 && (L-sum < S*D-index) ))  // Can still satisfy load
                 && ((S==2)
                       ||((S==3) && (index%3!=2))
                       ||((S==3 ) && (index%3==2) )
                          && !(Possible[index-2]==1 && Possible[index]==1))
                              generateLTable(index + 1, Possible , sum, L);
             }
      }
}
```

Having generated the "Load" table of size countL, the next step is to calculate the penalty each nurse would incur if she is assigned each row of the load array. This is done to avoid having to calculate the penalty multiple times when generating the schedule table. We create an array "Pen" with the same number of rows as "Load" and N column, each representing the penalty a nurse would have if assigned to that schedule, based on the preferences in the "Pref" table. Figure 2 shows a sample "Pen" table when D = 7, S = 2, C = 2, and N = 6. It corresponds to the "Load" table in Fig. 1. So if Schedule "00000001010101" is assigned to nurse 0 her total penalty would be 4, if assigned to nurse 1, her penalty would be 0 and so on. Algorithm 3 shows how to generate the "Pen" table.

```
4 0 4 100 103 100
3 1 3 100 102 101
2 2 2 101 102 102
....
....
....
....
....
....
100 104 100 104 0 104
101 103 101 103 100 4
100 104 100 104 100 4
```

Fig. 2. The "Pen" table at D = 7, S = 2, C = 2, and N = 6

```
Algorithm 3: Pseudocode for generating the "Pen" table
generatePen() {
   for (int i=0; i< countL; i++)
      for (int n=0; n<N; n++) {
           pen_sum=0;
           for ( int j=0; j< S*D; j++)
                pen_sum  += Load[i][j]* pref[n][j];
          Pen[i][n] = pen_sum;
      }
}
```

With the "Load" and "Pen" table ready, generating the schedule is a simple but time-consuming operation. Picking schedules from the "Load" table guarantees that they satisfy all the hard constraints except the coverage constraint. The schedules are generated top to bottom by recursively picking schedules from "Load" and adding them to the "Sch" table. So, in case of D = 7, S = 2, C = 2, and N = 6, we assign nurse 0 to the first row of "Load" and then assign nurse 1 to all possible 582 rows and so on. The number of possibilities is enormous ($582^6 \approx 3.9 \times 10^{16}$); however most of these combinations will be rejected early on due to the Coverage constraint or the Penalty excess. Algorithm 4 shows the Top-to-Bottom recursive algorithm to generate the optimal schedule table.

```
Algorithm 4: Generating the schedule table
generateSchedule(int index, int[][] current, int[] SumSoFar, int totsofar )
{
   if ( index==N)  // found a solution, is it optimal?
      {    max = 0;
          for (int i=0; i<N;i++)
             if ( PenSoFar[i] > max)
                max = PenSoFar[i];
            if (( max < MAX)|| ( (max==MAX) && (totsofar <TOT)  ) )
              {  MAX = max;                         // Optimal Schedule so far
                 TOT = totsofar;
                 PrintSchedule(current);
              }
      }
   else { int[] SumSoFar1 = new int[S*D];  // total of assignments per shift
          int totsofar1 = 0;                // total penalties so far
          first:
          for (int j=0; j<countL; j++) {
            if ( Pen[j][index] > MAX )
                 continue;
             else {   // Pen <= MAX  , so there is hope of finding optimal
sol.
                  totsofar1 = totsofar +  Pen[j][index];
                  if ((Pen[j][index]== MAX) && ( totsofar1>= TOT))
                       continue;    // not good enough
                  else { // Check for the coverage constraint now
                          for (int r=0; r<S*D; r++){
                             SumSoFar1[r] = SumSoFar[r] + Load[j][r];
                             if (SumSoFar1[r] > C )   // bigger than C
                                continue first;
                             if (  SumSoFar1[r] <= (C*N+index))
                                continue first;   // will be less than C
                             current[index][r] = Load[j][r];
                          }
                          PenSoFar[index] = Pen[j][index];
                       }
                  }
            generateSchedule(index+1, current, SumSoFar1, totsofar1);
          }
      }
}
```

4.2 Left-to-Right Recursive Solution to the NSP

In this solution, we build up the schedule shift by shift, in a left-to-right fashion. We first create two arrays, the "Cov" and the "Nbr", and then run the main procedure for generating the schedule. Algorithm 5 shows an overview of the solution.

Algorithm 5: The Left-to-Right recursive solution to the NSP

```
main
{ generateCtable();                              //Algorithm 6
  generateNbrtable();                            //Algorithm 7
  Loadmin = Math.floor(D*S*C/N);
  Loadmax = Math.ceil(D*S*C/N);

  int[] SumSoFar = new int[N];
  int[] PenSoFar = new int[N];
for ( int i = 0 ; i< countC; i++) {     // fill the first column from "Cov"
     System.out.println( (int)(100.0*i/countC) + " percent done ");
     for (int j=0;j<N;j++) {
        SumSoFar[j]=Cov[i][j];
        PenSoFar[j]=Cov[i][j] * pref[j][0];
        current[j][0]= Cov[i][j];
     }
     generateSTable(1,Sch, SumSoFar, PenSoFar, i);  //Algorithm 8
  }
}
```

We start by generating all the possible columns of the Schedule table and save them in an array "Cov", the array has only binary values of 0(unassigned) or 1(assigned). It has N columns and Combination(N,C) rows. The column contains all possible combinations of choosing C nurse from the N nurse. Table 1 shows a sample of the array when N = 6 and C = 2, which has 15 rows and 6 columns. Algorithm 6 shows how to recursively generate the array using simple pseudocode. All rows in the "Cov" array satisfy constraint 1 (Table 1).

Table 1. Sample "Cov" array when N = 6 and C = 2

Variation number	Nurse 0 1 2 3 4 5
0	0 0 0 0 1 1
1	0 0 0 1 0 1
2	0 0 0 1 1 0
3	0 0 1 0 0 1
4	0 0 1 0 1 0
5	0 0 1 1 0 0
6	0 1 0 0 0 1
7	0 1 0 0 1 0
8	0 1 0 1 0 0
9	0 1 1 0 0 0
10	1 0 0 0 0 1
11	1 0 0 0 1 0
12	1 0 0 1 0 0
13	1 0 1 0 0 0
14	1 1 0 0 0 0

Algorithm 4: Pseudocode for Generating the "Cov' table

```
generateCTable(int index, int[] current, int sum ) {
      if  (index==N)                    // generated a full "solution"
            if  (sum==C)  {              // copy the solution to "cov"
                for(int i=0; i<N; i++)
                     Cov[countC][i] = current[i];
                countC++;              // How many rows in "Cov"
            }
      else {  for (int j=0; j<2 ; j++) {
                current[index] = j;
                if ((sum<=C) &&(C-sum < N-index))
                  generateCTable(index + 1, current ,sum);
                sum+=j;
              }
           }
}
```

Next, we work on satisfying constraint 2: no consecutive shifts. Having decreases the possibilities of each column from 2^N to Combination(N,C) by using the "Cov" table, it is possible to decrease it even further by considering that a nurse cannot work two back to back shifts. This is implemented by creating a neighbor table "Nbr", which includes each of the possible neighbors of a certain column, so that there is no adjacent 1's in the final scheduling table. Each column will have Combination(N-C,C) neighbors. So if N = 6 and C = 2 the column possibilities are dropped from 2^6 (256) to just 6. The neighbor table is used to choose the next column when building the schedule table. Only the first column will iterate among all the different possibilities provided in "Cov". Table 2 shows a sample of the neighbor array when N = 6 and C = 2. Algorithm 7 shows the pseudocode of how to generate the "Nbr" array.

Table 2. Sample "Nbr" table when N = 6 and C = 2

Row in "Cov"	Possible neighbor rows in "Cov" 0 1 2 3 4 5
0	5 8 9 12 13 14
1	4 7 9 11 13 14
2	3 6 9 10 13 14
3	2 7 8 11 12 14
4	1 6 8 10 12 14
5	0 6 7 10 11 14
6	2 4 5 11 12 13
7	1 3 5 10 12 13
8	0 3 4 10 11 13
9	0 1 2 10 11 12
10	2 4 5 7 8 9
11	1 3 5 6 8 9
12	0 3 4 6 7 9
13	0 1 2 6 7 8
14	0 1 2 3 4 5

Algorithm 7: Pseudocode for generating the "Nbr" table

```
GenerateNbrTable() {
   int j=0;
   boolean adj;
   for (int i=0; i<countC; i++) {
      j=0;
      for(int k=0; k<countC; k++) {
            adj = false;
            for (int r=0; r<N; r++)
                 if (Cov[i][r]==1 && Cov[k][r]==1)
                     {adj = true;break;}
              if (adj==false)
                {Nbr[i][j] = k; j++;}
         }
      }
    countN=j;    // Size of the Nbr table
}
```

Having generated the "Cov" and "Nbr" tables that satisfies the first two hard constraints, it is now possible to generate all possible schedules that satisfies constraints three (one shift per day), four (fairness) and the soft constraints. The Sch table is filled column by column, left to right, and at each new column, we do a feasibility test to check if we should abandon this solution. The first column will iteratively contain all the rows of the Cov table, and the second will iteratively contain all the possible neighbors of the data in the first column, and so on. With the addition or change in each column, we check for the third constraint, one shift per day. If it is violated, we abandon this column and get the next column from the neighbor table.

We also maintain a vector of the number of the shifts that each nurse acquires. If one of them gets more than $Load_{max}$, or the number of remaining columns will not guarantee $Load_{min}$, we abandon this column. We also maintain the maximum and sum of the penalties acquired by the nurses as we go, and one the Maximum is bigger than the global maximum, we abandon that solution. Otherwise, we start on the next column.

A valid solution is found if all the columns are filled up. This means that all hard constraints were satisfied and we have a schedule with a maximum schedule penalty equal or less than the global maximum penalty. If the maximum penalty is less than the global maximum, the schedule is saved and declared the optimal schedule so far, and the global maximum is updated. If the maximum penalty is equal to the global maximum penalty and the total schedule penalty is less than the global total penalty, we save the schedule and it becomes the new optimal schedule, and the global total penalty is updated. This is the main algorithm of our solution and its time complexity depends on the values of N, D, S, and C. Algorithm 8 shows how the schedule table is generated when S = 2.

Algorithm 8: Left-to-right Optimal Solution to the NSP

```
generateSTable(int index,   int[][] Sch, int[] SS, int[] Pen, int prev )
{
       if (index == S*D )
         {    // found a solution, calculate maximum & total penalty
                         max=0;
                         tot=0;
                         for ( ii=0; ii<N ; ii++){
                             tot= tot +Pen[ii];
                             if (Pen[ii] > max)
                                  max = Pen[ii];
                         }
                         if ( ( max < MAX) || ((max==MAX)&&(tot <TOT) ) ) {
                             TOT=tot;
                             MAX = max;
                             PrintSchedule();
                           }
         }
     else {
         int[] SumSoFar = new int[N];
         int[] PenSoFar = new int[N];
         int totsofar;
         int max = 0;
         first:
         for (int j = 0 ;j<countN; j++) {   // All possible neighbors
             totsofar=0;
             for (int i=0; i<N ;i++){
               sum = Cov [Nbr[prev][j]][i];
               SumSoFar[i]  =  SS[i]  + sum;
               if  ( SumSoFar[i] > Loadmax )          //Too many shifts
                        continue first;
              if (sub+index >  S*SumSoFar[i])       // Too few shifts
                        continue first;
              PenSoFar[i]  =  Pen[i] + sum * pref[i][index];
              if (PenSoFar[i] > MAX)          // Too much penalty so far
                        continue first;
              if (PenSoFar[i] > max )
                   max = PenSoFar[i];
              totsofar = totsofar +  PenSoFar[i] ;
              Sch[i][index] = sum  ;
             }
          if ( (max == MAX) && ( totsofar >= TOT)) // too much penalty
                  continue first;
          generateSTable(index+1,Sch,SumSoFar,PenSoFar, Nbr[prev][j] );
       }
    }
}
```

4.3 Long-Term Optimal Scheduling

An advantage of our technique is the ability to extend it to multiple scheduling periods, not just repeat the same algorithm each month. An inherent problem of independent scheduling is that certain nurses may always get relatively large or small penalties compared to the others. To guard against this, and to provide long-term fairness, we can easily modify our technique by simply creating a 1-dimensional array "OldPen" of length N. It contains the penalties each nurse has acquired over the previous scheduling periods. That penalty is to be added to the "Pen" array at the beginning of scheduling. If there is no previous history, then we start "Pen" with all zeros. Therefore we will obtain the optimal and most fair schedule for that scheduling period.

5 Analysis and Experimental Results

We implemented our technique using the Java programming language and the Netbeans IDE. We used an IBM Lenovo ultrabook with an Intel i5 processor and 8 GB of memory. We tested the code with different values of N, S, D, and C. We used penalties of 0 for preferred shift, 1 for un-preferred shift, and 100 for days-off and shifts-off requests. However, any values can be used to emphasize the importance of each preference.

For testing and comparison purposes, we generated two standard Preference table format: a "Best" format where the nurses preferences distribute evenly, so there is as little overlap as possible between their needs. So their shift preferences will be divided equally, so if S = 3, one third will prefer the morning shift, one third will prefer the afternoon shift and one third will prefer the night shift. Their days-off and shifts-off will have as little overlap as possible. Figure 3 shows a sample "Best" Preference table at D = 7, S = 2, C = 2, and N = 6. We assume each nurse gets to request one day off and one shift off per week. This format should have the least possible penalties. Figure 4 shows one of the optimal solutions for this case, as there are usually multiple optimal schedules. The algorithm can be easily modified to show all the optimal schedules with the same Maximum and total penalties for the head nurse to pick from them.

```
Nurse 0: 100 100 0   1   0   1   0   100 0   1   0   1   0   1
Nurse 1: 1   0   100 100 1   0   1   0   1   100 1   0   1   0
Nurse 2: 0   1   0   1   100 100 0   1   0   1   0   100 0   1
Nurse 3: 1   0   1   0   1   0   100 100 1   0   1   0   1   100
Nurse 4: 0   100 0   1   0   1   0   1   100 100 0   1   0   1
Nurse 5: 1   0   1   100 1   0   1   0   1   0   100 100 1   0
```

Fig. 3. "Best" nurse preference table at D = 7, S = 2, C = 2, and N = 6

The second standard format is the "Worst" format, where all the nurses have identical preferences in shift preference, days-off requests and shifts-off requests. Figure 5 shows a sample "Worst" Preference table at D = 7, S = 2, C = 2, and N = 6. This format should have the highest penalties. We also use randomly generated preference table for testing, which would fall between the "Best" and "Worst" preferences tables in performance. Figure 6 shows one of the optimal solutions for this case.

We tested several different values of the input parameters over the "Best", "Worst", and 10 random nurse preferences table. Table 3 shows the execution time for each test

Nurse 0:	0	0	1	0	1	0	0	0	1	0	1	0	1	0	Penalty: 0
Nurse 1:	0	0	0	0	0	1	0	1	0	0	0	1	0	1	Penalty: 0
Nurse 2:	1	0	0	1	0	0	1	0	1	0	0	0	1	0	Penalty: 1
Nurse 3:	0	1	0	1	0	0	0	0	0	1	0	1	0	0	Penalty: 0
Nurse 4:	1	0	1	0	1	0	1	0	0	0	1	0	0	0	Penalty: 0
Nurse 5:	0	1	0	0	0	1	0	1	0	1	0	0	0	1	Penalty: 0

Fig. 4. An optimal schedule for the "Best" case

Nurse 0:	0	1	100	1	0	1	0	1	0	1	0	1	100	100
Nurse 1:	0	1	100	1	0	1	0	1	0	1	0	1	100	100
Nurse 2:	0	1	100	1	0	1	0	1	0	1	0	1	100	100
Nurse 3:	0	1	100	1	0	1	0	1	0	1	0	1	100	100
Nurse 4:	0	1	100	1	0	1	0	1	0	1	0	1	100	100
Nurse 5:	0	1	100	1	0	1	0	1	0	1	0	1	100	100

Fig. 5. "Worst" preference table at D = 7, S = 2, C = 2, and N = 6

Nurse 0:	0	0	0	0	0	1	0	0	1	0	0	1	0	1	Penalty: 102
Nurse 1:	0	0	0	0	0	1	0	0	1	0	0	1	0	1	Penalty: 102
Nurse 2:	0	1	0	1	0	0	1	0	0	0	1	0	1	0	Penalty: 102
Nurse 3:	0	1	0	1	0	0	1	0	0	0	1	0	1	0	Penalty: 102
Nurse 4:	1	0	1	0	1	0	0	1	0	1	0	0	0	0	Penalty: 102
Nurse 5:	1	0	1	0	1	0	0	1	0	1	0	0	0	0	Penalty: 102

Fig. 6. An optimal schedule for the "Worst" case

for both the Top-to-Bottom and the Left-to-Right techniques. Each time measurement for the Left-to-Right technique is the average time of 100 runs.

Table 3. Performance results the Top-to-Bottom vs. the Left-to-Right techniques

D	S	C	N	Search space size	Top-to-Bottom			Left-to-Right		
					Best Pref.	Random Pref.	Worst Pref.	Best Pref.	Random Pref.	Worst Pref.
7	2	1	5	1.2×10^{21}	1.2 s	3 s	49 s	0.1 s	0.3 s	4.1 s
7	2	2	6	1.9×10^{25}	22 min	13 min	19 days	1.1 s	1.2 s	1384 s
7	3	1	5	1.8×10^{44}	7 min	18 min	30 days+	1 s	5 s	5486 s
14	2	1	3	1.9×10^{25}	10 h	8.9 h	30 days+	0.4 s	0.6 s	14 s

Clearly, the performance time of the Top-to-Bottom is very large, even for small values of N. This is because the recursive technique abandons possible paths based on the coverage constraints only, while the Left-to-right technique abandons paths based on the other three hard constraints so there is more pruning of schedules, so their search space is much smaller. Therefore, this technique was dropped in the experimental phase for larger values of N. Table 4 shows additional results for the Left-to-Right technique.

Table 4. Execution times for the Left-to-Right technique

D	S	C	N	Search space size	Left-to-Right scheduling		
					New Search Space size	Best Pref.	Random Pref.
7	2	6	13	6.13×10^{54}	1.66×10^{14}	26 s	79 s
7	2	6	14	1.004×10^{59}	1.95×10^{22}	6129 s	9808 s
7	2	7	15	1.64×10^{63}	3.5×10^{15}	48632 s	21573 s
7	3	1	6	8.5×10^{37}	5.7×10^{14}	7 s	239 s
7	3	3	9	7.8×10^{56}	8.8×10^{27}	42872 s	39243 s
14	2	1	3	1.9×10^{25}	4×10^{8}	0.4 s	.6 s
14	2	2	5	1.4×10^{42}	7.6×10^{13}	1 s	4 s
14	2	3	7	1.004×10^{59}	2.3×10^{9}	52 s	18 s
14	3	2	7	3.2×10^{88}	2.1×10^{21}	46413 s	53523 s
14	3	1	5	1.64×10^{63}	5.5×10^{12}	120 s	86 s

An interesting observation in Table 3 is in the "Worst" Schedule, which takes the longest time to run, but converges the quickest, and finds only one optimal solution, as all the preferences are identical and it finds that all schedules produces the same optimal results as the first one found. In this case, it makes sense to stop the technique as soon as the first optimal schedule is found, rather than iterate needlessly.

Another interesting observation was the number of optimal solution that were found, that all have the same maximum nurse penalty and the same average nurse penalty. The more similar the nurse requests are, the greater the number of optimal schedules to choose from, the longer the total running time. Figure 7 shows the results of several runs using parameters D = 7, S = 2, C = 2, and N = 6. Each dot represents the results of 1000 runs using random schedules. The x-axis is the range of standard deviations of the nurse preferences, calculated per shift. The first dot represents the most different nurse preferences (close to the "best" schedule), and the last one represent the most similar nurse preferences (the "Worst" schedule). The running time exponentially depends on the standard deviation, as well as the distribution of the nurse preferences.

To test the effect of long-term scheduling, we ran the Left-to-Right technique over 13 scheduling periods, once independently and once by adding old penalties to the technique. Figure 8 shows the results for D = 7, S = 2, C = 2, and N = 6 over 13 scheduling periods. The long-term scheduling technique clearly satisfies more fairness over the long run, as the difference between the nurse with the least penalty and the

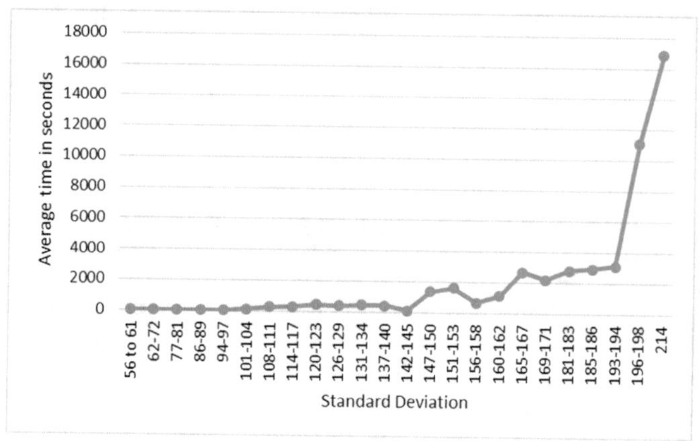

Fig. 7. Relationship between the variations of the nurse preferences and the running time.

nurse with the maximum penalty is kept as small as possible. Figure 8(a) and 8(b) show the results when the nurses have the same preferences in each scheduling period, and Fig. 8(c) and 8(d) shows the results when the nurses have random preferences in every scheduling period.

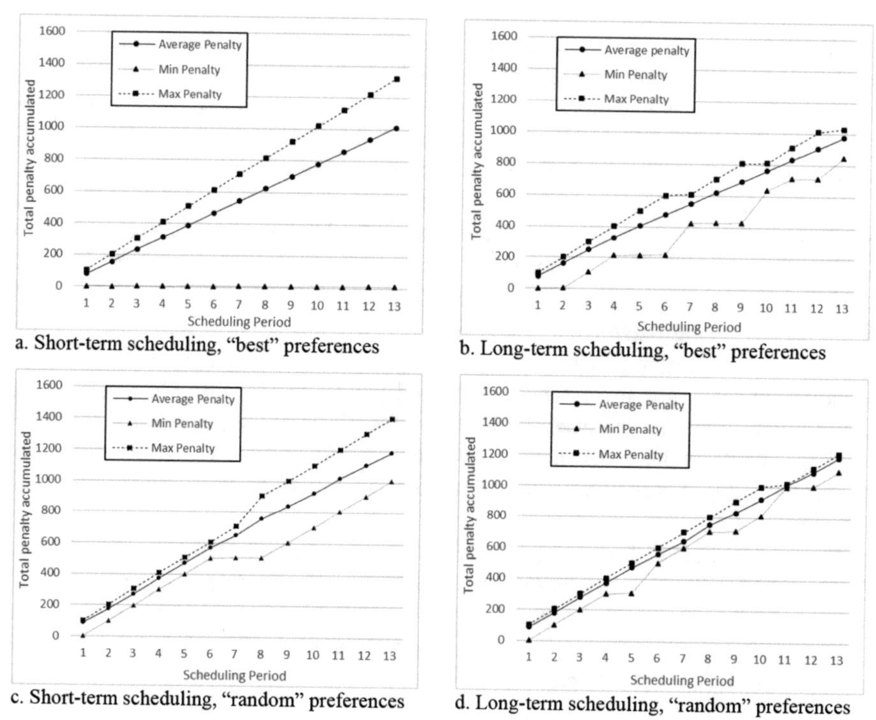

Fig. 8. Long term scheduling performance analysis

6 Conclusion and Future Work

In this paper, we present two simple techniques to solve the nurse scheduling problem by decreasing the search space and recursively finding the optimal solution. The first technique recursively builds the scheduling table top-to-bottom, nurse-by-nurse and the second one recursively builds it left-to-right, shift-by-shift. We prove that the left-to-right technique is much more efficient in finding the optimal solution. The technique can also be extended to provide better fairness over multiple scheduling periods. We are currently experimenting with using Amazon Web services to obtain faster results and well as developing a parallel solution to the problem.

References

1. Lu, K., Lin, P., Wu, C., Hsieh, Y.: The relationships amongst turnover intentions, professional commitment and job satisfaction of hospital nurses. J. Professional Nursing **18** (4), 214–219 (2002)
2. Warner, D.: Scheduling nurse personnel according to nursing preference: a mathematical programming approach. Oper. Res. **24**, 842–856 (1976)
3. Burke, E., De Causmaecker, P., Vanden Berghe, G., Van Landeghem, H.: The state of the art of nurse rostering. J. Sched. **7**, 441–499 (2004)
4. Bai, R., Burke, E.K., Kendall, G., Li, J., McCollum, B.: A hybrid evolutionary approach to the nurse rostering problem. Trans. Evolutionary Computing **14**, 580–590 (2010)
5. Oughalime, A., Ismail, W.R., Yeun, L.C.: A Tabu search approach to the nurse scheduling problem. In: Proceeding of the International Symposium on Information Technology, Kuala Lumpur, Malaysia, pp. 1–7 (2008)
6. Kundu, S., Mahato, M., Mahanty, B., Acharyya, S.: Comparative performance of simulated annealing and genetic algorithm in solving nurse scheduling problem. In: Proceeding of the International Multi Conference on Engineers and Computer Sciences, pp. 19–21 (2008)
7. Lü, Z., Hao, J.-K.: Adaptive neighborhood search for nurse rostering. Eur. J. Oper. Res. **218**, 865–876 (2012)
8. Youssef, A., Senbel, S.: A bi-level heuristic solution for the nurse scheduling problem based on shift-swapping, IEEE 8th Annual Computing and Communication Workshop and Conference (2018)
9. Gunther, M., Nissen, V.: Particle swarm optimization and the agent-based algorithm for a problem of staff scheduling. Lect. Notes Comput. Sci. **6025**, 451–461 (2010)
10. Gutjahr, W.J., Rauner, M.S.: An ACO algorithm for a dynamic regional nurse-scheduling problem in Austria. Comput. Oper. Res. **34**, 642–666 (2007)

Override Control Based on NARX Model for Ecuador's Oil Pipeline System

Williams R. Villalba[1], Jose E. Naranjo[2], Carlos A. Garcia[2], and Marcelo V. Garcia[2,3](✉)

[1] Escuela Politecnica del Chimborazo, ESPOCH, 60155 Riobamba, Ecuador
wvillalba5@hotmail.com

[2] Universidad Tecnica de Ambato, UTA, 180103 Ambato, Ecuador
{jnaranjo0463,ca.garcia,mv.garcia}@uta.edu.ec

[3] University of Basque Country, UPVEHU, 48013 Bilbao, Spain
mgarcia294@ehu.eus

Abstract. Override control systems based on Adaptive Neuro-Fuzzy Inference System (ANFIS) allows to optimize the work of a process by series of predictions of its possible inputs and outputs. It is fully compatible with linear and non-linear systems, but the requirement of a controller with the enough processing capabilities to ensure that all the involved operations are done in real time. For extraction of the dynamic properties and identification of a plant model this research work uses a Nonlinear Autoregressive Exogenous (NARX) neural networks. With this model the authors develop the simulation of the Neuro-Fuzzy Controllers trained by ANFIS, which are part of an override control structure for an Oil & Gas pipeline process. Finally, the behavior of neuro-fuzzy controllers is analyzed by effects of disturbances and human-errors because of the non-correlation in the setpoints of the main variables.

Keywords: Adaptive Neuro-Fuzzy · Inference System (ANFIS) · Override control · Nonlinear Autoregressive Exogenous (NARX) · Slug-Flow and oil pipeline process

1 Introduction

One of the most economical means of transporting hydrocarbons is through an oil pipeline system whose goal is to take petroleum products from refineries passing through re-pumping stations to the points of distribution found in dispatch terminals. The re-pumping stations are responsible for raising pressure to cover the hydraulic gradient necessary to reach the next pumping station. The number of pumping stations and the type of pipes to be used will depend on the altimetric profile and hydraulic design of each pipeline.

The pumping units, that are part of the pipelines stations, consist of three main sections: (i) Variable Frequency Drive (VFD); (ii) three-phase induction motor; and (iii) multi-stage centrifugal pump. These three elements make up a pumping unit of a hydrocarbon transport system whose purpose is to provide the

K. Arai et al. (Eds.): CompCom 2019, AISC 997, pp. 376–389, 2019.
https://doi.org/10.1007/978-3-030-22871-2_26

necessary energy to counteract the friction losses along the pipeline path and to control the changes of the hydraulic parameters of the mountain pipeline. Here the main problem of control lies in keeping the Maximum Operating Pressure (MAOP) within the permitted limits as well as eliminate the Slug-flow phenomenon which is the presence of vapor-liquid fuel at the highest point of the altimetric profile.

Thus far, a large part of the petroleum products transport industry employs conventional Proportional-Integral-Derivative (PID) control whose tuning methods have improved through optimization algorithms as expressed in [1]; however, the PID from its conception is defined for linear plants, reason why [2] and [3] have raised hybrid mechanisms for motor control in order to take advantage of a classic controller PID and a Fuzzy controller. Additionally, [4] proposes an ANFIS training as well as the use of neural networks for plant identification.

This work seeks to incorporate the benefits of a Neuro-fuzzy System to a rotating machine that controls a hydrocarbon transport system. The structure of the proposed work considers the acquisition of plant data, finding a model that represents the pumping unit through Nonlinear Autoregressive Exogenous Neural Networks (NARX) [5], the cloning of an Integral Proportional (PI) controller and finally an advanced structure of neuro-fuzzy control for the control of the main parameters of the pump and consequently of the hydraulic parameters of the duct.

This article is divided into six sections including the introduction. Section 2 shows the State of Technology that has been used as a starting point for this research, Sect. 3 illustrates the case study in Petroecuador Ecuadorian Oil company, Sect. 4 presents the hardware and software control proposal implemented under the concepts of NARX and ANFIS, Sect. 5 shows the results of the system; and finally, Sect. 6 discusses the conclusions and future work.

2 State of the Technology

The aim of this section is to present the methodology which other authors have developed similar systems in the industry. In this context, a set of related works is presented in addition to a compendium of terms for a better understanding of this research.

2.1 Literature Review

Artificial intelligence has changed the way in which industries are handled proposing not only a linear control of production but an immersion and greater control of rotating machines. Jadhav and Jaladi [6] propose a Control based on Artificial Intelligence for a Space Vector Modulated Direct Torque Control (SVM DTC) which has succeeded in decimating the effects of non-linearities and especially the effects of disturbances.

On the other hand, Chaouali et al. [7] present an improvement of the indirect measurement of the magnetic flow of the IFOC (Indirect Field Oriented Control) rotor. Here, techniques of fuzzy logic for speed control for a three-phase pump were used. Furthermore, they propose a control for a specific vector that makes it possible to modify the motor torque independently.

Because fuzzy driver training requires a high degree of experience, an adaptation mechanism has been considered as proposed by Hydayat [2] and AL-Saedi et al. [3] which is based on an Adaptive Diffuse Inference model for fuzzy systems (ANFIS). Finally, Malamura and Murata [8] propose a gas transport optimization model using combined methods of intelligent control to reduce the complexity of the formulation due to a system with non-linear restrictions, continuous-discrete hybrid parameters and impact of the discontinuities by the automatic switching of the network of components.

These works have been the basis for the present research which demonstrates the adaptation of the neuro-fuzzy controller, trained by an ANFIS network, to the dynamic parameters of the mountain transport pipes for oil products.

2.2 Non-linear Autoregressive Exogenous Neural Networks (NARX)

According to what is expressed in [5,9] and [10] it is established that the NARX neural networks are recursive networks which are based on time series, where future values are estimated from the present and past values whose Mathematical support is manifested in 1:

$$Y_{n+1} = F(Y_n, ...Y_{n-q+1}; U_n.....U_{n-p+1}) \tag{1}$$

Equation (1) determines that the present output data depend on the previous output and input values; F is a non-linear function, Y_n and U_n are the present

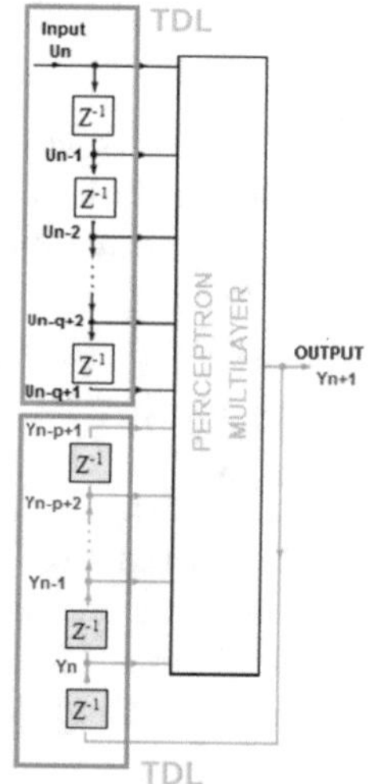

Fig. 1. Model of the neural network NARX

output and input values, the sub-indices p/q are the number of inputs/outputs that should be considered to calculate the output prediction data. Additionally, Fig. 1 shows the structure of the NARX network where the time series enters on the left side and is transformed into vectors of size p/q passing through the delays (Z_{-1}); these memories are called tapped-delay-line (TDL) whose content feeds the input layer of a multi-layer perceptron network.

2.3 General Description of ANFIS Systems

A brief review of concepts on ANFIS is carried out in [3] and [11] whose main difference is the ability to adapt the parameters of the antecedent and consequent, where the Hybrid training model is deployed forward with the least squares (LS) method. The output of the ANFIS is a linear combination of the parameters of the consequent; consecutively, after adjusting the parameters of the consequent, the approximation error is retro-propagated to update the parameters of the antecedent using the principle of decreasing gradient; this cycle is repeated for each cycle until obtaining an acceptable error.

The diffuse inference mechanism of five layers according to [11] (see Fig. 2), where each layer represents an operation is detailed below:

Layer 1: Input variables are fuzzified by determining the degree of membership that satisfy the linguistic term associated to each adaptive node. As indicated in the Eq. (2).

$$O_i^1 = A_i\left(x\right) \tag{2}$$

Layer 2: In this layer, the operation process of Standard T is carried out, being $U_{Ai \cap Bi} = min\left[U_{Ai}(x), U_{Bi}\left(y\right)\right]$, where each fixed node calculates the degree of activation of the rule associated with that node. The output of each node is calculated with Eq. (3).

$$O_i^2 = w_i(x) = A_i(x) \cdot B_i(x) \tag{3}$$

where $i = 1, 2.....\infty$

Layer 3: The static nodes of this layer are represented by a letter N whose meaning is the normalization of the degrees of activation, obtaining at the output the degree of activation normalized. Its equation is the following:

$$O_i^3 = \bar{\omega}_i(x) = \frac{\omega_i}{\omega_1 + \omega_2} \tag{4}$$

where $i = 1, 2.....\infty$

Layer 4: The output of the adaptive nodes of layer 4 is the product between the degree of normalized intensity $\bar{\omega}_i(x)$ of layer 3 and the parametrization $[p_i; q_i; r_i]$ of the output polynomial Z_i. The output equation is:

$$O_i^4 = \bar{\omega}_i(x) \cdot z_i = \bar{\omega}_i(p_i x + q_i y + r_i) \tag{5}$$

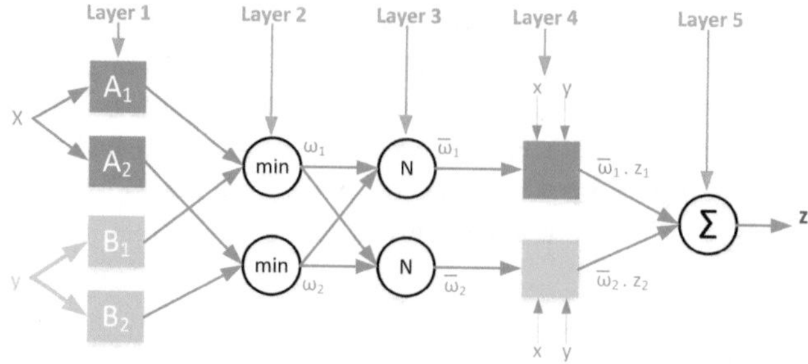

Fig. 2. 5-layer ANFIS structure

Layer 5: This layer only has a single static node whose functionality is to deliver the sum of all the inputs. Its equation is:

$$O_i^5 = \sum_i \bar{\omega}_i \cdot z_i = \frac{\sum_i \omega_i \cdot z_i}{\sum_i \omega_i} \tag{6}$$

3 Case Study

The transportation process of petroleum products in Ecuador mostly is through pipelines that cross the different geographical regions; these products are extracted from the refineries and delivered to the different pipelines that have the mission of transporting them to the different domestic dispatch points in the country.

A Mountain Poliduct is comprised by a station header; re-pump stations that increase pressure by means of pumping units to compensate the loss of energy due

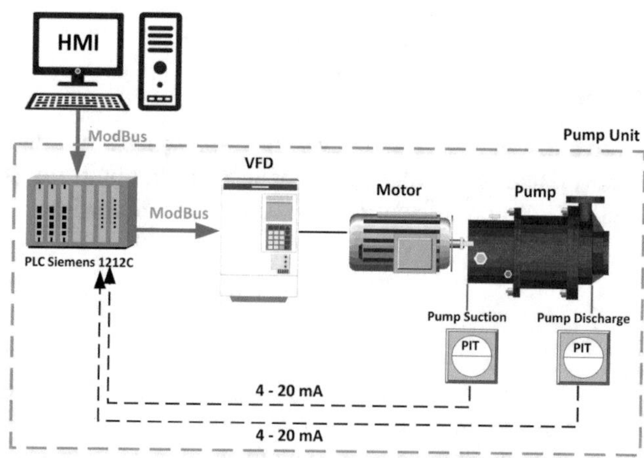

Fig. 3. Block diagram of the pumping unit.

to the friction of the liquid in the walls of the pipeline; and a reducing pressure station of arrival. The shipment of the products depends on the programmed demand and they travel between the stations by batches or separate lots with a product of different API density which is called interface. The sequence of the bumps depends of its density compatibility [12].

According to Fig. 3, each station has pumping units that consist of a variable frequency drive (VFD); a motor; and the multistage centrifugal pump that is responsible for increasing pressure to overcome the highest point of the pipeline altimetric profile. Here, the physical factors to control within a pumping system are the pressures of suction, the discharge of the pump and consequently the pressure of the highest point of the altimetric profile to relieve the Slug-Flow phenomenon. These pressures are related to the rotation speed of the pump whose magnitude depends on the milliamps supply received by the variable frequency drive (VFD) coming from the output card of the Basic Process Control System (BPCS). In this way, manual control and monitoring is carried out by a plant operator through a Man-Machine Interface (HMI).

The goal pursued by this case study is the design of the override control structure presented in Fig. 4. The plant to be controlled has the input variable in milliamperes (mA), that is transmitted to the VFD.

Figure 4 shows the P&ID in which is observed that the Pressure Transmitters (PIT) deliver the signal to the BPCS where the pressure magnitudes pass through a selection element (LS). This performs the neuro-fuzzy control to maintain the set points both in the suction and in the discharge by varying the rotation speed through controlling the lower value of milliamperes delivered to the VFD. This proposal was developed and simulated using Matlab-Simulink to obtain the dynamic model of the plant that represents the data flow of Fig. 3, to carry out simulations that validate the expected control.

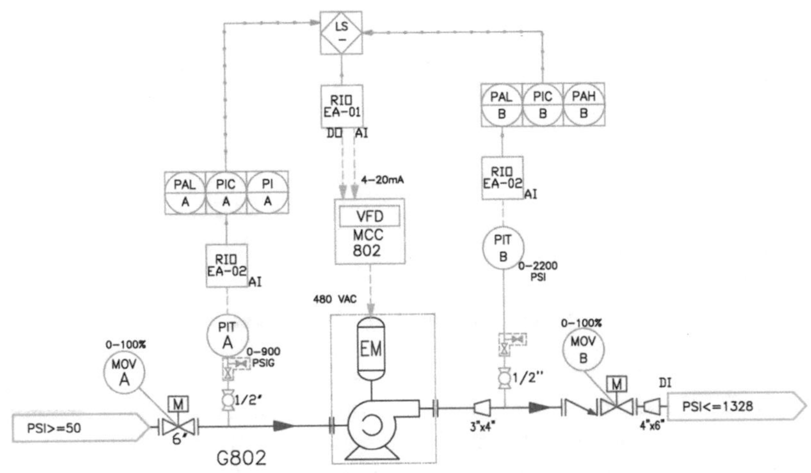

Fig. 4. P&ID and architecture of the override control

Table 1. Variable ranges of the plant to be modeled

Variable	Real range	Standardize range
Miliamperes	14.4–20 (mA)	0–1
Suction pressure	68.69–290.17 (psig)	0–1
Discharge pressure	1056–1328 (psig)	0–1

4 Proposed Solution

4.1 Plant Training Using NARX Networks

Under the conceptualization of NARX and heuristic procedure [13], the training of the plant is carried out with the normalized control signal as the input pattern and the normalized pressures as output targets as can be observed in Table 1, where the Mean Quadratic Error (MSE) objective was established in 1e−6. The MSE obtained were 9.5972e−07 and 9.9098e−07 for the suction and discharge pressure respectively, these are shown in Fig. 5.

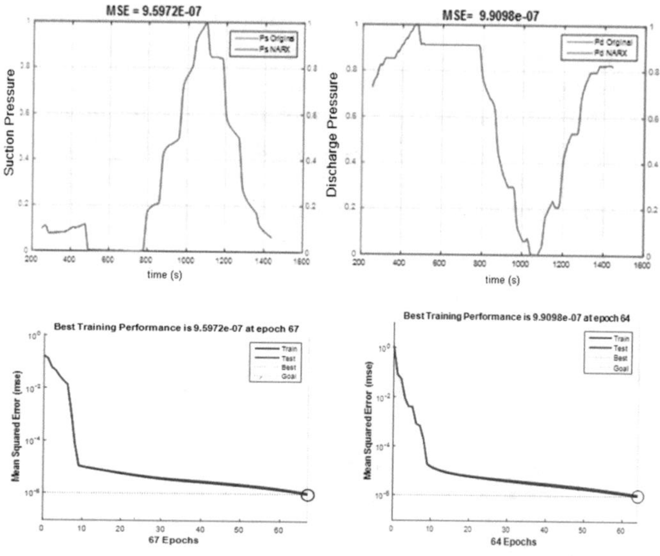

Fig. 5. Supervised training with NARX plant networks

Consequently, the estimated standard error (S_e) and the determination coefficient (Sy^2) ease the extraction of the correlation coefficient (R^2) which is the magnitude and direction of the relationship between the real output of the plant and the output of the Narx network that is being used for training.

4.2 ANFIS Training for Controllers

ANFIS is a supervised training, reason why it is essential to consider as a first step the tuning of a classic PID from where the error signal $e(k)$ and the control signal $u(k)$, that will serve as patterns and as the system control curve, can be acquired [14].

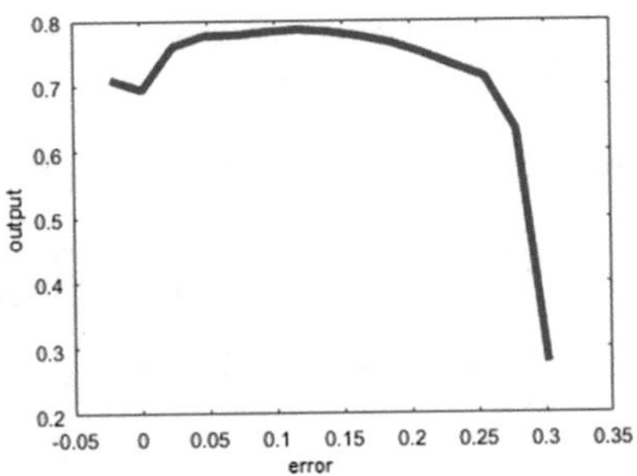

Fig. 6. Control curve for the controller by discharge pressure

With the control curve of Fig. 6, a neuro-fuzzy controller for the discharge pressure has been trained. This is represented by a TSK (Takagi-Sugeno-Kan) diffuse system of order 1 (one input and one output) with 5 rules generated by the combinations of the antecedents that causes its respective consequent rule.

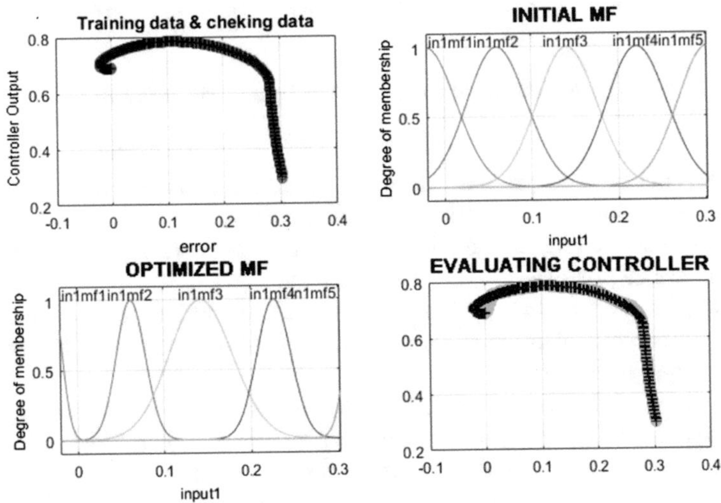

Fig. 7. Controller ANFIS by discharge pressure

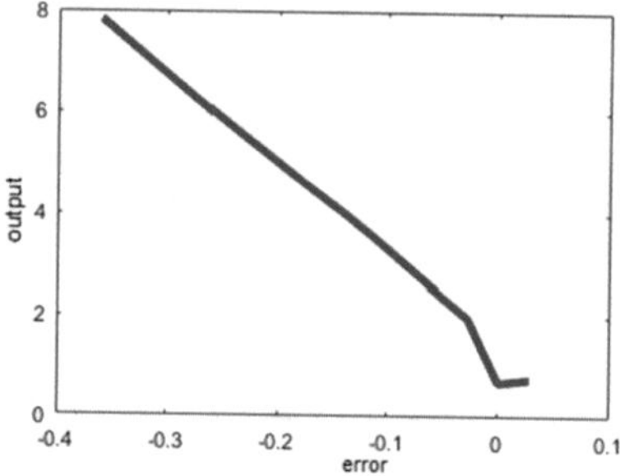

Fig. 8. Control curve for the controller by suction pressure

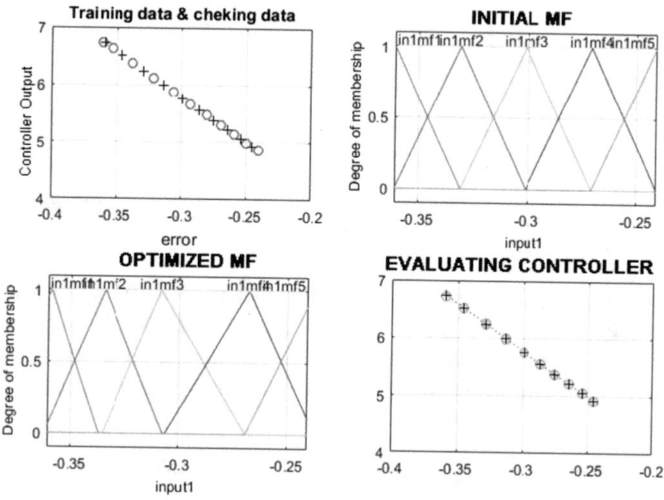

Fig. 9. ANFIS of the controller by suction pressure

Figure 7 details the process. Here the training data can be observed, the five Gaussian membership functions at the beginning and optimization after the training process; ending in the evaluation of the neuro-fuzzy controller trained for the discharge pressure.

In a similar way to the discharge pressure, the suction pressure test is formulated, whose control curve of error variables $e(k)$ and output of the controller $u(k)$ is detailed in Fig. 8.

The training data (x) and validation (o) are shown in Fig. 9; here the suction pressure required five triangular 'trimf' type start member functions. For the precedent of the fuzzy rules and as an optimization methodology for adaptive networks, hybrid learning was used to adjust the parameters of the ANFIS architecture, that is, the application of the least squares estimator. With this, what is intended is to correct the error measurement for each pair of training data making the convergence time faster than a backpropagation training.

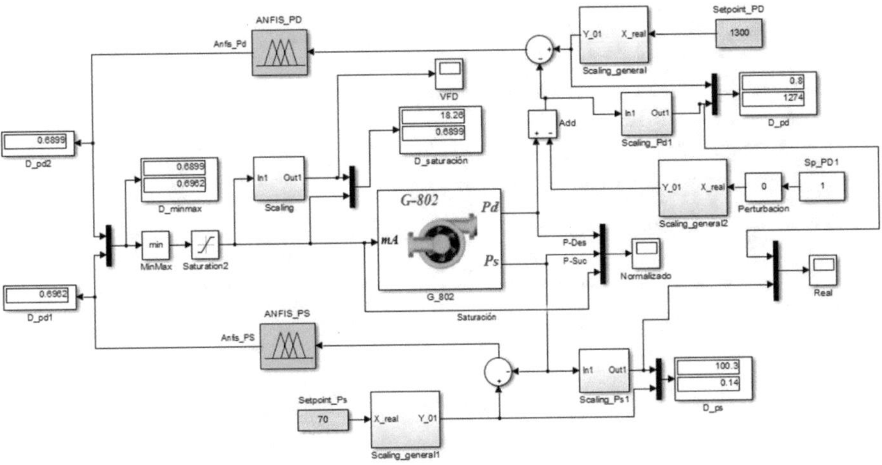

Fig. 10. Architecture override of the neuro-fuzzy controllers

4.3 Neuro-Fuzzy Control with Override Structure

The override control structure, detailed in Fig. 10 shows the two neuro-fuzzy controllers trained by Adaptive Neuro-Fuzzy Inference System-ANFIS using Matlab Simulink™.

In normal operation, the pumping unit will require the two setpoints of pressures mentioned above, whose narrow relationship depend on the speed of rotation of the pump and of the "mA" that is supplied to the VFD. These setpoints of the pressures are taken from the records of the pipeline operation and its hydraulic study as it is presented in Table 2.

After the correct configuration of the setpoints of the pressures that govern the pump, the ANFIS_PD will always have priority when it comes to controlling the discharge pressure, on the other hand the ANFIS_PS controller will act in case of disturbances in the discharge. The selector element of the neuro-fuzzy controllers will allow control of the lower control signal value. This value will pass through a saturation element to limit the values of the input signal to the VFD.

The variables of the pressures PS and PD that the plant delivers go through a new scaling to bring from normalized values to real values to know the effect of the control signal on the output variables of plant.

Table 2. Set-point in normal operation

Variable	Unit	Value
Suction pressure	psig	70
Discharge pressure	psig	1300

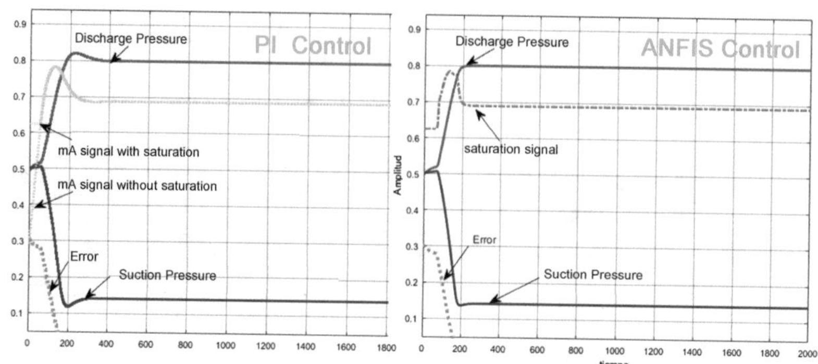

Fig. 11. Controllers comparison for discharge pressure

5 Simulation and Discussion of Results

5.1 Step Signal Response in Neuro-Fuzzy Controllers

By applying a step type signal to the plant input, the responses obtained at the PD and PS outputs are shown in Fig. 11, where the conventional Proportional-Integral (PI) controller on the left shows an overshoot of approximately 10% of the plant output signals.

Contrary to this effect, with the input-output control curve of the PI, the overshoot is eliminated maintaining the same rise time of 200s and reducing the time to reach the steady-state error in 0.

On the other hand, the response of the neuro-fuzzy controller for the suction pressure of Fig. 12 was made by means of a negative step type input because the suction pressure has an inverse effect to the angular speed of the pump, i.e., when the rpms decreases the suction pressure increases. For this reason, the step

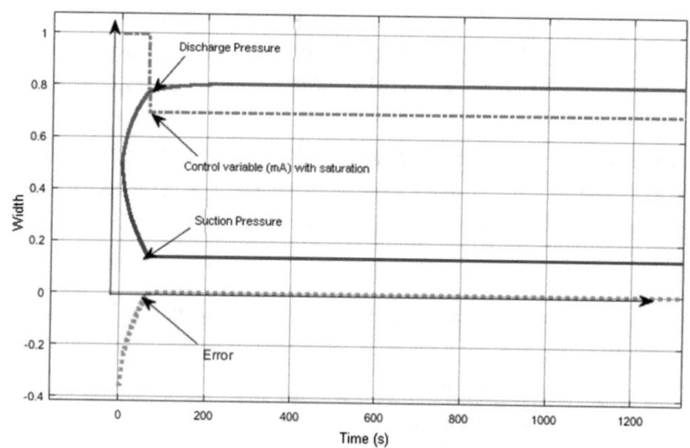

Fig. 12. Simulation of the neuro-fuzzy controller for suction pressure

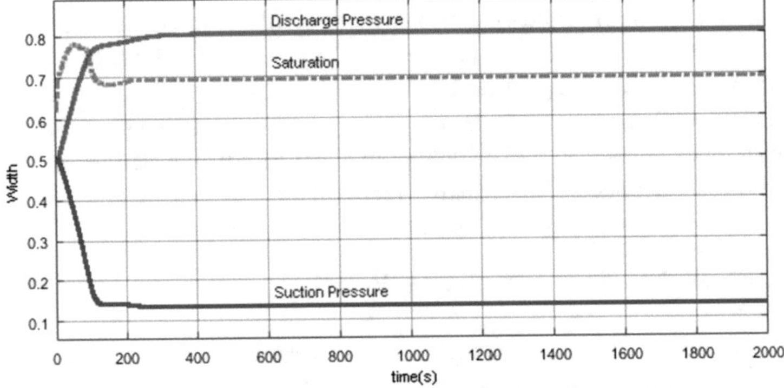

Fig. 13. Suction control curve to modification of discharge pressure setpoint without correlation with suction.

input was taken inside the normalized scale of 0.2 (15.52 mA) to 0.14 (15.18 mA) giving as a result a suction pressure output that stabilizes at approximately 75 s. This causes a mild effect on the discharge pressure with its stabilization time of 200 s, the difference between these times of stabilization is that the packaging of fluid in the suction is in a short section of pipe, however, the discharge must overcome all the hydrostatic pressure of the altimetric profile of the pipeline so that the stabilization time can be slower. The controller on the suction side also has a faster saturation signal to meet the desired setpoint.

5.2 System Override Structure Response

The neuro-fuzzy drivers overlap with respect to plant operability; for this case, the optimal operational continuity, according to the hydraulic study of this pipeline, is to maintain around 1300 psig the discharge pressure of the pump in such a way that the highest point of the altimetric profile can be maintained with a pressure greater than 15 psig. Decreasing the "Slug-Flow" effect at piping system.

Normally, the control priority is to use the discharge data through the ANFIS_PD controller of the pumping unit, where the override structure keeps up the response of the neuro-fuzzy controller; as the discharge Setpoint_PD is reduced without correlating with the suction pressure, in other words that it must reduce the speed of rotation of the machine, will take control by suction because the control signal of the ANFIS_PS has the lowest value, so the selector allows to take control of the plant.

An important consideration in the pumping units is that the suction and discharge pressures depend directly on the revolutions per minute (rpm) and maintain an inverse relation between them, in such a way that the modifications of the set points of pressures must have correlation and cannot be modified randomly.

Despite this recommendation, discharge control supports variation or human adjustment errors between 1285 psi to 1300 psi without affecting the discharge pressure response; however, below 1280 psi at the discharge setpoint the response would tend to deform as shown in Fig. 13 with a discharge setpoint of 1270 psi and a setpoint of 70 psig in the suction; that is, the control passes to the suction without relation between the pressure variables.

6 Conclusions and Ongoing Work

In the case of plants with more complexity, finding a mathematical model is a real challenge, and from the industrial and practical point of view a waste of human resources. The study and understanding of artificial neural networks fulfills these purposes, that is, a model of plant that contributes with the dynamic properties of the process for the simulations using different control and automation software is obtained.

This paper presents an approach to modeling using the NARX algorithm of a mountain pipeline for the transport of petroleum products in Ecuador. Additionally, the design and simulation of ANFIS controllers has been carried out both for the suction and the discharge of the pumping system of the process.

By using a plant modeling based on NARX training whose current outputs depend on the past input-output values concatenated with the delay units (TDL), an MSE of 1e−6 was reached, validating the dynamic properties of the pumping process.

The neuro-fuzzy controllers in the simulation environment had a robust behavior to uncorrelated changes of the discharge pressure setpoints, transferring the control to the neuro-fuzzy suction pressure controller. It was shown that the presence of disturbances that reduce the output pressure of the pumping unit is absorbed by the controller up to (−30 psig); however, below this limit, there is no control since it may be due to a break in the pipeline along its path.

Future work focuses on the integration of other control algorithms based on the Machine Learning paradigm. In addition, the implementation of the new control systems developed in real industrial processes should be carried out to verify the benefits with respect to the traditional control architecture based on the use of Programmable Controllers (PLCs) and its IEC 61131 standard.

Acknowledgments. This work was financed in part by Universidad Tecnica de Ambato (UTA) under project CONIN-P-0167-2017, EP Petroecuador and Government of Ecuador through grant INEDITA.

References

1. Ribeiro, J.M.S., Santos, M.F., Carmo, M.J., Silva, M.F.: Comparison of PID controller tuning methods: analytical/classical techniques versus optimization algorithms. In: 2017 18th International Carpathian Control Conference (ICCC), Sinaia, Romania, pp. 533–538 (2017)
2. Hidayat, Pramonohadi, S., Suharyanto, S.: A comparative study of PID, ANFIS and hybrid PID-ANFIS controllers for speed control of brushless DC motor drive. In: 2013 International Conference on Computer, Control, Informatics and Its Applications (IC3INA), Jakarta, Indonesia, pp. 117–122 (2013)
3. AL-Saedi, M.I., Wu, H., Handroos, H.: ANFIS and fuzzy tuning of PID controller for trajectory tracking of a flexible hydraulically driven parallel robot machine. J. Autom. Control Eng. **1**(3), 70–77 (2013)
4. Farid, A.M., Barakati, S.M., Seifipour, N., Tayebi, N.: Online ANFIS controller based on RBF identification and PSO. In: 2013 9th Asian Control Conference (ASCC), Istanbul, Turkey, pp. 1–6 (2013)
5. Liu, H., Song, X.: Nonlinear system identification based on NARX network. In: 2015 10th Asian Control Conference (ASCC), Kota Kinabalu, pp. 1–6 (2015)
6. Jadhav, S., Jaladi, K.: Advanced VSC and intelligent control algorithms applied to SVM_DTC for induction motor drive: a comparative study. In: 2016 12th World Congress on Intelligent Control and Automation (WCICA), Guilin, China, pp. 2694–2698 (2016)
7. Chaouali, H., Othmani, H., Mezghani, D., Mami, A.: Enhancing classic IFOC with fuzzy logic technique for speed control of a 3 Ebara Pra-50 moto-pump. In: 2016 17th International Conference on Sciences and Techniques of Automatic Control and Computer Engineering (STA), Sousse, Tunisia, pp. 423–428 (2016)
8. Malamura, E., Murata, T.: Hybrid system modeling and operation schedule optimization for gas transportation network based on combined method of DE, GA and hybrid petri net. In: 2016 5th IIAI International Congress on Advanced Applied Informatics (IIAI-AAI), Kumamoto, Japan, pp. 1032–1035 (2016)
9. Haykin, S.S., Haykin, S.S.: Neural Networks and Learning Machines, vol. 3. Pearson, Upper Saddle River (2009)
10. Yusuf, Z., Wahab, N.A., Sahlan, S.: Modeling of submerged membrane bioreactor filtration process using NARX-ANFIS model. In: 2015 10th Asian Control Conference (ASCC), Kota Kinabalu, pp. 1–6 (2015)
11. Kusagur, A., Kodad, S.F., Ram, B.V.S.: Modeling, design and simulation of an adaptive neuro-fuzzy inference system (ANFIS) for speed control of induction motor. Int. J. Comput. Appl. **6**(12), 29–44 (2010)
12. Lipu, M.S.H., Hussain, A., Saad, M.H.M., Ayob, A., Hannan, M.A.: Improved recurrent NARX neural network model for state of charge estimation of lithium-ion battery using pso algorithm. In: 2018 IEEE Symposium on Computer Applications & Industrial Electronics (ISCAIE), Penang, pp. 354–359 (2018)
13. Benabdelwahed, I., Mbarek, A., Bouzrara, K., Garna, T.: Non-linear system modelling based on NARX model expansion on Laguerre orthonormal bases. IET Signal Process. **12**(2), 228–241 (2018)
14. Olabe, X.B.: Redes neuronales artificiales y sus aplicaciones, 101 p. Publicaciones Esc. Ing. (1998)

MatBase E-RD Cycles Associated Non-relational Constraints Discovery Assistance Algorithm

Christian Mancas(✉)

Mathematic and Computer Science Department, Ovidius University, Constanta, Romania
christian.mancas@gmail.com

Abstract. *MatBase* is a prototype data and knowledge base management expert intelligent system based on the Relational, Entity-Relationship, and (Elementary) Mathematical Data Models. Entity-Relationship diagrams are lattice-type oriented graphs that may have cycles, with object sets as nodes and functions as edges. Most of these cycles are generally uninteresting, some of them should be broken by eliminating computable functions, but others have associated non-relational constraints that must be added to corresponding database schemes and enforced to guarantee database instances plausibility. This paper presents and discusses all four types of such cycles and an algorithm that assists database designers in discovering all non-relational constraints associated to them.

Keywords: Expert system · e-Learning tools · Data modelling · Entity-Relationship models · Database constraints theory · Relational constraints · Non-relational constraints · Data structures and algorithms for data management · Commutative function diagrams · (Elementary) mathematical data model · *MatBase*

1 Introduction

In the Entity-Relationship Data Model (E-RDM) [1–3], the major contribution is the powerful Entity-Relationship Diagrams (E-RDs), most probably the only data modelling formalism understandable by both customers, business analysts, database (db) designers, and software developers, as it uses labelled graphs made from only a few graphical symbol types: rectangles, diamonds, ellipses, lines, and, in our extension, arrows.

Any db scheme is a triple $<S, M, C>$, where S is a non-void finite collection of sets, M a finite non-void set of mappings defined on and taking values from sets in S, and C a similar one of constraints (i.e. closed first-order predicate calculus with equality formulas) over the sets in S and mappings in M. Sets and mappings constitute the structure ("skeleton") of any db, while constraints, which are formalizing business rules, are meant to only allow storing plausible data ("flesh") into dbs.

In the Relational Data Model (RDM) [3–5], sets of S are tables and views, mappings of M are their columns, and constraints of C are incorporated in the table schemes.

Unfortunately, both RDM and E-RDM have only a handful of constraint types, which are not at all enough to guarantee data plausibility. The (Elementary) Mathematical Data

K. Arai et al. (Eds.): CompCom 2019, AISC 997, pp. 390–409, 2019.
https://doi.org/10.1007/978-3-030-22871-2_27

Model ((E)MDM) [6–8] provides 61 types of constraints. *MatBase* [3, 7–12] is a prototype data and knowledge base management expert intelligent system providing both (E) MDM, E-RDM, RDM, and Datalog [5, 8] user interfaces, intensively used as an e-Learning tool in our labs and projects of undergraduate Databases and M.Sc. Advanced Database Systems, as well as by a couple of software developing companies.

Just like for object sets and mappings between them, constraints can be discovered only by humans. However, computer science and math can assist in this process: e.g. the keys discovery assistance algorithms [11], the constraint sets coherence and minimality ones [12], etc. This paper introduces and discusses an algorithm for assisting discovery of non-relational constraints associated to E-RD cycles.

This first section presents our E-RD variant notation (see [3] for more details and examples), recalls the difference between relational and non-relational constraints [9], and syntactically characterizes the four types of E-RD cycles [8, 10].

1.1 E-R Diagrams

In its original version, atomic (*entity*-type) object sets are represented by rectangles, compound (*relationship*-type) ones by diamonds, and their *attributes* (properties) by ellipsis. We advocate drawing first atomic ones, having only one rectangle or diamond and the associated ellipsis, and then only *structural* E-RDs without any ellipsis. The latter are directed graphs whose edges are so-called "*roles*". We are using a slightly different notation [3]: only non-functional relationships are represented by diamonds, but functional ones are represented as arrows, just like in math. Hence, in our version, structural E-RD (from now on abbreviated as E-RD) edges are (structural) functions. In relational dbs, structural functions are foreign keys.

Obviously, as in any lattice-type graph, E-RD ones may have cycles (polygons).

1.2 Relational and Non-relational Constraint Types

All business rules of any data sub-universe to be modelled, be them explicit or implicit, should be discovered, added to our data models, and enforced to be able to guarantee that only plausible data is stored in the corresponding dbs. In db schemes, business rules are incorporated as constraints. RDM provides only five types of constraints (that are embedded in almost all Relational Database Management Systems (RDBMS)), namely, *domain* (range), *totality* (not null), *key* (uniqueness), *referential integrity* (foreign keys), and *tuple* (check; e.g. $StartDate \leq EndDate$).

As these relational constraint types are not enough for accurate data modelling, both extended and embedded SQL provide means to also enforce non-relational constraints, namely triggers and trigger-type event driven methods, respectively.

For example, an obvious non-relational constraint exists even in an extremely simple db only consisting of the two tables *COUNTRIES*(*ID*, *Country*, *CapitalCity*) and *CITIES* (*ID*, *City*, *Country*): "any country has as capital a city of its own" (or, dually, "no country may have as its capital a city of another country"). Considering functions $Country : CITIES \rightarrow COUNTRIES$ and $CapitalCity : COUNTRIES \rightarrow CITIES$, this constraint can be formalized as $Country \circ CapitalCity = \mathbf{1}_{COUNTRIES}$ (where $\circ$ denotes function composition and $\mathbf{1}_{COUNTRIES}$ is the unity mapping of *COUNTRIES*, $\mathbf{1}_{COUNTRIES}(x) = x$, $\forall x \in COUNTRIES$) or, equivalently, as $Country \circ CapitalCity$ *reflexive*.

This constraint is associated to the E-RD cycle of length 2 from Fig. 1. Failing to enforce this constraint could result, for example, in letting users store in this db instance the fact that New York is the capital of Romania and/or that Craiova is the capital of the U.S.! As shown in this paper, E-RD cycles are privileged places to search for non-relational constraints.

Fig. 1. An example of an E-RD cycle having an associated non-relational constraint.

1.3 E-RD Cycle Types

Apart from the ones of length one (i.e. autofunctions), there may be only the following three types of such cycles [8, 10] (where a *source node* is a node that is the domain of both functions that connects it to the cycle, a *destination node* is the co-domain of both functions that connects it to the cycle, whereas all other nodes are *intermediate* ones: the domain of one function and the codomain of the other):

✓ *commutative*: any cycle having only one source and one destination nodes;
✓ *circular*: any cycle having only intermediate nodes;
✓ *general*: any other type of cycle than commutative and circular.

The general ones have length greater than 3, at least two source and two destination nodes, and may have attached only *generalized commutativity constraints* (i.e. formulas only involving the sets and mappings of an E-RD cycle), which, dually, may also be associated to commutative and circular types cycles. The commutative and circular ones may also "hide" constraints of math well-known types.

1.4 Related Work

Generally, there are very few db research results on non-relational constraints, except for (E)MDM related ones [3, 6–8, 13, 14]. For example, [1] only presents enforcement of some particular ones (by using the SQL functions COUNT, IS_EMPTY, and its inclusion operator).

Constraints associated to E-RD cycles were investigated in [14], which also includes a first much simpler version of the algorithm presented in this paper, as, at that time, (E)MDM provided far less constraint types than today.

MatBase provides its users with the ability to detect and classify all cycles in any db scheme structural E-RD it manages, by using an extended DFS-type algorithm [10].

Autofunction and dyadic relations, as well as commutative function diagram properties were extensively studied (e.g. [15–19]).

1.5 Paper Outline

Single autofunctions are analysed in Sect. 2. Sections 3 to 5 are devoted to the three types of E-RD cycles, respectively. Section 6 discusses constraints associated to adjacent cycles. Section 7 presents the algorithm for assisting discovery of non-relational constraints associated to E-RD cycles. Section 8 discusses its complexity, optimality, and usefulness. The paper ends with conclusions and references.

2 Autofunctions

Autofunctions (i.e. functions defined on and taking values from a same set, hence circular cycles of length 1) are very privileged places to search for non-relational constraints. Besides general function properties, like one-to-oneness and ontoness, as autofunctions are dyadic relations, we should always wonder whether they are acyclic, reflexive, irreflexive, symmetric, asymmetric, idempotent, or/and anti-idempotent (as Euclideanity and connectivity do not make sense for functions).

For example, all autofunctions that model trees (e.g. *Folder* : *FILES* → *FILES*, *ReportsTo* : *EMPLOYEES* → *EMPLOYEES*, *Mother* : *PEOPLE* → *PEOPLE*, *Father* : *PEOPLE* → *PEOPLE*) are acyclic (which, in this context, means that they are also irreflexive, asymmetric, and anti-idempotent [8]), as trees do not contain cycles (e.g. no folder may be its subfolder, neither directly, nor indirectly; no employee may report to her/himself, neither directly, nor indirectly; nobody may be his/her ancestor, neither directly, as parent, nor indirectly).

For example, *ElectedRepresentative* : *PEOPLE* → *PEOPLE* is idempotent, as any elected representative is also representing him/herself. Generally, any canonical surjection is not only onto, but also idempotent.

For example, *Spouse* : *PEOPLE* → *PEOPLE* is trivially irreflexive and symmetric (as nobody can get married to her/himself and as whenever y is a spouse of x, then x is a spouse of y too).

The only interesting cases of single autofunction reflexivity constraints are those for mappings that are not totally defined (i.e. that may take null values, but, whenever for a given x they are not null, they should always be x): what would be the use of storing in a db table a column whose values should always be equal to those of the primary surrogate key of that table? On the contrary, reflexivity is very often encountered for composed autofunctions, as proved by Sect. 4 below, as well as by the example in Fig. 1. Any reflexive autofunction is also symmetric, idempotent, and, when totally defined, bijective too [8].

3 Commutative-Type E-RD Cycles

3.1 Commutative Function Diagrams

Math studies *function diagram commutativity* [15, 16]: although most of the time exemplified with commutative type cycles of length three, for which a diagram made out of mappings $f : A \rightarrow B$, $g : B \rightarrow C$, and $h : A \rightarrow C$ commutes *iff* (i.e. *if and only if*) $h = g \circ f$, they can be generalized to any cycle length greater than 1; given any mappings $f_1 : A_1 \rightarrow A_2, f_2 : A_2 \rightarrow A_3, \ldots, f_n : A_n \rightarrow A_{n+1}, g_1 : A_1 \rightarrow B_1, g_2 : B_1 \rightarrow B_2, \ldots, g_m : B_{m-1} \rightarrow A_{n+1}$, for some strictly positive naturals n and m, such a diagram commutes *iff* $f_n \circ \ldots \circ f_2 \circ f_1 = g_m \circ \ldots \circ g_2 \circ g_1$.

Recall that the technique used for deciding whether a diagram commutes is the following: you consider any element of the source, compute for it both elements of the destination for each path and wonder whether they should always be the same; if they should, then the diagram commutes.

Let us consider, for example, the commutative-type diagram from Fig. 2 (having $n = m = 2$, length $n + m = 4$, *STATES* as source and *CONTINENTS* as destination). As, for any state, both the country and the region to which it belongs must belong to a same continent, this diagram commutes and the corresponding constraint (C_1) should be added to this db scheme: *C1*: *Continent* ° *Country* = *Continent* ° *Region*

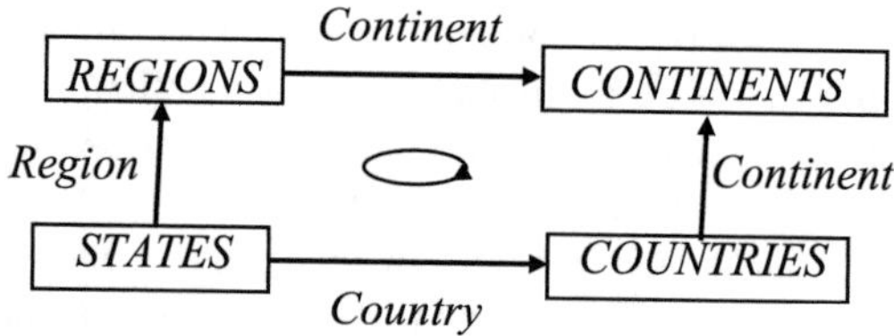

Fig. 2. An example of a commutative diagram of length four

Failing to enforce this constraint might allow storing implausible data (e.g. Texas, which is a state of the U.S.A., which belongs to North America, belongs to the South America's Andes region).

A particular case of commutative diagrams that needs a special type of treatment was already discussed in [3] (see best practice rule R-DA-7 from Sect. 2.10.7): those for which either n or m is equal to one. Whenever the corresponding computable mappings are not canonical ones and null values are not interfering with them either, they might and should be broken (by eliminating the mapping that is computable from the other ones forming the cycle or by replacing it with a computed one).

For example, the commutative-type diagram of Fig. 3 ($n = 1$, $m = 3$, length 4, source *CUSTOMERS*, destination *COUNTRIES*) commutes (i.e. *CCountry* = *Country* ° *State* ° *CCity*), as, for any customer, the country where it has its headquarters is the country containing the state that includes the city where its headquarters are located.

The cycle in this E-RD should be broken, by dropping *CCountry* (which may be computed anytime when needed), implying that no extra constraint is needed anymore.

Whenever querying speed is critical or updates to *CCountry* are much less frequent than queries involving it (and additional needed db space is affordable), it may be added as the computed function **CCountry* = *Country* ° *State* ° *CCity* (with *controlled redundancy*, i.e. being read-only to end-users and automatically updated whenever changes occur in *CCity*, *State*, and/or *Country* corresponding values, by trigger-like methods); note that this solution (see Fig. 4) is as costly as the initial one with the commutativity constraint added too, since plausibility is always guaranteed.

Moreover, in such cases, even if costlier, in fact it is often better that this redundant function does not store pointers to the corresponding countries, but plain names for customer cities, states, and countries (i.e. **CCity* = *CityName* ° *CCity*, **CState* = *StateName* ° *State* ° *CCity*, and **CCountry* = *CountryName* ° *Country* ° *State* ° *CCity*),

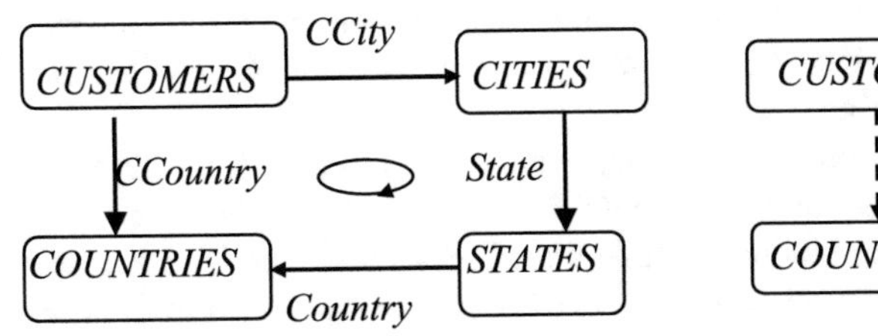

Fig. 3. An example of a breakable commutative diagram

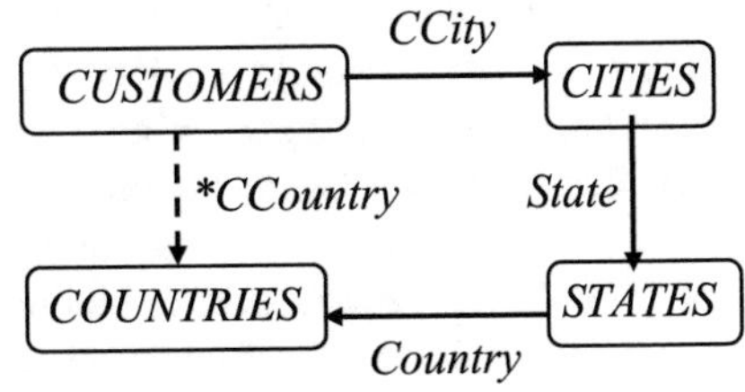

Fig. 4. Resulting E-RD after replacing in Fig. 3 the computable *CCountry* with the correspondingly computed **CCountry*

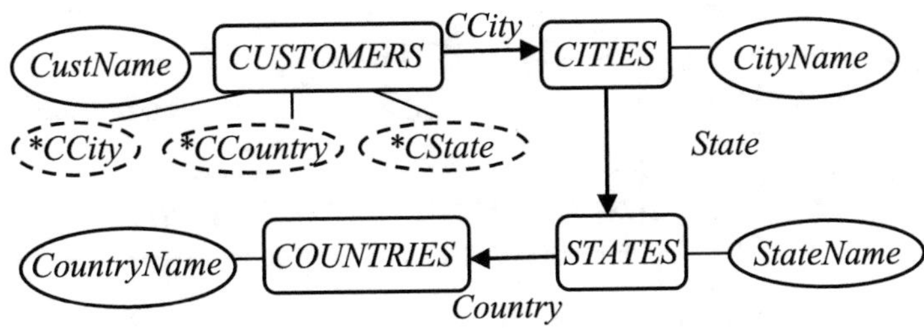

Fig. 5. Resulting E-RD after optimizing the controlled redundancy of the one in Fig. 4

like in Fig. 5: with this solution, when querying data about customers, only *CUSTOMERS* have to be accessed (whereas even in the one from Fig. 4, for getting corresponding city, state, and country names too, all four underlying tables have to be accessed).

Theoretically, cycle breaking can be done in some cases even if both n and m are greater than one: for example, if $f \circ g = e \circ h$ and, say, f is bijective, then g is obviously computable ($g = f^{-1} \circ e \circ h$, where $f^{-1} : B \rightarrow A$ is the inverse of $f : A \rightarrow B$, i.e. $f \circ f^{-1} = \mathbf{1}_A$ and $f^{-1} \circ f = \mathbf{1}_B$). However, bijective functions are not at all interesting in conceptual data modelling, as equipotent object sets should always be merged into only one set [3, 8]. It is true that ontoness is easily achievable by restricting the co-domains to the corresponding function images, so one-to-oneness might be sometimes enough to obtain at least partially reversible functions. For example, we could declare $CAPITAL\_CITIES = \{x \in CITIES \mid \exists y \in COUNTRIES,\ Capital(y) = x\}$ and the restricted $Capital : COUNTRIES \leftrightarrow CAPITAL\_CITIES$ becomes bijective. Generally, however, this is very rare; for example, in C_1 above there is no one-to-one mapping, as, generally, continents include both several regions and countries, whereas both countries and regions include several states.

Moreover, there are cases when commutative-type diagrams cannot be broken even when $n = 1$ or $m = 1$. The first one is whenever the syntactically computable mapping is not also semantically computable because it is a canonical one. For example, the cycle from Fig. 6 is obviously of commutative-type (with $n = 1$, $m = 2$, length 3, source *ISLANDS_SHARING*, and destination *COUNTRIES*), but you should never wonder whether it should be broken, as *SharingCountry* is a canonical Cartesian projection, so it is not computable.

In this example, *Country* stores the country which either fully owns the corresponding island or the biggest part of it; if an island is fully owned by only one country, then *ISLAND_SHARING* does not store anything on it; for example, nothing is stored for Britain; for shared islands, all other countries (except for the one stored in *Country*, as no data should be stored more than once in any db) are stored in it; for example, *Country* stores for Ireland the Irish Republic and *ISLAND_SHARING* stores for it the U.K.

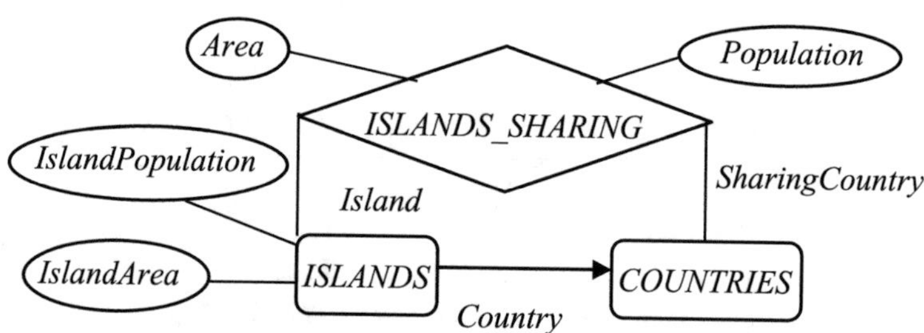

Fig. 6. An example of an unbreakable commutative type E-RD cycle, because canonical Cartesian projections are not computable

Obviously, this cycle should not commute, but if no constraint is associated with it, then users might violate the above-mentioned axiom, for example by also adding in *ISLAND_SHARING* a pair <Ireland, Irish Republic>. To prevent it, the following anti-commutativity constraint (corresponding to the business rule "no country may own more than one share of any island") should be added to this db scheme:

$$C'\text{: } (\forall x \in ISLANDS)(\forall y \in ISLANDS\_SHARING)$$
$$(Country(x) = SharingCountry(y) \Rightarrow Island(y) \neq x)$$

Moreover, another constraint must be added to this E-RD cycle too in order to prevent storing implausible data, namely: "For any shared island, the sum of the areas owned by all corresponding sharing countries should be at most equal to the total island area; similarly, the sum of the populations of the corresponding areas should also be at most equal to the total island population". Its formalization is:

$$C''\text{: } (\forall x \in ISLANDS)(\forall y \in ISLANDS\_SHARING,\ Island(y) = x)$$
$$(IslandArea(x) \geq sum(Area(y)) \wedge IslandPopulation(x) \geq sum(Population(y)))$$

Obviously, C' and C'' above may be merged into the following equivalent constraint: C: $(\forall x \in ISLANDS)(\forall y \in ISLANDS\_SHARING,\ Island(y) = x)$

$$(Country(x) \neq SharingCountry(y) \wedge IslandArea(x) \geq sum(Area(y)) \wedge$$
$$IslandPopulation(x) \geq sum(Population(y)))$$

In (E)MDM, because it is associated to an E-RD cycle, such an object constraint is considered of type *generalized commutativity* and the corresponding cycle is said to *unconventionally commute* according to that constraint. Obviously, any E-RD cycle of length at least 2 might have such an associated constraint.

To conclude, whenever $f = f_1$ or $g = g_1$, you should not ever wonder whether $f = g_m \circ \ldots \circ g_2 \circ g_1$ or $g = f_n \circ \ldots \circ f_2 \circ f_1$ conventionally commute if f or g are a canonical projection, inclusion, or surjection, but only if the corresponding commutative-type cycle unconventionally commutes.

The second case in which commutative cycles cannot be broken is due to interactions with null values. For example, considering the commutative-type diagram from Fig. 7 ($n = 1$, $m = 5$, length 6, source *RIVERS*, destination *CONTINENTS*), let us first note that *Mountain* may take null values too (as there are rivers that do not take their sources from mountains). Obviously, whenever it doesn't, this diagram is commuting (*Continent* = *Continent* ∘ *Range* ∘ *Subrange* ∘ *Group* ∘ *Mountain*), because any river that springs from a mountain belongs to the same continent as that mountain.

For example, the Mississippi has its source at only 450 m altitude in the Itasca lake; as its source is not in the mountains, *Mountain* has for it a null value, which makes impossible computing its continent from *Continent* ∘ *Range* ∘ *Subrange* ∘ *Group* ∘ *Mountain*; dually, the Danube springs from the South-eastern Black Forest mountains, a massive of the European Alps, so the db should enforce the constraint that the Danube cannot belong to any other continent but Europe.

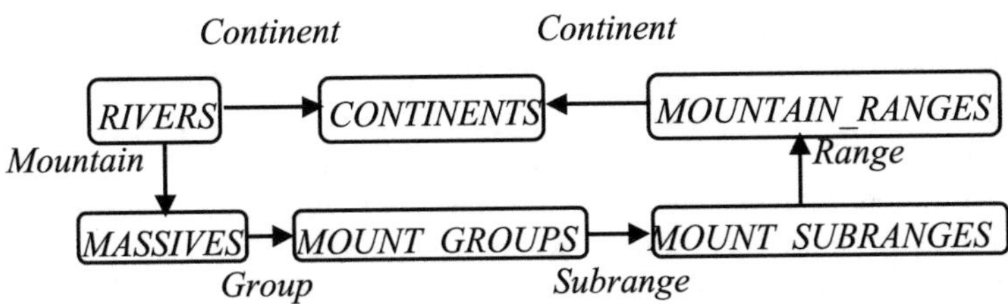

Fig. 7. An example of a commutative diagram with a non-total mapping

To conclude with, no commutative cycle containing at least one non-totally defined mapping should be broken either. *MatBase* does not allow its users to break unbreakable cycles.

3.2 Anti-Commutative Function Diagrams

Sometimes, as already seen with constraint *C'* associated to the cycle from Fig. 6 above, commutative-type diagrams are anti-commuting: $(\forall x \in A_1)((f_n \circ \ldots \circ f_2 \circ f_1)(x) \neq (g_m \circ \ldots \circ g_2 \circ g_1)(x))$. Let us consider, for example, the commutative-type diagram from Fig. 8 (having $n = 3$, $m = 2$, length $n + m = 5$, *ISLANDS_SHARING* as source and *CONTINENTS* as destination), in which *OCEANS* is a subset of *CONTINENTS* (because data for them is identical, except for some dualisms –e.g. *MaxAltitude* stores for oceans the corresponding maximum depth-, and a Boolean *Ocean?* is distinguishing between them: *OCEANS* = {*x*∈*CONTINENTS* | *Ocean?*(*x*)} and i_{OCEANS} is the associated canonical inclusion mapping (i.e. $i_{OCEANS}(x) = x$, ∀*x*∈*OCEANS*).

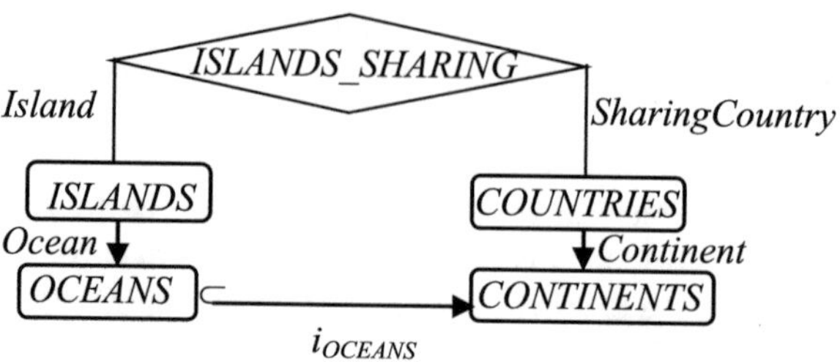

Fig. 8. An example of an anti-commutative E-RD cycle

Trivially, as, for any country sharing a part of an island, the continent to which it belongs must be distinct from the ocean to which that island belongs, this diagram anti-commutes and the corresponding constraint (C_2) should be added to this db scheme:

C_2: $(\forall x \in ISLANDS\_SHARING)((i_{OCEANS} \circ Ocean \circ Island)(x) \neq Continent \circ SharingCountry)(x))$

The technique used for deciding whether a diagram anti-commutes is almost identical to the one employed for commutative ones: you consider any element of the source, compute for it both elements of the destination for each path and wonder whether they are never the same; if they are, then the diagram anti-commutes.

Please note that anti-commutativity is a much stronger constraint than function inequality; for example, let us denote by $f = f_n \circ \ldots \circ f_2 \circ f_1$ and by $g = g_m \circ \ldots \circ g_2 \circ g_1$, both of them being functions defined on A_1 and taking values from A_{n+1}; by function equality definition, as they have same domains and codomains, $f = g$ *iff* $f(x) = g(x)$, $\forall x \in A_1$; when negating this definition for such mappings (i.e. having same domains and codomains), it trivially follows that $f \neq g$ *iff* $\exists x \in A_1$ such that $f(x) \neq g(x)$, which is much weaker than anti-commutativity (that requires this inequality to hold for any value of x, not for just one or some of them).

3.3 Commutative-Type Function Diagrams that Commute Unconventionally

As constraint *C* proves (see Fig. 6 above), commutativity-type E-RD cycles may also have associated generalized commutativity constraints.

The technique used for deciding whether a commutative-type diagram unconventionally commutes is almost identical to the one employed for (conventional) commutative ones: you consider any element of the source, compute for it both elements of the destination for each path and wonder whether there is any other kind of connection between them (other than equality and inequality); if there is, then the diagram has the corresponding associated generalized commutativity constraint.

3.4 Peculiarities of Commutative-Type Function Diagrams of Length 2

"Degenerated" commutative-type diagrams of length two need special treatment:

First, any commutative one should always be broken, by eliminating any of the two involved mappings: what would be the sense of storing in a table two columns that must always be identical?

Anti-commutativity should always be enforced, just like for corresponding cycles of length greater than two. For example, to the *BOARDING_PASSES* set from Fig. 9, the following constraint should be added:

($\forall x \in$*BOARDING_PASSES*)(*Departure*(x) $\neq$ *Destination*(x)).

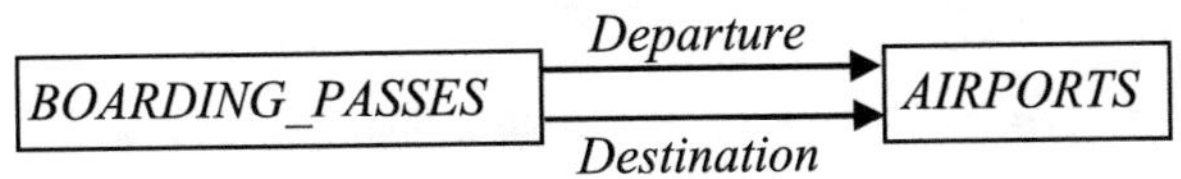

Fig. 9. An example of an anti-commutative diagram of length two

Note that, in this particular case of commutative-type diagrams of length two, made out of mappings $f : D \to C$ and $g : D \to C$, commutativity corresponds to $f \bullet g$ reflexivity and anti-commutativity to $f \bullet g$ irreflexivity.

Generally, according to the equivalence between E-R and functional data modelling [8, 13], entity-type object set D is equivalent to a dyadic relationship-type object set D', whose canonical Cartesian projections are f and g; hence, all homogeneous binary function products (i.e. products of two functions having same co-domain) may also have any dyadic relation properties. Consequently, for such diagrams it is natural to also investigate $f \bullet g$ for symmetry, asymmetry, transitivity, intransitivity, Euclideanity, inEuclideanity, acyclicity, and connectivity. Let us consider, for example, the diagram from Fig. 10, in the context of any kind of championship where each team plays against any other one twice (once home and once away).

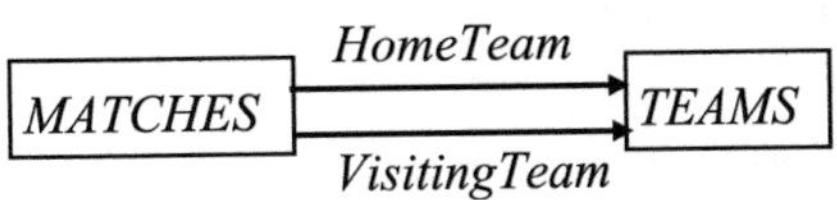

Fig. 10. An example of a length two diagram having several associated dyadic type constraints

As no team may play against itself, this diagram is anti-commuting, i.e. *HomeTeam* • *VisitingTeam* is irreflexive; as there are always two matches between any two teams *HomeTeam* • *VisitingTeam* is symmetric too; as it is a championship (i.e. any team must play against all other ones) *HomeTeam* • *VisitingTeam* is transitive, Euclidean, and connected too; as it is symmetric, it cannot be acyclic. Consequently, the following constraints should be added to this *MATCHES* scheme: *HomeTeam* • *VisitingTeam irreflexive, symmetric, transitive, Euclidean, connected.*

Similar to autofunctions, dyadic relation properties are related by several implications [8], out of which in this context the following are relevant: acyclicity implies asymmetry; asymmetry, intransitivity, as well as inEuclideanity imply irreflexivity.

3.5 Uninteresting Commutative-Type Function Diagrams

Generally, most of the commutative-type diagrams are uninteresting, as they do not have any associated constraint, like, for example, the one from Fig. 11 (having $n = 2$, $m = 2$, length 4, *STATES* as source and *CITIES* as destination; *Capital* are one-to-one as no city may simultaneously be the capital of more than one country and/or region).

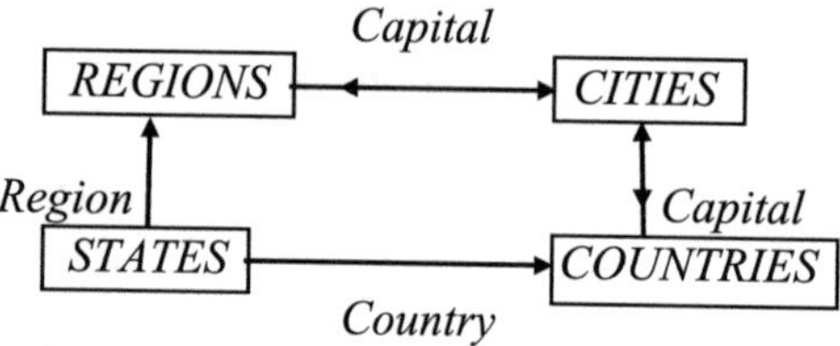

Fig. 11. An example of an uninteresting commutative-type diagram

As there is only one state per country to which the country capital belongs (and even fewer ones for which the capitals of the regions to which they belong is also the corresponding country's capital), this diagram does not commute. Dually, it does not anti-commute either: for example, Bucharest is the capital of both Romania and the Bucharest state. Moreover, it does not have any other associated constraint, so it is uninteresting: considering any state *s* belonging to a region *r* that has capital *c* and to a country *t* that has capital *c'* there is no relation whatsoever between *c* and *c'* (as sometimes $c' = c$, sometimes $c' \neq c$, and sometimes *c* and/or *c'* may even be nulls).

4 Circular-Type E-RD Cycles

To any node of a circular-type cycle a corresponding composed autofunction is associated. As such, we should always investigate what properties such auto-functions have (among reflexivity, irreflexivity, symmetry, asymmetry, idempotency, anti-idempotency, and acyclicity).

4.1 Reflexivity and Irreflexivity

Most frequently, circular-type cycles have associated reflexivity constraints (e.g. Fig. 1). For example, let us consider the diagram from Fig. 12 (length 3, all nodes are intermediate ones):

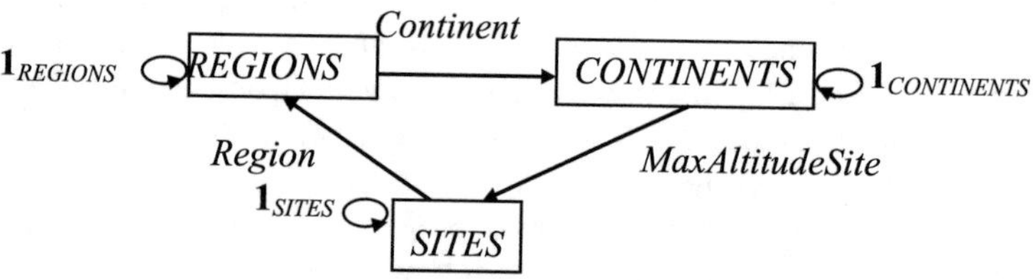

Fig. 12. Another example of a locally commutative circular-type diagram

1. Considering any region *r* belonging to a continent *c* having the maximum altitude site *s* belonging to region *r'* it is obvious that *r* and *r'* are not always equal (and not always unequal either, as there is only one maximum altitude site per continent, but there are several regions in a continent). Consequently, this diagram does not either commute or anti-commute in its *REGIONS* node.

2. Considering any continent *c* having the maximum altitude site *s* belonging to region *r'* of continent *c'* it is obvious that *c* and *c'* are always equal. Consequently, this diagram commutes in its *CONTINENTS* node and the corresponding constraint *Continent* ∘ *Region* ∘ *MaxAltitudeSite reflexive* should be added to the *CONTINENTS* scheme. If failing to enforce it, users might store such implausible data as the Death Valley, which contains the lowest altitude point of North America, belongs to the South America's Andes region.

3. Considering any site *s* belonging to region *r* belonging to a continent *c* having the maximum altitude site *s'* it is obvious that *s* and *s'* are not always equal (and not always unequal either). Consequently, this diagram does not either commute or anti-commute in its *SITES* node either.

The technique employed for checking whether a circular-type diagram locally commutes in one of its nodes is obviously a particularization of the one used for (conventional) commutative ones: you consider any element of the current source node, compute for it the value of the corresponding composed mapping and wonder whether they are always the same; if they are, then the diagram locally commutes in that node (as the composed function is reflexive, i.e. equal to that node's unity mapping); if they should always be distinct, then the diagram anti-commutes.

Composed function reflexivity is also called in the (E)MDM *local commutativity*, because it is a particular case of (conventional) commutativity, where one member of the equality is a unity mapping. For example, the diagram from Fig. 12 above locally commutes in *CONTINENTS*, as *Continent* ∘ *Region* ∘ *MaxAltitudeSite* = $\mathbf{1}_{CONTINENTS}$.

Similarly, corresponding irreflexivities are called *local anti-commutativities*. For example, let us consider the circular cycle from Fig. 13 (where $i_{EMPLOYEES}$ is the canonical injection associated to the inclusion *EMPLOYEES* ⊂ *PEOPLE* and *MD* stores for any person his/her doctor of medicine): as nobody may be his/her own MD, it is anti-commuting in *EMPLOYEES* (i.e. $MD(i_{EMPLOYEES}(x)) \neq x, \forall x \in EMPLOYEES$).

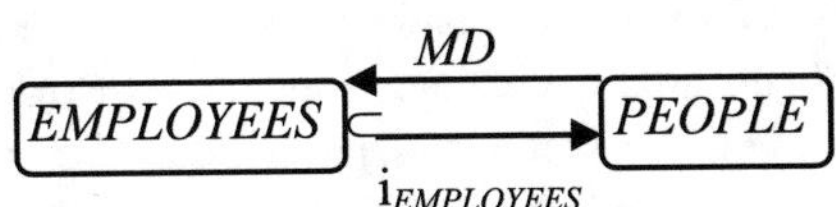

Fig. 13. An example of a locally anti-commutative circular-type diagram

4.2 Other Compound Autofunction Constraint Types

Let us consider, for example, the circular diagram (length 2 and *RootFolder* is one-to-one, as no folder may be the root one for more than one logic drive) from Fig. 14.

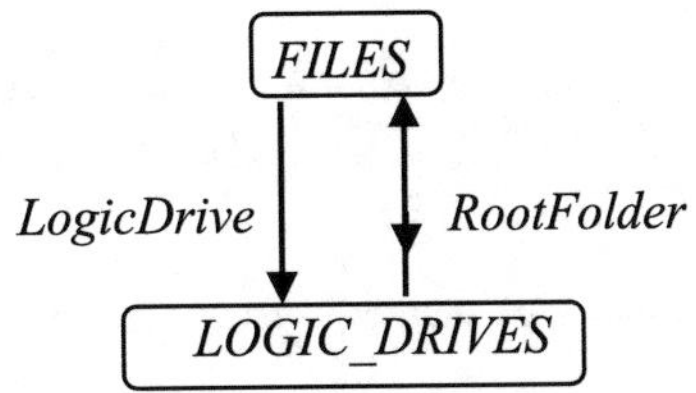

Fig. 14. A circular-type cycle that has an associated idempotency constraint too

Let us analyse first whether it is locally commuting in node *LOGIC_DRIVES*, i.e. *Logic-Drive* ° *RootFolder reflexive*?; considering any logic drive x, x is either not formatted (i.e. *RootFolder*(x) ∈ NULLS) or it is, in which case it has a root folder y = *RootFolder*(x) ∈ *FILES*; as any file, y should belong to a logic drive and it is obvious that it has to belong to x, the logic drive whose root folder it is, i.e. x = *Logic-Drive*(y), which means that this mapping composition is reflexive.

Being reflexive, as autofunction reflexivity implies both symmetry and idempotency, it is also symmetric and idempotent, but these constraints should not be added to this db scheme, as it would become not minimal. Being reflexive, it cannot be acyclic (as its graph is made out only of length one cycles). Consequently, only the constraint *LogicDrive* ° *RootFolder reflexive* must be added to *LOGIC_DRIVES*.

Let us analyse now whether it is locally commuting in node *FILES*, i.e. *RootFolder* ° *LogicDrive reflexive*?; considering any file x, x belongs to a logic drive y = *LogicDrive* (x); as it has at least one file, y has to be formatted, so there is a folder z belonging to *FILES* such that z is the root folder of y: z = *RootFolder*(y) ∈ *FILES*; as not all files in a logic drive may be its root folders, generally $x \neq z$, i.e. this cycle is not locally commuting in *FILES*, that is *RootFolder* ° *LogicDrive* $\neq \mathbf{1}_{FILES}$. Hence, we should analyse whether it is locally anti-commuting in node *FILES*, i.e. *RootFolder* ° *LogicDrive*(x) *irreflexive*?, which is not true either, as any formatted logic drive has to have a root folder, for which $x = z$.

Then, we should analyse whether f = *RootFolder* ° *LogicDrive* is symmetric (i.e. $f^2 = \mathbf{1}_{FILES}$): using above notations, for any file x, z = *RootFolder*(y) ∈ *FILES*, y = *LogicDrive*(z) (as any file stored on a logic drive is stored on the same drive as its root folder), so $f^2 = f$, which obviously means that f is not symmetric, but it is idempotent.

It is very easily provable that f is neither asymmetric (as equality holds for any root folder), hence nor acyclic either. Consequently, constraint *RootFolder* ° *LogicDrive idempotent* must be added to this db scheme too.

The techniques used to discover any other (but reflexivity) constraint types that are associated to circular-type cycles are based on the corresponding type definitions and are very similar to the ones employed for the corresponding autofunctions ones: the only difference is that instead of dealing with an atomic function, in such cases we deal with composed ones.

Generally, in each node of any circular-type diagram the corresponding composed autofunction should be investigated just like explained in Sect. 2 above. Moreover, as circular-type diagrams may also have associated generalized constraints, their existence should be considered too. For example, the diagram from Fig. 15 (where *SOLAR_SYSTEMS* is modelled as a subset of *CELESTIAL_BODIES*, with *Sun* as the corresponding canonical inclusion) has associated not only the constraint "the sun of any solar system should belong to its solar system" (i.e. *SolarSystem* ° *Sun reflexive*), but also the generalized (unconventional) commutativity one "solar systems may not contain solar systems" (i.e. (∀*x*,*y*∈*SOLAR_SYSTEMS*)(*SolarSystem*(*Sun*(x)) $\neq y \wedge$ *SolarSystem*(*Sun*(y)) $\neq x$)).

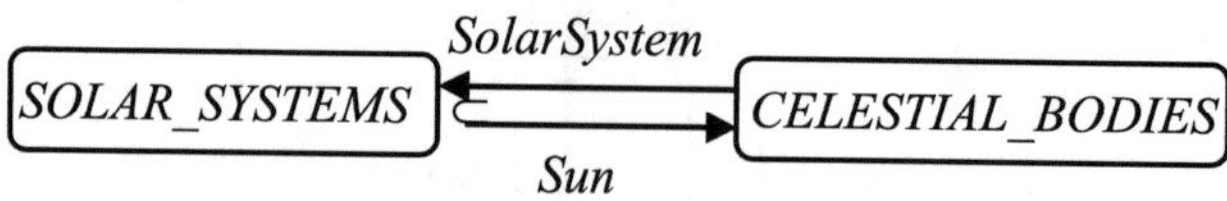

Fig. 15. An example of a circular-type diagram that has an associated generalized commutativity constraint

In this case, fortunately, this constraint should not be enforced, as it is implied by the reflexivity of *SolarSystem* ° *Sun*, hence it is redundant.

5 General-Type E-RD Cycles

Let us consider, for example, the general-type cycle from Fig. 16 (length 7, source nodes *NEIGHBOR_SEAS* and *NEIGHBOR_OCEANS*, destination nodes *COUNTRIES* and *CONTINENTS*).

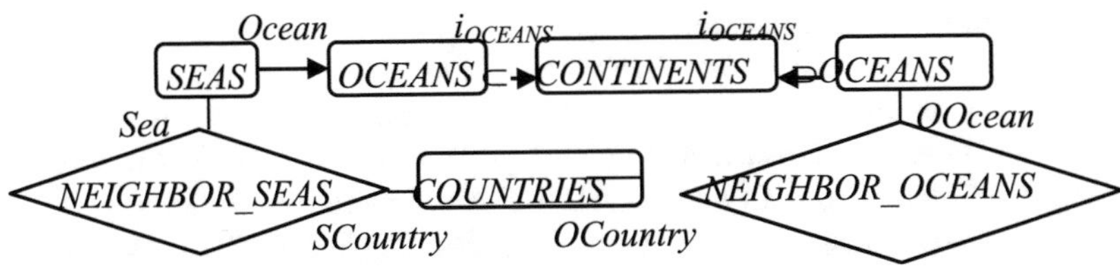

Fig. 16. An example of general-type diagram that has an associated generalized commutativity constraint

Consider any country *c* that is neighbour to a sea *s* which belongs to an ocean *o* (for example *c* = France, *s* = English Channel; *o* = Atlantic); by transitivity, it follows that *c* is also neighbour to *o*, so that this diagram has the following associated generalized commutativity constraint ("any country that is neighbour to a sea embedded in an ocean is also neighbour to that ocean"): $(\forall x \in NEIGHBOR\_SEAS, Ocean(Sea(x)) \notin \text{NULLS})$

$$(\exists y \in NEIGHBOR\_OCEANS)(SCountry(x) = OCountry(y) \land OOcean(y) = Ocean(Sea(x)))$$

Failing to enforce this constraint might allow users to store in the corresponding db such implausible data as France is neighbour to the English Channel, which is embedded into the Atlantic Ocean, but France is not neighbour to the Atlantic as well.

The technique used for deciding whether or not a general-type diagram unconventionally commutes is of the same type to the one employed for (conventional) commutative ones: you consider any element of a source, compute for it both elements of the destination for each path and wonder whether there is any other kind of connection between them and those similarly obtained for any or at least for one element of another source; if there is, then the diagram has the corresponding associated generalized commutativity constraint.

To conclude, for any general-type diagram you should check whether it has an associated generalized commutativity constraint and whenever this is the case you should add it to the corresponding db scheme and enforce it.

6 Constraints Associated to Adjacent E-RD Cycles

Two cycles are *adjacent* if they share at least one node. Sometimes, adjacent cycles have associated constraints too.

For example, the constraint "any two neighbour countries should belong either to a same continent or to neighbour continents" is associated to the three adjacent cycles from Fig. 17 (all of them of type commutative, two having length 2 and the one between them length 3), where *Continent* : *COUNTRIES* → *CONTINENTS* stores, for each country, the continent where their capital is located (e.g. North America for the U. S.A., Europe for Russia, Asia for Turkey, etc.), while the *COUNTRIES_CONTINENTS* binary relationship stores the other continents on which countries may also span, if any (e.g. Australia and Oceania for the U.S.A., with Hawaii, Asia for Russia, Europe for Turkey, etc.). Obviously, failing to enforce it might allow users to store in *NEIGHBOR_COUNTRIES* such implausible data as South Africa being neighbour to Brazil.

Formalization of this constraint is:

$$(\forall x \in NEIGHBOR\_COUNTRIES)$$
$$(Continent(Country(x)) = Continent(NeighborCountry(x)) \vee (\exists y \in COUNTRIES\_CONTINENTS)(Country(x)) = Country(y)) \wedge Continent(NeighborCountry(x)) = Continent(y) \vee NeighborCountry(x)) = Country(y) \wedge Continent(Country(x)) = Continent(y)) \vee (\exists z \in NEIGHBOR\_CONTINENTS)(Continent(z) = Continent(Country(x)) \wedge NeighborContinent(z) = (NeighborCountry(y)) \vee Continent(z) = Continent(NeighborCountry(x)) \wedge NeighborContinent(z) = Continent(Country(y))))$$

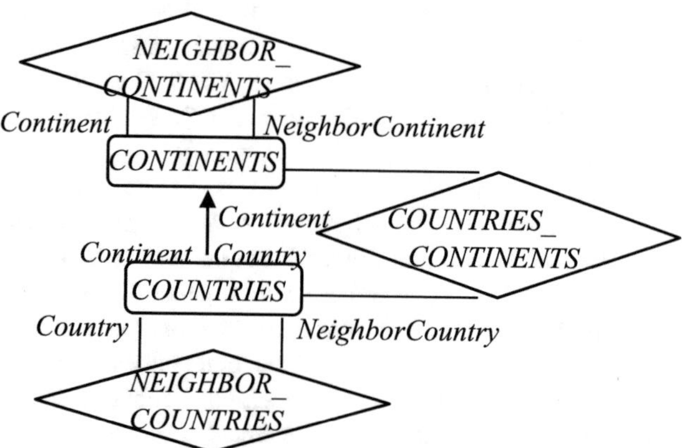

Fig. 17. An example of three adjacent cycles having an associated constraint

7 *MatBase* E-RD Cycle Analysis Assistance Algorithm

Figures 18 and 19 show the pseudocode algorithm for assisting analysis of E-RD cycles and discovery of all their associated constraints, which summarizes all the above and that is implemented in both current *MatBase* versions:

```
ALGORITHM A1. MatBase E-RD Cycles Analysis Assistance
Input: a db scheme S and its associated set of E-RDs
Output: S augmented with the newly discovered constraints associated to its E-RD
        cycles (if any).
  loop for all E-RDs d
   loop for all autofunctions a of d
      investigateAutoFunction(a);
   end loop;
   loop for all cycles c of d having n = length(c) > 1
    case type(c):
     commutative:
      if length(c) > 2 then
       if c anti-commutes then S = S ∪ {c anti-commutes};
       else
        if c commutes then
         if c can be broken by eliminating edge f from it then
          if it is not desired that computed mappings be added
           then drop f;
           else replace f by the corresponding computed ones;
          end if;
         else S = S ∪ {c commutes};
         end if;
        end if;
       end if;
      else        // cycle of length 2 (mappings f and g)
       if f = g then remove redundant f or g;
       else if c anti-commutes then S = S ∪ {f • g irreflexive};
          if f • g is acyclic then S = S ∪ {f • g acyclic};
          else if f • g symmetric then S = S ∪ {f • g symmetric};
               else if f • g asymmetric then
                        S = S ∪ {f • g asymmetric} — {f • g irreflexive};
               end if;
          end if;
          if f • g transitive then S = S ∪ {f • g transitive};
          else if f • g intransitive then
                  S = S ∪ {f • g intransitive} — {f • g irreflexive};
          end if;
          if f • g Euclidean then S = S ∪ {f • g Euclidean};
          else if f • g inEuclidean then
                 S = S ∪ {f • g inEuclidean} — {f • g irreflexive};
          end if;
          if f • g is connected then S = S ∪ {f • g connected};
          end if;
       end if;
     circular:
       loop for all edges f = f_{i+1} ° ... ° f_n ° f_1 ° ... ° f_i of c (1 ≤ i ≤ n)
         investigateAutoFunction(f);
       end loop;
    end case type(c);
    if c unconventionally commutes according to constraint G then S = S ∪ {G};
    if c and some adjacent cycle(s) unconventionally commute according to
    constraint G' then S = S ∪ {G'};
   end loop (for all cycles c of d);
  end loop (for all E-RDs d);
 End ALGORITHM A1;
```

Fig. 18. Algorithm *A*1 (*MatBase* assistance for analysing E-RD cycles)

```
ALGORITHM A2. MatBase investigateAutoFunction(f) procedure
if f is acyclic then
  S = S ∪ {f acyclic};
  if f is idempotent then S = S ∪ {f idempotent};
  else if f is anti-idempotent then S = S ∪ {f anti-idempotent};
  end if;
else
  if f is reflexive then
    if f is atomic and totally defined then remove f;
    else S = S ∪ {f reflexive};
    end if;
  else if f is symmetric then S = S ∪ {f symmetric};
       else if f is asymmetric then S = S ∪ {f asymmetric};
       end if;
       if f is idempotent then S = S ∪ {f idempotent};
       else if f is anti-idempotent then S = S ∪ {f anti-idempotent};
       end if;
       if not (f asymmetric or f anti-idempotent) then
         if f is irreflexive then S = S ∪ {f irreflexive};
       end if;
  end if;
end if;
End ALGORITHM A2;
```

Fig. 19. Algorithm A2 (*MatBase investigateAutoFunction* procedure)

8 Results and Discussion

8.1 Algorithm's Complexity and Optimality

It is very easy to check that this algorithms is very fast, as its time complexity is $O(|AF| + |C_{>1}|)$ (i.e. linear in the sum of the cardinals of the set of single autofunctions and compound ones associated to the nodes of its circular-type cycles (denoted AF) and the one of the greater than 1 length non-circular cycles subset (denoted $C_{>1}$)): each single autofunction (i.e. length 1 cycle) is considered only once; each cycle of length 2 is considered at most twice (which is the number of compound autofunctions associated to its node, for circular ones; for commutative ones it is considered only once); for any cycle of length greater than 2, if it is commutative or general is considered only once, but if it is circular every compound autofunction associated to its nodes is considered.

Moreover, this algorithm is also optimal, as, using math (e.g. the second algorithm from the previous section), it asks db designers the minimum number possible of questions for any cycle type and subtype. For example, if something is reflexive, it cannot be irreflexive (and the same goes for any dual pair of properties); if a dyadic relation is acyclic, then it is also irreflexive, asymmetric, and anti-idempotent; autofunctions cannot be Euclidean or connected; transitivity for autofunctions is called idempotency.

8.2 Algorithm's Usefulness

First, analysing E-RD cycles is interesting even *per se*, within the study of sets, functions, and relations semi-naïve algebra, as it helps getting a better understanding on function diagram commutativity and anti-commutativity, as well as on dyadic relation properties of both single and compound autofunctions and homogeneous binary function products. Moreover, the generalized commutativity constraints (e.g. the one attached to the cycles presented in Figs. 6 and 15 to 17) are excellent real-world examples of closed first order predicate calculus with equality formulas.

The main utility of this algorithm is, of course, in the realms of data modelling, db constraints theory, db and db software applications design and development: all constraints (business rules) that are governing the sub-universes modelled by dbs, be them relational or not, should be enforced in the corresponding dbs' schemas; otherwise, their instances might be implausible. Structural E-RD cycles have sometimes associated non-relational constraints that may be much more easily discovered by applying the assistance algorithm presented in this paper.

Generally, almost all autofunctions have associated non-relational constraints. Very many circular-type cycles have reflexivity (but not only) type constraints associated. A significant number of commutative-type cycles also have associated constraints, while some of them should be broken by eliminating computable functions (or sometimes replacing them with redundant computable ones). Although infrequently, even general-type cycles have associated constraints, especially when their length is small.

For example, in a geography and astronomy db (from which we've extracted almost all examples presented in this paper) that has in its E-RD 63 nodes and 352 edges there are 21,806 cycles of length at most 16, out of which 36 are circular, 850 commutative, and the rest are of the general type. Analysing them with the help of this algorithm, we discovered 7 circular cycles that have a reflexivity constraint associated to one of their associated compound autofunctions, 4 having reflexivity and 7 irreflexivity ones, 4 commutative-type cycles that commute and are unbreakable, 8 which are anti-commutative, 9 irreflexivities, 8 asymmetries, and one acyclicity for compound autofunctions, 16 generalized commutativity constraints (out of which 6 are associated with adjacent cycles), and plus 7 acyclic autofunctions and one that is irreflexive, as well as one acyclic, 8 asymmetric, and 9 irreflexive dyadic relations.

Obviously, except for single autofunctions, this algorithm may be used only after all cycles of a db E-RD have been detected and classified, which is done by *MatBase* too [10], as a prerequisite of applying it.

9 Conclusion

In summary, we have designed, implemented, and successfully tested in both *MatBase* latest versions (for MS Access and C# and SQL Server) an algorithm for assisting detection (and enforcement) of all non-relational constraints associated to E-RD cycles, analysed its complexity and optimality, as well as outlined its usefulness for both sets, functions, and relations algebra, and, especially, for data modelling, db constraints theory, db and db software application design and development practices.

Very many non-relational db constraint types may be attached to E-RD cycles. The algorithm presented in this paper helps users to analyse each such cycle exhaustively and intelligently, such that they discover all non-relational constraints associated to them in the minimum possible time. Moreover, *MatBase* also includes automatic enforcement of these constraints, through automatic code generation.

This algorithm is successfully used both in our lectures and labs on Advanced Databases (for the postgraduate students of the Mathematics and Computer Science Department of the Ovidius University, Constanta and the Computer Science Taught in English Department of the Bucharest Polytechnic University) and by two Romanian IT companies developing db software applications for many U.S. and European customers in the Fortune 100 ones.

References

1. Chen, P.P.: The entity-relationship model: toward a unified view of data. ACM TODS **1**(1), 9–36 (1976)
2. Thalheim, B.: Fundamentals of Entity-Relationship Modeling. Springer, Berlin (2000)
3. Mancas, C.: Conceptual Data Modeling and Database Design: A Completely Algorithmic Approach. Volume I: The Shortest Advisable Path. Apple Academic Press / CRC Press (Taylor & Francis Group), Waretown, NJ (2015)
4. Codd, E.F.: A relational model for large shared data banks. CACM **13**(6), 377–387 (1970)
5. Abiteboul, S., Hull, R., Vianu, V.: Foundations of Databases. Addison-Wesley, Reading, MA (1995)
6. Mancas, C.: A deeper insight into the mathematical data model. In: Draghici, M. (ed.) ISDBMS'90, pp. 122–134. ICI Press, Bucharest, Romania (1990)
7. Mancas, C.: On knowledge representation using an elementary mathematical data model. In: Boumedine, M. IKS'02, pp. 206–211. Acta Press, Calgary, Canada (2002)
8. Mancas, C.: Conceptual Data Modeling and Database Design: A Completely Algorithmic Approach. Volume II: Refinements for an Expert Path. Apple Academic Press / CRC Press (Taylor & Francis Group), Waretown, NJ (2019, in press)
9. Mancas, C., Dorobantu, V.: On enforcing relational constraints in *MatBase*. London J. Res. in Comp. Sci. and Technol. **17**(1), 39–45 (2017)
10. Mancas, C., Mocanu, A.: *MatBase* DFS detecting and classifying E-RD cycles algorithm. J. Comput. Sci. Appl. Inf. Technol. **2**(4), 1–14 (2017)
11. Mancas, C.: Algorithms for key discovery assistance. In: Repa, V., Bruckner, T. (eds.) BIR 2016, LNBIP, vol. 261, pp. 322–338. Springer, Cham (2016)
12. Mancas, C.: *MatBase* constraint sets coherence and minimality enforcement algorithms. In: Benczur, A., Thalheim, B., Horvat, T. (eds.) ADBIS 2018, LNCS, vol. 11019, pp. 263–277. Springer, Cham (2018)
13. Mancas, C.: On the equivalence between entity-relationship and functional data modeling. In: Hamza, M.H. (ed.) SEA'03, pp. 335–340. Acta Press, Calgary, Canada (2003)
14. Mancas, C.: On modeling closed E-R diagrams using an elementary mathematical data model. In: Manolopoulos, Y., Navrat, P. (eds.) ADBIS 2002, vol. 2, pp. 65–74. Slovak Technology University Press, Bratislava, Slovakia (2002)
15. Adámek, J., Herrlich, H., Strecker, G.E.: Abstract and Concrete Categories (The Joy of Cats). John Wiley & Sons, Hoboken, NJ (1990)

16. Purdea, I., Pic, G.: Modern Algebra Treaty (Volume 1, in Romanian). Editura Academiei, Bucharest, Romania (1977)
17. Lippe, E., ter Hofstede, A.H.M.: A category theory approach to conceptual data modeling. RAIRO-Theor. Inform. Appl. **30**(1), 31–79 (1996)
18. Fredericks, P.J.M., ter Hofstede, A.H.M., Lippe, E.: A unifying framework for conceptual data modelling concepts. Inf. Softw. Technol. **39**(1), 15–25 (1997)
19. Johnson, M., Rosebruck, R.: Sketch data models, relational schema and data specifications. Elsevier Science B.V., http://www.elsevier.nl/locate/entcs/volume61.html (2002). Last accessed 1 Dec 2002

Heuristic Search for Tetris: A Case Study

Giacomo Da Col(✉) and Erich C. Teppan

Alpen-Adria Universität Klagenfurt, 9020 Klagenfurt, Austria
{giacomo.da,erich.teppan}@aau.at

Abstract. Games represent important benchmark problems for AI. One-player games, also called puzzles, often resemble real world optimization problems and, thus, lessons learned on such games are also important for such problems. In this paper we focus on the game of Tetris, which can also be seen as a packing problem variant. We provide an empirical evaluation of a heuristic search approach for Tetris with the following goal: Having an effective heuristic function at hand, we want to answer the question how much the additional tree search pays off. We are especially interested if there is a so called sweet spot that represents the best ratio between score achieved and time invested in the search. This knowledge is crucial in order to be able to implement deep-learning approaches in the light of limited computing resources, i.e. to produce many good games to be learned from in rather short time. Our experiments reveal that such a sweet spot exists and hence, using this knowledge, only a fraction of time is needed for producing the same amount of learning data with similar quality.

Keywords: Heuristics · Search · Tetris · Beam-search · AI

1 Introduction

Since the late nineties, the discipline of artificial intelligence (AI) could come up with many major success stories perceived not only by computer scientists but also by a broad mass of non experts. For example, driver assistance systems increase safety of your car rides [30], smart home applications configure your living space conforming to automatically identified needs and preferences [21], navigation systems direct you to your goal as fast as possible [18,20] and recommender systems let you find products and places you are interested in [16].

In the industrial context, image processing systems automatically identify production errors [12], scheduling systems optimize workflows in production lines [9,19,27], big data applications identify hidden relationships in giant business data sets [15] and AI equipped tools are used for designing tailored solutions for customers [10,22].

Different developments have led to the positive progression of AI, which is on the one hand improved hardware [1,17], and on the other hand these are advancements in classic fields like heuristic search [13] or machine learning [8,25].

K. Arai et al. (Eds.): CompCom 2019, AISC 997, pp. 410–423, 2019.
https://doi.org/10.1007/978-3-030-22871-2_28

Systems like Apple's personal assistance software Siri [2] clearly reveal that we currently might approach the original goal of strong artificial intelligence that is comparable with human intelligence.

In this context, popular games like Chess, Backgammon or Go represent good benchmarks because they are designed to be challenging for humans, and have a compact set of rules that defines the environment and the winning conditions. A milestone in the field is certainly the defeat of the Chess champion Garry Gasparov by IBM's Deep Blue, in 1997 [7]. Almost 20 years later, Deepmind's Alpha-Go scores a clear victory over the grand master Lee Sedol in the game of Go [23].

If both games represented a big challenge due to the great number of possible game states, Go presented the additional issue of the lack of effective heuristic functions to estimate the goodness of the states. To overcome the problem, Silver *et al.* implemented a deep neural network architecture to learn such a heuristic function from millions of example games. In the last version, named Alpha-Go Zero, the whole learning is done using just games generated by self-play [24]. Like with a given non-learned heuristic function, the neural network is then used to guide a tree-search based approach based on the minimax principle towards the most promising moves.

Hence, also in the Alpha-Go systems the neural net alone is not doing the whole job. Even after the network is fully trained, Alpha-Go Zero would not be able to beat the previous version using just the raw output of the neural network. Similarly, using only the neural net without any additional search Alpha-Go Lee would probably not have been able to beat grandmaster Lee Sedol. The search guided by the neural net is crucial to ensure the master-defeating capabilities of the system. For example, in Alpha-Go zero there are four TPUs (tensor processing units) concerned with the search. Although this does not sound much, one has to know that such units are able to perform 180 tera flops per second.

According to game and decision theory, games like Chess or Go represent zero-sum perfect information two-player games. Though, many real world problems that could be tackled by AI applications like optimization problems rather resemble one-player games, i.e. puzzles. For example, the famous game Tetris could be seen as a form of packing problem, which occurs in many different variants in many different industrial environments.

Interestingly, there has not been much research on AI approaches applied on Tetris. One of the early results in this field is due to Bertsekas and Tsitsiklis [4], which used their λ-policy iteration (a variant of the standard Bellman policy iteration [3]) and scored 3200 lines cleared on average. A more recent achievement used a policy iteration technique based on classification (CBMPI) that was able to reach 51 millions of lines cleared on average [14].

In terms of heuristics, the approach followed by the majority of researchers are classic evaluation functions that are based on feature vectors for estimating the goodness of search states. These heuristic methods are divided in one-piece controllers, when the information about the next piece is not used (i.e. no look-ahead), and two-piece controllers, which take advantage of the knowledge of the next piece (like in the classic version of Tetris). Various techniques have been used to optimize the weights of these features in the heuristic function.

Dellacheire, creator of one of the best one-piece controller still known, tuned the weights by hand. While there is no official publication by Dellacheire, other authors re-implemented his algorithm and confirmed the result of 660 000 lines cleared [28]. Better results are achieved when taking advantage of the next piece information. Szita and Lörincz used a noisy-cross entropy method and were able to reach 300 000 lines cleared on an average on 30 games, with 800 000 pick on the best game [26]. Böhm *et al.* used an evolutionary algorithm to fine-tune the weights [5]. They both tried a linear sum of the features and an exponential sum, showing that the linear weighted sum works better, being able to reach more than 480 million removed lines in the best game.

Taking Tetris as an appealing version of important industrial optimization problems, we contribute an in-depth analysis of costs and benefits of heuristic search applied to two versions of the Tetris game. We particularly focus on the question whether and how much additional search pays off in the light of a predefined state-of-the-art heuristic. For purposefully applying deep learning approaches, this knowledge is absolutely crucial out of the following reasons.

Clearly it is important to be able to produce good learning data, i.e. good games, for the majority of cases. At the same time it is also important to be able to perform the games quickly in order to produce enough learning data in the light of limited computing resources. The computing power that is at Deep Mind's disposal is not realistic for most research groups or companies. Although Deep Mind has not provided this information in the famous Nature paper [23], different sources in the internet talk about up to 5000 TPUs (i.e. 5000×180 tera flops per second) to be used for producing the learning data [29].

Hence, with respect to the invested amount of search, we are particularly interested in identifying a 'sweet spot' at which already good games are produced but at the same time more search does not improve the outcome significantly. In our approach, we use the Böhm heuristic in combination with beam search and allow different beam widths in order to answer the question whether there is such a sweet spot. To the best of our knowledge, there is no comparable work described in literature.

2 Background

Tetris is a puzzle videogame created by Alexey Pajitnov in the mid eighties. The game consists of:

- a board of cells of fixed width and height. The most common notation used to describe the board is *height* $\times$ *width*, e.g. the original settings are 20×10 cells.
- A set of pieces (Fig. 1) which appear sequentially on the top of the board and gradually fall to the bottom until they collide with an obstacle, i.e. another piece or the bottom of the board. The piece can be rotated or moved horizontally by the player as it falls. When a collision happens, the piece is marked

as placed and can not be moved any more. The placement of a piece in the board is called action. After each action, the next piece spawns at the top center of the board.

- A game state is defined by all the pieces that are present in the cells of the board, the current piece that is about to fall and a certain number of future pieces. The number of future pieces given is called look-ahead l. In the original settings, $l = 1$.
- If the player completely fills a row of the board with pieces, this row disappears and all the stones above it fall down of one position to fill the gap. This mechanic is the core of the game, because otherwise the board will be filled with pieces after very few actions.
- The game ends when the current piece can not spawn because it collides with at least one other piece.

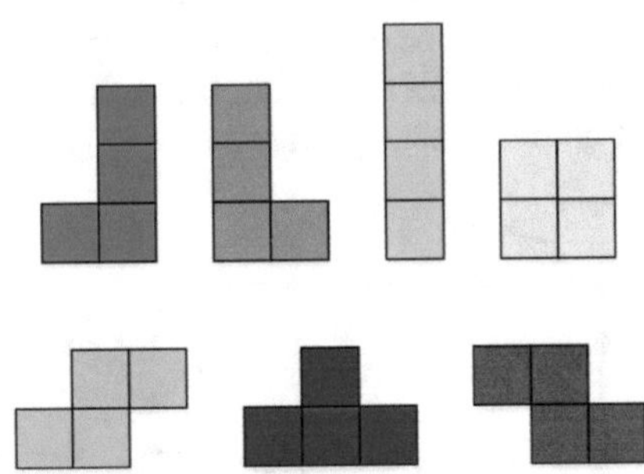

Fig. 1. The 7 pieces of the standard Tetris game. They are commonly referred to (from top left): J, L, I, O, S, T, Z

The goal of the player is to "survive" as long as possible. The pace increases with the proceeding of the game, making it difficult to place the pieces in an effective way. Every Tetris game comes to an end eventually, as it exists at least one sequence of pieces that makes the pile height grow, no matter the placement of the pieces [6].

It has been proven that Tetris is an NP-hard optimization problem if the sequence of pieces is part of the input [11]. This fact, in conjunction with the small set of rules for game definition and the fame of the game, makes Tetris a valuable challenge for intelligent systems.

Similarly to other works in the literature [4,5,26], we are considering a special version of Tetris, where the rotation and the position of the piece is decided after the piece spawns but before it starts to drop. This makes the list of possible actions faster to compute with no significant influence on the gameplay.

2.1 Tetris Heuristic

For Tetris there exists a number of effective heuristic functions that directly lead to highly performing heuristic algorithms. The most effective of these functions are built upon a linear combination of weighted state features. For the experiments discussed in this paper, we used the famous heuristic by Böhm *et al.* [5], which operates using the features summarized below:

1. *Max height:* Height of the tallest column in the board.
2. *Holes:* Number of all free cells with at least one occupied cell above them.
3. *Removed lines:* Number of lines removed in the last step to get in the current board.
4. *Height difference:* Difference between the tallest and the smallest column.
5. *Max well depth:* The depth of the deepest well (with a width of one) on the board.
6. *Sum well:* The sum of the depths of all the wells.
7. *Weighted Cells:* Weighted sum of occupied cells, whereby the weight of a cell is defined as its height on the board.
8. *Row transition:* Row-wise sum of all the transition between occupied/free cell. The right/left borders of the board count as occupied.
9. *Column transition:* Same as row transition, but column-wise. The top/bottom borders of the board count as occupied.

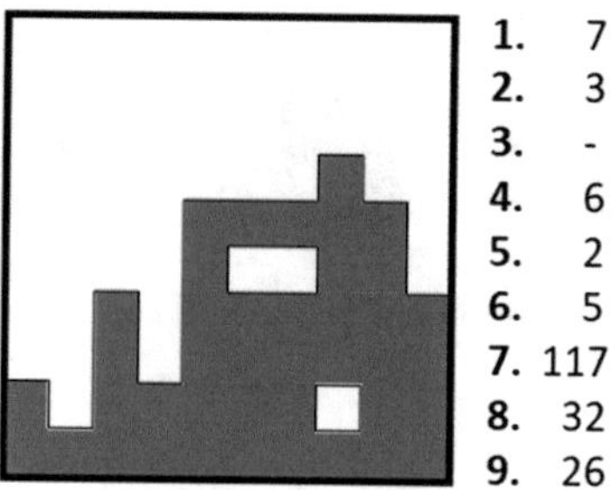

Fig. 2. Calculation of features values conforming to the Böhm *et al.* heuristic function

Figure 2 shows an example of all the features calculated for the given state on a 10×10 board. Features are weighted conforming to the vector $[-62709, -30271, -48621, 35395, -12, -43810, -4041, -44262, -5832]$.

Based on our observations, we would describe the heuristic behavior as follows. At first, it aims to build dense blocks along the left and the right border. At each iteration, the current piece is only put in the middle if it fits perfectly, otherwise is used to enlarge one of the outer blocks. The blocks are maintained compact by the row/column transition features, while *Max height* and *Weighted Cells* control the growth of the right and left blocks.

2.2 Search Algorithm

Our search approach is based on beam search for two main reasons. First, beam search possesses linear space complexity. Second, the search can be tuned by just one parameter, the beam width b, which determines how many nodes to expand at each level of the search tree. Thus, beam search represents a perfect candidate for the purpose of this paper.

In the case of Tetris, every node in the search tree corresponds to a game state. When a node is expanded, for every possible action (*i.e.* for every positioning of the current piece), a child node is generated with the resulting state. Consecutively, the states are ranked conforming to the Böhm heuristic function, and just the best b nodes are expanded further. Starting with the current piece, search proceeds until there are no more look-ahead pieces.

Example. Figure 3 offers a graphical representation in a small example with beam width $b = 2$ and number of look-ahead pieces $l = 1$. The states are ordered from left to right conforming to the Böhm heuristic function. For every level, just the best 2 nodes are expanded further, the rest is pruned. Note that, if we would just consider the placement of the current piece, the first node at the first level would be selected. By considering also the look-ahead piece, we can determine that the second node leads to a situation where a line can be removed, and therefore is preferable to the first one.

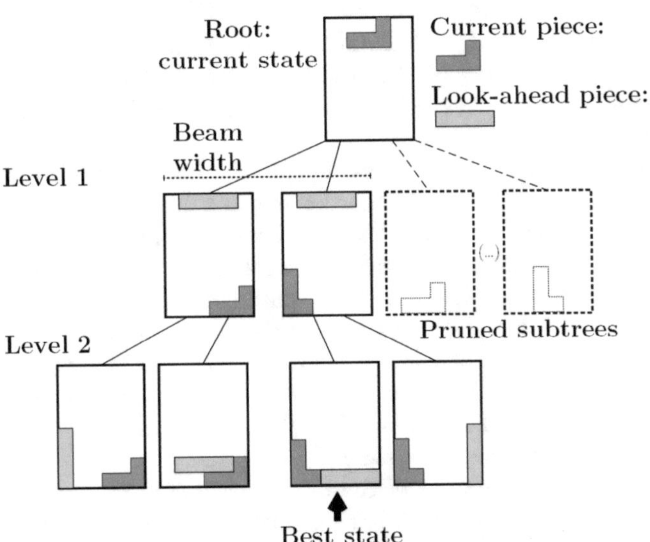

Fig. 3. Small example of a beam search tree for Tetris

3 Experimental Setup

The goal of the experiments is to analyze the impact of the search procedure in combination with the best evaluation function found in literature. The analysis is conducted with respect to the beam width b, which determines the number of search nodes that are maximally expanded at each level of the search tree. Our aim is the empirical determination of the beam width that represents the best compromise between performance and time invested in the search.

Definition 1. *Given the range $W = [1 \dots maxWidth]$ of possible beam widths, we define the* ***sweet spot*** *as the beam width $b \in W$ such that $\frac{score}{search\,time}$ is maximum.*

Conforming to other works in literature [4,5,26], we measure the score in terms of the number of removed lines in a game until the end condition is reached (game over). By search time we denote the time invested in the decision on the best action for a current piece. In our experiments, we will measure the search time in milliseconds.

Thus, we can formulate our main hypothesis as follows:

- H1: With respect to Tetris, a sweet spot can be found at a relatively small beam width.

This would allow to reach acceptable performance within a fraction of the time compared to the complete search.

The experiments are carried out against a small board configuration (12×6) and a standard board configuration (20×10). In both cases, we test games with look-ahead of 1 conforming to the standard version of Tetris, and look-ahead of 2, a benchmark that was rarely tested due to the higher number of possible combinations ($\tilde{4}0\,000$ per action).

Please note that for the small board configuration it is harder to reach high scores in comparison with the standard board. In fact, the set of pieces is the same as for the standard case, but the board is almost four times smaller, therefore the games end much more quickly. Furthermore, the smaller search space allows to conduct a wider analysis on the parameters and provides an approximation of the behavior we can expect from the standard board.

The implementation of the search algorithm and the Tetris player has been developed in C. The experiments ran on a system equipped with Intel i7-3930K processor (3.20 GHz), 64 GB of RAM. Every game was run in a separate parallel thread, in batch of 12 games at a time.

To mitigate the randomness of the piece generation, we use the arithmetic mean of the scores of 36 games. All the games are randomly initialized and are different from each other. Each game is played following the action suggested by our search approach until the end condition is reached. To calculate the search time we take the arithmetic mean of 100 calculations of the best action, for each different value of b.

4 Results

4.1 Small Board Test

We use a board size of 12×6, 12 rows tall and 6 columns wide. This configuration was also used by [5] in their preliminary evaluation. By testing all the beam widths between 1 and 24 (because for each piece there are at most 6 positions and 4 rotations), we can evaluate all the different degrees of search completeness, from greedy to complete. We can lower this upper-bound considering that some of the placements closer to the borders are not available because of collisions with the edge. Therefore, by counting the possible actions for pieces without rotation symmetries (T, J and L), we can derive that a beam width of 18 is sufficient to ensure the completeness of the search. Note that these completeness considerations are valid when testing games with look-ahead of 1. With look-ahead of 2, a much larger beam width is needed to make the search complete.

Table 1. Results of the small board test, look-ahead 1 and 2. Search time measured in ms. Sweet spots are in bold

Beam width	Look-ahead 1			Look-ahead 2		
	Score	Time	Score/time	Score	Time	Score/time
1	39.3	0.06	628.82	36.9	0.09	410.37
2	85.7	0.09	919.63	105.7	0.16	675.67
3	202.3	0.13	1576.64	539.9	0.23	2374.12
4	218.2	0.16	1346.30	525.4	0.30	1769.85
5	357.5	0.19	1845.99	1977.2	0.36	5464.58
6	532.8	0.22	2375.55	3532.9	0.43	8242.83
7	634.3	0.26	**2447.13**	5723.9	0.50	11538.51
8	543.8	0.29	1852.86	5261.1	0.56	9328.59
9	550.8	0.32	1695.36	9928.1	0.63	15728.62
10	803.4	0.35	2279.49	9702.8	0.70	13933.39
11	687.3	0.39	1760.91	12191.3	0.76	15969.17
12	525.1	0.41	1274.51	12712.3	0.83	15312.04
13	781.5	0.45	1718.14	17212.0	0.90	19064.34
14	633.5	0.48	1315.74	21687.4	0.97	22429.85
15	732.4	0.52	1418.25	25449.1	1.03	24666.32
16	921.6	0.55	1689.24	23860.5	1.09	21903.86
17	940.3	0.58	1615.15	35822.5	1.17	**30673.31**
18	747.1	0.62	1214.16	21229.4	1.23	17231.96

Table 1 summarizes the results for the small board test. As expected, the search time grows linearly with the beam width, and in the case of look-ahead of 2 is roughly double respect to the correspondent look-ahead of 1. The search time is measured in milliseconds. Since the values are pretty small, one may think that the difference between two beam widths in terms of search time is insignificant. However, please note that a good controller can easily play *millions* of actions in a single game, lasting for several hours.

Figure 4 offers a representation of the data for the look-ahead of 1 experiment. The bars represent the score reached (axis on the left side) and the line represents the score/search time ratio (axis on the right side). The score is steeply increasing at small beam widths (<8), but then the growth becomes less prominent. While in theory a larger beam width should always lead to a better or equal score, there is still a small fluctuation due to the random nature of the game. The score/search time curve highlights how good is the performance compared to the time invested in the search. The sweet spot is the pick of this curve, which is reached at beam width of 7. In particular, using such beam width it is possible to achieve 67.5% of the best mean score (940.3) by sparing almost 60% of search time required by the complete search (0.25 instead of 0.61 ms per action).

Concerning the look-ahead of 2, we checked beam widths up to 18. Note that, unlike the look-ahead of 1 case, 18 is not a sufficient beam width to consider the search complete. In fact, based on the example in Fig. 3, we would need to expand *all* the states at level 2 in order to find the best state at level 3 (which corresponds to the look-ahead of 2 states). The total number of states at level 2 in this case is $18^2 = 324$, which is the beam width needed for the complete search. However, since games with look-ahead of 2 last longer, it will be not feasible to test all the possibilities, so we limit to search a sweet spot within the same beam width range used with look-ahead of 1.

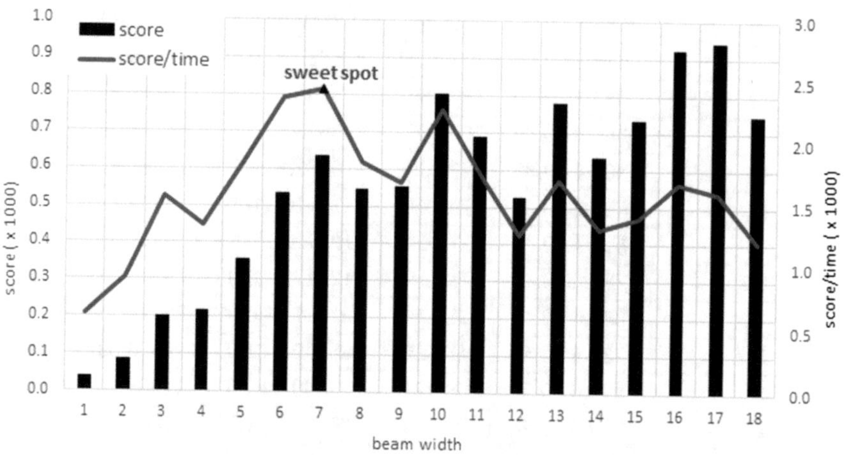

Fig. 4. Representation of the score and the score/search time ratio of the small board test, look-ahead of 1

Figure 5 shows the results for look-ahead of 2. First, there is a notable increase in performance compared to the look-ahead of 1 case (35 822.5 against 940.3), while the search time is just doubling. This increase means that the search algorithm makes good use of the additional information provided by the extra piece of look-ahead, which allows a better optimization of piece placements. We can see that a sweet spot is present at beam width 17. Although we can not

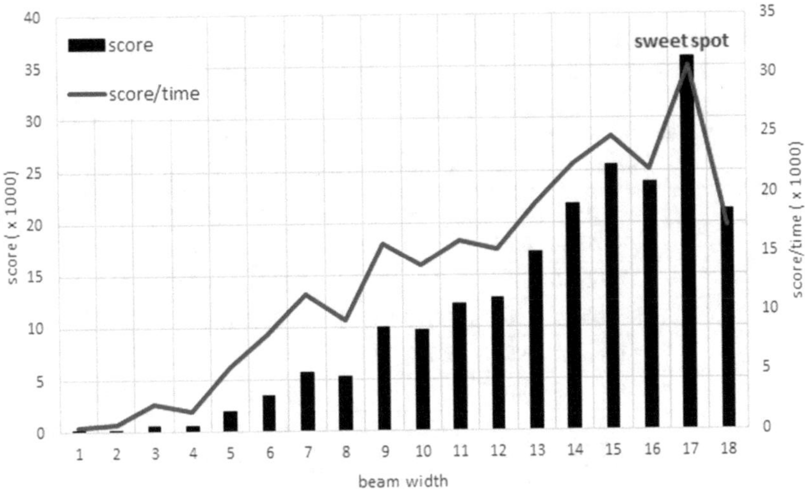

Fig. 5. Representation of the score and the score/search time ratio of the small board test, look-ahead of 2

exclude that a better one would be found at a larger beam width, we can still conclude that a sweet spot exists within the beam width range of the look-ahead of 1.

4.2 Standard Board Test

The experimental setup and the results presentation for the standard board test (20×10) follows the same considerations of the small one, the only significant difference is that in this case the beam width required to test the complete search for the look-ahead of 1 is 34, due to the bigger board.

Table 2 summarizes the results for the standard board test. Clearly, also in this case the search time increases linearly with the beam width. The mean scores reached are much higher than the small board case, in particular in the look-ahead of 2. For this reason we had to stop our investigation at beam width of 10, when the mean score of 1 722 081 was reached. Note that more than 4 millions actions are needed to reach such a score, meaning that several hours are spent to calculate just one of these games. Note also that the reported results are mean values over 36 games, but the best scores achieved by some of the games are much higher. For example, several games reached more than 1 million lines cleared in the look-ahead of 1 case. The best game overall is reached by the look-ahead of 2 case at beam-width of 10, with 12 790 074 lines cleared in total.

Figure 6 shows the results for look-ahead of 1. Compared to the small board, where the sweet spot was found on the first half of the beam widths, in this case the sweet spot is closer to the complete search, at beam width 22. There are also several other good candidates, such as beam width 17, 25, 30 and 32. At beam width 17 it is possible to achieve 71.2% if the best mean score (293 560.3) by

Table 2. Results of the standard board test, look-ahead 1 and 2. Search time measured in ms. Sweet spots are in bold

Beam width	Look-ahead 1			Look-ahead 2		
	Score	Time	Score/time	Score	Time	Score/time
1	80.9	0.27	297.51	68.4	0.40	169.85
2	205.8	0.41	507.42	251.4	0.68	369.60
3	2653.3	0.54	4875.50	10994.4	0.96	11468.18
4	2957.2	0.69	4310.62	13997.5	1.23	11423.66
5	12530.7	0.82	15312.09	206053.9	1.51	136415.92
6	17002.9	0.96	17658.03	275428.7	1.78	154662.32
7	27119.3	1.09	24822.42	455791.0	2.05	222188.68
8	30746.6	1.23	24977.18	593727.9	2.33	255334.98
9	47442.4	1.37	34631.20	1799754.5	2.61	**690123.79**
10	52171.9	1.51	34528.86	1722081.6	2.88	598252.86
11	69253.2	1.64	42210.43	-	-	-
12	63769.8	1.78	35863.34	-	-	-
13	82992.5	1.92	43267.56	-	-	-
14	108700.1	2.05	53000.28	-	-	-
15	88904.8	2.19	40580.24	-	-	-
16	105017.7	2.33	45033.48	-	-	-
17	149921.3	2.46	60857.04	-	-	-
18	144120.9	2.61	55264.19	-	-	-
19	138611.5	2.74	50663.26	-	-	-
20	158969.0	2.87	55303.66	-	-	-
21	152562.5	3.01	50617.99	-	-	-
22	208968.7	3.16	**66137.27**	-	-	-
23	178196.8	3.29	54207.51	-	-	-
24	180857.1	3.42	52905.48	-	-	-
25	229853.9	3.58	64228.47	-	-	-
26	205471.4	3.70	55607.17	-	-	-
27	192488.7	3.83	50280.26	-	-	-
28	179093.8	3.97	45090.08	-	-	-
29	166287.5	4.10	40598.02	-	-	-
30	251980.5	4.24	59439.87	-	-	-
31	188231.8	4.38	42940.84	-	-	-
32	293560.3	4.52	64983.70	-	-	-
33	215709.8	4.65	46377.12	-	-	-
34	253823.0	4.79	52979.96	-	-	-

sparing more than a third of the search time required by complete search (3.16 instead of 4.79 ms per action).

Concerning look-ahead of 2, just the first 10 beam width are explored, but the performance in terms of score is already high. This means that, despite the small beam width, the additional information of one look-ahead pieces in deeply exploited by the search algorithm. Another interesting point is that the identified sweet spot, at beam width 9, shows an astonishing improvement compared to the previous beam width (the mean score of beam 9 is *triple* the mean score of beam 8). This is the steepest improvement among all the other sweet spots in this study.

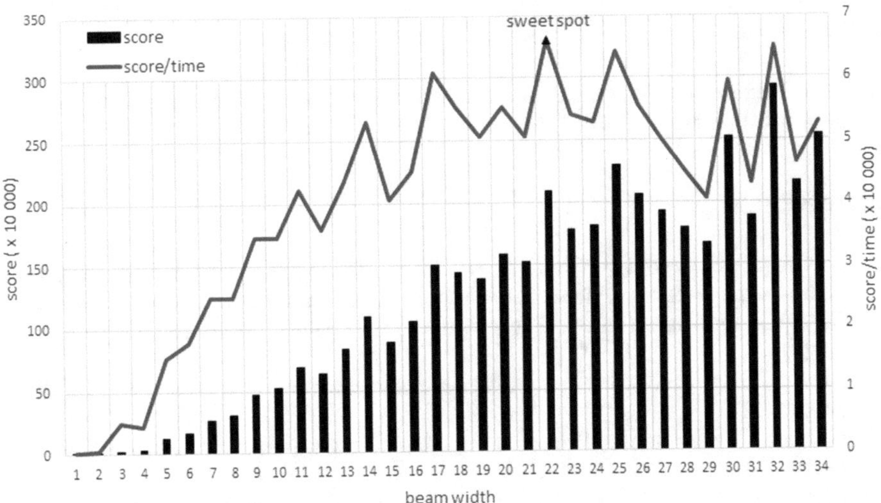

Fig. 6. Representation of the score and the score/search time ratio of the standard board test, look-ahead of 1

5 Conclusion

In this paper we investigated the applicability of an heuristic search method on Tetris, with the aim of reaching comparable scores with respect to a complete search method, but in a fraction of the time. Using a combination of beam search with a strong Tetris heuristic, we run our experiments against a small board configuration and the standard one. By tuning the beam width, in the small board we were able to find a sweet spot that reached up to 67.5% of the best score in less than half of the time compared to the complete search. We reached a similar result in the standard board, where the identified sweet spot reached 71.2% of the best score in two thirds of the time of the complete search. With such sweet spot we reached an average of 208 968 lines cleared, with a pick on the best game at almost one million lines cleared (932 798), results that are at least comparable with others in literature. By exploiting the linear space complexity of beam search, we were to tackle games with 2 pieces of look-ahead, a benchmark that, to the best of our knowledge, was never addressed by other researchers due to high computational costs. In both the small board and the standard one, by limiting the beam width, we were able to reach scores much higher than the look-ahead of 1 case, in comparable time.

Such an approach can scale well also in real world optimization problems scenarios, like online packing or scheduling problems, where typically the number of look-ahead pieces (or jobs) is much bigger than 1. For such cases, complete search approaches are typically out of reach, due to the exponential increase in memory consumption. Thus, the knowledge about pruning strategies in terms of beam width is absolutely crucial for a successful application.

Acknowledgement. Work has been conducted in the scope of the research project Productive4.0 (H2020-ECSEL-GANo.: 737459).

References

1. Al-Omari, A., Arnold, J., Taha, T., Schüttler, H.B.: Solving large nonlinear systems of first-order ordinary differential equations with hierarchical structure using multi-GPGPUs and an adaptive Runge Kutta ODE solver. IEEE Access **1**, 770–777 (2013)
2. Aron, J.: How innovative is apple's new voice assistant, siri? New Sci. **212**(2836), 24-24 (2011)
3. Bellman, R.: Dynamic Programming. Courier Corporation, North Chelmsford (2013)
4. Bertsekas, D.P., Tsitsiklis, J.N.: Neurodynamic Programming. Athena Scientific, Belmont (1996)
5. Böhm, N., Kókai, G., Mandl, S.: An evolutionary approach to tetris. In: The Sixth Metaheuristics International Conference (MIC2005), p. 5 (2005)
6. Burgiel, H.: How to lose at tetris. Math. Gaz. **81**(491), 194–200 (1997)
7. Campbell, M., Hoane, A.J., Hsu, F.H.: Deep blue. Artif. Intell. **134**(1–2), 57–83 (2002)
8. Ciregan, D., Meier, U., Schmidhuber, J.: Multi-column deep neural networks for image classification. In: 2012 IEEE Conference on Computer Vision and Pattern Recognition, pp. 3642–3649, June 2012
9. Da Col, G., Teppan, E.C.: Declarative decomposition and dispatching for large-scale job-shop scheduling. In: Joint German/Austrian Conference on Artificial Intelligence (Künstliche Intelligenz), pp. 134–140. Springer (2016)
10. Da Col, G., Teppan, E.C.: Learning constraint satisfaction heuristics for configuration problems. In: 19th International Configuration Workshop, vol. 8 (2017)
11. Demaine, E.D., Hohenberger, S., Liben-Nowell, D.: Tetris is hard, even to approximate. In: COCOON, pp. 351–363. Springer (2003)
12. Demant, C., Garnica, C., Streicher-Abel, B.: Industrial Image Processing: Visual Quality Control in Manufacturing. Springer, Heidelberg (2013)
13. Edelkamp, S., Schroedl, S.: Heuristic Search: Theory and Applications. Elsevier, Amsterdam (2011)

14. Gabillon, V., Ghavamzadeh, M., Scherrer, B.: Approximate dynamic programming finally performs well in the game of tetris. In: Advances in Neural Information Processing Systems, pp. 1754–1762 (2013)
15. Hameed, M., Qamar, U., Akram, U.: Business intelligence: self adapting and prioritizing database algorithm for providing big data insight in domain knowledge and processing of volume based instructions based on scheduled and contextual shifting of data. In: 2016 Future Technologies Conference (FTC), pp. 1168–1175, December 2016
16. Jannach, D., Resnick, P., Tuzhilin, A., Zanker, M.: Recommender systems – beyond matrix completion. Commun. ACM **59**(11), 94–102 (2016)
17. Mahajan, T., Masih, S.: Enhancing blowfish file encryption algorithm through parallel computing on GPU. In: 2015 International Conference on Computer, Communication and Control (IC4), pp. 1–4, September 2015
18. Nakajima, S., Kitayama, D., Sushita, Y., Sumiya, K., Chandrasiri, N.P., Nawa, K.: Route recommendation method for car navigation system based on estimation of driver's intent. In: 2012 IEEE International Conference on Vehicular Electronics and Safety (ICVES 2012), pp. 318–323, July 2012
19. Nicolo, G., Salido, M.A., Ferrer, S., Giret, A., Barber, F.: A multi-agent approach using dynamic constraints to solve energy-aware unrelated parallel machine scheduling problem with energy-dependent and sequence-dependent setup time. In: COPLAS 2017, p. 31 (2017)
20. Otsuka, R., Kitagami, S., Terashima, Y., Saito, M., Kiyohara, R.: Improved location estimation based on the pressure sensor for car navigation system. In: 2014 IEEE International Conference on Consumer Electronics (ICCE), pp. 276–277, January 2014
21. Park, H., Hwang, S., Won, M., Park, T.: Activity-aware sensor cycling for human activity monitoring in smart homes. IEEE Commun. Lett. **21**(4), 757–760 (2017)
22. Shen, J., Wu, B.: Service configuration knowledge representation and reasoning based on a hybrid approach. In: 2013 International Conference on Service Sciences (ICSS), pp. 224–229, April 2013
23. Silver, D., Huang, A., Maddison, C.J., Guez, A., Sifre, L., Van Den Driessche, G., Schrittwieser, J., Antonoglou, I., Panneershelvam, V., Lanctot, M., et al.: Mastering the game of Go with deep neural networks and tree search. Nature **529**(7587), 484–489 (2016)
24. Silver, D., Schrittwieser, J., Simonyan, K., Antonoglou, I., Huang, A., Guez, A., Hubert, T., Baker, L., Lai, M., Bolton, A., et al.: Mastering the game of Go without human knowledge. Nature **550**(7676), 354 (2017)
25. Szegedy, C., Liu, W., Jia, Y., Sermanet, P., Reed, S., Anguelov, D., Erhan, D., Vanhoucke, V., Rabinovich, A.: Going deeper with convolutions. In: 2015 IEEE Conference on Computer Vision and Pattern Recognition (CVPR) (2015)
26. Szita, I., Lörincz, A.: Learning Tetris using the noisy cross-entropy method. Neural Comput. **18**(12), 2936–2941 (2006)
27. Teppan, E.C., Da Col, G.: Automatic generation of dispatching rules for large job shops by means of genetic algorithms. In: 8th International Workshop on Combinations of Intelligent Methods and Applications (CIMA 2018) (2018)
28. Thiery, C., Scherrer, B.: Building controllers for tetris. ICGA J. **32**, 3–11 (2009)
29. Wikipedia: Alphazero (2018). https://en.wikipedia.org/w/index.php?title=AlphaZero&oldid=822188094. Accessed 31 Jan 2018
30. Yenikaya, S., Yenikaya, G., Düven, E.: Keeping the vehicle on the road: a survey on on-road lane detection systems. ACM Comput. Surv. **46**(1), 2:1–2:43 (2013)

A Comparative Study of the Kinematic Response and Injury Metrics Associated with Adults and Children Impacted by an Auto Rickshaw

A. J. Al-Graitti[1](✉), G. A. Khalid[2], P. R. Berthelson[3], R. K. Prabhu[3], and M. D. Jones[1]

[1] Cardiff School of Engineering, Cardiff University, Wales, UK
[2] Middle Technical University, College of Electrical and Electronic Engineering Techniques, Baghdad, Iraq
[3] Department of Agricultural and Biological Engineering and the Centre for Advanced Vehicular Systems, Mississippi State University, Starkville, USA

Abstract. Child pedestrian-vehicle collisions are a leading public health issue and road safety challenge worldwide with injury risk four times greater in developing countries. The auto rickshaw is a common mode of urban transport in many developing countries and poses a significant risk to pedestrian safety. A comparative study was conducted of the kinematic responses and injury metrics associated with both adults and children impacted by an auto rickshaw. A Finite Element model of an auto rickshaw simulated the impact with two pedestrian Anthropometric Test Devices (ATDs), a 50th percentile adult male Hybrid III and a six-year-old child ATD at velocities between 5 and 40 km/h at front, side and rear pedestrian positions, relative to the vehicle centreline. The simulation outcomes were correlated against reported injury metrics, Head Injury Criterion (front, side and rear) and Neck Injury Criterion in front impacts and compared with pedestrian size. The results suggest that child pedestrians are subject to a relatively high risk of neck injury at velocities in excess of 15 km/h and head injury at velocities in excess of 20 km/h.

Keywords: Child pedestrian · Auto rickshaw · Head and neck injury criterion

1 Introduction

Children are particularly vulnerable road users as passengers, bicyclists and pedestrians. As pedestrians they may walk, play or even work on roads in close proximity to vehicles. All these interactions, together with a variety of other risk factors associated with childhood, increase the vulnerability of children to road traffic injury [1]. Trauma is the major cause of paediatric injury [2] and statistically, child pedestrians are the group most likely to be involved in vehicle-pedestrian impacts. Males aged between 5 and 14 years and females between 10 and 19 years have the highest fatality rates amongst all pedestrians. In developed countries between 5% and 10% of children suffering road traffic injuries are pedestrians, however, in developing countries the

K. Arai et al. (Eds.): CompCom 2019, AISC 997, pp. 424–443, 2019.
https://doi.org/10.1007/978-3-030-22871-2_29

proportion ranges between 30% and 40% [1]. A high fatality rate is associated with bad practice, such as a walking in roads [3, 4]. The World Bank predicts that 1.2 million pedestrians die in road accidents annually, of which, 35% are children [5]. Deaths and injuries from road traffic impacts are predicted to increase across the world by 67% between 1990 to 2020 [6], largely as a result of a growth in motorisation and associated trends towards urbanisation and globalisation. Roads have always been unsafe places for children, however, the rapid increase in traffic and a shift towards transportation on roads makes the issue particularly pressing [5].

Childhood injury is a leading public health issue [7] and although the consequence of child pedestrian injuries has been reported for many years [8], there still is insufficient progress in the post kinematic response of the child pedestrians, injury metrics and injury prevention measures [4, 9]. Significant effort has been focused on the adult pedestrian kinematics and injury mechanics cause by four-wheel vehicle collisions [9, 10].

Many aspects, such as post kinematic response, injury distribution and injury severity, vary, according to the pedestrian age [11, 12] and vehicle type [13–16]. Other variations may include anatomical or biomechanical aspects, such as body proportion, mass distribution, skin resilience and skull bone elasticity. These variations make the child less resilient to impact trauma [17–20] and result in adults and children having different injury thresholds.

The auto rickshaw whilst being a common urban transport mode in many developing countries, has received relatively little consideration [21, 22] with respect to pedestrian injury risk. Those limited efforts have focused on adult pedestrian-auto rickshaw collisions [23], no research has been conducted on child pedestrians. Accident data of vehicle-adult pedestrian impacts suggest that head and neck injuries are the most common and potentially life threating [24, 25]. Vehicle-child pedestrian road accident statistics, based on hospital data in four developing countries (Bangladesh, Colombia, Pakistan and Egypt) indicate that children under 12 years suffered concussion or other head injury, cuts, bruises, open wounds, fractures and sprains in more than 25% of all cases [26]. Head injuries have a high incidence and frequently, the worst neurological outcomes in pedestrian accidents, compared with other types of road accidents [27, 28]. Furthermore, the European Enhanced Vehicle-Safety Committee (EEVC) carried out a study of 321 child pedestrian accidents and found that 36% of the reported AIS1 + injuries and 72% of the AIS4 + injuries were head injuries [29]. It was reported that the vehicle windscreen and windscreen screen frame were the most frequently impacted automotive components [30, 31].

It has been previously concluded that impact velocity and location has the greatest influence on pedestrian injury risk [32–34]. Road traffic accident data suggests that 14% of pedestrians are impacted to their anterior surface, 8% to their posterior face, almost 73% are impacted laterally and 5% are impacted under unknown conditions [35]. In addition, 6% of the pedestrians were impacted whilst standing with their legs together, 5% were impacted with legs apart laterally, 65% were found to be in a gait stance and 24% were in an unknown stance [36].

As a result of regulatory and ethical concerns, paediatric cadavers are infrequently used for biomechanical experiments. Therefore, numerical techniques, especially the use of finite element (FE) models are applied to understanding the response and injury mechanisms of children in road traffic accidents and can provide a detailed biomechanical response and improve injury prediction in children [37]. Therefore, a six-year-old child ATD (6YO-child) was employed in pedestrian-auto rickshaw impact tests [38].

This study investigates the influence of auto rickshaw frontal geometry, impact velocity, impact positions and pedestrian size on the kinematic and injury mechanics at the vehicle centreline using two different ATDs, the 50% percentile Hybrid III adult male and the 6YO-child Hybrid III.

2 Methodology

2.1 Simulation Setup

Auto rickshaw-pedestrian impact simulations and analyses were conducted with two Hybrid III ATDs, an adult male and a 6YO-child model in LS-DYNA software (Livermore Software Technology Corporation (LSTC)). The auto rickshaw FE model was adopted from AL-Graitti et al. 2017 [30]. The ground surface was modeled as a rigid body. The contact between the auto rickshaw and the ATD was defined by AUTOMATIC_SURFACE_TO_SURFACE. AUTOMATIC_GENERAL, which was utilised to define the interaction between the auto rickshaw tyre and the ground along the moving path [39]. In addition, the contact friction coefficient, between the pedestrian ATD parts and auto rickshaw was 0.65 and the coefficient between the ATD feet/ground and vehicle was 0.7 [10, 11, 28], shown in Fig. 1.

The auto-rickshaw-pedestrian impact simulations were conducted at the vehicle centreline with three major pedestrian impact positions, front, rear and a side walking posture (facing laterally with the right leg forward, without a walking velocity and the left arm positioned backward), see Fig. 2. Vehicle impact velocities were 5, 10, 15, 20, 25, 30, 32, 35 and 40 km/h. In addition, LS-DYNA software was used to collect the post-collision output data. To reduce the uncertainty in time shifts, when comparing the simulations, a Butterworth pre-filter with a frequency of 180 Hz was applied to collect the head and neck loading response for adult pedestrian, as recommended by the National Highway Traffic Safety Administration (NHTSA) [40, 41]. Whilst a frequency of 1000 Hz and 60 Hz were applied to collect the head and neck loading respectively for 6YO-child pedestrian as defined by the LSTC and National Crash Analysis Centre (NCAC) [42].

2.2 Comparative Study

A comparative study was conducted on the post kinematic response, injury criterion and injury risk levels of adult and paediatric pedestrians impacted by the auto-rickshaw. Primary impacts were investigated utilising the adult and 6YO-child, impacted at the centreline of an auto rickshaw. The influences of impact velocity, pedestrian size and position were examined at velocities between 5 and 40 km/h and injury parameters calculated using LS-DYNA software from primary impacts. Head injury criteria (HIC_{15}) for the front, rear and side impacts and neck injury criteria (N_{ij}) were calculated for frontal impacts. The Abbreviated Injury Scale (AIS) and injury thresholds were compared against risk level, to indicate the threat to life.

2.3 Selected Injury Parameters and Injury Risk Level

The Head Injury Criterion (HIC_{15}) and Neck Injury Criterion (N_{ij}) were applied to assess the risk of pedestrian head and upper neck injury using the injury thresholds shown in Table 1. HIC injury thresholds for adults are 1000 for frontal and rear impacts and 800 for side impacts, corresponding to a risk of severe head injury (AIS4+), which is associated with a 16% and 18% risk, respectively, of bone structure deformation soft tissue damage, skull fractures, brain contusions and lacerations and/or brain bleeding [43–46]. The injury risk level could predict, for example, whether a skull fracture is a simple linear crack or a depressed fracture with a bone fragment larger than the skull thickness, which can be depressed into the cranial cavity and risking damage to the brain [47].

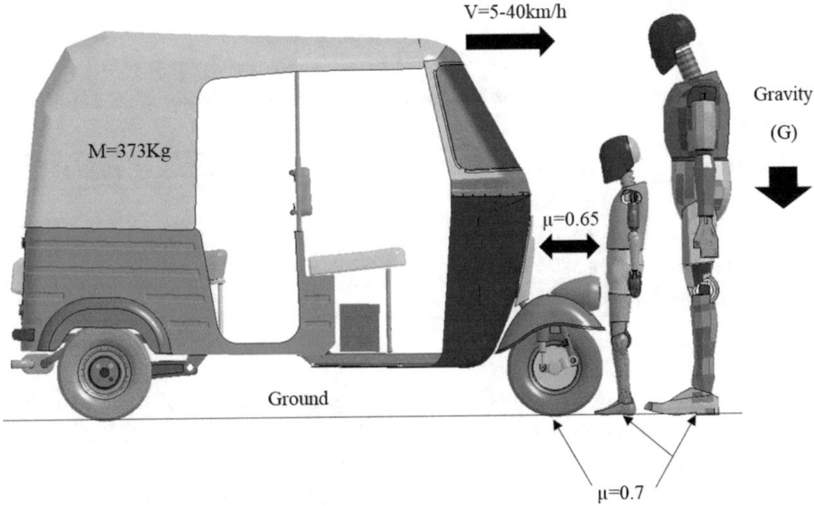

Fig. 1. Vehicle-pedestrian impact simulation setup.

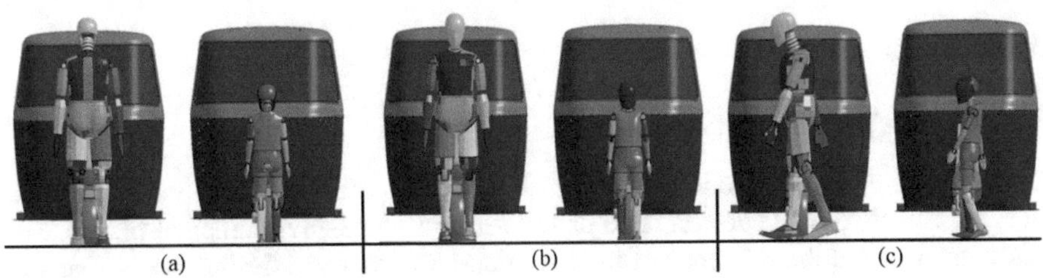

Fig. 2. Auto-rickshaw–adult and 6YO-child pedestrian impact simulations at the vehicle centreline at different positions; (a) front (b) rear and (c) side.

Table 1. Head and upper neck injury thresholds.

Impact position	Adult head injury thresholds (HIC_{15})	6YO-child head injury thresholds (HIC_{15})	Adult and child upper neck injury thresholds (N_{ij})
Front	1000	700	1
Rear	1000	700	–
Side	800	700	–

Whilst an (HIC_{15} = 700) for 6YO-child indicates a 23% risk of serious head injury (AIS3+) [48] or a 5% risk of severe head injury (AIS4+), corresponding with brain injury or skull fracture [49], it is also associated with head injuries such as injuries occurring at the base of the skull, or compound, comminuted or depressed skull fractures [48, 50]. Upper neck injury criteria (N_{ij}) in the frontal impacts for the adults and six-year-old child were calculated, based on the combination of the axial force and bending moment at the occipital condyles. The critical intercept values for adults and child are different, as shown in Table 2.

Table 2. Intercept load values for the 50th percentile hybrid III male and 6YO-child ATDs for frontal impacts

Load case	Intercept values of adult [51]	Intercept values of child [51]
Extension	−135 Nm.	−37 Nm.
Flexion	+310 Nm.	+93 Nm.
Tension force	+6806 Nm.	+2800 Nm.
Compression force	−6160 N.	−2800 N.

The upper neck injury thresholds (N_{ij} = 1), for adults and a 6YO-child indicate a 22% risk of serious neck injury (AIS3+) [40]. It is associated with a risk of rupture of small blood vessels of the occipital condylar joints, alar ligament rupture with symptoms of neck pain and stiffness, shoulder weakness, dizziness, headache, memory loss and damage to the spinal cord, resulting from cervical vertebrae compression (disc rupture and nerve root damage, damage to the brainstem and even death [48, 52–54].

3 Results

3.1 Pedestrian Kinematic Response

Various factors may influence pedestrian kinematics, injury patterns and injury risks such as, vehicle type, pedestrian and driver behaviour and environmental factors. This study investigated the influence of the auto rickshaw front-end design, pedestrian size, impact velocity and impact positions on post kinematic injury metrics and injury risk. The dynamic response of the adult and 6YO-child shows that the windscreen and windscreen frame are the most frequently contacted area with the pedestrian head, see Fig. 3.

Head Contact Locations, Angles, and Contact Duration

Head contact location and head contact angle, the angle between the direction of the head during the impact and the ground reference level [55], varied with pedestrian size and impact position, see Fig. 3 and Table 3. Head contact period is the time duration of the first contact of the head with vehicle components [56]. The head contact period for both the adult and child decreased with impact velocity in all impact positions, as shown in Fig. 4. However, in most cases, the head contact time for the 6YO-child was less than the adult. During all impact simulations, no contact occurred between the head and vehicle components at 5 km/h.

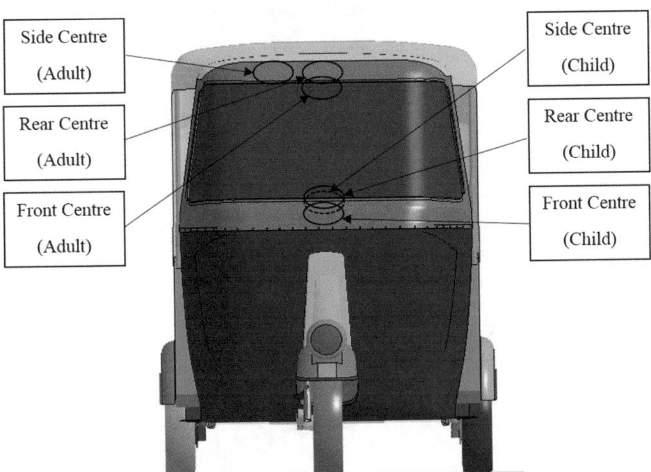

Fig. 3. Head contact locations of adult and 6YO-child at the auto rickshaw centreline.

Table 3. Head contact angles during various impact positions.

Impact position	Head contact angles of adult (degrees)	Head contact angles of 6YO-child (degrees)
Front	31	20
Rear	18	19
Side	45	20

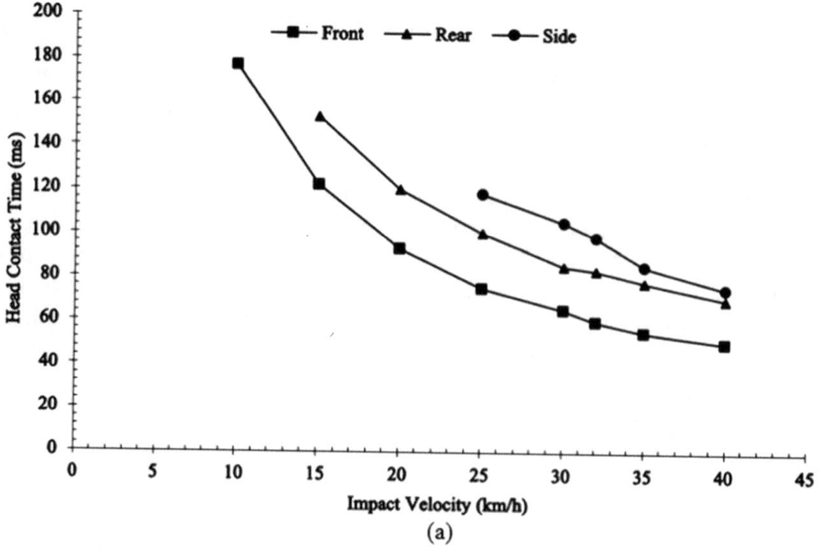

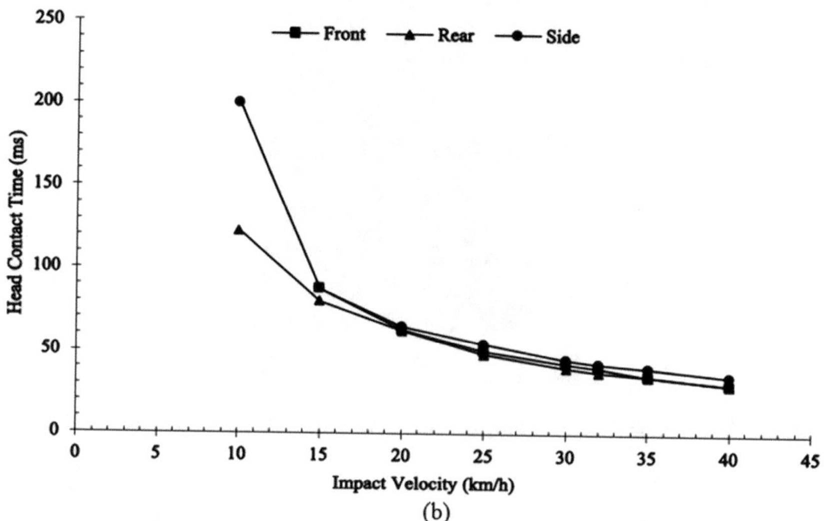

Fig. 4. Head contact time at the vehicle centreline in different impact positions (a) adult and (b) 6YO-child.

Pedestrian kinematics of adult and 6YO-child ATDs were simulated at the vehicle centreline for the front, rear and side impact positions. Dynamic responses of 30 km/h impacts are provided at the vehicle centreline in Fig. 5. Tables 4 and 5 show pedestrian-auto rickshaw interactions and contact time sequences during the impact simulations. Figure 5(a) shows the kinematic response of impacts at the vehicle centreline for front, rear and side adult pedestrian impact orientations and Fig. 5(b) shows the kinematic response of impacts at the vehicle centreline for front, rear and side child pedestrian impact orientations. AL-Graitti et al. 2017 reported adult pedestrian-vehicle impacts [30] with initial contacts occurring between the adult pedestrian knees and the front headlamp in the front impact position, followed by lower leg contact with the mudguard and the lower torso impacting with the lower part of the windscreen frame and the chest with the windscreen; the head simultaneously collided with the upper windscreen and upper part of the windscreen frame, see Fig. 5(a). The 6YO-child's first interaction was between both upper legs and the mudguard and the lower torso with the headlamp. Lower leg contact was produced at the front tyre. Finally, the head impacted with the lower region of the windscreen frame, as shown in Fig. 5(b).

During the adult rear impact, the first contact was between the back of both knees and the headlamp. Lower leg contact was produced at the mudguard and the lower torso impacted the frontal vehicle edge. The right and left hands produced glancing interactions with the lower frontal sheet plate and the upper torso contacted the windscreen. The final contact was between the head and the upper vehicle windscreen frame, see Fig. 5(a). During the child rear impact, the lower torso struck the headlamp, followed by the knees contacting with the mudguard. The lower leg impacted with the front tyre, followed by the upper leg with the headlamp and the upper torso with the lower front sheet plate. Finally, the head contacted with the lower part of the windscreen and windscreen frame components, see Fig. 5(b).

During the adult side impact, the first contact was between the right knee and the headlamp, producing ankle pro and supination. Subsequently, the right arm pushed behind the upper torso and the upper leg rotated, consistent with the frontal vehicle geometry and the upper right arm contacted with the windscreen. Finally, the head impacted with the upper side of the windscreen frame, as shown in Fig. 5(a). During the child side impact, the initial contact occurred between the right upper leg and the mudguard, followed by the upper part of the right upper leg and the lower torso contacting with the headlamp. Simultaneously, the right lower leg struck the front tyre and the right arm pushed behind the upper torso and collided with the frontal sheet plate. Finally, the head collided with the lower part of the windscreen and windscreen frame, as shown in Fig. 5(b).

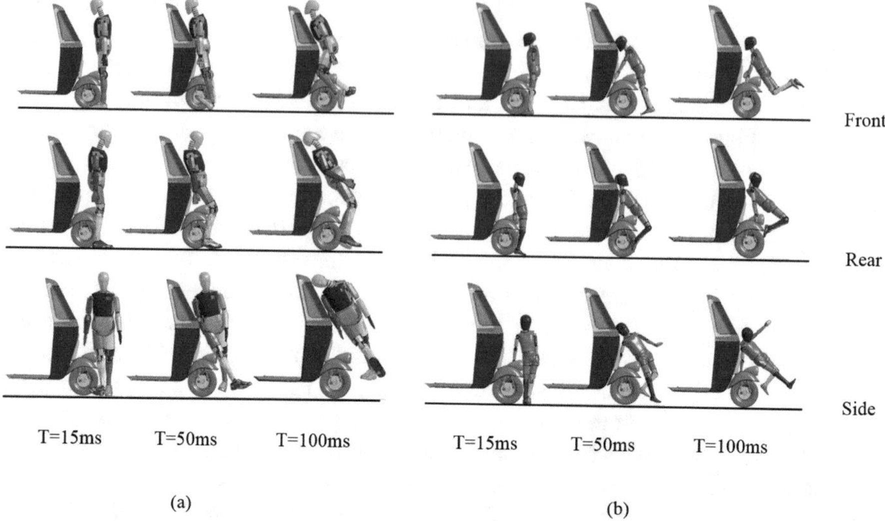

Fig. 5. The kinematic response of (a) adult and (b) 6YO-child pedestrian ATDs impacted at the vehicle centreline at the front, rear and side positions at 30 km/h.

Table 4. The sequence of adult pedestrian-vehicle interaction and contact time at vehicle centreline.

Impact position	Vehicle contact region	Pedestrian contact region	Contact period (ms)
Front	- Front headlamp - Mudguard - (Windscreen frame) - Windscreen - Upper windscreen/upper frame	- Knees - Lower leg - Lower torso - Upper torso, chest - Head	0–20 0–35 30–75 35–75 50–65
Rear	- Front headlamp - Mudguard - (windscreen frame) - Windscreen - Upper windscreen frame	- Knees - Lower leg - Lower torso - Upper torso - Head	0–17 2–47 27–54 62–72 82–92
Side	- Headlamp - Lower front plate - Windscreen - Windscreen	- Right knee - Right hand and arm - Upper right arm - Head	0–49 22–27 52–85 100–105

Table 5. Sequence of 6YO-child pedestrian-vehicle interaction and contact period at the vehicle centreline.

Impact position	Vehicle contact region	Pedestrian contact region	Contact period (ms)
Front	- Mudguard	- Lower upper leg	0–47.5
	- Front headlamp	- Upper leg	5–40
	- Front headlamp	- Lower torso	5–100
	- Frontal tyre	- Lower leg	7.5–27.5
	- Lower windscreen frame	- Head	42.5–55
Rear	- Front headlamp	- Lower Torso	0–75
	- Mudguard	- Knees	7.5–32.5
	- Frontal tyre	- Lower leg	7.5–35
	- Headlamp	- F- Upper legs	15–32.5
	- Lower front sheet plate	- Upper torso	40–95
	- Lower windscreen/frame	- Head	40–50
Side	- Mudguard	- Right upper leg	0–20
	- Headlamp	- Right upper leg	2.5–27.5
	- Headlamp	- = - Lower torso	5–95
	- Frontal tyre	- Right lower leg	5–15
	- Lower windscreen/frame	- Head	45–60

3.2 Head Injury and Injury Risk Level

The HIC values for both adult and paediatric contacts increase considerably with vehicle impact velocity. Figure 6(a) suggests a relationship between the head injury criteria (HIC) and vehicle impact velocity in the front impact position for adult and child ATDs. The HIC values for the adult exceeded the threshold (HIC_{15} = 1000) at 30 km/h. Whilst the 6YO-child exceeded the threshold (HIC_{15} = 700) at 20 km/h.

Figure 6(b) demonstrates an increase in HIC with vehicle impact velocity in the rear impact position for adult and child ATDs. The adult HIC values exceeded the HIC_{15} threshold (1000) at 25 km/h and the 6YO-child exceeded the threshold (700) at 20 km/h.

Figure 6(c) shows the HIC and injury risk increase with vehicle impact velocity in the side impact position for adult and child pedestrians. The HIC values for an adult did not exceeded the threshold (HIC_{15} = 800) between 5 and 40 km/h, while the 6YO-child exceeded the HIC_{15} threshold (700) at 20 km/h.

3.3 Upper Neck Injury and Injury Risk Level in Front Impacts

Neck injury risk for both adult and 6YO-child is represented by the Neck Injury Criteria (N_{ij}). N_{ij} values were determined by selecting the worst load condition, from each of the different collision velocities, see Fig. 7. For the adult pedestrian, compression-extension (N_{ce}) was the worst case at 5 km/h and 10 km/h, as shown in Fig. 7(a). Tension-extension (N_{te}) was selected as the worst load case at impact velocities between 15 km/h and 48 km/h. While, for the 6YO-child, the compression-extension (N_{ce}) was the worst case at 5 km/h. Tension-extension (N_{te}) was selected as the worst possible load case at impact velocities from 10 km/h to 40 km/h, see Fig. 7(b).

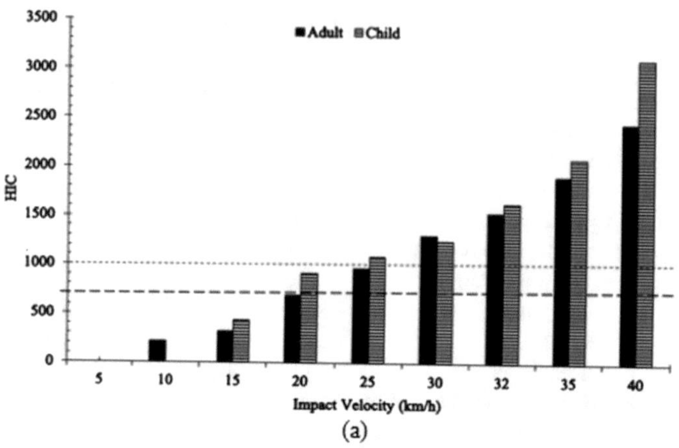

(a)

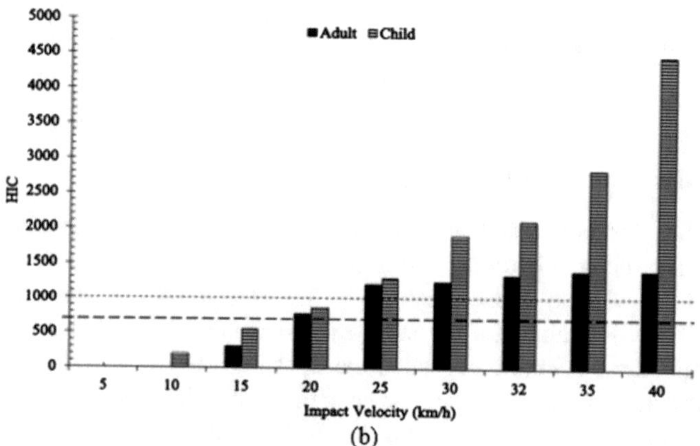

(b)

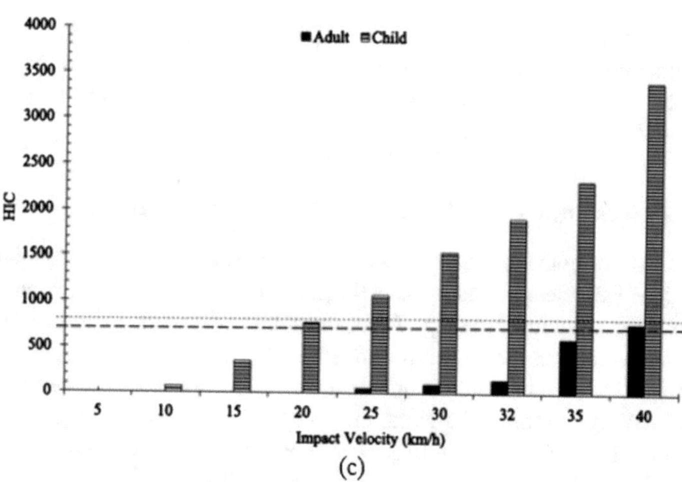

(c)

Fig. 6. HIC for impacts at the vehicle centreline in (a) front, (b) rear and (c) side impact positions for adult and 6YO-child pedestrians.

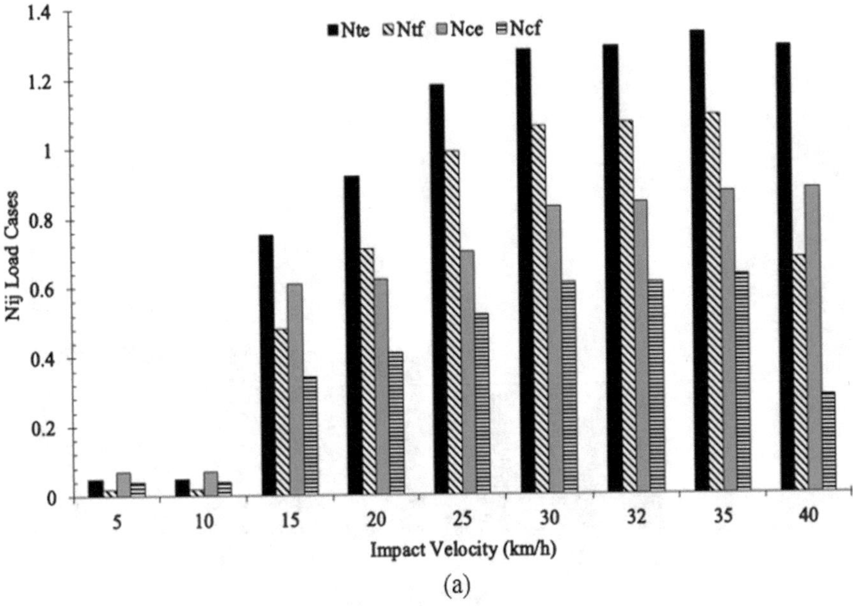

(a)

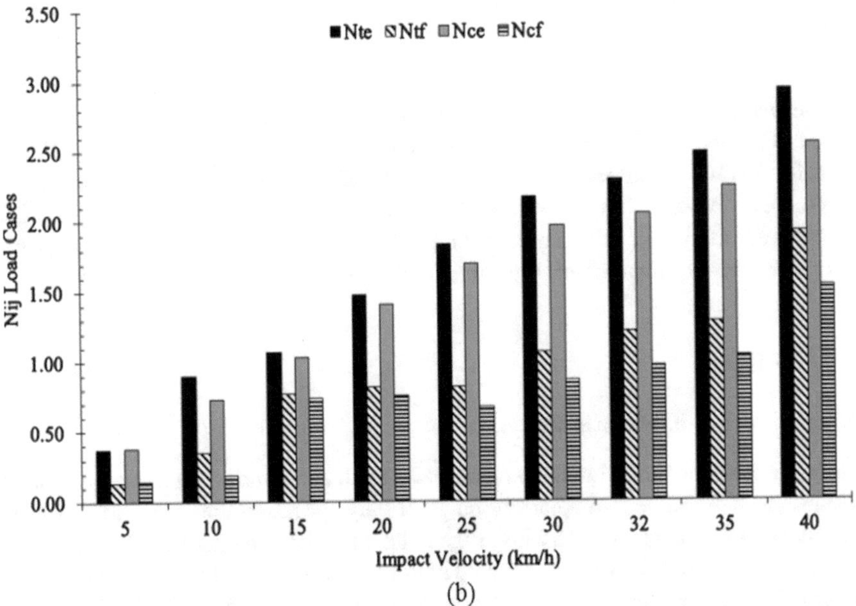

(b)

Fig. 7. Upper neck load conditions for pedestrian (a) adult and (b) 6YO-child, impacted at the vehicle centreline.

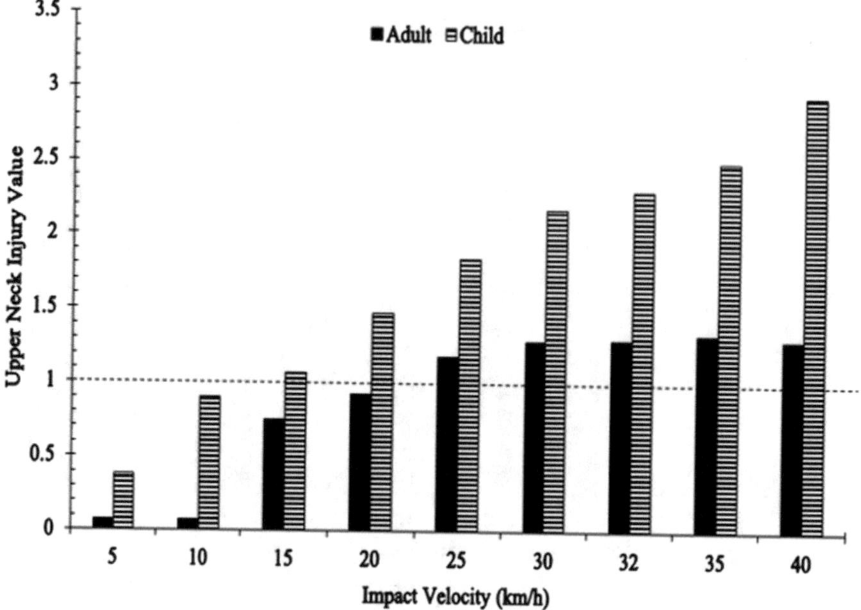

Fig. 8. Upper neck injury values for adult and 6YO-child pedestrians impacted at the vehicle centreline.

The maximum load conditions for the front impact for the adult and child were chosen to provide a worst case upper neck load scenario, see Fig. 8. N_{ij} for both adult and child ATDs was observed to increase noticeably with vehicle impact velocity. For the adult, Nij exceeded the threshold, (N_{ij} = 1) at 25 km/h, whilst the N_{ij} for the child exceeded the threshold at 15 km/h, which indicates a significantly higher upper neck injury vulnerability than an adult during frontal impacts.

4 Discussion

4.1 Pedestrian Kinematic Response

This study examined the kinematic response of adult and child pedestrians impacted by an auto rickshaw, subject to a narrow range of very specific initial conditions, since its objective was to computationally investigate the unfamiliar impact biomechanics of the auto rickshaw. The 50th percentile male Hybrid III and 6YO-child were integrated with an auto rickshaw-human impact FE model at a range of pre-impact velocities between 5 km/h and 40 km/h. The effect of the pedestrian impact position and pedestrian size were investigated by varying three orientations (front, rear and side; relative to the vehicle) with two standing and gait postures, for a total of six simulated mechanisms. Pedestrian kinematics and injury risks were evaluated and compared across all simulations.

Generally, significant momentum was transferred to the pedestrians during impacts at the auto rickshaw centreline in the forward direction. Whilst the first interaction with the pedestrian occurred most frequently at the lower limbs, noticeably below the pedestrian's centre of gravity and the lower frontal vehicle components, nevertheless, pedestrian size still had a significant influence on the dynamic response and injury mechanics resulting from the lower centre of gravity compared to the adult. In addition, different mass, length and anatomical features produced numerous kinematic perturbations at the vehicle centreline, thus, influencing the subsequent dynamic response of the pedestrian.

In summary, contact patterns of the pedestrian body region with the vehicle components producing from the dynamic response of the ATD impacted by the auto rickshaw varied with impact position, pedestrian size and impact velocity, see Tables 4 and 5.

Head Contact Locations, Angles, and Contact Duration

With reference to head impacts across all impact simulations, the most frequently impacted auto rickshaw regions were the windscreen and windscreen frame for both adult and 6YO-child pedestrians. Variation in pedestrian impact position and pedestrian size changed head contact location and head contact angle, as shown in Fig. 3 and Table 3. Head contact duration was investigated at different impact velocities and different impact positions with adult and child pedestrian ATDs, see Fig. 4. Three pedestrian orientations were investigated for impacts at the centreline of the auto rickshaw, frontal, rear and side impacts produced no head impacts at 5 km/h for adult and 6YO-child pedestrians.

Head contact duration was observed to decrease with increasing impact velocity for all impact positions as a result of increasing head accelerations for both adult and child pedestrians, see Fig. 4. The numerical results of the head contact duration illustrate a near agreement with the velocity-head impact duration relationship, outlined in previous studies [9, 10, 30, 55, 57]. Although, in most cases, the child pedestrian head contact period was less than that of the adult pedestrian. This outcome is consistent with a previous study, which investigated the influence of different vehicles, such as Sport Utility Vehicles (SUV) and Sedans, on adult and child pedestrians during impacts [11]. These findings may be a result of the lower height and centre of gravity of the child compared to that of the adult, which leads to a reduced head contact duration. Thus, the head contact duration of the auto rickshaw front was observed to vary widely as a result of the vehicle impact velocity, impact position and pedestrian size. Additionally, these parameters may be useful for automotive designers to improve passive safety and active safety, such as pop up panels and/or airbag technologies [22, 55, 58].

4.2 Head Injury and Injury Risk Level

Variations in pedestrian impact position and pedestrian size produced differences in head impact angles. HIC and head injury risk level changed significantly with impact position, pedestrian size and impact velocity.

The 6YO-child pedestrian impacted by an auto rickshaw produced high HIC values compared to the adult impacts, which exceeded the injury risk threshold for adults (HIC_{15} = 1000 for front and rear impacts and 800 for side impacts). The thresholds correspond with a risk of severe head injury (AIS + 4) between 16% and 18% and are associated with bone structure deformation, soft tissue injury, skull fractures, brain contusions, lacerations and/or brain bleeding [43–46]. Increasing injury severity can be used to predict whether a skull fracture may be linear or depressed and or comminuted, producing ingress into the cranial cavity and subsequent damage to the brain [47]. A HIC_{15} of 700 for a 6YO-child indicates a 23% risk of serious head injury (AIS3+) [48] or a 5% risk of severe head injury (AIS4+), corresponding with brain injury [49], it associates with head injuries, such as skull injuries occurring at the base of the skull, or compound, comminuted or depressed skull fractures [48, 50].

Head and upper neck injury parameters and injury risk increase with the vehicle impact velocity. HIC values were produced from the pedestrian impacts. For the adult pedestrian, during frontal impacts, no head contact occurred at 5 km/h. Head contact was produced at 10 km/h and the corresponding HIC values were observed to increase between 10 and 40 km/h to produce HIC values of between 222 and 2449, see Fig. 6 (a). The values exceed the head injury threshold (HIC_{15} = 1000) between the velocities of 30 and 40 km/h. Whilst for the 6YO-child pedestrian, no head contact occurred at 5 and 10 km/h. Head impacts did occur at 15-40 km/h with corresponding HIC values of between 433 and 3108, as shown in Fig. 6(a). The results highlight that the HIC values for the adult exceed the threshold (HIC_{15} = 1000) at 30 km/h and the 6YO-child exceeded the threshold (HIC_{15} = 700) at 20 km/h.

Rear impacts produced no adult head contacts at 5 and 10 km/h, whilst impacts at 15-40 km/h, produced corresponding HIC values of 309-1693, as shown in Fig. 7. The outcomes emphasise that the HIC values exceed the threshold (HIC_{15} = 1000) at 25 km/h or greater, as shown in Fig. 6(b). However, no head contact occurred for the 6YO-child at 5 km/h, though, impacts at 10-40 km/h produced HIC values of 201–4470, as shown in Fig. 6(b). Hence, the results emphasise that the HIC values for the adults exceed the HIC_{15} threshold (1000) at 25 km/h, whilst the HIC values for the child exceed the threshold (700) at 20 km/h.

For side impacts, no head contact was produced for the adult pedestrian between 5 km/h and 20 km/h. The HIC values were lower compared to the front and rear impacts, though increased steadily. Impact velocities of 25–40 km/h produced HIC values of 47–772, see Fig. 6(c). While, no head interaction occurred for the 6YO-child at 5 km/h, impacts at 10–40 km/h corresponded with HIC values of between 70–3418, see Fig. 8. Consequently, the results emphasise that the HIC values for the adult did not exceed the threshold (800) between 5 and 40 km/h, whilst the HIC values for the child exceeded the threshold (700) at 20 km/h.

Thus, a 6YO-child may sustain a serious head injury at an impact velocity of 20 km/h or greater in all impact positions about the vehicle centerline, a finding which is in agreement with the outcomes of field studies [25, 59]. In addition to the vehicle impact velocity influencing the response, impact position and pedestrian size were observed to influence the kinematics, head contact duration, location, injury severity and body region impacted by the vehicle; a finding that is consistent with Liu and Yang, 2003 [9]. There is, however, a significant difference between the HIC values and injury risk between the 6YO-child and the adult; findings that show agreement with Venkatason et al. 2014 [12]. The reason for the injury risk variation between children and adults may be as a result of the disproportionately larger head of the child, the relatively soft skin or elastic skull bones. In addition, the heavier head mass coupled with weaker neck supporting structures, may form the basis for the higher incidence of head injury. Compared to the adult, these features make the head of the child less resilient to impact trauma [17, 19]. In addition, the child's head may be more physiologically vulnerable to injury than adults. It is noteworthy that the frontal geometry of the auto rickshaw, which is almost similar to flat-front vehicles, light truck vehicles, vans, etc. produces a higher injury risk than passenger cars [14–16].

4.3 Upper Neck Injury and Injury Risk Level in Front Impacts

The upper neck is a vulnerable region for injury which is associated with head movement. Upper neck injury risk for the adult and child is represented by the Neck Injury Criterion (N_{ij}), which is produced by considering the combination of forces and moments, measured at the occipital condyles. Neck injury criteria (N_{ij}) is applied using neck injury thresholds (N_{ij} = 1) [40]. This specifies a 22% risk of serious neck injury (AIS3+) [40], which corresponds to the rupture of small blood vessels of the occipital condylar joints, alar ligament rupture, damage to spinal cord (disc rupture and nerve root damage), brainstem injury and even death [40, 54]. N_{ij} values for the adult and child were calculated based on the combination of axial force and moment. For the adult pedestrian it was observed that the compression-extension load case (N_{ce}) represents the worst load case of N_{ij} at both 5 km/h and 10 km/h; producing values of 0.16 and 0.20, respectively, see Fig. 7(a). The tension-extension load case (N_{te}) denotes the maximum neck load, as shown in Fig. 7(a); with N_{ij} values at 15, 20, 25, 30, 32, 35 and 40 km/h corresponding to 0.65, 0.93, 1.09, 1.12, 1.14, 1.20 and 1.30, respectively. The neck injury load cases (compression-flexion (N_{cf}) and tension-flexion (N_{tf})) have been excluded from the upper neck injury assessment of the frontal impact, since their values were lower than the maximum-case injury loads (N_{ce}) and (N_{te}). The collision velocity exceeded the upper neck injury threshold (N_{ij} = 1) at 25 km/h, see Fig. 8. Whilst for the 6YO-child, it appeared that the compression-extension load case (N_{ce}) represents the worst load case of N_{ij} at 5 km/h, corresponding to a value of 0.38, see Fig. 7(b). The tension-extension load case (N_{te}) represents the worst neck load, with N_{ij} values at 10–40 km/h of between 0.90 and 2.93, as shown in Fig. 8. The upper neck injury load conditions (compression-flexion (N_{cf}) and tension-flexion (N_{tf})) again have been omitted from the neck injury valuation of the frontal impact, since their values were less than the worst-case neck injury loads, (N_{ce}) and (N_{te}). The velocity of impact exceeds the injury threshold (N_{ij} = 1) between 15 and 40 km/h. Therefore, the upper neck load cases and N_{ij} values for the frontal impacts are considerable, varying with impact velocity and pedestrian size at the vehicle centreline. In addition, the simulation results in Figs. 7 and 8 illustrate that the upper neck injuries (N_{ij}) will be more predominant than head injuries. What represents a harmless impact velocity for upper neck injury will not be the same as a head injury. In addition, in most impact velocities, head and upper neck injury values for the 6YO-child are higher than the adult pedestrian, thus a child may be seriously injured at impact velocities below those expected to produce injury when impacted by the auto rickshaw at the frontal position at the vehicle centreline.

5 Limitations and Future Work

In all of the auto rickshaw-to-pedestrian impact simulations, both adult and child pedestrians were located at the vehicle centreline. In an unpublished study, when the adult pedestrian was impacted at a position offset from the vehicle centerline, the kinematic responses and injuries were significantly different [30]. Thus, it is apparent that the effects of other vehicle contact regions on the kinematic responses and injury metrics requires investigation.

6 Conclusion

This study provides an improved understanding of auto rickshaw–pedestrian impacts related pedestrian kinematics and associated head and neck injury risks for impacts with the vehicle centreline for both adult and 6YO-child ATDs. Key findings are summarised as follows:

1. The kinematic response is remarkably sensitive and influenced by many factors, such as pedestrian size, impact velocity and impact position.
2. Reducing the traveling velocity of the auto rickshaw in urban areas appears to be confirmed as effective for mitigating the injury severity of pedestrians in vehicle impacts.
3. The simulation results show that the injury severity of the head and upper neck of pedestrians vary significantly with body size, vehicle impact velocity and impact position. Additionally, the upper neck load cases vary with impact velocity and pedestrian size.
4. Vehicle windscreen and windscreen frame provide the greatest risk of injury.
5. Child pedestrians are exposed to higher injury risks to the head and upper neck than adults.
6. Head and neck injuries could occur independently of each other. Even in the absence of a head impact, neck injury risk values were observed to exceed the injury threshold, as a result of vehicle frontal geometry and pedestrian kinematics response.
7. Velocities below 20 km/h are associated with a low head injury risk, whilst velocities below 15 km/h are required to provide the same low risk for neck injuries.

To decrease the injury risk level and increase pedestrian safety during primary impacts an auto rickshaw traveling velocity regulation for urban areas requires implementation. In addition, engineering solutions need to be developed, enhancing the vehicle frontal geometry, material response for the frontal vehicle components and the retro-fitting of injury mitigation technologies to the relevant auto rickshaw impact regions, which are the subject of the highest injury risk to pedestrians.

Acknowledgements. The authors wish to thank The Higher Committee for Education Development (HCED) in Iraq and the Ministry of Construction, Housing, Municipalities and Public works for their financial support of this work.

References

1. World Health Organization (W.H.O).: Youth and Road Safety. Geneva-Switzerland (2007)
2. Rivara, F.P.: Injuries in the United States: current status of the problem, potential interventions, and future research needs. J. Am. Med. Assoc. **144**(985), 25–29 (1990)
3. Hyder, A.A., et al.: A new challenge to child and adolescent survival in Urban Africa: an increasing burden of road traffic injuries. Traffic Inj. Prev. **7**(4), 381–388 (2006)
4. Linnan, M., et al.: Child mortality and injury in Asia: survey results and evidence. Florence. 3 (2007)
5. World Bank.: World Bank Grouper, Road Safety [Online]. Available: http://www.worldbank.org/transport/roads/safety.htm (2002). Accessed 10 Sept 2016
6. Peden. M., et al.: World Report on Road Traffic Injury Prevention. Geneva-Switzerland (2004)
7. World Health Organization (W.H.O).: The Global Burden of Disease: 2004 Update. Geneva-Switzerland (2008)
8. Ashton. S.J., et al.: Child Pedestrian Injuries. In: Int. Meeting on Biomech. Trauma in Children, pp. 159–170 (1974)
9. Liu, X.J., Yang, J.K.: Effects of vehicle impact velocity and front-end structure on dynamic responses of child pedestrians. Traffic Inj. Prev. **4**(4), 337–344 (2003)
10. Han, Y., et al.: Finite element analysis of kinematic behaviour and injuries to pedestrians in vehicle collisions. Int. J. Crashworthiness **17**(2), 141–152 (2012)
11. Yang, J.: Effects of vehicle front design parameters on pedestrian head-brain injury protection. In: Proceeding of 18th International Technical Conference on the Enhanced Safety of Vehicles (ESV), pp. 1–8. Nagoya, Japan (2003)
12. Venkatason, K., et al.: The head injury mitigation of an adult and child pedestrian in a frontal vehicle impact using response surface methodology. Appl. Mech. Mater. **575**, 952–955 (2014)
13. Henary, B.Y., et al.: Child and adult pedestrian impact: the influence of vehicle type on injury severity. In: Annual Proceedings/Association for the Advancement of Automotive Medicine. Association for the Advancement of Automotive Medicine, pp. 105–126, 47 (2003)
14. Roudsari, B.S., et al.: An evaluation of the association between vehicle type and the source and severity of pedestrian injuries. Traffic Inj. Prev. **6**(2), 185–192 (2005)
15. Tanno, K., et al.: Patterns and mechanisms of pedestrian injuries induced by vehicles with flat-front shape. Leg. Med. **2**(2), 68–74 (2000)
16. Crocetta, G., et al.: The influence of vehicle front-end design on pedestrian ground impact. Accid. Anal. Prev. **79**, 56–69 (2015)
17. Huelke, D.F.: An overview of anatomical considerations of infants and children in the adult world of automobile safety design. In: 42nd Annual Proceedings of Association for the Advancement of Automotive Medicine, pp. 93–113 (1998)
18. Sumchi, A., Sternback, G.: Hangman's fracture in a 7-week old infant. Ann. Emerg. Med. **20**, 86–89 (1991)
19. Wiart, J., et al.: Analysis of RF exposure in the head tissues of children and adults. Phys. Med. Biol. **53**(13), 3681–3695 (2008)
20. Cassan, F.B., et al.: Comparative study of restrained child dummies and cadavers in experimental crashes. In: The 37th Stapp Car Crash Conference, pp. 243–260 (1993)
21. Mani, A., et al.: Sustainable Urban Transport Policy in India-Role of the Auto-rickshaw Sector. INDIA, Mumbai (2012)

22. Civitas Consultancies Pvt Ltd for City Connect Foundation Chennai (CCCF).: Study on the Autorickshaw Sector in Chennai, pp. 1–220 (2010)
23. Mohan, D., et al.: Impact modelling studies for a three-wheeled scooter taxi. Accid. Anal. Prev. **29**(2), 161–170 (1997)
24. Han, Y., et al.: Effects of vehicle impact velocity, vehicle front-end shapes on pedestrian injury risk. Traffic Inj. Prev. **13**(5), 507–518 (2012)
25. Otte. D.: Severity and mechanism of head impacts in car to pedestrian. In: International Conference on the Biomechanics of Impact (IRCOBI), pp. 329–341 (1999)
26. Peden, M., et al.: World Report on Child Injury Prevention. Geneva-Switzerland (2008)
27. Viano, D., et al.: Serious brain injury from traffic-related causes: priorities for primary prevention. Accid. Anal. Prev. **29**(6), 811–816 (1997)
28. Yao, J.F., et al.: Reconstruction of head-to-hood impact in an automobile-to-child-pedestrian collision. Int. J. Crashworthiness **11**(4), 387–395 (2006)
29. European Enhanced Vehicle-Safety Committee (EEVC).: Pedestrian injury accidents. In: 9th International Technical Conference on the Enhanced Safety of Vehicles (ESV), pp. 638–671 (1982)
30. Al-Graitti, A.J., et al.: Auto rickshaw impacts with pedestrians: a computational analysis of post-collision kinematics and injury mechanics. Int. J. Biomed. Biol. Eng. **11**(11), 568–587 (2017)
31. Yang, J.: Review of injury biomechanics in car-pedestrian collisions. Int. J. Veh. Saf. **1** (1/2/3), 100–117 (2005)
32. Anderson, R.W.G., et al.: Vehicle travel speeds and the incidence of fatal pedestrian crashes. Accid. Anal. Prev. **29**(5), 667–674 (1997)
33. Cuerden, R., et al.: Pedestrians and their survivability at different impact speeds. In: Enhanced Safety Vehicle (ESV), pp. 1–12 (2007)
34. Rosén, E., Sander, U.: Pedestrian fatality risk as a function of car impact speed. Accid. Anal. Prev. **41**, 536–542 (2009)
35. Jarrett, K.L., Saul, A.: Pedestrian injury - analysis of the PCDS field collision data. In: 16th International Technical Conference on the Enhanced Safety of Vehicles (ESV), pp. 1204–1211 (1998)
36. Chidester, B., Isenberg, R.: Final report - the pedestrian crash data study. In: Proceeding in the 17th International Conference on the Enhanced Safety of Vehicles (ESV) 248, pp. 1–12 (2001)
37. Jiang, B., et al.: Development of a 10-year-old paediatric thorax finite element model validated against cardiopulmonary resuscitation data. Comput. Methods Biomech. Biomed. Engin. **17**(11), 1185–1197 (2014)
38. Meng, Y., et al.: A 6 year-old pediatric finite element model for simulating pedestrian impacts. In: 14th International LS-DYNA Users Conference, pp. 1–10 (2016)
39. LS-DYNA.: What is the difference between *CONTACT_AUTOMATIC_ SINGLE_SURFACE and *CONTACT_AUTOMATIC_GENERAL? [Online]. Available: http://www.dynasupport.com/faq/contact/what-is-the-difference-between-contact-automatic (2016). Accessed 2 Apr 2017
40. Eppinger, R., et al.: Development of Improved Injury Criteria for the Assessment of Advanced Automotive Restraint Systems-II. National Highway Traffic Safety Administration (NHTSA), USA (1999)
41. Guha, S., et al.: LSTC Hybrid III Dummies Positioning and Post-processing. Livermore Software Technology Corporation (LSTC), Michigan (2008)
42. Mahadevaiah, U., et al.: LSTC Hybrid III 6 Year-Old Finite Element Model Documentation. Livermore Software Technology Corporation (LSTC), USA (2013)

43. Mertz, H.J., et al.: Head injury risk assessment for forehead impacts. SAE technical paper (1996)
44. Prasad, P., Mertz, H.J.: The Position of the United States Delegation to the ISO Working Group 6 on the Use of HIC in the Automotive Environment. SAE Tech. Pap (1985)
45. Kikuchi, A., et al.: Human head tolerance to lateral impact deduced from experimental head injuries using primates. In: The 8 Enhanced Safety Vehicle Conference (ESV), pp. 251–261 (1982)
46. Mcintosh, A.S., et al.: Head impact tolerance in side impacts. In: Proceedings: 15th International Technical Conference on the Enhanced Safety of Vehicles, pp. 1273–1280 (1996)
47. Genneralli, T.A.: The state of the art of head injury biomechanics – a review. In: Proc. of the 29th Annual American Association for Automotive Medicine (AAAM) Conf (1985)
48. Klinich, K.D., et al.: Techniques for developing child dummy protection reference values. General Motors, pp. 1–7–5, U.S.A (1996)
49. Mertz, H.J.: Injury risk assessments based on dummy responses. Accidental Injury: Biomechanics and Prevention. 2nd ed. A.M. Nahum and J.W. Melvin. Springer Verlag, New York pp. 89–102 (2002)
50. Association for the Advancement of Automotive Medicine (AAAM).: The Abbreviated Injury Scale-2005 Revision, Update 2008. Des Plaines, IL (2008)
51. Kleinberger, M.: Development of Improved Injury Criteria for the Assessment of Advanced Automotive Restraint Systems. National Highway Traffic Safety Administration (NHTSA), USA (2000)
52. Muser, M.H., et al.: Biomechanical significance of the rebound phase in low speed rear end impacts. In: International IRCOBI Conference on the Biomechanics of Impact, pp. 393–410 (2000)
53. Hai-bin, et al.: Biomechanics of whiplash injury. Orthopade **27**(5), 813–819 (1998)
54. Parr, J.C., et al.: Evaluation of the nij neck injury criteria with human response data for use in future research on helmet mounted display mass properties. In: Proceedings of the Human Factors and Ergonomics Society Annual Meeting **56**(1), pp. 2070–2074 (2012)
55. Peng, Y., et al.: A study of adult pedestrian head impact conditions and injury risks in passenger car collisions based on real-world accident data. Traffic Inj. Prev. **14**(6), 639–646 (2013)
56. Peng, Y., et al.: A study of pedestrian and bicyclist exposure to head injury in passenger car collisions based on accident data and simulations. Saf. Sci. **50**(9), 1749–1759 (2012)
57. Liu, X.J., et al.: A study of influences of vehicle speed and front structure on pedestrian impact responses using mathematical models. Traffic Inj. Prev. **3**(1), 31–42 (2002)
58. Rosén, E., et al.: Pedestrian injury mitigation by autonomous braking. Accid. Anal. Prev. **42** (6), 1949–1957 (2010)
59. Leaf, W.A., Preusser, D.F.: Literature Review of Vehicle Travel Speeds and Pedestrian Injuries. National HighwayTraffic Safety Administration (NHTSA), U.S. Department of Transportation, Washington, DC (1999)

A Discrete Cosine Transform Based Evolutionary Algorithm and Its Application for Symbolic Regression

Quanchao Liu[1(✉)] and Yue Hu[1,2]

[1] Institute of Information Engineering, Chinese Academy of Science, Beijing 100093, China
liuquanchao@iie.ac.cn

[2] University of Chinese Academy of Science, Beijing 100049, China

Abstract. A discrete cosine transform (DCT) expresses a finite sequence of data points in terms of a sum of cosine functions oscillating at different frequencies, the energy-compaction property of which makes itself very suitable for symbolic regression problems with noisy data samples. In this paper, we propose a DCT based genetic programming (DCT-GP) for symbolic regression or other optimization problems with noisy experimental observations. Firstly, a sequence of experimental samples was changed from time domain into frequency domain with DCT. Secondly, the boundary between the low frequency components and high frequency components was decided, by which most of the noise inserted into the sample data can be discarded. The experimental sample sequence was rebuilt with the low frequency components. Finally, the new obtained samples were looked on as another fitness function and were used to direct the evolutionary procedure of GP. In this way, DCT was integrated into the GP algorithm. The major advantage of the DCT-GP algorithm is that, on the one hand, it can deal with the regression problem with noisy data samples and obtain accurate solutions. On the other hand, if the sample scale is very small, it can deal with the over-fitting problem in the regression progress and can avoid losing of structure information when the available samples were departed into training set and validating set. The performance of the proposed DCT-GP algorithm with several regression problems indicates that the proposed algorithm can find the solution more efficiently and effectively than traditional sample-partitioning based GP algorithms.

Keywords: Discrete cosine transform (DCT) · Symbolic regression · Over-fitting · Genetic Programming (GP)

1 Introduction

Given a set of data samples and a class of functions, the problem of regression is to find a function in the given class that can fit the data samples accurately [1–3]. For a classical regression, the structure of the function is predefined and the task left is to find appropriate coefficients for the function. If both of the structure and the parameters are unknown, the optimization procedure is defined as symbolic regression. Many

K. Arai et al. (Eds.): CompCom 2019, AISC 997, pp. 444–462, 2019.
https://doi.org/10.1007/978-3-030-22871-2_30

techniques have been used to construct regression models (linear regression, nonlinear regression, radial basis functions, neural networks, support vector machines and genetic programming) [4–7]. Among them, GP, neural networks and SVMs are suitable for symbolic regression. The reason is that without a prior information, the search space is a set of all possible symbolic models representing valid operations from the fixed set. The classical numerical optimization techniques will become ineffective. If we want find an interpretable analytic expression with an explicit expression manner, GP is a more suitable algorithm for its non-parametric and non-linear features [8, 9].

To deal with the problem of symbolic regression, different versions of genetic programming are proposed, for example, linear genetic programming (LGP), Cartesian genetic programming (CGP), etc. In [10], a grammar guided genetic programming systems was proposed, in which several syntactical constraints were set on programs. These constraints have the capabilities of overcoming the closure requirement in canonical genetic programming, which cannot always be fulfilled. In [11], a tree adjunct grammar guided genetic programming (TAG3P) was presented, in which tree adjunct grammars along with context free grammars were used to set syntactical constraints as well as search bias for the evolution of programs. The main idea of TAG3P is to evolve the derivation sequence in genotype rather than evolve the derivation tree. In [12], gene expression programming (GEP) was introduced by Ferreira, which incorporates simple, fixed length linear chromosomes, similar to the ones used in GA. The chromosomes were called genotype in GEP and changed as the ramified structures of different sizes and shape. The phenotype, as a ramified expression tree of different sizes and shapes, is translated from chromosome according to certain criteria. Many symbolic regression and time series prediction problems were dissolved by GEP effectively [13, 14].

Despite the plethora of improved genetic programming, there are still two difficult problems left in symbolic regression via GP. The first one is the slow evolutionary, it is an inefficient procedure when the number of the sample points is very large and the infinite size of the search space is considered [15–17]. The other one is the difficulty in avoiding over-fitting, which will face the dilemma of balancing the performance and complexity of the model [18–22]. For the first issue, a Hoeffding based evolutionary algorithm has been proposed to deal with it in [23]. For the second issue, the two commonly used methods to deal with it are hold-out method and N-fold cross validation method [19, 24]. The former divides the obtained data into two disjoint sets (training set and test set). Training data set is used for developing a model and the test data set is used for examining the generalization ability. The later divides the obtained data into N disjoint parts and the N-1 parts were used for building a model and the remaining part was used for testing. Though many experiments have shown the effectiveness of the two kinds of algorithms for large scale of data sets, there still have several problems that they cannot deal with. For example, when the scale of the given data set is very small, many structure information will be discarded after partitioning. That will lead to an inaccuracy result [25–29].

In this paper, we further propose a DCT based genetic programming (DCT-GP) to deal with the second problem. Firstly, the sequence of experimental samples was changed from time domain to frequency domain with DCT. Secondly, the boundary between the low frequency components and high frequency noise was decided, by which most of the noise inserted into the sample data can be discarded. Then, we rebuild the experimental sample sequence with the low frequency components. Finally, the obtained experimental samples were looked on as another fitness function and were used to direct the evolutionary procedure of GP. In this way, DCT was integrated into the proposed GP algorithm and the major advantage of the proposed DCT-GP algorithm is that: (1) Unlike the hold-out method and N-fold cross validation method, the DCT-GP algorithm can avoid deleting structure information when the sample number is very small; (2) The DCT-GP algorithm can effectively delete the noise inserted in the training data set and describe the whole trend of it.

The rest of this paper is organized as follows. We first review the related work on Discrete Cosine Transform, symbolic regression and genetic programming. Next, the DCT based genetic programming algorithm was introduced and some theorem analyses were represented. Thirdly, we implemented several experiments to examine the parameters and the efficiency of the algorithm. Finally, we conclude this paper by highlighting the key contributions of this work.

2 Related Work

2.1 Discrete Cosine Transform (DCT)

A discrete cosine transform is a finite sequence of data points in terms of a sum of cosine functions oscillating at different frequencies. It was first defined by AHMED in 1974 and widely applied in science and engineering, especially in lossy compression of audios (e.g. MP3) and images (e.g. JPEG) [30]. It is very similar to the discrete Fourier transform (DFT), but using only real numbers. There are 8 standard DCT variants and only 4 of them are commonly used. The mainly feature of it is the property of energy compaction: most of the information in a given signal was concentrated in a few low-frequency components.

Formally, the discrete cosine transform is a linear, invertible function f: $R^N \rightarrow R^N$ (where R denotes the set of real numbers), or equivalently an invertible $N \times N$ square matrix. There are several variants of the DCT with slightly modified definitions and four commonly used forms are DCT-I, DCT-II, DCT-III, DCT-IV. In each of them, N real numbers $x_0,\ldots, x_{N-1}$ are transformed into N real numbers $X_0,\ldots, X_{N-1}$ according to the formulas below:

DCT-I:
$$X_k = \frac{1}{2}(x_0 + (-1)^k x_{N-1}) + \sum_{n=1}^{N-2} x_n \cos[\frac{\pi}{N-1} nk] \quad k = 0, 1, \cdots, N-1$$

DCT-II:
$$X_k = \sum_{n=0}^{N-1} x_n \cos[\frac{\pi}{N}(n + \frac{1}{2})k] \quad k = 0, 1, \cdots, N-1$$

DCT-III:
$$X_k = \frac{1}{2} x_0 + \sum_{n=1}^{N-1} x_n \cos[\frac{\pi}{N} n(k + \frac{1}{2})] \quad k = 0, 1, \cdots, N-1$$

DCT-IV:
$$X_k = \sum_{n=0}^{N-1} xn \cos[\frac{\pi}{N}(n + \frac{1}{2})(k + \frac{1}{2})] \quad k = 0, 1, \cdots, N-1$$

2.2 Genetic Programming and Symbolic Regression

GP is a soft computing search technique, it deals with a tree-structured program, called GP tree, whose structure is evolved toward minimizing its fitness value by using genetic operators. The structure of the GP tree is modified and optimized so as to it is more appropriate for model approximation. Cramer [31] introduced GP in 1985 and Koza improved it and enabled it to be applicable to wider areas, i.e., data mining, artificial intelligence and machine learning [32–34]. Now a day, GP is becoming a popular modeling approach and a large amount of research works regarding GP with regression problems and system identifications were carried out.

Earlier, the Minimum Description Length (MDL) method was inserted into GP algorithm to improve the generalization abilities of the solutions. However, further studies have shown that over-fitting and bloat are two independent phenomena. In [35], Silva reviewed most of the past theories about bloat phenomenon and introduced a new approach "Dynamic Limits" to control the bloat. In fact, over-fitting refers to the phenomenon of excessively fitting model to samples so as to the noise hided in the training samples were presented by the model. Many approaches have been proposed to avoid obtaining an over-fitted model by GP. The Interval Arithmetic algorithm [36, 37] was used to evolve reliable models having no undefined behaviors in its output range. The Partitioning Data algorithm was used to protect the generalization ability by dividing the available data into different sample sets, some for evolving the model and others for approximating the generalization ability. An excessively complex model than needed has little practical usage and was hard for interpreting. So, the Reducing Complexity algorithms were proposed to simply the models with complex structures [17, 28, 38]. To deal with multiple and conflicting objectives, the Multi-Objective Optimization algorithms has also been proposed in symbolic regression [39].

3 Discrete Cosine Transform Based Fitness Function

3.1 The Fitness Function for Regression

In general, regression models involve three kinds of variables, the unknown parameters β, the independent variables X and the dependent variable Y. If we relate Y to a function of X and β, a regression model can be expressed by $Y \approx f(X, \beta)$. Let k be the length of the parameter vector β, N be the number of the observed data. If N < k, the regression model is called an under-determined model for no enough data can be used to recover β.

In fact, evolutionary algorithm is one of most important methods for symbolic regression, in which the possible and acceptable solutions is called population, the offspring creation way contains crossover, selection, mutation and the aim of it is to synthesize the best formula to fit the given observed data. To evaluate or measure the quality of an obtained model (individual), the definition of prediction error should be determined in advance. Sum of absolute errors (SAE), mean squared error (MSE) and root mean squared error (RMSE) are popular tools.

Let (x_i, y_i), i = 1:N be the training set of N input-output records, let $(x_i, \tilde{f}(x_i))$, i = 1:N be the calculated result obtaining by the formula $\tilde{f}$. The sum of absolute errors (SAE) of a solution can be expressed by

$$SAE_i = \sum_{i=1}^{N} |y_{(i)} - \tilde{f}(x_{(i)})| \tag{1}$$

The mean absolute error (MAE) of a solution can be expressed by

$$MAE_i = \frac{\sum_{i=1}^{N} |y_{(i)} - f(x_{(i)})|}{N} \tag{2}$$

The root mean squared error (RMSE) can be expressed by

$$RMSE_i = \sqrt{\frac{\sum_{i=1}^{N} |y_{(i)} - f(x_{(i)})|^2}{N}} \tag{3}$$

3.2 Changing the Experimental Observation Sequence into Frequency Area

Let $(x_{(i)}, y_{(i)})$, i = 1:N be the sequential data points in a given training set, $y_{(i)}$ be the observed sequence of dependent variable y, $dy_{(k)}$ be the coefficients of frequency components, then, the observed sequence $y_{(i)}$ can be changed into a series of components with different frequency coefficients $dy_{(k)}$.

$$dy_{(k)} = w_{(k)} \sum_{i=0}^{N-1} y_{(i)} \cos(\frac{\pi(2i+1)k}{2N}) \quad k = 0, 1, \cdots, N-1 \tag{4}$$

Where

$$w_{(k)} = \begin{cases} \frac{1}{\sqrt{N}} & k = 0 \\ \sqrt{\frac{2}{N}} & 1 \leq k \leq N-1 \end{cases}$$

$dy_{(k)}$ return the unitary discrete cosine transform of $y_{(i)}$, $y_{(i)}$ and $dy_{(k)}$ are the same size. The series is indexed from i = 0 and k = 0.

In fact, these frequency components are very useful for constructing a regression model. Generally speaking, the low-frequency components were suitable to be used to describe the global variation trend, the high-frequency components were suitable for describing the detail and noise.

3.3 Deleting the Noisy Components

One of the main features of DCT is the property of energy compaction, by which the most information in a given signal was concentrated in a few low-frequency components. According to this property, we can devise a GP algorithm whose evolutionary direction was guided by the fitness function building with the low-frequency components in the signal. However, how to decide the number of the low-frequency components and how to delete the high-frequency noise components were still a hard problem left to us.

Definition 1 (maximal noise coefficient, $dy_{(n)}$): Let $dy_{(1)}$, $dy_{(2)}$,…, $dy_{(INT(N*70\%))}$,…, $dy_{(n)}$,…, $dy_{(N-1)}$, $dy_{(N)}$ be a frequency sequence dY, maximal noise coefficient, $dy_{(n)}$, is a coefficient whose absolute value, $|dy_{(n)}|$, is the maximal in the later 70% coefficient sequence of dY.

Definition 2 (frequency balance point, $dy_{(b)}$): Let $dy_{(1)}$, $dy_{(2)}$,…, $dy_{(b)}$,…, $dy_{(n)}$,…, $dy_{(N)}$ be a frequency sequence dY, the frequency balance point $dy_{(b)}$ is a coefficient before maximal noise coefficient $dy_{(n)}$, who has the minimal absolute value in sequence $\{dy_{(1)}, dy_{(2)},\ldots, dy_{(b)},\ldots, dy_{(n)}\}$. The frequency balance point $dy_{(b)}$ can be used to partition low or high frequency components, before which the low frequency components can be used to describe the trend of the function, after which the high frequency components can be used to delete noise in dY.

Definition 3 (relative energy $E^{\%}[dy_{(k)}]$): Let $dy_{(1)}$, $dy_{(2)}$,…, $dy_{(k)}$,…, $dy_{(N-1)}$, $dy_{(N)}$ be a frequency sequence dY, let $dy_{(k)}$ be a frequency coefficient, the relative energy of point $dy_{(k)}$, $E^{\%}[y_{(k)}]$, can be calculated by

$$E^{\%}[dy_{(k)}] = \frac{\sum_{i=k+1}^{N} |dy_{(i)}|}{\sum_{i=1}^{k} |dy_{(i)}|} \tag{5}$$

The procedure for rebuilding a new DCT-based fitness function was presented below:

(1) Searching for the maximal noise coefficient $dy_{(n)}$;
(2) Searching for the frequency balance point $dy_{(b)}$;
(3) Calculating the relative energy of point $dy_{(b)}$, $E^{\%}[dy_{(b)}]$;
(4) If $E^{\%}[dy_{(b)}] < 1\%$, let k = b and go (5). Otherwise, go (7).
(5) Replace the coefficients $\{dy_{(k+1)},\ldots, dy_{(n)},\ldots, dy_{(N)}\}$ with zero and rebuild the DCT-based fitness function with sequence $\{dy_{(1)}, dy_{(2)},\ldots, dy_{(k)}, 0, 0, 0, \ldots\}$ (shown in Sect. 3.4).
(6) Terminate the procedure.
(7) From $dy_{(b+1)}$ to $dy_{(n)}$, search for another point $dy_{(a)}$ whose relative energy $E^{\%}[dy_{(a)}]$ is less than 1%. If such a point was found, let k = a and go (5); otherwise, let k = N and go (6).

3.4 Rebuilding the DCT-Based Fitness Functions

Let $(x_{(i)}, y_{(i)})$, i = 1:N be the sequential data points in a given training set, $y_{(i)}$ be the observed sequence of dependent variable y, $dy_{(k)}$ be the new obtained frequency coefficient of sequence dY. Then, the re-building data points $(x_{(i)}, y^*_{(i)})$, i = 1:N can be obtained by the formula below.

$$y*(i) = \sum_{i=0}^{N-1} w(i) dy_{(k)} \cos\left(\frac{\pi(2k+1)i}{2N}\right) \quad i = 0, 1, \cdots, N-1 \tag{6}$$

Where

$$w(i) = \begin{cases} \frac{1}{\sqrt{N}} & i = 0 \\ \sqrt{\frac{2}{N}} & 1 \leq i \leq N-1 \end{cases}$$

4 DCT-Based Genetic Programming

4.1 The DCT-GP Algorithm

Genetic Programming (GP) is an automatic learning methodology, which breeds a population of trial solutions with several biologically inspired operators, such as reproduction, selection, crossover and mutation. The main difference between algorithm DCT-GP and GP is the fitness function. The most popular method for calculating the fitness value in symbolic regression is to compare the difference between the evolved model and the given experimental samples. However, in DCT-GP, we first change the raw dependent variable sequence Y into its frequency expression, dY. Then, we delete the noise in dY and rebuilding a new dependent variable sequence Y*. Finally, we calculate the errors between sequence Y* and the evolved model. The proposed DCT-GP algorithm for symbolic regression is presented below and is illustrated in Fig. 1:

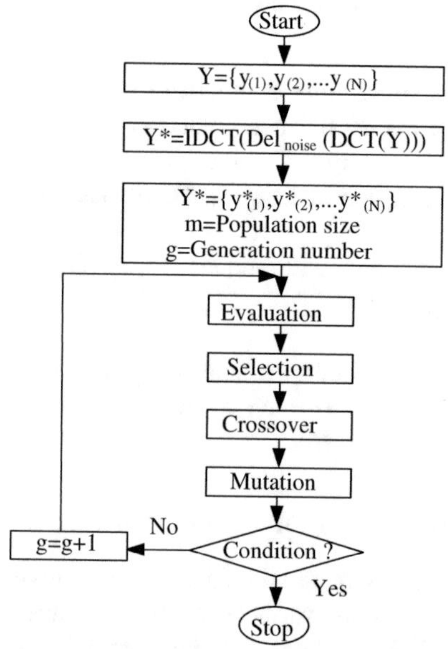

Fig. 1. The DCT-GP algorithm for symbolic regression.

(1) Initialization: Initialize the raw dependent variable sequence $Y = \{y_{(0)}, y_{(1)}, \ldots y_{(K)}\}$.
(2) Fitness function calculation: Calculate the DCT-based fitness function with the method presented in Sects. 3.2, 3.3 and 3.4.
(3) Population initialization: Initialize the generation counter g (g = 0), the individual counter i (i = 1 and i <= I), the gth generation population X^g, $X^g = \{X^{g,1}, X^{g,2}, \ldots X^{g,I}\}$, where I is the population scale.
(4) Fitness calculation: Calculate the fitness values for individuals in population X^g.
(5) Selection: Do a traditional selection operation.
(6) Crossover: Do a traditional crossover operation.
(7) Mutation: Do a traditional mutation operation.
(8) Check convergence: Repeat the steps (4) to (8) until the stopping criterion is met.

4.2 Algorithm Analysis

Definition 4 (independent variable sequence, X): The independent variable sequence X, $X = \{x_{(0)}, x_{(1)}, \ldots x_{(K-1)}\}$ is a set of independent variable x_i separated from the set of experimental observations (x_i, y_i).

Definition 5 (dependent variable sequence, Y): The dependent variable sequence Y, $Y = \{y_{(0)}, y_{(1)}, \ldots y_{(K-1)}\}$ is a set of dependent variable y_i separated from the set of experimental observations (x_i, y_i).

Definition 6 (frequency coefficient sequence of Y, dY). The frequency coefficient sequence dY, $dY = dct(Y) = \{dy_{(0)}, dy_{(1)}, \ldots, dy_{(k)}, \ldots, dy_{(N-1)}\}$, is the set of frequency coefficients of Y. $dy_{(k)}$ is the value of the k_{th} coefficient, the value of it can be calculated by formula below:

$$dy_{(k)} = w_{(k)} \sum_{j=0}^{N-1} y_{(i)} \cos\left(\frac{\pi(2i+1)k}{2N}\right) \quad k = 0, 1, \ldots N-1 \tag{7}$$

Where,

$$w_{(k)} = \begin{cases} \frac{1}{\sqrt{N}} & k = 0 \\ \sqrt{\frac{2}{N}} & 1 \le k \le N-1 \end{cases}$$

Definition 4 (The DCT-based dependent variable sequence of Y, Y*): The DCT-based dependent variable sequence Y*, $Y^* = \{y^*_{(0)}, y^*_{(1)}, \ldots, y^*_{(i)}, \ldots, y^*_{(N-1)}\}$ is a set of dependent variable obtained from the frequency coefficient sequence dY with the method presented in Sects. 3.2, 3.3 and 3.4, $Y^* = IDCT(Del_{noise}(DCT(Y)))$.

Theorem 1: Let $Y = \{y_{(i)}, i \in [0, N-1]\}$ be the dependent variable sequence coming from experimental observations, $y_{(i)}$ ($y_{(i)} = y^0_{(i)} + Z_{(i)} \sim N(y_{(i)}, n)$) is the sum of the low frequency variable $y^0_{(i)}$ and high frequency noise $Z_{(i)}$ ($Z_{(i)} \sim N(0, n)$), $dY = \{dy_{(0)}, dy_{(1)}, \ldots, dy_{(K-1)}\}$ be the frequency coefficient sequence of Y,

Then,

(A) $dy^0_{(k,k\to N-1)} = 0,\ \left(y^0_{(i)} \approx y^0_{(i+1)},\ i\epsilon[0,\ N-1]\right);$
(B) $dy^Z_{(k,k\to N-1)} \neq 0,\ \left(Z_{(i)} \approx N(0,n),\ i\epsilon[0,\ N-1]\right).$

Proof:
According to the character of DCT, we can get

$$\begin{aligned} dy_{(k)} = dy^0_{(k)} + dy^z_{(k)} = w_{(k)} \cdot \sum_{i=0}^{N-1} (y^0_{(i)} + Z_{(i)}) \cos(\frac{\pi(2i+1)k}{2N}) \\ = w_{(k)} \cdot \sum_{i=0}^{N-1} y^0_{(i)} \cdot \cos(\frac{\pi(2i+1)k}{2N}) + w_{(k)} \cdot \sum_{i=0}^{N-1} Z_{(i)} \cdot \cos(\frac{\pi(2i+1)k}{2N}) \end{aligned} \tag{8}$$

In (8),

$$\begin{aligned} & dy^0_{(k)} = w_{(k)} \cdot \sum_{i=0}^{N-1} y^0_{(i)} \cdot \cos(\tfrac{\pi(2i+1)k}{2N}) \\ & = w_{(k)} \cdot \left(y^0_{(0)} \cdot \cos(\tfrac{\pi k}{2N}) + y^0_{(1)} \cdot \cos(\tfrac{3\pi k}{2N}) + \ldots + y^0_{(N-1)} \cdot \cos(\tfrac{(2N-1)\pi k}{2N})\right) \\ & = w_{(k)} \cdot \left(y^0_{(0)} \cdot \cos(\tfrac{\pi}{2} - \tfrac{\pi}{2N}) + y^0_{(1)} \cdot \cos(\tfrac{3\pi}{2} - \tfrac{3\pi}{2N}) + \ldots + y^0_{(N-1)} \cdot \cos(\tfrac{(2N-1)\pi}{2} - \tfrac{(2N-1)\pi}{2N})\right) \\ & (k \to N-1) \end{aligned}$$

$$\begin{aligned} & \approx w_{(k)} \cdot \left(y^0_{(0)} \cdot \cos(\tfrac{\pi}{2} - \tfrac{\pi}{2N}) - y^0_{(1)} \cdot \cos(\tfrac{\pi}{2} - \tfrac{3\pi}{2N}) + y^0_{(0)} \cdot \cos(\tfrac{\pi}{2} - \tfrac{5\pi}{2N}) - y^0_{(1)} \cdot \cos(\tfrac{\pi}{2} - \tfrac{7\pi}{2N}) + \ldots + y^0_{(N-1)} \cdot \cos(\tfrac{\pi}{2} - \pi)\right) \\ & \approx \left(k \to N-1;\ y^0_{(i)} \approx y^0_{(i+1)},\ i\ \epsilon[0,\ N-1]\right) \end{aligned}$$

$$\begin{aligned} & \approx w_{(k)} \cdot y^0_{(0)} \cdot \left(\cos(\tfrac{\pi}{2} - \tfrac{\pi}{2N}) - \cos(\tfrac{\pi}{2} - \tfrac{3\pi}{2N}) + \cos(\tfrac{\pi}{2} - \tfrac{5\pi}{2N}) - \cos(\tfrac{\pi}{2} - \tfrac{7\pi}{2N}) + \ldots + \cos(\tfrac{\pi}{2} - \pi)\right) \\ & \approx w_{(k)} \cdot y^0_{(0)} \cdot \left(\sin(\tfrac{\pi}{2N}) - \sin(\tfrac{3\pi}{2N}) + \sin(\tfrac{5\pi}{2N}) - \sin(\tfrac{7\pi}{2N}) + \ldots + \sin(\pi)\right) \end{aligned}$$

$$\approx w_{(k)} \cdot y^0_{(0)} \cdot 0 \qquad \left(k \to N-1;\ y^0_{(i)} \approx y^0_{(i+1)},\ i\ \epsilon[0,\ N-1],\ N \to \infty\right)$$

$$\begin{aligned} & dy^Z_{(k)} = w_{(k)} \cdot \sum_{i=0}^{N-1} Z_{(i)} \cdot \cos(\tfrac{\pi(2i+1)k}{2N}) \\ & = w_{(k)} \cdot \left(Z_{(1)} \cdot \cos(\tfrac{\pi k}{2N}) + Z_{(2)} \cdot \cos(\tfrac{3\pi k}{2N}) + \ldots + Z_{(N-1)} \cdot \cos(\tfrac{(2N-1)\pi k}{2N})\right) \\ & \neq 0 \qquad (k \to N-1,\ Z_{(i)} <> Z_{(i+1)}, Z_{(i)} \sim N(0,\ n),) \end{aligned}$$

So, for the low frequency coefficient $dy^0(k)$, the low frequency components were counteracted at the end of the sequence dy ($k \to N-1$). For the high frequency coefficient $dy^Z_{(k)}$, the high frequency noise were accumulated. If we rebuild a new DCT-based fitness function by removing the noise from sequence dy and direct the evolution with it, we can obtain a more accurate solution.

5 Performance Evaluation

The experiments were conducted on a 3.0 GHz Pentium4 with 1.0 GB of memory running Microsoft Windows XP. All code was compiled using Microsoft Visual C++6.0.

5.1 Sample Number

Sample scale is an important factor for most approaches in area of symbolic regression, especially for sampling based algorithms, the accuracy and generalization ability of which were largely decided by it. To demonstrate the influence of the sample number, we give a function defined by y = min{2/x, sin(4x) + 1} and the variable x was uniformly distributed in (1,15). We symmetrically select 9, 18 and 150 sample points to draw three sketches. From Fig. 2(a) and (b), we can see that when the unrepresentative points (9/18 points) were extracted from the original data set and the sketches represented by them will lost the primary shape of the given function. Naturally, the regression model constructed with them will be an incorrect one.

In fact, it is very hard to decide which point in the data set is a representative point carrying the key structural information before obtains a model. So, regression with sampling based algorithm and insufficient data points will bear the risk of losing structural information. Adequate samples are necessary for acquiring an accuracy model. However, in the field of engineering, regressing with insufficient samples is a

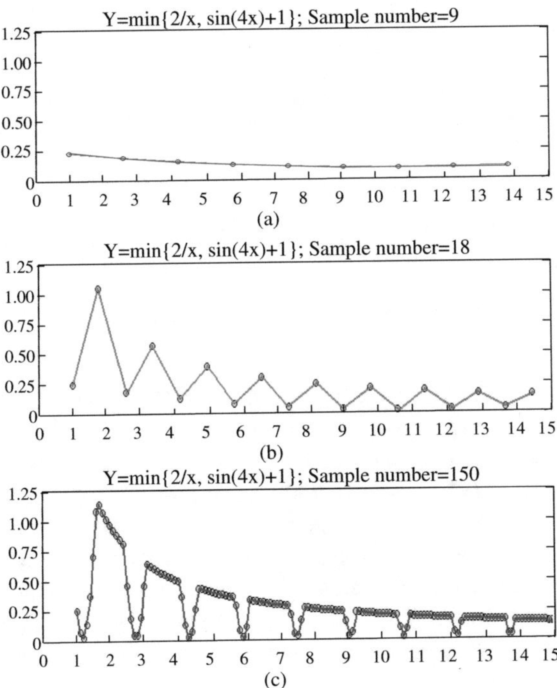

Fig. 2. The sketches of different sample numbers.

frequently encountered problem. The traditional partitioning-data based algorithms cannot deal with this problem for the sampling action of them (appeared in the partitioning progress) makes the rare samples more scarce. On the contrary, the DCT-GP algorithm searches for variation tendency of the given data set by DCT transformation and avoid the over-fitting not by portioning the samples into different subsets, and testing the model by subsets, but by comparing the models with the tendency calculated by all of the points in the data set. So, the DCT-GP algorithm is more suitable for this kind of regression problems.

5.2 How to Avoid Over-Fitting

In this section, we designed a simple toy problem to show how does the proposed DCT-GP algorithm avoids over-fitting. The function $Y = 2x + 1 + \delta_{Rnd}$ with 11 equidistant points in the interval −1 to +1 was shown in Fig. 3(b) and was looked on as the object function of symbolic regression. For each point, the output value of it was added a random number δ_{Rnd}, the range of it was set to $[-0.1 \sim +0.1]$. For traditional GP algorithm, we performed 40 independent runs for the given regression problem and in each one we used the primary 11 data points as objective functions. After 50 generation running, most of them obtained the over-fitting models, one of the models was shown in Fig. 3(a).

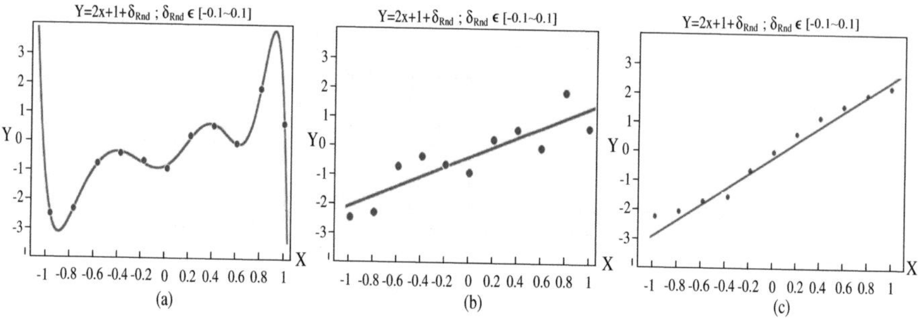

Fig. 3. How to avoid over-fitting.

For the proposed DCT-GP algorithm, we firstly changed the primary data set $\{(y_{(1)}, x_{(1)}), (y_{(2)}, x_{(2)}), \ldots\ldots, (y_{(11)}, x_{(11)})\}$ into frequency domain. Ten frequency coefficients $dy_{(1)}, dy_{(2)}, \ldots, dy_{(11)}$ were obtained. And then, we constructed a new data set $\{(y^*_{(1)}, x_{(1)}), (y^*_{(2)}, x_{(2)}), \ldots\ldots, (y^*_{(11)}, x_{(11)})\}$ with the former 6 frequency coefficients $dy_{(1)}, dy_{(2)}, \ldots, dy_{(6)}$ and changed them into time domain. 95% energy of them was persisted in the new obtained sequence $\{y^*_{(1)}, y^*_{(2)}, \ldots, y^*_{(11)}\}$. Finally, we make the new obtained data set as the fitness function to evaluate the fitness of the individuals appeared in the proposed DCT-GP algorithm. The obtained result was shown in Fig. 3(c), from which we can see that the obtained solution is more similar with the primary one.

5.3 Avoiding the Interference of Noise

In this section, the function $Y = x^4 + x^3 + x^2 + x + 100$ with 100 equidistant points in the interval −5 to +5 was looked on as the regression objective. For each point, the output value of it was added a random number δ_{Rnd}, the range of which was set to $[-\delta \sim +\delta]$. To analyze the interferences from noises, we give two groups of experiments. In group-1, the noise parameter $\delta_{Rnd\text{-}1}$ of function-1 was set to δ_0 ($\delta_0 \in [-10 \sim +10]$), the noise parameter $\delta_{Rnd\text{-}2}$ and $\delta_{Rnd\text{-}3}$ of function 2 and 3 were the same as 1, but amplified 5 and 10 times. In group-2, we use different parameters δ_1, δ_2 and δ_3, but the range of which are the same. In Fig. 4(a) and (c), we show the original six objective functions. In Fig. 4(b) and (d), we show the corresponding sequence of them in frequency domain.

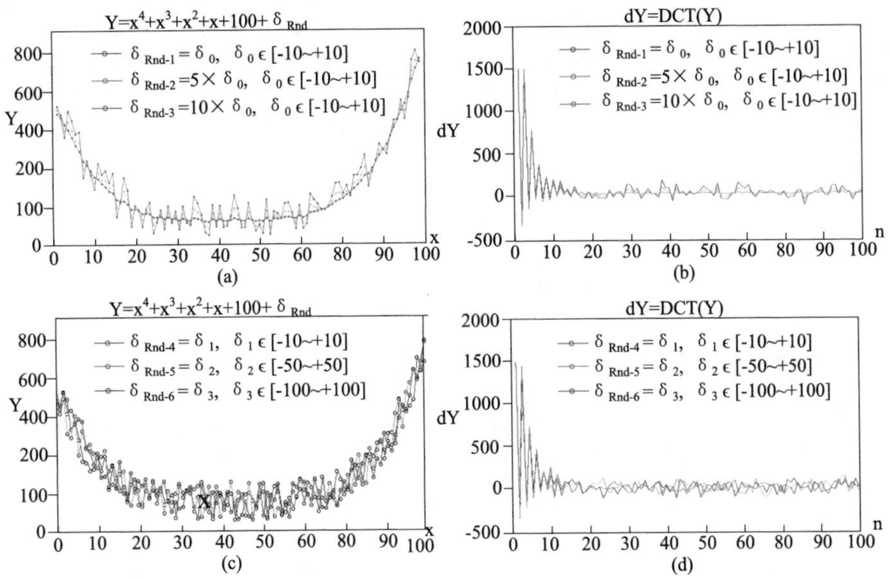

Fig. 4. How to avoid the interference coming from noise.

From Fig. 4(a) and (b) we can find that for the three functions Y1, Y2 and Y3 (which have the same noise parameter δ_0, but different amplification values 1, 5 and 10), the low frequency components of them ($n < 10$) were approximately equal. But the high frequency components ($n > 10$) were very different. From Fig. 4(c) and (d), we can find the same trend. The only differences of the two groups of experiments were that the high frequency components in groups-1 were proportional, but the high frequency components in groups-2 were irregular.

So, if we rebuild the sample points with low frequency components and make them as the fitness function in the proposed GP algorithm, we will avoid the interferences coming from noises and obtain a robust solution.

5.4 How to Select the Low Frequency Components?

In fact, one of the main features of DCT is the property of energy compaction, by which the most information in a given signal was concentrated in a few low-frequency components. According to this property, we can devise a GP algorithm whose evolutionary direction was guided by the fitness function building with the low-frequency components of the signal. However, how to decide the number of the low-frequency components and how to delete the high-frequency noise components were still a hard problem left to us.

In this section, we give a method to determine the range of noise and to build a mildly fitness function for avoiding over-fitting. Firstly, we search for a point "T" whose vertical coordinate has the largest value in the later 70% components. We can see the found point "T" in Fig. 5(a). And then, we can find the frequency balance point "T′" on the left side of "T". Finally, the horizontal ordinate n1 of point "T′" is the boundary of trend data and noise data. After doing that, we rebuild the fitness function sequence with dY`, whose low frequency components 0−n1 were the same as dY and the high frequency components n1−n were filled with zero. We can see the result from Fig. 5(b).

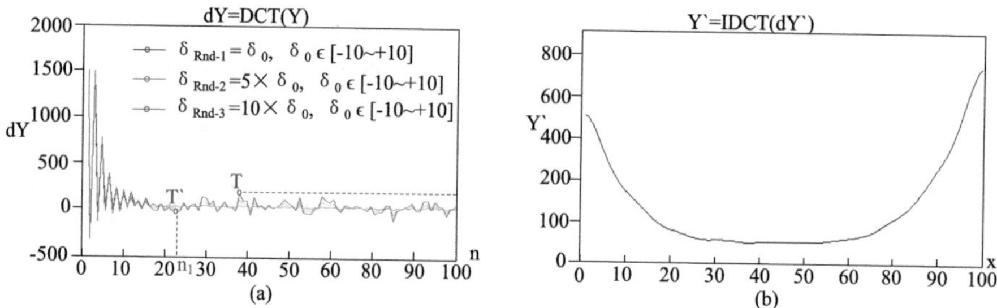

Fig. 5. How to select the low frequency component.

5.5 Comparisons-1

In fact, the generalization ability of a given regression approach refers to the ability to find solutions that perform well not only on samples for training samples but also on samples for testing. Developing an unknown model from finite training data will carry the risk of excessively fitting the model. The over-fitted model always tries to model the noises in training samples rather than expressing the whole trend. In complex and sparse search spaces, GP has the tendency of finding solutions that were biased towards the training set. So, in [24], Gagne divided the available data into two disjoint data sets, training set and test set. The former was used to evolve the model and the latter was used to approximate the generalization ability. In [19], Zavoianu proposed the N-Fold Cross Validation method to find generational solutions. In [20], Vanneschi gives a measure to compute the over-fitting of a given model.

In this section, we compare the algorithm DCT-GP with partitioning based algorithm [24] on test data set $\{Y = x^4 + x^3 + x^2 + x + 100 + \delta_{Rnd}\}$ and examine performances of them when different parameters were used. The mean absolute error (MAE) was looked on as the fitness functions. The results were shown in Fig. 6, in which (a1), (a2), (a3) and (a4) are four primary functions whose sample number and noise scope are respectively initialized to 280, 280, 28, 28 and [−1000, 1000], zero, [−1000, 1000], zero. The corresponding frequent sequence dY of them after DCT transformation are shown in Fig. 6(b1), (b2), (b3), (b4). The implementation procedures of the two algorithms were shown in Fig. 6(c1), (c2), (c3), (c4). What calls for special attention is that, in algorithm [24], the available data was divided into three parts, training data set, validation data set and test data set. The training set was used to evaluate the fitness of the individuals, the validation set was used to select the best individual and the test set was used to estimate the accuracy on unseen data. In algorithm DCT-GP, the available data was divided only into two parts, evolving data set and test data set. The evolving data set contains all of the available data except the data in test set. So, more available structure information will be used for building the regression model in DCT- GP than in algorithm [24].

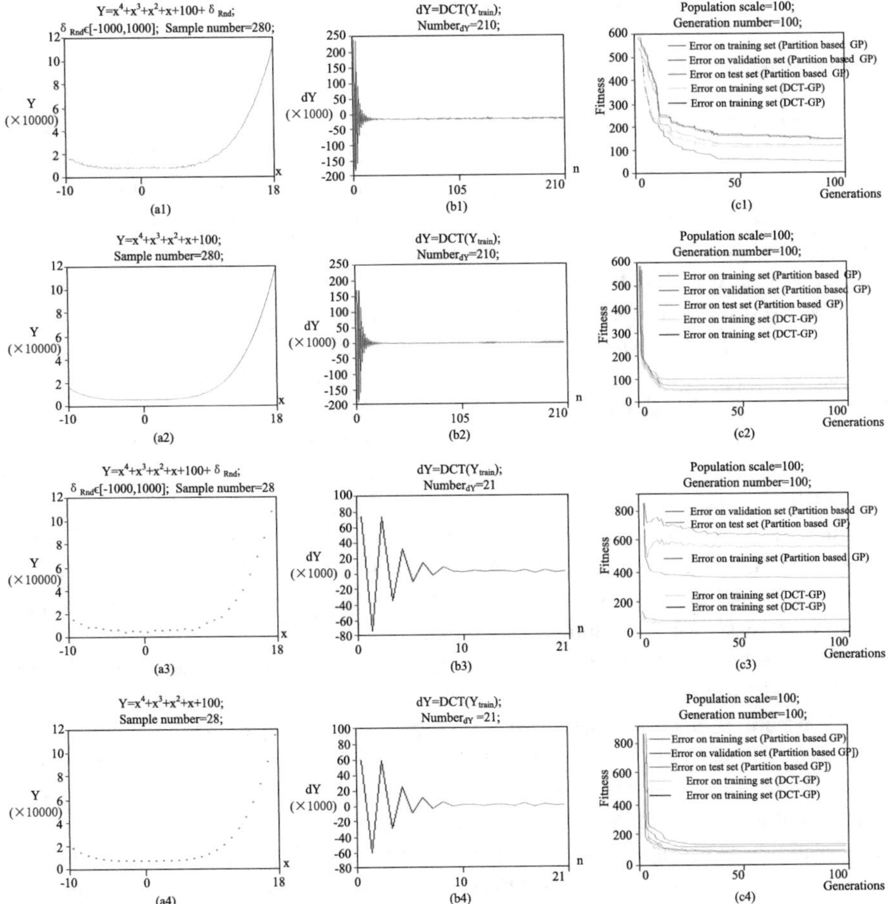

Fig. 6. The influences of different parameters.

From Fig. 6(a1), (a2), (c1) and (c2) we can see that the noise δ_{Rnd} has great influence on the mean absolute error (MAE). In c1 ($\delta_{Rnd} \in [-1000, 1000]$), the MAE of algorithm [24] on training set, validation set and test set were respectively greater than that in (c2), where no noise was inserted into the available data. However, for algorithm DCT-GP, the difference of MAE on evolving set and test set were not very clear in Fig. 6(c1) and (c2). It is because that in the rebuilt sample set, most of the noise was deleted by DCT-GP.

The sample number is another important factor in symbolic regression. For traditional partition based algorithm, such as algorithm [24], partitioning the available data set into training set and validation set will lose some structure information. So, the accuracy of the obtained model will be less than the proposed DCT-GP algorithm, which make use of all the available data and rebuild them for regression. We can see this from Fig. 6(c1) and (c3), or from Fig. 6(c2) and (c4). In fact, when the sample number is very large and no noise was inserted into the samples, the performances of the DCT-GP algorithm and partitioning based algorithms were very similar. However, when the sample number is very small, the partition based algorithms will lost some useful structure information in the regression progress and so as to bad performances.

5.6 Comparisons-2

In order to further examine the performance of the proposed algorithm, we implement it on several test data sets which are computer generated data sets from the functions listed in Table 1. The number of the pairs (x_i, y_i) and their distribution range are listed in it too. The parameters of algorithm GP, used in the regression process were listed in Table 2. The best mean absolute errors (MAE), absolute errors (SAE) and root mean squared errors (RMSE) of the two algorithms were represented in Table 3. In Sample 1–3, the maximum number of nodes in the trees was set to 25. In Sample 4–6, the maximum number of nodes in the trees was set to 50. In all test samples, the elitism was applied to the individuals ranked in the top 10/100th of the population. For all functions we use MAE as the fitness function. We define protected division as the operator that returns one if the denominator is exactly zero. The ephemeral random constant R appeared in Table 1 is uniformly distributed over a specified interval of [−10, 10]. The parameter of relative energy $E^{\%}[dy_{(a)}]$ was set to 1%. To assess the accuracy of the solutions, we considered the fitness of the best individual found within given generations. We computed the sum of absolute errors (SAE), the mean absolute error (MAE) and the root mean squared error (RMSE) across all runs. Table 3 shows the average result for 20 running.

From the analysis in Sect. 4.2, we know that the accuracy of the model obtained with DCT-GP should be better than that with traditional fitness functions. In fact, it is partly the truth. In Sample 5, we can find that, when the sample number is very small, all of the three errors (SAE, MAE, and RMSE) of DCT-GP and PA-GP were very similar. The explanation is that when no such a point, $dy_{(a)}$ was found (the relative energy $E^{\%}[dy_{(a)}]$ of it is not less than 1%), the DCT based fitness function will degenerate into a traditional fitness function. The generational ability of itwill disappear. This can be seen in Sect. 3.3 and Table 3. So, reasonably estimating the scope of noise and selecting suitable parameters for DCT-GP before implementation is very important.

Table 1. Sample functions used in experiments.

No	Function	Range	Noise	Function set	Terminal set
Sample 1	$f(x) = x^4 + x^3 + x^2 + x + 100$	x∈ [0, 14]	δ ϵ [−100 ~+100];	+, −, *,/	X, 1.0, 2.0, 3.0
Sample 2	$f(x) = \cos(2x)$	x∈ [−3.14, 3.14]	δ ϵ [−0.25 ~+0.25];	+, −, *,/, sin,cos,sqr	X, 1.0, 2.0, 3.0
Sample 3	$f(x) = e^{-(\sin 3x+2x)}$	x∈ [−1.57, 1.57]	δ ϵ [−0.25 ~+0.25];	+, −, *,/, sin,cos, sqr,exp	X, 1.0, 2.0, 3.0
Sample 4	$f(x) = \min\{2/x, \sin(x) + 1\}$	x∈ [1, 15]	δ ϵ [−0.5 ~+0.5];	+, −, *,/ ,exp	X, R
Sample 5	$f(x) = 2.718x^2 + 3.1416x$	x∈ [−3.14, 3.14]	δ ϵ [−2 ~+2];	+, −, *,/	X, R
Sample 6	$f(x) = \min\{\sin(x), \cos(x)\}$	x∈ [−3.14, 3.14]	δ ϵ [−0.5 ~+0.5];	+, −, *,/, sin,cos	X, R

Table 2. General settings.

Parameter	Sample-1	Sample-2	Sample-3	Sample-4	Sample-5	Sample-6
Max. depth	10	10	10	10	10	10
Population size	600	600	1000	500	500	600
Number of generations	100	100	100	200	300	500
Crossover rate	0.96	0.96	0.96	0.96	0.96	0.96
Mutation rate	0.25	0.25	0.25	0.25	0.25	0.25
Selection	Tournament of size 10					

Table 3. Performance measures for DCT-GP and partition based GP(PA-GP).

Algorithm	Fitness function	Sample 1		Sample 2		Sample 3		Sample 4		Sample 5		Sample 6	
		δ	No noise	δ	No noise	δ	No noise	δ	No noise	δ	No noise	δ	No noise
SAE	DCT-GP	5305	662	4.42	0.0098	2.01	0.025	7.56	6.99	18.98	0.039	6.21	1.37
	[PA-GP]	5622	681	4.89	0.012	2.56	0.032	8.02	7.44	18.60	0.035	6.87	1.41
MAE	DCT-GP	647	81	0.34	0.0007	0.23	0.003	0.88	0.81	2.11	0.016	0.44	0.098
	[PA-GP]	716	202	0.65	0.0022	0.58	0.042	3.12	1.18	1.98	0.012	1.02	0.132
RMSE	DCT-GP	982	151	0.45	0.0021	0.34	0.005	1.76	1.11	4.64	0.007	0.87	0.188
	[PA-GP]	1623	383	1.72	0.0052	1.12	0.078	5.54	2.43	4.41	0.006	2.14	0.212

6 Discussions and Conclusion

In fact, the major advantage of the proposed DCT-GP algorithm is the capability of deleting the noise components from the primary sample data set and rebuilding it to deal with the over-fitting problem. When the sample data set was very large and no noise was inserted in it, there exists a lot of structure information can be used. The

DCT-GP algorithm and the traditional GP algorithm are very similar. On the other hand, the DCT-GP algorithm is not only an algorithm, but a framework. Any kinds of GP algorithm can be inserted into this framework and works with the rebuilt fitness function. Connecting the DCT based regression framework with more practical GP algorithms is a good way to evolve regression model.

Acknowledgements. We are grateful to the anonymous referees for their invaluable suggestions to improve the paper. This work is financially supported by The National Key Research and Development Program of China (No. 2018YFC0704304).

References

1. Sen, A., Srivastava, M.: Regression Analysis – Theory, Methods, and Applications. Springer-Verlag, Berlin (2011, 4th printing)
2. Armstrong, J.S.: Illusions in regression analysis. Int. J. Forecast. (forthcoming) **28**(3), 689 (2012). https://doi.org/10.1016/j.ijforecast.2012.02.001
3. Carroll, R.J.: Transformation and weighting in regression. Routledge (2017)
4. Sacks, J., Welch, W.J., Mitchell, T.J., Wynn, H.P.: Design and analysis of computer experiments. Statist. Sci. **4**, 409–435 (1989)
5. Haykin, S.: Neural Networks: A Comprehensive Foundation. Prentice Hall PTR (1994)
6. Vapnik, V.N.: The support vector method. In: International Conference on Artificial Neural Networks, pp. 261–271. Springer, Berlin, Heidelberg (1997)
7. Breiman, L.: Classification and Regression Trees. Routledge (2017)
8. Cano, A., Ventura, S., Cios, K.J.: Multi-objective genetic programming for feature extraction and data visualization. Soft. Comput. **21**(8), 2069–2089 (2017)
9. Bannister, C.A., Halcox, J.P., Currie, C.J., et al.: A genetic programming approach to development of clinical prediction models: a case study in symptomatic cardiovascular disease. PLoS One **13**(9), e0202685 (2018)
10. Tsakonas, A., Dounias, G., Doumpos, M., Zopounidis, C.: Bankruptcy prediction with neural logic networks by means of grammar-guided genetic programming. Expert Syst. Appl. **30**, 449–461 (2006)
11. Hoai, N.X., McKay, R.I., Essam, D., et al.: Solving the symbolic regression problem with tree-adjunct grammar guided genetic programming: The comparative results. In: Proceedings of the 2002 Congress on Evolutionary Computation, CEC 2002, vol. 2, pp. 1326–1331. IEEE (2002)
12. Ferreira, C.: Gene expression programming: a new adaptive algorithm for solving problems. Complex Syst. **13**(2), 87–129 (2001)
13. Ferreira, C.: Gene Expression Programming: Mathematical Modeling by an Artificial Intelligence. Springer (2006)
14. Peng, Y.Z., Yuan, C.A., Qin, X., et al.: An improved gene expression programming approach for symbolic regression problems. Neurocomputing **137**, 293–301 (2014)
15. Jin, Y., Olhofer, M., Sendhoff, B.: On evolutionary optimization with approximate fitness functions. In: Proceedings of the 2nd Annual Conference on Genetic and Evolutionary Computation. Morgan Kaufmann Publishers Inc., 786–793 (2000)
16. Vladislavleva, E.Y.: Model-based problem solving through symbolic regression via pareto genetic programming. CentER, Tilburg University (2008)

17. Vladislavleva, E.J., Smits, G.F., Den Hertog, D.: Order of nonlinearity as a complexity measure for models generated by symbolic regression via pareto genetic programming. IEEE Trans. Evol. Comput. **13**(2), 333–349 (2009)
18. Silva, S.G.O.: Controlling bloat: individual and population based approaches in genetic programming. (2008)
19. Zavoianu, A.C.: Towards solution parsimony in an enhanced genetic programming process. Master's thesis, International School Informatics: Engineering & Management, ISI-Hagenberg, Johannes Kepler University, Linz (2010)
20. Vanneschi, L., Castelli, M., Silva, S.: Measuring bloat, overfitting and functional complexity in genetic programming. In: Proceedings of the 12th Annual Conference on Genetic and Evolutionary Computation, pp. 877–884. ACM (2010)
21. Mousavi Astarabadi, S.S., Ebadzadeh, M.M.: Avoiding overfitting in symbolic regression using the first order derivative of GP trees. In: Proceedings of the Companion Publication of the 2015 Annual Conference on Genetic and Evolutionary Computation, pp. 1441–1442. ACM (2015)
22. Chen, Q., Xue, B., Niu, B., et al.: Improving generalisation of genetic programming for high-dimensional symbolic regression with feature selection. In: 2016 IEEE Congress on Evolutionary Computation (CEC), pp. 3793–3800. IEEE (2016)
23. Zhao, Li, Wang, Lei: Hoeffding bound based evolutionary algorithm for symbolic regression. Eng. Appl. Artif. Intell. **25**, 945–957 (2012)
24. Gagné, C., Schoenauer, M., Parizeau, M., et al.: Genetic programming, validation sets, and parsimony pressure. In: European Conference on Genetic Programming, pp. 109–120. Springer, Berlin, Heidelberg (2006)
25. Liu, Y., Khoshgoftaar, T.: Reducing overfitting in genetic programming models for software quality classification. In: Eighth IEEE International Conference on High Assurance Systems Engineering, pp. 56–65. IEEE Computer Society (2004)
26. Foreman, N., Evett, M.: Preventing overfitting in GP with canary functions. In: Proceedings of the 7th Annual Conference on Genetic and Evolutionary Computation, pp. 1779–1780. ACM (2005)
27. Gonçalves, I., Silva, S.: Experiments on controlling overfitting in genetic programming. In: 15th Portuguese Conference on Artificial Intelligence (EPIA 2011), pp. 10–13 (2011)
28. Trujillo, L., Silva, S., Legrand, P., et al.: An empirical study of functional complexity as an indicator of overfitting in genetic programming. In: European Conference on Genetic Programming, pp. 262–273. Springer, Heidelberg (2011)
29. Austel, V., Dash, S., Gunluk, O., et al.: Globally Optimal Symbolic Regression (2017)
30. Ahmed, N., Natarajan, T., Rao, K.R.: Discrete cosine transform. IEEE Trans. Comput. 90–93 (1974)
31. Cramer, N.L.: A representation for the adaptive generation of simple sequential programs. Proceedings of the First International Conference on Genetic Algorithms, pp. 183–187 (1985)
32. Zameer, A., Arshad, J., Khan, A., et al.: Intelligent and robust prediction of short term wind power using genetic programming based ensemble of neural networks. Energy Convers. Manag. **134**, 361–372 (2017)
33. Enríquez-Zárate, J., Trujillo, L., de Lara, S., et al.: Automatic modeling of a gas turbine using genetic programming: an experimental study. Appl. Soft Comput. **50**, 212–222 (2017)
34. Pimenta, A., Nametala, C.A.L., Guimarães, F.G., et al.: An automated investing method for stock market based on multiobjective genetic programming. Comput. Econ. **52**(1), 125–144 (2018)
35. Silva, S., Costa, E.: Dynamic limits for bloat control in genetic programming and a review of past and current bloat theories. Genet. Program. Evolvable Mach. **10**(2), 141–179 (2009)

36. Keijzer, M.: Improving symbolic regression with interval arithmetic and linear scaling. In: European Conference on Genetic Programming, pp. 70–82. Springer, Heidelberg (2003)
37. Kotanchek, M., Smits, G., Vladislavleva, E.: Trustable symbolic regression models: using ensembles, interval arithmetic and pareto fronts to develop robust and trust-aware models. Genetic programming theory and practice V, pp. 201–220. Springer, US (2008)
38. Stinstra, E., Rennen, G., Teeuwen, G.: Metamodeling by symbolic regression and Pareto simulated annealing. Struct. Multidisc. Optim. **35**(4), 315–326 (2008)
39. Ngatchou, P., Zarei, A., El-Sharkawi, A.: Pareto multi objective optimization. In: Proceedings of the 13th International Conference on Intelligent Systems, pp. 84–91. IEEE (2005)

Similarity Measurement of Handwriting by Alignment of Sequences

Katalin Erdélyi(✉) and Bálint Molnár(✉)

Information Systems Department, Eötvös Loránd University,
1117 Pázmány Péter sétány 1/C, Budapest, Hungary
{l5gtow,molnarba}@inf.elte.hu

Abstract. In court proceedings, the question often arises: how much correspondence is there exactly between disputed writing and comparative writing patterns? According to handwriting examiners, handwriting cannot be measured in numbers. One of the reasons is that the handwriting experts use only a small part of the full feature complex, it is impossible to find and compare all of them manually. The selection is based on the expert's experience, although contradictory opinions on the same test material can be given by different experts. Thus, the enhancement of expert objectivity in the court processes is highly desirable, my goal is to search for opportunities of application of computer science and approaches. The Soviet-type handwriting expertise methodology that is currently used in Hungary primarily interpreted the writing as a movement. In fact, traditional writing comparison treats handwriting as a moving stream. This observation led to the analysis of handwriting using the sequence matching method. This paper provides an example of how a sequence alignment algorithm for comparing DNA sequences can be used to compare handwriting. The suitability of the algorithm is illustrated by the fitting of fake and genuine signatures, and the expert opinion is based on numerical results on the value of the alignment.

Keywords: Handwriting · Signature · Similarity · Sequence alignment · Bioinformatics

1 Development of Handwriting Research

Handwriting expertise in criminology sciences is always developed by means and techniques taken from other sciences. It has similar evolution to bioinformatics although handwriting analysis lags far behind bioinformatics.

The exploration, modelling and influence of biological processes is nowadays done with the most advanced IT tools and sophisticated methods [1] to analyze the gene stock with Markov models [2], with fuzzy systems for DNA sequencing [3], clustering and classification procedures for data analysis [4], neural networks [5] are applied too. These algorithms have the proven capability to employ in handwriting examination, as evidenced by countless scientific attempt and studies [6–10]. There are several approaches to the automatic extraction of computable handwriting characteristics, but all writers of investigation agree that exploration of most of the classic writing

K. Arai et al. (Eds.): CompCom 2019, AISC 997, pp. 463–473, 2019.
https://doi.org/10.1007/978-3-030-22871-2_31

characteristics, nevertheless—in agreement with Professor Srihari—"While it could be argued that all conventional features could eventually be computational features—when the correct algorithms have been defined the fact remains that most of the conventional features are not yet computable." [8].

2 Bioinformatics Background

When bioinformatics investigates the homology of DNA sequences, they first determine the order of the nucleotides constructing the molecules and then represent it with a code line [12]. The goal is to find the best match between code lines. The popular algorithms [13, 14] for this purpose use patches, inserts, deletions, or exchanges to find the best fit and record the costs of these operations. We consider the greatest similarity between two strings if the operations to transform the two strings into each other is at the lowest cost.

The DNA chain can be defined by a finite string over an alphabet *{A, C, G, T}*. Two arbitrary sequences could be as follows: S_1 = *CAGTGGATCA* and S_2 = *CAGTTGGACAC* (of course, in reality, sequences are much longer and less consistent). The goal is to align the two strings so that the character pairs are monotonous to follow the indexes of the two strings (S_1 and S_2) that is, the string sequence completely delivers sequences $S_{1,1\ldots n}$ and $S_{2,1\ldots m}$, if the gaps are ignored.

Table 1. Improving the quality of alignment

In part match, the value of the alignment is 7										
S_1	C	A	G	T	G	G	A	T	C	A
	|	|	|	|	|	|	|			
S_2	C	A	G	T	G	G	A	C	A	C
Alignment is enhanced by a gap insertion										
S_1	C	A	G	T	G	G	A	T	C	A
	|	|	|	|	|	|	|			
S_2	C	A	G	T	G	G	A	–	C	A
After reorder, value 9										
S_1	C	A	G	T	G	G	A	T	C	A
	|	|	|	|	|	|	|		|	|
S_2	C	A	G	T	G	G	A	–	C	A

The fitting can be evaluated by a point system. For the previous S_1 = *CAGTGGATCA* and S_2 = *CAGTGGACAC* sequences, the value of the assignment is 7 if each character match is worth 1 point. If a gap is inserted in position 8th of S_2, then S_2 9–10th positioning elements move to the right to create two more matches (Table 1).

Similar two strings if one is the subset of the other, or after riding, a partial match is created. In the further interpretation of similarity, it is necessary to determine the cost of transferring one sequence to another by changing, deleting or inserting characters. Costs can be defined for example this way: *deletion = 1*, *insertion* = 2, *change* = 3. The slightest change means the greatest similarity of the sequences. If S_1 = *C-GTGGATCT* and S_2 = *CATT-GACC-*, then aligning and patching them as well as the actions performed are shown in Table 2. At a cost of 11 units, the original fit increased by 4 units.

Table 2. Scores for operations and costs

	1.	2.	3.	4.	5.	6.	7.	8.	9.	10.
S_1	C	–	G	T	G	G	A	T	C	T
	|			|		|	|		|	
S_2	C	A	T	T	–	G	A	C	C	–
The score of alignment = 5										
S_1	C	–	G	T	G	G	A	T	C	T
	|		|	|	|	|	|	|	|	|
S_2	C	–	G	T	G	G	A	T	C	T
The score of alignment = 9										

Sequence	Position	Item	Operation	Costs
S_2	2.	A	deletion	1
S_2	3.	T	change for G	3
S_2	5.	-	insertion (G)	2
S_2	8.	C	change for T	3
S_2	10.	-	insertion (T)	2
$\sum$ **costs:**				**11**

3 Examining the Similarity of Writing Sequences

In writing, the properties of individuality information appear in the writing features, and the purpose of sequence alignment is to determine the similarity of two handwritings. As in the case of DNA sequencing, the sequencing elements of handwriting must first be determined, their correct selection is crucial for comparison and efficiency.

3.1 Defining the Sequence Elements

Let the sequence elements be the writing attributes and their values. Attributes (features) of handwriting are distinguishing character attributes of letters, more precisely the drawing elements of the letters. The Soviet-type handwriting examination method uses a wide variety of writing features (shape, direction, aspect ratio, relative position, complexity, continuity, pen pressure), each of them can have multiple values (e.g. relative position: below, above, near, far). Exploring the features and setting the values can be a task of human handwriting experts or algorithms. If a full investigation is possible, the similarity test should be extended to the full feature complex (Fig. 1), i.e. all the letters of the writing, all items of a letter, and all their categories of features.

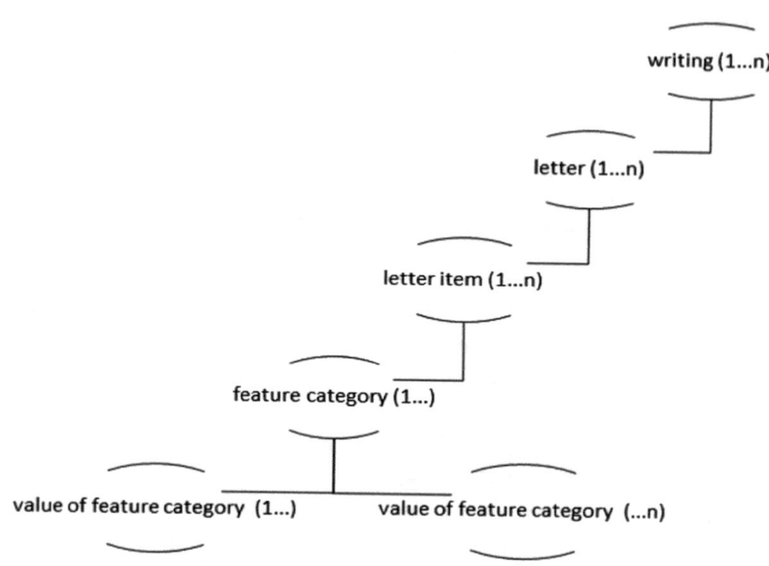

Fig. 1. Full feature complex of handwriting

The elements of the feature chain can be defined by a finite string over an alphabet or numbers:

{letter(a,A...z,Z), letterElement(a...z), letterElementSection(f, m, l), featureCategory(a ... z), featureCategoryValue(1...10)}

It is desirable to create "verbose" character codes for markup of feature values, for example, *Bbfw5*, it means "B"/baseline/first part/weight/very strong.

3.2 Aligning the Writings, the Value of the Matching

The purpose of sequence matching is therefore to determine the degree of similarity between two writing, which will then form the basis for the appropriate expert judgment. The greater the similarity between the two handwriting, the greater the likelihood that the two handwriting comes from the same person.

For the sake of simplicity, consider the writing feature that counterfeiters are less able to imitate. This is the pen pressure appearing on the elements of the letters. The local pen pressure is a local deviation from the average pen pressure generated when the pen is pressed on the writing paper, and on the writing,the substrate is manifested in different thicknesses (dye) of the lines. Therefore, the pixel density of the digitally recorded signature lines refers to the degree of dying on the line. The DBSCAN algorithm can be used to automatically detect pixel density, which is "…locates regions of high density that are separated from one another by regions of low density" [10]. By the help of some sorting algorithm, the line can be classified into the appropriate pressure category. The value of local pressure can be *very weak*, *weak*, *medium*, *strong*, *very strong*.

Apply a unique score system to evaluate the fit. The scores are determined by reference to the same person. A high score indicates the same writer (Table 3).

Table 3. Scoring system for assessment of the similarity measure

Alignment	Point	Comment
full match	3	The writers are certainly the same
similarity	2	The adjacent categories do not necessarily indicate another writer, but some outstanding extraordinary writing situation
gap (-)	1	A missing feature refers to a lack of a letter, which is a natural consequence of simplification, does not refer to different persons
difference	−1	There exist two categories of difference
big difference	−2	There exist three categories of difference
extreme big difference	−3	Creators of handwriting texts must be different if two features fall into two opposite categories (four categories of difference)

3.3 The Type of Opinion

The types of opinions shown in Table 4 are used for traditional writing identification. A percentage of the degree of identity can be given for these. The table explains the meaning of the type of opinion. The authority prefers only categorical and probable opinion. The greatest expert uncertainty is when about half and half of the identities and differences appear (contingency statement).

Table 4. Degree in percentage for identity and their explanation for sequence coding of handwritings

Degree of identity	Type of opinion	Explanation
90–100%	categorical identity statement	absolutely same person
80–90%	probable identity statement	the very likely same person
65–80%		likely the same person
40–65%	contingency statement	it cannot be decided; the similarity may be random
less than 40%	probable exclusionary statement	unlikely same person
less than 20%	categorical exclusionary statement	absolutely different person

3.4 Align Genuine Signatures

The sign-type signatures in Fig. 2 are from the same person (both are genuine), and the characteristics in Table 5 are the values of the pen pressure characteristics of the two signs.

Table 5. Values of the pen pressure characteristics of the two genuine signs

Pen pressure			**Sign 1**	**Sign 2**
Letter	**Letter element**	**Letter element section**	**Value**	
1. B	baseline(b)	first (f)	very strong	very strong
2. B	baseline(b)	last (l)	very weak	very weak
3. B	additional line (a)	first (f)	very strong	medium
4. B	additional line (a)	middle (m)	strong	medium
5. B	additional line (a)	last (l)	medium	medium
6. I	starting line (s)	first (f)	weak	strong
7. I	baseline(b)	middle (m)	very strong	-
8. I	baseline(b)	last (l)	weak	weak
9. I	dot (d)	first (f)	strong	strong
10. I	dot (d)	middle (m)	medium	-
11. I	dot (d)	last (l)	very weak	very weak

Fig. 2. The two signatures from the same person

Write the contents of Table 5 as a sequence chain:
S_1 = *Bbf5Bbl1Baf5Bam4Bal3Isf2Ibm5Ibl2Idf4Idm3Idl1*
S_2 = *Bbf5Bbl1Baf3Bam3Bal3Isf4Ibm-Ibl2Idf4Idm-Idl1*,
Then match them (Table 6).

Table 6. Scores of alignment and the alignments itself of two genuine signs

	1.	2.	3.	4.	5.	6.	7.	8.	9.	10.	11.	$\sum$
S_1	Bbf5	Bbl1	Baf5	Bam4	Bal3	Isf2	Ibm5	Ibl2	Idf4	Idm3	Idl1	
score	0003	0003	000-**1**	0002	0003	000-1	0001	0003	0003	0001	0003	**20**
S_2	Bbf5	Bbl1	Baf3	Bam3	Bal3	Isf4	—	Ibl2	Idf4	—	Idl1	

3.5 Align Fake and Genuine Signatures

The two sign-type signatures in Fig. 3 are not from the same person. Table 7 shows the values of the detected pen pressure characteristics.

Fig. 3. The fake and the genuine signature (the forged is made of this real pattern)

Table 7. Values of the pen pressure characteristics of the fake and genuine sign

Pen pressure			**Fake sign**	**Genuine sign**
Letter	**Letter element**	**Letter element section**	**Value**	
1 B	baseline(b)	first (f)	very strong	very strong
2 B	baseline(b)	last (l)	very weak	medium
3 B	additional line (a)	first (f)	very strong	weak
4 B	additional line (a)	middle (m)	strong	medium
5 B	additional line (a)	last (l)	medium	medium
6 I	starting line (s)	first (f)	weak	medium
7 I	baseline(b)	middle (m)	weak	strong
8 I	baseline(b)	last (l)	weak	strong
9 I	dot (d)	first (f)	strong	very strong
10 I	dot (d)	middle (m)	medium	very strong
11 I	dot (d)	last (l)	very weak	very weak

Write the contents of (Table 7) as a sequence chain:
$S_1 = Bbf5Bbl1Baf5Bam4Bal3Isf2Ibm2Ibl2Idf4Idm3Idl1$
$S_2 = Bbf5Bbl3Baf2Bam3Bal3Isf3Ibm4Ibl4Idf5Idm5Idl2$
Then match them (Table 8).

Table 8. The alignment between the two sequences of features and the Score of alignments

	1.	2.	3.	4.	5.	6.	7.	8.	9.	10.	11.	$\sum$
S₁	Bbf5	Bbl1	Baf5	Bam4	Bal3	Isf2	Ibm2	Ibl2	Idf4	Idm3	Idl1	
score	000**3**	000-1	000-**1**	0002	000**3**	000**2**	000-**1**	000-1	000**2**	000-**1**	000**3**	**10**
S₂	Bbf5	Bbl3	Baf2	Bam3	Bal3	Isf3	Ibm4	Ibl4	Idf5	Idm5	Idl1	

3.6 Value of the Align

The sum of the two genuine signatures shown in Table 5 is 20 points, using the points system in Table 3. By examining the 11-letter local pressure characteristic, the maximum match would be 11 * 3 = 33 points. Thus, the 20 points correspond to approximately 61% similarity to Table 4, and it is also apparent that the 61% result suggests a contingency statement. This means that the decision on the origin of the two signatures shown in Fig. 2 is based on a single feature, based on the local pen pressure: the two signatures can come from the same person, but can also come from two different persons. The result is that the origin cannot be decided.

The fake and genuine signature alignment value is 10 points, using the scoring system of Table 4, 10/33, approximately 30%. This is a probable negative opinion in Table 4.

3.7 Applying Sequence Matching Operations

Writing variation is the reason why deletion-insert-replacement operations can be interpreted for signatures. With these changes, the letters that we want to compare can be aligned with each other, making comparison easier (Table 9). Operations can also determine the cost of transferring one signature to another. Operational execution or change costs can be seen as differences at the expense of similarity, so you can measure the extent to which controversial writing could be put together from the writing pattern.

(1) Gap insertion between sequence elements: the letters to compare can be a shift from one character to another, making comparison easier (Table 9, Column A).

(2) Deletion of the sequence element: also allows sliding one character to another (Table 9/Column B).

(3) Changing from one sequence element to another sequence element: the costs of operations designate the scope of possible operations that can be applied to the controversial handwriting for the reason that whether the handwriting can be compiled together from patterns of handwriting provided in samples. If there is a large number of patterns of handwritings for comparative analysis, it is possible to reconstruct the controversial writing in all details from the available patterns, of course, in the case the samples of handwritings originated from the same person. If different persons created the samples of handwriting, the disputed handwriting cannot be constructed from samples of handwriting that are dedicated for comparison so that the investigated handwriting cannot be reconstructed by any permitted operations on the sequences of feature codes, or if even there is a chain of operations providing a satisfactory reconstruction results it would be associated with an unreasonably high cost.

Table 9. Operations of alignment on handwritings

Detail of an original writing	9/A Alignment operations Insertion	9/B Deletion
 In position 6th there are different letter elements (upper oval, bottom start line)	 Inserting a gap in the top 6th position, the two "ó" letters are placed under one another in position 7th	 By deleting the bottom start line "ó", the two "ó" letters are placed under one another in the 6th position

4 Conclusion and Results Evaluation

The key step in applying sequence alignment to handwriting is to correctly select the sequence elements (handwriting attributes and their values). The exploration of features can be a task of a handwriting expert or algorithm(s).

The goodness of the fitting value can be judged by the experience of the human scribe: the quantification of the sequence alignment is consistent with the result of the traditional expert examination.

The examples presented can be evaluated with the help of experience on human handwritings: whether the matching between the aligned sequences in quantified forms how correlates with the result of the traditional examination of experts.

As a handwriting expert, the authors can claim that the matching score 61% of the two genuine signatures and the fake-genuine matching score 30% are acceptable. Although an incompetent person is likely to think that the degree of the similarity on the signatures in Fig. 3 assessed as 30% is not correct as they seem much closer to each other, but it must be borne in mind that a layman's eyes are captured by easy-to-imitate formatting that is less relevant in identification than a more hidden (such as pen-pressure) feature. Another handwriting expert could find 30% more exaggerated, but this value can be explained by physiological determination: people usually press the pen stronger on paper at typical places, less at other places, because this is the way for convenient and effective handwriting.

In any case, there are two important conclusions:

- A single feature does not give a precise answer to the origin of handwriting, a 61% match for the same person only allows a contingency statement that the authorities do not like, furthermore such an opinion gives a bad impression of the capability of the expert. It is therefore strongly advised to make decisions according to Soviet methodology based on several types of handwriting features.

- The sequence alignment method for comparing texts, DNA sequences, etc. works well for comparing of human handwritings (for signatures and longer textual handwritings, too) but accurate results can only be obtained by examining the full spectra of features in their complexity.

Acknowledgment. The project has been supported by the European Union, co-financed by the European Social Fund (EFOP-3.6.3-VEKOP-16-2017-00002).

References

1. Pearson, W.R.: An Introduction to Sequence Similarity ("Homology") Searching. Univ. of Virginia, Charlottesville, VA (2014). https://www.ncbi.nlm.nih.gov/pmc/articles/PMC3820096/. Last accessed 21 Jan 2018
2. Yoon, B-J.: Hidden Markov Models and Their Applications in Biological Sequence Analysis. Dep. of Electrical&Computer Engineering, Texas A&M Univ (Ch. 3.) (2009). https://www.ncbi.nlm.nih.gov/pmc/articles/PMC2766791/. Last accessed 21 Jan 2018
3. Cordón, O., Gomide, F., Herrera, F., Hoffmann, F. Magdalena L.: Ten years of genetic fuzzy systems: current framework and new trends. Univ. of Granada, Spain (2004). https://www.sciencedirect.com/science/article/pii/S0165011403001118. Last accessed 03 Dec 2017
4. Negi, S.S., Schein, C.H., Ladics, G.S., Mirsky, H., Chang, P. Rascle, J-B., Kough, J.,… Braun, W.: Functional Classification of Protein Toxins as a Basis for Bioinformatic Screening. Univ. of Texas, Sealy Center for Structural Biology and Molecular Biophysics, Galveston, USA (2017) https://doi.org/10.1038/s41598-017-13957-1. Last accessed 15 June 2018
5. Serra, A., Galdi, P., Tagliaferri, R.: Machine Learning for Bioinformatics and Neuroimaging. Overview. MRC Centre for Reproductive Health, Edinburgh, UK (2018). https://www.researchgate.net/publication/322626573_Machine_Learning_for_Bioinformatics_and_Neuroimaging. Last accessed 21 Aug 2018
6. Igarza, J.J., Goirizelaina, I., Espinosa, K., Hernáez, I., Méndez, R., Sánchez, J.: Online Handwritten Signature Verification Using Hidden Markov Models. Univ. of Basque Country, Spain (2017). https://link.springer.com/chapter/10.1007/978-3-540-24586-5_48. Last accessed 29 May 2018
7. Shaout, A., Sterniak, J.: Neural and Fuzzy Methods for Handwritten Character Recognition. Univ. of Michigan, Dearborn, MI. (2010). https://pdfs.semanticscholar.org/0ef8/5b102c1df561e3e69627eaba8b2cf2966d1f.pdf. Last accessed 21 Aug 2018
8. Xu, B., Lin, D., Chao, H., Li, W., Liao, Q.: Off-line Signature Verification Using Linear Regression Classification. Univ of Tsinghua, China (2015). https://www.researchgate.net/publication/301444543_Off-line_Signature_Verification_Using_Linear_Regression_Classifier. Last accessed 15 June 2018
9. Kumar, P., Singh, S., Grag, A., Prabhat, N.: Hand Written Singature Recognition & Verification using Neural Network. Research Paper. PIET, Samalkha (2013). https://s3.amazonaws.com/academia.edu.documents/36795392/Online_Signature_verification_algorithm.pdf?AWSAccessKeyId = AKIAIWOWYYGZ2Y53UL3A&Expires = 1540668324&Signature = JlZZIWrrY0oCKVq7Z%2BIkhu6QwUc%3D&response-content-disposition=inline%3B%20filename%3DOnline_Signature_verification_algorithm.pdf. Last accessed 29 May 2018

10. Hafemann, L.G., Sabourint, R., Oliveora, L.S.: Learning Features for Offline Handwritten Signature Verification using Deep Convolutional Neural Networks. Univ. Of Quebec, Montreal, Canada, Brazil (2017) https://arxiv.org/abs/1705.05787. Last accessed 10 April 2018
11. Srihari, N.S., Cha, S-H., Arona, H., Lee, S.: Individuality of Handwriting, Univ. of Buffalo, USA (2001). https://cedar.buffalo.edu/papers/articles/Individuality_Handwriting_2002.pdf. Last accessed 12 April 2018
12. Autenrith, F., Isralewitz, B., Suthey-Schulten, Z., Sethi, A., Pogorelov, T.: Bioinformatics and Sequence Alignment. Univ. Of Illinois, San Francisco Workshop (pp. 6–7) (2005). https://www.ks.uiuc.edu/Training/Tutorials/science/bioinformatics-tutorial/bioinformatics.pdf. Last accessed 15 April 2018
13. NCBI Homepage. https://blast.ncbi.nlm.nih.gov/Blast.cgi. Last accessed 18 Mar 2018
14. UVaFASTA Server. https://fasta.bioch.virginia.edu/fasta_www2/fasta_list2.shtml. Last accessed 18 Mar 2018
15. Tan, P-N., Steinbach, M., Kumar, V.: Introduction to data mining (Chapter 8, p. 526) cluster analysis: basic concepts and algorithms. http://www-users.cs.umn.edu/~kumar/dmbook/ch8.pdf

Multi-channel Speaker Separation Using Speaker-Aware Beamformer

Conggui Liu[1(✉)] and Yinhua Liu[2]

[1] Tokyo Institute of Technology, Tokyo, Japan
conggui@ks.cs.titech.ac.jp
[2] Institue for Future, Qingdao University, Qingdao, China
liuyinhua@qdu.edu.cn

Abstract. In this work, we address the problem of multi-channel speech separation. We use a localization network to estimate delay times to compute steering vectors and derive spatial filters using these vectors and mixtures, in a similar way as a recently proposed method. The beamformer has difficulties in speech separation when speakers are close to each other or their locations are estimated inaccurately. To overcome this problem, we propose to inform beamforming about speakers so that it tracks speakers using not only locations but also speaker characteristics through utterances. We investigate and compare different methods of using the speaker information in beamforming such as multiplying steering vectors with speaker weights. Experiments on simulated data demonstrate that the proposed method can improve the performance of both speech separation and speech recognition.

Keywords: Speech separation · i-vector · Multi-channel · Deep neural network · Speech recognition

1 Introduction

Although automatic speech recognition (ASR) has recently advanced greatly in noisy conditions, it performs poorly for overlapped speech, where multiple speakers are active simultaneously. The overlapped speech is frequent in many scenarios such as automatic meeting translation. Conventionally, speech separation (SS) is required to extract each speaker's voice before ASR. Speech separation has been investigated using both single and multi-channel techniques, including non-negative matrix factorization (NMF) [1], independent component analysis (ICA) [2] or Computational Auditory Scene Analysis (CASA) [3].

This work addresses the task of separating all speakers from multi-channel recordings. Recently, there have been great advances in separating speech from noise by a scheme combining mask estimation neural networks with beamformer [4–6]. Such a scheme has been suggested by recent works and seems to be effective. The mask estimation neural network extracts a time-frequency mask from a recording. The mask is then used to compute covariance matrices of speech and

K. Arai et al. (Eds.): CompCom 2019, AISC 997, pp. 474–484, 2019.
https://doi.org/10.1007/978-3-030-22871-2_32

noise that are then employed in estimating spatial filters. This scheme enables to accurately estimate the mask of speech because speech is very different from noise in time-frequency characteristics.

Using the above-mentioned scheme in separating speech from speech is, however, not straight-forward. Several works have been developed to address speech separation by combining neural networks with beamformer [7–9]. In [9], the neural networks are applied for mask estimation that is then used to compute spatial filters. In contrast to mask estimation, neural networks are applied for estimating the time difference of arrival (TDOA) in [8]. Because of using multi-channel recordings, these works are able to use spatial cues for separating speech. It has difficulties in separating close speakers, e.g. if speakers' locations are close to each other, it may fail to separate these speakers.

In this work, we explore an alternative approach where we inform each spatial filter about the speaker so that it still enables to separate close speakers. This is accomplished by using additional information in each spatial filter about the speaker such as speaker weights learned from i-vectors through an auxiliary network. This scheme enables to separate close speakers as the proposed beamformer can distinguish speakers by using speaker characteristics. Furthermore, it is also robust to estimation errors of speakers' directions.

Different mechanisms for using speaker information in neural networks were previously explored for the task of speaker extraction. The most common methods include using speaker representation to adapt one layer [10] or using speaker representations as additional input feature [11]. They can obtain good performance on mask estimation and are able to be applied for speech separation. However, they need reference signals which are not available in many real conditions. In contrast, our scheme doesn't demand reference signals as it allows us to extract speaker representations from separated speech.

The rest of the paper is structured as follows. Section 2 describes neural network based beamforming. In Sect. 3, we describe our proposed method in detail. Section 4 describes data, experimental settings, and discuss the results.

2 Neural Network Based Beamforming

In this section, we review a previous scheme proposed in [8] for multi-speaker speech recognition and its difficulty. The scheme is shown in Fig. 1.

2.1 Overview of the Method

The observed signal at the m-th microphone in the frequency domain can be modeled as:

$$X_m(t,f) = \sum_n H_{nm}(t,f) S_n(t,f) + N(t,f), \tag{1}$$

Where t denotes the index of frames, f is frequency bin, $S_n(t,f)$ is the n-th speaker's signal, and $N(t,f)$ is the noise signal. $H_{nm}(t,f)$ is the transform function from speaker n to microphone m.

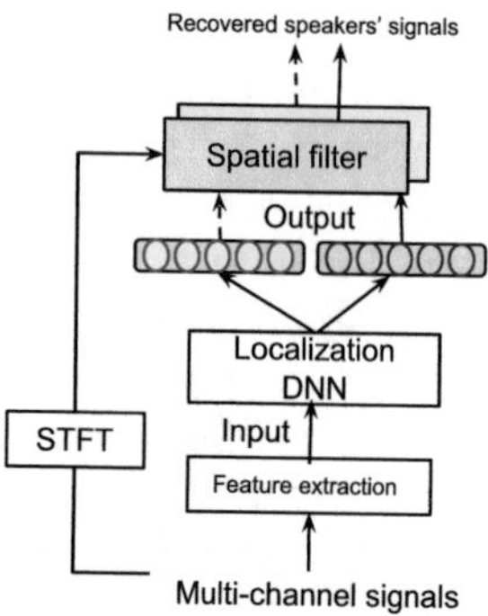

Fig. 1. The scheme of neural network based beamforming for two-speaker cases [8]. The output of the localization DNN includes 8 TDOAs for each speaker, e.g. the output with blue color belongs to one speaker, and the output with red color belongs to the other speaker. Each speaker' TDOAs are used in the corresponding spatial filter, e.g. the TDOAs with blue color pass into the spatial filter block with blue color for recovering the speaker's signal. STFT refers to short-time Fourier transform.

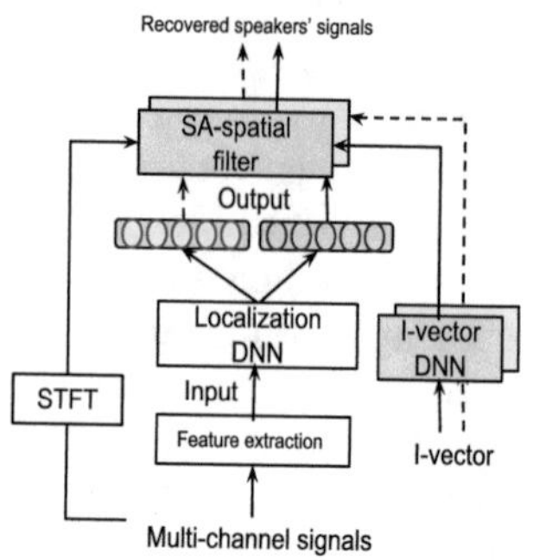

Fig. 2. The scheme of the proposed speaker-aware method. Taking two-speaker cases as an example. The i-vector DNN transforms utterance-level i-vector into speaker weights. Different colors denote the processing for different speakers. Each speaker has an i-vector DNN, which shares parameters with that of other speakers. The SA-spatial filter refers to speaker-aware spatial filter. The i-vector DNN can be jointly optimized with the localization DNN or independently optimized.

The beamforming typically aims to recover each speaker's signal by using a spatial filter. For the n-th speaker, the signal $S_n(t,f)$ at frame t and frequency bin f is estimated by:

$$\widehat{S}_n(t,f) = \mathbf{w}_n^H(f)\mathbf{X}(t,f), \tag{2}$$

where $\mathbf{X}(t,f) = [X_1(t,f), .., X_m(t,f), .., X_M(t,f)]^T$, and $\mathbf{w}_n(f)$ denotes the spatial filter for speaker n. Following [6], we use Minimum Variance Distortionless Response beamformer (MVDR), where spatial filters are computed by:

$$\widehat{\mathbf{w}}_n(f) = \frac{(\mathbf{R}(f))^{-1}\mathbf{a}_n(f)}{\mathbf{a}_n^H(f)(\mathbf{R}(f))^{-1}\mathbf{a}_n(f)}, \tag{3}$$

where $\mathbf{R}(f)$ is the spatial covariance matrix of the noise-plus-interference signal. Usually, this matrix is approximately estimated by computing the covariance matrix of observed signals. $\mathbf{a}_n(f)$ expresses the n-th speaker's steering vector, which is computed by using the time difference of arrival (TDOA).

$$\mathbf{a}_n(f) = [1, .., e^{-2i\pi f \Delta\tau_{n,m}}, .., e^{-2i\pi f \Delta\tau_{n,M}}]^T \tag{4}$$

where $\Delta\tau_{n,m}$ denotes the TDOA of speaker n between microphone m and the reference microphone.

The TDOA $\Delta\tau_{n,m}$ is estimated by neural network. In particular, the neural network estimates frame-level TDOAs and these TDOAs are then combined using an average operation to get an utterance level TDOA:

$$\Delta\tau_{n,m}(t) = g(\mathbf{x}(t)), \tag{5}$$

$$\Delta\tau_{n,m} = mean(\Delta\tau_{n,m}(t)). \tag{6}$$

Here, g is the transform of the neural network, and $\mathbf{x}(t)$ denotes the input feature of the neural network. The neural work is trained by using ideal TDOAs as targets, which are obtained by using microphone configuration and speakers' locations.

2.2 The Problem

In the previous work [8], speech separation is accomplished by combining a localization neural network with beamforming. While speakers' locations are not close to each other and speakers' TDOAs can be accurately estimated from the neural network, speakers' signals can be effectively separated from observed signals. Such conditions may however not be fulfilled in some real scenarios. The problem can be avoided if we use mask-based methods [12] instead of beamforming for recovering speaker. However, the estimated mask may cover more than one speakers. This results in failing to separate speakers.

3 Proposed Method

There are different ways to solve the above-mentioned problem. In this paper, we propose to inform each spatial filter of the beamformer about the speaker. The processing of speech separation can be expressed as:

$$\widehat{S}_n(t, f) = \mathbf{w}_n^H(f, g_s(\lambda_n))\mathbf{X}(t, f), \tag{7}$$

where λ_n denotes a vector representing speaker n, and g_s is the transform of an auxiliary network. In this way, the beamformer can distinguish signals using not only speakers' directions but also speaker-specific characteristics. Informing each spatial filter about the speaker allows our system to learn to track the speaker's characteristics and thus solve the above-mentioned problem.

Different ways can be applied for extracting speaker representations. In this paper, we explore to use an i-vector as speaker representation for the following two reasons. The i-vector as a representation of speaker characteristics [13] has been successfully applied for speaker verification [14] and speaker recognition [15] recently. In addition, it can be extracted without using speaker identities and thus allows our system to generalize in open-speaker conditions, e.g. test speakers are unseen in the training set.

Previous works have explored to modify the behavior of neural networks for adapting an acoustic model or for estimating a mask in speaker extraction. Different from them, we aim to modify the behavior of beamforming for speech separation. To modify the behavior of beamforming, we have to solve two challenges: (1) using the i-vector for speech separation, and (2) modifying the spatial filters. In this paper, an auxiliary neural network is trained to learn to transform the i-vector to a vector with speaker weights. The speaker weights are then applied to modify the spatial filters. The proposed scheme is shown in Fig. 2. We explore two ways to use the speaker weights in spatial filters: (1) beamformer with speaker-aware weights and (2) beamformer with speaker-aware steering vectors. In addition, we investigate two loss functions as training criterion: (1) square error (SE) and (2) cross-entropy error (CE).

3.1 Beamformer with Speaker-Aware Weights

One straightforward way is to use speaker-aware weights on the spatial filters. We refer this method to SW-Beam. In beamforming, the spatial filters are assumed to be speaker-specific. Thus, it is reasonable to multiply each spatial filter by speaker-specific weights. For speaker n, the processing of MVDR can be modified as:

$$\widehat{S}'_n(t,f) = \alpha_n(f)\frac{\mathbf{a}_n^H(f)\mathbf{R}(f)^{-1}\mathbf{X}(t,f)}{\mathbf{a}_n^H(f)\mathbf{R}(f)^{-1}\mathbf{a}_n(f)}. \tag{8}$$

Here, $\alpha_n(f)$ denotes the speaker-specific weight of speaker n. The weight can be obtained using an auxiliary neural network which has a speaker representation λ_n as its input.

This approach is suitable for the case of a training set, which has a lot of speakers. In this case, effective speaker-aware weights can be learned through the auxiliary neural network and enable to track speakers. However, such conditions may not be satisfied in real scenarios as it is expensive to collect such training data.

3.2 Beamformer with Speaker-Aware Steering Vector

To overcome the drawback mentioned in the beamformer with speaker-aware weights and be able to effectively use speaker weights in beamforming, we utilized a speaker-aware steering vector in the beamformer. We refer this method to SV-Beam.

In contrast to the previous approach, speaker weights are not used to directly weigh the spatial filters. Instead, the speaker weights weigh a steering vector, which is then used in computing spatial filters. We can express this approach as:

$$\widehat{S}'_n(t,f) = \frac{\mathbf{h}_n^H(f)\mathbf{R}(f)^{-1}\mathbf{X}(t,f)}{\mathbf{h}_n^H(f)\mathbf{R}(f)^{-1}\mathbf{h}_n(f)}. \tag{9}$$

Here, $\mathbf{h}_n(f)$ is a speaker-aware steering vector, which is computed by:

$$\mathbf{h}_n(f) = \mathbf{a}_n(f) \odot \alpha_n(f), \tag{10}$$

where $\odot$ denotes inner product operation. Note, we modify the steering vector rather than the spatial matrix $\mathbf{R}(f)$ because only the steering vector is assumed to be speaker-specific.

In beamforming, the steering vector usually has a norm constraint. Thus, adapting the steering vector can be learned from a training set without lots of speakers.

3.3 Training Criterion

In our system, there are three neural networks: localization DNN, acoustic-model DNN, and i-vector DNN. Following [8], we firstly pre-train localization DNN and acoustic-model DNN. Then, we train the i-vector DNN using clean speech as targets or alignments as targets. With the clean speech, the parameters in the i-vector DNN are optimized by minimizing the following squared error (SE):

$$\mathcal{L}_{SE} = \frac{1}{TF} \sum_{t,f} (|\widehat{S}'_n(t,f)| - |S_n(t,f)|)^2. \tag{11}$$

For speech recognition purpose, the i-vector DNN is trained by minimizing frame-level cross entropy error (CE):

$$\mathcal{L}_{CE} = -\frac{1}{T} \sum_{t}^{T} \log p_n(\widehat{\mathbf{S}}'_n(t)), \tag{12}$$

$$\widehat{\mathbf{S}}'_n(t) = [\widehat{S}'_n(t,1), .., \widehat{S}'_n(t,f), .., \widehat{S}'_n(t,F)]^T, \tag{13}$$

Here, $p_n(\widehat{\mathbf{S}}'_n(t))$ is the output of the acoustic model DNN. Note other parameters in Eqs. (11) and (12) are fixed during training.

3.4 Speech Reconstruction

The speaker-aware beamformer aims to reduce interfering speakers and keep the specific speaker. However, this behavior may lead to producing significant artifacts due to a limited number of training speakers. To address this, we use the following formula to recover speech in the test phase:

$$\tilde{S}_n(t,f) = \eta \widehat{S}'_n(t,f) + (1-\eta)\widehat{S}_n(t,f), \tag{14}$$

where η is a control parameter. In this paper, $\eta = 0.5$.

Given estimated spectrum and phase information in mixtures, we can reconstruct time-domain signals using inverse short-time Fourier transform. For each test speaker, assuming that the clean speech is unavailable, we extract the speaker's i-vector from the separated signal using the neural network based beamforming.

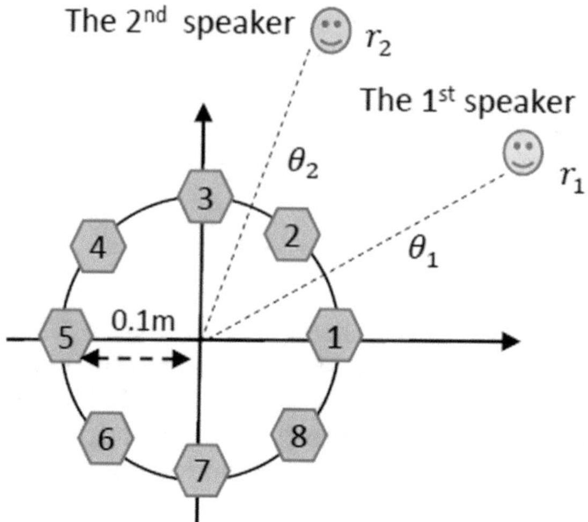

Fig. 3. The configuration of a circular microphone array with 8 microphones. Two speakers locate at r_1 and r_2 meter distance from the array in angle θ_1 and θ_2, respectively.

4 Experiments

We carried out all experiments on simulated data. The accuracy is evaluated with measures: Signal-to-interference ratio (SIR) [16] and Word error ratio (WER). In this section, we described generated data, experimental settings, and discuss results.

4.1 Data

The simulated data is based on WSJCAM0 corpus. In particular, we simulated multi-channel mixtures by mixing multi-channel observed signals from one speaker into that from another speaker and adding noise from the REVERB Challenge corpus. For each speaker, the multi-channel observed signals were generated by convolving each clean speech utterance with artificial room impulse responses (RIRs). The RIRs are generated using *Roomsimove Toolbox* with a circular microphone array with 8 microphones, 20 cm in diameter and RT60 ranging from 0.2 s to 1.0 s. The configuration of the array is shown in Fig. 3. The signal-to-noise ratio (SNR) ranges from 0 dB to 30 dB. For training data, two speakers are located at 1, 2, or 4 m distance from the array in randomly selected angles from 0 to 360°.

4.2 Experimental Settings

We used the same configuration for the localization network as in [8], which consists of three fully connected layers with ReLu activations and a fully connected

Table 1. SIR (dB) results for different speech separation methods on simulated data at different SNRs. Beam ≡ the baseline, SW-beam ≡ the beamformer with speaker-aware weights, SV-Beam ≡ the beamformer with speaker-aware steering vectors.

SNR(dB)	Beam	SW-Beam	SV-Beam
0	3.84	4.62	4.67
5	4.15	4.78	4.89
10	4.38	4.88	5.08
20	4.80	5.00	5.16
30	5.12	5.04	5.17

layer. For these layers, the numbers of units are 644-1024-1024-16, respectively. The input of the network includes generalized cross-correlations (GCCs) and utterance-level TDOAs extracted from Generalized cross-correlations phase transform (GCC-PHAT) [17]. The output is the corresponding target TDOAs for all speakers between 8 microphone pairs. Here, we used utterance-level TDOAs for estimating target TDOAs because the utterance-level TDOAs cannot be estimated accurately for some cases, e.g. two speakers are too close to each other, and allow the GCCs feature to include global information. For the training, we used Adam optimization scheme to minimize mean square error on the simulated data.

The i-vector DNN consists of three fully connected layers with ReLu activations and a fully connected layer. The numbers of units are 100-512-512-513, respectively. The input of the network is i-vector extracted from each utterance using an i-vector extractor. The i-vector extractor is trained on the clean speech of WSJCAM0 corpus. The output is speaker weights, which is used in beamforming. The network is trained to minimize cross-entropy error with target alignments or mean square error with clean spectra.

Following [8], the acoustic model is first trained on clean speech, called as C-AM, then adapted on separated speech, called as R-AM, and further adapted jointly with the localization DNN on the simulated data, called as J-AM. We used the combination of localization DNN, the MVDR beamforming [18], and C-AM as a baseline method.

4.3 Speech Separation

Table 1 summarizes the results of speech separation using different methods. First, we carried out experiments using the beamformer with speaker-aware weights. The results show an SIR improvement in low-SNR conditions, which confirms the ability of speaker-aware beamformer to separate speakers' speech. We also observe a lower SIR compared to the baseline when the SNR is 30dB. This suggests that using speaker weights directly on spatial filters is beneficial for reducing noise but has difficulties in suppressing interfering speakers due to a limited number of training speakers.

Table 2. Speech recognition results of different methods on simulated data. LF ≡ loss function: (1) square error (SE) and (2) cross-entropy error (CE). AM ≡ the acoustic model: (1) C-AM obtained on clean data, (2) R-AM obtained by adapting C-AM on separated data, and (3) J-AM obtained by jointly adapting R-AM and Localization DNN. The simulated data has SNRs ranging from 0dB to 30 dB.

Method	LF	AM	WER(%)
Beam	-	C-AM	83.1
SW-Beam	SE	C-AM	82.7
SW-Beam	CE	C-AM	80.9
SW-Beam	CE	R-AM	60.5
SW-Beam	CE	J-AM	58.6
SV-Beam	SE	C-AM	82.5
SV-Beam	CE	C-AM	81.5
SV-Beam	CE	R-AM	52.8
SV-Beam	CE	J-AM	48.8

Then, we investigated the beamformer with speaker-aware steering vectors. In this experiment, speaker-weights obtained from an i-vector DNN are used to inform the steering vector about the speaker. We can see an improvement compared to the baseline and the beamformer with speaker-aware weights. This suggests that informing steering vectors about speakers and then using the speaker-aware steering vectors in beamforming, is beneficial.

4.4 Speech Recognition

Table 2 presents speech recognition results on word error rate (WER). For speech recognition purpose, there are different options for training our models, e.g. the loss function can be a square error or cross-entropy error, the acoustic model can be C-AM, R-AM, or J-AM, and the training target can be clean spectrum or alignments. We investigated our methods in several settings. First, we carried out our experiments using C-AM with different loss functions. We can see an improvement compared to the baseline and an improvement when comparing that with a cross-entropy error to that with a square error. This means that training models with WER-related loss functions are more beneficial for speech recognition than with spectrum-related functions.

Then, we investigated different acoustic models for different methods. In the following experiments, models are trained using a cross-entropy error because changing acoustic model has no effect on the square error loss. We can see a significant improvement compared to these with a clean acoustic model (C-AM), and a consistent improvement for different methods when comparing that with a jointly trained acoustic model to that with a retrained acoustic model. This suggests that combing robust acoustic model with speaker-aware beamformer is beneficial for speech recognition.

Finally, compared to the beamformer with speaker-aware weights, the method with speaker-aware steering vectors has consistently better performance in different setups because speaker-aware steering vectors work better on reducing interfering speakers. This suggests that not only encoding speaker characteristics in beamforming but also using such characteristics in a good way is important for speech recognition.

5 Conclusions

In this paper, we proposed a method for informing beamforming about speakers so that it can effectively separate speakers from mixtures. This enables to use previously proposed neural network based beamformer for separating speakers when speakers are close to each other or their locations are estimated inaccurately. The experimental results show that speaker-aware beamformer allows better tracking speakers. Following the findings in our investigations, we plan to explore different ways to utilize speaker representations in future work.

References

1. Schmidt, M.N., Olsson, R.K.: Single-channel speech separation using sparse non-negative matrix factorization. In: INTERSPEECH, pp. 2614–2617 (2006)
2. Lee, T.-W.: Independent Component Analysis, pp. 27–66. Springer, Heidelberg (1998)
3. Cooke, M.: Modelling Auditory Processing and Organisation, vol. 7. Cambridge University Press, Cambridge (2005)
4. Erdogan, H., Hershey, J.R., Watanabe, S., Mandel, M.I., Le Roux, J.: Improved MVDR beamforming using single-channel mask prediction networks. In: Interspeech, pp. 1981–1985 (2016)
5. Heymann, J., Drude, L., Haeb-Umbach, R.: Neural network based spectral mask estimation for acoustic beamforming. In: 2016 IEEE International Conference on Acoustics, Speech and Signal Processing (ICASSP), pp. 196–200. IEEE (2016)
6. Higuchi, T., Ito, N., Araki, S., Yoshioka, T., Delcroix, M., Nakatani, T.: Online MVDR beamformer based on complex gaussian mixture model with spatial prior for noise robust ASR. IEEE/ACM Trans. Audio Speech Lang. Process. **25**(4), 780–793 (2017)
7. Drude, L., Haeb-Umbach, R.: Tight integration of spatial and spectral features for BSS with deep clustering embeddings. In: Proceedings of Interspeech, pp. 2650–2654 (2017)
8. Liu, C., Inoue, N., Shinoda, K.: A unified network for multi-speaker speech recognition with multi-channel recordings. In: Accepted to APSIPA (2017)
9. Yoshioka, T., Erdogan, H., Chen, Z., Alleva, F.: Multi-microphone neural speech separation for far-field multi-talker speech recognition. In: 2018 IEEE International Conference on Acoustics, Speech and Signal Processing (ICASSP), pp. 5739–5743. IEEE (2018)
10. Zmolikova, K., Delcroix, M., Kinoshita, K., Higuchi, T., Ogawa, A., Nakatani, T.: Speaker-aware neural network based beamformer for speaker extraction in speech mixtures. In: Interspeech (2017)

11. Wang, Q., Muckenhirn, H., Wilson, K., Sridhar, P., Wu, Z., Hershey, J., Saurous, R.A., Weiss, R.J., Jia,Y., Moreno, I.L.: Voicefilter: targeted voice separation by speaker-conditioned spectrogram masking. arXiv preprint arXiv:1810.04826 (2018)
12. Hershey, J.R., Chen, Z., Le Roux, J., Watanabe, S.: Deep clustering: discriminative embeddings for segmentation and separation. In: ICASSP, pp. 31–35 (2016)
13. Dehak, N., Kenny, P.J., Dehak, R., Dumouchel, P., Ouellet, P.: Front-end factor analysis for speaker verification. IEEE/ACM Trans. Audio Speech Lang. Process. **19**(4), 788–798 (2011)
14. Nautsch, A., Darmstadt, H.: Speaker verification using i-vector. University of Applied Science Hochschule Darmstadt, Germany (2014)
15. Kanagasundaram, A., Vogt, R., Dean, D.B., Sridharan, S., Mason, M.W.: I-vector based speaker recognition on short utterances. In: INTERSPEECH, pp. 2341–2344 (2011)
16. Vincent, E., Gribonval, R., Févotte, C.: Performance measurement in blind audio source separation. IEEE/ACM Trans. Audio Speech Lang. Process. **14**(4), 1462–1469 (2006)
17. Knapp, C., Carter, G.: The generalized correlation method for estimation of time delay. IEEE/ACM Trans. Acoust. Speech Signal Process. **24**(4), 320–327 (1976)
18. Mestre, X., Lagunas, M.A.: On diagonal loading for minimum variance beamformers. In: ISSPIT, pp. 459–462 (2003)

Revisiting Skip-Gram Negative Sampling Model with Rectification

Cun (Matthew) Mu(✉), Guang Yang, and Yan (John) Zheng

Jet.com/WalmartLabs, Hoboken, NJ 07030, USA
{matthew.mu,guang,john}@jet.com

Abstract. We revisit skip-gram negative sampling (SGNS), one of the most popular neural-network based approaches to learning distributed word representation. We first point out the ambiguity issue undermining the SGNS model, in the sense that the word vectors can be entirely distorted without changing the objective value. To resolve the issue, we investigate the intrinsic structures in solution that a good word embedding model should deliver. Motivated by this, we rectify the SGNS model with quadratic regularization, and show that this simple modification suffices to structure the solution in the desired manner. A theoretical justification is presented, which provides novel insights into quadratic regularization. Preliminary experiments are also conducted on Google's analytical reasoning task to support the modified SGNS model.

Keywords: Word embedding · SGNS model · Quadratic regularization

1 Introduction

Distributed word representations, a.k.a. word embeddings, represent each word with a real-valued vector as an approximation to its linguistic meaning. Different from the traditional discrete and sparse one-hot encoding, such continuous and dense representations are shown to better capture syntactic and semantic regularities in language, and have been successfully applied in various natural language processing tasks, such as document classification [1], information retrieval [2,3], question answering [4,5], named entity recognition [6,7], and parsing [8].

One of the main approaches to learning distributed word representation is the neural-network based one [9–17], in which word vectors are trained to maximize the likelihood of word-context occurrences observed from large text corpus (e.g., news collections, Wikipedia and Web Crawl) based on probabilistic models. In particular, a series of recent papers by Mikolov et al. [18–22] culminated in and popularized the skip-gram model with negative-sampling training scheme (a.k.a. the SGNS model), which together with its variants [23,24] is shown to achieve state-of-the-art results on a variety of linguistic tasks.

Despite the empirical success of the SGNS model, in this paper, we will first point out an observation that the optimization problem introduced by the SGNS

K. Arai et al. (Eds.): CompCom 2019, AISC 997, pp. 485–497, 2019.
https://doi.org/10.1007/978-3-030-22871-2_33

model is essentially an *ill-posed* one. In specific, we can easily distort the output solution without changing its objective value. To fix this issue, we investigate solution structures that a good word embedding model should deliver, and argue that a meaningful word embedding model should allow and only allow the ambiguities introduced by *orthogonal transformations*. Motivated by this goal, we rectify the SGNS model by appending quadratic regularization terms to the original objective of SGNS, and show this simple modification *suffices* in enforcing the solution to be structured in the desired manner. A theoretical justification is presented, which provides novel insights into quadratic regularization. Preliminary experiments are conducted to evaluate word vectors on Google's analytical reasoning task, which shows the modified SGNS model outperforms the original SGNS model in a consistent manner.

2 SGNS Model

The SGNS model is essentially the skip-gram word neural embedding model introduced in [20] trained using the negative-sampling procedure proposed in [21]. In this section, we will briefly review the SGNS model together with its related notation. Although the SGNS model is initially proposed and described in the language of neural network, we find the explanation provided by Goldberg and Levy [25] is more transparent and could better disclose the rationale behind the model. Therefore, in the following, we adopt their approach in formulating the SGNS model.

Let $\mathcal{W}$ be the word vocabulary of our interest with $n := |\mathcal{W}|$. The training data $\mathcal{D}$, normally collected based on some text corpus, consists of word-context pairs $(w, c) \in \mathcal{W} \times \mathcal{W}$ in both positive and negative sense. For a word w, its positive context word c is often sampled from the neighborhood centering around the locations where w shows up in the text corpus, while its negative context word c is normally sampled from $\mathcal{W}$ randomly according to certain predefined distribution [26]. For each word $w \in \mathcal{W}$, its *center-word embedding* and *context-word embedding* are assumed to exist and represented as $\mathcal{U}[w]$ and $\mathcal{V}[w]$, where

$$\mathcal{U} : \mathcal{W} \to \mathbb{R}^d \quad \text{and} \quad \mathcal{V} : \mathcal{W} \to \mathbb{R}^d. \tag{1}$$

The center-word embedding $\mathcal{U}[\cdot]$ is normally outputted as word representation, which will be used either by itself or as an important ingredient in subsequent natural language processing and machine learning applications.

The SGNS model learns the embeddings by solving the following optimization problem,

$$\max_{\mathcal{U}:\mathcal{W}\to\mathbb{R}^d,\ \mathcal{V}:\mathcal{W}\to\mathbb{R}^d} \quad \sum_{(w,c)\in\mathcal{D}^+} \log\sigma(\mathcal{U}[w]^\top\mathcal{V}[c]) + \sum_{(w,c)\in\mathcal{D}^-} \log\sigma(-\mathcal{U}[w]^\top\mathcal{V}[c]), \tag{2}$$

where $\mathcal{D}^+$ and $\mathcal{D}^-$ denotes the positive and negative pairs in $\mathcal{D}$, and $\sigma(\cdot)$ denotes the usual sigmoid function, i.e. $\sigma(x) = 1/(1+\exp(-x))$. For simplicity, we denote the center-word embedding matrix $\boldsymbol{U}$ (resp. context-word embedding matrix $\boldsymbol{V}$) as the matrix in $\mathbb{R}^{n\times d}$ whose row vectors are stacked by the center-word embeddings (resp. context-word embedding) of all words from the vocabulary. We will use $\boldsymbol{u}_i$, $\boldsymbol{v}_i \in \mathbb{R}^d$ to denote the i-th row of $\boldsymbol{U}$ and $\boldsymbol{V}$. With a slight abuse of notation, we will also use interchangeably $\boldsymbol{U}[w]$ and $\mathcal{U}[w]$, $\boldsymbol{V}[w]$ and $\mathcal{V}[w]$, i.e.

$$\boldsymbol{U}[w] := \mathcal{U}[w] \quad \text{and} \quad \boldsymbol{V}[w] := \mathcal{V}[w] \tag{3}$$

to represent the center-word and the context-word embeddings of the word $w \in \mathcal{W}$. Then clearly we can rewrite (2) equivalently as a maximization problem over the matrices $\boldsymbol{U}$ and $\boldsymbol{V}$ in $\mathbb{R}^{n\times d}$,

$$\max_{\boldsymbol{U},\boldsymbol{V}\in\mathbb{R}^{n\times d}} \mathcal{L}(\boldsymbol{U},\boldsymbol{V}) := \sum_{(w,c)\in\mathcal{D}+} \log\sigma(\boldsymbol{U}[w]^\top \boldsymbol{V}[c]) + \sum_{(w,c)\in\mathcal{D}-} \log\sigma(-\boldsymbol{U}[w]^\top \boldsymbol{V}[c]). \tag{4}$$

The SGNS model models how words are interacted with their contexts, which is rooted deeply in the distributional hypothesis of Harris [27], stating that *words sharing similar contexts possess similar meanings.* Intuitively, the SGNS model attempts to find embeddings $\{\boldsymbol{U}[w]\}_{w\in\mathcal{W}}$ and $\{\boldsymbol{V}[c]\}_{c\in\mathcal{W}}$ in a way such that their inner-products are encouraged to be large for good context pairs, but to be small for bad ones. Several insightful interpretations–e.g., implicit matrix factorization [28], representation learning [29], weighted logistic PCA [30], to just name a few–have been further proposed to better understand the underlying principles of the model. However, as we will point out in the next section, the SGNS model is essentially an ill-posed problem from the perspective of optimization.

3 Ambiguity in the SGNS Model

In this section, we will address a fundamental ambiguity issue undermining the SGNS model (4). Specifically, we will show that the solution from SGNS can be easily distorted without affecting the objective value.[1]

Suppose $(\boldsymbol{U}^\star, \boldsymbol{V}^\star)$ is one optimal solution to (4). Then for any invertible matrix $\boldsymbol{M} \in \mathbb{R}^{d\times d}$, $(\boldsymbol{U}^\star\boldsymbol{M}, \boldsymbol{V}^\star\boldsymbol{M}^{-\top})$ is another optimal solution to SGNS as the objective value remains the same:

$$\begin{aligned}
&\mathcal{L}(\boldsymbol{U}^\star\boldsymbol{M},\, \boldsymbol{V}^\star\boldsymbol{M}^{-\top}) \\
&= \sum_{(w,c)\in\mathcal{D}+} \log\sigma\left(\left\langle \boldsymbol{M}^\top\boldsymbol{U}^\star[w], \boldsymbol{M}^{-1}\boldsymbol{V}^\star[c]\right\rangle\right) + \sum_{(w,c)\in\mathcal{D}-} \log\sigma\left(-\left\langle \boldsymbol{M}^\top\boldsymbol{U}^\star[w], \boldsymbol{M}^{-1}\boldsymbol{V}^\star[c]\right\rangle\right) \\
&= \sum_{(w,c)\in\mathcal{D}+} \log\sigma(\langle \boldsymbol{U}^\star[w], \boldsymbol{V}^\star[c]\rangle) + \sum_{(w,c)\in\mathcal{D}-} \log\sigma(-\langle \boldsymbol{U}^\star[w], \boldsymbol{V}^\star[c]\rangle) \\
&= \mathcal{L}(\boldsymbol{U}^\star, \boldsymbol{V}^\star).
\end{aligned} \tag{5}$$

[1] In addition to the SGNS model, following the same logic, the fundamental ambiguity issue is shared by many other prevailing word embedding models (e.g., the CBOW model with negative sampling [20–22], and the GloVe model [17].

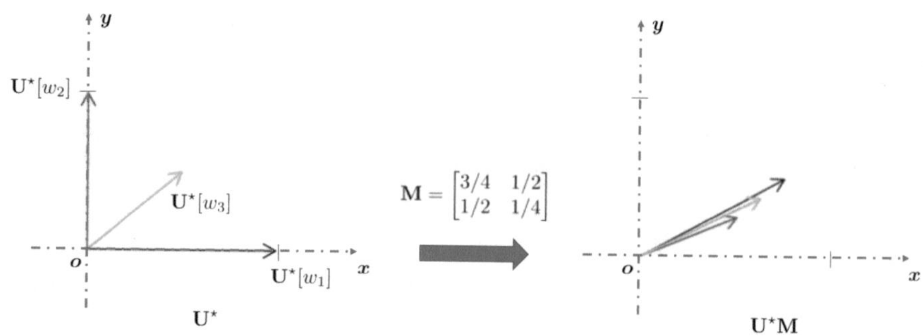

Fig. 1. Illustration of Example 1. Here we choose $\boldsymbol{M}$ as specified in (6) with $\varepsilon = 1/4$. Although both embedding matrices $\boldsymbol{U}^\star$ and $\boldsymbol{U}^\star\boldsymbol{M}$ are solutions to the SGNS model, we can clearly observe that their word vectors are quite different in terms of encoded linguistic properties.

Therefore, there is an extremely large amount of freedom to manipulate $(\boldsymbol{U}^\star, \boldsymbol{V}^\star)$ without affecting the optimality, which could lead to entirely different embeddings in terms of encoded semantic and syntactic properties (i.e., vector lengths and angles). To better understand the severity of this ambiguity, let us think about the following toy example.

Example 1. Suppose we have $\mathcal{W} = \{w_1, w_2, w_3\}$, and

$$\boldsymbol{U}^\star = \begin{bmatrix} \boldsymbol{U}^\star[w_1] \\ \boldsymbol{U}^\star[w_2] \\ \boldsymbol{U}^\star[w_3] \end{bmatrix} = \begin{bmatrix} 1 & 0 \\ 0 & 1 \\ 1/2 & 1/2 \end{bmatrix},$$

whose row vectors are pretty spread out in $\mathbb{R}^2$. However, by choosing

$$\boldsymbol{M} = \begin{bmatrix} 1/2+\varepsilon & 1/2 \\ 1/2 & 1/2-\varepsilon \end{bmatrix}, \tag{6}$$

where $0 \neq \varepsilon \in \mathbb{R}$, as argued above, $\boldsymbol{U}^\star\boldsymbol{M}$ is also an optimal solution to (4) with

$$\boldsymbol{U}^\star\boldsymbol{M} = \begin{bmatrix} 1 & 0 \\ 0 & 1 \\ 1/2 & 1/2 \end{bmatrix} \begin{bmatrix} 1/2+\varepsilon & 1/2 \\ 1/2 & 1/2-\varepsilon \end{bmatrix} = \begin{bmatrix} 1/2+\varepsilon & 1/2 \\ 1/2 & 1/2-\varepsilon \\ 1/2+\varepsilon/2 & 1/2-\varepsilon/2 \end{bmatrix},$$

whose row vectors now become almost parallel as ε approaches 0 (see Fig. 1).

To sum up, even though $\boldsymbol{U}^\star$ and $\boldsymbol{U}^\star\boldsymbol{M}$ have entirely different word representations in essence, the SGNS model makes no differentiation among them. In order to ensure intrinsic embeddings being learned, we have to avoid those $\boldsymbol{M}$'s that distort the linguistic properties of the word vectors. As the linguistic properties of the word vectors are mostly reflected by their *lengths* and *inner products*, we should allow and only allow linear transformations that preserve these quantities. For arbitrary $\boldsymbol{u}, \boldsymbol{v} \in \mathbb{R}^d$, we are guaranteed to have $\|\boldsymbol{M}\boldsymbol{u}\| = \|\boldsymbol{M}\boldsymbol{u}\|$ and $\langle \boldsymbol{M}\boldsymbol{u}, \boldsymbol{M}\boldsymbol{v}\rangle = \langle \boldsymbol{u}, \boldsymbol{v}\rangle$ if and only if $\boldsymbol{M} \in \mathbb{R}^{d\times d}$ is orthogonal, i.e., $\boldsymbol{M}^\top \boldsymbol{M} = \boldsymbol{I}$. So the only *innocuous* ambiguities are the ones resulting from orthogonal transformation. Geometrically, this means that the rows of the embedding matrix $\boldsymbol{U}$ are transformed through *rotation* and *reflection*. Therefore, an ideal word embedding model should be expected in general to have unique optimal solutions up to orthogonal transformation, i.e.,

[*] $(\boldsymbol{U}^\star \boldsymbol{M}, \boldsymbol{V}^\star \boldsymbol{M}^{-\top})$ *is optimal if and only if* $\boldsymbol{M}$ *is orthogonal.*

We will elaborate how we are able to achieve this in the next section.

4 SGNS Model with Quadratic Regularization

In this section, we will work towards the goal stated in [*] by modifying the SGNS model.

Let us consider the extended SGNS model with regularization,

$$\max_{\boldsymbol{U},\boldsymbol{V}\in\mathbb{R}^{n\times d}} \quad \mathcal{L}(\boldsymbol{U},\boldsymbol{V}) - \mathcal{R}(\boldsymbol{U},\boldsymbol{V}), \tag{7}$$

where $\mathcal{R} : (\mathbb{R}^{n\times d}, \mathbb{R}^{n\times d}) \to \mathbb{R}\cup\{+\infty\}$ is some regularizer. The aim is to leverage the regularization term $\mathcal{R}$ to enforce the solution to be unique up to orthogonal transformation without (on the other hand) making the model too hard to be optimized. In the following, we will choose $\mathcal{R}$ to be a simple quadratic form, and show this slight modification is sufficient to achieve the goal stated in [*] and thus resolve the ambiguity issues undermining the SGNS model (2).

Consider the following SGNS model with quadratic regularization (named as the SGNS-qr model thereafter):

$$\begin{aligned}\max_{\boldsymbol{U},\boldsymbol{V}\in\mathbb{R}^{n\times d}} f(\boldsymbol{U},\boldsymbol{V}) := &\sum_{(w,c)\in\mathcal{D}+} \log\sigma(\boldsymbol{U}[w]^\top \boldsymbol{V}[c]) + \sum_{(w,c)\in\mathcal{D}-} \log\sigma(-\boldsymbol{U}[w]^\top \boldsymbol{V}[c]) \\ &- \frac{\lambda}{2}\|\boldsymbol{U}\|_F^2 - \frac{\lambda}{2}\|\boldsymbol{V}\|_F^2,\end{aligned} \tag{8}$$

where $\lambda > 0$ is the regularization parameter and $\|\cdot\|_F$ denotes the matrix Frobenius norm. A similar model has been proposed in [31] in the context of collaborative filtering, which falls into the general framework of low-rank models [32] with the logistic loss function and the quadratic regularization. The quadratic regularizer $\mathcal{R}(\boldsymbol{U},\ \boldsymbol{V}) := \frac{\lambda}{2}\|\boldsymbol{U}\|_F^2 + \frac{\lambda}{2}\|\boldsymbol{V}\|_F^2$ explicitly encourages entries in both $\boldsymbol{U}$ and $\boldsymbol{V}$ to be small in magnitude, which (perhaps surprisingly) has the effect of penalizing the non-orthogonal transformation. We will state this novel insight regarding quadratic regularization in the following theorem.

Theorem 1. *Let* $(\boldsymbol{U}^\star, \boldsymbol{V}^\star)$ *be an optimal solution to* (8). *Suppose* $\boldsymbol{U}^\star$ *and* $\boldsymbol{V}^\star$ *are both full rank. Then* $(\hat{\boldsymbol{U}}, \hat{\boldsymbol{V}}) := (\boldsymbol{U}^\star \boldsymbol{M}, \boldsymbol{V}^\star \boldsymbol{M}^{-\top})$ *is an optimal solution if and only if* $\boldsymbol{M}$ *is orthogonal.*

Proof. Let us first prove the *if* direction. Since $\boldsymbol{M}$ is orthogonal,

$$\|\boldsymbol{U}^\star\|_F = \|\boldsymbol{U}^\star\boldsymbol{M}\|_F, \quad \|\boldsymbol{V}^\star\|_F = \|\boldsymbol{V}^\star\boldsymbol{M}\|_F = \left\|\boldsymbol{V}^\star\boldsymbol{M}^{-\top}\right\|_F, \tag{9}$$

and therefore

$$\begin{aligned} f(\boldsymbol{U}^\star, \boldsymbol{V}^\star) &= \mathcal{L}(\boldsymbol{U}^\star, \boldsymbol{V}^\star) + \frac{\lambda}{2}\|\boldsymbol{U}^\star\|_F^2 + \frac{\lambda}{2}\|\boldsymbol{V}^\star\|_F^2 \\ &= \mathcal{L}(\boldsymbol{U}^\star\boldsymbol{M}, \boldsymbol{V}^\star\boldsymbol{M}^{-\top}) + \frac{\lambda}{2}\|\boldsymbol{U}^\star\boldsymbol{M}\|_F^2 + \frac{\lambda}{2}\|\boldsymbol{V}^\star\boldsymbol{M}^{-\top}\|_F^2 \\ &= f(\hat{\boldsymbol{U}}, \hat{\boldsymbol{V}}), \end{aligned}$$

which implies the optimality of $(\hat{\boldsymbol{U}}, \hat{\boldsymbol{V}})$.

In the rest of the proof, we will focus on the *only if* direction.

Let $\boldsymbol{U}\boldsymbol{\Sigma}\boldsymbol{V}^\top$ be the reduced singular value decomposition (SVD) [33] of $\boldsymbol{U}^\star(\boldsymbol{V}^\star)^\top$, i.e., $\boldsymbol{U}^\star(\boldsymbol{V}^\star)^\top = \boldsymbol{U}\boldsymbol{\Sigma}\boldsymbol{V}^\top$ where $\boldsymbol{U} \in \mathbb{R}^{n\times d}$ and $\boldsymbol{V} \in \mathbb{R}^{n\times d}$ have orthonormal columns, and $\boldsymbol{\Sigma} = \operatorname{diag}(\sigma_1, \sigma_2, \ldots, \sigma_d)$ with $\sigma_1 \geq \sigma_2 \geq \cdots \geq \sigma_d > 0$. Here we write $\sigma_d > 0$ since $\boldsymbol{U}^\star$ and $\boldsymbol{V}^\star$ are full rank, and by Sylvester inequality [34]

$$d = \operatorname{rank}(\boldsymbol{U}^\star) + \operatorname{rank}(\boldsymbol{V}^\star) - d \leq \operatorname{rank}(\boldsymbol{U}^\star(\boldsymbol{V}^\star)^\top) \leq \min\{\operatorname{rank}(\boldsymbol{U}^\star), \operatorname{rank}(\boldsymbol{V}^\star)\} = d.$$

Now we will first derive a upper bound for $f(\boldsymbol{U}^\star, \boldsymbol{V}^\star)$:

$$\begin{aligned} f(\boldsymbol{U}^\star, \boldsymbol{V}^\star) &= \mathcal{L}(\boldsymbol{U}^\star, \boldsymbol{V}^\star) - \mathcal{R}(\boldsymbol{U}^\star, \boldsymbol{V}^\star) \\ &= \mathcal{L}(\boldsymbol{U}^\star, \boldsymbol{V}^\star) - \frac{\lambda}{2}\|\boldsymbol{U}^\star\|_F^2 - \frac{\lambda}{2}\|\boldsymbol{V}^\star\|_F^2 \\ &\leq \mathcal{L}(\boldsymbol{U}^\star, \boldsymbol{V}^\star) - \lambda \cdot \|\boldsymbol{U}^\star\|_F \cdot \|\boldsymbol{V}^\star\|_F \\ &\leq \mathcal{L}(\boldsymbol{U}^\star, \boldsymbol{V}^\star) - \lambda \cdot \left\|\boldsymbol{U}^\top\boldsymbol{U}^\star\right\|_F \cdot \left\|\boldsymbol{V}^\top\boldsymbol{V}^\star\right\|_F \\ &\leq \mathcal{L}(\boldsymbol{U}^\star, \boldsymbol{V}^\star) - \lambda \cdot \operatorname{trace}(\boldsymbol{U}^\top\boldsymbol{U}^\star(\boldsymbol{V}^\star)^\top\boldsymbol{V}) \\ &= \mathcal{L}(\boldsymbol{U}^\star, \boldsymbol{V}^\star) - \lambda \cdot \|\boldsymbol{\sigma}\|_1, \end{aligned} \tag{10}$$

where the third and the fifth lines uses Cauchy-Schwartz inequality, the fourth line holds as the operator norms $\|\boldsymbol{U}\| \leq 1$, $\|\boldsymbol{V}\| \leq 1$, and $\|\boldsymbol{A}\boldsymbol{B}\|_F \leq \|\boldsymbol{A}\|\,\|\boldsymbol{B}\|_F$ for any compatible matrices $\boldsymbol{A}$ and $\boldsymbol{B}$, and the last line follows directly from the definition of SVD.

But on the other hand, we can also derive the following lower bound for $f(\boldsymbol{U}^\star, \boldsymbol{V}^\star)$:

$$\begin{aligned} f(\boldsymbol{U}^\star, \boldsymbol{V}^\star) &\geq f(\boldsymbol{U}\boldsymbol{\Sigma}^{\frac{1}{2}}, \boldsymbol{V}\boldsymbol{\Sigma}^{\frac{1}{2}}) \\ &= \mathcal{L}(\boldsymbol{U}^\star, \boldsymbol{V}^\star) - \frac{\lambda}{2}\left\|\boldsymbol{U}\boldsymbol{\Sigma}^{\frac{1}{2}}\right\|_F^2 - \frac{\lambda}{2}\left\|\boldsymbol{V}\boldsymbol{\Sigma}^{\frac{1}{2}}\right\|_F^2 \\ &= \mathcal{L}(\boldsymbol{U}^\star, \boldsymbol{V}^\star) - \frac{\lambda}{2}\left\|\boldsymbol{\Sigma}^{\frac{1}{2}}\right\|_F^2 - \frac{\lambda}{2}\left\|\boldsymbol{\Sigma}^{\frac{1}{2}}\right\|_F^2 \\ &= \mathcal{L}(\boldsymbol{U}^\star, \boldsymbol{V}^\star) - \lambda\|\boldsymbol{\sigma}\|_1, \end{aligned} \tag{11}$$

where $\boldsymbol{\Sigma}^{\frac{1}{2}} := \operatorname{diag}(\sqrt{\sigma}_1, \sqrt{\sigma}_2, \ldots, \sqrt{\sigma}_d)$.

Combining (10) and (11), one can easily derive that

$$f(\boldsymbol{U}^\star, \boldsymbol{V}^\star) = \mathcal{L}(\boldsymbol{U}^\star, \boldsymbol{V}^\star) - \lambda \|\boldsymbol{\sigma}\|_1, \quad \text{and} \tag{12}$$

$$\frac{1}{2}\|\boldsymbol{U}^\star\|_F^2 + \frac{1}{2}\|\boldsymbol{V}^\star\|_F^2 = \|\boldsymbol{U}^\star\|_F \|\boldsymbol{V}^\star\|_F = \|\boldsymbol{U}^\star\|_F^2 = \|\boldsymbol{V}^\star\|_F^2 = \|\boldsymbol{\sigma}\|_1. \tag{13}$$

Now we are ready to show that $\boldsymbol{U}^\star = \boldsymbol{U}\boldsymbol{\Sigma}^{\frac{1}{2}}\boldsymbol{Q}$ for some orthogonal matrix $\boldsymbol{Q} \in \mathbb{R}^{d\times d}$.

As $\boldsymbol{U}\boldsymbol{\Sigma}\boldsymbol{V}^\top$ is the SVD of $\boldsymbol{U}^\star(\boldsymbol{V}^\star)^\top$, there exist full rank matrices $\boldsymbol{S} \in \mathbb{R}^{d\times d}$ and $\boldsymbol{T} \in \mathbb{R}^{d\times d}$ such that $\boldsymbol{U}^\star = \boldsymbol{U}\boldsymbol{S}$, $\boldsymbol{V}^\star = \boldsymbol{V}\boldsymbol{T}$ and $\boldsymbol{S}\boldsymbol{T}^\top = \boldsymbol{\Sigma} = \operatorname{diag}(\boldsymbol{\sigma})$. Then from (13), one has

$$\|\boldsymbol{U}^\star\|_F = \|\boldsymbol{U}\boldsymbol{S}\|_F = \|\boldsymbol{S}\|_F = \|\boldsymbol{\sigma}\|_1^{1/2}, \tag{14}$$

$$\|\boldsymbol{V}^\star\|_F = \|\boldsymbol{V}\boldsymbol{T}\|_F = \|\boldsymbol{T}\|_F = \|\boldsymbol{\sigma}\|_1^{1/2}. \tag{15}$$

Now let us write

$$\boldsymbol{X} := \begin{bmatrix}\boldsymbol{S}\\ \boldsymbol{T}\end{bmatrix}\begin{bmatrix}\boldsymbol{S}^\top \boldsymbol{T}^\top\end{bmatrix} = \begin{bmatrix}\boldsymbol{S}\boldsymbol{S}^\top & \boldsymbol{S}\boldsymbol{T}^\top\\ \boldsymbol{T}\boldsymbol{S}^\top & \boldsymbol{T}\boldsymbol{T}^\top\end{bmatrix} = \begin{bmatrix}\boldsymbol{S}\boldsymbol{S}^\top & \boldsymbol{\Sigma}\\ \boldsymbol{\Sigma}^\top & \boldsymbol{T}\boldsymbol{T}^\top\end{bmatrix} \succeq \mathbf{0}. \tag{16}$$

Define

$$s^\star \in \arg\min_{i\in[d]}\left\{(\boldsymbol{S}\boldsymbol{S}^\top)_{ii} - \sigma_i\right\} \quad \text{and} \quad t^\star \in \arg\min_{i\in[d]}\left\{(\boldsymbol{T}\boldsymbol{T}^\top)_{ii} - \sigma_i\right\}. \tag{17}$$

Due to the facts that

$$\sum_{ii}(\boldsymbol{S}\boldsymbol{S}^\top)_{ii} = \|\boldsymbol{S}\|_F^2 = \sum_{i\in[d]}\sigma_i \quad \text{and} \quad \sum_{ii}(\boldsymbol{T}\boldsymbol{T}^\top)_{ii} = \|\boldsymbol{T}\|_F^2 = \sum_{i\in[d]}\sigma_i, \tag{18}$$

we must have

$$(\boldsymbol{S}\boldsymbol{S}^\top)_{s^\star s^\star} - \sigma_{s^\star} \le 0 \quad \text{and} \quad (\boldsymbol{T}\boldsymbol{T}^\top)_{t^\star t^\star} - \sigma_{t^\star} \le 0. \tag{19}$$

Since $\boldsymbol{X}$ is positive semidefinite [34],

$$(\boldsymbol{e}_{s^\star} - \boldsymbol{e}_{t^\star})^\top \boldsymbol{X}(\boldsymbol{e}_{s^\star} - \boldsymbol{e}_{t^\star}) = (\boldsymbol{S}\boldsymbol{S}^\top)_{s^\star s^\star} + (\boldsymbol{T}\boldsymbol{T}^\top)_{t^\star t^\star} - \sigma_{s^\star} - \sigma_{t^\star} \ge 0. \tag{20}$$

which together with (19) leads to

$$(\boldsymbol{S}\boldsymbol{S}^\top)_{s^\star s^\star} = \sigma_{s^\star} \quad \text{and} \quad (\boldsymbol{T}\boldsymbol{T}^\top)_{t^\star t^\star} = \sigma_{t^\star}. \tag{21}$$

Combining (17) and (19), it can be easily verified that

$$\operatorname{diag}\left(\boldsymbol{S}\boldsymbol{S}^\top\right) = \boldsymbol{\sigma} = \operatorname{diag}\left(\boldsymbol{T}\boldsymbol{T}^\top\right), \tag{22}$$

which implies that for any $i \in [d]$, $\boldsymbol{s}_i$ and $\boldsymbol{t}_i$ (the i-th row of $\boldsymbol{S}$ and $\boldsymbol{T}$) satisfies $\|\boldsymbol{s}_i\|^2 = \|\boldsymbol{t}_i\|^2 = \sigma_i$. In addition, since $\boldsymbol{S}\boldsymbol{T}^\top = \boldsymbol{\Sigma}$, the inner-product $\langle \boldsymbol{s}_i, \boldsymbol{t}_i\rangle = \sigma_i$. Due to Cauchy-Schwartz inequality, $\boldsymbol{s}_i = \boldsymbol{t}_i$. Therefore, $\boldsymbol{S} = \boldsymbol{T}$, $\boldsymbol{\Sigma} = \boldsymbol{S}\boldsymbol{T}^\top =$

$\boldsymbol{S}\boldsymbol{S}^\top = \boldsymbol{T}\boldsymbol{T}^\top$. Then it can be easily verified that $\boldsymbol{S} = \boldsymbol{T} = \boldsymbol{\Sigma}^{\frac{1}{2}}\boldsymbol{Q}$ for some orthogonal matrix $\boldsymbol{Q}$. Therefore, we have proved that $\boldsymbol{U}^\star = \boldsymbol{U}\boldsymbol{\Sigma}^{\frac{1}{2}}\boldsymbol{Q}$ for some orthogonal matrix $\boldsymbol{Q} \in \mathbb{R}^{d\times d}$.

Next, as $(\hat{\boldsymbol{U}}, \hat{\boldsymbol{V}})$ is also optimal and $\hat{\boldsymbol{U}}\hat{\boldsymbol{V}}^\top = \boldsymbol{U}^\star(\boldsymbol{V}^\star)^\top$, we can follow exactly the same argument to show that $\hat{\boldsymbol{U}} = \boldsymbol{U}\boldsymbol{\Sigma}^{\frac{1}{2}}\hat{\boldsymbol{Q}}$ for anther orthogonal matrix $\hat{\boldsymbol{Q}} \in \mathbb{R}^{d\times d}$. Therefore, in order to satisfy

$$\hat{\boldsymbol{U}} = \boldsymbol{U}\boldsymbol{\Sigma}^{\frac{1}{2}}\hat{\boldsymbol{Q}} = \underbrace{\boldsymbol{U}\boldsymbol{\Sigma}^{\frac{1}{2}}\boldsymbol{Q}}_{\boldsymbol{U}^\star}\boldsymbol{M}, \tag{23}$$

one must have $\boldsymbol{M} = \boldsymbol{Q}^\top\hat{\boldsymbol{Q}}$, which is also orthogonal. That completes our proof.

Theorem 1 states that optimal solutions to (8) are not unique, but are essentially all equivalent in terms of their encoded linguistic properties, as a result of the quadratic regularization removing all the adversarial ambiguities (e.g. the one described in Example 1) undermining the original SGNS model (4).

5 Experiment

In this section, we will conduct some preliminary experiments to compare the SGNS model with our SGNS-qr model.

Algorithm. We use the popular toolbox word2vec [20,21] with its default parameter setting to solve the SGNS model, which leverages the standard *stochastic gradient method* (SGM) [35,36] to optimize the objective. We solve the SGNS-qr model by modifying the SGM in word2vec to accommodate the additional quadratic terms.

Dataset. We use a publicly accessible dataset Enwik9[2] as our text corpus, which contains about 128 million tokens collected from English Wikipedia articles. The vocabulary $\mathcal{W}$ is constructed by filtering out words that appear less than 200 times. The positive and negative word-context pairs are generated in exactly the same manner with the one implemented in word2vec using its default setting. We adopt Google's analogy dataset to evaluate word embeddings on analytical reasoning task.

Evaluation. In Google's analogy dataset, 19,544 questions are presented with the form "a is to $a^\star$ as b is to $b^\star$", where $b^\star$ is hidden and to be inferred from the whole vocabulary $\mathcal{W}$ based on the input $(a, a^\star, b)$. Among all these analogy questions, around half of them are syntactic ones (e.g., "think is to thinking as code is to coding"), and the other half are semantic ones (e.g., "man is to women as king is to queen"). The questions are answered using the 3CosMul scheme [37]:

$$\mathcal{B}^\star = \arg \max_{x\in\mathcal{W}/\{a,a^\star,b\}} \frac{\cos(\mathcal{U}[x],\mathcal{U}[a^\star]) \cdot \cos(\mathcal{U}[x],\mathcal{U}[b])}{\cos(\mathcal{U}[x],\mathcal{U}[a]) + \varepsilon} \tag{24}$$

[2] http://mattmahoney.net/dc/textdata.html.

where $\mathcal{U} : \mathcal{W} \to \mathbb{R}^d$ is the word embedding to evaluate and $\varepsilon = 1\mathrm{e}-3$ is set to avoid zero-division. The performance is measured as the percentage of questions answered correctly, i.e., $b^\star \in \mathcal{B}^\star$.

Experiment Result. We evaluate the SNGS model (4) and the SGNS-qr model (8) with different choices of λ. The performance of each model is reported in Table 1 in terms of the analytical reasoning accuracy. As presented in Table 1, within a wide and stable range of choices in λ, the SGNS-qr model outperforms the SGNS model ($\lambda = 0$) in a consistent manner, and the improvement becomes more and more non-trivial with the growth in the embedding dimension d. To better visualize this, we plot in Fig. 2 the prediction accuracies of the SGNS model and the SGNS-qr ($\lambda = 250$) model over d. As we can see clearly, the improvement rate rises from (nearly) 0% to more than 3% quickly as d increases. This suggests that the ambiguity issue undermining the SGNS model becomes substantially more severe when the optimization problem (2) is solved over larger ambient space. Remarkably, our simple rectification through quadratic regularization is capable of boosting the prediction accuracy by around 3%.[3]

Table 1. Evaluation of SGNS ($\lambda = 0$) and SGNS-qr models on Google's analytical reasoning task.

d	λ						
	0	10	50	100	250	500	1000
100	0.5642	0.5652	0.5666	**0.5665**	0.5645	0.5570	0.5397
200	0.6618	0.6617	0.6640	0.6656	**0.6668**	0.6605	0.6355
300	0.6768	0.6772	0.6798	0.6848	**0.6909**	0.6869	0.6593
400	0.6851	0.6860	0.6902	0.6938	**0.7005**	0.6952	0.6658
500	0.6909	0.6920	0.6947	0.6971	**0.7035**	0.6965	0.6554
600	0.6755	0.6763	0.6825	0.6888	**0.6973**	0.6926	0.6508
700	0.6781	0.6798	0.6835	0.6885	**0.6981**	0.6901	0.6399
800	0.6736	0.6744	0.6808	0.6848	**0.6926**	0.6860	0.6328
900	0.6713	0.6731	0.6785	0.6818	**0.6903**	0.6829	0.6275
1000	0.6622	0.6631	0.6689	0.6738	**0.6820**	0.6716	0.6181

6 Future Work

In this paper, we rectify the SGNS model with quadratic regularization, and prove that this simple modification cures ambiguity issues undermining the

[3] We note that similar empirical observation of the use of quadratic regularization being capable of improving the performance of the SGNS model has also been made by Vilnis and McCallum (2014) for a different NLP task: word similarity task.

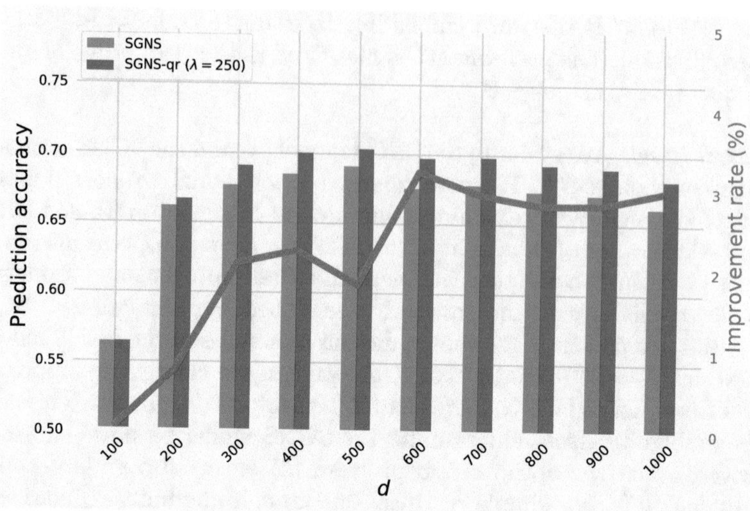

Fig. 2. Comparison between SGNS and SGNS-qr model on Google's analytical reasoning task. When the embedding dimension is small (e.g., $d = 100$), the SGNS-qr model is almost on a par with the SGNS model. But when d becomes larger, the SGNS-qr model soon surpasses the SGNS model. Remarkably, the improvement is increasingly enlarged with the growth in d and culminates in a boost of around 3% in prediction accuracy.

SGNS model. Formulating the appropriate optimization to solve is an important but first step towards learning word vectors in a robust and efficient manner. We believe a (possibly) larger gain from this rectification comes from the perspective of optimization algorithm. Numerical methods, which perform poorly on machine learning tasks related with the SGNS model, might be solely due to the ill-posedness of the model rather than the inefficacies of the algorithms. In the future, we will tailor some recently designed numerical optimization methods (e.g., [38–43]) beyond SGM to solve our SGNS-qr model. Another interesting research direction is to resolve the ambiguity issue by leveraging higher-order relations among words and estimating underlying word embeddings through tensor decompositions [44–48].

References

1. Kim, Y.: Convolutional neural networks for sentence classification. In: Proceedings of the 2014 Conference on Empirical Methods in Natural Language Processing (EMNLP), pp. 1746–1751 (2014)
2. Grbovic, M., Djuric, N., Radosavljevic, V., Silvestri, F., Bhamidipati, N.: Context- and content-aware embeddings for query rewriting in sponsored search. In: International ACM SIGIR Conference on Research and Development in Information Retrieval (2015)

3. Nalisnick, E., Mitra, B., Craswell, N., Caruana, R.: Improving document ranking with dual word embeddings. In: International Conference Companion on World Wide Web (2016)
4. Iyyer, M., Boyd-Graber, J., Claudino, L., Socher, R., Daumé III, H.: A neural network for factoid question answering over paragraphs. In: Conference on Empirical Methods in Natural Language Processing (2014)
5. Shih, K., Singh, S., Hoiem, D.: Where to look: focus regions for visual question answering. In: Conference on Computer Vision and Pattern Recognition (2016)
6. Sienčnik, S.: Adapting word2vec to named entity recognition. In: Nordic Conference of Computational Linguistics (2015)
7. Lample, G., Ballesteros, M., Subramanian, S., Kawakami, K., Dyer, C.: Neural architectures for named entity recognition. In: Proceedings of the 2016 Conference of the North American Chapter of the Association for Computational Linguistics: Human Language Technologies, pp. 260–270 (2016)
8. Socher, R., Bauer, J., Manning, C., Ng, A.: Parsing with compositional vector grammars. In: Annual Meeting of the Association for Computational Linguistics (2013)
9. Bengio, Y., Ducharme, R., Vincent, P., Jauvin, C.: A neural probabilistic language model. J. Mach. Learn. Res. **3**(Feb), 1137–1155 (2003)
10. Morin, F., Bengio, Y.: Hierarchical probabilistic neural network language model. In: International Conference on Artificial Intelligence and Statistics (2005)
11. Bengio, Y., Schwenk, H., Senécal, J.-S., Morin, F., Gauvain, J.-L.: Neural probabilistic language models. In: Innovations in Machine Learning. Springer (2006)
12. Collobert, R., Weston, J.: A unified architecture for natural language processing: deep neural networks with multitask learning. In: International Conference on Machine Learning (2008)
13. Mnih, A., Hinton, G.: A scalable hierarchical distributed language model. In: Advances in Neural Information Processing Systems (2009)
14. Collobert, R., Weston, J., Bottou, L., Karlen, M., Kavukcuoglu, K., Kuksa, P.: Natural language processing (almost) from scratch. J. Mach. Learn. Res. **12**(Aug), 2493–2537 (2011)
15. Le, H., Oparin, I., Allauzen, A., Gauvain, J.-L., Yvon, F.: Structured output layer neural network language model. In: International Conference on Acoustics, Speech and Signal Processing (2011)
16. Baroni, M., Dinu, G., Kruszewski, G.: Don't count, predict! A systematic comparison of context-counting vs. context-predicting semantic vectors. In: Annual Meeting of the Association for Computational Linguistics (2014)
17. Pennington, J., Socher, R., Manning, C.: GloVe: global vectors for word representation. In: Conference on Empirical Methods in Natural Language Processing (2014)
18. Mikolov, T., Karafiát, M., Burget, L., Černockỳ, J., Khudanpur, S.: Recurrent neural network based language model. In: Annual Conference of the International Speech Communication Association (2010)
19. Mikolov, T., Yih, W., Zweig, G.: Linguistic regularities in continuous space word representations. In: Conference of the North American Chapter of the Association for Computational Linguistics: Human Language Technologies (2013)
20. Mikolov, T., Chen, K., Corrado, G., Dean, J.: Efficient estimation of word representations in vector space. arXiv preprint arXiv:1301.3781 (2013)
21. Mikolov, T., Sutskever, I., Chen, K., Corrado, G.S., Dean, J.: Distributed representations of words and phrases and their compositionality. In: Advances in Neural Information Processing Systems (2013)

22. Mikolov, T., Grave, E., Bojanowski, P., Puhrsch, C., Joulin, A.: Advances in pre-training distributed word representations. arXiv preprint arXiv:1712.09405 (2017)
23. Sun, F., Guo, J., Lan, Y., Xu, J., Cheng, X.: Sparse word embeddings using ℓ_1 regularized online learning. In: Proceedings of the Twenty-Fifth International Joint Conference on Artificial Intelligence (2016)
24. Yang, W., Lu, W., Zheng, V.: A simple regularization-based algorithm for learning cross-domain word embeddings. In: Proceedings of the 2017 Conference on Empirical Methods in Natural Language Processing, pp. 2898–2904 (2017)
25. Goldberg, Y., Levy, O.: Word2vec explained: deriving Mikolov et al.'s negative-sampling word-embedding method. arXiv preprint arXiv:1402.3722 (2014)
26. Levy, O., Goldberg, Y., Dagan, I.: Improving distributional similarity with lessons learned from word embeddings. Trans. Assoc. Comput. Linguist. **3**, 211–225 (2015)
27. Harris, Z.: Distributional structure. Word **10**(2–3), 146–162 (1954)
28. Levy, O., Goldberg, Y.: Neural word embedding as implicit matrix factorization. In: Advances in Neural Information Processing Systems (2014)
29. Li, Y., Xu, L., Tian, F., Jiang, L., Zhong, X., Chen, E.: Word embedding revisited: a new representation learning and explicit matrix factorization perspective. In: International Joint Conference on Artificial Intelligence (2015)
30. Landgraf, A.J., Bellay, J.: Word2vec skip-gram with negative sampling is a weighted logistic PCA. arXiv preprint arXiv:1705.09755 (2017)
31. Johnson, C.: Logistic matrix factorization for implicit feedback data. In: NIPS Distributed Machine Learning and Matrix Computations Workshop (2014)
32. Udell, M., Horn, C., Zadeh, R., Boyd, S.: Generalized low rank models. Found. Trends® Mach. Learn. **9**(1), 1–118 (2016)
33. Trefethen, L.N., Bau III, D.: Numerical Linear Algebra, vol. 50. SIAM (1997)
34. Horn, R., Johnson, C.: Matrix Analysis. Cambridge University Press, Cambridge (1990)
35. Robbins, H., Monro, S.: A stochastic approximation method. Ann. Math. Stat. **22**(3), 400–407 (1951)
36. Bertsekas, D.P.: Incremental gradient, subgradient, and proximal methods for convex optimization: a survey. Optim. Mach. Learn. **2010**(1–38), 3 (2011)
37. Levy, O., Goldberg, Y.: Linguistic regularities in sparse and explicit word representations. In: Conference on Computational Natural Language Learning (2014)
38. Reddi, S.J., Hefny, A., Sra, S., Poczos, B., Smola, A.: Stochastic variance reduction for nonconvex optimization. In: International Conference on Machine Learning, pp. 314–323 (2016)
39. Wang, X., Ma, S., Goldfarb, D., Liu, W.: Stochastic quasi-newton methods for nonconvex stochastic optimization. SIAM J. Optim. **27**(2), 927–956 (2017)
40. Goldfarb, D., Mu, C., Wright, J., Zhou, C.: Using negative curvature in solving nonlinear programs. Comput. Optim. Appl. **68**(3), 479–502 (2017)
41. Fonarev, A., Grinchuk, O., Gusev, G., Serdyukov, P., Oseledets, I.: Riemannian optimization for skip-gram negative sampling. In: Proceedings of the 55th Annual Meeting of the Association for Computational Linguistics (2017)
42. Reddi, S.J., Kale, S., Kumar, S.: On the convergence of adam and beyond. In: International Conference on Learning Representations (2018)
43. Chen, R., Menickelly, M., Scheinberg, K.: Stochastic optimization using a trust-region method and random models. Math. Programm. **169**(2), 447–487 (2018)
44. Anandkumar, A., Ge, R., Hsu, D., Kakade, S.M., Telgarsky, M.: Tensor decompositions for learning latent variable models. J. Mach. Learn. Res. **15**(1), 2773–2832 (2014)

45. Mu, C., Hsu, D., Goldfarb, D.: Successive rank-one approximations for nearly orthogonally decomposable symmetric tensors. SIAM J. Matrix Anal. Appl. **36**(4), 1638–1659 (2015)
46. Mu, C., Hsu, D., Goldfarb, D.: Greedy approaches to symmetric orthogonal tensor decomposition. SIAM J. Matrix Anal. Appl. **38**(4), 1210–1226 (2017)
47. Bailey, E., Aeron, S.: Word embeddings via tensor factorization. arXiv preprint arXiv:1704.02686 (2017)
48. Frandsen, A., Ge, R.: Understanding composition of word embeddings via tensor decomposition. In: ICLR (2019)

LANA-I: An Arabic Conversational Intelligent Tutoring System for Children with ASD

Sumayh Aljameel[1(✉)], James O'Shea[2], Keeley Crockett[2], Annabel Latham[2], and Mohammad Kaleem[2]

[1] Department of Computer Science, College of Computer Science and Information Technology, Imam Abdulrahman Bin Faisal University, Dammam, Saudi Arabia
saljameel@iau.edu.sa

[2] Department of Computing, Math and Digital Technology, Manchester Metropolitan University, Manchester, UK
{j.d.oshea,k.crockett,a.latham,m.kaleem}@mmu.ac.uk

Abstract. Children with Autism Spectrum Disorder (ASD) share certain difficulties but being autistic will affect them in different ways in terms of their level of intellectual ability. Children with high functioning autism or Asperger syndrome are very intelligent academically but they still have difficulties in social and communication skills. Many of these children are taught within mainstream schools but there is a shortage of specialised teachers to deal with their specific needs. One solution is to use a virtual tutor to supplement the education of children with ASD in mainstream schools. This paper describes research to develop a novel Arabic Conversational Intelligent Tutoring System, called LANA-I, for children with ASD that adapts to the Visual, Auditory and Kinaesthetic learning styles model (VAK) to enhance learning. This paper also proposes an evaluation methodology and describes an experimental evaluation of LANA-I. The evaluation was conducted with neurotypical children and indicated promising results with a statistically significant difference between user's scores with and without adapting to learning style. Moreover, the results show that LANA-I is effective as an Arabic Conversational Agent (CA) with the majority of conversations leading to the goal of completing the tutorial and the majority of the correct responses (89%).

Keywords: Autism · Intelligent tutoring system · String similarity · Arabic language

1 Introduction

The number of children being diagnosed with autism spectrum disorder (ASD) is increasing [1]. Children with high functioning Autism (HFA) or Asperger's Syndrome (AS) (i.e. those with higher verbal IQ) are usually offered education in the mainstream schools. However, many mainstream schools are not able to include students with ASD because of the lack of skilled teachers and the poor training and provisions from the responsible institutions [2]. In addition, traditional education using human tutors is a challenge for students with autism, who have difficulties in communication and social

K. Arai et al. (Eds.): CompCom 2019, AISC 997, pp. 498–516, 2019.
https://doi.org/10.1007/978-3-030-22871-2_34

interactions. Many researchers have reported that using a virtual tutor with students with ASD could meet the individual student's needs [3, 4].

Conversational Intelligent Tutoring System (CITS) is a software system which uses natural language interfaces to allow users to learn topics through discussion as they would in the classroom. Many CITS have been developed for different domains. To our knowledge, no academic research exists on Arabic CITS developed specifically for Autistic children. LANA-I [5] is an Arabic CITS, which engages autistic children with a science tutorial where the curriculum material is mapped to VAK model. One challenge in building such a system is the requirement to deal with the Arabic language grammatical features and its morphological nature. The research into Conversational Agent (CA) development techniques revealed that hybrid approach was the most appropriate approach to develop an Arabic CA [6]. The engine of LANA-I is based on the two main CA development strategies, A Pattern Matching (PM) engine and a Short Text Similarity (STS) algorithm that calculate the matching strength of a pattern to the user utterance. The two parts of the engine work together to overcome some of the unique challenges of the Arabic language and to extract responses from resources in a particular domain (Science topic). The main contributions in this paper are:

- A novel architecture for an Arabic CITS using VAK model for an appropriate education scenario.
- The results of designing an experimental methodology to validate the educational tutoring scenario in the Arabic CITS.

In order to evaluate LANA-I, an initial pilot study was conducted on the general population. This study took place in UK with neurotypical children from the target age group (10–12) years whose first language is Arabic. It is important to test LANA-I, with the general population before testing it with Autistic children to avoid any problems and issues that may occur in the tutorial or confusion in the presentation of the tutorial material. The study used the learning gain measurement (defined in Sect. 3) to evaluate the ability of CITS to adapt to a child's learning styles.

This paper is organized as follows: Sect. 2 describes the architecture and methodology of implementing LANA-I. Section 3 explains the experimental methodology and the experiments. Section 4 provides the results and discussion. Finally, Sect. 5 presents the conclusions and future works.

2 LANA-I Architecture and Implementation

LANA-I was developed based on two phases. The first phase involved designing and implementing an Arabic CA, and the second phase focused on development of an educational tutorial on science, mapping the tutorial to the VAK model and introducing the Arabic ITS interface to the CA. In the first phase, a new architecture for developing an Arabic CA was developed using both PM and STS. The key features are summarised as follows:

- Ability to control the conversation through context.
- Ability to personalise the lesson with the user's learning style and provide suitable supporting material to the user.
- A scripting language to provide Arabic dialogue for LANA-I.
- A novel CA engine that manages the response using a combination of the PM technique and the STS technique.
- Managing the response when the context is changed. For example, creating the right response when the user writes something that is not related to the tutorial topic.

The proposed framework for the Arabic CA consists of five components as shown in Fig. 1:

1. Graphical User Interface: Responsible for the communication between the user and CA (in this case the CITS tutor) through a web interface with panels to display supporting material.
2. Controller: The controller manages the conversation between the user and the Arabic CA, as well as cleaning the user utterance by removing diacritics and other illegal characters (e.g. ! £ $).
3. Conversation Agent Manager: Responsible for controlling and directing the conversation through contexts. In addition, the manager ensures that the discussion is directed towards completion of the tutorial.
4. Conversation Agent engine: Responsible for pattern matching and calculating the similarity strength between the user utterance and the scripted patterns.
5. The knowledge base: The knowledge base is responsible for holding the tutorial domain knowledge in a relational database, which includes CA scripts, Log file, and General contexts such as weather, agreement and rude words.

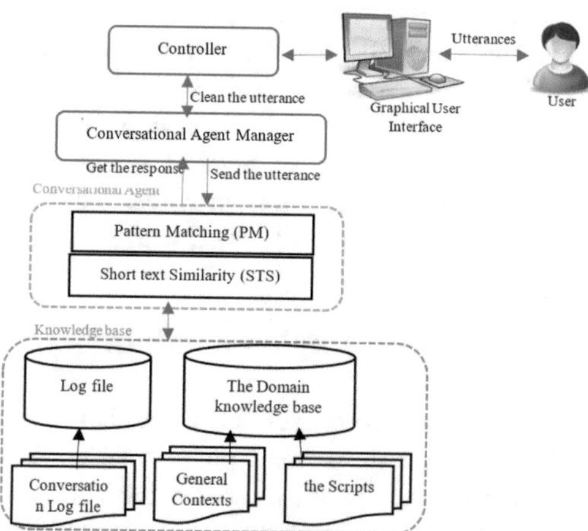

Fig. 1. The LANA-I CA architecture

In the second phase the ITS architecture is designed to adapt to the user's learning style within the Arabic CA. Based on the typical ITS architecture [7], LANA-I ITS architecture consists of four main components as shown in Fig. 2:

1. User Interface Model: responsible for the interaction between the user and the ITS components.
2. Tutor Model: the ITS manager, which is the main component that interacts with the user through the GUI, and personalises the tutorial based on the user's learning style.
3. Student Model: a temporal memory structure, which records the user's responses during the tutoring session and the student's profile such as user ID, user's age, gender, user's learning style, and Pre-test and Post-test scores.
4. Domain Model: the Tutorial Knowledge Base, which contains structured topics that are presented to the user.

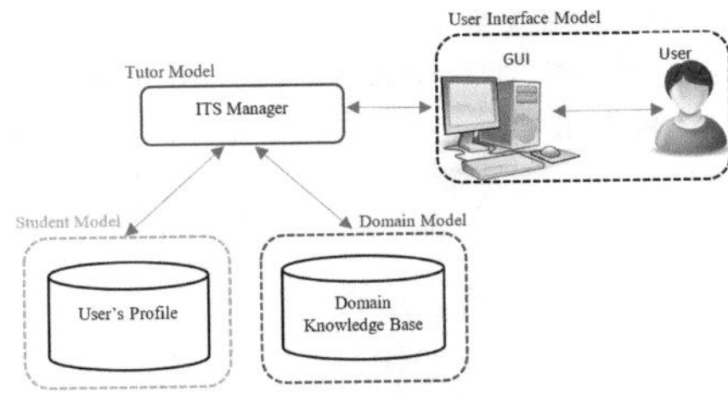

Fig. 2. LANA-I ITS Architecture.

The LANA-I architecture combines the Arabic CA (Fig. 1) and ITS architecture (Fig. 2). The proposed LANA-I CITS architecture, shown in Fig. 3 contains three main components that are described in the following sections:

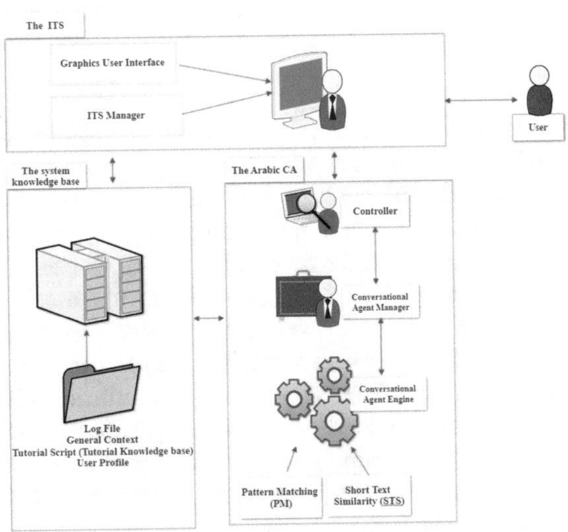

Fig. 3. LANA-I CITS architecture

2.1 The Knowledge Base

The knowledge base consists of four sub-components: (1) the Tutorial Knowledge base, (2) Arabic general contexts (e.g. weather, and greetings), (3) user's profile, and (4) the log file. The tutorial domain CA was from the Science curriculum for ages 10-12. The material was provided by the Ministry of Education in Saudi Arabia [8]. Script contexts were structured according to the topics in the science book taught in schools in Saudi Arabia, such as the Earth, Moon, Solar System, eclipse, etc. A short interview was conducted to design the knowledge base with three primary school teachers in Saudi Arabia, who teach Science. The aim of this interview was to extract knowledge of lesson design and delivery of tutorials similar to the traditional classroom delivery model. The teachers evaluated the designed tutorial in order to give feedback and approval. The tutorials were implemented within LANA-I using a relational database MySQL to store and retrieve the resources. The LANA-I knowledge base consists of two contexts: the domain, which is the science tutorials, and a general context, illustrated in Fig. 4.

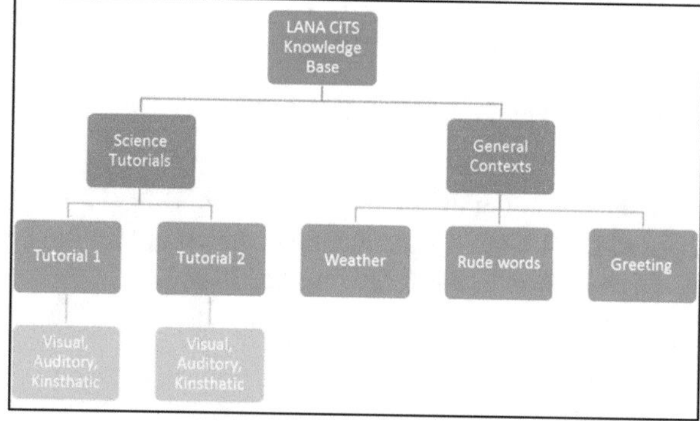

Fig. 4. Knowledge base contexts

In Fig. 4, each context has all related sub contexts mapped to it. The science tutorials hold the tutorials sessions that are related to the science subject such as solar system, earth, and moon. General contexts deal with general conversation that is not related to the domain, such as weather, greeting, rude words, and user leaving words (any words or sentences means that the user will leave the conversation). The general context makes LANA-I seem more aware by responding to the user's utterances, which are not related to the main domain. However, only a few general sub-contexts have been implemented as a proof of concept within the scope of this research.

Once the tutorial was designed and approved by the primary school teachers then the tutorial questions were mapped to the VAK learning style model. Each question in the tutorial is mapped to Visual learning style through pictures and videos, Auditory learning style through sound recording, Kinaesthetic learning style through models and instructions as shown in Table 1.

Table 1. Part of the tutorial and how it is mapped to the VAK model

Question: What cause the four seasons? Answer: Seasons result from the yearly orbit of the Earth around the Sun and the tilt of the Earth's rotational axis relative to the plane of the orbit	V: video A: Audio K: solar system model
Question: How long the earth takes to complete the rotation around the sun? Answer: 365 days, which is one year	V: video A: Audio K: earth-moon model
Question: The moon is the nearest object to the earth but it is different from the earth in many ways. What are the differences between the earth and the moon? Answer: The moon doesn't have atmosphere and water whereas the earth has	V: video A: Audio K: earth-moon model

2.2 Arabic Scripting Language

The three different approaches to develop an Arabic CA and a number of challenges faced by Arabic language were discussed in survey paper [6]. It was concluded that there is a lack of Arabic NLP resources leading to limited capabilities within Arabic CAs. Consequently, this research needs a new scripting language to enable the Arabic CA combined with an ITS to deliver Arabic tutoring sessions using the modern standard Arabic language.

The foundation of LANA-I scripting language is based on the InfoChat scripting language [9]. InfoChat was designed using English scripting language and based on the pattern matching (PM) technique, where the domain is organised into a number of contexts and each context contains rules, each rule in the domain contains a number of patterns and a response that forms the CA output to the user. In the LANA-I CA engine, the similarity strength is calculated through combining the use of the PM, which is based on InfoChat method, and the Cosine algorithm [10] to compute the similarity strength between the utterance and the scripted patterns in order to improve the CA accuracy. The highest matching strength pattern will generate the response back to the user. Based on the scripting methodology reported by Latham [11] the procedures to create the scripts within the Knowledge base are:

1. The methodology for scripting each context is:
   - Create a context table, which has a record with a unique name to represent that context (topic).
   - When the context is invoked, an initialisation rule is fired.
   - All rules and patterns are scripted for the associated context.
   - Test each context to check that the expected rule is fired for a sample of user utterances.

2. The methodology for scripting each rule is:
   - Create a rule table, which has a unique rule name to represent the tutorial question.
   - For each rule, create patterns that match user utterances. Extract the important words and use the wildcards to replace unimportant words.
   - Create patterns from each different utterance phrase. E.g. saying the same things using alternative words.
   - Create the CA response to the utterance.
   - Add Image, Audio, and instructions to the rule in order to map the learning style.

2.3 Scripting Arabic CA for LANA-I

In LANA-I, the tutorial topics were represented as the contexts and the agent's questions for such topic were represented as the rules, while the pattern represent the user's utterances, which belong to such a rule. The scripting language in LANA-I includes the following features:

- Provide supporting material to the user depending on the user's learning style (Visual: images and videos – Auditory: Sounds – Kinaesthetic: Instructions and objects). The learning style will be determined using a child friendly-customised VAK learning style preferences questionnaire be-fore the tutorial.
- This material is stored in the scripting database and once a rule is fired, the corresponding material is provided to the user through the interface. All images, videos and audios provide the right answer. All of them have the same knowledge that the written text provides.
- The scripting language works with the ITS manager to check the user's learning style and provide the suitable material with the fired rule.

Table 2 shows an example of the scripting language and how the tutorial mapping the VAK model. When the user is visual learner, the rule is fired with the video or image material. In this example (Table 2), the rule has video that will be executed

Table 2. Scripting language (translated)

Context: 1 Context Name: Topic 1 (solar system)	Context: 1 Context Name: Topic 1 (الـ نظام الـ شمـسي)
Rule Number: 1 Rule Name: The earth rotates around the sun. what the result of this rotation? Pattern: * four seasons Pattern: * 4 seasons Pattern: four seasons * Response: *Well done,* the earth rotates around the sun every year and this cause the four seasons. Image: yes Audio: yes Instruction: Look at the solar system model and its object	Rule Number: 1 Rule Name: تدور الأرض حول الشمس كل سنه. ماهي نتيجة هذا الدوران؟ Pattern: *الفصول الأربعة Pattern: * المواسم الأربعة Pattern: * أربع فصول24 ساعه Response: أحسنت بارك الله فيك. تدور الارض حول الشمس وينتج عنه الفصول الأربعة Image: yes Audio: yes Instruction: أنظر إلى نموذج النظام الشمسي والأجرام المحيطه بها

along with the rule. When the user is an Auditory learner, the rule will be fired with the audio material. The same thing with the kinaesthetic learning style, the rule will be fired with instructions to use the models. Figure 5 shows the models that are used with the Kinaesthetic learning style.

Fig. 5. The solar system model and earth model

2.4 The Arabic Conversational Agent

The second component of LANA-I (Fig. 3) is the Arabic CA. The main components within the Arabic CA architecture (Fig. 1) are: (i) Controller, (ii) Conversational Agent Manager, and (iii) Conversational Agent Engine.

The controller directs and manages the entire conversation by working with several other components, which are ITS manager, GUI and CA manager to achieve the conversation goal. In the beginning, the controller finds the student's learning style by communicating with the ITS manager. Then, the controller receives the user's utterance and provides an utterance checking process based on the following constraints:

- Filter the utterance: The controller filters the utterance to remove any special characters (i.e. $, &, *, !, ?, "", £, (), ^). For example (as shown in English), if the user writes: (Hi, how are you Lana???) the system will convert it to (Hi how are you Lana).
- Check for rude and offensive words: if the utterance contains any of rude or offensive words, the system will warn the user three times before terminating the session.

The controller then checks if the conversation is within the tutorial scenario or not by communicate with the CA. The response will be delivered back to the user with providing supporting material such as video, picture, audio, or instructions depending on the user's learning style as shown in Fig. 6.

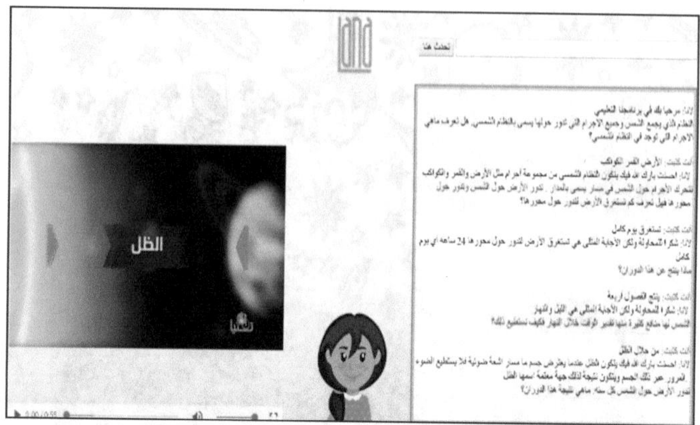

Fig. 6. Conversation between LANA-I and visual learner

The Conversation Manager is responsible for controlling the conversation to make sure that the goal is achieved. The goal is completing the tutorial by checking whether the user stays on the tutorial topic, or the user switches the context i.e. asks about something irrelevant to the tutorial. The Arabic CA implemented in LANA-I adopts a goal-oriented methodology [11].

The Conversational Agent Engine contains a combination of methods of string similarity and PM approaches to determine the similarity between two sets of strings within CA's, while traditional CA's used only a PM approach that involves a strength calculation through different aspects of the user utterance and the scripted pattern such as activation level and number of words, etc. It is important to use string similarity methods to overcome some of the challenges in the Arabic language (described in Sect. 2.3). In the field of CA development, the scripting is the most time consuming part of CA development [12]. The biggest challenge of scripting CAs is the coverage of all possible user utterances [11]. The engine handles the challenge of Arabic scripting by combining the Wildcard PM Function with the string similarity algorithm that calculates similarity strength and overcomes the scripting length challenge. Consequently, strings with minor differences should be recognized as being similar. The LANA-I CA engine has two main components:

- Pattern matching approach (Wildcard PM [9]).
- String similarity algorithm (Cosine algorithm [10])

These components were used to calculate the similarity between the user utterance and the scripted patterns in order to reduce the need to cover all possible utterances when scripting the domain.

Pattern matching is based on similar method to InfoChat [9]. In PM technique, the user utterance will be matched to the stored patterns; these patterns contain wildcard characters to represent any number of words of characters. An example of that, the wild card symbol (?) matches a single character, where the wild card symbol (*) matches any number of words. An example for the PM approach in LANA-I is illustrated in the following table:

Table 3. Example of matching the utterance with the patterns script

Rule: Although the moon can appear a very bright white, its surface is actually dark. So from where the moon reflects the light?	**Rule:** القمر لايضئ بنفسه فهو انعكاس لضوء آخر. ماهو؟
Pattern 1: the moon reflects * sun **Pattern 2**: * the sunlight **Pattern 3**: * the sunshine	**Pattern 1:** القمر يعكس * الشمس **Pattern 2**: * ضوء الشمس **Pattern 3**: * نور الشمس
Utterance 1: it is the sunlight **Utterance 2:** reflects the moon the sun.	**Utterance 1:** هو ضوء الشمس **Utterance 2:** يعكس القمر الشمس

In Table 3, the utterance#1 will match pattern#2 (* the sunlight), where the wild card symbol (*) will match any number of words. The results from this matching will be 1 whereas the utterance#2 will not match any patterns because of the word order. In this case the matching result will be 0.

The second component of the Arabic CA engine uses the STS algorithm, which for the purpose of this work is the Cosine similarity measure [10]. The similarity between two pieces of text is determined by representing each piece of text in the form of word vector. A word vector is a vector of length *N* where *N* is the number of different tokens in the text. The similarity is computed as the angle between the word vectors for the two sentences in vector space.

For two texts t_1 and t_2 the cosine similarity is illustrated in the following equation:

$$\mathrm{SIM}(t_1.t_2) = \frac{\sum_{i=1}^{n} t_1 t_2}{\sqrt{\sum t_{1i}^2} \times \sqrt{\sum t_{2i}^2}} \tag{1}$$

The cosine similarity result is nonnegative and bounded between [0,1] where 1 indicates that the two texts are identical.

2.5 The Workflow of LANA-I CA Engine

In the beginning, the PM Wildcard will be used to match the user utterance with the patterns stored in the database. If the match is not found, the STS Cosine similarity will be applied. For example, assume the pattern stored in the LANA-I database was (S1), while the user utterance was (S2), as shown in Table 4:

The utterance is not recognised by the PM approach because of the word order or minor differences from the pattern. Therefore, the system applies the Cosine similarity, which is illustrated in the following steps:

Table 4. Example of calculating the similarity

Stored pattern in LANA-I: S1= يختلف القمرعن الأرض بأن ليس له غلاف جوي	S1: The moon is different from the earth that it does not have an atmosphere.
User Utterance: S2= القمر ليس له غلاف جوي لذلك يختلف عن الأرض	S2: The moon does not have atmosphere therefore it is different from the earth.

- Create Matrix[][] where the columns are the unique words from S1 and S2, and the rows are the words sequence of S1 and S2.
- Calculate the similarity between each word, 1 means the two words are identical, 0 means the two words are different.

The result obtained from Cosine similarity when applied to S1 and S2 is equal (0.88), which is greater than the threshold (0.80). This threshold is empirically determined using a small set of sentence pairs as in [13]. When the user's utterance is recognized by the similarity measure, the corresponding response will be generated and delivered to the user.

2.6 LANA-I ITS

The third component in the LANA-I architecture (Fig. 3) contains: The Graphics User Interface (GUI), and the ITS manager.

Fig. 7. LANA-I GUI and its character

The GUI is the point where LANA-I and the user interact with each other. A character called LANA (shown in Fig. 7) is used in LANA-I. This character appears in all the system interfaces to make the conversation more natural and engaging for the users. The LANA character was designed by the author and then evaluated by primary school teachers in Saudi Arabia in order to make the tutorial more engaging.

The ITS manager adapts the tutorial depending on to the user's learning style VAK, which is determined at the beginning of the tutorial through a questionnaire. The questionnaire was developed on the basis of a widely disseminated version of Smith's visual, auditory, and kinaesthetic styles [15] to be completed by pupil with parents or teacher help. There were three questions focused on Smith's visual, auditory, and kinaesthetic styles (VAK). For each question, the pupils had to respond 'yes' or 'no' to each question. The total questions in the questionnaire are nine and the student's learning style result will be based on the highest score in one area.

2.7 LANA-I Workflow

This section describes the LANA-I workflow from perspectives of teacher and the pupil in order to understand how each activity communicates with others. Figure 8 presents the workflow.

Initially LANA-I starts from the learning style questionnaire, which is taken by the teacher. In the registration screen, the system asks the teacher to enter the pupil ID number (provided by the researcher), age, gender, and the result of the learning style questionnaire, conducted before using the system. There are four options for the learning style: Visual, Auditory, Kinaesthetic, and No-learning style, which used with the control group in the testing stage. After completing this stage, the system shows the pre-test interface, and asks the user to complete the test. The next interface after submitting the test is the CA tutorial, The ITS manager is responsible in this stage for personalising the tutoring session according to the user's learning style by providing the CA components, through the controller, the suitable materials from the Knowledge base component. The ITS manager also saves the user's registration information and the pre-test score in the log file/student's profile. In the CA screen, the user is led through the topic from one question to another until the end of the tutoring session. The user has only one attempt to answer. After completing the tutorial, the user is asked to fill the post-test questions and the score is saved by the ITS manager.

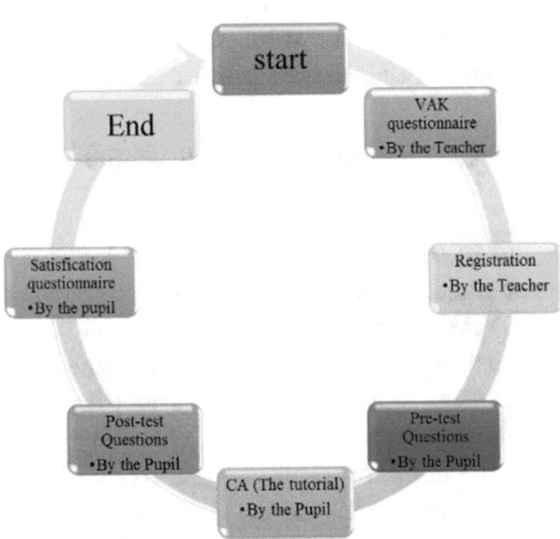

Fig. 8. LANA-I workflow

3 Experimental Methodology

The LANA-I prototype was tested through two main experiments to evaluate the system. The evaluation was based on a set of objective and subjective metrics [16]. The objective metrics were measured through the log file/temporal memory and the pre-test and post-test score. The subjective metrics were measured using the user feedback questionnaire. The aim of capturing the subjective and objective metrics is to test the hypotheses, which relate to the effectiveness of the conversation and using the VAK learning style, end user satisfaction, usability and system robustness. The main hypothesis of the experiments is:

- HA_0: LANA-I using VAK model cannot be adapted the student learning style.
- HA_1: LANA-I using VAK model can be adapted the student learning style.
- HB_0: LANA-I is not an effective Arabic CA.
- HB_1: LANA-I is an effective Arabic CA.

The aim of the first experiment was to test Hypothesis A (LANA-I using VAK model can adapt the tutorial to the student's learning style). The second experiment was conducted to test the Hypothesis B (LANA-I is an effective Arabic CA).

3.1 Participants

The total size of the sample was 24 neurotypical participants within the age group (10–12) years and their first language was Arabic. The participants recruited for the evaluation were residents of the Greater Manchester area within the UK and none of them had previous experience of using LANA-I. All participants' parents received an information sheet about the project and its aims, and a consent form to obtain their permission before conducting the experiment. The participants were divided into two groups. The first group is a control group (n = 12), who used the LANA-I without adapting to the learning style VAK model as basis comparison. The second group is an experimental group (n = 12), who used LANA-I with adapting to the learning style VAK model. From 15 participants, 12 were selected based on their results on the learning style questionnaire in order to divide them into 3 groups of 4.

3.2 Experiment 1: LANA-I Tutoring with and Without VAK Learning Style

Subjective and objective metrics were used to answer the two questions related to Hypothesis A. The experiment is based on the pre-test and post-test scores with and without adapting to the VAK learning Style model. Participants were divided into 2 groups. Each group of participants was asked first to register into the system and complete the pre-test questions. The Control Group used LANA-I without adapting the VAK learning style model. They started the tutorial without the VAK questionnaire, whereas the experimental Group, who used the LANA-I with adapting to VAK learning style model, were asked to fill the VAK learning style questionnaire in order for LANA-I to be adapted to the learning style. After adapting the learning style, the tutorial provided the most suitable material during the session such as videos, images or instructions and physical resources. The Control Group did the tutorial session based on a text conversation without any additional materials. When the session ended, both groups did the post-test questions in order to measure their learning gain. Finally, both groups completed the user feedback questionnaire.

3.3 Experiment 2: LANA-I CA System Robustness

The data for this experiment was gathered from the LANA-I log file and the user feedback questionnaire whilst participants were completing experiment 1. The subjective and objective metrics were used to answer questions related to Hypothesis B. The data gathered from the log file allows assessment of the performance of LANA-I CA and the algorithms deployed in the architecture. This data will measure success using objective metrics. The data from user feedback questionnaire will be analysed in order to measure success using subjective metrics.

4 Results and Discussion

The learning gain was measured using a pre-test and post-test approach [17–19]. The same test questions were completed before and after the tutoring conversation. The test scores were compared to establish whether there is any improvement as follows:

$$Learning\ gain = post - test\ score - pre - test\ score \tag{2}$$

The results from experiment 1, illustrate that there is a statistically significant difference (as shown in Table 5) between the user's test scores with and without adapting to the VAK learning style. The mean rank of the differences between the pre-test and post scores in each case are without VAK (M = 9.50) and with VAK (M = 15.50). The P value of the mean's variation between the two groups is 0.03 (significant at p <= 0.05), that meaning the users performed better when they are using their preferred VAK learning style.

Table 5. Mann-Whitney U test for the differences in learning gain between the control group and the experimental group.

	Group	N	Mean rank	Sum of ranks
Learning gain	Control	12	9.50	114.00
	Experimental	12	15.50	186.00
	Total	24		

Test statistics[a]		
		Learning gain
Mann-Whitney U		36.000
Wilcoxon W		114.000
Z		−2.173
Asymp. Sig. (2-tailed)		.030
Exact Sig. [2*(1-tailed Sig.)]		.039[b]

[a]Grouping Variable: VAK
[b]Not corrected for ties.

The effect size (r) is calculated using the Eq. (3) and found that r = 0.4 indicating a medium to large effect size using [20] criteria of 0.1 = small effect, 0.3 = medium effect, 0.5 = large effect. It can be concluded that there is a statistically significant improvement in the learning gain in the Experimental Group comparing to the Control Group.

$$r = Z / \textit{square root of } N \textit{ where } N = \textit{total number of cases} \quad (3)$$

Relative learning gain was also measured in this experiment. Relative learning gain is a measure that calculate the average improvement in test scores as a percentage of the possible improvement [21]. This measure additionally takes into account the opportunity for improvement. Average test score improvements were calculated and compared using the following formula:

$$\textit{Relative learning gain} = (\textit{PostTest} - \textit{PreTest})/(\textit{TotalScore} - \textit{PreTest}) \quad (4)$$

For example, if student get 8/10 in pre-test and only improves by 1, this is different to another getting 3/10 in pre-test and only improving by 1 – improvement is 50% in first case but only 14.2% in the second case. Table 6 illustrates the results of the Mann-Whitney U test conducted in order to measure the relative learning gain between Control Group and Experimental Group. It shows that there is a difference in the mean value between Control Group and Experimental Group. The average improvement in test scores in the Experimental Group was improved more than the Control Group. The ranks in each case are: Control Group (M = 9.04) and Experimental Group (M = 15.96). The relative learning gain between the Control Group and the Experimental Group is statistically significant different, p-value less than 0.05 (P = .015). The effect size (r) is calculated using the Eq. (3) and found that r = 0.5 indicating a large effect size using [20] criteria.

Table 6. Mann-Whitney U test for the relative learning gain between control group and experimental group.

	Group	N	Mean rank	Sum of ranks
Relative Learning gain	Control	12	9.04	108.50
	Experimental	12	15.96	191.50
	Total	24		

Test Statistics[a]		
		Relative learning gain
Mann-Whitney U		30.500
Wilcoxon W		108.500
Z		−2.428
Asymp. Sig. (2-tailed)		.015
Exact Sig. [2*(1-tailed Sig.)]		.014[b]

This result highlights that the participants performed better when they have adapted to their preferred VAK learning style. It can therefore be concluded that there is statistically a significant difference between the pre-test and post-test scores with and without adapting to VAK learning style, indicating that the HA_1 can be accepted.

Another test was conducted to find out whether there was a significant difference in the scores for participant's perception of remembering what they have learned from LANA-I. The users who learned with adapting to VAK learning style are much happier with remembering what they have learned from LANA-I (91.7%) than users who learned without adapting to VAK (50%) as shown in Fig. 9. This result indicates that HA_1 can be accepted.

Further tests have been carried out to find out whether there was a significant difference in the scores for participant's satisfaction with adapting the tutorial to the VAK learning style. The data was gathered from the user feedback questionnaire (Question 4: I think I can learn better with adapting to my learning style (Visual – Auditory -Kinaesthetic). The results illustrated that participants are happy with (83.3%), **indicating that HA_1 can be accepted.**

The results in Table 7 reveal that in general LANA-I performed well as an Arabic CA based on the number of corrected and uncorrected responses. The results show from the total conversations there are 281 correct responses (about 89.8% from the total utterances) and 32 incorrect responses, when the algorithm failed to detect the students answer because of spelling mistakes (about 10.2% from the total utterances). The spelling mistakes affected the strength of similarity of the utterances, this cause the incorrect responses. Moreover, from the conversations that did reach the end of the tutorial, the majority (95.8%) of them reached the goal, meaning the user stayed within the tutorial and did not interrupt the conversation before the end of the tutorial. This provides evidence that the users did enjoy using LANA-I as a tutoring system; this is

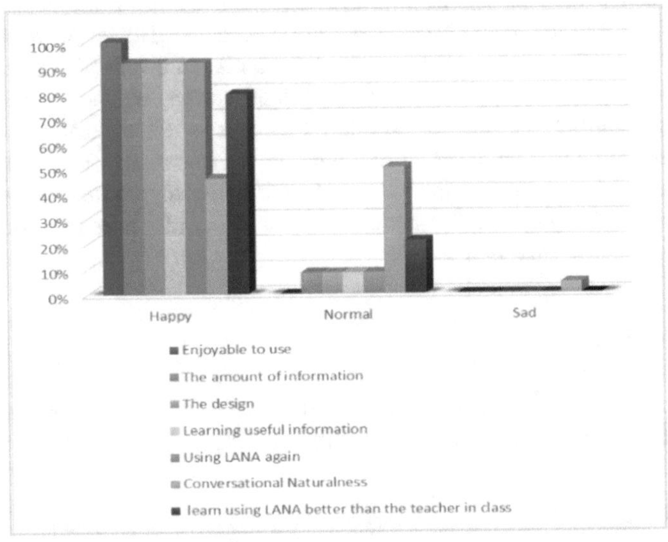

Fig. 9. Frequency analysis LANA-I questionnaire

also reflected in the opinion of the participants in the end user questionnaire who thought that LANA-I is enjoyable (100%) and a vast majority stated they would use the system again (91.6%).

Table 7. Data analysis from the log file

Conversation analysis	
METRIC	
Total number of conversations	24
Total number of utterances in all conversations	313
Average number of words per user utterance	3.1
Average number of utterances per conversation	31
Number of correct response from LANA	89.8%
Number of incorrect response from LANA	10.2%
Number of unrecognized utterances	38%
Number of utterances using Similarity Strength	57.18%
Number of rules fired	170
Average conversation duration (mins)	11.5
Percentage of conversations leading to goal	95.8%

An additional insight provided from these results is that the LANA-I engine is allowing the reduction of scripted patterns. The results illustrated that (57.18%) of all the utterances input by the users were actually different from the scripted patterns and in this case the system used the Short Text Similarity algorithm (Cosine algorithm). The log file shows that 34 unique utterances relating to 15 different rules were correctly recognised and dealt with by the LANA-I algorithm by firing the appropriate rule. Based on these findings Hypothesis HB_1 can be accepted. However, the results have identified some weaknesses in the LANA-I architecture, mainly the number of unrecognised utterances. LANA-I failed to recognise some utterances from the users (38%). Upon further analysis of the log file it was found that some of these unrecognised utterances were due to the gaps in the knowledge base, which can be easily addressed by expanding the patterns in the knowledge base. However, further research is needed to improve the similarity measure in order to reduce the number of patterns in the knowledge base.

5 Conclusion

This paper outlined a novel Arabic CITS called LANA-I, which used the VAK learning style model to enhance the learning of children. Our findings provide evidence for theses novel features:

1. LANA-I can be adapted to the VAK learning style for the tutorial.

2. LANA-I is able to converse in the Arabic language to teach children age (10-12) the science subject.
3. The LANA-I engine used PM with STS similarity algorithm was effective in recognising the user utterance.

The pilot study illustrated some key information with regards to the effectiveness of using the VAK learning style with the tutorial and with regards the functionality and robustness of LANA-I CA. To summarise the main findings of the pilot study are as follows:

- Using VAK learning style with LANA-I improves the participants (neurotypical) learning.
- LANA-I can be adapted to the user's learning style and enhance their learning.
- LANA-I engine managed to reduce the number of scripted patterns by an average of 11% using the STS algorithm comparing to PM approach.
- LANA-I engine managed to respond correctly to the user by an average of 89%
- LANA-I is able to lead the user towards the goal of the tutorial, with 95.8% success of completing the tutorial.

However, the first evaluation highlighted areas of weakness within LANA-I architecture. Further research is required to make components and algorithms within LANA-I more robust and to achieve the main objective, which is developing an Arabic CITS for people with Autism. Additional research is required as follows:

- Further improvement to the knowledge base and the CA engine will be made based on the results of the first pilot study.
- New methodologies will be researched and developed to overcome the spelling variations in the Arabic language, which affect the performance of the similarity algorithm.
- Further enhancement to the similarity algorithm to improve the matching of Arabic text much more efficiently in order to reduce the number of incorrect responses.

These weakness and further enhancements will be addressed by further research and development, to make the system ready for use with autistic pupils.

References

1. The National Autistic Society: Autism (2015). Available: http://www.autism.org.uk/about-autism/introduction/what-is-autism.aspx
2. Alotaibi, A.A.: Knowledge and Use of Applied Behavior Analysis Among Teachers of Students with Autism Spectrum Disorder in Saudi Arabia. Washington State University (2015)
3. Fletcher-Watson, S., Hammond, S., O'Hare, A.M., Petrou, M., Helen, Pain, H.: Click-East: evaluating the impact of an Ipad app on social communicative abilities in young children with autism. In: Presented at the 2013 International Meeting for Autism Research (2013)
4. Bernard-Opitz, V., Sriram, N., Nakhoda-Sapuan, S.: Enhancing social problem solving in children with autism and normal children through computer-assisted instruction. J. Autism Dev. Disord. **31**(4), 377 (2001)

5. Aljameel, S.S., O'Shea, J.D., Crockett, K.A., Latham, A., Kaleem, M.: Development of an Arabic conversational intelligent tutoring system for education of children with ASD. In: Computational Intelligence and Virtual Environments for Measurement Systems and Applications (CIVEMSA), 2017 IEEE International Conference, pp. 24–29. IEEE (2017)
6. Aljameel, S.S., O'Shea, J.D., Crockett, K.A., Latham, A.: Survey of string similarity approaches and the challenging faced by the Arabic language. In: Computer Engineering & Systems (ICCES), 2016 11th International Conference, pp. 241–247. IEEE (2016)
7. Ahuja, N.J., Sille, R.: A critical review of development of intelligent tutoring systems: Retrospect, present and prospect. Inter. J. Comput. Sci. Issues **10**(4), 39–48 (2013)
8. M. o. Education: Ministry of education (2017). Available: https://www.moe.gov.sa/en/Pages/default.aspx
9. Michie, D., Sammut, C.: Infochat scripter's manual. Convagent Ltd., Manchester (2001)
10. Qian, G., Sural, S., Gu, Y., Pramanik, S.: Similarity between Euclidean and cosine angle distance for nearest neighbor queries. In: Proceedings of the 2004 ACM Symposium on Applied Computing, pp. 1232–1237. ACM (2004)
11. Latham, A., Crockett, K., McLean, D., Edmonds, B.: A conversational intelligent tutoring system to automatically predict learning styles. Comput. Educ. **59**(1), 95–109 (2012)
12. O'Shea, J., Bandar, Z., Crockett, K., McLean, D.: A comparative study of two short text semantic similarity measures. In: KES International Symposium on Agent and Multi-Agent Systems: Technologies and Applications, pp. 172–181. Springer (2008)
13. Li, Y., McLean, D., Bandar, Z.A., O'Shea, J.D., Crockett, K.: Sentence similarity based on semantic nets and corpus statistics. IEEE Trans. knowl. Data Eng. **18**(8), 1138–1150 (2006)
14. Massaro, D.W., Bosseler, A.: Read my lips: the importance of the face in a computer-animated tutor for vocabulary learning by children with autism. Autism **10**(5), 495–510 (2006)
15. Smith, A.: Accelerated Learning in Practice. A&C Black, London (1998)
16. O'Shea, J., Bandar, Z., Crockett, K.: Systems Engineering and Conversational Agents, vol. 10, pp. 201–232 (2011)
17. Kelly, D., Tangney, B.: Adapting to intelligence profile in an adaptive educational system. Interact. Comput. **18**(3), 385–409 (2006)
18. Person, N.K.: AutoTutor improves deep learning of computer literacy: Is it the dialog or the talking head? In: Artificial Intelligence in Education: Shaping the Future of Learning Through Intelligent Technologies, vol. 97, p. 47 (2003)
19. Lee, J.-L., Orwig, G., Gunter, G., Witta, L.: The effect of cognitive styles on a visually-oriented task in online learning environments. In: Society for Information Technology & Teacher Education International Conference, pp. 3489–3494. Association for the Advancement of Computing in Education (AACE) (2006)
20. Cohen, J.: Statistical Power Analysis for the Behavioral Sciences, 2nd edn. Erlbaum Associates, Hillsdale (1988)
21. Latham, A., Crockett, K., McLean, D.: An adaptation algorithm for an intelligent natural language tutoring system. Comput. Educ. **71**, 97–110 (2014)

A Formal Model for Robot to Understand Common Concepts

Yuanxiu Liao, Jingli Wu, and Xudong Luo(✉)

Guangxi Key Laboratory of Multi-source Information Mining and Safety, Guangxi Regional Multi-source Information Integration and Intelligent Processing Collaborative Innovation Center, Guangxi Normal University, Guilin 541004, Guangxi, China
luoxd@mailbox.gxnu.edu.cn

Abstract. Can the robot think? This is a major topic of artificial intelligence and has always attracted the attention and exploration of many researchers. Since Searle proposed the *Chinese House Thought Experiment* in 1980, researchers have conducted extensive research on whether or not a formal symbolic system itself can understand the meaning of symbols, and have obtained many insights. From the perspective of cognitive science, this paper studies how robots understand knowledge, and proposes a general model that expresses robots' understanding of common concepts (the simplest form of knowledge) to describe the process by which robots understand some simple knowledge. We actually find a new way for robots to understand complex knowledge. In this paper, we also illustrate our method by an example of robot's understanding the concept of "pen".

Keywords: Concept understanding · Formal method · Knowledge representation and reasoning · Logic · Situation calculus

1 Introduction

In recent years, great progress has been made in the research and application of Artificial Intelligence (AI). AI products (such as language translation, face recognition, self-driving car, smartphones, and many more) have shown many of the machine's intelligent features. The AlphaGo program defeated one of the top Go players in the country and demonstrated the machine's superior ability in Go games. However, these characteristics and talents do not definitely mean that artificial intelligence has exceeded human intelligence. Many AI researchers think only in the situation where the relevant rules are well-defined and contain all the necessary information, the computer system can surpass human beings. For example, calculators are more powerful than human computing, and the logistics separation system is much faster than manual sorting, and AplpaGo also belongs to this category. When facing unclear rules or rules that do not

K. Arai et al. (Eds.): CompCom 2019, AISC 997, pp. 517–526, 2019.
https://doi.org/10.1007/978-3-030-22871-2_35

contain all the necessary information, human intelligence far exceeds current artificial intelligence.

Thus, how to enhance the intelligence of computer systems is one of the major issues in the field of artificial intelligence and has always attracted the attention and exploration of many researchers. And since Searle [1] put forward the *Chinese House Ideological Experiment* in 1980, researchers have conducted in-depth studies on whether or not the formal symbol system can understand the meaning of symbols and obtained many beneficial results. However, most existing understanding models are not based on formal systems, so there is no direct response to the question of whether or not formal symbol systems can understand the meaning of symbols themselves. To this end, according to he theories and methods proposed in recent years in the fields of artificial intelligence, philosophy, and cognitive science, we study how robots understand knowledge by proposing a formal model of the concept of robot understanding (one of the simplest forms of knowledge). Although the model just describes the process that the robot's understanding of some simple knowledge, it actually finds a new way for robot to understand complex knowledge.

The rest of this paper is organised as follows. Section 2 discusses the related work. Section 3 presents the logical framework for describing robot sensors, actions and knowledge. Section 4 proposes our model of understanding concepts. Section 5 gives an example to show how our model works, Finally, Sect. 6 concludes the paper with future work.

2 Related Work

Recently there are many studies on knowledge comprehension. Yoshimura et al. [2] used a computer-generated intuition and sensory association method to study the meaning of analogy. The sensory association method outputs the feelings and impressions of human beings in responding to a given noun, and the association between words is implemented through the application of a concept library and a large-scale knowledge base associative mechanism. The goal of the system is to make the computer system understand the metaphor. Zhou, Li, and Bai [3] devised a deep understanding framework for implementing description generators that relate to ordinary images of human behaviour. They proposed a syntax tree model for implementing a deep model of image recognition and a realization of natural language generation. This framework can identify the interaction between human and object in the image based on 3D spatial distribution co-occurrence analysis, and generate natural language description according to what actually happened in the image. Terai and Nakagawa [4] constructed a computational model based on corpus statistical analysis and metaphor understanding. Their model includes a classification process and a dynamic interaction process, which is implemented on a recurrent neural network. The model can highlight the characteristics of metaphorical expression. Doboli et al. [5] studied the construction of semantic knowledge structure for creative problem solving and proposed a semantic network representation model. The model can quickly and accurately understand the characteristics and structure of the solution and then make a reliable prediction of their outcomes and returns. Roy et al. [6] studied the discovery and comprehension of the user's intention on the level of words in the web search query. The words in the inquiry can be divided into the content and the intention. The content words indicate the subject of the inquiry, while the users and the intentions Words make their inquiry more obvious. Their research focuses on the discovery and understanding of intention words. In their opinion, intelligent processing of intent words is crucial to improving the quality of the query results. Sheng et al. [7] proposed a new robot semantic mapping method based on human activity recognition in the environment where humans and robots coexist. This method gives a new perspective of robot semantic mapping and can greatly reduce the difficulties in the traditional object-based visual classification algorithms. Semantic information can help robots to better understand unknown environments. Ramirez-Amaro, Beetz, and Cheng [8] studied the extraction of semantic representations through observation of human activities to transfer skills to humanoid robots and proposed a method of semantically representing a high-level understanding that allows the robot to acquire and determine the behavior of a demonstrator. Thus, a meaningful semantic description of human motion and object properties is obtained. Khalifa and Gadomski [9] view understanding as an explanatory knowledge. That is, a computer system S understands why p if and only if S knows p; and for some q, S produces a true belief that q correctly interprets p, or maintains a correct interpretation of p truth by credible explanatory evaluations.

More researchers in the fields of philosophy, cognitive science, psychology, and computer science also have conducted discussions and studies in different ways and from different perspectives on whether or not a computer system can understand the meaning of symbols. For example, Zhong [10] proposed a new mechanistic theory of mechanism-based artificial intelligence. Different from classical structuralism, functionalism, and behaviourism, institutionalist artificial intelligence considers the conversion process of "object information → perception information → knowledge → intelligent strategy → intelligent behaviour" as the common core mechanism of intelligent generation of all the intelligent systems (human intelligence systems and artificial intelligence systems). In the process of the interaction between the subject and the object, the perception information is generated from the object information, the knowledge is generated from the perception information, and then the intelligence strategy, and the intelligent behaviour are derived from the knowledge so as to form the common core mechanism of intelligence generation. This is a new methodology, namely, *Methodology of Information Ecology*. She et al. [11] proposed a new theoretical framework of thinking activities represented by network nodes, propositions by network connection, knowledge representation by network circuits, and deductive reasoning stimulated by space-time dynamics among network nodes. The basis of a new generation of expert system design. He et al. [12] argued that "understanding" is intuitively equivalent to "knowing the meaning of A" or attributing it to "knowing how" and attempting to describe the objectivity of understanding with knowing objectivity. Understanding can be used to state that the subject understands a proposition if and only if the subject knows the truth or knows the proposition. Xia et al. [13] discussed the problem of symbol foundation and its solution strategy.

There are several major solutions to Searle's Chinese-language problem, including internal foundation laying, causal foundation, semiotics foundation and action-based semantics. However, their solutions all have different degrees of defects. How to blend the artificial intelligence structures with different approaches is not only free from the shortcomings of various approaches but also gives full play to its advantages in order to better solve the problem of symbolic foundation and is worthy of study in artificial intelligence science. Jiang [14] discussed whether or not a computer can have the problem of semantic comprehension. It is assumed that the semantic comprehension of the computer must have the semantic content of the formal syntax. The semantic meaning given to the formal syntax refers to whether or not the intentionality can be given. Therefore, from the ideological experiment in Chinese House to the ultimate intentions caused by the problem, there is an inner logic clues to change ideas. By revealing each of these ideas in depth and detail, it helps us to have a clearer understanding of the origin and position of the Chinese-language ideological experiment in artificial intelligence theory and finally to some enlightenment on the development of artificial intelligence theory. Fan [15] discussed the understanding of what is a biological phenomenon or language phenomenon. He pointed out that Searle thinks the machine cannot think because it cannot understand, because understanding depends on the specific biochemical reaction. Wittgenstein's theory shows that understanding is an external language phenomenon. That is, understanding is the social phenomenon produced by the language, the product of the language and its context. These ideas for the study of machine thinking have some inspiration. Chen [16] developed a new theory of knowledge understanding, which states that the purpose (behaviour events) constitutes the unique research field of "epistemic theory of understanding", and the understanding of the purpose, meaning and value of behaviour is the goal of understanding. Comprehension is carried out under the framework of space of understanding. It consists of the following three parts: (i) a knowledge system that we already have provided a knowledge background for understanding; (ii) it contains deep thinking rules and value system, which provide normative reasons for cognitive activities; and (iii) perceptions and testimony provide information that provides factual evidence or justification. Understanding epistemology is not descriptive, but normative. And fundamental principles, restoration principles, and rationality principles constitute the three basic principles of understanding.

3 Logical Framework

This section establishes a logical framework L_R for describing robot sensors, actions, and knowledge. The logical framework is an extension to the classical theory of situational calculus, in which we added two types of sensors, and define a semantic based on the sensor output value. Most of the concepts and notations used in this paper about action theory and situation calculus can be found in [17].

3.1 L_R Syntax

The grammar of L_R is defined as follows:

1. L_R alphabet:
   - standard first-order logic alphabet;
   - infinite individual meta-symbols that represent actions: $a_1, a_2, \ldots$;
   - infinite individual meta-symbols are used to represent a situation: $s_1, s_2, \ldots$;
   - infinite individual argument symbols used to represent an object: any available symbol;
   - finite number of special function constant symbols: $f, g, h, \ldots$, corresponding to the functional sensors mounted on a robot;
   - finite number of specific predicate constants symbols: $P, Q, R, \ldots$, corresponding to relational sensors mounted on the robot.
2. The composition rules of terms and well-defined formulas are the same as those of the standard first-order logic and the classical situation calculus.
3. Inference rules in L_R are the same as inference rules in standard first-order logic and classical situation calculus.

3.2 L_R in the Sensor and Perception of Action

It is assumed that the robot is equipped with multiple sensors. A sensor here is a mechanical device that is attached to the robot. Each sensor has a specific function. For example, they can identify a bed, a toy, or the relative position between two objects, and so on. There are two types of sensors in L_R:

- The relational sensors, denoted as $\overline{TOY}(x), \overline{BED}(x), \ldots$, have the function of determining whether or not there is a specific relationship among given objects. These sensors have two output values "YES" and "NO". If the relation is true, the output is "YES"; otherwise, it is "NO". For example, sensor $\overline{TOY}(x)$ is used to detect whether or not x is a toy; sensor $\overline{BED}(x)$ is used to detect whether or not x is a bed; and sensor $\overline{IN}(x, y)$ is used to check whether or not x is in y.
- Functional sensors, denoted as $\overline{dist}(x, y), \overline{temp}(x), \ldots$, can determine certain properties of the object. The output values of these sensors vary in a range like mathematical functions. For example, the sensor $\overline{dist}(x, y)$ is used to measure the distance between x and y and its value changes in interval $[0, +\infty)$; sensor $\overline{temp}(x)$ is used to detect the temperature of x, the value of which varies in the range of $(-270, +\infty)$.

Perceptional action refers to the actions that a robot takes according to external information that it uses its own sensor to obtain. For example, when a robot wants to determine whether or not an object a is a toy, it needs to perform a perceptual action: start sensor $\overline{TOY}(x)$ and apply it to a. If a is a toy, the output value of $\overline{TOY}(a)$ is "YES"; otherwise, it is "NO". Similarly, when a robot needs to understand whether or not two objects b and c have a relation between them of b in c, the robot needs to perform a perceptual action: start the sensor $\overline{IN}(x, y)$ and apply it upon b and c. If b is in c, the output of $\overline{IN}(b, c)$ is "YES"; and otherwise, it is "FALSE".

3.3 L_R semantics

The semantic interpretation of a formula in L_R is defined as follows:

1. Let $P(x_1, \ldots, x_n)$ be an atom predicate formula in L_R, $\overline{P}(x_1, \ldots, x_n)$ be a sensor corresponding to the formula, and $a_1, \ldots a_2$ be individual objects in a robot environment. Then
   - $P(a_1, \ldots, a_n)$ is true, and if the output value of $\overline{P}(a_1, \ldots, a_n)$ is "YES"; and
   - $P(a_1, \ldots, a_n)$ is false if the output of $\overline{P}(a_1, \ldots, a_n)$ is "NO".
2. The interpretation of formulas

$$\begin{aligned}
&\neg P(a_1, \ldots, a_n),\\
&P(a_1, \ldots, a_n) \wedge Q(a_1, \ldots, a_n),\\
&P(a_1, \ldots, a_n) \vee Q(a_1, \ldots, a_n),\\
&P(a_1, \ldots, a_n) \rightarrow Q(a_1, \ldots, a_n)
\end{aligned}$$

   are the same as the ones in the first-order logic.
3. $\exists x P(x)$ is true if and only if $p(a)$ is true for some object a in the robot environment.
4. $\forall x P(x)$ is true if and only if $P(a)$ is true for all objects a in the robot environment.

By the above definition, we have the following theorem:

Theorem 1. *Let $P(x_1, \ldots, x_n)$ be an atomic predicate formula in L_R, and $a_1, \ldots, a_n$ be individual objects in a robot environment. Then the robot can recognise whether $P(a_1, \ldots, a_n)$ is true or not.*

Proof. By the definition of the semantic interpretation of a formula in L_R, we know $P(x_1, \ldots, x_n)$ corresponds to sensor $\overline{P}(x_1, x_2, ..., x_n)$. When it is necessary to determine whether or not $P(a_1, \ldots, a_n)$ is true, sensor $\overline{P}(x_1, \ldots, x_n)$ is started and applied to $a_1, \ldots, a_n$. If the output value of $\overline{P}(a_1, a_2, ..., a_n)$ is "YES", $P(a_1, \ldots, a_n)$ is true; otherwise, it is false. □

4 The Model of Understanding Concepts

In this section, we construct a model for a robot to understand a concept.

Different people may have different opinions on how to understand a concept, so there are all kinds of descriptions. However, *connotation* and *extension* are commonly recognised as two basic characteristics. Here *connotation* refers to the essential attribute(s) of the object reflected by a concept, and *extension* refers to all the objects with the essential attribute(s) reflected in the concept. Therefore, understanding a concept means understanding the connotation and denotation of the concept, which is the minimum requirement. Then, how to make sure that the robot understood a concept? Researchers have not reach an agreement yet.

However, we argue that the following three conditions can be used to test whether or not a robot understands a given concept:

1. The robot can recognise the extension of a given concept, *i.e.*, the robot can recognise the object of the essential attribute(s) reflected by a given concept. In other words, for an object in a robot's environment, the robot has the ability to identify whether or not the object is in an extension of a given concept. This task can be accomplished by a computer vision system or other identification system installed on the robot.
2. A robot knows the essential attributes of a given concept. Specifically, it is the basic property that a robot can query a given concept in its own knowledge base.
3. The robot grasps the application of a given concept, *i.e.*, the robot can use the connotation and extension functions of a given concept to solve related problems or accomplish related tasks.

That is, meeting the above three conditions means that a robot understands a given concept. Formally, we have:

Definition 1. *Let P be a special predicate in L_R, corresponding to the robot's sensor $\overline{P}(x)$, and the atomic predicate formula $P(x)$ represent the concept p. If for any object a in the robot environment, sensor $\overline{P}(x)$ can identify whether or not object a is in the extension of the concept p, then concept p is said to be (robot) epitaxially recognisable, denoted as $ExId(p)$.*

Definition 2. *Let p be a concept. The properties of p may be expressed by the formula in L_R. A is the attribute set of concept p. If A is contained in the robot's knowledge base, then concept p is (robot) attribute-awarable, denoted as $AttAwa(p)$.*

Definition 3. *Let p be a concept. If a robot's knowledge base contains common sense and application experience about concept p, and these common sense and experience are expressed by formulas in L_R, then concept p is called (robot) masterable, denoted as $Mast(p)$.*

To facilitate the discussion, we introduce an operator U in L_R. Let $U(R, \phi)$ represent "R understands ϕ", where R is a robot and ϕ is a formula in L_R. And we add $U(R, \phi)$ to L_R as a formula.

Definition 4. *Let R be a robot and formula $P(x)$ in L_R represent concept p. We say that robot R understands $P(x)$, denoted as $U(R, P)$, if p is epitaxially recognisable, attribute-awarable, and masterable.*

Definition 5. *The model of a robot's concept understanding is a triple tuple* $M(p) = (ExId(p), AttAwa(p), Mast(p))$*, where*

- *p represents a concept;*
- *ExId (p) represents concept p is (robot) epitaxially recognisable;*
- *AttAwa(p) represents concept p is attribute-awarable; and*
- *Mast(p) represents concept p is masterable.*

Theorem 2. *Let R be a robot and formula* $P(x)$ *in* L_R *be concept p. Then* $U(R, P(x))$ *holds if and only if* $ExId(p) \wedge AttAwa(p) \wedge Mast(p)$.

Proof. ($\Rightarrow$) Assuming that $U(R, P(x))$ holds, by Definition 4, p is epitaxially recognisable, attribute-awarable, and masterable. Then $ExId(p) \wedge AttAwa(p) \wedge Mast(p)$ is established by Definitions 1–3.

($\Leftarrow$) Suppose that $ExId(p) \wedge AttAwa(p) \wedge Mast(p)$ holds. By Definitions 1–3, concept p is epitaxially recognisable, attribute-awarable and masterable. Then by Definition 4, the robot can understand $P(x)$. So $U(R, P(x))$ holds. □

5 Examples

This section constructs a formal model of a robot's understanding of concept "pen", illustrating it is possible to represent the concept of a robot's understanding in a formal system.

Suppose that a robot has the following facilities:

1. A computer vision system, which is capable of recognising whether the object in the robot environment is a pen or not.
2. A knowledge base that contains the attributes of a pen: a writing instrument; its body is made of metal or synthetic fibre; the writing tip is made of metal; and it can hold and supply ink continuously.
3. A knowledge base that contains the common sense about pen and its usage: Pens can be used with ink; pens can be refilled with ink many times; pens can be used to write on paper; pens can be used to draw symbols on paper; pens can be used to drawn pictures on paper; pens can be used to fill forms, and so on.

We can use the following formulas in L_R to represent the above things:

1. $PenId(x)$ represents a computer vision system that identifies "pens".
2. $WriteTool(x)$ represents that x is a writing instrument; $Pen(x)$ represents that x is a pen; $PenBody(x)$ represents that x is a pen body; $MadeMetal(x)$ represents that x is made of metal; $Has(x, y)$ represents that x has y; $MadeSynFib(x)$ represents that x is made of synthetic fibers; $Ink(x)$ represents that x is ink; PenInkContainer(x) represents x is the ink container of a pen; $Inside(x, y)$ represents that x is in y; and $SustSup(x, y)$ represents that x can sustainably supply y.
3. $Cooperate(x, y)$ represents that x is used in conjunction with y; $Paper(x)$ represents x is paper; $Symbol(x)$ represents x is a symbol; $DrawOn(x, y, z)$ represents that x draws z on y, $Form(x)$ represents that x is a form, and $UsedToFill(x, y)$ represents that x is used to fill y.

Thus, the formal model of the robot understanding concept "pen" can be constructed as follows:

$$M(p) = (ExId(p), AttAwa(p), Mast(p)),$$

where

$$\begin{aligned}
ExId(p) =& PenId(x);\\
AttAwa(p) =& \{WriteTool(x),\\
& Has(x,y) \wedge PenBody(y) \wedge (MadeMetal(y) \vee MadeSynFib(y)),\\
& Has(x,y) \wedge Nib(z) \wedge MadeMetal(z),\\
& Has(x,z) \wedge RubTube(z) \wedge Ink(y) \wedge Inside(y,z),\\
& SustSup(x,y) \wedge RubTube(x) \wedge Ink(y)\};\\
Mast(p) =& \{Cooperate(x,y) \wedge Pen(x) \wedge Ink(y),\\
& DrawOn(x,y,z) \wedge Pen(x) \wedge Paper(y) \wedge Symbol(z),\\
& DrawOn(x,y,z) \wedge Pen(x) \wedge Paper(y) \wedge Pictures(z),\\
& UsedToFill(x,y) \wedge Pen(x) \wedge Form(y)\}
\end{aligned}$$

Of course, the above example is only a simple model. A real robot system may need to add more attributes, common sense and application experience. Actually, human beings have different understandings of concept "pen". A primary school pupil's understanding of pens may not be the same as that of a university student. In addition, a pen also has many attributes and usages, so robots with the different functions may need different knowledge bases.

6 Conclusion

It is well known that knowledge is the foundation of intelligence. An intelligent system must master certain knowledge, and understanding knowledge is a necessary condition for mastering knowledge. Therefore, enabling a computer system understand knowledge is one of the starting points for realising intelligence on a machine. To this end, based on the viewpoints and theories of understanding from artificial intelligence, cognitive science, and philosophical research [13–16], in this paper we propose a formal model of the concept of robot comprehension, and illustrate a model of robot understanding of "pen". The model proposed in this paper emphasises the following three main elements of understanding: the robot can recognise the extension of a given concept; the robot knows the basic attributes of a given concept; and the robot can master common sense and apply experience about a given concept. Using our formal model, we can implement the function of understanding simple knowledge for robot systems.

Although the model presented in this paper is relatively simple, after constructing the model of understanding a single concept, it is possible to construct a model of understanding multiple concepts. Therefore, further work can be considered from a number of simple conceptual model to construct complex conceptual model, and then further consider the construction of the sentence understanding model, and even the model of understand more complicated knowledge.

Acknowledgments. This paper is supported by the National Natural Science Foundation of China (Nos. 61662007, 61762015, and 61762016).

References

1. Searle, J.R.: Minds, brains, and programs. Behav. Brain Sci. **3**, 417–457 (1980)
2. Yoshimura, E., Imono, M., Tsuchiya, S., Watabe, H.: A simile recognition system using a commonsense sensory association method. Procedia Comput. Sci. **60**, 55–62 (2015)
3. Zhou, Z., Li, K., Bai, L.: A general description generator for human activity images based on deep understanding framework. Neural Comput. Appl. **28**(8), 2147–2163 (2017)
4. Terai, A., Nakagawa, M.: A corpus-based computational model of metaphor understanding consisting of two processes. Cogn. Syst. Res. **19–20**, 30–38 (2012)
5. Doboli, A., Umbarkar, A., Doboli, S., Bets, J.: Modeling semantic knowledge structures for creative problem solving: studies on expressing concepts, categories, associations, goals and context. Knowl.-Based Syst. **78**, 34–50 (2015)
6. Roy, R.S., Katare, R., Ganguly, N., Laxman, S., Choudhury, M.: Discovering and understanding word level user intent in Web search queries. Web Semant.: Sci. Serv. Agents World Wide Web **30**, 22–38 (2015)
7. Sheng, W., Du, J., Cheng, Q., Li, G., Zhu, C., Liu, M.: Robot semantic mapping through human activity recognition: a wearable sensing and computing approach. Robot. Auton. Syst. **68**, 47–58 (2015)
8. Ramirez-Amaro, K., Beetz, M., Cheng, G.: Transferring skills to humanoid robots by extracting semantic representations from observations of human activities. Artif. Intell. **247**, 95–118 (2017)
9. Khalifa, K., Gadomski, M.: Understanding as explanatory knowledge: the case of Bjorken scaling. Stud. Hist. Philos. Sci. **44**, 384–392 (2013)
10. Zhong, Y.: Artificial intelligence: concepts, methods and opportunities. Chin. Sci. Bull. **62**(22), 2473–2479 (2017)
11. She, Z.-S., Ni, Z.: On complex system model of thinking and design of a new generation of expert s ystem. Acta Sci. Natur. Univ. Pekinensis **47**(5), 960–968 (2011)
12. He, C.-X., Li, X.-W.: Logical characterization: a new perspective of studying "understanding". J. Xiamen Univ. (Philos. Soc. Sci. Edn.) **4**, 28–34 (2010)
13. Xia, Y.-H., Li, J.: The problem of symbol laying foundation and its solving strategy. Philos. Res. **2**, 102–110 (2017)
14. Jiang, J.-T.: From the Chinese room thought experiment to the problem of intentionality. Philos. Sci. Technol. **30**(3), 40–44 (2013)
15. Fan, Y.-H.: Understanding: a linguistic phenomenon - the revision of Chinese Room thought experiment from Wittgenstein's point of view. J. Dialect. Nat. **36**(4), 44–50 (2014)
16. Chen, J.-M.: Knowledge theory of "understanding". Philos. Trends **11**, 69–75 (2016)
17. Retire, R.: Knowledge in Action: Logical Foundations for Specifying and Implementing Dynamical Systems. MIT Press, Cambridge (2001)

Structure for Knowledge Acquisition, Use, Learning and Collaboration Inter Agents over Internet Infrastructure Domains

Juliao Braga[1,2(✉)], Joao Nuno Silva[1], Patricia Endo[3,4], and Nizam Omar[2]

[1] INESC-ID, Instituto Superior Técnico, Universidade de Lisboa, Lisboa, Portugal
juliao.braga@tecnico.ulisboa.pt
[2] Mackenzie Presbiterian University, Sao Paulo, SP, Brazil
[3] Universidade de Pernambuco, Recife, PE, Brazil
[4] Dublin City University, Dublin, Ireland

Abstract. This article describes an environment for knowledge acquisition, learning, use and collaboration inter agents over Internet Infrastructure. Four agent types are used in a previously applied four-tier model, such as the use case on the Internet Routing Registry. This model, which can be implemented in each Autonomous System domain of the Internet infrastructure, is integrated into an environment with (a) capturing information from unstructured databases, (b) creating and updating training bases appropriate to machine learning algorithms, and (c) creation and feeding of a knowledge base. Such resources become readily available to agents in each domain and to agents in all other domains with the aim of making them autonomous. The agents collaborate and interact with each other, through individual blockchain structures that also take care of operational security and integration aspects. In addition, a test bed to validate the entire model, including the functionalities of the agents, is also proposed and characterized.

Keywords: Artificial intelligence · Agents collaboration · Machine learning · Knowledge acquisition · Knowledge use

1 Introduction

Autonomous System (AS) (or routing domain) is the name given to the networks that making up the Internet [29]. The collection of all ASes interconnected, represent the Internet as shown in Fig. 1 [21,28].

The ASs are interconnected by the *Border Gateway Protocol* (BGP) [48]. BGP is a complex protocol requiring expert AS administrator knowledge to be configured correctly. This complexity is in turn enhanced by the over imposing Internet governance, whose main stakeholders and relationships are represented in Fig. 2 [15].

K. Arai et al. (Eds.): CompCom 2019, AISC 997, pp. 527–547, 2019.
https://doi.org/10.1007/978-3-030-22871-2_36

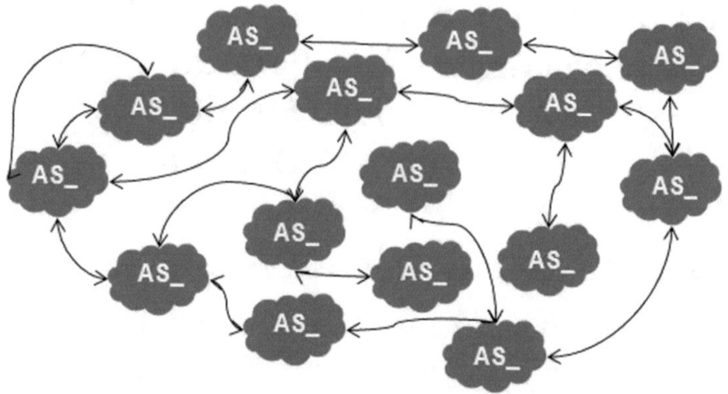

Fig. 1. How AS build the Internet: _ is some i, where $0 < i < 4.294.967.296 \simeq (2^{32})$

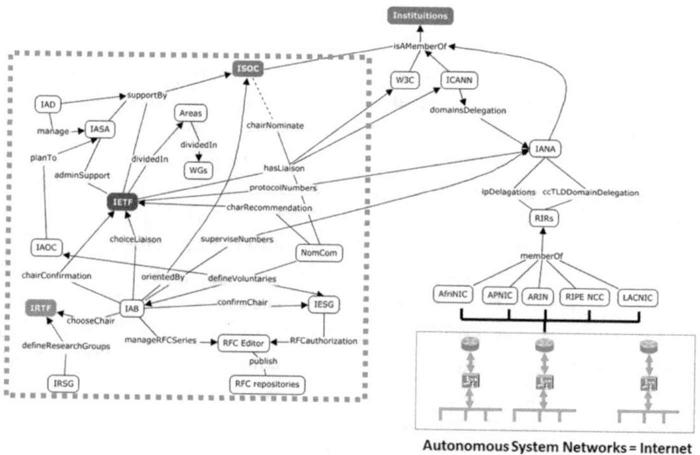

Fig. 2. The internet infrastructure ecosystem

The Internet Society[1] (ISOC) is a non-profit organization created in 1993 to achieve two results: aggregate and support the initiatives of stakeholders and other participants and, to remove all non-technical discussion out of the *Internet Engineering Task Force*[2] (IETF) [31]. Hence, ISOC manages the infrastructure necessary for the operation of the IETF and develops supporting frameworks such as the Internet Architecture Board[3] (IAB). In short, ISOC aims to promote

[1] https://www.internetsociety.org/.

[2] https://ietf.org/.

[3] https://www.iab.org/.

the development of standard[4] on the Internet, as well as to promote educational initiatives and public policies linked to the Internet.

Since the IETF was created in 1986, before ISOC, its members were strongly attached to the IETF with the role of managing various actions to improve the Internet. After the emergence of ISOC, the IETF stoped being funded by the US government. With this unlinking and under the management of ISOC, the IETF became popular worldwide as a reference in defining and discussing standards for the Internet. Figure 3 briefly shows this historical evolution.

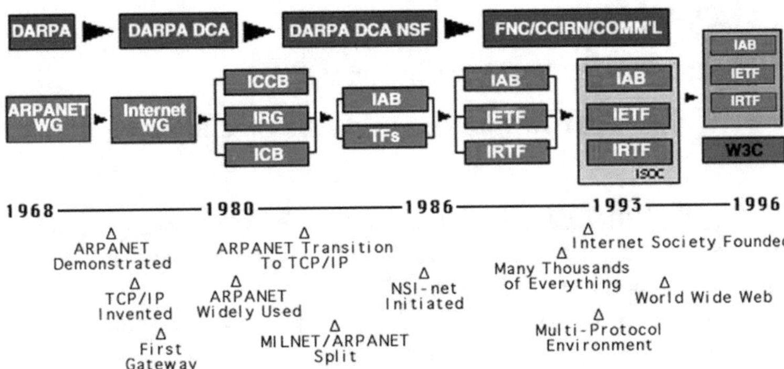

Fig. 3. Timeline and the main movements that defined the future of the Internet. Source: [36]

Traffic engineering and routing policy are key AS information that should be updated often. This information is hosted on important servers such as the such as *Internet Routing Registry*[5] (IRR) [3,4,13,37,61]. The IRR is a distributed database of route and route-related information [14]. Despite its importance, it is vulnerable to the capacity of the AS administrator to add, maintain and update IRR objects. Hence, there was a pressing need to develop a functional model of agents capable of replacing human interventions on such objects. So the *Autonomous Architecture Over Restricted Domains* (A2RD) was integrated into the domain of an AS [16]. A2RD replaces the human with its agents, *Intelligent Elements* (IEs), establishing a new IRR model, namely, *IRR revised* (IRRr) (Fig. 4).

A specialized A2RD IE automatically create objects as defined by the *Route Policy Specification Language* (RPSL) [2,12]. The objects unable to be automatically created receive support from AS administrators through a human-computer cooperation mechanism. With the application of the A2RD IEs, the IRR and IRRr models remain the same, characterized by the expectations recommended by the stakeholders to the IETF and *Internet Research Task Force*[6] (IRTF)

[4] These standards allow us to orchestrate the complexity of the Internet in such a way that it always works very well.

[5] http://www.irr.net/.

[6] https://irtf.org/.

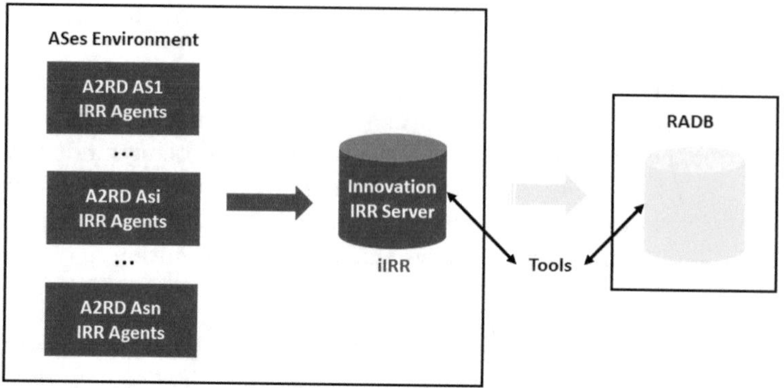

Fig. 4. The *IRR revised model* established by A2RD

[34,37,38,43,53]. A2RD IEs will also not affect the security concerns surrounding the IRR and Internet governance and, similarly, tools that use IRR databases can be used without any modification [35].

This article presents with the theoretical framework that guarantees the perfect understanding of the possible applications of A2RD on the Internet Infrastructure. It is divided into six sections. The first is this introduction. The second section is a presentation of the theoretical foundations and related works. The third is a description of the A2RD abstract model. The fourth section shows the A2RD implementation model. The fifth is the with the conceptual model of the A2RD development and the sixth section ends the article with the description of the current stage of the project.

2 Theoretical Foundations and Related Work

2.1 Theoretical Foundations

The notion of intelligent agents [50], was already perceived as feasible to be applied in the various areas of knowledge, in particular, those that affect the resources and facilities that make the Internet work. As stated above, the interests were to apply intelligent techniques over agents, called IE, in restricted domains of the Internet Infrastructure, that is, in the so-called routing domains represented by AS, was presented in 2014 [17].

Many terms used or need to be used to form the set of concepts necessary to apply the specific notion of IEs are still not well defined or clear in the literature. To avoid ambiguities and therefore to leave the understanding clear, it is necessary to present some definitions that will be used in this article.

Self-organization. The notion of *self-organization* was defined in the early 1970s by Ilya Prigogine, Nobel Prize in Chemistry in 1977, in relations with natural systems [22]. In the context of this work, when dealing with intelligent agents, by **self-organization** is meant the ability of a agent to react readily when it perceives in some way a threat (or instability) in its environment indicating the possibility of a deviation in its functional objectives. This reaction returns the functionality of the agent to its stable condition existing in the previous state (self-organized). Figure 5 is a simplified view of this definition.

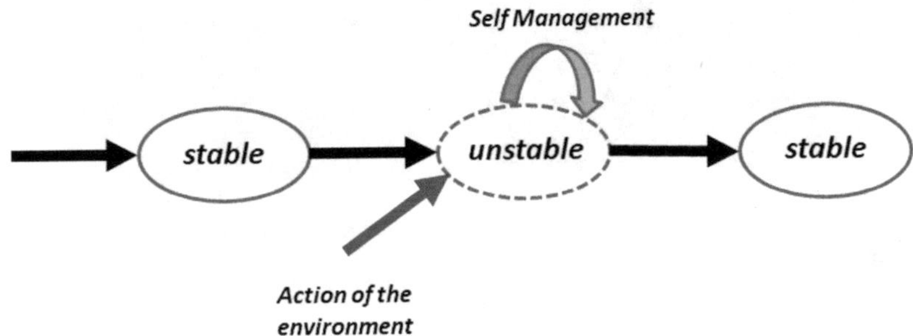

Fig. 5. Self-organization. Adapted from [22, page 2]

For a **intelligent agent** to be able to self-organize, it must have certain properties and appropriate functional characteristics, which will be seen later. For the purpose of this text, if an **intelligent system** exercises its capacity for self-organization without absolutely no human intervention, it will be recognized as **autonomous**, with definite freedom to act [1]. If, however, to exercise the function of self-organization, the intelligent system depends on a human orientation, not directly, but through pre-defined parameters indicating how to react, then the given denomination will be **autonomic**. Put differently, a **intelligent system** is **autonomic** if there is a human orientation (a plan) about how it should behave to self-organize, when reacting to an action of the environment. If an **intelligent system**, in order to exercise its capacity to self-organize, performs functions integrally put by the human being, then it is considered **automatic**. Finally, if the intelligent system has not the conditions to self-organize, that is, under an unusual action of the environment, it eventually fails, then it is said to be **legacy**. Figure 6 shows the relationship between such **intelligent systems** indicating the degrees of **independence** and **intelligence** aggregates and how it will be interpreted in this article.

The notion of **independence** is linked to human participation. The more intense that participation, the less autonomous the system is. Already, the notion of **intelligence** is associated with the ability of **self-organization** of the system. The greater the ability to effectively use the properties and functionality of self-organization, the more autonomous the system is. On the other hand, the notions of intelligence and independence are directly associated with the learning ability of the intelligent elements. This learning will be effective if there is **cooperation** between the intelligent elements, without distinction. The project is not interested in the functional aspects of **legacy systems** but, whenever

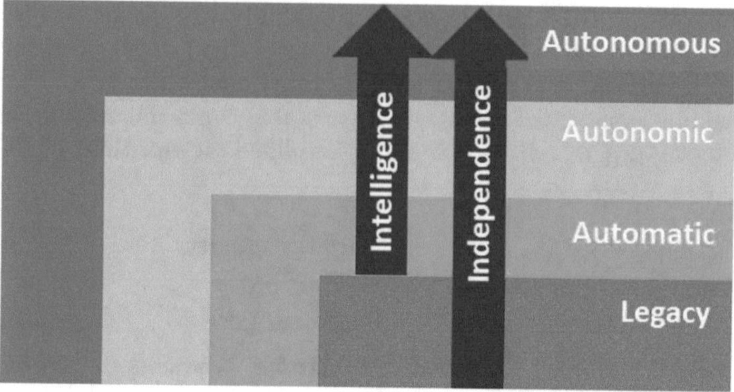

Fig. 6. Characterization of intelligence and independence of Intelligent Systems in the A2RD project

necessary, will use results or configuration parameters of these systems, in order to maintain balance between all the elements involved.

Domains. A *domain* represents a collection of things (actors, entities, etc.) that are aligned and united around common goals, within the specific limits of a particular area of interest[7]. In the context of the A2RD project, the area of interest is the environment outlined by the activities associated with the ASs which together, as we know, represent the Internet.

ASs, in turn, have subsets of interests that are identified as *sub-domains.* In general, the environment where *Internet of Things* (IoT) techniques are applied is a real example of a sub-domain.

Interoperability and Ontology. *Interoperability* is the term used in this context to designate the ability to cooperate between domains and/or subdomains through its respective IEs, in the achievement of common objectives. There are two types of interoperabilities: *syntactic interoperability* and *semantic interoperability.*

Syntactic interoperability refers to information that is exchanged between IEs during the connection. Such information, in general, is inserted in the context of the protocols that allow such connectivity. Eventually, additional information, which ascends to the upper layers of the TCP/IP model, uses additional information to maintain the connection that does not need interpretation of meaning, or are just pure data. In this case, some features such as *eXtensible Markup Language* (XML) or *Structured Query Language* (SQL) in formats recommended and described in this document enable interoperability.

The IEs, after the interconnection, must maintain the process of communication and cooperation between them. This is what represents *semantic inter-*

[7] http://www.ncoic.org/what-is-interoperability.

operability. The exchange of information between them begins to produce data with understandable meaning, whose interpretation is appropriate to produce the expected results, throughout the process. Such data, in appropriate repositories, are accompanied by a special formatting called *ontology*. Ontology is the resource used to represent knowledge. The ontology (the knowledge associated with data in the pure state), has adequate languages to serve its purposes. Such languages vary depending on its ability to clearly express desired knowledge. Although details are discussed later in this text, Fig. 7 displays the main languages available in a comparison of its formalism and its ability to express knowledge. Note that the most powerful of languages is the natural language, which has such a restriction of interpretation by non-human procedures, which will be avoided in this work.

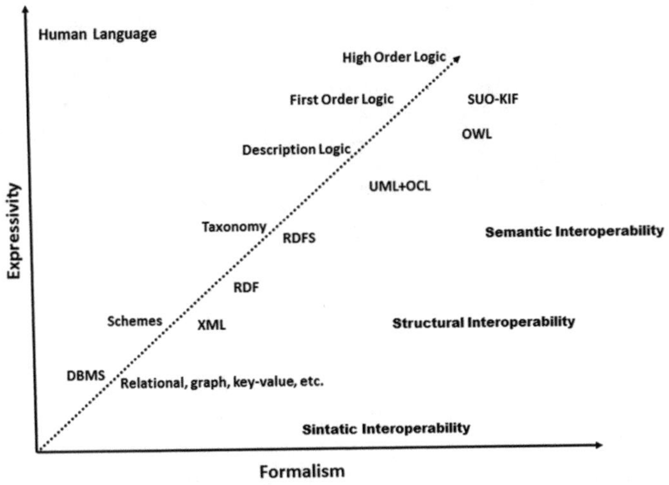

Fig. 7. Languages to represent knowledge and its expressivity. Source: [46]

Intelligent Agents. In the early 1990s, Michael Wooldridge from his doctoral thesis in 1992 made a considerable effort in the formalization of intelligent agents [60]. Also he established the theoretical and practical concepts involving intelligent agents, its architecture, mathematical models, logic and semantics, as well as considerations about the issues surrounding software projects and development, with approaches on the multi-disciplinarity around of intelligent agents [58]. This same work recalls that a cluster of intelligent agents form a **agency**, characterization perfectly adequate to the set of IEs. An agency actually consists of *multi-agents*. On two occasions, Wooldridge consolidated the ideas on *multi-agents*, [56,57]. In the second edition, using an article he defines, in free translation [58]:

> "An agent is a computer system that is situated in some environment, and that is capable of autonomous action in this environment in order to meet its delegated objectives" [57, Chapter 2].

On the other hand, Russel and Norvig [50], more simply, illustrates the concept of the relationship between the agent and the environment in which it is associated (Fig. 8). For them, *an agent is something that has perception of its environment through sensors and acts on the environment through actuators.*

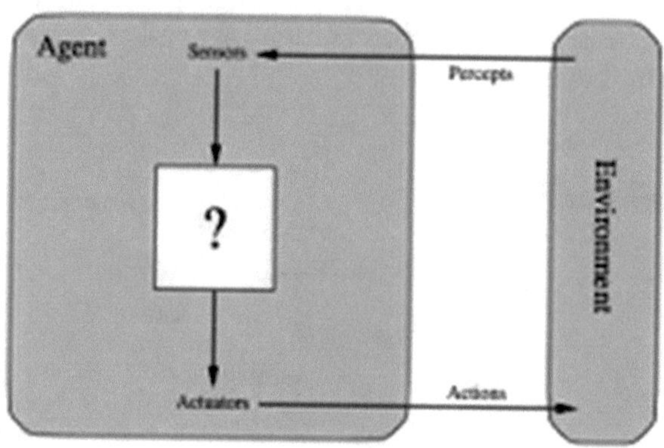

Fig. 8. How the agents interact with the environment. Source: [50]

[50,56,57] are appropriate texts for the learning and application of agents and multi-agents. The first, with its broad description of Artificial Intelligence techniques and the second a specific text about multi-agents. To them, join a book, also, essential and complementary [59]. In addition, one can not forget the text of [54]. Everyone at the end of each chapter makes appropriate and convincing approaches to the issues discussed above, including a literature review. Naturally, [50] present a more up-to-date text.

Multi-agents. In addition to the definitions of agents, in the previous sections, a very clear multi-agent model proposed by [33] is represented in Fig. 9.

This model is suitable to establish the main motivation of the proposal of the A2RD model. Before proceeding with this analysis of the peculiarities and specific properties of an IE, it is worth noting the definition given by this author:

> An agent is an encapsulated computer system that is situated in some environment and that is capable of flexible, autonomous action in that environment in order to meet its design objectives.

Immediately, the same author clarifies some points associated with the definition of his canonical model. According to him, agents are:

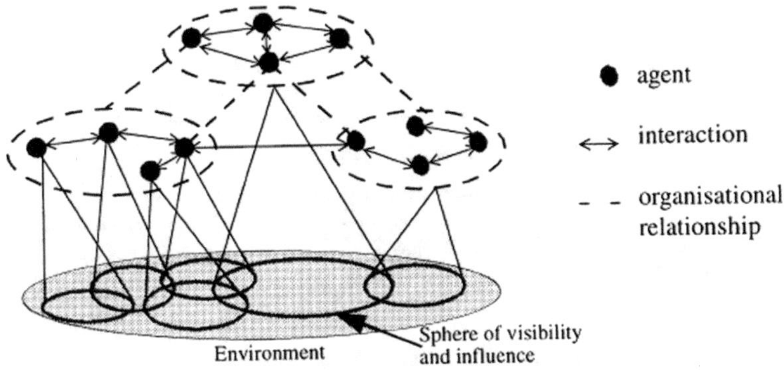

Fig. 9. Canonical view of an agent-based system. Source: [33, page 281]

(a) *Clearly identifiable*: As entities that solve problems, with well-defined interfaces and boundaries.
(b) *Encapsulated in a particular environment*: They receive inputs related to the state of its environment, through sensors and act on the environment through actuators (which he calls *effectors*).
(c) *Designed to meet a specific goal*: They have particular goals to meet.
(d) *Autonomous*: They have control over both its internal state and its own behavior. The control characteristic about its own behavior is what distinguishes agents from objects.
(e) *Able to exhibit flexible problem solver behavior*: In addressing its goals they need to be both reactive (able to respond in time to changes that occur in its environment) and proactive (empowered to act ahead of your future goals).

These observations, complemented by [55,58] make evident the definitions of multi-agents that continues a very active and effervescent area of research. Thus, the analysis of Fig. 9 can proceed, on which two aspects stand out:

(i) From the point of view of interconnection there is not a complete mesh or full mesh, i. e., some agents do not communicate directly with other agents.
(ii) In relation to the environment, the agents act in specific sub-domains and more than one agent can act on the same sub-domain.

Communication Between Intelligent Agents. There is a huge research effort in the direction of defining the communication properties between agents. One such initiative is the *Foundation for Intelligent Physical Agents*[8] (FIPA). Founded in 1995 with the aim of developing standards aimed at systems of agents. These patterns, specified in categories are grouped[9] according to Fig. 10.

FIPA standards systematically use Software Engineering and, in particular, Unified Modeling Language (UML), as meta-language [23,24]. In fact, FIPA

[8] http://www.fipa.org.
[9] http://www.fipa.org/repository/bysubject.html.

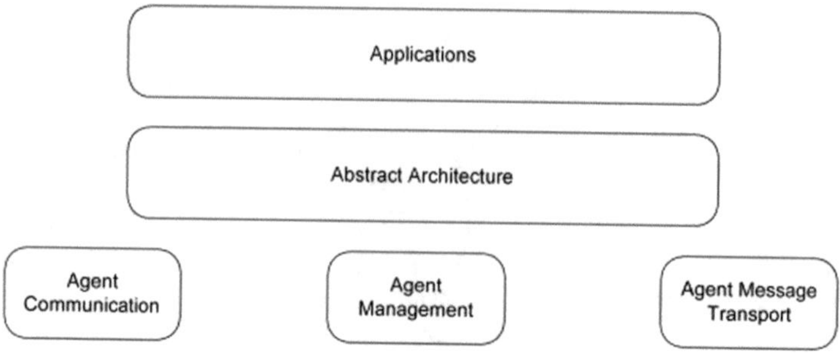

Fig. 10. FIPA specification category groups.

recommends extensions for UML dealing with agents. Author in [11] present a comparison of several meta-models used in some known methodologies, in search of common aspects between them. In addition, in a UML approach, [6] provides a very expressive overview and presents an extension of the UML, called Agent Unified Modeling Language (AUML). [45] describe the UML meeting the specification [26]. In [5], the AUML extension is described comprehensively. Object Management Group (OMG) discusses agents[10], and recommends a specific environment for AUML[11].

The Abstract Architecture [25] is the prerequisite for moving on to other specifications. In [27], the proposal for an inter-agent communication language (ACL) that gave rise to Java Agent Development Framework (JADE), whose best-known original document is [9] followed by a complementary article [8] and a much more complete text in [10]. The importance of the environment, in which the agents interact, is characterized in a very lucid manner in [44].

Although the FIPA proposal is not the only one, whether it is an attempt to standardize or not, for example, in cases of agents in *peer-to-peer* environments, described in [40], This project deviates considerably from FIPA's proposal, making the desired scenario for the IES Agency more flexible and closer to the characteristics of the Internet Infrastructure if this concept is necessary.

2.2 Related Work

Table 1 shows the main works and its respective characteristics involving intelligent agents, which had strong influence during the development of A2RD.

The origin of the associated ideas arose from the proposal of Autonomic Computing [30]. [41] display details of MAPE-K model with its respective control cycles and compare several other proposals of autonomic architectures with emphasis on networks. [52] proposes changes in the MAPE-K model, simplifying it, for elements of autonomic networks. [7] started in the Internet Research Task Force (IRTF) NMRG group, proposals that continued in the IETF ANIMA

[10] http://www.objs.com/agent/.
[11] http://www.auml.org/.

Table 1. Related works

Characteristics	MAPE-k	Others	Schmid	ANIMA
Reference	IBM [30]	[41]	[52]	[7] (IRTF + IETF)
Domain	Application. IBM Products	Application. Autonomic Architecture	Networks	Networks: autonomic nodes with the same intention
Integration between elements	Through an executor	Follow the MAPE-K: executor equivalent	Variation of MAPE-K	If necessary, use the Feedback cycle
ID	Undefined	Undefined	Undefined	IPv6 (host interface)
Human Interference	High Level Objective	High Level Objective	High Level Objective. Deterministic Behavior	Intention. Autonomic Control Plane
Specific	Self management	Self management	Self management and self-adaptation	Self-management, network knowledge, self knowledge (self-awareness)
Architecture	Centralized, restricted scalability	Hierarchical, peer, restricted scalability	Similar to FIPA's proposal: Agency	Scalable in the domain

group, with studies that are in full activity, with specific recommendations for protocols, indicating the most recent and active studies on autonomic networks. From the models described in Table 1, only ANIMA proposes an identification (ID) for its autonomic functions, associating them with an IPv6 address, indicating that the autonomic functions are aggregated to the host interface.

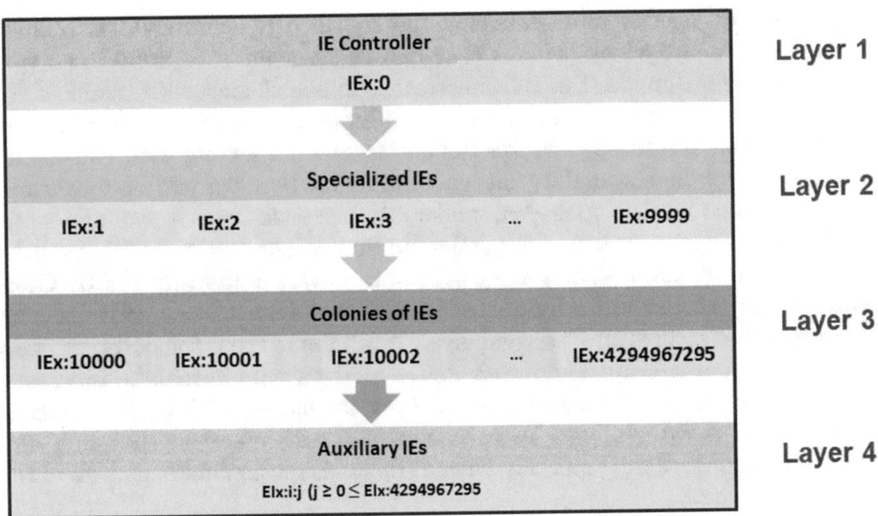

Fig. 11. A2RD layer model for the ASx domain, where x is the AS number.

3 The A2RD Abstract Model

A2RD is a project that proposed a model of agents divided into layers as can be seen in Fig. 11.

The model serves the interest of establishing an architecture of IEs on the administrative domain of ASs. It may exist in any of the 2^{32} possible ASs. However, on 07/11/2018 there were only **62351** ASs, originating traffic on the Internet, according to CIDR-Report[12]. An AS is identified by a unique number. This uniqueness is controlled by the *Public Technical Identifiers*[13] (PTI) and is named *Autonomous System Number* (ASN). Thus, the largest possible value of **x** is **61612**, corresponding to **AS61612**, at the date above. A2RD implementations are independent and restricted to an AS, but with a high degree of interoperability and, of course, intensive cooperation because AS administrators depend on the behavior of all others. The PTI has reserved two contiguous ranges of AS numbers for private use: **64512-65534** and **4200000000-4294967294** [39]. Conveniently, these AS numbers can be used to designate IE domains.

The first of the four layers hosts the IE called the Controller. Its identification is unique and fixed: **x:0**, that is, the number **0** placed to the right side of the **:** symbol, following the ASN hosting the model. Sometimes, to make clear which IE is being referenced, **IE** is used before identification, for example, by stating that the **IE Controller** is **IEx:0**. Thus, if **ASn** is the host domain of the model, then the controller element is **IEn:0**. No IE from the lower layers may exist without the prior consent of the **IE Controller**. It has the property of keeping himself organized (self-organization) and of ensuring the self-organization of any IE from the lower layer.

The second layer is represented by the so-called **Specialized IEs**. These elements are identified by suffixes that can range from **1** to **9999**. The specialized elements support the **IE Controller** in specific activities required for functionalities ranging from ensuring the interoperability of the entire system of implemented IEs to specific functionalities such as servers with end-to-end characteristics[14] that stimulate the understanding between two architectures: the layer model and the topological model [51], access features to bank semantic repositories, proprietary software (similar to SouthBound SDN APIs), facilities required for lower-tier IEs, and many others. However, support for the **IE Controller** is the primary objective of the **Specialized IEs**. This objective is what determines the features of the second layer. It is assumed that some **Specialized IEs** may be **Autonomic Elements** or IEs that execute automatic processes, such as proprietary software and procedures associated with legacy systems, among others. A **Specialized IE** can be created with functions that only concern the **IE Controller**, especially when it depends on the functionalities of IEs of the third layer.

---

[12] http://thyme.rand.apnic.net/current/data-summary.

[13] https://pti.icann.org/.

[14] Recognized as end-to-end arguments.

In the third layer lies the largest agglomeration of IEs, which is why it is called the **IE Colonies**. Elements of this layer can be **autonomous**, **autonomic** or **automatic**, and are directly responsible for the most important activities of the application, including software reuse. They act under the influence of a high degree of interoperability and cooperation between them IEs of other layers and other domains/sub-domains. They do not directly participate in interconnections or exchange messages with other IEs outside the domain, but they do so through IEs in the upper tiers. There is intense semantic interoperability activity on the part of these IEs, which have a high capacity for self-learning due to continuous interactions with the domain environment, and produce improvement effects on the knowledge of other IEs of the *colony* itself and the IEs of the layers the **IE Controller**. In other words, these IEs favor the learning of the entire cluster of IEs of the layer model, which is being described. The IEs of the colonies receive an identification with numerical suffixes, ranging from **10000** to **4294967295**.

The fourth layer is the **Auxiliary IEs**. This layer exists, in order to allow the transfer of computing demands to a new set of IEs (A2RD successiveness). It reproduces, successively, the first, second, third and fourth layers. This new IEs sequence has an additional suffix **:j:0** for a new **IE Controller** responsible for the next four layers. In the second, third and fourth new layers, the IEs identifications are post-fixed with **:j:id** where, **j** is the colony IE number that originated the new fourth layer and **id** is a number with the above specifications. A typical application for the fourth layer are sub domains, such as home networks (**home-net**).

The use case for the A2RD was the addition and update of objects in IRR server. The application was considered useful mainly because the tasks of the AS administrator did not guarantee the accuracy in its completion nor the permanent need to update the objects making the IRR an unreliable system from the point of view of its content [18].

4 The A2RD Implementation Model

According to Fig. 12, the IEs are arranged and distributed between layers, similar to what was said in the previous section and are implemented in the domain of any ASN.

It is also observed, in the same figure, that the IEs is important functionally in inter-domain operations reside in the upper layers. For example, a classification of relevance is the intensity of aggregation that an IE possesses, in relation to the auto-* (or self-*) properties. If an IE, however, has some self-organizing capability, it must participate directly linked to the **IE Controller**. Even if it participates in the **Auxiliary IEs** layer, an **IE Controller** can logically construct a new layer architecture. And so on.

On the other hand, the representation of the model is logical (abstraction of the physical implementation). Physically, the locating if an IE in the domain environment is essential. The best alternative is IP addressing, preferably IPv6, for reasons of availability. The IE Controller must maintain a table associating

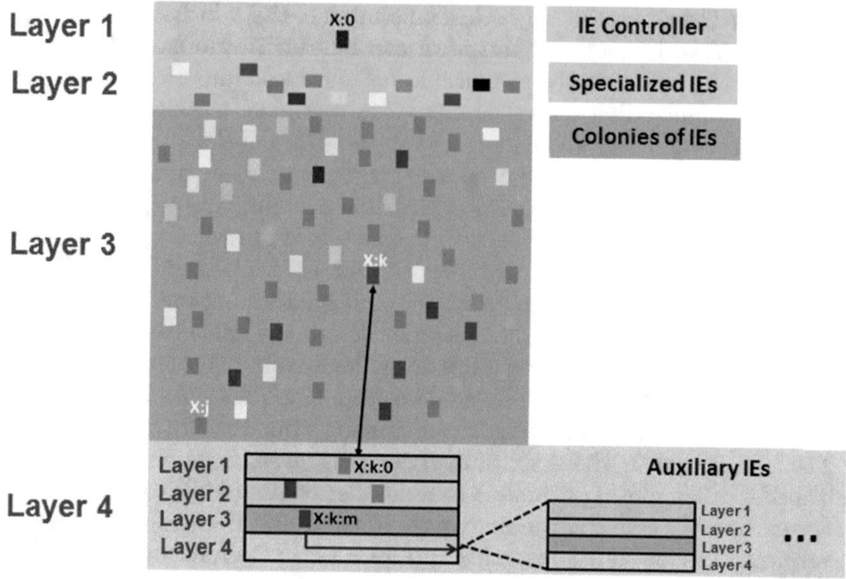

Fig. 12. A2RD implementation model, where x is any ASN.

the reference logic with the IP designated by the IE Controller itself, from the premise that an IPv6 block should be available at the beginning of the implementation. In the implementation of prototypes related to the case study, the

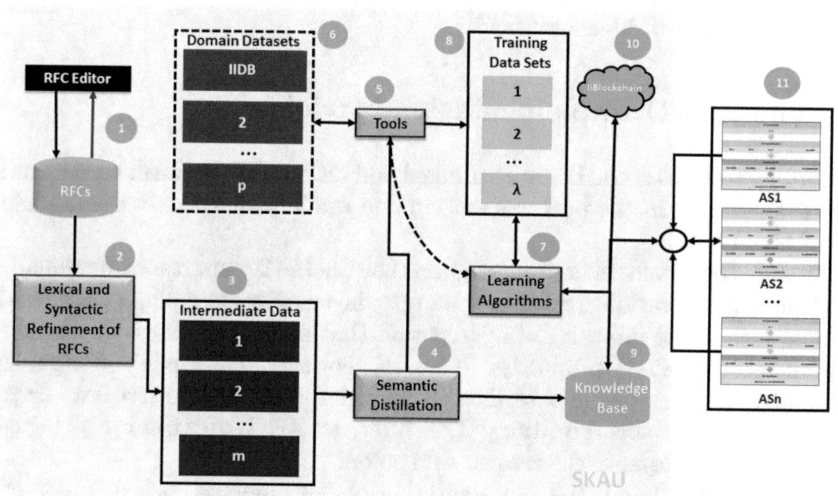

Fig. 13. Structure for Knowledge Acquisition, Use, Learning and Collaboration model (SKAU)

Python language will be used. When needed, features closer to the operating system will be used ("scripts" and other inherent facilities).

5 The A2RD Environment Conceptual Model

Figure 13 shows the *A2RD Environment Conceptual Model*, namely, *Structure for Knowledge Acquisition, Use, Learning and Collaboration Inter A2RD Agents* (SKAU) in which each implementation of A2RD, into an AS, is represented as an agglomeration of IEs in a four layers model (11).

The other components of the SKAU are dynamically constructed from non-structured databases. For this project, only the Request for Comments (RFCs) repository will be used. These repository contains documents authored by network operators, engineers and computer scientists, documentary methods, behaviors, research, or innovations applicable to the Internet, all of them resulting from working groups of the IETF and IRTF, and maintained by RFC-Editor[15].

The SKAU model components can be described as follows:

- RFCs are captured/updated and stored locally (1);
- A set of tools responsible for acting lexical and syntactically on RFCs (2), transforming them into intermediary databases (3);
- Other tools (4), like *Semantic Distillation*, that act on the intermediary databases producing inputs for the construction of *Domain Datasets* (6) among them the *Internet Infrastructure Data Base* (IIDB), and so these into *Training Data Sets* (TDS) (8). Also, these tools will support for provide part of the **knowledge base** (9) [32];
- *Learning algorithms* (7) that support the construction and use of TDS to renew the knowledge base and meet the demand of agents of A2RD models in the process of developing and applied intelligent actions. The efficient use of TDS will respond to the classic algorithms of Machine Learning (ML): (a) supervised learning, (b) unsupervised learning, (c) reinforcement learning and (d) semi-supervised learning, that combines (a) and (b). ML has a vast reference about its techniques, entertaining, but [42] in Section II has an appropriate overview, with close approach, i.e., optical networks.
- A database, namely, *IIBlockchain* (10) built by each implemented A2RD model and stored together in the Git Hub (so, in cloud), that serves as support for the process of collaboration and effective interaction, inter/intra agents of the models [20]. The *IIBlockchain* cloud interacts with the *learning algorithm* and *knowledge base* allowing agents to exercise *offline and online computation*[16].

Each AS can implement an A2RD, which is controlled by the IE, namely, *IE Controller*, and receives the identification *x:0*, where *x* is the *AS Number* (ASN).

[15] https://www.rfc-editor.org.

[16] *Offline computation is the computation done by the agent before it has to act, and online computation is the computation done by the agent between observing the environment and acting in the environment* [47].

6 Current Stage, Conclusions and Results

6.1 IIDB

IIDB is a data set that covers the Internet Infrastructure domain and can be used for quick access not only by IEs but also by third parties and will help build the KB and support to update the KB and will be used as learning content for ML algorithms. The first three data bases which initially make up the IIDB – *iidb.rfc* (3,903 MBytes), *iidb.person* (4,961 MBytes) and *iidb.acronym* (9,590 MBytes) – are available in the repositories Open System Foundation (OSF), include the JSON templates [19].

The final execution result of the acronym algorithm, implemented in Python is represented by the numbers in Table 2.

Table 2. Statistics of the execution of the acronyms extraction algorithm

#	Representation	Meaning
1	RFCs processed (files)	8,340
2	Processing time (seconds)	14,163.1
3	Mean size of acronyms (chars)	3.36
4	Total number of acronyms extracted	69,198
5	Total number of acronyms extracted (no repetition)	12,273
6	Acronyms (no repetition) automatically confirmed	11,098
7	Precision	90.42%

The number of RFCs processed is less than the number of the last RFC processed, because there are numbers without RFCs. (1) The processing time is the result of the execution time given by the Sublime Text.[17] (2) The acronyms average size, in number of characters. (3) The total number of acronyms extracted, with repetition. (4) Number of non-repeating acronyms from a non-human point of view. (6) Representing 90.42% of (5), (7).

6.2 IIBlockchain

IIBlockchain routines are already implemented and ready for use by IEs in test bed. Preliminary tests demonstrate that the mechanism is effective for the process of collaboration and interaction between agents with appropriate security.

6.3 Test Bed

Currently we are focusing on the implementation of the A2RD and SKAU test bed, whose design can be seen in Fig. 14, that will validate the theoretical model.

[17] https://www.sublimetext.com/.

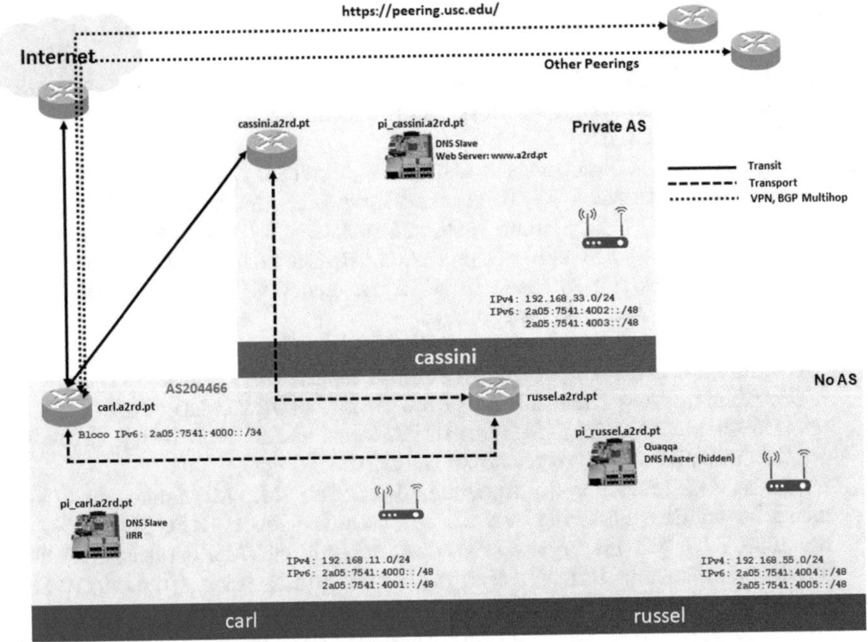

Fig. 14. A2RD and SKAU testbed

It involves three environments and a set of appropriate resources to test all the requirements of the proposed model:

(i) One local environment with public ASN (AS204466).
(ii) One public IPv6 block will be announce: 2a05:7541:4000::/34.
(iii) Two local environment with private ASN.
(iv) Private IPv4 in all environment.
(v) Python language for all required code [49].
(vi) Availability of the code in the GitHub[18].
(vii) Use of the domain a2rd.pt and DNS servers (one in each ASN and one of them an hidden master).
(viii) Use of the Zope[19] for the web server of the project and as hosting environment for the code will be developed.

Acknowledgment. This work is supported by CAPES – Brazilian Federal Agency for Support and Evaluation of Graduate Education within the Brazil's Ministry of Education and was also supported by national funds through Fundacao para a Ciencia e a Tecnologia (FCT) with reference UID/CEC/50021/2019.

[18] https://github.com/juliaobraga/a2rd.

[19] http://www.zope.org/en/latest/.

References

1. Agoulmine, N.: Introduction to autonomic concepts applied to future self-managed networks. In: Autonomic Network Management Principles: From Concepts to Applications, p. 1 (2010)
2. Alaettinoglu, C., Villamizar, C., Gerich, E., Kessens, D., Meyer, D., Bates, T., Karrenberg, D., Terpstra, M.: Routing policy specification language (RPSL). Technical report, RFC Editor, June 1999, rFC2622. https://www.rfc-editor.org/info/rfc2622. (Obsoletes RFC2280) (Updated-By RFC4012, RFC7909) (Status: PROPOSED STANDARD) (Stream: IETF, Area: ops, WG: rps). https://doi.org/10.17487/RFC2622. Accessed 03 Feb 2018
3. Awduche, D., Chiu, A., Elwalid, A., Widjaja, I., Xiao, X.: Overview and principles of internet traffic engineering. Technical report, RFC Editor, December 2002, rFC3272. http://www.rfc-editor.org/info/rfc3272. (Updated-By RFC5462) (Status: INFORMATIONAL) (Stream: IETF, Area: subip, WG: tewg). https://doi.org/10.17487/RFC3272. Accessed 10 Nov 2018
4. Awduche, D., Malcolm, J., Agogbua, J., O'Dell, M., McManus, J.: Requirements for traffic engineering over MPLS. Technical report, RFC Editor, September 1999, rFC2702. http://www.rfc-editor.org/info/rfc2702. (Status: INFORMATIONAL) (Stream: IETF, Area: rtg, WG: mpls). https://doi.org/10.17487/RFC2702. Accessed 10 Nov 2018
5. Bauer, B., Muller, J.P., Odell, J.: An extension of UML by protocols for multi-agent interaction. In: Proceedings of the Fourth International Conference on MultiAgent Systems, pp. 207–214. IEEE (2000)
6. Bauer, B., Müller, J.P., Odell, J.: Agent UML: a formalism for specifying multiagent software systems. Int. J. Softw. Eng. Knowl. Eng. **11**(03), 207–230 (2001)
7. Behringer, M., Pritikin, M., Bjarnason, S., Clemm, A., Carpenter, B., Jiang, S., Ciavaglia, L.: Autonomic networking - definitions and design goals. Technical report, RFC Editor, December 2014
8. Bellifemine, F., Caire, G., Pogg, A., Rimassa, G.: JADE: a software framework for developing multi-agent applications. Lessons learned. Inf. Softw. Technol. **50**(1), 10–21 (2008)
9. Bellifemine, F., Poggi, A., Rimassa, G.: JADE–a FIPA-compliant agent framework. In: Proceedings of PAAM, London, vol. 99, p. 33 (1999)
10. Bellifemine, F.L., Caire, G., Greenwood, D.: Developing Multi-Agent Systems with JADE, 1st edn. Wiley, Hoboken (2007)
11. Bernon, C., Cossentino, M., Pavón, J.: Agent-oriented software engineering. Knowl. Eng. Rev. **20**(02), 99–116 (2005)
12. Blunk, L., Damas, J., Parent, F., Robachevsky, A.: Routing policy specification language next generation (RPSLng). Technical report, RFC Editor, March 2005. https://www.rfc-editor.org/info/rfc4012. (Updates RFC2725, RFC2622) (Updated-By RFC7909) (Status: PROPOSED STANDARD) (Stream: IETF, WG: NON WORKING GROUP). https://doi.org/10.17487/RFC4012. Accessed 03 Feb 2018
13. Boyle, J., Gill, V., Hannan, A., Cooper, D., Awduche, D., Christian, B., Lai, W.: Applicability statement for traffic engineering with MPLS. Technical report, RFC Editor, August 2002, rFC3346. http://www.rfc-editor.org/info/rfc3346. (Status: INFORMATIONAL) (Stream: IETF, Area: subip, WG: tewg). https://doi.org/10.17487/RFC3346. Accessed 10 Nov 2018

14. Braga, J.: Curso IRR – Parte I a Parte X (2010). https://ii.blog.br/2010/03/18/curso-irr-parte-i-introducao/, Internet Infrastructure Blog: Accessed 19 Jan 2018
15. Braga, J., Granville, L.Z., O'Flaherty, C., Moreiras, A.M. (eds.): O Livro do IETF/El Libro del IETF/The Tao of the IETF. CGI.br, São Paulo (2014). https://cgi.br/media/docs/publicacoes/1/o-livro-do-ietf.pdf
16. Braga, J., Omar, N., Granville, L.Z.: Uma proposta para o uso de elementos inteligentes em domínios restritos da infraestrutura da internet. In: Anais CSBC 2015 - WPIETFIRTF. Recife, Pernambuco, Brasil, July 2015
17. Braga, J., Omar, N.: Semantic repository in internet infrastructure knowledge domain: methodology. In: Anais do CSBC 2014 - IWPIETF LAC, Brasília, DF, pp. 2179–2184, July 2014
18. Braga, J., Omar, N., Thome, L.F.: Acquisition and use of knowledge over a restricted domain by intelligent agents. In: Proceedings of the SouthEast Conference, ACM SE 2017, pp. 203–207. ACM, New York (2017). http://doi.acm.org/10.1145/3077286.3077293
19. Braga, J., Silva, J.N., Endo, P.T., Omar, N.: Autonomous Architecture Over Restricted Domains (A2RD), January 2019. https://doi.org/10.17605/OSF.IO/TKA9U. https://osf.io/tka9u/. Accessed 19 Mar 2019
20. Braga, J., Silva, J.N., Endo, P.T., Ribas, J., Omar, N.: Blockchain to improve security, knowledge and collaboration inter-agent communication over restrict domains of the internet infrastructure. In: Proceeding of CSBC 2018 - V Workshop pre IETF, Natal, RN Brazil, p. 13, July 2018. https://arxiv.org/abs/1805.05250
21. Colel, R., Callon, R., Gardner, E., Rekhter, Y.: Guidelines for OSI NSAP allocation in the internet. Technical report, RFC Editor, May 1994, rFC1629. (Obsoletes RFC1237) (Status: DRAFT STANDARD) (Stream: IETF, WG: osinsap). https://doi.org/10.17487/RFC1629. Accessed 07 Nov 2018
22. Ebeling, W., Feistel, R.: Physics of Self-organization and Evolution. Wiley, Hoboken (2011)
23. Eriksson, H.E., Penker, M.: Business Modeling with UML. OMG Press, New York (2000)
24. Eriksson, H.E., Penker, M.: UML 2 Toolkit. OMG Press, New York (2004)
25. FIPA: Sc00001l: Abstract architecture specification (2002). Standard. Disponível em: http://www.fipa.org/specs/fipa00001/index.html. Acessado em: 12 Sept 2014
26. FIPA: Sc00036h: Propose interaction protocol specification (2002). Standard. Disponível em: http://www.fipa.org/specs/fipa00036/index.html. Acessado em: 12 Sept 2014
27. FIPA: Sc00023k: Agent management specification (2004). Standard. Disponível em: http://www.fipa.org/specs/fipa00023/index.html. Acessado em 12 Sept 2014
28. Hares, S., Katz, D.: Administrative domains and routing domains: a model for routing in the internet. Technical report, RFC Editor, December 1989, rFC113. (Status: INFORMATIONAL) (Stream: Legacy). https://doi.org/10.17487/RFC1136. Accessed 06 Sept 2014
29. Hawkinson, J., Bates, T.: Report on MD5 performance. Technical report, RFC Editor, March 1996, rFC1930. https://tools.ietf.org/rfc/rfc1930.txt. (Updated-By RFC6996, RFC7300) (Also BCP0006) (Status: BEST CURRENT PRACTICE) (Stream: IETF, Area: rtg, WG: idr). Accessed 6 Sept 2014
30. Horn, P.: Autonomic Computing: IBM's Perspective on the State of Information Technology. IBM Research (2001)
31. Huizer, E.: IETF-ISOC relationship. Technical report, RFC Editor, October 1996, rFC2031. (Status: INFORMATIONAL) (Stream: IETF, Area: gen, WG: poised95). https://doi.org/10.17487/RFC2031. Accessed 08 Nov 2018

32. Isotani, S., Bittencourt, I.I.: Dados abertos conectados. Novatec Editora, São Paulo, SP, Brasil (2015)
33. Jennings, N.R.: On agent-based software engineering. Artif. Intell. **117**(2), 277–296 (2000)
34. Kisteleki, R., Haberman, B.: Securing routing policy specification language (RPSL) objects with resource public key infrastructure (RPKI) signatures. Technical report, RFC Editor, June 2016, rFC7909. http://www.rfc-editor.org/rfc/rfc7909.txt. (Updates RFC2622, RFC4012) (Status: PROPOSED STANDARD) (Stream: IETF, Area: rtg, WG: sidr). https://doi.org/10.17487/RFC7909. Accessed 29 July 2017
35. Kuerbis, B., Mueller, M.: Internet routing registries, data governance, and security. J. Cyber Policy **2**(1), 64–81 (2017). https://doi.org/10.1080/23738871.2017.1295092
36. Leiner, B.M., Cerf, V.G., Clark, D.D., Kahn, R.E., Kleinrock, L., Lynch, D.C., Postel, J., Roberts, L.G., Wolff, S.: A brief history of the internet. ACM SIGCOMM Comput. Commun. Rev. **39**(5), 22–31 (2009)
37. McPherson, D., Amante, S., Osterweil, E., Blunk, L., Mitchell, D.: Considerations for internet routing registries (IRRs) and routing policy configuration. Technical report, RFC Editor, December 2015, rFC7682. http://www.rfc-editor.org/rfc/rfc7682.txt. (TXT = 47996) (Status: INFORMATIONAL) (Stream: IETF, Area: ops, WG: grow). https://doi.org/10.17487/RFC7682. Accessed 29 July 2017
38. Meyer, D., Schmitz, J., Orange, C., Prior, M., Alaettinoglu, C.: Using RPSL in practice. Technical report, RFC Editor, August 1999, rFC2650. https://tools.ietf.org/rfc/rfc2650.txt. (Status: INFORMATIONAL) (Stream: IETF, Area: ops, WG: rps). https://doi.org/10.17487/RFC2650. Accessed 29 July 2017
39. Mitchell, J.: Autonomous system (AS) reservation for private use. Technical report, RFC Editor, July 2013, rFC6996. https://tools.ietf.org/rfc/rfc6996.txt. (Updates RFC1930) (Also BCP0006) (Status: BEST CURRENT PRACTICE) (Stream: IETF, Area: rtg, WG: idr). Accessed 03 Mar 2015
40. Moro, G., Sartori, C., Singh, M.P.: Agents and Peer-to-Peer Computing. Springer, Heidelberg (2005)
41. Movahedi, Z., Ayari, M., Langar, R., Pujolle, G.: A survey of autonomic network architectures and evaluation criteria. IEEE Commun. Surv. Tutor. **14**(2), 464–490 (2012)
42. Musumeci, F., Rottondi, C., Nag, A., Macaluso, I., Zibar, D., Ruffini, M., Tornatore, M.: A survey on application of machine learning techniques in optical networks. IEEE Commun. Surv. Tutor. 1 (2018). https://ieeexplore.ieee.org/document/8527529/. http://arxiv.org/abs/1803.07976
43. Newton, A.: Cross registry internet service protocol (CRISP) requirements. Technical report, RFC Editor, February 2004. https://www.rfc-editor.org/info/rfc3707. (Status: INFORMATIONAL) (Stream: IETF, Area: app, WG: crisp). https://doi.org/10.17487/RFC3707. Accessed 03 Feb 2018
44. Odell, J., Parunak, H.V.D., Fleischer, M.: Modeling agents and their environment: the communication environment. J. Object Technol. **2**, 39–52 (2003)
45. Odell, J.J., Parunak, H.V.D., Bauer, B.: Representing agent interaction protocols in UML. In: Agent-Oriented Software Engineering, pp. 121–140. Springer (2001)
46. Peace, A.: Ontology: A Practical Guide, 1st edn. Articulate Software Press, Angwin (2011)
47. Poole, D.L., Mackworth, A.K.: Artificial Intelligence: Foundations of Computational Agents. Cambridge University Press, Cambridge (2010)

48. Rekhter, Y., Li, T., Hares, S.: A border gateway protocol 4 (BGP-4). Technical report, RFC Editor, January 2006, rFC4271. http://www.rfc-editor.org/rfc/rfc4271.txt. (Obsoletes RFC1771) (Updated-By RFC6286, RFC6608, RFC6793) (Status: DRAFT STANDARD) (Stream: IETF, Area: rtg, WG: idr). Accessed 07 Sept 2014
49. Rhodes, B., Goerzen, J.: Foundations of Python Network Programing, 3rd edn. Apress, New York (2011)
50. Russel, S., Norvig, P.: Artificial Intelligence, 3rd edn. Prentice Hall, New York (2010)
51. Saltzer, J.H., Reed, D.P., Clark, D.D.: End-to-end arguments in system design. ACM Trans. Comput. Syst. (TOCS) **2**(4), 277–288 (1984)
52. Schmid, S., Sifalakis, M., Hutchison, D.: Towards autonomic networks. In: Autonomic Networking, pp. 1–11. Springer (2006)
53. Villamizar, C., Alaettinoglu, C., Meyer, D., Murphy, S.: Routing policy system security. Technical report, RFC Editor, December 1999, rFC2725. http://www.rfc-editor.org/info/rfc2725. (Status: PROPOSED STANDARD) (Stream: IETF, Area: ops, WG: rps). https://doi.org/10.17487/RFC2725. Accessed 03 Feb 2018
54. Weiss, G.: Multiagent Systems: A Modern Approach to Distributed Artificial Intelligence. MIT Press, Cambridge (1999)
55. Wooldridge, M.: Agent-based software engineering. IEEE Proc. Softw. **144**(1), 26–37 (1997)
56. Wooldridge, M.: An Introduction to Multiagent Systems, 1st edn. Wiley, Hoboken (2002)
57. Wooldridge, M.: An Introduction to Multiagent Systems, 2nd edn. Wiley, Hoboken (2009)
58. Wooldridge, M., Jennings, N.R.: Intelligent agents: theory and practice. Knowl. Eng. Rev. **10**(02), 115–152 (1995)
59. Wooldridge, M.J.: Reasoning About Rational Agents. MIT Press, Cambridge (2000)
60. Wooldridge, M.J.: The logical modelling of computational multi-agent systems. Ph.D. thesis, Citeseer (1992)
61. Xiao, X., Hannan, A., Bailey, B., Ni, L.M.: Traffic engineering with MPLS in the internet. IEEE Netw. **14**(2), 28–33 (2000)

The Space Between Worlds: Liminality, Multidimensional Virtual Reality and Deep Immersion

Ralph Moseley(✉)

School of Science and Technology, Middlesex University, London, UK
r.moseley@mdx.ac.uk

Abstract. Normal, everyday consciousness is a mixture of mental states which change, dependent upon both internal and external factors, ordinarily without any particular focused direction. When deliberate change is desired, however, there are mechanisms of transition available. These gateways allow smooth induction between stable, sustained, brain states. Building upon the concept of Deep Immersion, multidimensional virtual reality is explored utilizing personal narrative, portals and symbolic anchoring links as mechanisms of transit. Together, these target specific states within the individual, using biofeedback and fluid, self-adapting environments. This work is informed by Buddhist, yogic and other spiritual traditions, employing the latest technologies in Virtual and Augmented Reality, combined with mobile sensors and EEG equipment.

Keywords: Virtual reality (VR) · Deep immersion · Meditation · Biofeedback · Neuroscience

1 Introduction

There are many ways we can deliberately induce particular states of consciousness. It may be that our physical and mental condition at a point in time, naturally flows to a specific mental condition. For example, when we are tired, we are predisposed toward sleep and delta brainwave coherence. Our busy minds calculating abudget, or some scientific formula, would naturally tend toward a dominance of beta in specific areas of the brain.

Cultures developed their own technologies for deliberate induction to target states. Earliest attempts could be that of the shaman creating ecstatic trance through dance, pharmacological input or other types of physical arousal.

These specific states were explored and catalogued by many spiritual traditions and cultures until finer discernment was achieved over time, which maps to scientific understanding of brainwave patterns.

Previous work [1, 2] looked at how brain entrainment can be used and more recently Virtual Reality was utilized to engage Buddhist and Yogic techniques, leading to specific states of consciousness, inducing particular brainwave patterns. Furthermore, the latter work involved utilizing virtual, augmented and mixed reality to form a Deep Immersion which included analysis of brainwave patterns in real-time to form a

K. Arai et al. (Eds.): CompCom 2019, AISC 997, pp. 548–562, 2019.
https://doi.org/10.1007/978-3-030-22871-2_37

biofeedback loop to modify techniques being used "on-the-fly". Such techniques included Buddhist, yogic and hybrid forms of induction into meditative states.

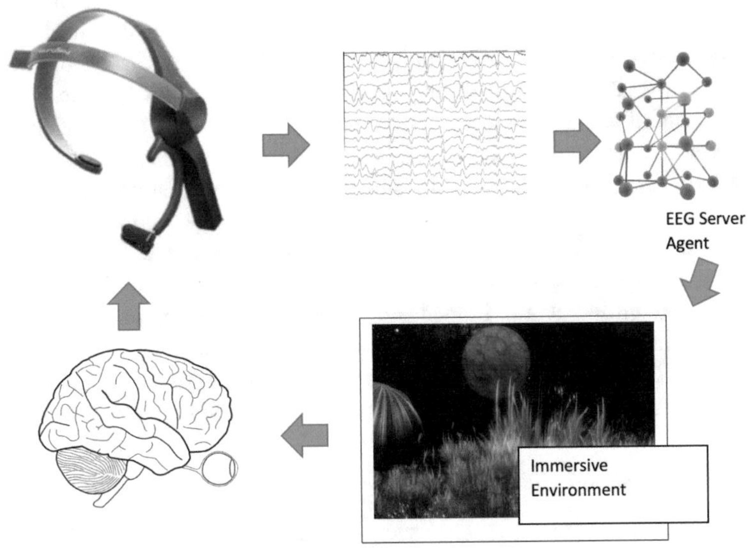

Fig. 1. The Deep Immersion feedback loop

In *Deep Immersion with Kasina* [3], this aspect was looked at in depth, using a particular technique, known as Kasina, to induce and train an individual in meditative states using biofeedback loops (see Fig. 1). The method involved focusing on objects having particular attributes associated with specific qualities of the mind.

1.1 Transition States and Presence

This paper concerns the process of transition between mental states, the induction from one to another and the ability of new technologies to facilitate the transference. Not only this, but the question of how smooth transfer can be achieved into mixed reality environments [4] through portals and the use of symbolic persistent gateways, known here as *anchors*.

2 The Doorway Effect (Localized Memory)

The doorway effect is the concept that the mind associates information with the surrounding environment. A common example being that you want something from one room and when you reach there, you can't remember why you are there. This is also known as 'event perception' by psychologists [5]. Here the process frees information related to previous environments when a new environment is presented; the passage through a door signals to the mind to release space for new thoughts. Studies have determined that this effect applies to both real and virtual environments [6, 7].

There have been many studies investigating the process of transition between real and virtual worlds, in order to enhance the VR experience such as Slater et al. [9], Interrante [10], Steinicke [11] and Sproll [12]. Milgram and Kishino [13] formulated the idea of a 'mixed reality continuum'. This work investigated the in-between transition phases of reality and VR, through total immersion and back again.

The concept of starting the journey from real to virtual within the local environment was proposed for various reasons, including enhancing distance estimation within a virtual space [11]. It was believed that appearing in a known environment first would permit a more gradual transition from reality to the virtual. Studies concluded that participants who visited the transitional environment first and then used a portal to go into the virtual world felt an increased sense of presence, compared to participants who appeared in the virtual world immediately. Further Steinicke, proposed the use of portals to the virtual world in order to help users navigate the environment and improve the experience. An interesting idea here was the use of transparent portals which allow the user the ability to get a glimpse of the environment they are about to enter.

2.1 Multidimensional Virtual Reality

Proponents of deeper immersive states such as Slater, suggest the concept of "stacking depths" that is, the idea that additional environment layers could be included within a virtual space. Here it was discovered that participants who had difficulties immersing in the first virtual layer were able to feel a sense of presence after greater depths were joined.

The idea of localized memory and transitional spaces, together with the idea of virtual environments, would tend to suggest that spaces can be created to preserve and explore specific memories, linking to the idea of the doorway effect.

There have been attempts to provide conceptual models which split the immersion depth by levels. Sproll et al. proposed a 5-stage transition model that divided into physical and mental levels. The physical level was composed of direct transposition of the physical via cameras; the following two layers represent the mental level, capturing the user's attention, before passing them through a liminal space (actually in this case, labelled as a limbo), followed by full immersion. This limbo is described as a space between worlds, that breaks the connection to the real world and prepares the user to enter the virtual environment.

It could be asserted that the first question to ask, would be, why this is necessary? In normal everyday reality our brains have become accustomed to logical passage through the world – it simply makes sense. There is no disjuncture – the laws of physics are consistent and are upheld wherever there is travel. A kitchen will always lead to the living room, if indeed this was the case when it was built – barring some logical change, an extension, or major disaster. Humans also travel through a landscape of meaning and connect symbolically with our world. This symbolic passage through world includes attached emotional content, for example, we can walk from our kitchen to our living room physically but on route there may be a memory attached to a particular chair that is passed. In other words, our mental and emotional content can be changed in the process, our state is changing in a multi-dimensional, multi-layered world.

3 Depth of Presence

3.1 Personal and Fictional Narratives

Stories provide not only a break from daily life but a basis for belief and communication at unconscious levels [14], as well as reinforcement of memories. Fiction in effect mimics this episodic approach. In myths, fantasy and stories we see how a common device used is that of some kind of gateway which is an inductive passage to some new land and hence a shift from the mundane. One of the more classic examples of this is the story of *The Lion, The Witch and The Wardrobe* [15] where the entry point for a whole fantasy world is the back of a cupboard. The cupboard becomes the transfer area. Likewise, the story of *Alice in Wonderland* [16] provides a similar template of experience, in which the heroine disappears into an alternate reality populated by anthropomorphic entities and normal rules of physics are suspended.

Myths often have transition mechanisms from here-and-now to some alternate world. In Norse mythos, for example, there is the Bifrost, also known as the "Rainbow Bridge" as the only permanent portal leading to Midguard, complete with another feature that re-occurs – the guardian, daemon, or psychopomp, which in this case is Heimdall. This guide often leads the person who experiences the journey from their normal state of mind into accepting, more smoothly, an entirely different world, with different rules.

In Irish mythology there is the Sidhe, the mound portal which gives access to an Otherworld populated by the Faye, where many travelers have accidently stumbled at their peril. Many other cultures include entrances to alternate worlds; the Greeks with portals to Hades, Romans similarly with the Elysian fields.

There are several common features between myths, and the fictions that are created, expressing some need for this transition state to alternate worlds and our preparation to enter them. There is an experience of the normal here and now; there is some symbol or invitation to enter; the portal; then there is the experience of the "other" state or world that is entered.

Personal narrative and epistemology drive an individual to explore new experiences or re-live old ones. Forms within stories and myths echo not only parts of a cultural experience but the individual's personal story, archetypes and belief which has been known to encourage neuroplasticity [17]. To promote the ability to be trained to produce desirable states, systems need to encapsulate these ideas, creating worlds which echo an individual's belief system and resonate at a fundamental level.

3.2 Lucidity, Focus and Presence of Awareness

Daily life presents us with what would seem to a continuous, logical progression with well-known physical rules. The brain accepts this and has come to understand it. However, there is an experience of some elasticity – the perception of time, for example. Pleasurable events pass by too quickly, whereas somethings less so – a visit to the dentist, make minutes appear to be experiencedin terms of hours.

There are too, within our own era and modern-day culture, experiences of more alternate states. There are stories of near-death experiences which often echo deeper cultural roots – the Tibetan BardoThodol [18] and the experiences leading to death.

Probably the most common altered state that is encountered is that of sleep – which may be full of strange landscapes, people and entities that don't necessarily conform to the usual understanding of physics. These dream states consist of unusual narratives that encompass the whole being, that is, there is an emotional component as well as physical. There appears often fragments of meaning, conflict and resolution.

A special case of dreaming is that of lucidity where the dreamer gains an awareness within the dream state itself. The usual mechanisms of dreaming at this point become more malleable to the dreamer – there would appear to be experience of the dream and self-creation within the same domain. Lucid dreaming as a term was coined by Van Eeden in 1913 [19] to denote a dream in which "… the sleeper remembers day life and his own condition, reaches a state of perfect awareness, and isable to direct his attention, and attempt different acts of free volition". Intriguingly, early scientific experiments revealed that it was possible within lucid states to signal to the outside world while within their lucidity using eye movements, in this case two consecutive left-right turns. This is possible because eye muscles are able to move freely and under control during REMS.

Less dramatic but present all the same, are experiences found in sleep, meditation, exercise and arousal, where we are shifted, albeit sometimes subtly, away from our usual day to day state of mind.

3.3 Utilizing Narrative as a Means of Engaging Depth

A narrative in this context, one of virtual and Deep Immersion, constitutes a layer of meaning present at several levels, acting on the subject in ways beyond the lab construct developed in the Kasina experiments. It implies symbolic, emotive and intellectual constructs, which deeply embed the user beyond the bare techniques applied. In short, this aspect engages a human multidimensionally.

The Kasina Deep Immersion experiments applied techniques, noted above, without regard to components essential to human experience – indeed some of which may be hard to emulate but other aspects may be possible to enrich and embed any experience.

There are many factors which contribute to the journey into a meditative or even restful state. A strong factor is belief, or a backdrop in which the techniques themselves hang. While it is possible to "sterilize" techniques and bring them to "barebones", it is often the case that belief components act as strong motivators and bolster regular practices when all else fails. A belief set often precipitates a particular world, social or individual view, or in fact, no view at all. This belief set constitutes a paradigm or narrative which dictates the interactions that a person makes with both their external world and internal life. Belief factors high in understanding a holistic view of the individual and their capacity to function, biologically and mentally to their optimum capability. This matrix of belief affects at a biological level and can enhance or stress the immune system and the ability to heal [20].

Imbuing objects with history and a degree of spiritual weight is a common component which drives regular spiritual practices such as yoga. It is often the case that an artefact or relic can create a symbolic link with past teachers or an event of significance. While the symbol or object can be religious, in context, this is not always the case. Particularly resonant experiences, lead to association with a particular artefact.

While some form of object or symbol may imply meaning, there are whole symbolic contexts which give rise to altering the state of mind of the individual. Here, the symbolic context may be a set of actions performed over time – essentially a ritual. The narrative here creates human meaning-making and ordering of experience.

The Kasina experiments dealt with techniques derived from Buddhist practices, essentially hundreds of years old. Here the elements included symbolic forms which are essentially elemental in nature, along with aspects of spaciousness, color and light. These objects are used as concentration tools, leading to gateway states of consciousness, known as jnanas [21]. An object and its attributes are concentrated upon to bring the element into the meditator's experiential world – rather than any losing of self, quite the reverse is implied.

In sequence the symbolic objects form successive stepping stones to rarefied states and prolonged concentration.

Not all spiritual traditions rely in this way on objects as tools to access states other than usual consciousness. As noted in the previous papers, there is rough split between object and non-object focus. Open focus techniques, seen in Zazen and Shi-ne (Tibetan Buddhism) allow for the idea of non-referentiality that is, where there is a sense of being but lacking in perceptive duality.

Within Buddhism, particularly Tibetan, this can be broken into an approach of view (*tawa*), meditation (*gompa*) and action (*chodpa*) [22]. The view here is not philosophical but an uncharacterized way of seeing ourselves within our current context. To explain further, there is a removal of the conceptually constructed way of seeing the world. It is directly seeing the world in an effortless and uncontrived way. This lack of preconceived ideas is used as part of a methodology that can be used to investigate the situation in which the meditator finds themselves. It is the recognition that logical analysis, in this context is limited and that it is direct experience that is key.

Meditation in this particular approach is used to develop this view, the discovery of space. Here meditation is seen not as some artificially induced state but rather it is a natural state which is discovered. It is this mechanism of discovery that in itself is discovered. The quality of this meditation is effortlessness and the approach is led by the discovery of space (Shi-ne). The two re-enforce each other and form the basis of the third aspect of the approach, that of action, which is the dynamic of relationship within the world (in this context). It is the response when view and meditation are present in the moment. In this sense, action is not simply a way of acting but rather being in an unrestricted, uncontrived manner.

3.4 The Guide

As we have initially seen, the guide, known also as the psychopomp or daemon forms an important part of myth and fictional stories about journeys into other worlds. This particular archetypal form, as noted by Jung, is resident in popular culture, also in personal stories of encounters with the afterlife. Indeed, the principal role of such an entity is often seen as the provider of safe passage between worlds and the intermediary between Conscious and Unconscious, thanks to the integration of Anima (each man's feminine nature) and Animus (each woman's male principle) in form of the "Self". From a neuroscience point of view the origin of such archetypes is intriguing. The possibilities include in some sense being hard-wired via genetically inherited structures or self-organizing patterns, which emerge in response to the development of meaning by the human brain. The latter option seemingly the better to fit to Jung's explanation in early work which stressed the emergence of archetypes as fundamental dichotomies of self-experience, the character of these being enantiodromia (superabundance of any phenomenon leads to its opposite, a restoration of balance). Here there is the potential for Jungian archetypes as primarily distributed phenomena, sets of pairwise partitions of the conscious brain. Later works of Jung, offer a different interpretation, labelled sparse, in which small populations of neurons code for a given semantic character. In this representation, hypothetical "Grandmother cells" could code for whole semantic classes and therefore the archetypes themselves. Some experiments suggest this may be the case, implicating the Medial Temporal Lobe in representing semantic classes across modalities in single neurons [23]. This is, however, an area of some debate.

In the context of the yoga teacher, guru or guide in relaxation states such as Savasana (corpse pose) and Yoga Nidra (literally, yogic sleep) likewise, just like the archetype, there is a directing of awareness which may be to sensation, observation of phenomenon, which could be internal or external. A guide seemingly knows and lights the way, effortlessly, sometimes only hinting at what the experiencer should do, other times guiding explicitly.

4 Development

4.1 Toward Implementation

This holistic view of the human experience taken here has, so far, realized the following important concepts:

- *Narrative* engages the human experience at a fundamental level, with effects on biological and psychological mechanisms.
- The usage of symbolic objects or gateways which can be underpinned with sensory feedback association – *an anchor*. The anchor is the instigator of the new experience or heralds the return to 'safe ground'.
- *Portals*, or tunnels of experience which link one domain to another, the transition itself.
- There is the *guide*, a way of drawing the presence of awareness to specific points within the experience.

This paper now considers how to bring these points together to explore possibilities and implications within virtual environments and Deep Immersion.

4.2 Yoga Nidra

A suitable project, based on a yogic practice known as Yoga Nidrawas chosen which can investigate how the various presented ideas can function as a whole. In yoga practice there are examples of how a student is led progressively into deeper states of relaxation; this is present in both the practice of Savasana (corpse pose) and Yoga Nidra. In Savasana, usually in a prone state, the individual is allowed the opportunity to "body scan", where body parts are visited mentally, led by the teacher. This may include visualization and breathing practices, concurrently. In Yoga Nidra, again, there is the leading of the individual, but this tends to follow a more particular format and stages such as those enumerated here:

1. Preliminary Relaxation Technique
2. Creation of a Sankalpa (Intention or Resolve)
3. Stage 1: Rotation of Awareness through the body
4. Stage 2: Awareness of the subtle body
5. Stage 3: Awareness of Feeling and Emotion
6. Stage 4: Visualization
7. Final Steps

This journey allows a release of the physical body, a quietening of the brain waves, tapping into the emotional body and the intellect.

The process of Yoga Nidra from a brainwave point of view is an intriguing one. The initial breathing exercises trigger the relaxation response. This balances the sympathetic and parasympathetic nervous systems, as well the left and right brain. In this process the brain shifts from being in the usual beta awakened state with heightened activity to alpha, where the mood-regulating hormone serotonin is released. From this initial alpha state, deep alpha is then gained and also a high theta brain-wave state which corresponds to REM sleep. It is at this point where super-learning can occur and is also present in artists, for example. Emotional integration and release occur here and structures in the brain can change in response. Visions, colors and auditory hallucinations are all possible at this point. At this stage there can be the experience of "nothingness". Theta leads into delta, a restorative state in which organs can regenerate and stress hormones such as cortisol is removed from the system (Table 1).

Table 1. Brainwave description for Yoga Nidra

Frequency range	Name	Related attributes and states
> 40 Hz	Gamma waves	Higher mental activity, perception, problem solving, fear, and consciousness. Appears in specific meditative states, relating to Buddhist compassion meditations in the Tibetan tradition
13–39 Hz	Beta waves	The most usual state for normal everyday consciousness. Active, busy or even anxious thinking. Also appears in active concentration, arousal, cognition, and or paranoia
7–13 Hz	Alpha waves	Relaxed wakefulness, pre-sleep and pre-wake drowsiness, REM sleep. Considered as the brainwaves of meditation. These waves also appear in the relaxation process before sleep. Occuring in initial stages of Yoga Nidra, leading to deep Alpha

(continued)

Table 1. (*continued*)

Frequency range	Name	Related attributes and states
4–7 Hz	Theta waves	Appears in deep meditation/relaxation, NREM sleep. A theta prominent individual may be awake but lose their sense of bodily location for example. High theta occurs following deep Alpha in Yoga Nidra
< 4 Hz	Delta waves	Deep dreamless sleep with loss of body awareness. Maintaining consciousness while delta present is difficult. Occurs in Yoga Nidra following on from Theta states achieved

After delta, the Yoga Nidra experience can take the individual to even slower brainwave states that cannot be reached in conventional sleep. Here the brain is thoughtless – but the experiencer is still present and awake. Not all Yoga Nidra practitioners experience this but regular guided journeys here make it more likely, at least as initial brief encounters which slowly extend.

Finally, the Yoga Nidra experience ends with the guide taking the practitioner safely back to wakefulness.

Previous work looked at forms of stimulation which essentially aim at manipulating the physical components of the human body through mechanisms existing at purely a biological level. For example, techniques which rely on entrainment – the frequency following response of the brain when exposed to two signals that are set apart by a specific amount, locking to the difference of the two. This allows a way of guiding the brain through successive brainwave patterns to arrive at some target state.

In *Deep Immersion with Kasina* [3], this was taken further, and Buddhist meditation techniques were applied within a virtual environment. Software was developed which allowed the system to create optimum circumstances for the meditative state to develop. A biofeedback loop was incorporated which adapted the system depending on how the subject was progressing. Several variations were added to the system:

- The Kasina technique, with biofeedback in the virtual environment
- Introduction of modulated light and sound – brain entrainment
- Introduction of a vocal (leading) stream
- A virtual mind machine

The virtual environment provides a way of introducing the perfect controllable setting, which can react and adapt to the person undergoing the meditative experience. It provides a narrative that the mind can link the physical affects to the mind, just as would happen in a real-world setting but with the added benefit of an environment which adapts as a session is underway.

In a minor sense, an issue raised here could be: How is it best for an individual to transition to a virtual world regardless what their current state of mind is? There is an inevitable jarring effect in transition. This is one aspect looked at here. A deeper aspect is the combination of how transfer can be gained smoothly, in both a narrative and biofeedback/arousal techniques, combining to form a powerful induction into specific targeting states of mind.

Fig. 2. The Oculus Rift VR equipment

4.3 Adapting Yoga Nidra to Deep Immersion

The construction of a suitable virtual environment was developed from the Kasina experiment. Developed primarily in Unity [24] as before, with upgraded software and equipment. The targeted platform in this instance was the Oculus Rift (see Fig. 2) [25] with added Bluetooth EEG, Mindwave Mobile by Neurosky [26]. This system allows for Deep Immersion utilizing biofeedback to adapt the environment according to the current brainwave pattern of the individual.

Construction

The environments produced, along with the guide, facilitate the narrative, that is, the story which is being related at various levels to the user's mind toward some target state. In this case matching a particular point in the phases of Yoga Nidra.

Prelude phase – building a narrative and the anchor mechanism

As noted, presence in the virtual world can be enhanced by using stacks of 'folded' or occluded environments. To make this part of the experiment, a symbolic form was created allowing access to the next state; it seems the case that such a symbol is best to be consistent, in some way personal, and also to act as an anchor with external stimulus attached to it when utilized, thus creating a link. In this case, the individual travels into the Yoga Nidra chamber from three successive prelude narratives, progressively going deeper into the virtual domain. Narrative one involved the transition to a virtual half way portal (level 0) encompassing aspects of the real environment, narrative two, acclimatizing within the virtual level one and finally level two where the Yoga Nidra lab can be accessed from. Between each successive phase, the symbol is used as an entrance point to the next, briefly.

The Simulacrum – a symbolic body in the virtual world

The subject's actual physical body is laid prone for this experiment. A simulacrum body is produced which guides the individual to use their awareness through particular states. This simulacrum is best situated, in this instance, above the subject but equally could be in front.

Where the body states are explored the simulacrum is used; abstract symbolic forms, as used in the Kasina experiment, as symbols – particularly as elemental forms.

(1) ***The Relaxation Phase***

This is normally composed of a brief body scan and guidance to situate the body in a way which will allow a restful state in body and mind. A body scan at its most simple, is usually a quick bringing of the awareness of the mind over the body searching for areas of tension.

(2) ***Sankalpa (Intention or resolve)***

A Sankalpa can be described as a statement of deeply held fact – a vow that is true in the present moment. It is not a petition or prayer; it is a short phrase clearly expressed, focusing on a chosen goal. It could be to make a positive change in life, reforming a habit or changes in personality.

(3) ***Rotation of Awareness***

This involves bringing attention to various body parts in sequence. For example, the teacher will say, something along the lines of "bringing attention to the fingertips of the right hand, visit each finger, thumb…fill the palm with awareness…the whole hand…wrist…lower arm…" and so on. The sequence may start with small detail then repeat with larger scenes focusing on whole limbs, torso etc. An important element here is the ability to involve imagination and creative use of language. Note here that the sequence used relates directly to the motor homunculus (the symbolic person embedded within the brain matter). The sensory motor cortex is therefore accessed during this stage of Yoga Nidra.

In a virtual world context, the rotation of awareness is mirrored on the simulacrum body, bringing the attention to that part of the body, then lowered to the subject to mimic a 'joining' of the two ideas into the physical form.

(4) ***Awareness of the Subtle Body (Breath)***

The next stage relates to breath and a descent from beta (busy mind) to alpha (relaxed mind). Bringing attention to the breath makes breathing a function of higher brain, cerebral cortex from its normal functioning within the brain stem. The body at this point is known to produce endorphins, the body's natural painkillers. In the virtual experiment there is guidance through lengthening and timing the breath, utilizing their anchor and the simulacrum.

(5) ***Awareness of Feeling and Emotion***

This stage is said to target the limbic system (reptilian brain). It uses combinations of opposite emotions or sensations while practicing non-attachment. This particular phase is aimed at producing will-power, emotional control and equanimity. Again, similar to the Kasina experiment, elemental forms were produced in succession to match the format of the technique.

(6) ***Visualization***

This stage utilizes simple imagery. In Buddhist and yogic texts this relates to cleansing of painful and disturbing material (samskaras), or deep-rooted conditioning. The idea here is to work with the contents of the unconscious mind including mental and emotional patterns. The guide at this point directs the individual through a series of archetypal images which will evoke responses in the relaxed mind.

(7) ***Final Phase***

There are two main steps to the final phase. The first is the revisiting of the intent – the Sankalpa – to plant this in the field of the unconscious mind, in its now relaxed and altered state.

The final step is the slow return to embodiment within the normal physical reality. When Yoga Nidra is done within the context of a purely physical environment the attention and therefore, awareness, is moved from the internal space back to the sense of physical embodiment.

In the VR counterpart to this, there must be a moving of the awareness in a similar way but back through the stack of virtual domains via the narrative, portals and anchor. At the final level there is a re-acclimatization with the body, as simple slowly bringing awareness to physical aspects and embodiment.

4.4 Outcomes

This experiment used several concepts explored in this paper. There is the idea of Deep Immersion; adaptive, responsive, environments and the idea of utilizing symbolic elements to move from one VR environment to another, as embedded stacks. Another concept used here was that of the guide – presenting a narrative of targeted content which diverts the attention from being lost and keeps presence focused from one stage to the next, and onward.

This paper mainly concerns the synthesis of concepts and technology to drive further tests into this area. However, some useful information was collected from these initial experiments.

A yoga practitioner and a complete novice were the first to be both exposed to the described system. They both were not used to the technology involved – virtual reality. So, a period of adjustment was given to them utilizing simple scenes and programs. The complete Yoga Nidra system runs for 23 min, which is comparable with real world sessions.

Both subjects quickly learned the way the system works and entered the deeper 'stack' within the program. Each user chose a personal anchor as a persistent link to travel between states, each time the token was used, vibration feedback was applied through hand units.

The yoga practitioner seemed to adapt quicker to the initial phase – and the identification with the Simulacrum. Induction into initial stages of relaxation were much quicker and an alpha brainwave pattern was visible much earlier than anticipated. The novice began a dominant alpha wave pattern later, at the end of Stage 3.

Both subjects reported the use of the anchor as a satisfying technique of moving between virtual domains.

The system was then utilized for sessions over the period of a week with good reports from a further 10 users. The recorded EEG patterns showed good matches between the stage achieved and the expected pattern for that point. It was noticeable that both novices and yoga practitioners became quicker at developing these patterns showing perhaps some training aspect taking place.

4.5 Future Work

Much more experimentation needs to be completed with this work but the use in this project shows potential toward many unique applications. This maybe orientated toward general health care, personal development or treatment of particular conditions.

There may be arole for TMS (transcranial magnetic stimulation) within the multi-modal framework of VR for individuals suffering from MDD (Major Depressive Disorder) and PTSD (Post-Traumatic Stress Disorder). Individuals with MDD and PTSD exhibit hyperactivity in brain regions associated with fear and rumination and hypoactivity in regions correlated with reappraisal, resiliency, and self-regulation. Within the scope of multi-dimensional VR, a unique landscape may be created to invoke safety and stabilization while jointly focusing on influencing states of consciousness. Accordingly, when used in conjunction with VR, targeted TMS may support remission from MDD and PTSD by increasing neuronal activity in executive regions and reducing stimulation in limbic, that would likely interfere with positive states of consciousness offered by VR. Often underactivity of executive network and increased firing in fear circuits negates the capacity for reframing of experience and correlated neuroplasticity. By providing personalized VR during more optimal brain activity, via TMS, there may be a potential for markedly shifting conditional psychological experience and entrenched circuitry, in a manner that has not previously been explored.

5 Conclusions

This paper explored several aspects of liminality within Deep Immersion – the idea of rebuilding a virtual environment, continuously, according to target states using biofeedback as a mechanism which informs this process over time.

More generally, there is the idea that mixed reality states require an induction in themselves, a way of acclimatization, ending in a particular environment being reached, which is then accepted.

At a deeper level, a person can be taken to greater presence within virtual domains by creating stacks of embedded environments, each a level further in. To reach those more present states inside a virtual reality, portals and entry points can be used – this may be as simple as utilizing symbols, or anchors, or depending on the desired effect, to form a complete narrative.

Several mechanisms of transit have been explored in this work acting on different levels of the human experience. Narratives provide a pathway of meaning, not necessarily to the rational mind. Portals form a tunnel of transition between 'here' and 'there', a plausible link. The anchor acts as a symbolic, persistent link to a transition, a safety point that is logical and consistent. Finally, there is the guide, a way of bringing attention and presence of awareness to desired internal, or external dimensions of the experience.

Perhaps another novel of this aspect of the work was the inclusion of the Simulacrum, a virtual symbolic body which the user identifies with and forms a link between the somatic, physical form and the 'other' world.

In terms of Deep Immersion, there is a relationship between the scene or environment and the brainwave pattern produced, as there is in real-life. In this paper, the idea presented is that narrative, persistent symbolic anchors and portals enhance Deep Immersion and also can be applied in a wider context to enhance studies in virtual presence and the exploration of training systems or health care applications.

References

1. Moseley, R.: Inducing targeted brain states utilizing merged reality systems. In: Proceedings of Science and Information Conference (SAI) (2015)
2. Moseley, R.: Immersive brain entrainment in virtual worlds: actualizing meditative states, emerging trends and advanced technologies for computational intelligence. Studies in Computational Intelligence, vol. 647, pp. 315–346, 7 June 2016. ISBN 9783319333533
3. Moseley, R.: Deep Immersion with Kasina: An exploration of meditation and concentration within Virtual Reality Environments. Published by IEEE in the Proceedings of Science and Information Conference (SAI) (2017). ISBN 978-1-5090-5443-5
4. Georgiou, M.: A smooth transition between the real and a virtual reality world. MSc dissertation, Middlesex University (2017)
5. Radvansky, G.A., Zacks, J.M.: Event Perception, Wiley Interdisc. Rev. **2**, 608–620 (2011)
6. Radvansky, G.A., Copeland, D.E.: Walking through doorways causes forgetting: situation models and experienced space. Mem. & Cogn. **34**(5), 1150–1156 (2006)
7. Radvansky, G.A., Krawietz, S.A., Tamplin, A.K.: Walking through doorways causes forgetting: further explorations. Q. J. Exp. Psychol. **64**(8), 1632–1645 (2011)
8. Lawrence, Z., Peterson, D.: Mentally walking through doorways causes forgetting: the location updating effect and imagination. Memory. **24**(1), 12–20 (2016)
9. Slater, M., Usoh, M., Steed, A.: Depth of presence in virtual environments, presence: teleoperators and virtual environments. MIT Press **3**(2), 130–144 (1994)
10. Interrante, V., et al.: Elucidating factors that can facilitate veridical spatial perception in immersive virtual environments. In: IEEE Virtual Reality Conference, pp. 11–18. IEEE (2007)
11. Steinicke, F., et al.: Transitional environments enhance distance perception in immersive virtual reality systems. In: Proceedings of the 6th Symposium on Applied Perception in Graphics and Visualization (APGV 2009), pp. 19–26. ACM
12. Sproll, D., et al.: Poster: Paving the way into virtual reality - A transition in five stages. In: 2013 IEEE Symposium on 3D User Interfaces (3DUI), pp. 175–176. IEEE (2013)
13. Milgram, P., Kishino, F.: A taxonomy of mixed reality visual displays. In: IEICE transactions on information and systems. search.ieice.org. (1994). https://search.ieice.org/bin/summary.php?id=e77-d_12_1321
14. Jung, C.G.: The Archetypes and the Collective Unconscious (2nd edn.). Routledge, London. ISBN 978-0415058445

15. Lewis, C.S.: The Lion, the Witch and the Wardrobe. HarperCollins, New York (2009). ISBN 978-0007323128
16. Carroll, L.: Alice's Adventures in Wonderland. Ostrich Books. ISBN 978-1772261189
17. SZak, P: How Stories Change the Brain. https://greatergood.berkeley.edu/article/item/how_stories_change_brain. Accessed 1 Nov 2019
18. Coleman, G.: The Tibetan Book of the Dead. Penguin Classics, London (2006)
19. Eeden, F.V.: A Study of Dreams, Proceedings of the Society for Psychical Research, vol. 26. Society for Psychical Research (1913)
20. Koenig, H.G., Cohen, H.J.: The Link Between Religion and Health Psychoneuroimmunology and the Faith Factor. Oxford University Press, Oxford. (2002). ISBN 978-0195143604
21. Gunaratana, H.: Beyond Mindfulness in Plain English: An Introductory Guide to Deeper States of Meditation. Wisdom Books (2009). ISBN 978-0861715299
22. Chogyam, N.: Spectrum of Ecstasy. Shambhala (2003). ISBN 978-1590300619
23. Quiroga, R.Q., Reddy, L., Kreiman, G.C., Koch, I.: Fried, Invariant visual representation by single neurons in the human brain. Nature **435**, 1102–1107 (2005). http://www.nature.com/nature/journal/v435/n7045/full/nature03687.html
24. Unity, Software development. https://unity3d.com/. Accessed 1 Nov 2018
25. Oculus, Virtual Reality equipment. https://www.oculus.com/. Accessed 1 Nov 2018
26. NeuroSky, EEG & Biosensors. http://neurosky.com/. Accessed 1 Nov 2018

Predicting Endogenous Bank Health from FDIC Statistics on Depository Institutions Using Deep Learning

David Jungreis(✉), Noah Capp, Meysam Golmohammadi, and Joseph Picone

Neural Engineering Data Consortium, Temple University, Philadelphia, PA 19122, USA
dbjungreis@gmail.com

Abstract. The Federal Deposit Insurance Corporation (FDIC) keeps records of banking data in its Statistics on Depository Institutions (SDI) going back to the fourth quarter of 1992. The data are reported quarterly on approximately 1,050 variables, such as total assets, liabilities, and deposits. We hypothesized that impending failure could be predicted from these data. We restricted the database to 60 quantitative variables that had no missing data for any bank in any quarter during the epoch under consideration: the first quarter of 2000 through the second quarter of 2017. Deep learning approaches based on multilayer perceptrons and convolutional neural networks were evaluated and failed to accurately predict failures better than guessing based on priors. These baseline experiments, particularly the inability to overfit to the training data, show the challenges of finding failure-predicting trends that are strictly intrinsic to a bank. Future deep learning work would have to include exogenous factors to link quarterly observations between banks.

Keywords: Data analysis · Bank health · Convolutional neural networks

1 Introduction

The frequent turbulence of financial institutions in recent years has generated significant interest in an automated tool to assess bank health. For example, an equity trader might be interested in identifying if any of the publicly traded banks are due to fail like Lehman Brothers and Bear Stearns did during the Global Financial Crisis (GFC) of 2007–2009 [1]. For the investor in more exotic instruments, bank health could be of interest for the trader of derivatives, particularly credit default and interest rate swaps. Lenders would want any advantage in

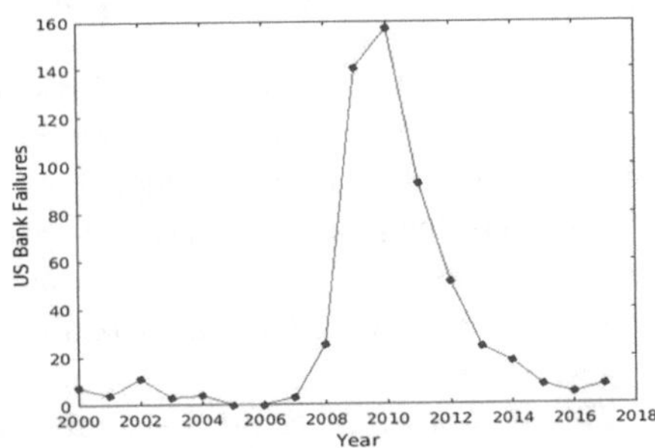

Fig. 1. US Bank failures spiked in 2009, in the wake of the Global Financial Crisis, peaking in 2010 at 157.

K. Arai et al. (Eds.): CompCom 2019, AISC 997, pp. 563–572, 2019.
https://doi.org/10.1007/978-3-030-22871-2_38

determining risk and appropriate interest to charge on interbank loans. Regulators would want to know which banks are in danger of failing, as many did in the wake of the GFC [2].

In Fig. 1, we show the number of bank failures per year. Following a scarcity of failed banks in 2000–2007, bank failures began to rise as the effects of the GFC set in, causing a surge of bank failures starting in 2009, peaking at 157 in 2010, and only returning to pre-crisis levels in 2015. Bank failures during the GFC were driven by a combination of leverage, concentration of risk (particularly in mortgages and mortgage derivatives) and a focus on short-term gains at the expense of long-term risk [3].

Automated bank health analysis has been able to achieve strong performance using support vector machines [4–6], fully-connected neural networks [7], and logistic regression [8]. These studies, however, suffer from various shortcomings that we address in this paper. For example, many of these studies (e.g., Erdogan [4, 5, 6, 9] and Boyacioglu [7]) used small data sets (fewer than 100 samples). Second, all six studies considered factors exogenous to the banks. Erdogan and Boyacioglu considered banks from roughly the same eras, limiting variability due to differences in the general health of the economy, while Zheng [8] explicitly included macroeconomic random effects. We address these issues in this work by considering 1,000 banks using variables strictly endogenous to the banks.

State-of-the-art machine learning is based on neural networks, particularly multi-layer networks known as deep learning [10]. These networks have the ability to discover hidden relationships within the data. This is a clear advantage over regression with interaction terms, where investigators must determine which interactions to explore. For instance, Lopez et al. [11, 12] investigated EEG normal/abnormal binary classification using traditional forms of machine learning, such as random forests (RF) and k-nearest neighbors (kNN), yet she achieved superior performance when deep learning, in particular, Convolutional Neural Networks (CNN), was used.

Deep learning has been used in finance [13], but for bankruptcy prediction, less advanced techniques are emphasized in the literature [4–9]. Moreover, publications seem to address bankruptcy in general, not endogenous bank health for a specific bank. Addo [14] showed that deep learning could be used for determining credit risk (essentially the goal of this paper), but he did not demonstrate the technique on bank data. Tam [15, 16], Le [17] and Abdou [18] developed neural network-based models to determine bank bankruptcy risk, but these were not based on modern deep learning architectures such as CNN.

The goal in this work is to assess baseline performance of deep learning systems on operational banking data provided by the U.S. Federal Government [2]. Operational data contain many forms of imperfection that make it extremely difficult for the application of traditional deep leaning systems. Given a consistent number of reporting periods, our goal was to predict if the bank will fail before the end of the next quarter, only considering the fixed effects of the data from the bank, not random effects of exogenous variables.

This is in contrast to previous work on bank failure and bankruptcy prediction, which has incorporated, either explicitly or implicitly [4–9], exogenous macroeconomic and time effects. That work has had some success in predicting bank failures but predicting within the same time period seems more in the spirit of descriptive rather than inferential statistics, and it certainly is not a form of forecasting. Using explicit macroeconomic conditions [8] presents the problem of having to predict the economy at large.

Further, our work does not develop an early warning system of any particular duration [9], only whether or not the bank will fail in the next quarter. The desired amount of warning depends on the user of the system. While an early warning system may be ideal for a regulator who would like to address bank mismanagement and avoid a failure, the trader of credit default swaps on banks might find this less helpful.

Indeed, to avoid excessively paying the spread on the swaps (a premium, in insurance terms), the ideal system for such a trader would give a tighter bound on the expected failure date, not a vague notion that a bank is likely to fail within the next two years. However, that does not account for liquidity and the availability of credit default swaps at the moment the trader wants them, and purchase and pricing of the swaps would be a judgment call by the particular trader. For these reasons, we developed a baseline system that performs a binary classification of failure or success.

2 Description of the Data

The Federal Deposit Insurance Corporation (FDIC) in the United States keeps records of banking data in its Statistics on Depository Institutions (SDI) [2]. This was selected for use in this study because this data was analogous to the official Turkish statistics used in several comparable studies (e.g., [4–7, 9]).

SDI record-keeping goes back to the 1992Q4 and continues through to the present. Data are reported for the end of each fiscal quarter. Data analysis began after the release of the 2017Q2 data, which was taken as the cutoff point for this work. The number of variables was dynamic: 1,040 in 1992Q4, and 1,076 in 2017Q2. The number of banks followed a strictly decreasing trend throughout the entire epoch, with fewer banks every quarter since 1992Q4.

It is interesting to note that data for the post-2000 epoch did capture numerous macroeconomic events. The epoch spanned the end of the Clinton years, the post-9/11 recession, the Global Financial Crisis, the post-Crisis bull market, and the "Trump run" bull market acceleration. Hence, the data was suitable for investigating the extent to which failures could be predicted without explicit knowledge of these events.

Organizing and preprocessing this data set to make it suitable for machine learning experiments proved to be quite a challenge. For example, tracking an institution by name does not always work. Many banks change names over time (e.g., Chase Manhattan acquired JPMorgan and renamed the company JPMorgan Chase). Acquisitions further muddy the waters (e.g., Chemical Bank purchased Chase Manhattan but continued to use the Chase Manhattan name, even though Chase was the smaller bank in terms of assets). Nonetheless, the merger and acquisition activity had to be included in the data.

An alternative to the bank name is to use the FDIC Certificate Number, which is labeled "cert" in the FDIC database. Each bank is assigned a unique cert by the FDIC for issuing insurance certificates [19]. Prior to the Chase acquisition, Chemical Bank had cert 628. After the acquisition, the combined company still used cert 628. Likewise, the entirety of JPMorgan Chase uses cert 628, even after the acquisition of

JPMorgan by Chase Manhattan. Unfortunately, these kinds of transformations are extremely disruptive to machine learning algorithms attempting to model relationships between inputs and outputs since there are discontinuities in the input data. There are no input variables that describe or represent these changes explicitly, so external knowledge is required to fully interpret the data.

Table 1. Aggregate statistics for a subset of the data extending from 2000Q1 to 2017Q2 are shown.

Variable	Quantity
Banks	11,576
Quarters	70
Variables	1,076
Complete variables	60
Bank failures	558

Perhaps the most challenging aspect of working with SDI data is the abundance of missing data. There are quarters where variables are completely missing for all banks (e.g., trading liabilities, labeled "tradel", is completely missing for 1992Q4, though the variable is included in the header). For example, as summarized in Table 1, out of 1,076 variables that occurred in 70 quarters of data, there were only 60 variables that were complete.

Due to the abundance of missing data, we had to focus on deep learning techniques and training strategies that are robust to missing data. Alternatively, we could have focused on techniques that attempt to synthesize missing data points, but these techniques are much more experimental in nature and hence not suitable for a baseline study. Two approaches were taken to combat the issue of missing data. First, the time period considered was restricted to 2000Q1 to the present (2017Q2). Record-keeping improved over time, and this has an added effect of not modeling old data from the time series. Second, we defined a "complete variable" as a quantitative variable which had no missing data for the period under consideration. Only 60 variables for the post-2000 epoch qualified as complete variables out of over 1,000 possible variables. The others were discarded either for having missing data somewhere in post-2000 epoch, or because they were categorical variables like address.

Formatting data to be amenable to processing by approaches such as CNN also proved challenging due to the way the SDI data were organized. The FDIC organized all data by quarter, rather than by bank, as is necessary for predicting classifications of the banks. Further, each quarter is divided into approximately 65 categories in separate spreadsheets organized by topics such as "assets & liabilities", "derivatives", or "securities". Consequently, data had to be merged between files within each quarter. These files contained duplicated variables, such as finding derivatives ("obsdir") in both the "assets & liabilities" spreadsheets and the "derivatives" spreadsheets. Further complicating the issue was the fact that variables did not correspond to the same column indices in every quarter.

The process of condensing these data so they were organized by bank was not trivial since the post-2000 epoch contains 4,278 files of data with 11,536 cert numbers. Each quarter had its spreadsheets condensed into one master spreadsheet for the entire quarter, containing no duplicated variables. There were 70 such files, each of which had to be opened 11,536 times for a total of 807,520 file openings. This still resulted in prohibitively long run times, even spread over 12 parallel jobs of 1,000 banks each. Since software was scripted in Python, this added an additional level of inefficiency.

Once the data were properly preprocessed, they still needed to be organized into training and testing data so that pilot experiments could be run. We also needed to integrate failure data. We did not consider blind evaluation data in our pilot experiments due to the limited amount of data and the fact that we were not going to do excessive parameter tweaking. In this study, we were mainly interested in assessing the suitability of the data for this type of research.

The FDIC provides a list of bank failures and the date of failure. Exactly 500 failed banks existed for 20 or more quarters during the post-2000 epoch, and thousands of successful banks existed for 20 quarters. Those 500 failed banks were partitioned into 400 for training and 100 for testing. The successful banks were allocated at random into a training set of 400 and a testing set of 100, giving both the training and testing sets an equal number of failed and healthy banks.

To run our deep learning algorithms, which generally prefer to see data of the same duration, the data for each bank were cropped to span equal lengths of time. For the failed banks, cropping was done at the end of the last quarter for which the bank reported data. For successful banks, the starting dates were selected at random. Sliding frames of bank data were designed to span 20 consecutive quarters. This duration was optimized by series of informal experiments using closed-loop training and evaluation.

3 Approach

For this study we leveraged a deep learning architecture that we have been developing for a variety of applications including speech processing, EEG analysis, and digital pathology [11, 20, 21]. The front end of these systems uses a recursive structure based on CNN and Long Short-Term Memory (LSTM) networks. Since the CNN approach, which captures temporal and spatial correlations, was pioneered for image processing applications, the time series for each bank will be referred to as its "bank image," as each is a multichannel time series. The technique has been successful in the assessment of multichannel time series and produces results superior to more traditional techniques such as k-nearest neighbors and random forests [11, 12]. We also include results for a pilot baseline system we developed based on logistic regression in which vectors were flattened to a length 1,200.

Baseline deep learning performance was evaluated using a fully-connected artificial neural network with one hidden layer with 512 nodes, 50% dropout, and "ReLU" activation. The batch size was 32, and 10 training epochs were used. The training data set consisted of 20×60 bank images. This approach was stepped up to a 20-layer multilayer perceptron neural network using steps of 2, 3, 4, 5, 6, 7, 8, 9, 10, 15, 20.

A CNN was implemented to inspect the data for correlations. Filtering layers used 3×3 filters. Three rounds of 3×3 filtering were performed, each followed by 2×2 maxpooling. This approach, depicted in Fig. 2, follows our work in other types of deep learning research [20, 21]. Finally, a fully-connected layer of 512 nodes was used after all CNN steps, and the 512 nodes mapped to the outcomes of a successful or failed bank. A dropout of 25% was used after all three maxpooling steps.

Marginal time series differences and second-order differences were applied to the bank images, with the differenced data run through the 3×3-filtering convolutional and fully-connected neural networks. Throughout all experiments, a binary cross-entropy loss function was used along with the Adam optimizer [20].

4 Results

Logistic regression failed to separate the distributions to any significant ($\alpha < 0.05$) extent, even on the training data. In-sample results were 56% accuracy, 65% sensitivity, and 47% specificity, while out-of-sample results were 51% accuracy, 57% sensitivity, and 44% specificity. Because of the poor performance on the training data, these results suggested that the logistic regression was an underfit, high-bias model. The opportunity to fit hundreds of thousands or even millions of parameters in neural networks motivated us to take a closer look at deep learning approaches, as we searched for bias reduction to correct this underfitting.

Baseline deep learning performance with a one hidden layer neural network was unable to beat random guessing based on the prior distribution of 50% successful and 50% failed banks ($\alpha < 0.05$), even on closed-set experiments on the training data. The system was able to find two distributions, but seemingly only because the input data gave the two groups of successful and failed banks. Performance did not significantly improve as the neural network was extended to 20 hidden layers. This network had over 24 million parameters when dropout was set to 0% to induce overfitting. However, the network could not accurately classify the training data under this overfitting condition.

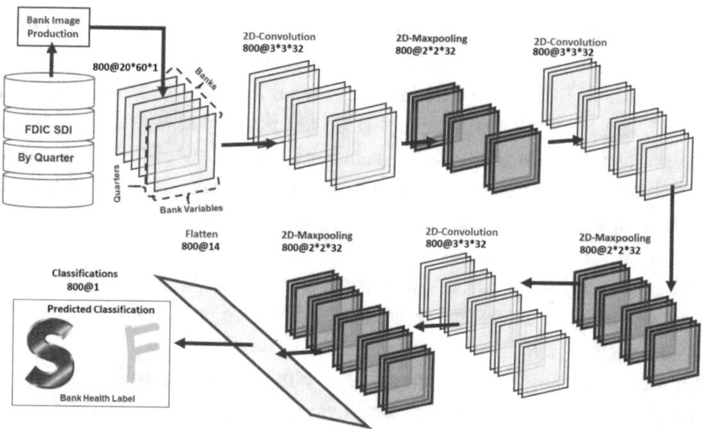

Fig. 2. A convolutional neural network (CNN) of three layers of 3 × 3 filtering, each followed by a layer of 2 × 2 maxpooling, was used as a baseline system.

Most surprisingly, the multilayer perceptron (MLP) networks were unable to produce two distributions. As larger and more parameter-heavy networks were explored, results tended to give zero values along an entire row of the confusion matrix: 100% sensitivity with 0% specificity, or vice versa, as shown in Fig. 3. Further, the same model setup could be run multiple times and not give the same result. Some runs gave 100% sensitivity and 0% specificity, while the next run of the same model would give the reverse. CNN gave this same lack of performance. This indicates that the models found nothing in the data onto which they could latch and separate the data.

5 Discussion

The inability even to overfit to the training data was particularly surprising, especially since previous work [4–9] was able to produce strong accuracy in predicting bank failures, both in terms of sensitivity and specificity. These results raise a number of questions. The first is a criticism of the techniques used, particularly given the somewhat small amount of data used to build the fully-connected and convolutional neural networks. Why use such complicated modeling techniques? That the neural networks were unable even to overfit indicates that an overly complicated modeling technique for the data size is not the culprit of poor performance. Moreover, logistic regression was unable to separate the successful and failure distributions. Since all classifiers failed on training data, we conclude that the FDIC data [2] are inadequate to solve this bank failure question.

The second question concerns the time sensitivity of bank health. In particular, do exogenous factors contribute to bank failures? Intuitively, this seems like it would be the case. Before and after the Global Financial Crisis, bank failures were rare. During the GFC, however, bank failures spiked. It is hard to dismiss this as coincidence. This brings up the frightening prospect that the generally good economy since the end of the GFC has masked problems at banks that simply have yet to experience another exogenous trigger of failure.

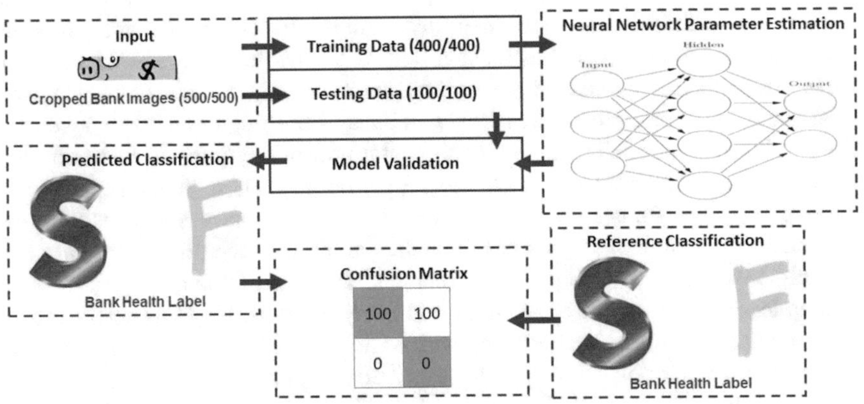

Fig. 3. The results of closed-loop analysis using an MLP network are shown.

The third question is if the variables in this study are adequate to capture the health of a bank. Indeed, the variables were chosen due to limitations of the SDI and the requirements of CNNs that preclude the use of variables with missing data. Further work could allow for some missing data but not having such a hard restriction of excluding any variables that had even one missing datum for any bank across the entire time span. For variables with the occasional missing datum, various techniques could be used to fill in the gaps, ranging from simple interpolation to more complex techniques such as autoencoders [22].

Once the 20-layer fully-connected neural network showed no ability to separate the training data, the prospects for the other techniques looked rather grim. Indeed, high parameter counts for models of the data produced no separation of the data. The lack of even producing two classifications suggests that the data for both failed and successful banks are generated by the same process, the opposite of the expectation that the successful banks have their processes of good management while the failed banks have their processes of mismanagement.

6 Conclusion

It appears that the information necessary to classify a bank as healthy or unhealthy is not endogenous to the signals used in this study. High-parameter neural networks were unable to give performance significantly ($\alpha < 0.05$) better than random guessing based on the 50/50 prior distribution, even on the training data.

Endogenous bank health predicted solely from the SDI data seems to be an intractable problem. Introduction of data capturing exogenous factors seems critical. The data used in this study come from the reputable source of the American Federal Government, yet existing state-of-the-art classification techniques did not work to make the ever-critical prediction of banks in danger of failure.

Regardless of data set and the problems present (such as missing data), perhaps the biggest challenge of a neural network approach to bank health is that failed banks necessarily have shorter time series than successful banks. The limited amount of data makes it difficult to build complex models. This is further complicated by the fact that the data spans different periods in time and the durations vary. Signals of uneven length are routine in some disciplines like speaker and language recognition. Techniques have been developed to normalize the inputs to neural network systems so that each input has the same number of elements [23, 24]. Such normalization techniques are worth exploring in future studies on bank health. Similarly, we have yet to explore data augmentation procedures for bank health prediction though such procedures have been extremely effective in disciplines like speech and image processing.

Deep learning has shown great promise in many fields, making this a very exciting time in the history of machine learning and artificial intelligence. However, even the best algorithms cannot overcome problems created by poor data. Future research needs to address improvements in the quantity and quality of the data, particularly including quantitative measures for exogenous factors.

References

1. Greenspan, A.: Never saw it coming: why the financial crisis took economists by surprise. Foreign Aff. **92**, 88–96 (2013)
2. Federal Deposit Insurance Corporation: Failures and Assistance Transactions - Historical Statistics on Banking, https://www5.fdic.gov/hsob/SelectRpt.asp?EntryTyp=30&Header=1&tab=bankFailures. Accessed 8 Dec 2018

3. Financial Crisis Inquiry Commission: The Financial Crisis Inquiry Report: Final Report of the National Commission on the Causes of the Financial and Economic Crisis in the United States. U.S. Government Printing Office, Washington, DC, USA (2011)
4. Erdogan, B.E.: Prediction of bankruptcy using support vector machines: an application to bank bankruptcy. J. Stat. Comput. Simul. **83**, 1543–1555 (2013)
5. Erdogan, B.E., Akyüz, S.Ö.: A weighted ensemble learning by SVM for longitudinal data: Turkish bank bankruptcy. In: Tez, M., von Rosen, D. (eds.) Trends and perspectives in linear statistical inference, pp. 89–103. Springer, Istanbul (2018)
6. Erdogan, B.E., Egrioglu, E.A.: Support vector machines vs multiplicative neuron model neural network in prediction of bank failures. Am. J. Intell. Syst. **7**, 125–131 (2017)
7. Boyacioglu, M.A., Kara, Y., Baykan, Ö.K.: Predicting bank financial failures using neural networks, support vector machines and multivariate statistical methods: a comparative analysis in the sample of savings deposit insurance fund (SDIF) transferred banks in Turkey. Expert Syst. Appl. **36**, 3355–3366 (2009)
8. Zheng, C., Cheung, A., Cronje, T.: Bank liquidity, bank failure risk and bank size. In: The 2016 ECU Business Doctoral and Emerging Scholars Colloquium. pp. 1–32, Edith Cowan University, Perth, Western Australia (2016)
9. Erdogan, B.E.: Bankruptcy prediction of Turkish commercial banks using financial ratios. Appl. Math. Sci. **2**, 2973–2982 (2008)
10. LeCun, Y., Bengio, Y., Hinton, G.: Deep learning. Nature **521**, 436–444 (2015)
11. Lopez, S.: Automated Identification of Abnormal Adult Electroencephalograms, Department of Electrical and Computer Engineering, Temple University, Philadelphia, Pennsylvania, USA (2017)
12. Lopez, S., Suarez, G., Jungreis, D., Obeid, I., Picone, J.: Automated identification of abnormal EEGs. In: IEEE Signal Processing in Medicine and Biology Symposium, pp. 1–4, Philadelphia, Pennsylvania, USA (2015)
13. Bao, W., Yue, J., Rao, Y.: A deep learning framework for financial time series using stacked autoencoders and long short-term memory. PLoS One **12**, 25 (2017)
14. Addo, P.M., Guegan, D., Hassani, B.: Credit risk analysis using machine and deep learning models. Risks **6**, 38 (2018)
15. Tam, K.: Neural network models and the prediction of bank bankruptcy. Omega Int. J. of Mgmt. Sci. **19**(5), 429–445 (1991)
16. Tam, K., Kiang, M.: Managerial applications of neural networks: the case of bank failure predictions. Manage. Sci. **38**, 926–947 (1992)
17. Le, H.H., Viviani, J.-L.: Predicting bank failure: an improvement by implementing a machine-learning approach to classical financial ratios. Res. Int. Bus. Financ. **44**, 16–25 (2018)
18. Abdou, H., Abdallah, W., Mulkeen, J., Ntim, C., Wang, Y.: Prediction of financial strength ratings using machine learning and conventional techniques. Invest. Manag. Financ. Innov. **14**, 194–211 (2017)
19. FDIC: Bankfind Glossary, https://research.fdic.gov/bankfind/glossary.html. Accessed 1 Oct 2017
20. Shah, V., Golmohammadi, M., Ziyabari, S., von Weltin, E., Obeid, I., Picone, J.: Optimizing channel selection for seizure detection. In: Obeid, I., Picone, J. (eds.) Proceedings of the IEEE Signal Processing in Medicine and Biology Symposium, pp. 1–5. IEEE, Philadelphia, Pennsylvania, USA (2017)
21. Golmohammadi, M., Ziyabari, S., Shah, V., Obeid, I., Picone, J.: Gated recurrent networks for seizure detection. In: Obeid, I., Picone, J. (eds.) Proceedings of the IEEE Signal Processing in Medicine and Biology Symposium, pp. 1–5. IEEE, Philadelphia, Pennsylvania, USA (2017)

22. Beaulieu-Jones, B.K., Moore, J.H.: Missing data imputation in the electronic health record using deeply learned autoencoders. In: Pacific Symposium on Biocomputing, pp. 207–219, Puako, Hawaii (2017)
23. Ghahremani, P., Sankar Nidadavolu, P., Chen, N., Villalba, J., Povey, D., Khudanpur, S., Dehak, N.: End-to-end deep neural network age estimation. In: Interspeech, pp. 277–281, Hyderabad, India (2018)
24. Snyder, D., Garcia-Romero, D., Sell, G., Povey, D., Khudanpur, S.: X-vectors: robust DNN embeddings for speaker recognition. In: 2018 IEEE International Conference on Acoustics, Speech and Signal Processing (ICASSP), pp. 5329–5333, Alberta, Canada (2018)

A Novel Ensemble Approach for Feature Selection to Improve and Simplify the Sentimental Analysis

Muhammad Latif[1,2] and Usman Qamar[1,2(✉)]

[1] Department of Computer Engineering, College of Electrical & Mechanical Engineering (E&ME), National University of Sciences and Technology (NUST), Islamabad, Pakistan
latif_awang001@yahoo.com, usmanq@ceme.nust.edu.pk
[2] National Centre for Big Data and Cloud Computing (NCBC), Lahore, Pakistan

Abstract. Text Classification is a renowned machine learning approach to simplify the domain-specific investigation. Consequently, it is frequently utilized in the field of sentimental analysis. The demanding business requirements urge to devise new techniques and approaches to improve the performance of sentimental analysis. In this context, ensemble of classifiers is one of the promising approach to improve classification accuracy. However, classifier ensemble is usually done for classification while ignoring the significance of feature selection. In the presence of right feature selection methodology, the classification accuracy can be significantly improved even when the classification is performed through a single classifier. This article presents a novel feature selection ensemble approach for sentimental classification. Firstly, the combination of three well-known features (i.e. lexicon, phrases and unigram) is introduced. Secondly, two level ensemble is proposed for feature selection by exploiting Gini Index (GI), Information Gain (IG), Support Vector Machine (SVM) and Logistic Regression (LR). Subsequently, the classification is performed through SVM classifier. The implementation of proposed approach is carried out in GATE and RapidMiner tools. Furthermore, two benchmark datasets, frequently utilized in the domain of sentimental classification, are used for experimental evaluation. The experimental results prove that our proposed ensemble approach significantly improve the performance of sentimental classification with respect to well-known state-of-the-art approaches. Furthermore, it is also analyzed that the ensemble of classifiers for the improvement of classification accuracy is not necessarily important in the presence of right feature selection methodology.

Keywords: Feature selection ensemble · Classifiers ensemble · Sentimental classification · SVM

1 Introduction

The concept of World Wide Web (WWW) leads to the exceptional increase in online data from the last decade. Facebook and Twitter are the true example of this trend where large amount of raw data is available in the form of user's posts/comments.

K. Arai et al. (Eds.): CompCom 2019, AISC 997, pp. 573–592, 2019.
https://doi.org/10.1007/978-3-030-22871-2_39

Similarly, the prominent online selling groups like Amazon are also dealing with the large amount of online data e.g. product reviews etc. In this context, sentimental classification is a field of machine learning where the available raw data is analyzed to evaluate the operational business strategies by utilizing the current people's trends about some issue or product. Particularly, sentimental classification usually categorizes the raw textual data into three classes [1] i.e. positive, negative and neutral. This automated classification facilitates business groups and other stakeholders to take right decision to meet some particular business objective.

The sentimental classification comprises three major steps i.e. feature extraction, feature selection and classification. In first step, features are extracted for further evaluation. However, it is not usually required to keep all extracted features due to several reasons e.g. irrelevance, computational complexity, etc. Therefore, in second step, feature selection is performed to keep the most significant features for further processing. Finally, in third step, classification is performed to categorize the given textual data into positive, negative and neutral classes.

As sentimental classification shows promising outcomes in various domains [13, 19] to achieve certain business objectives, it is frequently researched in the past decade [4] to improve the classification accuracy. In this context, ensemble of classifiers [3] is one of the promising approach to improve classification accuracy. However, classifier ensemble is usually done for classification while ignoring the significance of feature selection. In the presence of right feature selection methodology, the classification accuracy can be significantly improved even when the classification is performed through a single classifier. Therefore, in this article, a novel feature selection ensemble approach is presented for sentimental classification. The overview of approach is shown in Fig. 1.

It is essential to build the strong foundation for the proposed feature selection ensemble. Therefore, preliminary groundwork details are summarized in Sect. 2. Subsequently, the ensemble approach for feature selection is proposed (Sect. 3) by utilizing Gini Index (GI), Information Gain (IG), Support Vector Machine (SVM) and Logistic Regression (LR) algorithms as shown in Fig. 1. The feature selection ensemble approach is implemented in Rapid miner [33], Visual Studio .NET and GATE tool [34]. Particularly, proposed combination of features is implemented in GATE and .Net framework. Subsequently, proposed feature selection ensemble approach and classification is performed through Rapid Miner tool. The implementation details are summarized in Sect. 3.1. The experimental evaluation is performed (Sect. 4) through two benchmark datasets i.e. multi domain datasets [35] and Polarity [36]. The comparative analysis of our proposed ensemble approach with state-of-the-art is presented in Sect. 5. The comparison results prove that our proposed feature selection ensemble approach outperforms the existing state-of-the-art approaches as shown in Fig. 1. The significant aspects of the research are discussed in Sect. 6. Finally, the article is concluded in Sect. 7.

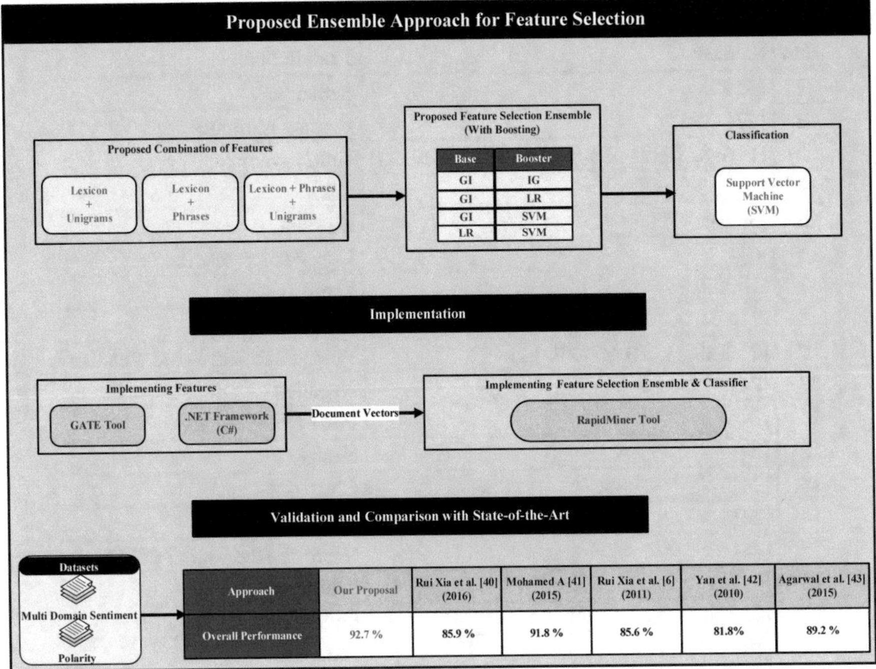

Fig. 1. Overview of research

2 Preliminaries

In this section, we briefly describe the primary concepts like leading features, ensemble techniques and classifiers in the domain of sentimental classification. Particularly, a Systematic Literature Review (SLR) is performed [4] to investigate the latest trends (2008–2016) in sentimental classification. This provides the basis to select features, classification algorithms, ensemble approach and benchmark datasets for the proposed feature selection ensemble approach. The details are presented in subsequent sections.

2.1 Features for Proposed Approach

We investigate the leading features in the domain of sentimental classification. The results are summarized in Table 1. The feature name is given in third column and corresponding research works are given in second column.

It can be seen from the Table 1 that Uni/N gram feature is frequently utilized in the field of sentimental classification. Similarly, Part-Of-Speech (POS) and SentiWordNet are also renowned features. However, these features are usually considered individually. We believe that the right combination of these features certainly improve the classification performance. Therefore, in our proposed approach, we select and combine lexicon, unigram and phrases features as follows:

Table 1. Prominent features for sentimental analysis

Sr.#	Studies	Feature Name
1	[9, 29]	Semantic
2	[17]	Product attributes
3	[1, 6–8, 10, 11, 13–16, 18–26, 28, 30, 32]	Uni/N-gram
4	[2, 6, 29]	POS
5	[3, 10]	Hashing
6	[6]	Dependency relations
7	[16, 28, 29]	SentiWordNet
8	[11, 24, 28]	Length
9	[2, 3, 6, 11, 20, 27, 29, 31]	Word Pairs and Word Relation
10	[24]	Negation
11	[24]	Polarity Dictionary
12	[24]	Stems
13	[24]	Clustering
14	[24]	ALLCAPS
15	[25]	Vector Space Model (VSP)
16	[9, 12]	Sentiment
17	[9]	Stylometric
18	[10]	Numbers

- Firstly, lexicon and unigram features are combined to provide more meaningful feature.
- Secondly, lexicon and phrases features are combined.
- Finally, all three (lexicon, phrases and unigram) are combined to provide the most powerful representation.

2.2 Classification Algorithms and Ensemble Approaches

We analyze various classification algorithms that are frequently used in the domain of sentimental classification. It has been found that Support Vector Machine (SVM) is the foremost algorithm for sentimental classification. Furthermore, it is frequently ensemble with other algorithms to improve sentimental classification as shown in Table 2.

It can be seen from Table 2 that SVM is frequently ensemble with other 23 algorithms during 2008–2016 for sentimental classification. However, SVM is mostly ensemble with other algorithms for classification rather than feature selection. Although such ensemble improves the classification accuracy, it is generally complex and computational intensive. Therefore, we have proposed the ensemble of feature selection to simplify the process while significantly improving the classification accuracy.

In addition, it is also investigated that the bagging (bootstrap) [1], boosting [2], stacking [9], daggling [2], majority voting [2], ada boost [10], weighted combination [6], meta classifiers combination [6], bagging random space [1] and meta cost [9] are the leading ensemble approaches for sentimental classification. However, these ensemble approaches are usually applied for classification rather than feature selection.

Table 2. Algorithms ensemble with SVM for sentimental classification

Sr. #	Studies	Classifier
1	[23, 29]	Radial Base Function Neural Network (RBF NN)
2	[1–3, 5, 6, 8–10, 15, 19, 20, 22, 26, 27, 29, 32]	Naïve Bayes (NB)
3	[1, 9, 15, 19]	Decision trees
4	[2]	Bayesian Logistic Regression (BLR)
5	[22]	Senti Strength
6	[3, 19]	Random Forests (RF)
7	[5, 11, 15]	Conditional Random Fields (CRF)
8	[7, 31]	Back Propagation Neural Network (BPNN)
9	[1, 10, 20, 23]	Probabilistic Neural Network (PNN)
10	[1, 10, 20, 23]	K neighbors
11	[14]	General Inquirer Based Classifier (GIBC)
12	[2, 7]	Linear Discriminant Analysis (LDA)
13	[14]	RBC
14	[2, 3, 9, 23]	Logistic Regression (LR)
15	[15, 19, 27, 29]	Bayesian Network (BN)
16	[16]	Hidden Markov Model (HMM)
17	[32]	Scoring
18	[14]	Selective Bayesian Classifier (SBC)
19	[20]	Classification Base (CB)
20	[1, 5, 6, 12, 15, 18, 20, 24, 26, 32]	Maximum Entropy
21	[23]	Multi-Layer Perceptron (MLP)
22	[31]	Extreme Learning Machine (ELM)
23	[12]	ANN (Artificial Neural Networks)

On the basis of above investigation, we only select SVM for classification in the proposed approach. This leads to simplify the classification process. On the other hand, we propose the ensemble of Gini Index (GI), Information Gain (IG), Support Vector Machine (SVM) and Logistic Regression (LR) for feature selection. Furthermore, we choose boosting as an ensemble technique. The selection of GI and IG algorithms can be argued as these are unavailable in Table 2. Basically, we select both (GI & IG) algorithms due to their promising results for feature selection as presented in state-of-the-art [4].

2.3 Datasets

We investigated the leading datasets, used to performed validation, in the field of sentimental classification. It has been found that Movie [1, 5], Product [10, 21], Twitter [3, 12] and medical [1, 2] are the leading open source benchmark datasets in the domain of sentimental classification. Therefore, we have selected movies and polarity datasets. The description of these datasets is given below

(1) *The Multi Domain Sentiment Data Set*

The multi domain sentiment dataset of product reviews (book, DVD, electronic, and kitchen appliances) taken from amazon was first used by [37]. Data set can be downloaded at [35]. The data set contains 1000 negative and 1000 positive labeled reviews for each domain i.e. book, DVD, electronic, and kitchen appliances. Consequently, overall size of dataset is 8000. Each review have the information consisting of a rating (0–5 stars), a reviewer name and location, a product name, a review title and date, and the review text. Reviews with ratings >3 are labeled positive; those with rating <3 are labeled negative.

(2) *The Polarity Data Set*

This dataset is available at [36]. The polarity dataset v2.0 is a set of film review documents available for research in sentiment analysis and opinion mining. It was first introduced as a research data set for sentiment classification presented in [38] with 700 positive and 700 negative reviews. Revised dataset is presented in 2004 and comprises of 1000 positive and 1000 negative labeled film reviews extracted from the Internet Movie Database Archive.

3 Proposed Approach and Implementation

On the basis of the detailed investigation (Sect. 2), we propose novel feature selection ensemble approach to provide the best classification performance while restricting the stress on system resources at acceptable level. Firstly, we propose the combination of three leading features (Sect. 2.1) i.e. lexicon + unigram, lexicon + phrases and lexicon + unigram + phrases. In next step, we propose the feature selection ensemble by exploiting Gini Index (GI), Information Gain (IG), Support Vector Machine (SVM) and Logistic Regression (LR) algorithms as shown in Fig. 2. Finally, SVM is used to perform the classification.

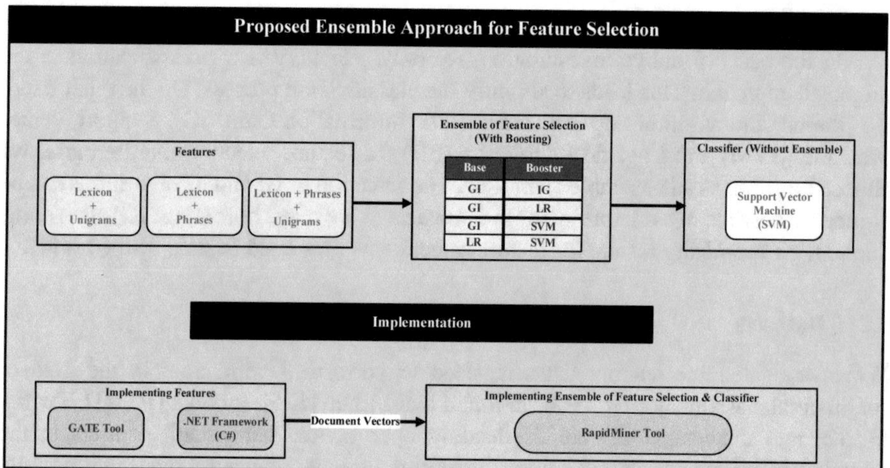

Fig. 2. Proposed ensemble for feature selection with implementation architecture

3.1 Implementation

We selected Multi Domain Sentiment and polarity benchmark datasets (Sect. 2.3) for the evaluation of proposed approach. Both datasets contains raw unannotated text in the form of positive and negative reviews. Therefore, it is required to perform pre-processing steps for extraction/selection of required features i.e. lexicon + unigram, lexicon + phrases and lexicon + unigram + phrases. Consequently, we have used GATE tool [34] for text pre-processing as shown in Fig. 2. Particularly, we use ANNIE framework to perform tokenization, POS tagging and sentence split. Moreover, we have used Morphological analyzer for root finding instead of traditional strimmer's. Furthermore, we have also utilized Noun Phrase (NP) and Verb Phrase (VB) build-in chucker of GATE tool. Finally, we got annotated XML document against each plain text review of datasets. The sample document is shown in Fig. 3.

It can be seen from the figure that we annotate three types of features NP, VP and tokens. Tokens contains other information like POS and term root etc. Finally, we got 10,000 xml documents for the reviews of both datasets. Although we have annotated our desired features (Fig. 3), we cannot extract these features through XML queries/paths for further utilization as there format is incompatible for RapidMiner tool [33]. Therefore, we develop Gate2CVS tool [39] to transform GATE xml file with desired features to CSV files which are compatible with RapidMiner. The Gate2CVS tool is able to read the XML document, extract features and finally save the features in CVS file. The screenshot of CVS file for book dataset, generated by Gate2CVS tool, is shown in Fig. 4.

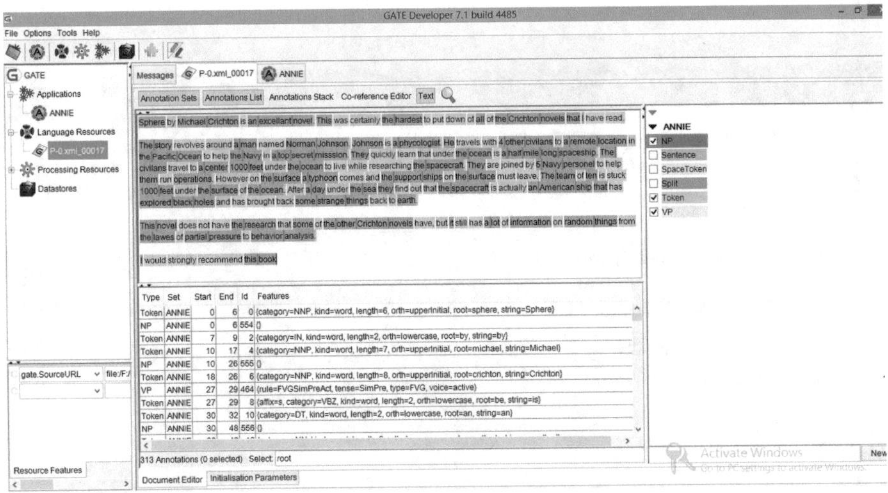

Fig. 3. Sample annotated XML document for plain text review

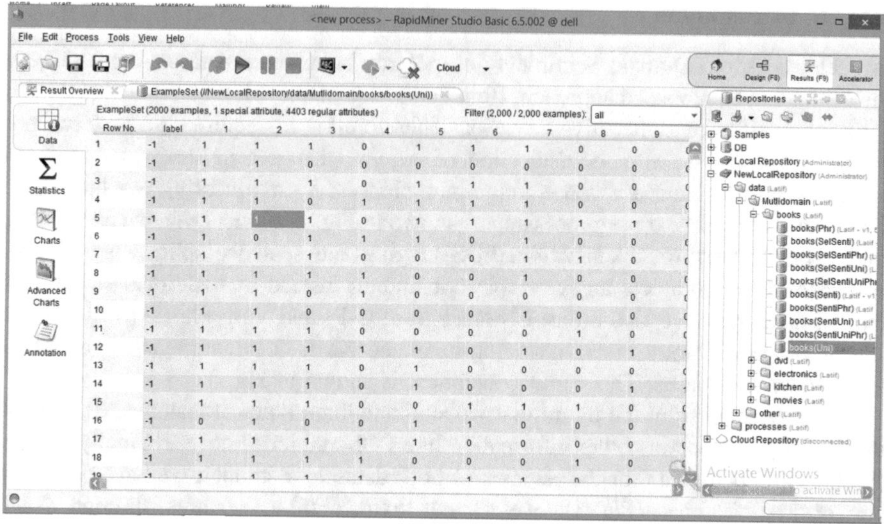

Fig. 4. CVS file generated through GATE2CVS tool for books dataset

It can be seen from Fig. 4 that dataset contains 4403 features with 2000 examples. The row no. represents examples (2000) and label represents positive (1) and negative −1 labels. 4403 features ids are shown against each sample. The above document vectors are generated on the basis of extracted feature space as shown in Fig. 5.

ID	POS	Root	Isphrase
1	v	be	FALSE
2	n	book	FALSE
3	v	have	FALSE
4	r	not	FALSE
5	v	do	FALSE
6	r	n't	FALSE
7	v	read	FALSE
8	v	can	FALSE
9	v	will	FALSE
10	r	so	FALSE
11	r	just	FALSE
12	n	time	FALSE
13	v	write	FALSE
14	n	story	FALSE
15	v	make	FALSE
16	v	get	FALSE

Fig. 5. Extracted feature space for books dataset

Gate2CVS tool is able to generate feature space and document vectors as shown in Figs. 4 and 5. The sample generated CVS files for books dataset and source code of Gate2CVS tool can be found at [39] for further evaluation.

Once we generate document vectors of datasets through Gate2CVS tool, we use RapidMiner to implement the proposed feature selection ensemble. Furthermore, SVM is used for classification by utilizing 10X validation approach. The screenshot of RapidMiner process for book dataset is shown in Fig. 6. It can be seen from the figure that Logistic Regression (LR) and SVM ensemble is applied for feature selection and SVM is used for classification. The RapidMiner processes for feature selection and classification can be found at [39] for further evaluation.

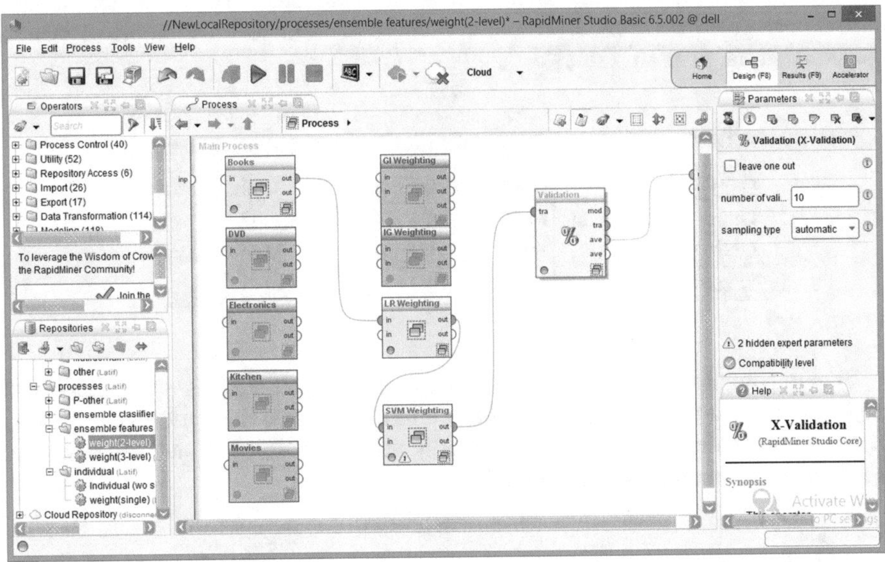

Fig. 6. Screenshot of RapidMiner process (Books) for feature selection ensemble and classification

4 Experimental Evaluation

In this section, we perform experiment evaluation to investigate the results of proposed approach (Sect. 3). We selected two benchmark datasets (Sect. 2.3) i.e. Multi Domain Sentiment [35] and Polarity [36]. The multi domain sentiment dataset of product reviews (book, DVD, electronic, and kitchen appliances) is taken from amazon and polarity dataset v2.0 is a set of film review documents. The number of original features in both datasets is given in Table 3. This facilitates to accurately investigate the performance of proposed approach.

Table 3. Total number of original features in both datasets

Features	Books	DVD	Electronics	Kitchen	Movies
Lexicon (Baseline)	935	855	312	302	1182
Unigram	4403	3938	2498	2361	6452
Phrasal	4408	4607	2601	2132	5269

4.1 Evaluating Combination of Features

Primarily, we select three types of features (i.e. lexicon, phrases and unigram) and propose their combination i.e. lexicon + unigram, lexicon + phrases and lexicon + unigram + phrases. The total number of individual features in the selected datasets are given in Table 3. The total number of features, after proposed features combination, is given in Table 4.

Table 4. Total number of features after proposed combination

Features	Books	DVD	Electronics	Kitchen	Movies
Lexicon + Unigrams	4878	4354	2649	2502	7382
Lexicon + Phrases	5278	5392	2889	2420	6122
Lexicol + Uni + Phr	7951	7679	4432	3952	9285

It is important to evaluate the effectiveness of proposed features combination with respect to individual features. Therefore, we have evaluated the performance of individual and combined features by individually utilizing SVM, LR and NB classifiers. The results are summarized in Table 5.

Table 5. Performance of original features and ensemble of features with individual classifiers

Features	Books			DVD			Electronics			Kitchen			Movies		
	SVM	LR	NB	SVM	LR	NB	SVM	LR	NB	SVM	LR	NB	SVM	LR	NB
Accuracy with original features															
Lexicon	66.10	61.60	65.85	71.30	66.85	71.00	69.50	70.90	71.85	72.80	72.95	73.20	82.20	79.20	79.80
Uni	77.75	75.70	72.55	77.45	77.40	70.20	74.20	74.95	73.50	80.45	78.25	77.10	83.40	82.70	82.40
Phr	74.80	71.45	72.10	73.05	69.70	67.00	71.25	73.10	75.55	74.10	72.60	74.10	79.90	79.10	79.80
Accuracy with Ensemble Features															
Lexicon + Uni	77.40	75.70	72.45	78.75	77.35	69.15	74.10	75.25	73.25	79.35	79.10	76.05	83.30	84.55	81.35
Lexicon + Phr	76.90	74.65	73.05	75.45	75.15	69.95	73.40	75.65	76.65	77.10	76.30	76.60	82.90	82.80	82.30
Lexicol + Uni + Phr	79.45	78.25	75.00	79.75	76.65	69.40	74.80	76.95	76.25	80.55	80.10	76.55	84.10	85.20	82.50

It can be seen from the table that our proposed features combination certainly improves the performance as compared to individual features. Furthermore, it is also analyzed that SVM classifier provides best results as compared to LR and NB. On the basis of this evaluation (Table 5), we use proposed combined features and lexicon features as a base line for all further evaluation.

4.2 Evaluating Feature Selection Ensemble

In this section, we evaluate the performance of the proposed feature selection ensemble. For accurate evaluation with simplicity, we fix the number of features by utilizing GI algorithm with (weight > 0.0015) as shown in Table 6.

Table 6. Fixed number of features (GI weight > 0.0015) to evaluate feature selection ensemble

Features	Books	DVD	Electronics	Kitchen	Movies
Lexicon (Baseline)	44	65	44	49	89
Lexicon + Uni	282	323	297	283	330
Lexicon + Phr	265	302	316	257	305
Lexicol + Uni + Phr	404	442	435	390	533

To comprehensively investigate the performance of proposed feature selection ensemble, we perform this evaluation in two steps. In first step, we evaluate the performance by employing all four classifiers (i.e. GI, IG, SVM and LR) individually for feature selection. Furthermore, SVM is used for classification. This facilitates to thoroughly investigate the performance of proposed feature selection ensemble. The results are summarized in Table 7.

Table 7. Performance of GI, IG, SVM and LR classifiers individually for feature selection

	Books	DVD	Electronics	Kitchen	Movies
	Feature selection by GI (weight $\geq$ 0.0015)				
Lexi (Baseline)	67.6	70.85	70.9	73.9	79.2
Lexicon + Uni	82.2	85.05	82.45	86.25	87.1
Lexicon + Phr	78.85	82.8	80.5	84	85.4
Lexi + Uni + Phr	82.35	86	83.4	87.8	86.55
	Feature selection by IG				
Lexi (Baseline)	67.5	70.55	70.6	74.05	82.8
Lexicon + Uni	81.15	85.65	83.05	85.7	86.4
Lexicon + Phr	79.1	83.6	80.4	84.5	86.4
Lexi + Uni + Phr	83.05	87	83.4	87.6	91
	Feature selection by LR				
Lexi (Baseline)	59.25	69.4	67.3	72.05	79.3
Lexicon + Uni	80.7	84	84.55	87.25	88.55
Lexicon + Phr	77.75	81.15	83.15	84.65	86.55
Lexi + Uni + Phr	82.75	85.4	87.45	89.95	91.45
	Feature selection by SVM				
Lexi (Baseline)	67.85	73.5	71.55	73.05	82.45
Lexicon + Uni	84.35	86.2	84.65	88.3	93.25
Lexicon + Phr	80.95	83.8	83	85.75	91.55
Lexi + Uni + Phr	85.45	88.3	86.35	90.55	95.1

It can be seen from the table that features selected by SVM classifier provide best performance as compared to IG, GI and LR. Now, in second step, we evaluate the proposed ensemble with boosting (Fig. 2). Particularly, four combinations have been evaluated where GI is used as a base in first three combinations and IG, LR and SVM are used as a booster, respectively. In last combination, LR is used as a base and SVM is used as booster. The classification is performed through SVM. The evaluation results are summarized in Table 8.

Table 8. Performance results of proposed ensemble approach

Features	Books	DVD	Electronics	Kitchen	Movies
Feature selection by GI-IG					
Lexi (Baseline)	67.50	70.55	70.60	74.05	79.35
Lexicon + Uni	81.15	85.65	83.05	85.70	86.55
Lexicon + Phr	79.50	83.60	80.40	84.50	85.90
Lexi + Uni + Phr	83.15	87.00	83.40	87.60	86.35
Feature selection by GI-LR					
Lexi (Baseline)	65.90	70.15	69.65	72.70	81.45
Lexicon + Uni	85.25	86.60	85.80	88.05	88.75
Lexicon + Phr	85.10	84.95	83.15	86.30	88.20
Lexi + Uni + Phr	88.45	88.60	86.85	89.70	93.10
Feature selection by GI-SVM					
Lexi (Baseline)	68.10	72.00	70.25	73.50	80.75
Lexicon + Uni	87.75	88.20	86.15	89.65	90.95
Lexicon + Phr	85.25	86.00	85.05	87.75	89.65
Lexi + Uni + Phr	90.60	91.20	88.15	91.05	92.55
Feature selection by LR-SVM					
Lexi (Baseline)	65.50	70.30	70.30	72.45	81.85
Lexicon + Uni	88.20	89.55	87.00	90.75	95.70
Lexicon + Phr	84.95	85.95	87.10	88.65	94.45
Lexi + Uni + Phr	91.85	91.45	89.70	93.05	97.60

It can be seen from the Table 8 that proposed feature selection ensemble approach certainly improves the performance as compared to individual feature selection algorithm (Table 7). Particularly, the ensemble of LR and SVM provides the best performance as shown in Table 8. It is also analyzed that improper ensemble of classifiers for feature selection not only degrades the performance but also increase computational complexities. For example, the ensemble of GI and IG classifiers notably reduces the performance (Table 8) as compared to the feature selection performed through SVM alone (Table 7). Therefore, it is always essential to select the right classifiers and ensemble approaches for feature selection.

4.3 Performance Comparison

This section summarizes the performance comparison of proposed approach. Firstly, the comparative analysis for proposed feature combination is performed. The classification is performed through SVM. The results are summarized in Table 9.

Table 9. Performance comparison for proposed features combination

Data Set	Lexi (Baseline)	Lexicon + Uni	Lexicon + Phr	Lexi + Uni + Phr	Increase %
Books	66.10	77.40	76.90	79.45	13.35
DVD	71.30	78.75	75.45	79.75	8.45
Electronics	69.50	74.10	73.40	74.80	5.30
Kitchen	72.80	79.35	77.10	80.55	7.75
Movies	82.20	83.30	82.90	84.10	1.90

It can be seen from the table that the proposed features combination significantly improves the performance. Particularly, the combination of all three features (i.e. lexicon + unigram + phrases) provides the best classification results. There is 13.35% increase in the performance for books datasets where lexicon feature is used as a baseline. Similarly, there is a significant increase in the performance for DVD, Electronics and Kitchen datasets as shown in the Table 9.

We have also performed the performance comparison for proposed feature selection ensemble approach. For this comparison, we only consider the combination of all three features (i.e. lexicon + unigram + phrases). Firstly, SVM is used for classification without employing any feature selection algorithm. Secondly, SVM is used for both feature selection and classification. Finally, the proposed ensemble of LR and SVM is used for feature selection and classification is performed through SVM. The results are summarized in Table 10.

Table 10. Performance comparison for proposed feature selection ensemble

Data set	SVM classifier (without feature selection)	SVM classifier (weight by SVM feature selection)	SVM classifier (LR-SVM feature selection)	Increase %
Books	79.45	85.45	91.85	12.40
DVD	79.75	88.30	91.45	11.70
Electronics	74.80	86.35	89.70	14.90
Kitchen	80.55	90.55	93.05	12.50
Movies	84.10	95.10	97.60	13.50

It can be seen from the Table 10 that the performance without feature selection is significantly low as compared to proposed feature selection ensemble (LR & SVM). Even feature selection through SVM only has less classification accuracy as compared to the ensemble of LR and SVM for feature selection. Particularly, there is 12.40, 11.70, 14.90, 12.50 and 13.50 increase in the performance through proposed feature selection ensemble for books, DVD, Electronics, Kitchen and Movies datasets, respectively.

The outcomes of the experimental evaluation (Sects. 4.1, 4.2, 4.3) demonstrate that the proposed features combination and feature selection ensemble significantly improves the performance of sentimental classification. To further verify the novelty and applicability of proposed approach, the detailed comparative analysis with high impact state-of-the-art approaches is performed in subsequent section.

5 Comparison with State-of-the-Art

We perform comparative analysis of our proposed ensemble approach with existing ensemble techniques for further confirmation. In first step, we perform methodology comparison as given in Table 11.

Table 11. Comparison with state-of-the-art in the context of approach / methodology

Research work	Feature extraction		Feature selection		Classifiers ensemble	
	Combination	Types	Ensemble	Methods	Ensemble	Classifier
Rui Xia et al. [40] (2016)	YES	Unigram and bigram	NO	–	Yes (Stacking)	Linear SVM, Logistic Regression and Naive Bayes
Mohamed Abdel Fattah [41] (2015)	NO	Unigrams	NO	TF-IDF, TF-ICF, MI,OR, WLLR, CHI	Yes (Voting, Borda Count)	SVM, Neural Network, Gaussian Mixture Model
Rui Xia et al. [6] (2011)	Yes POS, Word Relation	Unigrams, Bigrams and Dependencies	NO	–	Yes (Fixed, Weighted, Meta-classifier)	Naive Bayes, Maximum Entropy, SVM
Yan Dang et al. [42] (2010)	YES	Content free, Unigrams, Bigrams and Sentiment	NO	IG	NO	SVM
Basant Agarwal et al. [43] 2015	YES Dependency and Semantic	Unigrams and bigrams, Bi-tagged, Dependency Features	NO	mRMR	NO	SVM
Our approach	YES	Lexicon, Unigrams and Phrase	YES	GI, IG, LR, SVM	NO	SVM

It can be analyzed from the Table 11 that the main advantage of our approach is the feature selection ensemble which is not supported by state-of-the-art approaches. Furthermore, the combination of phrases, unigram and lexicon features is hard to find in the literature review. Consequently, our proposal is novel and significant contribution in the domain of sentimental classification. Furthermore, our proposal significantly improves the performance as compared to the state-of-the-art approaches. The comparison of performance results is given in Table 12.

Table 12. Performance comparison of proposed approach with state-of-the-art

Datasets	Our approach	Rui Xia et al. [40] (2016)	Mohamed Abdel Fattah [41] (2015)	Rui Xia et al. [6] (2011)	Yan Dang et al. [42] (2010)	Basant Agarwal et al. [43] (2015)
Books	91.85	82.9	89.8	81.8	78.50	88.5
DVD	91.45	83.7	91.4	83.8	80.75	89.2
Electronics	89.70	86.6	91.9	85.95	83.75	88.9
Kitchen	93.05	89.1	93.7	88.65	84.15	–
Movies	97.60	–	92.3	87.70	–	90.1
Overall	92.70	85.9	91.8	85.6	81.8	89.20

It can be seen from Table 12 that our proposal has improved overall accuracy as compared to state-of-the-art. However, the performance of Electronics and Kitchen datasets is slightly less as compared to Mohamed Abdel Fattah [41]. Other than that, the results of our proposed approach are significantly better than all other state-of-the-art approaches as shown in Table 12. It is important to note that Rui Xia et al. [40] and Yan Dang et al. [42] do not include Movies dataset in their evaluation results. Therefore, it is not possible for us to compare the results of Movies dataset. Similar is the case with Basant Agarwal et al. [43] for Kitchen dataset.

6 Discussion

In this article, we propose straightforward feature selection ensemble approach by utilizing four algorithms i.e. GI, IG, SVM and LR. The experimental evaluation prove that the proposed approach significantly improve the classification performance. However, it can be argued that the nested boosting of these classifiers for feature selection might further improve the performance. We believe that the nested boosting of classifiers for feature selection increases the complexity and deviates the spirit of our proposal. The main theme behind our proposal is to simplify the sentimental classification with optimum performance. However, to investigate such ensemble possibilities, we perform the nested boosting of GI, IG, SVM and LR classifiers for feature selection as shown in Fig. 7.

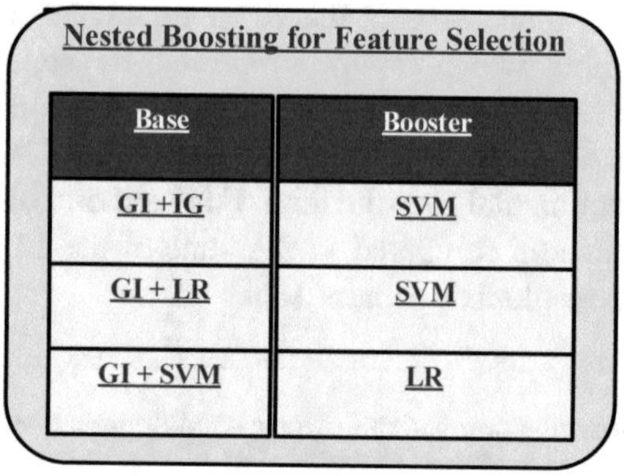

Nested Boosting for Feature Selection

Base	Booster
GI +IG	SVM
GI + LR	SVM
GI + SVM	LR

Fig. 7. Structure of nested boosting for feature selection ensemble

The results are summarized in Table 13.

Table 13. Performance of feature selection ensemble with nested boosting

	Books	DVD	Electronics	Kitchen	Movies
	Feature selection by GI-IG-SVM				
Lexi (Baseline)	67.8	73.1	70.9	73.65	80.5
Lexicon + Uni	86	88.45	85.35	88.6	88.2
Lexicon + Phr	83.85	85.9	83.45	86.45	88.05
Lexi + Uni + Phr	88.35	90.65	87.25	90.75	92.05
	Feature selection by GI-SVM-LR				
Lexi (Baseline)	65.9	70.5	69.9	72.7	81.55
Lexicon + Uni	86.45	87.35	84.75	88.5	92.1
Lexicon + Phr	85.45	85	83.7	87.05	90.35
Lexi + Uni + Phr	88.15	90.6	88.35	90.2	93.4
	Feature selection by GI-LR-SVM				
Lexi (Baseline)	68.1	72.2	70.25	73.5	80.8
Lexicon + Uni	87.7	89	85.9	88.8	93.75
Lexicon + Phr	84.75	85.8	84.75	86.95	92.75
Lexi + Uni + Phr	90.05	91.3	88.15	91.45	96

It has been analyzed from the table that the results of feature selection ensemble with nested boosting (Table 13) are not as good as that of proposed feature selection ensemble with simple boosting (Table 8). Consequently, it can be concluded that the ensemble of features selection can only improve the performance results up to certain level. Thereafter, performance is starting to decrease gradually as analyzed from Tables 13 and 8. Furthermore, such ensemble approach is computational intensive.

Another investigable point of this research is the application of proposed feature selection ensemble with the ensemble of classifiers for classification. Therefore, we have evaluated the proposed approach with the ensemble of classifiers for classification. Particularly, we have used SVM, LR and Naïve Bayes (NB) algorithms for this evaluation as shown in Fig. 8. Furthermore, Bayesian Boosting (BB), Stacking (ST) and Voting (VT) ensemble techniques are considered. The performance results are summarized in Table 14.

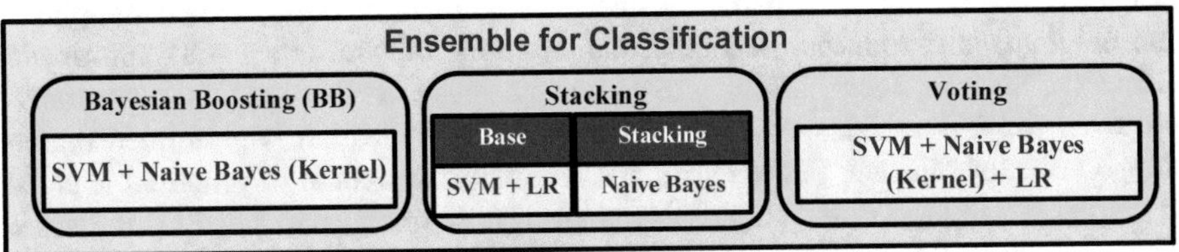

Fig. 8. Structure of classifiers ensemble for evaluation

It can be seen from Table 14 that in first stage, we evaluate the ensemble of classifiers (Fig. 8) where only SVM is used for feature selection. In second stage, we evaluate the ensemble of classifiers where feature selection is performed by the proposed ensemble (Sect. 3) of LR and SVM. Finally, the evaluation of classifiers ensemble is evaluated where nested boosting of GI, LR and SVM is used (Table 13) for feature selection. From the results of classifier ensemble (Table 14), it can be analyzed that classifier ensemble techniques partially improve the performance results of proposed approach while significantly increasing the complexity and computational stress of classification problem. Consequently, it can be concluded that proposed feature selection ensemble provides the optimum performance while significantly limiting the complexity because it is not required to ensemble classifiers for classification.

Table 14. Performance of proposed feature selection ensemble with classifiers ensemble for classification

	Books			DVD			Electronics			Kitchen			Movies		
	BB	VO	ST	BB	VO	ST	BB	VO	ST	BB	VO	ST	BB	VO	ST
	Feature selection by SVM														
Lexi (Baseline)	68.80	68.75	68.65	73.25	74.05	74.10	72.80	72.00	72.85	75.15	75.90	76.50	76.90	73.35	78.80
Lexicon + Uni	83.25	84.60	85.05	84.60	86.40	86.35	87.25	87.85	89.25	86.50	88.25	88.75	91.35	89.20	95.20
Lexicon + Phr	79.95	81.35	81.70	82.90	84.35	84.25	83.95	84.60	85.90	84.80	86.20	86.65	88.05	85.95	91.85
Lexi + Uni + Phr	84.70	85.70	85.90	86.35	88.65	87.70	88.70	88.95	90.10	88.25	90.50	90.10	92.80	90.30	96.05
	Feature selection by LR-SVM weighting														
Lexi (Baseline)	67.65	68.20	67.95	72.15	74.65	74.80	67.55	67.15	68.30	74.50	76.55	76.35	72.85	71.65	75.45
Lexicon + Uni	87.10	87.30	87.85	85.80	89.15	88.20	87.00	86.25	88.20	88.15	91.05	89.75	92.30	90.75	95.35
Lexicon + Phr	83.85	84.35	85.15	85.85	86.95	87.95	83.75	83.30	85.50	88.20	88.85	89.50	89.05	87.80	92.65
Lexi + Uni + Phr	89.40	91.20	90.00	87.70	90.55	90.00	89.30	90.15	90.35	90.05	92.45	91.55	94.60	94.65	97.50
	Feature selection by GI-LR-SVM weighting														
Lexi (Baseline)	68.35	68.65	69.55	72.30	74.15	74.70	68.30	67.35	68.60	72.60	74.15	75.45	73.80	68.00	74.05
Lexicon + Uni	86.90	87.90	88.30	88.05	89.40	89.25	86.85	86.60	87.35	88.35	89.40	90.00	92.35	87.25	92.80
Lexicon + Phr	85.65	85.85	87.30	86.00	87.80	88.75	85.60	84.55	86.35	86.30	87.80	89.50	91.10	85.20	91.80
Lexi + Uni + Phr	89.05	90.55	91.10	89.65	91.80	91.40	89.00	89.25	90.15	89.95	91.80	92.15	94.50	89.90	95.60

7 Conclusions and Future Work

This article presents a novel methodology to ensemble classifiers for feature selection in order to simplify the sentimental classification while attaining the optimum performance. Firstly, the combination of three well-known features (i.e. lexicon, phrases and unigram) is introduced. Secondly, the ensemble of Gini Index (GI), Information Gain (IG), Support Vector Machine (SVM) and Logistic Regression (LR) algorithms is proposed for feature selection. Subsequently, the classification is performed through Support Vector Machine (SVM) classifier. The implementation of proposed approach is carried out in GATE and RapidMiner tools. Furthermore, two benchmark datasets, frequently utilized in the domain of sentimental classification, are used for experimental evaluation. The experimental results prove that our proposed ensemble approach significantly improves the performance of sentimental classification with respect to well-known state-of-the-art ensemble approaches. Furthermore, it is also analyzed that the ensemble of classifiers for the improvement of classification accuracy is not necessarily important in the presence of right feature selection methodology.

In this research, we highlight the significance of features combination and feature selection ensemble to improve the performance of sentimental analysis with simplicity. This research can be extended in multiple directions. For example, one probable direction is to add more features and their different combinations in the proposed methodology for further performance improvements. Similarly, different other classifiers (e.g. Max. Entropy etc.) and ensemble approaches (e.g. meta cost etc.) can be investigated for feature selection in the given methodology.

References

1. Wang, G., et al.: Sentiment classification: the contribution of ensemble learning. J. Decis. Support Syst. **57**, 77–93 (2013)
2. Onan, A., Korukoğlu, S., Bulut, H.: A multiobjective weighted voting ensemble classifier based on differential evolution algorithm for text sentiment classification. JESA **62**, 1–16 (2016)
3. da Silva, N.F.F., et al.: Tweet sentiment analysis with classifier ensembles. J. Decis. Support Syst. **66**, 170–179 (2014)
4. Athar, A., Butt, W.H., Anwar, M.W., Latif, M., Azam, F.: Exploring the ensemble of classifiers for sentimental analysis—a systematic literature review. In: 9th International Conference on Machine Learning and Computing, Singapore 2017
5. Fersini, E., Messina, E., Pozzi, F.A.: Sentiment analysis: Bayesian ensemble learning. DSS **68**, 26–38 (2014)
6. Xia, R., Zong, C., Li, S.: Ensemble of feature sets and classification algorithms for sentiment classification. Inf. Sci. **181**(6), 1138–1152 (2011)
7. Vinodhini, G., Chandrasekaran, R.M.: A comparative performance evaluation of neural network based approach for sentiment classification of online reviews. J. King Saud Univ. – Comput. Inf. Sci. **28**(1), 2–12 (2016)
8. Catal, C., Nangir, M.: A sentiment classification model based on multiple classifiers. Appl. Soft Comput. **50**, 135–141 (2017)

9. King, M.A., Abrahams, A.S., Ragsdale, C.T.: Ensemble learning methods for pay-per-click campaign management. ESA **42**(10), 4818–4829 (2015)
10. Lochter, J.V., Zanetti, R.F., Reller, D., Almeida, T.A.: Short text opinion detection using ensemble of classifiers and semantic indexing. ESA **62**, 243–249 (2016)
11. Ekbal, A., Saha, S.: Combining feature selection and classifier ensemble using a multi objective simulated annealing approach: application to named entity recognition. Soft. Comput. **17**(1), 1–16 (2013)
12. Wang, Y., Rao, Y., Zhan, X., Chen, H., Luo, M., Yin, J.: Sentiment and emotion classification over noisy labels. KBS **111**, 207–216 (2016)
13. Vinodhini, G.: A sampling based sentiment mining approach for e-commerce applications. JIPM **53**(1), 223–236 (2017)
14. Prabowo, R., Thelwall, M.: Sentiment analysis: a combined approach. Informatics **3**(2), 143–157 (2009)
15. Saha, S., Ekbal, A.: Combining multiple classifiers using vote based classifier ensemble technique for named entity recognition. JD&KE **85**, 15–39 (2013)
16. Abbasi, A., Chen, H., Thoms, S., Fu, T.: Affect analysis of web forums and blogs using correlation ensembles. IEEE Trans. Knowl. Data Eng. **20**(9), 1168–1180 (2008)
17. Vinodhini, G., Chandrasekaran, R.M.: Sentiment mining using SVM-based hybrid classification model. In: Computational Intelligence, Cyber Security and Computational Models, vol. 246, pp. 155–162. Springer, New Delhi (2013)
18. Lin, Y., Wang, X., Zhang, J., Zhou, A.: Assembling the optimal sentiment classifiers. In: Proceedings of 13th International Conference, Paphos, Cyprus, November 28–30, 2012, vol. 7651, pp. 271–283 (2012)
19. Wan, Y., Gao, Q.: An ensemble sentiment classification system of Twitter data for airline services analysis. In: IEEE 15th Data Mining Workshops (2015)
20. Su, Y., Zhang, Y., Ji, D., Wang, Y., Wu, H.: Ensemble learning for sentiment classification. In: Workshop on Chinese Lexical Semantics, vol. 7717, pp 84–93. Springer, Berlin, Heidelberg (2013)
21. Whitehead, M., Yaeger, L.: Sentiment mining using Ensemble classification model. In: Innovations and advances in computer sciences and engineering, pp. 509–514. Springer, Dordrecht (2010)
22. Chalothom, T., Ellman, J.: Simple approaches of sentiment analysis via ensemble learning. In: Information science and applications, vol. 339, pp 631–639. Springer-Verlag, Berlin, Heidelberg (2015)
23. Prusa, J., Khoshgoftaar, T.M., Dittman, D.J.: Using ensemble learners to improve classifier performance on tweet sentiment data. In: IEEE 16th ICIRI 2015
24. Hagen, M., Potthast, M., Büchner, M., Stein, B.: Twitter sentiment detection via ensemble classification using averaged confidence scores. In: European Conference on Information Retrieval, pp. 741–754. Springer, Cham (2015)
25. Dai, L., Chen, H., Li, X.: Improving sentiment classification using feature highlighting and feature bagging. In: 11th IEEE ICDMW 2011, pp. 61–66
26. Wang, Z., Li, S., Zhou, G., Li, P., Zhu, Q.: Imbalanced sentiment classification with multi-strategy ensemble learning. In: Proceedings Asian Language Processing (2011)
27. Wang, W.: Heterogeneous Bayesian ensembles for classifying spam emails. In: Proceedings on Neural Networks, 2010
28. Kumar, V., Minz, S.: Multi-view ensemble learning for poem data classification using SentiWordNet. In: Advanced Computing and Informatics Proceedings of ICACNI 2014, vol. 27, pp. 57–66
29. Hassan, A., Abbasi, A., Zeng, D.: Twitter sentiment analysis: a bootstrap ensemble framework. In: International Conference on Social Computing (2013)

30. Vinodhini, G., Chandrasekaran, R.M.: Sentiment mining using SVM-based hybrid classification model. Comput. Intell. Cyber Secur. Comput. Models **246**, 155–162 (2013)
31. Wang, F., Zhang, Y., Rao, Q., Li, K., Zhang, H.: Exploring mutual information-based sentimental analysis with kernel-based extreme learning machine for stock prediction. Soft Comput. 2016, PP 1–13
32. Lu, B., Tsou, B.K.: Combining a large sentiment lexicon and machine learning for subjectivity classification. In: Proceedings of the Ninth International Conference on Machine Learning and Cybernetics, 11–14 July 2010
33. RapidMiner Tool. https://rapidminer.com/
34. GATE Tool. https://sourceforge.net/projects/gate/
35. Multi Domain Sentiment Data Set. https://www.cs.jhu.edu/~mdredze/datasets/sentiment/
36. Polarity Data Set. http://www.cs.cornell.edu/people/pabo/movie-review-data/
37. Blitzer, J., Dredze, M., Pereira, F.: Biographies, bollywood, boom-boxes and blenders: domain adaptation for sentiment classification. In: ACL, vol. 7, pp. 440–447, Jun 2007
38. Pang, B., Lee, L., Vaithyanathan, S.: Thumbs up? sentiment classification using machine learning techniques. In: Proceedings of the ACL-02 Conference on Empirical Methods in Natural Language Processing, vol. 10, pp. 79–86 (2002)
39. GATE2CVS Tool. http://ceme.nust.edu.pk/ISEGROUP/gate2cvs.html. Accessed Apr 2017
40. Xia, R., Xu, F., Yu, J., Qi, Y., Cambria, E.: Polarity shift detection, elimination and ensemble: A three-stage model for document-level sentiment analysis. J. Inf. Process. Manage. **52**, 36–45, 2016
41. Fattah, M.A.: New term weighting schemes with combination of multiple classifiers. J. Neurocomput. **167**, 434–442 (2015)
42. Dang, Y., Zhang, Y., Chen, H.: A lexicon-enhanced method for sentiment classification: an experiment on online product reviews. IEEE Intell. Syst. **25**(4), 46–53 (2010)
43. Agarwal, B., Poria, S., Mittal, N., Gelbukh, A., Hussain, A.: Concept-level sentiment analysis with dependency-based semantic parsing: a novel approach. J. Cogn. Comput. **7**(4), 487–499 (2015)

High Resolution Sentiment Analysis by Ensemble Classification

Jordan J. Bird(✉), Anikó Ekárt, Christopher D. Buckingham, and Diego R. Faria

School of Engineering and Applied Science, Aston University, Birmingham B4 7ET, UK
{birdj1,a.ekart,c.d.buckingham,d.faria}@aston.ac.uk

Abstract. This study proposes an approach to ensemble sentiment classification of a text to a score in the range of 1–5 of negative-positive scoring. A high-performing model is produced from TripAdvisor restaurant reviews via a generated dataset of 684 word-stems, gathered by information gain attribute selection from the entire corpus. The best performing classification was an ensemble classifier of RandomForest, Naive Bayes Multinomial and Multilayer Perceptron (Neural Network) methods ensembled via a Vote on Average Probabilities approach. The best ensemble produced a classification accuracy of 91.02% which scored higher than the best single classifier, a Random Tree model with an accuracy of 78.6%. Other ensembles through Adaptive Boosting, Random Forests and Voting are explored with ten-fold cross-validation. All ensemble methods far outperformed the best single classifier methods. Even though extremely high results are achieved, analysis documents the few mis-classified instances as almost entirely being close to their real class via the model's given error matrix.

Keywords: Sentiment analysis · Opinion mining · Machine learning · Ensemble learning · Classification

1 Introduction

The applications of Sentiment Analysis are increasingly growing in importance in both the sciences and industry, for example, through human-robot interaction [19] and as a business tool in terms of user feedback to products [9], giving more prominence to the field of Affective Computing. Affective Computing [20] is the study of systems capable of empathetic recognition and simulation of human affects including but not limited to sentimental and emotional information encapsulated within human-sourced data.

In this work, various methods of Sentiment Classification are tested on top of a generated set of word-stem attributes that are selected by their ranking of information gain correlating to their respective class. The best model is then analysed in terms of its error matrix to further document the classification results. The main contributions of this work are as follows:

K. Arai et al. (Eds.): CompCom 2019, AISC 997, pp. 593–606, 2019.
https://doi.org/10.1007/978-3-030-22871-2_40

- Effective processing of text to word-stems and information gain based selection suggests a set of 684 attributes for effective classification of high resolution sentiment.
- Single and ensemble Models are presented for the classification of sentiment score on a scale of 1–5 as opposed to the standard three levels of classified sentiment (Positive-Neutral-Negative). In this study, 1 is the most negative result, and 5 is the most positive.
- Methods of Sentiment Classification are based entirely on text and correlative score rather than taking into account metadata (user past behaviour, location, etc.), enabling a more general application to other text-based domains.

This paper will document related works in sentiment analysis modeling, a proposed approach to the experiment, the preprocessing and acquisition of sentiment-based data and the training of various single and ensemble models with analysis. Finally, the impact will be reviewed as well as noting the future works enabled by a general text-to-score sentiment classifier.

2 Related Work

Sentiment analysis, or opinion mining, is the study of deriving opinions, sentiments, and attitudes from lingual data such as speech, actions, behaviours, or written text. Sentiment Classification is an approach that can class this data into nominal labels (e.g. *'this remark has a **negative** valence'*) or continuous polarities or score which map to their overall sentiment. With the rise of online social media, extensive amounts of opinionated data are available online, which can be used in classification experiments which require large datasets. Negative polarity was used to analyse TripAdvisor reviews [34] on a scale of negative-neutral-positive, findings show that each review rating of one to five stars each have unique distributions of negative polarity. This unique pattern suggests the possibility of further extending the two polarity three-sentiment system to a further five levels. A similar three-class sentiment analysis was successfully trained on Twitter data [29]. Related work with Twitter Sentiment Analysis found that hashtags and emoticons/emoji were very effective attributes to train classifiers [17]. Exploration of TripAdvisor reviews found that separation via root terms, 'food', 'service', 'ambiance' and 'price' provided a slight improvement for machine learning classification [22].

Human-Robot Interaction has increasingly become concerned with Sentiment Analysis as an extra dimension of interaction with a robotic agent. A chatbot architecture was constructed that analysed input and output sentiment of messages and provided it as meta-information within the chatbot's response [4]. The usage of a robot's perception of sentiment is prominent in multiple applications of classification such as mental state [3,5], facial expressions [11,12], voice/speech [25], and observed physical activities [1,35] (Fig. 1).

With increasing availability of computing resources for lower costs, accurate classification is enabled on increasingly larger datasets over time, giving rise to cross-domain application through more fine-tuned rules and complex patterns

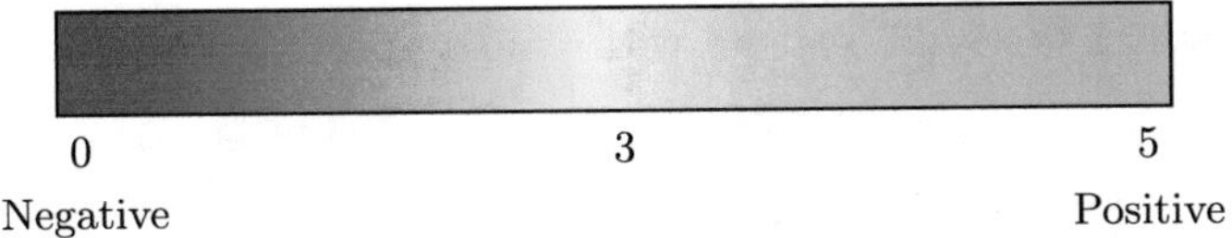

Fig. 1. A diagram to show sentiment gradient, 1 is the most negative score and 5 is the most positive score.

of attribute values. That is, a model trained on dataset A can be used to classify dataset B. This has been effectively shown through multiple-source data to produce an *attribute thesaurus* of sentiment attributes [6]. Researchers also found that rule-based classification of large datasets are unsuited to cross-domain application, but machine-learning techniques on the same data shows promising results [10].

Observing the results of the related works into social media sentiment analysis (Twitter, TripAdvisor, IMDB, etc.) shows the sheer prominence of three level sentiment classification, with only one class for negativity and one for positivity along with a neutral class, with an overall result being calculated with derived polarities. With the large amount of data available correlating to a user's specification of class outside of this range of three, this paper suggests a more extensive sentiment classification paradigm to co-ordinate with user's review scores, to better make use of human-sourced data. To engineer this, the number of data classes in this experiment will be equal to the range of scores available to a user. The end goal would be more points of polarities to give a finer measurement of sentiment. Moreover, many of the state-of-the-art studies experiment with single classifiers, many strong models are produced with Bayesian, Neural Network, and Support Vector Machine approaches but they have not been taken further to an ensemble and explored.

3 Data Acquisition and Processing

3.1 Data Acquisition

A dataset of 20,000 user reviews of London based restaurants was gathered from TripAdvisor[1], in which a review text was coupled with a score of 1 to 5, where 1 is the most negative and 5 is the most positive review. All reviews were in English, and all other meta information such as personal user information was removed, this was performed for the more general application of the classifier to all text based data containing opinions. All restaurants from the Greater London Area were chosen randomly as well as the reviews themselves selected at random.

3.2 Preprocessing

Resampling was performed with a 0.2 weighting towards the lower reviews due to the prominence of higher reviews, to produce a more balanced dataset. The

[1] TripAdvisor - http://tripadvisor.co.uk.

resulting dataset of 17,127 reviews with their respective scores can be seen in Table 1.

Table 1. Reduced dataset

Score	*No. of reviews*
1	2960
2	2983
3	3179
4	3821
5	4283

It is worth noting that even after weighted re-sampling, there remains a higher frequency of positive reviews which will be factored into the analysis of results, specifically in analysis of the classification accuracy of low review scores by way of error matrix observation.

With unprocessed text having few statistical features, feature generation was performed via a filter of word vectors of the string data, based on the statistics of word-stem prominence. Firstly, worthless stopwords were removed from the text using the Rainbow List [24] (i.e. words that hold no important significance), and then the remaining words were reduced to their stems using the Lovins Stemmer algorithm [21]. Stopword removal was performed to prevent misclassification of class based on the coincidental prominence of words with no real informative data, and stemming was performed to increase the frequency of terms by removing their formatting e.g. time based suffixes and clustering them to one stem.

The process of word vectorisation with the aforementioned filtering produced 1455 numerical attributes mapped to the frequency of the word stem. Further attribute selection was required to remove attributes that had little to no influence of class, which would reduce the computational complexity of classification.

3.3 Attribute Selection

Attribute selection was performed on the 1455 numerical word stem attributes to produce a reduced dataset of 684 attributes. The information gain (IG) of each attribute was calculated and sorted using a simple Ranker algorithm (where higher IG = more correlation to a class).

Information Gain or Relative Entropy is the expected value of the Kullback-Leibler divergence where a univariate probability distribution of a given attribute is compared to another [18]. This is a measure of a comparison of states given as follows:

$$IG(T, a) = H(T) - H(T|a) \tag{1}$$

This denotes the measured change in entropy, where IG is the Information Gain of Class T and Attribute a.

A cutoff point of 0.001 Information Gain (the lowest measure) was implemented and this removed 771 attributes (word-stems) that were considered to have no impact on the class. This meant that all remaining attributes had a measurable classification ability when it came to sentiment. Of the highest information gain were the word-vector attributes "disappointing" (0.08279), "worst" (0.06808), "rude" (0.0578) and "excellent" (0.05356) - which, regardless of domain, can be observed to have high sentimental polarity.

The dataset to result from this processing was taken forward for classification experiments.

4 Classification Model Background

A range of models were selected following distinctly different methods of deriving a classification, this was for a range of performances as well as for a future ensemble of non-correlative models.

4.1 Rule Based Models

Rule Based Classification is performed via an 'if-then' approach to labelling [36] (e.g. *'if it is* $-15°$*' then 'it is Winter'*), where a defined number of rules are generated and refined using their classification entropy (see Eq. 4).

Zero Rules (ZeroR) Classification is used as a benchmark, it is the simple process of labelling all data with the most common class, i.e. if the most common binary class encompasses 51% of the dataset then ZeroR will have a 51% accuracy. One Attribute Rule (OneR) is a common example of effective classification based on selecting the single best rule, such as classifying the Season by temperature in the aforementioned example. Experimentation of multiple rules are then compared based on their minimum-error, and the best one chosen to classify objects.

4.2 Bayes Theorem

Bayesian Models are classification models based on Bayes Theorem. Bayes Theorem [2] is the comparable probability that data point d will match to Class C. Bayes Theorem is given as follows:

$$P(A|B) = \frac{P(B|A)P(A)}{P(B)} \tag{2}$$

Where P is 'probability of', A and B are the evidence i.e. the probability of *P(A)* being true is related to the probability of the H with evidence $P(A|B)$. In terms of this work, this would take inputs of word-stem parameters and classify the text via its highest probability as calculated by the formula.

Naive Bayes classification is given as follows:

$$\hat{y} = \underset{k \in (1,\ldots,K)}{\arg\max}\, p(C_k) \prod_{i=1}^{n} p(x_i \mid C_k) \tag{3}$$

Where class label y is given to data object k. The naivety in Bayesian algorithms concerns the assumed independence of attribute values (or existence), whether or not the assumption holds true for a data. Naive Bayes (NB) is the application of Bayes' theorem which selects the class based on the lowest risk, i.e. based on the evidence it will select the most likely label for the data. Naive Bayes Multinomial (NBM) is a classification method that differs from NB by defining the distribution of each attribute $p(f_i|c)$ as a multinomial distribution [23] e.g. a word count within a text.

4.3 Decision Trees

Decision Tree classification is the derivation of conditional control statements based on attribute values, which are then mapped to a tree. Classification is performed by cascading a data point down the tree through each conditional check it meets until an end node is reached, which contains a Class. The growth of the tree is based on the entropy of its end node, that is, the level of disorder in classes found on that node. E.g. if a certain ruleset on a trained tree ends on a node that has 90% sentiment level 4 and 10% sentiment level 5. The entropy of a node with c Class results is calculated as follows:

$$E(S) = -\sum_{i=1}^{c} P_i \times log_2 P_i \tag{4}$$

A tree will continue to grow either until the level of disorder of the node is 0 (i.e. 1 Class), or it has been stopped by other means such as a set stopping length (pre-pruning) or reduction after the entire tree is grown (post-pruning).

A random tree classifier is a decision tree generated in which k-random attributes are selected at each node [27]. The model is simple since no pruning is performed and thus an overfitted tree is produced to classify all input data points. Therefore, cross-validation is used to create an average of the best performing random trees, or with a testing set of unseen data.

J48 is a decision tree operating the C4.5 algorithm [28]. Whilst growing the decision tree, information entropy is used to calculate a best split at each node, i.e. that which enriches both splits with differing classes (or each class in a binary classification problem). If features are worthless for the problem with zero information gain, higher nodes on the tree are used to classify the current instance, though, this results in the waste of computational resources.

4.4 Neural Networks

An Artificial Neural network (ANN) is a system loosely inspired by the natural brain for weighted classification or regression [32]. In observation of Fig. 2, a simplified diagram of a Neural Network - input neurons (attributes) are generated for layer 1, where n is the attributes available to the classifier. A number of hidden layers for classification are generated, in the figures' case, two layers of three and two nodes respectively. Finally, at the end of the diagram, c output nodes that map to c number of classes. Each link between neurons carries a weight, which is trained. The network diagram is considered a Deep Neural Network due to the existence of more than one hidden layer of neurons [33]. For regression problems (e.g. prediction of house price), a single output node of real numbers is used.

A Multilayer Perceptron (MLP) is a neural network that through optimisation (or learning) updates the weights between neurons via the process of backpropagation, i.e. to derive a gradient which is used to calculate connection weighting values [31]. Backpropagation is the process of calculating the error at the output (the green layer in Fig. 2) and fed backwards throughout the network to distribute new weightings, to refine the system and reduce the errors or loss.

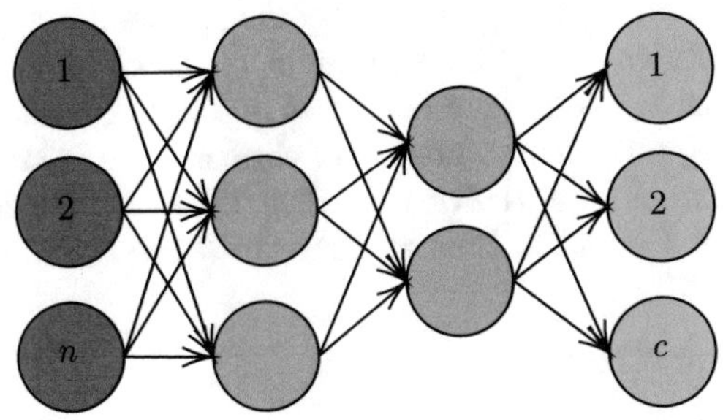

Fig. 2. A simplified Deep Neural Network diagram with n input attributes, two hidden layers of 3, 2 neurons and c output classes.

4.5 Support Vector Machines

A Support Vector Machine (SVM) is the classification of entities by calculation of an optimised separator between groups, formed via an n-dimensional hyperplane within the given problem space [8]. An optimal solution considers the margin between the separating hyperplane and the classifier points, with a larger average margin defining a best possible classification vector.

Sequential Minimal Optimization (SMO) is an algorithm to implement an SVM with several classes by reducing the quadratic function to a linear constraint [26]. This is performed by breaking the optimisation problem into the smaller subproblems, which are then analytically solved [7].

5 Ensemble Method Background

An ensemble is a fusion of classifiers at each prediction, where a result is calculated based on the results of the models being fused. This section will detail the theory behind ensemble methods as well as those chosen for this experiment.

5.1 Voting

Voting is a simple democratic process that takes into account predictions of the models encompassed. Each model will subsequently vote on each class or a regression result in turn, and the final decision is derived through a selected operation:

- **Average of Probabilities** - Models vote on all classes with a vote equal to each of its classification accuracies of said class. E.g. if a model can classify a binary problem with 90% and 70% accuracies, then it would assign those classes 0.9 and 0.7 votes respectively if voting for them. The final output is the class with the most votes.
- **Majority Vote** - All models will vote on the class it predicts the data to be, and the one selected is the class with the most votes.
- **Min/Max Probability** - The minimum or maximum probabilities of all model predictions are combined and class is selected based on this value.
- **Median** - For regression, all models will vote on a value, and the one selected will be the median of all of their values. E.g. if two models in a voting process vote for values of 1.5 and 2, then the output of the classifier will be 1.75.

5.2 Adaptive Boosting

Adaptive Boosting (AdaBoost) is a Gödel Prize winning meta algorithm for the improvement of classifiers [13]. AdaBoost follows an iterative approach by, at each iteration, selecting a random training subset to improve on the previous iteration's results, and further uses those results to produce a weighting of classification in a combination. This is given by:

$$F_T(x) = \sum_{t=1}^{T} f_t(x) \tag{5}$$

where F is the set of t models classifying the data object x [30].

5.3 Random Forest

A Random Forest is an ensemble of Random Trees through Bootstrap Aggregating (bagging) and Voting [15]. Training is performed through a bagging process where multiple random decision trees are generated, a random selection of data is gathered and trees are grown to fit the set. Once training is completed, the generated trees will all vote, and the majority vote is selected as the predicted class (see Section V, subsection A, *'Voting'*). Random Forests tend to outperform Random Trees due to their decreasing of variance without increasing of the model bias.

5.4 Adopted Approach

Manual tuning noted the performance of Random Forests, Adaptive Boosting, and Vote (average probabilities) and thus they were used in this experiment. Adaptive Boosting was performed on Random Tree as well as Random Forest (*ensemble of ensemble*). Voting on average probabilities were tested with models that had techniques unique to one another; firstly, Naive Bayes Multinomial, Random Tree and Multilayer Perceptron. Secondly, Random Forest, Naive Bayes Multinomial, and Multilayer Perceptron.

6 Results

Training and evaluation of models were performed using 10-fold cross validation ($k = 10$), in which one fold of the data is used for testing a classifier trained on the remaining nine folds of data. This is performed ten times, thus testing and training encompass all ten folds. This is done to prevent a model being overfitted to the dataset in question and therefore having no useful application. Leave-one-out cross-validation ($k = n\ data\ points$) was not performed due to computational resource requirement of such validation. Where required, all random seeds for model training were set to 1. Random numbers are generated by the Java Virtual Machine (JVM).

6.1 Single Classifiers

Results of single classifiers can be seen in Table 2. The two best models were both Decision Tree algorithms, with the best being Random Tree with an accuracy of 78.6%.

The selected methods had the following parameters set:

- **MLP** - Three layers of 5, 10, and 15 neurons with training at 2000 epochs.
- **RT** - No limits imposed on the number of random attributes selected or the tree depth, a minimum total weight of instances on a node set to 1.
- **J48** - As for RT but with tree depth pruning at a confidence factor of 0.25.
- **SMO** - A complexity parameter of 1.0 with a Logistic calibrator.

6.2 Ensemble Classifiers

The models applied in the ensemble experiments (Table 3, column 2) were sourced from the training experiments performed previously, seen in Table 2, and are therefore more directly comparable as a sum to their parts.

Results of ensemble methods and their classifiers can be seen in Table 3. The best model was a Vote of Average Probabilities by the three previously trained models of Random Forest, Naive Bayes Multinomial, and a Multilayer Perceptron.

The selected methods had the following parameters:

Table 2. Classification accuracy of single classifier models

Classifier	Classification accuracy
OneR	29.59%
MLP	57.91%
NB	46.28%
NBM	59.02%
RT	**78.6%**
J48	75.76%
SMO SVM	68.94%

Table 3. Classification accuracy of ensemble models

Ensemble	Classifiers	Classification accuracy
RF	N/A	84.9%
Vote	NBM, RT, MLP	80.89%
Vote	RF, NBM, MLP	**91.02%**
AdaBoost	RT	79.36%
Adaboost	RF	84.93%

- **Random Forest** - Bag size was 100% of the provided data, tree maximum depth was unlimited, selected features were not limited, and 100 iterations were performed for each forest.
- **Vote** - Class voting was decided upon via the highest average of probabilities.
- **AdaBoost** - 10 iterations of each model were performed for the Adaptive Boost.

6.3 Analysis

ZeroRules benchmarking resulted in an accuracy of 24.88%, all of the models far outperformed the benchmark with the exception of OneR which had an only slightly better accuracy of 29.59% (+4.71), this is due to One Rule Classification having diminishing returns on higher dimensionality datasets, the one in this particular experiment taking place in 684-dimension space.

All ensemble approaches to classification outperformed the single classifiers. Interestingly an 'ensemble of ensemble' approach produced better results when it came to AdaBoost of a Random Forest (+0.03%), and most importantly factoring in the Random Forest within a vote model along with Naive Bayes Multinomial and a Multilayer Perceptron, which produced a classification accuracy of 91.02%.

In terms of the error matrix (Table 4), it is observed that the best model put forward misclassified predictions at a gradient around the real class, due to the crossover of sentiment based terms. Most prominently, classes 4 and 5 were the most difficult to predict, and further data analysis would give concrete examples of lingual similarity between reviews based on these two scores.

Table 4. Error matrix for the classifications given by the vote (RF, NBM, MLP) model

Predicted class					Real class
1	*2*	*3*	*4*	*5*	
2919	26	7	3	5	*1*
42	**2887**	25	19	10	*2*
55	71	**2873**	126	54	*3*
14	25	139	**3090**	544	*4*
17	13	63	288	**3902**	*5*

7 Conclusion

To conclude, this work presented results from models for classification of multi-level sentiment at five distinct levels after performing effective feature extraction based on lingual methods. The best single classifier model was a Random Tree with a classification accuracy of 78.6%, which was outperformed by all applied ensemble methods and their models. The best overall model was an ensemble of Random Forest, Naive Bayes Multinomial, and a Multilayer Perceptron through a Vote of Average Probabilities, with a classification accuracy of 91.02%.

The findings suggest future work in the development of text-based ensemble classifiers as well as their single classification parameters, due to the trained models in this experiment successfully being improved when part of an ensemble. The effectiveness of Neural Networks for sentiment classification is well documented [14] implying that further work with more computational resources than were available for this experiment is needed due to the low results achieved. Furthermore, leave-one-out cross validation has been observed to sometimes be more effective than k-fold cross validation [16] but proved too computationally expensive for the dataset in this experiment, therefore exploration of this method of training validation is needed. Voting had very promising results, which could be refined further through adding new trained models as well as experimentation with different democratic processes.

Successful experiments were performed purely on a user's message and no other meta information (e.g. previous reviews, personal user information) which not only shows effectiveness in the application in the original domain of user reviews, but also a general application to other text-based domains such as chatbots and keyword-based opinion mining. The application of the classifiers put forward in this paper are useful in the aforementioned domains, though future work should encompass a larger range of sources to smooth out some of the remaining domain-specific information.

Table 5. Indirect comparison of this study and state-of-the-art sentiment classification work (different datasets)

Work	Resolution	Accuracy
This study (ensemble - vote)	5	**91.02%**
Read [29]	3	84.6%
Bollegala et al. [6]	2	83.63%
Denecke [10]	2	82%
This study (single - RT)	5	78.6%
Kouloumpis et al. [17]	3	75%

In terms of contribution, a comparison of the results of this study and the state of the art can be seen in Table 5. With a far higher resolution than the 3 (Pos-Neu-Neg) or 2 (Pos-Neg) observed in many works, a high accuracy of 91.02 was still achieved through the method of ensemble. It can be observed that the best single classifier fits within related works whereas an ensemble approach with the same dataset far outperforms related works.

Acknowledgments. This work was supported by the European Commission through the H2020 project EXCELL (https://www.excell-project.eu/), grant No. 691829.

This work was also partially supported by the EIT Health GRaCEAGE grant number 18429 awarded to C.D. Buckingham.

References

1. Adama, D.A., Lotfi, A., Langensiepen, C.: Key frame extraction and classification of human activities using motion energy. In: UK Workshop on Computational Intelligence, pp. 303–311. Springer (2018)
2. Bayes, T.: An essay towards solving a problem in the doctrine of chances (1763). By the late Reviewed: R. Price, J. Canton
3. Bird, J.J., Ekárt, A., Buckingham, C.D., Faria, D.R.: Mental emotional sentiment classification with an EEG-based brain-machine interface. In: The International Conference on Digital Image and Signal Processing (DISP 2019). Springer (2019)
4. Bird, J.J., Ekárt, A., Faria, D.R.: Learning from interaction: an intelligent networked-based human-bot and bot-bot chatbot system. In: UK Workshop on Computational Intelligence, pp. 179–190. Springer (2018)
5. Bird, J.J., Manso, L.J., Ribiero, E.P., Ekárt, A., Faria, D.R.: A study on mental state classification using EEG-based brain-machine interface. In: 9th International Conference on Intelligent Systems. IEEE (2018)
6. Bollegala, D., Weir, D., Carroll, J.: Using multiple sources to construct a sentiment sensitive thesaurus for cross-domain sentiment classification. In: Proceedings of the 49th Annual Meeting of the Association for Computational Linguistics: Human Language Technologies, vol. 1, pp. 132–141. Association for Computational Linguistics (2011)
7. Chang, C.-C., Lin, C.-J.: LIBSVM: a library for support vector machines. ACM Trans. Intell. Syst. Technol. (TIST) **2**(3), 27 (2011)

8. Cortes, C., Vapnik, V.: Support-vector networks. Mach. Learn. **20**(3), 273–297 (1995)
9. Cui, H., Mittal, V., Datar, M.: Comparative experiments on sentiment classification for online product reviews. In: AAAI, vol. 6, pp. 1265–1270 (2006)
10. Denecke, K.,: Are SentiWordNet scores suited for multi-domain sentiment classification? In: 2009 Fourth International Conference on Digital Information Management, ICDIM 2009, pp. 1–6. IEEE (2009)
11. Faria, D.R., Vieira, M., Faria, F.C.C., Premebida, C.: Affective facial expressions recognition for human-robot interaction. In: 2017 26th IEEE International Symposium on Robot and Human Interactive Communication (RO-MAN), pp. 805–810. IEEE (2017)
12. Faria, D.R., Vieira, M., Faria, F.C.C.: Towards the development of affective facial expression recognition for human-robot interaction. In: Proceedings of the 10th International Conference on PErvasive Technologies Related to Assistive Environments, pp. 300–304. ACM (2017)
13. Freund, Y., Schapire, R.E.: A decision-theoretic generalization of on-line learning and an application to boosting. J. Comput. Syst. Sci. **55**(1), 119–139 (1997)
14. Ghiassi, M., Skinner, J., Zimbra, D.: Twitter brand sentiment analysis: a hybrid system using n-gram analysis and dynamic artificial neural network. Expert Syst. Appl. **40**(16), 6266–6282 (2013)
15. Ho, T.K.: Random decision forests. In: 1995 Proceedings of the Third International Conference on Document Analysis and Recognition, vol. 1, pp. 278–282. IEEE (1995)
16. Kohavi, R., et al.: A study of cross-validation and bootstrap for accuracy estimation and model selection. In: Ijcai, Montreal, Canada, vol. 14, pp. 1137–1145 (1995)
17. Kouloumpis, E., Wilson, T., Moore, J.D.: Twitter sentiment analysis: the good the bad and the OMG!. Icwsm **11**(538–541), 164 (2011)
18. Kullback, S., Leibler, R.A.: On information and sufficiency. Ann. Math. Stat. **22**(1), 79–86 (1951)
19. Lee, C.-W., Wang, Y.-S., Hsu, T.-Y., Chen, K.-Y., Lee, H.-Y., Lee, L.-S.: Scalable sentiment for sequence-to-sequence chatbot response with performance analysis. arXiv preprint arXiv:1804.02504 (2018)
20. Lisetti, C.L.: Affective computing. Pattern Anal. Appl. **1**(1), 71–73 (1998)
21. Lovins, J.B.: Development of a stemming algorithm. Mech. Transl. Comput. Linguist. **11**, 22–31 (1968)
22. Lu, B., Ott, M., Cardie, C., Tsou, B.K.: Multi-aspect sentiment analysis with topic models. In: 2011 11th IEEE International Conference on Data Mining Workshops, pp. 81–88. IEEE (2011)
23. McCallum, A., Nigam, K., et al.: A comparison of event models for naive bayes text classification. In: AAAI 1998 Workshop on Learning for Text Categorization, vol. 752, pp. 41–48. Citeseer (1998)
24. McCallum, A.K.: Bow: a toolkit for statistical language modeling, text retrieval, classification and clustering (1996). http://www.cs.cmu.edu/mccallum/bow
25. Nogueiras, A., Moreno, A., Bonafonte, A., Mariño, J.B.: Speech emotion recognition using hidden Markov models. In: Seventh European Conference on Speech Communication and Technology (2001)
26. Platt, J.: Sequential minimal optimization: a fast algorithm for training support vector machines (1998)
27. Prasad, A.M., Iverson, L.R., Liaw, A.: Newer classification and regression tree techniques: bagging and random forests for ecological prediction. Ecosystems **9**(2), 181–199 (2006)

28. Quinlan, J.R.: C4.5: Programs for Machine Learning. Elsevier, Amsterdam (2014)
29. Read, J.: Using emoticons to reduce dependency in machine learning techniques for sentiment classification. In: Proceedings of the ACL Student Research Workshop, pp. 43–48. Association for Computational Linguistics (2005)
30. Rojas, R.: AdaBoost and the super bowl of classifiers a tutorial introduction to adaptive boosting. Technical report, Freie University, Berlin (2009)
31. Rosenblatt, F.: Principles of neurodynamics. Perceptrons and the theory of brain mechanisms. Technical report, Cornell Aeronautical Lab Inc., Buffalo, NY (1961)
32. Schalkoff, R.J.: Artificial Neural Networks, vol. 1. McGraw-Hill, New York (1997)
33. Schmidhuber, J.: Deep learning in neural networks: an overview. Neural Netw. **61**, 85–117 (2015)
34. Valdivia, A., Luzón, M.V., Herrera, F.: Sentiment analysis in TripAdvisor. IEEE Intell. Syst. **32**(4), 72–77 (2017)
35. Vieira, M., Faria, D.R., Nunes, U.: Real-time application for monitoring human daily activity and risk situations in robot-assisted living. In: Robot 2015: Second Iberian Robotics Conference, pp. 449–461. Springer (2016)
36. Zhang, C., Zhang, S.: Association Rule Mining: Models and Algorithms. Springer, Berlin (2002)

Modelling Stable Alluvial River Profiles Using Back Propagation-Based Multilayer Neural Networks

Hossein Bonakdari[1(✉)], Azadeh Gholami[2], and Bahram Gharabaghi[1]

[1] School of Engineering, University of Guelph, Guelph, ON NIG 2W1, Canada
hbonakda@uoguelph.ca

[2] Department of Civil Engineering, Razi University, Kermanshah, Iran

Abstract. Modelling stable alluvial river profile is one of the most important and challenging issues in river engineering that several studies have been dedicated to it. The main objective of this study is to evaluate the back propagation-based multilayer neural network (BP-MLNN) performance in predicting stable alluvial river profile. We used eighty-five observational datasets to train and test, three separate models to predict each of the channel width (w), flow depth (h) and longitudinal slope (s) of stable channels. The network input parameters are the flow discharge (Q), mean sediment size (d_{50}) and affecting Shields parameter (τ^*) and w, h and s parameters are the output. It is concluded from the results that the proposed models to predict the width, depth, and slope of stable channels, with a correlation coefficient (R) of 0.96, 0.886, and 0.870 respectively, perform well. The mean absolute relative error (*MARE*) value of 0.063 related to the width estimation model in comparison with the depth and slope estimation model with *MARE* value of 0.077 and 0.518 shows the superior accuracy of the BP-MLNN model. The presented BP-MLNN models in this study are therefore recommended in river engineering projects to estimate the cross-sectional dimensions of stable alluvial channels as simple and robust design tools.

Keywords: Back propagation · Channel dimension · Regime state · Multilayer perceptron neural network · Stable channel

1 Introduction

Alluvial channels and rivers regarding flow discharge value and sediment that should be transported, regulate the dimensions as do not have sedimentation and pure scour. A river that is reached to a stable state (equal sedimentation and erosion rates) is known as regime state. Regime means dynamic equilibrium that the channel dimension remains constant over time [1]. The regime or equilibrium geometry of alluvial natural channels remains a topic of fundamental scientific and engineering interest in river engineering [2]. So that, the appropriate design of channel dimensions under specific discharge values and sediment load causes of decreasing restoration and maintenance costs and increasing life time of channels. The main hydraulic and geometric

K. Arai et al. (Eds.): CompCom 2019, AISC 997, pp. 607–624, 2019.
https://doi.org/10.1007/978-3-030-22871-2_41

parameters of stable alluvial channels are the width (w), depth (h), velocity (v), and channel slope (s) [3].

There are several methods to study and estimate the stable channel dimensions as (i) empirical methods, (ii) semi-analytical or theoretical methods, (iii) extremal methods, and (iv) rational or mechanical methods [4]. In general, these methods are based on two approaches: (i) the methods that are based on statistical and regression relations and without application of theoretical concepts pf fluid mechanics (empirical methods) and (ii) the methods that used the fundamental principles of fluid mechanics and thermodynamic studies (analytical methods). Some researchers were presented different regime relationships based on regression, and analytical methods on different datasets related to different rivers which they can be used as a good review work about previous traditional methods [3, 5–14].

Some relationships based on analytical and regression methods have been suggested for estimating hydraulic geometry which presents the channel width, depth, and slope based on flow discharge (Q) parameter [6, 15–17]. Most of these relationships are based on flow discharge, and a few have chosen mean sediment size (d_{50}) [3, 5, 8, 18], and very few are based on Q, d_{50} parameters and Shields parameter (τ*) as an independent variable [2, 11, 12, 19]. Significant studies on dimensions of alluvial channels under equilibrium conditions are carried out by several researchers. In 1930, Lacey [20] and in 1953, Leopold and Maddock [21] were the first ones that suggested relationships based on empirical methods to estimate dimensions of stable channels.

In 1980, Chang [22] was the first who estimated channel dimensions using extremal methods based on the resistance flow and low power relations. Cheema et al. [23] using the extremal method defined that an alluvial channel was reached a stable width when the rate of change of unit stream power respectfully to its width is a minimum. After that, Singh [24], Eaton et al. [25] and Nanson and Huang [26] based on extremal methods measured the stable channel dimensions by different approaches. Julien and Wargadalam [19] using semi-analytical methods determined bed-load motion, numerical modeling of bed transition and self-forming modeling in rivers based on different datasets. They referred to the significant influence of the Shields parameter (τ*) on the estimation of depth and slope parameters. Also, they declared that the proposed theoretical model was more accurate than previous empirical methods in estimation of stable channel dimensions.

Using rational methods for the first time was done by Glover and Florey [5] who first applied the tractive force approach in order to estimate stable channel dimensions and shape. They referred to no movement of sediment in anywhere of the channels in stable states. In the following, Vigilar and Diplas [27], Dey [28] proposed a numerical scheme for estimating dimensions and shapes of stable channels by considering lateral momentum induced by turbulence and hence the non-uniform distribution of shear stress.

Kolberg and Howard [29] examined the variability in exponents of hydraulic geometry relations for different rivers. They analyzed active channel geometry and discharge relations using data from alluvial channels. Their analysis showed that the discharge-width exponents were distinguishable, depending on the variations in materials forming the bed and banks of alluvial channels. Afzalimehr et al. [12] using non-linear regression analysis obtained the geometry of the stable channels and declared that the d_{50} and τ^* parameters are not considered as effective variables in the channel width prediction and only depends on the flow discharge.

Generally, regime relationships are obtained based on observed dataset related to special rivers with different hydraulic, geographic and geomorphologic conditions [12]. It is evident that for other rivers or artificial channels with different conditions, these relationships have less precision and yield results with high error values compared to the observed datasets [8].

Artificial intelligence (AI) methods are an alternative means of reducing the inaccuracies of regression-based models and decreasing exhausting calculations of the analytical methods. AI methods have been widely utilized in diverse hydraulic and hydrology engineering sciences, such as: water surface elevation prediction [30], scour depth prediction [31, 32], discharge capacity estimation [33], sediment transport [34], aeration efficiency of the stepped weir [35], open channel bend channel [36, 37] wave prediction on beaches [38], friction factor prediction [39].

Using of AI methods in stable channel geometry prediction is so applicable so that Khadangi et al. [40], Tahershamsi et al. [41] and Bonakdari and Gholami [42] used an artificial neural network (ANN) to model the width, depth, and slope of stable alluvial channels based on input parameters of Q and d_{50}. Their results were compared with previous empirical equations, and the results demonstrated the high ability of AI methods compared to other empirical methods.

Gholami et al. [43] examined the performance of group method of data handling (GMDH) neural network in the estimation of width, depth and slope of stable channels, and also they examined the sensitivity analysis of models respect to three input parameters of Q, d_{50} and τ^*. They referred to the high accuracy of the proposed model in channel dimension prediction. Also, they declared that the input parameter combinations of (Q, d_{50}), (Q, d_{50}, τ^*) and (d_{50} and τ^*) are the influential input combinations in the prediction of width, depth and slope parameters of stable channels, respectively.

Shaghaghi et al. [44] compared the two different optimization methods in combination with GMDH based on genetic algorithm and particle swarm optimization in the prediction of the stable channel width. They referred to the acceptable adaption of GMDH in combination with genetic algorithm results with corresponding observed values. Shaghaghi et al. [45] evaluated the application of GMDH model in comparison with gene expression programming (GEP) in order to estimate width, depth, and slope of stable channels. They declared that both GMDH and GEP models have acceptable accuracy in the prediction of sable channel dimensions.

In the following, the application of M5 Model Tree (M5Tree), Multivariate Adaptive Regression Splines (MARS) and Least Square Support Vector Regression (LSSVR) were assessed by Shaghaghi et al. [46]. Their results declared the high ability of M5Tree model compared to other methods. Also, Gholami et al. [47–49] used GEP and different adaptive neuro-fuzzy inference system (ANFIS) models combined with various optimization and evolutionary methods in the estimation of shape profiles formed on stable channel banks. They declared that the proposed AI methods had able to predict the stable shape profile on channel banks with more accuracy than other previous traditional methods.

The main advantage of the multilayer perceptron (MLP) is its ability to create clustering shapes that are highly nonlinear which are arranged in a feed-forward layered structure [50, 51]. Therefore, the application of the MLP model used in several tasks, e.g. information processing, coding and pattern recognition in phenomena [52–54]. Furthermore, one of this application is reducing the dimension of feature space. For this especial type of operation, auto-association, identity mapping or encoding, the output pattern is similar to the input variable patterns [50]. Hence, the output layer does not contain any nonlinear function, in particular in real input values. These advantages make the MLP as a robust method. Furthermore, in this paper, the main aim is to construct an appropriate relation between input and output values. In the other hands, the MLP model associated with the BP algorithm allow on-line learning. This issue is an important advantage of the BP algorithm notably when the number of training patterns is somewhat large [50].

In this study, the capability of MLP neural network model based on back-propagation (BP) training algorithm (BP-MLNN) to predict the important geometry parameters of width (w), depth (h) and slope (s) in the stable channels designing are evaluated. Also, the notable aim of the present paper is the consideration and evaluation of Shields parameter (τ*) as an effective input parameter in estimating channel dimensions, in addition, to flow discharge and mean sediment size parameters (Q and d_{50}). Hence, the parameters of Q, $d_{50,}$ and τ* are as input parameters, and the parameters of w, h, and s of the stable channels are the output values. Observational data related to Afzalimehr et al. [12] measured in the 85 river cross sections were used to train and test the models. Three individual BP-MLNN neural networks to predict the width, depth and slope of are designed. The various statistical indices are used to determine the model's accuracy.

2 Methodology

2.1 Case Studies

In this study, data from three rivers (Kaj and Beheshtabad river (southwest of Iran), Gamasyab River (West of Iran)) in the 85 cross-section with gravel bed rivers which are located in regime conditions (stability) are collected by Afzalimehr et al. [12] was used as observation data in width, depth and slope measurement. For each of the mentioned 85 cross sections, 3–5 velocity profiles were measured, and in each velocity profile, 13–16 point velocity measurements were taken. Therefore, in total, there were 85, 425 and 6000 cross sections, velocity profiles, and point velocity measurements.

In each reach using the current-meter bar, the flow depth was measured at 0.5-m intervals across each cross section, and then the top width was recorded. As Afzalimehr et al. [12] stated, they divided the water-surface width into the five parts (vertical intervals) to mark the locations of measuring profiles of velocity. A butterfly meter with horizontal axes was used to gauge the velocities at each site across the depth direction from the bed to water surface.

A cylindrical leg (diameter = 15 cm and thickness = 5 cm) was linked to the current meter bar during measurements and in this way, the nearest point to the bed was measured 5 cm above the bed as in 20% of the depth the intervals were 1–2 cm and were 3–5 cm in the upper 80% of the depth. This issue was because of the velocity gradient in the upper 80% depth is smaller than 20% of the depth near the bed. The "50 s" was considered as time step for measuring each point velocity measurement. Surveying was done to measure the longitudinal slope in selected stations at the central river axis.

The bed material size was determined by Wolman's method [55]. A rectangular cross-section was assumed as an average cross-section of the channel with the same width equal to the main cross-section in order to calculate the average flow depth. In this case, by dividing the area of rectangular cross-section (A) into the corresponding flow width, the flow depth (h) was obtained as $h = A/w$ [2, 8, 12].

Observational measurements in the selected river reaches were conducted during different seasons that represented the sections were generally stable [12]. Indeed, despite erosion and sedimentation in each section but the rates of erosion and sedimentation are in equilibrium. In this case, after reaching the stability state, the average dimension (depth, slope and notably width) of reaches don't change in different time periods (one season to another along a year) which this criterion is approved by many researchers [56, 57].

In the study of Afzalimehr et al. [12], the Shields parameter was computed based on boundary layer theory. In this case, the shear velocity (u^*) was calculated by velocity value located on an inner region of boundary layer theory (20% depth) (log-law). Therefore, the Shields parameter was directly calculated without the presence of channel depth and slope (h and s). Certainly, such estimation will remove the risk of any correlation in practice and will prepare a better picture of flow behavior [58]. Table 1 represents the range of used data gathered by Afzalimehr et al. [12].

Table 1. The range of used datasets related to Afzalimehr et al. [12].

Parameters	Data range	Average value
Discharge (m^3/s)	1.25–6.41	3.83
Mean sediment size (m)	0.0039–0.13	0.06695
Shields parameter	0.0001–0.8141	0.4071
Width (m)	5.5–27	16.25
Depth (m)	0.18–0.57	0.375
Slope	0.00001–0.0284	0.014205

2.2 Multilayer Artificial Neural Network Model

The artificial neural network is an idea inspired by biological neural networks and is used to process data [59]. The MATLAB R2011b software was utilized to design an optimum neural network model. The structure of feed forward artificial neural network that is called multilayer perceptron is completed with back-propagation (BP) algorithm [60].

Three separate multilayer neural networks (MLNN) were used in this study: one to predict width, one to predict the depth and the other to predict slope. All cases are multilayer perceptron (MLP) neural networks. Figure 1 provides a general view of these networks, which are organized into three layers, namely input, hidden, and output layers (there may be one or more hidden layers). Each layer is made up of a number of neurons, which are connected to the neurons of the adjacent layer through some weights.

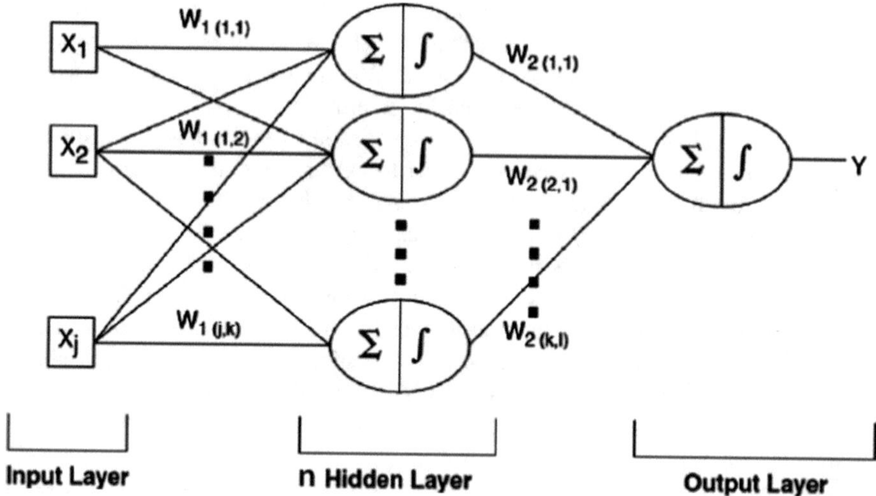

Fig. 1. Schematic diagram of a multilayer perceptron neural network model [61, 66].

This kind of network is known as feed forward network. In Fig. 1, each neuron k receives the input signals from each node j in the previous layer. In this case, there is a weight ($W_{(j,k)}$) associated with each input signal (x_j). In this case, the effective output signal (E_k) for node k, is the summation of weights of all input signals as [61]:

$$E_k = \sum_{j=1}^{N} W_{jk} x_j + W_{0k} \tag{1}$$

In Eq. (1), there is an extra term as W_{0k} which is named *bias*. The activation function for E_k value is applied to induce the ultimate output from the neuron. The network training process entails correcting the connective weights between neurons through the best training method. The activation function can be a linear, discrete or other distribution functions.

However, in order to use back propagation algorithm for training network, the activation function should be continues to function in all spaces. The sigmoid function can fulfill this issue and is generally used in the application of feed-forward neural network [61, 62]. Therefore, the "Sigmoid" activation function is used in this paper. In order to model ANN, the training network requires using pair input-output parameters. Indeed, a neural network trains by modifying bias and weights that are connected with neurons.

Before training starts, the biases and weights randomly are initialized. An especial case is that the initialization is done between ranges of ($-2/a$, $2/a$) which a is the input numbers to neuron. If the initialization was not limited in this range, the network training might be done slowly [63]. After initializing, the network is trained by some samples (the training pairs of input-output related to training datasets). Each training pairs have special input and output values which the network produces the output value based on the input parameter. Therefore, the network is trained repeatedly (specific time numbers) based on training datasets until the corresponding output is produced (or the output value is close to acceptable value). Criteria "stop" training in the present study is considered 100 epoch which occurs when the model converges completely [64–66]. Generally, in each training algorithm, the goal is reducing network error according to Eq. (2):

$$E_p = \frac{1}{2}\sum_k (o_{pk} + y_{pk})^2 \tag{2}$$

where E_p is the linear value of pattern p, o_{pk} the ideal corresponding output and y_{pk} the real output in the kth node (neuron) of the network. The number of neurons within the hidden layer(s) is obtained by trial and error method [67–69].

Therefore, the trial and error method was used, and different models were examined with various numbers of neurons considered within the hidden layer [70]. Then the model that presented the best results was selected as the optimal neural network model. In this study, 4, 10, 10 and 1 neurons were utilized in the input layer, two hidden layers, and the output layer respectively, in all cases. The input variables are Q, d_{50}, and τ^* parameters and the output or target variable are the corresponding width, depth and slope (w, h, and s) of the stable channel. A total of 85 observational data were used to predict width, depth, and slope of the stable channel in the present study. In all cases, the data were divided into two groups: 60 data (70% of whole data) for training the network and 25 data (30% of whole data) for testing.

2.3 Statistical Indexes for Model Evaluation

The results of the BP-MLNN models are explained in this section along with the statistical indices. Three indices of Root Mean Squared Error (*RMSE*), Mean Absolute Error (*MAE*), Mean Absolute Relative Error (*MARE*), Correlation coefficient (R) and average absolute deviation (δ) and determination coefficient (R^2) are tabulated in Table 2 as Eqs. (3–8).

In the equations in Table 2, O_i is the observed parameter, P_i is the parameter predicted by the models, $\overline{O_i}$ and $\overline{P_i}$ is the mean observed and predicted parameters respectively and N is the number of parameters. R and R^2 provides a measure of how well the observed outcomes are replicated by the model. Higher model accuracy leads to *RMSE, MAE, MARE* values closer to zero [71–74]. The index of δ is as a dimensionless parameter to measure the relative error between the predicted and observed values and also can be used as a criterion to evaluate the model performances [75, 76].

Table 2. Different statistical indices for evaluation of BP-MLNN model proposed in this study in order to estimate stable channel dimensions.

Statistical index	Equation	Equation number
RMSE	$RMSE = \sqrt{\frac{1}{N}\sum_{i=1}^{N}(O_i - P_i)^2}$	(3)
MAE	$MAE = \frac{1}{N}\sum_{i=1}^{N}\lvert O_i - P_i\rvert$	(4)
MARE	$MARE = \frac{1}{N}\sum_{i=1}^{N}\left(\frac{\lvert O_i - P_i\rvert}{O_i}\right)$	(5)
δ	$\delta = \frac{\sum_{i=1}^{N}(O_i - P_i)^2}{\sum_{i=1}^{N}O_i} \times 100$	(6)
R	$R = \frac{\sum_{i=1}^{N}(O_i - \overline{O_i}).(P_i - \overline{P_i})}{\sqrt{\sum_{i=1}^{N}(O_i - \overline{O_i})^2\sum_{i=1}^{N}(P_i - \overline{P_i})^2}}$	(7)
R^2	$R^2 = 1 - \frac{\sum_{i=1}^{N}(O_i - P_i)^2}{\sum_{i=1}^{N}(O_i - \overline{O_i})^2}$	(8)

3 Results

Geometry and hydraulic parameters of stable channels (width, depth, and slope) are estimated using BP-MLNN models in the present study. Scatter plots of width, depth and slope prediction by models compared with observational data [12] are showed in Fig. 2 and R^2 values between the BP-MLNN models show in the plots in train and test datasets. The BP-MLNN model is present by ANN model in graphs. *R*, *MARE*, *RMSE*, *MAE*, δ error indexes between ANN models with observed data are shown in Table 3. In Fig. 2, more data compression has been around the exact line and also R^2 values in three models (value close to 1) shows high accuracy of ANN models in prediction of all three parameters (R^2 = 0.9159, 0.8 and 0.9 for predicting width, depth, and slope respectively in train datasets).

The linear equation of the fitted line on both graphs at the top part are shown as "$y = C_1x + C_2$," and C_1 values is close to 1, and the C_2 values close to zero for all three models in both the train and test which present the accuracy and performance of ANN models. Table 3 shows the statistical error indices between predicted variables by ANN model in comparison with the observed data in all train and test datasets. It can be seen from the table that the ANN model in all parameters of width, depth, and slope parameters prediction has high accuracy (the *R* is close to 1 in all three models). In all three models, the error indices have fewer values. However, it seems that the ANN model in *w*, *h* and *s* prediction in training dataset is more accurate than testing datasets. ANN model in width prediction has a less relative error (*MARE*) and greater *R*-value than the other models (*R* = 0.96, *MARE* = 0.063). Because the *RMSE* and *MAE*

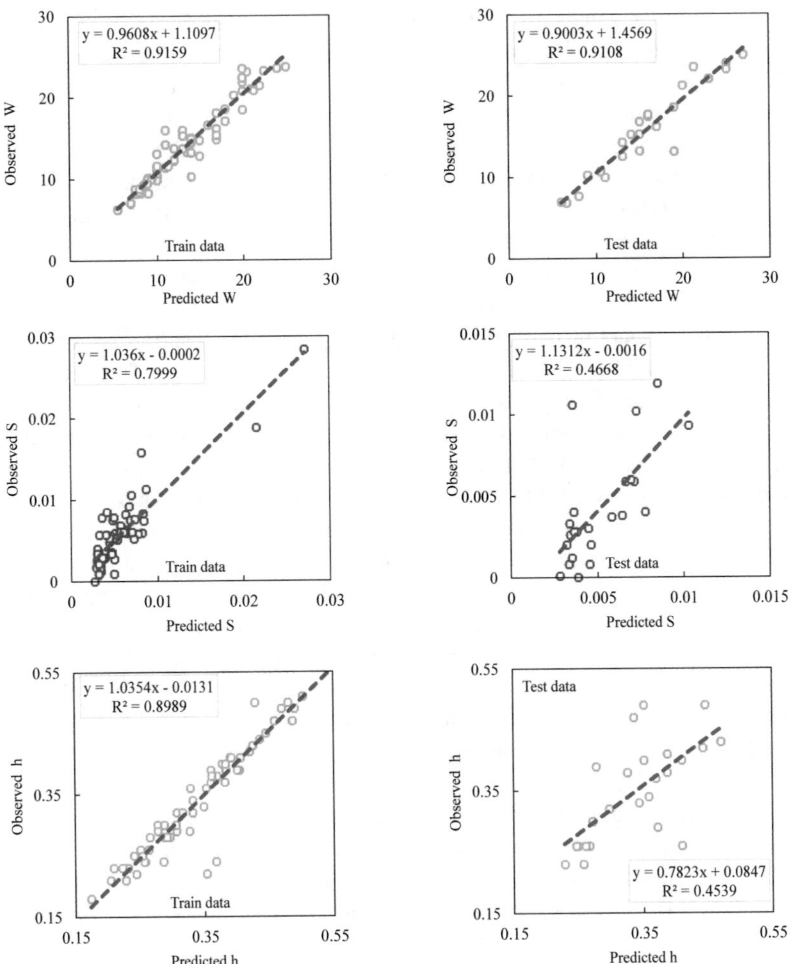

Fig. 2. The scatter plot diagrams to predict the width, depth, and slope by ANN model in comparison with observational data.

Table 3. The different error indexes to compare and evaluate ANN models in prediction width, depth and slope of stable channels for all datasets.

Index	Width prediction	Depth prediction	Slope prediction
R	0.96	0.886	0.870
MARE	0.063	0.077	0.518
RMSE	1.408	0.041	0.0022
MAE	0.364	0.025	0.0017
δ	0.135	0.005	0.00086

absolute error index values are high in width prediction model that in this model, the data value is greater than 1, while the depth and slope values in 85 cross points are less than 1. In other hands, the *RMSE* and *MAE* values are absolute error values, hence they only show the model difference with observed values and cannot detect the model performance with high accuracy (they are not divided by the observed values, and they are a dimensional values similar to dimension of *w*, *h* and *s* parameters). Hence, the great values of *RMSE* and *MAE* in width prediction (1.408 and 0.364 respectively) is not demonstrated the less efficiency of ANN model in prediction, and they are only because of the large value of width parameters unlike depth and slope parameters. The large width value may cause deviation in model accuracy evaluation as well, so in width prediction model the δ value is greater than other models (0.135). On the other hand, the small values of *RMSE* and *MAE* in depth (*h*) (*RMSE* = 0.041 and *MAE* = 0.025) and notable slope prediction (*s*) (*RMSE* = 0.0022 and *MAE* = 0.0017) is because of less values of depth and slope parameters unlike width parameter. So, it can be said that the proposed ANN model in width prediction is more accurate than depth and slope prediction based on two rational R^2 and *MARE* indices. As seen in Fig. 2, the ANN model is more compressed around the exact line in width prediction than depth and slope prediction in both training and testing datasets. Also, according to Fig. 2, it can be seen that the ANN model in slope prediction is more scattered than other parameter predictions in addition to less value of R^2 value in test datasets for slope prediction (R^2= 0.4539). Furthermore, the high value of *MARE* (0.518) for slope prediction according to Table 2 confirms the less efficiency of the ANN model.

Also, Table 4 presents the width, depth and slope values of stable channel predicted by the ANN model in some points of test datasets and corresponding observed values which are compared to each other. In this able, the so small values of slope parameters is removed to better comparisons. In this table, the values of each *w*, *h* and *s* parameter can be seen. Higher width values than depth and slope values are obvious. Furthermore, the dimensionless error index of *MARE* is calculated and gathered for each data and averaged *MARE* value for better evaluations. As seen, the *MARE* values in width prediction is so smaller than the *MARE* values in depth and slope predictions. In width prediction, the *MARE* values are located in ranges of 0–0.17 which is less than this range for depth (0–0.57) and slope predictions (0–1.346). In the other hand, in these points the averaged *MARE* value for width, depth and slope prediction is 0.083, 0.118 and 0.2205 respectively which the less value of *MARE* value in width prediction approves the more efficiency of ANN model than rest parameters. In slope prediction, the high values of *MARE* almost for all points is clearly seen which is represented the high difference of predicted values by ANN model with the observed values.

It can be said that the proposed ANN model in this study is able to detect the width value accurately. Width parameter is one of the main factors in regime rivers [8]. During reaching to stability state, the channels adjusted its width, depth, and longitudinal slope. In this way, in the channel cross-section the bank slope is milder, and the channel width is more than the initial form of a channel. Therefore, the channel widening occurs in the process of stability which is a critical phenomenon [76]. In this paper, the proposed BP-MLNN can carefully estimate this parameter in almost all reaches. Moreover, it can be said that the input parameters of Q, $d_{50,}$ and τ^* is selected as a good input combination so that the Q is the main influential parameter in width

Table 4. Comparison of predicted width, slope and depth values by the ANN model and the corresponding measured values and an *RMSE* value of each data.

Width prediction			Depth prediction			Slope prediction		
EXP	ANN	*MARE*	EXP	ANN	*MARE*	EXP	ANN	*MARE*
16	17.452	0.091	0.4	0.350	0.125	0.0037	0.004	0.213
15	16.80	0.120	0.26	0.249	0.042	0.004	0.006	0.551
15	13.2	0.120	0.26	0.245	0.058	0.0102	0.009	0.093
17	16.26	0.044	0.42	0.442	0.052	0.0059	0.005	0.080
14	15.253	0.090	0.29	0.371	0.279	0.0059	0.005	0.135
13	14.259	0.097	0.39	0.277	0.290	0.0038	0.005	0.247
13	12.563	0.034	0.26	0.266	0.023	0.002	0.002	0.159
19	13.142	0.308	0.41	0.387	0.056	0.006	0.005	0.227
11	10.006	0.090	0.23	0.257	0.117	0.004	0.004	0.022
16	17.704	0.107	0.43	0.469	0.091	0.0033	0.003	0.139
15	15.256	0.017	0.32	0.296	0.075	0.0026	0.002	0.050
6.6	6.883	0.043	0.37	0.369	0.003	0.003	0.003	0.102
17	16.2	0.047	0.26	0.260	0.000	0.002	0.002	0.007
19	18.570	0.023	0.4	0.409	0.024	0.0028	0.003	0.035
8	7.69	0.039	0.34	0.357	0.050	0.0028	0.003	0.033
6	7.013	0.169	0.23	0.228	0.009	0.0008	0.001	0.242
9	10.258	0.140	0.49	0.3507	0.284	0.0012	0.001	0.288
10.5	10.710	0.020	0.47	0.335	0.287	0.0106	0.025	1.346
23	22.10	0.039	0.38	0.387	0.018	–	–	–
25	24	0.040	0.38	0.324	0.147	–	–	–
25	23.25	0.070	0.26	0.409	0.573	–	–	–
27	25	0.074	0.49	0.446	0.090	–	–	–
21.3	23.52	0.104	0.33	0.342	0.036	–	–	–
20	21.256	0.063	0.3	0.2707	0.098	–	–	–
Averaged *MARE* value		0.083	Averaged *MARE* value		0.118	Averaged *MARE* value		0.2205

prediction [2, 12] whereas the effective parameter in depth and slope prediction is the mean sediment size. Also, the Shields parameter has an important role in slope prediction [19]. According to the result in this paper, it can be said that the variation ranges of observed Q values (according to Table 1) as an input parameter is acceptable and enough to predict width values by ANN model. In the depth prediction it seems that the applied variation range of d_{50} parameter accompany by Q rang values are good and can satisfy the ANN model in order to yield acceptable prediction for depth parameter compared to observed values. However, it seems that the considered Shields parameters ranges can be controlled in order to get better modeling by ANN and increase the ANN model sensitivity to this parameter.

In Fig. 3, the width, depth and slope of channel values are compared to experimental values at test datasets until the difference can be seen well. It can be seen that the maximum difference between the ANN model and measured data is related to the slope prediction model. ANN model in the most peak points predict more slope values, and it can be said that the model is overestimating. On the opposite, in the depth prediction models in the peak points, the model predicts lower values. Compliance between data in the width model prediction is seen as well, and the difference is less than the slope and depth prediction models. Furthermore, it can be seen that the ANN

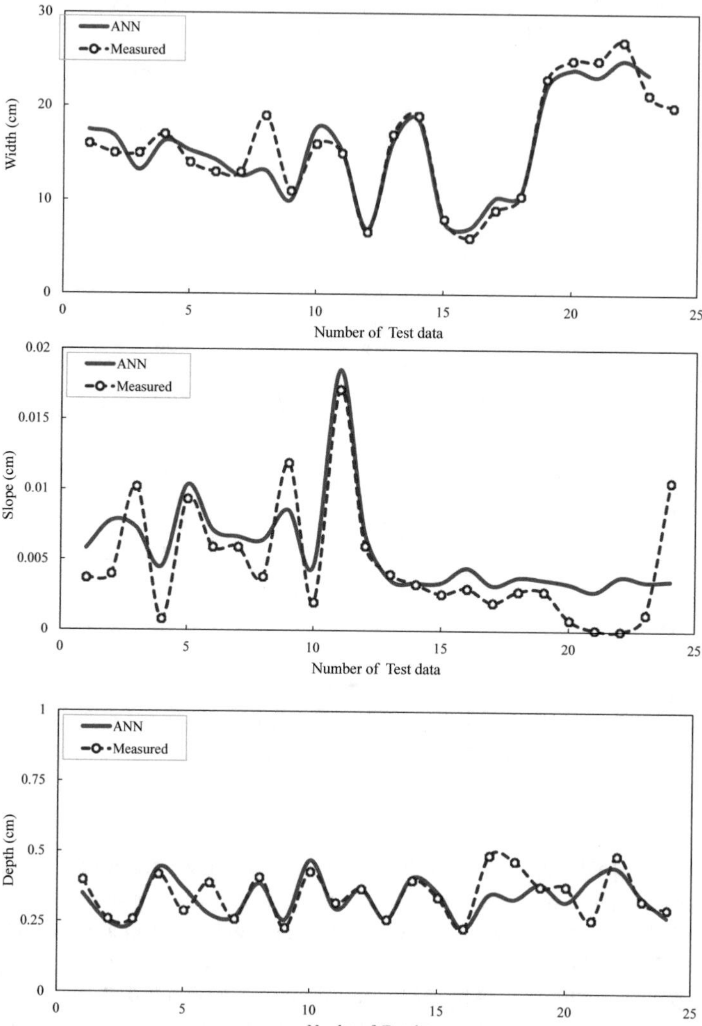

Fig. 3. Comparison of predicted width, slope and depth values by the ANN model and the corresponding measured values in test datasets.

model in the estimation of maximum and minimum width values have high compliance with the corresponding observed values unlike the model performances in depth and slope prediction.

In Fig. 4, the bar graphs of *RMSE, MAE, MARE* error value in two training and testing datasets of both slope and flow depth prediction is drawn (for comparison of this two models). It can be seen from these graphs (*RMSE, MAE*) that depth prediction model in the training datasets performs better than the testing datasets. Furthermore, the slope prediction model in the two training and testing datasets has been performed almost the same.

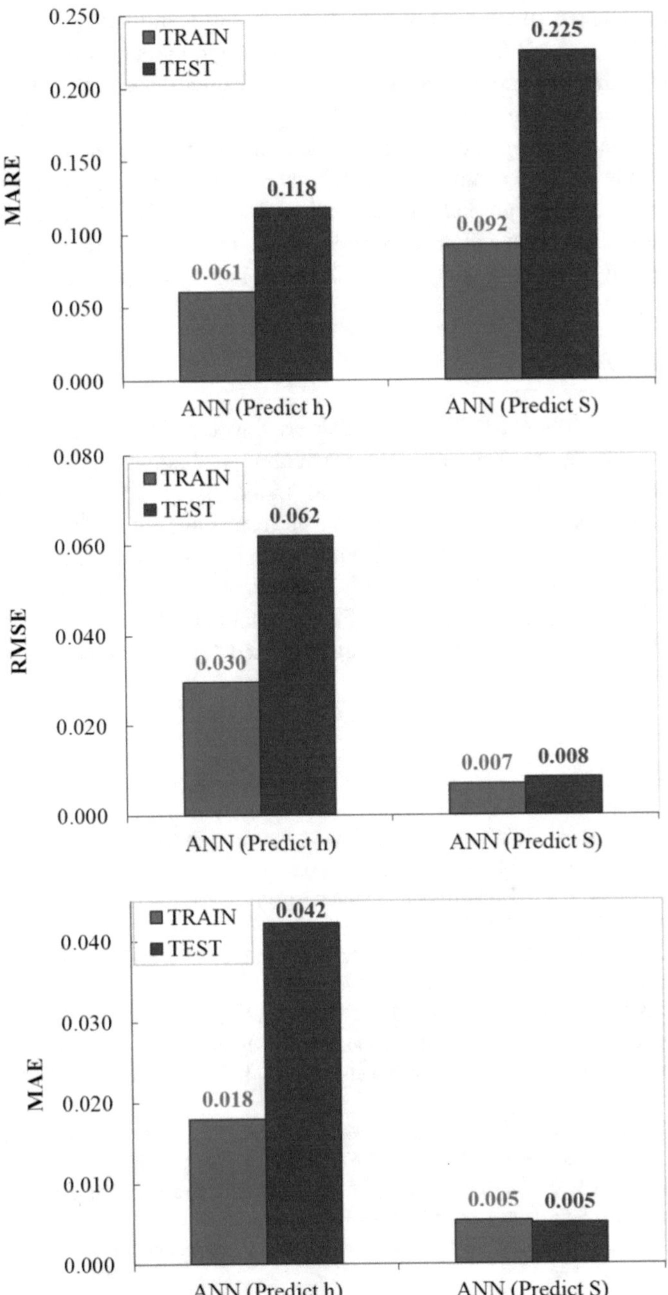

Fig. 4. The error bar graphs for slope and flow depth prediction models in two testing and training datasets for *MARE*, *RMSE* and *MAE* error indices.

Also with comparison of bars can be found that in-depth prediction model, the bars related to *RMSE* and *MAE* values are longer than slope model, thus is shown that the error value in the flow depth prediction model is greater than the slope prediction model. Moreover, the error value in the slope prediction model, especially in the test mode, is less than the depth prediction model. But as it was mentioned, this issue is because of the fewer slope values than depth values which cause to decrease the *RMSE* and *MAE* values in slope prediction. Also, the high values of *MARE* notably in training datasets approve this issue and also according to the mentioned section the ANN model in the prediction of depth model is more accurate than slope prediction.

4 Conclusions

In this study, three robust BP-MLNN models are developed that can accurately predict the width (w), the depth (h) and the slope (s) of alluvial channels in the stable state. The three BP-MLNN models were trained and tested using eighty-five observational datasets related to stable state of channels. The results show that the new BP-MLNN models can accurately predict observed values of w, h, and s of stable channels with R value of 0.96, 0.886 and 0.870 and *MARE* values of 0.063, 0.077 and 0.518 respectively for width, depth and slope of stable channel. Therefore, the proposed BP-MLNN models developed in this study can be used for the design of stable alluvial channel profiles.

References

1. Singh, V.P., Zhang, L.: At-a-station hydraulic geometry relations, 1: theoretical development. Hydrol. Process. **22**, 189–215 (2008)
2. Lee, J.S., Julien, P.Y.: Downstream hydraulic geometry of alluvial channels. J. Hydraul. Eng. **132**(12), 1347–1352 (2006)
3. Parker, G.: Self-formed straight rivers with equilibrium banks and mobile bed. Part 2. The gravel river. J. Fluid Mech. **89**(1), 109–146 (1978)
4. ASCE Task Committee on Hydraul. Bank Mech., and Model. of River Width Adjust.: River width adjustment. I: processes and mechanisms. J. Hydraul. Eng. **124**(9), 881–902 (1998)
5. Glover, R.E., Florey, Q.L.: Stable channels profiles. US Bur Reclam, Hydraul Rep No 325 (1951)
6. Bray, D.L.: Regime equations for gravel bed rivers. In: Hey, R.D., Bathurst, J.C., Thorne, C. R. (eds.) Gravel bed rivers, pp. 242–245. Wiley, New York (1982)
7. Andrews, E.D.: Bed material entrainment and hydraulic geometry of gravel-bed rivers in Colorado. Geol. Soc. Am. Bull. **95**, 371–378 (1984)
8. Hey, R.D., Thorne, C.R.: Stable channels with mobile gravel beds. J. Hydraul. Eng. **112**(8), 671–689 (1986)
9. Farias, H.D., Pilian, M.T., Matter, M.T., Pece, F.J.: Regime width of alluvial channels. ICHE Conference, pp. 1–21. Cottbus (1998)
10. Millar, R.G.: Theoretical regime equations for mobile gravel-bed rivers with stable banks. Geomorphology **64**, 207–220 (2005)
11. Afzalimehr, H., Singh, V.P., Abdolhosseini, M.: Effect of nonuniformity of flow on hydraulic geometry relations. J. Hydrol. Eng. **14**(9), 1028–1034 (2009)

12. Afzalimehr, H., Abdolhosseini, M., Singh, V.P.: Hydraulic geometry relations for stable channel design. J. Hydrol. Eng. **15**(10), 859–864 (2010)
13. Davidson, S.K., Hey, R.D.: Regime equations for natural meandering cobble-and gravel-bed rivers. J. Hydraul. Eng. **137**(9), 894–910 (2011)
14. Mohamed, H.I.: Design of alluvial Egyptian irrigation canals using artificial neural networks method. Ain Shams Eng. J. **4**, 163–171 (2013)
15. Nixon, M.: A study of bankfull discharges of the rivers in England and Wales. Proc. Inst. Civil Eng. **12**(2), 157–174 (1959)
16. Simons, D.B., Albertson, M.L.: Uniform water conveyance channels in alluvial material. Transactions-ASCE **128**(3399), 65–167 (1963)
17. Langbein, W.B.: Geometry of river channels. J. Hydraul. Div. ASCE **90**(HY2), 301–311 (1964)
18. Neill, C.R., Yaremko, E.K.: Regime aspects of flood control channelization. In White, W.R. (ed.), International Conference on River Regime. Hydraulics Research Ltd., Wallingford, England, and John Wiley, New York (1988)
19. Julien, P.Y., Wargadalam, J.: Alluvial channel geometry: theory and applications. J. Hydraul. Eng. **121**, 312–325 (1995)
20. Lacey, G.: Stable channels in alluvium. Minutes of the Proceedings of the Institution of Civil Engineers, Thomas Telford-ICE Virtual Library **229**, 259–292 (1930)
21. Leopold, L.B., Maddock, T.: The hydraulic geometry of stream channels and some physiographic implications. US Government Printing Office 252 (1953)
22. Chang, H.H.: Geometry of gravel streams. J. Hydraul. Div. **106**, 1443–1456 (1980)
23. Cheema, M.N., Marifio, M.A., DeVries, J.J.: Stable width of an alluvial channel. J. Irrig. Drain. Eng. **123**(1), 55–61 (1997)
24. Singh, V.P.: On the theories of hydraulic geometry. Int. J. Sedim. Res. **18**(3), 196–218 (2003)
25. Eaton, B.C., Church, M., Millar, R.G.: Rational regime model of alluvial channel morphology and response. Earth Surf. Proc. Land. **29**(4), 511–529 (2004)
26. Nanson, G.C., Huang, H.Q.: Least action principle, equilibrium states, iterative adjustment and the stability of alluvial channels. Earth Surf. Proc. Land. **33**, 923–942 (2008)
27. Vigilar, G., Diplas, P.: Stable channels with mobile bed: model verification and graphical solution. J. Hydraul. Eng. ASCE **124**(11), 1097–1108 (1998)
28. Dey, S.: Bank profile of threshold channels: a simplified approach. J. Irrig. Drain. Eng. ASCE **127**(3), 184–187 (2001)
29. Kolberg, F.J., Howard, A.D.: Active channel geometry and discharge relations of U.S. Piedmont and midwestern streams: the variable exponent model revisited. Water Resour. Res. **31**(9), 2353–2365 (1995)
30. Adib, A., Banetamem, A., Navaseri, A.: Comparison between results of different methods of determination of water surface elevation in tidal rivers and determination of the best method. Int. J. Integr. Eng. **9**(1) (2017)
31. Firat, M.: Scour depth prediction at bridge piers by Anfis approach. Proc. Inst. Civil Eng. Water Manag. **162**(4), 279–288 (2009)
32. Najafzadeh, M., Saberi-Movahed, F., Sarkamaryan, S.: NF-GMDH-Based self-organized systems to predict bridge pier scour depth under debris flow effects. Mar. Resour. Geotech. **36**(5), 589–602 (2018)
33. Kisi, O., Bilhan, O., Emiroglu, M.E.: Anfis to estimate discharge capacity of rectangular side weir. Proc. Inst. Civil Eng. Water Manag. **166**(9), 479–487 (2013)
34. Gharabaghi, B., Bonakdari, H., Ebtehaj, I.: Hybrid evolutionary algorithm based on PSOGA for ANFIS designing in prediction of no-deposition bed load sediment transport in sewer pipe. In: Science and Information Conference, pp. 106–118. Springer, Cham (2018)

35. Sattar, A.A., Elhakeem, M., Rezaie-Balf, M., Gharabaghi, B., Bonakdari, H.: Artificial intelligence models for prediction of the aeration efficiency of the stepped weir. Flow Meas. Instrum. **65**, 78–89 (2018)
36. Gholami, A., Bonakdari, H., Zaji, A.H., Akhtari, A.A., Khodashenas, S.R.: Predicting the velocity field in a 90° open channel bend using a gene expression programming model. Flow Meas. Instrum. **46**, 189–192 (2015)
37. Gholami, A., Bonakdari, H., Ebtehaj, I., Akhtari, A.A.: Design of an adaptive neuro-fuzzy computing technique for predicting flow variables in a 90° sharp bend. J. Hydro. Inform. **19** (4), 572–585. jh2017200 (2017)
38. Power, H.E., Gharabaghi, B., Bonakdari, H., Robertson, B., Atkinson, A.L., Baldock, T.E.: Prediction of wave runup on beaches using gene-expression programming and empirical relationships. Coast. Eng. **144**, 47–61 (2018)
39. Milukow, H.A., Binns, A.D., Adamowski, J., Bonakdari, H., Gharabaghi, B.: Estimation of the Darcy-Weisbach friction factor for ungauged streams using gene expression programming and extreme learning machines. J. Hydrol. **568**, 311–321 (2018)
40. Khadangi, E., Madvar, H.R., Kiani, H.: Application of artificial neural networks in establishing regime channel relationships. In: 2nd International Conference on Computer, Control and Communication, pp. 1–6. IEEE (2009)
41. Tahershamsi, A., Majdzade, M.R.T., Shirkhani, R.: An evaluation model of artificial neural network to predict stable width in gravel bed rivers. Int. J. Environ. Sci. Technol. **9**(2), 333–342 (2012)
42. Bonakdari, H., Gholami, A.: Evaluation of artificial neural network model and statistical analysis relationships to predict the stable channel width. In: River Flow 2016, vol. 417, Iowa City, USA (2016)
43. Gholami, A., Bonakdari, H., Ebtehaj, I., Shaghaghi, S., Khoshbin, F.: Developing an expert group method of data handling system for predicting the geometry of a stable channel with a gravel bed. Earth Surf. Proc. Land. **42**(10), 1460–1471 (2017)
44. Shaghaghi, S., Bonakdari, H., Gholami, A., Ebtehaj, I., Zeinolabedini, M.: Comparative analysis of GMDH neural network based on genetic algorithm and particle swarm optimization in stable channel design. Appl. Math. Comput. **313**, 271–286 (2017)
45. Shaghaghi, S., Bonakdari, H., Gholami, A., Kisi, O., Shiri, J., Binns, A.D., Gharabaghi, B.: Stable alluvial channel design using evolutionary neural networks. J. Hydrol. **566**, 770–782 (2018)
46. Shaghaghi, S., Bonakdari, H., Gholami, A., Kisi, O., Binns, A.D., Gharabaghi, B.: Predicting the geometry of regime rivers using M5 model tree, multivariate adaptive regression splines and least square support vector regression methods. Int. J. River Basin Manag. 1–67 (2018)
47. Gholami, A., Bonakdari, H., Zeynoddin, M., Ebtehaj, I., Gharabaghi, B., Khodashenas, S.R.: Reliable method of determining stable threshold channel shape using experimental and gene expression programming techniques. Neural Comput. Appl. 1–19 (2018)
48. Gholami, A., Bonakdari, H., Ebtehaj, I., Mohammadian, M., Gharabaghi, B., Khodashenas, S.R.: Uncertainty analysis of intelligent model of hybrid genetic algorithm and particle swarm optimization with ANFIS to predict threshold bank profile shape based on digital laser approach sensing. Measurement **121**, 294–303 (2018)
49. Gholami, A., Bonakdari, H., Ebtehaj, I., Gharabaghi, B., Khodashenas, S.R., Talesh, S.H.A., Jamali, A.: A methodological approach of predicting threshold channel bank profile by multi-objective evolutionary optimization of ANFIS. Eng. Geol. **239**, 298–309 (2018)
50. Bourlard, H., Kamp, Y.: Auto-association by multilayer perceptrons and singular value decomposition. Biol. Cybern. **59**(4–5), 291–294 (1988)

51. Mazroua, A.A., Salama, M.M.A., Bartnikas, R.: PD pattern recognition with neural networks using the multilayer perceptron technique. IEEE Trans. Electr. Insul. **28**(6), 1082–1089 (1993)
52. Chen, X.Y., Chau, K.W., Wang, W.C.: A novel hybrid neural network based on continuity equation and fuzzy pattern-recognition for downstream daily river discharge forecasting. J. Hydroinform. **17**(5), 733–744 (2015)
53. Wen, W., Wu, C., Wang, Y., Chen, Y., Li, H.: Learning structured sparsity in deep neural networks. In Advances in Neural Information Processing Systems 2074–2082 (2016)
54. Gholami, A., Bonakdari, H., Zaji, A.H., Ajeel Fenjan, S., Akhtari, A.A.: Design of modified structure multi-layer perceptron networks based on decision trees for the prediction of flow parameters in 90° open-channel bends. Eng. Appl. Comput. Fluid Mech. **10**(1), 194–209 (2016)
55. Wolman, M.G.: A method of sampling coarse river bed material. EOS, Transactions American Geophysical Union **35**(6), 951–956 (1954)
56. Mikhailova, N.A., Shevchenko, O.B., Selyametov, M.M.: Laboratory of Investigation of the formation of stable channels. Hydro Tech. Constr. **14**, 714–722 (1980)
57. Abdelhaleem, F.S., Amin, A.M., Ibraheem, M.M.: Updated regime equations for alluvial Egyptian canals. Alex. Eng. J. **55**(1), 505–512 (2016)
58. Afzalimehr, H., Anctil, F.: Accelerating shear velocity in gravel-bed channels. Hydrol. Sci. J. **7**, 37–44 (2000)
59. Menhaj, M.B., Ray, S.: Neuro-based adaptive controller for longitudinal flight control. In Intelligent Control. 2003, IEEE International Symposium on, pp. 158–163. IEEE (2003)
60. Mehrabian, A.R., Menhaj, M.B.: A real-time neuro-adaptive controller with guaranteed stability. Appl. Soft Comput. **8**(1), 530–542 (2008)
61. Dawson, C.W., Wilby, R.: An artificial neural network approach to rainfall-runoff modelling. Hydrol. Sci. J. **43**(1), 47–66 (1998)
62. Ebert, T., Bänfer, O., Nelles, O.: Multilayer perceptron network with modified sigmoid activation functions. In International Conference on Artificial Intelligence and Computational Intelligence, pp. 414–421. Springer, Berlin, Heidelberg (2010)
63. Gallant, S.I.: Neural Network Learning and Expert Systems. Cambridge, MIT Press, (1993)
64. Yuhong, Z., Wenxin, H.: Application of artificial neural network to predict the friction factor of open channel flow. Commun. Nonlinear Sci. Numer. Simul. **14**(5), 2373–2378 (2009)
65. Bilhan, O., Emiroglu, M.E., Kisi, O.: Application of two different neural network techniques to lateral outflow over rectangular side weirs located on a straight channel. Adv. Eng. Softw. **41**(6), 831–837 (2010)
66. Fenjan, S.A., Bonakdari, H., Gholami, A., Akhtari, A.A.: Flow variables prediction using experimental, computational fluid dynamic and artificial neural network models in a sharp bend. Int. J. Eng. Trans. A: Basics **29**(1), 14–22 (2016)
67. Kisi, O.: Multi-layer perceptrons with Levenberg-Marquardt training algorithm for suspended sediment concentration prediction and estimation. Hydrol. Sci. J. **49**(6), 1025–1040 (2008)
68. Gholami, A., Bonakdari, H., Zaji, A.H., Akhtari, A.A.: Simulation of open channel bend characteristics using computational fluid dynamics and artificial neural networks. Eng. Appl. Comp. Fluid Mech. **9**(1), 355–361 (2015)
69. Kalteh, A.M.: Rainfall-runoff modelling using artificial neural networks (ANNs): modelling and understanding. Casp. J. Environ. Sci. **6**(1), 53–58 (2008)
70. Kalteh, A.M., Hjorth, P.: Monthly runoff forecasting by means of artificial neural networks (ANNs). Desert **13**(2), 181–191 (2008)
71. Chiteka, K., Enweremadu, C.C.: Prediction of global horizontal solar irradiance in Zimbabwe using artificial neural networks. J. Clean. Prod. **135**, 701–711 (2016)

72. Almonacid, F., Fernandez, E.F., Mellit, A., Kalogirou, S.: Review of techniques based on artificial neural networks for the electrical characterization of concentrator photovoltaic technology. Renew. Sustain. Energy Rev. **75**, 938–953 (2017)
73. Gholami, A., Bonakdari, H., Zaji, A.H., Fenjan, S.A., Akhtari, A.A.: New radial basis function network method based on decision trees to predict flow variables in a curved channel. Neural Comput. Appl. **30**(9), 2771–2785 (2018)
74. Afram, A., Janabi-Sharifi, F., Fung, A.S., Raahemifar, K.: Artificial neural network (ANN) based model predictive control (MPC) and optimization of HVAC systems: A state of the art review and case study of a residential HVAC system. Energy Build. **141**, 96–113 (2017)
75. Gholami, A., Bonakdari, H., Zaji, A.H., Michelson, D.G., Akhtari, A.A.: Improving the performance of multi-layer perceptron and radial basis function models with a decision tree model to predict flow variables in a sharp 90° bend. Appl. Soft Comput. **48**, 563–583 (2016)
76. Pizzuto, J.E.: Numerical simulation of gravel river widening. Water Resour. Res. **26**, 1971–1980 (1990)

Towards Adaptive Learning Systems Based on Fuzzy-Logic

Soukaina Ennouamani(✉) and Zouhir Mahani

Ibn Zohr University, Agadir, Morocco
soukaina.ennouamani@edu.uiz.ac.ma, z.mahani@uiz.ac.ma

Abstract. E-learning systems have the ability to facilitate the interaction between learners and teachers without being limited by temporal and/or spatial constraints. However, the high number of students at universities, the huge number of available learning in the web, the differences between learners in term of characteristics and needs make the traditional e-learning systems more limited. For this purpose, adaptive learning has been recently explored in order to cope with these limitations and to meet the individual needs of learner. In this context, many artificial intelligence methods and approaches have been integrated in such computer-based systems in order to create effective learner models, structured domain models, adaptive learning paths, personalized learning format, etc. Such methods are highly recommended for designing adaptive e-learning and m-learning systems with good quality. In this paper, we focus only on one of these methods, called fuzzy logic, which is widely used in educational area. We present the integration of fuzzy logic as a valuable approach that has the ability to deal with the high level of uncertainties and imprecision related to learners' characteristics and learning contexts.

Keywords: E-learning · Adaptive learning · Personalized education · Artificial intelligence · Fuzzy logic · Fuzzy inference systems

1 Introduction

In recent years, computer-based learning is considered as one of the research areas that have attracted the attention of many researchers, and particularly e-learning that has had an impressive revolution in the field of smart learning. In this context and based on the use of Information and Communication Technologies (ICT), e-learning environments have the potential to be integrated in heterogeneous groups of learners in order to enhance their experiences and to provide them with a myriad of learning resources. In fact, smart learning platforms support the flexible teaching-learning process through mitigating temporal and spatial constraints [1].

In this context, teachers who used to know everything become teachers who must be continuously learning and reflective on their knowledge and skills [2]. It becomes more and more difficult to satisfy every learner's needs, and to achieve the desired quality of teaching and learning which is continuously changing over time. This is because of instructors who can not adjust their teaching strategies for every single learner, especially at universities where the number of students is high. In addition to

K. Arai et al. (Eds.): CompCom 2019, AISC 997, pp. 625–640, 2019.
https://doi.org/10.1007/978-3-030-22871-2_42

this, the total number of remote learners makes the diagnosing of their interests and the determination of the best instructional actions, a difficult task for instructors [1].

In addition, e-learning is challenged by a considerable number of limitations related to the technical aspects, viz. crowded learning platforms, various user profiles, huge number of learning resources, etc. These limitations have encouraged researches in the field of smart learning to move forward to adaptive e-learning, which is an appropriate approach for heterogeneous learning groups [3]. Therefore, the integration of new methods and tools becomes an obligation in order to map with the new learning requirements. Artificial intelligence based methodologies are one of the considerable techniques that have been used to design intelligent educational systems [4].

Various artificial intelligence approaches (Bayesian Network, Machine Learning, Heuristic Methods, Ontology Construction, Fuzzy Logic) have been integrated to solve an important number of problems related to e-learning. Each solution has been designed regarding to the objective of each proposal, such as recommendation, content adaptation, format personalization, learning path generation, learning collaboration, etc. In this paper, we introduce the use of fuzzy logic in e-learning as a significant method to create practical learner models and/or to design learning device parameters in order to develop adaptive e-learning systems.

In order to help researchers to understand the artificial-intelligence-based learning systems, this paper focuses only on fuzzy logic and describes its integration trends into some adaptive educational systems. Furthermore, we aim to analyze the importance of this integration to enhance the effectiveness of different adaptive e-learning solutions, and to review up to date application developments of these solutions. To achieve this objective, this paper is structured as follows: The second section draws on the concept and the definition of smart learning as well as the adaptation in learning systems. The third section presents the fuzzy logic by introducing Zadeh's fuzzy theory, fuzzy If-Then rules as well as fuzzy inference systems. The next section discusses the fuzzy-based adaptive learning systems by reviewing how the some previous related works have integrated this logic to design adaptive e-learning systems. After all, we conclude our paper and we present our future work and perspectives.

2 Adaptation in Smart Learning Environments

2.1 Smart Learning

Authors in [5], claimed that "there is no clear and unified definition of smart learning so far". However, and in spite of the difficulty of forming a definition of smart learning that has a high usage in our daily life, many researches attempt to define it. In [6], authors reported that smart learning environments not only employ digital technologies in supporting learning, education as well as training but also provide a significant design of the future learning environments. In fact, the word "smart" is now continuously used in the field of educational research in order to create new concepts, say, Smart Education, Smart University, Smart Learning, Smart Classroom, and Smart Learning Environment [7].

This drives us to mention the meaningful definition suggested by [8] who stated that: "the essence of smarter education is to create intelligent environments by using smart technologies, so that smart pedagogies can be facilitated to provide personalized learning services and enable learners to develop talents of wisdom that have better value orientation, higher thinking quality, and stronger conduct ability."

One of the important instances of smart learning is e-learning that utilizes computer, internet and network technologies as an integral part to facilitate learning anytime and anywhere. In this perspective, authors in [9] indicated that internet has become a central core of the educative environment experienced by learners, and that explains the quick adoption of e-learning all over the world. Furthermore, e-learning becomes a prominent alternative to traditional learning paradigms and the predominant method of delivering teaching materials to a learner supported by an instructor with the aim of being evaluated [6, 10].

"E-Learning is an umbrella term that describes learning done at a computer, usually connected to a network, giving us the opportunity to learn almost anytime and anywhere" [11]. It is known that e-learning takes the advantages of computer networks to provide leaning resources at/outside classrooms. It also operates internet technologies to insure the storage as well as the information sharing. Therefore, e-learning can be considered as a rich experience that generates new skills, understanding and knowledge for any teacher, instructor, trainer, or manager hoping to offer a suitable learning for their learners.

On the other hand, and to be accurate, we have observed a fast transition in learning methodologies. It has been started with d-learning (distanced learning), then the progress of Information and Communication Technologies has led to e-learning (electronic learning) and recently, m-learning (mobile learning) has become the latest progress in this area. M-learning is usually associated to the use of handheld devices, namely, mobile phones, game consoles, laptops, and tablet computers in order to perform training and teaching activities in a dynamic learning environment [12].

Mobile technologies are becoming more ubiquitous, pervasive and networked, with important capabilities for context awareness and internet connectivity [13]. As a result, learning has jumped from classrooms environment to real and virtual learning environments, and has become more personal, adapted and collaborative. Thus, the integration of mobile learning is seen, by the majority of researches, as an important alternative to enhance learner's interest and motivation [14].

Authors in [15] and [16] indicated that m-learning cross multiple contexts including space and time borders, and involve social and effective interactions through using personal electronic devices. These devices enable the mobile users to take the advantage of virtual learning activities in order to improve their learning abilities without being tethered to a specific location, say, university environment [17].

Other researchers [18, 19] pointed out that m-learning has the strength of ubiquity, availability as well as technical functions such as geospatial technologies, social networking, sensors detection, multimedia tools, visual identification and audio recorder. These features have promoted e-learning to m-learning as a wireless learning that combines learning resources from the current real environment with the numerical one.

The above definitions of e-learning and m-learning have been frequently cited by the majority of academic researches. This allow us to describe how ICT can change the traditional paradigms of teaching and learning, and consequently to discover how to use these technologies to transform learning from a limited academic activity into a flexible and enjoyable daily experience.

2.2 Adaptation in Learning Systems

At the turn of the century, the considerable growth in the field of computer-based learning including e-learning and m-learning have performed further concepts, viz. adaptive/personalized learning systems, recommendation in learning systems, collaborative learning systems and so on. The main core of this progress is the integration of artificial intelligence techniques to obtain the desirable solution regarding the users' needs, objectives and motivations. In this paper, we pay a considerable attention to adaptive and personalized learning systems only.

Researches in adaptive systems can be traced back to the early 1990s. At that period of time, teachers started to make considerable efforts to evaluate the ability of their students, including their knowledge level, skills, background, learning styles, interests, etc. The principal objective was to take these characteristics into consideration in order to provide courses in a way that fits every student's needs. This adjustment can be done inside classrooms with smaller number of participants [20] or for one-to-one teaching [21]. However, and since the majority of traditional classrooms consist of a huge number of students, teachers definitely can not consider each learner's needs to afford the tailored learning.

In order to solve the above problems, and since the ultimate goal of any smart learning system is to maximize the student's quality of learning, several researchers have begun to study the different possibilities to adapt the educational systems and to facilitate the learning process in an individual way. In this context, learner modeling techniques that have been introduced in Intelligent Tutoring Systems (ITS) are also integrated into the learning applications aiming to be adaptive and personalized [22].

As indicated in [23], adaptive e-learning systems have the power to provide students with personalized online learning materials and services. Adaptive educational systems attempt to enhance student learning experiences by modeling the ideal learning environment based on their specific needs [24]. Therefore, this kind of systems are getting more attention due to their potential of making learning available, more flexible and in a dynamic design.

Generally, the target of any adaptive learning system is to personalize the different learning approaches in order to achieve the students' satisfaction, and to facilitate the learning process [25]. In other words, learners must be modelled in a way to assimilate not only their needs but also their individual characteristics that can be classified in different categories. Thus, and through this kind of software, many connections and links are created between learner characteristics and available learning resources.

As mentioned in [10], the development of any adaptive learning system involves three components that dynamically interact with each other, namely, learner model, domain model and adaptation model (Fig. 1). The learner model is the core of these systems that provides a structured presentation of the learner characteristics [26]. The domain model is a structured design of the existing knowledge based on an abstract representation [27]. Finally, the adaptation model is the bridge between the learner and domain models by combining the learners' needs and characteristics with the relevant learning materials [28].

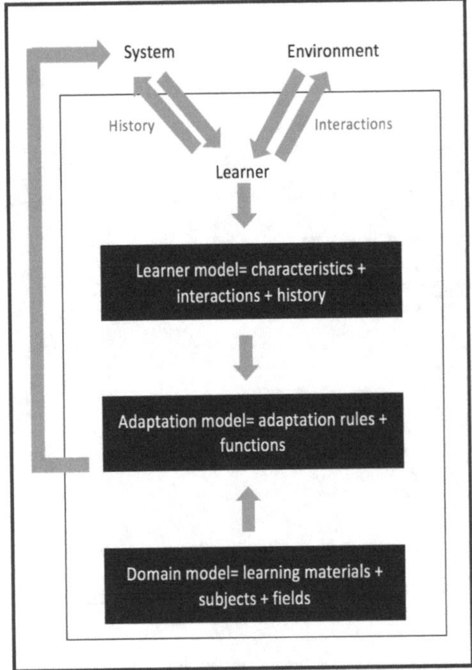

Fig. 1. Adaptive e-learning system components [10]

According to [22] and [29], an adaptive system must be able to manage a structured learning navigation in order to generate suitable learning resources, to control each learner activity, to estimate learner's behavior and interactions, to deduce user's needs and preferences and to create better links between learner model and domain model. In fact, every adaptive e-learning or m-learning system has a specific target, namely, adapting learning path, contents, support/instruction, and presentation. The target can not be reached without having a basic source of adaptation that can be classified as learner characteristics and learner interactions. A predetermined pathway that can be followed to achieve that target is also required in such educational systems (Fig. 2). These three elements are considered to determine what information about the learner should be collected, and how it will be used in order to provide the desired adaptation [30, 31].

In addition, m-learning systems can be considered more adaptive than e-learning systems. This is due to the available sensor detectors in mobile devices that allow to provide more information about the learner's context. In other words, mobile sensors can determine the learner's motions via the accelerometer, the distance between the learner and the handled device using the proximity sensor, the environmental noise through the microphone, etc. Consequently, the user's context can be predicted with more accuracy allowing the system to build a conception about the appropriate presentation of the learning content.

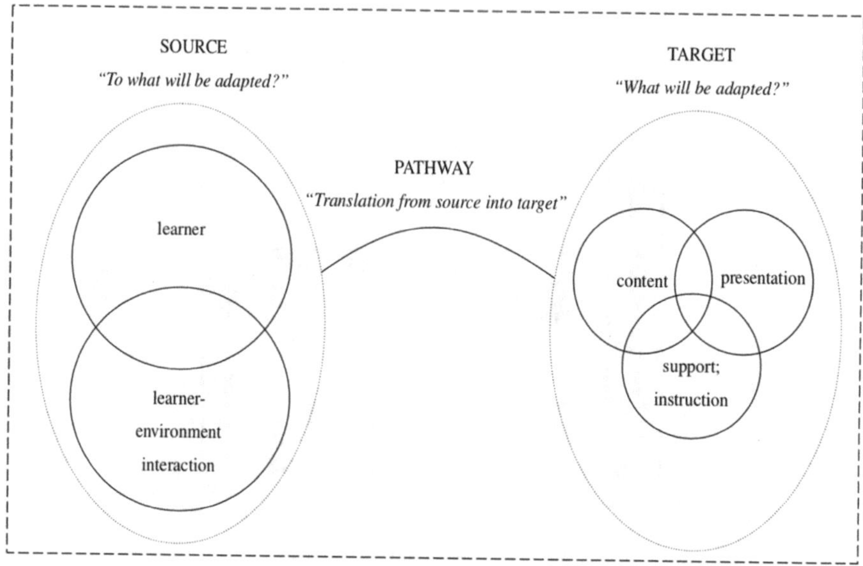

Fig. 2. The tripartite structure of adaptive learning systems [38]

3 Fuzzy Logic Theory and Fuzzy Inference Systems (FIS)

3.1 Fuzzy Logic Theory

Obviously, our brains have the capacity to deal with the facts which have a lack of exactness. Thus, we can answer a question about how much a car is near to another, using the expressions which are open to more than one interpretation, for example, very near, not very near, far away, quite away, etc. In contrast, machines in general and computers in particular can not give such precision, because it is based on binary set (0,1) membership. Therefore, and based on this issue, fuzzy logic was introduced by Zadeh [32] as a further theory to the classical set theory. The fuzzy set theory was developed to generalize the classical crisp sets, and to answer such uncertain questions using the concept of partial membership.

In literature, it has been proved that the fuzzy set theory offers a considerable rang of methods to clarify the ambiguous and uncertain situations. It is a convenient solution to manage the precision of information as well as the fuzziness of different states [33]. Fuzzy sets deal with vague terminologies, viz. young, rich, tall, and others. It represents a transition from the rigid membership of a class of objects inside the crisp set (1 if it belongs to the set, and 0 otherwise), to the fuzzy set that gives the opportunity to have a normalized partial grade of membership (a value between 0 and 1) [34]. Figure 3 represents an example of the degree of membership of three fuzzy sets related to the determination of the age.

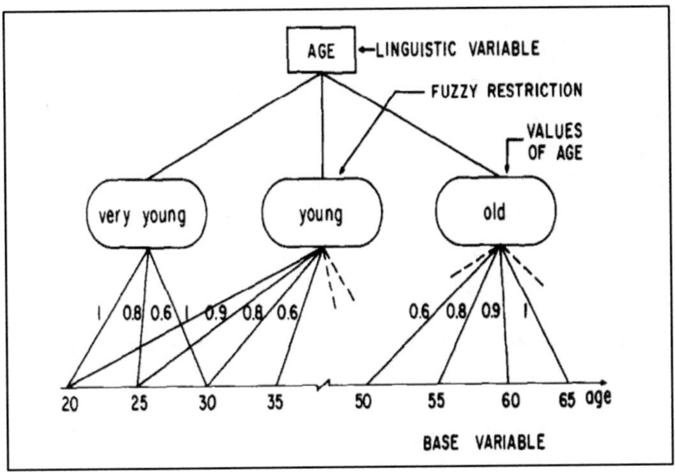

Fig. 3. Hierarchical structure of a linguistic variable [39]

Zadeh introduced fuzzy set theory to solve our daily problems of uncertainty caused by incomplete data as well as human subjectivity [35]. It can be considered as the most compatible method with the human-being decision-making process [36], using linguistic variables as well as the degrees of membership which symbolize the grade of stability of an item inside a fuzzy set [37]. A fuzzy set is defined as an ordered set (x, μA(x)), where x belongs to X and $\mu_A(x)$ belongs to [0, 1], equipped with a membership function $\mu_A(x) : X \rightarrow [0, 1]$, where:

$$\mu_A(x) = \begin{cases} 1 & , x\, belongs\, to\, A \\ [0,1] & , x\, is\, partially\, in\, A \\ 0 & , x\, does\, not\, belongs\, to\, A \end{cases} \quad (1)$$

3.2 Fuzzy If-Then Rules and Fuzzy Inference Systems (FIS)

– *Fuzzy If-Then rules:*

Zadeh's theorem has become more useful when Mamdani [40] employed it in a practical application to control an automatic steam engine. Mamdani's method is a relational model known as a linguistic method because the premise and the outcome are computed with words. It is a manually developed method that uses defined rules in order to generate the adequate results via a fuzzy membership function [41].

Mamdani's fuzzy If-Then rules are the expressions of the form "IF A THEN B", where A and B are labels of fuzzy sets [32] associated with convenient membership functions. Considering the flowing example:

If the speed is high, then apply the brake a little

In this example, "speed" is a linguistic variable, and "brake" is also a linguistic term characterized by a membership function. This example represents a reasoning situation in an environment of imprecision guided by the human ability of making decisions.

On the other hand, Takagi and Sugeno [42] introduced another form of if-then rules that follow a data driven model. In contrast to Mamdani's rules, this method is based on historical data computed with weighted average [41]. Another example suggested by [43] describes the resistant force on a moving object:

$$\textit{If velocity is high, then force} = k * (\textit{velocity})^2$$

Where, "high" is a linguistic proposition described by a membership function. After then, the inferred conclusion is described by a non-fuzzy equation of the input (velocity).

The above types of fuzzy if-then rules have been employed in a myriad number of application domains. They represent an essential block in the Fuzzy Inference Systems (FISs) that provides a description of the system's workflow. FIS are explained in the following paragraphs.

There are many application domains of fuzzy logic and fuzzy sets, namely, Fuzzy Inference Systems(FIS), Fuzzy Decision Trees (FDT) and so on. In this paper, we focus on FIS because it has been integrated in the majority of adaptive computer-based educational systems that we reviewed in this paper (Sect. 4).

– *Fuzzy Inference Systems (FIS):*

FIS or fuzzy-rule-based systems introduced by Mamdani consist of four blocks, namely, Fuzzifier, Fuzzy Rule Base, Inference Engine, and Defuzzifier (Fig. 4). These blocks are also known as the basic stages of any FIS, described as follows:

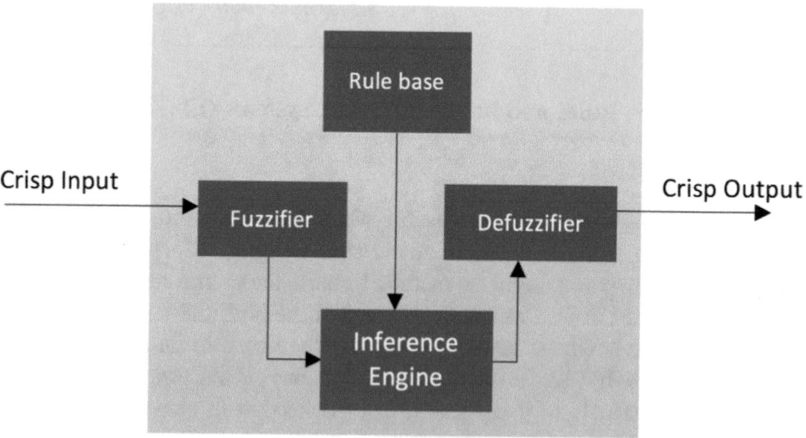

Fig. 4. The Mamdani's fuzzy inference system

- Fuzzifier: Based on the crisp sets of input data, the Fuzzifier converts this input to fuzzy sets using membership functions as well as fuzzy linguistic variables. In this block, the performed process is known as Fuzzification (crisp to fuzzy) [41]. Linear triangular and linear trapezoid membership functions are frequently used in FIS [44].

- Fuzzy Rule Base: It represents the key block in any FIS by containing a number of fuzzy if-then rules (Mamdani and/or Sugeno types). As claimed by [37] and [41], these rules are determined through training and/or historical data, or defined by experts. Each fuzzy if-then rule can be considered as a native description of the scheduled system that applies the membership values for prospective inferences.
- Inference Engine: It applies the fuzzy rules on the fuzzy input generated by the Fuzzifier in order to establish the desired results. This engine represents the kernel of decision making process [45] which is designed to dynamically integrate the different system resources in a way to run a series of fuzzy-based outputs.
- Defuzzifier: This block is responsible of transforming the fuzzy output generated by the inference engine into a crisp output. It allows to map each fuzzy set with one crisp set in order to create a transition from linguistic values to numerical values using, again, the membership functions. This process is called Defuzzification, it consists of performing the inverse transformation (fuzzy to crisp) and it has the most computational complexity [37]. Different Defuzzification methods and approaches have been introduced in the literature, namely, the center of area method, bisector of area method, mean of maximum method, smallest of maximum method, and the largest of maximum method [46].

As concluded by [43], a fuzzy inference system that involves fuzzy if-then rules can model the qualitative human knowledge aspects and reasoning processes, without using accurate quantitative values as well as numerical sources and analyses.

4 Fuzzy-Based Adaptive Learning Systems: Related Works

Due to its popularity, fuzzy logic has become successfully applied in a variety of complicated systems and applications that are involving a lack of exactness [47, 48]. In recent years, fuzzy logic has been highly employed in computer-based learning in order to cope with imprecision as well as the variation of the users' characteristics, device features, and domain knowledge structure. Fuzzy logic does not need a lot of data to be executed, it has predictable processes [49], and these are the main reasons why it has been and still used for developing e-learning systems. Fuzzy logic has been used for achieving various goals in computer-based educational systems, namely, for creating suitable assessment tests [50], predicting the learning styles in web environments [51], evaluating a web based LMS [52], diagnosing students' cognitive profiles and comprehension [53], creating a system that delivers personalized tips to students [54], and so on. In this paper, we focus only on the use of fuzzy logic in adaptive learning systems through presenting and studying a list of related works. Moreover, we present the different techniques used in this area as well as how FIS is employed to create such kind of learning systems.

In [45], the authors introduced and designed an online educational module based on fuzzy set theory. They aim, through this proposal, to generate adaptive learning activities in a virtual environment. The basic motivation of this solution is that a learner model is known by the uncertainty of knowledge acquisitions as well as decision making process. Thus, fuzzy logic has been used to model learner profiles in order to present the suitable learning activity in an individual way. This system starts with a collection of all data that establish a learner profile as a crisp input for the FIS. After that, the fuzzification step transforms the crisp input to a fuzzy input using the membership functions. Half trapezoidal, two triangular and half right trapezoidal functions are used in this model. Always regarding the FIS stages, the inference engine starts working in the third stage by the use of rules base related to the learner's age, gender, level of knowledge, level of difficulty of the learning activity as well as the learning session duration. The inference engine uses if-then rules that allow to deduce the appropriate learning activity. At the end, the defuzzification is performed via the extraction of the suitable learning activity from the related data base, based on the value given by the inference engine.

A personalized system for learning English proposed by [55] is based on learner profiles and aims to help students improve their English language competency through an extensive reading environment. The proposal uses learner preferences, fuzzy inference process, memory cycle updates, and analytic hierarchy process in order to provide learners with the most suitable English articles. The techniques of fuzzy inferences and personal memory cycle updates allow the system under consideration to extract the appropriate articles for the targeted learner's ability as well as the desired vocabulary. The authors employed fuzzy inference mechanism respecting the four blocks described in the previous section, including the Input, the Fuzzifier, the Inference Engine and the Defuzzifier in order to determine the suitable articles from the available database for every single learner. In accordance with this objective, the considerable characteristics in this proposal are related to the article's content and the learner's language ability, namely, the Average Difficulty of Vocabulary (ADV), the Average Length of Sentence (ALS), the Total Length of Article (TLA), the Average Ability of Vocabulary of the learner (AAV), and the Article Correlation (AC). These characteristics represent the input of the FIS that are used in the following stage of Fuzzification in order to calculate the degree of membership using a trapezoidal type for each linguistic variable (low, medium, high). The inference step in this algorithm is fulfilled via 243 rules based on combining three linguistic terms and five fuzzy input items. In the final step, the Defuzzification involves the discrete center of area method in order to generate an output that represents how appropriate an article is for a determined learner based on a value between 0 and 1. The final decision is taken based on the maximum value that describes the most convenient article for a learner. This proposal represents a significant illustration of the integration of fuzzy logic in an adaptive e-learning system.

Another work was suggested by [56] aims to use fuzzy sets to represent the learner's knowledge level as a subset of the domain knowledge. The proposed approach is implemented and evaluated in a Mental Architecture Digitized model which is based on Fuzzy Cognitive Maps (FCM) in order to represent the dependence among the domain concepts. The use of fuzzy logic in this proposal performs the user modeling through the identification and the update of each learner's knowledge level related to the concepts of the domain knowledge under consideration. Consequently, the system provides users with the significant adaptive tips through an automatic modeling of the learning or forgetting process of each learner regarding the FCMs.

Fuzzy logic was also used to model the student profile presented in [57] for personalizing online educational systems. In order to meet with this objective, the authors have developed a fuzzy epistemic logic to present the learner's knowledge state through a multi-agent based student profiling system. This system stores the learning activities and interaction history of each learner into the student profile database, and then the profiling data is abstracted into a learner model. Afterward, dynamic learning plans for each learner are established based on the student model and content model. Therefore, every single learner gets personalized learning materials, personalized quiz, and personalized advice. This fuzzy-based educational system have proven that the personalization has the strength of increasing learning motivation, and consequently enhancing learning effectiveness.

In [58], a personalized competence-based instructional system called "Inter-MediActor" has been introduced. The system tends to generate an individualized navigation path for each learner based on the graph of dependencies between competences and student model. This proposal adopts fuzzy logic to deal with uncertainty of the student's assessment, including marks as well as prerequisite knowledge. Like all fuzzy-logic-based systems, the first step of FIS is the Fuzzification where the system transforms the student's mark value (crisp value between 0 and 1) into a linguistic value (negative, positive, or no mark) using the membership functions. Concerning prerequisite knowledge level, it has five linguistic terms: not, little, enough, well, and very well. The estimation of this characteristic is based on the marks obtained during the final-assessment tests. The Fuzzification of the level of difficulty is calculated for each competence of the course, and it is unchangeable during the session. Three membership functions are used to describe the level of difficulty, namely, easy, normal, and difficult. After Fuzzification, fuzzy if-then rules of the Inference Engine are applied for describing relations between the level of difficulty of the competence, the marks obtained in the final-assessment test for the competence, and the estimated prerequisites of this competence. At the end and regarding FIS stages, the linguistic values of the level of recommendation are defuzzified in order to be transformed into crisp values (step 4). To be accurate, more recommended, recommended, less recommended, and forbidden become 4, 3, 2, 1, and 0, respectively.

A fuzzy-based framework for learning path personalization was suggested by [59]. It takes into consideration the most preferred learning material and the least preferred learning material. This by using the learning characteristics and pedagogical models that are based on the learner personality factors determined through Myers-Briggs Type Indicator (MBTI). Since Fuzzy logic is most suitable for working on imprecise input data and for supporting natural description of knowledge, fuzzy logic techniques are used in this proposal for enabling the system to classify learning material structure, and then to adapt the selection of possible learning which are suitable for the student's fuzzy learning styles membership. For this purpose, the algorithm under consideration runs in a way to combine and mix two approaches, namely, learning style approach and fuzzy logic theory. It consists of three divisions; learning style (MBTI) approach, fuzzy logic approach and dynamic course adaptation. The four stages of FIS are respected, starting by Fuzzification stage that transforms the crisp learner's personality characteristics into the appropriate linguistic values. On the one hand, the input linguistic variables are the learner's personality scores: extrovert, introvert, sensing and intuitive. On the other hand, the fuzzy outputs illustrate the learner's acceptance level of the learning material: theory, example, exercise and activities. Afterward, the inference system based on 76 rules (if-then statements) and various inputs combined with AND operator, generate multiple consequents. At the end, every student receives the appropriate structure of learning materials which match his/her learning style and personality.

The authors in [60] present the design and implementation of a fuzzy-logic expert system for adaptation in e-learning. This model aims to introduce an adaptive system for learning English as a second language, based on the student's knowledge and characteristics. The proposed approach follows the process of detecting the learner's sensory preferences in order to decide the appropriate learning content form. This is by the use of an expert system which detects the learner's knowledge level, helps to decide the needed part of courses, adapts and presents the tailored learning materials based on the previous characteristics. Regarding FIS, the Fuzzification of the student's knowledge, assessment tests as well as his/her preferences is always followed by the performance of the Inference Engine. At this level, if-then rules determined by a language education expert are employed to deal with the input related to the number of correct answers within the category (V1), weight of correct answers within the category (V2), importance of the category for further studies (V3), and time spent over answers of the category (V4). The output linguistic variable from the fuzzy expert system represents the need of further studies of the given category (V5). At the end, appropriate learning materials are selected for each category through an algorithm that calculates its importance (ISM). As a result, based on the length of the study time, the study variant is adapted and a reduced one is generated when the study time is less. It includes only the necessary learning materials.

In this section, we presented how fuzzy-logic can be integrated and involved in adaptive learning systems. The related works described above follow the steps of creating a Fuzzy Inference System in order to generate adaptive and personalized e-learning platforms. This is by operating Fuzzification, Inference Engine based on rule base, and Defuzzification. These stages are the core of the adaptation process in any fuzzy-based adaptive learning system.

5 Conclusion and Future Work

The adaptation of learning has a positive effect on the learning process, leading to an increased efficiency, effectiveness as well as learner satisfaction [61]. In this context, the integration of pedagogical approaches as well as artificial intelligence methodologies becomes necessary to meet with the learner's need and to deliver a more interactive learning environment. For this purposes, we presented in this paper one of the artificial intelligence methods called fuzzy logic. As claimed Mamdani, fuzzy logic allows to "compute with words" which cannot be accomplished using other methods, and it has the ability to represent human conceptualizations in a realistic way.

Therefore, fuzzy logic can contribute to create an big number of useful solutions in the modelling of intelligent systems [57] including computer-based educational systems.

In this perspective, we studied the integration of fuzzy logic in adaptive learning systems by reviewing some previous research in this field. We can deduce that fuzzy logic is one of the significant ways to improve the system performance and to make it able to generate serious decisions about the adaptation of learning materials that must be delivered. This is because of the learner characteristics' fuzziness that can be handled with such method of resolving uncertainty problems. Consequently, this integration increases the learner's performance, comprehension, skills as well as his/her satisfaction. In addition to this, authors in [62] share the same opinion by indicating that an algorithm based on fuzzy logic helps to select the ideal model based on a set of inputs and model specifications.

The main motivation of our next research is to develop a new fuzzy-based adaptive learning system for higher education. We aim to incorporate the fuzzy logic in our previous model of m-learning systems [29] in order to deal with the uncertainty of the learner characteristics as well as context parameters. This will enable the adaptation engine to run in a dynamic way allowing the system to be more effective. It will also enhance the learning content adaptation as well as the format adaptation which is more and more required especially in mobile environments. Therefore, the learning experiences of every single student in higher education will be more individual in the transmission of the maximum crucial knowledge.

References

1. Almohammadi, K., Hagras, H., Yao, B., Alzahrani, A., Alghazzawi, D., Aldabbagh, G.: A type-2 fuzzy logic recommendation system for adaptive teaching. Soft Comput. **21**, 965–979 (2017). https://doi.org/10.1007/s00500-015-1826-y
2. Mergler, A.G., Spooner-Lane, R.S.: What pre-service teachers need to know to be effective at values-based education. Aust. J. Teach. Educ. **37**, 66–81 (2012)
3. Schiaffino, S., Garcia, P., Amandi, A.: eTeacher: Providing personalized assistance to e-learning students. Comput. Educ. **51**, 1744–1754 (2008). https://doi.org/10.1016/j.compedu.2008.05.008
4. Almohammadi, K., Hagras, H.: An adaptive fuzzy logic based system for improved knowledge delivery within intelligent e-learning platforms. In: 2013 IEEE International Conference on Fuzzy Systems (FUZZ-IEEE). pp. 1–8 (2013)
5. Zhu, Z.-T., Yu, M.-H., Riezebos, P.: A research framework of smart education. Smart Learn. Environ. **3**, 4 (2016). https://doi.org/10.1186/s40561-016-0026-2
6. Hoel, T., Mason, J.: Standards for smart education – towards a development framework. Smart Learn. Environ. **5**, 3 (2018). https://doi.org/10.1186/s40561-018-0052-3
7. Uskov, V.L., Bakken, J.P., Heinemann, C., Rachakonda, R., Guduru, V.S., Thomas, A.B., Bodduluri, D.P.: Building smart learning analytics system for smart university. In: Uskov, V.L., Howlett, R.J., and Jain, L.C. (eds.) Smart Education and e-Learning 2017. pp. 191–204. Springer, Berlin (2018)
8. Zhu, Z.T., Bin, H.: Smart education: a new paradigm in educational technology. Telecommun. Educ. **12**, 3–15 (2012)

9. Zhao, C., Wan, L.: A shortest learning path selection algorithm in e-learning. In: Sixth IEEE International Conference on Advanced Learning Technologies (ICALT'06). pp. 94–95 (2006)
10. Ennouamani, S., Mahani, Z.: An overview of adaptive e-learning systems. In: 2017 Eighth International Conference on Intelligent Computing and Information Systems (ICICIS). pp. 342–347 (2017)
11. Tan, H., Guo, J., Li, Y.: E-learning recommendation system. In: 2008 International Conference on Computer Science and Software Engineering. pp. 430–433 (2008)
12. Cavus, N., Bicen, H., Akcil, U.: The Opinions of Information Technology Students on Using Mobile Learning (2008)
13. Naismith, L., Lonsdale, P., Vavoula, G.N., Sharples, M.: Mobile Technologies and Learning (2004)
14. Rahamat, R.B., Shah, P.M., Din, R.B., Aziz, J.B.A.: Students' readiness and perceptions towards using mobile technologies for learning the english language literature component. Engl. Teach. **8**, 16 (2017)
15. Traxler, J., Kukulska-Hulme, A.: Mobile Learning: A Handbook for Educators and Trainers. Routledge, Abingdon (2007)
16. Chan, T.-W., Roschelle, J., Hsi, S., Kinshuk, Sharples, M., Brown, T., Patton, C., Cherniavsky, J., Pea, R., Norris, C., Soloway, E., Balacheff, N., Scardamalia, M., Dillenbourg, P., Looi, C.-K., Milrad, M., Hoppe, U.: One-to-one technology-enhanced learning: an opportunity for global research collaboration. Res. Pract. Technol. Enhanc. Learn. **01**, 3–29 (2006). https://doi.org/10.1142/s1793206806000032
17. Norris, C.A., Soloway, E.: Learning and schooling in the age of mobilism. Educ. Technol. **51**, 3–10 (2011)
18. Wu, W.-H., Jim Wu, Y.-C., Chen, C.-Y., Kao, H.-Y., Lin, C.-H., Huang, S.-H.: Review of trends from mobile learning studies: a meta-analysis. Comput. Educ. **59**, 817–827 (2012). https://doi.org/10.1016/j.compedu.2012.03.016
19. Kukulska-Hulme, A.: How should the higher education workforce adapt to advancements in technology for teaching and learning? Internet High. Educ. **15**, 247–254 (2012). https://doi.org/10.1016/j.iheduc.2011.12.002
20. Kidd, T.T.: Online education and adult learning: new frontiers for teaching practices. Information Science Reference (2010)
21. BLOOM, B.S.: The 2 sigma problem: the search for methods of group instruction as effective as one-to-one tutoring. Educ. Res. **13**, 4–16 (1984). https://doi.org/10.3102/0013189x013006004
22. Chrysafiadi, K., Virvou, M.: Student modeling approaches: a literature review for the last decade. Expert Syst. Appl. **40**, 4715–4729 (2013). https://doi.org/10.1016/j.eswa.2013.02.007
23. Pandey, H., Singh, V.K.: A fuzzy logic based recommender system for e- learning system with multi-agent framework. Int. J. Comp. Appl. **122**(17), 0975–8887 (2015)
24. V. J. Shute, D.Z.-R.: Adaptive Educational Systems (2012)
25. Boticario, J., Santos, O., Van Rosmalen, P.: Issues in developing standard-based adaptive learning management systems. Presented at the EADTU 2005 working conference: Towards Lisbon 2010: Collaboration for innovative content in lifelong open and flexible learning. (2005)
26. Kass, R.: Building a user model implicitly from a cooperative advisory dialog. User Model. User-Adapt. Interact. **1**, 203–258 (1991). https://doi.org/10.1007/bf00141081
27. Moore, M.G.: Editorial: three types of interaction. Am. J. Distance Educ. **3**, 1–7 (1989). https://doi.org/10.1080/08923648909526659

28. Alshammari, M., Anane, R., Hendley, R.J.: Adaptivity in e-learning systems. In: 2014 Eighth International Conference on Complex, Intelligent and Software Intensive Systems, pp. 79–86 (2014)
29. Ennouamani, S., Mahani, Z.: Designing a practical learner model for adaptive and context-aware mobile learning systems. IJCSNS Int. J. Comput. Sci. Netw. Secur. **18**, 84–93 (2018)
30. Millán, E., Loboda, T., Pérez-de-la-Cruz, J.L.: Bayesian networks for student model engineering. Comput. Educ. **55**, 1663–1683 (2010). https://doi.org/10.1016/j.compedu.2010.07.010
31. Nguyen, L., Do, P.: Combination of Bayesian network and overlay model in user modeling. In: Allen, G., Nabrzyski, J., Seidel, E., van Albada, G.D., Dongarra, J., and Sloot, P.M.A. (eds.) Computational Science – ICCS 2009. pp. 5–14. Springer, Heidelberg (2009)
32. Zadeh, L.A.: Information and control. Fuzzy Sets. **8**, 338–353 (1965)
33. Zenebe, A., Norcio, A.F.: Representation, similarity measures and aggregation methods using fuzzy sets for content-based recommender systems. Fuzzy Sets Syst. **160**, 76–94 (2009). https://doi.org/10.1016/j.fss.2008.03.017
34. Al-Shamri, M.Y.H., Bharadwaj, K.K.: Fuzzy-genetic approach to recommender systems based on a novel hybrid user model. Expert Syst. Appl. **35**, 1386–1399 (2008). https://doi.org/10.1016/j.eswa.2007.08.016
35. Drigas, A.S., Argyri, K., Vrettaros, J.: Decade Review (1999–2009): Artificial intelligence techniques in student modeling. In: Lytras, M.D., Ordonez de Pablos, P., Damiani, E., Avison, D., Naeve, A., and Horner, D.G. (eds.) Best Practices for the Knowledge Society. Knowledge, Learning, Development and Technology for All, pp. 552–564. Springer, Heidelberg (2009)
36. Shakouri G., H., Tavassoli N., Y.: Implementation of a hybrid fuzzy system as a decision support process: a FAHP–FMCDM–FIS composition. Expert Syst. Appl. **39**, 3682–3691 (2012). https://doi.org/10.1016/j.eswa.2011.09.063
37. Amindoust, A., Ahmed, S., Saghafinia, A., Bahreininejad, A.: Sustainable supplier selection: a ranking model based on fuzzy inference system. Appl. Soft Comput. **12**, 1668–1677 (2012). https://doi.org/10.1016/j.asoc.2012.01.023
38. Vandewaetere, M., Desmet, P., Clarebout, G.: The contribution of learner characteristics in the development of computer-based adaptive learning environments. Comput. Hum. Behav. **27**, 118–130 (2011). https://doi.org/10.1016/j.chb.2010.07.038
39. Zadeh, L.A.: The concept of a linguistic variable and its application to approximate reasoning—I. Inf. Sci. **8**, 199–249 (1975). https://doi.org/10.1016/0020-0255(75)90036-5
40. Aessilan, S., Mamdani, E.: An experiment in linguistic synthesis of fuzzy logic controllers. Int. J. Man-Mach. Stud. **7**, 1–13 (1974)
41. Khan, F.A., Shahzad, F., Altaf, M.: Fuzzy based approach for adaptivity evaluation of web based open source learning management systems. Clust. Comput. (2017). https://doi.org/10.1007/s10586-017-1036-8
42. Takagi, T., Sugeno, M.: Derivation of fuzzy control rules from human operator's control actions. IFAC Proc. **16**, 55–60 (1983). https://doi.org/10.1016/s1474-6670(17)62005-6
43. Jang, J.-R.: ANFIS: adaptive-network-based fuzzy inference system. IEEE Trans. Syst. Man Cybern. **23**, 665–685 (1993). https://doi.org/10.1109/21.256541
44. Chen, C.-M.: A fuzzy-based decision-support model for rebuy procurement. Int. J. Prod. Econ. **122**, 714–724 (2009). https://doi.org/10.1016/j.ijpe.2009.06.037
45. Mohamed, F., Abdeslam, J., Lahcen, E.B.: Personalization of learning activities within a virtual environment for training based on fuzzy logic theory. In: International Association for the Development of the Information Society (2017)
46. Sivanandam, S.N., Sumathi, S., Deepa, S.N.: Introduction to fuzzy logic using MATLAB. Springer, Berlin (2007)

47. Wallace, M., Ioannou, S., Karpouzis, K., Kollias, S.: Possibilistic rule evaluation: a case study in facial expression analysis. Int. J. Fuzzy Syst. **8** (2006)
48. Lin, C.-T., Fan, K.-W., Yeh, C.-M., Pu, H.-C., Wu, F.-Y.: High-accuracy skew estimation of document images. Int. J. Fuzzy Syst. **8** (2006)
49. Gomathi, C., Rajamani, V.: Skill-based education through fuzzy knowledge modeling for e-learning. Comput. Appl. Eng. Educ. **26**, 393–404 (2018). https://doi.org/10.1002/cae.21892
50. Goyal, M., Yadav, D., Choubey, A.: Fuzzy logic approach for adaptive test sheet generation in e-learning. In: 2012 IEEE International Conference on Technology Enhanced Education (ICTEE), pp. 1–4 (2012)
51. Deborah, L.J., Sathiyaseelan, R., Audithan, S., Vijayakumar, P.: Fuzzy-logic based learning style prediction in e-learning using web interface information. Sadhana **40**, 379–394 (2015). https://doi.org/10.1007/s12046-015-0334-1
52. Cavus, N.: The evaluation of learning management systems using an artificial intelligence fuzzy logic algorithm. Adv. Eng. Softw. **41**, 248–254 (2010). https://doi.org/10.1016/j.advengsoft.2009.07.009
53. Tsaganou, G., Grigoriadou, M., Cavoura, T., Koutra, D.: Evaluating an intelligent diagnosis system of historical text comprehension. Expert Syst. Appl. **25**, 493–502 (2003). https://doi.org/10.1016/s0957-4174(03)00090-3
54. Kosba, E., Dimitrova, V., Boyle, R.: Using fuzzy techniques to model students in web-based learning environments. In: Palade, V., Howlett, R.J., Jain, L. (eds.) Knowledge-Based Intelligent Information and Engineering Systems, pp. 222–229. Springer, Heidelberg (2003)
55. Hsieh, T.-C., Wang, T.-I., Su, C.-Y., Lee, M.-C.: A fuzzy logic-based personalized learning system for supporting adaptive english learning. J. Educ. Technol. Soc. **15**, 273–288 (2012)
56. Guimarães, R. dos S., Strafacci, V., Tasinaffo, P.M.: Implementing fuzzy logic to simulate a process of inference on sensory stimuli of deaf people in an e-learning environment. Comput. Appl. Eng. Educ. **24**, 320–330 (2016). https://doi.org/10.1002/cae.21707
57. Xu, D., Wang, H., Su, K.: Intelligent student profiling with fuzzy models. In: Proceedings of the 35th Annual Hawaii International Conference on System Sciences. pp. 8 (2002)
58. Kavcic, A.: Fuzzy student model in InterMediActor platform. In: 26th International Conference on Information Technology Interfaces, 2004. pp. 297–302, vol. 1 (2004)
59. Salim, N., Haron, N.: The construction of fuzzy set and fuzzy rule for mixed approach in adaptive hypermedia learning system. In: Pan, Z., Aylett, R., Diener, H., Jin, X., Göbel, S., and Li, L. (eds.) Technologies for e-Learning and Digital Entertainment. pp. 183–187. Springer, Heidelberg (2006)
60. Bradac, V., Walek, B.: A comprehensive adaptive system for e-learning of foreign languages. Expert Syst. Appl. **90**, 414–426 (2017). https://doi.org/10.1016/j.eswa.2017.08.019
61. Popescu, E., Badica, C., Moraret, L.: Accommodating learning styles in an adaptive educational system. Informatica **34** (2010)
62. Shakouri G., H., Menhaj, M.B.: A systematic fuzzy decision-making process to choose the best model among a set of competing models. IEEE Trans. Syst. Man Cybern. - Part Syst. Hum. **38**, 1118–1128 (2008). https://doi.org/10.1109/tsmca.2008.2001076

Discrimination of Human Skin Burns Using Machine Learning

Aliyu Abubakar[(✉)] and Hassan Ugail

Centre for Visual Computing, University of Bradford, Bradford, UK
a.abubakar6@bradford.ac.uk

Abstract. Burns become a serious concern issue affecting thousands of lives worldwide and subjecting victims to physical deformities which usually led them to discrimination in the society due to the scary looks. High mortality rates are being reported annually and is associated with lack of healthcare facilities in most of the remote locations such as towns and villages, as well as unavailability or inadequate experienced burn surgeons. Moreover, studies have shown that experienced burn surgeons have drawback in their assessment due to visual fatigue. Therefore, we propose this study to determine whether Machine Learning (ML) can be used to discriminate between burnt skin and normal skin images. We expect to minimize the unwanted hospital delays and render service delivery improvement when conducted with ML algorithms. As such, we employed one of the variant architectures of Residual Network (ResNet) - ResNet101 for the operation. The kernels of this model were used in extracting useful features and Support Vector Machine (SVM) along with 10-fold cross-validation technique was used in classifying the images and obtained a recognition accuracy of 99.5%.

Keywords: Skin burn · Machine learning · Convolutional neural network classification · Cross-validation

1 Introduction

Burns become a serious concern issue and affecting thousands of lives worldwide [1, 2]. Report by researchers [1] shows 265,000 or thereabout deaths are recorded globally every year. Precise identification of burn injury is important in order to give a right medication. Burn assessment is vitally important, as this will provide an insight of whether the burn will heal spontaneously or may require surgery. Visual inspection (known as clinical assessment) is the most common adopted means of identifying burns but concern has been raised due to the poor accuracy of the technique. Moreover, lack of medical professionals in most burn centres and unavailability of aiding diagnostic tools has subjected thousands in danger. Study by [3] shows that wrong identification of burn severity leads to poor assessment. The authors [3] stated that when a positive burn injury is falsely classified based on visual perception of the surgeon may lead to unnecessary surgery, or prolong the patient hospital stay. Similarly, the authors [4–6]7 have described clinical assessment as ineffective means of diagnosing burn as it gives a maximum accuracy less or equal to 75%, in addition of been subjective.

K. Arai et al. (Eds.): CompCom 2019, AISC 997, pp. 641–647, 2019.
https://doi.org/10.1007/978-3-030-22871-2_43

2 Related Literatures

According to [7], accurate burn depth assessment is a challenging task even with experienced surgeons, but very important. Burn depth assessment with experienced surgeons was reported to be 60–75% accurate [7, 8]. In view of this, [7] reported that Laser Doppler Imaging is the best technique to adopt for burn diagnosis over thermal imaging. Thermal imaging operates by establishing contact with the burn wound which as result may subject patient to uncomfortable situation such as pain, in addition to confined measurement of a sample site which might not be the actual burn representation to be assessed. Laser Doppler imaging operates by shining laser light on the wound surface without contact. The light is reflected and frequency change occurs which is picked up and translated by a computer into images with different colours that correspond to the degree of burn present. However, the underpinning factor in adopting LDI is the cost and portability issues.

In another development towards improving medical care for burn patients, [9] conducted a study for the burn depth assessment using LDI. Utilizing LDI in this study was the fact that clinical means doesn't give a satisfied result as expected. 40 patients were diagnosed with both clinical methods and LDI on different days after sustaining burn injury. Assessment was carried out on 0,1,3,5, and 8 days using LDI where 54%, 79.5%, 95%, 97% and 100% accuracies were recorded. Clinical assessment provided 40.6%, 61.5%, 52.5%, 71.4% and 100% accuracies for the same number of days' assessments. This shows that Laser Doppler imaging is unsuitable for the accurate diagnosis of burn in first 24 h after injury. Patients are likely to be misdiagnosed or risk waiting long for the accurate assessment.

Authors in [10] were motivated and conducted a comparative study to determine the accuracies of LDI and clinical assessment in discriminating superficial and deep partial thickness burn wounds. The study sample comprised of 34 patients in which 92 injuries were identified from March 2015 to November 2016. Features such as colour, dislodgement of hair follicles, fade, and pain are the parameters used to differentiate the different classes of burn depth. This was analysed statistically using SPSS package. Depth of 57 wounds was correctly classified using clinical approach and provides an accuracy of 81.52% and sensitivity of 81%. 83 wounds were classified correctly using LDI with accuracy of 90.21% and sensitivity of 92.75%. Both methodologies have 82% specificity. LDI in burn depth assessment outperforms clinical assessment but it takes much time and very expensive.

However, an attempt to use machine learning in burn diagnosis was conducted by [11]. A database of 611 pairs of Colour and infrared images was acquired through the usage of infrared camera. Features were extracted using Histogram of Oriented Gradient (HoG), Local Binary Pattern (LBP) and Topographic features, which are then fed into a pre-trained models - LeNet and 34-layered architecture of Residual Network to detect normal skin and burned skin, normal skin or light burn and serious burn, normal skin, light burn, and serious burn. In a nutshell, LeNet was used for binary classification while both were applied in multi-class classification. The overall precision for binary classification is 75.41% with high false positive rate of up to 29.34%.

The contribution of this paper is to investigate whether machine learning can be used to discriminate between human skin burn injury and normal skin with high precision and accuracy that superseded experienced burn surgeon using very deep convolutional neural network. Additionally, we will compare our finding with what was obtained using a very light (shallow) architecture in [11].

3 Method

In this section, detail of how we form the image database is explained, in addition to how the images were processed.

3.1 Image Acquisition

Burn images were acquired through searches on internet using key words such as 'burn images' and 'skin burns'. Then the acquired images were cropped, rotated, and flipped in order to increase the size of the database. We obtained 1360 burnt skin and normal skin images via the aforementioned processes known as augmentation. A sample of example of burn and healthy skin images is shown in Fig. 1.

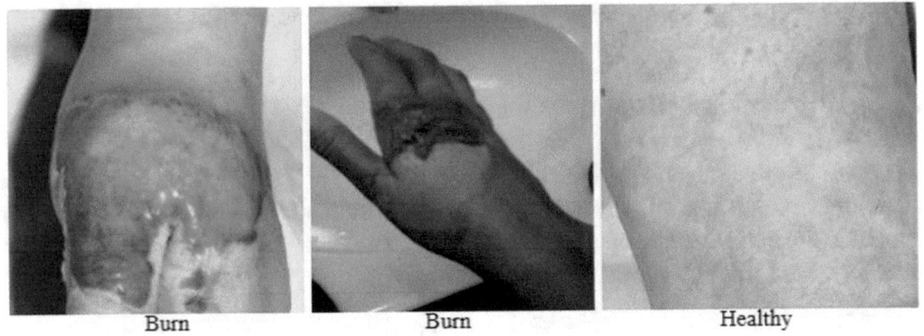

Fig. 1. Sample example of burn and healthy skin images

3.2 Feature Extraction and Classification

For the feature extraction, we used fully trained convolutional neural networks in performing the operation and the features were subsequently fed in to SVM classification algorithm. Features were extracted from the 2D coloured images using deep convolutional neural network architectures.

In recent times, Convolutional Neural Network (CNN) has been used as a powerful tool for extracting features from images [12]. This development has consequently led to a new paradigm in computer vision community, because is has obviously improved state-of-the-art results on many machine vision task [13]. However, the performance of CNN is underpinned by many factors such as the data size available for the training and the depth of the network [13]. Both [13] and [12] have shown the possibility of using

CNN model which was trained on certain datasets and the use the lower layers as learned filters, a process called '*transfer learning*'.

Research conducted by [12] shows that results obtained by using off-the-shelf ConvNet outperforms the work carried out using the ConvNet that was trained from the scratch in a situation where data size is not enough to train the model. Similarly, [13] achieved excellent results via the same technique of feature extraction. For this reason, we used the 101-layered achitecture of Residual Networks.

ResNet has recently achieved a remarkable performance by providing means of developing CNN with more layers with increasing accuracy as against earlier versions of its counterparts that suffered poor performances due to gradient fading. Input of the previous layer (lower layer) is made available to subsequent layer (upper layer), this unique feature of ResNet makes it a good successfully very deep learning algorithm.

SVM is one of the techniques in machine learning used to solve a classification problem [14]. It a type of supervise machine learning algorithm that works based on the known input and output. The algorithm is trained to learn how to map a known input data to the corresponding known output. It achieves the required result by creating an optimum separating hyperplane that separates data points into different classes. Hyperplane separates the data points in such a way that the margin between the data points belonging to different classes is at least a maximum in order to obtain good generalization ability. The term generalization refers to the ability of the system to give an intense predictive accuracy when presented with an unseen data [15].

4 Result and Discussion

The result of our study are presented here and depicted in Table 1 which displayed the result obtained using 101-layered residual network and SVM classifier. We also computed the performance metrics in order to evaluate the accuracy of the classification algorithm used by the following equations:

$$\text{Accuracy} = \frac{\text{TP} + \text{TN}}{\text{TP} + \text{TN} + \text{FP} + \text{FN}} \tag{1}$$

$$\text{Precision} = \frac{\text{TP}}{\text{TP} + \text{FP}} \tag{2}$$

$$\text{Sensitivity (Recall)} = \frac{\text{TP}}{\text{TP} + \text{FN}} \tag{3}$$

$$\text{Specificity} = \frac{\text{TN}}{\text{TN} + \text{FP}} \tag{4}$$

Table 1. Result obtained using ResNet-101 layered architecture

Output Class	Target Class	
	1	2
1	676 49.7%	3 0.2%
2	4 0.3%	677 49.8%

In Table 1, the predicted classes are the rows (Output Class) and the actual classes are represented by the column (Target Class). The correctly classified instances are represented by the leading diagonal, while the off-diagonal cells present misclassified observations. Using the equations in Sect. 3, we obtained the following corresponding values of each metric as shown in Table 2.

Table 2. Performance metrics obtained by ResNet-101 as feature extractor and SVM for the classification

	Precision	Sensitivity	Specificity	Accuracy
ResNet101_SVM	99.56%	99.41%	99.56%	99.49%

We have presented in Table 2 the performance metrics with their corresponding values for the employed algorithms in this study. A total accuracy of approximately 99.5% was recorded.

Precision which provides how correctly the algorithm predicted an accurate result, and it achieved approximately 99.6%. Sensitivity on the other hand, gives the correctly positive predicted result, and in this case, 99.4% was recorded. When we consider specificity on the other side, it tells us the true negative predicted outcome obtained by the classification algorithm, which is 99.7%. The overall accuracy is 99.5%

However, when we compared our outcome with what was obtained by the proposed work in [11] using LeNet, which has a total of five layers, as presented in Table 3, deep convolutional neural network with SVM classifier outperformed the LeNet architecture used in extracting features and SVM for the classification.

Table 3. Performance comparison of ResNet-101 and LeNet architectures

	ResNet101	LeNet [11]
Precision	99.56%	81.83%

5 Conclusion

This study has shown that machine learning can be used to discriminate between skin burn injury and normal skin with high accuracy. The obtained result has evidently shows that machines can be used to give first assessment in identifying human skin burn injury with high accuracy as compared to clinical assessment. With that, we believe that this can significantly minimize the waiting times in hospitals by improving the efficiency of service delivery. The results show evidently machines can perform binary classification with maximum accuracy that superseded human experts. However, this shows that convolutional neural network with multiple stacked layers can be utilized to assist medical personnel in making decision objectively. Moreover, since we were able to present a research work which ultimately provides an avenue of using deep learning to discriminate burn skin and healthy skin with high accuracy, we believe this can be applied to discriminate burns of different category, therefore the further work is expected to investigate classifying burn degree using machine learning algorithm.

References

1. Log, T.: Modeling skin injury from hot rice porridge spills. Int. J. Environ. Res. Pub. Health **15**(4), 808 (2018)
2. Abraham, J., Hennessey, M., Minkowycz, W.: A simple algebraic model to predict burn depth and injury. Int. Commun. Heat Mass Transf. **38**(9), 1169–1171 (2011)
3. Zhao, Y., Maher, J.R., Kim, J., Selim, M.A., Levinson, H., Wax, A.: Evaluation of burn severity in vivo in a mouse model using spectroscopic optical coherence tomography. Biomed. Optics Exp. **6**(9), 3339–3345 (2015)
4. Chatterjee, J.S.: A critical evaluation of the clinimetrics of laser Doppler as a method of burn assessment in clinical practice. J. Burn Care Res. **27**(2), 123–130 (2006)
5. Ye, H., De, S.: Thermal injury of skin and subcutaneous tissues: a review of experimental approaches and numerical models. Burns **43**(5), 909–932 (2017)
6. Singla, N., Srivastava, V., Mehta, D.S.: In vivo classification of human skin burns using machine learning and quantitative features captured by optical coherence tomography. Laser Phys. Lett. **15**(2), 025601 (2018)
7. Monstrey, S., Hoeksema, H., Verbelen, J., Pirayesh, A., Blondeel, P.: Assessment of burn depth and burn wound healing potential. Burns **34**(6), 761–769 (2008)
8. Singer, A.J., Boyce, S.T.: Burn wound healing and tissue engineering. J. Burn Care Res. **38** (3), e605–e613 (2017)
9. Hoeksema, H., Van de Sijpe, K., Tondu, T., Hamdi, M., Van Landuyt, K., Blondeel, P., Monstrey, S.: Accuracy of early burn depth assessment by laser Doppler imaging on different days post burn. Burns **35**(1), 36–45 (2009)
10. Jan, S.N., Khan, F.A., Bashir, M.M., Nasir, M., Ansari, H.H., Shami, H.B., Nazir, U., Hanif, A., Sohail, M.: Comparison of Laser Doppler Imaging (LDI) and clinical assessment in differentiating between superficial and deep partial thickness burn wounds. Burns **44**(2), 405–413 (2017).
11. Badea, M.-S., Vertan, C., Florea, C., Florea, L., Bădoiu, S.: Severe burns assessment by joint color-thermal imagery and ensemble methods. In: e-Health Networking, Applications and Services (Healthcom), 2016 IEEE 18th International Conference on, 2016, pp. 1–5

12. Razavian, A.S., Azizpour, H., Sullivan, J., Carlsson, S.: CNN features off-the-shelf: an astounding baseline for recognition. In: Computer Vision and Pattern Recognition Workshops (CVPRW), 2014 IEEE Conference on, 2014, pp. 512–519.
13. Bukar, A.M., Ugail, H.: Automatic age estimation from facial profile view. IET Comput. Vision **11**(8), 650–655 (2017)
14. Vapnik, V., Guyon, I., Hastie, T.: Support vector machines. Mach. Learn. **20**(3), 273–297 (1995)
15. Xue, H., Yang, Q., Chen, S.: SVM: Support vector machines. The Top Ten Algorithms in Data Mining **6**(3), 37–60 (2009)

Sometimes You Want to Go Where Everybody Knows Your Name

Reuben Brasher(✉), Justin Wagle(✉), and Nat Roth(✉)

Microsoft, Redmond, USA
{rebrashe,justiwag,narot}@microsoft.com

Abstract. We introduce a new metric for measuring how well a model personalizes to a user's specific preferences. We define personalization as a weighting between performance on user specific data and performance on a more general global dataset that represents many different users. This global term serves as a form of regularization that forces us to not overfit to individual users who have small amounts of data. In order to protect user privacy, we add the constraint that we may not centralize or share user data. We also contribute a simple experiment in which we simulate classifying sentiment for users with very distinct vocabularies. This experiment functions as an example of the tension between doing well globally on all users, and doing well on any specific individual user. It also provides a concrete example of how to employ our new metric to help reason about and resolve this tension. We hope this work can help frame and ground future work into personalization.

Keywords: Personalization · Artificial intelligence

1 Introduction

In a wide range of fields, such as music recommendations, advertising and healthcare, learning users' personal tendencies and judgments is essential. Many current approaches demand a centralized data store to aggregate and learn globally. Models trained in these centralized datasets then leverage specific features known about a given user, such as their location or click histories, to make predictions that appear personalized to that user. While such global models have proven to be widely effective, they inherently conflict with user privacy concerns. User data must leave their device, and training a central model requires regular communication between a given user and the remote model. Further, if users are in some way truly unique, and exhibit different preferences than seemingly similar users, large centralized models may have trouble quickly adapting to this behavior.

With these disadvantages in mind, we present a definition of personalization that allows for no direct sharing or centralization of user data. We see personalization as the balance between generalization to global information and specialization to a given user's quirks and biases.

K. Arai et al. (Eds.): CompCom 2019, AISC 997, pp. 648–658, 2019.
https://doi.org/10.1007/978-3-030-22871-2_44

To make this definition concrete, we show how a simple baseline model's performance changes on a sentiment analysis task as a function of user bias and the way information is shared across models. We hope this work can contribute to framing the discussion around personalization and provide a metric for evaluating in what ways a model is truly providing a user personal recommendations.

We also discuss related areas such as differential privacy, and federated learning, which have been motivated by similar considerations. Our work could easily fit into the frameworks of federated learning or differential privacy.

1.1 Personalized Models

There has been a long history of research into personalization within machine learning. There is a wealth of work on using Bayesian hierarchical models to learn mixes of user and global parameters from data. These works have achieved success in areas from health care [1], to recommendation systems [2], to generally dealing with a mix of implicit and explicit feedback [3]. There has also been increasing work on helping practitioners to integrate these Bayesian techniques with deep learning models [4].

Many approaches to personalization within deep learning have relied on combining personal features, hand written or learned, with some more global features to make predictions. For example, in deep recommender systems, a feature might be whether a user is a certain gender, or has seen a certain movie. A model may learn to embed these features, and combine them with some linear model as in [5] in order to make recommendations for a specific user. It is also common to learn some vector describing the user end to end for a task, rather than doing this featurization by hand. In such scenarios your input might be a sentence and a user id and the prediction would be the next sentence as in [6], in which the user is featurized via some learned vector. Similarly, Chunseong Park et al. [7], learn a vector representation of a user's context to generate image captions that are personal to the user, and the work in [8] learns a user vector alongside a language model to determine if a set of answers to a question will satisfy a user. These approaches have the benefit of not requiring any manual description of the important traits of a user.

Here, when we discuss personalization, we focus more on personalization work within deep learning. In general, deep learning models are large, complicated, and very non-linear. This makes it hard to reason about how incorporating a new user, or set of training examples will affect the state of the model at large, a phenomenon known as catastrophic forgetting [9], a topic which itself has seen a large amount of research [10–12]. In general, this means that if we add a new user whose behavior is very different from our previous training examples, we need to take extra steps to preserve our performance on previous users. This makes online personalization of models to outlier users an open problem within deep learning.

1.2 Federated Learning

Our other key personalization constraint is privacy related; to get users to trust a model with extremely personal data, it is our belief that it is becoming increasingly necessary, and even legally mandated, to guarantee them a degree of privacy [13,14]. Research on federated learning has demonstrated that intelligence from users can be aggregated and centralized models trained without ever directly storing user data in a central location, alleviating part of these privacy concerns. The field focuses on training models when data is distributed on a very large number of devices, and further assumes each device does not have access to a representative sample of the global data [15,16].

Federated learning is concerned with training a central model that does well on users globally. However, since model updates are based on an average of all users' behavior, the contribution from an individual user tends to be washed out after each update to the global model. Konečnỳ et al., [16] admit as much, explicitly saying that the issues of personalization are separate from federated learning. Instead, much of the current research focuses on improving communication speed [17] and how to maintain stability between models when communication completely drops or lags [18]. As [19] comes closest to our concerns, it hypothesizes a system in which each user has a personal set of knowledge and some more global mechanism aggregating knowledge from similar users. They do not propose an exact mechanism for how to do this aggregation and how to determine which users are similar. We hope to contribute to the conversation on how to best minimally compromise the privacy and decentralization of learning, while not enforcing all models to globally cohere and synchronize.

Finally, it is important to note that federation itself does not guarantee privacy. While in practice this aggregation of gradients, in the place of storing of raw data, will often obscure some user behavior, it may still leak information about users. For example, if an attacker observes a non-zero gradient for a feature representing a location, it may be trivial to infer that some of the users in the group live in that location. Making strong guarantees about the extent to which data gives us information about individual users is the domain of differential privacy [20,21]. In future work, we hope to incorporate these stronger notions of privacy into our discussion as well, but believe that federated learning is a good first step towards greater user privacy.

2 Personalization Definition

With these problems in mind, we define personalization as the relative weighting between performance of a model on a large, multi-user, global dataset, and the performance of that model on data from a single user. This definition implies several things. In particular, the extent to which a model can be personalized depends both on the model itself, and the variance in user behavior. On a task in which users always behave the same, there is little room for personalization, as a global model trained on all user data will likely be optimal globally and locally. However, on any task where user behavior varies significantly between individuals, it is possible a model trained on all users may perform poorly on a specific user. Nonetheless, a specific user may benefit from some global data; for example, a user with less training data may see better performance if they use a model trained with some global data. Therefore, the best personalization strategy will have some ability to incorporate global knowledge, while minimally distorting the predictions for a given user.

In addition, we add the constraint that user specific data be private and not explicitly shared between models. In particular, this means that even if all user data is drawn from the same distribution, we cannot simply aggregate and then train on all the data. Instead we must determine other ways to share this knowledge, such as federating or ensembling user models.

In this paper, we establish some simple benchmarks for evaluating how well a model respects this definition of personalization.

Formally, suppose we have number of users, N, and for each user we have some user specific data, $\{X_i : i = 1 \dots N\}$, as well as some user specific models, $\{M_i : i = 1 \dots N\}$. Let the global data be $D = \bigcup X_i$. And suppose we have a loss function, $\mathcal{L}$, which is a function of both X_i and M_i, with $\mathcal{L}_i = \mathcal{L}(X_i, M_i)$. We define our success at personalization as:

$$\alpha \mathcal{L}(X_i, M_i) + (1 - \alpha)\mathcal{L}(D, M_i), \tag{1}$$

where α is between 0 and 1, and determines how much we weight local user and global data performance. In the case where X_i follows the same distribution as all D, this definition trivially collapses to approximately optimizing $\mathcal{L}(D, M_i)$, the familiar, non personal objective function on a dataset. However, as α increases and X_i diverges from D, we introduce a tension between optimizing for the specific user, while still not ignoring the whole dataset. Finally, to enforce our constraint of privacy, each model, M_i, has access only to X_i, and the weights of all the other models, M_j for $j = 1 \dots N$, but does not have access to the other datasets, X_j, $j = 1 \dots N$, $j \neq i$.

3 Personalization Motivation and Implications

One question might be why we bother at all with adding global data to the equation, since it is more intuitive to think about personalization as just using the model that does best on a single user's data, and that data alone. However, that intuition ignores the fact that we may have only observed a small amount of behavior from any given user. If we only fit optimally to a specific user's data, we risk overfitting and performing poorly on new data even from that same user.

A Bayesian interpretation of our definition is to view the global data term as representing our prior belief about user behavior. Another interpretation is to view α as how much catastrophic forgetting we will allow our model to do in order to personalize to a user.

From the Bayesian perspective, the global data serves as a type of regularization that penalizes the local model from moving too far away from prior user data in order to fit to a new user. We can think about α as a hyperparameter representing the strength of our prior belief. The smaller α is, the less we allow the model to deviate from the global state. There may be no perfect rule for choosing α, as it depends on the task and rate at which we want to adapt to the user.

One strategy could be to slowly increase α for a given user as we observe more data from them. With this strategy, data rich users will have large α and data poor users will have small α. Thus, data rich users will be penalized less for moving further away from the global state. This is close to treating our loss as the maximum a posteriori estimate of the users data distribution, as we observe more data. The rate of changing α could be chosen so as to minimize the loss of our approach on some held out user data, following the normal cross validation strategy for choosing hyperparameters. Alternatively, we may have domain specific intuition on how much personalization matters, and α provides an easy way to express this.

From the catastrophic forgetting perspective, our definition is similar to the work in [12], which penalizes weights from moving away from the values that were optimal for a previous task. That work upweights the penalty for weights that have a high average gradient on the previous task, reasoning that such weights are likely to be most important. We directly penalize the loss of accuracy on other users, rather than indirectly penalizing that change, as the gradient based approach does. The indirect approach of [12] has the benefits of being scalable, as it may be expensive to recalculate total global loss, and can potentially adapt to unlabeled data. Still, we see a common motivation, as in both cases, we have some weighting for how much we want to allow our model to change in response to new examples.

To calculate $\mathcal{L}(D, M_i)$ we do not need to gather the data in a central location (which would violate our privacy constraint). It is enough to share M_i with each other user, or some sampling of other users, and gather summary statistics of how well the model performs. We could then aggregate these summary statistics to evaluate how well M_i does on D. However, sharing a user's model with other users still compromises the original user's privacy, since model weights potentially offer insight into user behavior.

In practice, we often have a subset of user data that we can centralize from users who have opted in, or a large, public, curated dataset that is relevant to our task of interest. We can treat such a dataset as a stand in for how users will generally behave. This approach does not compromise user privacy. Alternately, since our $\mathcal{L}(D, M_i)$ is meant to regularize and stabilize our local models, there may be other approaches that achieve this global objective without directly measuring performance on global data. In future work, we will more deeply explore how best to measure this global loss without violating user privacy.

4 Experiments

We run an experiment with a simple model to demonstrate the trade-offs between personal and global performance and how the choice of α might affect the way we make future user predictions.

4.1 Setup and Data

We use the Stanford Sentiment Treebank (SSTB) [22] dataset and evaluate how well we can learn models for sentiment. As a first step, we take the 200 most positive and 200 most negative words in the dataset, which we find by training a simple logistic regression model on the train set. We then run experiments simulating the existence of 2, 5, or 8 users. In each experiment, these words are randomly partitioned amongst users, and users are assigned sentences containing those words for their validation, train, and test sets. Sentences that contain no words in this top 400 are randomly assigned, and sentences that contain multiple of these are randomly assigned to one of the relevant users. This results in a split of the dataset in which each model has a dataset in which a subset of words are significantly enriched, but are very underrepresented in all splits of the datasets.

This split is meant to simulate a pathological case of user style; we try to simulate users in our train set that are very biased and almost non-overlapping in terms of the word choice they use to express sentiment. While this may not be the case for this specific review dataset, in general there will be natural language tasks in which users have specific slang, inside jokes, or acronyms that they use that may not be used by others. For such users, an ideal setup would adapt to their personal slang, while still leveraging global models to help understand the more common language they use.

4.2 Architecture

For each user, we train completely separate models with the same architecture. Roughly following the baseline from the original SSTB paper, we classify sentences using a simple two-layer neural network, and use an average of word embeddings as input and a tanh non-linearity. We use 35 dimensional word embeddings, dropout of 0.5 [23], and use ADAM [24] to optimize the models. It is important to note that while this approach is not the state of the art, a model this size could run on low powered and memory constrained devices, and therefore could realistically be deployed. We start with an initial learning rate of 0.001, which we slowly decay, by multiplying by 0.95, if the validation accuracy has not decreased after a fixed number of batches. Finally, we use early stopping in the case validation accuracy does not decrease after a fixed number of batches. Once all the models are trained, we evaluate two ways of combining our networks: averaging model predictions, and simply taking the most confident models, where confidence is defined as the absolute difference between 0.5 and the models prediction.

4.3 Evaluation Metrics

To evaluate, we use the train, validation, test splits as provided with the dataset, and use pytreebank [25] to parse the data. We only evaluate on the sentence level data for test and validation sets. In all tables, we report accuracy, and test accuracy for user specific data is evaluated solely on sentences that contain only their assigned words and none of the other users' specific words. Global data scores represent the whole test set. We report all results averaged over 15 independent trials.

5 Results and Analysis

5.1 Single User Performance on User-Specific Data vs. Single User Performance on Global Data

Unsurprisingly, as the second and third columns of Table 1 show, single user models perform much better on their own heavily biased user-specific test set than on the global data. This makes sense as each model has purposely been trained on more words from their biased test set. Those words were also specifically selected to be polarizing, but the gap makes concrete the extent to which varying word usages can hurt model performance on this task.

Table 1. Accuracy by number of users. The second column reports accuracy of the single user model (SUM) on the user-specific datasets. The last three columns report performance on the global dataset for the single user model and the two ensemble models.

Num. users	SUM (user-specific dataset)	SUM (global dataset)	Average aggregation (global dataset)	Confidence aggregation (global dataset)
2	0.826	0.797	0.816	0.803
3	0.824	0.783	0.813	0.789
5	0.806	0.739	0.794	0.746
8	0.795	0.697	0.772	0.704

5.2 Single User Performance on User-Specific Data vs. Ensembled Models on User-Specific Data

As the number of users increases, the single user model outperforms both aggregation methods on user-specific data (Table 2). This is particularly pronounced for the confidence aggregation method: ensembling hurts performance across all experiments, with this effect increasing as we add users. As the number of users increases, for any given prediction we are less likely to choose the specific user's model, which performs best on their own dataset. The averaging aggregation method outperforms the confidence aggregation method and is competitive with the single user model for up to five users. However, for more than five users, the averaging approach starts to perform worse on the user's own data, again suggesting that we start to drown out much of the personal judgment and rely on global knowledge.

Table 2. Comparison of ensemble model performance to single user model performance on user-specific data. "Difference" columns denote the single user model accuracy minus the ensemble model accuracy. As the number of users increases, user-specific models outperform ensemble models by increasingly wide margins.

Num. users	Difference (average aggregation)	Difference (confidence aggregation)
2	−0.001	0.013
3	−0.001	0.026
5	0.005	0.059
8	0.022	0.096

5.3 User Performance on Global Data vs. Ensembled Models on Global Data

While it might be easy to conclude that we should just use a single user model, Table 3 demonstrates that the average-aggregated ensembled models outperform the single user model on global data, particularly as the number of users increases. Again, this is what we would expect, since the aggregated models have collectively been trained on more words in more examples, and ought to generalize better to unbiased and unseen data. This global knowledge is still important, as it may contain insights about phrases a user has only used a few times. This may be especially true for a user who has little data. Recall we divide the whole dataset amongst all users, so as the number of users increases, each user-specific model is trained on less data. In this case the lack of a word in the training set may not indicate that a user will never use that word. It may be that the user has not interacted with the system enough for their individual model to have fully learned their language.

Table 3. Comparison of ensemble model performance to single user model performance on global data. "Difference" columns denote single user model accuracy minus ensemble model accuracy. As the number of users increases, the average-aggregated ensemble model increasingly outperforms the single user model.

Num. users	Difference (average aggregation)	Difference (confidence aggregation)
2	−0.019	−0.006
3	−0.029	−0.005
5	−0.055	−0.007
8	−0.075	−0.006

5.4 Choosing an Approach Based on α

These experiments demonstrate the tensions between performing well on global and user data, the two terms in our loss in Eq. (1). We can apply Eq. (1), vary α, and see at what point we should prefer different strategies.

Specifically, suppose we have two approaches we can choose from, with personalized losses of p_0 and p_1, and global losses of g_0 and g_1 respectively. If $\mathcal{L}_0 - \mathcal{L}_1 < 0$ the first approach is superior. We can solve for the α such that $\mathcal{L}_0 = \mathcal{L}_1$, where our loss term again comes from Eq. 1. Plugging our definition in, we see that

$$\alpha p_0 + (1-\alpha)g_0 = \alpha p_1 + (1-\alpha)g_1,$$

or equivalently,

$$\alpha p_0 + (1-\alpha)g_0 - \alpha p_1 - (1-\alpha)g_1 = 0.$$

Rearranging this yields

$$\alpha = \frac{g_1 - g_0}{(p_0 - p_1) - (g_0 - g_1)}, \tag{2}$$

as our break even personalization point. For this value of α, we ought to see our two models as equally valid solutions to the problem of personalization.

$\mathcal{L}_0 - \mathcal{L}_1$ is linear with respect to α, so if $\mathcal{L}_0 - \mathcal{L}_1 \geq 0$ for any α above our cutoff, it will be greater everywhere above the cutoff, and vice versa. This yields a rule for how to approach making a decision between multiple types of models. It only requires choosing a single hyperparameter α between 0 and 1, representing one's belief on how much personalization matters to the task at hand.

We illustrate how to apply these ideas to our experimental results. We see from Tables 2 and 3 that when we compare using a single model to averaging predictions from 5 models, we have:

$$g_{\text{average}} - g_{\text{single}} = -0.054$$
$$p_{\text{single}} - p_{\text{average}} = -0.00523$$

So, the cutoff value for our hyperparameter α, where the single model and averaged models yield equivalent personalization losses is $-0.0545/(-0.0545 - 0.00523) = 0.9124$. We can also compute the ranges of α where we should prefer each model. Because, as explained above, $\mathcal{L}_{\text{single}} - \mathcal{L}_{\text{average}}$ is linear in α, evaluating a single point above the cutoff suffices: we choose $\alpha = 1$ for computational convenience, and have $\mathcal{L}_{\text{single}} - \mathcal{L}_{\text{average}} = p_{\text{single}} - p_{\text{average}} = -0.00523 \leq 0.0$. Consequently, we prefer the single model for values of α above the cutoff of 0.9124, and the averaged model for values of α below the cutoff. Again, recall that to determine what α is appropriate for our task, we can draw upon some expert or domain knowledge. For example, we may know from user research that personalization is 4 times more important than global performance for a specific feature, so we should choose $\alpha = .8$. Alternatively, we could try various values of α and run experiments to see which best maximizes accuracy on some group of held out users. Once α is chosen, the arithmetic above quickly allows us to best choose amongst different strategies for aggregating models.

6 Conclusion

Our definition of personalization allows for a decoupling of models at train time, while only requiring aggregate knowledge of other models inference in order to potentially benefit from this global knowledge. In addition, it gives a practitioner a simple, one parameter way, of deciding how to choose amongst models that may have different strengths and weaknesses. We have shown how this approach might look on a simplified dataset and model, and why the naïve approach of only using a single model, or always aggregating all models, may not be optimal in all situations.

In the future, we will work to develop better methods for combining this aggregate global knowledge, while not hurting user performance. To better protect user privacy, we will also consider alternate methods for regularizing our models outside of the global loss term, $\mathcal{L}(D, M_i)$. We hope that this work will provide a useful framing for future work on personalization, learning in decentralized architectures, such as Ethereum and Bitcoin [26,27], and serve as a guideline for situations in which the normal single loss and centralized server training paradigm cannot be used.

References

1. Fan, K., Aiello, A.E., Heller, K.A.: Bayesian models for heterogeneous personalized health data. arXiv preprint arXiv:1509.00110 (2015)
2. Zhang, Y., Koren, J.: Efficient Bayesian hierarchical user modeling for recommendation system. In: Proceedings of the 30th Annual International ACM SIGIR Conference on Research and Development in Information Retrieval, pp. 47–54. ACM (2007)
3. Zigoris, P., Zhang, Y.: Bayesian adaptive user profiling with explicit & implicit feedback. In: Proceedings of the 15th ACM International Conference on Information and Knowledge Management, pp. 397–404. ACM (2006)
4. Shi, J., Chen, J., Zhu, J., Sun, S., Luo, Y., Gu, Y., Zhou, Y.: Zhusuan: a library for Bayesian deep learning. arXiv preprint arXiv:1709.05870 (2017)
5. Cheng, H.-T., Koc, L., Harmsen, J., Shaked, T., Chandra, T., Aradhye, H., Anderson, G., Corrado, G., Chai, W., Ispir, M., et al.: Wide & deep learning for recommender systems. In: Proceedings of the 1st Workshop on Deep Learning for Recommender Systems, pp. 7–10. ACM (2016)
6. Al-Rfou, R., Pickett, M., Snaider, J., Sung, Y.-H., Strope, B., Kurzweil, R.: Conversational contextual cues: the case of personalization and history for response ranking. arXiv preprint arXiv:1606.00372 (2016)
7. Chunseong Park, C., Kim, B., Kim, G.: Attend to you: personalized image captioning with context sequence memory networks. arXiv preprint arXiv:1704.06485 (2017)
8. Chen, Z., Gao, B., Zhang, H., Zhao, Z., Liu, H., Cai, D.: User personalized satisfaction prediction via multiple instance deep learning. In: Proceedings of the 26th International Conference on World Wide Web, pp. 907–915. International World Wide Web Conferences Steering Committee (2017)
9. French, R.M.: Catastrophic forgetting in connectionist networks. Trends Cogn. Sci. **3**(4), 128–135 (1999)

10. Kirkpatrick, J., Pascanu, R., Rabinowitz, N., Veness, J., Desjardins, G., Rusu, A.A., Milan, K., Quan, J., Ramalho, T., Grabska-Barwinska, A., et al.: Overcoming catastrophic forgetting in neural networks. Proc. Nat. Acad. Sci. **114**(13), 3521–3526 (2017)
11. Kemker, R., Abitino, A., McClure, M., Kanan, C.: Measuring catastrophic forgetting in neural networks. arXiv preprint arXiv:1708.02072 (2017)
12. Aljundi, R., Babiloni, F., Elhoseiny, M., Rohrbach, M., Tuytelaars, T.: Memory aware synapses: learning what (not) to forget. arXiv preprint arXiv:1711.09601 (2017)
13. The European Parliament and the Council of the European Union. Regulation (eu) 2016/679 of the European parliament and of the council of 27 April 2016 (2016). http://eur-lex.europa.eu/legal-content/EN/TXT/PDF/?uri=CELEX:32016R0679&from=en. Accessed 23 Jan 2018
14. State of California Department of Justice Office of the Attorney General. Privacy laws—state of California - department of justice - office of the attorney general. https://oag.ca.gov/privacy/privacy-laws. Accessed 22 Jan 2018
15. McMahan, H.B., Moore, E., Ramage, D., y Arcas, B.A.: Federated learning of deep networks using model averaging (2016)
16. Konečný, J., McMahan, H.B., Ramage, D., Richtárik, P.: Federated optimization: distributed machine learning for on-device intelligence. arXiv preprint arXiv:1610.02527 (2016)
17. Konečný, J., McMahan, H.B., Yu, F.X., Richtárik, P., Suresh, A.T., Bacon, D.: Federated learning: strategies for improving communication efficiency. arXiv preprint arXiv:1610.05492 (2016)
18. Smith, V., Chiang, C.-K., Sanjabi, M., Talwalkar, A.: Federated multi-task learning. arXiv preprint arXiv:1705.10467 (2017)
19. Malle, B., Giuliani, N., Kieseberg, P., Holzinger, A.: The more the merrier-federated learning from local sphere recommendations. In: International Cross-Domain Conference for Machine Learning and Knowledge Extraction, pp. 367–373. Springer (2017)
20. Dwork, C.: Differential privacy: a survey of results. In: International Conference on Theory and Applications of Models of Computation, pp. 1–19. Springer (2008)
21. Abadi, M., Chu, A., Goodfellow, I., McMahan, H.B., Mironov, I., Talwar, K., Zhang, L.: Deep learning with differential privacy. In: Proceedings of the 2016 ACM SIGSAC Conference on Computer and Communications Security, pp. 308–318. ACM (2016)
22. Socher, R., Perelygin, A., Wu, J., Chuang, J., Manning, C.D., Ng, A., Potts, C.: Recursive deep models for semantic compositionality over a sentiment treebank. In: Proceedings of the 2013 Conference on Empirical Methods in Natural Language Processing, pp. 1631–1642 (2013)
23. Srivastava, N., Hinton, G.E., Krizhevsky, A., Sutskever, I., Salakhutdinov, R.: Dropout: a simple way to prevent neural networks from overfitting. J. Mach. Learn. Res. **15**(1), 1929–1958 (2014)
24. Kingma, D.P., Adam, J.B.: A method for stochastic optimization. arXiv preprint arXiv:1412.6980 (2014)
25. Raiman, J.: Stanford sentiment treebank loader in python. https://github.com/JonathanRaiman/pytreebank. Accessed 05 Jan 2018
26. Nakamoto, S.: Bitcoin: a peer-to-peer electronic cash system (2008)
27. Wood, G.: Ethereum: a secure decentralised generalised transaction ledger. Ethereum Project Yellow Paper, p. 151 (2014)

Fuzzy Region Connection Calculus and Its Application in Fuzzy Spatial Skyline Queries

Somayeh Davari[1,2(✉)] and Nasser Ghadiri[1]

[1] Department of Electrical and Computer Engineering, Isfahan University of Technology, 84156-83111 Isfahan, Iran
somayehdvr@gmail.com

[2] Department of Computer Science, National Research University Higher School of Economics, Moscow 125319, Russia
nghadiri@cc.iut.ac.ir

Abstract. Spatial data plays a pivotal role in decision-making applications in a way that nowadays we witness its ever-growing and unprecedented use in both analyses and decision-making. In between, spatial relations constitute a significant form of human understanding of spatial formation. Regarding this, the relationships between spatial objects, particularly topological relations, have recently received considerable attention. However, real-world spatial regions such as lakes or forests have no exact boundaries and are considered fuzzy. Therefore, defining fuzzy relationships between them would yield better results. So far, several types of research have addressed this issue, and remarkable advances have been achieved. In this paper, we propose a novel method to model the "Part" relation of fuzzy region connection calculus (RCC) relations. Furthermore, a method based on fuzzy RCC relations for fuzzification of an important group of spatial queries, namely the skyline operator, is proposed in spatial databases that can be used in decision support, data visualization, and spatial databases applications. The proposed algorithms have been implemented and evaluated on real-world spatial datasets. The results of the carried out evaluation demonstrate more flexibility in comparison with other well-established existing methods, as well as the appropriateness of the speed and quality of the results.

Keywords: Fuzzy spatial reasoning · Spatial skyline query · Spatial data analysis · Fuzzy spatial skyline query

1 Introduction

1.1 Background

Spatial data represents locations and characteristics of spatial phenomena including points, lines, and areas. As an example, urban areas or dispersion of factors leading to diseases, such as the distribution of lead, can be displayed with these data. Many studies have been conducted on the relationship between spatial factors and diseases, and the results have been proved to be beneficial. One of the major causes of malignant diseases is environmental pollution. Instances are industrial plants and the distribution

K. Arai et al. (Eds.): CompCom 2019, AISC 997, pp. 659–677, 2019.
https://doi.org/10.1007/978-3-030-22871-2_45

of lead element which has an effective role in increasing the infections [1]. By looking at the map of the distribution of the infection and disease and their relationship, fruitful results can be achieved. Geographic Information Systems play a key role in this investigation. The topological relations between spatial phenomena can be applied in these systems to obtain the relationships between diseases and spatial factors.

Topological relations figure an essential aspect of how people understand spatial configurations. As a result, a significant portion of the spatial information is transmitted in natural language discourse related to topology: for example, a specific geographical area may be close to, inside, or can overlap the other one. The Region Connection Calculus (RCC; [2]) is provided as a tool for modeling the topological relations. It is also applicable in the reasoning of available topological data (e.g., if *A* is nearby *B* and *B* is a part of *C*, then *C* cannot be a part of *A*, regardless of how *A*, *B*, and *C* regions are defined). Generality is one of the main features of the calculus which distinguishes it from related approaches such as 9-intersection model [3]. Starting with an arbitrary universal set U of areas, topological relationships are defined with conditions of a reflective symmetrical relationship C in U, called connection. Visual definition of some RCC relationships is demonstrated in Fig. 1. In particular, note that EC (externally connected) model the *adjacency* relation, while the *including* relation is modeled by Tangential Proper Part (TPP), Non-Tangential Proper Part (NTPP) and Equality (EQ). In different applications, areas can be modeled in various ways, and connection can be defined accordingly.

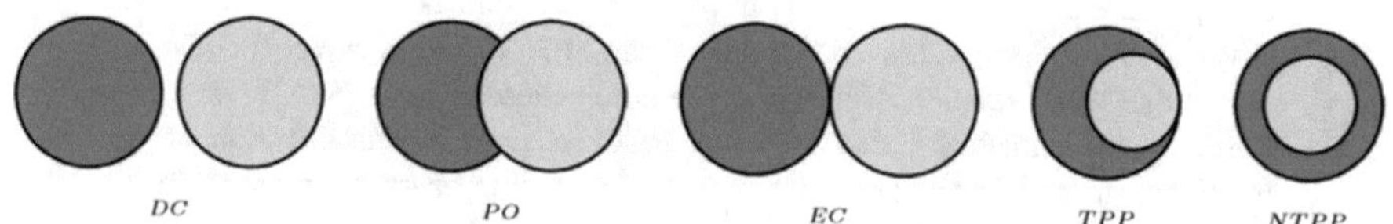

Fig. 1. Visual meaning of some of RCC relations [3]

When using the RCC in applications, it is usually assumed that regions are distinct entities; for example, they are determined by exact boundaries. On the other hand, many geographical areas, for instance, are inherently ill-defined. As an illustration, although administrative areas, such as countries, states, and provinces, have officially defined -and then exact- borders, many places where people refer to in everyday communication (e.g., local places) have not exact boundaries. Fuzzy topological relations can be a proper solution to work with this uncertainty. Fuzzy topological relations are applicable in many fields such as route tracking algorithms based on fuzzy relations [4] and medical diagnosis from patient records [5]. It also has applications in mining topological relations from the web [6], image interpretation [7], robot navigation and manipulation control [8], brain MRI segmentation [9], and representation and reasoning of moving entities [10].

Many types of research have been undertaken on fuzzy spatial topological relations, and considerable advances have been recently made. A variety of methods have been both represented and tested for modeling the fuzzy spatial phenomena and their fuzzy relations [11–15]. Moreover, some systems have been designed and implemented for employing fuzzy topological relations. More definitions of fuzzy topological relations have been developed with either Region Connection Calculus [2, 16–20] or 4-intersection and 9-intersection matrices. Implementation of fuzzy RCC relations in geographical information systems (GIS) helps to improve the user interface compared to most current GIS systems [33]. We have studied and implemented fuzzy RCC relations in our previous work [21]. Also, we utilized this implementation to find the spatial relationships between diseases. As another application, it can be used in fuzzification of the Skyline Operator [22]. So far, some approaches have been developed to make database systems more flexible to support the needs of the users. One of the most advantageous approaches is benefiting from skyline operator. Receiving a database D of tuples or n-dimension points, this operator returns a set of dominated tuples or points in D. A tuple *u* dominates another tuple *u'* if *u* is at least as good as *u'* in all dimensions and is better than *u'* at least at one point. Using spatial skyline, the user can get the nearest points or areas to his given spots or areas. Issues such as spatial analysis of diseases can be better solved by taking advantage of this type of the query.

1.2 The Research Objective of the Paper

The RCC relations in database management systems (DBMS) have not been developed in a fuzzy form. In this work, we have implemented the fuzzy RCC on PostGIS since not only is it useful, but also it is proved to be powerful. Furthermore, existing fuzzy sky operators have shortcomings regarding the execution time, the quantity or quality of their results. We also have studied some of these operators whose shortcomings in the execution time, results, and non-fuzziness of input tuples are shown. Moreover, we have tried to offer ways to overcome the shortcomings of previous approaches, for this is essential and vital in many applications. An appropriate example would be the use of the proposed fuzzy operator to evaluate the spatial relationship between diseases. One of the significant features of the proposed methods can be the possibility of using fuzzy regions as inputs of queries that can be applied in the study of the spatial relationship of disease in the real world where regions have not exact boundaries.

1.3 The Paper Organization

The rest of the paper is organized as follows. The second section presents an overview of the literature, and the related works are reviewed. In the third section, the proposed method for implementing the Part relation of fuzzy RCC relations in PostGIS and fuzzification of skyline operator is presented. In the fourth section, the proposed methods are investigated, evaluated, and compared with existing methods. Eventually, an overview of the achievements, general conclusions, and recommendations to continue the work is given.

2 Background

Many attempts have been made towards developing optimized algorithms and introduction of several models for skyline queries. Skyline queries had been studied as the maximum vectors before Borzsony and colleagues [22] introduced skyline queries for database applications. Various algorithms are provided for skyline computations, including calculation of the progressive skyline using auxiliary structures, the nearest neighbor algorithm for processing the skyline query. Other implementations include branch and bound skyline (BBS) algorithm, the sort-filter skyline (SFS) algorithm that exploits pre-arranged lists, and linear erase-sorting (LESS) algorithm for the skyline. Recently, there has been a growing interest in addressing the problem of "large dimensions" for skyline queries [23] using the intrinsic properties of skyline points, including the skyline frequency, k-dominants skyline, and k-representative skyline. Meanwhile, for a case, when the reason of data objects movement varies, some studies have been suggested to track skyline changes. These works make use of spatial-temporal solidarity in changes to enhance the skyline query processing further. All these efforts, however, have not considered the spatial domination relationships between data points.

Most spatial studies on spatial query mechanism include the ranking of neighboring points based on the distance to a query point. This research direction widely contains the problem of continuous nearest neighbor query processing in moving objects as well. For multiple points of a query, Papadias and his colleagues [24] have investigated the ranking with a total distance for a set of coordinated functions which add distances to multiple points of the query. As nearest neighbor queries requiring a distance function, which is often difficult to define, another research path has studied concepts of the skyline query. For a spatial skyline query with a query point, Huang and Jensen [25] have studied the problem of finding locations that have been dominated with regards to the network distance to the query point. For this type of query with multiple query points, Sharifzadeh and Shahabi [26] have suggested two algorithms that obtain skyline positions for the given query points so that no other position is more close to all of the query points. In [27], it is shown that the proposed method is not correct. Sharifzadeh and colleagues [28] have proposed another version of the method that eliminates the flaws of the previous version, but it is computationally costly.

Other methods for calculating skyline have been suggested as well. A multi-objective optimized process for skyline has been presented based on the genetic algorithm by Özyer et al. [29]. Fuzzy skyline has been proposed to make the skyline more flexible. In particular, fuzzy skyline in [30] is considered and a flexible overcome relation is provided. This relationship makes it possible to extend skyline with points that are not highly dominated by any other points (even if they are strongly dominated). Yet, other methods arise to improve the skyline. A variety of fuzzy skyline methods are studied in [31]. In the paper, five incentives are raised to fuzzify the skyline. The first motivation is to refine the skyline with the introduction of a sequence of points to find the best skyline points. The second is to make the skyline more flexible. This means that where points' conditions are close to the skyline points, they are also considered as members of the skyline. In the third case, the skyline could be made more manageable with different methods. For example, by weighting the criteria in a hierarchy, this simplicity is achieved. Alternatively, larger scales can be used in the calculation of the skyline. From another point of view, it is possible to place those points that are somewhat similar in categories and calculate the skyline. In the last mode, context-dependent and incomplete priorities can be applied in calculating the skyline.

In the field of disease analysis, among the five incentives, making the skyline more flexible and using larger scales are applicable. With making the skyline more flexible, it will be possible to gain skylines that are huge to the user request in a way that the user can do intended analysis and research on the obtained tuples. Larger scales can help the diversity of obtained tuples from the skyline and make the skyline more understandable. Spatial factors in disease studies have not been given proper attention. However, spatial factor has a significant impact in this area. To investigate the effect of the location, maps can be used. To increase the speed, accuracy, and quality of the results, geographic information systems are a good choice considered for the last half-century in the review [32]. In another work [33], geographic information systems applications in human health studies have been divided into two parts. An overview of the applications of geographic information systems in epidemiology and public health have been made as well.

3 Proposed Model

3.1 The Proposed Approach Regarding the Implementation of Fuzzy Part Relation of the Fuzzy RCC Relations

Fuzzy RCC theory has been suggested by Schockaert [3]. In our previous work [21], we analyzed an estimate based on limiting areas, presented divisions of regions in polynomial time, and proposed its algorithm. To implement fuzzy RCC, the implementation of fuzzy connection is required. Then, using the fuzzy connection, other fuzzy RCC relations can also be implemented.

DC relation is the complement of C relation and easily computable. To enter the relation P to PostGIS, P function is proposed in Appendix A. Receiving two areas from the input, this function returns the extent to which the first region is part of the second one as output, which will be a fuzzy binary number. In this function, firstly it is checked that regions are not empty. Afterwards, the dimensions of each area are calculated, and the number 3 will be taken as the number of divisions for each dimension. This number is subject to change, and its increase to some extent can lead to more accurate results. After marking the beginning of the regions, array G will be initialized. After that the formula for the calculation of P on these sections is implemented.

3.2 The Fuzzy Skyline Proposed Method

The proposed method for using Fuzzy RCC in skyline and fuzzification of the skyline has two phases which will be explained in the following. In the first phase, consider attributes as linguistic variables. For example, each distance measure is divided into four linguistic variables as far away, away, close, and too close. Afterward, check whether there are any tuples in each area or not. For instance, in Fig. 2, one or a set of tuples are close to attribute A1 and are away from attribute A2. This share range that is close to toward A1 and is away from A2 here, is called a granule G. Table 1 illustrates the features of example tuples each of which would be members of one or more granules. For example, the tuple u is too close to A1 and away from A2.

A granule G is in a skyline if and only if there is a tuple member of the granule, and its center dominates other granules:

$$G \in S \Leftrightarrow \exists u \in D | \forall i\ u_i \in G, \forall \acute{G} \in D, \exists \acute{u} \in \acute{G} | center(G) >_{dom} center(\acute{G}) \quad (1)$$

In the second phase, the tuples that are fuzzily connected to the skyline tuples are added to the skyline:

$$\acute{u} \in S \Leftrightarrow \exists u \in S, 0 < C(\acute{u}, u) \leq 1 \quad (2)$$

For this stage, we use the fuzzy connect which was defined in [21]. We can change this connection measure as a parameter to increase or decrease the number of output tuples. Moreover, the degree of connectivity can be considered as part of the response to the query. Thus, the user will consider tuples depending on the needed number based on their connection measure's order.

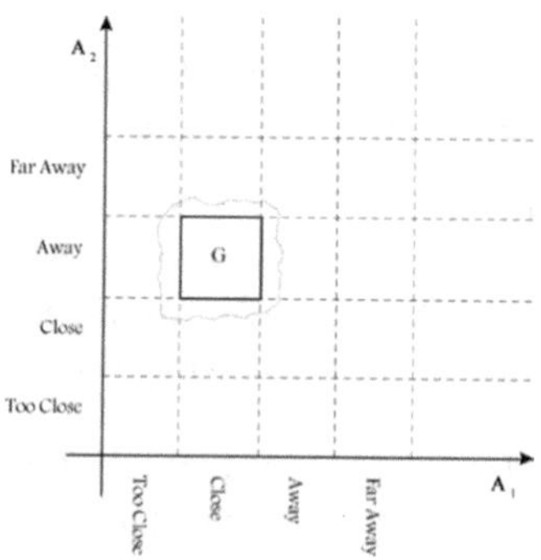

Fig. 2. Considering attributes as linguistic variables, a granule and its fuzzy boundaries

Table 1. Attributes of example tuples

	A_1	A_2
u	Too close	Away
$\overline{u}$	Away	Close
$\overline{\overline{u}}$	Far away	Too close

In Fig. 2, the line around granules G shows the tuples involving this connection boundary. The tuples can be regarded as fuzzy areas in this method. Therefore, skyline results can be expanded fuzzily to the entire database.

3.3 The Results Assessment Methods

The results obtained from the proposed skyline are not necessarily the best choices of data collections because they are extensions of the obtained skylines. There are some ways to assess the selected results. The following criteria are suggested for measuring the results efficiency:

(1) Nearest skylines: A method to assess the efficiency of the results is to measure the sums of distances between each of the points or regions of the results and points or regions of the query. As the average of this sums become less than the average of the sums of other points or regions that are not in the skyline, the results of the skyline get more optimal. For example, if we measure the sum of distances between obtained regions from the fuzzy skyline and query points and compare them with the sum of the distances between other areas of the data set and query points, it is expected that the values obtained initially be lower than secondary values [24].
(2) Representative skylines: In particular, Lin and colleagues [23] have studied how to choose a subset of k size that maximizes the number of points which become dominated by at least one of these skyline points. With this method, the better members dominate, the greater number of points or areas will be obtained.
(3) Area measurement: Measuring and surveying the area between approximate skyline and exact skyline [34] is another way to measure the quality of the results.
(4) The proposed method of accuracy measurement: To evaluate the results and their accuracy measurement, there are methods such as Precision, Recall, and F-measure. But these methods do not consider the order of output results and only consider gained results. Therefore, we suggest a method for evaluating the results that, in addition to the output results, attends their order. In this method, we obtain a correct arrangement for the output results based on the sum of distances to query tuples or based on the number of dominated tuples by each tuple. Afterward, we perform the preferred method for the evaluation and obtain its derived order of output results. To evaluate the results, suppose x is the number of output results and y is the total number of tuples whose order we have achieved. If each tuple of x is a member of x first tuples of y, its rank is zero; otherwise, its rank is equal to sum of number of tuples that are between this tuple and x first tuples of y plus one. Accordingly, we calculate the total ranks of tuples member of x. Next, we divide the resulting number by x. Then we add it by one and divide it by y. The result will be the results percentage of inaccuracy whose complement shows the accuracy of the results of the method that is under assessment.

For example, if we have the numbers one to four as the correct order, and the obtained numbers of the assessment method are two and three in order, the number two is one of the two correct order numbers, and its rank will be zero. However, the number three is not a component of the first two correct order numbers, and it is located after the two first numbers with no distance. Then the number three will have the rank one. The total rank of these two numbers will be one. We will divide the number one by the number of members which are under assessment, too. We add the number 0.5, which is the division result, with one and next divide it by the number of members of the correct order, i.e., four. The resulted number is 0.37; therefore, the percentage of the results inaccuracy is 37%. Its complement which is 63% shows the percentage of accuracy of the results obtained from the method which we are evaluating.

Using the described method, the results of the proposed skyline and their order can be assessed. We use three of the methods, i.e., the first, second and fourth ones, in this paper to show the efficiency of the selected points. These methods are more appropriate for evaluating the results of spatial skylines and are more straightforward and more compelling for analysis. The first two methods provide approximately the same results on spatial skylines, but with the implementation of the proposed method, we can assess the results of our proposed fuzzy skyline more accurately.

Also, to check the speed and the number of query results of the proposed method, we implemented and carried out a comparison with some of the previously offered methods for the skyline calculation which their quality have been approved. One of these methods is the basis skyline which is previously mentioned. Other methods, including Derived and Enhanced methods, and the proposed fuzzy skyline method will be presented below. Then, the FlexPref which is a framework for assessing extensible excellence in database systems is expressed to be also used in evaluating the methods.

The Basis Skyline's Metrics. To apply this method, it is predicted that a change in the number of responses is observed by changing the dimensions.

The Derived Method's Metrics. In this method, it is expected that changes in the results be observed by taking into account different figures and different limits for the level of fuzziness. Also, by changing the dimensions' number, the speed of execution of the algorithm is checked. As a result of increasing the dimensions' number, an increase in the runtime is expected.

The Enhanced Method's Metrics. In this approach, all obtained points of the spatial skyline are more efficient than when we use only the exact skyline by combining the exact and approximate skylines algorithms. Instead of overcoming the test for all data points, we can find random skylines by approximate skyline effectively and then obtain the rest of the skyline points by the exact skyline. The speed of execution is analyzed by changing the dimensions' number in this method as well. It is expected that the runtime is scaled up with growth in the dimensions' number.

The Proposed Fuzzy Skyline Metrics. This technique employs fuzzy RCC according to the priorities and preferences of the user, and the user's intended number of output tuples as output is achieved. In the proposed method, the speed of implementation and results number can be changed by varying parameters. For example, the user can increase the number of output results by adding the connection range.

The FlexPref Assessment Framework. The personalized database systems respond to the users according to their priorities. The FlexPref framework is for extensible priority assessment in database systems and includes three functions. A function is to compare two tuples and if the first tuple dominates another one, it returns one. If the second tuple dominates the first tuple, then this function will have the negative one as output. Except for the two modes, the output of the function is equal to zero. Another function is to check the superiority of a tuple to a set of top tuples. If the tuple is preferred to other superior tuples, the output of the function will be equivalent to one; otherwise, it will be zero. The third function is to add the preferred tuple to a set of preferred tuples and sort or delete it from the elements of preferred tuples set. These three functions are implemented for each one of the skyline calculation methods and are integrated for each method in another function that is to evaluate the method. With the implementation of the final function for each method, the methods can be evaluated and compared with each other [35].

4 Implementation and Evaluation

4.1 Design of the Experiments

The described methods are implemented with PostGIS and PostgreSQL in the pgAdmin software. The obtained results are due to running these methods in the system with CPU 2.0 Core i7 and RAM 6.0 GB. To evaluate the methods in this paper, four data collections are used which will be analyzed in the following.

Data Collections of Isfahan Counties and Distribution of Lead in Isfahan Province. The geometric dataset of Iran counties is downloaded from 'DIVA-GIS'. http://diva-gis.org, and then the required data, i.e. Isfahan counties is extracted (Fig. 3a). We created the distribution of lead in geometry dataset of Isfahan province manually in accordance with article [1] using the QGIS software.

Data Collection of Paris Hotels. This dataset includes the geometry of 1183 hotels in Paris. Paris locations Collection is available on 'GeoNames'. http://geonames.org which is employed with hotel filtering in this work.

Data Collection of Urban Areas in Isfahan Province. This dataset contains 31 urban areas of Isfahan province. The complete earth data collection is downloaded from 'Natural Earth'. http://naturalearthdata.com and the data of Isfahan province is separated and used (Fig. 3b).

Data Collection of Urban Areas on the Earth. This dataset includes the geometry of 11878 densely populated areas of the planet. The collection is downloaded from 'Natural Earth' and used in this paper.

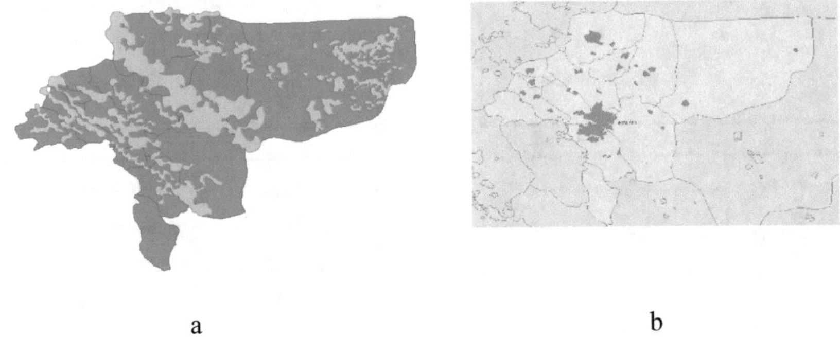

Fig. 3. (a) Map of distribution of lead in Isfahan province. (b) The distribution map of urban areas in Isfahan province.

4.2 The Impact of the Parameters of the Proposed Fuzzy Skyline Method

In this subsection, the influence of each parameter of the proposed method on the number of output tuples and runtime will be investigated. These investigations have been carried out on the urban areas dataset of Isfahan province. In each of these implementations, parameters are considered to clarify the underlying impact of the parameter. Using these results, the user can adjust the number of output results and the runtime of the method as he desires.

Figure 4(a) shows the effect of dividing the number of each area's dimension on the number of output tuples. In accordance with the chart, as the number of divisions increases, the number of output tuples rises and from the number four of the divisions, the number of tuples in the output will be equal. The reason for this is that the higher the number of divisions, the more the accuracy in measuring the amount of areas' connection. Figure 4(a) to (d) and Fig. 5(a) to (f) show the impact of other factors. The rest of the results are interpretable in the same manner.

4.3 Paris Hotels Dataset Results

In this study, we intend to find a skyline of hotels which have the shortest distance to the Eiffel Tower and Charles de Gaulle Airport. The basis skyline (Sect. 1.1) returns 43 hotels as output in four seconds. If the user wants more hotels as output or provides a variety in results, he can implement the proposed methods on this data. With the implementation, it can be seen that as the Grid number changes, the runtimes and obtained results change as well, but these changes do not have a direct or inverse

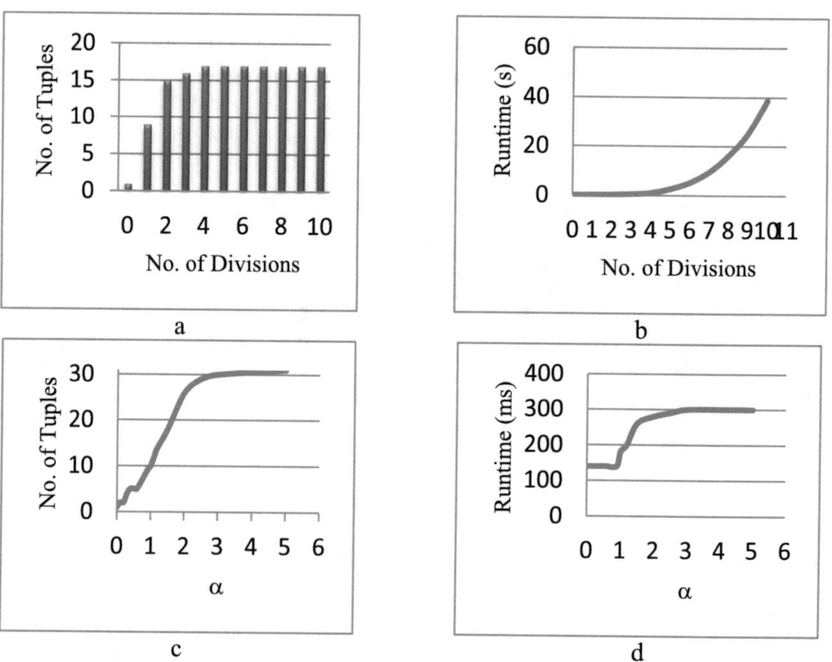

Fig. 4. (a) Effect of the number of divisions on the number of tuples ($M = 0$, $N = 1$, $R = 1$, $0.01 < C < 1$). (b) effect of the number of divisions on the runtime ($M = 0$, $N = 1$, $R = 1$, $0.0 < C < 1$). (c) Impact of α on the number of tuples ($N = 0.2$, $R = 1$, $0.1 < C < 1$). (d) Impact of α on the runtime ($N = 0.2$, $R = 1$, $0.1 < C < 1$)

relationship with the Grid number. Therefore, different results can be seen by varying this number. If we run the algorithm on hotel points, it applies only R and doesn't apply C; in this case, we are still expanding skyline, but with less execution time. However, it does not show the Connect on Fuzzy RCC. As a result, hotels are not like areas, and the number of skyline points is equal to the point case. Those types of data are required so that Fuzzy RCC application can be shown better. For example, those in which the hotels are in the form of areas. Thus, the data of the urban areas of Isfahan province have been employed in Sect. 4.4.

4.4 Results of Isfahan Province Urban Areas' Dataset

Implementation of the Method. Suppose you want to investigate the influence of factories on people or climate of urban areas closer to the Oil Refinery and Steel Company of Isfahan. For instance, the amount of each type of disease in these areas and the relationship between these diseases and proximity to refineries can be examined. To obtain these areas, we applied the basic skyline on the urban areas of Isfahan province dataset, and it returned one urban area closer to Isfahan oil refinery and Steel Company in 93 ms, but when we apply the proposed methods on this dataset as well, we can obtain more areas based on the desired value and needs of the subject to be used in the research.

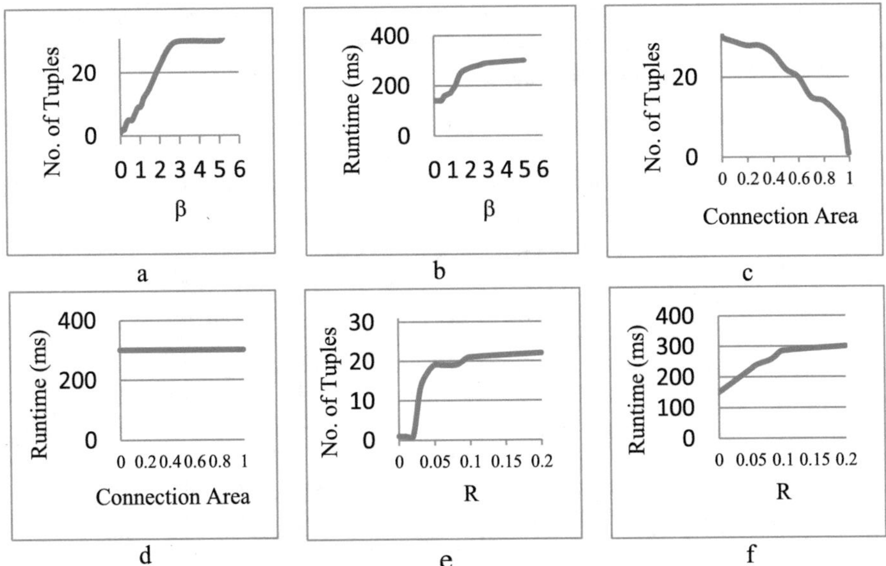

Fig. 5. (a) Impact of β on the number of tuples (M = 0.2, R = 1, 0.1 < C < 1). (b) Impact of β on the runtime (M = 0.2, R = 1, 0.1 < C < 1). (c) Influence of connection area on the number of tuples (M = 1, N = 2, R = 1). (d) Impact of connection area on the runtime (M = 1, N = 2, R = 1). (e) Effect of R on the number of tuples (M = 1, N = 1, 0.01 < C < 1). (f) Impact of R on the runtime (M = 1, N = 1, 0.01 < C < 1)

In such data, the grid method with any value for the number of divisions returns all the data. So, the fuzzy method has only been tested on these data, and the combined method has not been used. The headlines in Table 2 have been defined in accordance with what is stated in Appendix B. In all of the results presented in the table, except those of the two last rows, the amount of fuzziness is intended as one, but, in the last two rows, 0.5 and 0.01 have been set, respectively. In the rows whose letters are bold, only R is applied, and R and C are applied to the next rows. As it is visible in the table, any number of areas that we want to examine can be obtained through the proposed method. In these implementations, the high number of divisions is intended to increase the accuracy of results in applying fuzzy connection. If the runtime is more important, we can reduce the number of divisions and achieve a dramatic increase in execution speed. According to the table, with an increase in M and N and reduction in Bet, an increase in the number of outputs, as well as a slight increase in runtime has been observed. Also, the results' number and the runtime increase by applying C. Using C, you can distinguish between areas with different fuzzy boundaries. As observed, however, the fuzzy skyline runtime will increase, too.

Evaluating the Results of the Fuzzy Skyline Proposed Method. According to the assessment methods discussed in Subsect. 3.3, the four methods can be used to evaluate the output results. Since the spatial skyline is being studied, the first two methods and the proposed method are more suitable and faster to evaluate the results. Applying the first two methods produced similar results on the second dataset.

Applying the method revealed that, based on the proposed evaluation method and using the order which is obtained from the two methods of evaluation, the proposed fuzzy skyline method on these data has 0.03455% error and, as a result, has the accuracy of 99.9654% indicating the high accuracy of the method.

Table 2. Results of implementation of the second phase of the proposed fuzzy skyline on the dataset of urban areas of Isfahan province (as we represented earlier [21])

Tfuzzy (ms)	TSum (ms)	NFuzzy	NSum	M	N	Bet.
141	141	2	2	0	0.3	0.01
171	171	4	4	0	0.5	0.01
181	181	6	6	0	0.8	0.01
191	191	9	9	0	1	0.01
195	195	6	6	0	1	0.2
285	285	30	30	2	2	0.01
16901	16901	4	4	0	0.3	0.01
19882	19882	10	10	0	0.5	0.01
21926	21926	15	15	0	0.8	0.01
23480	23480	17	17	0	1	0.01
21864	21864	14	14	0	1	0.2
25197	25197	20	20	0	1.5	0.2
28284	28284	28	28	0.5	1.5	0.2

(continued)

Table 2. (*continued*)

Tfuzzy (ms)	TSum (ms)	NFuzzy	NSum	M	N	Bet.
31345	31345	30	30	1	1.5	0.2
28615	28615	28	28	1	1	0.2
26217	26217	23	23	1	0.5	0.2
22390	22390	16	16	1	0	0.2
28458	28458	26	26	1	1	0.2
16702	16702	4	4	1	1	0.2

4.5 Results of the Dataset of Urban Areas on the Earth

Implementations of the methods on the data collection of Isfahan province urban areas did not lead to very different and comparable runtimes. To compare the runtimes of the methods, we required large datasets so that the differences between runtimes could be revealed tangibly. Hence, we employed the dataset of urban areas on the Earth since these data could be considered as regions on which all the methods could be verified and applied. Furthermore, due to the plurality of the data, the runtimes could increase and the difference between the methods could be revealed.

In Fig. 6, the results of the implementation of the three methods and the proposed method are visible. In this figure, it is shown that the runtimes of the proposed method and the two methods of Skyline and Derived are close together while the Efficient method execution time is much higher. Moreover, the fact that the runtime increases with an increase in the dimensions has been shown.

For a more detailed comparison of the three methods whose runtimes are close to each other, we remove the Efficient method of comparison (Fig. 7). As it is visible, the runtime of the proposed approach against these two methods is reasonable and acceptable. In the lower divisions, it is also less than the Derived method. In the proposed method, the runtime is near the basic skyline regardless of fuzzy connection and only with regard to R, while the number of tuples of the output can increase to the maximum number of data. According to the results, the proposed algorithm has a

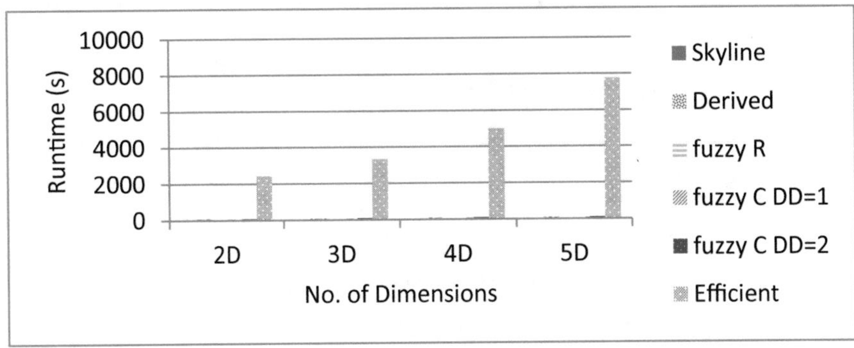

Fig. 6. Comparison of the execution time of four methods on the dataset of urban areas of the earth in terms of the number of dimensions

reasonable running time as well. In its implementation, furthermore, less running time can be achieved considering the lower number of divisions. Although the lower number of divisions slightly decrease the algorithm accuracy, it can highly increase the implementation speed.

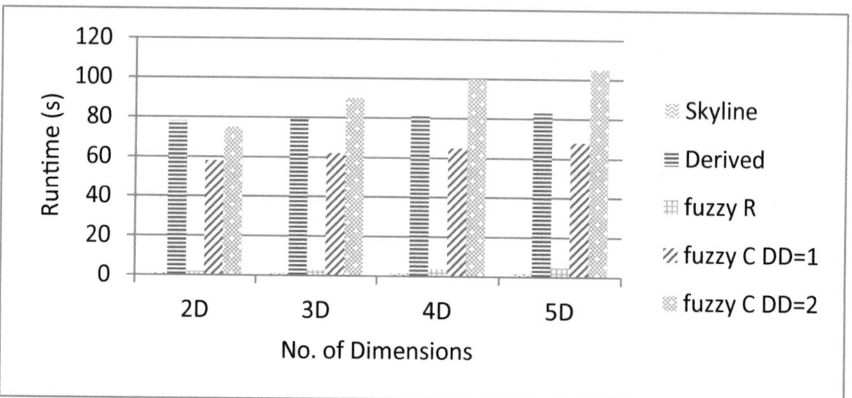

Fig. 7. Comparison of the execution time of three methods on data collection of urban areas of the earth in terms of the number of dimensions

In Fig. 8 the number of output tuples in the implementation of the basic and Efficient skyline is compared. In both methods, as the dimensions increase, the number of tuples of output grow, and the tuples obtained from the Efficient method are much more than those from the basic skyline. Moreover, the number of output tuples compared with the size of the dataset is very insignificant, and this number cannot be changed, while in the proposed and Derived methods, this number can decrease to zero and increase to all the tuples. In these two methods, increasing the dimensions also had no significant impact on the number of output tuples.

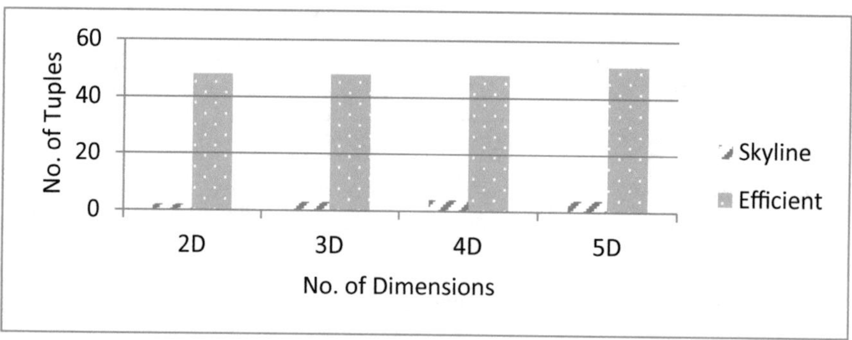

Fig. 8. Comparison of output tuples of running two methods on the dataset of urban areas of the earth regarding the number of dimensions

To evaluate the proposed method, the effect of the number of input tuples on the runtime of each method is also investigated. The result of this study is visible in Fig. 9. According to the study, in all methods, as the number of tuples increased, the runtime rose as well. This impact in the proposed method and the basic skyline is much smaller than that in other two methods. Also, in the Derived method, the effect of the number of input tuples on the runtime is much more than that in the Efficient method.

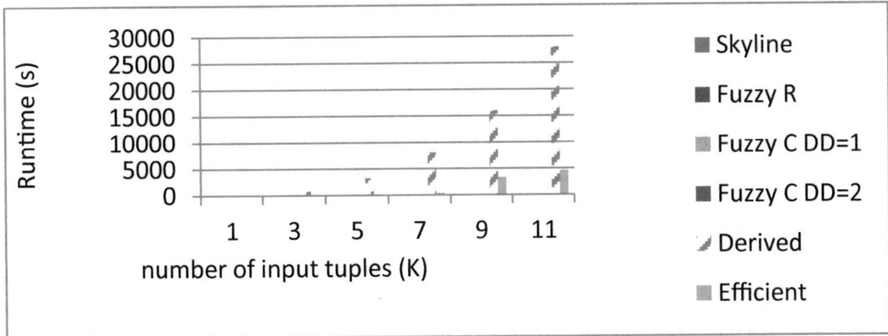

Fig. 9. Comparison of the impact of the number of input tuples on the runtime of the three methods on the dataset of urban areas of the earth

The Results of Running the FlexPref Evaluation Method. In this part, with the implementation of the FlexPref framework, the runtimes and number of output tuples of the methods are compared. Figure 10 represents the results of the implementation of the FlexPref framework for each of the methods. In this figure, the numbers of output tuples are compared by varying the number of input tuples. Obviously, the number of input tuples does not have a direct impact on the number of output tuples, but as the number of input tuples changes, the number of output tuples has also changed. As can be seen in the figure, the numbers of output tuples in the methods are close together, but in the proposed method and the Derived method, according to the parameters, this number is subject to change from zero to the number of all tuples.

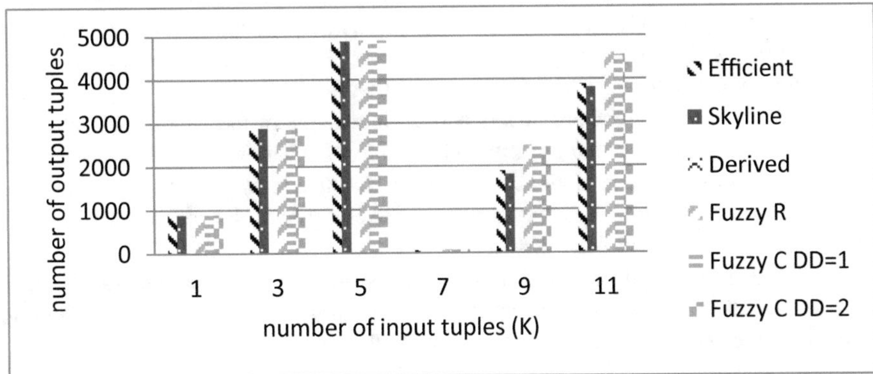

Fig. 10. Comparison between the number of input tuples on the runtime of the FlexPref framework in three methods on the dataset of urban areas of the Earth

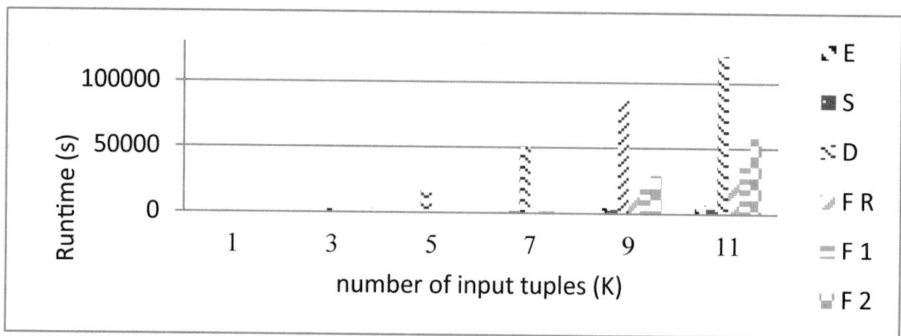

Fig. 11. Comparison of the effect of the number of input tuples on the number of output tuples in the execution of the FlexPref framework in three methods on the dataset of urban areas of the Earth

To compare the runtimes, the methods are implemented with the FlexPref framework and tested with the dataset of urban areas of the earth. The results of this assessment are shown in Fig. 11. According to the figure, the runtimes of all the methods increase with a rise in input tuple numbers. It has also been found that the runtime of the proposed method compared to other methods is acceptable. Compared to other methods, the number of tuples is according to user preferences in the Derived method, but its runtime is high. In the proposed method, the number of output tuples can be changed based on user preferences, whereas its execution is much less than the Derived method.

5 Summary and Conclusion

In this paper, we presented a method to import the Part relationship of the fuzzy RCC relations into PostGIS spatial database that has been neglected before. Afterward, these relationships were used to fuzzify the skyline and made the basis skyline more flexible. The proposed skyline fuzzification method showed to be effective, fast, and flexible regarding the number of resulting outputs so that a user (e.g. a researcher in the medical field) will be able to fit the needs of his specific issue by changing the parameters and observing the results in the output areas. Moreover, the proposed approach contributes to obtaining the skyline of fuzzy areas for the first time.

Observing the accomplished works and achievements of the paper, the importance of the issue and the potential for researching the related fields are characterized. As observed, various methods have been proposed and tested for modeling fuzzy spatial objects and fuzzy relationships between them, the analysis of the spatial relationship of disease, and skylines and fuzzy skylines.

Even though the method proved to be effective and flexible in skyline fuzzification, there will be still room for further improvement. There is a wide area of research in this field. Examples of routes can be taken in the following:

- Presenting and improving the fuzzy indexing structures
- Presenting methods or improving existing methods for modeling the fuzzy objects in databases
- Providing methods or improving existing methods to obtain fuzzy relationships between spatial objects
- Studying the relationship between diseases and various spatial factors
- Studying the spatial relationship of distribution of a variety of diseases

Appendix A. The Function P to Calculate the Fuzzy Relationship Part

```
CREATE OR REPLACE FUNCTION P(G1 Geometry, G2 Geometry)
RETURNS  fuzzy_boolean
AS
$$
DECLARE
height float; width float; h float; w float;
G geometry[3][3];
min float;
A float[3][3];
x float; y float;
i integer; j integer;
BEGIN
if ST_IsEmpty(G1) or ST_IsEmpty(G2) then return 0;
else
height := ST_YMax (G1) - ST_YMin (G1);
width := ST_XMax (G1) - ST_XMin (G1);
h := height / 3 ; w := width / 3 ;
x := ST_XMin (G1); y := ST_YMin (G1);
G:=array_fill('point(0 0)'::geometry, array[3,3]);
for i IN array_lower(G, 1)..array_upper(G, 1) LOOP
for j IN array_lower(G, 1)..array_upper(G, 1) LOOP
G[i][j]  :=  ST_GeomFromText  ('POLYGON  (('||x||'  '||y||',  '||x||'  '||y+h||',
'||x+w||' '||y+h||', '||x+w||' '||y||', '||x||' '||y||'))', 0 );
x:=x+w;
END LOOP;
y:=y+h;
END LOOP;
A:=array_fill('0'::float, array[3,3]);
for i IN array_lower(A, 1)..array_upper(A, 1) LOOP
for j IN array_lower(A, 1)..array_upper(A, 1) LOOP
A[i][j] := least (1, 1- C(G[i][j], G1) + C(G[i][j], G2));
END LOOP;
END LOOP;
SELECT min(elem) INTO min
FROM   unnest(A) AS elem;
return min; end if;
End
$$
LANGUAGE 'plpgsql' STABLE STRICT;
```

Appendix B. The Meaning of Obrivations

The meaning of each obrivations in the paper are as follow:

TGrid:	skyline runtime by Grid method
TFuzzy:	skyline runtime by fuzzy method with or without applying Grid method
TSum:	Total runtime
NGrid:	the number of skyline members after applying Grid method
NFuzzy:	the number of skyline members with fuzzy method with or without applying Grid method
NSum:	Total skyline members after applying the method or methods
M:	alpha number in the fuzzy equation.
N:	beta number in the fuzzy equation.
Bet.:	the minimum acceptable value for a new tuple connection to the previous one to be added to the skyline in fuzzy method
Grid:	The number of divisions of a range of attributes, such as distance to the airport from zero to maximum in Grid method

References

1. Rashidi, M., Ghias, M., Roozbahani, R., Ramesht, M.H.: Investigated the relationship between the spatial distribution of malignant disease and lead in Isfahan province. J. Isfahan Med. Sch. In Persian. 29(135) 2012
2. Egenhofer, M.J., Franzosa, R.D.: Point-set topological spatial relations. Int. J. Geogr. Inf. Syst. **5**(2), 161–174 (1991)
3. Schockaert, S., De Cock, M., Kerre, E.E.: Spatial reasoning in a fuzzy region connection calculus. Artif. Intell. **173**(2), 258–298 (2009)
4. Obradović, Đ., Konjović, Z., Pap, E., Rudas, I.J.: Linear fuzzy space based road lane model and detection. Knowl.-Based Syst. **38**, 37–47 (2013)
5. Norris, D., Pilsworth, B.W., Baldwin, J.F.: Medical diagnosis from patient records—a method using fuzzy discrimination and connectivity analyses. Fuzzy Sets Syst. **23**(1), 73–87 (1987)
6. Schockaert, S., Smart, P.D., Abdelmoty, A.I., Jones, C.B.: Mining topological relations from the web. In: IEEE DEXA Workshops, pp. 652–656 (2008)
7. Hudelot, C., Atif, J., Bloch, I.: Fuzzy spatial relation ontology for image interpretation. Fuzzy Sets Syst. **159**(15), 1929–1951 (2008)
8. Tan, J., Ju, Z., Hand, S., Liu, H.: Robot navigation and manipulation control based-on fuzzy spatial relation analysis. Int. J. Fuzzy Syst. **13**(4), 292–301 (2011)
9. Colliot, O., Camara, O., Bloch, I.: Integration of fuzzy spatial relations in deformable models —application to brain MRI segmentation. Pattern Recognit. **39**(8), 1401–1414 (2006)
10. Wu, J.: A qualitative spatio-temporal modelling and reasoning approach for the representation of moving entities. Doctoral dissertation, Université de Brest (2015)
11. Muñoz-Velasco, E., Burrieza, A., Ojeda-Aciego, M.: A logic framework for reasoning with movement based on fuzzy qualitative representation. Fuzzy Sets Syst. **242**, 114–131 (2014)
12. Salamat, N., Zahzah, E.-H.: On the improvement of combined fuzzy topological and directional relations information. Pattern Recognit. **45**(4), 1559–1568 (2012)

13. Salamat, N., Zahzah, E.-H.: Two-dimensional fuzzy spatial relations: a new way of computing and representation. Adv. Fuzzy Syst. **2012**, 5 (2012)
14. Du, H., Alechina, N.: Qualitative spatial logics for buffered geometries. J. Artif. Intell. Res. **56**, 693–745 (2016)
15. Chen, J., Cohn, A.G., Liu, D., Wang, S., Ouyang, J., Yu, Q.: A survey of qualitative spatial representations. Knowl. Eng. Rev. **30**(01), 106–136 (2015)
16. Liu, W., Li, S.: On standard models of fuzzy region connection calculus. Int. J. Approx. Reason. **52**(9), 1337–1354 (2011)
17. Schockaert, S., De Cock, M., Cornelis, C., Kerre, E.E.: Fuzzy region connection calculus: an interpretation based on closeness. Int. J. Approx. Reason. **48**(1), 332–347 (2008)
18. Schockaert, S., De Cock, M., Cornelis, C., Kerre, E.E.: Fuzzy region connection calculus: representing vague topological information. Int. J. Approx. Reason. **48**(1), 314–331 (2008)
19. Schockaert, S., Li, S.: Realizing RCC8 networks using convex regions. Artif. Intell. **218**, 74–105 (2015)
20. Bjørke, J.T.: Topological relations between fuzzy regions: derivation of verbal terms. Fuzzy Sets Syst. **141**(3), 449–467 (2004)
21. Davari, S., Ghadiri, N.: Spatial database implementation of fuzzy region connection calculus for analysing the relationship of diseases. In: 23rd Iranian Conference on Electrical Engineering (ICEE), pp. 734–739. IEEE (2015)
22. Borzsony, S., Kossmann, D., Stocker, K.: The skyline operator. In: 17th International IEEE Conference, pp. 421–430 (2001)
23. Lin, X., Yuan, Y., Zhang, Q., Zhang, Y.: Selecting stars: the k most representative skyline operator. In: ICDE 2007, IEEE 23rd International Conference on Data Engineering, pp. 86–95 (2007)
24. Papadias, D., Tao, Y., Mouratidis, K., Hui, C.K.: Aggregate nearest neighbor queries in spatial databases. ACM Trans. Database Syst. **30**(2), 529–576 (2005)
25. Huang, X., Jensen, C.S.: In-route skyline querying for location-based services, in web and wireless. In: Geographical Information Systems, pp. 120–135. Springer (2005)
26. Sharifzadeh, M., Shahabi, C.: The spatial skyline queries. In: Proceedings of the 32nd International Conference on Very Large Data Bases, VLDB Endowment, pp. 751–762 (2006)
27. Lee, M.W., Son, W., Ahn, H.K., Hwang, S.W.: Spatial skyline queries: exact and approximation algorithms. GeoInformatica **15**(4), 665–697 (2011)
28. Sharifzadeh, M., Shahabi, C., Kazemi, L.: Processing spatial skyline queries in both vector spaces and spatial network databases. ACM Trans. Database Syst. (TODS) **34**(3), 14 (2009)
29. Özyer, T., Zhang, M., Alhajj, R.: Integrating multi-objective genetic algorithm based clustering and data partitioning for skyline computation. Appl. Intell. **35**(1), 110–122 (2011)
30. Goncalves, M., Tineo, L.: Fuzzy dominance skyline queries. Database and Expert Systems Applications, pp. 469–478. Springer, Berlin, Heidelberg (2007)
31. Hadjali, A., Pivert, O., Prade, H.: On different types of fuzzy skylines. In: Foundations of Intelligent Systems, pp. 581–591. Springer (2011)
32. Glass, G.E.: Update: spatial aspects of epidemiology: the interface with medical geography. Epidemiol. Rev. **22**(1), 136–139 (2000)
33. Rezaeian, M.: Use of geographical information systems in epidemiology. J. Qazvin Uni. Med. Sci. **10**(38), 115–123 (2006)
34. Li, H., Tan, Q., Lee, W.-C.: Efficient progressive processing of skyline queries in peer-to-peer systems. In: Proceedings of the 1st International Conference on Scalable Information Systems, p. 26 (2006)
35. Levandoski, J.J., Mokbel, M.F., Khalefa, M.E.: FlexPref: a framework for extensible preference evaluation in database systems. In: ICDE, 26th International IEEE Conference on Data Engineering, pp. 828–839 (2010)

Concerning Neural Networks Introduction in Possessory Risk Management Systems

Mikhail Vladimirovich Khachaturyan[1(✉)] and Evgeniia Valeryevna Klicheva[2]

[1] PhD in Economics, Associate Professor of Department of Organizational and Managerial Innovations, Plekhanov Russian University of Economics, Moscow, Russia
mike-hach@mail.ru

[2] PhD in Economics, Associate Professor of Department of Restaurant Business, Plekhanov Russian University of Economics, Moscow, Russia
evklicheva@mail.ru

Abstract. The increasing rate of implementation of machine learning and artificial intelligence is currently a key component of development of organization's possessory risk management systems. Owners and managers of major, medium, and small entities strive to have improved and more efficient analytical mechanisms to improve management systems as well as systems for collection, structuring, and analysis of the increasing volumes of data in statutory regulation and of other unstructured data for compliance with the requirements of legislation and financial risk management. It is also obvious that the use of neural networks both in core business processes and in organization management systems has become an important means of economic competition. In terms of the innovative advantage created using machine learning in possessory risk management systems, two preliminary conclusions can be made. First, an important competitive advantage is the fact that machine learning methods enable analysis of large data volumes providing a high level of detail and the depth of predictive analysis, which makes it possible for possessors to obtain additional opportunities for analysis in risk management and compliance with statutory regulation in finance. This article is devoted to the analysis of such opportunities and advantages as well as trends of neural networks implementation in entity's possessory risk management systems.

Keywords: Issues · Implementation · Neural networks · Management system · Possessory risk · Machine learning

1 Introduction

Recently, the rate of implementation of machine learning and artificial intelligence methods in entities' possessory risk management systems has increased significantly. This is due to a conviction in management science and practice that use of neural networks can significantly improve analytical potential, optimize and automate all types of management business processes, including client underwriting, compliance methods, client interaction, and risk management.

K. Arai et al. (Eds.): CompCom 2019, AISC 997, pp. 678–687, 2019.
https://doi.org/10.1007/978-3-030-22871-2_46

Recently, data volumes collected by entities' management systems have increased significantly. This increase is since the regulators both in Russia and in other countries started to set new requirements to the level of reporting detail, while services digitalization results in huge volumes of high-frequency unstructured consumer data. As a result, entity management systems have a greater need for more powerful analytical tools for work with large data volumes of all kinds and formats with simultaneous preservation or even increase in the level of detail of such data for subsequent analysis.

According to the authors, the use of neural networks and machine learning in possessory risk management is a promising field that can provide the required level of analytical capability. This statement is confirmed with empirical data, the analysis of which reveals that, despite a history of using some components of machine learning in management stems from as early as the beginning of XX century, it was not until recent decades that the increase in using such tools as computational innovations and greater availability of high-frequency data made it possible to model complex nonlinear relations, which simplified significantly the use of machine learning methods in entity management systems in general, particularly, in possessory risk management.

This article is devoted to analysis of the concept of machine learning and trends of its use in entity's possessory risk management an attempt is made to trace the relation of machine learning with other types of statistical analysis as well as to analyze the potential and limits for such use of neural networks in entity management systems. In addition, the authors will briefly touch upon the issues of study and use of deep learning, a form of artificial intelligence directly related to machine learning. In addition, issues of implementation of neural networks in possessory risk management systems will be discussed in the article using three examples of machine learning: risk modeling, identification of fraud and laundering of illegally obtained proceeds as well as misconduct and abuse in financial risk management.

2 Prerequisites for Using Neural Networks in Possessory Risk Management Systems

Machine learning includes a broad range of analytical tools that may be divided into controlled and uncontrolled learning tools. Controlled machine learning includes making a statistic model for forecasting or evaluating data based on one or several resources used (e.g. GDP growth forecast based on several variables). During learning, the data set is activated without the dependent variable for assessment or forecasting. Instead, the data is analyzed to demonstrate patterns and structures in the data set.

In the modern context, use of neural networks and machine learning is among the most promising tools for forecasting possessory risks. Through identification of interrelations or regularities in a data sample, it may create a model incorporating the interrelations that lead to the most significant changes beyond the sample [1, 2].

Such model is developed through the launch of variables and the model using data subsets for identification of the strongest predictors with subsequent testing of the model on many different data subsets.

This can be done thousands of times so that the model could learn using the data sets available and improve its forecast indicators. It should be noted that using large data sets and large computational power, an increase of machine learning and neural networks involvement in possessory risk management systems is closely related to the big data revolution in the modern world [3, 4]. Jointly, acceleration and sophistication of the use of computational methods based on big data in recent years, together with significant theoretical achievements in the development of machine learning algorithms, have led to a renaissance in computational modeling about analysis of entity management systems in general management systems and in possessory risk management, in particular.

The accuracy of certain approaches of controlled machine learning increases significantly due to their ability to perform nonparametric analysis of data on risks that may flexibly comply with any model for estimation of information on risk events. This contrasts with some traditional statistic approaches that start from an assumption of the interrelation between the dependent and independent variables. For instance, linear regression assumes that this relation is linear while it is not always so. On the contrary, certain machine learning approaches can identify nonlinear relations, which makes them fit for data analysis in possessory risk management.

3 Machine Learning Methods and the Range of Their Use in Possessory Risk Management Systems

The range of machine learning comprises many different analytical methods, the applicability of which depends on the type of statistic problems, which is of special importance in risk management system in general and in possessory risk management, in particular. Sensu lato, machine learning may apply to three classes of statistic problems: regression, classification, and clustering.

Regression and classification problems may be solved through controlled machine learning, while clustering represents an uncontrolled approach to machine learning.

Regression issues include forecasting of a quantitative continuous dependent variable, such as GDP growth or inflation.

Linear learning methods aim to solve regression problems, including partial least squares and principal component analysis; nonlinear learning methods include penalty regression approaches such as LASSO (least absolute shrinkage and selection operator) and elastic nets. Under penalty approaches to the model, the factor for penal sanctions for complexity is normally added, which should increase the forecast productivity [5–7].

Classification issues normally include forecast of qualitative (discrete) dependent variable, which takes on values within a class, such as blood group (A/B/AB/O). Spam filtering is an example where the dependent variable may take on values SPAM/NO SPAM. Such issues may be resolved using a decision tree, the purpose of which is to provide a structured set of Yes/No questions, which may quickly sort out a broad range of functions and thus perform accurate forecast of specific results. Support vector machines also classify observations, but they use and optimize the stock that separates different classes more efficiently [5–7].

Finally, in clustering, only incoming variables are observed, while the respective dependent variable is missing. As an example, data examination by the entity's owner for fraud or opportunistic behavior of managers may be considered, provided that the owner does not know, which of the observed activities are fraudulent and which are not. However, analysis of aggregate information grouped in clusters according to the observable properties may provide an understanding of the required data. It can allow an analyst or the owner to understand, which transactions or managers' actions resemble the target ones. In some case, uncontrolled learning is used to study the data set; then, the results of this approach are used as input for controlled learning methods [8].

Within the framework of this paper, the authors should note that the use of machine learning and neural networks in possessory risk management systems has disadvantages, too. For instance, the ability of machine learning to make forecasts beyond the sample does not necessarily facilitate formulating exhaustive explanations or conclusions, which is especially important in risk management system. In addition, it is important to emphasize that statistic methods forming the basis of most machine learning systems and neural networks are usually subject to compromise solutions between the explanatory and the forecast productivity. A good prognostic model may be very complex hence difficult to interpret.

Another specific feature of machine learning and neural networks to be considered when integrating them in possessory risk management systems is that for forecasting purposes only correlations between variables should be indicated, not causal relationships. In the event of assessment of receivables level, a good inference model should explain why some debtors do not repay their debts. The debtor's performance may be estimated through its statistic relevance and fitness in the data sample. On the other hand, a good prognostic model will select indicators that demonstrate most fully and properly the prerequisites for the debtor's default. In this respect, it does not matter if any of the indicators reflect the causative factor of the debtor's insolvency or the symptoms that come with insolvency. It is important that it contains information on the possibility of debt repayment.

Another important aspect of the use of machine learning methods and neural networks in possessory risk management systems is that the use of over-sophisticated models for risk assessment may lead to the network's 'overtraining' when it describes an occasional mistake or noise instead of basic relations in the data set. The model's complexity may be motivated by a too great number of parameters in relation to the number of observations. Overtraining in machine learning is most widespread in nonparametric nonlinear models, which have complex structures so that they are difficult to interpret.

Where a model describes the noise in the data set, it may fit well for a data sample but may not work for testing the data beyond the sample.

There are several methods to cope with overtraining and improve the predictive power of machine learning models, including:

- bootstrap;
- enhancement;
- bootstrap aggregating (also known as data bundling).

In the first case, concerns are raised about the preponderance of observations due to the limitation of the training set to ensure more intensive preparation of the model based on it. For instance, in training of a model to identify fraudulent transactions, it may be required to reweight the fraudulent observations due to their relatively low number.

The data bundling algorithm is performed by the model hundreds and thousands of times, each time in a separate data subsample to increase the forecasting productivity. The final model is the average value of each of the launch models. As that average model was tested on many different data samples, it should be more tolerant of basic data modifications. Random forest is an example of a model comprising a set of different models powered by a decision tree.

When using econometrics methods in the possessory risk management system, a resulting model may be combined with a model based on different machine learning method. The result is the so-called ensemble, i.e. a model comprising a group of models, the results of which are combined through weighted average or voting. It should be noted that for many small models' averaging normally grants better forecasting for the data beyond the sample than one model selection.

It should be noted that, due to lack of significant explanatory power and immanent complexity of machine learning models, the use of such methods in risk management systems is criticized as a method of theoretical analysis of simple correlations that is inevitably fragile. It is also obvious that machine learning is based on identified correlations (past) in the sample to forecast correlations beyond the sample (future) not necessarily understanding the analyzed relations. In this respect, it is just as a questionable forecasting method, in fact—like any other statistic approach. In this case, forecast accuracy may be increased if these correlations are incorporated into the model. In addition, we should remember that if the observer has no idea about what a correlation stands for, the observer also has no idea as to what may cause a disturbance of such correlation.

4 Deep Learning and Neural Networks: From Machine Learning to Artificial Intelligence as a Basis for Possessory Risk Management Systems

Thus far, the discussion focused on the possibility of using traditional machine learning methods within the framework of possessory risk management systems, which are applied to statistic issues with clearly defined and structures datasets [9]. Under current conditions, machine learning approaches were refined and united to solve all types of complex issues functioning as artificial intelligence or neural networks.

One of the dominant approaches is deep learning, a learning approach that may be based on both controlled and uncontrolled methods, which are nonlinear by nature [10].

In deep learning, several algorithm levels are combined to imitate neurons during the multilayer process of human brain learning.

Each algorithm is equipped to extract a certain function from the data. Then, this so-called representation or abstraction is transferred to the next algorithm, which again touches on a different aspect of the data.

The stack of learning representation algorithms enables the deep learning approaches to study all data types, including low-quality unstructured data and the algorithms' ability to create respective data abstractions makes it possible for the system to generally perform the relevant analysis. It should be noted that these function levels are not developed by human engineers but rather extracted from data using the procedure for general purpose learning.

Currently, deep learning methods are widely used in different areas. This is determined by the possibility of processing large volumes of unstructured data and identification of complex patterns. Thus, deep learning was used most widely in big data analysis such as user data sets of technology giants like Google, Microsoft, and Amazon. However, the authors believe that these opportunities currently make it possible to use more actively deep learning mechanisms in possessory risk management systems.

It should be noted that in recent years, the volumes of data on different types of risks, including financial, accumulated by entities' management systems increased significantly, as the requirements to reporting expanded, and services digitalization creates a vast volume of high-frequency unstructured consumer data. As a result, owners and managers of entities need to establish and use the more powerful analytical instruments for work with big volumes of data of all types and formats subject to preservation and in some cases, to increase the level of detail of data sets for analysis.

After the economic crisis of 2014–2016, regulators both in Russia and abroad introduced many new rules and surveillance measures requiring entities' management systems to provide more detailed and frequent data on the greater number of aspects of their business models and balances. Due to this, entities' management systems have to detail the data on the main types of risks, liquidity indicators, and equity levels.

Thus, nearly all aspects of the business model of a contemporary entity are regulated and controlled through such risk indicators, in fact, the very management system often is consigned to matters of business processes optimization in accordance with certain limitations. Consequently, to compete effectively, entity management systems should find a way to search for and collect consumer data that should be optimal for regulators and efficient for owners, for a detailed understanding of the customers' preferences and behavior.

Obviously, the most efficient mechanism to solve the aforesaid issue is using machine learning methods taking into account the capabilities for collection and detailed analysis of vast data sets that they provide to the entities' owners and managers.

In addition, the authors believe that it is required to distinguish the machine learning methods that may be integrated into possessory risk management systems. More traditional machine learning methods could be used for intelligent analysis—mining—of high-quality structures control data. Also, Google-like methods of deep learning and neural networks, which—due to their ability of learning representation—cope better with high-frequency low-quality big data sources, should be used to collect and analyze such data.

Below, we consider three cases of use of machine learning in possessory risk management systems: risk modeling, identification of fraud and laundering of illegally obtained proceeds as well as misconduct and abuse in financial risk management.

5 Peculiarities of the Use of Machine Learning in Possessory Risk Management Systems

Since the early 2000s, the management science developed a vast base of research devoted to the use of machine learning methods for model entity's risks.

For instance, Russian and foreign authors pay utmost attention to the analysis of the theory and practice of neural networks use for modeling the main types of risks accompanying the activities of small and medium businesses. A significant number of papers are devoted to the assessment of company solvency using support vector methods, using which demonstrates that they provide more accurate forecasts for the data beyond the samples than the existing methods [3–5, 10–12].

Another significant aspect of theoretical understanding of possessory risk management systems development based on machine learning is studying of application of generalized classification and regression trees (CART) to large data set of a commercial bank to build consumer credit risk models. They combine traditional credit factors, such as debt to income ratio and bank consumer transactions, which increases significantly the forecasting power of the model [10].

It should be noted that the said studies used linear, logit and probit regression to model entities' risks, including financial risks [3–5, 8]. The authors believe that currently, use of machine learning methods within the frameworks of possessory risk management systems to increase the quality of financial risk forecasts. For data studies, uncontrolled methods are normally used while regression and classification methods (decision trees, support vector methods) can predict key variables of financial risks such as the probability of default or default with due regard to losses. It is also evident that forecast quality may be improved through establishing the mechanisms for interaction between entities' possessory risk management systems and bank's risk management systems, which normally have vast data sets on their borrowers and may be used as inputs for analysis of risk of interaction between entities. However, the creation of such databases may be complicated by the fact that the methods used may be sophisticated and the models may be sensitive to data overtraining. Thus, the quality of the data available in banks' risk management systems is not sufficient for deep statistical analysis within the framework of operation of entities' possessory risk management systems, especially where banks' risk management systems not always can consolidate the data in accordance with the logic required for efficient operation of an entity's possessory risk management system due to controversial data definitions in different aspects of risk assessment and absence of a uniform risk assessment system.

Another area of interaction for entities' possessory risk management systems and banks' risk management systems where machine learning has been used for over ten years - and rather successfully - is fraud identification in non-cash payments between legal entities for goods and services. Banks equipped their clearing and settlement operations infrastructure with monitoring systems (so-called document flow mechanisms) that track incoming payment orders for potentially fraudulent activities. Fraudulent transactions may be blocked in real time. It should be noted that the model examples of fraudulent activities with payment orders of legal entities were collected jointly by entities' possessory risk management systems and banks' risk management systems for a certain period. High frequency of payment order transactions of entities being the customers of banks ensures large data sets required for training of the algorithm forming the basis of interaction of entities' possessory risk management systems and banks' risk management systems as well as testing and functionality check of such interaction. Besides, as banks may precisely check, which transactions were fraudulent, they are able to create clear data sets with relevant tags that allow differentiating between fraud and everyday transactions. It should be noted that the least developed aspect of the use of machine learning mechanisms and neural networks in the development of entities' possessory risk management systems and banks' risk management systems is the identification of laundering of illegally obtained proceeds and financing of terrorism. When interacting with banks' risk management system, many entities' possessory risk management systems still rely on traditional systems, which are powered by rules focused on analysis and tracing of individual transactions or simple schemes of assessment of transaction sets. In most cases, such systems are incapable of timely identification of complex transaction schemes or obtaining a holistic picture of transaction behavior in the bank's payment infrastructure or in the accounting system within the framework of entity's possessory risk management system. The use of such schemes of possessory risk management results in many false positives, which reduces confidence in the possessory risk management systems of both the management of the entity and the owners. In its turn, assessment, prevention, and filtering of such false positives from suspicious transactions require significant human resources both in banks' risk management systems and entities' possessory risk management systems. Evidently, as new digitalization standards are implemented both in Russia and abroad, the scale of use of such obsolete systems will decrease, so the quality and efficiency of possessory risk management systems will increase as will their interaction with banks' risk management systems.

6 Conclusion

Evidently, use of machine learning, neural networks, and artificial intelligence is studied and analyzed now in all areas of life of society, including in entity management. In the environment of changing the technological paradigm, owners and managers of entities both in Russia and abroad search for the most efficient theoretical and practical approaches to identification, analysis, assessment, and management of ever-increasing volumes of data related to both statutory reporting and everyday operation of an entity and, consequently, strive to improve the efficiency and performance of the entity and create competitive advantages.

It should be noted that in the digital economy, every single aspect of an entity's business model that can be improved using machine learning, neural networks, and artificial intelligence. These technologies are used now in such areas of operation as customer preferences analysis, risk management, fraud identification, and violation preventing as well as automation of customer support systems or implementation of automated processes of counterparty check.

This paper attempts to start the analysis of the implementation of machine learning and neural networks for entities' possessory risk management systems in modeling credit risks, fraud identification in settlement and clearance service in banks, and combating the laundering of illegally obtained proceeds.

Evidently, a preliminary conclusion can be made based on this paper, whereas the scale of applying machine learning and neural networks in entities' possessory risk management systems increases and many management systems are at the experimental stage.

Machine learning includes several tools of statistic learning, which usually can analyze very big volumes of data, thereby ensuring a high level of detail and depth of analysis, which is currently the most promising method of data structuring for the purposes of risk situation forecasting. The ability to identify nonlinear relations and perform data analysis without making assumptions on the form or relations among variables (i.e. nonparametric) implemented in many machine learning methods and neural networks increases the level of detail, with which risk data can be analyzed and the results of risk event release can be predicted.

Such improved and often automated analytical capabilities make it possible for entity management systems to better understand the logic of functioning of basic business processes such as receivables formation, possessory risk management, customer interaction, and counterparty payments.

Due to ever-increasing data volumes received during those processes, machine learning may identify richer, more complex models and interrelations in the analysis of transactions with counterparties or credit risks or, by combining different data sets, to receive the more accurate comprehensive conclusions in the monitoring of fraud in settlements with customers and counterparties.

References

1. Goncharenko, L.P., Sybachin, S.A., Khachaturyan, M.V.: Peculiarities of Organizational Economic Mechanism Development in correspondence with State Strategic Management in Russia. In: Proceedings of Conference Trends of Technologies and Innovations in Economic and Social Studies (TTIESS 2017), Advances in Economics, Business and Management Research, vol. 38. Atlantis Press (2017)
2. Khachaturyan, V.M.: Organizational-economic mechanism of formation and realization of the industrial policy within the framework of the CMEA and the EU: experience and prospects for Russia. Int. Bus. Manag. **10**(14), 2677–2686 (2016)
3. Badvan Nemer, L., Blazhenkova Natalia, M., Klicheva Evgeniia, V., Karaev Alan, K., Yarullin Raul, R.: Increasing the efficiency of the state fiscal and budgetary policy in modern conditions. Int. J. Appl. Bus. Econ. Res. **15**(23), 125–138 (2017)
4. Aiginger, K., Davies, St.: Industrial Specialization and Geographic Concentration: Two Sides of the Same Coin, p. 235. World Bank, Washington (2011)
5. Baldwin, R., Martin, P.: Handbook of Regional and Urban Economics, p. 2671. Palgrave Macmillan, London (2010)
6. Krugman, P., Venables, A.J.: Integration, specialization, and adjustment. Eur. Econ. Rev. **40,** 959 (2010)
7. Leeder, E., Sysel, Z., Lodl, P.: Cluster - Basic Information, p. 56. Cambridge University Press, Cambridge (2011)

8. Marshall, A.: Elements of the Economics of Industry, p. 145. Palgrave Macmillan, London (2011)
9. Marshall, A.: The Economics of Industry, p. 134. Palgrave Macmillan, London (2011)
10. McKee, D., Dean, R., Leahy, W.: Regional Economics Theory and Practice, p. 93. Free Press, New York (2010)
11. Kochetkov, V.N., Shipova N.: Economic risk and methods of measurement: tutorial. K.: European University of Finance, Informational Systems, Management and Business (2014)
12. Bychkova, S.M., Rastamanov, L.N.: Risks in Auditing, Bychkova, S.M. (ed.). Finance and statistics (2013)

Size and Alignment Independent Classification of the High-Order Spatial Modes of a Light Beam Using a Convolutional Neural Network

Aashima Singh[1], Giovanni Milione[1], Eric Cosatto[2], and Philip Ji[1](✉)

[1] Optical Networking and Sensing Department, NEC Laboratories America, Inc., Princeton, NJ 08540, USA
{gmiilione,pji}@nec-labs.com

[2] Machine Learning Department, NEC Laboratories America, Inc., Princeton, NJ 08540, USA

Abstract. The higher-order spatial modes of a light beam are receiving significant interest. They can be used to further increase the data speeds of high speed optical communication, and for novel optical sensing modalities. As such, the classification of higher-order spatial modes is ubiquitous. Canonical classification methods typically require the use of unconventional optical devices. However, in addition to having prohibitive cost, complexity, and efficacy, such methods are dependent on the light beam's size and alignment. In this work, a novel method to classify higher-order spatial modes is presented, where a convolutional neural network is applied to images of higher-order spatial modes that are taken with a conventional camera. In contrast to previous methods, by training the convolutional neural network with higher-order spatial modes of various alignments and sizes, this method is not dependent on the light beam's size and alignment. As a proof of principle, images of 4 Hermite-Gaussian modes (HG_{00}, HG_{01}, HG_{10}, and HG_{11}) are numerically calculated via known solutions to the electromagnetic wave equation, and used to synthesize training examples. It is shown that as compared to training the convolutional neural network with training examples that have the same sizes and alignments, a $\sim 2\times$ increase in accuracy can be achieved.

Keywords: Convolutional neural network · Machine vision · Optical communication · Spatial modes

1 Introduction

A light beam's transverse (i.e., perpendicular to the light beam's direction of propagation) spatial shape can be made up of one or a superstition of multiple spatial modes. A spatial mode is an electric field whose complex amplitude is described by a mathematical function that is a solution to an electromagnetic wave equation. For example, the light beam of a conventional laser pointer typically comprises a spatial mode that is referred to as the fundamental spatial mode, i.e., the lowest order solution to such a wave equation. For the fundamental mode, the electric field is characterized by being strongest at the light beam's center and becoming gradually less strong farther from the light beam's center. The electric fields of higher-order spatial modes, being high-order

K. Arai et al. (Eds.): CompCom 2019, AISC 997, pp. 688–696, 2019.
https://doi.org/10.1007/978-3-030-22871-2_47

solutions to the wave equation, have more complex spatial dependencies. For example, there are higher-order spatial modes referred to as Hermite-Gaussian (HG) modes, as shown in Fig. 1, which have "lobe"-like spatial dependencies. Higher-order spatial modes can propagate over free space (e.g. Earth's atmosphere, outer space) and waveguides (e.g. optical fibers).

Higher-order spatial modes are receiving significant interest for various applications. For example, higher-order spatial modes can be used to further increase the data speeds of high speed optical communications over free space and optical fibers, i.e., each higher-order spatial mode can be used as a channel over which data is encoded [2–4]. Or, each higher-order spatial mode can be used as a data state with which to encode data [5]. Also, higher-order spatial modes can be used for novel imaging modalities. For example, using higher-order spatial modes, the lateral motion and, rotational orientation of a remote object can be remotely detected [6, 7]. And, using higher-order spatial modes, the shape of an object can be remotely detected [8].

The classification of higher-order spatial modes is ubiquitous, especially with respect to the applications above. For fundamental spatial modes, classification comprises characterization of the spatial modes' quality via the so-called M^2 factor, i.e., a product of the beams' measured size and divergence. However, higher-order spatial modes are more various and, the complex amplitude of the electric field of each has more complex spatial dependence. Therefore, classification of higher-order spatial modes requires a more complex spatial analysis including, differentiation of the high-order spatial modes from each other. Measurement of the M^2 factor is not sufficient.

Canonical systems and methods to classify higher-order spatial modes comprise measurement of the complex amplitude of a light beam's electric field. Typically, the complex amplitude of a light beam's electric field is indirectly measured using holographic techniques via unconventional optical devices. Such optical devices must emulate the complex spatial dependencies of the complex amplitudes of the electric fields of higher-order spatial modes. Such optical devices include, liquid crystal on silicon based spatial light modulators [6–9]. While effective, complex holographic techniques via unconventional optical devices are dependent on a light beam's alignment and size. Additionally, such optical devices require quality of fabrication that depends on how well the complex spatial dependencies of the complex amplitudes of the electric fields of higher-order spatial modes can be emulated. As such, measuring the complex amplitude of a light beam's electric field using holographic techniques via unconventional optical devices may have prohibitive complexity, cost and efficacy. Therefore, a method to classify the high-order spatial modes of a light beam is required that: (1) does not require measurement of the complex amplitude of a light beam's electric field using holographic techniques via unconventional optical device; (2) is independent on a light beam's alignment and size.

Recently, methods to classify the higher-order spatial modes of a light beam that do not require the measurement of the complex amplitude of a light beam's electric field using holographic techniques via unconventional optical devices were demonstrated. In those methods, the higher-order spatial modes of a light beam have been classified using a conventional camera and, a convolutional neural network (CNN) [10, 11]. The conventional camera records the transverse, spatial dependencies of the intensities of the higher-order modes, which are then classified by the CNN. The CNN is trained via training examples of the transverse, spatial dependencies of the intensities of the higher-order spatial modes, which are recorded using the camera. While effective, those methods are dependent on the light beam's alignment and size because, the training examples are higher-order spatial modes that have the same sizes and alignments. While CNNs boast translation (alignment) independence, that independence is constrained in great part by the size of the convolution kernels and, the use of pooling layers. As such, a method to classify higher-order spatial modes that is independent on a light beam's alignment and size is needed.

In this work, a method to classify the higher-order spatial modes of a light beam that is not dependent on a light beam's alignment and size is proposed. In this method, a CNN is used to classify the transverse, spatial dependencies of the intensities of higher-order spatial modes. However, in contrast to previous work, the transverse, spatial dependencies of the intensities of higher-order spatial modes of various sizes and alignments are used as training examples, with which the CNN is trained. As a proof of principle, 4 HG modes (HG_{00}, HG_{01}, HG_{10}, and HG_{11}) are classified using a CNN that is based on the LeNet architecture [12]. The transverse, spatial dependencies of the intensities of the HG modes of various sizes and alignments are numerically calculated via known solutions to the electromagnetic wave equation, and used to synthesize training examples. It is shown that, as compared to training the CNN with training examples that have the same sizes and alignments, a $\sim 2\times$ increase in accuracy can be achieved.

In Sect. 2, higher-order spatial modes as higher-order solutions to the electromagnetic wave equation are described. In Sect. 3, the convolutional neural network that is used to classify the higher-order spatial modes is described. Additionally, the training of the convolutional neural network for misaligned higher-order spatial modes is described. And, the results of the classification of the higher-order spatial modes are presented. In Sect. 4, the results are summarized and prospects for future work are discussed.

2 Higher-Order Spatial Modes

Spatial modes are the mathematical functions that can be used to describe the transverse (i.e., perpendicular to a light beam's direction of propagation) spatial dependence of the complex amplitude of a light beam's electric field [1]. The mathematical functions are solutions to an electromagnetic wave equation. For example, the Helmholtz wave equation is given by [1]:

$$\nabla^2(x,y)u(x,y) + k^2 u(x,y) = 0, \tag{1}$$

where $\nabla^2(x,y)$ is the Laplacian in rectangular coordinates and, $k = 2\pi/\lambda$, where λ is the light beam's wavelength.

HG modes are solutions to Eq. (2) in rectangular coordinates. Hermite-Gaussian modes are given by the equation:

$$HG_{m,n}(x,y) = c_{m,n} H_m\left(\frac{x\sqrt{2}}{w}\right) H_n\left(\frac{y\sqrt{2}}{w}\right) \exp\left(\frac{-x^2-y^2}{w^2}\right), \tag{2}$$

where $H_m(\ldots)$ and $H_n(\ldots)$ are Hermite polynomials, w is the waist size of the higher-order spatial modes, $c_{m,n}$ are complex coefficients, and $m, n = 0,1,2,\ldots$

The transverse, spatially dependent intensities of HG modes are given by the equation ($|HG_{mn}(x,y)|^2$):

$$I_{m,n}(x,y) = |c_{m,n}|^2 H_m^2\left(\frac{x\sqrt{2}}{w}\right) H_n^2\left(\frac{y\sqrt{2}}{w}\right) \exp\left(\frac{-2(x^2 - y^2)}{2w^2}\right), \tag{3}$$

Now, consider the transverse, spatial dependencies of HG modes that have varying sizes and alignments, with respect to, for example, the pixelated sensor area of a camera. The sizes of HG modes will vary if their waist sizes vary, i.e., $w + \delta w$, where δw is the variation of the waist size. The alignments of HG modes will vary if the position of the center of the light beam varies, i.e., $x + \delta x$ and $y + \delta y$, where δx and δy are the variations of the x and y coordinates, respectively. The variation of an HG mode's size and alignment are shown schematically in Fig. 1.

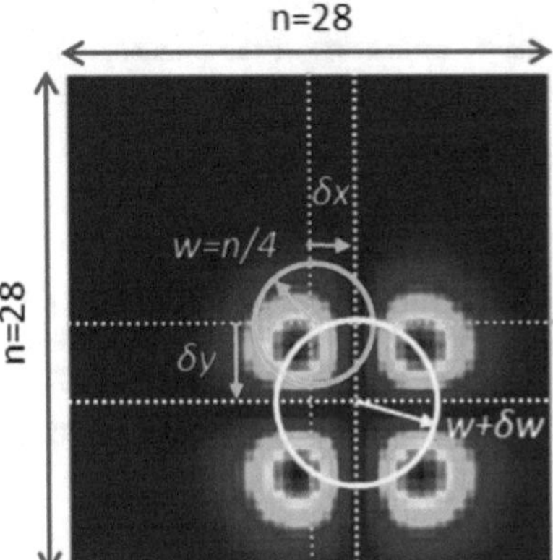

Fig. 1. The variation of an HG mode's waist size, i.e., $w + \delta w$, where δw is the variation of the waist size, and alignment, i.e., $x + \delta x$ and $y + \delta y$, where δx and δy are the variations of x and y coordinates, respectively.

The transverse, space dependent intensities of HG modes whose sizes and alignments vary are given by the equation:

$$I_{m,n}(x,y)(x + \delta x, y + \delta y; w + \delta w) = |c_{m,n}|^2 H_m^2\left(\frac{(x + \delta x)\sqrt{2}}{w + \delta w}\right) H_n^2\left(\frac{(y + \delta y)\sqrt{2}}{w + \delta w}\right) \times \exp\left(\frac{-2\left((x + x\delta)^2 - (y + y\delta)^2\right)}{2(w + \delta w)^2}\right) \tag{3}$$

The transverse, spatial dependencies of the intensities of numerically calculated HG_{00}, HG_{01}, HG_{01}, and HG_{11} modes from Eq. (3) that have the same sizes ($\delta w = 0$) and, are not misaligned ($\delta x = 0,\ \delta y = 0$) are shown in Fig. 2.

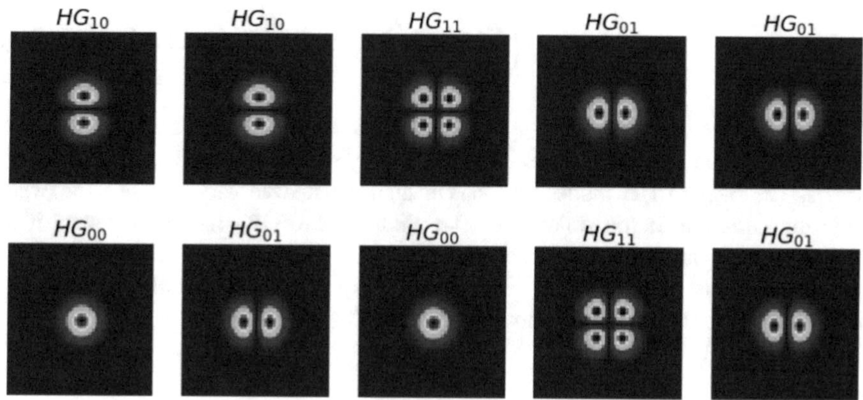

Fig. 2. The transverse, spatial dependencies of the intensities of numerically calculated HG_{00}, HG_{01}, HG_{01}, and HG_{11} from Eq. (3) that have the same sizes ($\delta w = 0$) and, are not misaligned ($\delta x = 0,\ \delta y = 0$).

The transverse, spatial dependencies of the intensities of numerically calculated HG_{00}, HG_{01}, HG_{01}, and HG_{11} modes from Eq. 3 that have varying sizes $\delta w = 0$; $\delta x, \delta y = 0$, Test examples: $-w/2 \leq \delta w \leq +w/2$; $-2w \leq \delta x, \delta y \leq +2w$ (Case III) Training examples: $-w/2 \leq \delta w \leq +w/2$; $-2w \leq \delta x, \delta y \leq +2w$, Test examples: $-w/2 \leq \delta w \leq +w/2$; $-2w \leq \delta x, \delta y \leq +2w$ are shown in Fig. 3.

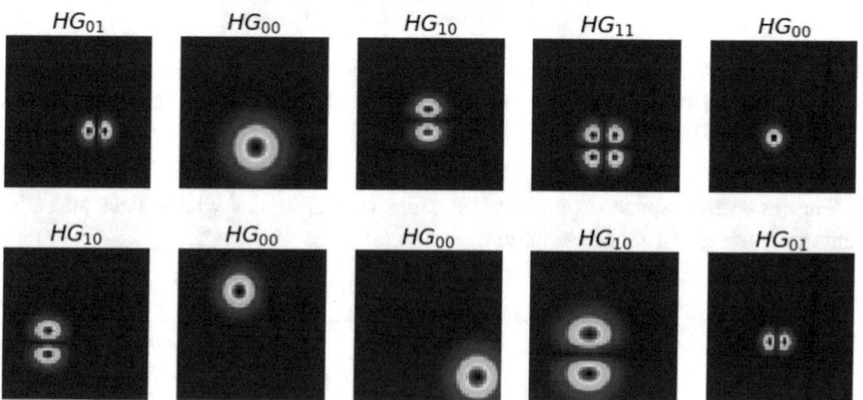

Fig. 3. The transverse, spatial dependencies of the intensities of numerically calculated HG_{00}, HG_{01}, HG_{01}, and HG_{11} from Eq. 3 that have varying sizes (δw: $\{-w/2, \ldots, +w/2\}$) and, misalignments (δx: $\{-3w/2, \ldots, +3w/2\}$, δy: $\{-3w/2, \ldots, +3w/2\}$).

3 Convolutional Neural Network

A block flow diagram of the CNN architecture that was used is shown in Fig. 4. The CNN is based on the LeNet architecture [12]. The CNN comprised 3 convolutional layers, where the 1st convolutional layer had 32 3 × 3 filters, the 2nd layer had 64 3 × 3 filters and, the 3rd layer had 128 3 × 3 filters (i.e., the kernel size of all filters was 3 × 3). After each convolutional filter, there was a leaky rectifier linear unit (ReLU) then, a max pooling layer of size 2 × 2. Finally, after the last max pooling layer, there was a flatten layer, then a dense (fully connected) layer that comprised 128 units and, an output layer that comprised 4 units. In total, the CNN comprised 896,132 trainable parameters.

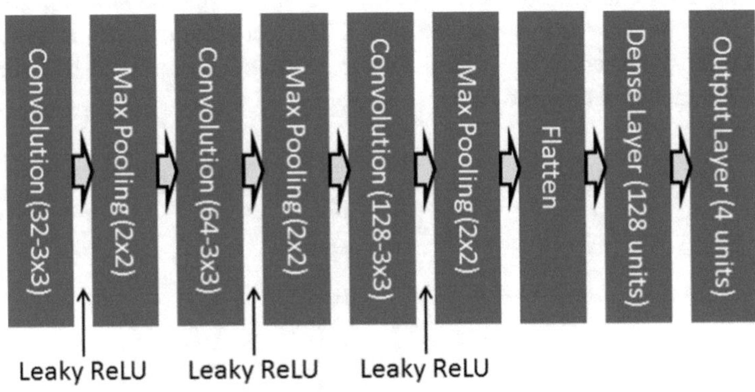

Fig. 4. CNN architecture: 3 convolutional layers, where the 1st convolutional layer had 32 3 × 3 filters, the 2nd layer had 64 3 × 3 filters and, the 3rd layer had 128 3 × 3 filters. After each convolutional filter, there was a leaky ReLU then, a max pooling layer of size 2 × 2. Finally, after the last max pooling layer, there was a flatten layer, then a dense (fully connected) layer that comprised 128 units and, an output layer that comprised 4 units.

The CNN was trained using an Adam optimizer, which used categorical cross entropy loss. 1000 training examples were used with a batch size of 64, where 800 of the training examples were used for training and, 200 of the training examples were used for validation. The size of each training example was 28 × 28 × 1. The CNN trained over 25 epochs. After each epoch, training accuracy and training loss and, validation accuracy and validation loss were calculated. Finally, after 25 epochs, 200 more training examples were passed through the CNN, with which a testing accuracy and a testing loss were calculated.

As a proof of principle, the CNN was trained on and made to classify 4 HG modes, i.e., 4 classes: HG_{00}, HG_{01}, HG_{10}, and HG_{11}. Training examples and test examples of HG modes were synthesized by numerically calculating their transverse, spatially dependent intensities using Eq. (3). The sizes of the training examples and, the testing examples were 28 × 28 × 1, as shown in Fig. 1. The sizes and alignments of the HG modes were varied by varying δw, δx, and δy in Eq. 3, as shown in Fig. 1.

The classification of the HG modes by the CNN was evaluated for three cases, where each case comprised different combinations of training examples and test examples that had various combinations of alignments and sizes, as shown in Table 1: (Case I) Training examples: $\delta w = 0$; $\delta x, \delta y = 0$, Test examples: $\delta w = 0$; $\delta x, \delta y = 0$. (Case II) Training examples: $\delta w = 0$; $\delta x, \delta y = 0$, Test examples: $-w/2 \leq \delta w \leq +w/2$; $-2w \leq \delta x, \delta y \leq +2w$ (Case III) Training examples: $-w/2 \leq \delta w \leq +w/2$; $-2w \leq \delta x, \delta y \leq +2w$, Test examples: $-w/2 \leq \delta w \leq +w/2$; $-2w \leq \delta x, \delta y \leq +2w$.

First, consider case I. For case I, a plot of the resulting training accuracy and validation accuracy and a plot of the resulting training loss and validation loss, as a function of epoch number are shown in Fig. 5. As can be seen, as expected, the training accuracy and validation accuracy both reach 1.0 and, the training loss and validation loss reach 10^{-7} as the number of epochs increase. This is expected as all of the training examples for each class are the same. The test accuracy and the test loss were calculated and, are shown in the first row of Table 1. The test accuracy and the test loss were 100% and 10^{-7}, respectively. Again, this is expected because, all off the testing examples for each class are the same.

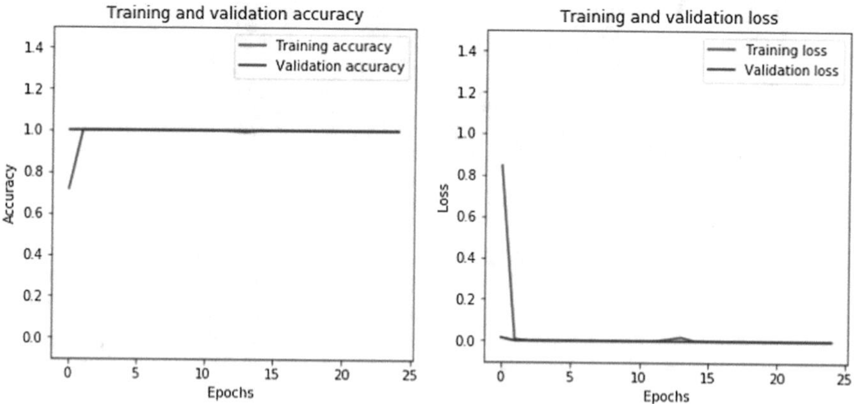

Fig. 5. A plot of the resulting training accuracy and validation accuracy and a plot of the resulting training loss and validation loss, as a function of epoch number, for cases I and II: Training examples: $\delta w = 0$; $\delta x, \delta y = 0$.

Next, consider case II. For case II, a plot of the resulting training accuracy and validation accuracy and a plot of the resulting training loss and validation loss, as a function of epoch number, are shown in Fig. 5, being the same as for case II. Again, as can be seen, as expected, the training accuracy and validation accuracy both reach 1.0 and, the training loss and validation loss reach 10^{-7} as the number of epochs increase. Again, this is expected as all of the training examples for each class are the same. The test accuracy and the test loss were calculated and, are shown in the second row of Table 1. In contrast to case I, the test accuracy and the test loss were 41.5% and 8.3976, respectively. This is due to the fact that the training examples are HG modes that have the same sizes and alignments but, the testing examples have varying sizes and alignments. As such, there is a $\sim 2\times$ reduction in test accuracy.

Finally, consider case III. For case III, a plot of the resulting training accuracy and validation accuracy and a plot of the resulting training loss and validation loss, as a function of epoch number, are shown in Fig. 6. The training accuracy and validation accuracy both reach 1.0 and, the training loss and validation loss reach 0.0038 as the number of epochs increase. The test accuracy and the test loss were calculated and, are shown in the third row of Table 1. In contrast to case II, the test accuracy and the test loss were 100.0% and 0.0032, respectively. This is due to the fact that the training examples are HG modes that have varying sizes and alignments and, the testing examples have varying sizes and alignments. As such, there is a ~2× increase in test accuracy, being similar to case I.

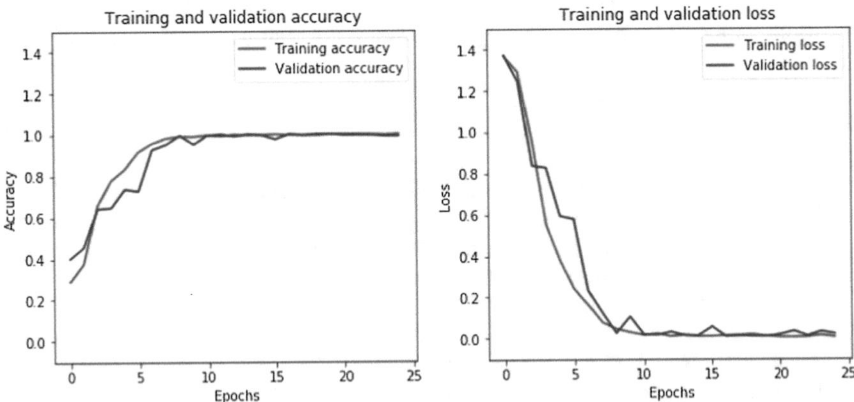

Fig. 6. A plot of the resulting training accuracy and validation accuracy and a plot of the resulting training loss and validation loss, as a function of epoch number, for case III: Training examples: $-w/2 \leq \delta w \leq +w/2$; $-2w \leq \delta x, \delta y \leq +2w$.

Table 1. Test accuracy and test loss for cases I, II, and III.

Case	Training examples	Test examples	Test accuracy	Test loss
I	$\delta w = 0$; $\delta x, \delta y = 0$	$\delta w = 0$;; $\delta x, \delta y = 0$	100.0%	10^{-7}
II	$\delta w = 0$, $\delta x, \delta y = 0$	$-w/2 \leq \delta w \leq +w/2$, $-2w \leq \delta x, \delta y \leq +2w$	41.5%	8.3976
III	$-w \leq \delta w \leq +w$, $-2w \leq \delta x, \delta y \leq +2w$	$-w/2 \leq \delta w \leq +w/2$, $-2w \leq \delta x, \delta y \leq +2w$	100.0%	0.0035

4 Conclusion

In conclusion, a method to classify the higher-order spatial modes of a light beam that is not dependent on a light beam's alignment and size was proposed. In this method, a CNN was used to classify the transverse, spatial dependencies of the intensities of higher-order spatial modes. However, in contrast to previous work, the transverse, spatial dependencies of the intensities of higher-order spatial modes of various sizes and alignments were used as training examples, with which the CNN was trained. As a proof of principle, 4 HG modes (HG_{00}, HG_{01}, HG_{10}, and HG_{11}) were classified using a CNN that was based on the LeNet architecture [12]. The transverse, spatial dependencies of the intensities of the HG modes of various sizes and alignments were numerically calculated via known solutions to the electromagnetic wave equation, and used to synthesize training examples. It was shown that, as compared to training the CNN with training examples that have the same sizes and alignments, a $\sim 2\times$ increase in accuracy was achieved.

Higher-order spatial modes can be reliably simulated numerically. As such, it is expected that an experimental evaluation of this work will yield comparable results. Future work will comprise experimentally generating higher-order spatial modes of various alignments and sizes and using a convolutional neural network to classify them.

References

1. Sale, B.E.A., Teich, M.C.: Fundamentals of photonics, 2nd edn. Wiley, Hoboken, NJ (2007)
2. Huang, H., Milione, G., Lavery, M.P.J., Xie, G., Ren, Y., Cao, Y., Ahmed, N., Nguyen, T. A., Nolan, D.A., Li, M.-J., Tur, M., Alfano, R.R., Willner, A.E.: Mode division multiplexing using an orbital angular momentum mode sorter and MIMO-DSP over a graded-index few-mode optical fibre. Sci. Rep. **5**, 14931 (2015)
3. Milione, G., Lavery, M.P.J., Huang, H., Ren, Y., Xie, G., Nguyen, T.A., Karimi, E., Marrucci, L., Nolan, D.A., Alfano, R.R., Willner, A.E.: 4 × 20 Gbit/s mode division multiplexing over free space using vector modes and a q-plate mode (de)multiplexer. Opt. Lett. **40**(9), 1980–1983 (2015)
4. Ip, E., Milione, G., Li, M.-J., Cvijetic, N., Kanonakis, K., Stone, J., Peng, G., Pinto, X., Montero, C., Moreno, V., Linares, J.: SDM transmission of real-tome 10Gbe traffic using commercial SFP + transceivers over 0.5 km elliptical-core few-mode fiber. Opt. Express **23**, 17120–17126 (2015)
5. Milione, G., Nguyen, T.A., Nolan, Leach, J.D.A., Alfano, R.R.: Using the nonseparability of vector beams to encode information for optical communication. Opt. Lett. **40**, 4887–4890 (2015)
6. Cvijetic, N., Milione, G., Wang, T.: Detecting lateral motion using light's orbital angular momentum. Sci. Rep. **5**, 15422 (2015)
7. Milione, G., Wang, T., Han, J., Bai, L.: Remotely sensing an object's rotational orientation using the orbital angular momentum of light. Chin. Opt. Lett. **15**, 030012 (2017)
8. Xie., G., Song, H., Zhao, Z., Milione, G., Ren, Y., Liu, C., Zhang, R., Changjing, B., Li, L., Wang, Z., Pang, K., Starodubov, D., Lynn, B., Tur., M., Willner, A.E.: Using a complex optical orbital-angular-momentum spectrum to measure object parameters. Opt. Lett. **42**(21), 4482–4485 (2017)
9. Forbes, A., Dudley, A., Mclaren, M.: Creation and detection of optical modes with spatial light modulators. Adv. Opt. Photon. **8**, 200–227 (2016)
10. Lohani, S., Knutson, E.M., O'Donnell, M., Huver, S.D., Glasser, R.T.: On the use of deep neural networks in optical communications. App. Opt. **57**(15), 4180 (2018)
11. Doster, T., Watnik, A.T.: Machine learning approach to OAM beam demultiplexing via convolutional neural networks. App. Opt. **56**, 3386–3396 (2017)
12. Lecun, Y., Bottuo, L., Bengio, Y., Haffner, P.: Gradient-based learning applied to document recognition (1998)

Automatic Induction of Neural Network Decision Tree Algorithms

Chapman Siu[1,2(✉)]

[1] Faculty of Engineering and Information Technology, School of Electrical and Data Engineering, University of Technology Sydney, Sydney, Australia
chapman.siu@student.uts.edu.au
[2] Suncorp Group Limited, Brisbane, Australia

Abstract. This work presents an approach to automatically induction for non-greedy decision trees constructed from neural network architecture. This construction can be used to transfer weights when growing or pruning a decision tree, allowing non-greedy decision tree algorithms to automatically learn and adapt to the ideal architecture. In this work, we examine the underpinning ideas within ensemble modelling and Bayesian model averaging which allow our neural network to asymptotically approach the ideal architecture through weights transfer. Experimental results demonstrate that this approach improves models over fixed set of hyperparameters for decision tree models and decision forest models.

Keywords: Decision tree · Neural network · Oblique decision trees

1 Introduction

Decision trees and their variants have had a rich and successful history in machine learning in general, and their performance has been empirically demonstrated in many competitions and even in automatic machine learning settings.

Various approaches have been used to enable decision tree representations within a neural network setting, in which this paper will consider non-greedy tree algorithms which are built on top of oblique decision boundaries through probabilistic routing [7]. In this way, decision tree boundaries and the resulting classification is treated as a problem which can be learned through back propagation in a neural network setting [8].

On the neural network component, it has been further demonstrated that highway networks can viewed as an ensemble of shallow neural networks [9]. As ensembles of classifiers are related to the Bayesian Model Averaging in an

This is a pre-print of a contribution "Chapman Siu, Automatic Induction of Neural Network Decision Tree Algorithms". To appear in Computing Conference 2019 Proceedings. Advances in Intelligent Systems and Computing.

K. Arai et al. (Eds.): CompCom 2019, AISC 997, pp. 697–704, 2019.
https://doi.org/10.1007/978-3-030-22871-2_48

asymptotic manner [5], thus, creating a decision tree model within a neural network setting over a highway network can be used to determine the optimal neural network architecture and by extension the optimal hyperparameters for decision tree learning.

As our contribution, we aim to provide an automated way to induce decision tree whilst retaining existing weights in order to progressively grow or prune decision trees in an online manner. This simplifies the hyperparameters required in choosing models, instead allowing our algorithm to automatically search through the ideal neural network architecture. As such in this work, we modify existing non-greedy decision tree algorithms through stacking our models through modifying the routing algorithm of decision trees. Thus whilst previously, a single induced decision tree may have only one set of training for the leaf nodes, in our approach, a single decision tree can have different set of leaf nodes stacked in order determine the ideal neural network architecture.

This paper seeks to identify and bridge two commonly used Machine Learning techniques in the form of tree models and neural networks, as well as identifying some avenues for future research.

2 Background

Within decision tree algorithms, research has been done to grow tree and prune trees in a post-hoc manner, greedy trees are limited in their ability to fine tune the split function once a parent node has already been split [7]. In this section, we will briefly outline related works for non-greedy decision trees, approaches to extending non-greedy tree to ensemble models and finally mechanisms for performing model choice through Bayesian model determination.

2.1 Inducing Decision Trees in Neural Networks

Inducing non-greedy decision trees has been done through construction of oblique decision boundaries [7,8]. This has been done through soft-routing of the decision tree wherein the contribution of each leaf node to the final probability is determined probabilistically. One of the contributions by Kontschieder et al. [8] compared with other approaches the separating of training the probabilistic routing of the underlying binary decision tree and the training of the leaf classification nodes which need not be binary classification. The decision tree algorithm was also modified in through a shallow ensemble manner to a decision forest through bagging the classifiers.

In early implementation of decision trees, algorithms used were often using recursive partitioning methods, which aim to perform partitions in the form of $X_i > k$ where X_i is one of the variables in the dataset and k is a constant which is the split decision. These decision trees are also called *axis-parallel* [6], because each node produces a axis-parallel hyperplane in the attribute space. These trees are often considered greedy trees, as they grow a tree one node at and time with no ability to fine tune the splits based on the results of training at lower levels of the tree [7].

In contrast, recent implementations of decision trees focus instead on the ability to update the tree in an *online* fashion leading to non-greedy optimizations typically based on *oblique decision trees* [7,8]. the goal of oblique decision trees is to change the partition decisions instead to be in the form $\sum_{i=1}^{p} a_i X_i + a_{p+1} > 0$ where $a_i, ..., a_{p+1}$ are real-valued coefficients. Theses tests are equivalent to hyperplanes at an oblique orientation relative to the axis hence the name *oblique decision trees.* From this setting, one could convert oblique decision trees to the axis-parallel counterpart by simply setting $a_i = 0$ for all coefficients except one.

2.2 Ensemble Modelling and the Model Selection Problem

Ensemble modelling within the neural networks has also been covered by Veit et al. [9], who demonstrated the relationship between residual networks (and by extension Highway Networks) and the shallow ensembling of neural networks, in the form $y_i^{(1)} \cdot t(y_i^{(1)}) + y_i^{(2)} \cdot t(y_i^{(2)}) + ... + y_i^{(n)} \cdot t(y_i^{(n)})$. Furthermore, in this setting as we are interested in stacking models of the same class, Le and Clarke [5] have demonstrated the asymptotic properties in stacking and Bayesian model averaging. Approaches like sequential Monte Carlo methods [3] can be used in order in order to change state and continually update the underlying model.

A simple approach to consider an ensemble approach to the problem. In this setting we would simply treat the new data *independent* of the old data and construct a separate learner. Then we can combine it together using a stacking approach. In this setting, we aim to combine models as a linear combination together with a base model which might represent any kind of underlying learner [10].

More recently there have been attempts at building ensemble tree methods for online decision trees, including the use of bagging techniques in a random forest fashion [8]. Furthermore it has been demonstrated that boosting and ensemble models have connections with residual networks [4,9], giving the rise to the possibility of constructing boosted decision tree algorithms using neural network frameworks.

These approaches to ensemble models have a Bayesian parallel. In the Bayesian model averaging algorithms. These models are related to stacking [5], where the marginal distribution over the data set is given by $p(\mathbf{x}) = \sum_{h=1}^{H} p(\mathbf{x}|h)p(h)$. The interpretation of this summation over h is that just one model is responsible for generating the whole data set, and the probability distribution over h reflects uncertainty as to which model that is. As the size of the data set increases, this uncertainty reduces and the posterior probabilities $p(h|\mathbf{x})$ become increasingly focused on just one of the models [1].

3 Our Approach

In this section we present the proposed method in which we describe our approach to automatically grow and prune decision trees. This section is divided into the following parts: decision routing, highway networks and stacking.

3.1 Decision Routing

In our decision routing, a single neural network can have multiple routing paths for the same decision tree. If we start from the base decision tree, we could have two additional variations, a tree with one node pruned and a tree with one additional node grafted. In all three scenarios, the decision tree would share the same set of weights; the only alteration is that the routing would be different in each case as shown in Fig. 1.

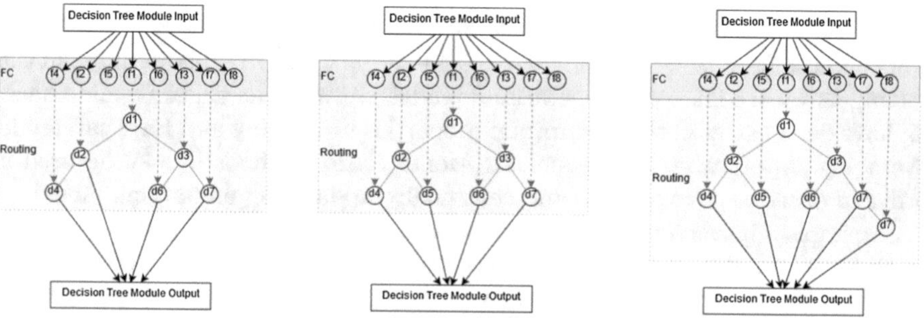

Fig. 1. Multiple decision routing for the same shared fully connected layers. Left to right: tree of depth two with one node pruned, tree of depth two with no perturbation, tree of depth two with one node grafted. The fully connected (FC) block (in blue) is shared among all the trees, through the parameters are not updated equally if they are not routed as part of the routing algorithm. When constructed in a neural network and stacked together, the network weights would only comprise of the rightmost structure (with the additional node pruned), with multiple outputs representing each of the routing strategies. At each of the leaf nodes, there would be a separate classifier layer that is built in accordance with the number of classes which the decision tree is required to train against.

In this scenario, all trees shared the same underlying tree structure and were connected in the same way. I it is in this manner which weights can be shared among all the trees. The routing layer determines whether nodes are to be pruned or grafted. The decision to prune or graft a node was done through $p(x_{t+1}|x_t, \theta)$. In the simpliest case, we simply pick a leaf node uniformly at random to prune or to graft. Additional weighting could be given depending on the past history of the node and updated using SMC approaches with a uniform prior.

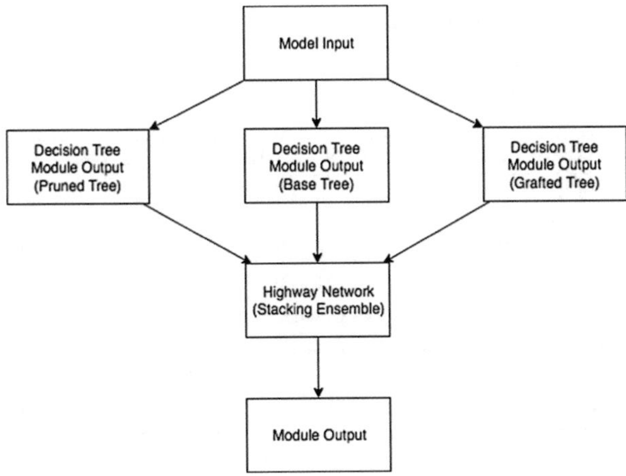

Fig. 2. A stacked ensemble model rearranged to resemble a highway network

3.2 Highway Networks and Stacked Ensembles

To demonstrate a stacked ensemble model can be rearranged to represent a highway network, the function t would be weights of one dimension, that is that that it is a scalar of one dimension, that is $t^j(y_i, \theta^j)$ is $\theta^j \in \mathbb{R}$ for all j where $\sum_{\forall j} \theta^j = 1, \theta^j \geq 0, \forall \theta^j$.

In this manner, the different decision trees which are perturbed as a stacked ensemble (Fig. 2). Using this information, the corresponding weights can be interpreted in the Bayesian model sense.

The construction of such a network differs from the usual highway network in the sense that the underlying distribution of data does not alter how it is routed, instead all instances are routed in the same way which is precisely how the stacking ensembles operate, as oppose to the usage of other ensemble methods. The advantages of using a simple stacking in this instance, is primarily the interpretation of the weights as posterior probability of the Bayesian model selection problem.

3.3 Model Architecture Selection

Finally the model architecture selection problem can be constructed through combining the two elements above. In this setting, at every iteration we would randomly select nodes to prune and grow. At the end of each iteration, we would perform weighted random sampling based on the model posterior probabilities.

After several iterations we would expect that $p(h|x_t)$ will eventually converge to a particular depth and structure of the decision tree. In order to facilitate this, the slope annealing trick $\pi = \text{softmax}_\tau(y/\tau)$ [2], where π is the modified weighted samples, and y is the output from the highway network and tau is the temperature. This is introduced to the highway net weights in order to progressively reduce the temperature so that base model selected to perturb becomes more deterministic in the end.

Algorithm 1. Automatic Induction of Neural Decision Trees

1: randomly initialize θ
2: **for** $t = 1$ to T **do**
3: Based on $p(h|x_t, \theta_t)$ (i.e. the weights inferred by the Highway Network), draw a weighted sample to perturb the model x_t based on softmax_τ
4: Compute and update the weights θ using back propagation
5: Lower temperature of softmax to $\tau = \tau \times \delta$, where δ is the discount rate
6: **end for**
7: Return final model x_t with probabilistic routing and parameters θ_t

Furthermore, this can be extended to ensemble approach through construction of such trees in parallel leading to the decision forest algorithms. In this scenario, each tree in the forest will have its own set of parameters and will induce different trees randomly. As they would be induced separately and randomly, we may yield more diverse set of classifiers leading to stronger results which may optimize different portions of the underlying space.

4 Experiments and Results

In our experiments we use ten publicly available datasets to demonstrate the efficacy of our approach. We used training and test datasets where provided to compare performance as shown in the table below. Where not provided, we performed a random 70/30 split into training and testing respectively. The following table (Tables 1 and 2) reports the average and median, over all the datasets, the relative improvement in log-loss over the respective baseline non-greedy decision tree.

Table 1. Average Error Improvement Across compared with baseline

Model	Avg Impr. (Train)	Avg Impr. (Test)	Median Impr. (Train)	Median Impr. (Test)
Tree	8.688%	6.276%	1.877%	0.366%
Forest	7.351%	7.351%	0.247%	0.397%

Table 2. Average Error Improvement Across compared with baseline with fine tuning

Model	Avg Impr. (Train)	Avg Impr. (Test)	Median Impr. (Train)	Median Impr. (Test)
Tree	22.987%	12.984%	11.063%	4.461%
Forest	23.223%	15.615%	11.982%	5.314%

In all instances, we began training our decision tree with a depth of 5, with τ initialized at 1.0 with a discount rate of 0.99 per iteration. Our decision forest was also set to have 5 decision trees, and combined through average voting. For all datasets we used standard data preparation approach from the recipes R library whereby we center, scale, and remove near zero variance predictors from our datasets. All models for baseline and our algorithm were built and trained using Python 3.6 running Keras and Tensorflow. In all models we use decision tree of depth 5 as a starting point with benchmark models trained for 200 epochs. With out automatic induced decision trees we train our models for 10 iterations, each with 20 epochs. We further train the final selected models to fine tune the selected architecture with the results as shown.

From the results, we notice that both approaches improve over the baseline where the tree depth is fixed to 5. With further fine tuning, it become apparent that the decision forest algorithm outperforms the vanilla decision tree approach. Even without fine tuning, it is clear that the forest approach is more robust in its performance against the testing dataset, demonstrating the efficacy of our approach.

5 Conclusion

From the results above, and compared with other benchmark algorithms, we have demonstrated an approach for non-greedy decision trees to learn ideal architecture through the use of sequential model optimization and Bayesian model selection. Through the ability to transfer learning weights effectively, and controlling the routing, we have demonstrated how we can concurrently train strong decision tree and decision forest algorithms, whilst inducing the ideal neural network architecture.

References

1. Bishop, C.M.: Pattern Recognition and Machine Learning. Information Science and Statistics. Springer, Heidelberg (2006)
2. Chung, J., Ahn, S., Bengio, Y.: Hierarchical multiscale recurrent neural networks. In: ICLR (2017)
3. Hastie, D.I., Green, P.J.: Model choice using reversible jump Markov chain Monte Carlo. Statistica Neerlandica **66**(3), 309–338 (2012)
4. Huang, F., Ash, J.T., Langford, J., Schapire, R.E.: Learning deep resnet blocks sequentially using boosting theory. In: International Conference of Machine Learning 2018, vol. abs/1706.04964 (2018)
5. Le, T., Clarke, B.: On the interpretation of ensemble classifiers in terms of Bayes classifiers. J. Classifi. **35**(2), 198–229 (2018)
6. Murthy, S.K., Kasif, S., Salzberg, S.: A system for induction of oblique decision trees. J. Artif. Int. Res. **2**(1), 1–32 (1994)
7. Norouzi, M., Collins, M., Johnson, M.A., Fleet, D.J. Kohli, P.: Efficient non-greedy optimization of decision trees. In: Advances in Neural Information Processing Systems (2015)

8. Kontschieder, P., Fiterau, M., Criminisi, A., Bulo, S.R.: Deep neural decision forests. In: Proceedings of the Twenty-Fifth International Joint Conference on Artificial Intelligence, IJCAI 2016, New York, NY, USA, 9–15 July 2016, pp. 4190–4194 (2016)
9. Veit, A., Wilber, M.J., Belongie, S.: Residual networks behave like ensembles of relatively shallow networks. In: Advances in Neural Information Processing Systems (2016)
10. Wolpert, D.H.: Stacked generalization. Neural Netw. **5**(2), 241–259 (1992)

Reinforcement Learning in *A Marketing Game*

Matthew G. Reyes(✉)

Ann Arbor, MI 48105, USA
matthewgreyes@yahoo.com
https://amarketinggame.com, http://www.matthewreyes.com

Abstract. This paper discusses a reinforcement learning interpretation of *A Marketing Game*, a model of socially-contingent decision-making that includes marketing by companies, which enables optimization of influence on a social network. Specifically, we consider a simple cycle network with asymmetric social biases, where Companies A and B each get a single unit of marketing allocation. We illustrate the steps in Company B's allocation decision, following allocation by Company A. The paper shows numerically that minimum conditional description length (MCDL) can be used to track convergence of asymmetric Glauber dynamics and correctly estimate the asymmetric social biases. The paper then demonstrates Company B's use of the learned network biases to simulate the network under its own candidate allocations in order to select the optimal one.

Keywords: Reinforcement learning · Social networks · Marketing · Glauber dynamics · Gibbs distributions · Minimum conditional description length

1 Introduction

In this paper, we propose a reinforcement learning interpretation of *A Marketing Game* (AMG), a recently introduced approach for Companies A and B to optimize influence over consumer preference in a market consisting of alternatives Product A and Product B. Reinforcement learning is a machine learning technique in which agents take *actions* to maximize some *reward* in a dynamic interactive environment learned through *sensing* and observing the results of previous or simulated actions [22]. The *environment* is the network of social connections through which influence is transmitted by observation and direct communication, and the manner in which individual consumers make decisions in response to such influence. *Actions* taken by Companies A and B are selections of marketing allocations, which are respective subsets of consumers to whom the companies market, together with specific types of marketing targeted to individual consumers. *Sensing* is inference of consumer preferences from data, and

M. G. Reyes—Independent Researcher and Consultant.

K. Arai et al. (Eds.): CompCom 2019, AISC 997, pp. 705–724, 2019.
https://doi.org/10.1007/978-3-030-22871-2_49

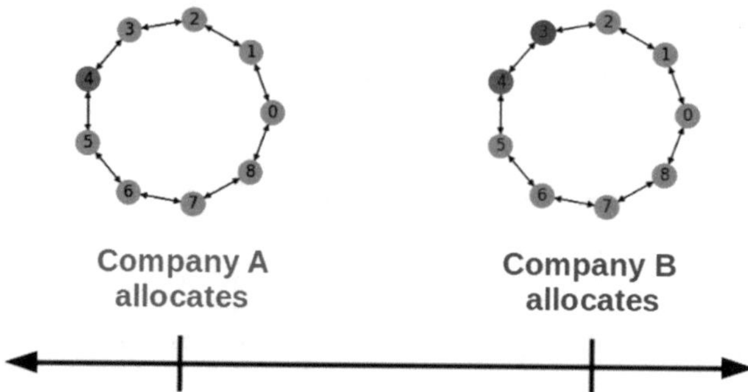

Fig. 1. Timeline indicating when Companies A and B implement respective marketing allocations on a cycle network of consumers. Red indicates marketing allocation by Company A, blue allocation by Company B. In this paper we consider asymmetric interactions $\theta_{i+1 \to i} = 1$, $\theta_{i-1 \to i} = .6$, and no inherent biases, i.e., $\alpha_i = 0$ for all i. Company A allocates to consumer 4 with marketing bias $m_A^4 = 2$, Company B will form estimates $\hat{\theta}_i$ and $\hat{\theta}_{j \to i}$ of the direct and social biases, then use these estimates to evaluate expected market share resulting from targeting consumer i with marketing bias m_B^i.

from these inferred preferences, the influences consumers exert on one another. The *reward* is market share, revenue, or some other measure of Company performance. This ongoing dynamic of inference from data and allocation based on anticipated reward is called *A Marketing Game* [17].

There are many social environments for the sharing of information. In practice, one might aggregate several such social networks to form a more comprehensive view of the influences upon consumer decision-making. In this paper, we focus the discussion to a single social *network* whose parametrization can be adapted to different social media platforms. A *network* is denoted by a *graph* $G = (V, E)$, where V is the set of *sites*, or consumers, and E is the set of *edges*, or social relationships connecting consumers, referred to as *neighbors*. Different social media platforms will have different rules regarding what defines a neighbor. For example, in the *Facebook* platform, neighbors are undirected: if consumer i is a neighbor of consumer j, then consumer j is a neighbor of consumer i. On the other hand, neighbors on *Twitter* are directed: if i is a neighbor of j when i "follows" j, then it is not necessarily the case that j is a neighbor of i. In this paper we assume an undirected network so that E consists of unordered pairs of sites $\{i, j\}$. Nevertheless, the *social bias* $\theta_{i \to j}$ of consumer i on consumer j will in general be distinct from the social bias $\theta_{j \to i}$ that consumer j exerts on consumer i.

While there are different social networks, they do have certain general properties that researchers have been able to identify and characterize. For example, many social networks have *small-world* [26] and *scale-free* [1] topologies. Moreover, recent research indicates that so-called *spanning trees* of scale-free networks are themselves scale-free [12], which means analysis and algorithm development with respect to analytically and computationally tractable models can still be informative and useful on real-world networks. In this paper we consider an extremely simple network topology, a cycle, illustrated in Fig. 1. While topologically speaking, this is an unrealistic network to consider, the simplicity affords analysis and a gradual building up of insight, that can then be carried over to more realistic networks, that would not likely be possible by beginning analysis on a more complicated topology. For instance, sites in a cycle are topologically uniform. As a result, we can isolate the influence that marketing to a particular consumer has on the preferences of other consumers, or how changes in social biases affect the overall influence of marketing.

The consumer decision-making model in *A Marketing Game* is based on *random utility* [13]. *Utility* can be viewed as a scale for measuring differences in a consumer's perception of the respective values provided by a set of alternatives [10,23], which determines consumers' preference for and selection among the alternatives. The theory of *random* utility states that consumers *maximize* utility, but that in any selection, there are factors influencing the decision that the company cannot explicitly account for. As such, consumer decisions are *modeled* as random, where utility is viewed as a parametrization of observed frequencies of consumer choice. The particular decomposition of utility will depend on knowledge about how consumers make decisions with respect to a given market, as well as data that can be collected or is under the control of the company.

A socially-contingent version of random utility selection was developed by making utility contingent upon the choices of one's neighbors within a social network [3]. The simplest parametrization of consumer i's socially-contingent choice includes the *social biases* $\{\theta_{j\to i}\}$ introduced above, and *inherent biases* $\{\alpha_i\}$ of individual consumers towards the respective alternatives. We let $x_i = 1$ indicate preference for Product A, and $x_i = -1$ a preference for Product B. Under a standard assumption on the *unknown sources* of utility in selection between Products A and B [13], this leads to the following *Glauber dynamics* [3,6,14,17] for consumer choice on the network:

$$p(x_i|\mathbf{x}_{\partial i}^{(t)}) = \frac{\exp\left\{\sum_{j\in\partial i} \theta_{j\to i} x_i x_j^{(t)} + \alpha_i x_i\right\}}{Z_{i|\mathbf{x}_{\partial i}^{(t)}}}, \tag{1}$$

where $Z_{i|\mathbf{x}_{\partial i}^{(t)}}$ is the normalizing constant referred to as the (local) partition function at i conditioned on $\mathbf{x}_{\partial i}$ at time t. The set ∂i consists of the *neighbors* of consumer i, and $\mathbf{x}_{\partial i}^{(t)}$ are the preferences of these neighbors at time t.

Versions of socially contingent decision-making have been explored by others, for example, Kempe et al. [11], Watts and Dodds [26], and Montanari and Saberi [14], in addressing the question of how preference spreads on a social network. As discussed in [17], from the perspective of marketing to influence preference on the network, these approaches suffer from two mathematically subtle yet operationally significant drawbacks. The first is that they model consumer choice as best-response, or deterministic maximization of some hypothesized utility.

The second shortcoming is that they do not include marketers for the respective companies. For example, analyses in [11,14,26] view marketing as *altering the preference* of a select group of consumers, then examining contagions resulting from a best-response scaling of the dynamics in (1). Free trials and other such promotional seeding mechanisms likely have an impact and should be considered. However, companies do not just seed their products, but rather continue to market them for the purpose of influencing the decisions that consumers make. Therefore, it is essential to explicitly factor in the influence of marketing in order to evaluate the effect that a particular marketing strategy can have on consumer preference throughout a social network.

To enhance the perception that consumer i has of their respective products, Companies A and B *market* their products to consumer i with respective *marketing biases* $m_A^i > 0$ and $m_B^i > 0$. The Glauber dynamics for consumer choice on the network now become [17]

$$p(x_i|\mathbf{x}_{\partial i}^{(t)}) = \frac{\exp\left\{\sum_{j\in\partial i} \theta_{j\to i} x_i x_j^{(t)} + (\alpha_i + m_A^i - m_B^i)x_i\right\}}{Z_{i|\mathbf{x}_{\partial i}^{(t)}}}. \quad (2)$$

We refer to $\theta_i = \alpha_i + m_A^i - m_B^i$ as the *direct bias* of consumer i.

Additional approaches to modeling the spread of information on social networks include affinity-based models [9,20] in which individuals share information based upon their *affinity* for the information, and after sharing the information, undergo a *refractory* period during which their affinity declines and they eventually stop sharing the information. In some respects, such a model can be likened to that of *A Marketing Game* where the inherent bias in (1) is time-varying, increasing with further exposure, then declining after having selected the Product. However, in a AMG, there are *competing* Products, for example, advertisements for two political candidates. To incorporate the affinity into AMG, one would employ a *nested* version of the logit model (1). For instance, the first logit model would concern the decision of whether or not to select *either* candidate. Then, conditioned on the decision to select *one* of them, a second logit model would concern the decision of *which* candidate to select. The refractory period would then be tantamount to losing interest in the candidate after a period of time.

The paper is organized as follows. Section 2 provides a preliminary discussion with contributions of the present paper. Section 3 discusses the distinction between transient and steady-state dynamics. Section 4 provides a reinforcement learning interpretation of *A Marketing Game*. Section 5 discusses the use of minimum conditional description length to track Glauber dynamics. Section 6 discusses Company B's allocation. Section 7 concludes with a discussion of limitations and future directions. The appendix includes additional plots illustrating parameter estimates for individual consumers.

2 Preliminary Discussion and Contributions

Companies A and B will each launch a *marketing campaign* consisting of a marketing *allocation* and marketing *horizon*. The marketing allocation by Company A is specified by a subset $M_A \subset V$ of consumers, a set of *types of marketing* $\{t^i_A : i \in M_A\}$, and a corresponding set of *investment levels* $\{d^i_A : i \in M_A\}$. The marketing bias m^i_A applied to consumer i due to to targeting consumer i with marketing type t^i_A at investment level d^i_A depends on consumer i's *marketing response* $m^i_A = f_i(\alpha_i, t^i_A, d^i_A)$. Consumer i's marketing response is a function that indicates the change in consumer i's perception of the value of Product A in response to marketing type t^i_A at investment level d^i_A. Likewise for Company B's marketing allocation consisting of the subset $M_B \subset V$ of consumers to target with marketing types $\{t^i_B : i \in M_B\}$, at respective investment levels $\{d^i_B : i \in M_B\}$, which result in marketing strengths $\{m^i_B : i \in V\}$ as a result of consumer marketing responses $\{f_i(\alpha_i, t^i_B, d^i_B)\}$. The well-known Weber-Fechner and Stevens' Laws [21] regarding perception in response to stimulus intensity have been used to model such marketing response functions. A Company's marketing horizon will be the duration for which the marketing allocation is implemented. The timing and horizon of a marketing campaign are understood to be important aspects of marketing strategy [8].

Consumer i's preference $x^{(t)}_i$ at time t will be influenced by the marketing strengths m^i_A and m^i_B, his inherent bias α_i, and the social biases $\{\theta_{j\to i} : j \in \partial i\}$ exerted by his neighbors. This preference will influence material that consumer i posts on social media. Suppose that $y^{(t)}_i$ is a post made by consumer i at time t on a particular social networking platform. The post $y^{(t)}_i$ will be correlated with the preference $x^{(t)}_i$ by expressing either positive sentiment with respect to the Product that consumer i prefers, or negative sentiment with respect to the other Product. In either case, a Company will form an estimate $\hat{x}^{(t)}_i$ of consumer i's preference by applying machine learning algorithms. For example, if $y^{(t)}_i$ is a Tweet, the Company may first apply a topic model to infer the subject matter of consumer i's post, and then apply sentiment analysis to subsequently infer consumer i's sentiment with respect to the subject matter. The essence of *A Marketing Game* [17] is a framework for using inferred preferences to further infer social influences on a network, which play a significant role in influencing consumer preference. By understanding the influences that consumers exert upon one another, Companies can leverage this information to make more strategic decisions regarding selection of their respective marketing allocations.

In order to make sense of concepts, we consider a simplified scenario in this paper. We assume that each consumer makes a social media post when they update their preference, and furthermore, we assume that Companies' machine learning algorithms are perfect, so that they are able to observe the true preferences for consumers. Moreover, while Companies are ultimately interested in purchases or votes stemming from consumer preferences, here the objective of Companies A and B is to increase overall preference for their respective Products. Given this setup, a simple setting to consider is the cycle network in Fig. 1.

The social biases are asymmetric, with $\theta_{i+1\to i} = 1$ and $\theta_{i-1\to i} = 0.6$. There are no inherent biases, the only direct biases are due to marketing. Companies A and B will each allocate to a single consumer on the network, of equal strengths $m_A = m_B = 2$. More specifically, consider the situation where Company A has allocated to consumer 4 at time $T = 0$. At each time $t \geq T$, Company B uses minimum conditional description length (MCDL) to form estimates $\{\hat{\theta}_i^{(t)}\}$, $\{\hat{\theta}_{j\to i}^{(t)}\}$ of direct and social biases. At a time t_0, Company B uses these estimates to assess anticipated gains in market share resulting from its own allocation to one of the sites on the cycle, selecting the allocation that results in the greatest expected market share. We demonstrate numerically that MCDL estimates converge to the true values and that Company B's optimal allocation is to consumer 3. That is, given asymmetric social biases, a Company should market to consumers who are the most influenced by those consumers preferring the other Product, thus blocking preference for the other Product from spreading on the network. See [17] for a derivation of the socially-contingent model of consumer choice (2). Additionally, [16] considers the effect of marketing response on allocation selection on a cycle with symmetric social biases.

3 Transient and Steady-State Preference Dynamics in a Marketing Game

After Company A has marketed to consumer 4, the system of network preferences will evolve according to (2). We assume that each consumer updates his preference according to a Poisson process with the same rate. We make the standard assumption that time is discretized according to those times at which consumers update their preferences. This results in a *time-series* of preference configurations $\mathbf{x}^{(0)}, \mathbf{x}^{(1)}, \ldots, \mathbf{x}^{(t)}, \ldots$, with one-step conditional distribution $p(\mathbf{x}^{(t)}|\mathbf{x}^{(t-1)})$ given by

$$p(\mathbf{x}^{(t)}|\mathbf{x}^{(t-1)}) = \begin{cases} p(x_i^{(t)}|\mathbf{x}_{\partial i}^{(t-1)}) & \mathbf{x}_{V\setminus i}^{(t)} = \mathbf{x}_{V\setminus i}^{(t-1)} \\ 0 & \text{otherwise} \end{cases},$$

where $V\setminus i$ is the set of all consumers not including consumer i. That is, the configurations $\mathbf{x}^{(t)}$ and $\mathbf{x}^{(t+1)}$ at successive time points differ in at most a single site. For each t, $\mathbf{x}^{(t)}$ is a *cross-section* of the network preferences. The conditional distribution of cross-section $\mathbf{x}^{(t)}$ at time t given cross-section $\mathbf{x}^{(t_0)}$ at time t_0 is

$$p\left(\mathbf{x}^{(t)}|\mathbf{x}^{(t_0)}\right) = \sum_{\mathbf{x}^{(t_0+1)}} \cdots \sum_{\mathbf{x}^{(t-1)}} p\left(\mathbf{x}^{(t_0+1)}|\mathbf{x}^{(t_0)}\right) \cdots p\left(\mathbf{x}^{(t)}|\mathbf{x}^{(t-1)}\right). \quad (3)$$

We say that the cross-sectional distribution $p(\mathbf{x}^{(t)})$ on preference configurations is *stationary* if $p(\mathbf{x}^{(t)}) = p(\mathbf{x}^{(t+1)})$ for sufficiently large t. For instance, Blume [3] showed that if for all pairs of neighbors i and j, the social biases $\theta_{i\to j}$ and $\theta_{j\to i}$ are the same, i.e., $\theta_{i\to j} = \theta_{j\to i} = \theta_{ij}$, then the dynamics (2) converge to a stationary distribution given by the *Gibbs equilibrium*

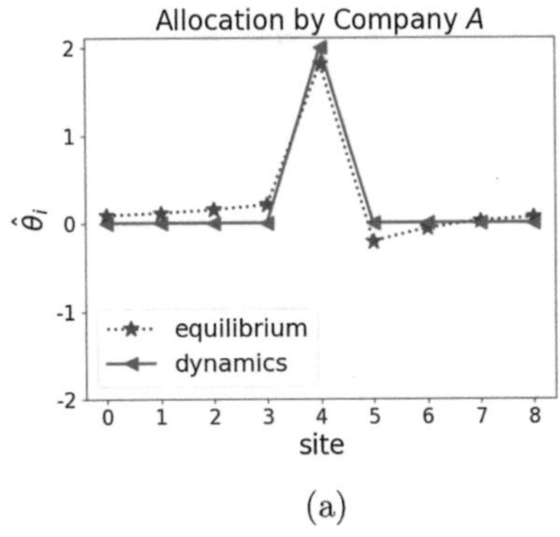

(a)

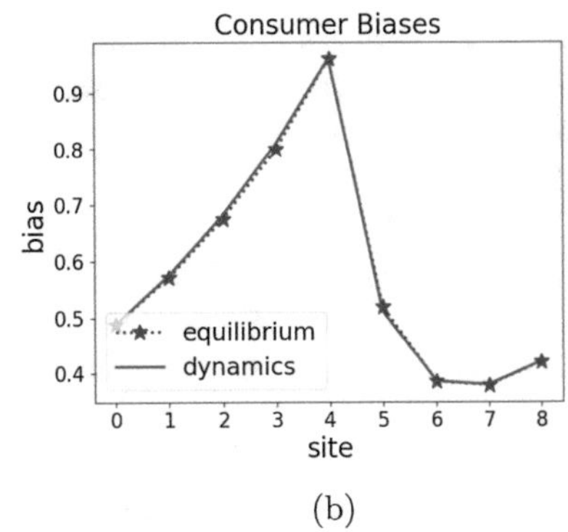

(b)

Fig. 2. (a) Estimated direct biases $\{\hat{\theta}_i\}$ for stationary equilibrium, and true direct biases $\{\theta_i\}$ of dynamics; and (b) stationary consumer biases $\{\mu_i = p_i(1) - p_i(-1)\}$, computed both empirically with respect to the asymmetric dynamics, and exactly with respect to the symmetric equilibrium with direct biases given in (a). Note that if the allocation were instead placed by Company B, the estimated direct biases and consumer biases would simply be the negative of what they are in (a) and (b).

$$p(\mathbf{x}) = \frac{1}{Z} \exp\{\sum_{\{i,j\}} \theta_{ij} x_i x_j + \sum_{i \in V} \theta_i x_i\}, \tag{4}$$

where Z is the normalizing constant referred to as the *partition function.* A *Gibbs equilibrium* [3,5,6], in addition to stationarity, satisfies the *detailed balance* condition, i.e., $p(x_i'|\mathbf{x}_{\partial i})p(\mathbf{x}) \neq p(x_i|\mathbf{x}'_{\partial i})p(\mathbf{x}')$, which is often referred to as *reversibility.* Reversibility is in fact a stronger condition that implies stationarity. On the other hand, if $\theta_{i\to j} \neq \theta_{j \to i}$ for some pair of neighbors i and j, then the dynamics (2) will not converge to a Gibbs equilibrium, in that the dynamics are not reversible. Nevertheless, Godreche [7] showed that for a cycle with uniformly asymmetric social biases, i.e., $\theta_{i\to j} = \theta_s \pm \Delta$ for all neighbors i, j, and uniform direct biases, $\theta_i = \theta_d$ for all i, the Glauber dynamics of (2) *do* converge to a stationary distribution that coincides with the Gibbs equilibrium for the dynamics with symmetric social biases θ_s and the same uniform direct biases θ_d.

For t not sufficiently large, the dynamics are said to be *transient.* That is, after Company A allocates marketing to consumer 4, the full effects of that marketing will not be observed immediately [8]. During the transient phase, the influence of Company A's marketing is distributed throughout the network, through the social biases that neighboring consumers exert upon one another. While dynamics in the transient phase can be described using (3), in practice such analysis often becomes unwieldy, and depends on knowledge of the initial configuration $\mathbf{x}^{(t_0)}$. However, if Company A's marketing horizon is sufficiently long, recent numerical analysis [18] suggests that the dynamics on this network converge to a stationary distribution coinciding with the Gibbs equilibrium for dynamics with symmetric social biases, similar to Godreche's analytical solution [7]. For instance, Fig. 2 (a) shows the direct biases for the time-series dynamics, and the direct biases estimated for a stationary cross-sectional Gibbs model. Figure 2(b) shows expected consumer preferences generated empirically from simulated dynamics and with respect to the estimated stationary Gibbs model.

Suppose that at time t_0 Company B selects its marketing allocation. Company B will have formed estimates $\{\hat{\theta}_i^{(t_0)}\}$ and $\{\hat{\theta}_{j\to i}^{(t_0)}\}$ of direct and social biases. In evaluating each of its candidate marketing allocations, it will add the corresponding marketing strength m_B^i to the appropriate $\hat{\theta}_i$, then simulate the dynamics (2) beginning from configuration $\mathbf{x}^{(t_0)}$. As in the case following Company A's allocation, the dynamics proceed through a transient phase in which the probability of a configuration $\mathbf{x}^{(t)}$, $t > t_0$, will be given by (3). If Company B maintains its allocation sufficiently long, the dynamics will eventually reach a stationary distribution of the form (4), assuming, of course, that Company A maintains its original marketing allocation. If Company A changes its allocation, and Company B continues with the allocation it makes at time t_0, the dynamics will settle into a *different* steady-state. And if Company A changes its allocation frequently enough, even if Company B persists with the allocation it implements at time t_0, the dynamics will lurch from one transient phase to another, without ever settling into a stationary phase.

These, however, are considerations for future work. In what follows we assume that Company A indeed maintains its original allocation, and that Company B makes its own allocation selection based on steady-state computations.

4 A Reinforcement Learning Interpretation

Company B's allocation selection can be viewed within the rubric of *reinforcement learning*, in which an *agent* seeks to optimize some *reward* through its *actions*. Actions are selected based upon partial information gleaned through *sensing* the *environment*. In our setting, the state of the environment includes the collection of inherent biases $\{\alpha_i : i \in V\}$, marketing biases $\{m_A^i : i \in V\}$ from Company A, social biases $\{\theta_{j\to i} : \{i,j\} \in E\}$, and the configuration of preferences $\mathbf{x}^{(t_0)}$ on the network at the time t_0 when Company B makes its allocation selection.

Company B gains partial information about its environment by *sensing*. In *A Marketing Game*, this amounts to computing estimates $\{\hat{\mathbf{x}}_i^{(t)} : i \in V\}$ of consumer preference by applying machine learning algorithms to data posted on social media. Furthermore, from such preference estimates, Company B computes estimates $\{\hat{\theta}_i^{(t_0)} : i \in V\}$ of direct biases and estimates $\{\hat{\theta}_{j\to i}^{(t_0)} : i \in V\}$ of social biases. We let $\hat{\boldsymbol{\theta}}^{(t_0)}$ denote the collection of these estimated bias parameters computed by Company B at time t_0. Company B will use its estimate $\hat{\mathbf{x}}^{(t_0)}$ of the current preference configuration, and its estimate $\hat{\boldsymbol{\theta}}^{(t_0)}$ of the direct and social biases, in selecting the marketing allocation that maximizes reward.

The reward in *A Marketing Game* is market share, or equivalently, the difference in market share between Companies A and B. Recall that at time t, $x_i^{(t)}$ is the preference of consumer i. The *total preference* for configuration $\mathbf{x}^{(t)}$ is

$$r\left(\mathbf{x}^{(t)}\right) = \sum_{i \in V} x_i^{(t)},$$

and similarly, for a sequence of configurations $\mathbf{x}^{(t_1)}, \ldots, \mathbf{x}^{(t_K)}$,

$$r\left(\mathbf{x}^{(t_1)}, \ldots, \mathbf{x}^{(t_K)}\right) = \sum_{j=1}^{K} r\left(\mathbf{x}^{(t_j)}\right) = \sum_{j=1}^{K} \sum_{i \in V} x_i^{(t_j)}.$$

The *expected* total preference on the network at time t is then

$$\begin{aligned} R^{(t)} &= \mathbf{E}\left[r\left(\mathbf{X}^{(t)}\right)\right] = \mathbf{E}\left[\sum_{i \in V} X_i^{(t)}\right] = \sum_{i \in V} \mathbf{E}\left[X_i^{(t)}\right] \\ &= \sum_{i \in V} p_i^t(A) - p_i^t(B) \\ &\triangleq \sum_{i \in V} \mu_i^{(t)}, \end{aligned}$$

where $p_i^t(A)$ is the *market share* for Company A with respect to consumer i at time t [2]. Likewise, $p_i^t(B)$ is the market share of Company B with respect to consumer i at time t. The quantity $\mu_i^{(t)} = p_i^t(A) - p_i^t(B)$ is the expected preference, or *bias*, for consumer i at time t. The expected total preference is the difference in market share of the two companies with respect to the entire network. The goal of Company A is to maximize expected total preference, whereas for Company B the goal is to minimize expected total preference.

As mentioned, we assume that Company B directly observes the true preferences $\{\mathbf{x}^{(t_0)}\}$. Moreover, as we discuss in Sect. 5, Company B will form estimates $\{\hat{\theta}_i\}$ and $\{\hat{\theta}_{j\to i}\}$ that converge to the true direct and social biases. We also assume that Company B knows the inherent biases $\{\alpha_i\}$. With this information, Company B will compute the expected total preference on the network for some *horizon* extending from t_0 for a duration T. In optimizing over expected total preference, Company B may include a *discount factor* γ to emphasize rewards earned closer in time to when the marketing allocation is implemented.

We let $Q_{\hat{\mathbf{x}}^{(t_0)}}(\hat{\boldsymbol{\theta}}^{(t_0)}, i_B, \gamma, T)$ be the expected total preference that Company B earns through time duration T, by allocating to consumer i_B at time t_0 with estimated parameters $\hat{\boldsymbol{\theta}}^{(t_0)}$ and discount factor γ. That is, Company B will compute

$$Q_{\hat{\mathbf{x}}^{(t_0)}}(\hat{\boldsymbol{\theta}}^{(t_0)}, i_B, \gamma, T) \triangleq \mathbf{E}\left[\sum_{\tau=0}^{T} \gamma^{\tau} r\left(\mathbf{X}^{(t_0+1+\tau)}\right) \middle| \hat{\mathbf{x}}^{(t_0)}\right] \tag{5}$$

for each $i_B \in V$, and will then allocate to $i_B^* = \arg\max_{i_B} Q_{\hat{\mathbf{x}}^{(t_0)}}(\hat{\boldsymbol{\theta}}^{(t_0)}, i_B, \gamma, T)$.

Suppressing notation in the interest of space, we can express (5) recursively in the familiar Bell's equation [22]

$$\begin{aligned}
Q_{\hat{\mathbf{x}}^{(t_0)}} &= \mathbf{E}\left[\sum_{\tau=0}^{T} \gamma^{\tau} r\left(\mathbf{X}^{(t_0+1+\tau)}\right) \middle| \hat{\mathbf{x}}^{(t_0)}\right] \\
&= \mathbf{E}\left[r\left(\mathbf{x}^{(t_0+1)}\right) + \sum_{\tau=1}^{T} \gamma^{\tau} r\left(\mathbf{X}^{(t_0+1+\tau)}\right) \middle| \hat{\mathbf{x}}^{(t_0)}\right] \\
&= \mathbf{E}\left[r\left(\mathbf{x}^{(t_0+1)}\right) + \gamma \sum_{\tau=0}^{T} \gamma^{\tau} r\left(\mathbf{X}^{(t_0+2+\tau)}\right) \middle| \hat{\mathbf{x}}^{(t_0)}\right] \\
&= \sum_{\mathbf{x}^{(t_0+1)}} p\left(\mathbf{x}^{(t_0+1)} \middle| \mathbf{x}^{(t_0)}\right) \left[r\left(\mathbf{x}^{(t_0+1)}\right) + \gamma Q_{\hat{\mathbf{x}}^{(t_0+1)}}\right].
\end{aligned}$$

This recursive decomposition can be extended to an intermediate time $T_1 < T$, $T_1 > t_0 + 1$. Then, rather than the reward $R^{(t_0+1)} = \mathbf{E}\left[r\left(\mathbf{x}^{(t_0+1)}\right)\right]$ gained by moving to the next point in time, we would consider the expected reward

$$\begin{aligned}
R_{\hat{\mathbf{x}}^{(t_0)}}^{\mathbf{x}^{(T_1)}} &= \mathbf{E}\left[r\left(\mathbf{X}^{(t_0+1)}, \mathbf{X}^{(t_0+2)}, \ldots, \mathbf{X}^{(T_1)}\right) \middle| \hat{\mathbf{x}}^{(t_0)}\right] \\
&= \mathbf{E}\left[\sum_{\tau=t_0+1}^{T_1} r\left(\mathbf{X}^{(\tau)}\right) \middle| \hat{\mathbf{x}}^{(t_0)}\right] \\
&= \sum_{\tau=t_0+1}^{T_1} \mathbf{E}\left[r\left(\mathbf{X}^{(\tau)}\right) \middle| \hat{\mathbf{x}}^{(t_0)}\right] \\
&= \sum_{\tau=t_0+1}^{T_1-1} \sum_{\mathbf{x}^{(\tau)}} p\left(\mathbf{x}^{(\tau)} \middle| \hat{\mathbf{x}}^{(t_0)}\right) r\left(\mathbf{x}^{(\tau)}\right) + p\left(\mathbf{x}^{(T_1)} \middle| \hat{\mathbf{x}}^{(t_0)}\right) r\left(\mathbf{x}^{(T_1)}\right)
\end{aligned}$$

gained in transitioning from configuration $\hat{\mathbf{x}}^{(t_0)}$ to configuration $\mathbf{x}^{(T_1)}$ over some $T_1 - t_0$ duration of time. This would be useful, for example, if we wanted to evaluate expected reward by discounting the transient phase differently than the steady-state phase. Moreover, it could be used to incorporate residual gains in expected total preference that remain after Company B discontinues its marketing allocation [8].

5 Company B Tracks Network Influences

In this section, we discuss a method that Company B uses to compute estimates $\{\hat{\theta}_i^{(t_0)} : i \in V\}$ and $\{\hat{\theta}_{j \to i}^{(t_0)} : i \in V, j \in \partial i\}$ of direct biases and social biases. In Sect. 6, we illustrate Company B's use of this model to simulate the dynamics in (2) as part of its allocation selection.

Minimizing description length is a well-known, well-founded framework for both estimating parameters of a known model and model selection [19]. For example, Company B can estimate direct and social biases by minimizing the *conditional* description length (CDL)

$$\bar{D}(x_i^{(t_0-T:t_0)}|\mathbf{x}_{\partial i}^{(t_0-T:t_0)};\boldsymbol{\theta}_{\bar{i}}) = -\sum_{\tau=0}^{T}\log p(x_i^{(t_0-\tau)}|\mathbf{x}_{\partial i}^{(t_0-\tau)};\boldsymbol{\theta}_{\bar{i}}) \qquad (6)$$

of consumer i's choices *conditioned* on the choices $\mathbf{x}_{\partial i}$ of his neighbors, where $\boldsymbol{\theta}_{\bar{i}} = \theta_i \cup \{\theta_{j\to i} : j \in \partial i\}$ is the *neighborhood parameter* directly influencing consumer i's preference. Minimizing (6) at time t_0 yields estimates $\{\hat{\theta}_i^{(t_0)}\}$, $\{\hat{\theta}_{j\to i}^{(t_0)}\}$ of the network biases. Figure 3(a) shows the average errors, over 40 Monte Carlo runs, in the estimated direct bias $\hat{\theta}_0$ and the estimated social biases $\hat{\theta}_{8\to 0}$ and $\hat{\theta}_{1\to 0}$, as a function of sample complexity. Figure 3(b) and (c) show the parameter estimates for two different Monte Carlo runs.

The gradient of $\bar{D}(x_i^{(t_0-T:t_0)}|\mathbf{x}_{\partial i}^{(t_0-T:t_0)};\tilde{\boldsymbol{\theta}}_{\bar{i}})$ with respect to neighborhood parameter $\tilde{\boldsymbol{\theta}}_{\bar{i}}$ is

$$\nabla_{\tilde{\boldsymbol{\theta}}_{\bar{i}}} D = \frac{1}{T+1}\sum_{\tau=0}^{T}\left(\mathbf{E}\left[X_i|\mathbf{x}_{\partial i}^{(t_0-\tau)};\tilde{\boldsymbol{\theta}}_{\bar{i}}\right] - x_i^{(t_0-\tau)}\right)\mathbf{x}_{\bar{i}}^{(t_0-\tau)}, \qquad (7)$$

where

$$\mathbf{x}_{\bar{i}}^{(t_0-\tau)} = \begin{bmatrix} 1 \\ x_{j_1}^{(t_0-\tau)} \\ \vdots \\ x_{j_n}^{(t_0-\tau)} \end{bmatrix}$$

is the *empirical neighborhood* of site i at time $t_0-\tau$. Setting (7) to zero, yields the (conditional) *moment-matching* conditions

$$\sum_{\tau=0}^{T}\mathbf{E}\left[X_i \;\middle|\; \mathbf{x}_{\partial i}^{(t_0-\tau)};\hat{\boldsymbol{\theta}}_{\bar{i}}\right] = \sum_{t=0}^{T} x_i^{(t_0-\tau)},$$

for each site i, and for $j \in \partial i$,

$$\sum_{\tau=0}^{T}\mathbf{E}\left[X_i X_j \;\middle|\; \mathbf{x}_{\partial i}^{(t_0-\tau)};\hat{\boldsymbol{\theta}}_{\bar{i}}\right] = \sum_{\tau=0}^{T} x_i^{(t_0-\tau)} x_j^{(t_0-\tau)},$$

for minimization of CDL. The Hessian of CDL with respect to neighborhood parameter $\tilde{\boldsymbol{\theta}}_{\bar{i}}$ is

$$H_{\tilde{\boldsymbol{\theta}}_{\bar{i}}} = \frac{1}{T+1}\sum_{\tau=0}^{T}\operatorname{cov}\left(X_i, X_i|\mathbf{x}_{\partial i}^{(t_0-\tau)};\tilde{\boldsymbol{\theta}}_{\bar{i}}\right)\mathbf{x}_{\bar{i}}^{(t_0-\tau)}\mathbf{x}_{\bar{i}}^{(t_0-\tau)}.$$

As a sum of covariance matrices, $H_{\tilde{\theta}_{\bar{i}}}$ is positive semi-definite, so that CDL is a convex function of the neighborhood parameter $\tilde{\theta}_{\bar{i}}$ and is therefore amenable

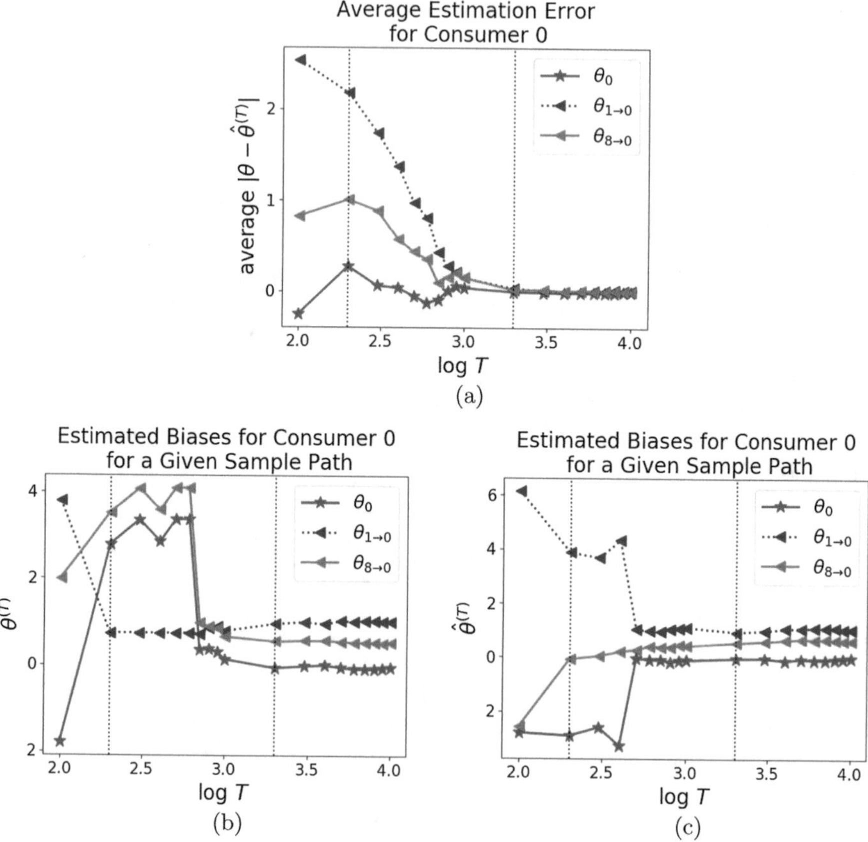

Fig. 3. For Glauber dynamics with uniformly asymmetric social biases, i.e., $\theta_{i+1\to i} = 1$ and $\theta_{i-1\to i} = .6$, for all i, $\theta_4 = 2$, $\theta_i = 0$ for $i \neq 4$, (a) estimation error of direct and social biases for consumer 0; (b) and (c) direct and social bias estimates for two different sample paths.

to standard convex optimization algorithms[1]. From Fig. 3(a), one can see that the errors in the estimated biases converge to zero with more samples. Figures 5 and 6 in the appendix give estimates from a single Monte Carlo run, as well as average estimation errors, for additional sites on the cycle.

One can alternatively derive the right hand side of (6) under the guises of both maximum *pseudo*-likelihood (MPL) [4] and logistic regression (LR) [25]. However, MPL involves the *pseudo* step

$$p\left(\mathbf{x}^{(t_0)}; \boldsymbol{\theta}\right) \approx \prod_i p\left(x_i^{(t_0)} \middle| \mathbf{x}_{\partial i}^{(t_0)}; \boldsymbol{\theta}\right),$$

[1] For the numerical analysis in this paper, the author used the 'Newton-CG' method.

whereas with LR, one requires that $x_i^{(t)}$ and $\mathbf{x}_{\partial i}^{(t-\tau)}$ are independent so that

$$p\left(x_i^{(t_0-T)}, \ldots, x_i^{(t_0)} \middle| \mathbf{x}_{\partial i}^{(t_0-\tau)}, \ldots, \mathbf{x}_{\partial i}^{(t_0)}; \boldsymbol{\theta}\right) = \prod_{\tau=0}^{T} p\left(x_i^{(t_0-\tau)} \middle| \mathbf{x}_{\partial i}^{(t_0-\tau)}; \boldsymbol{\theta}\right).$$

That is, the MPL and LR based approaches seek a factorizing distribution, from which the logarithm simplifies the objective function into a sum. However, by instead focusing on (conditional) description length, we can derive the same objective function in a more principled manner that does not require independent samples or a heuristic approximation. We thus refer to the minimization of (6) as minimum conditional description length (MCDL) estimation [15].

In addition to computing estimates of the direct and social biases for time-series dynamics, MCDL can also be used to compute direct and social biases for cross-sectional steady-state Gibbs equilibria. Note that in the case of the latter, social biases are necessarily symmetric, as indeed there is a single social bias θ_{ij} for each pair of neighbors. The difference in using MCDL to estimate dynamics versus equilibrium parameters is as follows. For estimating equilibrium parameters, one would use configurations at all time points. For estimating dynamics parameters, one would only use configuration time points in (6) corresponding to those times at which consumer i updates his choice. This is because when one of consumer i's neighbors updates his preference, the resulting preference configuration will not reflect preference selections *made by* consumer i.

A reason for wanting to estimate equilibrium parameters is that with an equilibrium model, one can determine probabilities without the extensive computing often required for Monte Carlo simulation. Indeed, it has been shown [24] that in resource-limited settings, suboptimal variational methods with respect to an equilibrium model can outperform exact Monte Carlo. In the next section, however, we consider Monte Carlo simulation with respect to an estimated dynamics model.

6 Company B's Steady-State Allocation

In this section we illustrate Company B's use of Monte Carlo simulation to select its marketing allocation. For each candidate $i_B \in V$, we simulate the dynamics of (2) with bias parameters given by the sum of the estimate $\hat{\boldsymbol{\theta}}^{(t_0)}$ obtained in the previous section, and the marketing strength $m_B^{i_B} = 2$ applied to site i_B. As can be observed in Figs. 3, 5, and 6, the MCDL estimates converge to the correct values for the direct and social biases.

For each candidate i_B, we perform $r = 100$ Monte Carlo runs beginning from initial configuration $\mathbf{x}^{(t_0)}$. Each run consists of $T = 100000$ time points. This generates a collection $\left(\mathbf{x}^{(t,\rho)}\right)$ of preference configurations, where $\mathbf{x}^{(t,\rho)}$ is the configuration at simulation time point t in the ρ-th Monte Carlo run. For time t in the simulation, we compute the empirical bias estimates

$$\hat{\mu}_i^{(t)} = \frac{1}{r} \sum_{\rho=1}^{r} x_i^{(t,\rho)},$$

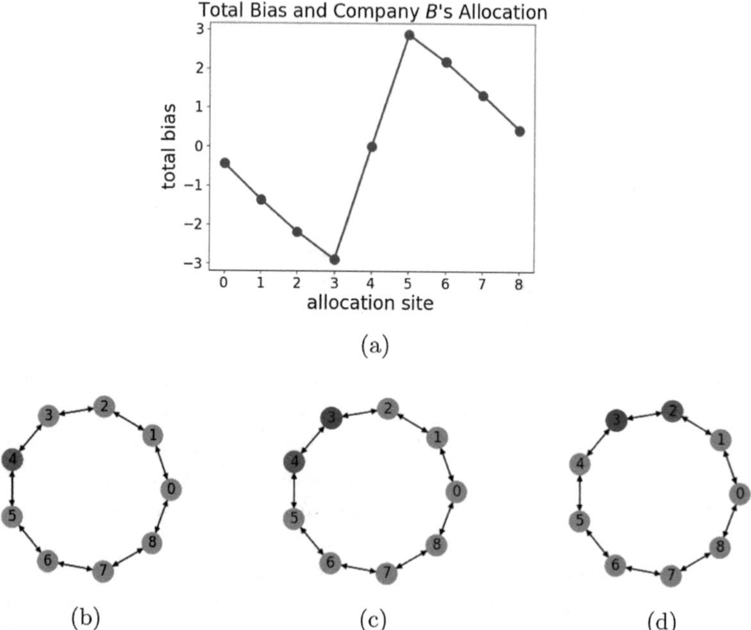

Fig. 4. (a) Total expected preference as a function of allocation site i_B for Company B, for the case $\theta_{i+1\to i} = 1$, $\theta_{i-1\to i} = .6$, $\theta_4 = 2$, $\theta_{i_B} = -2$, and $\theta_i = 0$ for $i \neq 4, i_B$, computed using Monte Carlo simulation of the asymmetric dynamics. In (b), (c), and (d), we show the sequence of allocations beginning with Company A's initial allocation, then Company B's optimal stationary allocation, followed by Company A's subsequent optimal stationary allocation.

which are estimates of the expected preferences $\mu_i^{(t)} = \mathbf{E}\left[X_i^{(t)}\right]$ of each consumer i at each time t. The expected total preference on the network under allocation i_B is then

$$\begin{aligned}\hat{Q}_{\mathbf{x}^{(t_0)}} &= \sum_{\tau=0}^{T} \gamma^\tau \frac{1}{r} \sum_{\rho=1}^{r} \sum_{i\in V} x_i^{(t_0+1+\tau,\rho)} = \sum_{\tau=0}^{T} \gamma^\tau \sum_{i\in V} \frac{1}{r} \sum_{\rho=1}^{r} x_i^{(t_0+1+\tau,\rho)} \\ &= \sum_{\tau=0}^{T} \gamma^\tau \sum_{i\in V} \hat{\mu}_i^{(t_0+1+\tau)}.\end{aligned}$$

which can be thought of as total expected preference.

We observed that the estimates $\hat{\mu}^{(t_0+1+\tau)}$ converged with respect to the number of Monte Carlo iterations. Therefore, we now see that expected total preference is

$$\hat{Q}_{\mathbf{x}^{(t_0)}} = \sum_{\tau=0}^{T} \gamma^{\tau} \sum_{i \in V} \mu_i^{(t_0+1+\tau)}$$

where $\mu^{(t_0+1+\tau)}$ is the statistical bias for consumer i for the dynamics (3), τ time steps from $\mathbf{x}^{(t_0)}$. We further observed that the $\mu_i^{(t_0+1+\tau)}$ converged with respect to numerical time index τ. Therefore, using a steady-state assumption $\mu_i^{(t_0+1+\tau)} \approx \mu_i$, the expected reward is essentially

$$\hat{Q}_{\mathbf{x}^{(t_0)}} = \sum_{\tau=0}^{T} \gamma^{\tau} \sum_{i \in V} \mu_i = \left(\frac{\gamma^{T+1} - 1}{\gamma - 1} \right) \sum_{i \in V} \mu_i. \tag{8}$$

In other words, if Company B assumes a steady-state model in its reward calculation, the discount factor γ simply provides a positive scaling of $\sum_{i \in V} \mu_i$.

We now discuss the results of Company B's Monte Carlo based allocation selection based on expected total preference

$$\sum_{i \in V} \hat{\mu}_i^{(t_0+1+T)} \approx \sum_{i \in V} \mu_i^{(t_0+1+T)} \approx \sum_{i \in V} \mu_i,$$

shown in Fig. 4(a). Recall that, due to our convention, Company B seeks to *minimize* expected total preference. As one can see from Fig. 4(a), Company B optimizes its market share by placing its unit of allocation at consumer 3. Figure 4(b) depicts Company A's allocation at consumer 4. Due to the directionality of the social biases, i.e., $\theta_{i \to i-1} > \theta_{i \to i+1}$, there is a directionality to Company A's influence, i.e., the influence from Company A's marketing allocation is greatest clockwise from consumer 4, which one can observe in the consumer biases in Fig. 2(b). Figure 4(c) illustrates Company B's optimal allocation. By allocating to consumer 3, Company B effectively blocks the strongest influence from Company A's allocation, and is able to propagate its own strongest influence in the direction of Company A's weakest influence. Conversely, the worst allocation for Company B's is to consumer 5, the consumer in the counterclockwise direction. If Company A in turn follows the same strategy as Company B, that is, estimating direct and social biases using a sufficient amount of data, and bases its allocation decision on a sufficiently long horizon, Company A will then allocate to consumer 2, as depicted in Fig. 4(d).

7 General Discussion: Limitations and Future Directions

In this paper we have introduced a reinforcement learning interpretation of *A Marketing Game* [17]. Specifically, we followed Company B through the process of estimating direct and social biases, using Monte Carlo simulation to compute total expected preference for different allocations, then selecting the allocation that optimized this reward. In the interest of exposition, the setting considered in this paper was by necessity extremely simple. As such, there remain many further explorations to better understand this problem, not only in realm of theoretical models, but perhaps more importantly, how these principles can be applied in the real world with real data.

One direction for further investigation is the case where preference configurations $\mathbf{x}^{(t)}$ are not known perfectly, and instead have to be estimated. This amounts to the realistic case that the machine learning algorithms processing consumers' social media posts $\mathbf{y}^{(t)}$ will make misclassification errors, both with regard to inferred subject matter, as well as with respect to assigned sentiment.

Another important point that requires examination is the fact that, given estimated bias parameter $\hat{\boldsymbol{\theta}}^{(t_0)}$, Company B will in general not be able to disambiguate the direct biases $\{\theta_i = \alpha_i + m_A^i - m_B^i\}$. For example, in Sect. 5, Company B estimates direct and social biases. In reality it does not know whether the estimated direct bias $\hat{\theta}_4$ is due to marketing from Company A or the inherent bias of consumer i. Moreover, in simulating preference dynamics under candidate marketing allocations, candidate allocations will be in terms of not only the subset M_B, but also the marketing types $\{t_B^i\}$ and investment levels $\{d_b^i\}$. Since both the inherent biases $\{\alpha_i\}$ and marketing responses $\{f_i(\alpha_i, t_B^i, d_B^i)\}$ will be unknown, Company B's allocation selection will be with respect to an expected total preference of the form

$$\mathbf{E}\left[\sum_{\tau=0}^{T} \gamma^\tau r\left(\mathbf{X}^{(t_0+1+\tau)}\right) \middle| \mathbf{x}^{(t_0)}\right] = \int_\theta \mathbf{E}\left[\sum_{\tau=0}^{T} \gamma^\tau r\left(\mathbf{X}^{(t_0+1+\tau)}\right) \middle| \theta, \mathbf{x}^{(t_0)}\right] p\left(\theta \middle| \hat{\theta}\right),$$

where the conditional distribution $p\left(\theta \middle| \hat{\theta}\right)$ will take into account not only measurement error in $\hat{\theta}$, but also uncertainty in the inherent biases and marketing responses. The construction of this conditional distribution will in practice involve input from market research, where consumers will be profiled, ideally allowing a Company to infer the likelihoods that consumers have certain inherent biases and marketing responses.

8 Appendix: Additional MCDL Bias Estimates

Here we include results for estimates of direct and social biases for additional consumers.

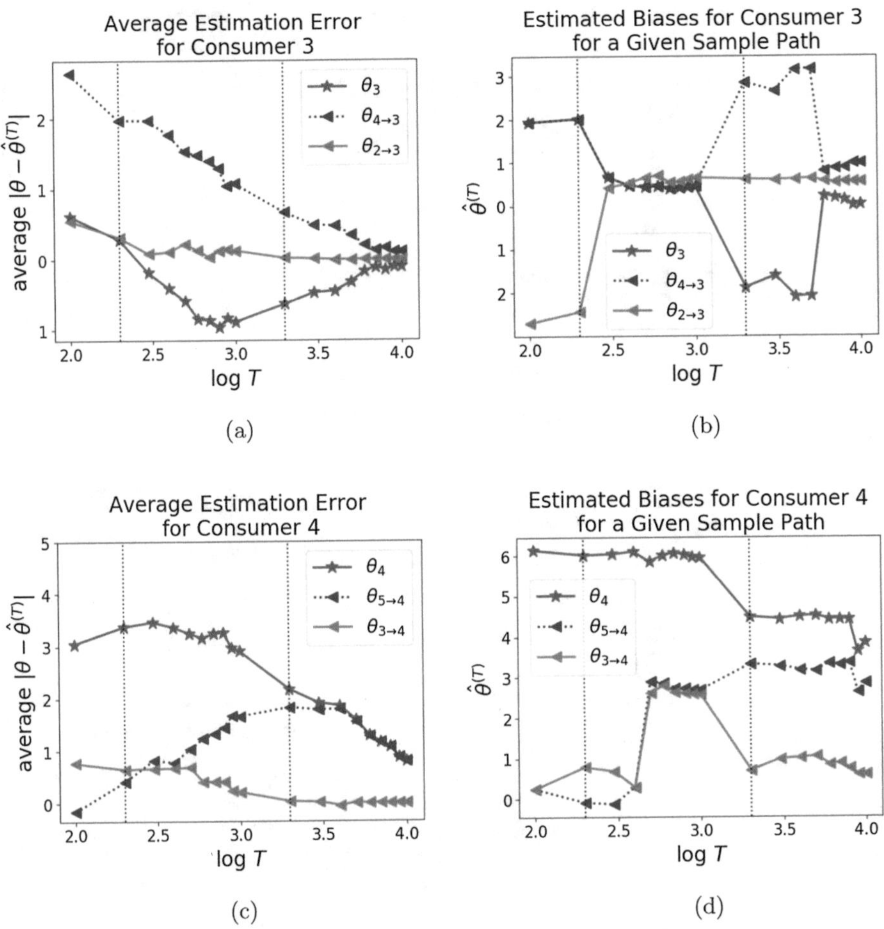

Fig. 5. For Glauber dynamics with uniformly asymmetric social biases, i.e., $\theta_{i+1\to i} = 1$ and $\theta_{i-1\to i} = .6$, for all i, $\theta_4 = 2$, $\theta_i = 0$ for $i \neq 4$, (a) and (c) estimation error; (b) and (d) direct and social bias estimates for a sample path.

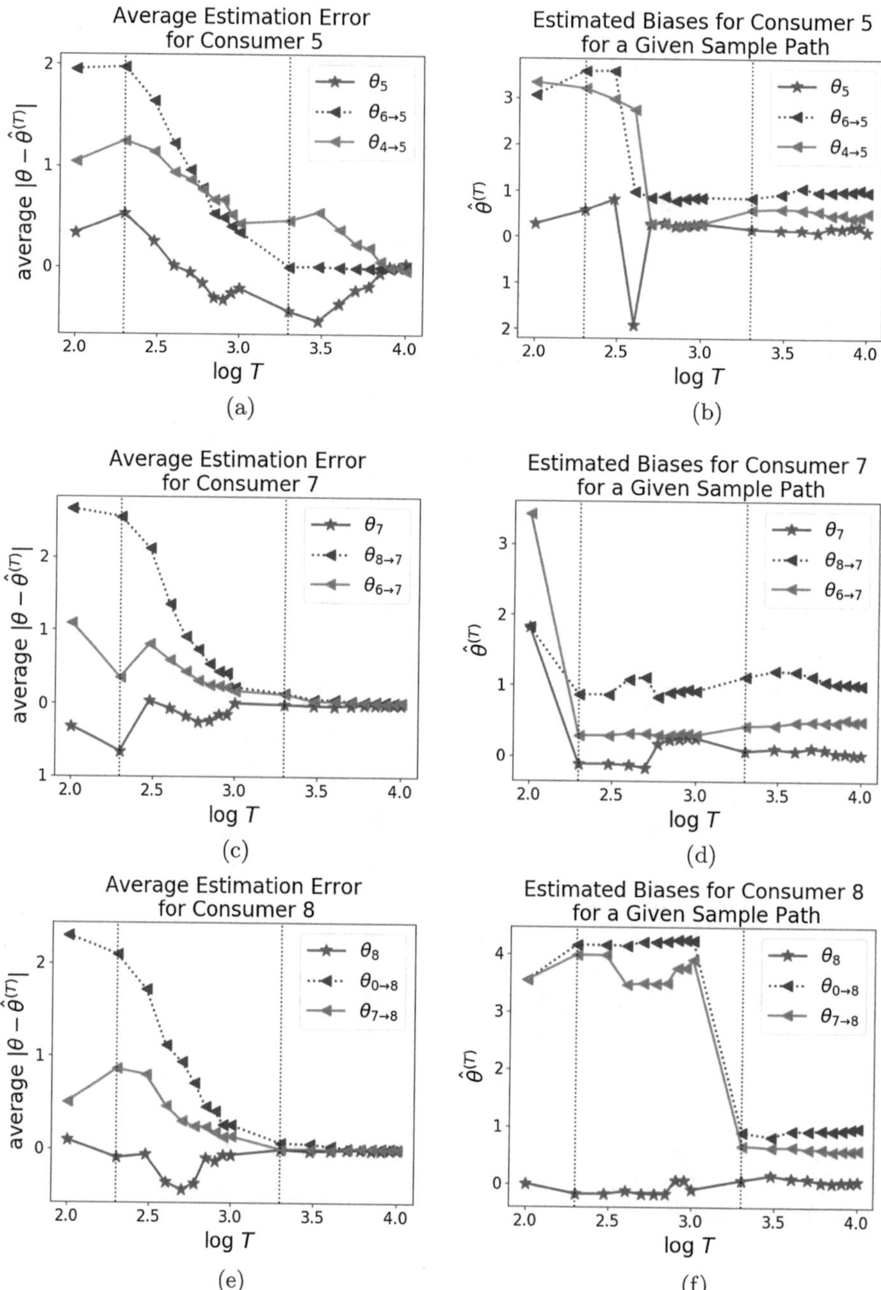

Fig. 6. For Glauber dynamics with uniformly asymmetric social biases, i.e., $\theta_{i+1\to i} = 1$ and $\theta_{i-1\to i} = .6$, for all i, $\theta_4 = 2$, $\theta_i = 0$ for $i \neq 4$, (a), (c), and (e) estimation error; (b), (d), and (f) direct and social bias estimates for a sample path.

References

1. Barabasi, A.-L., Bonabeau, E.: Scale-free networks. Sci. Am. **288**(5), 60–69 (2003)
2. Bell, D.E., Keeney, R.L., Little, J.D.C.: A market share theorem. J. Mark. Res. **12**(2), 136–141 (1975)
3. Blume, L.E.: Statistical mechanics of strategic interaction. Games Econ. Behav. **5**(3), 387–424 (1993)
4. Csiszar, I., Talata, Z.: Consistent estimation of the basic neighborhood structure of a Markov random field
5. Georgii, H.O.: Gibbs Measures and Phase Transitions. De Grutyer, Berlin (1988)
6. Glauber, R.J.: Time-dependent statistics of the Ising model. J. Math. Phys. **4**, 294–307 (1963)
7. Godreche, C.: Dynamics of the directed Ising chain. J. Stat. Mech. Theory Exp. (2011). https://doi.org/10.1088/1742-5468/2011/04/P04005
8. Hanssens, D.M., Parsons, L., Schultz, R.L.: Market Response Models: Econometric and Time Series Analysis. Kluwer Academic Publishers, Norwell (2002)
9. Liu, H., Xie, Y., Hu, H., Chen, Z.: Affinity based information diffusion models in social networks. Int. J. Modern Phys. C, **25** (2014)
10. Luce, D.: Individual Choice Behavior. Dover, New York (1959)
11. Kempe, D., Kleinberg, J., Tardos, E.: Influential nodes in a diffusion model for social networks. In: Proceedings of the 32nd International Colloquium on Automata, Languages, and Programming (ICALP), pp. 1127–1138 (2005)
12. Kim, D.-H., Noh, J.D., Jeong, H.: Scale-free trees: the skeletons of complex networks. Phys. Rev. E, **70** (2004)
13. McFadden, D.: Conditional logit analysis of qualitative choice behavior. In: Frontiers in Econometrics. Academic Press, New York (1974)
14. Montanari, A., Saberi, A.: The spread of innovations in social networks. PNAS **107**(47), 20196–20201 (2010)
15. Reyes, M.G., Neuhoff, D.L.: Minimum conditional description length estimation of Markov random fields. In: Information Theory and Applications Workshop, February 2016
16. Reyes, M.G.: A marketing game: a rigorous model for strategic resource allocation. In: ACM Workshop on Machine Learning in Graphs, 19–23 August 2018, London, UK (2018)
17. Reyes, M.G.: A marketing game: a model for social media mining and manipulation. Accepted to Future of Information and Communication Conference, 14 March 2019, San Francisco, CA (2019)
18. Reyes, M.G.: Frustrated Equilibrium of Coordinating Dynamics in A Marketing Game. In preparation
19. Rissanen, J.: A universal prior for integers and estimation by minimum description length. Ann. Stat. **11**(2), 416–431 (1983)
20. Shang, Y.: Lie algebraic discussion for affinity based information diffusion in social networks. Open Phys. **15** (2017)
21. Stevens, S.S.: To honor fechner and repeal his law. Science **133**(3446), 80–86 (1961)
22. Sutton, R.S., Barto, A.G.: Reinforcement Learning: An Introduction. MIT Press, Cambridge (1998)

23. Thurstone, L.L.: Psychological analysis. Am. J. Psychol. **38**(3), 368–389 (1927)
24. Wainwright, M.J.: Estimating the "wrong" graphical model: benefits in the computation-limited setting. J. Mach. Learn. Res. **7**, 1829–1859 (2006)
25. Ravikumar, P., Wainwright, M., Lafferty, J.: High-dimensional ising model selection using l_1-regularized logistic regression. Ann. Stat. **38**(3), 1287–1319 (2010)
26. Watts, D.J., Dodds, P.S.: Influentials, networks, and public opinion formation. J. Consum. Res. **34**, 441–458 (2007)

Pedestrian-Motorcycle Binary Classification Using Data Augmentation and Convolutional Neural Networks

Robert Kerwin C. Billones(✉), Argel A. Bandala, Laurence A. Gan Lim, Edwin Sybingco, Alexis M. Fillone, and Elmer P. Dadios

De La Salle University, Manila, Philippines
robert.billones@dlsu.edu.ph

Abstract. One common problem in vehicle and pedestrian detection algorithms is the mis-classification of motorcycle riders as pedestrians. This paper focused on a binary classification technique using convolutional neural networks for pedestrian and motorcycle riders in different road context locations. The study also includes a data augmentation technique to address the un-balanced number of training images for a machine learning algorithm. This problem in un-balanced data sets usually cause a prediction bias, in which the prediction for a learned data set usually favors the class with more image representations. Using four data sets with differing road context (DS0, DS3-1, DS4-3, and DS4-3), the binary classification between pedestrian and motorcycle riders achieved good results. In DS0, training accuracy is 96.96% while validation accuracy is 81.52%. In DS3-1, training accuracy is 93.17% while validation accuracy is 86.58%. In DS4-1, training accuracy is 94.42% while validation accuracy is 97.00%. In DS4-3, training accuracy is 95.94% while validation accuracy is 88.59%.

Keywords: Binary classification · Convolutional neural networks · Data augmentation · Pedestrian activity analysis

1 Introduction

Traffic monitoring in urban environments typically involves the detection of both vehicles [1, 2] and pedestrians, and their interaction with each other. Computer vision techniques can be used to detect and recognize these road traffic agents. Computer vision involves the extraction of relevant visual features to represent moving and non-moving objects [3] in road traffic scenes. Along with pattern recognition techniques, several traffic monitoring objectives can be attained such as congestion analysis [4], traffic violations detection [5], license plate recognition [6], and vehicle detection [7, 8] and classification [9]. Other research areas for road traffic scene understanding are pedestrian detection [10, 11], human behavior recognition [12, 13], and crowd activity analysis [14–16]. Pedestrian detection posed a challenge because of the variations in human poses, as well as, lighting conditions, occlusion, and even camera views [17]. Road user types such as motorcycles, bicycles, and even three-wheeled vehicles that have similar features to a pedestrian adds to the confusion of people or pedestrian detection.

K. Arai et al. (Eds.): CompCom 2019, AISC 997, pp. 725–737, 2019.
https://doi.org/10.1007/978-3-030-22871-2_50

This study focused on the binary classification of pedestrians and motorcycles. Most of the time, motorcycle riders (even bicycles) are mis-classified as pedestrians. In applications such as behavior monitoring of boarding and alighting passengers, as well as, people crossing a pedestrian lane, the mis-classification problem contributes to the number of false positive detections. To correctly differentiate between a pedestrian and motorcycle riders, a convolutional neural network [18, 19] is used as the learning algorithm to classify pedestrians and motorcycles. Also, in some instances where there is un-balanced number of training images, a data augmentation technique was used to generate a balanced data set. An un-balanced data set usually caused problems involving prediction bias in which the prediction for a learned data set usually favors the class with more image representations.

2 Pedestrian Detection

There has been an increased scope of urban traffic activity analysis using automatic video analytics. The richness of information from video cameras provided a suitable platform for understanding these traffic activities. In urban environments, we can include pedestrians, as well as bicycles, as road users [20, 21]. This prompted several studies on the effects of human activities in road traffic. Social interactions are a natural part of a crowded [14–16] traffic scene, especially in urban traffic scenarios. Pedestrian activity analysis [10, 11] is important in understanding the complex interactions of road users in urban intersections. In visual systems, pedestrian or people must first be recognized as a distinct object. Figure 1 shows a sample of a people detector.

Fig. 1. Sample of people detection [10]

Several models for people or pedestrian feature extraction have been proposed such as Haar cascades, and histogram of gradients (HOG) [22]. These features are then trained using learning algorithms such as support vector machines (SVMs) [22], boosting classifiers [23], and artificial neural networks [8]. Multi-classifier models [24–26] were also proposed to improve the learning capabilities of single classifiers. In the study in [17], K-means algorithm was used to prepare a diverse training data sets for a deep multi-classifier model. Equation (1) below was used extract features for each image sample.

$$X = \{x_i\}_{i=1}^{n} \tag{1}$$

Equation (2) shows how cluster centroids are selected at random from k images. While Eq. 3, calculates the distance between the cluster centroids and feature images using Eq. (1).

$$\{c_j\}_{j=1}^{k} \tag{2}$$

$$d_{ij} = \left\| x_i - c_j \right\|_2 \tag{3}$$

Equation (4) calculates the minimal distance by dividing the images into corresponding clusters.

$$C_j = \left\{ i | \forall m \neq j, d_{im} > d_{ij} \right\} \tag{4}$$

where $1 \leq m \leq k$.

Equation (2) can be updated by calculating the mean of images within each cluster. This can be computed with Eq. (5).

$$c_j = \frac{1}{|c_j|} \sum\nolimits_{i \in C_j} x_i \tag{5}$$

This process is repeated until clustering results become constant.

3 Convolutional Neural Networks

According to the recent findings in neuroscience, information representation in the neocortex of a mammalian brain does not pre-process sensory signals [27, 28]. This leads to the development of new design architectures that does not rely on conventional computer vision feature extraction techniques. It is important to note that sensory signals propagate through the brain in a hierarchical manner that allows for learning of any exhibited regularities in the signal representations or observations. The emergence of deep learning architectures is inspired by these key findings [29]. Convolutional neural networks (CNNs) [19] are the first successful implementation of a robust deep learning architecture. Elementary features of an image [30] such as oriented corners and edges can be access by the neurons, thus providing invariance to scale, rotation, and shift [31, 32].

The convolution process [33–35] involves convolving an input image I to a filter K. A grayscale image I can be defined by Eq. (6), while filter K can be defined by Eq. (7).

$$I : \{1,\ldots,n_1\} \times \{1,\ldots,n_2\} \rightarrow W \subseteq \mathbb{R}, (i,j) \rightarrow I_{i,j} \tag{6}$$

$$K \in \mathbb{R}^{2h_1+1 \times 2h_2+1} \tag{7}$$

Image I can be represented by $n_1 \times n_2$ array. For a color image, this array can be represented by three (RGB) color channels, $n_1 \times n_2 \times 3$. Equivalently, Eq. (8) also describes filter K.

$$K = \begin{pmatrix} K_{-h1,-h2} & \cdots & K_{-h1,h2} \\ \vdots & K_{0,0} & \vdots \\ K_{h1,-h2} & \cdots & K_{h1,h2} \end{pmatrix} \tag{8}$$

The discrete convolution of image I and filter K is calculated using Eq. (9).

$$(I * K)_{r,s} := \sum\nolimits_{u=-h_1}^{h_1} \sum\nolimits_{v=-h_2}^{h_2} K_{u,v} I_{r+u,s+v} \tag{9}$$

After discrete convolution, we need to define the borders of an image by using a smoothing filter. A Gaussian filter $K_{G(\sigma)}$ which is calculated using Eq. (10) is commonly used.

$$\left(K_{G(\sigma)}\right)_{r,s} = \frac{1}{\sqrt{2\pi\sigma^2}} exp \frac{r^2 + s^2}{2\sigma^2} \tag{10}$$

where σ is Gaussian distribution's standard deviation [36].

Figure 2 shows a single convolutional layer with output feature maps derived from the input image through convolution process.

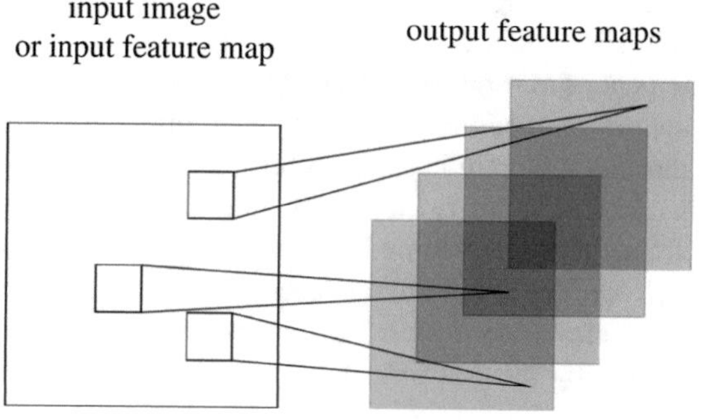

Fig. 2. Single convolutional layer [33]

4 Methodology

4.1 Image Data Sets

The image data set used for this study is composed of four different locations. DS0 is composed of traffic videos captured using low-altitude cameras which are used for traffic violations monitoring (number coding and illegal loading/unloading) of a T-type road intersection. It is further composed of LS1, LS2, LS3, TA1, TA2, and TA3 video data sets. DS3-1 is a traffic video captured using high-altitude camera which is usually used for traffic congestion monitoring of a wide intersection. DS4-1 is a traffic video captured using medium-altitude camera which is usually used for traffic violations (number coding and illegal loading/unloading) and congestion monitoring of a bus stop area. DS4-3 is the same camera used in DS4-1, but the video capture is done during night-time. Figure 3 shows the sample video capture images from the different context locations.

Fig. 3. Context locations: DS0, DS3-1, DS4-1, and DS4-3

Table 1 shows the number of training and validation samples used for each context locations (DS0, DS3-1, DS4-1, DS4-3). Figures 4, 5, 6, and 7 show the sample images of pedestrians and motorcycles in different context locations.

Table 1. Training and validation samples

Location	No. of training samples	No. of validation samples
DS0	600	200
DS3-1	1400	600
DS4-1	1200	400
DS4-3	600	200

Fig. 4. Sample of pedestrian and motorcycle for DS0

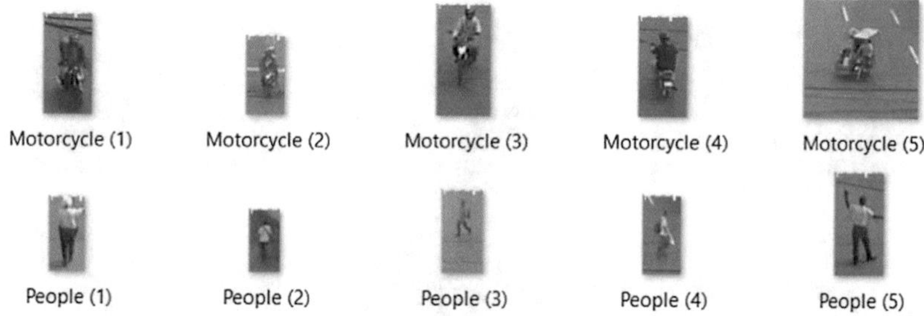

Fig. 5. Sample of pedestrian and motorcycle for DS3-1

Fig. 6. Sample of pedestrian and motorcycle for DS4-1

Fig. 7. Sample of pedestrian and motorcycle for DS4-3

4.2 Data Augmentation

In cases when there are very few training and validation samples available, data augmentation can be use so that the learning model would not process the exact same image repeatedly. This can be achieved by random transformations such as rotation, width shift, height shift, rescale, shear, zoom, and horizontal flip. Data augmentation also prevent overfitting and helps the learning model to have prediction bias.

The study used the Keras library to pre-process image. Table 2 lists the parameters used in the data augmentation process. Data augmentation process is used for the DS4-3 motorcycle dataset. Only 10 motorcycle images are available for this dataset.

Table 2. Data augmentation parameters

Parameter	Value
Rotation range	20
Width shift range	0.2
Height shift range	0.2
Shear range	0.2
Zoom range	0.2
Horizontal flip	False
Fill mode	Nearest

4.3 Convolutional Neural Network Parameters

The study used a CNN model with three (3) layers of 2D convolutional (CONV) layers and an output layer of densely-connected neural network. The input to CNN is the images from the training and validation samples, and the binary output is either people or motorcycle. The CNN model and its parameters are listed in Table 3.

Table 3. CNN model and parameters

CNN model	Parameters
Input layer (2D convolutional layer)	32 filters 3×3 kernel size
Activation	Rectified linear unit (relu)
2D max pooling layer	2×2 pool size
2D convolutional layer	32 filters 3×3 kernel size
Activation	Rectified linear unit (relu)
2D max pooling layer	2×2 pool size
2D convolutional layer	64 filters 3×3 kernel size
Activation	Rectified linear unit (relu)
2D max pooling layer	2×2 pool size
Core layer (Densely connected neural network layer)	64 filters
Activation	Rectified linear unit (relu)
Dropout	0.5
Densely connected neural network layer	2 filters (output)
Activation	Sigmoid

5 Results and Discussions

5.1 Data Augmentation Results

In the DS4-3 dataset, there is only 10 raw images for motorcycles. To augment the motorcycle dataset, the data augmentation described in section IV-B was used. With input of 10 raw images, the process generated 390 images. These images are then used for training and validation of the DS4-3 dataset. Figure 8 shows samples of raw motorcycle images, while Fig. 9 shows samples of generated motorcycle images.

Fig. 8. Sample of raw motorcycle images for DS4-3

Fig. 9. Sample of generated motorcycle images for DS4-3

5.2 CNN Training and Validation Results

The DS0 dataset used a total of 600 training and 200 validation images. The DS0 learning model was simulated for 723 s with 30 epochs. DS0 results showed that the training accuracy is 96.96% with a loss function of 0.0987, while the validation accuracy is 81.52% with a loss function of 0.6819. Figure 10 shows the plot of accuracy versus number of epochs for DS0 model.

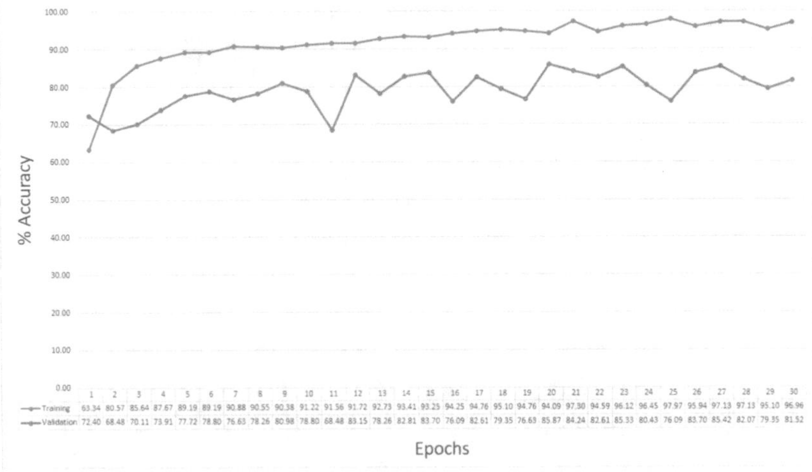

Fig. 10. CNN results on pedestrian-motorcycle binary classification for DS0

The DS3-1 dataset used a total of 1400 training and 600 validation images. The DS3-1 learning model was simulated for 1197 s with 20 epochs. DS3-1 results showed that the training accuracy is 93.17% with a loss function of 0.1864, while the validation

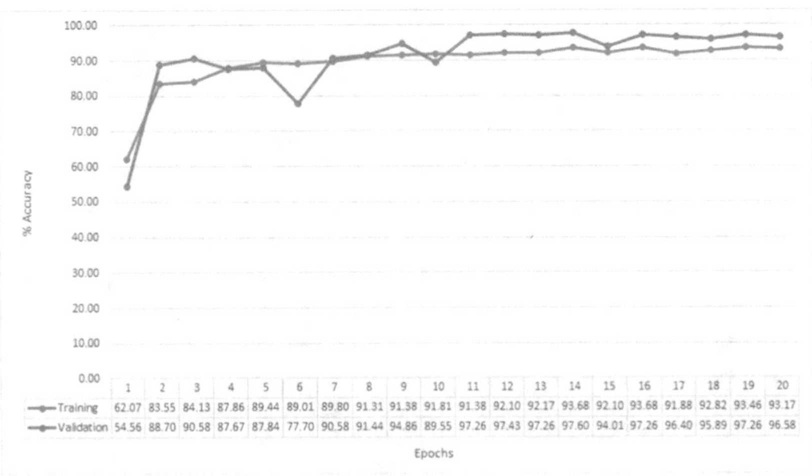

Fig. 11. CNN results on pedestrian-motorcycle binary classification for DS3-1

accuracy is 96.58% with a loss function of 0.1468. Figure 11 shows the plot of accuracy versus number of epochs for DS3-1 model.

The DS4-1 dataset used a total of 1200 training and 400 validation images. The DS4-1 learning model was simulated for 965 s with 20 epochs. DS4-1 results showed that the training accuracy is 94.42% with a loss function of 0.1567, while the validation accuracy is 97.00% with a loss function of 0.521. Figure 12 shows the plot of accuracy versus number of epochs for DS4-1 model.

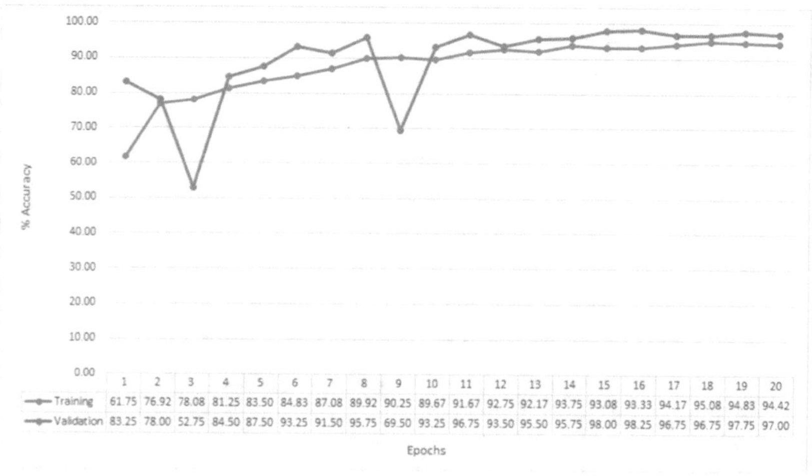

	1	2	3	4	5	6	7	8	9	10	11	12	13	14	15	16	17	18	19	20
Training	61.75	76.92	78.08	81.25	83.50	84.83	87.08	89.92	90.25	89.67	91.67	92.75	92.17	93.75	93.08	93.33	94.17	95.08	94.83	94.42
Validation	83.25	78.00	52.75	84.50	87.50	93.25	91.50	95.75	69.50	93.25	96.75	93.50	95.50	95.75	98.00	98.25	96.75	96.75	97.75	97.00

Fig. 12. CNN results on pedestrian-motorcycle binary classification for DS4-1

The DS4-3 dataset used a total of 600 training and 200 validation images. The DS4-3 learning model was simulated for 700 s with 20 epochs. DS4-3 results showed that the training accuracy is 95.94% with a loss function of 0.1291, while the validation

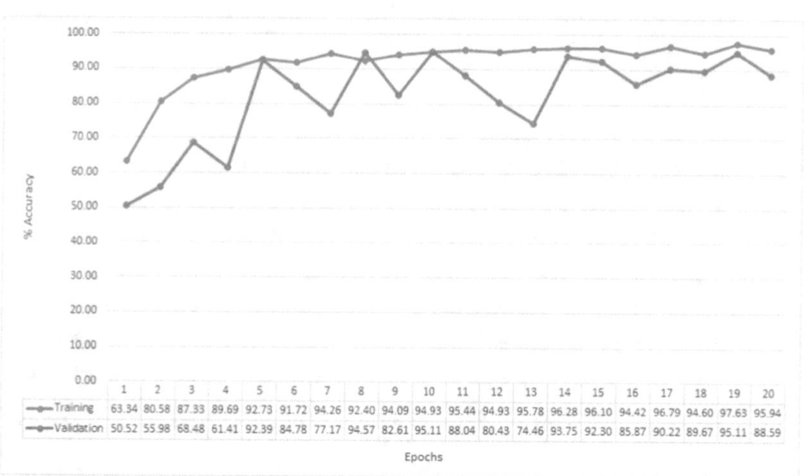

	1	2	3	4	5	6	7	8	9	10	11	12	13	14	15	16	17	18	19	20
Training	63.34	80.58	87.33	89.69	92.73	91.72	94.26	92.40	94.09	94.93	95.44	94.93	95.78	96.28	96.10	94.42	96.79	94.60	97.63	95.94
Validation	50.52	55.98	68.48	61.41	92.39	84.78	77.17	94.57	82.61	95.11	88.04	80.43	74.46	93.75	92.30	85.87	90.22	89.67	95.11	88.59

Fig. 13. CNN results on pedestrian-motorcycle binary classification for DS4-3

accuracy is 88.59% with a loss function of 0.4234. Figure 13 shows the plot of accuracy versus number of epochs for DS4-3 model.

5.3 Summary of Pedestrian-Motorcycle Binary Classification Results

Table 4 lists the summary results of pedestrian-motorcycle binary classification using convolutional neural networks. The table includes the training and validation accuracy, loss function, no. of epochs, and simulation time for each context locations (DS0, DS3-1, DS4-1, DS4-3).

Table 4. Summary of CNN results

Results	DS0	DS3-1	DS4-1	DS4-3
Training accuracy	96.96%	93.17%	94.42%	95.94%
Validation accuracy	81.52%	96.58%	97.00%	88.59%
Training loss function	0.0987	0.1864	0.1567	0.1291
Validation loss function	0.6819	0.1468	0.521	0.4234
No. of epochs	30	20	20	20
Simulation time (seconds)	723	1197	965	700

6 Conclusion

The study aims to address a mis-classification problem in pedestrian detection in urban environment. Typically, motorcycles and bicycles (even three-wheeled vehicles) are mis-classified as people or pedestrians. In some applications, this mis-classification contributes to false positive detections, and thus lowering the accuracy of the system. This study implemented a pedestrian-motorcycle binary classification using convolutional neural networks to specifically differentiate between a pedestrian (people) and motorcycles (or bicycles, and even three-wheeled vehicles). This learning model was used in four different locations with differing road conditions (DS0, DS3-1, DS4-1, and DS4-3). Also, a data augmentation technique was used to address the scarcity of training and validation samples of motorcycles for DS4-3 (bus stop area, during night-time). Using the DS0 dataset, the training accuracy is 96.96%, while validation accuracy is 81.52%. Using the DS3-1 dataset, the training accuracy is 93.17%, while validation accuracy is 96.58%. Using the DS4-1 dataset, the training accuracy is 94.42%, while validation accuracy is 97.00%. Using the DS4-3 dataset, the training accuracy is 95.94%, while validation accuracy is 88.59%. These experiments showed that we can further differentiate between pedestrians and motorcycle riders in different road conditions.

Acknowledgment. The authors highly appreciate the Department of Science and Technology - Philippine Council for Industry, Energy, and Emerging Technologies for Research and Development (DOST-PCIEERD) for providing funds for this research study. The authors also acknowledged Dr. Alvin B. Culaba, and Dr. Marlon D. Era for sharing their valuable knowledge for the completion of this research.

References

1. Sivaraman, S., Trivedi, M.: Integrated lane and vehicle detection, localization, and tracking: a synergistic approach. In: IEEE Transactions on Intelligent Transportation Systems (2013)
2. Bedruz, R.A., Sybingco, E., Bandala, A., Quiros, A.R., Uy, A.C., Dadios, E.: Real-time vehicle detection and tracking using a mean-shift based blob analysis and tracking approach. In: 2017 IEEE 9th International Conference on Humanoid, Nanotechnology, Information Technology, Communication and Control, Environment and Management (HNICEM), Manila, pp. 1–5 (2017)
3. Buch, N., Velastin, S.A., Orwell, J.: A review of computer vision techniques for the analysis of urban Traffic. IEEE Trans. Intell. Transp. Syst. **12**(3), 920 (2011)
4. Billones, R.K.C., Bandala, A.A., Lim, L.A.G., Sybingco, E., Fillone, A.M., Dadios, E.P.: Microscopic road traffic scene analysis using computer vision and traffic flow modelling. J. Adv. Comput. Intell. Intell. Inform. **22**(5), 704–710 (2018)
5. Billones, R.K.C., Bandala, A.A., Sybingco, E., Lim, L.A.G., Dadios, E.P.: Intelligent system architecture for a vision-based contactless apprehension of traffic violations. In: 2016 IEEE Region 10 Conference (TENCON), pp. 1871–1874 (2016)
6. Bedruz, R.A., Sybingco, E., Quiros, A.R., Uy, A.C., Vicerra, R.R., Dadios, E.: Fuzzy logic based vehicular plate character recognition system using image segmentation and scale-invariant feature transform. In: 2016 IEEE Region 10 Conference (TENCON), Singapore, pp. 676–681 (2016)
7. Barth, A., Franke, U.: Tracking oncoming and turning vehicles at intersections. In: 13th International IEEE Conference on Intelligent Transportation Systems (ITSC), pp. 868–881 (2010)
8. Billones, R.K.C., Bandala, A.A., Sybingco, E., Lim, L.A.G., Fillone, A.D., Dadios, E.P.: Vehicle detection and tracking using corner feature points and artificial neural networks for a visionbased contactless apprehension system. In: Computing Conference 2017, pp. 688–691 (2017)
9. Buch, N., Orwell, J., Velastin, S.A.: Three-dimensional extended histograms of oriented gradients (3-DHOG) for classification of road users in urban scenes. In: Proceedings BMVC, London, UK. (2009)
10. Tuzel, O., Porikli, F., Meer, P.: Pedestrian detection via classification on riemannian manifolds. IEEE Trans. Pattern Anal. Mach. Intell. **30**(10), 1713–1727 (2008)
11. Escolano, C.O., Billones, R.K.C., Sybingco, E., Fillone, A.D., Dadios, E.P.: Passenger demand forecast using optical flow passenger counting system for bus dispatch scheduling. In: 2016 IEEE Region 10 Conference (TENCON), pp. 1–4 (2016)
12. Candamo, J., Shreve, M., Goldgof, D.B., Sapper, D.B., Kasturi, R.: Understanding transit scenes: a survey on human behavior-recognition algorithms. IEEE Trans. Intell. Transp. Syst. **11**(1), 206–224 (2010)
13. Shah, M.: Understanding human behavior from motion imagery. Mach. Vis. Appl. **14**(4), 210–214 (2003)
14. Shao, J., Kang, K., Loy, C.C., Wang, X.: Deeply learned attributes for crowded scene understanding. In: IEEE Conference Publication, pp. 4657–4666 (2015)
15. Zhang, C., Li, H., Wang, X., Yang, X.: Cross-scene crowd counting via deep convolutional neural networks. In: IEEE Conference Publication, pp. 833–841 (2015)
16. Zhou, B., Wang, X., Tang, X.: Understanding collective crowd behaviors: learning a mixture model of dynamic pedestrian-agents. In: IEEE Conference Publication, pp. 2871–2878 (2012)

17. Song, H., Wang, W., Wang, J., Wang, R.: Collaborative deep networks for pedestrian detection. In: IEEE Third International Conference on Multimedia Big Data, pp. 146–153 (2017)
18. Wang, L., Wang, Z., Du, W., Qiao, Y.: Object-scene convolutional neural networks for event recognition in images. In: IEEE Conference Publication, pp. 30–35 (2015)
19. Huang, F.-J., LeCun, Y.: Large-scale learning with SVM and convolutional nets for generic object categorization. In: Proceedings Computer Vision and Pattern Recognition Conference (CVPR 2006) (2006)
20. Wang, X., Ma, X., Grimson, W.E.L.: Unsupervised activity perception in crowded and complicated scenes using hierarchical bayesian models. IEEE Trans. Pattern Anal. Mach. Intell. **31**(3), 539–555 (2009)
21. Shao, J., Loy, C.C., Wang, X.: Scene-independent group Profiling in crowd. In: 2014 IEEE Conference on Computer Vision and Pattern Recognition, pp. 2227–2234 (2014)
22. Dalal, N., Triggs, B.: Histograms of oriented gradients for human detection. CVPR **1**, 886–893 (2005)
23. Cosma, C., Brehar, R., Nedevschi, S.: Pedestrians detection using a cascade of lbp and hog classifiers. In: ICCP, pp. 69–75 (2013)
24. Guo, W., Xiao, Y., Zhang, G.: Multi-scale pedestrian detection by use of adaboost learning algorithm. In: ICVRV, pp. 266–271 (2014)
25. Dai, D., Gool, L.V.: Ensemble projection for semisupervised image classification. In: ICCV, pp. 4321–4328 (2013)
26. Veni, C.V.K., Rani, T.S.: Ensemble based classification using small training sets: a novel approach. In: CIEL, pp. 1–8 (2014)
27. Lee, T., Mumford, D.: Hierarchical Bayesian inference in the visual cortex. J. Opt. Soc. Amer. **20**(7), 1434–1448 (2003)
28. Lee, T., Mumford, D., Romero, R., Lamme, V.: The role of the primary visual cortex in higher level vision. Vision. Res. **38**, 2429–2454 (1998)
29. Arel, I., Rose, D.C., Karnowski, T.P.: Deep machine learning—a new frontier in artificial intelligence research. In: IEEE Computational Intelligence Magazine, pp. 13–18 (2010)
30. LeCun, Y.: Generalization and network design strategies. In: Connectionism in Perspective (1989)
31. Girshick, R.: FastR-CNN. In: 2015 IEEE International Conference on Computer Vision, vol. 8, no. 1, pp. 1440–1448 (2015)
32. Szegedy, C., Liu, W., Jia, Y., Sermanet, P., Reed, S., Anguelov, D., Erhan, D., Vanhoucke, V., Rabinovich, A.: Going deeper with convolutions. In: IEEE Conference Publication, pp. 1–9 (2015)
33. Stutz, D.: Seminar report: understanding convolutional neural networks. Fakultät für Mathematik, Informatik und Naturwissenschaften Lehr- und Forschungsgebiet Informatik VIII (2014)
34. Jarrett, K., Kavukcuogl, K., Ranzato, M., LeCun, Y.: What is the best multi-stage architecture for object recognition? In: International Conference on Computer Vision, pp. 2146–2153 (2009)
35. LeCun, Y., Kavukvuoglu, K., Farabet, C.: Convolutional networks and applications in vision. In: International Symposium on Circuits and Systems, pp. 253–256 (2010)
36. Forsyth, D., Ponce, J.: Computer Vision: A Modern Approac. Prentice Hall Professional Technical Reference, New Jersey (2002)

Performance Analysis of Missing Values Imputation Methods Using Machine Learning Techniques

Omesaad Rado(✉), Muna Al Fanah, and Ebtesam Taktek

Department of Computer Science, University of Bradford, Bradford BD7 1DP, UK
{o.a.m.rado,m.m.s.alfanah, e.a.m.taktek}@bradford.ac.uk

Abstract. Real world data often contain missing values. Data mining techniques have been actively used to overcome this problem by using methods of imputing the missing values. In particular, before applying any classification model, handling missing data in the dataset is an important task in pre-processing stages for ensuring the quality of classification results. Using the appropriate method of missing value imputation can help to generate complete datasets for improving the classifier's performance. Many approaches have been proposed in the field of machine learning and data mining for handling missing values. Techniques used for imputing missing values can be divided into single and multiple methods. Some techniques, namely random forests, CART, k-NN imputation method and mean method, remove attributes and observations, predicting missed values by Multivariate Imputation method by Chained Equations (MICE) for example. In this study various approaches of treating missing values were applied on different decision trees algorithms to investigate how these techniques can be used effectively to improve the performance of selected classifiers. The Stroke data set was used in these experiments to check how well the methods of handling missing values work. Moreover, the paper reports how using data imputation methods affect classification results. The best results are obtained from the classifiers with removing variables that have missing data more than the rest of attributes. This work presents an attempt to analyse the chosen techniques with the purpose to investigate their strengths and weaknesses in handling missing values, and reports that both imputation methods (MIM and MICE) are efficient and yield similar accuracy.

Keywords: Machine learning · Data imputation methods · Classification model

1 Introduction

Missing values is one of the many problems associated with real-life data sets that affect the quality of data mining results. The goal of data mining techniques is to provide fast and accurate decision making in several domains by analysing the

K. Arai et al. (Eds.): CompCom 2019, AISC 997, pp. 738–750, 2019.
https://doi.org/10.1007/978-3-030-22871-2_51

gathering data. However, many techniques have been used to impute missing values in the data based on data types and the missingness pattern. Particularly for data used in healthcare, these data may contain missing values that occur due to measurement errors, lack of information or because the patient may not wish to record or share certain personal information. Consequently data cleaning and exploratory analysis are necessary stages in model building processes to produce powerful models.

The most common problem in this stage is handling missing values of the real-world data effectively. The most commonly used methods of data imputation are deleting the observations or attributes or both. The most common methods of missing (numerical) data replacement use mean or median or mode. However, the results of a study in [1, 2] showed that there is no best method of imputation which can provide a good performance for all classifiers.

Scholars in the health sector such as Mallinckrodt [3] used a comprehensive strategy for preventing and treating missing data. Other machine learning scholars replaced missing data using various techniques such as Schafer schemes [4, 5]. The same scheme was used in [5]. Moreover, the Multi-layer perceptron (MLP), Self-organisation maps (SOM), K-nearest neighbours, K-NN prognosis model, and K-nearest neighbours are also used [6].

There are many reasons to select the Stroke dataset. Firstly, dealing with missing data in the health sector is expensive. According to [7], missing data analysis within a randomised clinical trial is a current problem in cost-effectiveness. Therefore, the choice of method should be grounded in the hypothesised missing data mechanism, which as a result should be informed by the available evidence [7].

Secondly, stroke disease is a challenge disease to diagnosis and high-priced. For instance, an early nineties report study [8] indicated that young adults having stroke disease were uncommon and often a diagnostic challenge. A survey conducted at that time named as "The National Survey of Stroke" declared that only 3.7% of all strokes occurred in patients aged 15–45 years. The review of this study has the medical records of 113 young patients during (1982–1987) aged 15–45 years who were recorded to the Medical Center Hospital of Vermont with a diagnosis of stroke. In recent years, a study by [9] estimated that by 2013 an approximate 5% of young adults died of stroke. Most of young adults who were discharged routinely to stay at home is at an average cost associated with their hospital stay of $34,886 for ischemic stroke, $146,307 for sub-arachnoid haemorrhage, and $94,482 for intracerebral haemorrhage [9].

In addition, health sector data is a critical source of enrichment for both, tmedical studies and patients care. In fact, stroke is a disease resulting from several factors, such as interruption of the blood supply to the brain. According to World Health Organisation (WHO) [10], stroke is one of the common chronic diseases and are of the leading causes of death. Therefore, treating the missing values in stroke data records using the chosen techniques will improve the quality of the dataset and consequently their mining outputs.

Typically, there are three types of missing data patterns [11]: (1) missing completely at random or MCAR; (2) missing at random or MAR; and (3) Missing Not At Random (MNAR).

The contributions of this study are twofold: the first part is the method of completing the missing values during the pre-processing phase; the second contribution is using the completed data that is obtained in the pre-processing phase to build decision trees and random forests models. Finally, comparing the performance of the chosen classifiers will provide insights in the effect of missing data and imputation methods to the classifiers.

The potential of this research targets the problem of handling missing values in the Stroke dataset, by exploring different options of how to deal with missing values or how to impute these values. A comparison of classifiers performance was done to verify the best imputation methods with tree-structured classifiers which include decision trees and random forests.

Many approaches have been proposed in the field of machine learning and data mining for dealing with missing values, mainly, including remove attributes, remove observations, predicting by MICE [12]. This paper has been organized in the following way. The second section of this paper contains the literature review. The third part describes the methodology used for this study, methods and the model construction. The forth section includes analyses the results of the study. Finally, conclusion and areas for further research are identified.

2 Related Work

In 1988, Cheeseman noticed the importance of treating missing data in his early study on Machine Learning using autoclass [2]. Also, authors of [13] did a comparative study between Autoclass and Decision Trees (C4.5) to impute data. This study highlighted the need for predicting missing data and led to series of developments in algorithms to impute missing data.

In recent years, there has been an increasing amount of literature on handling missing data. In a current major study [14], the authors examined two multilevel multiple imputation approaches: Fully Conditional Specification (FCS) imputation and Joint Model (JM) imputation were examined for replacing incomplete variables. Another study involved applying data imputation to replace missing information in the dataset of the Samho gauging station at Taehwa River [15]. In [16] authors presented a comparative study to evaluate four models of imputation of environmental data. Surveys, such as [17] were conducted and had analysed the use of data imputation methods on different classifiers based on clinical Heart Failure dataset in an investigation into imputation methods. In [18], the authors found that chosen strategies of variable selection methods should consider missing data mechanisms utilized for dealing with missing information. Another study [19] identified that the incorrect assignment of disease can lead to misclassification bias; therefore bootstrapping methods were used to measure misclassification bias.

Recent studies that addressed the methods for missing data use tensor method in Breast cancer data to reduce missing values in order to provide more accuracy for a diagnostic accuracy [20]. This study use the particle swarm optimization algorithm with adaptive adjustment (RAPOS) which is developing by chaotic search applied by solve the problem of imbalance data [20]. The results are an easy diagnosis prediction through using different classifiers. For RAPOS method, Mean Square Error (MSE), accuracy level and sensitivity are all superior to other method [20].

Another study conducted by Yu et al. [21] focused in eliminating MSE by using advance modification of the original extreme learning machine with new tools for solving missing data problem. The applied method uses five data sets, and shows fast computational speed, including no parameter need to be tuned and it appears more reliable and stable by using two cascade of; L1 penalty (LARS) and L2 penalty (Tikhonov regularization) on ELM(TROP-ELM) to regularize [21].

In contrast, various scholars referred to missingness in data treating by dummy codes for each variable with missing data. These missing variables were encoded as a predictor in a regression model [22]. However, a study by [22] found out the drawbacks of applying different method for imputation. One of the most important advantage is Imputation minimises Bias, less costly to collect the missing values. It also allows for analysis using a rectangular dataset, since using regular software and techniques, so that standard analysis can then proceed. In contrast, the drawbacks of imputed data are: influence the type of imputation can affect the completed data, the second drawback is the uncertainty due to imputation [22]. However, there are many criteria for replacing missing data as addressed by White in [23]. Studies carried out in different sector investigate replacing missing values as pointed out by [24].

The study conducted by Walczak and Massart in two parts [25, 26]. The first study depends on iterative approached and named as Imputation Approach (IA) algorithm. It highlights that the validation of the final models must be performed very carefully, since the attenuation of data variability. This attenuation depends on the correlation of the studied variables and on the percentage of missing elements. While, in the second study, used the EM algorithm that apply maximum likelihood, in order to estimate the missing values [26].

3 Methodology

Suppose the dataset has $x_1, x_2, \ldots, x_n$ features and x_n contains missing values. x_n can be replaced or predicted based on other features. Data imputation refers to the process for replacing missing values and for predicting values by using algorithms of imputation data [24].

The mentioned methods in the previous section allow us to highlight the terminologies used within concerning literature predicting missing data. Before deciding about the way of handling missing values in dataset, it is important to consider the pattern of missingness. Most of literature refers to mechanisms of missing data can be classified into categories based on the missing value. According to [27], EM algorithm was used for replacing missing values and went further to analysis EM algorithm with missing values. Typically, majority of scholars as agreed that types of patterns of missing values in relation to randomness of missing values are categorized as follows:

First the Missing Completely At Random (MCAR)

Missing values are unrelated to any feature or observed data. It might occur in situations where the data is missing due to some random phenomena. Skip pattern in a questionnaire might appear when the person did not answer a part of the questions [26].

Second the Missing At Random (MAR)
The missing data can be incompletely at random and related to observed data. Otherwise, when an outcome variable is missing at random and accepted to exclude the missing cases. A study in [28] approved that for every the Missing Not At Random (MNAR) model an MAR counterpart can be constructed that has exact the same fit to the observed data, in the sense that it produces exact the same predictions to the observed data.
Third the Missing Not At Random (MNAR)
This includes another case of data missingness that depends on the missing value itself. The missingness is not related to observed or available data.

3.1 Imputation Methods

Generally, techniques for handling the missing data can be categorised to single imputation and multiple imputation. Precisely, several approaches can be used in both single and multiple imputation. The methods used by this work are as follows:

- **Single Imputation**

The first approach for dealing with missing values is called a single imputation. This approach handles missing values by using only the cases that have complete observed data and exclude the rest that have missing values. When number of variables have missing values, this approach can be achieved by mean, mode or median imputation, deleting the variable, removing observations of the original data.

However, keeping or losing variables is an important decision in the analysing process. For numerical data, the treatment of missing values can be done by replacing the missing values with the mean, mode or median to gain better results. These are considered as the most frequently used methods to replace the missing value for a given dataset. In some cases, a specific variable has a higher rate of missing values than the rest of the variables in the data. Therefore, eliminating that attribute can be better solution to save information. Furthermore, removing that variable cannot affect the meaning of the problem. In another hand, removing observations can be a solution of missing values particularly if the missing data is not a big number of observations of big dataset. Bias in data can be reduced by using a simple method is called a weighting method. However, the model may lose information by deleting these observations. Additionally, the bias is another issue can be brought here and it will affect the classes in the training data.

- **Multiple Imputation**

The second method is multiple imputation. It works based on Multivariate Imputation method by Chained Equations (MICE) for dealing with missing data. MICE is an approach to impute based on a set of imputation methods. MICE consider the missing data that are Missing at Random (MAR) [29]. MICE is helpful for large imputation procedures as addressed by White [23]. A few machine learning algorithms are used as mechanisms of missing value imputation.

3.2 Classification Algorithms

3.2.1 Datasets

The original dataset used in this experiment is Stroke dataset which contains training data set and tested data set. The training dataset has 43400 instances with 12 attributes. The test dataset contains 10 attributes and 18601 instances for testing dataset it contains 10 attributes. Table 1 shows the description of the columns of the data. The data is available on Kaggle the datasets source [30].

Table 1. Data description.

No	Feature name	Data type	Description
1	ID	Numeric	id-Patient
2	gender	Categorical	0%
3	Age	Numeric	Age of Patient
4	hypertension	Categorical	0 - no hypertension, 1 - suffering from hypertension
5	heart_disease	Categorical	0 - no heart disease, 1 - suffering from heart disease
6	ever_married	Categorical	Yes/No
7	work_type	Categorical	Type of occupation
8	Residence_type	Categorical	Area type of residence (Urban/Rural)
9	avg_glucose_level	Numeric	Average Glucose level (measured after meal)
10	Bmi	Numeric	Body mass index
11	smoking_status	Categorical	patient's smoking status
12	stroke	Categorical	0 - no stroke, 1 - suffered stroke

The entire stroke dataset includes the following: There are almost 96% of the values in the dataset with no missing information and 0.03% are missing. Figure 1 shows that from the total instances of the whole stroke dataset is 43,400 instances. The missing data (uncompleted instances verses the completed data) as 20% variables representing variables, that is, 2 variables. For instance, there is 3.31% cases with missing values of Bmi variable, and the most missing values is 30.71% of Smoking Status feature. The ratio of 33.01% cases are missing, that is, 14,328 cases, and approximately 15% (14,754) values were missing.

Classification tree are included within the tree-based models. The decision tree classifier is well-known supervised learning technique. A decision tree works based on recursive partition of the space of data points [31]. It separates points by choosing the best partition of the data via measures. J48 and PART algorithms are considered as decision trees. In general, common measures of a split point are Entropy and Gini index [14]. They are defined as follows:

$$Entropy(p) = -\sum\nolimits_{i=1}^{k} p_i log_2 p_i \tag{1}$$

Where *pi* denotes class probability in *p*, and *i* is class number

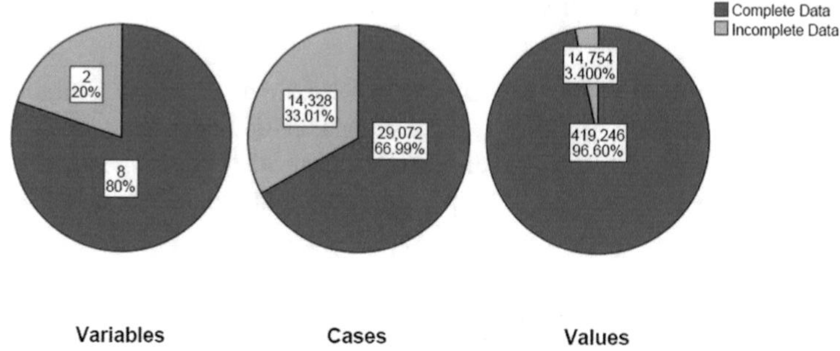

Fig. 1. The influence of missing values in the variables.

$$Gini\ index = 1 - \sum\nolimits_{i=1}^{k} p_i^2 \qquad (2)$$

The measure of Classification and Regression Trees (CART) is defined as follows:

$$CART(H_Y, H_N) = 2\frac{nY}{n}\frac{nN}{n}\sum\nolimits_{i=1}^{k} |p(c_i/H_Y) - p(c_i/H_N)| \qquad (3)$$

where n, nN and nY indicate points number in H, H_Y, and H_N [31].

This study deployed four classifiers: the first Random Forests (RF) is ensemble learning classifier that contains a collection of trees [32]. The second classifier, Support Vector Machines (SVM) classifier, formally works which is based on finding the optimal hyperplane and separates the boundary between classes in a d – dimensional space. Suppose given two label classes in 2 – dimensional space the hyperplane divides the plan into two sections for the classes [33]. The third, Naive Bayes algorithm (NB) is a supervised learning algorithm for classification. Several studies used this classifier to complete missing values see [34].

3.2.2 Tools

The R packages (MICE), Weka are used to run the experiments for missing value imputation in this study. These packages include functions to impute missing data for continuous and categorical variables.

4 The Model Construction

Classification stage contains conventional methods are pre-processing, special-purpose learning, post-processing, and hybrid approaches [35]. To solve the problem of imbalance data, there are four elementary techniques: sampling, cost-sensitive and ensemble approach. In this study, sampling technique Therefore, this research is achieved in three stages: pre-processing contains completing the missing data by removing observation, removing variables, and classifiers: Random forest and CART, along with ROSE.

The pre-processing phase obtains the data and applying different imputed methods to deal with missing values. It is an important to recognize the number of positive and negative cases for class variable to see whether the data are imbalanced data. In this data, the number of positive values of stork is 42617 and 783 for negative value. This means the data are imbalanced data set as was pointed out in the introduction to this paper. Therefore, ROSE method is used to correct imbalanced data.

The second stage is to use a classification algorithm to build classifiers for each obtained data from previous step. The primary distinction between this study and previous study is by He [36] is that this study use ROSE to balance the classes.

The potential of this study is to enhance the small dataset to reach high accuracy and precision. Missing values can be handled before or during the modelling [37]. According to [38], the methods are categorized into pre-replacing method and embedded method which can be performed during the classification. Missing values are estimated in the pre-process stage before creating the model. The model construction is displayed in Fig. 2. We have tried different missing value imputation mechanisms with stroke dataset and built classifiers for each method.

The third stage is to evaluate the performance based on the accuracy, precision and F-measure of the classifiers. The data preparation stage and Single imputation lead to find imbalance dataset. The imbalanced data was treated to build a model and finally evaluation stage was done.

Notably, the smoking status attribute has 0.306 missing values, BMI attribute has only 0.034 missing values and no missing values for the rest of features. BMI missing values are imputed with mean but for Smoking Status variable, it is not suitable to replace the values. Roughly, there are different ways to treat missing values such as deleting or predicting the variable before feeding the data to a model.

Figure 2 below indicated the pre-processing of data set: there are certain actions, such as, dealing with imbalance data and as a treatment of the missing data in stroke dataset.

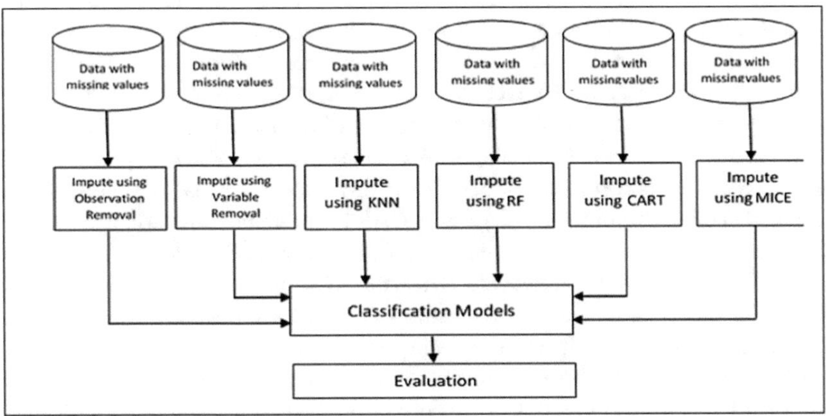

Fig. 2. The model construction

5 Results and Discussion

This study compares the performance of imputed methods on the Stroke dataset as the construction for the model is shown in Fig. 2. A comparison of several imputation methods were done and analysed their performance when applied to different classification algorithms the data set contains missing values in some columns with different ratios. The applied imputation methods replaced the missingness before building the model. The chosen algorithms have been implemented for all experiments using R programming language. The performance of chosen techniques is compared according to the accuracy level and precision.

Table 2 illustrated the accuracy level according to the replacement method and classifiers.

Table 2. The accuracy of different imputed methods with classification algorithms

No	Imputation method	J48	RF	NB	SVM	PART
1	Removing observation	84.40%	85.42%	74.62%	74.36%	83.23%
2	Removing variables	86.65%	86.65%	78.46%	77.47%	83.23%
3	Imputed data by Rf	86.01%	87.03%	77.67%	77.20%	85.47%
4	Imputed data by CART	86.36%	87.01%	77.58%	77.30%	85.02%
5	Imputed data by Mice	85.95%	86.82%	77.63%	77.29%	85.20%
6	Imputed data by KNN	86.10%	87.06%	77.53%	77.03%	84.95%

Some imputed methods (the six methods in Table 2 above) are used in preprocessing phase before building the classifiers for handling missing values namely the removing observation, removing variables, imputed by using RF, imputed by using CART, imputed by using MICE and imputed by using K-nn.

The classical methods of imputation, namely (Removing observation and Removing variables) are low in accuracy levels and merely reached 86.65% with two classifiers: J 4.8 and RF. While, the six remaining methods of imputation (as shown in Table 2) indicates that the best accuracy level was performed by 87.06% when implementing RF. Despite the fact that around 85.42% was obtained from the RF classifier with removing variables that has missing data more than the rest of attributes. This method performs the best on the Stroke data set. While, MICE imputation method shows an accuracy percentage of 85.95% with J4.8 classifier that is slightly less than RF method. J4.8 algorithm with RF, KNN, CART, and removing variables imputation method shows a very similar performance with accuracy of 86.01%, 86.10%, 86.36%, and 86.65% respectively. However, removing observation shows lower accuracy compared to other imputations. The accuracy of removing observation is 84.40%.

Selecting a method for handling missing values is not important when the rate of missing values is small. The type of randomness should be considered for selecting appropriate method for a given data. According to [39], argue that the rate of missing data less than 5%, a chosen method cannot make big different to the results.

The results obtained from the model by using only completed record (not to include observations with missing values) were with the accuracy of 74.36%, 74.62%, 83.23%, 84.40% and 85.42% for SVM, NB, PART, J4.8 and RF classifiers respectively. However, removing observation way may cause lose power of the model and present bias of the classes of the dataset.

The result of applying Decision Trees (J4.8) algorithm is shown in Tables 2 and 3. First, the accuracy level is high for the Imputed data by CART method while, the precision term showing around 0.841 for removing observation. F-measure is counted for almost 0.867 which is quite similar for imputation by all methods.

Table 3. Precision of different imputed methods with classification algorithms

No	Imputation method	J4.8	RF	NB	SVM	PART
1	Removing observation	0.841	0.855	0.746	0.747	0.832
2	Removing variables	0.867	0.867	0.785	0.777	0.865
3	Imputed data by Rf	0.86	0.871	0.777	0.776	0.855
4	Imputed data by CART	0.864	0.871	0.776	0.777	0.851
5	Imputed data by Mice	0.86	0.869	0.776	0.777	0.852
6	Imputed data by KNN	0.861	0.872	0.775	0.774	0.851

In addition, precision statistic measure is used for evaluating the prediction performance of classifiers as is shown in Fig. 2. The the highest influence of imputed data on the data was gathered from Removing variables with J4.8.

MICE use multivariate imputations to estimate the missing values for MAR values. MICE is a common package used by R users. Figure 3 shows 10 completed data sets obtained by MICE. Mice has several methods available for estimating missing values different features such as Bayesian polytomous regression (polyreg), logistic regression (logreg) or (Predictive Mean Matching) (PMM) for numeric variables. A certain variable has a specific method for imputation of missing values.

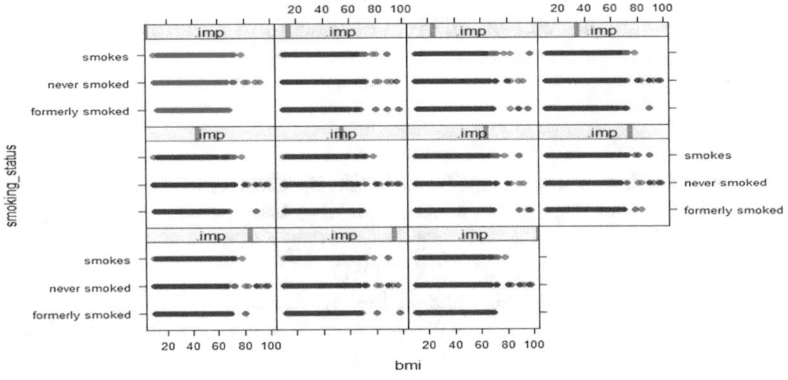

Fig. 3. 10 imputed data sets obtained by MICE

From the obtained results are presented in Table 4 and by comparing these results, the F-measure of SVM was in between 0.743 and 0.769 with all imputed methods.

Table 4. F-Measure of different imputed methods with classification algorithms

No	Imputation method	J48	RF	NB	SVM	PART
1	Removing observation	0.84	0.854	0.746	0.743	0.832
2	Removing variables	0.867	0.867	0.785	0.774	0.856
3	Imputed data by Rf	0.86	0.87	0.777	0.771	0.855
4	Imputed data by CART	0.864	0.87	0.776	0.772	0.85
5	Imputed data by Mice	0.859	0.868	0.776	0.772	0.852
6	Imputed data by KNN	0.861	0.87	0.775	0.769	0.849

Any model cannot always work better than any other model without gaining knowledge described the addressing problem [40]. Different performance metrics were used including the accuracy, precision and F-measure for evaluating yielded results.

6 Conclusion and Future Work

Missing values are common problem in real-world datasets. Different solutions for data imputation applied on stroke dataset in this study. The datasets processed by different imputation methods and popular classifiers were used to predict whether a patient can have stroke or not. ROSE method was used to balance the data. The results of this study have a few suggestions for future practice. Other method that might be possible is using MissForest [41].

Further experimental investigations are needed to use other machine learning techniques to predict missing values in stroke dataset. Examination of how using data imputation methods affect classification algorithm results with embedded imputation methods that can be replaced missing values during the classification. We are going to study the possibility of predicting and replacing missing values, despite the huge missing data in some attributes on different datasets form different domains.

References

1. Zhang, Y., Kambhampati, C., Davis, D.N., Goode, K., Cleland, J.G.F.: A comparative study of missing value imputation with multiclass classification for clinical heart failure data. In: Proceedings - 2012 9th International Conference Fuzzy Systems and Knowledge Discovery FSKD 2012, no. May, pp. 2840–2844 (2012)
2. Cheeseman, P., Kelly, J., Self, M., Stutz, J., Taylor, W., Freeman, D.: AutoClass: a Bayesian classification system. In: Machine Learning Proceedings 1988, pp. 54–64, Jan 1988
3. Schafer, J.L., Graham, J.W.: Missing data: our view of the state of the art. Psychol. Methods **7**(2), 147–177 (2002)

4. Schafer, J.: Analysis of Incomplete Multivariate Data, vol. 72. Chapman & Hall, London (1997)
5. Hron, K., Templ, M., Filzmoser, P.: Imputation of missing values for compositional data using classical and robust methods. Comput. Stat. Data Anal. **54**(12), 3095–3107 (2010)
6. Jerez, J.M., et al.: Missing data imputation using statistical and machine learning methods in a real breast cancer problem. Artif. Intell. Med. **50**(2), 105–115 (2010)
7. Faria, R., Gomes, M., Epstein, D., White, I.R.: A guide to handling missing data in cost-effectiveness analysis conducted within randomised controlled trials. Pharmacoeconomics **32** (12), 1157–1170 (2014)
8. Bevan, H., Sharma, K., Bradley, W.: Stroke in young adults 1, pp. 1–5 (2004)
9. Ellis, C.: Stroke in young adults. Disabil. Health J. **3**(3), 222–224 (2010)
10. WHO|Stroke, Cerebrovascular accident. WHO (2015)
11. Donders, A.R.T., van der Heijden, G.J.M.G., Stijnen, T., Moons, K.G.M.: Review: a gentle introduction to imputation of missing values. J. Clin. Epidemiol. **59**(10), 1087–1091 (2006)
12. van Buuren, S., Groothuis-Oudshoorn, K.: MICE: multivariate imputation by chained equations in R combustion. J. Stat. Softw. **VV**(Ii), 1–68 (2010)
13. Lakshminarayan, K., Harp, S., Goldman, R., T. S.- KDD, and U.: Imputation of missing data using machine learning techniques. aaai.org, pp. 140–143 (1996)
14. Enders, C., Mistler, S., B. K.-P. Methods, and U.: Multilevel multiple imputation: a review and evaluation of joint modeling and chained equations imputation. psycnet.apa.org, p. 222 (2016)
15. Kim, M., Baek, S., Ligaray, M., Pyo, J., Park, M., Cho, K.H.: Comparative studies of different imputation methods for recovering streamflow observation. Water (Switzerland) **7** (12), 6847–6860 (2015)
16. Williams, D.A., Nelsen, B., Berrett, C., Williams, G.P., Moon, T.K.: A comparison of data imputation methods using Bayesian compressive sensing and Empirical Mode Decomposition for environmental temperature data. Environ. Model. Softw. **102(**C), 172–184 (2018)
17. T. K., et al.: Validation of clinical classification schemes for predicting stroke: results from the National Registry of Atrial Fibrillation. J. Am. Med. Assoc. **285**(22), 2864–2870 (2001)
18. Faris, P.D., Ghali, W.A., Brant, R., Norris, C.M., Diane Galbraith, P., Knudtson, M.L.: Multiple imputation versus data enhancement for dealing with missing data in observational health care outcome analyses. J. Clin. Epidemiol. **55**(2), 184–191 (2002)
19. van Walraven, C.: Bootstrap imputation minimized misclassification bias when measuring Colles' fracture prevalence and its associations using health administrative data. J. Clin. Epidemiol. **96**, 93–100 (2018)
20. Nekouie, A., Moattar, M.H.: Missing value imputation for breast cancer diagnosis data using tensor factorization improved by enhanced reduced adaptive particle swarm optimization. J. King Saud Univ. - Comput. Inf. Sci. (2018)
21. Yu, Q., Miche, Y., Eirola, E., van Heeswijk, M., Séverin, E., Lendasse, A.: Regularized extreme learning machine for regression with missing data. Neurocomputing **102**, 45–51 (2013)
22. Scheffer, J.: Dealing with missing data (2002)
23. White, I.R., Royston, P., Wood, A.M.: Multiple imputation using chained equations: issues and guidance for practice. Stat. Med. **30**(4), 377–399 (2011)
24. Pigott, T.D.: A review of methods for missing data. Educ. Res. Eval. **7**(4), 353–383 (2001)
25. Walczak, B., Massart, D.L.: Dealing with missing data: part I. Chemom. Intell. Lab. Syst. **58**, 15–27 (2001)
26. Walczak, B., Massart, D.L.: Dealing with missing data: part II. Chemom. Intell. Lab. Syst. **58**(1), 29–42 (2001)

27. Ghahramani, Z., Jordan, M.I.: Supervised learning from incomplete data via an EM approach. Adv. Neural Inf. Process. Syst. **VI**, 81–87 (1994)
28. Beunckensc, C., Molenberghs, G., Kenward, M.G.: Direct likelihood analysis versus simple forms of imputation for missing data in randomized clinical trials (2005)
29. Rubin, D.B.: Multiple imputation after 18+ years (with discussion). J. Am. Stat. Assoc. **91** (434), 473–489 (1996)
30. Healthcare Dataset Stroke Data|Kaggle. https://www.kaggle.com/asaumya/healthcare-dataset-stroke-data. Accessed 17 Dec 2018
31. Alpaydın, E.: Introduction to machine learning. Methods Mol. Biol. **1107**, 105–128 (2014)
32. Wu, X., et al.: Top 10 algorithms in data mining. Knowl. Inf. Syst. **14**(1), 1–37 (2008)
33. Bramer, M.: Principles of Data Mining. Springer, London (2016)
34. Ramoni, M., Sebastiani, P.: Robust learning with missing data. Mach. Learn. **45**(2), 147–170 (2001)
35. Haibo He, H., Garcia, E.A.: Learning from imbalanced data. IEEE Trans. Knowl. Data Eng. **21**(9), 1263–1284 (2009)
36. He, Y.: Missing data analysis using multiple imputation: getting to the heart of the matter. Circ. Cardiovasc. Qual. Outcomes **3**(1), 98–105 (2010)
37. Garciarena, U., Santana, R.: An extensive analysis of the interaction between missing data types, imputation methods, and supervised classifiers. Expert Syst. Appl. **89**, 52–65 (2017)
38. Aljuaid, T., Sasi, S., Sasi, S.: Proper imputation techniques for missing values in data sets Predictive Modeling View project FIR Digital Filter View project Proper Imputation Techniques for Missing Values in Data sets (2016)
39. Saunders, J.A., et al.: Imputing missing data : a comparison of methods for social work researchers. Soc. Work. Res. **30**(1), 19–31 (2016)
40. Wolpert, D.H., Macready, W.G.: No free lunch theorems for optimization. IEEE Trans. Evol. Comput. **1**(1), 67–82 (1997)
41. Stekhoven, D.J., Bühlmann, P.: Missforest-non-parametric missing value imputation for mixed-type data. Bioinformatics **28**(1), 112–118 (2012)

Evolutionary Optimisation of Fully Connected Artificial Neural Network Topology

Jordan J. Bird(✉), Anikó Ekárt, Christopher D. Buckingham, and Diego R. Faria

School of Engineering and Applied Science, Aston University, Birmingham B4 7ET, UK
{birdj1,a.ekart,c.d.buckingham,d.faria}@aston.ac.uk

Abstract. This paper proposes an approach to selecting the amount of layers and neurons contained within Multilayer Perceptron hidden layers through a single-objective evolutionary approach with the goal of model accuracy. At each generation, a population of Neural Network architectures are created and ranked by their accuracy. The generated solutions are combined in a breeding process to create a larger population, and at each generation the weakest solutions are removed to retain the population size inspired by a Darwinian 'survival of the fittest'. Multiple datasets are tested, and results show that architectures can be successfully improved and derived through a hyper-heuristic evolutionary approach, in less than 10% of the exhaustive search time. The evolutionary approach was further optimised through population density increase as well as gradual solution max complexity increase throughout the simulation.

Keywords: Neural networks · Evolutionary computation · Neuroevolution · Hyperheuristics · Computational intelligence

1 Introduction

With the increasing growth of computational resources available to both people, private companies and governments, complex deep neural networks are growing in prominence to solve problems and provide predictions through the employment of Artificial Intelligence. Deep Neural Network Topology is a heuristic problem and therefore an optimal (best) solution is unique for individual problems, and thus tuning of topology is required. This paper provides a method to optimise deep neural networks for classification via evolutionary computation, saving a considerable amount of time and resources when compared to exhaustive search processes, autonomously with little to no human input.

The main contributions of this work are twofold:

- A hyper-heuristic evolutionary algorithm to derive an ANN solution through visualisation and analysis of the full problem space for comparison

K. Arai et al. (Eds.): CompCom 2019, AISC 997, pp. 751–762, 2019.
https://doi.org/10.1007/978-3-030-22871-2_52

- Algorithmic dynamism inspired by nature, through growing population density, gradual increase of possible complexities, and random chance solution generation both globally and for individual neuron layers.

2 Related Work

Famous research resulted in the *'No Free Lunch'* [16] theorem for optimisation problems. That is, for those problems that can only be completely solved in polynomial time, there is no single best algorithm to find the global best solution in non-polynomial time [7]. Optimisation algorithms must be employed that are more efficient than a random search within the problem space, and of course a complete exhaustive search of the entire problem space.

Evolution of Neural Networks through Augmenting Topologies (NEAT) is an algorithm for the genetic improvement of neural networks [13]. The algorithm proposes the evolution of network layers in a non-fully connected neural network topology, that is, neurons in layer n are not all necessarily connected to layer $n+1$. The algorithm has been particularly effective in the domain of real time problems of user input to a desired action, most notably for an evolving ANN that learns to play Super Mario in real time [14] (though became overfitted after completing the first level) and a similar study involving the evolution of an ANN that autonomously plays NERO [12].

Evolution of Recurrent Neural Networks (RNN) showed very promising results, arguing that an evolutionary algorithm is far superior to the traditional method of training within the constraints of a defined topology [1]. Improvement of Neural Networks has also been successfully experimented on with Particle Swarm Optimisation (PSO) [10] (velocity based agent swarm search of the problem space).

3 Background

3.1 Evolutionary Algorithms and Neuroevolution

An evolutionary algorithm is a population-based meta-heuristic optimisation method inspired by biological evolution [15]. That is, the optimisation of an organism due to levels of reproduction beyond that which can be supported by an environment, i.e. 'survival of the fittest' [4]. An organism, in this case a solution, forms a member of a generation, and will breed with other solutions to create offspring inspired by both parents. At each generation, strengths and weaknesses of population members are considered, and the environment will cause the death of the weakest (varying on implementation).

The general outline of an evolutionary optimisation algorithm is given as the following:

1. Create a 0th (initial) population by generating a set of random solutions.
2. Begin biological simulation until a specified termination requirement is met:

(a) Select parents and breed, creating an offspring solution (a chance of random mutation may be present).
(b) Evaluate the fitness of each individual that has not yet been evaluated
(c) Kill off solutions based on a chosen selection method (e.g. weakness), beyond the maximum size of the population.

Neuroevolution is the application of evolutionary algorithms to Artificial Neural Networks (ANN), in which parameters and/or topology are treated as inheritable traits [5]. The implementation of an Evolutionary Algorithm treats a neural network as a member of the solution population, in which classification accuracy, or low cost (cost-based error matrices and/or training time) are treated as the fitness metric.

3.2 Fully Connected Multilayer Perceptron Topology

An Artificial Neural Network (ANN) is a system of computing inspired by the biological brain [9], in that a brain's neuron (nerve cell) takes data input to the dendrites and produces an output at nerve endings [6].

Task specificity is not programmed, and thus learning must be performed. For example, in a dataset of images of 'cat' and 'dog', a neural network will autonomously learn minute differences between the two different sets of data and consider these rules when attempting to classify an image it has not seen before [11]. Learning to process data to output(s) is derived through considering examples and finding the best fit to correctly classify those examples, or produce a best distance-based single output in terms of regression problems (i.e. prediction of stock market prices).

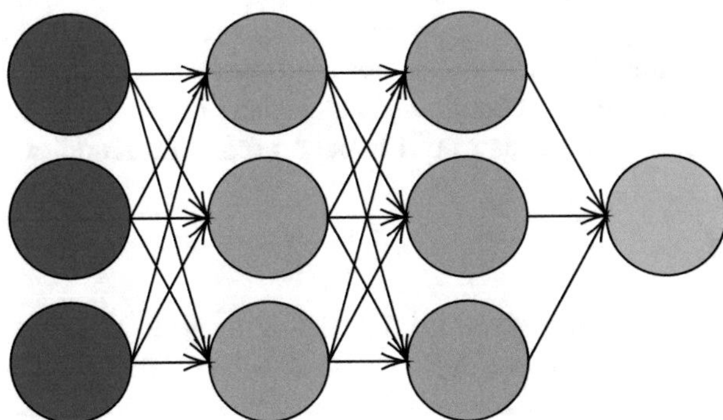

Fig. 1. A simplified diagram of a fully connected neural network. Three blue input nodes form the input layer, six grey hidden nodes form two hidden layers of three neurons, and one green node forms the regression output layer.

A fully connected topology can be seen in Fig. 1, in which all nodes of layer n are connected to all nodes of layer $n+1$. With fully connected topology assumed, this experiment is to optimise the search process of hidden layer topology, i.e. the number of neurons contained within the 'grey' layers. Secondly, the experiment is also to optimise the search process of the number of hidden layers themselves (the amount of 'grey' layers in Fig. 1). Note that this is also a feedforward neural network, since all connections pass to following layers and do not form any cycles.

Backpropagation. In addition to topology, the weights within an ANN are also a heuristic problem. That is, there is no 'free lunch' [16] and an optimisation algorithm must be used to fine tune weights. The training process for an Artificial Neural Network is backpropagation, a type of automatic differentiation which is a process of gradient calculation for the further derivation of weights to be used within the network [2]. Calculation of the gradient of the loss function (cost) is performed at the output compared to the real value of the data, and is passed backwards through the network starting from said output, i.e. 'backpropagation'.

Loss is calculated dependent on the class of problem that the network is designed to solve. Absolute euclidean distance is used for a single real number output for a regression problem e.g. if a house price is predicted to be £200,000 and the real value is £250,000 then the error of the output was 50,000 i.e. the distance the prediction was away from the real value.

For classification, entropy is considered. Entropy is the calculation of disorder, or randomness and is given as follows:

$$E(S) = -\sum_{i=1}^{c} P_i \times log_2 P_i \tag{1}$$

For example, if a ruleset were to classify a set of 10 instances in a binary problem with an equal distribution of 5 class A and 5 class B, entropy would be 1 as the results are completely random. Lower randomness is an indication of more order within a set of rules, but without cross-validation or a separate training set, can also indicate a dataset completely memorised by an ANN that cannot be transferred to other data. This would make a prediction model worthless.

4 Method

The evolutionary process is given by Algorithms 1 and 2. Random solutions are initialised and for a set number of generations, each solution will be bred with a randomly chosen member of the population resulting in a child solution (Algorithm 2). Random chance for complete mutation of child (i.e. return a random solution) and also for individual neuron layers (i.e. a random number between 1 and maximum neurons) takes place, and a child is created from the two parents, which is added to the population. After the breeding process, all untested solutions are tested, and the weakest below the maximum population are deleted. Maximum population and Maximum neuron counts are increased by the defined step, until the limit is reached.

```
Result: Array of best solutions at final generation
initialise Random solutions;
for Random solutions : rs do
    test accuracy of rs;
    set accuracy of rs;
end
while Simulating do
    for Solutions : s do
        parent2 = random Solution;
        child = breed(s, parent2); See Algorithm 2
        test accuracy of child;
        set accuracy of child;
    end
    Sort Solutions best to worst;
    for Solutions : s do
        if s index > population size then
            delete s;
        end
        increase maxPopulation by growth factor;
        increase maxNeurons by growth factor;
    end
end
Return Solutions;
```

Algorithm 1. Evolutionary Algorithm for ANN optimisation

The breeding process given in Algorithm 2 will take the length (number of layers) of one of the two parents at a 50/50 chance, and then the layers are filled in at an equal chance between the two parents, unless a parent does not have a layer at the index (i.e. it is too short) in which it is taken from the parent that does. Finally a random chance occurs in which, instead of a neuron count at a layer from either parent 1 or parent 2 is returned, a random number between 1 and the maximum neuron count is returned. This process increases variability in the population by both providing a complete random solution, or a slight random solution still inspired by both parents. A final logic check is performed on whether the child happened to be identical to parent 1 or parent 2, in which case a random solution is generated.

Complexity is scaled throughout since maximum neuron and population counts increase between each generation, and thus models with very high complexity and introduced later into the algorithm. This was implemented due to the computational resources required for very large networks that had low classification accuracy. The scale and maximum are manually tuned for each individual problem.

For fully-connected neural networks (solutions), 5 fold cross validation is used to create an averaged solution and thus prevent memorisation of the dataset, each ANN is allowed 500 epochs to train. The algorithm and all training/classification was performed on an AMD FX-8320 3.5 GHz 8-Core Processor with only core

```
Result: Child solution child of parents s and parent2
random number r [1-100];
random mutation chance c;
if r ≤ c then
  | child = random solution;
else
  | if r ≤ 50 then
  |   | child layerCount = s layerCount;
  | else
  |   | child layerCount = parent2 layerCount;
  | end
  | for Layers : l do
  |   | random number r2 [1-100];
  |   | if r2 ≤ 50 then
  |   |   | if s layerCount ≥ l then
  |   |   |   | child neuron at l = s neuron at l;
  |   |   | else
  |   |   |   | child neuron at l = parent2 neuron at l;
  |   |   | end
  |   | else
  |   |   | if parent2 layerCount ≥ l then
  |   |   |   | child neuron at l = parent2 neuron at l;
  |   |   | else
  |   |   |   | child neuron at l = s neuron at l;
  |   |   | end
  |   | end
  |   | if r2 ≤ c then
  |   |   | random number randomNeuron [1-maxNeurons];
  |   |   | child neuron at l = randomNeuron;
  |   | end
  | end
end
if child = s OR child = parent2 then
  | return random solution;
end
return child;
```

Algorithm 2. Breeding process for solution *s* and *parent2* derived from *Algorithm 1*

applications allowed to run, the Operating System used was Windows 10. The algorithm was implemented with Java and all random numbers were generated by the Java Virtual Machine with a seed of 0. For these experiments, hyper parameters were manually tuned to:

- A starting max population size of 4
- Population growth of 2 per generation
- Population capped at 30
- A starting max neuron count of 5 per layer

- Neuron grown of 5 per generation
- Neurons capped at 50

Two tests were performed on two different datasets. Glass Identification was a relatively simple dataset, whereas the Wine quality dataset was very complex. Due to the focus being on the search for better results, rather than the specific value of the result (high accuracy), a percentile is measured ie. where the solution exists in order of best to worst solutions. For example, if the best results were to be 2 of 100 results, they would exist within the 98th percentile. This is due to the low training times of the networks and resources available, and a benchmark of optimised search rather than the results themselves. The algorithm could then be applied to more complex neural networks (e.g. 2000 epoch training time) that would be impossible to exhaustively search.

It is important to note that many problems would require far too many computational resources to viably exhaustively solve. It is for this reason that testing is performed to prove the application of the algorithm alongside a comparison of exhaustive search, with the goal that the algorithm could be further applied to problems that cannot be solved in a brute-force manner, and thus, could not be compared to a global best solution.

5 Preliminary Results

5.1 Glass Identification Dataset

A dataset of Glass Identification was acquired from the UCI Repository of Machine Learning Databases [3]. The dataset contains nine numerical attributes of the corresponding chemical compounds of the glass, along with seven classes of the glass usage (e.g. 'headlamps').

The size of the dataset along with the few attributes allowed for exhaustive search to benchmark the evolutionary algorithm. Figure 2 shows the 3D problem space for two layer neural network architectures, observation shows many local maxima for solutions. Of all solutions, the three best had identical fitness measurements and thus contributed to the top 0.12% of all solutions.

Figure 3 and Table 1 show the results of the evolutionary algorithm for the Glass Identification Dataset. Within 6.69% of the time to exhaustively search all combinations, the algorithm found the second best solution within the top 99.57 percentile of all possible results. None of the three best results were found during simulation, of which contribute 0.12% of the total number of results. Interestingly, evolutionary improvement seemed to occur at similar times during the simulations, even though the starting populi were vastly different due to random initialisation. Further work is required to study this pattern, or coincidence, of improvement.

5.2 Wine Quality Dataset

A dataset of Wine Quality was acquired from the UCI Repository of Machine Learning Databases [3]. The dataset is comprised of eleven numerical attributes

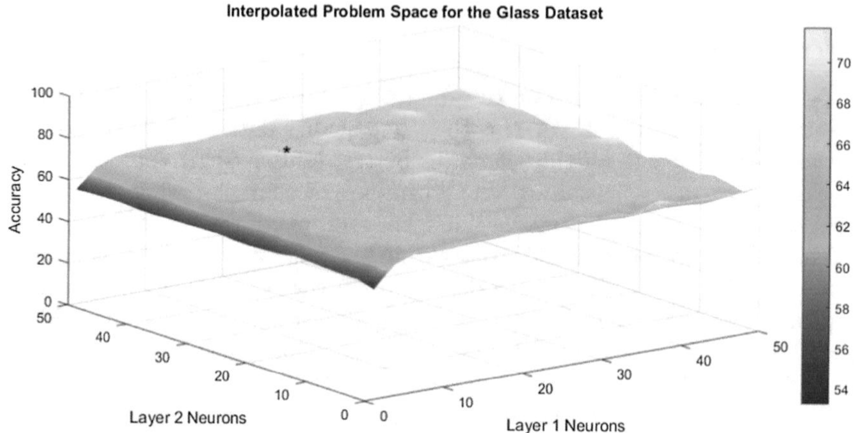

Fig. 2. 3D interpolated problem space for the glass dataset. X and Y data are layer 1 and layer 2 neuron counts respectively. Z height shows the model accuracy, and an asterisk shows the global best solution.

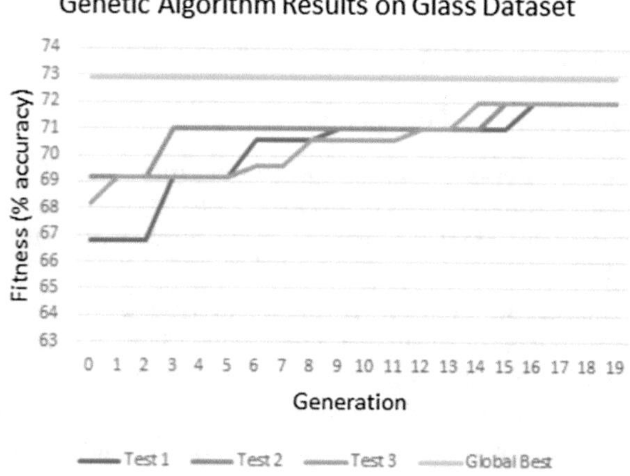

Fig. 3. Graph to show the strongest solution during three evolutionary simulations on the glass dataset.

which are measurements of wine contents, such as acidity and alcohol content. The data are linked to one of eleven classes, a score rating of 0–10 awarded to the wine. Figure 4 shows the 3D problem space for two layer neural network architectures derived from an exhaustive search taking over 5 days to complete, whereas Fig. 5 shows the genetic simulations on the wine dataset per generation.

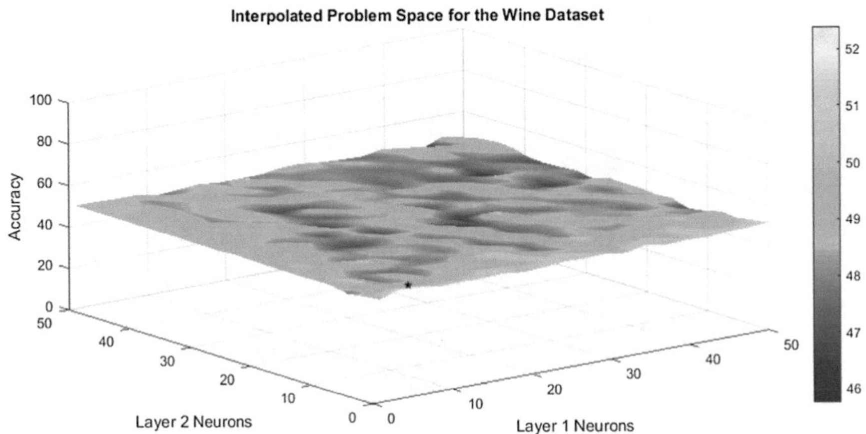

Fig. 4. 3D interpolated problem space for the wine dataset. X and Y data are layer 1 and layer 2 neuron counts respectively. Z height shows the model accuracy, and an asterisk shows the global best solution.

5.3 Discussion

The wine dataset was far more complex than the Glass dataset, as observed by the exhaustive search time having an increase of 417,547 s (4 days, 19:59:07), to a total of 482681 s (5 days, 14:04:41). Although this was true, a very slight percentage time increase for the wine dataset still produced very good results. This shows that the algorithm could be used for a more complex dataset, i.e. a problem that cannot be realistically exhaustively searched and thus could not be verified in the same manner of the two problems in this paper.

Results of the two experiments are presented in Table I. For the glass dataset, a second best (99.57 percentile) result was found, whereas for the wine dataset, the best (99.96 percentile) was found; within 6.69% and 7.59% of the exhaustive search time respectively. Further exploration should be performed on the glass dataset due to the late improvements at generations 13, 14, and 15, as this may possibly be due to the simulation not running for enough time.

The algorithm produced impressive results in both cases, in the case of the wine dataset saving over five days worth of computing time. The promising results presented point towards employing a genetic approach in future optimisation of artificial neural networks, due to both the time and resources saved from not performing an exhaustive search (which may be impossible). Also, the algorithm follows a logical process, presenting an autonomous system that requires little to no human intervention.

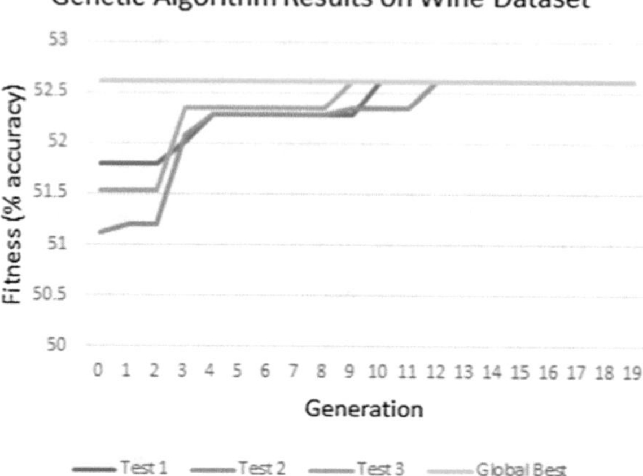

Fig. 5. Graph to show the strongest solution during three evolutionary simulations on the wine dataset.

Table 1. Results of three genetic simulations and their averages compared to exhaustive search on two separate datasets

Experiment	Dataset	
	Glass	*Wine*
Genetic time 1 (S)	4352	35997
Genetic percentile 1	99.57	99.96
Genetic time 2 (S)	4897	36933
Genetic percentile 2	99.57	99.96
Genetic time 3 (S)	4675	37024
Genetic percentile 3	99.57	99.96
Genetic average time (S)	4641	36651
Genetic average percentile	99.57	99.96
Exhaustive time (S)	65134	482681
Exhaustive best percentile	99.88	99.96
Genetic vs exhaustive time (%)	6.69%	7.59%
Genetic vs exhaustive percentile	−0.31	0

6 Next Steps

This section defines the limitations of the studies carried out and suggestions for contribution in further experiments.

This study focused on the optimisation of hidden layer neuron counts for Multilayer Perceptron Neural Networks, future work should concern the performance of optimisation in consideration of other types of Neural Networks, including the viability of this algorithm with said other Neural Network types. Additionally, only neuron and layer counts were optimised at each generation whereas parameters such as the amount of training time, momentum, and learning rate etc. should also be heuristically refined for each problem to possibly create a better solution.

The single-objective approach of this study was only concerned with the classification accuracy of a model (though more simple topology were enticed rather than their more complicated yet identical counterparts through complexity scaling), factors such as training time should be taken into account in a multi-objective optimiser to rate identical accuracy models by the computing resources they require to train, where the consumption of fewer resources would be better. This would result in a further optimised model, and would require either second comparison or a fitness score.

The evolutionary algorithm measured solution fitness by the general accuracy of the model, ie. comparing the number of correctly classified instances to the total instances in the testing dataset. For networks with an additional goal-based approach (such as medical prediction models with a focus on preventing misdiagnoses of healthy patients), a cost-based learning approach must be optimised to further achieve the model's goal. This would be implemented by the fitness value corresponding to a classification cost, and the compare operation sorting solutions lowest first rather than highest.

Due to the computational resources required for training a large volume of Multilayer Perceptrons, 5-fold cross validation was chosen for model averaging. Related work shows that the more complex leave-one-out cross validation techniques tend to be superior in attaining higher accuracy scores [8], and therefore it would be logical to perform this experiment with those techniques. This technique was not possible with the resources available for this experiment.

Finally, the breeding algorithm is unsuitable for single hidden layer neural networks since a child produced would always be random, since it would produce a child identical to either one of its parents, therefore other breeding processes should be explored such as mid-point (average) values due to the peaks and troughs observed in Figs. 2 and 4 having relatively large areas and therefore smooth gradient descent.

7 Conclusion

To conclude, this study presented a high-performing genetic algorithm for the optimisation of Artificial Neural Networks which, through preliminary results, had time savings of over 90% for both a simple and a complex dataset. The algorithm has original features such as the increasing growth of the environment and thus population, as well as a cap on solution complexity that increases at each solution towards a maximum value. These were introduced to closer simulate Darwinian evolution in nature.

Future work is suggested through the further optimisation of the algorithm's hyper-parameters, as well as more complex data comparisons with their exhaustive search statistics - these experiments would be enabled with access to more computational resources.

Acknowledgments. This work was supported by the European Commission through the H2020 project EXCELL (https://www.excell-project.eu/), grant No. 691829.

This work was also partially supported by the EIT Health GRaCEAGE grant number 18429 awarded to C.D. Buckingham.

References

1. Angeline, P.J., Saunders, G.M., Pollack, J.B.: An evolutionary algorithm that constructs recurrent neural networks. IEEE Trans. Neural Netw. **5**(1), 54–65 (1994)
2. Bengio, Y., Goodfellow, I.J., Courville, A.: Deep learning. Nature **521**(7553), 436–444 (2015)
3. Blake, C., Merz, C.J.: UCI repository of machine learning databases (1998). http://www.ics.uci.edu/mlearn/mlrepository.html, department of information and computer science. University of California, Irvine, CA
4. Darwin, C.: On the Origin of Species, 1859. Routledge, London (2004)
5. Floreano, D., Dürr, P., Mattiussi, C.: Neuroevolution: from architectures to learning. Evol. Intell. **1**(1), 47–62 (2008)
6. Hopfield, J.J.: Neurons with graded response have collective computational properties like those of two-state neurons. Proc. Nat. Acad. Sci. **81**(10), 3088–3092 (1984)
7. Donald Ervin Knuth: Postscript about NP-hard problems. ACM SIGACT News **6**(2), 15–16 (1974)
8. Kohavi, R., et al.: A study of cross-validation and bootstrap for accuracy estimation and model selection. In: Ijcai, Montreal, Canada, vol. 14, pp. 1137–1145 (1995)
9. Rosenblatt, F.: Principles of neurodynamics. Perceptrons and the theory of brain mechanisms. Technical report, Cornell Aeronautical Lab Inc., Buffalo, NY (1961)
10. Shi, Y., et al.: Particle swarm optimization: developments, applications and resources. In: Proceedings of the 2001 Congress on Evolutionary Computation, vol. 1, pp. 81–86. IEEE (2001)
11. Simonyan, K., Zisserman, A.: Very deep convolutional networks for large-scale image recognition. arXiv preprint arXiv:1409.1556 (2014)
12. Stanley, K.O., Bryant, B.D., Miikkulainen, R.: Real-time neuroevolution in the NERO video game. IEEE Trans. Evol. Comput. **9**(6), 653–668 (2005)
13. Stanley, K.O., Miikkulainen, R.: Evolving neural networks through augmenting topologies. Evol. Comput. **10**(2), 99–127 (2002)
14. Togelius, J., Karakovskiy, S., Koutník, J., Schmidhuber, J.: Super mario evolution. In: 2009 IEEE Symposium on Computational Intelligence and Games, CIG 2009, pp. 156–161. IEEE (2009)
15. Vikhar, P.A.: Evolutionary algorithms: a critical review and its future prospects. In: 2016 International Conference on Global Trends in Signal Processing, Information Computing and Communication (ICGTSPICC), pp. 261–265. IEEE (2016)
16. Wolpert, D.H., Macready, W.G.: No free lunch theorems for optimization. IEEE Trans. Evol. Comput. **1**(1), 67–82 (1997)

State-of-the-Art Convolutional Neural Networks for Smart Farms: A Review

Patrick Kinyua Gikunda[1,2(✉)] and Nicolas Jouandeau[2]

[1] Department of Computer Science, Dedan Kimathi University of Technology, Nyeri, Kenya
patrick.gikunda@dkut.ac.ke

[2] Le Laboratoire d'Informatique Avancée de Saint-Denis (LIASD), University of Paris 8, Saint-Denis, France
n@ai.univ-paris8.fr

Abstract. Farming has seen a number of technological transformations in the last decade, becoming more industrialized and technology-driven. This means use of Internet of Things (IoT), Cloud Computing (CC), Big Data (BD) and automation to gain better control over the process of farming. As the use of these technologies in farms has grown exponentially with massive data production, there is need to develop and use state-of-the-art tools in order to gain more insight from the data within reasonable time. In this paper, we present an initial understanding of Convolutional Neural Network (CNN), the recent architectures of state-of-the-art CNN and their underlying complexities. Then we propose a classification taxonomy tailored for agricultural application of CNN. Finally, we present a comprehensive review of research dedicated to applications of state-of-the-art CNNs in agricultural production systems. Our contribution is in two-fold. First, for end users of agricultural deep learning tools, our benchmarking finding can serve as a guide to selecting appropriate architecture to use. Second, for agricultural software developers of deep learning tools, our in-depth analysis explains the state-of-the-art CNN complexities and points out possible future directions to further optimize the running performance.

Keywords: Convolutional Neural Network · SmartFarm · State-of-the-art

1 Introduction

The global population is set to touch 9.6 billion mark by year 2050 [4]. The continous population growth means increase in demand for food to feed the population [5]. Agriculture is the practice of cultivation of land and breeding of animals & plants to provide food and other products in order to sustain and enhance life [6]. Due to the extreme weather conditions, rising climate change and environmental impact resulting from intensive farming practices [8], farmers are now forced to change their farming practices. To cope with the new farming

K. Arai et al. (Eds.): CompCom 2019, AISC 997, pp. 763–775, 2019.
https://doi.org/10.1007/978-3-030-22871-2_53

challenges, farmers are forced to practice smart farming [9], which offers solutions of farming management and environment management for better production. Smart farming focuses on the use of information and communication technology (ICT) in the cyber-physical farm management cycle for efficient farming [10].

Curent ICT technologies relevant for use in smart farming include IoT [11], remote sensing [12], CC [13] and BD [14]. Remote sensing is the science of gathering information about objects or areas from a distance without having physical contact with objects or areas being investigated. Data collected through remote sensing and distributed devices is managed by cloud computing technology, which offers the tools for pre-processing and modelling of huge amounts of data coming from various heterogeneous sources [15]. These four technologies could create applications to provide solutions to todays's agricultural challenges. The solutions include real time analytics required to carry out agile actions especially in case of suddenly changed operational or environmental condition (e.g. weather or disease alert). Figure 1 summarises the concept of smart farming with various distributed sensor nodes connected through a network. The continous monitoring, measuring, storing and analysing of various physical aspects has led to a phenomena of big data [16]. To get insight for practical action from this large type of data requires tools and methods that can process multidimensional data from different sources while leveraging on the processing time.

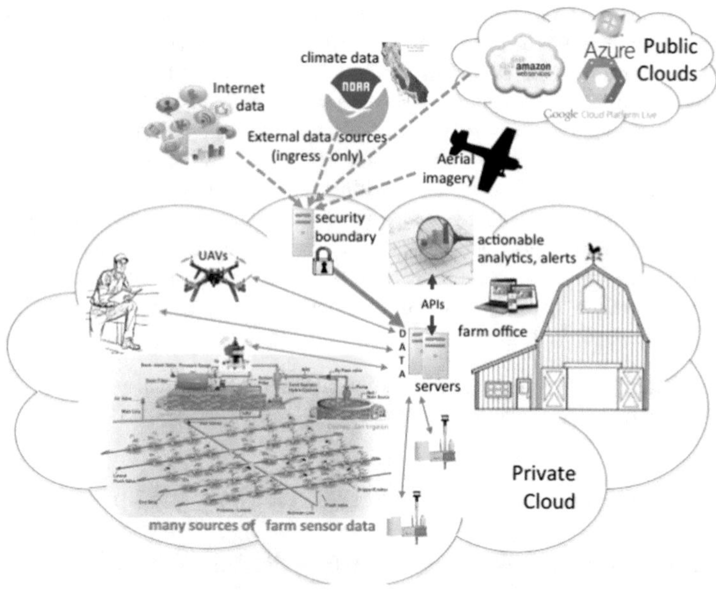

Fig. 1. SmartFarm made up of IoT nodes connected through a network [9]

One of the successful data processing tool applied in this kind of large dataset is the biologically inspired Convolutional Neural Networks (CNNs), which have achieved state-of-the-art results [17] in computer vision [18] and data mining [19]. As deep learning has been successfully applied in various domains, it has recently entered in the domain of agriculture [10]. CNN is a subset method of Deep Learning (DL) [20], defined as deep, feed-forward Artificial Neural Network (ANN) [21]. The CNN covolutions allow data representations in a hierarchical way [22]. The common characteristics of CNN models is that they follow the same general design principles of successive applying convolutional layers to the input, periodically downsampling the spatial dimensions while increasing the number of feature maps. These architectures serve as rich feature extractors which can be used for image classification, object detection, image segmentation and many more other advanced tasks. This study investigates the agricultural problems that employ the major state-of-the-art CNN architectures that have participated in the ImageNet Large Scale Visual Recognition Challenge (ILSVRC) [23] with highest accuracy in a multi-class classification problem. ImageNet [26] classification challange has played a critical role in advancing the CNN state-of-the-art [17]. The motivation for carrying out the study include: (a) CNNs has better precision compared to other popular image-processing techniques in the large majority of problems [27]. (b) CNN has entered in the agricultural domain with promising potential [28]. (c) All the CNN models that have achieved the top-5 error are successful when applied in other computer vision domain with remarkable results [27]. This review aims to provide insight on use of state-of-the-art CNN models in relation to smart farming and to identify smart farming research and development challenges related to computer vision. Therefore the analysis will primarily focus on the success on use of state-of-the-art CNN models in smart farms with a intention to provide this relevant information to the future researchers. From that perspective the research questions to be addressed in the study are:

(a) What is the role of CNN in smart farming?
(b) What type of the state-of-the-art CNN architecture should be used?
(c) What are the benefits of using state-of-the-art CNN in IoT based agricultural systems?

The rest of the paper is organized as follows: in Sect. 2, we present an overview of existing state-of-the-art CNNs architectures including their recent updates. Then we propose a taxonomy to provide a systematic classification of agricultural issues for CNN application. In Sect. 3, we present the existing state-of-the-art CNNs and their scope of application in agri-systems. We conclude the paper in Sect. 4.

2 Methodology

In order to address the research questions a bibliographic analysis in the domain under study was done between 2012 and 2018, it involved two steps: (a) collection of related works and, (b) detailed review and analysis of the works. The choice of the period is from the fact that CNN is rather a recent phenomenon.

In the first step, a keyword-based search using all combinations of two groups of keywords of which the first group addresses CNN models (LeNet, AlexNet, NIN, ENet, GoogLeNet, ResNet, DenseNet, VGG, Inception) and the second group refers to farming (i.e. agriculture, farming, smart farming). The analysis was done while considering the following research questions: (a) smart farm problem they addressed, (b) dataset used, (c) accuracy based on author's performance metric, (d) state-of-the-art CNN model used. Its important to note that use of state-of-the-art deep learning has great potential, and there have been recent small comparative studies to analyse and compare the most efficient architecture to use in agricultural systems. They include: Comparison between LeNet, AlexNet, and VGGNet on automatic identification of center pivot irrigation [29] and comparison between VGG-16, Inception-V4, Resnet and DenseNet for plant disease identification [30].

3 State-of-the-Art CNN: An Overview

CNNs typically perform best when they are large, meaning when they have more deeper and highly interconnected layers [31]. The primary drawback of these architectures is the computational cost, thus large CNNs are typically impractically slow especially for embedded IoT devices [32]. There are recent research efforts on how to reduce the computation cost of deep learning networks for everyday application while maintaining the prediction accuracy [33]. In order to understand the application of the state-of-the-art CNN architectures in agricultural systems, we reviewed the accuracy and computational requirements from relevant literature including recent updates of networks as shown in Fig. 2. The classical state-of-the-art deep network architectures include; LeNet [34], AlexNet [35], NIN [36], ENet [37], ZFNet [38], GoogleLeNet [39] and VGG 16 [40]. Modern architectures include; Inception [41], ResNet [42], and DenseNet [43].

LeNet-5 is a 7-layer pioneer convolutional network by LeCun et al. [34] to classify digits, used to recognise hand-written numbers digitized in 32×32 pixel greyscale input images. High resolution images require more convolutional layers, so the model is constrained by the availability of the computing resources.

AlexNet is a 5-layer network similar to LeNet-5 but with more filters [35]. It outperformed Lenet-5 and won the LSVRC challenge by reducing the top-5 error from 26.2% to 15.3%. Use Rectified Linear Unit (Relu) [1] instead of Hyperbolic Tangent (Tanh) [2] to add non-linearity and accelerates the speed by 6 times. Dropout was employed to reduce over-fitting in the fully-connected layers. Overlap pooling was used to reduce the size of the network while reducing top-1 error by 0.4% and top-5 error by 0.3%.

Lin et al. [36] created a Network in Network (NIN) which inspired the inception architecture of googlenet. In their paper, they replaced the linear filters with nonlinear multi linear perceptrons that had better feature extraction and accuracy. They also replaced the fully connected layers with activation maps and global average pooling. This move helped reduce the parameters and network complexity.

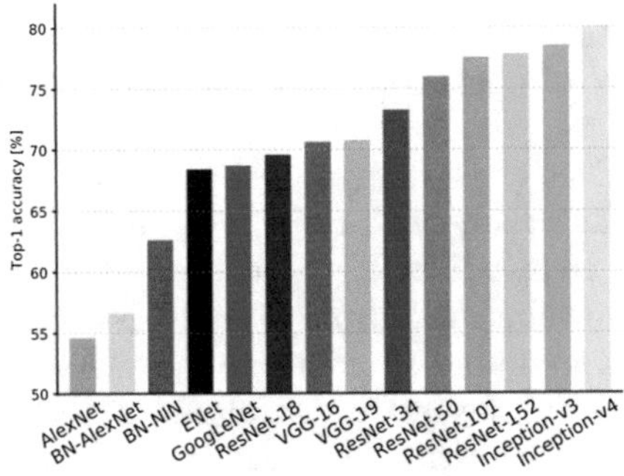

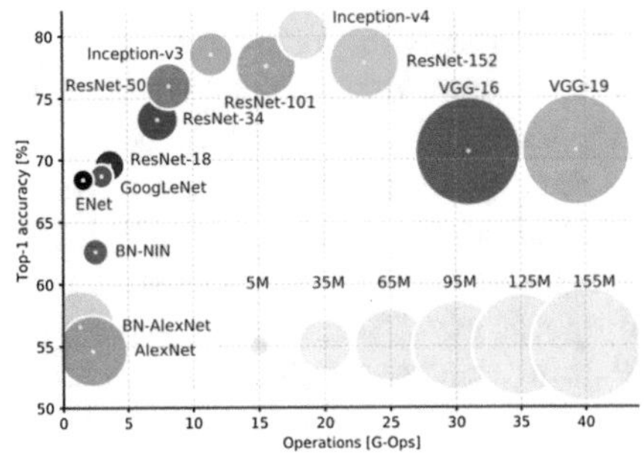

Fig. 2. Top-1 accuracy vs the computational cost. The size of the circles is proportional to number of parameters. Legend; the grey cirles at the bottom right represents number of parameters in millions. [32]

In their article Paszke et al. [37] introduced an Efficient Neural Network (ENet) for running on low-power mobile devices while achieving state-of-the-art results. ENet architecture is largely based on ResNets. The structure has one master and several branches that separate from the master but concatenate back.

In their work Zeiler and Fergus [38] created ZFNet which won a ILSVRC 2013 [25] image classification. It was able to achieve a top-5 rate of 14.8% an improvement of the AlexNet. They were able to do this by tweaking the hyper-parameters of AlexNet while maintaining the same structure with additional deep learning elements. There is no record observed of use of ZFNet in agricultural systems despite the accuracy improvement. Each branch consists of three convolutional layers. The 'first' 1×1 projection reduces the dimensionality while the latter 1×1 projection expands the dimensionality. In between these convolutions, a regular (no annotation/asymmetric X), dilated (dilated X) or full convolution (no annotation) takes place. Batch normalization [60] and PReLU [61] are placed between all convolutions. As regularizer in the bottleneck, Spatial Dropout is used. MaxPooling on the master is added only when the bottleneck is downsampling which is true.

GoogleNet, a 2014 ILSVRC image classification winner, was inspired by LeNet but implemented a novel inception module. The Inception cell performs series of convolutions at different scales and subsequently aggregate the results. This module is based on several very small convolutions in order to drastically reduce the number of parameters. There has been tremedious efforts done to improve the performance of the architecture: (a) Inception v1 [39] which performs convolution on an input, with 3 different sizes of filters (1×1, 3×3, 5×5). Additionally, max pooling is also performed. The outputs are concatenated and sent to the next inception module. (b) Inception v2 and Inception v3 [41] factorize 5×5 convolution to two 3×3 convolution operations to improve computational speed. Although this may seem counterintuitive, a 5×5 convolution is 2.78 times more expensive than a 3×3 convolution. So stacking two 3×3 convolutions infact leads to a boost in performance. (c) In Inception v4 and Inception-ResNet [44] the initial set of operations were modified before introducing the Inception blocks.

Simonyan and Zisserman created VGGNet while doing investigation on the effect of convolutional network depth on its accuracy in the large-scale image recognition setting. The VGGNet took the second place after GoogLeNet in the competition. The model is made up of 16 convolutional layers which is similar to [35] but with many filters. There have been a number of update to the VGGNet architecture starting with pioneer VGG-11 (11 layers) which obtained 10.4% error rate [40]. VGG-13 (13 layers) obtains 9.9% error rate, which means the additional convolutional layers helps the classification accuracy. VGG-16 (16 layers) obtained a 9.4% error rate, which means the additional $3 \times 1 \times 1$ conv layers help the classification accuracy. 1×1 convolution helps increase non-linearity of the decision function, without changing the dimensions of input and output, 1×1 convolution is able to do the projection mapping in the same high dimensionality. This approach is used in NIN [36] GoogLeNet [39] and ResNet [42]. After updating to VGG-16 it obtained 8.8% error rate which means the deep learning network was still improving by adding number of layers. VGG-19 (19 layers) was developed to further improve the performance but it obtained 9.0% showing no improvement even after adding more layers.

When deeper networks starts converging, a degradation problem is exposed: with the network depth increasing, accuracy gets saturated and then degrades rapidly. Deep Residual Neural Network (ResNet) created by He et al. [42] introduced a norvel architecture with insert shortcut connections which turn the network into its counterpart residual version. This was a breakthrough which enabled the development of much deeper networks. The residual function is a refinement step in which the network learn how to adjust the input feature map for higher quality features. Following this intuition, the network residual block was refined and proposed a pre-activation variant of residual block [45], in which the gradients can flow through the shortcut connections to any other earlier layer unimpeded. Each ResNet block is either 2 layer deep (used in small networks like ResNet 18, 34) or 3 layer deep (ResNet 50, 101, 152). This technique is able to train a network with 152 layers while still having lower complexity than VGGNet. It achieves a top-5 error rate of 3.57% which beats human-level performance on this dataset. Although the original ResNet paper focused on creating a network architecture to enable deeper structures by alleviating the degradation problem, other researchers have since pointed out that increasing the network's width (channel depth) can be a more efficient way of expanding the overall capacity of the network.

In DenseNet which is a logical extension of ResNet, there is improved efficiency by concatenating each layer feature map to every successive layer within a dense block [43]. This allows later layers within the network to directly leverage the features from earlier layers, encouraging feature reuse within the network. For each layer, the feature-maps of all preceding layers are used as inputs, and its own feature-maps are used as inputs into all subsequent layers, this helps alleviate the vanishing-gradient problem, feature reuse and reduce number of parameters.

3.1 Proposed Agricultural Issues Classification Taxonomy

Many agricultural CNN solutions have been developed depending on specific agriculture issues. For the study purpose, a classification taxonomy tailored to CNN application in the smart farming was developed as shown Fig. 3. In this section, we categorize use of state-of-the-art CNN based on the agricultural issue they solve:

(a) Plant management includes solutions geared towards crop welfare and production. This includes classification (species), detection (disease and pest) and prediction (yield production).
(b) Livestock management address solutions for livestock production (prediction and quality management) and animal welfare (animal identification, species detection and disease and pest control).
(c) Environment management addresses solutions for land and water management.

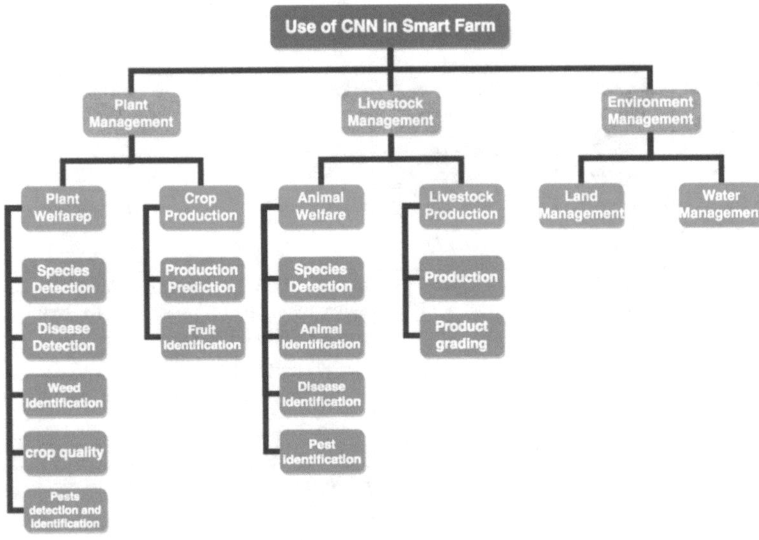

Fig. 3. Proposed classification taxonomy for CNN use in smart farm

3.2 Use of State-of-the-Art CNN in Smart Farms

Table 1, shows use of state-of-the-art CNN in agriculture and in particular the areas of plant and leaf disease detection, animal face identification, plant recognition, land cover classification, fruit counting and identification of weeds. It consist of 5 columns to show: the problem description, size of data used, accuracy according to the metrics used, the state-of-the-art CNN used and reference literature.

Table 1. Use of state-of-the-art CNN in smart farm

No.	Smartfarm problem description	Data used	Accuracy	CNN framework used	Article
1	Fruit detection	Images of three fruit varieties: apples (726), almonds (385) and mangoes (1154)	F1 (precision score) of 0.904 (apples) 0.908 (mango) 0.775 (almonds)	VGGNet	[46]
2	Detection of sweet pepper and rock melon fruits	122 images	0.838 (F1)	VGGNet	[47]
3	Recognize different plant species	Data set of 44 classes	99.60% (CA - correct prediction)	AlexNet	[48]
4	Recognize different plant	91 759 images	48.60% (LC-correct species classification)	AlexNet	[49]
5	Identify obstacles in row crops and grass mowing	437 images	99.9% in row crops and 90.8% in grass mowing (CA)	AlexNet	[50]
6	Identify crop species and diseases	54 306 images	0.9935 (F1)	AlexNet + GoogLeNet	[7]
7	Detect obstacles that are distant, heavily occluded and unknown	48 images	0.72 (F1)	AlexNet + VGG	[51]
8	Leaf disease detection	4483 images	96.30% (CA)	CaffeNet	[3]
9	Identify thistle in winter wheat and spring barley images	4500 images	97.00% (CA)	DenseNet	[52]
10	Predict number of tomatoes in images	24 000 images	91% (RFC-Ratio of total fruits counted) on real images, 93% (RFC) on synthetic images	GoogLeNet + ResNet	[53]
11	Classify banana leaf diseases	3700 images	96% (CA), 0.968 (F1)	LeNet	[54]
12	Identify pig face	1553 images	96.7% (CA)	VGGNet	[55]
13	Classify weed from crop species based on 22 different species in total	10413 images	86.20% (CA)	VGGNet	[56]
14	Detecting and categorizing the criticalness of Fusarium wilt of radish based on thresholding a range of color features	1500 images		GoogLeNet	[57]
15	Fruit counting	24 000 images	91% accuracy	Inception-ResNet	[53]
16	Automatic plant disease diagnosis for early disease symptoms	8178 images	overall improvement of the balanced accuracy from 0.78 to 0.87 from previous 2017 study	Deep ResNet	[58]

In their paper, Amara et al. [54] use the LeNet architecture to classify the banana leaves diseases. The model was able to effectively classify the leaves after several experiments. The approach was able to classify leaves images with different illumination, complex background, resolution, size, pose, and orientation. We also reviewed use of CaffeNet architecture [59] in agricultural application, which is a 1-GPU version of AlexNet. The success of this model at LSVRC 2012 [24] encourage many computer vision community to explore more on the application of deep learning in computer vision. Mohanty et al. [7] combined both AlexNet and GoogLeNet to identify 14 crop species and 26 diseases (or absence thereof) from a dataset of 54,305 images. The approach records an impressive accuracy of 99.35% demonstrating the feasibility of the state-of-the-art CNN architectures. Other areas AlexNet has been used with high accuracy record include; identify plants using different plant views [49], identify plant species [48], identify obstacles in the farm [50] and leaf disease detection [3]. Because of its achievement to improve utilization of the computing resources GoogLeNet has been used in fruit count [53] and plant species classification [7]. VGGNet has been used in classifying weed [56], detect obstacles in the farm [51], fruit detection [47] and animal face recognition [55]. Like ResNet, DenseNet is a recent model that explains why it has not been employed significantly in farming, nevertheless it has been used in thistle identification in winter wheat and spring barley [52]. Since ResNet is a such a recent model, it have only been used by one author in fruit counting [53]. Many of the CNN developed for agricultural use depend on the problem or challenge they solve.

4 Conclusions and Recommendations

Despite remarkable achievement in use state-of-the-art CNN in agriculture in general, there exist grey areas in relation to smart farm that future researchers may look at. These areas may include; real-time image classification, interactive image classification and interactive object detection. State-of-the-art CNN is relatively a new technology that explain why the finding of the study about their use in smart farm is relatively small. However, its is important to note that models built from state-of-the-art architectures have a impressive record of better precision performance. In this paper, we aimed at establishing the potential of state-of-the-art CNN in IoT based smart farms. In particular we first discussed the architectures of state-of-the-art CNNs and their respective prediction accuracy at the ILSVRC challenge. Then a survey on application of the identified CNNs in Agriculture was performed; to examine the particular application in a smart farm, listed technical details of the architecture employed and overall prediction accuracy achieved according to the author precision metrics. From the study its evident of continuous accuracy improvement of the state-of-the-art CNN architectures as computer vision community put effort to perfect the methods. The findings indicate that state-of-the-art CNN has achieved better precision in all the cases applied in the agricultural domain, scoring higher accuracy in majority of the problem as compared to other image-processing techniques.

Considering that the state-of-the-art CNN has achieved state-of-the-art results in prediction in general and high precision in the few farming cases observed, there is great potential that can be achieved in using the methods in smart farming. It has been observed that many authors apply more than one architecture in order to optimize the performance of the network without compromising the expected accuracy. This approach is very efficient in the observed cases, and we recommend similar hybrid approach when building robust IoT based networks which are computationally fair to the mobile devices. This study aims to motivate researchers to experiment and apply the state-of-the-art methods in smart farms problems related to computer vision and data analysis in general.

References

1. Xu, B., Wang, N., Chen, T., Li, M.: Empirical evaluation of rectified activations in convolutional network. CoRR, abs/1505.00853 (2015)
2. Luo, P., Li, H.: Research on quantum neural network and its applications based on tanh activation function. Comput. Digit. Eng. **16**, 33–39 (2016)
3. Sladojevic, S., Arsenovic, M., Anderla, A., Culibrk, D., Stefanovic, D.: Deep neural networks based recognition of plant diseases by leaf image classification. Comput. Intell. Neurosci. (2016)
4. Godfray, H.C., Beddington, J.R., Crute, I.R., Haddad, L., Lawrence, D., Muir, J.F., Pretty, J.N., Robinson, S., Thomas, S.M., Toulmin, C.: Food security: the challenge of feeding 9 billion people. Science **327**(5967), 812–818 (2010)
5. Lutz, W.L., Sanderson, W.C., Scherbov, S.: The coming acceleration of global population ageing. Nature **451**, 716–719 (2008)
6. Bruinsma, J. (ed.): World Agriculture: Towards 2015/2030: An FAO Perspective. Earthscan, London (2003)
7. Mohanty, S.P., Hughes, D.P., Salathé, M.: Using deep learning for image-based plant disease detection. Front. Plant Sci. (2016)
8. Gebbers, R., Adamchuk, V.I.: Precision agriculture and food security. Science **327**(5967), 828–831 (2010)
9. Krintz, C., Wolski, R., Golubovic, N., Lampel, B., Kulkarni, V., Sethuramasamyraja, B.B., Roberts, B.: SmartFarm: improving agriculture sustainability using modern information technology (2016)
10. Kamilaris, A., Prenafeta Boldú, F.: A review of the use of convolutional neural networks in agriculture. J. Agric. Sci. 1–11 (2018). https://doi.org/10.1017/S0021859618000436
11. Weber, R.H., Weber, R.: Internet of Things: Legal Perspectives. Springer, Heidelberg (2010). https://doi.org/10.1007/978-3-642-11710-7
12. Anindya, S.: Remote sensing in agriculture. Int. J. Environ. Agric. Biotechnol. (IJEAB) **1**(3), 362–367 (2016)
13. Jinbo, C., Xiangliang, C., Han-Chi, F., et al.: Cluster computing. https://doi.org/10.1007/s10586-018-2022-5
14. Chi, M., Plaza, A., Benediktsson, J.A., Sun, Z., Shen, J., Zhu, Y.: Big data for remote sensing: challenges and opportunities. Proc. IEEE **104**, 2207–2219 (2016)
15. Waga, D., Rabah, K.: Environmental conditions' big data management and cloud computing analytics for sustainable agriculture. World J. Comput. Appl. Technol. **2**, 73–81 (2017)

16. Chen, M., Mao, S., Liu, Y.: Big data: a survey. MONET **19**, 171–209 (2014)
17. Russakovsky, O., Deng, J., Su, H., Krause, J., Satheesh, S., Ma, S., Huang, Z., Karpathy, A., Khosla, A., Bernstein, M.S., Berg, A.C., Fei-Fei, L.: ImageNet large scale visual recognition challenge. Int. J. Comput. Vis. **115**, 211–252 (2015)
18. Liang, M., Hu, X.: Recurrent convolutional neural network for object recognition. In: 2015 IEEE Conference on Computer Vision and Pattern Recognition (CVPR), pp. 3367–3375 (2015)
19. Poria, S., Cambria, E., Gelbukh, A.F.: Aspect extraction for opinion mining with a deep convolutional neural network. Knowl.-Based Syst. **108**, 42–49 (2016)
20. Goodfellow, I.J., Bengio, Y., Courville, A.C.: Deep learning. Nature **521**, 436–444 (2015)
21. Bhandare, A., Bhide, M., Gokhale, P., Chandavarkar, R.: Applications of convolutional neural networks. Int. J. Comput. Sci. Inf. Technol. **7**(5), 2206–2215 (2016)
22. Schmidhuber, J.: Deep learning in neural networks: an overview. Neural Netw. **61**, 85–117 (2015)
23. http://image-net.org/challenges/LSVRC/2017 . Accessed 02 Sept 2018
24. http://image-net.org/challenges/LSVRC/2012/ . Accessed 02 Sept 2018
25. http://image-net.org/challenges/LSVRC/2013/ . Accessed 02 Sept 2018
26. ImageNet. http://image-net.org. Accessed 21 Oct 2018
27. Alom, M.Z., Taha, T.M., Yakopcic, C., Westberg, S., Sidike, P., Nasrin, M.S., Van Essen, B.C., Awwal, A.A.S., Asari, V.K.: The history began from AlexNet: a comprehensive survey on deep learning approaches. https://arxiv.org/pdf/1803.01164
28. Kamilaris, A., Prenafeta-Boldu, F.X.: Deep learning in agriculture: a survey. Comput. Electron. Agric. **147**, 70–90 (2018)
29. Zhang, C., Yue, P., Liping, D., Zhaoyan, W.: Automatic identification of center pivot irrigation systems from landsat images using convolutional neural networks. Agriculture **8**(10), 1–19 (2018)
30. Chebet, E., Yujian, l., Njuki, S., Yingchun, L.: A comparative study of fine-tuning deep learning models for plant disease identification. Comput. Electron. Agric. https://doi.org/10.1016/j.compag.2018.03.032
31. Andri, R., Cavigelli, L., Rossi, D., Benini, L.: Hyperdrive: a systolically scalable binary-weight CNN inference engine for mW IoT end-nodes. In: 2018 IEEE Computer Society Annual Symposium on VLSI (ISVLSI), pp. 509–515 (2018)
32. Canziani, A., Paszke, A., Culurciello, E.: An analysis of deep neural network models for practical applications. CoRR, abs/1605.07678 (2016)
33. HasanPour, S.H., Rouhani, M., Fayyaz, M., Sabokrou, M., Adeli, E.: Towards principled design of deep convolutional networks: introducing SimpNet. CoRR, abs/1802.06205 (2018)
34. LeCun, Y., Bottou, L., Bengio, Y.: Gradient-based learning applied to document recognition. Proc. IEEE **86**(11), 2278–2324 (1998)
35. Krizhevsky, A., Sutskever, I., Hinton, G.E.: ImageNet classification with deep convolutional neural networks. Commun. ACM **60**, 84–90 (2012)
36. Lin, M., Chen, Q., Yan, S.: Network in network. CoRR, abs/1312.4400 (2013)
37. Paszke, A., Chaurasia, A., Kim, S., Culurciello, E.: ENet: a deep neural network architecture for real-time semantic segmentation. CoRR, abs/1606.02147 (2016)
38. Zeiler, M.D., Fergus, R.: Visualizing and understanding convolutional networks. In: European Conference on Computer Vision, pp. 818–833. Springer (2014)
39. Szegedy, C., Liu, W., Jia, Y., Sermanet, P., Reed, S.E., Anguelov, D., Erhan, D., Vanhoucke, V., Rabinovich, A.: Going deeper with convolutions. In: IEEE Conference on Computer Vision and Pattern Recognition (CVPR), pp. 1–9 (2015)

40. Simonyan, K., Zisserman, A.: Very deep convolutional networks for large-scale image recognition. CoRR, abs/1409.1556 (2014)
41. Szegedy, C., Vanhoucke, V., Ioffe, S., Shlens, J., Wojna, Z.: Rethinking the inception architecture for computer vision. In: IEEE Conference on Computer Vision and Pattern Recognition (CVPR), pp. 2818–2826 (2016)
42. He, K., Zhang, X., Ren, S., Sun, J.: Deep residual learning for image recognition. In: 2016 IEEE Conference on Computer Vision and Pattern Recognition (CVPR), pp. 770–778 (2015)
43. Huang, G., Liu, Z., Maaten, L.V., Weinberger, K.Q.: Densely connected convolutional networks. In: 2017 IEEE Conference on Computer Vision and Pattern Recognition (CVPR), pp. 2261–2269 (2017)
44. Szegedy, C., Ioffe, S., Vanhoucke, V.: Inception-v4, inception-ResNet and the impact of residual connections on learning. In: AAAI (2017)
45. He, K., Zhang, X., Ren, S., Sun, J.: Identity mappings in deep residual networks. In: ECCV (2016)
46. Bargoti, S., Underwood, J.P.: Deep fruit detection in orchards. In: 2017 IEEE International Conference on Robotics and Automation (ICRA), pp. 3626–3633 (2017)
47. Sa, I., Ge, Z., Dayoub, F., Upcroft, B., Perez, T., McCool, C.: DeepFruits: a fruit detection system using deep neural networks. Sensors **16**(8), 1222 (2016)
48. Lee, S.H., Chan, C.S., Wilkin, P., Remagnino, P.: Deep-plant: plant identification with convolutional neural networks. In: 2015 IEEE International Conference on Image Processing (ICIP), pp. 452–456 (2015)
49. Reyes, A.K., Caicedo, J.C., Camargo, J.E.: Fine-tuning deep convolutional networks for plant recognition. In: CLEF (2015)
50. Steen, K.A., Christiansen, P., Karstoft, H., Jørgensen, R.N.: Using deep learning to challenge safety standard for highly autonomous machines in agriculture. J. Imaging **2**, 6 (2016)
51. Christiansen, P., Nielsen, L.N., Steen, K.A., Jørgensen, R.N., Karstoft, H.: DeepAnomaly: combining background subtraction and deep learning for detecting obstacles and anomalies in an agricultural field. Sensors **16**(11), 1904 (2016)
52. Sørensen, R.A., Rasmussen, J., Nielsen, J., Jørgensen, R.N.: Thistle detection using convolutional neural networks. In: EFITA WCCA 2017 Conference, Montpellier Supagro, Montpellier, France, 2–6 July 2017 (2017)
53. Rahnemoonfar, M., Sheppard, C.: Deep count: fruit counting based on deep simulated learning. Sensors **17**(4), 905 (2017)
54. Amara, J., Bouaziz, B., Algergawy, A.: A deep learning-based approach for Banana Leaf diseases classification. In: BTW (2017)
55. Hansen, M.F., Smith, M.L., Smith, L.N., Salter, M.G., Baxter, E.M., Farish, M., Grieve, B.: Towards on-farm pig face recognition using convolutional neural networks. Comput. Ind. **98**, 145–152 (2018)
56. Dyrmann, M., Karstoft, H., Midtiby, H.: Plant species classification using deep convolutional neural network. Biosyst. Eng. **151**, 72–80 (2016)
57. Hyun, J., Ibrahim, H., Irfan, M., Minh, L., Suhyeon, I.: UAV based wilt detection system via convolutional neural networks. Sustain. Comput. Inform. Syst. https://doi.org/10.1016/j.suscom.2018.05.010
58. Picona, A., Alvarez, A., Seitz, M., Ortiz, O., Echazarra, J., Johannes, A.: Deep convolutional neural networks for mobile capture device-based crop disease classification in the wild. Comput. Electron. Agric. https://doi.org/10.1016/j.compag.2018.04.002

59. Jia, Y., Shelhamer, E., Donahue, J., Karayev, S., Long, J., Girshick, R.B., Guadarrama, S., Darrell, T.: Caffe: convolutional architecture for fast feature embedding. In: ACM Multimedia (2014)
60. Ioffe, S., Szegedy, C.: Batch normalization: accelerating deep network training by reducing internal covariate shift. In: ICML (2015)
61. He, K., Zhang, X., Ren, S., Sun, J.: Delving deep into rectifiers: surpassing human-level performance on ImageNet classification. In: 2015 IEEE International Conference on Computer Vision (ICCV), pp. 1026–1034 (2015)

Neural Networks to Approximate Solutions of Ordinary Differential Equations

Georg Engel(✉)

Christian Doppler Laboratory for Quality Assurance Methodologies for Autonomous Cyber-Physical Systems, Institute for Software Technology, Graz University of Technology, Graz, Austria
engel@ist.tugraz.at

Abstract. We discuss surrogate data models based on machine learning as approximation to the solution of an ordinary differential equation. The surrogate model is designed to work like a simulation unit, i.e. it takes a few recent points of the trajectory and the input variables at the given time and calculates the next point of the trajectory as output. The Dahlquist test equation and the Van der Pol oscillator are considered as case studies. Computational demand and accuracy in terms of local and global error are discussed. Parameter studies are performed to discuss the sensitivity of the method.

Keywords: Ordinary differential equations · Machine learning · Surrogate model · Neural network

1 Introduction

1.1 Motivation and Related Work

Many physical systems in this world can be described effectively using differential equations. Solving these equations can be expensive in practice, in particular when solutions are required repeatedly, like in optimization problems. When combing these models with the real world, like in cyber-physical systems, the real-time requirement puts stringent upper bounds on the computational costs of the models. It is hence of interest to reduce the computational costs, taking into account some limited reduction of accuracy. This is the purpose of surrogate models, which usually apply model order reduction in some way.

Data surrogate models exploit that wealth of data which is generated during simulations. A prominent method to do so is proper orthogonal decomposition, which was successfully applied already many years ago, e.g. to analyze turbulent flows [1]. In the present work we consider a slightly different approach, looking for surrogate models from the perspective of co-simulation. Co-simulation denotes the dynamic coupling of various simulation units, which exchange information

K. Arai et al. (Eds.): CompCom 2019, AISC 997, pp. 776–784, 2019.
https://doi.org/10.1007/978-3-030-22871-2_54

during simulation time, for a review see [2]. Several co-simulation interfaces have been defined and implemented, e.g. the Functional Mock-Up Interface standard [3]. A discussion and comparison was performed e.g. in [4]. In such a setup the computational costs are often dominated by one (or few) of the individual simulation units. Furthermore, the communication between the simulation units introduces a further source of errors, which might outweigh the errors of surrogate models in some cases. Hence, it would be interesting to exploit the framework of co-simulation to replace costly simulation units by cheaper surrogate models.

Machine learning is becoming more and more popular in recent years (and decades) for many applications. In the context of differential equations, some work dates back more than twenty years, but a considerable boost appeared very recently. Artificial neural networks have been proposed to derive analytical solutions for differential equations by reformulating them as optimization problem [5]. Deep reinforcement learning was suggested for general non-linear differential equations, where the network consists of an actor that outputs solution approximations policy and a critic that outputs the critic of the actor's output solution [6]. An approximation model for real-time prediction of non-uniform steady laminar flow based on convolutional neural networks are proposed in [7]. Considering the simulation of buildings, several efforts have been made to reduce the computation costs, in particular within the IBPSA 1 project [8]. The input and output data of simulators is used to train neural networks in a co-simulation setting (referred to as "intelligent co-simulation"), and the model is compared to model order reduction by proper orthogonal decomposition [9]. A similar approach called "component-based machine learning", for the same field of application, is pursued in [10]. The same setting was investigated using deep-learning neural-networks [11], where a considerable computational speedup over the physical model at similar accuracy is claimed.

1.2 Background of This Work

The present work is motivated mainly by the following two points: On one hand, a wealth of data is generated during a simulation through various function calls, which in particular for long time trajectories (e.g. annual simulations for building performance) receive repeatedly very similar input. It is very tempting to replace costly function calls by cheap surrogate models which ideally could be trained on the fly using data generated in early phases of the simulation time. On the other hand, in practice, such surrogate models would be very useful when applicable in a modular way. Hence, they should comply during training and predictive phase to a standard interface. Furthermore, they should ideally work with minimal knowledge about the system, e.g. requiring only the input variable which would otherwise be fed to the original costly function call.

Set up in such a way, surrogate models for components or subsystems could be used in a modular way and trained on the fly during simulation, such that the user may only barely be aware when the surrogate model actually replaces the original simulation unit. Co-simulation interfaces appear to share many properties with an interface needed for such a surrogate model. An intriguing long-term goal is thus given by developing a modular surrogate model framework complying to a wide-spread co-simulation interface standard like FMI.

The present work is a first step in this direction, examining the method on simple case studies where extensive parameter studies are feasible and insights related to the structure of the method can be achieved. The main contribution of this work is as follows:

- A neural network is demonstrated as surrogate model for two different case studies.
- Accuracy and computational costs are discussed.
- Parameter studies are performed to discuss the sensitivity of the method.

2 Method

A neural network shall be trained as surrogate model for a simulation unit. The multi-layer perceptron of the package scikit-learn is used in this respect [12]. It is trained by the solution to differential equations generated by a traditional solver. For the latter, we employ the routine *odeint* from the python package *scipy* [13].

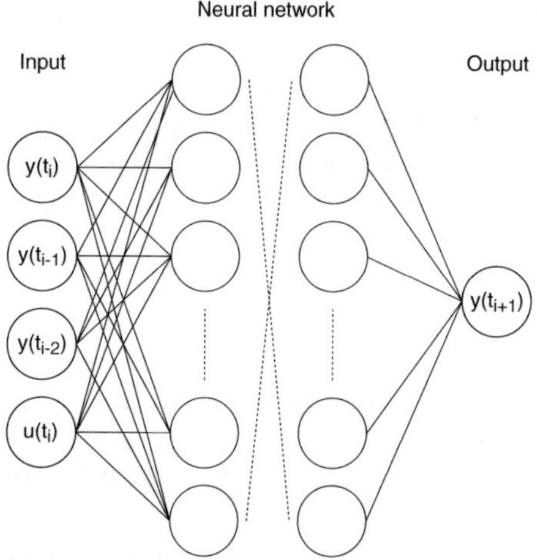

Fig. 1. Schematics of the introduced method. A neural network serves to predict the value of the state variable $y(t)$ at the next time step t_{i+1}. The input is given by the present and previous (here: two) values of the state variable plus an explicit function of time $u(t)$ evaluated at the respective time step t_i. The dots indicate that an arbitrary number of neurons and hidden layers can be used.

The surrogate model has the same input and output variables as the original simulation unit. A schematics of the design is shown in Fig. 1. The input of the surrogate model is given by the present and previous (here: two) values of the state variable plus an explicit function of time $u(t)$ evaluated at the respective time step. The neural network can have an arbitrary number of neurons and hidden layers. For the simple cases considered here, the output coincides with the state variable of the model. Note that in practice (e.g. through a co-simulation interface), the previous values of the state variable would not be available as input. Instead, a framework or a wrapper needs to be implemented, where these values are stored. Alternatively, such functionality could be integrated in the neural network itself, e.g. considering recurrent neural networks.

The method is set up in a way to allow easily for generalization in several respects to be considered in future. Other machine learning models can be implemented instead of the multi-layer perceptron, like multi-variate regression based on suitable basis functions or more specific types of neural networks, such as recurrent ones. When using a wrapper for simulation units in a co-simulation, the method can provide a surrogate model in the wrapper straightforwardly. While we consider here samples for training which are consecutive in simulation time, minor modifications of the method allow to take more arbitrary snapshots of the data.

3 Model

For testing the method, we consider two models. The first one is the Dahlquist test equation,

$$\dot{y}(t) - \lambda y(t) = 0, \tag{1}$$

with $\lambda = 0.1$, which gives an exponentially decaying solution. The second one is the the Van Der Pol oscillator, which is a second order non-linear ordinary differential equation,

$$\ddot{y}(t) - \mu \left[1 - y^2(t)\right] \dot{y}(t) + y(t) = 0, \tag{2}$$

with $\mu = 0.1$. The parameters are chosen such to give a trajectory in the range considered suitable for the present discussion.

4 Results

As initial condition we define $y(t = 0) = 1$ (and $\dot{y}(t = 0) = 0$ in case of the second order differential equation). We choose the interval $t \in [0, 100)$ and 1000 time steps. $t \in [0, 30)$ is used for training, $t \in [30, 50)$ as testing for potential model selection and $t \in [50, 100)$ for validation. The model selection is used to apply 100 different random seeds in the optimizer of the neural network, which is crucial to obtain a satisfactory validation. Unless stated otherwise, we use 4 input variables (the present and 2 previous values of the state variable plus a potential explicit function of time), one hidden layer with 5 neurons and a tolerance requirement of 10^{-7} for the training of the neural network. The figures shown in this section

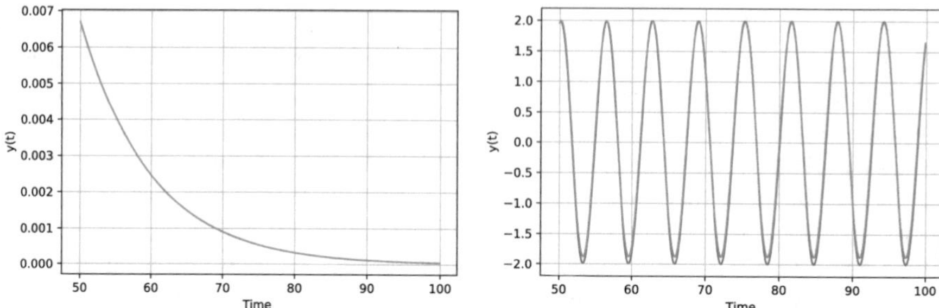

Fig. 2. Left: Results for the validation of the neural network for the Dahlquist test equation corresponding to Eq. (1). Right: Same as left, but for the Van der Pol oscillator corresponding to Eq. (2). The blue curve shows the "exact" solution generated by the traditional solver, the orange curve shows the prediction of the neural network. If only one curve is visible, the predicted one lies one top of the "exact" solution.

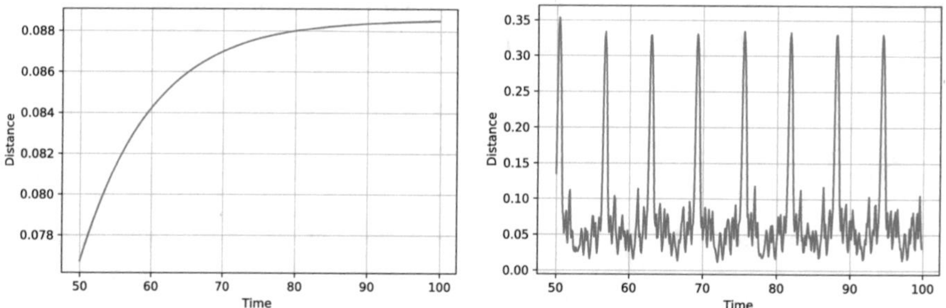

Fig. 3. The distance in theory space from input variables during validation compared to training data is shown for validation of the neural network. The distance is defined as Euclidean distance from the nearest point from the training data set. Left: Results for the neural network for the Dahlquist test equation. Right: Same as left, but for the Van der Pol oscillator.

always show results for the Dahlquist test equation corresponding to Eq. (1) on the left hand side and results for Van der Pol oscillator corresponding to Eq. (2) on the right hand side.

Figure 2 shows the results for the validation. For the chosen parameters, the validation works very well. Since the accuracy of a data model becomes worse the more it relies on extrapolation, we show a measure for that in Fig. 3. A distance is calculated as Euclidean distance of the input variables during the validation from the nearest point of the input variables during the training. The global error of the model can easily be estimated considering the deviation of the predicted trajectory from the trajectory generated by the traditional solver, shown in Fig. 4. As usual, the global error increases with time due to accumulation. Also, a correlation of its increase with the distance shown in Fig. 3 is verified. We note further that the global error for the Dahlquist test equation is considerably larger than the one for the Van der Pol oscillator, which is expected according to the different complexity of these models.

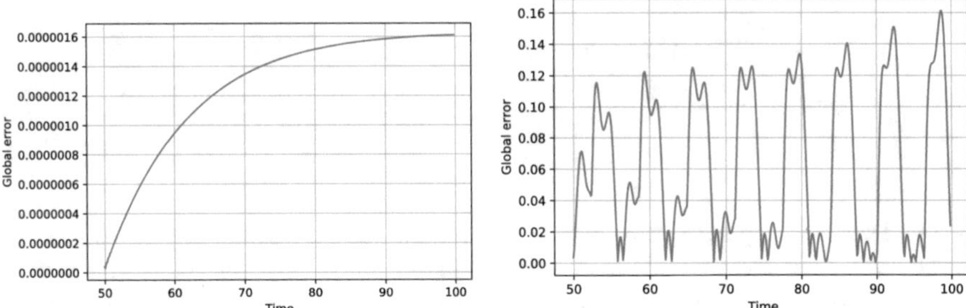

Fig. 4. The global integration error generated by the data model along the trajectory, determined by comparison to a trajectory generated by a traditional solver. Left: Results for the neural network for the Dahlquist test equation. Right: Same as left, but for the Van der Pol oscillator.

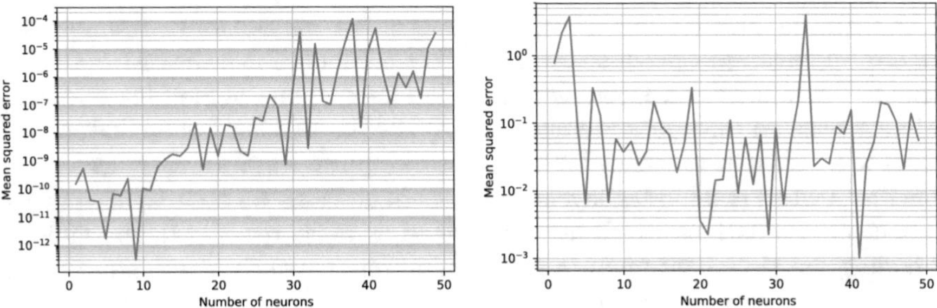

Fig. 5. The mean squared error during the validation, shown versus the number of neurons for one hidden layer. Left: Results for the neural network for the Dahlquist test equation. Right: Same as left, but for the Van der Pol oscillator.

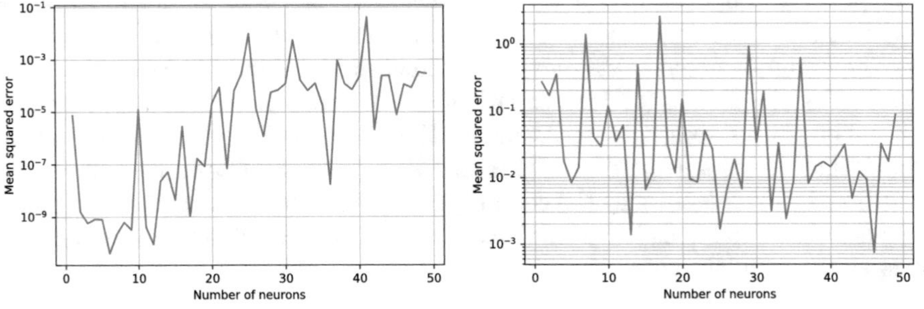

Fig. 6. The mean squared error during the validation, shown versus the number of neurons for two hidden layers. Left: Results for the neural network for the Dahlquist test equation. Right: Same as left, but for the Van der Pol oscillator.

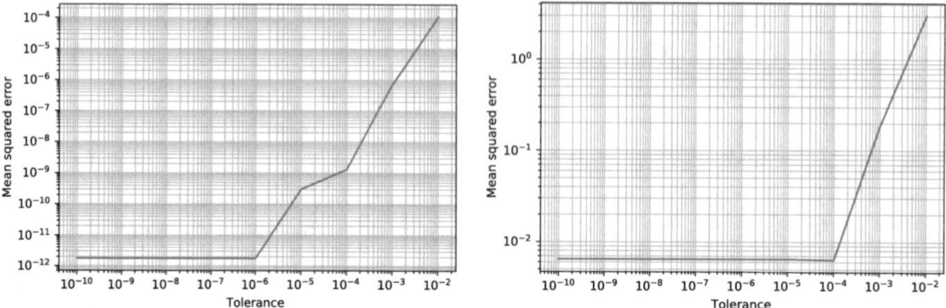

Fig. 7. The mean squared error during the validation, shown versus the required tolerance during the training of the neural network. Left: Results for the neural network for the Dahlquist test equation. Right: Same as left, but for the Van der Pol oscillator.

Parameter studies have been performed to discuss the sensitivity of the method. The mean square global error during the validation is shown versus the number of neurons in one hidden layer in Fig. 5 and in two hidden layers (each with the same number of neurons) Fig. 6. For the Dahlquist test equation, the error appears to worsen towards a higher number of neurons. Also, the performance of two hidden layers is worse than that of one hidden layer (not the different scale on the y-axis in Figs. 5 and 6) For the Van der Pol oscillator, there seems to be a slight tendency for the error to improve towards a higher number of neurons when two hidden layers are used. In all cases we observe a significant amount of fluctuations which are interpreted as noise relating to the difficulty of finding the global optimum in the training phase, which manifests itself also in the need for several random seeds.

The mean square global error during the validation is shown versus the required tolerance of the optimizer during the training in Fig. 7. In case of the Dahlquist test equation, the error can be improved by choosing a suitable tolerance, in particular noting that the default value in the training of the multi-layer perceptron of scikit-learn is 10^{-4}. Recall also that the error improves considerably when using several random seeds, which is essential for the method to work.

The computational costs have been measured using the python package *timeit*. For the Dahlquist test equation, the neural network surrogate model took about 30 ms compared to 0.2 ms for the traditional solver; for the Van der Pol oscillator, the surrogate model took about 35 ms compared to 4 ms for the traditional solver. Note that the computational costs for training the data model are neglected here. This is reasonable, as the training could be done offline or at least easily parallelized.

5 Summary/Discussion

Neural network surrogate models for ordinary differential equations have been investigated considering the Dahlquist test equation and the Van der Pol oscillator. It was shown that a fairly small neural network can describe these simple models to fairly good approximation. It was found crucial to apply various random seeds for the optimizer of the neural network during the training phase. A more stringent tolerance requirement than the default one might bring some improvement, while the number of neurons plays a minor role.

The computational costs of the surrogate model are higher than the one of the traditional solver, depending on the difficulty of the problem. This can be expected as the code is far from optimized and the model order was actually not reduced for these simple cases. Hence, the surrogate model actually implements a reverse engineering of the model solved by an explicit solver. An improvement in computational costs is expected when applying the method for more complicated models, where the order of the model is effectively reduced.

6 Outlook/Open Issues

The present work reports a first step in the development of a modular surrogate model framework suited for co-simulation scenarios. Hence, the method presented here shall be tested further for more complicated models, where model order reduction is desired. A realistic case study will be considered in the field of thermal energy engineering. It will be discussed which problems are particularly suited for this kind of surrogate models. An extension to other machine learning models will be considered. The method shall be applied in the context of co-simulation, replacing a simulation unit after sufficient training on the fly.

Acknowledgments. The financial support by the Austrian Federal Ministry for Digital and Economic Affairs and the National Foundation for Research, Technology and Development is gratefully acknowledged. We further acknowledge fruitful discussions with Gerald Schweiger, Claudio Gomes and Philip Ohnewein.

References

1. Berkooz, G., Holmes, P., Lumley, J.L.: The proper orthogonal decomposition in the analysis of turbulent flows. Annu. Rev. Fluid Mech. **25**(1), 539–575 (1993)
2. Gomes, C., Thule, C., Broman, D., Larsen, P.G., Vangheluwe, H.: Co-simulation: state of the art. CoRR, abs/1702.0, February 2017
3. Blochwitz, T., Otter, M., Arnold, M., Bausch, C., Clauß, C., Elmqvist, H., Junghanns, A., Mauss, J., Monteiro, M., Neidhold, T., Neumerkel, D., Olsson, H., Peetz, J.V., Wolf, S.: The functional mockup interface for tool independent exchange of simulation models. In 8th International Modelica Conference 2011, pp. 173–184 (2009)
4. Engel, G., Chakkaravarthy, A.S., Schweiger, G.: A general method to compare different co-simulation interfaces: demonstration on a case study. In: Kacprzyk, J. (ed.) Simulation and Modeling Methodologies, Technologies and Applications, Chap. 19. Springer (2018)
5. Lagaris, I.E.E., Likas, A., Fotiadis, D.I.I.: Artificial neural networks for solving ordinary and partial differential equations. IEEE Trans. Neural Netw. **9**(5), 1–26 (1997)

6. Wei, S., Jin, X., Li, H.: General solutions for nonlinear differential equations: a deep reinforcement learning approach. Technical report (2018)
7. Guo, X., Li, W., Iorio, F.: Convolutional neural networks for steady flow approximation. In: Proceedings of the 22nd ACM SIGKDD International Conference on Knowledge Discovery and Data Mining, San Francisco, CA, USA, 13–17 August 2016, pp. 481–490 (2016)
8. IBPSA. IBPSA Project 1. https://ibpsa.github.io/project1/
9. Berger, J., Mazuroski, W., Oliveira, R.C.L.F., Mendes, N.: Intelligent co-simulation: neural network vs. proper orthogonal decomposition applied to a 2D diffusive problem. J. Build. Perform. Simul. **11**(5), 568–587 (2018)
10. Geyer, P., Singaravel, S.: Component-based machine learning for performance prediction in building design. Appl. Energy **228**, 1439–1453 (2018)
11. Singaravel, S., Suykens, J., Geyer, P.: Deep-learning neural-network architectures and methods: using component-based models in building-design energy prediction. Adv. Eng. Inform. **38**(May), 81–90 (2018)
12. Pedregosa, F., Varoquaux, G., Gramfort, A., Michel, V., Thirion, B., Grisel, O., Blondel, M., Prettenhofer, P., Weiss, R., Dubourg, V., Vanderplas, J., Passos, A., Cournapeau, D., Brucher, M., Perrot, M., Duchesnay, E.: Scikit-learn: machine learning in Python. J. Mach. Learn. Res. **12**, 2825–2830 (2011)
13. Jones, E., Oliphant, E., Peterson, P., et al.: SciPy: open source scientific tools for Python (2001). http://www.scipy.org/. Accessed 06 Jun 2019

Optimizing Deep Learning Model for Neural Network Topology

Sara K. Al-Ruzaiqi[1(✉)] and Christian W. Dawson[2]

[1] Computer Science Department, Loughborough University, Muscat, Oman
s.k.s.al-ruseiqi@lboro.ac.uk
[2] Computer Science Department, Loughborough University, Loughborough, UK

Abstract. In this work, a method of tuning deep learning models using H2O is proposed, where the trained network model is built from samples of selected features from the dataset, in order to ensure diversity of the samples and to improve training. A successful application of deep learning requires setting its parameters in order to get better accuracy. The number of hidden layers and the number of neurons are the key parameters in each layer of a deep machine-learning network, which have great control on the performance of the algorithm. Hyper-parameter, grid search and random hyper-parameter approaches aid in setting these important parameters. In this paper, a new ensemble strategy is suggested that shows potential to optimize parameter settings and hence save more computational resources throughout the tuning process of the models. The data are collected from several airline datasets to build a deep prediction model to forecast airline passenger numbers. The preliminary experiments show that fine-tuning provides an efficient approach for tuning the ultimate number of hidden layers and the number of neurons in each layer when compared with the grid search method.

Keywords: Deep learning · H2O · Optimizing

1 Introduction

Deep learning has been applied in many contexts, for example, health (pattern recognition), education (machine translation) and vision interpretation. The optimization of such models involves calibration with large data sets, which are then used to make future predictions [1]. We frequently make use of different optimization techniques in deep learning to improve performance. For instance, grid search is one such technique that is used to enhance the calibration of deep neural network models. The challenge still remains to train deep learning neural networks with large data sets in an efficient manner [2].

Researchers have applied deep neural networks in a number of studies with promising outcomes [3]. Learning methods have been developed that can teach deep network variants such as conviction networks presented by Fukushima [4]. This development reestablished enthusiasm for deep neural networks. Neural networks with numerous layers will frequently encounter an issue of some transfer functions approaching zero. Hochreiter was the first researcher to summarize this disappearing gradient issue within his Ph.D. thesis [5]. Prior to deep learning, the majority of neural

K. Arai et al. (Eds.): CompCom 2019, AISC 997, pp. 785–795, 2019.
https://doi.org/10.1007/978-3-030-22871-2_55

networks used a simple quadratic error performance measure on the output layer [6]. De Boer et al. [6] introduced the cross-entropy error feature, and it frequently achieves much better outcomes than the quadratic function previously used as it deals with the vanishing gradient issue by permitting errors to change weights even if a neuron's gradient saturates (their results are close to zero).

Additionally, it offers a more irregular way of error representation as opposed to the quadratic error performed for classification neural networks. Thus, the analysis provided in this paper is going to utilize the cross-entropy error measure for classification as well as the more commonly used root mean square error (RMSE) for regression. Neural networks usually begin with random weights [9]. These random weights are often sampled within a range, such as (−1,1). This range initialization can sometimes create a set of weights that prove difficult for the back propagation training (for example, being caught in a local minima). The author in [8] unveiled the rectified linear unit (ReLU) transfer function to deal with this issue. The ReLU transfer function typically achieves better training benefits for deep neural networks as opposed to the sigmoidal transfer functions, which are often used. Based on recent studies [1, 7], the type of transfer function to use for deep neural networks is identified for each layer type.

The hidden layers of deep neural networks make use of the ReLU transfer function. For their output layer, the majority of deep neural networks employ a linear transfer function for regression, along with a softmax transfer function for classification. No transfer function is required for the input layer. Moreover, over-fitting is a regular issue for neural networks [10]. A neural network is believed to be over-fit when it has been trained to a stage such that the network starts to master the outliers in the data set. This neural network is learning how to commit to memory, not generalize [8]. Since there are many variables compared with the total number of training examples, feature selection, and regularization techniques are used to combat the over-fitting problem. Data are partitioned into training, validation, and testing sets to ensure robust statistics.

2 Optimizing Deep Prediction Model

The objective of this study is to train and build a deep prediction model. The focus is on feed-forward neural networks. We start by exploring the data, simplifying them and after that, applying the model and investigating the results. In these steps the R language is used along with the h2o R package. Monthly data (1989 to 2016) are sourced from the Oman Management Airport Company (OMAC) which contains 51,983 observations with 9 variables (2 are categorical explanatory variables; and the variable to predict passenger data (Pax) are also categorical with 13 levels).

The focus of this work is not on the data set as such, but on the analysis that should be done before making predictions and the features of H2O in R; and why it is important in deep learning. The primary reason behind selecting the H2O implementation over various other neural networks libraries is that it has plain hyper parameter tuning which makes it rapid. In addition, H2O possesses Java as a backend, which makes it easier to run over several cores of the processor.

In order to train models with the H2O engine, the datasets should be linked to the H2O cluster first, then run the Deep Learning model on the dataset, and the Deep

Learning model will be tasked to perform (multi-class) classification. The data are imported into R in the conventional way. Figure 1 presents the monthly distribution of the passenger data set.

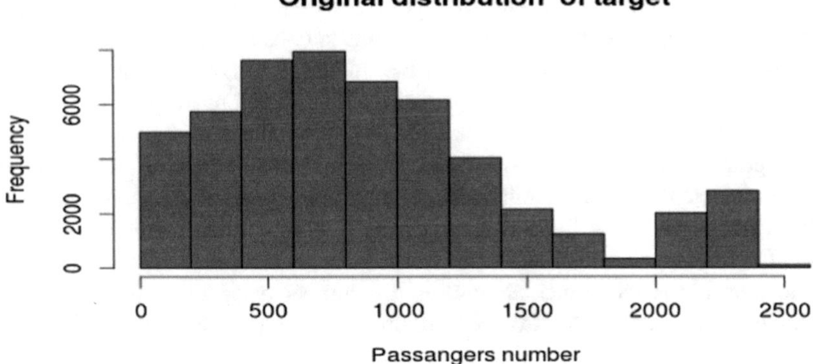

Fig. 1. The original distribution of target

Because of the size of the data set it was possible to carry out multiple runs to observe the variation in prediction performance and to investigate the impact of model regularization by tuning the 'Dropout' parameter in the ***h2o.deeplearning(…)*** function. This was undertaken using the following steps:

- Set-up and connect to a local H2O cluster from R.
- Train a deep neural networks model.
- Use the model for predictions.
- Ensemble models.
- Consider the memory usage.

A local cluster with 4 GB memory allowance was built to ensure that the memory utilization was sufficient over the period of the model training process. Only one thread was created to initiate deep learning in this project.

3 Methods

3.1 Training a Deep Neural Network Model

The dataset was split randomly into training and test sets (75% and 25%, respectively). In addition, after this we double-checked using a dropout of 25% data from the original training file. A dropout is a 1-fold cross validation taken to its extreme: the testing set contained 9840 observations while the training set was composed of all the remaining observations. Note that in a dropout 1 = number of observations in the dataset. It appears as the data lends itself to the Neural Network. Later analyse will determine if this is, in fact, the situation or perhaps if the default parameters for the Neural Network have been a lucky-hit.

3.2 Testing the Model by Using the Model for Prediction

The experiments were started by adding L1 and L2 regularization (and boosting the number of training rounds/epochs from 1 to 100). This seems to reduce the errors down to 292.5256 (errors before 308.0929), which makes sense as we have made our model less prone to catching 'noise' and to generalize better. We experiment with a **wide range of hidden-layers and number of neurons** to construct a list of potentials so that the chosen parameter was somewhere in the middle. The model results in less error, which was even lower than on the validation. RMSE before dropout is 308.09 passengers; after 292.525 – so dropout has led to improvement.

The graph in Fig. 2 on Y-axis shows the actual total values of passengers along with the predicted values from our predictor on x-axis. Some of the points near 2000 to 2500 on the y-axis show there is a miss match on actual and prediction count. The value of it can be confirmed as 0–500 from y-axis at last points. Hence, there was a need to train the ANN further. Regarding the number of epochs, the neural network should iterate the best number from the possible states of the greedy search 100 epochs.

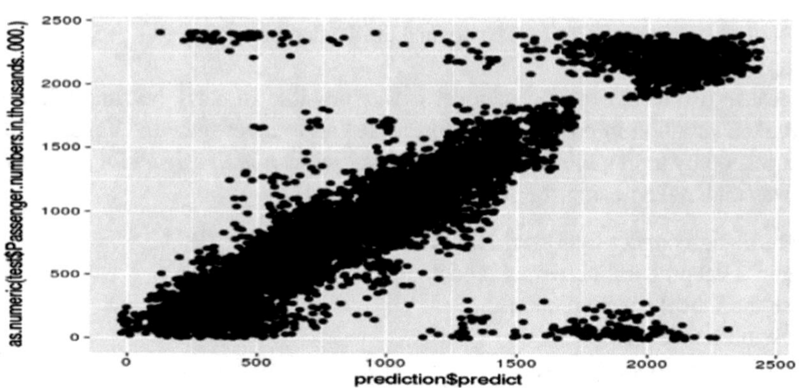

Fig. 2. Predicted values on the unseen test set against ground truth-values

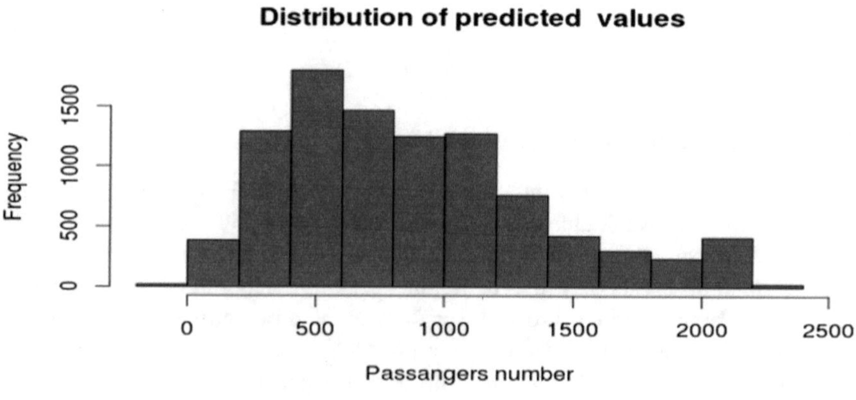

Fig. 3. The distribution of predicted values

A dropout of 25% of data from original train file was performed in order to have a 1-fold cross validation with 75% data for train and 25% data for evaluation to have some initial intuition about how the tuned deep learning model is perform by plotting predicted values on the unseen test set against ground truth values. This result is presented as the plot in Fig. 2 of the predicted distribution followed by plot Fig. 3 of the predicted values.

Figure 2 shows a scatter plot showing patterns in passenger flight list. It looks like there is enough variation in the data to make this a good predictor.

3.3 Fine Tuning Using Hyper-Parameter, Grid Search and Random Hyper-Parameter Search

It is possible to achieve less than 10% test set error rate in few seconds using some tuning. Hyper-parameter tuning is very important for Deep Learning, as it could influence model accuracy. The very first 10,000 rows of the training dataset will be trained. Hyper parameter optimisation can usually be accomplished better with random parameter search than with grid search.

Within the following model, Maxout, Tanh, and Rectifier, are utilized as activation operators. The Tanh function is a rescaled and shifted logistic function - with symmetry roughly zero it enables the training algorithm to converge more quickly. Rectifier has two benefits: it is quick and does not suffer through the vanishing gradient condition.

3.4 Extra Grid-Search to Optimise Parameters

When a sizable feed-forward neural network is trained on a tiny training set, it usually performs badly on the held out test data. This “over-fitting” is significantly decreased by randomly omitting one half of the characteristic detectors on each training situation.

This stops complex co-adaptations in which a function detector is just useful in the context of other certain element detectors. Rather, each neuron learns to identify a feature, which is usually of great help for creating the appropriate answer provided the combinatorial large variety of inner contexts in which it must operate.

Random grid searches a favorite option to locate the greatest parameters. Another grid search on the model was run since it scored perfectly, in the beginning, to find out if it will be better and overtake the NN.

3.5 Fine Tuning the Hyper-Parameters

Since the dataset is not considerably different in context from the initial dataset, which the pre-trained model is trained on, it must go for fine-tuning. As a result, preparing to apply a bit of topology, meaning dealing with neural network with 128 neurons in very first hidden layer, 63 neurons in next hidden layer as well as 32 neurons on third hidden level. Epochs (passes with the data) per iteration on N nodes are 100. Additional epochs were utilized for higher predictive accuracy, but just when in the position to afford the computational cost.

The training error value was based on the parameter ***training_frame = train***, which specifies the selection of randomly sampled training points to be utilized for scoring;

the default utilizes 10,000 points. The validation error is based on the parameter ***validation_frame = valid_frame***, which regulates the identical value on the validation set and it is set by default to become the whole validation set. Setting either of the parameters to zero instantly uses the whole corresponding dataset for scoring.

Here we have fed the data into our deep learning module ANN with predictors, target and train data as well as into the deep learning module with L1 and L2 for regularization. Deep Learning is based on a multi-layer feed forward artificial neural network that was trained with stochastic gradient descent using back-propagation. Below is the fine-tuning implementation, after performing hyper-parameter optimization using grid search the best neural network topology found is the following:

```
tuned_model <- H2O.deeplearning(
  x=predictors,
  y=target,
  training_frame=train,
  hidden=c(128,63,32),
  epochs=100,
  nfolds=5,
  fold_assignment="Modulo",
  l1=5.6e-05,
  l2=7.4e-05,
  input_dropout_ratio=0.05
)
```

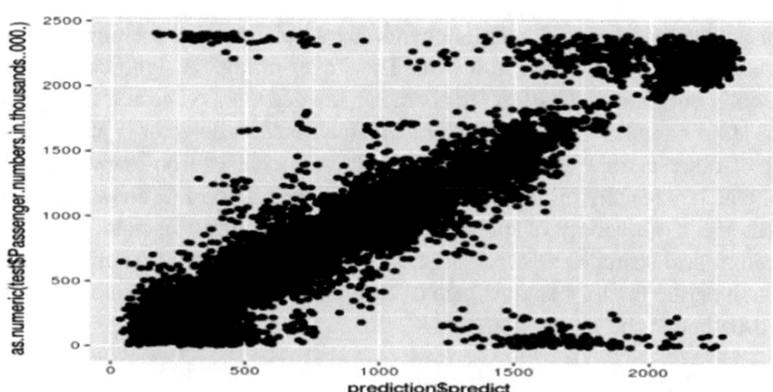

Fig. 4. A newly predicted values on the unseen test set against ground truth-values.

Changing input dropout ratio rate improved the training dataset to 0.8560758. Changing it definitely changes the results. Once again start predicting with the help of data, which already trained.

The plot in Fig. 4 is of newly generated predicted data with the help of gplot. The graph above on Y-axis shows the actual total values of passengers along with the predicted values from our predictor on x-axis. Some of the points near 2000 to 2500 on y-axis show there is a miss match on actual and prediction count. It can be confirmed from 0–500 from y-axis at last points. These are fewer as compared with the last plot, as the hidden layers increased in the ANN.

4 Improving Deep Neural Network Model Performance

A new ensemble strategy is suggested in this paper to enhance the overall performance of neural network classifiers. The suggested solution brings together a number of neural network classifiers, exactly where every classifier utilizes a unique distance function as well as likely another group of characteristics (feature engineered). These features engineered wish consuming a mix of grid the search engines (at the amount of this ensemble). Demonstrating that instead of optimizing the feature establish on their own for every distance metric, it's better to co adapt them, such that every feature set is enhanced within the context of this ensemble as entire.

Ensemble learning of neural network is a learning paradigm in which ensembles of a number of neural networks indicate increased generalization abilities, which outshine individuals of individual networks. For deep learning of multi-layer neural networks, ensemble learning remains relevant. Additionally, qualities of deep neural networks are able to offer possible chances to enhance the overall performance of conventional neural network ensembles. Within this paper, looking at this method having the best performing model and then trained the model. Furthermore, getting rid of the intense values has somewhat beneficial impact on the model effect on each training as well as validation datasets. A particular need to remember that even some enhancement within the metrics would result in a big leap of the model.

4.1 Split the Data According to Initial Train Test Split (1-Fold Cross Validation)

Half of the outlier labels were removed from the training set and considered unlabeled data (which treated as a contaminated normal class in the setup). At this point, no regularization has been applied for any method so that the results are more easily comparable. For feature transformation, trying to make use of a variety of established unsupervised outlier detection techniques: choosing arbitrary subsets of options, as well as k-NN outlier (compute sum of distances to k nearest neighbors). These functions are based upon a rough search for unsupervised algorithms, which function moderately effectively on the training set. It is a simple starting point along with options ought to be investigated.

4.2 Getting the Best Three Features

In the random forest approach, a large number of decision trees were created. The package "randomForest" has the function ***randomForest()*** which is used to create and analyze random forests, used for prediction. The below Fig. 5 shows that when use different features with ensemble functions, errors get reduced. So, using same technique to improve the ANN.

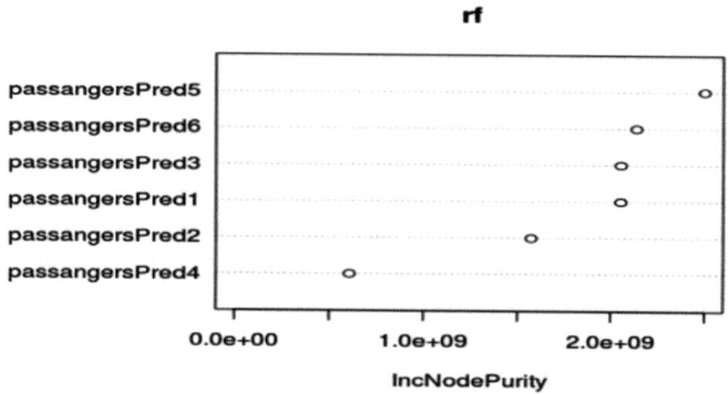

Fig. 5. Variable importance on generated features.

The outlier threshold was calculated by multiplying standard deviation (SD) passenger of tempTr with 1.5 and added to Passenger.numbers.in.thousands..000.of tempTr, after that take mean of that entire data for prediction of tr. If the prediction entry of tree not in between 0 and 1 that entry will be eliminated.

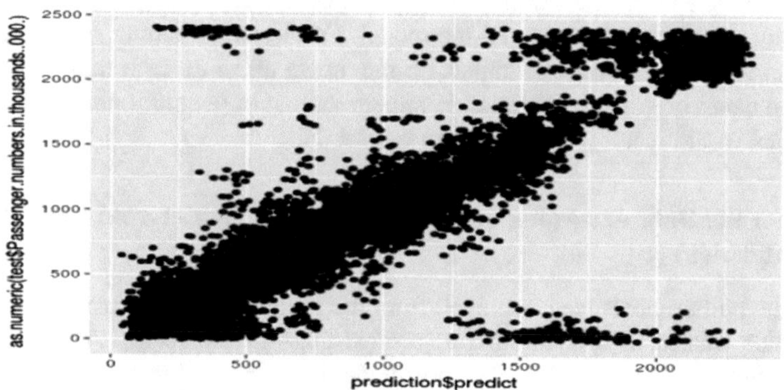

Fig. 6. A last predicted values on the unseen test set against ground truth-values.

Above plot in Fig. 6, is newly generated predicted data with the help of ***gplot***. The graph above on Y-axis shows the actual total values of passengers along with the predicted values from our predictor on x-axis. Some of the points near 2000 to 2500 on y-axis show there is a miss match on actual and prediction count. Similar can be confirmed from 0–500 from y-axis at last points. These are very less as compare to our last plot, as we have increased the hidden layers along with MSE i.e. mean square error methods in our ANN.

4.3 Features Obtained from Deep Learning Hidden Layers

The number of features extracted from the deep learning model was the same number of neurons in each hidden layer. For a better selection of only important non-linear features, the correlations were computed of each feature that was corresponding to each neuron in the hidden layer with the target variable (passenger number in thousands per month). Having computed the correlations, we then kept only one feature from each neuron, where is the maximum correlation compared with other features in the same hidden layer, resulting in final 4 non-linear features.

Figure 7 below shows that the features are very important for number of passenger's prediction from most important to least important. As it is observed from the plot

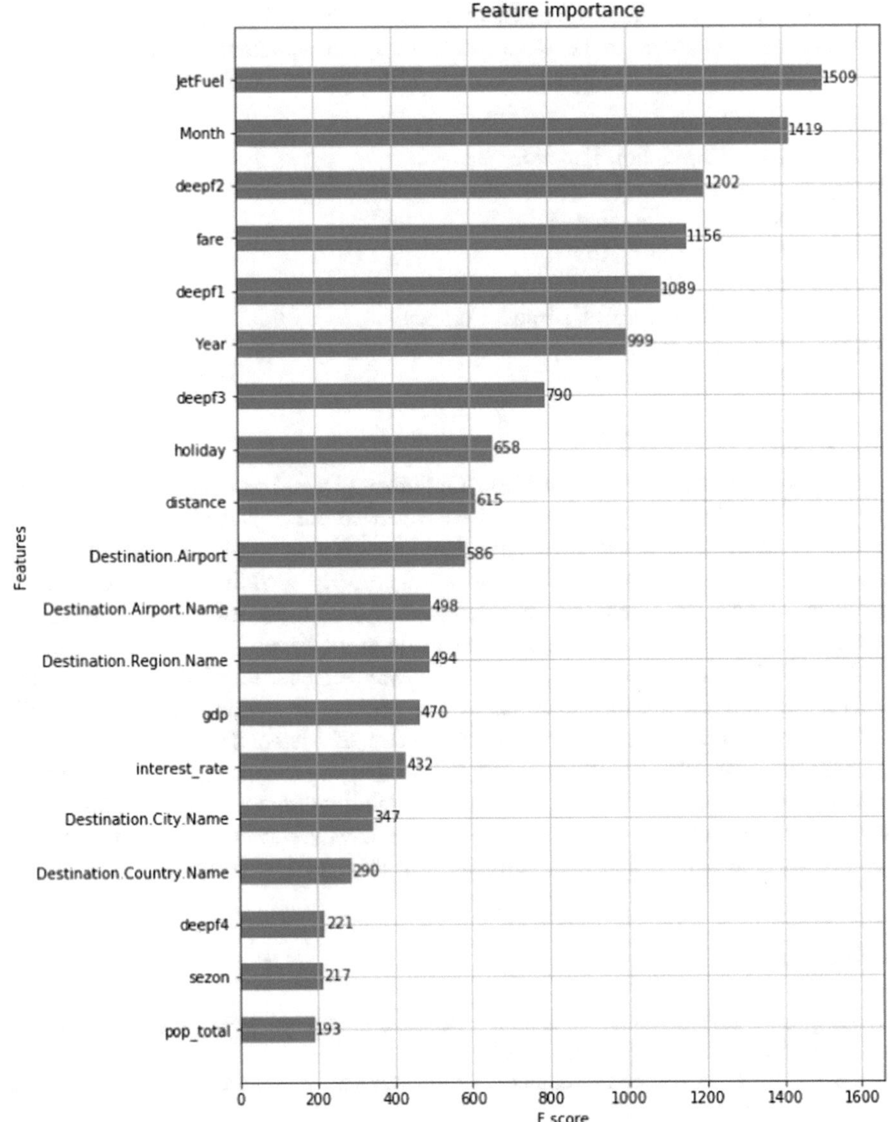

Fig. 7. Features importance obtained from deep learning hidden layers.

the nonlinear features are, deepf1, deepf2, deep3, and deepf4 obtained by the method described above are very influential and are the ones with highest influential impact captured by XGBOOST feature importance.

5 Conclusion

In this work a method of tuning deep learning models using h2o is proposed, where the trained network is built from samples of selected features from the dataset, in order to ensure diversity of the samples. The results show that the final predictions for the test set are better than predictions in the training data set, as the histogram of final predictions in test set seems to reassemble better histogram of the predictions in training set.

List of Abbreviations

ACF Auto Correlation Function
ANN Artificial Neural Network Model
RMSE Root Mean Squared Error
ReLU Rectified Linear Unit
PCA Principal Components Analysis
OMAC Oman Management Airport Company
H2O Library Packages
SD Standard Deviation

Declaration

I hereby declare that this paper entitled, "Optimizing Deep Learning Model for Neural Network Topology" is my own original work carried out as part of my PhD research except for the extent, which are from other researches and duly acknowledged. All sources used for this article have been fully and properly cited. It contains no material, which to a substantial extent has been accepted or used at any educational institution, except where the due acknowledgement is made in the paper.

I would like to address my special acknowledgements to all those people who provide me with data for my experiments. My warm appreciation is due to the Public Authority for Civil Aviation, Directorate General of Meteorology, and Ministry of Tourism in Oman. The data that support the findings of this study are available from these authorities but restrictions apply to the availability of these data, which were used under license for the current study, and data are however available from the authors upon reasonable request and with permission of these two authorities.

References

1. Bengio, Y.: Representation learning: a review and new perspectives. IEEE Trans. Pattern Anal. Mach. Intell. **35**(8), 1798–1828 (2013)
2. Hornik, K.: Approximation capabilities of multilayer feedforward networks. Neural Netw. **4**, 251–257 (1991)

3. Fukushima, K.: Neocognitron: a self-organizing neural network model for a mechanism of pattern recognition unaffected by shift in position. Biol. Cybern. **36**, 193–202 (1980)
4. Hochreiter, S.: Untersuchungen zu dynamischen neuronalen Netzen. Diploma, Technische Universität München (1991)
5. Bishop, C.M.: Neural Networks for Pattern Recognition, 1st ed. Oxford University Press, Oxford (1995)
6. De Boer, P.-T., Kroese, D.P., Mannor, S., Rubinstein, R.Y.: A tutorial on the cross-entropy method. Ann. Oper. Res. **134**, 19–67 (2005)
7. Glorot, X., Bengio, Y.: Deep sparse rectifier neural networks. In: International Conference on Artificial Intelligence and Statistics (2011)
8. Bastien, F., et al.: Theano: new features and speed improvements (2012). arXiv preprint arXiv:1211.5590
9. Russell, S., Norvig, P.: Artificial Intelligence: A Modern Approach, 3rd ed. Artificial Intelligence (1995)
10. Masters, T.: Practical Neural Network Recipes in C++, 1st edn. Morgan Kaufmann, San Mateo (1993)

Detecting Traces of Bullying in Twitter Posts Using Machine Learning

Caroline Jin[1(✉)], Harpreet Kaur[2], Amena Khatun[3], and Sitara Uppalapati[4]

[1] MIT Beaver Works Summer Institute, Chantilly, VA, USA
cj.jin@hotmail.com

[2] MIT Beaver Works Summer Institute, Bronx, NY, USA
Hkaur0745@bths.edu

[3] MIT Beaver Works Summer Institute, Queens, NY, USA
khatuna@bxscience.edu

[4] MIT Beaver Works Summer Institute, Herndon, VA, USA
sitarauppalapati@gmail.com

Abstract. There are over 100,000 bullying-related tweets sent through Twitter every week. These social media posts show the prevalence of bullying both online and in the physical world. Researchers find that victims of bullying are more likely to have symptoms of depression. Oftentimes, these tweets are unreported, leaving the victim to face mental consequences when kept in a bullying situation. Natural language processing (NLP) and machine learning, however, can efficiently detect bullying-related tweets and bring help to those in need. The purpose of this research was to explore various methods of identifying bullying related tweets posted by various groups, including perpetrators and victims, through NLP and machine learning. We produced word embeddings (word vectors) and used the term frequency-inverse document frequency (TF-IDF) algorithm. Then, we trained various machine learning models, including multinomial Naive Bayes classifier, convolutional neural networks (CNN) and recurrent neural networks (RNN), to improve the predictions for the labels of tweets and observe patterns among them. For binary classification (bullying traces vs. no bullying traces), our CNN model produced the highest area under receiver operating curve (AUROC) of 0.924. AUROC was a more comprehensive metric which accounted for our skewed dataset. For multiclass classification, the tweets were separated into six classes: NA (the tweet was not bullying-related and was not annotated), self-disclosure, report, accusation, denial, and cyberbullying. Our RNN model produced AUROC value for each class, ranging from .854 to .936, for multiclass classification. We aim to implement a script to Twitter to help identify users who are being impacted by bullying and may require mental health assistance. This research has important real world applications because as social media usage increases so do the number of malicious posts online.

Keywords: Social media · Bullying · Tweets · Machine learning · Neural networks

K. Arai et al. (Eds.): CompCom 2019, AISC 997, pp. 796–803, 2019.
https://doi.org/10.1007/978-3-030-22871-2_56

1 Introduction

Depression is a major issue among youth mainly due to bullying. High school and middle school students surveyed by the Cyberbullying Research Center report having higher rates of self-deprecating and suicidal thoughts after experiencing bullying (Hinduja and Patchin 2018). Bullying emerges from the negative emotions that rise from relationship problems including breakups, intolerance, and ganging up on one individual. These problems result in teens and adolescents experiencing anxiety, fear, depression, and low self-esteem in their everyday lives.

One key factor contributing to the expansion of bullying and thus depression is social media. As technology and social media infiltrate our lives, more than 1/3 of the youth become victims of bullying [1–3]. 71% of young people say that they are concerned about cyberbullying [1–3]. Furthermore, through research conducted by the University of Wisconsin in 2011, 250 million public tweets are sent daily during the study—a number almost 10 times the population of the state of Texas [7]. This problem must be tackled in order to prevent further cases of bullying.

Previously conducted research related to our topic gives us some insight on how to proceed. The research, "Using Machine Learning to Detect Cyberbullying," focused on Formspring.me, a question-and-answer website, and implemented a decision tree and instance based learner model to determine cyberbullying content [6]. Researchers have also worked on detecting bullying content in gaming chat rooms like the World of Tanks. Using SQL database queries and AI-based sentiment text analysis services, they found the prevalence of offensive language and racist sentiment [4].

Few research papers in machine learning have focused on bullying in both the real and virtual world. Furthermore, few research has considered the specifics of bullying, particularly on the different roles involved in a bullying situation. Researchers from the University of Wisconsin considered bullying in both worlds and these complex relationships while analyzing tweets containing bullying traces. They utilized a simple Naive Bayes classifier and a support vector machine (SVM) to classify whether a tweet was bullying-related and what kind of response the tweet was [7]. The research was mostly a broad overview of different NLP techniques and machine learning models that could be used in studying bullying. Our research focuses specifically on which models are more effective in examining various tweets and detecting traces of bullying among them.

2 Materials

We utilized a labeled public dataset from the University of Wisconsin. The dataset had 7,322 data points, each one containing the tweet ID, user ID, and labels including 'Bullying Traces?,' 'Type,' 'Form,' 'Teasing?,' 'Author Role,' and 'Emotion.' The labels we focused on in this research were 'Bullying Traces?' for our binary classification and 'Type' for our multiclass classification. 'Bullying Traces?' indicated whether or not a tweet contained content related to bullying and used 'n' or 'y' as its values. In the 'Type' column, tweets were labeled as either 'NA' (the tweet is not a bullying trace and was not annotated), 'self-disclosure,' 'report,' 'accusation,' 'denial,' or 'cyberbullying.'

We extracted the raw tweets using the provided tweet IDs. The first part was checking if the tweet existed; we did this by checking if the url (https://twitter.com/statuses/TweetID) for each tweet was valid. The second part in-volved removing tweets that were private using Twitter Developer Platform through tweepy. Through this process, 5,443 tweets were left in the first round, and 3,755 tweets remained in the second round. Figure 1 shows a sample of the dataset we obtained after the second round. Using the remaining tweets, we applied NLP techniques to extract specific features and trained various machine learning models.

	Unnamed: 0	Tweet ID	User ID	Bullying_Traces?	Type	Form	Teasing?	Author_Role	Emotion	Text
0	0	107688644067856384	185389094	1	1	1	1	1	1	@bellathorne143 i herd that you got bullied wh...
1	1	102206417217392640	226320672	0	0	0	0	0	0	Bullying: O gesto mais idiota, estpido e irrac...
2	2	102779484725448704	297557032	0	0	0	0	0	0	The Bully at School Goes High Tech Part 1: \n...
3	3	108676972149874688	157724561	0	0	0	0	0	0	Esse @Felipemath fazendo bullying comigo! Haha'
4	4	106590106873372672	62179998	0	0	0	0	0	0	AUISHUAHS eu e o @wallace_mancha tiramos o dia...
5	5	109034091743154176	177913822	0	0	0	0	0	0	@Looweihao @haoyangg @sleepybed Hello, since w...
6	6	102533497637437441	70412906	1	1	1	1	1	3	For those keeping score at home- cousin who ju...
7	7	103679008855691264	11363462	0	0	0	0	0	0	BETTER ANTI-BULLYING AD SLOGANS: Hey, Bullies....
8	8	105202476927549441	293593896	0	0	0	0	0	0	@luttylutz bahh lo kaga demen dia juga? Gue ki...
9	9	107278521805713408	150964152	0	0	0	0	0	0	cara3: bully iam gara2 uda di ucpin tpi gk bls'
10	10	107478741424414720	360086723	0	0	0	0	0	0	http://t.co/nbBZDxf Please help to stop workpl...
11	11	103471384973418497	219025244	0	0	0	0	0	0	@lucy_robbo yes on thur, she said she might, w...
12	12	106840582453870592	244569315	0	0	0	0	0	0	@bbuk @brianofficial bobby needs threw out tha...
13	13	101324602789208064	210803457	1	1	1	1	1	3	don't you get it? This is me getting rid of yo...
14	14	105799405784936448	264606687	0	0	0	0	0	0	No bullying se eu jogar um tijolo na sua cara...

Fig. 1. Sample of Filtered Bullying Dataset

3 Methods

With the tweets, we formatted the data into both binary and multiclass by changing the textual labels to numerical representations. Afterwards, we ran various natural language processing algorithms such as TF-IDF, stop words, and word-vectors. We used a multinomial Naive Bayes, recurrent neural network, and convolutional neural network to determine which model(s) would best identify the tweets that hint toward bullying.

3.1 Multinomial Naive Bayes with TF-IDF and Stop Words

TF-IDF was a simple NLP algorithm we used to process our text. This algorithm is similar to bag-of-words, which considers only the frequency of each word in the tweets. TF-IDF extends this idea by also considering the number of tweets that contains the word. In doing so, TF-IDF weighs words like 'the' and 'a' less heavily than words that better differentiate the bullying-related tweets from the non-bullying-related tweets.

We applied this algorithm to our data and inputted the processed data into a multinomial Naive Bayes classifier. We also considered removing these common words, also known as stop words, before using TF-IDF algorithm. We took out the stop words from tweets but kept words such as 'I' and 'you' as a point of view could also be useful for classification. We then fed the cleaned tweets into TF-IDF and a multinomial Naive Bayes classifier.

3.2 Word Embedding: Word2Vec

Word embeddings map each word to its corresponding vector. We used Word2Vec, which took raw text as input and learned a word by considering its surrounding context or predicted a word given its surrounding context using gradient descent with randomly initialized vectors. Word2Vec used different vectors for word embeddings depending on whether it was the word we conditioned on or the word we tried to predict. The probability we maximized then was:

$$P(V_{out}|V_{in}), \text{ where } V_{out} \text{ is the output word and } V_{in} \text{ the input.}$$

The interesting property of word vectors obtained from this equation was that they encoded not only syntactic but also semantic relationships between words as shown in Fig. 2. Not only were similar words close to each other in the vector space (as measured by some norm), but word analogies were reflected by the difference between word vectors. This property known as 'additive compositionality' referred to the linear structure in the vector space that allows analogical reasoning. Word vectors thus can be seen as representing the distribution of the context in which a word appears, and the sum of vectors roughly represents an AND concatenation.

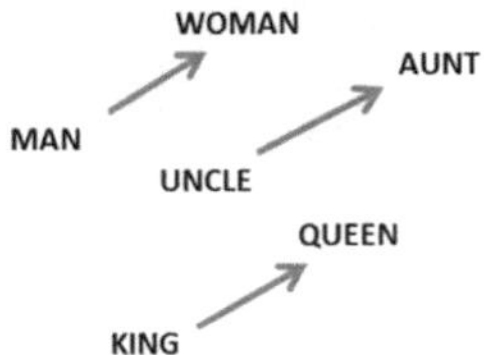

Fig. 2. Example vectors with corresponding words

For our research, we used Stanford's pretrained global vectors for word representation dictionary, which held 400,000 words with their corresponding vectors of length 100 [5]. Before converting the tweets into vectors, we made the words all lowercase as the dictionary contained only lowercase words. Further-more, we removed words that were not in the dictionary; these words were often long strings of repeated letters like 'hahahahaha,' references to other users like '@bellathorne,' or html links. After preprocessing, we converted the words in the tweets into word vectors using the Stanford dictionary. We stored each tweet into a vector of size 30 by 100 (30 being the max number of words per tweet and 100 being the length of each word vector). For tweets with less than 30 words, zero vectors were placed. The resulting vector for all of our data was of size 3755 by 30 by 100.

3.3 Neural Networks

We divided the Word2Vec data into training, validation, and test of 2403, 601, and 751 tweets respectively. Before applying our data into our models, we transformed our labels from strings to numerical values. For our binary class, we relabeled 'n' and 'y' as 0 and 1, respectively. The multiclass labels changed from 'NA,' 'self-disclosure,' 'report,' 'accusation,' 'denial,' and 'cyberbullying' to 1-D vectors of length 6 using one-hot encoding. Using this transformed data, we developed two different neural network models for each of our binary and multiclass classification.

For binary, we used a RNN model containing a long-short term memory (LSTM) layer of 100 perceptrons and a fully-connected layer. LSTM can improve the classification of sequential data like text as these units enable the model at arbitrary points of time to remember the order of the sequence. The second model was a CNN of four layers. CNN models are often used in image processing but also can be used in NLP. Our model contained 2-D convolutional kernels, 2-D max pooling kernel, a flattening layer, and a fully-connected layer. This model is shown in Fig. 3.

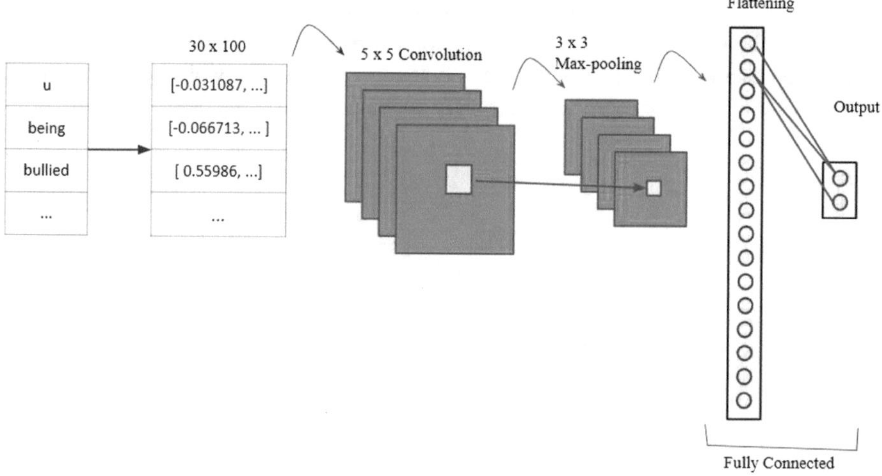

Fig. 3. CNN model used for binary classification

We used similar models for multiclass changing just the input and output shape to our desired dimensions. For each model, we trained for 40 epochs with a batch size of 64. After 40 epochs, the learning curve of the model followed an asymptotic behavior, so 40 epochs was a reasonable period to train our models.

4 Results

Since our task was identifying potential tweets with bullying traces, a classifier that would classify actual bullying tweets as non-bullying would not be optimal. Thus, we considered the precision rate through the confusion matrix and the area under the curve

(AUC) as our evaluation metrics. For our binary classification, our RNN performed slightly worse than the multinomial Naive Bayes and CNN based on a lower AUC of .772. Figure 4 shows similarities between binary classification of the multinomial Naive Bayes and CNN based on our confusion matrix. From top to bottom and left to right, the confusion matrix referred to the number of cases that were true positives, false positives, false negatives, and true negatives. In both cases, there were 606 tweets that were non-bullying traces and 147 tweets that were bullying traces. This disparity in the number of tweets contributed to the shade representing the true negative being much darker than the shade representing the true positive in both confusion matrices.

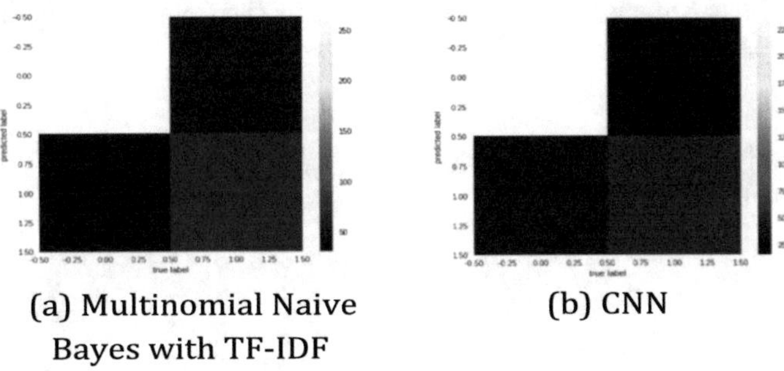

Fig. 4. Color map of confusion matrix for binary classification. Main diagonal from top to bottom refers to the number of true positives and true negatives. Lighter color means more data is being classified in that class

The multinomial Naive Bayes classifier tended to have a higher number of cases that were false positives rather than false negatives whereas the CNN classifier tended to have a higher number of false negatives rather than false positives. For our classification task, we wanted to capture all potential bullying-related tweets rather than miss tweets that were actually bullying-related. In this aspect, the CNN was the better choice. In Fig. 5, the receiver operator curve (ROC) curve showed that the Multiple Naive Bayes classifier had an AUC of .832 while the CNN had had an AUC .924. This metric further revealed our CNN model as the better classifier for binary classification.

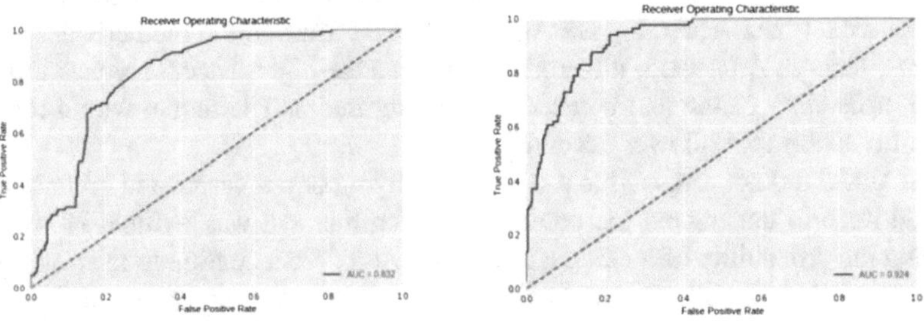

Fig. 5. ROC for different binary classifiers

For the multiclass, the bullying traces were separated into six classes: NA, self-disclosure, report, accusation, denial, and cyberbullying. Our dataset had a lack of tweets that indicated cyberbullying. Out of the 3755 tweets, only eight were labeled as 'cyberbul-lying' with four tweets in the training set, two in the validation set, and two in the test set. Due to the lack of data, neither multiclass models could classify the few cyberbullying cases. Thus, for our comparison, we focused on the remaining five classes.

Unlike our binary classification, our RNN classifier tended to perform better in the five classes as shown in Fig. 6. The AUROC for RNN was between .854 and .936 for all the classes whereas the AUROC for the CNN was between was .806 and .871. Each class for the RNN had a higher AUROC than that for the CNN, showing that our RNN model was a stronger classifier for this task.

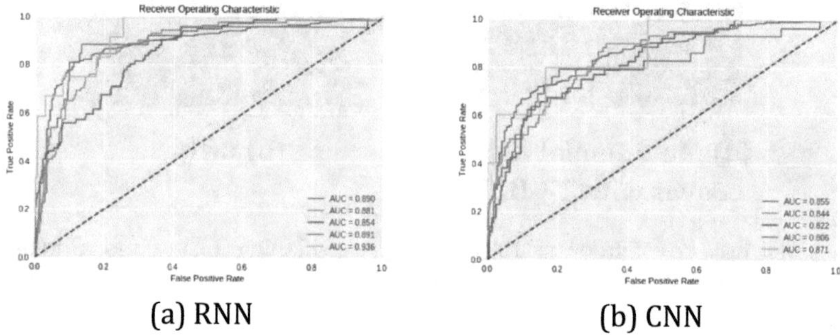

Fig. 6. ROC curve for each of the multiclass classifiers. The key from top to bottom, the color corresponds to NA, self-disclosure, report, accusation, and denial respectively. *Note: cyberbullying ROC curve was not considered due to lack of data.*

5 Conclusion

To demonstrate how our models worked, we created a dummy twitter account Medlytics_5Test. The account was followed by us. We then posted some text that we thought had traces of bullying and some that did not. Every post was extracted by our model using tweepy. The model would get a timeline of the user's posts and pick out the most recent post. Then, the text was formatted based on the methods described in *Meth-ods* section. Afterward, the models that gave the highest accuracy was utilized to get the probability of the text containing bullying traces. The model would report the probability using the follower account.

The model did not predict that a post was bullying or not because in order to do that it would have to understand the context in which the post was written. However, by reporting the probability of it being a bullying post it drew attention to the post so other users could read it and determine if it counted as bullying or not and took the appropriate action.

6 Discussion

The problems faced while attempting to address cyberbullying on Twitter utilizing machine learning were multifaceted. Because the labeled data was from 2012, many of the posts were either deleted or made private. In the end, the data frame had about 4,000 data points which was a small sample size considering the number of interactions that took place on Twitter. Additionally, the Stanford dictionary was limited to common words and included only text emoticons, such as ':)' or ':p.' Many of the tweets had keyboard emojis and usernames that were hard to vectorize. The data set had few cases involving cyberbullying and was also biased because the labeling was very subjective to the researchers who had preprocessed the data. Lastly, there were contextual problems because it was difficult to identify a tweet as bullying without understanding the context of the situation.

The Stanford dictionary contained words from multiple languages including Spanish, meaning our model was able to detect bullying present in tweets of different parts of the world. But as mentioned before, the contextual problem was exacerbated when addressing different languages.

To take this project to the next level, more tweets needed to be extracted from Twitter, analyzed, and labeled as containing traces of bullying or not. A dictionary with the ability to turn abbreviations, emojis, and slang words to vectors, a way to account for spelling mistakes and contextualizing our data, and greater computing power would improve the functionality of the model. Lastly, adapting the algorithm to Twitter would allow us a chance to see how the algorithm can help with bullying.

References

1. Cyberbullying statistics: (n.d.). Retrieved from https://enough.org/statscyberbullying
2. Hinduja, S., Patchin, J.W.: Connecting adolescent suicide to the severity of bullying and cyberbullying. J. Sch. Violence (2018). https://doi.org/10.1080/15388220.2018.1492417
3. Hodges, E.V.E., Perry, D.G.: Personal and interpersonal antecedents and consequences of victimization by peers. J. Pers. Soc. Psychol. **76**(4), 677âĂß685 (1999). https://doi.org/10.1037//0022-3514.76.4.677
4. Murnion, S., Buchanan, W.J., Smales, A., Russell, G.: Machine learning and semantic analysis of in-game chat for cyberbullying. Comput. Secur. **76**, 197–213. http://www.sciencedirect.com/science/article/pii/S0167404818301597, https://doi.org/10.1016/j.cose.2018.02.016
5. Pennington, J., Socher, R., Manning, C.: Glove: global vectors for word representation. In: Proceedings of the 2014 Conference on Empirical Methods in Natural Language Processing (emnlp), pp. 1532–1543 (2014)
6. Reynolds, K., Kontostathis, A., Edwards, L.: Using machine learning to detect cyberbullying. In: 2011 10th International Conference on Machine Learning and Applications and Workshops, vol. 2, pp. 241–244. https://doi.org/10.1109/ICMLA.2011.152. December 2011
7. Xu, J.-M., Jun, K.-S., Zhu, X., Bellmore, A.: Learning from bullying traces in social media. In: Proceedings of the 2012 Conference of the North American Chapter of the Association for Computational Linguistics: Human Language Technologies, pp. 656–666. Association for Computational Linguistics, Stroudsburg, PA, USA (2012). http://dl.acm.org/citation.cfm?id=2382029.2382139

Credit Risk Analysis Applying Machine Learning Classification Models

Roy Melendez(✉)

Pontificia Universidad Católica del Perú, Lima 32, Lima, Peru
roy.melendez@pucp.edu.pe

Abstract. The credit risk analysis is a major problem for financial institutions, credit risk models are developed to classify applicants as accepted or rejected with respect to the characteristics of the applicants such as age, current account and amount of credit. In the present investigation, we will apply four classification models to evaluate their performance and compare it with other previous investigations.

Keywords: SVM · Neural networks · Logistic regression · Decision trees · German credit risk · Classification models · Machine learning

1 Introduction

Credit risk assessment decisions are a key determinant of the success of financial institutions due to the heavy losses associated with the wrong decisions. For this reason, credit score is a method to measure the potential risk of credit applications. In general, it uses historical data to produce a score that financial institutions can use to assess credit applicants in terms of risk. This model of credit score is developed to classify applicants as accepted or rejected with respect to the characteristics of the applicants such as age, credit amount or level of income to name a few.

Therefore, to examine credit applications, new techniques must be developed to help predict credit more accurately. The benefits of the credit rating include the reduction of the cost of the credit analysis, which allows faster credit decisions, closer monitoring of existing accounts and the prioritization of credit collections.

Various machine learning techniques have been adopted to build credit rating models. Most credit rating models have been extensively developed to improve their accuracy in recent years.

In the credit risk analysis, the objective is to apply models that can extract knowledge from the applicant's profile and apply it to assess the credit risk of people with a much broader scope.

The objective of the present research will be to apply automatic learning classification models on credit risk analysis, evaluating their performance and comparing the results with the studies already carried out on the articles used.

The machine learning models adopted for our evaluation are Logistic Regression, Support Vector Machine, Neural Networks and Decision Trees, capable of generating

K. Arai et al. (Eds.): CompCom 2019, AISC 997, pp. 804–814, 2019.
https://doi.org/10.1007/978-3-030-22871-2_57

simple results that can be easily interpreted, but this usually requires expert domain knowledge and an in-depth understanding of the data.

This article is organized as follows: Sect. 2 describes the State of the Art with a summary of the main articles related to the purpose of the research. Section 3, Design of the Experiment, details the set of data used and the methodology to be followed. Section 4, Experimentation and Results, shows the results of the experiment compared to the results of the baseline. Section 5, Discussion, interprets the results. Section 6, Conclusions and Future Works and finally References.

2 State of the Art

Below, we summarize some scientific articles that have contributed to this problem for classification algorithms.

One of the reference reports is that of Baesens [13], where it makes a comparison of all existing classification methods to date, using different data sets, including the German Credit data set. This report concludes that the algorithms of SVM, neural networks and logistic regression are very good options to use in the evaluation of credit risk.

In another comparative research, an empirical analysis of the results extracted from the KNN algorithms (k-nearest neighbors), SVM and Neural Networks [2] of Feng-Chia is performed. Using the non-parametric Wilcoxon test, to find the best classifier of the German and Australian data set. With an error rate of 5% it is concluded that the best algorithm for the German data set is SVM and for the Australian it is KNN.

In a similar analysis, two classification models were applied: vector support machine (SVM) and a neural network [3] of Zhang. The cross-validation procedure was performed. The result shows that the SVMs surpassed the neural network model; And the 10-fold cross-validation and exit procedures obtained comparable results. Where the algorithms were able to classify up to 79.25% of cases.

Further research deepens the use of neural networks. For example, in [1] of Al-shayea, they constructed a supervised learning neural network using forward propagation and backward propagation of 2 layers with 24 inputs, 20 hidden sigmoid neurons and continuous output neurons. For the algorithm to learn, 765 samples were used and the rest (235 samples) for the tests. Where the network was able to classify up to 80.4% of cases in the training set.

One of the important aspects of credit risk analysis is not only knowing how to predict whether the credit is risky or not, it is also interesting to know what criteria have been used to compare them. In [12] de Baesens, explain that the fundamental problem of the neural network is its behavior of black box, being important for financial institutions, to know what are the rules that were used when deciding whether the client is viable or not for credit, and thus comply with the standards of their respective countries. To achieve this, they used three techniques for extracting rules from neural networks: Neurorule, Trepan and Nefclass. In which it is concluded that Neurorule and Trepan give a good accuracy when classifying.

Finally, research was carried out on how to combine several classification models by means of assembly models, this is analyzed in Baesens [14]. Here they update the research done 10 years ago by adding several assembly models to their evaluation.

3 Experiment Design

The data source to be used is from the UCI Machine Learning Repository. The original data set called German Credit Data were extracted from those published by Professor Hoffmann, which contains 1000 instances (records) with 20 characteristics.

Next, we will show the structure of the most relevant data.

- Age (Numeric)
- Sex (Text: Male, Female)
- Work (Numeric: 0 - not trained and non-resident, 1 - not trained and resident, 2 - trained, 3 - highly trained.)
- House (Text: Owner, Rent, Free)
- Saving Accounts: (Text: moderate, quasi-rich, rich)
- Current Accounts: (Numeric, DM German frame)
- Amount of Credit: (Numeric, DM German frame)
- Duration (Numeric: months)
- Purpose (Text: car, furniture/equipment, radio/TC, domestic use, repair, education, business, holidays/others.

The data set has 700 instances with low risk credits and 300 with high risk credits. Given that we have 70% of instances with low risk credits, we can say that this data set is unbalanced.

3.1 Base Line

The strategies to be used will be aligned with the research carried out by Baesens [13].

3.2 Strategies

In order to use the data set in some of the classification models such as SVM, it is necessary to normalize them.

We will reserve 2/3 of the data as a training set and the remaining third as a set of tests. We will not perform balancing techniques such as oversampling or undersampling to align ourselves to the baseline.

For the selection of the best model and the choice of hyper parameters in each classification model, we will use 10-fold CV cross-validation.

For the performance measurement we will use the confusion matrix and the ROC curve under a threshold of 0.5. As a metric to make the comparisons with the results of the baseline we will use the accuracy, the exhaustivity, the specificity and the area under the curve (AUC).

3.3 Classification Models Chosen

For the following experiment we chose four classification models.

3.3.1 Support Vector Machine

We will use linear logistic regression models with a Gaussian kernel. The hyper parameters to be optimized will be C and gamma.

3.3.2 Logistic Regression

We will use the descendant gradient algorithm to find the model, in addition we will evaluate the possibility of using regularization. The hyper parameter to optimize will be C.

3.3.3 Decision Trees

To align ourselves to the baseline, two experiments were performed with the data set using normalization and discretization using the one-hot-encoding technique. The hyper parameters to be optimized will be the maximum depth and the minimum number of examples per branch. Although in the baseline they use algorithm C4.5, we will use the optimized algorithm CART.

3.3.4 Neural Networks

The initial strategy will be to use a neural network with backward propagation. The hyper parameters to be optimized will be Alpha and the type of activation. We will use a single hidden layer to align with the baseline.

3.4 Tools

We will carry out the experiments using Anaconda as a development environment and Python 3.6 as a programming language. We will use scikit-learn as a library to obtain the classification algorithms.

4 Experimentation and Results

Following the strategies defined in the design of the experiment, we will show the results obtained for each of the chosen classification models.

4.1 Choice of Hyper Parameters

To avoid a high computational cost, we have established maximum and minimum values for each hyper parameter, the range was established based on the experience and the previous experiments. Using 10-fold CV cross validation, we have obtained the best hyper parameters for each model in Table 1.

Table 1. Best hyper parameters

	Hyper parameters
SVM linear	C = 3.1623, gamma = 0.0001
SVM Gaussian	C = 3.1623, gamma = 0.10
Logistic regression	C = 3.1623
Continuous trees	max_depth = 6, min_samples_leaf = 19
Discrete trees	max_depth = 6, min_samples_leaf = 13
Neural networks	alpha = 1.053, activation = logistic

4.2 Performance

Next, the results obtained from our baseline for the experiments [13]. The experiments were performed under a threshold of 0.5. The best scores of each category are shown in bold (Table 2).

Table 2. Base line results

	Exactitud (PCC)	*Exhaustividad (Sens)*	*Especificidad (Spec)*	*AUC*
SVM linear	71.0	**95.6**	17.1	76.6
SVM Gaussian	74.0	92.6	33.3	77.2
Logistic regression	**74.6**	89.5	41.9	77.7
Continuous trees	72.2	88.2	37.1	74.7
Discrete trees	**74.6**	87.3	46.7	74.6
Neural networks	73.7	85.2	**48.6**	**78.7**

In Table 3, we will show the results obtained from the experiment. In bold you will find the best scores of each category.

Table 3. Results of the experiment

	Exactitud (PCC)	*Exhaustividad (Sens)*	*Especificidad (Spec)*	*AUC*
SVM linear	75.8	85.9	51.0	76.5
SVM Gaussian	73.3	**86.8**	40.6	**77.9**
Logistic regression	75.8	85.0	53.1	76.9
Continuous trees	72.7	85.5	41.7	72.1
Discrete trees	70.0	**86.8**	29.2	67.0
Neural networks	**76.7**	85.5	**55.2**	77.2

The difference of our experiment with the baseline of research, in most cases the difference does not exceed 5% and in many cases exceeding the results of the baseline as can be seen in Table 4. More differences were observed in the comparison of specificity.

Table 4. Comparative table (in percentage)

	Exactitud (PCC)	*Exhaustividad (Sens)*	*Especificidad (Spec)*	*AUC*
SVM linear	−4.8	9.7	−33.9	0.1
SVM Gaussian	0.7	5.8	−7.3	−0.7
Logistic regression	−1.2	4.5	−11.2	0.8
Continuous trees	−0.5	2.7	−4.6	2.6
Discrete trees	4.6	0.5	17.5	7.6
Neural networks	−3.0	−0.3	−6.6	1.5

The difference in accuracy between the training data data set and the test data set can be seen in Table 5. In all cases the accuracy in the test set is less than in the training set.

Table 5. Accuracy of the training and test set

	Entrenamiento	*Prueba*
SVM linear	78.21	75.76
SVM Gaussian	80.15	73.33
Logistic regression	77.76	75.76
Continuous trees	78.06	72.73
Discrete trees	78.36	70.00
Neural networks	77.46	76.67

5 Discussion

As part of this paper, based on Tables 1 to 5, we will discuss some topics that will serve as an interpretation of the experiments carried out and opportunities for improvement in the cases raised.

5.1 Empirical Analysis

5.1.1 Precision

For our validation set of 330 instances (33%) we have the following results. Using the linear SVM model, the financial institution will grant credit to 47 risk clients and will have an accuracy of 81.04% of not making a mistake. While using the Gaussian SVM algorithm the financial institution will grant credit to 57 risk clients and will have an accuracy of 78.1% of not making a mistake. Using the Regression Logistics algorithm, the financial institution will grant credit to 45 risk clients and will have an accuracy of 81.6% of not making a mistake. Using the Trees of Decision algorithm continues, the financial institution will grant credit to 56 risk clients and will have an accuracy of 78.1% of not making a mistake. While using the Discrete Decision Trees algorithm the financial institution will grant credit to 68 risk clients and will have an accuracy of 74.9% of not making a mistake. Finally, using the Neural Networks algorithm, the financial institution will grant credit to 43 risk clients and will have an accuracy of 82.3% in not making a mistake. The latter being the one with the highest percentage of precision in not giving credit to risk clients.

5.1.2 Overfitting

Some symptoms of overfitting are: having very large C parameters in the models that use it and having a significant difference in accuracy between the training set and the test set.

According to Table 1, the values of C are not greater than 3.16, so it would be an acceptable value and not an important risk.

In general, the accuracies between the data sets of training and tests are not very different (Table 5), except in the case of trees with discrete information and the SVM with Gaussian kernel, where they present variations of up to 8%, here we could have certain overfitting risk.

5.1.3 Matrix of Confusion

For our experiment we have that most models have a completeness index greater than 85% in all cases (see Table 3), while in the specificity we have no greater than 55% while in Table 2 of the baseline we have percentages of up to 17%. In both experiments, the percentage of specificity differs considerably among the different classification models.

5.1.4 ROC Curve

In Figs. 1, 2, 3, 4, 5, and 6 we see the ROC curve for each of the models used. It is important to note that all graphs have a similar shape, with the exception of decision trees. The ideal would have been to approach the point (0, 1) so we can say that the results are not the most optimal, but this is a behavior that is repeated in the baseline experiment.

5.2 Best Models

There was no model that was clearly superior, in the baseline (Table 2) the neural networks were better in specificity and AUC, while the logistic regression and the discrete trees were superior in accuracy.

In our experiment (Table 3), neural networks obtained the best results in accuracy and specificity. SVM Gaussiano was superior in AUC and completeness.

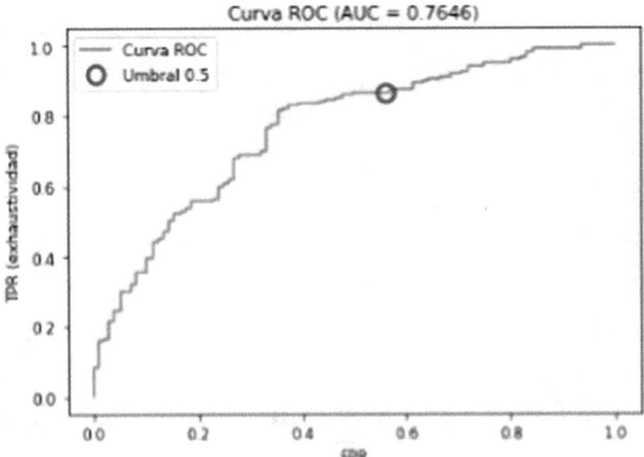

Fig. 1. ROC curve for Linear SVM

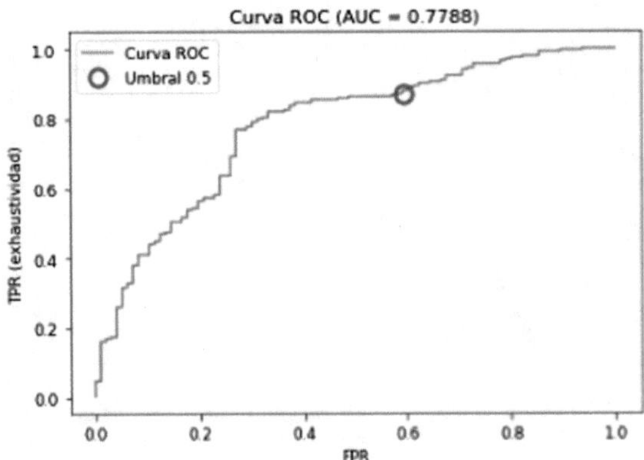

Fig. 2. ROC curve for SVM with Gaussian Kernel

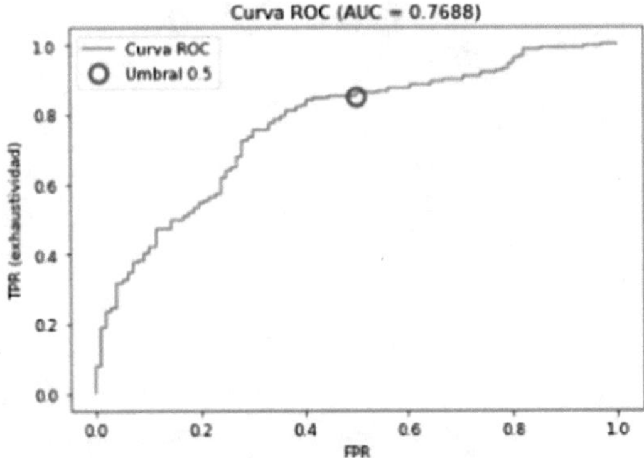

Fig. 3. ROC curve for logistic regression

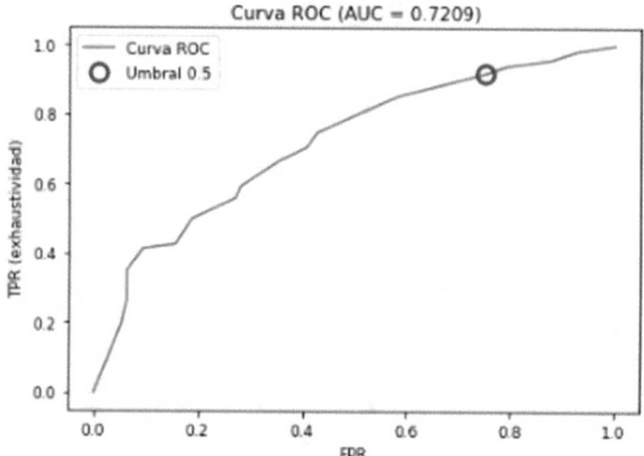

Fig. 4. ROC curve for decision tree

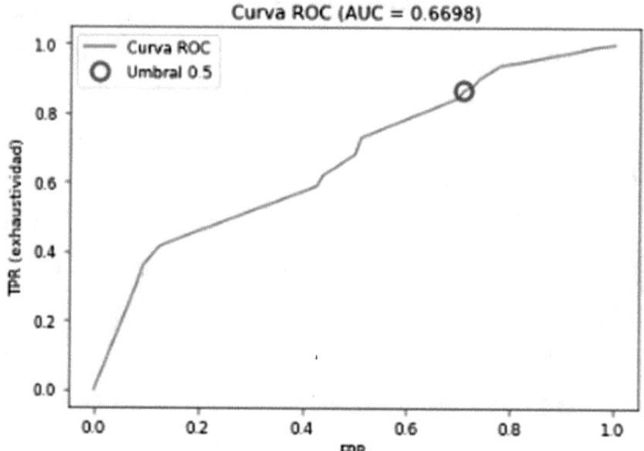

Fig. 5. ROC curve for discrete decision tree

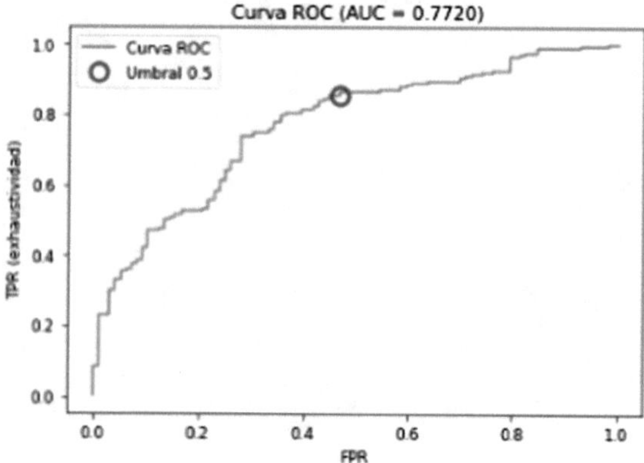

Fig. 6. ROC curve for neural networks

5.3 Improvement Opportunities

5.3.1 Umbral

During the experiment we used a threshold of 0.5 following the baseline, making a credit qualify as positive when the probability is greater than 0.5. However, it would be good to analyze what is most important for a financial institution, if the possibility that good credits are rated as bad or the possibility that bad credits are rated as good. This analysis must be carried out by each financial institution according to the credits it offers, the cost that each error implies and the level of accuracy that it wishes to reach.

5.3.2 Possible Noise Between the Data

After analyzing the confusion matrix and the ROC curve, it is important to note that none of the classification models stood out in all the evaluation criteria, this may be due to some noise in the credit risk data [15]. An improvement strategy for future experiments could be the elimination of certain characteristics that do not contribute to correct learning.

6 Conclusions and Future Work

We review that credit risk analysis with classification models can be an important competitive advantage for financial institutions, taking advantage of the expert judgment of existing credit analysts to support the creation of the data set and the techniques that learning provides us automatic.

The main objective of this work was to perform the experimentation of classification algorithms applied to credit risk assessment using logistic regression, SVM, neural networks and decision trees, having as base line an experiment already carried out [13]. We chose the exhaustivity and area under the curve (AUC) as main measures.

In general there was no classifier that was a clear winner in all evaluation categories, but the Neural Networks that were superior in more than one category for both our experimentation and the baseline were highlighted, on the other hand, the algorithm of Gaussian SVM also showed good performance.

It is a challenge to improve the performance of decision trees that are very useful, since they give us a descriptive level that other models do not provide us.

Evaluate the computational cost in subsequent investigations, since models such as SVM or decision trees could be very expensive if no restrictions are established on the hyper parameters.

Evaluate the cost of errors in future experiments, since it could give us an approximation of the economic impact that these classification models would have on a financial institution.

Finally, it is proposed to deepen the research on topics such as the balancing of data, reduction of dimensionality, choice of best threshold and assembly models in order to achieve better results. Perform the experiment with other classification models and different data sets.

References

1. Al-Shayea, Q., El-Refae, G.: Evaluation credit risk using artificial neural network. In: Global Engineers & Technologists Review, September 2011, pp. 21–28 (2011)
2. Feng-Chia, L.: Comparison of the primitive classifiers without features selection in credit scoring. In: International Conference on Management and Service Science, September 2009
3. Zhang, L., Hui, X., Wang, L.: Application of adaptive support vector machines method in credit scoring. In: International Conference on Management Science and Engineering, September 2009, pp. 1410–1415 (2009)
4. Feng-Chia, L., Wang, P., Wang, G.: Comparison of the primitive classifiers with extreme learning machine in credit scoring. In: IEEE International Conference on Industrial Engineering and Engineering Management, December 2009, pp. 685–688 (2009)
5. Beninel, F., Bouaguel, W., Belmufti, G.: Transfer learning using logistic regression in credit scoring. In: Université de Tunis, December 2012
6. Sudhamathy, G.: Credit risk analysis and prediction modelling of bank loans using R. Int. J. Eng. Technol., October 2016, **8**, pp. 1954–1966 (2016)
7. Khandaniy, A., Kim, A., Lo, A.: Consumer credit risk models via machine-learning algorithms. Massachusetts Institute of Technology, March 2010
8. Sitt, M., Wu, T.: Evaluation of credit risk. Stanford University
9. Charpignon, M., Horel, E., Tixier, F.: Prediction of consumer credit risk. Stanford University
10. Pacelli, V., Azzollini, M.: An artificial neural network approach for credit risk management. J. Intell. Learn. Syst. Appl., May 2011, **3**, 103–112 (2011)
11. Zhuravlev, M., Wang, C.: An analysis of profit and customer satisfaction in consumer finance. College of New Jersey (2009)
12. Baesens, B., Setiono, R., Mues, C., Vanthienen, J.: Using neural network rule extraction and decision tables for credit-risk evaluation. J. Manag. Sci., March 2003, **49**, 312–329 (2003)
13. Baesens, B., Van Gestel, T., Viaene, S., Stepanova, M., Suykens, J., Vanthienen, J.: Benchmarking state-of-the-art classification algorithms for credit scoring. J. Oper. Res. Soc., June 2003, **54**, 627–635 (2003)
14. Lessmann, S., Baesens, B., Seow, H., Thomas, L.: Benchmarking state-of-the-art classification algorithms for credit scoring: an update of research. Eur. J. Oper. Res. **247**, 124–136 (May 2015)
15. Baesens, B.: Developing intelligent systems for credit scoring using machine learning techniques. Katholieke Universiteit Leuven (2003)

Aligning Ground Truth Text with OCR Degraded Text

Jorge Ramón Fonseca Cacho(✉) and Kazem Taghva

Department of Computer Science, University of Nevada, Las Vegas, USA
{Jorge.FonsecaCacho,Kazem.Taghva}@unlv.edu

Abstract. Testable data sets are a valuable commodity for testing and comparing different algorithms and experiments. An important aspect of testing OCR error correction and post processing is the ability to match an OCR generated word with the original word regardless of errors generated in the conversion process. This requires string alignment of both text files, which is often lacking in many of the available data sets. In this paper, we cover relevant research and background on the String Alignment Problem and propose an alignment algorithm that when tested with the TREC-5 data set achieves an initial alignment accuracy average of 98.547% without zoning problems and 81.07% with. We then address the Zoning Problem, and other issues, and propose solutions that could increase both accuracy and overall performance up to an alignment accuracy of 98% in both scenarios.

Keywords: String alignment · OCR · Alignment · Ground truth generation · String matching · OCR post processing · Multiple sequence alignment problem

1 Introduction

While researching new techniques to understand and improve different parts of the OCR (Optical Character Recognition) post processing work flow, specifically correcting errors generated during the conversion of scanned documents into text, we required test data that would include both the original text and the converted copy with any OCR generated errors (which we will refer from now on as OCR'd text) that were generated in the process. This is usually referred to as the degraded text. Furthermore, in order to test the accuracy of our OCR text correction techniques we needed to be able to compare a word with an error to the equivalent 'correct' word found in the ground truth. This required having the correct text aligned with the degraded text for all words that were present. As we looked for public data we could use we found that most data sets that had both the correct and degraded texts did not have both files aligned. This forced us to seek a possible alignment program or algorithm we could use, or implement, to assists us in aligning the text before we could continue our research with error correction.

K. Arai et al. (Eds.): CompCom 2019, AISC 997, pp. 815–833, 2019.
https://doi.org/10.1007/978-3-030-22871-2_58

2 Ongoing Research

Historically, the Multiple Sequence Alignment Problem for n strings is an NP-Complete problem [1]. However, for only two strings, which is a common problem in bioinformatics, the dynamic programming algorithm given by Needlman and Wunsch [2] has a time complexity of $\mathrm{O}(\mathrm{k}^2)$ where k is the length of the string after the alignment and is bound by the sum of the lengths of the original strings. This time complexity can be misleading as the length of the string does matter when generating the pathways that must be evaluated. In bioinformatics these, can be long proteins that are hundreds of characters long. Therefore, the real time complexity is $\mathrm{O}((2\mathrm{k})^\mathrm{n})$ [3].

This should not discourage us from finding a better solution as, unlike molecular biology where each string given could be aligned to any location of the entire reference genome, for OCR the majority of the text should be in a similar order to the correct text aside from some insertions, deletions, or replacements that cause only minor shifting in the overall alignment. The only instances when it could be completely out of order is if the OCR'd text has multiple columns that were read wrong or an image caption in the middle of a page appears in a different location in the original file compared to the OCR'd text. These however are special cases that are not common, but exist and pose a real problem, and the initial generation of the OCR'd text will have dealt with the majority of these issues via OCR Zoning [4]. Using all this knowledge we can try and improve efficiency as we develop our own algorithm since we no longer need to check the entire string but can use information of the previously aligned word.

The old standard for aligning text was having human typists enter in the text by hand as it was done in the Annual Tests of OCR Accuracy [5] and the UNLV-ISRI document collection [6] where in both cases the words were matched manually with the correct word to create the ground truth. This is very laborious, expensive, and time consuming and not ideal for expanding to larger data sets, or automating the process.

A lot of the existing packages such as Google's Tesseract [7] are one-in-all packages that will align back to the original converted image that was used to generate the file; however this does not help for aligning to the ground truth because when the character recognition happens, it keeps a pointer to the location or cluster where that character came from in an image. This is not what we want, but could nevertheless be useful to add on top of our alignment to increase accuracy.

Matching documents with the ground truth using shape training and geometric transformations between the image of the document and the document description [8] has limited uses as it still requires a description of where the words appear [9] and, like Tesseract, involves matching the images with the text which is more aimed at issues with columns versus finding feature points whereas we are interested in aligning OCR'd text that was generated from the images and the original text.

Another solution mentioned in [9] involves using clustering for Context analysis by taking training data that is then fed to several modules that are known as

JointAssign (select triplets of common letters that are matched as part of existing words in training data), UniqueMatch (unlabeled clusters are given assignments to try and match them), MostMatch (continues guesswork to try and assign highest probability letters), VerifyAssign (tries to improve guess by assigning other letters to improve score) with the goal of matching a word in a corpus with the correct version [10]. This is a very effective matching algorithm that can produce the correct versions of words (which is ultimately what we want) but does not know where in the corpus it aligned to if the word appears more than once. So while this is an idea whose concept we use in our implementation for matching out of order text along with the basic idea of maintaining a score we are also interesting in taking advantage of the location where we are aligning to decrease having to search in an entire corpus; however should all else fail in aligning a word, using this method would be an effective backup, but the tidy data [11] preparation for it would be expensive in terms of time complexity. So for now we leave it to be implemented in the future once we can evaluate the cost-benefit since it is mentioned that this algorithm has trouble coping with digits, special symbols, and punctuation - something that could be solved if we did a soft pass with an OCR correction software like our past OCRSpell [12–14]; however doing this has the risk of biasing our data unless there is a 100% effectiveness in combination with our algorithm.

More recently, research on this area has focused more on creating ground truth for image oriented post processing such as in the case of camera-captured text [15]. The reason behind this shift is due to the ongoing demand of image recognition from different uses; however, the important aspect to take from this is the idea of understanding words as sets of letters and using that to help locate the most likely candidate for that alignment. Ultimately they are still using Levenshtein distance [16] as is our algorithm. So while the focus of research may have shifted to this area. The core alignment problem is still very relevant.

3 The TREC-5 Data Set

When we were searching for a data set we could use for our ongoing OCR Post Processing research, including the alignment part, we needed to have matching original and OCR'd versions of the text that was both a large corpus, but could be split into smaller parts we could test and compare. The U.S. Department of Commerce's National Institute of Standards and Technology (NIST) Text REtrieval Conference (TREC) TREC-5 Confusion Track [17] was just what we needed. The following is an excerpt from the README file contained in the TREC-5 Confusion Track file *confusion_track.tar.gz* [18]:

> This is the document set used in the TREC-5 confusion track. It consists of the 1994 edition of the Federal Register. The United States Government Printing Office (GPO) prints the Federal Register as a record of the transactions of the government. One issue is published each business day and contains notices to Federal agencies and organizations, executive orders

and proclamations, proposed rules and regulations, etc. The Federal Register was selected for these experiments because it is a large collection for which both hardcopy and electronic versions are readily available. The corpus contains 395MB of text divided into approximately 55,600 documents. There are three different text versions of the corpus. The "original" version is derived from the typesetting files provided by the GPO; this is regarded as the ground truth text version of the collection and was used to design the questions. The "degrade5" version is the output obtained by scanning the hardcopy version of the corpus. The estimated character error rate of this version of the corpus is 5%. The "degrade20" version was obtained by downsampling the original page images produced above and scanning the new images. The estimated character error rate of the downsampled images is 20%.

For our purposes we have selected to work with the original and degrade5 versions as they are the OCR generated errors that were naturally created when scanning the image and converting it to text without the additional downsampling that generated more errors in the degrade20 version. Each version is distributed into folders numbered 01–12 and within each folder are files named "fr940104.0" where each file contains approximately 100 documents, like "FR940104-0-00001", separated by tags. Each document varies in length, but they are approximately 1,000 words long, but with several larger ones with some being over 30,000 words long. This gives a great variety to experiment within the data set.

4 Preparing the Data Set

As part of our larger project with this data set [19,20] we have created a MySQL database where we insert the text files word by word without modifying much of the data aside from basic cleanup that includes removing all tag data. The only tag data we use before disposing of it is to separate individual documents within a file <DOCNO> so we can annotate where each document begins and ends in the database along with its document number. Along with the word we also store a numerical location that represents where in the file this word was found. So location 3 would indicate it is the third word in the text. We consider a word anything separated by white space. Using MySQL rather than direct text is not necessarily a key part of the alignment, but it does increase efficiency in the way our database was carefully curated. Because it is outside of the scope of this paper that is all we will delve into this subject. But we definitely gain efficiency from querying this way and we also have an expense when initially inserting and creating the database.

5 The Algorithm

Before delving into the algorithm, we have to understand what we are trying to accomplish. We need to match a word in the OCR'd text with the correct word and location from the original text. However, this word may or may not contain an OCR generated error. It is because of this that one cannot limit to finding exact matches and we have to be flexible and allow a certain edit distance (in our case Levenshtein edit distance) in the case that the word contains an error, otherwise we would never find a match. However, as soon as we introduce this leniency, we risk aligning to incorrect words or locations. Aligning to incorrect locations can also happen to words without an error when they are very common in a document. For example, the word "the" is known as stop words because of how common it is in search engines. What is more is that a combination of all of these errors can happen. Let us take a look at the following string to show examples of some of these errors,

```
The register in the building is next to the registrar's
office.
```

Suppose we were aligning the OCR'd text word *the* to the following string. It is possible we end up matching with any of the three *the* present. Even if we used the beginning of the next word it would still be ambiguous between the first and third case. In any case we have matched the word correctly, but we may have matched to the incorrect location. Furthermore, if instead of *the* we had an OCR generated error where h→n (h is misread as n) and we allowed an edit distance of 1 we could still match to any of the correct *the*. See Fig. 1.

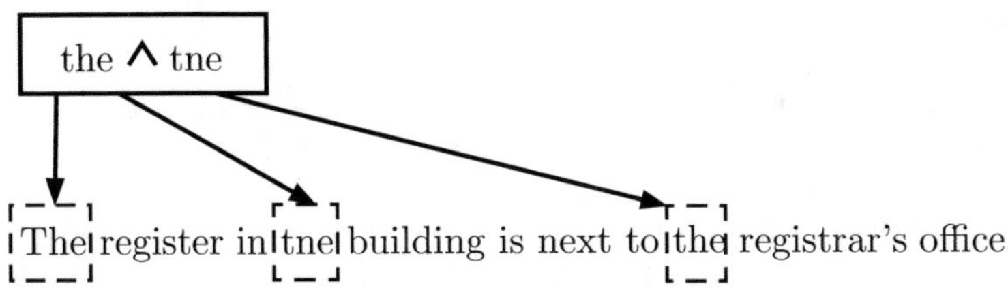

Fig. 1. Both `the` and `tne` can match to any of the three `the` occurrences.

However, if the error was an i_ such as s→g and we tried to match it with an edit distance of 1 to the string we would not know which of the two candidates *is* or *in* is the correct one. It is also possible that in the unlikely error that s→n in our OCR'd text that we would match to the incorrect word as *in* exists in the string as well as *is*. See Fig. 2.

Because of this it is important to align not just to the correct word, but to the correct location. This is a key factor not just for ensuring we do not run into issues of aligning to false positives, but also to greatly increase the alignment algorithm's speed from the non-deterministic polynomial time complete multiple string alignment problem to a more manageable polynomial time complexity. We can do this by taking advantage that statistically more words are correct than wrong in the OCR'd text. We know this because the OCR'd text has an

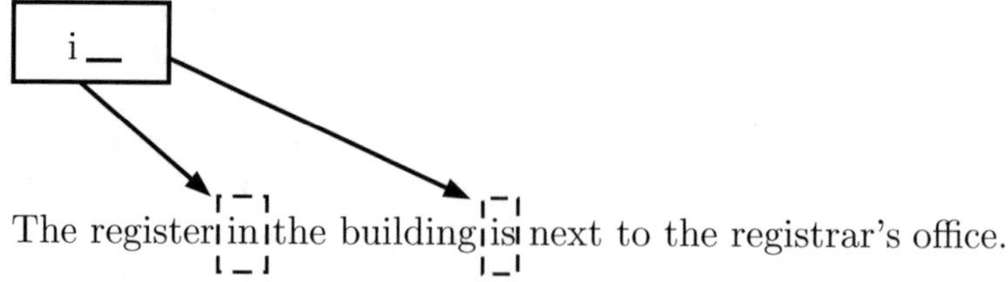

Fig. 2. With an edit distance of 1 *i_* can match to either *in* or *is*.

approximate error rate of 5% as given in the TREC-5 README file which is in line with state of the art OCR software. Furthermore, because the OCR'd text, aside from column errors or other out of order text segments, should still be in an overall same order and location as the correct text we can align words in the order they appear. For words that do not have an OCR generated error they can be used to help align ones that are not initially found due to errors. In the case that there are shifts in locations due to insertions or deletions, the relative location between neighboring words should remain similar as well allowing us to ensure we do not lose track of where the matches are happening. For example, if we have the following two strings we are matching. Assuming the initial locations are 100, and 102 respectively,

```
The hegister in the building is next to the registrar is
office.
```

```
The register in the building is next to the registrar's
office.
```

if we know that location 100 and 102 are a match for *the* then we can initially assume that location's 101 and 103 should match and so on. However location 108 and 109 are words that should both match to 110 which can pose a challenge of its own as well, but ultimately it is to our advantage to use preceding aligned words to assist in aligning the succeeding words.

With this we present the algorithm starting at the beginning of both corpus pointing to what we believe to be the first word of the OCR'd text and the original/correct text:

0. We assign a confidence weight to each word we will process that will store how confident we are that the current word was aligned to the correct word/location. This weight will vary whether it was an exact match (potentially no OCR generated errors) or a match after edit distance was applied (in which case the word has one or more OCR generated errors). Initialize the confidence for all words to 0.
1. Take the next OCR'd word and check against a dictionary to see if it is a real word. If it is then increase confidence by .5 and continue to step 2.

2. Check the previous word we aligned for the last location. (0 if this is first word). Using that location we check the immediate next word in the correct text to see if it is a match.
   - If it is a match we increase both this word confidence and the previous by .25 Then we are done aligning. Store the location found in the database and move on to the next word (start over in step 1).
   - If it is not a match we decrease the confidence of this word by .25 and if the current confidence of this word was .5 (was match in dictionary) then we decrease the confidence of the previous word by .25 and continue to step 3.
3. Since it was not a match we begin to look for the word in a nearby location: We search for exact matches after the word for X words (X for now is 20 words).
   - If we find a matching candidate, we increase the confidence by .25 and move on to the next word.
   - If we do not find a matching candidate we search X words before the previous word.
   - If we find a matching candidate we increase the confidence by .1 and decrease the confidence of the previous word by .4 and move on to the next word.
   - Else go to step 4.
4. No match for words either immediately after or in the neighboring X words: We are going to search for matches like in step 3 for neighboring words but rather than search for exact matches we are going to search for words that have an edit distance of Y. Y can be greater than 1 but we will iterate starting with 1 edit distance until Y. Great care must be taken to limit the size of Y to avoid false positives. For now we will set a maximum Y of 3:
   While edit distance checked is less than Y:
   - If we find a matching candidate with Y distance, we increase the confidence by .1 and move on to the next word.
   - If not then we increase the edit distance (Y++) and continue loop.
5. Once we exhausted our edit distance allowed, if we have not found a word we then search in a similar fashion for the previous X words (as we did in step 3, but in this case allowing an edit distance until Y).
6. If after all this we do not find a match. Then we give it a confidence of -1 and store the location of the previous word plus 1. We mark in the database as well that the word failed to be aligned so we can ensure that when we are testing OCR data we can handle either not using that word for testing or try to align it using something else. We also keep a counter of match failures and if we have more than Z in a document (so 20 in our case) we warn the user to take a closer look to identify the problem as most likely the alignment has lost its way because of out of order text or other unforeseen circumstances.

We continue this algorithm until we have reached the end of the document we are aligning. If we are aligning more than one document at a time, due to the way the data is stored we are always able to start at the beginning word of both documents which may or may not be a match but does guarantee that any alignment problems from one document do not bleed into the other. Lastly, the algorithm as implemented reads from the database by chunks in order to avoid huge number of queries and work efficiently with a database of any size, but does not commit changes until it is done.

6 Performance and Analysis of Our Algorithm

Using the first 100 documents in the TREC-5 Database, we insert 153,454 words in the database from the original/correct text. We then insert 155,282 words in the database from the corresponding OCR'd text. This already tells us that there are over 1,800 additional words in the OCR'd text. We then run the algorithm and compute the average success rate, which we define as

$$\frac{\text{Number of Words Aligned Successfully}}{\text{Total Number of Words Attempted}} = \textbf{Success Rate}$$

This returns us a success rate of 81.07% for the 155,282 words processed. With the total amount of match failures at 29,397. At first this may seem like a low number as this means that nearly 1 of every 5 words failed to be aligned. But we must understand that the theoretical limit due to the discrepancy in word count between the degraded and original texts is

$$\frac{153,454}{155,282} \times 100\% = 98.823\%$$

More importantly, analyzing the individual document success rate, see Fig. 3.

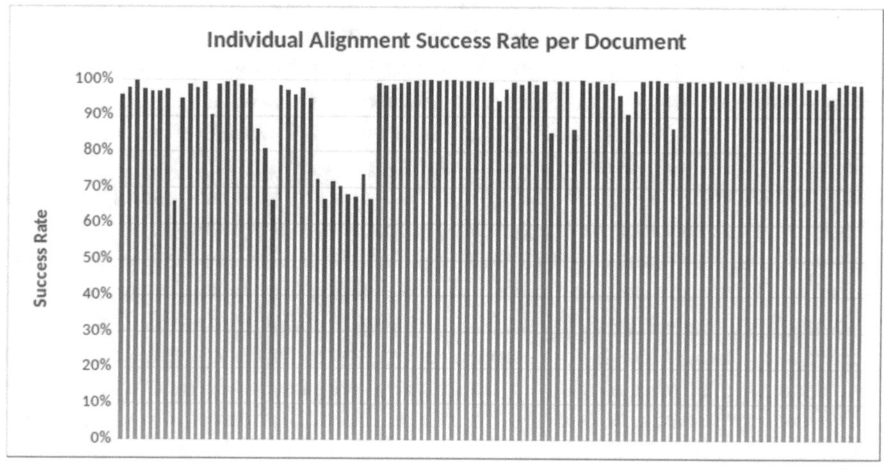

Fig. 3. Individual document success percentage rate.

Summary:

```
Total Match Failures: 29,397
Number of words processed:155,282
Average Success Rate: 81.07%
Lowest  Individual Success Rate:  66.12%
Highest Individual Success Rate: 100.00%
```

We can see that the majority of the documents had a very high success rate over 90.00%, the only documents with a success rate lower than that were the following 15 documents,

```
66.12%, 66.54%, 66.75%, 66.79%,
67.33%, 68.22%, 70.25%, 71.83%,
72.49%, 73.54%, 80.73%, 85.34%,
86.36%, 86.49%, 86.71%
```

Looking at several of these more carefully,

```
Processing Document: FR940104-0-00008
Degrade Range [3216,3892] and Original Range [3172,3868].
Match Failures: 229. Local Success Rate in Document: 66.12%
Processing Document: FR940104-0-00021
Degrade Range [14433,15210] and Original Range [14328,15092].
Match Failures: 260. Local Success Rate in Document: 66.54%
Processing Document: FR940104-0-00027
Degrade Range [18021,18479] and Original Range [17866,18219].
Match Failures: 126. Local Success Rate in Document: 72.49%
Processing Document: FR940104-0-00028
Degrade Range [18480,19235] and Original Range [18220,18773].
Match Failures: 251. Local Success Rate in Document: 66.75%
Processing Document: FR940104-0-00029
Degrade Range [19236,19914] and Original Range [18774,19242].
Match Failures: 191. Local Success Rate in Document: 71.83%
Processing Document: FR940104-0-00030
Degrade Range [19915,20668] and Original Range [19243,19782].
Match Failures: 224. Local Success Rate in Document: 70.25%
Processing Document: FR940104-0-00031
Degrade Range [20669,26255] and Original Range [19783,25111].
Match Failures: 1775. Local Success Rate in Document: 68.22%
Processing Document: FR940104-0-00032
Degrade Range [26256,49803] and Original Range [25112,48504].
Match Failures: 7692. Local Success Rate in Document: 67.33%
Processing Document: FR940104-0-00033
Degrade Range [49804,68849] and Original Range [48505,67470].
Match Failures: 5040. Local Success Rate in Document: 73.54%
Processing Document: FR940104-0-00034
Degrade Range [68850,105839] and Original Range [67471,104343].
Match Failures: 12283. Local Success Rate in Document: 66.79%
```

We see a set of them located near each other with some of the documents having a large amount of words in them. Document FR940104-0-00034 has 36,872 words of which 12,283 were failures. This single document accounts for over 41% of the failures found and along with the previous two documents with 7,692 and 5,040 match failures account for a total of 85.10% of the match failures and for a total of 79,231 words which is over half of the words. Analyzing these files we find that the top reasons for match failures were out of order text due to zoning problems such as a column misread or image captions and data being in a different order between both files, and misalignment where the threshold that we had for checking neighboring words became greater than the program allowed. This is bound to happen with larger files where not everything is aligned. The files also contain a lot of numerical and special character data that the OCR software appears to have had issues processing therefore creating a lot of OCR generated errors. In all of these cases the alignment algorithm failed to work when the errors created a greater discrepancy than the threshold allowed. At that point when the alignment was so misaligned it could not reset itself, would then cascade into more mismatches until the end of the document. The Alignment algorithm 'lost its way' midway. If we were to be able to solve the issue by only trying to search through the entire document after the algorithm is determined to have lost its way we could salvage the matching and greatly increase our success rate. As we can see in documents where the alignment algorithm didn't lose its way, those that were above 90% success rate, which accounts for an 85% of the documents, had an average success rate is `98.547%` Meaning that if we can solve the issues mentioned, primarily the occasional out of order text problem above we can get as close to the theoretical limit as 0.0276%.

Original Text for DOCNO FR940104-0-00001:

```
<DOC>
<DOCNO> FR940104-0-00001 </DOCNO>
<PARENT> FR940104-0-00001 </PARENT>
<TEXT>

<!-- PJG FTAG 4700 -->

<!-- PJG STAG 4700 -->

<!-- PJG ITAG l=90 g=1 f=1 -->

<!-- PJG /ITAG -->

<!-- PJG ITAG l=90 g=1 f=4 -->
Federal Register
<!-- PJG /ITAG -->

<!-- PJG ITAG l=90 g=1 f=1 -->
&blank;/&blank;Vol. 59, No. 2&blank;/&blank;Tuesday, January 4,
1994&blank;/&blank;Rules and Regulations
```

```
<!-- PJG 0012 frnewline -->

<!-- PJG /ITAG -->

<!-- PJG ITAG l=01 g=1 f=1 -->
Vol. 59, No. 2
<!-- PJG 0012 frnewline -->

<!-- PJG /ITAG -->

<!-- PJG ITAG l=02 g=1 f=1 -->
Tuesday, January 4, 1994
<!-- PJG 0012 frnewline -->

<!-- PJG 0012 frnewline -->

<!-- PJG /ITAG -->

<!-- PJG /STAG -->

<!-- PJG /FTAG -->
</TEXT>
</DOC>
```

Cleaned Up File Inserted to Database for DOCNO FR940104-0-00001:

```
Federal Register
&blank;/&blank;Vol. 59, No. 2&blank;/&blank;Tuesday, January 4,
1994&blank;/&blank;Rules and Regulations
Vol. 59, No. 2
Tuesday, January 4, 1994
```

OCR'd Text for DOCNO FR940104-0-00001:

```
<DOC>
<DOCNO> FR940104-0-00001 </DOCNO>
<TEXT>

Federal hegister ) Vol. 391 No. 2 )
Tuesday1 'anuary 41 1994 )           hules and

hegulations
Vol. 391 No. 2
Tuesday1 'anuary 41 1994

</TEXT>
</DOC>
```

Cleaned Up File Inserted to Database for DOCNO FR940104-0-00001:

```
Federal hegister ) Vol. 391 No. 2 )
Tuesday1 'anuary 41 1994 )          hules and

hegulations
Vol. 391 No. 2
Tuesday1 'anuary 41 1994
```

The following queries show the database tables where the algorithm reads the text files from and writes the alignment output to. These samples are all based on the same DOCNO FR940104-0-00001 for both Original and OCR'd text:

```
USE TRECSample;
SELECT * FROM doctext LIMIT 24; -- (See Table 1 for output)
SELECT * FROM doctextORIGINAL LIMIT 19; -- (See Table 2)
SELECT * FROM matchfailList LIMIT 1; -- (See Table 4)
SELECT * FROM docErrorList LIMIT 14; -- (See Table 3)
SELECT * FROM doctextSolutions LIMIT 24; -- (See Table 5)
```

The above Table 5 is the updated table after the alignment algorithm is completed. The column `word` is OCR'd Text word in the location shown, and the column `wordOR` is the word from the Original Text file that it was aligned to. The column `location` shows the OCR'd Text location for that word and the column `locationOR` shows the matching location from the Original file. As we can see the majority of the words were matched to the correct words regardless if they had an OCR error that was generated during the conversion. The only match failure was `hules` not being matched with `Rules` because it was not inserted into the database correctly. In this case the word `Unknown` is written into wordOR and this location is added to the matchfailList table. See Table 4. Words that are aligned but have an edit distance of 1 or more are also inserted into the docErrorList table, see Table 3, as they contain at least one OCR generated error since they were not found in the database. As we can see some of the words aligned to an incorrect instance of the word due to invalid tag data being read into the database. This shows that the algorithm will search for and try to latch on to nearby groups of words that could match the OCR'd text before giving up. In this case it successfully did this. If we wanted to avoid the original issue we would have to clean the text further by considering `␣` as tag data and automatically replace it with whitespace so text could be read properly into the database. Finally, Table 6 provides a visual representation of the doctextSolutions table and aids in showing how the alignment of both text files jumps around into two major contigs, 1 small one and 3 individual words aligned.

Table 1. doctext

word	location	locationOR	docsource
Federal	0	0	FR940104-0-00001
hegister	1	1	FR940104-0-00001
)	2	14	FR940104-0-00001
Vol	3	11	FR940104-0-00001
391	4	12	FR940104-0-00001
No	5	13	FR940104-0-00001
2	6	14	FR940104-0-00001
)	7	17	FR940104-0-00001
Tuesday1	8	15	FR940104-0-00001
'anuary	9	16	FR940104-0-00001
41	10	17	FR940104-0-00001
1994	11	18	FR940104-0-00001
)	12	14	FR940104-0-00001
hules	13	15	FR940104-0-00001
and	14	9	FR940104-0-00001
hegulations	15	10	FR940104-0-00001
Vol	16	11	FR940104-0-00001
391	17	12	FR940104-0-00001
No	18	13	FR940104-0-00001
2	19	14	FR940104-0-00001
Tuesday1	20	15	FR940104-0-00001
'anuary	21	16	FR940104-0-00001
41	22	17	FR940104-0-00001
1994	23	18	FR940104-0-00001

7 Improvements for Special Cases

The main issue we are facing in our alignment is when there is out of order text. Typically such problem comes when something like two columns are read as one while generating the OCR'd text file. Another example is when a caption from an image is read in as part of the regular text. This problem is exactly what we are facing, but in a different situation. In this case the OCR tool used to create the TREC-5 database did a good job at reading columns correctly. The problem lies in the original document. The original TREC-5 Files have tags for FOOTNOTES, captions and other special text locations that allow the parser that created the prints to place this text in the correct places. Nevertheless, because those tags allow the writer to place these tags anywhere in the source files, they can place a footnote tag with the according text in the middle of

Table 2. doctextORIGINAL

word	location	docsource
Federal	0	FR940104-0-00001
Register	1	FR940104-0-00001
& blank;/& blank;Vol	2	FR940104-0-00001
59,	3	FR940104-0-00001
No	4	FR940104-0-00001
2& blank;/& blank;Tuesday,	5	FR940104-0-00001
January	6	FR940104-0-00001
4,	7	FR940104-0-00001
1994& blank;/& blank;Rules	8	FR940104-0-00001
and	9	FR940104-0-00001
Regulations	10	FR940104-0-00001
Vol	11	FR940104-0-00001
59,	12	FR940104-0-00001
No	13	FR940104-0-00001
2	14	FR940104-0-00001
Tuesday,	15	FR940104-0-00001
January	16	FR940104-0-00001
4,	17	FR940104-0-00001
1994	18	FR940104-0-00001

Table 3. docErrorList

word	location	docsource
hegister	1	FR940104-0-00001
)	2	FR940104-0-00001
391	4	FR940104-0-00001
)	7	FR940104-0-00001
tuesday1	8	FR940104-0-00001
'anuary	9	FR940104-0-00001
41	10	FR940104-0-00001
)	12	FR940104-0-00001
hules	13	FR940104-0-00001
hegulations	15	FR940104-0-00001
391	17	FR940104-0-00001
tuesday1	20	FR940104-0-00001
'anuary	21	FR940104-0-00001
41	22	FR940104-0-00001

Table 4. matchfailList

word	location	docsource
hules	13	FR940104-0-00001

Table 5. doctextSolutions

word	wordOR	location	locationOR	docsource
federal	federal	0	0	FR940104-0-00001
hegister	register	1	1	FR940104-0-00001
)	2	2	14	FR940104-0-00001
vol	vol	3	11	FR940104-0-00001
391	59,	4	12	FR940104-0-00001
no	no	5	13	FR940104-0-00001
2	2	6	14	FR940104-0-00001
)	4,	7	17	FR940104-0-00001
tuesday1	tuesday,	8	15	FR940104-0-00001
'anuary	january	9	16	FR940104-0-00001
41	4,	10	17	FR940104-0-00001
1994	1994	11	18	FR940104-0-00001
)	2	12	14	FR940104-0-00001
hules	Unknown	13	15	FR940104-0-00001
and	and	14	9	FR940104-0-00001
hegulations	regulations	15	10	FR940104-0-00001
vol	vol	16	11	FR940104-0-00001
391	59,	17	12	FR940104-0-00001
no	no	18	13	FR940104-0-00001
2	2	19	14	FR940104-0-00001
tuesday1	tuesday,	20	15	FR940104-0-00001
'anuary	january	21	16	FR940104-0-00001
41	4,	22	17	FR940104-0-00001
1994	1994	23	18	FR940104-0-00001

the page. Because we ignore tag information when cleaning up the data in the beginning, we end up with such text out of order. The simple solution would be to add a tag exception when cleaning up the text or to tag that data in our database as a potential caption or table name in order to process it more carefully and check against mismatched text. This would work perfectly, but this would overspecialize our algorithm for just this data set and would not work with other data sets that may actually have out of order text causing zoning problems. We could use a heuristic approach to generate our own tags such as in Autotag [21]

Table 6. First number represents location in OCR'd text, second number represents location in original text.

Alignment Output

Correct Text from Source File (columns); OCR Generated Conversion Text (rows)

	word	Federal	Register	&blank	59,	No.	2␣/␣Tuesday,	January	4,	1994&blank	and	Regulations	Vol.	59,	No.	2	Tuesday,	January	4,	1994
word																				
Federal		0,0																		
hegister			1,1																	
)																2,14				
Vol.													3,11							
391														4,12						
No.															5,13					
2																6,14				
)																			7,17	
Tuesday1																	8,15			
'anuary																		9,16		
41																			10,17	
1994																				11,18
)																12,14				
hules										?							X			
and											14,9									
hegulations												15,10								
Vol.													16,11							
391														17,12						
No.															18,13					
2																19,14				
Tuesday1																	20,15			
'anuary																		21,16		
41																			22,17	
1994																				23,18

which is part of MANICURE designed at ISRI [22–24], but to use this we would have to generate data for the entire document, and while this may give us better results, it is possible we would lose efficiency, so while we can take some ideas from these past projects, another solution must be found that will work in all cases where text is out of order regardless of if either version of the text has tags all while maintaining efficiency and speed in our algorithm.

8 Finding a Solution to Zoning Problems

There are many options we could take to try and match text that was not originally found. The majority of these are text that is out of order by more than the distance we check. However, as mentioned earlier, some of these can be very time expensive operations that would defeat the efficiency of our algorithm. So a balance must be striken. The main problem is that once we lose our way the rate of errors increases exponentially due to a domino effect. So, all we really need to solve this is to 'find our way' and then re-activate the matching algorithm from there. Using a method like this only requires us to search the entire document for each unmatched section. How we do this varies, but trying to search the initial word that we could not align can create false positives especially if it is a word that appears many times, such as if it is a stop-word or a very common word in the document. The other case can be a word that does not appear at all, as is the case if the word contains an OCR generated error. Because of this we want to search a set of words instead of a single word in order to find a small amount of high quality matches. There are many ways to do this but the goal is to match a set of a few words ideally with the correct set of words in the correct text. This is similar to matching gene sequences. To do this We could be lenient and allow permutations as described in [25]; however, while effective with columns, this is not effective with pages containing tables. Using this method we will still have issues with sets of words that have OCR generated errors. For this reason, we have to be flexible with the words in order to allow for OCR generated errors, but being careful that we do not have false positives by aligning to a wrong set of words. Because of this we would have to keep the edit distance to a low threshold. We could also split a word into a beginning, middle and end, and only match the beginning of each word. This would solve OCR generated words not matching except for when the error is at the beginning of the word. Which should be a lower occurrence than an error appearing overall in any part of the word. However, we risk mismatching to words that start with the same few letters but are different words, such as Register and Regional both starting with *Reg*.

9 Conclusion

The ability to test the accuracy of OCR post processing software by having an available ground truth that can easily be matched to the corresponding OCR generated text is important when trying to test improvements between different available OCR Post Processing Software including the one we are currently developing. Doing this requires string alignment of both text files, which is something that a lot of the data sets lack. In this paper we discussed some of the background and relevant research that involves string alignment and the Multiple Sequence Alignment Problem and the special case of it that we are dealing with. We then introduced our own alignment algorithm along with the TREC-5 Data set we used to test it, followed by examples of the algorithm's performance. We then took these results and analysized them in order to identify the primary issues that prevented 100% alignment accuracy in our algorithm. Furthermore, we discussed the zoning problem as the main issue of the alignment decreasing the accuracy from 98.547% to 81.07%. We then proposed several solutions to this problem that could improve the overall performance to nearly 100%. Because reproducible research is a keystone of advancing our field and collaboration between researchers [26,27], the implementation of the alignment algorithm will be available along with the entire OCR Post Processing Workflow, from Creating the Database to running a Linear regression experiment [20], on multiple repositories including Docker's Hub (Search: UNLVCS), Github (https://github.com/UNLVCS), and zenodo (See DOI: https://doi.org/10.5281/zenodo.2536524).

Acknowledgment. Dr. Jan B Pedersen and Ben Cisneros for all of their contributions.

References

1. Wang, L., Jiang, T.: On the complexity of multiple sequence alignment. J. Comput. Biol. **1**(4), 337–348 (1994)
2. Needleman, S.B., Wunsch, C.D.: A general method applicable to the search for similarities in the amino acid sequence of two proteins. J. Mol. Biol. **48**(3), 443–453 (1970)
3. Barigozzi, M., Pin, P.: Multiple string alignment, vol. 6, November 2006
4. Kanai, J., Rice, S.V., Nartker, T.A., Nagy, G.: Automated evaluation of OCR zoning. IEEE Trans. Pattern Anal. Mach. Intell. **17**(1), 86–90 (1995)
5. Rice, S.V., Jenkins, F.R., Nartker, T.A.: The Fifth Annual Test of OCR Accuracy. Information Science Research Institute, Los Angeles (1996)
6. Taghva, K., Nartker, T.A., Borsack, J., Condit, A.: UNLV-ISRI document collection for research in OCR and information retrieval. In: Document Recognition and Retrieval VII, vol. 3967, pp. 157–165. International Society for Optics and Photonics (1999)
7. Tesseract OCR. https://opensource.google.com/projects/tesseract. Accessed 15 Sept 2018
8. Hobby, J.D.: Matching document images with ground truth. Int. J. Doc. Anal. Recognit. **1**(1), 52–61 (1998)
9. Kornfield, E.M., Manmatha, R., Allan, J.: Text alignment with handwritten documents. In: 2004 Proceedings of the First International Workshop on Document Image Analysis for Libraries, pp. 195–209 (2004)
10. Ho, T.K., Nagy, G.: OCR with no shape training. In: 2000 Proceedings of the 15th International Conference on Pattern Recognition, vol. 4, pp. 27–30. IEEE (2000)
11. Wickham, H., et al.: Tidy data. J. Stat. Softw. **59**(10), 1–23 (2014)
12. Taghva, K., Stofsky, E.: OCRSpell: an interactive spelling correction system for OCR errors in text. Technical report, Citeseer (1996)
13. Taghva, K., Stofsky, E.: OCRSpell: an interactive spelling correction system for OCR errors in text. Int. J. Doc. Anal. Recognit. **3**(3), 125–137 (2001)
14. Taghva, K., Nartker, T., Borsack, J.: Information access in the presence of OCR errors. In: Proceedings of the 1st ACM Workshop on Hardcopy Document Processing, pp. 1–8. ACM (2004)
15. Ahmed, S., Malik, M.I., Afzal, M.Z., Kise, K., Iwamura, M., Dengel, A., Liwicki, M.: A generic method for automatic ground truth generation of camera-captured documents. arXiv preprint arXiv:1605.01189 (2016)
16. Levenshtein, V.I.: Binary codes capable of correcting deletions, insertions, and reversals. Sov. Phys. Dokl. **10**(8), 707–710 (1966)
17. Kantor, P.B., Voorhees, E.M.: The TREC-5 confusion track: comparing retrieval methods for scanned text. Inf. Retr. **2**(2–3), 165–176 (2000)
18. TREC-5 confusion track. https://trec.nist.gov/data/t5_confusion.html. Accessed 10 October 2017
19. Cacho, J.R.F., Taghva, K., Alvarez, D.: Using the Google web 1T 5-gram corpus for OCR error correction. In: Information Technology-New Generations. Springer (2019, in press)
20. Fonseca Cacho, J.R., Taghva, K.: Using linear regression and MySQL for OCR post processing (to appear)

21. Taghva, K., Condit, A., Borsack, J.: Autotag: a tool for creating structured document collections from printed materials. In: Electronic Publishing, Artistic Imaging, and Digital Typography. Springer, pp. 420–431 (1998)
22. Taghva, K., Condit, A., Borsack, J., Kilburg, J., Wu, C., Gilbreth, J.: Manicure document processing system. In: Document Recognition V, vol. 3305, pp. 179–185. International Society for Optics and Photonics (1998)
23. Taghva, K., Borsack, J., Bullard, B., Condit, A.: Post-editing through approximation and global correction. Int. J. Pattern Recognit. Artif. Intell. **9**(06), 911–923 (1995)
24. Taghva, K., Borsack, J., Condit, A.: Expert system for automatically correcting OCR output. In: Document Recognition, vol. 2181, pp. 270–279. International Society for Optics and Photonics (1994)
25. Latifi, S.: Correcting OCR-generated text using permutations. In: Proceedings of the ICEE 1993 (1993)
26. Fonseca Cacho, J.R., Taghva, K.: Reproducible research in document analysis and recognition. In: Information Technology-New Generations, pp. 389–395. Springer (2018)
27. Fonseca Cacho, J.R., Taghva, K.: The state of reproducible research in computer science (to appear)

Incremental Alignment of Metaphoric Language Model for Poetry Composition

Marilena Oita(✉)

The Swiss AI Lab IDSIA, USI, SUPSI, Manno, Switzerland
marilena@idsia.ch

Abstract. The ability to automatically generate meaningful text with respect to a topic is an important AI mission. In particular, the automatic generation of content which is deemed creative is a great challenge. In this paper, poetry generation is approached through the lenses of a new architectural design for creativity, that leverages semantics for the creation of variance and the preservation of the content coherency throughout the generation process. The full implemented system is made available on github (https://github.com/marilenaoita/poetry-composition-resources).

Keywords: Creativity · Automatic poetry generation · Metaphor identification · Neural language model · Semantic alignment

1 Introduction

Understood as the generation of novelty that is goal-appropriate, creativity is the most valued of human abilities [1]. Modelled in artificial agents, creativity or *useful imagination* [1] is expected to bridge the gap to generic AI [2], with considerable implications towards speeding up the creation of value.

The composition of poetry is a unique artifact of human language, in itself a compelling manifestation of creativity. Similar to science, poetry decodes the reality, but in a more unconventional, subjective way.

> *One demands two things of a poem. Firstly, it must be a well-made verbal object that does honor to the language in which it is written. Secondly, it must say something significant about a reality common to us all, but perceived from a unique perspective.—W. H. Auden*

Expressing the perceptions of an unexpressed (encoded) reality, poetry decodes it by acting "as a shortcut to the truth"[1]. By making explicit a novel relation between observations of reality [3,4], poetry generation is a core creative process that can be reverse-engineered using the INNGenuity framework [5].

[1] Sarah Howe, https://www.newscientist.com/article/2073697-verse-in-the-universe-the-scientific-power-of-poetry/.

K. Arai et al. (Eds.): CompCom 2019, AISC 997, pp. 834–845, 2019.
https://doi.org/10.1007/978-3-030-22871-2_59

Building autonomous systems that automatically generate poetry which is both meaningful and original is a huge challenge in AI. Besides the practical interest in augmenting the human expertise in poetry composition, by speeding up the process and widening its reach, a system which generates meaningful and original poetry is however most interesting for the creative breakthrough artificial agents would demonstrate.

Machine learning models, by nature, are only retaining 'by heart' the patterns of the input corpus. Since creativity needs understanding [6], pure learning of sequences does not produce something truly new, i.e., *different*.

Current state-of-the-art deep learning models for natural language generation have perfected the mechanisms of generating next-best-sequence starting from a seed text. Despite that, the results are far from being recognized by humans as coherent, and even less as creative. Despite mimicking the patterns of expression of the emulated author or style, the generated text either does not have coherency and meaning, either reproduces too much of the original works.

An originality of this work is the use of neural networks in alignment with semantics with the main goals of preserving the overall meaning intended in the seed, and global coherency of the creation. Semantics encourages the creation of what we perceive as imaginative by activating the creativity conditions [7]: (1) the creation of variation (novelty), and (2) the assessment of meaning (coherency).

The contributions of this paper are:

1. identifying by using NLP techniques metaphoric expressions, and exploiting them with the goal of poetic imagination modelling;
2. training a figurative language model to predict the next best metaphor given a seed text sequence;
3. employing Universal Sentence Encoder (uSE)[2] in an incremental way on the output of the language model to semantically align 'imagined' sequence to verse; to this end, an index with more than 3 million uSE embeddings of the Gutenberg poetry corpus[3] has been built.

This work continues as follows: Sect. 2 outlines related work to automatic poetry generation. Further in Sect. 3, the collaboration between computational linguistic, semantics and deep learning is described. The system's architecture and pipeline output results are illustrated in Sect. 4. Section 5 wraps up with discussion, conclusion and future work.

2 Related Work

Automatically identifying and generating metaphoric constructs is an important step in the process of poetry composition. Metaphor models have been explored in [8], but from a pure computational linguistics perspective. A semantic approach to metaphor interpretation and generation is described in [9]. In this paper,

[2] Google module, tensorflow_hub, 2018.

[3] https://github.com/aparrish/gutenberg-poetry-corpus.

the metaphor model approached as a figurative language model, trained with a deep neural network.

By using Gutenberg poetry corpus, the work presented here classifies as corpus-based poetry-generation. A template method for this category is described in [10].

NLP techniques and semantics have been already used for poetry generation, but not in collaboration with neural networks. Neural networks, trained at word and/or character level to learn style and rhythm [11], model the task of poetry composition as text generation using sequence models [12]. More than putting a word or character after another, poems develop ideas that unfold incrementally as the discourse evolves [13]. More recent works on natural language generation do optimize their architecture by integrating some form of semantics [14] encoded directly into the model training.

More traditional approaches to poetry generation are represented by Hidden Markov Models, which employ a list hard-coded rules to add constraints on meter, word similarity and rhythm [15]. An alternative is the usage of evolutionary algorithms, where meaningfulness, grammaticality, and 'poeticness' are properties of a goal state, and poetry generation is reduced to a state space search problem [16,17].

Not surprisingly, semantics has been pervasively employed when the focus of the content generation task has been less the rhyme, and more the poem *coherency*, which is supposed to enhance 'feeling, insight and wit' [17]. Semantic features of poems, like type token ratio, concrete or abstract object tokens, etc. have been emphasized as characteristic of good poems [18]. Semantics integrated from Wordnet [19] is exploiting the relations between concepts present in the seed text [20]. PoeTryMe[4] builds verses with spans of text that contain the seed tokens involved in a semantic relation. The method presented in this paper uses semantics in two ways: indirectly through the uSE embedding model, and by using Wordnet synsets for the metaphoric expressions identification.

Related works on text-to-poem, e.g., plot-to-poem[5], or on meaning transfer using alignment[6], consider solely the direct correspondence of the initial text to the target (poetic verse) by means of concepts *similarity*, but in this "translation" no creativity is involved. The reason is first, there is no creation of variation: there will always be one single correspondence from the first text structure to the target one, and second, in this linear transformation no overall creative goal is enforced (global coherency of the generated content). These two shortcomings have been addressed in this paper. Preventing "mindlessness" [21] by creating (in our case, *linguistic*) variation and enforcing semantic coherency have been achieved with subtle architectural changes, but which lead to a non-linear impact on results. The metaphor language model creates variation, while the incremental alignment strategy takes care of the overall coherency. These are further discussed in the next section.

[4] http://poetryme.dei.uc.pt/.
[5] http://static.decontextualize.com/plot-to-poem.html.
[6] http://yknzhu.wixsite.com/mbweb.

3 Methodology

The present work follows INNGenuity architecture [5] for creativity by modelling, in the context of poetry composition, the following hubs: imagination, executive and salience (attention to goal). Having as input any type of seed text (e.g. phrase, story, poem, etc.), the goal is to create a corresponding poetic structure that preserves the intention communicated in the seed text, but expresses it in new ways.

3.1 Main Idea

Figure 1 outlines the fully unsupervised end-to-end method for poetry composition.

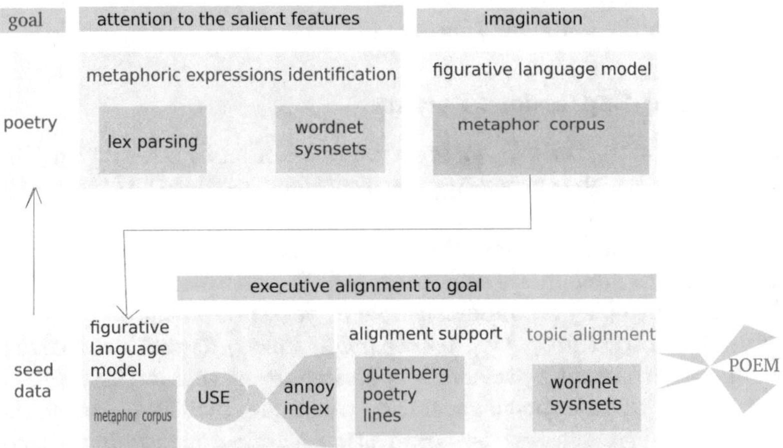

Fig. 1. The proposed pipeline for poetry composition

Typical deep learning approaches to poetry generation train a language model on the poetry corpus *itself*[7], which results in a close-to-zero-creativity text generator, since in these cases what the model actually mimicks is a 'plagiator'.

The strategy implemented in this paper is built on the following reasoning: the poet uses *imagination* for encoding the message of the poem, in the form of metaphors: succinct, semantically-rich, unexpected associations of concepts. Therefore, what we need is to model is metaphor 'imagination', not the poems' manners of expression themselves. The *metaphor generator* is implemented using a generative language model on a metaphor corpus. Imagined outputs will be fuzzy and imperfect, typical to what a language model can currently achieve, but

[7] http://karpathy.github.io/2015/05/21/rnn-effectiveness/.

inline with the nature of what imagination typically generates. The imagined output is next put in context, as a verse, using semantic alignment to a very large poetry corpus. Similar to the human process of creation, this latter step models the executive part of our brain that puts an idea in its most rightful form given the abilities (here, the expressivity of the semantic model), and the resources (the reach of available expression means, i.e., vocabulary).

Culture or 'experience', inherently semantic, comes therefore to complement and correct the imagination. Its presence is needed as a way to apprehend the universe to which poetry refers to, and distills "truth" from, in the form of poetic constructs.

In this paper, knowledge is modelled inherently using the Universal Sentence Embedding (uSE) model [22], which has been trained by Google on very large generic datasets like Wikipedia, web news, web question-answer pages and discussion forums. Inherent semantics helps achieving, through alignment, the best mapping of an idea to a poetic 'universe' denoting the existing manners of expression (here, the Gutenberg lines of poetry).

3.2 Metaphor Expressions Parsing

Language at its most distilled, *metaphors* are essential to poetry [23]. For their identification in text, the current approach considers parsing the text in a specific way.

Word associations are an important element of linguistic creativity, because of their direct link to metaphoric constructs [24]. In particular, lexical associations have been found to play an important role in poetic text [25].

Multi-token structures, e.g., *tender_light*, *stormy_flames*, are particularly important for the meaning preservation, an important goal in this paper.

In contrast with other poem generating techniques that simply parse the text token by token, here the text is parsed using a lexical parser first. Candidate constructs to metaphoric expressions are obtained by serializing the tokens linked by the `compound` or `relcl` relations, as identified by a dependency parser (with additional rules, aka POS tag). Using simple rules on the part-of-speech tag of tokens composing the candidates, non-metaphoric structures like *one_shade* or *these_lilies* can be pruned.

The advantages of segmenting the text at (possibly) multi-token are in this context the following: (I) it allows the training of a more appropriate model, since our goal is to generate metaphors which are multi-token expressions; (II) it prevents the loss of meaning: empirically, a language model loses sight of meaning very quickly when trained at smaller granularity, i.e. word or character level); (III) it optimizes the training since the sequence window has to be much smaller for the model to learn useful patterns, while capturing more useful dependencies.

This specific parsing also allows the assessment whether a text sequence may contain metaphors, or not at all. This can be important in the final step, where more metaphoric constructs can be given more attention, and verses aligned semantically, but containing no metaphors, can be ignored. The metaphoric potential of a sequence text is computed based on the existence, and the quality of metaphoric expressions. The parameters for quality are whether the expression contains highly ambiguous and semantically-rich token components. Ambiguity and rich semantic properties [18] (e.g., number of attributes, relations to other concepts in an ontology) are both qualities derived using Wordnet [26]. Ambiguity is measured based on the number of synsets a token has in Wordnet. For the semantic interpretation, we need to fix the joint meaning of a candidate metaphor. For this, the associated Wordnet synset definition sets to each component token are extracted. Each pair of synset definitions is next considered. Pair definitions are embedded using the uSE model, and best joint interpretation is chosen to be the pair closest in terms of semantic coherency.

3.3 A Figurative Language Model

The process of creation of metaphors has the merit of engaging both cognition and imagination [27]. A metaphor corpus[8] containing verses from Poetry Foundation, known to contain metaphors, has been selected for building the figurative language model. Other sources[9] of metaphors could be added to increase the expressivity and robustness of the model.

The training is done on metaphoric expressions parsed as described above. A neural network consisting of 4 stacked bidirectional LSTMs, with a dropout of 0.4 at each layer, and the hidden layers of 1024 dimensions has been considered. A sequence length of 8 is chosen, with a learning rate of 0.0001, and all is trained over 60 epochs. The considered vocabulary size of metaphor lexical expressions is 1624. This is not much, but since the training is done at expression level and not at word or character level, it is sufficient to create variation.

The general shortcoming of training at expression level is that the generated expressions will not be linked by typical stop words, etc. Here, the priority is given to the preservation of meaning, rather than the form. The rationale for this choice is also that the form of the generated sequences will be 'corrected', through alignment, in the next step.

3.4 Semantic Alignment of Imagined Text to Verse

In contrast with other works, the semantic similarity considered here does not work at token level, but at phrase level instead. Universal Sentence Encoder model, which has been trained with a deep averaging network, encodes given text into a 512-dimension vector. In addition to carrying semantics from external generic sources, uSE allows the transformation of multiple verses into a single "resuming" sentence, an appealing quality which is next leveraged in the incremental alignment.

The alignment proceeds between the 'imagined' sequence generated at each step by the figurative language model, and the Gutenberg poetry corpus. Note

[8] https://github.com/marilenaoita/nips-poetry-composition-resources/blob/master/metaphor_annotated.txt.

[9] http://ota.ahds.ac.uk/headers/2541.xml.

that the corpus on which the language model is trained is not the same as the corpus of alignment: this ensures the creation of meaningful variation: each time we run the process, we may obtain different results given a appropriate *temperature* parameter value to the generative model.

For efficiency reasons, an annoy[10] index of dimension 512, and built with 20 trees, is storing the 3,085.117 uSE embeddings corresponding to the Gutenberg poetry lines.

The transformation from the initial seed text to a poem is done *incrementally* as follows. The seed text is segmented into sentences (if various). A first seed sentence (or the title) is given to the metaphoric language model. The predicted next sequence of the model, which will be a metaphoric construct, is aligned to the closest verse from the Gutenberg poetry corpus by means of vector cosine similarity. Starting with the next seed sentence, at each step, the top alignment candidates to the imagination output will be ranked based on semantic coherency with the poem lines generated so far. That means that when searching for the best alignment there exists a retroactive check for balancing the variation (given by the imagination output), and the content consistency with respect to what already exists as poem 'draft' so far. This contributes at ensuring, at the best of the uSE model capabilities, that a *narrative flow* is present.

Finally, structural constraints can be integrated by means of mimicking the style of expression of the initial poem, using explicit pronunciation hints and the syllables number in verses from cmudict[11]. There exists two options: either when extracting the alignment candidates we select the one which mostly respects these constraints, either we replace the ending tokens of the verses such that the rhythm is present (using an index of semantically similar tokens that exhibit the desired structure for instance). Since the primary concern in this paper is not the rhyme (or structural constraints), this more straightforward part has not been currently addressed.

4 Experiments and Results

For illustrating the resulting creations of this pipeline, the seed text has been chosen to be either short stories (e.g. about the universe[12]), other poems, or the interpretation of the poems directly[13].

4.1 "The Big Bang"

short story about the beginning of the universe[14]

[10] https://github.com/spotify/annoy.
[11] http://www.speech.cs.cmu.edu/cgi-bin/cmudict.
[12] https://www.esa.int/esaKIDSen/StoryoftheUniverse.html.
[13] http://classicalpoets.org/10-greatest-poems-ever-written/.
[14] https://www.esa.int/esaKIDSen/SEMSZ5WJD1E_OurUniverse_0.html.

Aligned to Gutenberg (current pipeline):
"Slow-syllabled of weed and bloom.
But time and earth case-harden us to live;
The Serpent sleeping, in whose mazie foulds
The clearest stream through Phrygias land which flows.
To print our poems, the propulsive cause,
Where sunset spreads serenest."

We can observe here creative transformation of the original, rather formal content, to a poem which has a flow and conveys the "big bang" idea and semantic threads about the genesis of the universe.

Aligned to the metaphor corpus itself (experiment)[15].

Unfathomable sea ! whose waves are years,
while what lurks below the surface is another story,
the ocean was salt before we crawled to tears.
the western wave was all a-flame
earth is a door i cannot even face
the dead man is the flywheel of the spinning planet
the breath of the moist earth is light
over why why. causation is sequence
hath guest fire-fledged as thine.. whose lord is love?

4.2 "Ozymandia"

by Percy Bysshe Shelley.
"I met a traveler from an antique land
Who said: 'Two vast and trunkless legs of stone
Stand in the desert . . . Near them, on the sand,
Half sunk, a shattered visage lies, whose frown,
And wrinkled lip, and sneer of cold command,
Tell that its sculptor well those passions read
Which yet survive, stamped on these lifeless things,
The hand that mocked them, and the heart that fed:
And on the pedestal these words appear:
'My name is Ozymandias, king of kings:
Look on my works, ye Mighty, and despair!'
Nothing beside remains. Round the decay
Of that colossal wreck, boundless and bare
The lone and level sands stretch far away.' "

Meaning of the Poem: the poem's interpretation as extracted from[16]. "In this winding story within a story within a poem, Shelley paints for us the image

[15] https://github.com/marilenaoita/nips-poetry-composition-resources/blob/master/metaphor_annotated.txt.
[16] http://classicalpoets.org/10-greatest-poems-ever-written/.

of the ruins of a statue of ancient Egyptian king Ozymandias, who is today commonly known as Ramesses II. This king is still regarded as the greatest and most powerful Egyptian pharaoh. Yet, all that's left of the statue are his legs, which tell us it was huge and impressive; the shattered head and snarling face, which tell us how tyrannical he was; and his inscribed quote hailing the magnificent structures that he built and that have been reduced to dust, which tells us they might not have been quite as magnificent as Ozymandias imagined. The image of a dictator-like king whose kingdom is no more creates a palpable irony. But, beyond that there is a perennial lesson about the inescapable and destructive forces of time, history, and nature. Success, fame, power, money, health, and prosperity can only last so long before fading into 'lone and level sands'."

Generated poem from the interpretation of the original poem directly (so that we can observe the meaning preservation):
"To the green doublet; bitter is the wind, as though it blew
The steepy rock, and frantic tide,
Winding and vague was the family road–
The fires are dead, the gold is stone."

4.3 Evaluation of Generated Poetry

The task of poetry generation is challenging, but even more is evaluating its results in terms of the creativity involved[17]. Current means to properly evaluate automatically generated content are very limited [28].

An aesthetic measure introduced by [29] formalizes a poem's beauty solely based on phonemic features, therefore this measure cannot be used for semantic coherency since it fails to quantify the meaning of poetic texts.

Common in language generation tasks, the perplexity [30] measure is used to evaluate language models. Capturing the likelihood of correctly predicting the next character/word/expression in a sequence, perplexity can however only measure the learning capabilities of a model. Indeed, the pipeline proposed in this paper does contain multiple modules, not only a language model, therefore the usage of perplexity to evaluate the generated poems is not suitable. To study the different results that are obtained using a typical language model, a language model has been trained on the Gutenberg corpus directly, with the Adam optimizer, and the following parameters: hidden_units of 128, batch_size of 512, num_epochs of 300, using an Embedding layer, two stacked bidirectional LSTMs, and dropouts in between. The training of this alternative for poetry generation, takes several days to train on the entire Gutenberg corpus. Nevertheless, even with this state-of-the-art model, the results mostly reproduce the context of the original poem. That is, the model correctly predicts the next possible sequence of words, given the seed, but without much variation from the original poem's content. This is not surprising, since a language model trained with deep networks is the result of a learning process which is not creative per-se.

[17] http://www.thepaintingfool.com/papers/pease_aisb11.pdf.

More semantic-oriented measures would be needed to capture the subtlety of the imagination or richness of meaning. For this, empyrical tests involving human interpretation are still the most trusted [31].

Besides the illustrated poem generated by the method, chosen short for practical reasons, more examples are given at[18]. The code and other ressources (alignment index and models) are available on indicated github for further, direct experiments.

5 Discussion and Further Work

This work demonstrates the potential of the interaction between a figurative language model and semantics, in order to enable meaningful poetry composition. Towards this goal, all main dimensions of text: linguistic, semantic and statistical have been leveraged in collaboration, in a fully unsupervised manner.

Although the automatic evaluation techniques are still in development for assessing the quality of generated content (be it video, image, or text), empyrical tests attest that poems generated with this pipeline show a preservation the global meaning, while developing threads of the ideas of the original poem. The resulting verses are also being expressed creatively, and not mimicking the original content, as it is the case for one-shot language models.

Further work includes the usage of a joint embedding with visual features on the metaphor structure representation. A multimodal [32] representation can enrich the semantic properties of the concepts involved in metaphors, and further improve the quality of the poetic composition. Illustrating the visual images the poems are creating by means of their metaphoric language is another promising direction that can be pursued by adapting the ideas presented in Chat-Painter [33], in order to create *PoemPainter*. Increasing the quality of generated metaphoric expressions by incorporating semantic relations in the learning process is another line of improvement.

Creative text generation needs an architecture that supports the integration of semantics to neural networks in a dynamic and flexible way [5]. The technical means towards creativity are being forged, and there exists a global effort towards the integration of meaning, as a way towards automatic machine understanding, a more flexible 'generator' (implementing *couldBe*), as opposed to the shallower, more rigid machine learning 'translator' (implementing *isA*).

References

1. Runco, M.A., Jaeger, G.J.: The standard definition of creativity. Creat. Res. J. **24**, 92–96 (2012)
2. Weisberg, R.W.: The creative mind versus the creative computer. Behav. Brain Sci. **17**(3), 555–557 (1994)
3. Gibbs, R.W., et al.: The Poetics of Mind (1994)

[18] https://github.com/marilenaoita/nips-poetry-composition-resources.

4. Willis, P.: Don't call it poetry. Indo-Pac. J. Phenomenol. **2**(1), 1–14 (2002)
5. Oita, M.: Reverse engineering creativity into interpretable neural networks. In: Proceedings of the Future of Information and Communication Conference, FICC (2019)
6. Dartnall, T.: Artificial Intelligence and Creativity: An Interdisciplinary Approach, vol. 17. Springer, Dordrecht (2013)
7. Campbell, D.T.: Blind variation and selective retention in creative thought as in other knowledge processes. Psychol. Rev. **67**, 380–400 (1960)
8. Shutova, E.: Models of metaphor in NLP. In: Proceedings of the 48th Annual Meeting of the Association for Computational Linguistics, ACL 2010. Association for Computational Linguistics (2010)
9. Veale, T., Hao, Y.: A fluid knowledge representation for understanding and generating creative metaphors. In: Proceedings of the 22nd International Conference on Computational Linguistics, vol. 1, pp. 945–952. Association for Computational Linguistics (2008)
10. Colton, S., Goodwin, J., Veale, T.: Full-face poetry generation. In: ICCC, pp. 95–102 (2012)
11. Xie, S.C.: Deep poetry: word-level and character-level language models for Shakespearean sonnet generation (2017)
12. Mikolov, T., Karafiát, M., Burget, L., Černockỳ, J., Khudanpur, S.: Recurrent neural network based language model. In: Eleventh Annual Conference of the International Speech Communication Association (2010)
13. Moss, J.: What is Imitative Poetry and Why Is It Bad? (2007)
14. Tran, V.-K., Nguyen, L.-M.: Neural-based natural language generation in dialogue using RNN encoder-decoder with semantic aggregation. CoRR, abs/1706.06714 (2017)
15. Hopkins, J., Kiela, D.: Automatically generating rhythmic verse with neural networks. In: Association for Computational Linguistics. Association for Computational Linguistics (2017)
16. Manurung, H.: An evolutionary algorithm approach to poetry generation (2004)
17. Veale, T.: Less rhyme, more reason: knowledge-based poetry generation with feeling, insight and wit. In: ICCC, pp. 152–159 (2013)
18. Kao, J., Jurafsky, D.: A computational analysis of style, affect, and imagery in contemporary poetry. In: Proceedings of the NAACL-HLT 2012 Workshop on Computational Linguistics for Literature, pp. 8–17 (2012)
19. Agirrezabal, M., Arrieta, B., Astigarraga, A., Hulden, M.: POS-tag based poetry generation with WordNet. In: Proceedings of the 14th European Workshop on Natural Language Generation, pp. 162–166 (2013)
20. Oliveira, H.G.: PoeTryMe: a versatile platform for poetry generation. Comput. Creat. Concept Invent. Gen. Intell. **1**, 21 (2012)
21. Langer, E.J., Piper, A.I.: The prevention of mindlessness. J. Pers. Soc. Psychol. **53**(2), 280 (1987)
22. Cer, D., Yang, Y., Kong, S., Hua, N., Limtiaco, N., John, R.St., Constant, N., Guajardo-Cespedes, M., Yuan, S., Tar, C., Sung, Y.-H., Strope, B., Kurzweil, R.: Universal sentence encoder. CoRR, abs/1803.11175 (2018)
23. Indurkhya, B.: Creativity in interpreting poetic metaphors. New Directions in Metaphor Research, pp. 483–501 (2007)
24. Krishnakumaran, S., Zhu, X.: Hunting elusive metaphors using lexical resources. In: Proceedings of the Workshop on Computational Approaches to Figurative Language, FigLanguages 2007. Association for Computational Linguistics (2007)

25. Netzer, Y., Gabay, D., Goldberg, Y., Elhadad, M.: Gaiku: generating Haiku with word associations norms. In: Proceedings of the Workshop on Computational Approaches to Linguistic Creativity, CALC 2009. Association for Computational Linguistics (2009)
26. Miller, G.A.: WordNet: a lexical database for English. Commun. ACM **38**(11), 39–41 (1995)
27. Ricoeur, P.: The metaphorical process as cognition, imagination, and feeling. Crit. Inq. **5**(1), 143–159 (1978)
28. Oliveira, H.G.: A survey on intelligent poetry generation: languages, features, techniques, reutilisation and evaluation. In: Proceedings of the 10th International Conference on Natural Language Generation, pp. 11–20 (2017)
29. Birkhoff, G.D.: Aesthetic Measure, vol. 38. Harvard University Press, Cambridge (1933)
30. Yan, R.: i, poet: automatic poetry composition through recurrent neural networks with iterative polishing schema. In: IJCAI, pp. 2238–2244 (2016)
31. Colton, S., Pease, A., Corneli, J., Cook, M., Llano, T.: Assessing progress in building autonomously creative systems. In: ICCC, pp. 137–145 (2014)
32. Liu, D., Guo, Q., Li, W., Lv, J.: A multi-modal Chinese poetry generation model. In: 2018 International Joint Conference on Neural Networks (IJCNN), pp. 1–8. IEEE (2018)
33. Sharma, S., Suhubdy, D., Michalski, V., Kahou, S.E., Bengio, Y.: Chatpainter: improving text to image generation using dialogue. CoRR, abs/1802.08216 (2018)

A Trie Based Model for SMS Text Normalization

Niladri Chatterjee(✉)

Indian Institute of Technology Delhi, New Delhi 110016, India
niladri@maths.iitd.ac.in

Abstract. Language takes varied forms as used on different platforms. The instinctive nature of humans to use shorter message length, facilitating faster typing while maintaining semantic clarity, shapes the structure of a non-standard form of written text known as the *texting language*. The present work focuses on developing a Trie-based technique to model words in texting language for normalizing SMS text to Standard English text. The model is conceived and developed through systematic analysis of training data on user behaviour of texting language. Trie as a data structure is not only compact, but also is easy to manipulate for performing operations related to text normalization. Although Trie is a well-known data structure, its application to model SMS text has so far been unexplored. The results obtained using this model, despite the scheme being computationally cheap, are comparable with existing HMM based schemes that are available in the literature.

Keywords: SMS · Text normalisation · Trie · Machine learning

1 Introduction

Text normalization is a process of translating *texting language* into plain English. The need for normalization arises due to certain deviations from grammar rules, and changes in spellings as is often practiced by the users of texting language (TL). The stipulated number of characters allowed per text message, the constraints of the small phone keypads, and mostly informal communication between friends and relatives, are the typical reasons that compel users to resort to typing in texting language. There has been a rapid increase in social text in the last few years, including the mobile phone text messages (SMS), comments from the social media websites, and real-time communication platforms, such as MSN and Gtalk. Usage of texting language in social media is natural and can also be explained by the increase in mobile usage of social media apps[1]. Unfortunately, traditional NLP tools often perform poorly when processing this kind of text [1]. One of the possible reasons is that social text, in general, is very informal, and contains many misspelled words, abbreviations and other non-standard tokens. As a consequence, SMS texts behave quite differently from normal written texts, and have

[1] 91% of mobile Internet access is for social activities vs. 79% on desktops, according to Microsoft (http://www.social4retail.com/the-growth-of-mobile-marketing-and-tagging-info-graphic.html).

K. Arai et al. (Eds.): CompCom 2019, AISC 997, pp. 846–859, 2019.
https://doi.org/10.1007/978-3-030-22871-2_60

some unique characteristics [2]. Thus, two opposing forces, viz., maintaining short message length, and yet keeping semantic unambiguity eventually shape the structure of this compressed non-standard form, of texting language. The Hidden Markov Model approach [3, 4], when constructed for each word in the corpus allows to model all possible normalized texts, and their occurrence possibilities. This allows efficient modelling of the frequency distribution of possible normalized words. Further, it can also be used to choose the syntactically highest ranked possibility, or to use the top ranked normalized word for the purpose of semantic correctness.

The objective of the current work is to study as well as model the characteristics of the texting language (TL) more formally. The aim is to solve the problem of SMS translation by using a Trie Data Structure to model a corpus of English words. More precisely the aim is to construct a sentence level decoder for TL so that given an input sentence in the TL the decoder would be able to generate the corresponding standard form. For example, the input text "Ill b thr 2morow in d hous" is expected to be mapped to "I'll be there tomorrow in the house".

A review [5] on behaviour of users of texting language concludes that the linguistic errors in language can mostly be one of the following forms:

1. Insertion error - Extra alphabet inserted in the word, For example, 'foood' → 'food', 'soooooo' → 'so'
2. Deletion error - Corresponding to any alphabet deleted in the query word. E.g. 'Botle' → 'Bottle'
3. Replacement Error - A particular alphabet in the given word is replaced by another alphabet from (a–z). For example, 'Talo' → 'Talk' or 'Tall'.
4. Swap error - Any two adjacent alphabets are swapped. For example, 'Laese' → 'Lease'
5. Phonetic error – An alphabet or a collection of alphabets are replaced with similar sounding word or number. For example, '2morow' → 'tomorrow'

Phonetic error is comparatively more difficult to handle. This happens because while typing a word people often switch between the graphemic and phonemic domains. For example, while establishing correspondence between the TL word 'transl8in' and standard English word 'translating', the strings 'transl' and 'in' are in the graphemic domain, whereas the '8' represents a phonetic substitution. However, the difficulty arises here is that phonemes are differently interpreted depending on the way a person pronounces the word. For example, the numeral '8' can be, phonetically interpreted as 'eight' or 'ate'. Similarly, the numeral '2' can stand for 'to' or 'two'. In order to address the above-mentioned problems with Texting Language, the present work proposes to use an efficient search based data structure that can store the database of Standard English words in a compact way from which the best match(es) can be retrieved. In order to address the above-mentioned problems with Texting Language, the requirement is to use an efficient search based data structure that can store the database of Standard English words in a compact way from which the best match(s) can be retrieved. This prompted us to use Trie as the underlying data structure for the above purpose.

The rest of the paper is organized as follows. Section 2 discusses some recent relevant work in this area. Section 3 describes the design and implementation of Trie data structure for SMS text normalization. Section 4 explains the training mechanism of the model and the probability assignment technique. Section 5 illustrates the algorithms involved in Trie based SMS Normalization. Section 6 discusses two series of experiments conducted for this work and the results. Section 7 presents the time complexity analysis. The concluding remarks are mentioned in Sect. 8.

2 Some Relevant Past Work

Several attempts have been made to address the problem of SMS text normalization. A majority of the existing work, as mentioned above, has used the method of statistical machine translation [3, 4] yielding a baseline accuracy of 74.7%. Pennell and Liu [6] have used statistical classifiers to learn the probability of deleting a given character using features based on character context, position in the word and containing syllable, and function within the word. They used statistical methods to determine whether or not to remove a character on the basis of contextual information, and created a list of possible abbreviations for English words. A reverse mapping has then been used to create a look-up table from text message lingo to proper English.

Dictionary substitution is another technique used in the domain of SMS text normalization. Jose and Raj [7, 8] have used four different databases (namely, Abbreviation Database, Non-noisy Database, Grapheme Database, and Phoneme Database) to convert SMS Texting Language to Standard English at word level.

There are, however, other hybrid techniques, e.g. [9], that use both spell checking, and machine translation techniques based on finite state framework. These methods have achieved an accuracy of 76.23%. Nocon et al. [10] use a permutation of SMT and dictionary substitution to find the best results converting Filipino Text Shortcuts. The best results (BLEU score of 0.807) were achieved when SMT was used before dictionary search.

3 Design and Implementation of the Trie

Trie is an efficient information retrieval data structure. The search complexity in a Trie is O(M), i.e. a linear function of the key length, while a well-balanced Binary Search Tree needs time proportional to M * log N, where M is the maximum string length, and N is the number of keys in the tree. The search can terminate due to end of string, or lack of key in the Trie. The Trie is initialized by storing a set of keys, which are valid words from Standard English language. Every character of an input key is inserted as an individual state or node in the Trie. The children of a node are stored in an array of pointers directed from the parent node. The key character under consideration ('a'–'z') acts as an index into the array of its children. If the input key is new or an extension of an existing key, there is a need to add the non-existing nodes of the key to the Trie, and to mark the leaf node. If the input key is a prefix of an existing key in the Trie, the last node of the key is marked as leaf. The length of the longest word in the Trie determines the depth of the data structure.

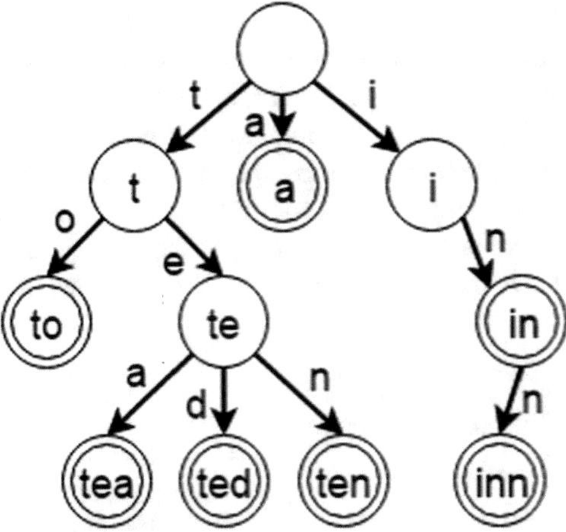

Fig. 1. Example of a Trie

The structure thus created represents a network of interlinked alphabets, and the frequencies of occurrence of individual words. Every node of the said Trie consists of possibly multiple branches. Each branch represents a possible character. In the Trie model, the branches emerging from a node represent the possible paths, or the collection of words following that path. A node in the path where a word terminates is either a leaf node, or is marked as a *special node*. This is required to capture the fact that a valid Standard English word may be a prefix of another word. For illustration, consider the words *in* and *inn*, as shown in Fig. 1. In order to treat both these words separately, for each word the terminating node is marked. For the initial experiments a corpus of 20,000 English Dictionary [11] words has been used in the Trie. Once insertion of all the words from the dictionary to the Trie is complete, one can query a particular TL string, and can receive a suggestion list from the Trie using the error checking algorithms as discussed in Sect. 5. The approach is self-learning as after each query the Trie is updated with the insertion of the highest ranked word from the suggestion list. This increases the frequency of occurrence of this word, and helps the model be more efficient in the subsequent runs, as explained in Sect. 4. The scheme for updating the probabilities is discussed in the next section. A flowchart explaining the working of this model is shown in Fig. 2.

4 Probability Assignment and Training of the Trie

So far the construction of the Trie assumed equi-probable emerging states from a given state. In order to have a more accurate prediction model, a probability assignment technique has been developed for the words, as discussed below. After correctly identifying the English word corresponding to a TL word, the correct word is inserted into the Trie, the choice for which is provided to the user.

The probabilities are assigned for each state (character) on the basis of the structure of the network of words created in the Trie. A small sample of the proposed Trie-based modelling of words is shown in Fig. 3. With each of the states in the Trie a counter variable is associated that indicates the number of times the particular state has been reached while training the Trie model using the training data. To get the probability for any state, the sum of the counter values for all the states at that level is obtained. The probability for any individual state is computed as the ratio of its counter value and the sum of all counter values corresponding to that level. Below we explain the scheme.

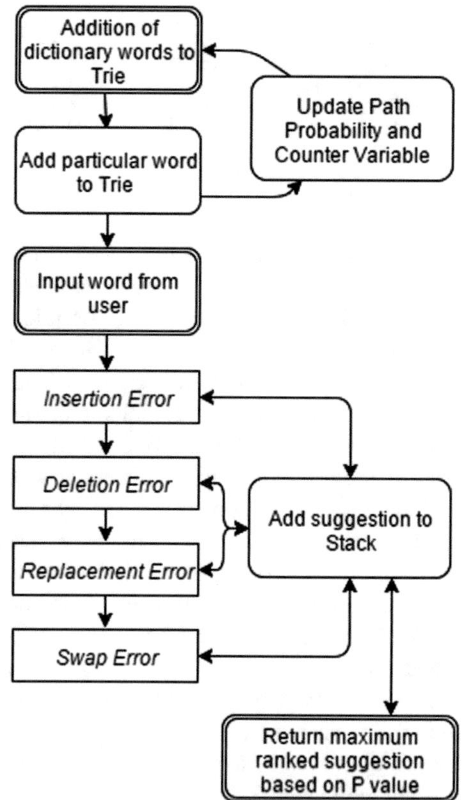

Fig. 2. Flowchart with working of the TRIE model

Let n be the number of words in the Trie model. Further, let binary variables t_{ij} represent whether the state i has been traversed during the insertion of the jth word. The *count* variable is calculated by summing the number of times the state i is traversed over the insertion of all words. The variable *total* is the sum of the *count* variable of state i and all its sibling states. The sibling states are all the states having the same parent, i.e. emerging out of the same node as state i. The probability of a state is the number of times the state is traversed divided by the number of times all states emerging from its parent node are traversed, shown in Eqs. (1–3).

$$\text{Count(i)} = \sum_{\text{j}=1}^{\text{n}} t_{ij} \tag{1}$$

$$\text{Total}(i) = \text{Count}(i) + \sum_{j \in Sibling(i)} \text{Count}(j) \tag{2}$$

$$\text{Probability}(i) = \frac{\text{Count}(i)}{\text{Total}(i)} \tag{3}$$

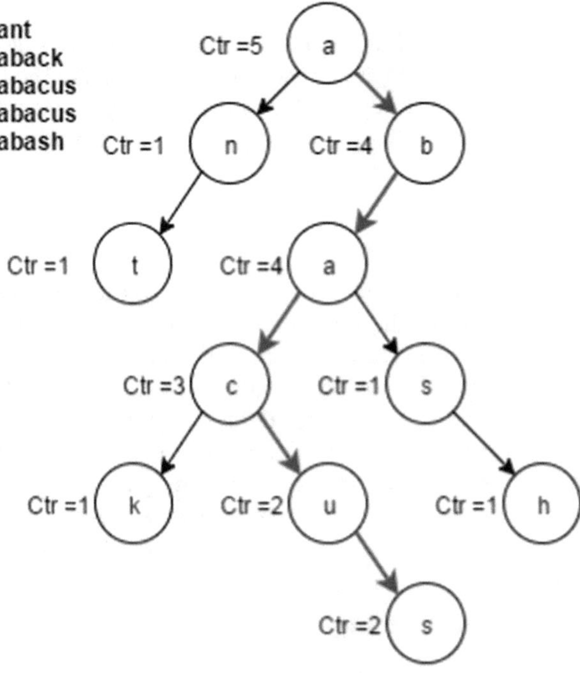

Fig. 3. Assignment and finding probability of a word in TRIE model

In the example shown in Fig. 3, five keys are inserted into an initially empty Trie, one by one. The keys are '*ant*', '*aback*', '*abacus*', '*abacus*' and '*abash*'. Note that the word '*abacus*' is input twice, in order to illustrate the change in values of the counter variable. Upon inserting these keys, the counter values for the states that are coming as a part of insertion process get modified. The counter value for the first state '*a*' takes the final value '5', which indicates that five strings starting with '*a*' have been inserted in the model. Since no other state is present at this level (for the Trie of Fig. 3), the probability for it is 1. Similarly, for the state '*c*', which comes as a part of insertion for '*aback*' and '*abacus*', the final value of the *counter* variable is '3' as *abacus* has been searched twice. As the total sum of all the counter values at that level comes out to be 4, the probability for state 'c' will thus be 0.75. The probabilities of all other states are calculated in a similar way.

In order to get the probability for any of the input string, it needs to traverse the path of that particular string and multiply the probabilities of all the states in the word model that occur in the path. To determine the probability of the input string '*abacus*', trace the path in the word model and multiply the probabilities of the states *'a', 'b', 'a', 'c', 'u', 's'*. Hence its probability will be:

$$((5/5) * (4/5) * (4/4) * (3/4) * (2/3) * (2/2)) = 0.4.$$

This can be verified by referring to the input data. Hence the probability of the word '*abacus*' is 0.4 as five words have been inserted in the Trie model, and the occurrence of '*abacus*' is 2.

The model can then be used to normalize words from English SMS texts to their Standard English forms with a high accuracy at the word level. The decoder may be used for automatic correction as well as information extraction and retrieval from noisy English documents, e.g. emails, blogs, wikis and chat logs that are written in TL.

5 SMS Normalization

Given a TL word, the following algorithm is employed to check whether the TL word has the specific error for which the algorithm corresponds to. The multiple algorithms for checking errors, as described ahead, run sequentially and check if the particular type of error exists. The order of algorithms in this research work is *Insertion, Swap, Replacement and Deletion*. Each algorithm returns a suggestion list containing possible Standard English translations with their respective probabilities. Since this work accounts for single errors in words, the order of running the algorithms is not significant. As an example, the algorithm for insertion error is shown below.

Insertion Error

```
For i= 0 to String Length, do
{
  Obtain Substring STR by removing ith character;
  If STR exists in TRIE
  {
    Add (STR, Probability(STR)) to Suggestions List.
  }
}
```

Algorithms corresponding to *Swap, Replacement* and *Deletion* have been designed in similar way. However, certain phonemic errors cannot be tackled by the above algorithms. Words such as *'fne'* for *'phone'*, or *'thx'* for *'thanks'* are a combination of graphemic errors and phonemic errors. A purely character-based Trie fails to capture these words by traversing the Trie from the top. In order to overcome the issue of a combination of errors, the following procedure is applied to reduce the complexity of errors to a single graphemic error form, for which the algorithms discussed so far can be employed. We add a layer of phonemic substitution and query the resulting word in the Trie. The substitutions performed are as follows:

- *'d'* is replaced with *'th'*
- *'x'* is replaced with *'aks'*
- *'f'* is replaced with *'ph'*
- *'z'* is replaced with *'se'*
- *'c'* is replaced with *'see'*

A Trie designed as above, however, has been found to generate incorrect results while translating TL words involving phonetic substitutions of numeric characters such as '2', '4' and '8'. In order to handle these words the basic Trie is extended by making the following transformations during the pre-processing of the SMS training data:

- '2' is replaced with *to*,
- '*4*' is replaced with '*for*',
- '*8*' is replaced with '*eight*'.

The suggestion list generated by querying the resultant word in the Trie is combined with the earlier suggestion list from the graphemic error handling. The suggestion list is ranked as per probabilities of occurrence and returned to user. The Trie is also updated to incorporate the new corrected word at a later time, to increase its probability of occurrence and adjust it as per rules discussed in Sect. 4.

6 Experiments and Results

Experiments have been conducted in two stages. In the first stage all the words have been considered, the translations for which are discussed with respect to the HMM Model as given by *Choudhury* et al. in [3]. The output using the Trie Model is compared to the output reported the above Model.

In the second experiment, we obtain the efficiency of the Trie-based scheme by obtaining the rank, highest frequency word, and the time taken to return the suggestion list. The same is tabulated and averaged over the size of input. This experiment further helps us in time complexity analysis as discussed in Sect. 7.

6.1 Experiment 1: Comparison with Benchmark

The phonemic substitution acts as a pre-processing layer before the word is queried in the Trie. It is responsible for converting the word into the form of a single graphemic error. Some observations from the above comparative analysis as shown in Table 1 are as follows:

Table 1. Comparison of output of Trie model with HMM based model [4]

Impure word (target word)	Suggestion list using Trie model	Rank using Trie model	Suggestion list using HMM model [4]	Rank using HMM model [4]
2day (today)	today, body, tory, tidy, toady, toy, tony	1	today, stay, away, play, clay	1
fne (phone)	phone, fee, foe, one, pine, fine pane	1	fine, phone, funny, fined, fines	2
thx (thanks)	thy, thanks, tax, the	2	the, tax, thanks, tucks, takes	3
m8 (mate)	mare, mete, hate, mane, male, mate, made, matte, date, maze, fate, gate, math, rate, mite, mace, late, ate, bate, tate, make, mute, pate	6	my, ms, mr, mate, me	4
ant (cannot/cant)	pant, ani, cant*, apt, aft, any, art, rant, aunt, act, ana, want, and	3	ant, aunt, ants, cannot, cant	4, 5
dem (them)	deem, dim, dew, dam, dey, diem, they, den, gem, hem, them, then, ahem, thee, hem	11	deem, deems, dec, dream, drum	10
orite (alright)	rite, orate, write, trite	–	write, omit, writes, writer, writers	–
cuz (cause)	cup, cur, cut, cud, cue, cub, cusp, cause, muse, fuse, use, curse, cube, case, cute, cue, cure, ruse	8	crews, cut, cup, occurs, acres	–
cin (seeing/seen)	min, con, can, coin, chin, fin, win, tin, yin, din, pin, gin, sin, kin, cit, bin, seeing, stein, seqin, seen	17, 20	coin, chin, clean, coins, china	–

Some of the interesting observations from Table 1 are as follows:

- The word *'fne'* with the phonetic substitution of *'f'* (being replaced with *'ph'*) gets converted to *'phne'* which is an *insertion* graphemic error for the word 'phone'. It is returned correctly by the existing *insertion* algorithm used in the Trie-based model.
- Similarly, the word *'thx'*, with the phonetic substitution of *'x'* by graphemic string *'aks'* gets converted to *'thaks'* which is captured easily by the same algorithm for *insertion*, for the word *'thanks'*.

- Interestingly, the word *'orite'*, even with a phonetic substitution of *'o'* (replaced with *'al'*) takes the form *'alrite'*, and is not transformed into a standard graphemic error of any form as discussed in Sect. 1, and hence is not reported in the Suggestion List by the algorithm.
- It is observed that the Trie Model performs better than the HMM Model, if the word can be brought into the standard form of a single graphemic error. For example, the texting language word *'cin'*, with a phonetic substitution of *'c'* being replaced with *'see'*, takes the form *'seein'* which leads to the translation *'seeing'* using the *insertion* algorithm, and the translation *'seen'* using the *deletion* algorithm, and hence both translations are reported. The same word is not correctly translated in the HMM model as given in [3].
- In case of multiple errors or a combination of phonemic or graphemic errors which cannot be reduced to a single graphemic error form, the Trie Model is presently unable to translate the word.
- The ranking of target word is seen to be better in the Trie Model for majority of words (*'2dy', 'fne', 'thx'*, etc.). However, it is observed that the method of merging the suggestion lists before and after phonetic substitution leads to a large distribution of words and their probabilities, because of which the target word in some cases ranked lower.

6.2 Experiment 2: Robustness of the Trie Based Scheme

Experiment 2.1.

In a more elaborated experiment, we have used 5000 TL word selected randomly from a corpus of SMS. In order to obtain the efficiency of the Tries-based scheme we obtained the rank, highest frequency word, and the time taken to return the suggestion list. The same is tabulated in Table 2. For this experiment, the average time taken over 5000 runs (One run corresponding to each TL word) of the program is 359 ms. The time includes the time taken to form the Trie by inputting dictionary words and then querying the model. The average number of words in the suggestion list for a large number of input words is $\sim$ 3 words. The average rank, indicating the position at which the actual target word occurs in the suggestion list is $\sim$ 2. The accuracy of finding the target word (correct translation) in the suggestion list came out to be 87.67%. The accuracy of the highest frequency word (Word with rank 1 in the suggestion list) being same as the expected word is 71.88%. The above results are with respect to an English dictionary available online with over 20,000 words [11].

Experiment 2.2.

In order to validate the above results, the algorithm has been on another data set of over 3,50,000 words obtained from a different online repository [12]. To be able to verify the results, a large data set containing TL words also had to be artificially generated for this experiment. The typical errors committed by users of QWERTY keyboards have been studied from an existing research [13]. A dataset of 30,000 impure words was obtained by using the same dataset and replacing *'e'* with *'r'*, *'y'* with *'u'*, *'o'* with *'oo'*, a total of 15 such graphemic errors that are common mistakes made by humans while using QWERTY keypads in mobile phones, and other such devices. The number of errors of each type were kept equal for consistency in the dataset. The focus was on alphabets/characters that are placed in proximity or adjacent to one another on a QWERTY keypad and are commonly interchanged by humans. An example of such an error is use of *'s'* in place of *'z'* due to a graphemic error – *"He sapped him with the laser"* instead of *"He zapped him with the laser"*. The alphabets *'j'* and *'k'* are also commonly interchanged while typing. The results obtained from this experiment are tabulated in Table 2.

Table 2. Sample output for TRIE model

Impure word	Target word	Normalized word (suggestion list)	Rank	Highest frequency word	Time taken (ms)
tran	train	ran, trag, tray, than, train, tram, trap, tan, bran	5	ran	389
lng	long	lung, leg, lag, long, lug, log	4	lung	308
aple	apple	pale, axle, aile, ape, apple, ample, able, ale, apse, maple	5	pale	330
beleive	believe	believe	1	believe	351
moble	noble	mole, noble, mobile	2	mole	342
gost	ghost	ghost, gust, gout, gosh, post, lost, host, goat, most, gist, cost, got	1	ghost	450
2night	tonight	tonight	1	tonight	349
2morow	tomorrow	tomorrow	1	tomorrow	330
v2	veto	veto	1	veto	331
8en	eighteen	eighteen	1	eighteen	345
h8	height	height, eight, weight	1	height	348
4evr	forever	forever	1	forever	339
4wrd	forward	forward	1	forward	348
a4d	afford	afford, ford	1	afford	365

The average time taken per word for a corpus of 30,000 impure words is 370 ms. The average number of suggestions is $\sim$ 4 words. The accuracy for finding the word in the suggestion list is 86.43% and the accuracy of highest ranked word matching the expected word is 69.25%.

7 Time Complexity Analysis

The discussion so far does not account for the running time of the algorithms and whether the method is scalable. In order to understand this an empirical analysis of Order of Time Complexity is presented in this section. The running time of the algorithm is recorded by varying the following two parameters:

7.1 Size of Dictionary or Dataset Used to Train the TRIE

The incorrect English word *'alin'* for the correct word *'alien'* was queried using a TRIE model formed with different number of words in the data set. In each run of the algorithm, the time taken to return the suggestion list and highest ranked word is noted for different sizes of the dictionary.

Table 3. Time taken for different sizes of dataset.

No of words	Time (ms)	Ratio (relative increase)
500	17	–
1000	37	2.17
2000	57	1.54
4000	97	1.70
8000	157	1.61
16000	260	1.65
32000	319	1.22

Subsequently, the size of dataset was doubled in each iteration in order to compare the time taken. The average ratio of time requirement on doubling the number of words in the dataset, is equal to 1.64. Hence asymptotically, the time complexity may be considered as $O(n^{1.64})$ where n is the number of words in the dataset. Table 3 summarizes the results.

7.2 Length of Impure Word Being Queried by User

To understand the dependency of time complexity of the algorithm, on the length of the TL words, the program was run for TL words of different lengths. The time taken to return the suggestion list and highest ranked correct English translation is noted as follows (Table 4):

Table 4. Time taken by words of different length

Word length	Time (ms)
2	1
3	1
4	3
5	3
6	3
7	4
8	5
9	6
10	7

On the average, the time varies as ~ (n-3) where n is the length of the word being queried in the Trie. Hence, the time complexity with respect to n, the length of the word, is O(n).

8 Conclusion

The present paper describes a TRIE-based conversion model between TL and the standard language. The model has been used to construct a decoder from English SMS texts to their Standard English forms with an accuracy of 86.43% at the word level. The decoder can be used for automatic correction as well as information extraction and retrieval from noisy English documents such as emails, blogs, wikis and chat logs that are written in TL. The novelty of the work resides in the construction of the TRIE to model the dictionary words. While the results of the TRIE based strategy are at par with the benchmark and the time complexity is significantly smaller, nevertheless, the structure of the current TRIE model can be improved in several ways, such as:

1. Addition of self-loops to graphemic states to capture emphasis as in *soooo* for *so*
2. Modelling of phonemic errors which have multiple interpretations, such as *two* or *to* for *'2'*, *eight* or *ate* for *'8'*.
3. Handling the translation of words having more than one error or a combination of graphemic and phonemic errors, even after pre-processing or adding a layer of phonetic substitution in order to reduce complexity. E.g.: As discussed in Sect. 4, the word *'orite'* cannot be reduced to a single graphemic error standard form even after the phonetic substitution of *'o'* with *'al'* to identify the word *'alright'*.
4. Modelling variations in proper nouns. For example, one can find several variations for the name '*Saurav Ganguly*' in the web ('*Saurabh Ganguly*', '*Sourav Ganguli*', '*Sourabh Ganguly*', etc.).
5. The scheme can presently model single words only, and TL strings that represent more than one word, e.g. *"IluvU"* for *"I love you"*, cannot be modelled using the method proposed at present

In the current research work the order of algorithms is not significant as we checked for only single graphemic error in words as discussed. However, algorithms for translation of words having multiple graphemic errors will vary in performance with respect to the order of error checking. The decoder may be further improved by incorporating language model and modules to handle abbreviations, deletion and fusion of words.

Acknowledgement. The author thankfully acknowledge the contribution of students Tushar Singla in conducting the experiments and preparing the draft.

References

1. Jurafsky, D., James, H.M.: Speech & Language Processing, 2nd edn. Pearson Education, India (2000)
2. Kobus, C., Yvon, F., Damnati, G.: Normalizing SMS: are two metaphors better than one? In: Proceedings of the 22nd International Conference on Computational Linguistics, vol. 1, pp. 441–448. Association for Computational Linguistics (2008)
3. Choudhury, M., Saraf, R., Jain, V., Mukherjee, A., Sarkar, S., Basu, A.: Investigation and modeling of the structure of texting language. Int. J. Doc. Anal. Recogn. (IJDAR) **10**(3–4), 157–174 (2007)
4. Bali, R.: A theoretical review on SMS normalization using hidden Markov models (HMMs). Int. J. Comput. Trends Technol. (IJCTT) **4**(7), 2387–2388 (2013)
5. Grinter, R.E., Eldridge, M.A.: y do tngrs luv 2 txt msg? In: Proceedings of ECSCW, pp. 219–238. Springer, Netherlands (2001)
6. Pennell, D., Liu, Y.: Toward text message normalization: modeling abbreviation generation. In: Proceedings of the International Conference on Acoustics, Speech, and Signal Processing (ICASSP), 5364–5367 (2011)
7. Jose, G., Raj, N.S.: Lexical normalization model for noisy SMS text. In: First International Conference on Computational Systems and Communications (ICCSC), pp. 57–62 (2014)
8. Jose, G., Raj, N.S.: Noisy SMS text normalization model. In: International Conference for Convergence of Technology, pp. 1–6, (2014)
9. Beaufort, R., et al.: A hybrid rule/model-based finite-state framework for normalizing SMS messages. In: Proceedings of the 48th Annual Meeting of the Association for Computational Linguistics, Uppsala, Sweden, pp. 770–779 (2010)
10. Nocon, N., et al.: Norm API: an API for normalizing Filipino shortcut texts. In: Asian Language Processing (IALP), pp. 207–210. IEEE (2014)
11. https://gist.github.com/h3xx/1976236
12. http://www.math.sjsu.edu/~foster/dictionary.txt
13. Al-Saudi, J.: Error analysis and spelling mistakes of EFL learners at Tafila Technical University: a case study. Eur. Sci. J. **8**(21), 56–75 (2013)

Word Topic Prediction Model for Polysemous Words and Unknown Words Using a Topic Model

Keisuke Tanaka[1] and Ayahiko Niimi[2(✉)]

[1] Graduate School of Systems Information Science, Future University Hakodate, 2-116 Kamedanakano, Hakodate, Hokkaido 041-8655, Japan
[2] Faculty of Systems Information Science, Future University Hakodate, 2-116 Kamedanakano, Hakodate, Hokkaido 041-8655, Japan
niimi@fun.ac.jp

Abstract. A prediction method for word topics is proposed. Sentences are sometimes difficult to read if the meaning of one or more words is unknown. However, they become easily readable if the topics of the words are known. Therefore, a topic model in which the words are divided by topic and several words are chosen from the available topics was used to predict the topics. However, the model performed poorly when dividing polysemous and unknown words by topic. To address this problem, a new method was formulated by adding two new elements to the conventional method. One of the elements comprised separating polysemous words according to their examples. The other element involved associating unknown words with sentences composed of known words. Several experiments were conducted using the proposed method, which achieved accuracies of ∼60% and ∼80% for dividing polysemous and unknown words, respectively, by topic. Such results imply the superiority of the proposed method for these two tasks.

Keywords: Text mining · Topic model · Latent Dirichlet Allocation · Document classification · Polysemous words · Unknown words

1 Introduction

We currently have the opportunity to read many Japanese sentences in media such as books, newspapers, and web pages. During reading a sentence, the meaning of the sentence cannot be understood and reading comprehension becomes difficult.

Among the problems that make reading comprehension difficult are unknown reasons of not knowing the meaning of a word, such as new words or unnamed words that are unknown words that make up sentences. In this case, it is ideal to solve the problem by examining the dictionary meanings of the unknown words, so it is possible to solve the problem by examining the definitions of indiscriminable words one by one, but there are dictionaries and web pages that

K. Arai et al. (Eds.): CompCom 2019, AISC 997, pp. 860–866, 2019.
https://doi.org/10.1007/978-3-030-22871-2_61

describe the words. It is assumed that the information for resources is prepared, so if you do not have such information at hand, you cannot expect a solution by this method. In addition, it takes time to finish and read all the definitions and commentaries of necessary words or everything necessary for understanding the sentence until you can understand the meaning of the sentences.

Even if you cannot know the dictionary definition of a word, if you can know the topic information to which a word belongs as a rough meaning or background, you can roughly grasp the meaning indicated by the whole sentence.

For one of the methods to help solve such problems, we proposed topics to which the word belongs without resorting to the dictionary meaning of the word, such as information on a web page describing a dictionary or a word. We proposed a method to show the information [1]. In the proposed method, based on the topic model, words in the sentence are classified into topics, and from among the classified groups of words, referring to the weights of the words, a representative word is selected and the word by estimating whether it is related to a topic, it shows which topic the word in the sentence belongs to. However, this method cannot distribute estimation results only to one topic for synonymous words that have multiple examples and are used for multiple subjects with the same notation, and the use of ambiguous words. There is a problem that the performance of topic estimation deteriorates with respect to unknown words, which are difficult to obtain for sufficient data compared with the known words.

The goal of this research is to solve the problems of the method in graduate research in the background and improve the performance of topic estimation for ambiguous words and unknown words.

In this research, we propose a method for improving topic estimation for polysemous languages by dividing polysemous languages for each example and handling them as different words and giving different estimation results, as well as a method for improving topic estimation for ambiguous words, we propose a new topic estimation method, which incorporates a method to improve topic estimation for unknown words by performing topic estimation in relation to the topics of the composed sentences.

Topics in this research consist of groups of words classified by topic to which the word belongs, known sentences consisted of known words, sentences with clear topics, unknown words consisted only of known words, which do not exist in sentences, unknown sentences indicate sentences containing unknown words.

2 Related Research

Shimizu et al. estimated the impression of adjectives and adjectival verbs from the co-occurrence frequencies of the words [2]. Co-occurrence was limited to the co-occurrence of the adjectives and adjectival verbs. Estimating the impression of a word is realized by obtaining the similarity from the measurement of frequency, quantifying it according to the impression from the degree of similarity, and arranging and displaying the word on the space.

In a previous study on a new adaptation that tunes the classifier learned from the training data of a certain area in the task of natural language processing to the test data of another area in the task of the new language [3], we prepared a mechanism to measure the similarities between words and used Latent Dirichlet Allocation (LDA), one of the topic models, to make it correspond to the clustering result of the word. It is carried out.

Shirai et al. studied the dynamic learning of the features of the labels of documents in document streams and using multi-label classification of document streams by using the correlation between labels for labeling [4]. In this current study, we propose a model with an extended topic model to learn the topic distribution of document set by estimating labels and topics, which are latent variables that generate words from label vectors and word sets of documents. We also estimated the label vector for an unknown document by finding a single label from the likelihood that the document would be generated from each label and from the set of multi-labels co-occurring with that single label. It is realized by selecting multi-labels.

Horiuchi et al. conducted automatic labeling using the article title of Wikipedia for question documents posted to the Apple Support Community [5]. In this research, questions to the Apple Support Community, we applied the mixture ratio of the topics of topics of each document obtained by applying LDA to the document and the word generation probability of each topic to the article set of Wikipedia, which is another corpus, and applied the article, which becomes the label from the generation probability of the topic. Automation of labeling is realized by selecting a title.

Sasaki et al. conducting cluster based on meaningful words as clues to automate the construction of a dedicated thesaurus aimed at solving the semantic discrimination problem [6]. In this study, by extracting the co-occurrence data of words having an origination relation, conducting latent meaning analysis and executing word clustering, clusters of semantically connected words were constructed.

3 Proposed Method

In the proposed method, the following process is performed on document data, thereby estimating to which topic a word in a sentence belongs:

1. Morpheme analysis is performed on document data and weighting is performed to obtain information of the words constituting sentences.
2. For ambiguous words, we create a chunk of nouns before and after clustering it and breaking polysemous words for each example.
3. Create classification models by LDA from word information and classify words by topic.
4. We associate unknown words with topics of sentences consisting only of known words and estimated topics.

Each step of this method is described in more detail.

3.1 Morphological Analysis and Word Weighting

Morphological analysis is performed on the document data to which the topic information of the sentence is linked to obtaining the word constituting the sentence and its weight and topic information of the sentence including the word. After that, referring to the topic information of the sentence including the word and linking the word and the topic information. When the word is used in plural sentences, the topic of the sentence has the largest weight value. Morphological analysis is performed using the morphological analysis software MeCab [7] and R language RMeCab [8] package to calculate the weight value by TF-IDF.

3.2 Division of Polysemous Words

We prepared a 2-g with each morpheme in the document data as a unit, extracted the elements that became the two previous words of the polysemous word from there, and the elements that become the latter two words and concatenate it with the polysemous word. For chunks created for the same polysemous word, by performing non-hierarchical clustering by the k-means method using the number of examples of polysemous words as the number of clusters, chunks can be used as examples of polysemous words. The polysemous words cut for each example were processed to treat them as different words even in the same notation in the future process and update the weight calculation of the words. The algorithm used for clustering was the Hartigan–Wong algorithm [9].

3.3 Creation of Classification Model by LDA

By classifying words by topic by creating a classification model that estimates topics possessed by sentences containing that word from words based on LDA [10] based on information on weights of each sentence of words obtained by morphological analysis, Estimate the topic of a word. To use the R language [11] MASS package [12] for LDA implementation.

3.4 Topic Estimation of Unknown Words

A sentence consisting only of known words whose cosine similarity to the sentence, including the word is the maximum, was selected for the unknown word and the topic information of the sentence was used as the estimation result of the topic of the unknown word. Overwrite the topic estimation result of the classification model of LDA.

4 Experiment and Results

To evaluate whether the accuracy of topic estimation performed on polysemous words or unknown words had actually improved on the basis of the integrated topic estimation method, we estimated the topic estimation result by integrated topic estimation method. Experiments were conducted to compare the topic information classified by topics and grading study methods.

We selected words that were assumed to be ambiguous words and unknown words from the words in the data to be the test part in the evaluation experiment [13]. In the case of polysemous words, the top 1,000 words used in more sentences were assumed to be polysemous words. It was divided into three examples according to the method of dividing ambiguous words, they were treated as different 3,000 words. Regarding unknown words, the books were published in 2005. Among the words included in the text, 3,681 words, which had not been included in the texts of the books published from 2001 to 2004, were assumed to be unknown words. Table 1 shows the number of polysemous and unknown words in the data.

In the experimental flow, morphological analysis first was performed on the document data in the corpus to extract words and weight information. Topic information obtained from the file in the corpus was used as topic information of a sentence including a word. After that, the entire word information was divided for training of the classification model, which is the 4/5 of all the words and for the test in the evaluation experiment which is the remaining 1/5 part, and the part for training and we created a classification model based on LDA based on LDA and classified the topic for the word of the test part. The topic based on the topic estimation method of the unknown word was additionally added to the word assumed as the unknown word. The topic estimation result by the classification model was overwritten and if the topic estimation result for the obtained ambiguous word or unknown word catches the topic information associated with the word, the classification is correct. Based on the standard, we evaluated and measured the accuracy of classification and compared it to the classification accuracy in the method of graduation research for the same ambiguous words and unknown words. In the case of polysemic words, the method in graduation research is a case of ambiguous words. Since it was not carving out, we treated all the same ambiguous words as giving the same estimation results.

Table 1. Numbers of polysemous and unknown words

Total number of words	Number of polysemous words	Number of unknown words
92,318	3,000	3,681

The experiment showed that the classification accuracy by the integrated topic estimation method for the total of 6,681 words of the polysemous words and the unknown words was 0.73. The classification accuracy by the LDA-only method was 0.46, among which the integrated topic estimation method for 3,000 ambiguous words. Classification accuracy was 0.64, classification accuracy by graduation research method was 0.41, classification accuracy by integrated topic estimation method for 3,681 unknown words was 0.81, and the classification accuracy by graduation research method was 0.52. The classification accuracy by the integrated topic estimation method for all words, including words other than the words assumed as ambiguous words and unknown words, was 0.71 and the classification accuracy by the LDA-only method was 0.62 The comparison of these results is shown in Table 2.

Table 2 shows that the integrated topic estimation method had improved by about 20% for polysemous words and about 30% for unknown words compared to LDA alone.

Table 2. Comparison of classification accuracy of integrated topic estimation and LDA-only methods

	Integrated method	LDA-only method	Increased value of classification accuracy
Polysemous and unknown words	0.73	0.46	0.27
Polysemous words only	0.64	0.41	0.23
Unknown words only	0.81	0.52	0.29
All Words	0.71	0.62	0.09

5 Conclusion

We proposed a word in sentences as one of the methods to help solve the problem of the difficulty of reading comprehension of a sentence due to the unknown meaning of the words constituting the sentence, which cannot be understood the topic estimation performance for polysemous words, and unknown words have a problem, which is the method of showing which topic belongs had been taken up. The problem of dividing polysemous words into sentences in the examples for this problem and proposing a new topic estimation method that incorporates a topic estimation method that associates unknown words with topics of sentences composed only of known words. The results of evaluation experiments on the proposed method respond to polysemous words. The method, which succeeds in the incorrect division of each example for polysemous word in sentences, with an accuracy of about 80%, the method according to unknown words is about 90% accurate to the unknown word in the sentence and the correct topic can be estimated. The new topic estimation method incorporating them improves the topic estimation accuracy for polysemous words by about 20% and the topic estimation accuracy for unknown words by about 30% compared with the conventional method. As a result, the proposed method that incorporates the method for dividing ambiguous words and the method for estimating topics of unknown words improves the problem of degrading the topic estimation performance of polysemous words and unknown words of conventional methods. It can be concluded that an estimate can be made.

As for future topic estimation of ambiguous words, as a future subject of this research, criteria on what kind of word is treated as a polysemous word and criteria for separate processing, as well as how many kinds of examples are used for polysemous words, clustering, it is necessary to clarify the criterion as to whether to define the number of clusters in the cluster and to set parameters in clustering automatically. Regarding the topic estimation of unknown words, it is necessary to realize topic estimation that spans sentences in multiple media in which topic definitions and classification methods are not consistent.

References

1. Tanaka, K., Niimi, A.: Topic classification and word extraction by word topic model. DEIM Forum 2016 (C2–6), pp. 1–6 (2016)
2. Shimizu, K., Hagiwara, M.: Image estimation of words based on adjective co-occurrences. The IEICE Transaction D: Information and Systems, J89-D (11), pp. 2483–2490 (2006)
3. Shinnou, H., Sasaki, M.: Domain adaptation for word sense disambiguation using k-nearest neighbor algorithm and topic model. J. Nat. Lang. Process. **20**(5), 707–726 (2013)
4. Shirai, M., Miura, T.: Multilabel classification of document streams based on topic model. DEIM Forum 2014 A9-1, pp. 1–5 (2014)
5. Horiuchi, Y., Wajima, K., Furukawa, T.: Labeling automation for help desk work efficiency. DEIM Forum 2015 D1-4, pp. 1–4 (2015)
6. Sasaki, M., Shinnou, H.: Word clustering for word sense disambiguation. IPSJ Natural Language Processing (NL) 154-21, pp. 142–145 (2003)
7. MeCab: Yet another part-of-speech and morphological analyzer. http://taku910.github.io/mecab/. Accessed 22 Dec 2017
8. RMeCab. Available: https://sites.google.com/site/rmecab/. Accessed 22 Dec 2017
9. Hartigan, J.A., Wong, M.A.: A K-means clustering algorithm. J. Royal Stat. Soc. Ser. C (Appl. Stat.) **28**(1), 100–108 (1979)
10. Blei, D.M., Ng, A.Y., Jordan, M.I.: Latent Dirichlet allocation. J. Mach. Learn. Res. **3**, 993–1022 (2003)
11. The R Project for Statistical Computing. http://www.rproject.org/. Accessed 22 Dec 2017
12. CRAN - Package MASS. https://cran.r-project.org/web/packages/MASS/index.html. Accessed 22 Dec 2017
13. The Balanced Corpus of Contemporary Written Japanese (BCCWJ). http://pj.ninjal.ac.jp/corpus_center/bccwj/. Accessed 22 Dec 2017

Improving Usability of Distributed Neural Network Training

Nathaniel Grabaskas(✉)

University of Washington, Bothell, WA 98011, USA
ngrab@uw.edu

Abstract. The recent success of Deep Neural Networks (DNN) has triggered a race to build increasingly large DNNs; however, training speed is a known limitation. To solve this limitation, distributed neural network training has become a large area of research. Usability, the complexity for a user to implement distributed neural network training, is an aspect rarely considered, yet critical. The impact of complexity on software development has been studied extensively and evidence shows increased complexity has a direct correlation on development effort, maintainability, and fault-proneness of software. In this work, we investigate if automation can greatly reduce the implementation complexity of distributing neural network training across multiple devices without loss of computational efficiency when compared to manual implementations. Experiments were conducted using convolution and fully-connected neural network layers to perform image classification on a four-node NVIDIA Jetson TX1 cluster. Our module, Torch Automatic Distributed Neural Network (TorchAD-NN) reduced the implementation complexity of data parallel neural network training by more than 90% and provided components to easily parallelize fully-connected neural layers.

Keywords: Machine learning · Distributed systems · Usability · Neural network · Torch · MPI

1 Introduction

Recently, Deep Neural Networks (DNNs) have seen tremendous success [1] in a wide range of applications. This success has triggered a race to build increasingly large DNNs [2]. These, in turn, need more and more data to solve complex problems within a quickly growing field of applications [3]. Increasing the scale of deep learning, with respect to the number of training samples, the number of model parameters, or both, can drastically improve ultimate classification accuracy [4]; however, a known scaling limitation is the training speed [5]. A high-accuracy DNN model can take weeks to train on a modern GPU [6] and is a key challenge affecting DNN development [7]. To solve this speed problem, distributed Neural Network (NN) training has become an increasingly large area of research [8, 9].

There are many research challenges with distributed neural network training, most commonly seen are increasing speed-up efficiency and maintaining network accuracy [10–12]. Usability, the complexity for a machine learning user or data scientist to

K. Arai et al. (Eds.): CompCom 2019, AISC 997, pp. 867–886, 2019.
https://doi.org/10.1007/978-3-030-22871-2_62

implement distributed neural network training, is an aspect rarely considered, yet we consider equally as critical. The problem is manual parallelization requires significant effort and knowledge. This effort (or implementation complexity) is a major concern as the impact of complexity on software development has been studied extensively and evidence shows increased complexity has a direct correlation on development effort, maintainability, and fault-proneness of software [13–15]. The terms usability and complexity are used interchangeably in our research. They refer to the difficulty of coding a distributed neural network training instance.

Table 1. Dataset and model parameter names for CPU training and the code necessary to convert the training to GPU.

CPU training	GPU training
Dataset	Dataset:cuda()
Model	Model:cuda()

Observing the ease in converting neural network training from CPU to GPU in Torch inspired our research. This ease is demonstrated in Table 1, where a Dataset and Model references can be converted to GPU training by simply adding the ":cuda()" function at the end. We sought to bring this high level of usability to distributed training across multiple devices to reduce development effort and mitigate fault proneness [16]. We framed our hypothesis: automation can greatly reduce the implementation complexity of distributing neural network training across multiple devices without loss of computational efficiency when compared to manual parallelization.

In this work, we present the Torch Automatic Distributed Neural Network (TorchAD-NN) module to improve usability, maintain speedup, and maintain the accuracy of distributed neural network training. This work aims to help users in small departments, scientists working in fields unrelated to computer science, and those working in machine learning as a hobbyist who lacks the funding and access to large-scale GPU clusters. TorchAD-NN also targets machine learning engineers and data scientists who do not have a strong foundation in software engineering and/or distributed computing. For those mentioned, access to current distributed neural network training comes with a high cost of implementation complexity. TorchAD-NN gives the powerful capabilities of distributed neural network training without requiring expertise and in-depth knowledge of distributed systems programming.

The rest of this paper is organized as follows: Sect. 2 clarifies background and discusses related work; Sect. 3 describes methods and metrics used; Sect. 4 illustrates and analyzes our experiment results; Sect. 5 concludes our discussion.

2 Related Work

Neural network parallelism comes in two flavors: data and model parallelism [3, 5, 9, 17]. Data parallelism is where each worker gets an equal subset of the data, a complete copy of the neural network, and then workers communicate by exchanging weight gradient updates throughout the training. Model parallelism is the case where each worker gets the complete dataset, a partition of the neural network, and workers communicate by synchronizing parameters and gradients from each partition of neurons. We chose to implement automation for both data and model parallel approaches to avoid limiting the applicability and usefulness of this work.

Throughout the paper, manual parallelism refers to manually adding the necessary code to take neural network training from a single node to multiple nodes. This uses the neural network libraries as they stand without modification. Automated parallelization refers to using our module to take neural network training from a single node to multiple nodes.

2.1 Data Parallelism Related Work

Data parallelism research has several similarities and differences between the various approaches. Research from Iandola et al. introduces FireCaffe [17] which used adjustable batch size to reduce the total quantity of communications required during training. Communication was conducted using a reduction tree. When training GoogLeNet [6] and Network-in-network [18] on ImageNet [19] they achieved 36% and 30% speedup efficiency, respectively, on a 128 GPU cluster. Their strategy for scaling up DNN training was to focus on reducing communication overhead. They use fast interconnects such as Infiniband or Cray Gemini to accelerate communication among the GPUs.

Sergeev et al. introduced the Horovod [20] framework. Building on the realization a ring-allReduce approach can improve over reduction trees they adopted Baidu's draft implementation of the TensorFlow ring-allReduce algorithm and built upon it. NCCL 2 introduced the ability to run ring-allReduce across multiple machines, enabling them to take advantage of its many performance boosting optimizations.

Gu et al. built the cNeural [10] framework for fast training of large-scale datasets with millions of training samples. To achieve this goal, firstly, cNeural adopts HBase for large-scale training dataset storage and parallel loading. Secondly, it provides a parallel in-memory computing framework for fast iterative training. Third, they choose a compact, event-driven messaging communication model instead of the heartbeat polling model for instant messaging delivery.

Horovod, FireCaffe, cNeural, and our research all conduct data parallelism in a similar manner. The dataset is split amongst the nodes and each node has an exact copy of the model being trained. Each node trains the model using its respective data and the models are synchronized given a set frequency and communication method. Our research does not focus on improving the communication scheme or training datasets with millions of instances, instead, we focus on small-scale implementation needs with lowered implementation complexity to the user. Our contribution is the automation module; it makes the parallelization invisible to the user by providing them with a single method call that handles the data distribution, synchronization frequency, and setting the training function synchronization override. Data parallelization is discussed further in Sect. 3.4.

2.2 Model Parallelism Related Work

DistBelief [5], the predecessor to TensorFlow, is a framework outlined by Dean et al. The user defines the computation to take place at each node in each layer of the model, and the messages to be passed during the forward and backward phases of computation. For large models, the user may partition the model across several machines, so responsibility for the computation on different nodes is assigned to different machines. When a model is split across multiple nodes vertically, horizontally or both, only the connections between the neurons on the edge of the graph are synchronized. This helps reduce the time needed for communication and allows flexibility in optimizing the network distribution for a specific topography. The downside is the user must define the computation and communication to take place at each node in each layer during forward and backward propagation. Our module provides components to automatically parallelize layers of the network. This allows the user to easily build their network just as they would normally and choose which layers are parallelized and which are not, enabling a high level of usability while still giving the user a large amount of control.

Commonly known from Google and outlined by Abadi et al., TensorFlow [9] is an interface for expressing machine learning algorithms and an implementation for executing such algorithms. The system has been used for conducting research and for deploying machine learning systems into production across computer science and other fields. They accomplish parallelism automatically by first analyzing the node topography. Given a computational graph, the input and output size, in bytes, of each layer is estimated, along with the computation time required. Simulated execution of the entire graph is then conducted, and the optimal device is chosen for each node using greedy heuristics. For model parallel training, this means each node is given a layer of the network and the input/output between layers is communicated. The downside is parallelization is dependent on the number of layers present in the network. A three-layer network cannot be parallelized on more than three nodes.

Our parallelization is not configured using an elaborate method of estimation and heuristics; we simply divide each layer to be parallelized across the given nodes, allowing parallelization to be implemented quickly. Horovod, FireCaffe, cNeural focus on data parallelism techniques and reducing communication overhead; therefore, there is no comparison between our model parallel approach and these libraries. Model parallelism is discussed further in Sect. 3.5.

3 Methods

This section starts with the high-level system overview, then outlines the communication protocol from TorchMPI and describes the custom synchronization function implemented. Next, data parallel and model parallel module designs are covered. Finally, metrics used for evaluation are defined.

3.1 High-Level System Design

Torch [21] and TorchMPI [22] (publicly available libraries) provide the necessary neural network and CUDA frameworks to conduct training on the GPU, and TorchMPI executes inter-node communication from within Torch and adds synchronization functions.

These automation modules are built on top of many existing layers (see Fig. 1), to provide extensive functionality with minimal complexity for the user. Our TorchAD-NN module is inserted between the user application on the top and existing Torch libraries such as cuNN (CUDA NN) and TorchMPI below. This enables the functionality of TorchMPI and cuNN libraries while abstracting the complexity from the user.

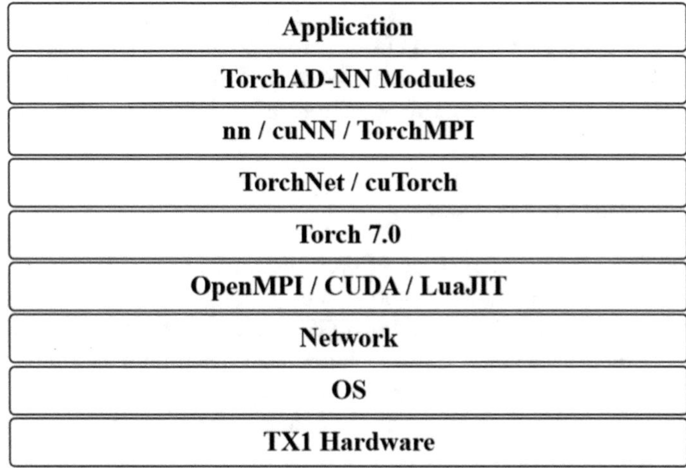

Fig. 1. High-level system architecture view. TorchAD-NN inserted at the top, right below the user application.

3.2 Communication Protocol

Torch is the library of choice for our research; however, the standard Torch installation does not support inter-node communication. TorchMPI, a library released by the Facebook research team around the beginning of 2017, provides many MPI communication paradigms as well as additional paradigms specific to NN training. Of the many methods provided by TorchMPI, the following are used for our research [22]:

- Collectives wrap a subset of MPI collectives useful for deep learning. These collectives operate on Torch tensors and scalar values with synchronous or asynchronous flavors; however, only synchronize calls are used in our work.
  - Allreduce Tensor – This function operates in the same manner as allReduce with MPI operation SUM. The input from each node is gathered and added together, every node is left with the exact summation result.
- TorchMPI NN extends Torch.nn with support for synchronous and asynchronous collectives to run on a distributed cluster of CPUs and GPUs. Only synchronize calls are used.
  - Synchronize Parameters – uses underlying Allreduce Tensor functionality to synchronize model parameters across all nodes during training.
  - Synchronize Gradients – uses underlying Allreduce Tensor functionality to synchronize model gradient weights across all nodes during training.

3.3 Synchronize Model

Initial tests showed Synchronize Parameters and Synchronize Gradients were insufficient to effectively distribute and train neural networks using Stochastic Gradient Descent (SGD) training. To overcome this, we wrote a custom model synchronize method which borrows functionality from the TorchMPI methods. This new method uses Synchronize Parameters with each node receiving the average value of the parameters, as opposed to the sum, from across all nodes. This method also uses Synchronize Gradients modified so that each node receives the average value of the gradients from across all nodes. This synchronization is used in all our data parallel results.

3.4 Data Parallel Module Design

We use Iandola's common definition of *data parallelism* [17]. Each worker gets an equal subset of the dataset, and then the workers communicate by exchanging weight gradient updates. For example, given the MNIST dataset with 60,000 samples and four nodes, each node gets 15,000 samples. Batch size determines synchronization frequency, at the end of each batch, the gradients and parameters are synchronized using the synchronize model method. At the end of training each node has a complete copy of the trained model (see Fig. 2).

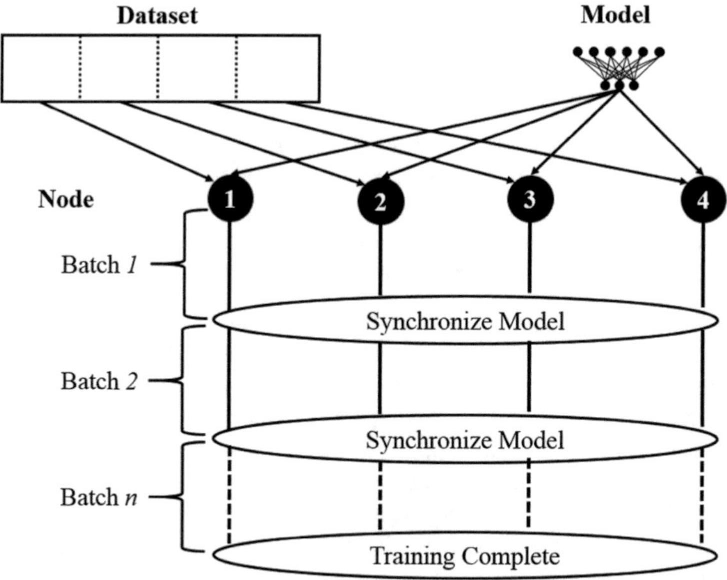

Fig. 2. In data parallelism, the data is partitioned equally across all nodes and each node gets a full copy of the NN model. During training, the models are synchronized at the end of each training batch. At the end of the training, each node contains a complete copy of the trained model.

Automated parallelization can be broken into four stages (see Fig. 3): initialize, partition dataset, determine synchronization frequency, and override. Stage (1) shows the main function of the data parallel module. The dataset and model being parallelized are passed in, communication and parallelization parameters are initialized.

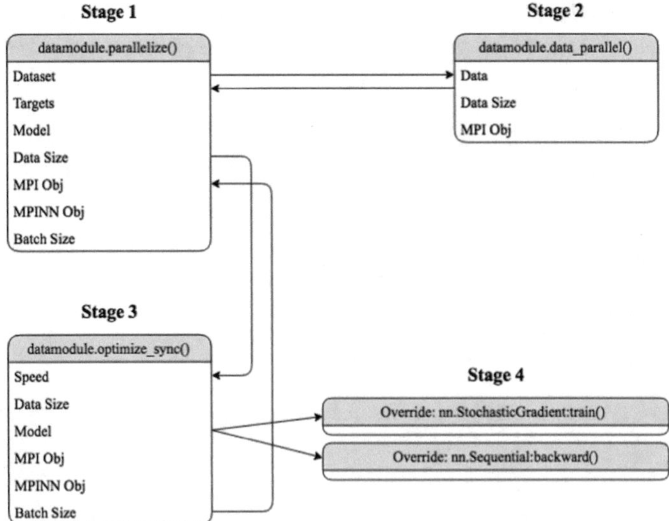

Fig. 3. Flow chart outlining the four stages of automated data parallelization. Parameters and objects used for each stage are included.

Stage (2) ensures the dataset is split evenly amongst all the nodes present. Once complete the new dataset is finalized.

Stage (3) performs optimization. If a batch size is provided this value is used to set the synchronization frequency Otherwise, one will be chosen using simple heuristics based on the size of the dataset.

Stage (4) overrides existing training functions with the necessary parallelization functionality. Training using the standard SGD engine from Torch 7.0 and any custom training using backward propagation are supported. This concludes stage 4 and continues execution along the initial training path. Example code for both automated and manual parallelization are provided below to demonstrate the reduction in complexity.

Automated parallelization example from experiments with SGD training. MPI, data partitioning, and synchronization are abstracted.

```
-- Import Libraries
automation = require 'torchad_nn.datamodule'

-- Define Model & Load DataSet
-- Parallelizing
trainData.data, trainData.labels, trsize = automation.parallelize
(trainData.data, trainData.labels, model, trsize)

-- Perform Training
trainer:train(trainData)
-- Evaluate
confusion = optim.ConfusionMatrix(classes)
```

Manual parallelization example. MPI must be controlled by the user, the dataset is partitioned onto each node, and training functions must be overridden to execute synchronization.

```
-- Start MPI
local mpi = require('torchmpi')
mpi.start(true)  --true equals use GPU

-- Define Model
-- Load DataSet

-- Synchronize Models
local mpinn = require('torchmpi.nn')
mpinn.synchronizeParameters(model)

-- Prepare Dataset for Parallelization
stripe = trsize / mpi.size()
newdata = {
             data = {},
             labels = {},
             size = function() return stripe end
       }

local start  = ( mpi.rank() * stripe ) + 1
local finish = start + (stripe - 1)
newdata.data = trainData.data[{ {start,finish} } ]
newdata.labels = trainData.labels[{ {start,finish} }

-- Override Training Function to Synchronize at end of Batch
function nn.Module:accUpdateGradParameters(input, gradOutput, lr)
       ...
       -- sync and shutdown when dataset is complete
       if (self.sync_counter == stripe) then
             mpinn.synchronizeGradients(model)
       elseif (self.sync_counter % opt.batchSize == 0) then
             mpinn.synchronizeGradients(model)
       end
       self.sync_counter = self.sync_counter + 1
       ...
end

-- Perform Training
trainer:train(newdata)
mpi.barrier()

-- Evaluate and Stope
confusion = optim.ConfusionMatrix(classes)
mpi.stop()
```

3.5 Model Parallel Module Design

For *model parallelism*, we again use Iandol's definition [17]. Each worker gets a subset of the model parameters, and the workers communicate to synchronize the output of a parallelized layer. Our model parallelism is accomplished by splitting each layer across available nodes. For a single fully-connected layer with size 1024 input and 10 output classes split across two nodes, each node takes 512 (1024 inputs/2 nodes) inputs and each generates output size 10, the output is synchronized across all nodes (see Fig. 4). This allows each node to share half of the input load and with correct synchronization minimizes accuracy loss. This method was inspired by Zhang et al. and their work with TorchMPI [23, 24].

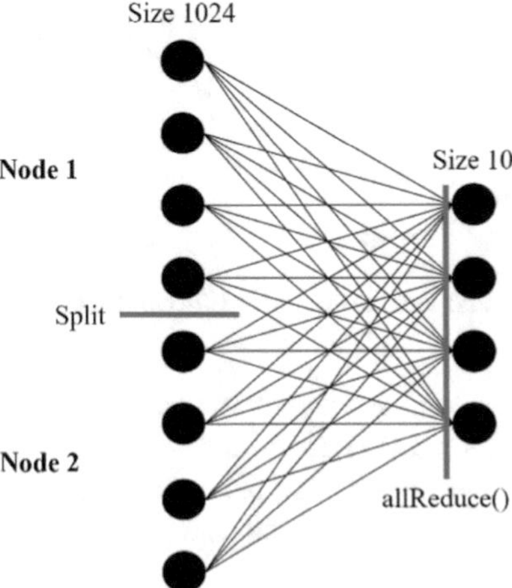

Fig. 4. This model demonstrates input of size 1024 split between two nodes, each node takes half (size 512). Each node generates its own output of size 10 and is then synchronized between the nodes using Allreduce Tensor.

Implementing this for a multi-layer network requires a deeper understanding of how the training takes places. For Torch's SGD training engine, a network trains through two phases: forward and backward propagation. Consider a MultiLayer Perceptron (MLP) network composed of two layers with the following components: Linear1-> Tanh-> Linear2. During forward propagation, the data moves through the network by entering Linear1, then moving to Tanh, and then entering Linear2. For backward propagation the output from Linear2 is then sent back through the network by entering Linear2, then Tanh, and finally Linear1 in reverse. As a result, multilayer parallelization is more difficult as the output and input dimensions must be coordinated through both propagation phases.

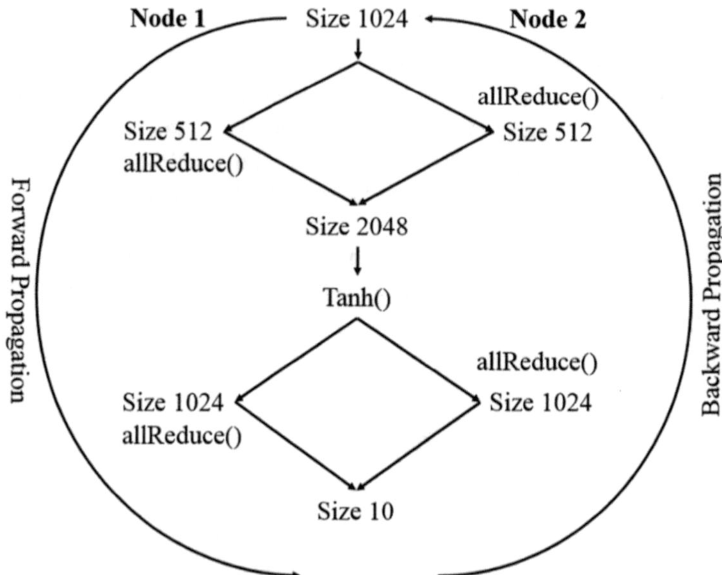

Fig. 5. Narrowing and expanding during forward and backward propagation.

The narrowing, expansion, and synchronization during forward and backward propagation using a simple MLP (1024-> 2048-> 10) are illustrated in Fig. 5. Input starts at size 1024, is narrowed to size 512 between two nodes, and each generates their respective size 2048 output. Each node then uses Allreduce to synchronize their output and proceed through the Tanh component. After Tanh, the input is again narrowed from 2048 to 1024 on each node. Allreduce is used again to synchronize output. This concludes the forward propagation. Backward propagation is the reverse, the output of 10 is then updated back to 1024 and this 1024 is synchronized between the nodes with Allreduce, which is fed into the Tanh component. Here the Tanh component must expand 1024 to 2048 since the next layer requires this size. The 2048 is then updated back up into each node's 512 and again synchronized using Allreduce. Each full cycle of forward and backward propagation requires four Allreduce calls.

Model parallel wrapper function example.

```
function MPInitialLinear.updateOutput(self, input)
      local input = narrowInput(input)
      self.output = nn.Linear.updateOutput(self, input)
      mpi.allreduceTensor(self.output)
      return self.output
end
```

This parallelization is implemented on each component using a wrapper function. For the forward propagation phase, updatOutput() is wrapped for each component, and during backward propagation, updateGradInput() is wrapped for each component. The code above shows an example of component wrapping. Line 2, shows the function input being narrowed using a custom narrowing function. This evenly splits the input neurons across all nodes in the MPI communicator. Line 3, shows the unmodified updateOutput() called with the narrowed input. Line 4, shows the output being synchronized using allreduceTensor(). In this way, any changes to the underlying neural network libraries will be reflected in TorchAD-NN model parallelism.

Linear and Reshape components vary slightly depending on if they are at the top of a neural network or on any layer below. Initial Reshape and Linear are used if they are the first or second layer, otherwise, Base Reshape and Linear are used. Three different operations are applied to variables within the components. Narrowed, is where values are evenly split amongst all nodes within the MPI communicator. Expanded, is where values are concatenated to reform their pre-narrowing size. Synchronized, is where the values are communicated across all nodes and each node is left with identical values. Each component is briefly described below:

- MPInitialReshape – This component narrows the input to be fed into the top layer of the network.
- MPBaseReshape – This component expands the output for the layer above during backward propagation.
- MPInitialLinear and MPBaseLinear – These components allow the input layer to be narrowed over multiple nodes within the network. The output is then synchronized. They perform the same function but in different ways depending on their location in the network.
- MPTanh – This component narrows the input and expands the output during backward propagation.

3.6 Usability Metrics

Two popular methods to measure complexity are McCabe's Cyclomatic complexity [25] and Halstead's software science [26]. Cyclomatic Complexity (CC) was first introduced by Thomas J. McCabe Sr. in 1976. Since then, it has become one of the most common methods to quantify the complexity of a program in a single number. CC is calculated by using the control flow graph of the program and describes the non-linearity of this graph. In summary, it counts the number of linearly independent paths through the source code [25].

Table 2. HCM measurements and their formulae [27].

Meaning	Symbol	Formula
Size of vocabulary	n	$n_1 + n_2$
Program length	N	$N_1 + N_2$
Program volume	V	N * log2 n
Difficulty level	D	$n_1/2 * N_2/n_2$
Effort to implement	E	V * D
Time to implement	T	E/18

Halstead's Complexity Measures (HCM) were first introduced by Maurice Howard Halstead in 1977. Much like McCabe's metric, HCM has become one of the most commonly used systems to quantify a program's complexity. HCM goes even further, measuring various metrics (see Table 2), such as time to program, programming effort, etc.

We chose to use HCM as our usability metric. CC only represents a program's complexity from the viewpoint of the computation to be performed and does not consider complexity from the viewpoint of the programmer. HCM delivers a quantitative measure of effort needed by the programmer to implement a given solution using unbiased metrics.

Operators are defined as keywords and symbols used to define a function. Examples include a function call, an if-then statement, parentheses, semicolons, require keyword, etc. In Table 2, n_1 is the number of unique operators and N_1 is the total number of operators.

Operands are defined as any variable or number used in the calculations. It can also refer to more complex entities such as the network model or keywords used to access a data instance, but not a function of these. In Table 2, n_2 is the number of operands and N_2 represent the total number of operands.

3.7 Usability Metrics

Below are listed the various performance metrics used and their associated definitions:

- Speed – Speed is measured using the built-in system tic and toc functions. Training time measures the amount of time needed to train the model. Time needed to test the trained model was not recorded or analyzed for this research.
- Speedup – Speedup is calculated by taking the training time from a (1) node implementation and dividing it by the time from a (n) node implementation (see Eq. 1). It represents how many times faster the program ran.

$$Speedup = \frac{Time\,Single\,Node}{Time\,n\,Nodes} \tag{1}$$

- Efficiency – Efficiency is calculated by taking the speedup and dividing it by the number of nodes used for the faster implementation (see Eq. 2). It is given as a percentage.

$$Efficiency = \frac{Speedup}{n\,Nodes} \tag{2}$$

- Accuracy – Accuracy is measured using a test data set and is calculated for each category using a confusion matrix and then aggregated to give the overall global accuracy of the trained model.

- Accuracy Loss – The amount of accuracy lost when parallelizing the model training (see Eq. 3).

$$Accuracy\,Loss = Accuracy\,Single - Accuracy\,n\,Nodes \tag{3}$$

4 Evaluation

First, we outline models used and the hardware setup. Next, the evaluation is covered in four main points broken down into the categories of complexity and performance. For each category data and model parallelism are analyzed.

4.1 Model and Hardware Setup

Models. Object detection and image classification use many convolution layers and are essential to applications such as robotics [28] and autonomous vehicles [29]. Increased training speed for these Convolutional Neural Networks (CNN) [2, 30] still often requires complex manual parallelization. Therefore, we focus our evaluation on image classification. We use publicly available datasets and networks directly related to current applications in image classification. Datasets used are: Canadian Institute For Advanced Research (CIFAR10) [31] and the Modified National Institute of Standards and Technology database (MNIST) [32]. Networks used are Network-in-Network (NiN) [33], MultiLayer Perceptron (MLP) (see code below), and CNN (see code below) with publicly available source code [34]. Three variations of the MLP are:

- MLP-Small = L1 (size = 1024) | L2 (size = 1024) | L3 (size = 1024)
- MLP-Medium = L1 (size = 1024) | L2 (size = 4096) | L3 (size = 2048)
- MLP-Large = L1 (size = 1024) | L2 (size = 8192) | L3 (size = 4096)

MultiLayer Perceptron architecture in Torch code. Three variations of this model were used with the layer sizes listed above.

```
Reshape(L1)
Linear(L1 -> L2)
Tanh
Linear(L2 -> L3)
Tanh
Linear(L3 -> 10)
```

Convolution Neural Network architecture in Torch code. Only one variation of this model was used.

```
SpatialConvolutionMM(3 -> 64, 5x5)
ReLU
SpatialMaxPooling(3x3, 3,3)
SpatialConvolutionMM(64 -> 64, 5x5)
ReLU
SpatialMaxPooling(3x3, 3,3)
View(64)
Dropout(0.500000)
Linear(64 -> 100)
ReLU
Linear(100 -> 10)
```

Hardware. All experiments were conducted on a cluster consisting of four Jetson TX1 developer boards, one 8-port 1gps switch, Ubuntu 16.04 LTS, OpenMPI 2.0.1, and Torch 7.0.

4.2 Data Parallel Complexity

Halstead's measures (HCM) demonstrate TorchAD-NN data parallelism requires significantly less effort to implement than manually coding a parallel solution. Table 3 measures the calculated effort needed for each implementation variation. These numbers show TorchAD-NN gives a 96% reduction in implementation complexity. GPU SGD training is also given as a comparison since the inspiration for the modules came from the ease of porting ANN training to the GPU.

Table 3. Effort and LoC measurements for data and model parallel implementations using HCM.

Parallel method	Implementation	Effort (HCM)	LoC
Data parallel	TF - multi GPU training	21,368.11	25
Data parallel	TF - distributed training	18,937.44	13
Data parallel	Torch manual SGD training	22,695.79	25
Data parallel	Torch GPU SGD training	1,423.01	6
Data parallel	Torch automated SGD training	1,008.31	2
Model parallel	Torch manual SGD training	N/A	131
Model parallel	Torch GPU SGD training	N/A	6
Model parallel	Torch automated SGD training	N/A	4

For manual implementation complexity comparison outside of Torch, we also looked at two TensorFlow (TF) distributed training examples (see Table 3). Multi-GPU Training is an example using a CNN on the CIFAR10 dataset and distributed across multiple GPUs on the same node. Distributed Training is an example of an MLP trained on the MNIST dataset across multiple nodes. When compared with TorchAD-NN, we see our module on average requires 95% less effort to implement than TensorFlow.

TorchAD-NN has low effort because it encapsulates nearly all the necessary parallelization techniques and reduces them to a single function call making the parallelization complexity invisible to the user. Manual complexity effort is high because it requires only a subset of the dataset be iterated over on each node, and this subset must be calculated specifically to its respective position in the node network to ensure the entire dataset is covered. These calculations have a high effort associated with them.

By encapsulating the necessary operations to hide data splitting and synchronization from the user to a single function call, TorchAD-NN significantly reduces coding effort, which increases usability. Supporting backward synchronization allows a broad range of training functions to be parallelized with TorchAD-NN.

4.3 Model Parallel Complexity

HCM was not used with model parallelism as the difference in implementation effort between manual and automated is misleading. In data parallel, the code between manual and automated differs. Therefore, HCM more accurately captured the difference in complexity. In contrast, model parallel manual and automated execution code are identical. The difference being, parallelized component code is included as part of the TorchAD-NN module and only requires the user to replace the component names (e.g., replace Linear with MPBaseLinear). HCM metrics could mislead into thinking there was a difference in implementation complexity because the code was different, but the only difference is abstraction away from the user. To represent this, we chose to use Lines of Code (LoC) as the metric.

LoC required for implementation show TorchAD-NN model parallelism requires significantly less effort to implement than manually coding a parallel solution. Table 3 illustrates the number of LoC required to implement each of the three variations. Only requiring 4 LoCs the TorchAD-NN model parallel module requires 33% fewer LoCs than GPU SGD Training and 97% fewer LoCs than manual model parallel SGD Training. GPU values are given as a baseline since the inspiration for the modules came from the ease of porting ANN training to the GPU.

The standard Torch NN library contains many more components than included within the TorchAD-NN library. Each layer component requires analysis and specific parallelization approach, which also depends on where it falls within the network structure. This cost led to only a limited number of layers being supported. The most common layer components were chosen for parallelization. The key to successful parallelization is the ability to hide communication overhead with calculations. By supporting individual component parallelization in TorchAD-NN, the power of a custom parallelized neural network is available with minimal complexity. This gives a high level of customization and flexibility to how a model is parallelized.

4.4 Data Parallel Performance

In Table 4, MLP-Small-Manual and MLP-Small-Auto are compared running on one, two, and four nodes. The automated version runs slightly faster and more efficiently than the manual version with equivalent accuracy. The slight increase in efficiency from manual to automated is because a small amount of time is saved by only pre-processing the node's partition of the dataset versus the entire dataset (in our results manual parallelization does not modify the dataset, only restricts which data is accessed).

Table 4. MLP-small on MNIST and network-in-network on CIFAR10 data parallelism results using SGD

Model	Nodes	Train time	Test acc.	Speedup	Efficiency
MLP-small	1	178.05	98.27%	1.00	–
MLP-small-manual	2	99.46	98.03%	1.69	84.65%
MLP-small-manual	4	52.01	97.40%	3.24	80.93%
MLP-small-auto	2	94.82	98.03%	1.78	88.79%
MLP-small-auto	4	51.90	97.40%	3.24	81.11%
NiN-auto	1	2,829.29	72.23%	1.00	–
NiN-auto	2	1,675.53	70.94%	1.69	84.43%
NiN-auto	4	966.81	71.67%	2.93	73.16%

Network-In-Network (NiN) results from training on the CIFAR10 dataset show automated parallelization can handle larger models and datasets while still demonstrating similar efficiency. Table 4 demonstrates for SGD training executed on the GPU with CIFAR10 the average speedup efficiency is 78.78% with an average model accuracy loss of 0.93%. Figure 6 visualizes the training time reduction from one to four nodes.

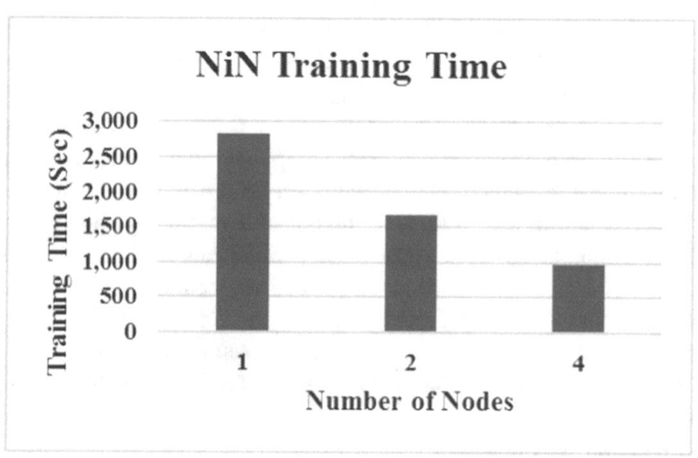

Fig. 6. Network in network training time comparison between 1–4 nodes

While performance is not the primary focus of our research, it is important to show automated parallelization achieves reasonable benchmarks. Network-In-Networks work well with parallelization because they are composed of the convolution layers from a CNN but without the fully-connected layers of the classifier. These fully-connected layers make up the bulk of the network size. A given CNN has a size of 787,788 kB while the comparable NiN has a size of only 7,612 kB. Less size means smaller communication size, and thus less communication overhead. Overall, TorchAD-NN achieved an average of 80% efficiency with less than 1.00% accuracy loss and demonstrates TorchAD-NN achieved a high level of efficiency.

4.5 Model Parallel Performance

Model parallel performance is the same between automated and manual model parallelization since the executed code is the same, the difference is the module abstraction. The focus for performance is instead on parallelization efficiency. From Table 5, model parallel variations train slower on two GPUs versus one GPU. With two nodes, an efficiency of less than 50% indicates the training took longer than the single node implementation.

Using MLP-Small from Table 5 as an example. By splitting the training across two nodes the time is halved (this assumes perfect parallelization for the sake of argument) from 9.64 s to 4.82 s, then take the total time for two nodes minus the estimated time for training, 29.77 s minus 4.82 s, and this leaves an estimated 24.95 s of communication overhead incurred. Because the original time was only 9.64 s, parallelization can only achieve a speedup if communication overhead is reduced or computation cost is increased.

Table 5 has two important observations. First, as model complexity increases GPU parallelization efficiency increases with it. Second, the limited resources environment restricted the size of the model trained and prohibited testing of larger sizes.

Table 5. MLP and CNN GPU SGD training speedup efficiency trained on MNIST dataset.

Model	Nodes	Time (sec)	Accuracy	Efficiency
MLP-small	1	9.64	77.30%	–
MLP-small	2	29.77	66.30%	16.19%
CNN	1	6.09	91.20%	–
CNN	2	16.59	91.60%	18.35%
MLP-medium	1	26.63	70.80%	–
MLP-medium	2	43.95	69.80%	30.30%
MLP-large	1	Not enough memory		
MLP-large	2	68.92	72.40%	No comparison

The GPU memory restricts larger models from training on a single node. Without this comparison, positive speedup efficiency was not demonstrated. It is estimated, further tests would show an efficiency greater than 50% if the current trend continued.

However, this failure highlights an important benefit of the parallelization. Models which require resources beyond what a single node can provide are able to be trained when model parallelism is used. Model parallelism reduces the requirements on a single node. By splitting the neurons across multiple nodes, each node is only required to store and compute a subsection of the total network. The model parallel approach allows for incredible flexibility and power to create networks split across multiple nodes with ease and precise control on which layers are modified.

5 Conclusion

In this work, we present a library for automated parallelization of NN training which greatly reduces human effort over existing approaches. TorchAD-NN data parallel results demonstrate implementation complexity (usability) is greatly reduced and achieves similar efficiency when the same training is manually parallelized. Our TorchAD-NN data parallel module successfully increased usability and maintained speedup efficiency of distributed neural network training.

TorchAD-NN model parallel also demonstrates a reduction in implementation complexity. TorchAD-NN model parallelism had the same performance as manual parallelization, yet overall was less efficient than single node training. The benefit is larger models can be trained which would otherwise be impossible on a single node and implementation complexity for this task is reduced.

In summary, TorchAD-NN automation greatly reduced the implementation complexity (Table 3) of distributing neural network training across multiple devices without loss of computational efficiency (Tables 4 and 5) when compared to manual parallelization. TorchAD-NN is released to the public and supports both data and model parallelism in Torch 7.0. As layers continue to be built upon other layers of abstraction the user becomes further removed from the underlying components. This allows machine learning engineers to train neural networks without worrying about low-level details. These layers of abstraction allow for further technological advances in every field from biology to farming and are critical to our development as a technological society.

6 Future Work

During this research, several areas were identified for continued work. Communication speed analysis and synchronization frequency optimization functionality could improve the efficiency of data parallel training. This would also eliminate the need for the user to specify batch size, further increasing the usability.

Additionally, complex models outside of our experiments could benefit from our model parallel approach. ResNet [35] and VGG16 [2] have incredibly large fully-connected classifier layers. VGG16, for instance, uses a three-layer classifier of 25088-> 4096-> 4096. This model is too large to be loaded on a single TX1 node but could be loaded and trained across four nodes.

References

1. LeCun, Y., Bengio, Y., Hinton, G.: Deep learning. Nature **521**, 436–444 (2015)
2. Simonyan, K., Zisserman, A.: Very deep convolutional networks for large-scale image recognition. arXiv:1409.1556 [cs] (2014)
3. Keuper, J., Preundt, F.J.: Distributed training of deep neural networks: theoretical and practical limits of parallel scalability. In: 2016 2nd Workshop on Machine Learning in HPC Environments (MLHPC), pp. 19–26 (2016)
4. Le, Q.V., Ngiam, J., Coates, A., Lahiri, A., Prochnow, B., Ng, A.Y.: On optimization methods for deep learning. In: Proceedings of the 28th International Conference on Machine Learning, pp. 265–272. Omnipress, USA (2011)
5. Dean, J., Corrado, G., Monga, R., Chen, K., Devin, M., Mao, M., Ranzato, M.A., Senior, A., Tucker, P., Yang, K., Le, Q.V., Ng, A.Y.: Large scale distributed deep networks. In: Pereira, F., Burges, C.J.C., Bottou, L., Weinberger, K.Q. (eds.) Advances in Neural Information Processing Systems 25, pp. 1223–1231. Curran Associates Inc., New York (2012)
6. Szegedy, C., Liu, W., Jia, Y., Sermanet, P., Reed, S., Anguelov, D., Erhan, D., Vanhoucke, V., Rabinovich, A.: Going deeper with convolutions. arXiv:1409.4842 [cs] (2014)
7. Dean, J.: CIKM '14: Proceedings of the 23rd ACM International Conference on Conference on Information and Knowledge Management. ACM, New York, NY, USA (2014)
8. Dean, J.: Large scale deep learning. In: CIKM 2014 Conference, Tsinghua University (2014)
9. Abadi, M., Agarwal, A., Barham, P., Brevdo, E., Chen, Z., Citro, C., Corrado, G.S., Davis, A., Dean, J., Devin, M., Ghemawat, S., Goodfellow, I., Harp, A., Irving, G., Isard, M., Jia, Y., Jozefowicz, R., Kaiser, L., Kudlur, M., Levenberg, J., Mane, D., Monga, R., Moore, S., Murray, D., Olah, C., Schuster, M., Shlens, J., Steiner, B., Sutskever, I., Talwar, K., Tucker, P., Vanhoucke, V., Vasudevan, V., Viegas, F., Vinyals, O., Warden, P., Wattenberg, M., Wicke, M., Yu, Y., Zheng, X.: TensorFlow: large-scale machine learning on heterogeneous distributed systems. arXiv:1603.04467 [cs] (2016)
10. Gu, R., Shen, F., Huang, Y.: A parallel computing platform for training large scale neural networks. In: 2013 IEEE International Conference on Big Data, pp. 376–384 (2013)
11. Li, S., He, J., Li, Y., Rafique, M.U.: Distributed recurrent neural networks for cooperative control of manipulators: a game-theoretic perspective. IEEE Trans. Neural Networks Learn. Syst. **28**, 415–426 (2017)
12. Teerapittayanon, S., McDanel, B., Kung, H.T.: Distributed deep neural networks over the cloud, the edge and end devices. In: 2017 IEEE 37th International Conference on Distributed Computing Systems (ICDCS), pp. 328–339 (2017)
13. Little, T.: Context-adaptive agility: managing complexity and uncertainty. IEEE Softw. **22**, 28–35 (2005)
14. Banker, R.D., Davis, G.B., Slaughter, S.A.: Software development practices, software complexity, and software maintenance performance: a field study. Manage. Sci. **44**, 433–450 (1998)
15. Antinyan, V., Staron, M., Derehag, J., Runsten, M., Wikström, E., Meding, W., Henriksson, A., Hansson, J.: Identifying complex functions: by investigating various aspects of code complexity. In: 2015 Science and Information Conference (SAI), pp. 879–888 (2015)
16. Gill, G.K., Kemerer, C.F.: Cyclomatic complexity density and software maintenance productivity. IEEE Trans. Softw. Eng. **17**, 1284–1288 (1991)
17. Iandola, F.N., Ashraf, K., Moskewicz, M.W., Keutzer, K.: FireCaffe: near-linear acceleration of deep neural network training on compute clusters. arXiv:1511.00175 [cs] (2015)
18. Lin, M., Chen, Q., Yan, S.: Network in Network. arXiv:1312.4400 [cs] (2013)

19. Deng, J., Dong, W., Socher, R., Li, L.J., Li, K., Fei-Fei, L.: ImageNet: A large-scale hierarchical image database. In: 2009 IEEE Conference on Computer Vision and Pattern Recognition, pp. 248–255 (2009)
20. Sergeev, A.: Meet Horovod: uber's open source distributed deep learning framework for TensorFlow, https://eng.uber.com/horovod/
21. Collobert, R., Bengio, S., Marithoz, J.: Torch: A Modular Machine Learning Software Library (2002)
22. TorchMPI: Implements a message passing interface (MPI). https://github.com/facebookresearch/TorchMPI
23. TorchMPI: Model parallel example. https://github.com/facebookresearch/TorchMPI
24. Zhang, S., Choromanska, A., LeCun, Y.: Deep learning with elastic averaging SGD. arXiv: 1412.6651 [cs, stat] (2014)
25. McCabe, T.J.: A complexity measure. IEEE Trans. Softw. Eng. **2**, 308–320 (1976)
26. Halstead, M.H.: Elements of Software Science (Operating and Programming Systems Series). Elsevier Science Inc., New York, NY, USA (1977)
27. IBM Knowledge Center - Halstead Metrics. https://www.ibm.com/support/knowledgecenter/SSSHUF_8.0.0/com.ibm.rational.testrt.studio.doc/topics/csmhalstead.htm
28. Coates, A., Ng, A.Y.: Multi-camera object detection for robotics. In: 2010 IEEE International Conference on Robotics and Automation, pp. 412–419 (2010)
29. Kim, B., Jeon, Y., Park, H., Han, D., Baek, Y.: Design and implementation of the vehicular camera system using deep neural network compression. In: Proceedings of the 1st International Workshop on Deep Learning for Mobile Systems and Applications, pp. 25–30. ACM, New York, NY, USA (2017)
30. Zeiler, M.D., Fergus, R.: Visualizing and understanding convolutional networks. arXiv: 1311.2901 [cs] (2013)
31. Krizhevsky, A.: Learning Multiple Layers of Features from Tiny Images. University of Toronto, Toronto (2012)
32. Lecun, Y., Bottou, L., Bengio, Y., Haffner, P.: Gradient-based learning applied to document recognition. Proc. IEEE **86**, 2278–2324 (1998)
33. Zagoruyko, S.: Network-in-Network on CIFAR10. https://github.com/szagoruyko/cifar.torch
34. Train-a-digit-classifier and train-on-cifar. https://github.com/torch/demos
35. He, K., Zhang, X., Ren, S., Sun, J.: Deep residual learning for image recognition. In: 2016 IEEE Conference on Computer Vision and Pattern Recognition (CVPR), pp. 770–778 (2016)

Towards a Better Model for Predicting Cancer Recurrence in Breast Cancer Patients

Nour A. AbouElNadar and Amani A. Saad(✉)

Computer Engineering Department, Arab Academy for Science and Technology, Alexandria, Egypt
nourabouelnadar@student.aast.com,
amanisaad2008@gmail.com

Abstract. Breast cancer is the most common type of cancer in Egyptian women. According to the International Agency for Research on Cancer, the causes of breast cancer are not yet fully known but some of the main risk factors have been identified and taken into consideration in researches performed on patients with high risk of breast cancer. In this paper various classification techniques are used to classify whether breast cancer is recurrent or non-recurrent for a number of patients. Classification techniques used are K-Nearest Neighbor (KNN), Decision Trees (DT), Naïve Bayes (NB), Support Vector Machines (SVM) and ensemble techniques Bagging, Voting and Random Forest (RF). The dataset is taken from the University of California Irvine (UCI) machine learning repository and experiments are conducted with Waikato Environment for Knowledge Analysis (WEKA) data mining tool. The research conducted goes through two phases, in the first phase the Random Forest classifier produced the best results with (84.3%) accuracy and the second phase, voting ensemble classifier produced the best results of 89.9% accuracy. The system model show an improvement in the overall accuracy compared to other researches done on the same dataset.

Keywords: Data mining · Classification · Breast cancer recurrence · Cancer prognosis · Prediction · Ensemble classifiers

1 Introduction

Breast Cancer is the most common type of cancer among Egyptian women according to The International Agency for Research on Cancer. In 2025, it is estimated that the number of breast cancer patients will increase by 15% [1]. One of the best ways to increase the level of survival of breast cancer is prediction and/or detection in its early stages. Essentially predicting the recurrence of the disease is one of the real-world medical obstacles [2].

Data mining techniques can be defined as a well-established way for efficient decision making where various approaches may generate valuable patterns of data. They are used to detect patterns in large data sets using machine learning, statistics and database systems to expose relevant and useful information from data. This is helpful for making important decisions. Data Mining can be efficiently used in health care

K. Arai et al. (Eds.): CompCom 2019, AISC 997, pp. 887–899, 2019.
https://doi.org/10.1007/978-3-030-22871-2_63

because healthcare industries are generating huge amounts of data and lacking intelligent decision tool for correct, timely and effective decision making. The data mining techniques play an active role in the healthcare industry. They are capable of collecting related patterns of a certain disease which can be beneficial in predicting prognosis and/or diagnosis and help medical practitioners in treatment decisions. These techniques can assist medical professionals in inspecting and identifying unexpected relationships within collected data that can be used in achieving quicker decision making and knowledge discovery [3].

This paper uses different classification techniques on data collected from University of Wisconsin Hospitals. The UCI machine learning repository dataset is used to predict the recurrence of breast cancer in patients. A chunk of the data is used in the learning phase of the model and the remaining is used for the testing phase. The rest of the paper is organized as follows: Sect. 2 presents the background and related work, Sect. 3 is the proposed model, Sect. 4 is the experiment and discussion and the final section is the conclusion.

2 Background and Related Work

2.1 Background

Classification. One of the most well-known techniques in data mining is classification. This technique is used for discovering a model that best differentiates and categorizes the classes and concepts of data. This model is used to anticipate the class label of objects where the class label is anonymous. Classification techniques can be divided into various structures.

Ensemble Techniques. The Ensemble learning is a technique, where a few distinct chosen classifiers are merged using combination rules to obtain a better result, to provide higher performance and accuracy [4]. Ensembles classifiers where proven to be more precise than the original classifiers that make them up [5]. The most widely used ensemble techniques are Bagging, Boosting, Voting and RF.

2.2 Related Work

Different research papers experimented on the UCI datasets using numerous data mining techniques such as KNN, DT, SVM, NB and many different ensemble methods in order to combine data mining techniques and help the model have better precision and more accurate results. In this section, we are going to review some of the papers that used the different UCI Breast Cancer datasets with different data mining techniques. The UCI machine learning repository contains 4 different breast cancer datasets. Wisconsin Breast Cancer Diagnostic (WBCD) and Wisconsin Breast Cancer Original (WBCO) datasets are solving the diagnostic problem (Class: malignant or benign) on the other hand, Wisconsin Breast Cancer Prognostic (WBCP) [6] and Wisconsin Breast Cancer (WBC) datasets are solving the prognostic problem (class: recurrent or non-recurrent). In this research we are focusing on the WBCP dataset.

Table 1 shows a summary of the most recent papers that used data mining and classification techniques in the past 5 years.

Table 1. Data mining techniques used on UCI breast cancer datasets

Reference	Year	Technique	Best accuracy %	Dataset	Pre-processing
[7] Kumar et al.	2013	DT, NB, SVM, KNN	SVM (97.59%)	WBCO	No
[5] Chaurasia et al.	2014	DT, KNN, SVM	SVM (96.2%)	WBCO	No
[8] Kumar et al.	2017	DT, NB, SVM	Voting with J48, NB, SVM SVM (97.13%)	WBCO	No
[9] Banu et al.	2017	DT, One R, Zero R	DT (75.52%)	WBC	No
[2] Pritom et al.	2016	NB, DT, SVM	SVM (75.75%)	WBCP	Yes
[10] Ojha et al.	2017	SVM, DT, NB, KNN	DT or SVM (81%)	WBCP	Yes

Diagnostic Datasets

As illustrated in Table 1, Kumar et al. [7] used WBCO dataset with 499 test sets and 200 patients. They tried to create a classification model to predict breast cancer NB, SVM and KNN using WEKA data mining tool. In this model, the SVM classifier outperformed all the other classification techniques with an accuracy of 97.59%.

Chaurasia et al. [5] performed an experiment on the WBCO using three classification techniques; DT Classifier, KNN and SVM using the WEKA software tool. The SVM classifier outperformed the other techniques with an accuracy of 96.2%.

Kumar et al. [8] compared performances of DT, NB, and SVM using the voting technique using the WEKA data mining tool on WBCO dataset with 699 instances. The Voting ensemble classifier approach combines the three algorithms for better prediction of breast cancer. SVM surpassed all the other techniques individually with an accuracy of 96.70% and the whole model outperformed a single classification technique with an increase of 0.28% in accuracy using voting technique 97.13%.

Prognostic Datasets

As shown in Table 1, Banu et al. [9] used the WBC with 286 instances and 10 attributes. Data was fed to DT, One R, Zero R algorithm and decision stump for classification. The classifier's performance was measured with the Confusion Matrix. The DT algorithm produced better performance than all the other classification methods with an accuracy of 75.52%.

Pritom et al. [2] used the WBCP which contains 198 instances and 34 attributes. Three different classification algorithms were applied on the dataset; NB, DT and SVM. Using feature selection technique in the process, unnecessary attributes were removed from the formula to help produce better results. SVM technique with feature selection Ranker technique produced better results than all the other techniques with an accuracy of 77.27%.

Ojha et al. [10] used WBCP with 198 instances. Four clustering data mining techniques were used; K means, Expectation Maximization EM, partition around medoids PAM and Fuzzy c-means and four classification algorithms SVM, DT, NB and KNN. The best result for the clustering techniques was 68% with EM. On the other hand, the DT and SVM classifiers achieved an 81% accuracy. Their conclusion was that classification algorithms are better predictors than clustering algorithms in this case.

3 Proposed Model

The main focus of this model is to predict the recurrence of breast cancer in cancer patients. When the system model is given an intended set of attributes it should output the proper class attribute that correctly classifies the patient provided. The proposed system applies data cleaning, pre-processing and various classification techniques in order to improve the prediction accuracy.

3.1 System Model

The model starts with the data preparation and cleaning, afterwards a preprocessing filter is applied to the dataset, then training and testing data are selected, subsequently different classification techniques are applied with 10-fold cross validation to overcome overfitting. Finally class prediction (recurrent or non-recurrent) is outputted and several performance metrics were generated for further analysis. Figure 1 shows a block diagram of the system model.

Data Preparation

The data preparation process is divided into two steps. The first step is the data transformation into required format and the second step is the data cleaning. The data cleaning step handles missing values inside the dataset by replacing them with the modes and means from the training data according to the attribute data type.

Missing values are handled before the classification phase to eliminate loss of efficiency, complications in handling and analyzing data and bias results that occur due to differences between missing and complete data fields. The data preparation is shown in the block diagram in Fig. 2.

Pre-Processing Filter. A resampling filter is used to improve the classifier's performance. Resampling is a pre-processing filtering technique (performed on the dataset before the classification takes place) for the datasets that are believed to have imbalanced classes. In our dataset 76% of the classes are non-recurrent and 23.8% only are recurrent. Since it is possible that this disproportion in classes causes biased or inaccurate results; therefore, the filtering resampling step is added before the classification phase.

Training and Testing. In this stage the dataset is split into a set for building the model "training" and a set for accrediting the accuracy of the model on new data records "testing". The training and testing part of the data should contain the outcome (target field or class field) to build and verify the model. This phase is done with 10-fold cross validation.

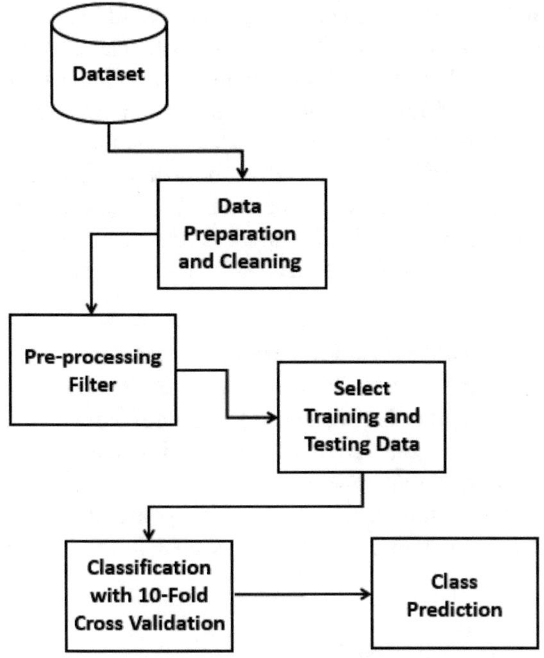

Fig. 1. Overall system model

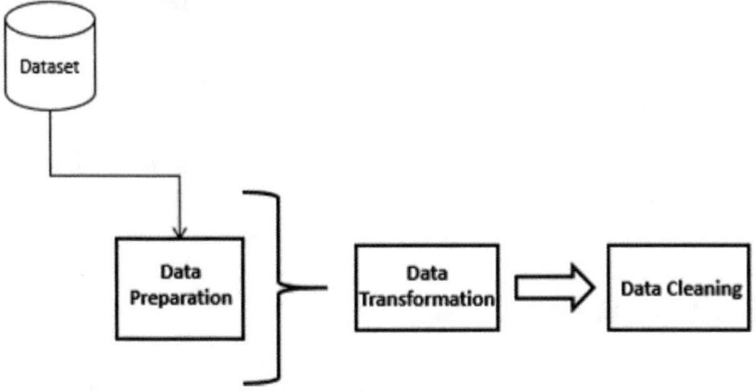

Fig. 2. Data preparation stage

10-Fold Cross Validation. 10-fold cross validation is chosen for more precise results and to avoid overfitting. Each classifier's training and testing phase is combined with the 10-fold cross validation. The dataset is split into 10 folds and each fold has is divided into 10 equal slices. One slice for testing and the 9 remaining are for training. This process is repeated k times. At the end, the recorded measures are averaged. It is common to choose k = 10 or any other size depending on the size of the original dataset. This decreases the chance of model overfitting.

Underfitting and Overfitting. Underfitting and overfitting are two of the main problems that should be avoided when dealing with classification/prediction techniques. Underfitting is when you had the opportunity to learn from a dataset but no learning was done as the dataset was not enough for training. For example, a student who did not study for their exam will underfit/fail it. On the other hand, overfitting is the opposite; it occurs when a model focuses too much on odd/idiosyncrasies features in the training data that do not describe the data. This means that the model is fitting into noise. An example for overfitting is a student who memorizes past exams answers without understanding, this student most likely will fail the exam (he overfits the training data) [10].

Classification with 10-Fold Cross Validation. The prediction process starts with the dataset format preparation, followed by the preprocessing and choosing the training dataset. Afterwards a classifier is chosen and the dataset is trained accordingly. At the end the model accuracy is measured through the test dataset and a conclusion is formulated from the overall experiment.

Following the data cleaning and pre-processing phase, the preprocessed dataset is passed to the selected classifiers. The classifiers used in this model are illustrated in Fig. 3.

3.2 The Proposed System Model

Different experiments were performed that will be explained thoroughly in the next section. The experimental process went through two different phases "with resampling filter" and "without resampling filter" and results were recorded with respect to the classifiers and ensemble classifiers. According to the presented results the best classifier is the ensemble classifier Voting (with KNN, DT and RF) which produced the most promising accuracy results than all the other classifiers. Figure 4 shows the proposed system model with data preparation, pre-processing and voting ensemble classifier to produce the best results.

4 Experiments and Discussion

This section presents the details of the conducted experiments such as; the tool, dataset, performance evaluation indicators that were calculated as well as the results of the experiments performed on the WBCP dataset.

4.1 Tool

The WEKA data mining tool is chosen for the experiments. It was developed by the University of Waikato [11] for data mining and classification techniques. WEKA, abbreviated from Waikato Environment for Knowledge Analysis, is a free and open-source machine learning and data mining software written in Java. WEKA provides different functions such as data processing, feature selection, classification, regression,

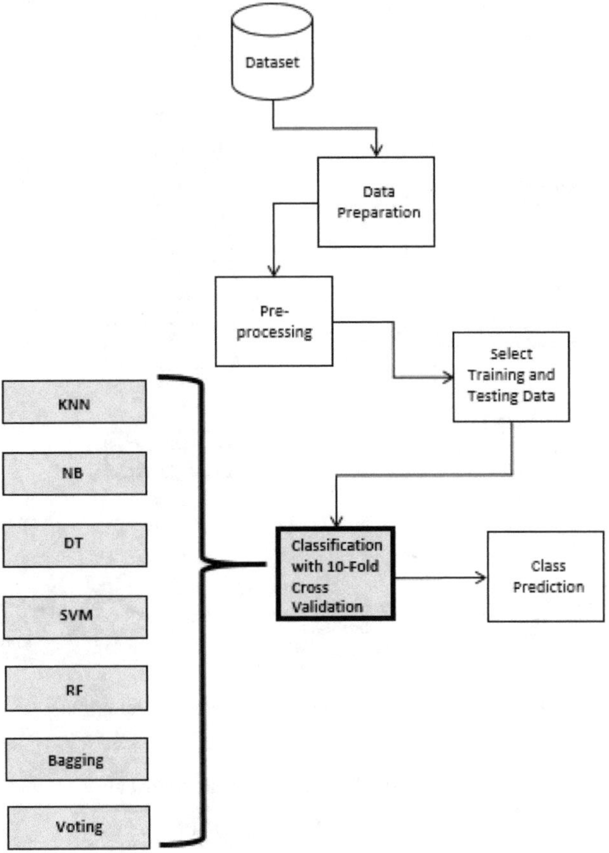

Fig. 3. Classification stage

clustering, association rule, and visualization, etc. All experiments were performed using libraries from WEKA machine learning tool version 3.6.

4.2 Experimental Setup

Dataset

The dataset used in the following experiment is the WPBC which has 198 instances, 34 attributes, two classes (recurrent and non-recurrent). The distribution of class attribute is shown in Table 2 while the attributes information is presented in Table 3.

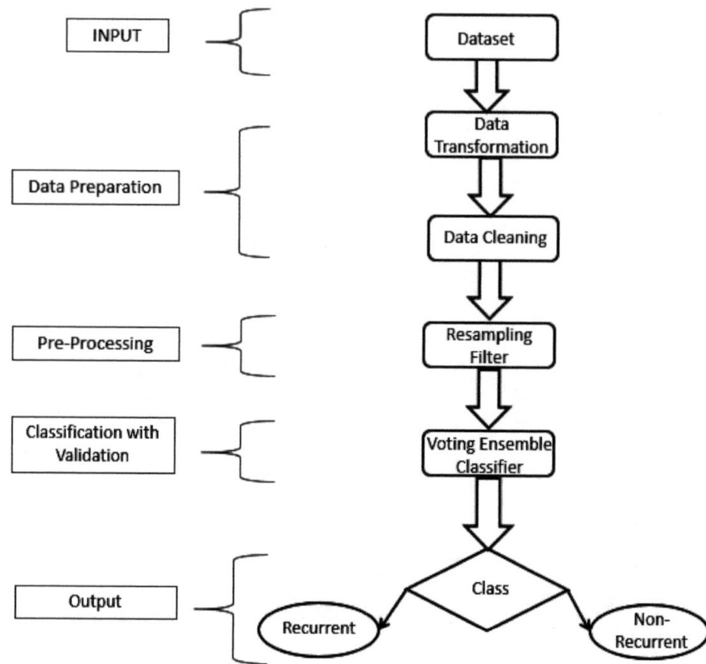

Fig. 4. Block diagram for the proposed system model

Table 2. The distribution of class attribute

Class	Frequency	Percentage
Non recurrent events	151	76.2%
Recurrent events	47	23.8%

4.3 Performance Evaluation and Results

Multiple experiments were carried out to measure and compare the performance of the proposed model with different approaches. The main performance evaluation formulas that were calculated on the results are the following:

$$\text{Accuracy: } (TP + TN)/TOTAL$$
$$\text{Sensitivity} = TP/(TP + FN)$$
$$\text{Specificity} = TN/(TN + FP)$$

Where TP = True Positive, TN = True Negative, FP = False Positive and FN = False Negative. Moreover the breakdown of the dataset's confusion matrix is shown below. Two classes: Positive class (a) and Negative class (b) where "a" is the non-recurrent class and "b" is the recurrent one.

Table 3. Attribute information

Attribute	Information
1 - ID Number	
2 - Outcome	R = recur N = nonrecur
3 - Time	Recurrence time if field 2 = R, disease-free time if field 2 = N
4 - 33	Ten real-valued features are computed for each cell nucleus: (a) radius (mean of distances from center to points on the perimeter) (b) texture (standard deviation of gray-scale values) (c) perimeter (d) area (e) smoothness (local variation in radius lengths) (f) compactness (perimeter^2/area - 1.0) (g) concavity (severity of concave portions of the contour) (h) concave points (number of concave portions of the contour) (i) symmetry (j) fractal dimension ("coastline approximation" - 1)

Confusion matrix	
a	b
TP	FN
FP	TN

The classification algorithms applied on the WPBC dataset after it was pre-processed produced promising and improved results as shown in Table 4. In phase one results, the Random Forest classifier produced higher accuracy than all other classifiers. On the other hand, in Ensemble classifiers, the RF produced the highest accuracy among all of them.

As shown in Tables 5 and 6 (phase two), the ensemble Voting classifier with KNN, DT and RF produced the best accuracy results of 89.899%.

Table 4. Phase one results

Classifier	Results
IBK	71.7172%
J48	75.7576%
NB	67.1717%
RF	84.3434%
SMO	76.2626%
Voting with J48 SMO and NB	77.7778%
Voting with IBK J48 and RF	79.798%
Bagging with J48	77.2727%
Bagging with RF	81.8182%

Table 5. Phase two results

Classifier	Results with resample filter
IBK	83.3333%
J48	83.8384%
NB	69.697%
RF	89.3939%
SMO	79.798%
Voting with J48 SMO and NB	83.3333%
Voting with IBK J48 and RF	89.899%
Bagging with J48	85.3535%
Bagging with RF	87.8788%

Table 6. Overall performance evaluation of all algorithms with resampling

Classifier	Confusion matrix		Accuracy	Sensitivity	Specificity
KNN	a	b	0.8333	0.904	0.634
	132	19			
	14	33			
DT	a	b	0.8383	0.899	0.653
	134	17			
	15	32			
NB	a	b	0.6969	0.832	0.393
	114	37			
	23	24			
RF	a	b	0.8939	0.882	0.964
	150	1			
	20	27			
SVM	a	b	0.7979	0.8006	0.705
	146	5			
	35	12			
Voting DTSVM and NB	a	b	0.8333	0.843	0.769
	145	6			
	27	20			
Voting KNNDT and RF	a	b	0.8989	0.911	0.846
	145	6			
	14	33			
Bagging with DT	a	b	0.8535	0.886	0.725
	140	11			
	18	29			
Bagging with RF	a	b	0.8787	0.862	1
	151	0			
	24	23			

Table 6 shows the results of the performance matrices that were applied to each classifier to evaluate the overall capability of the classifiers with regards to accuracy, sensitivity, and specificity and confusion matrix.

4.4 Validation Experiment

In this section an experiment for validation was performed on a new prognostic dataset. Wisconsin Breast Cancer (WBC) dataset contains 286 instances and 10 attributes. Table 7 shows attribute information for this dataset. WBC dataset was chosen for validation of the system model presented in Fig. 4. The same experiment performed on WBCP dataset in phase 2 was repeated on WBC dataset for validation and results were recorded for comparison with the results obtained from the previous experiments as shown in Table 7.

Table 7. WBC attribute information

Attribute	WBC with replacement
Class	No-recurrent, recurrent
Age	10–19, 20–29, 30–39, 40–49, 50–59, 60–69, 70–79, 80–89, 90–99
Menopause	lt40, ge40, premeno
Tumor-size	0–4, 5–9, 10–14, 15–19, 20–24, 25–29, 30–34, 35–39, 40–44,45–49, 50–54, 55–59
Inv-nodes	0–2, 3–5, 6–8, 9–11, 12–14, 15–17, 18–20, 21–23, 24–26, 27–29, 30–32, 33–35, 36–39.
Node-caps	Yes, no
Deg-malig	1, 2, 3
Breast	Left, right
Breast-quad	Left-up, left-low, right-up, right-low, central
Irradiant	Yes, no

As shown in Table 8 results shows that Voting with the best classifiers (KNN, RF and DT) produced the best results for the new dataset as well.

Table 8. WBC phase 2 results compared to WBCP

Classifier	WBC with resampling	WBCP with resampling
KNN	87.413%	83.333%
DT	80.420%	83.838%
SVM	78.322%	79.798%
NB	76.224%	69.697%
RF	84.965%	89.394%
Bagging with RF	85.664%	87.879%
Bagging with DT	81.119%	85.354%
Voting with KNN, RF and DT	88.112%	89.899

5 Conclusion

In this paper, a thorough study for different classification techniques was conducted. The objective is to provide a better system model for predicting breast cancer recurrence in cancer patients. The study started by reviewing the work done by different researchers that worked on the same dataset.

A system model has been proposed to serve as a second opinion for physicians to predict cancer recurrence for breast cancer patients and take the required medical precautions accordingly. Datasets has been collected from the UCI machine learning repository and the overall system model is divided into four stages as follows:

1. Data Preparation
2. Data Pre-Processing
3. Classification
4. Validation

Using different classification techniques the model presented delivered higher performance than existing state of the art models. These goals were achieved by using the pre-processing filter technique (resampling) and using the ensemble of classifiers "Voting Ensemble classifier". It was shown by experiments that the proposed model outperformed the results obtained by other researchers performed on the same dataset with an accuracy of 89.9%. In addition, the proposed model gave an insight into most of the top classification techniques and gave room for comparisons between different algorithms. Ensemble techniques produced very good results compared to one classification technique, which proved that combining classifiers is better than one classifier.

Finally for validation the same experiment was performed on a different prognostic dataset from the UCI machine learning repository. This dataset verified the proposed model and reinforced its results regarding the choice of the best classifier and pre-processing technique. The result of this research can help medical practitioners in making good prognostic decisions, thus treating patients in a more effective way.

References

1. International Agency for Research on Cancer Homepage. http://globocan.iarc.fr/. Last accessed 13 Oct 2018
2. Pritom, A.I., Munshi, M.A.R., Sabab, S.A., Shihab, S.: Predicting breast cancer recurrence using effective classification and feature selection technique. In: Proceedings of the 19th International Conference on Computer and Information Technology (ICCIT '16), pp. 310–314, December 2016
3. Rathore, N., Agarwal, S.: Predicting the survivability of breast cancer patients using ensemble approach. In: Proceedings of the International Conference on Issues and Challenges in Intelligent Computing Techniques (ICICT '14), pp. 459–464. IEEE, February 2014
4. Tseng, C.-J., et al.: Integration of data mining classification techniques and ensemble learning to identify risk factors and diagnose ovarian cancer recurrence. Artif. Intell. Med. **78**, 47–54 (2017)

5. Chaurasia, V., PalA, S.: Novel approach for breast cancer detection using data mining techniques. Int. J. Innov. Res. Comput. Commun. Eng. **2**(1), 17 pages (2014)
6. UCI Prognostic Dataset. https://bit.ly/2OWq7te. Last accessed 13 Oct 2018
7. Ravi Kumar, G., Ramachandra, G.A., Nagamani, K.: An efficient prediction of breast cancer data using data mining techniques. Int. J. Innov. Eng. Technol. (IJIET) **2**(4), 139–144 (2013)
8. Kumar, U.K., Nikhil, M.B.S., Sumangali, K.: Prediction of breast cancer using voting classifier technique. In: Proceedings of the IEEE International Conference on Smart Technologies and Management for Computing, Communication, Controls, Energy and Materials, Chennai, India, pp. 2–4 (2017)
9. Banu, G.R., Prakash, Bashier, I., Summera, : Applications of data mining classification techniques on predicting breast cancer disease. Int. J. Latest Trends Eng. Technol. **8**(2), 321–325 (2017)
10. Ojha, U., Goel, S.: A study on prediction of breast cancer recurrence using data mining techniques. In: 7th International Conference on Cloud Computing Data Science & Engineering-Confluence, pp. 527–530 (2017)
11. Weka Data Mining Tool Homepage. https://www.cs.waikato.ac.nz/ml/weka/. Last accessed 13 Oct 2018

Inducing Clinical Course Variations in Multiple Sclerosis White Matter Networks

Giovanni Melissari[1], Aldo Marzullo[1,2(✉)], Claudio Stamile[2], Francesco Calimeri[1], Françoise Durand-Dubief[2,3], and Dominique Sappey-Marinier[2,4]

[1] Department of Mathematics and Computer Science, University of Calabria, Rende, Italy
marzullo@mat.unical.it
[2] CREATIS, CNRS UMR5220, INSERM U1206, Université de Lyon, Université Lyon 1, INSA-Lyon, Villeurbanne, France
[3] Hôpital Neurologique, Service de Neurologie A, Hospices Civils de Lyon, Bron, France
[4] CERMEP - Imagerie du Vivant, Université de Lyon, Bron, France

Abstract. The incidence of neurological disorders is constantly growing; hence, the scientific community is intensifying the efforts spent in order to design approaches capable of determining the onset of such disorders. In this paper we focus on a specific neurological disorder, namely Multiple Sclerosis, a chronic disease of the central nervous system. We propose a method for identifying specific brain substructures that might underpin a worsening disease, thus allowing to delineate a number of potentially vulnerable brain regions. The task is addressed by means of a simulation procedure which iteratively disrupt brain regions. Experimental results show that the proposed simulation produces reliable graphs with respect to the used dataset.

Keywords: Multiple Sclerosis · Complex network analysis · Classification

1 Introduction

Multiple Sclerosis (MS) is a chronic autoimmune disease that affects the central nervous system; it can imply a significantly wide range of neurological symptoms, and can progress up to physical and cognitive disability. Despite the fact that pathological mechanisms still remain unknown, there is a common agreement on a description of an individual patient's disease course based on four

The work is partially funded by the Italian Ministry of University and Research under project "Dottorato innovativo a caratterizzazione industriale" PON R&I FSE-FESR 2014–2020.

K. Arai et al. (Eds.): CompCom 2019, AISC 997, pp. 900–917, 2019.
https://doi.org/10.1007/978-3-030-22871-2_64

clinical profiles that illustrate temporal information about the ongoing disease process [5,10]. The disease onset usually starts with a first acute episode, called Clinically Isolated Syndrome (CIS) which evolves into the Relapsing-Remitting (RR) course with a probability of 85%. RR patients will then evolve into the Secondary Progressive (SP) course after about 10–20 years [8]. With a probability of 15%, the disease starts directly with the Primary Progressive (PP) course. Unfortunately, such clinical courses are solely based on clinical observations and consensus; Defining automatic algorithms to determine the current clinical profile of a patient as well as understanding potential correlations among clinical courses is still an open challenge.

In the last decade, new promising approaches have been proposed to classify clinical profiles, based on different tools and techniques, from graph theory to machine learning [4,8,13]. In this context, a combination of Magnetic Resonance Images (MRI) acquisitions like $T1$ and Diffusion Tensor Imaging (DTI) [14,16] has been used to obtain structural connectivity matrices of the brain structure, thus paving the way to advanced complex network analysis. Such networks consist of nodes, corresponding to segmented cortical regions, and links, reconstructed by tractography from White Matter (WM) fibers-tracts.

In the brain connectivity domain, several measures variously capture functional integration and segregation, quantify centrality of individual brain regions or pathways, characterize patterns of local anatomical circuitry, and test resilience of networks to insult [13]. Different graph metrics can be estimated to measure brain network properties which can be analyzed to better characterize brain networks. In particular, in the MS context, several observations are obtained from the network analysis [2,8,13,15], providing useful biomarkers for the discrimination of each group. However, these metrics need to be further investigated in terms of their variations among clinical courses, since it is still unknown whether specific structures of the brain are more likely to be affected by the pathology, especially with respect to its evolution.

In this work, we take advantage from findings in recent network analysis [8] and propose an artificial deterioration algorithm of brain connectivity networks that induces a disease worsening from the RR to the SP clinical course. The main goal of the work is to find potential substructures of the brain that are more likely to be disrupted when a disease worsening occurs in RR subjects, thus allowing to delineate a small group of potentially vulnerable brain regions.

The remainder of the paper is structured as follows. We introduce the complex network metrics explored in previous works in Sect. 2. We discuss related works in Sect. 3 and then provide the reader with a detailed description of our approach in Sect. 4. We illustrate our experimental activities in Sect. 5 and discuss the results in Sect. 6. Eventually, we draw our conclusions in Sect. 7.

2 Complex Network Metrics

For the sake of readability, before illustrating our approach we briefly report the definition of the network metrics mainly explored in previous works, which guided the intuition behind the present proposal.

Let $G = \langle V, E, \omega \rangle$ be a weighted undirected graph, where V is the set of vertices with $|V| = q$, E is the set of edges and ω is a function ($\omega : \mathbb{N}_+^2 \rightarrow \mathbb{N}$) associating each edge to a weight so that $a_{i,j} = \omega(i, j)$.

Number of Edges. Since MS actually affects connections between brain regions, the *Number of Edges* for each graph is relevant. This is simply computed as:

$$l = \sum_{i,j \in V} a_{i,j}$$

Weighted Degree. The *Degree* of a node represents the number of edges connected to that node. For an undirected graph, it counts the number of neighbour nodes; for a directed graph, it is the number of outcoming/incoming edges; for a weighed graph, it represents the sum of weights corresponding to connected edges.

$$d_i = \sum_{j \in V} a_{i,j}$$

The mean network degree, or *Global Degree (GD)*, is commonly used as a measure of density.

Global Efficiency. The *Global Efficiency* of a graph is the average of the inverse Shortest Path Length for all pair of nodes. This measure is influenced by short paths; thus, the shortest the paths, the lower the measure. It is computed as:

$$GE = \frac{1}{q} \sum_{i \in V} \frac{\sum_{j \in V, j \neq i} k_{i,j}^{-1}}{q - 1}$$

where $q = |V|$ and $k_{i,j}$ is the shortest path from i to j. Global efficiency is maximum for a fully-connected network, while minimum for a totally disconnected one.

Modularity. The *Modularity* is the degree to which the network may be subdivided into clearly delineated and non-overlapping groups. Such groups are referred to as *community* or *modules*, and represent aggregated sets of highly interconnected nodes. Unlike most other network measures, the optimal modular structure for a given network is typically estimated by means of optimization algorithms; one of the most commonly used is based on Newman's Q-metric [12] coupled with an efficient optimization approach:

$$MOD = \sum_{u \in M} \left[p_{u,u} - \left(\sum_{v \in M} p_{u,v} \right)^2 \right]$$

where the networks is subdivided into M non-overlapping modules and $p_{u,v}$ is the proportion of edges that connect nodes belonging to module u with nodes belonging to module v.

Assortativity Coefficient. The *Assortativity Coefficient* (AC) of a network is the Pearson correlation coefficient between the degrees of all nodes on two opposite ends of a link [11]. Networks featuring a positive AC are likely to have a comparatively resilient core of mutually interconnected high-degree hubs (Fig. 1). On the other hand, networks featuring a negative coefficient are likely to have widely distributed high-degree hubs. AC is computed as:

$$AC = \frac{l^{-1}\sum_{(i,j)\in E} d_i d_j - [l^{-1}\sum_{(i,j)\in E} \frac{1}{2}(d_i + d_j)]^2}{l^{-1}\sum_{(i,j)\in E} \frac{1}{2}(d_i^2 + d_j^2) - [l^{-1}\sum_{(i,j)\in E} \frac{1}{2}(d_i + d_j)]^2}$$

where $l = \sum_{i,j\in V} a_{i,j}$ is the total number of links and $d_i = \sum_{j\in V} a_i, j$ is the degree of node i.

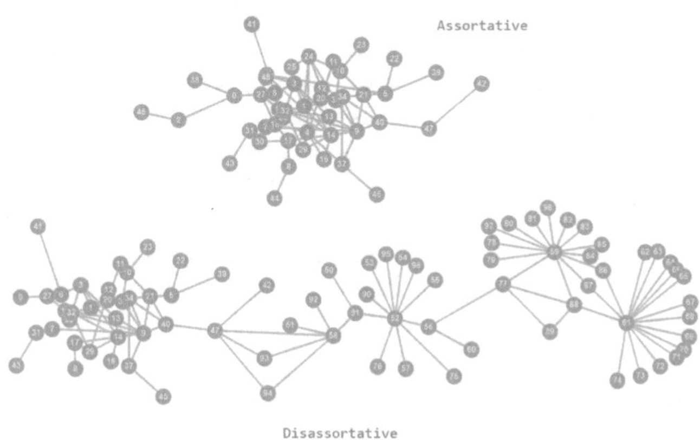

Fig. 1. "Assortative" graphs have a resilient core of interconnected hubs, and AC is positive. "Disassortative" graphs have a more widely distributed core of hubs that in turn connect low-degree nodes. The AC is negative.

3 Related Works

Complex Network Analysis is a new multidisciplinary approach to the study of complex systems aiming at characterizing brain networks with a small number of neurobiologically meaningful and easily computable measures [13]. Network metrics are often represented in terms of individual network elements (such as nodes or links). Measurement values of all individual elements comprise a distribution, which provides a more global description of the network.

Kocevar *et al.* in [8] proposed a graph theory based analysis to characterize the structural connectivity in every clinical profile of MS patients by estimating global network metrics. According to Rubinov *et al.* in [13], several graph metrics were estimated to measure global brain network properties, including density, assortativity, efficiency, among others. In particular, noteworthy findings are related to density, which significantly decreases in SP patients, showing that links (i.e., fibers connecting regions) are in fact disrupted. Global efficiency significantly decreases along with the progress of the pathology, and in particular from CIS-RR status to SP ($p < 0.001$). Modularity decreases from Healthy patients (HC) to CIS patients and increases in RR patients, suggesting a relevant presence of interconnected bunches of nodes that communicate each other through hubs. Finally, assortativity significantly decreases in SP and PP patients.

Among these results, it is interesting to focus on the significant differences in term of assortativity when comparing RR with SP groups. In particular, graphs in initial courses are likely to be *disassortative*, whilst graphs in SP and PP are *assortative* with a significant difference of trend on average. Despite the finding above is already established, it is still unclear why such kind of behaviour occurs and, in particular, what specific structures of the brain are likely to be affected in presence of a disease course worsening from RR to SP.

In this work, we describe a method that identifies potential substructures of graphs that may underpin a disease course worsening from RR to SP. In other words, we want to induce assortativity variation of RR graphs in order to obtain synthetic SP assortative graphs. Such a procedure would provide us with a chance to identify potential brain regions that are likely to be disrupted by the disease.

It is worth noting that the approach herein proposed is not meant to be a definitive simulation of the disease; however, it can suggest potentially vulnerable regions which could be further investigated.

4 Proposed Method

The approach herein proposed exploits structural representations of MR images. In particular, focusing on the already mentioned findings, we describe a method to select and disrupt edges of the graphs representing patients in the RR clinical course, such that they looks more similar to graphs representing patients in SP. In the following, in order to provide a detailed description of the proposed simulation strategy, we briefly recall the method used to extract graph from MRIs.

4.1 Brain Structural Connectivity Graph

The connectivity graph of each subject is obtained by applying the method originally proposed in Kocevar *et al.* [8]; for the sake of completeness, we briefly recall the mentioned approach. Starting from MR images, a brain structural connectivity matrix $A \in \mathbb{N}_+^{qxq}$ is produced, where each element is defined as $a_{i,j} = \Phi(i,j)$, and $\Phi : \mathbb{N}_+^2 \rightarrow \mathbb{N}$. In particular, each matrix A represents a weighted undirected graph $G = \langle V, E, \omega \rangle$ defined such that

$$E = \{\{i,j\} \mid \Phi(i,j) > 0 \wedge i,j \in \{1,\ldots,q\}\}$$

Weights in ω are related to the number of fibres connecting two nodes. Actually, higher weights between two regions imply stronger connectivity.

4.2 Artificial Deterioration Algorithm of Brain Connectivity Networks

As already stated above, the main goal of the present work is to find some structures of interest in the graphs that can represent a common meaningful pattern. The proposed strategy is mainly based on the observations related to the assortativity of the considered groups: given that graphs in RR are known to be disassortative while graphs in SP are known to be assortative, some sort of deterioration must affect graphs in the former group so that they eventually become assortative. Basically, we are going to induce assortativity variations to RR graphs in order to gradually transform them into synthetic SP assortative graphs.

By definition, disassortative graphs feature a broadly distributed high-degree hubs, meaning that a great portion of high-degree nodes is connected to low-degree nodes. We will refer to them as "disassortative" nodes, for simplicity. Assortative graphs, on the contrary, are likely to have a comparatively resilient core of mutually interconnected high-degree hubs; that is, a great proportion of high-degree nodes is connected to high degree neighbors. We will refer to them as "assortative" nodes, for simplicity.

Figure 2 illustrates an example of such definitions: *red circles* identify disassortative nodes, whilst *green circles* identify assortative nodes. Intuitively, disassortative nodes are likely to form a "bicycle wheel"; on the other hand, assortative ones are likely to form a cluster of interconnected nodes, constituting high-dense subcomponents of the graph.

Fig. 2. Red circles highlight high-degree nodes connected to low-degree nodes; green circles highlight high-degree nodes connected to other high-degree nodes.

Our strategy basically identifies specific edges linked to disassortative nodes and remove them from the original graph, so that the newly obtained graph is more likely to feature a more resilient core of assortative nodes. Deteriorate the edges of disassortative nodes causes a decrease in importance of the above-mentioned nodes, and an increase of global assortativity, accordingly. It is worth noting that the induced transformation must induce an obtained synthetic graph that still represents a brain network, with main metrics compatible with those of an SP graph.

The Strategy. In order to actually implement the idea, we need to define proper criteria for choosing nodes to be disrupted. To this aim, we inspect some relevant features revealed by local network metrics. *Weighted degree* is considered as a good measure for the density of a node, representing the intensity of connections (links) to it. The *average neighbour degree* [1] is a measure of density for the neighbourhood of a node and it is computed as:

$$nd_i = \frac{\sum_{j \in N(i)} d_j}{|N(i)|}$$

where $N(i)$ are the neighbours of node i and d_j is the degree of node j which belongs to $N(i)$. Since disassortative nodes are, by definition, high-degree nodes mostly connected with low-degree nodes, we identify the first k nodes with the lowest possible value of average neighbour degree coupled with the highest possible value of weighted degree. Now, we need to choose what edges will be disrupted. In particular, we want to disrupt links between disassortative nodes and neighbours featuring a non-high degree. The rationale comes from the observation that, even if selected nodes have a low neighbour degree, it might well be the case that some minor neighbour with high-degree exists. In this case, we clearly want to ignore the link with it. It is important to notice that, according to [4], where strong edges seem to be discriminatory for the classification, the strategy should avoid to select edges whose weight is *strong*.

More precisely, our strategy identifies the k most disassortative nodes in the following way:

1. take the nodes featuring a value of average neighbour degree lower than the 50% of its distribution;
2. among such nodes, identify the first k nodes with the highest value of weighted degree;
3. disrupt the edges between these nodes and their neighbours such that:
   (*i*) the neighbour node has a weighted degree lower than the minimum degree outlier,
   (*ii*) the edge weight is lower than the minimum edge weight outlier.

The parameter k is directly related with the quantity of damage to apply. Choosing a greater k will cause an excessive deterioration of links, while choosing smaller values would lead to non-significant variations of metrics. In this settings we choose $k = 10\%$ of the brain regions such that the resulting graphs are still fairly comparable with real brain networks.

Algorithm 1. Pseudo-code of the proposed simulated deterioration strategy.

```
function SIMULATEDDETERIORATIONSTRATEGY(graphs, K)
    Inputs: graphs: set of graphs to be deteriorated; k: number of
            disassortative nodes to deteriorate;
    Output: deteriorated_graphs: set of synthetic deteriorated graphs;

    Initialize:
        deteriorated_graphs ← {∅}

    for all g ∈ graphs do
        wd ← weightedDegree(g)
        nd ← averageNeighbourDegree(g)
        w ← edgeWeights(g)

        nd_median = percentile(nd, 50)
        F ← {node | ∀node ∈ node(g) ∧ nd[node] < nd_median}

        F ← sort(F, wd)
        F ← take(k)

        upper_edge_weight = min(MADoutliers(w))
        upper_weighted_degree = min(MADoutliers(wd))

        edges ← {∅}
        for all node ∈ F do
            for all nb ∈ neighbour(g, node) do
                if w[node][nb] < upper_edge_weight then
                    if nd[nb] < upper_weighted_degree then
                        e ← (node, nb)
                        append(edges, e)


        g0 ← g
        for all e ∈ edges do
            removeEdge(g0, e)

        append(deteriorated_graphs, g0)

    return deteriorated_graphs
```

The proposed method is illustrated by Algorithm 1. In particular, since the average neighbour degree is normally distributed (as confirmed by the normality test), we consider as *low* the points below the median, according to the *68-95-99.7 rule* [6]; such points will be lower than the 50% of the distribution. Furthermore, outliers in degree and weights distributions are identified by means of the *MAD (Median Absolute Deviation)* [9] outliers detection method. This method has been proven to be robust when the underlying distribution is not normal. Indeed, instead of using the average and the standard deviation as measures of centrality and dispersion respectively, MAD makes use of median and deviation around the median that, on the contrary, are not influenced by outliers. Since normality tests show that degree and edge-weight distributions are not normally distributed, the use of this method can be considered as sufficiently robust.

The goal of the proposed algorithm is to identify high-degree nodes that, at the same time, are likely to connect to low-degree nodes, as they are, by definition, those that make a graph disassortative. Moreover, the strategy chooses the edges with an inlier weight value. Removing these specific set of edges increases the expectation to obtain assortative graphs.

5 Experiments

In the following, we illustrate results of our experimental campaign, also providing a description of the used dataset.

5.1 Dataset Description

The dataset is a collection of 90 MS patients and 24 healthy control subjects (HC), distributed across the four aforementioned clinical courses, as shown in Table 1. Each patient underwent multiple brain MRI examinations over a different period for each patient (except for HC), ranging from 2.5 to 6 years. The minimum number of scans per patient is 3, while the maximum is 10. The gap between two consecutive scans is either 6 or 12 months. The total number of MRI detections in the dataset amounts to 602.

Table 1. MS patients on different clinical profiles and healthy control (HC).

Clinical courses	MRI examinations	Patients
HC	24	24
CIS	63	12
RR	199	30
SP	190	28
PP	126	20

In particular, we used the RR and the SP group of the described dataset to evaluate the proposed approach.

5.2 Evaluation Method

We evaluate the results by means of two different methods: First, we estimate the complex network metrics introduced in Sect. 2, both before and after the deterioration. We used non-parametric statistic tests in order to understand whether there are meaningful differences in distributions among clinical courses; furthermore, given the recent advances in MS clinical courses discrimination [3,4], we exploit three different Neural Network models in order to classify simulated graphs:

- Graph-based Neural Network (GB), [Calimeri et al. [4]].
- Convolutional Neural Network (CNN) [Calimeri et al. [3]].
- Brain Convolutional Neural Network (BN) [Kawahara et al. [7]].

Each network is trained to distinguish among the four clinical courses with high level of accuracy [3,4]. In particular, we used a slightly modified version of the mentioned models with comparable results. The expectation is that after the simulation, graphs classified as *RR* will be classified as *SP*.

We predict the course of all the obtained deteriorated graphs by using two main strategies: *(i)* an individual prediction made by each classifier; and *(ii)* a prediction based on *majority voting*. The latter simply assigns the class predicted by the majority of the classifiers. In this context, we have an overall of four classification responses (three individual classifiers + majority voting).

The average of the metric measurements and the classification results are shown in Table 2a and b, respectively.

Table 2. (a) reports average (± stdev) of the metric distributions calculated *before* (first column) and *after* (second column) the deterioration, as well as averages of SP graphs (third column). (b) reports, for each MS clinical course, the percentage of synthetic graphs classified within a certain group, with both individual classification and majority voting.

(a)

Metrics	Before	After	SP
assortativity	-0.03 ± 0.04	0.01 ± 0.03	0.01 ± 0.05
weighted degree	4229.90 ± 518.41	3484.29 ± 402.79	3667.65 ± 771.59
global efficiency	0.77 ± 0.03	0.70 ± 0.02	0.72 ± 0.05
modularity	0.52 ± 0.04	0.56 ± 0.03	0.56 ± 0.05
num of edges	1877.33 ± 191.63	1445.83 ± 144.25	1551.55 ± 314.76

(b)

MS course	Majority voting	GB	CNN	BN
CIS	1.01 %	0.00 %	1.01 %	0.00 %
RR	**53.53 %**	27.27 %	**62.12 %**	**65.15 %**
SP	36.88 %	**70.71 %**	31.82 %	24.24 %
PP	8.58 %	2.02 %	5.05 %	10.61 %

In order to assess the quality of the approach presented earlier, we compare the algorithm with a random-based simulated deterioration of graphs. This is crucial in order to understand whether the variations in results depend on the structure of the modified portion of connectome "attacked" with the proposed strategy, or simply depends on a random nature. More in detail, for each graph in the RR set, given the number n of edges to remove as identified by the proposed algorithm, we select n random edges to deteriorate. Choosing exactly the same number of removed edges as the previous strategy is important in order to maintain the same proportion of deteriorated edges.

Average measurements of network metrics and classification results are reported in Table 3a and b, respectively.

Table 3. (a) shows the average (± stdev) of the metric distributions calculated *before* (first column) and *after* (second column) the random deterioration, as well as the averages of SP graphs (third column). (b) shows for each MS clinical course, the percentage of synthetic graphs classified with a certain group, with both individual classification and majority voting.

(a)

Metrics	Before	After	SP
assortativity	-0.03 ± 0.04	-0.03 ± 0.04	0.01 ± 0.05
weighted degree	4229.90 ± 518.41	3252.98 ± 401.75	3667.65 ± 771.59
global efficiency	0.77 ± 0.03	0.70 ± 0.02	0.72 ± 0.05
modularity	0.52 ± 0.04	0.54 ± 0.04	0.56 ± 0.05
num of edges	1877.33 ± 191.63	1445.83 ± 144.25	1551.55 ± 314.77

(b)

MS course	Majority voting	GB	CNN	BN
CIS	0.51 %	0.00 %	1.01 %	0.51 %
RR	**94.44 %**	**93.43 %**	**90.91 %**	**74.75 %**
SP	1.01 %	5.56 %	1.01 %	9.59 %
PP	4.04 %	1.01 %	7.07 %	15.15 %

We perform the Wilcoxon rank-sum test [17] on the new distributions of measurements in order to understand whether the deteriorated graphs are reliable or not. We want to understand whether the obtained synthetic graphs preserve the characteristics of brain networks and their network metrics are similar to those of SP graphs. The significance is thus assessed on the distributions of synthetic graphs w.r.t. the graphs representing SP patients. Our null hypothesis is that the two distributions come from different population with 99% of confidence ($p < 0.01$), meaning they are different. Results are shown in Table 4a and b for proposed and random-based deterioration strategy, respectively.

In order to help the reader at appreciating differences among different approaches, we also propose the box-plots (Fig. 3) of each network metric estimated before and after the deterioration (with both random strategy and proposed method), as well as the SP distributions.

Table 4. (a) and (b) show results of the Wilcoxon runk-sum tests between the measurement distributions of deteriorated synthetic graphs and SP graphs, with the proposed deterioration strategy and the random strategy, respectively (with $p < 0.01$).

(a)

Metrics	p-value
assortativity	**0.1366**
weighted degree	**0.0248**
modularity	**0.7348**
global efficiency	0.0060
num of edges	0.0031

(b)

Metrics	p-value
assortativity	2.8788×10^{-17}
weighted degree	6.0176×10^{-9}
modularity	7.0000×10^{-6}
global efficiency	**0.0255**
num of edges	0.0031

5.3 Experimental Settings

As mentioned earlier, the parameter k is directly related with the quantity of damage to apply; intuitively, it is important to choose a proper value for k, in order to obtain deteriorated graphs that are fairly comparable with real SP brain networks. In our setting, we choose to attack 10% of the total brain regions; in the following, we describe how we made this decision.

In particular, we performed a significant amount of experiments, by running the proposed algorithm multiple times with increasing values of k. For each run, we computed the network metrics on the obtained graphs and inspected the classification performances, by ending with longitudinal measurements along with k. This experimental tuning gave us the possibility to understand how the deteriorated graphs vary in terms of network metrics and classification performances, along with an increasing number of nodes to be deteriorated (exactly k).

Figure 4 shows how each network metric estimated on resulting graphs varies along with k. Each plot also shows the best value for k w.r.t. each network metric.

Figure 5 shows how the accuracy of SP classification varies along with k. In particular, we consider only the percentage of deteriorated graphs discriminated as SP, given that, as already discussed, it is important to understand for which value of k resulting graphs look more similar to SP networks, w.r.t. our dataset.

As the results suggest, it is better to choose a low value for k as optimal, e.g., around 10% of the total number of brain regions. Indeed, even if classification performances show increasing accuracy when $k > 20\%$, each estimated metric suggest that such values for k induce an excessive deterioration of graphs. This means that the proper number of disassortative nodes to be deteriorated, in order to obtain synthetic graphs as close as possible to SP graphs, stands around 10% of the original brain regions (that is, 8.4–9 nodes).

6 Discussion

Results show interesting variations of assortativity when applying the proposed deterioration strategy as well as an interesting similarity with the SP distribu-

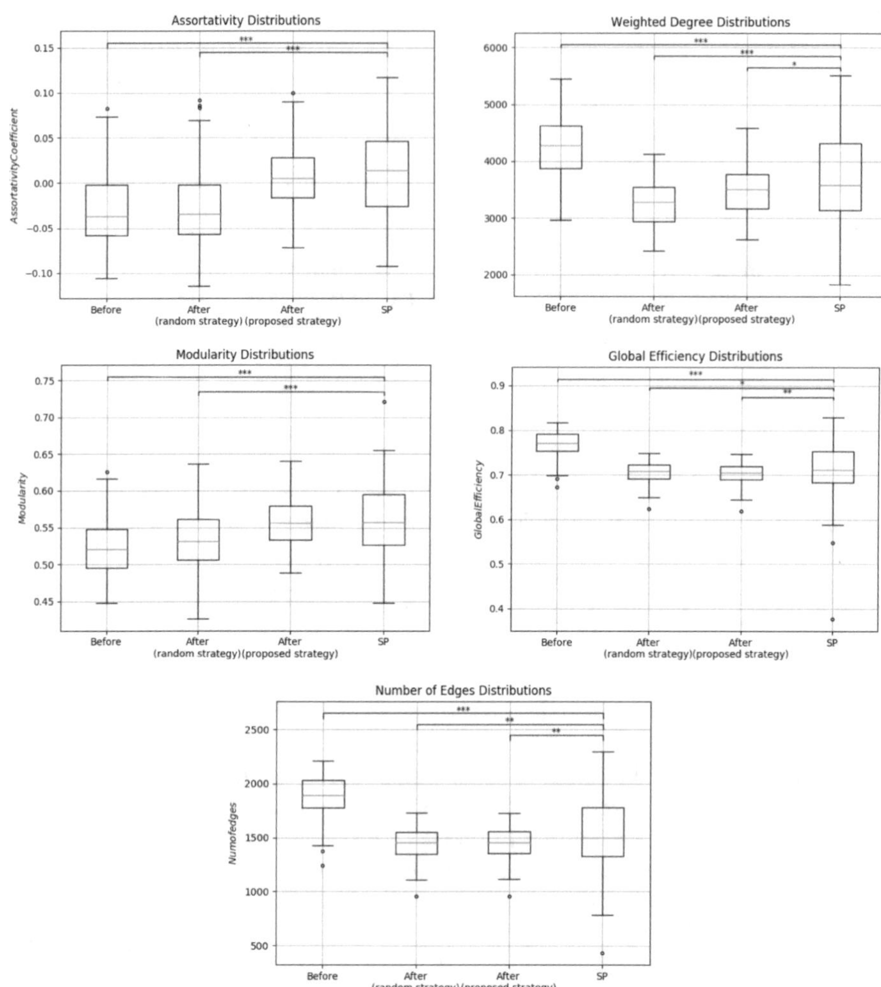

Fig. 3. Box-plots of the global network metrics (assortativity coefficient, weighted degree, global efficiency, modularity and number of edges) estimated. Differences between distributions were tested by means of a Wilcoxon runk-sum test [17] (*p < 0.05, **p < 0.01, ***p < 0.001).

tions. In particular, Table 2a shows how the averages of the proposed network metrics estimated on synthetic graphs are very close to the average measurements of original SP graphs. Statistic test of significance confirms the reliability of the results; indeed, Table 4a shows that assortativity ($p = 0.1366$), weighted degree ($p = 0.0248$) and modularity (0.7348) distributions are similar to SP distributions (with 99% of confidence). Furthermore, it is possible to exclude that deterioration occurs because of random events: Table 4b shows that the random

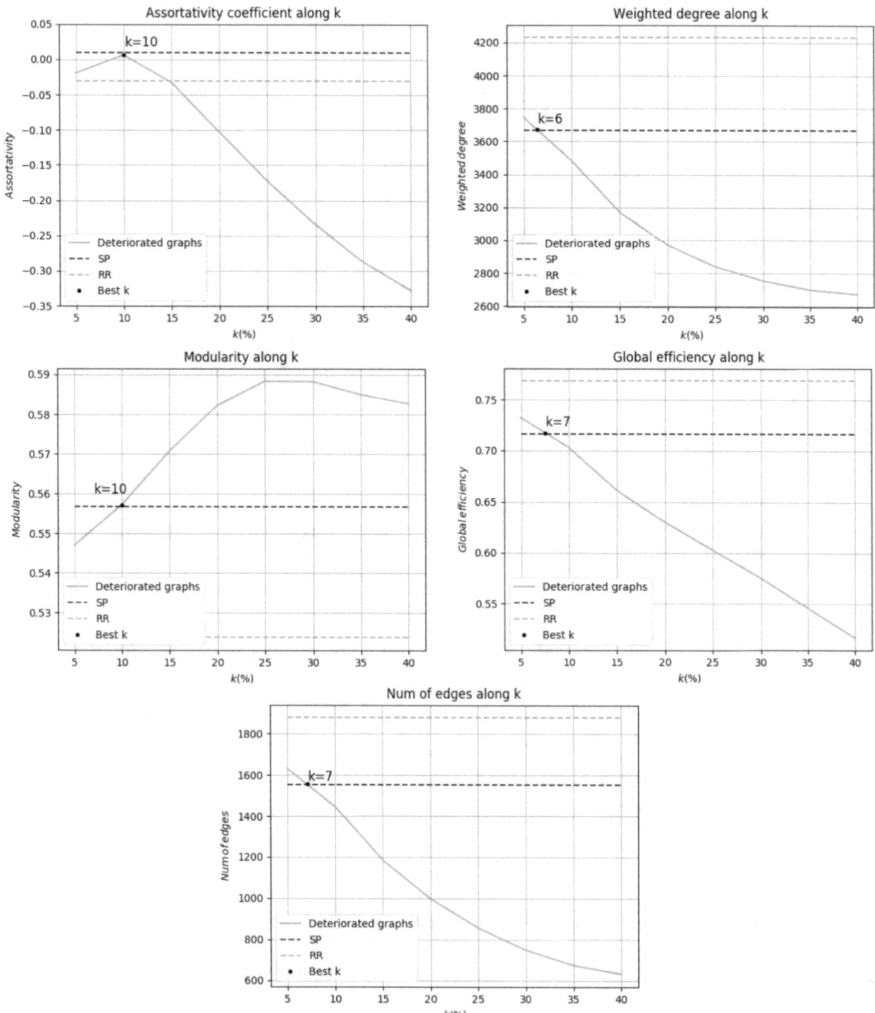

Fig. 4. Plot of the global network metrics (assortativity coefficient, weighted degree, global efficiency, modularity and number of edges) estimated along with different values of k.

approach does not induce noticeable variations in assortativity and, in addition, all others measure distributions are far different from SP ones.

Results of classification (Table 2b) show that, with majority voting, 53.55% of graphs deteriorated with the proposed method are classified as RR subjects and 36.88% as SP ones. Despite this, it is interesting to note that a significant decrease of accuracy occurs when comparing the majority voting of the random strategy with the proposed algorithm. In particular, Tables 2b and 3b show that accuracy of RR classification decreases from 94% to 53% while SP classifica-

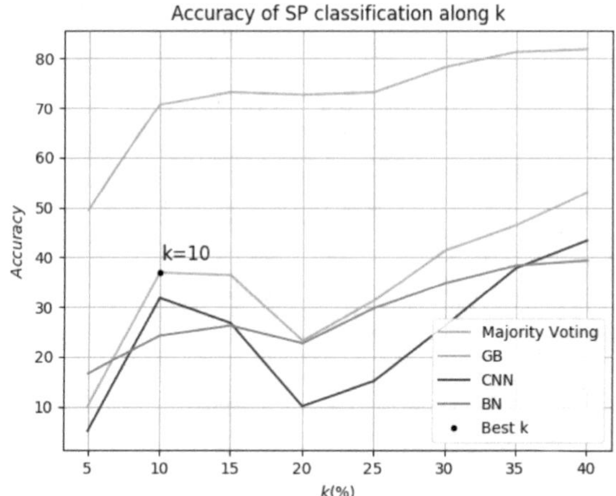

Fig. 5. This plot shows the accuracy of the SP classification, in particular the percentage of deteriorated graphs classified as SP with both individual classification and majority voting.

tion increases from 1% to 37%. In other words, disrupting the same amount of edges with the proposed strategy decreases the percentages of graphs classified as RR, while increasing the percentage of SP graphs, differently from the random-based approach. This suggests that, even if some other parameters that do not significantly change are quite discriminatory for classification purposes, regions deteriorated by the proposed approach play a crucial role in the disease worsening.

6.1 Analysis of the Most Disrupted Brain Regions

In the following, we focus on the study of the most disrupted regions. In particular, we want to highlight nodes that, during the deterioration process, have been chosen mostly as disassortative nodes to be deteriorated.

The bar chart in Fig. 6 shows, for each brain region, the number of times each node (neighbours included) is selected by the algorithm, on average, to be deteriorated. Blue bars represent outlier values identified by the MAD method; we consider these outliers as the most significant nodes in this context. A better visualization of this set of interesting regions can be observed in Fig. 7, where we plot the brain nodes together with their spatial coordinates in order to inspect their position within the brain network.

Interestingly, a strong symmetry can be noted. For instance, node 6 ("left inferior parietal cortex") is coupled with node 55, that is, its relative right region. Furthermore, the set of very important regions (the yellow-labelled nodes in the figure) are quite few w.r.t. the total number of brain regions (9/84, just 10% of total regions). Moreover, knowing that such this strong symmetry occurs, 4 or 5 regions per hemisphere are very likely to be chosen by the strategy.

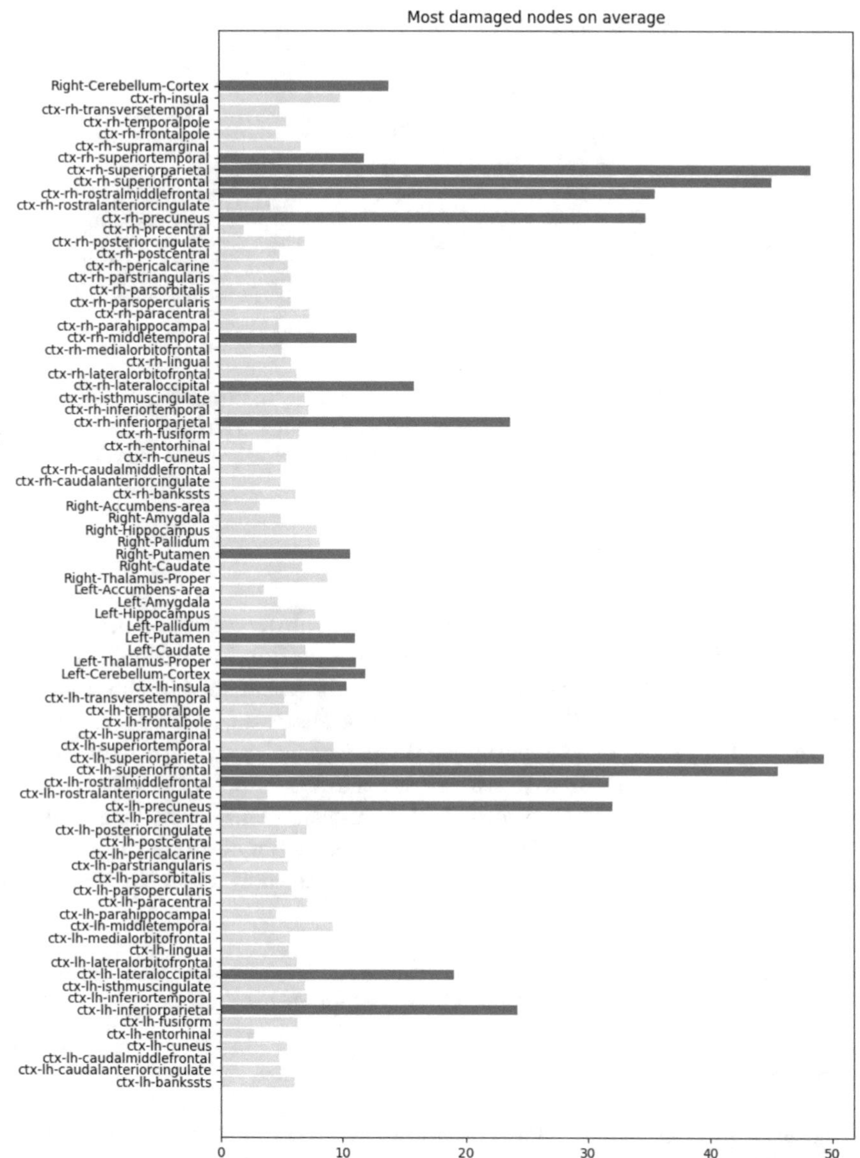

Fig. 6. Average number of times a node (neighbours included) has been selected to be deteriorated by the algorithm (x-axes), for each region (y-axes). Most significant ones, i.e., the most frequently selected, are highlighted in blue.

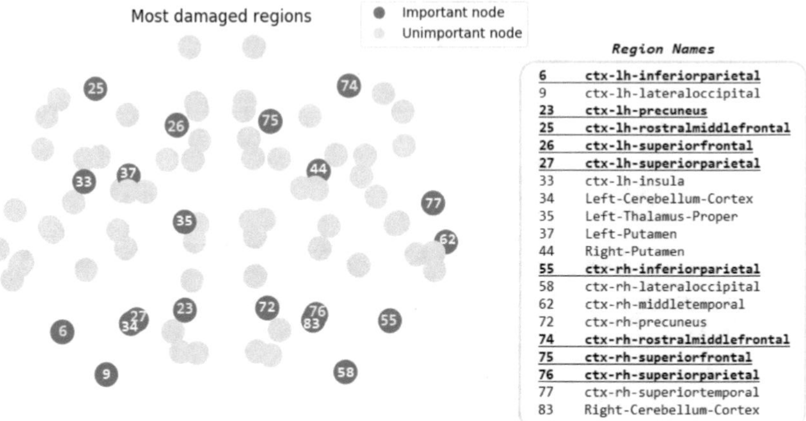

Fig. 7. Topological plot of the most damaged brain nodes and their names. Blue nodes are the most damaged. Among these, yellow-labelled nodes are those which relevance is above the average, thus representing regions that are likely to be disrupted much more than the others.

7 Conclusion

We proposed a method for identifying potential substructures of brain graphs that might lead to a disease worsening from RR to SP in MS patients. The proposed strategy is based on several findings already established in the literature, reached by means of a detailed complex network analysis of connectivity graphs in every clinical profile of MS patients. We performed a detailed experimental campaign; findings could allow to restrict the number of potential affected brain regions when a disease relapse occurs in RR subjects. This might be of interest, in the biomedical community, in order to focus on a small group of potentially vulnerable brain regions that are the ones that are most likely to be affected when a MS worsening occurs.

Acknowledgments. We gratefully acknowledge the support of NVIDIA Corporation with the donation of the Quadro $P6000$ GPU used for this research.

References

1. Barrat, A., Barthélemy, M., Pastor-Satorras, R., Vespignani, A.: The architecture of complex weighted networks. Neuroimage **101**(11), 3747–3752 (2004)
2. Betzel, R.F., Medaglia, J.D., Papadopoulos, L., Baum, G.L., Gur, R., Gur, R., Roalf, D., Satterthwaite, T.D., Bassett, D.S.: The modular organization of human anatomical brain networks: accounting for the cost of wiring. Quarterly **1**, 42–68 (2017)
3. Calimeri, F., Cauteruccio, F., Marzullo, A., Stamile, C., Terracina, G.: Mixing logic programming and neural networks to support neurological disorders analysis. In: International Joint Conference on Rules and Reasoning, pp. 33–47 (2018)

4. Calimeri, F., Marzullo, A., Stamile, C., Terracina, G.: Graph based neural networks for automatic classification of multiple sclerosis clinical courses. In: Proceedings of the European Symposium on Artificial Neural Networks. Computational Intelligence and Machine Learning (ESANN 2018) (2018)
5. Goldenberg, M.: Multiple sclerosis review. PT J **37**(3), 175–184 (2012)
6. Grafarend, E.W.: Linear and Nonlinear Models: Fixed Effects, Random Effects, and Mixed Models, p. 553. Walter de Gruyter, Berlin (2006)
7. Kawahara, J., Brown, C., Miller, S., Booth, B., Chau, V., Grunau, R., Zwicker, J., Hamarneh, G.: BrainNetCNN: convolutional neural networks for brain networks; towards predicting neurodevelopment. Neuroimage **146**, 1038–1049 (2017)
8. Kocevar, G., Stamile, C., Hannoun, S., Cotton, F., Vukusic, S., Durand-Dubief, F., Sappey-Marinier, D.: Graph theory-based brain connectivity for automatic classification of multiple sclerosis clinical courses. Front. Neurosci. **10**, 478 (2016)
9. Leys, C., Ley, C., Klein, O., Bernard, P., Licata, L.: Detecting outliers: do not use standard deviation around the mean, use absolute deviation around the median. J. Exp. Soc. Psychol. **49**(4), 764–766 (2013)
10. Lublin, F., Reingold, S., Cohen, J., Cutter, G., Sørensen, P., Thompson, A., Bebo, B.: Defining the clinical course of multiple sclerosis: the 2013 revisions. Neurology **83**(3), 278–286 (2014). https://doi.org/10.1212/WNL.0000000000000560
11. Newman, M.E.J.: Assortative mixing in networks. Phys. Rev. Lett. **89**, 208701 (2002)
12. Newman, M.E.J., Girvan, M.: Finding and evaluating community structure in networks. Phys. Rev. E. **69**, 026113 (2004)
13. Rubinov, M., Sporns, O.: Complex network measures of brain connectivity: uses and interpretations. Neuroimage **52**, 1059–1069 (2010)
14. Smith, R., Tournier, J.D., Calamante, F., Connelly, A.: Anatomically-constrained tractography: improved diffusion mri streamlines tractography through effective use of anatomical information. Neuroimage **62**(3), 1924–1938 (2012)
15. Sporns, O.: Structure and functions of complex brain networks. Neuroimage **15**(3), 247–262 (2013)
16. Tuch, D.S., Reese, T.G., Wiegell, M.R., Wedeen, V.J.: Diffusion MRI of complex neural architecture. Neurotechnique **40**, 885–895 (2003)
17. Wilcoxon, F.: Individual comparisons by ranking methods. Biom. Bull. **1**(6), 80–83 (1945)

The Efficacy of Various Machine Learning Models for Multi-class Classification of RNA-Seq Expression Data

Sterling Ramroach[1(✉)], Melford John[2(✉)], and Ajay Joshi[1(✉)]

[1] Department of Electrical and Computer Engineering, The University of the West Indies, St. Augustine Campus, Saint Augustine, Trinidad and Tobago
sramroach@gmail.com, ajay.joshi@sta.uwi.edu

[2] Department of Pre-Clinical Sciences, The University of the West Indies, St. Augustine Campus, Saint Augustine, Trinidad and Tobago
melford.john@sta.uwi.edu

Abstract. Late diagnosis and high costs are key factors that negatively impact the care of cancer patients worldwide. Although the availability of biological markers for the diagnosis of cancer type is increasing, costs and reliability of tests currently present a barrier to the adoption of their routine use. There is a pressing need for accurate methods that enable early diagnosis and cover a broad range of cancers. The use of machine learning and RNA-seq expression analysis has shown promise in the classification of cancer type. However, research is inconclusive about which type of machine learning models are optimal. The suitability of five algorithms were assessed for the classification of 17 different cancer types. Each algorithm was fine-tuned and trained on the full array of 18,015 genes per sample, for 4,221 samples (75% of the dataset). They were then tested with 1,408 samples (25% of the dataset) for which cancer types were withheld to determine the accuracy of prediction. The results show that ensemble algorithms achieve 100% accuracy in the classification of 14 out of 17 types of cancer. The clustering and classification models, while faster than the ensembles, performed poorly due to the high level of noise in the dataset. When the features were reduced to a list of 20 genes, the ensemble algorithms maintained an accuracy above 95% as opposed to the clustering and classification models.

Keywords: Machine learning · Cancer classification · Supervised learning

1 Introduction

Recent technological advances in molecular biology have resulted in cheaper and easier methods to perform RNA-seq expression analysis to extract the gene expression profile of a cell or tissue sample. These advances lead to the creation of large datasets of gene expression profiles for various diseases. Given these datasets, new fields emerged within bioinformatics to include diagnosis of diseases using gene expression data and the prediction of clinical outcomes with respect to treatment. RNA-seq expression analysis provides a quantitative measure of the expression levels of genes in a cell.

K. Arai et al. (Eds.): CompCom 2019, AISC 997, pp. 918–928, 2019.
https://doi.org/10.1007/978-3-030-22871-2_65

Current research estimates the existence of 19,000 genes per cell [1]. This data has been shown to contain the information necessary to diagnose cancer type.

Given the vast amount of data, it is not feasible for humans to understand relationships between samples and genes without assistance from a machine. Machine learning and other types of artificial intelligence have been used to build predictive models to classify and understand the relationships between gene expression levels and cancer type [2–8]. Previous works report wide variations in either accuracy, error rates, kappa coefficient, specificity, sensitivity, and so on. Prior studies also performed limited experiments using a small number of classes, classifiers, varying number of genes, small datasets, and a limited number of cancer types. A wealth of data arising out of expansive cancer genome projects such as The Cancer Genome Atlas (TCGA) now exists, which can be used to train machine learning algorithms to build decision models to diagnose cancer type, and possibly to identify causative genetic factors. A thorough understanding of the strengths and weaknesses of modern classifiers is necessary in order to build a successful RNA-seq expression diagnostic model. This study aims to provide sufficient information, to highlight the best classifiers for this type of data as well as similarly structured datasets, and to illustrate a structured comparison of the performance of various machine learning algorithms.

Podolsky et al. classified five cancer types using various gene expression datasets and found that K-nearest neighbor outperformed the support vector machine and an ensemble classifier [9]. Tarek et al. also found that the K-nearest neighbor achieves higher accuracy than ensemble methods for datasets of three types of cancer [10]. Azar et al. [3] and Uriarte et al. [11] showed that the Random Forest algorithm is preferable for gene selection using gene expression data. Al-Rajab et al. [1] and Tan and Gilbert [12] also highlighted the potential of using ensemble machine learning algorithms on gene expression data.

In this study, using a much larger number of tumours, a wider range of cancer types, and a larger number of genes, the possibility of achieving near 100% accuracy in the diagnosis of cancer type is investigated along with an investigation into the performances of the various models. The Random Forest (RF), along with other ensemble machine learning algorithms such as Gradient Boosting Machine (GBM), and Random Ferns (RFERN) are further analyzed in this work. The performance of these algorithms were compared with a classification and a clustering algorithm; Support Vector Machine (SVM) and K-Nearest Neighbor (KNN), respectively. Preliminary insight is also provided regarding the most important features or the genes used by the most successful models in relation to cancer type.

2 Methods

The dataset downloaded from the Catalogue of Somatic Mutations in Cancer (COSMIC) contained 5,629 samples with the expression levels of 18,019 genes each, resulting in an $n \times m$ matrix, where $n = 5{,}629$ and $m = 18{,}019$. The COSMIC data portal lists the gene expression dataset with the name CosmicCompleteGeneExpression.tsv. The version of the dataset downloaded is v80, with 101,406,435 rows in the format illustrated in Fig. 1.

Sample_id	Gene_name	Z_score
Sample_1	Gene_1	2.41
Sample_1	Gene_2	1.21
Sample_2	Gene_1	0.97
Sample_1	Gene_3	-0.20
...	...	...
Sample_*n*	Gene_*m*	-1.20

Fig. 1. Format of the cosmic_complete_gene_expression dataset from the COSMIC repository.

Inferring meaning from this raw dataset is difficult as the samples are not classified. Pre-processing is required before any machine learning models can be built. COSMIC also provides another dataset titled CosmicSample.tsv, which contains data on the primary site and the histology subtype for tumour samples as shown in Fig. 2.

Sample_id	Primary_site	Histology_Subtype
Sample_1	Lung	Adenocarcinoma
Sample_2	Lung	Squamous cell Carcinoma
Sample_3	Prostate	Adenocarcinoma
...	...	...
Sample_*n*	primary site	histology subtype

Fig. 2. Layout of the cosmic_sample dataset displaying the primary site and histology subtype of each sample.

Supervised machine learning models require the data to be labelled and thus, a combination of both datasets is needed for further analysis. The format of the processed dataset used in these experiments is illustrated in Fig. 3. There are 17 classes (types of cancer) in the dataset, ranging from 48 to 601 rows (samples).

Information on how to download the data is provided on the COSMIC website located at URL http://cancer.sanger.ac.uk/cosmic/download. All gene expression data provided by COSMIC were obtained from The Cancer Genome Atlas portal, which provides strict guidelines on the preparation of tumour samples. According to The Cancer Genome Atlas, gene expression (mRNA) data for all tumours were generated by the University of North Carolina at Chapel Hill, Chapel Hill, N.C. and all expression

			m genes			
Sample_id	Primary_site	Histology_subtype	Gene_1	Gene_2	...	Gene_m
Sample_1	Lung	Adenocarcinoma	2.41	1.21	...	0.99
Sample_2	Lung	Squamous cell Carcinoma	0.97	1.21	...	2.33
Sample_3	Prostate	Adenicarcinoma	2.39	2.05	...	-1.20
...	...	...	...	...	...	-0.55
Sample_*n*	primary site	histology subtype	1.23	1.21	...	-1.20

n samples

Fig. 3. Matrix representation of the pre-processed RNA-seq expression analysis, where rows represent samples, and columns represent the z-scores of each gene.

data are presented in a normalized form as z-scores [11]. All experiments were performed in the R environment [13] on an Asus Republic of Gamers G75 V laptop with an Intel Core i7 2.4 GHz processor and 16 GB RAM operating on Microsoft Windows 8.1 64-bit operating system.

RFbuilds classification trees are using a bootstrap sample of the dataset [6, 14, 15]. Each split is derived by searching a random subset (chosen by varying split points) of the given variables (16,718 genes) as the candidate set [16]. Although memory intensive, all trees are grown using all of the features in the training set to attain low-bias trees. Many deep uncorrelated trees are also grown to ensure low variance. This reduced bias and variance results in a low error rate. When classifying a sample, all trees in the forest output a vote declaring whether or not the new sample belongs to its class. The random forest classifies the new sample using the highest voted class. This ensemble is not prone to over-fitting since splitting points are randomly chosen (there will always be a random distribution). Another ensemble algorithm is the GBM. GBM is a nonparametric machine learning approach for classification based on sequentially adding weak learners to the ensemble [15, 17]. GBM reduces the error in its model by repeatedly combining weak predictors. One new weak learner is added to the ensemble with the sole purpose of complementing the existing weak learners in the model. If the new learner does not complement the ensemble, it is discarded. This process accounts for the usually slow training process of a GBM.

RFERN can be considered a constrained decision tree ensemble initially created for image processing tasks [18]. It is somewhat similar to a binary decision tree where all splitting criteria are identical making it a semi-naïve bayes classifier. The ensemble is built by randomly choosing different subsets of features for each fern. The SVM used a one-versus-rest technique where each class was separated into groups whereby it is positive compared to all other classes [19–21]. It finds hyper-planes that maximally separates classes. Due to the high dimensionality of the data, attributes are projected onto a high dimensional plane which make the data less likely to be linearly separable. This technique is known to perform poorly on high dimensional data. The KNN algorithm plots all attributes as points in a complex dimensional space (the number of genes). Using Manhattan distance, this algorithm classifies a new sample based on the votes of the nearest neighbors [21, 22]. KNN also performs poorly when tasked with classifying high dimensional data. Initial clusters fail to adapt to the training data. Attributes which were clustered incorrectly cannot be relocated at the end of modelling. Feature selections methods are commonly paired with SVM and KNN to avoid these issues.

3 Results and Discussion

The full dataset contained 5,629 samples comprised of 17 classes. The classes were unevenly distributed with the smallest class consisting of 48 samples, compared to the largest class of 601 samples. There were also NA, or NULL values for some attributes (genes) of various classes. The full dataset contained a total of 18,015 attributes, however, these were filtered to remove any NA or NULL values, resulting in 16,718 attributes per sample in the training set. After the models were built, they were then assessed with the task of classifying all samples in the test set, with the full 18,015 genes. Previous works focused on reducing the number of genes in training the prediction models [8, 11, 23, 24]. Feature reduction is performed after analysis of the results of utilizing the entire genome (or a larger number of genes than previously studied) when building prediction models. The dataset has the peculiar characteristic of having the number of attributes, orders of magnitude higher than the number of samples. High dimensional data often contains a high level of noise which is evident in the performance of some models.

Initially, each machine learning algorithm built its prediction model by examining a training set of gene expression data for which all 17 cancer types were visible. Models were built by finding relationships between the levels of expression of subsets of genes to a cancer type. The models were then assessed with a test set of gene expression data for which cancer types were withheld. Each algorithm was fine-tuned and trained on the full array of 18,015 genes per sample, for 4,221 samples (75% of the dataset). They were then tested with 1,408 samples (25% of the dataset). Table 1 shows the training and testing times for all models.

Table 1. Training and testing times of all models.

	RF	GBM	RFERN	SVM	KNN
Average training time (s)	15,859	21,445	1,332	3,271	1,524
Testing time (s)	441	294	338	749	411

Table 2 decomposes the overall classification accuracy by providing the accuracy for each class in the test set. The time taken for RF to be trained is 15,859 s. Although this is the second slowest algorithm, it attained the highest accuracy (99.89%) as shown in Table 2. The time taken to meticulously sift through the noise in the dataset lead to an overall better model. GBM accurately classified 99.68% of samples in the test set.

These results imply that greedy boosting can be a suitable direction in modelling gene expression data. RFERN classified 94.12% of test samples. It achieved the third best accuracy with the fastest training time of 1,332 s. This model was trained faster than RF and GBM because training time grows linearly with fern size, rather than exponentially with tree depth (as seen with the other ensemble techniques).

Table 2. Accuracy of all models.

Primary site Histology subtype	RF	GBM	RFERN	SVM	KNN
	Accuracy (%)				
Central Nervous System Astrocytoma Grade IV	100	100	100	16.67	64.29
Cervix Squamous cell Carcinoma	100	100	96.09	37.34	44.58
Endometrium Carcinosarcoma Malignant Mesodermal Mixed Tumour	100	100	98.25	0.00	5.56
Endometrium Endometrioid Carcinoma	100	100	80.92	72.86	78.57
Haematopoietic and Lymphoid Tissue Acute Myeloid Leukaemia	100	100	100	43.75	97.92
Haematopoietic and Lymphoid Tissue Diffuse Large B cell Lymphoma	97.92	97.92	95.83	0.00	45.45
Kidney Chromophobe Renal cell Carcinoma	100	100	100	0.00	86.96
Kidney Clear cell Renal cell Carcinoma	100	100	99.81	80.80	98.40
Large Intestine Adenocarcinoma	100	99.17	81.53	73.83	70.47
Liver Hepatocellular Carcinoma	100	100	100	54.12	95.29
Lung Adenocarcinoma	100	100	96.13	69.57	78.26
Lung Squamous cell Carcinoma	99.80	100	95.02	76.03	76.03
Ovary Serous Carcinoma	100	100	99.25	95.59	13.24
Pancreas Ductal Carcinoma	100	100	92.26	9.38	81.25
Prostate Adenocarcinoma	100	100	89.96	82.79	95.90
Stomach Adenocarcinoma	98.60	95.79	82.11	19.40	55.22
Upper Aerodigestive Tract Squamous cell Carcinoma	100	100	92.91	69.85	86.02
Average	**99.89**	**99.68**	**94.12**	**47.18**	**75.43**

There is a strong correlation between training times and prediction accuracy. The ensemble machine learning methods are all trained differently, with GBM and RF requiring the most time. Whereas the classification and clustering algorithms built their models relatively quickly. Ensemble algorithms outperformed SVM and KNN due to the nature of the problem. The number of genes per sample is much greater than the number of samples. There are 16,718 genes per sample, however, not all of these genes help differentiate between various cancer types. Ensemble algorithms are built to sift through the noise in large datasets to extract the core features.

The tumours belonging to the classes (cancer primary sites and histology subtypes): Central Nervous System Astrocytoma Grade IV, Cervix Squamous cell Carcinoma, Endometrium Carcinosarcoma Malignant Mesodermal Mixed tumour, Endometrium Endometrioid Carcinoma, Haematopoietic and Lymphoid Tissue Acute Myeloid Leukaemia, Kidney Chromophobe Renal cell Carcinoma, Kidney Clear cell Renal cell Carcinoma, Large Intestine Adenocarcinoma, Liver Hepatocellular Carcinoma, Lung Adenocarcinoma, Lung Squamous cell Carcinoma, Ovary Serous Carcinoma, Pancreas Ductal Carcinoma, Prostate Adenocarcinoma, and Upper Aerodigestive Tract Squamous cell Carcinoma were all classified with 100% accuracy by one of the ensemble algorithms. This implies that the genetic mutations which instigate the formation of these types of tumours share a similar distribution throughout all samples.

Given the stellar performances of RF and GBM, features were reduced based on a combination of the important variables selected by both ensembles. The top 80 genes were extracted and these were used to train the models again. This process was repeated for 60, 40, 20, and finally, 10 genes. Figure 4 illustrates the classification accuracy of all models built on these subsets of genes.

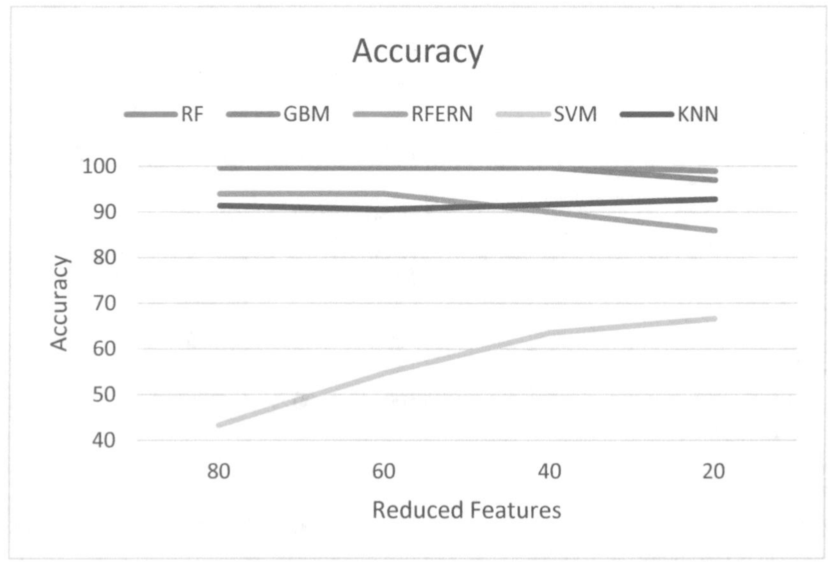

Fig. 4. Model accuracy based on a reduction of features

The accuracy dips when the feature count is 20. These 20 genes were extracted and displayed in Table 3. Defensin Beta, Interferon, and Keratin Associated Proteins are known to be strong cancer driver genes [25, 26]. Actin and Ribonuclease has been shown to play a significant role in some cancer types [27]. Further work is needed to determine the impact of olfactory genes on cancer as the current research is conflicted [28, 29].

Table 3. RF-GBM feature selection output

Name	Description
ACTL9	Actin Like 9
ACTRT2	Actin Related Protein T2
C10orf122	Testis
C17orf105	Chromosome 17
CLRN2	Clarin
DEFB104A	Defensin Beta
DEFB112	Defensin Beta
DEFB119	Defensin Beta
DEFB136	Defensin Beta
HS3ST3B1	Heparan Sulfate-Glucosamine
IFNA7	Interferon
KRTAP10.8	Keratin Associated Protein
KRTAP19.5	Keratin Associated Protein
KRTAP19.6	Keratin Associated Protein
KRTAP24.1	Keratin Associated Protein
OR10G9	Olfactory Receptor Family
OR2M2	Olfactory Receptor Family
OR52E4	Olfactory Receptor Family
OR5A2	Olfactory Receptor Family
RNASE12	Ribonuclease

Many tests and statistics were analyzed to ensure integrity of the results. The high levels of accuracy achieved are unlikely to be due to batch bias. The Cancer Genome Atlas gene expression values were generated by laboratory analysis at a single source, therefore errors in the equipment which were used for RNA-seq expression analysis would have been consistent throughout the dataset. Also, all expression values were normalized by conversion to z-scores. There is one major limitation to this study which arises from the data. There are no healthy samples in the dataset. Although the ensemble algorithms performed well at identifying cancer type, we are still unsure of their ability to identify healthy or non-cancerous tumours.

4 Conclusion

The use of machine learning in cancer diagnosis is becoming more feasible as algorithms become less prone to error and noise, and as the volume of training data increases. Using machine learning and data from RNA-seq expression, the potential exists for faster and more accurate diagnosis of cancer type. When trained with data derived from RNA-seq expression analysis, the random forest and gradient boosting machine were able to classify cancer type with an accuracy significantly greater than that of a support vector machine and K-nearest neighbors. The random forest and the gradient boosting machine were capable of diagnosing 17 types of cancer with accuracies of 99.89% and 99.68%, respectively. The random ferns algorithm was the third best performing algorithm with an accuracy of 94.12%. It was also the fastest algorithm. The support vector machine and the K-nearest neighbor algorithms classified cancer type with 47.18% and 75.13% accuracy, respectively. This difference in performance is attributed to the given task and the features of the dataset. RNA-seq expression produces a quantitative description of the levels of expression of genes in a cell. There are 18,015 genes and 5,629 samples in the dataset. The number of features (genes) is orders of magnitude higher than the number of samples, hence there is a high level of noise through which these algorithms need to sift. Pairing feature selection with all models lead to a significant improvement in the accuracy of the support vector machine, and K-nearest neighbors to 71.52% and 94.74%, respectively. Defensin beta, keratin associated protein, and the olfactory receptor family were found to be highly influential in the classification of cancer type.

References

1. Al-Rajab, M., Lu, J., Xu, Q.: Examining applying high performance genetic data feature selection and classification algorithms for colon cancer diagnosis. Comput. Methods Programs Biomed. **146**, 11–24 (2017)
2. Statnikov, A., Aliferis, C.F., Tsamardinos, I., Hardin, D., Levy, S.: A comprehensive evaluation of multicategory classification methods for microarray gene expression cancer diagnosis. Bioinformatics **21**, 631–643 (2005)
3. Azar, A.T., Elshazly, H.I., Hassanien, A.E., Elkorany, A.M.: A random forest classifier for lymph diseases. Comput. Methods Programs Biomed. **113**, 465–473 (2014)
4. Bartsch, G., Mitra, A.P., Mitra, S.A., Almal, A.A., Steven, K.E., Skinner, D.G., Fry, D.W., Lenehan, P.F., Worzel, W.P., Cote, R.J.: Use of artificial intelligence and machine learning algorithms with gene expression profiling to predict recurrent nonmuscle invasive urothelial carcinoma of the bladder. J. Urol. **195**, 493–498 (2016)
5. Breiman, L.: Random forests. Mach. Learn. **45**, 5–32 (2001)
6. Breiman, L., Friedman, J., Stone, C.J., Olshen, R.A.: Classification and Regression Trees. Wadsworth & Brooks. Cole Statistics/Probability Series (1984)
7. Ezkurdia, I., Juan, D., Rodriguez, J.M., Frankish, A., Diekhans, M., Harrow, J., Vazquez, J., Valencia, A., Tress, M.L.: Multiple evidence strands suggest that there may be as few as 19000 human protein-coding genes. Hum. Mol. Genet. **23**, 5866–5878 (2014)

8. Weinstein, J.N., Collisson, E.A., Mills, G.B., Shaw, K.R.M., Ozenberger, B.A., Ellrott, K., Shmulevich, I., Sander, C., Stuart, J.M., Network, C.G.A.R.: The cancer genome atlas pan-cancer analysis project. Nature Genet. **45**, 1113 (2013)
9. Podolsky, M.D., Barchuk, A.A., Kuznetcov, V.I., Gusarova, N.F., Gaidukov, V.S., Tarakanov, S.A.: Evaluation of machine learning algorithm utilization for lung cancer classification based on gene expression levels. Asian Pac. J. Cancer Prev. **17**, 835–838 (2016)
10. Tarek, S., Elwahab, R.A., Shoman, M.: Gene expression based cancer classification. Egypt. Inf. J. **18**, 151–159 (2017)
11. Díaz-Uriarte, R., De Andres, S.A.: Gene selection and classification of microarray data using random forest. BMC Bioinformatics **7**, 3 (2006)
12. Tan, Y., Shi, L., Tong, W., Hwang, G.G., Wang, C.: Multi-class tumor classification by discriminant partial least squares using microarray gene expression data and assessment of classification models. Comput. Biol. Chem. **28**, 235–243 (2004)
13. Team, R.C.: R Development Core Team R: a language and environment for statistical computing. R Foundation for Statistical Computing, Vienna, Austria (2014)
14. Khalilabad, N.D., Hassanpour, H.: Employing image processing techniques for cancer detection using microarray images. Comput. Biol. Med. **81**, 139–147 (2017)
15. Kursa, M.B.: rFerns: an implementation of the random ferns method for general-purpose machine learning. arXiv preprint (2012). arXiv:1202.1121
16. Meng, J., Zhang, J., Luan, Y.-S., He, X.-Y., Li, L.-S., Zhu, Y.-F.: Parallel gene selection and dynamic ensemble pruning based on affinity propagation. Comput. Biol. Med. **87**, 8–21 (2017)
17. Geurts, P., Ernst, D., Wehenkel, L.: Extremely randomized trees. Mach. Learn. **63**, 3–42 (2006)
18. Villamizar, M., Andrade-Cetto, J., Sanfeliu, A., Moreno-Noguer, F.: Bootstrapping boosted random ferns for discriminative and efficient object classification. Pattern Recogn. **45**, 3141–3153 (2012)
19. Zhi, J., Sun, J., Wang, Z., Ding, W.: Support vector machine classifier for prediction of the metastasis of colorectal cancer. Int. J. Mol. Med. **41**, 1419–1426 (2018)
20. Perez-Riverol, Y., Kuhn, M., Vizcaíno, J.A., Hitz, M.-P., Audain, E.: Accurate and fast feature selection workflow for high-dimensional omics data. PLoS One **12**, e0189875 (2017)
21. Li, X., Yang, S., Fan, R., Yu, X., Chen, D.: Discrimination of soft tissues using laser-induced breakdown spectroscopy in combination with k nearest neighbors (kNN) and support vector machine (SVM) classifiers. Opt. Laser Technol. **102**, 233–239 (2018)
22. Shang, Y., Bouffanais, R.: Influence of the number of topologically interacting neighbors on swarm dynamics. Sci. Rep. **4**, 4184 (2014)
23. Friedman, J., Hastie, T., Tibshirani, R.: The Elements of Statistical Learning. Springer Series in Statistics. Springer, Berlin (2001)
24. Cover, T., Hart, P.: Nearest neighbor pattern classification. IEEE Trans. Inf. Theory **13**, 21–27 (1967)
25. Ye, Z., Dong, H., Li, Y., Ma, T., Huang, H., Leong, H.S., Eckel-Passow, J., Kocher, J.-P.A., Liang, H., Wang, L.: Prevalent Homozygous deletions of type I interferon and defensin genes in Human Cancers Associate with Immunotherapy Resistance. Clin. Cancer Res. **24** (14), 3299–3308 (2018)
26. Rhee, H., Kim, H.-Y., Choi, J.-H., Woo, H.G., Yoo, J.E., Nahm, J.H., Choi, J.S., Park, Y.N.: Keratin 19 expression in hepatocellular carcinoma is regulated by fibroblast-derived HGF via a MET-ERK1/2-AP1 and SP1 axis. Cancer Res. **78**(7), 1619–1631 (2018)

27. Bram Ednersson, S., Stenson, M., Stern, M., Enblad, G., Fagman, H., Nilsson-Ehle, H., Hasselblom, S., Andersson, P.O.: Expression of ribosomal and actin network proteins and immunochemotherapy resistance in diffuse large B cell lymphoma patients. Br. J. haematol. **181**(6), 770–781 (2018)
28. Sanz, G., Leray, I., Dewaele, A., Sobilo, J., Lerondel, S., Bouet, S., Grébert, D., Monnerie, R., Pajot-Augy, E., Mir, L.M.: Promotion of cancer cell invasiveness and metastasis emergence caused by olfactory receptor stimulation. PLoS One **9**, e85110 (2014)
29. Lawrence, M.S., Stojanov, P., Polak, P., Kryukov, G.V., Cibulskis, K., Sivachenko, A., Carter, S.L., Stewart, C., Mermel, C.H., Roberts, S.A.: Mutational heterogeneity in cancer and the search for new cancer-associated genes. Nature **499**, 214 (2013)

Performance Analysis of Feature Selection Methods for Classification of Healthcare Datasets

Omesaad Rado(✉), Najat Ali, Habiba Muhammad Sani, Ahmad Idris, and Daniel Neagu

Faculty of Engineering and Informatics, University of Bradford, Bradford, UK
{o.a.m.rado,Nali50,hmsani,A.IdrisTambuwal, d.neagu}@bradford.ac.uk

Abstract. Classification analysis is widely used in enhancing the quality of healthcare applications by analysing data and discovering hidden patterns and relationships between the features, which can be used to support medical diagnostic decisions and improving the quality of patient care. Usually, a healthcare dataset may contain irrelevant, redundant, and noisy features; applying classification algorithms to such type of data may produce a less accurate and a less understandable results. Therefore, selection of optimal features has a significant influence on enhancing the accuracy of classification systems. Feature selection method is an effective data pre-processing technique in data mining, which can be used to identify a minimum set of features. This type of technique has immediate effects on speeding up classification algorithms and improving performance such as predictive accuracy. This paper, aims to evaluate the performance of five different classification methods including: C5.0, Rpart, k-nearest neighbor (KNN), Support Vector Machines (SVM), and Random Forest (RF), with three different feature selection methods, including: correlation-based feature selection method, Variables Importance selection method, and Recursive Feature elimination selection method on seven relevant numerical and mixed healthcare datasets. Ten-fold cross validation is used to evaluate the classification performance. The experiments showed that there is a variation of the effect of feature selection methods on the performance of classification techniques.

Keywords: Classification · Feature selection · Healthcare data

1 Introduction

Most of the real-world datasets usually contain useful information for understanding the data, at the same time the data may include several redundant or irrelevant features. Data with remarkably redundant variables have presented serious challenges to the existing machine learning algorithm including the curses of dimensionality, high computational complexity and storage. Such challenges can affect the performance of machine learning algorithms. However, methods of feature selection are often used to improve the efficiency of classification learning algorithms and to make the algorithms

K. Arai et al. (Eds.): CompCom 2019, AISC 997, pp. 929–938, 2019.
https://doi.org/10.1007/978-3-030-22871-2_66

computationally efficient. Feature selection is an important step of pre-processing stage in the process of building machine learning models [1]. Feature selection (FS) method refers to the process of identify and removing irrelevant features as much as possible from a dataset to obtain an optimal subset of features. FS method chooses a subset of variables from the original data set with no transformation of the original features space.

There are several experimental researches in the literature concerning the use of classification algorithms and feature selection on medical datasets. A primary concern of feature selection methods is for improving the accuracy of classifiers by selecting the most relevant features as possible from the data. The authors in [2] used Naïve Bayes, J48 and PART classifiers to examine the performance effect of the feature selection techniques. Another study presented in [3] has attempted to explore the challenge of using the methods of feature selection in imbalanced data with Bayesian learning. A comparative study to assess the level of consistency within selection algorithms outputs with regards to the task of high dimensional classification was presented in the study carried out by [4]. More recently, a study of analyzing the effectiveness of the filter, wrapper and fuzzy rough set-based feature selection methods [5] to produce high accuracy using KNN classifier based on microarray datasets. In addition, this paper also performs some comparative analysis between the filter, wrapper and fuzzy rough set-based feature selection methods. Similarly, in [3] a study of the analysis of the performance of different FS algorithms and classifiers on student data set in educational domain was performed. A considerable amount of literature has been published on feature selection techniques. One of these studies was to investigate some problems of online feature selection with feature streams has been discussed [5]. Surveys such as that conducted in [5] and [2] have shown a brief review of feature selection approaches used with machine learning in Neuroimaging and online feature selection with feature streams (OSFS).

Ideally, feature selection methods seek through a subset of features to detect the best subsets of features based on some feature evaluation procedures. The benefits of feature selection for learning can include a reduction in the amount of data needed to achieve learning, improved predictive accuracy of the learning algorithm, aid in achieving learned knowledge that is more compact and easily understood, and reduction in execution time [6, 7]. Feature selection methods for classification are broadly divided into three categories. Supervised Feature selection [8, 9], unsupervised Feature selection [10, 11], and semi-supervised feature selection [12]. Supervised Feature Selection is usually used for classification tasks. Three feature selection methods have been selected for this study. The first method is correlation-based feature selection (CFS) that works on the basis of correlation between features, the second method is variables importance (VImp) that ranks features based on certain criteria to handle redundant features, and the third method is recursive feature elimination (RFE) which work based on assumption of removing weakest features recursively [13, 14].

The purpose of this paper is to evaluate the performance of five different classification methods including: C5.0, Rpart, k-nearest neighbor (KNN), Support Vector Machines(SVM), and Random Forest (RF), with three different feature selection methods, including: correlation-based feature selection method, Variables Importance selection method, and Recursive Feature elimination selection method on seven relevant mixed and continuous healthcare datasets.

The rest of the paper is organized in as follows: Sect. 2 provides brief description of the chosen classification techniques and the feature selection methods as a common approach usually used for improving the accuracy of classification techniques. The experimental work and result analysis are represented in Sect. 3. Finally, Sect. 4 presents the conclusion and future work.

2 Background

This section briefly describes the techniques which areused in this study including:

2.1 Classification Techniques

Plenty classification techniques have been proposed for solving the classification problem, the most commonly used methods for data classification tasks include decision trees such as CART [15], C4.5 [16], C5.0 [17], K-nearest neighbor (KNN) [18], Support Vector Machines (SVMs) [19], Random forest (RF) [20], and Naïve Bayes [20]. In this study, five classification techniques have been selected for the experimental which are:

C5.0 classifier is a new decision tree algorithm developed based on C4.5 by Quinlan [17]. It contains all functionalities of C4.5 and apply a pack of new technologies [16]. The C5 classifier is able to anticipate which features are relevant and which are not relevant in classification [17]. The classifier can handle missing data values by estimating missing value as function of other features or apportion the missing one statistically among the results.

K-nearest neighbor (KNN) is a non-parameter and lazy algorithm used for both classification and regression tasks. Simply, the algorithm searches through all the dataset looking for the most similar instances using well-known distance such as Euclidian, Manhattan, and Minkowski distances for continuous data and Hamming distance for categorical data distance. The classifier performances better with low number of features then many features [18].

Recessive Partitioning and Regression Trees (Rpart) is an approach of classification, and it considers the basic of binary recursion splitting function. Mainly, Rpart is an implementation of Classification and regression trees (CART) approach [5]. Rpart builds models by using procedure of two stage. The obtained models can be represented as binary trees [15].

Support Vector Machine (SVMs) ware organically designed for binary classification problems [19], they have been extended to deal with, regression, and outlier detection problems with an intuitive model representation. Standard support vector machines (SVMs) have been adapted to work in metric space by building, a linear boundary in a high-dimensional similarity space. SVMs use similar way of kernel functions for defining similarity between samples, which depends on measuring the distance between the vector representations of the samples [21]. To apply SVM, the data should be scaled and represented as vector of real numbers.

Random forests (RF) is an ensemble learning method, constructed from some decision trees classifiers, and can be used for classification, regression, and feature selection tasks. The classifier can also be used to handle missing values. Moreover, it works well for small and medium data sets [5].

2.2 Feature Selection Methods

Feature selection methods have been applied on various datasets in several domains including healthcare. In recent years, there has been an increasing interest in analyzing healthcare datasets due to the fact that such data usually contains hidden information which needs to be extracted for right decision According to [22, 23], many potential benefits can be obtained with the use of feature selection, such as:

- Reducing the number of irrelevant features and reducing the measurement cost.
- Reducing redundant features and leads to increase in accuracy and efficiency of classification techniques.
- Facilitating data visualization and understanding of the data.

The most common feature selection methods include:

Correlation-based Feature Selection (CFS)

Feature selection for classification tasks in machine learning can also be accomplished on the basis of correlation between features, and such a feature selection procedure can be beneficial to machine learning algorithms [24–26]. CFS method generates ranks of the feature based on the correlation with the other features. All the features which give less correlation with the rest of the features will be selected and exclude the features that have high correlation from the data [27].

Variables Importance (VImp)

The variable importance can be quantified by using score of importance of given attributes. The use of the mean of misclassification rate for classification or mean square error (MSE) for regression [28].

Recursive Feature Elimination (RFE)

Recursive feature elimination (RFE) is a feature selection method that is used to remove weakest features. RFE seeks to improve generalization performance by ranking the features and recursively removing the least important features whose deletion will have the least effect on training errors [14, 29].

3 Experimental Evaluation

3.1 Datasets

Seven relevant healthcare datasets have been chosen from UCI Machine Learning Repository [30]. A brief description of these datasets is presented in Table 1.

Some datasets are pre-processed before we ran the experiments, inconsistent, erroneous, and missing values are removed. K-fold cross validation method is used for the classification performance.

Table 1. Description of the dataset used

SN.	Dataset	Data type	Instances	Features
1	Pima Indians	Numeric	768	9
2	Heart diseases	Numeric	303	14
3	Breast-cancer	Numeric	699	10
4	Indian liver patients	Mixed	583	11
5	SPECTF Heart	Numeric	267	44
6	Hepatitis	Mixed	115	17
7	Diabetes	Numeric	442	10

3.2 Experimental Setup

The experimental work is carried out in two phases. The first phase; includes applying the selected classification techniques directly to the data sets without the use of feature selection methods. The second phase, includes applying feature selection method as pre-processing step to select features subset, then applying the selected classification techniques to data subset. Figures 1 and 2 depicts the functional block diagram.

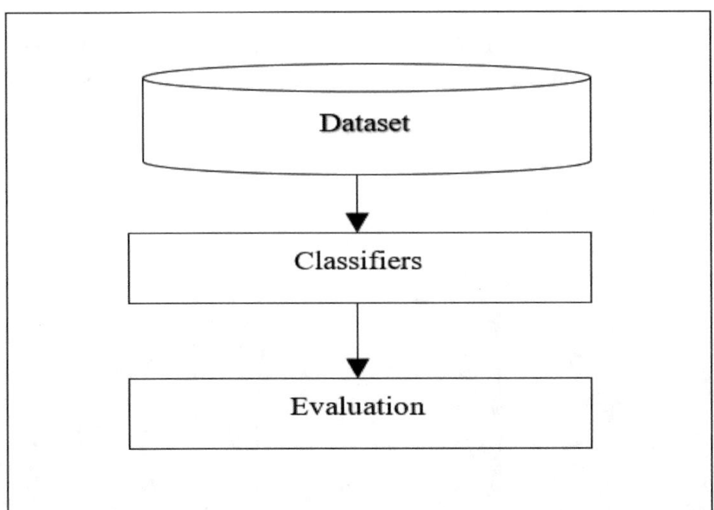

Fig. 1. Framework of data classification

3.3 Results and Discussion

Figure 3 illustrates the classification average accuracy which is obtained by applying five classification methods without feature selection methods to seven health care data sets. The results revealed that, the highest accuracy rate is obtained by applying KNN to Breast cancer data set with 97%. SVM obtained the highest accuracy rate reached Pima Indian diabetes with 77%, Statlog heart data set with 84%, and Hepatitis data set

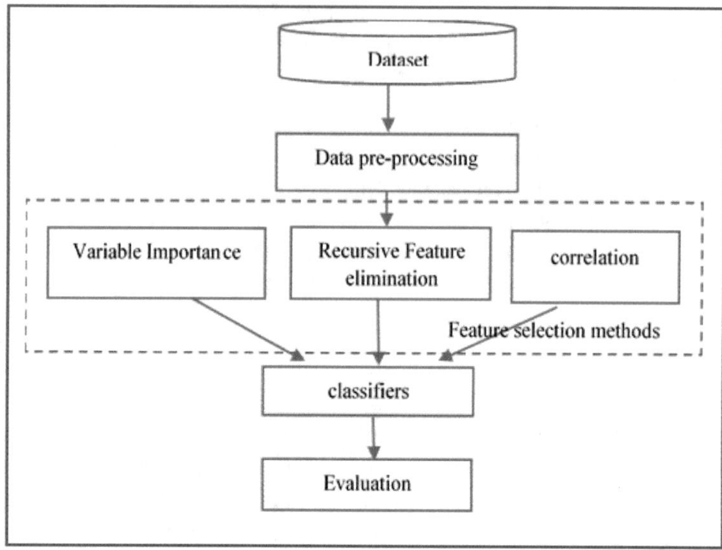

Fig. 2. Framework of data pre-processing and classification

with 86%. Conversely, Random forest obtained the highest accuracy rate reached 73% with Indian liver patient dataset, 82% with Spectf heart, and 71% with diabetes data sets.

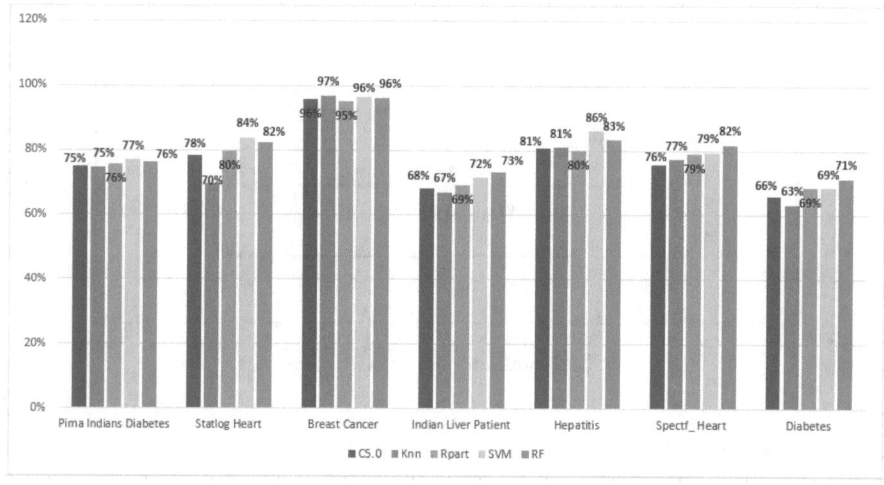

Fig. 3. Accuracy obtained by the chosen classifiers

Figure 4 shows the results obtained by applying the five chosen classification algorithms with correlation-based feature selection method. The highest accuracy among the seven data sets is obtained by applying KNN classifier to Breast Cancer dataset reached 96%. SVM to Pima Indian diabetes with 77%, Statlog heart dataset with 82%, Indian liver patient with 72%, and hepatitis dataset with 86%, and diabetes with 72%. RF gave highest performance reached 80% with spectf heart dataset. However, in some cases, such as applying SVM to Statlog heart dataset the classifier works well without applying correlation-based feature selection method, which indicted that all the features are very essential for efficient classification of the dataset. Other cases such as applying SVM to Pima Indian diabetes, hepatitis showed that there is no any affect for the feature selection method on the performance of the classifiers which indicated that the chosen features are not more important comparing to the whole features. Conversely, the performance of SVM increased to reaches 72% with Diabetes dataset after applying the feature selection method. The performance of KNN decreased after applying feature selection method comparing with the performance of the classifier which is obtained by apply KNN directly to Breast cancer dataset.

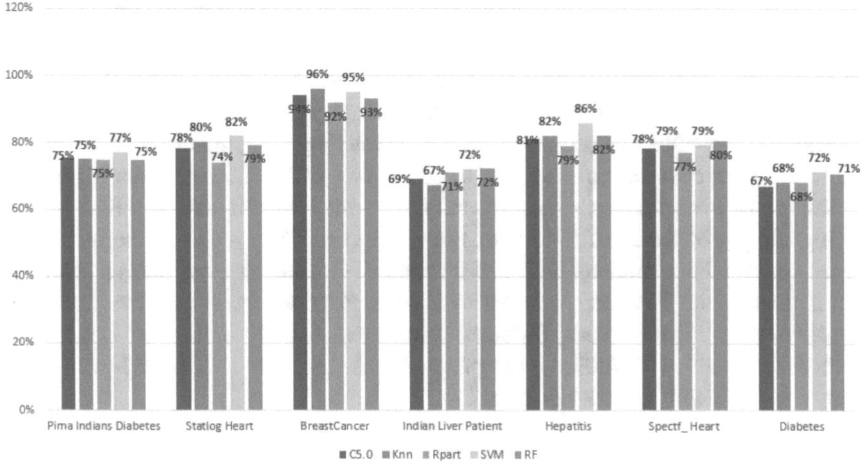

Fig. 4. Performance of chosen algorithms with correlation-based feature selection

Figure 5 shows the results obtained by applying the five chosen classification algorithms with Variables Importance feature selection method to seven data sets. The highest accuracy among all the data sets is obtained by applying KNN to Breast cancer data set achieved 97%. And SVM to Statlog heart with 82%, Hepatitis data set with 86%, and Spectf heart dataset with 81%. Random forest achieved highest performance with Pima Indian Diabetes dataset reached 78%, Indian liver Patient dataset reached 72%, and Diabetes dataset with 71%. The performance of RF classifier increased with applying Variables Importance feature selection method to Pima Indians Diabetes.

On the other hand, the performance of RF classifier decreased with applying feature selection method to Indian liver Patient dataset. Although, Rpart classifier did not achieve a highest performance with any dataset, but the performance of the classifier

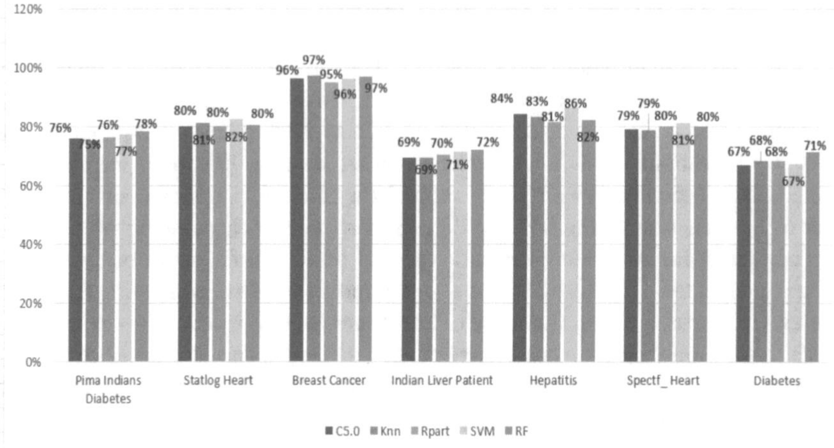

Fig. 5. Performance of chosen algorithms with variables importance selection

increased after applying it with Variables Importance feature toI ndian liver Patient, Hepatitis, and Spectf heart datasets reached 72%, 81%, and 80%, respectively.

Figure 6 illustrates the results which are obtained by applying the chosen classifiers with Recursive Feature elimination. Based on the results C5.0 has the highest performance of 97% with Breast cancer, and 85% with Hepatitis data set. Knn achieved the highest performance of 97% with Breast cancer, SVM achieved the highest performance with Pima Indians Diabetes reached 77%, Statlog heart dataset with 82%, Breast cancer with 97% and diabetes with 73%. Random forest also the highest performance with Pima Indians Diabetes dataset reached 77%, Statlog heart dataset reached 82%, and with Indian lever patient dataset achieved 73%, 83% with Spectf heart, and 73% with diabetes data sets.

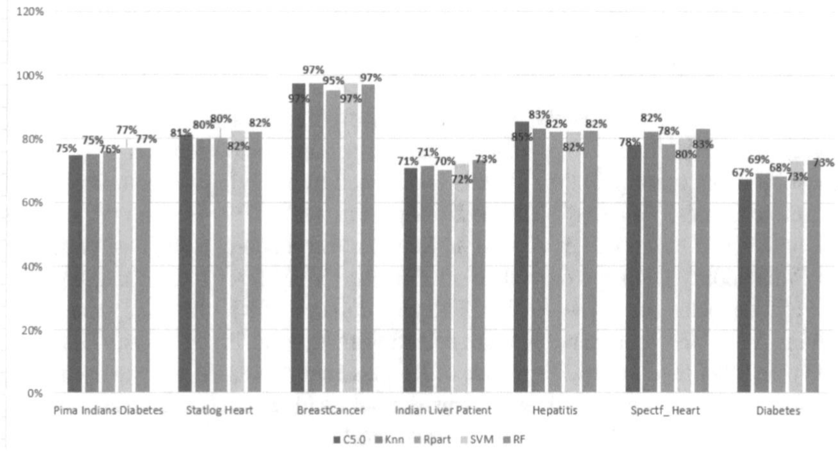

Fig. 6. Performance of chosen algorithms with recursive feature elimination

4 Conclusion and Future Work

This study, aims to highlight the influence of applying different feature selection methods (such as correlation-based Feature Selection, variables importance, and recursive feature elimination) towards five well-known classification techniques efficiency including; C5.0, Rpart, k-nearest neighbor (KNN), Support Vector Machines (SVM), and Random Forest (RF), for numerical and mixed healthcare datasets. The experimental work has been done in two phases: in the first phase; the classification techniques are directly applied to the healthcare datasets; in the second phase; each single feature selection method is applied as a pre-process step, and then the classification techniques are applied to the subsets. The results indicated that some classification techniques work well without applying feature selection methods and the classification techniques achieved better performance when all the features in some datasets are present. Some classification techniques did not yield promising results due to no influence of feature selection methods on the performance of classifiers. However, some classification techniques performed better after applying feature selection methods such as Rpart classifier.

Further work on this topic will include, investigating the effect of feature selection methods on classifying high-dimensional heterogeneous healthcare data sets using wide range of feature selection and classification techniques. Furthermore, the performance of hybrid combinations of features selection methods can be explored and evaluated on big healthcare datasets.

References

1. Chandrashekar, G., Sahin, F.: A survey on feature selection methods. Comput. Electr. Eng. **40**(1), 16–28 (2014)
2. Singh, R., Kumar, H., Singla, R.K.: Analysis of feature selection techniques for network traffic dataset. In: 2013 International Conference on Machine Intelligence Research Advancement, pp. 42–46 (2013)
3. Zaffar, M., Hashmani, M.A.: Performance analysis of feature selection algorithm for educational data mining, pp. 7–12 (2017)
4. Dessì, N., Pes, B.: Similarity of feature selection methods: an empirical study across data intensive classification tasks. Expert Syst. Appl. **42**(10), 4632–4642 (2015)
5. Arun Kumar, C., Sooraj, M.P., Ramakrishnan, S.: A comparative performance evaluation of supervised feature selection algorithms on microarray datasets. Procedia Comput. Sci. **115**, 209–217 (2017)
6. Khalid, S., Khalil, T., Nasreen, S.: A survey of feature selection and feature extraction techniques in machine learning. Undefined (2014)
7. Molina, L.C., Belanche, L., Nebot, À.: Feature selection algorithms: a survey and experimental evaluation
8. Weston, J.: Use of the zero-norm with linear models and Kernel methods André Elisseeff Bernhard Schölkopf Mike Tipping (2003)
9. Song, L., Smola, A., Gretton, A., Borgwardt, K.M., Bedo, J.: Supervised feature selection via dependence estimation

10. Dy, J.G., Brodley, C.E.: Feature selection for unsupervised learning. J. Mach. Learn. Res. **5**, 845–889 (2004)
11. Mitra, P., Murthy, C.A., Pal, S.K.: Unsupervised feature selection using feature similarity. IEEE Trans. Pattern Anal. Mach. Intell. PAMI. **24**(3), 301–312 (2002)
12. Xu, Z., King, I., Lyu, M.R.T., Jin, R.: Discriminative semi-supervised feature selection via manifold regularization. IEEE Trans. Neural Networks **21**(7), 1033–1047 (2010)
13. Ben Brahim, A., Limam, M.: A hybrid feature selection method based on instance learning and cooperative subset search. Pattern Recogn. Lett. **69**(C), 28–34 (2016)
14. Ramírez-Hernández, J.A., Fernandez, E.: Control of a re-entrant line manufacturing model with a reinforcement learning approach. In: Proceedings of the 6th International Conference on Machine Learning Applications, ICMLA 2007, pp. 330–335 (2007)
15. Wilkinson, L.: Classification and regression trees. Systat **11**, 35–56 (2004)
16. Quinlan, J.R., Ross, J.: C4.5 : Programs for Machine Learning. Morgan Kaufmann Publishers, San Mateo (1993)
17. Pang, S., Gong, J.: C5.0 classification algorithm and application on individual credit evaluation of banks. Syst. Eng. - Theory Pract. **29**(12), 94–104 (2009)
18. Cover, T., Hart, P.: Nearest neighbor pattern classification. IEEE Trans. Inf. Theory **13**(1), 21–27 (1967)
19. Vapnik, V.N.: The Nature of Statistical Learning Theory. Springer, New York (2000)
20. Zaki, M.J., Meira Jr., W.: Data Mining and Analysis: Fundamental Concepts and Algorithms. Cambridge University Press, New York (2014)
21. Osuna, E., Freund, R., Girosit, F.: Training support vector machines: an application to face detection. In: Proceedings of the IEEE Computer Society Conference on Computer Vision and Pattern Recognition, pp. 130–136, June 1997
22. Blum, A.L., Langley, P.: Selection of relevant features and examples in machine learning. Artif. Intell. **97**(1–2), 245–271 (1997)
23. Liu, H., Motoda, H.: Feature Selection for Knowledge Discovery and Data Mining. Springer, Boston, MA (1998)
24. Chou, T.-S., Yen, K.K., Luo, J., Pissinou, N., Makki, K.: Correlation-based feature selection for intrusion detection design. In: MILCOM 2007 - IEEE Military Communications Conference, pp. 1–7 (2007)
25. Michalak, K., Kwasnicka, H.: Correlation based feature selection method. Int. J. Bio-Inspired Comput. **2**(5), 319 (2010)
26. Hajek, P., Michalak, K.: Feature selection in corporate credit rating prediction. Knowl.-Based Syst. **51**(1), 72–84 (2013)
27. Yin, L., Ge, Y., Xiao, K., Wang, X., Quan, X.: Feature selection for high-dimensional imbalanced data. Neurocomputing **105**, 3–11 (2013)
28. Genuer, R., Poggi, J.M., Tuleau-Malot, C.: Variable selection using random forests. Pattern Recogn. Lett. **31**(14), 2225–2236 (2010)
29. Zeng, X., Chen, Y.-W., Tao, C., van Alphen, D.: Feature selection using recursive feature elimination for handwritten digit recognition. In: 2009 Fifth International Conference on Intelligent Information Hiding and Multimedia Signal Processing, pp. 1205–1208 (2009)
30. UCI Machine Learning Repository. https://archive.ics.uci.edu/ml/index.php. Accessed 18 Dec 2018

Towards Explainable AI: Design and Development for Explanation of Machine Learning Predictions for a Patient Readmittance Medical Application

Sofia Meacham[1], Georgia Isaac[1], Detlef Nauck[2(✉)], and Botond Virginas[2]

[1] Bournemouth University, Poole, UK
[2] British Telecom, Adastral Park, Ipswich, UK
detlef.nauck@bt.com

Abstract. The need for explainability of AI algorithms has been identified in the literature for some time now. However, recently became even more important due to new data protection act rules (GDPR 2018) and due to the requirements for wider applicability of AI to several application areas. BT's autonomics team has recognized this through several sources and identified the vitality of AI algorithms explainability to ensure their adoption and commercialization. In this paper, we designed and developed a system providing explanations for a prediction of patient readmittance using machine learning. The requirements and the evaluation were set by BT through their projects with real-customers in the medical domain. A logistic regression machine learning algorithm was implemented with explainability "hooks" embedded in its code and the corresponding interfaces to the users of the system were implemented through a web interface. Python-based technologies were utilized for the implementation of the algorithm (Scikit-learn) and the web interface (web2py), and the system was evaluated through thorough testing and feedback. Initial trade-off analysis of such an approach that presents the overhead introduced by adding explainability versus the benefits was performed. Lastly, conclusions and future work are presented, considering experimentation with more algorithms and application of software engineering methods such as abstraction to the aid of explainable AI, leading further along to "explainability by design".

Keywords: Explainable AI · Machine learning · Scikit-learn · Human-centered computing · Human–computer interaction

1 Introduction

This paper discusses the design and implementation of a medical-domain system for prediction of diabetic patient readmittance from the viewpoint of 'Explainable AI', a concept used to provide explanations and understandability of the corresponding machine learning algorithm's decisions. The dataset used to develop said system was provided by the UCI Machine Learning Repository [1, 2], first collected and donated by Strack et al. [3]. The dataset details diabetic patient admission encounters and

K. Arai et al. (Eds.): CompCom 2019, AISC 997, pp. 939–955, 2019.
https://doi.org/10.1007/978-3-030-22871-2_67

encourages the prediction of the binary attribute 'Readmitted' as either 'readmitted' or 'not readmitted'. The client for this research was BT's research labs at Adastral Park, Ipswich, UK. BT initiated the need for explainability in their machine learning algorithms and holds several clients in the health and other application sectors.

From extensive background study detailed in Sect. 2, it has been recognized that algorithms of the future, such as AI and classification/prediction algorithms, will need to explain themselves in order to ensure their adoption. Also, different types of algorithms are amenable to different approaches that need to be investigated.

In this paper, we took the bottom-up approach as regards to the explainability problem by starting from a specific case study that includes a specific classification algorithm (logistic regression) and attempts to make it explainable to users ranging from machine learning experts to domain experts (medical professionals in this case).

A systems engineering approach was taken to design the system at high-level of abstraction prior to detailed implementation. Specifically, UML semi-formal notations such as Class diagrams were used to define the system structure and functionality. Detailed implementation that consisted in integrating the web2py web framework with data science algorithms taken from the Scikit learn library followed. The developed web interface environment enabled experimentation with explainability aspects of the algorithm with promising results. It was observed that if "hooks" are positioned in specific points of the algorithm presenting relevant information to the user following would significantly improve explainability without compromising the overall algorithm performance.

The remainder of the paper will cover in Sect. 2.1 a background study on 'Explainable AI' research and in Sect. 2.2 explainability for several types of classification algorithms. In Sect. 3, the case study of an explainable algorithm for the medical domain is detailed. Specifically, in Sect. 3.1 the system requirements are presented, in Sect. 3.2 the system architecture and in Sects. 3.3 and 3.4 the design and implementation of the algorithm with the viewpoint of explainability are provided. Section 4 presents the evaluation of the proposed approach. Specifically, in Sect. 4.1 the classifier accuracy is detailed, in Sect. 4.2 the functional correctness is presented, in Sect. 4.3 the overhead to performance due to explainability is details and in Sect. 4.4 the client-machine learning expert feedback is presented. Section 5 offers conclusion and suggestions for future work and research directions.

2 Background Study: Explainable AI (XAI)

2.1 Explainable AI

'Explainable AI' (XAI) is a concept established in 2016 by DARPA, defining the practice of improving understandability, trustability, and manageability of emerging artificial intelligence systems [4]. Launchbury describes the evolution of artificial intelligence, starting with describing; initially, artificial intelligence was able to provide a description of data using sets of logic rules to represent knowledge in limited domains, though had no learning capability, only data perception and reasoning. Next, AI evolved to have the ability to predict and classify big data through training and testing of statistical models, however this is limited by its minimal ability to provide reasoning, leading to improvements in AI perception and learning, but little in terms of abstraction and reasoning. Launchbury introduces XAI as the next step in AI evolution; a stage in which developers construct explanatory models for use in real world situations, with the hope of facilitating natural communication between machines and people. In this evolution, systems can learn and provide reasoning, also improving perception and abstraction [5].

With the nature of machine learning classification algorithms being black-box and mostly uninterpretable, the integration of explainability through XAI into these algorithms should be imperative, especially in high-risk situations where the algorithm's result must be trusted enough to avoid repercussions. It was determined that two systems of XAI implementation can be used, being; ante-hoc and post hoc. Ante-hoc systems encompass machine learning algorithms that already hold a level of interpretability through 'glass-box' approaches, such as the decision tree or linear regression algorithms [6]. Post hoc systems alternatively provide explanations for uninterpretable, black-box systems such as neural networks, where their workings and decisions are undeterminable without the intervention of a separate tool such as LIME (Local Interpretable Model-Agnostic Explanations) [6, 7]. In both ante-hoc and post hoc systems, explainability is used to enable and inform domain experts to make their own decisions based on the outcomes of machine learning algorithms.

2.2 Types of Classification Algorithms and Their Explainability

Implementation of a classification system of any type requires thorough investigation into varying machine learning algorithms to ensure that the correct and most accurate algorithm is chosen for the task. In this case, as the system must not only classify but also explain and provide reasoning, there are further implications which must also be considered during investigation.

One such implication is the interpretability of the algorithm – or the ease at which the mapping of an output to descriptors can be understood by a user. These descriptors could be used to form the basis of explanations for the algorithm's decision [6]. A notable understanding though is that usually the more complex an algorithm, the less interpretable it is, even though the more complex algorithms tend to be more accurate [8]. This introduces a trade-off and begs the question: Do we require an incredibly accurate algorithm with limited to no explanations, or a reasonably accurate algorithm with sufficient explanations? PwC define a gap analysis approach for understanding and working with this trade-off. An algorithm's ability to explain and an organisation's readiness are measured on scales from low to high. A required level along with a current level are also defined on each scale. This method of analysis proves useful in determining which aspects of machine learning the organization should focus on. For instance, if they are comfortably past the required level of explainability, then this should indicate that they are able to trade-off some of their model explainability for accuracy instead [8]. In a business setting, a method such as gap analysis should be considered when working with minimizing negative impacts of the explainability-accuracy trade-off.

Beyond the main categorization of classification algorithms in ante-hoc and post hoc explainability types, current attempts in literature focus on reviewing (from the explainability perspective) the following commonly used algorithms:

Decision Tree Algorithms. The decision tree algorithm is considered one of the simplest and most interpretable white-box machine learning algorithms available, as it can be compared to the human decision-making process. Compared to the performance of a logistic regression model, it can be seen to underperform slightly, however due to its simplicity it can be used to provide helpful visual explanations for its results [9]. Though simple and interpretable, the decision tree algorithm is particularly susceptible to noisy data, for instance if two instances of data have the same values for some attributes but differing classification results and is also prone to overfitting.

Neural Networks. These are black-box algorithms, known to provide good accuracy but limited to no interpretability [10]. Neural networks require much lengthier training times in comparison to simpler models, owing to their complex nature. They contain considerably more connections and subtle properties when it comes to interactions between attributes. Mixing this with the workings of hidden layers alone being notoriously difficult to decipher, the overall interpretability of the algorithm suffers [8]. Due to the complexity of neural networks, extensive amounts of data are usually required to train a reliable model.

Logistic Regression. Similarly to decision trees, logistic regression is considered a simple, white-box algorithm which can provide interpretable explanations of results. Its interpretability stems from the ability to easily extract and understand the workings of the algorithm, for instance, through its coefficients [11]. Using these, the relationship between each attribute and the result can be visualized, providing an initial step towards explainability.

3 Case Study: Explainable Logistic Regression Algorithm for the Medical Domain

3.1 System Requirements

The system was designed and developed as a project at Bournemouth University in collaboration with an industrial partner, BT. BT's autonomics team provided the requirements for this project as they face the explainability problem in several of their projects with real clients. As part of their attempts to commercialize their autonomic algorithms and generally their classification algorithms adoption, explainability of the algorithm's decision making have arose several times. A medical application domain prediction algorithm was chosen for this project due to the criticality of the risk taken through classification decisions for these types of applications.

The main objective of the system was to predict patient readmission from the dataset suggested by the client and provided by UCI, and to give explanations for the given predictions.

As mentioned in the background study, there are two types of explainability in literature: ante-hoc where the explainability is inserted between the steps of the algorithm and post hoc where the algorithm is so complicated that it can't be stopped without serious compromising and explainability is inserted at the end of the algorithm, usually through applying external tools.

For this case study, an ante-hoc system with the aim of providing explainability of machine learning algorithm decisions for a medical application was designed and developed. Several technologies can be used towards this system with most cases incorporating interface design and web technologies. Interface design and web technologies are the most common approaches used to enable the communication between machines and users forming the backbone of any human–computer interaction. As explainability and specifically to domain experts in our case lies in this human–computer interaction point, it is apparent that interface design and web technologies should be incorporated.

The implementation decisions for interface design and web technologies were focused on the consistent use of Python as the backbone of the development. As Python is a strong language when it comes to compute-intensive tasks and is considered a leading language for machine learning [12], and the Scikit-learn library is both widely used and one of the top-performing machine learning libraries in terms of time efficiency [12], both technologies were chosen. Previous works have identified the ability of the Python web framework Web2py to work effectively with the Python machine learning library Scikit-learn [13], therefore this was the primary web technology chosen to develop the system.

The system described in this paper utilises a logistic regression algorithm in order to classify patients as regards to their probability of readmission. Several algorithms were evaluated from their capability to solve the case study problem and their explainability. The decision tree algorithms, however, simple and interpretable they maybe, were not chosen due to their susceptibility to noise data and proneness to overfitting. The neural networks are mostly categorized to post hoc explainability due to their complexity and their requirement for enormous amount of data. The logistic regression was the best choice as in terms of the medical dataset as enough data exists to train the logistic regression model efficiently, and resulting probabilities provide effective insight into how changes in their medical situation, for instance the number of inpatient procedures they have had, could affect their likelihood of being readmitted. Logistic regression also provides improved interpretability in comparison to more complex algorithms such as neural networks.

3.2 System Architecture

The high-level proposed system architecture is depicted in Fig. 1.

In this figure, the Model View Controller architectural pattern followed by web2py web framework is the center of the web interface design. In this model, the data stored in the model are manipulated by the functions of the Controller and are being viewed by the user.

In our proposed system architecture, this MVC model is enriched and extended with data science algorithms from the Scikit learn library which are interacting at two points: as part of the controller where the classification algorithm is being called to operate on data; as part of the model where we have integrated an existing data set and with the additional data stored/retrieved by the user.

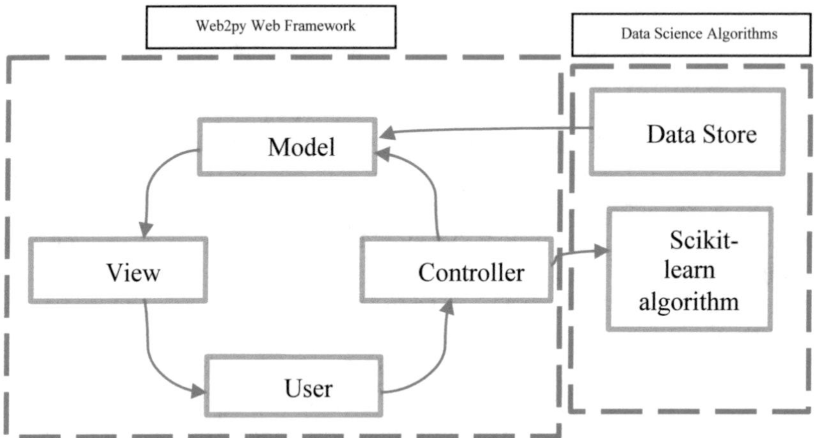

Fig. 1. System architecture diagram

This architecture is reproducible and can be applied to create web interfaces that use data science algorithms in several application contexts from explainable AI to intelligent interfaces.

3.3 Explainable Algorithm Ante-Hoc Design

The class diagram in Fig. 2 details the architecture of the classification system.

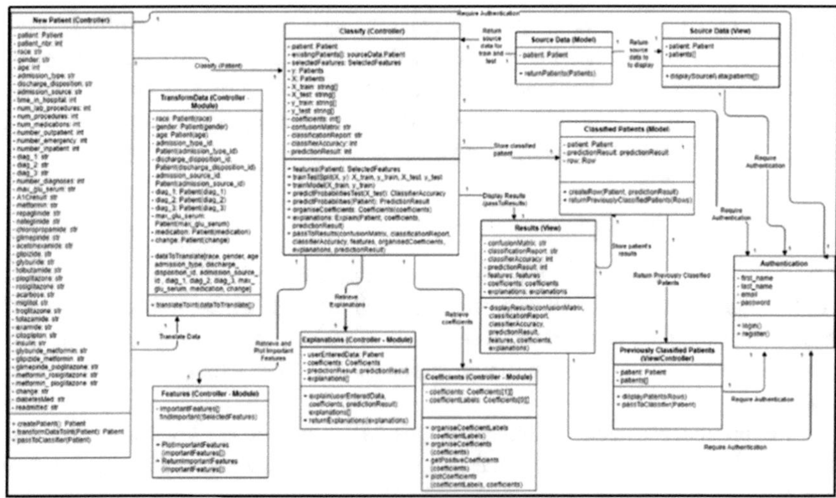

Fig. 2. UML class diagram for classification algorithm

Classification using the *Classify class* itself requires the user to be logged in, defined by the *Authentication class*, and a new patient to be created, defined by the *New Patient class*. Once the patient is created, the patient data is transformed to ensure all data points are integers, defined by the *Transform Data class*.

The classifier is then built by the *Classify class* using data returned by the *Source Data (model) class*, also displayed on a web page, defined by the *Source Data (view) class*. With a built classifier and transformed data, the *Classifier class* can classify the new patient.

During classification, a confusion matrix and classification report are produced. The *Features class* is used to find, plot, and return the important features. As well as this, logistic regression coefficients are retrieved, and both the label and value of each positive coefficient are retrieved, plotted, and returned, as defined by the Coefficients class.

Once classified, both the results and coefficients are used to determine the relevant explanation for each attribute, as defined by the *Explanations class*.

Finally, the patient's data and results are inserted into the database, defined by the *Classified Patients (Model)* all of which, when requested, can be viewed on the web page defined by the *Previously Classified Patients (View Controller) class*. The patient's classification results, along with the confusion matrix, classification report, plotted feature importances, plotted coefficients, and explanations for each coefficient are displayed on the results page, as defined by the Results class.

The following sections detail all major points where explainability was added and presented to the user and the corresponding interface will be depicted in figures.

3.4 Explainable Algorithm Ante-Hoc Implementation

Development of the machine learning classifier was carried out following the CRISP-DM methodology [14]. Both the data understanding and data preparation stages were carried out to ensure insufficiencies in the dataset, such as noise, missing data, and incorrect datatypes, were resolved. These stages led to a reduction in dataset size from 100,000 instances to 69,374.

With a prepared dataset, a logistic regression model was developed in Python using the Scikit-learn library and was chosen based on both its interpretability and suitability to the dataset. Due to its interpretability, it was possible to implement explainability using the ante-hoc method, therefore useful information could be extracted from the classifier wherever it was required. For instance, once the classifier was trained and tested using 70% and 30% of the dataset respectively, the following pieces of information were extracted:

- Probabilities of both Readmitted and Not readmitted
- A confusion matrix representation of model performance
- Positive coefficients determined by the model

These pieces of information could be extracted as and when required, owing to the ante-hoc implementation method used, allowing the creation of an "explainable" interface.

Displaying the Classifier's Results and Accuracy Metrics. As the final user's (domain expert's - doctors in this case) understanding of the overall result was paramount and considering interface design research, a progress bar-styled visualisation was used to display the results. As Wainer suggests, to display data accurately it is important to choose a method which clearly demonstrates both the order and magnitude of the resulting numbers [15]. Figure 3 demonstrates that this rule was followed, with each segment of the bar representing the magnitude of both the 'Readmitted' and 'Not Readmitted' results.

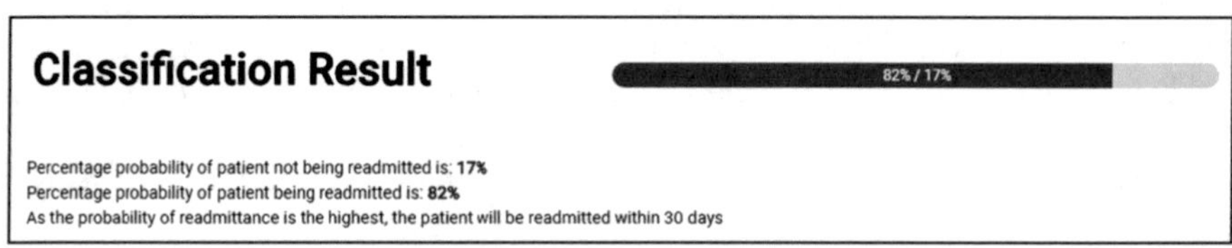

Fig. 3. Classification result progress bar

Additionally, the user is presented with a section detailing the classifier's accuracy. This provides an overall accuracy score achieved by the classifier when comparing classified test data against train data, and a confusion matrix to visualise the classifier's correct and incorrect predictions. This section can be seen in Fig. 4. This provides information explaining the performance of the algorithm and it refers to machine learning experts rather than the final users.

Accuracy & Confusion Matrix

Classifier Accuracy: 64.81%

	Predicted Readmitted	Predicted Not Readmitted
Actual Readmitted	60.22% / 12534	30.7% / 6390
Actual Not Readmitted	4.49% / 934	4.59% / 955

Correctly Predicted: 64.81% Incorrectly Predicted: 35.19%

Fig. 4. Confusion matrix for logistic regression classifier, as displayed on results interface

Understanding a classifier's performance is crucial when identifying methods of improvement requiring a suitable visualisation method, a simple and common one being the confusion matrix [11]. The confusion matrix provides the data needed to calculate various metrics; from a simple accuracy metric, to others such as precision, recall, and accuracy [16]. From the client's perspective, inclusion of the confusion matrix provides the user with a basic accuracy measurement, useful for users with limited machine learning experience, and more detailed metrics if required.

The confusion matrix in Fig. 4 demonstrates that the classifier is weak at predicting actual 'Not Readmitted' patients as 'Not Readmitted', and commonly predicts 'Readmitted' patients as 'Not Readmitted', therefore only receives an overall accuracy of ~64%. Reasoning for this is further detailed in Sect. 4.1.

Next, a graph displaying feature importance is presented to the user and produced using a Decision Tree algorithm to demonstrate a method for providing feature information. This graph can be seen in Fig. 5.

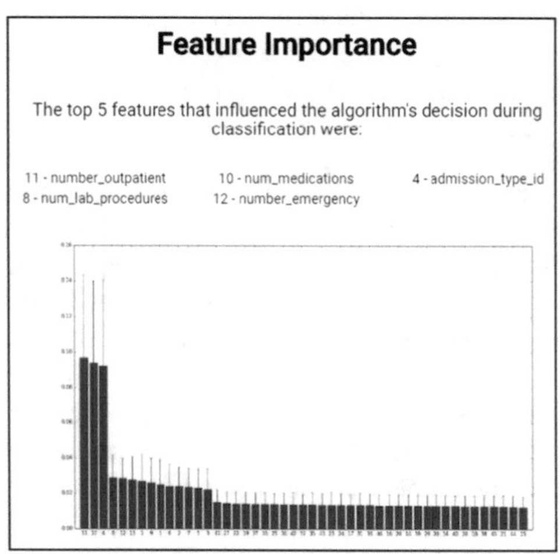

Fig. 5. Dataset feature importance, as displayed on results interface

Scikit learn's *ExtraTreesClassifier* was used to calculate a list of important features based on the impact each attribute from the dataset has on the overall prediction result. A similar activity was carried out by Saabas involving interpretability improvement of random forest classifications through the development of a new Python library, *treeinterpreter*. In his study, he extracted a figure for each attribute defining its contribution to the result, both for individual predictions and for the classifier as a whole [17]. Although produced through the *ExtraTreesClassifier* library instead of *treeinterpreter* a similar result is received, with an ordered list of each attribute's contribution to the overall classifier's result outputted. These results were plotted on a bar chart for easy visualisation of attribute contributions as shown in Fig. 5. This provided a level of explanation for the model as a whole, as the user can determine which features could have affected the results in what way, however it does not focus on specific explanations or reasoning for individual classifications.

Implementing Explainability for Individual Classifications. As the logistic regression algorithm uses coefficients to determine the influence of attributes on the result likelihoods, these coefficients must be retrieved and utilised to implement explainability for individual classifications. The coefficients were first extracted using the line of code: *coefficients = X.columns, np.transpose(model.coef_)*. Next, a *coefficientsHelper* module was developed which enabled the coefficients' attribute labels and coefficients to be received separately. It also contained a function used to retrieve only positive coefficients from the model, allowing the attributes which would most affect a likelihood of Readmitted to be easily identified and presented to the user. The *coefficientsHelper* module also contained a function which, using the retrieved coefficients' attribute labels and coefficients, would produce a bar chart plotted using the Matplotlib library.

Combining these functionalities, it was possible to develop a new module which would consider the coefficients and the user-entered patient data for each attribute. This module, *explainClassification*, determined sets of informative phrases, one of which, depending on the user-entered data for each positive coefficient, would be displayed upon hovering over bars on the coefficients bar chart.

Implementing the 'explainClassification' Module. Implementation of the *explainClassification* module made use of the algorithm's domain decisions, but also of domain knowledge where possible. For instance, in using its own knowledge on the dataset the algorithm can identify which supplied attributes influence the classification result and how. Figure 6 demonstrates the coefficients determined by the algorithm, plotted on a bar chart. Using this, we can see that the algorithm believes that some of the most influential attributes on a decision include; the number of times the patient has been admitted previously as an inpatient, whether the patient's diabetes medication/s has/have been changed, the number of times the patient has been admitted previously as an emergency, and the value of their last glucose serum test. The algorithm determines whether a patient will be readmitted or not based on the values of these coefficients. For instance, if a patient has had more inpatient visits than emergency and had an elevated glucose serum test result (>200 mg/dL or >8.59 A1C), the probability that they would be readmitted in the near future would increase. These most influential attributes are passed through to the *explainClassification* module, once the overall result is determined, in order to decide which set of statements inside the module to use to provide correct explanation statements. The code used to make this decision is shown in Fig. 6.

```
if notReadmitted > readmitted:
        resultStr = "As the probability of non-readmittance is the highest,
                     the patient will not be readmitted within 30 days"
        result = "Not readmitted within 30 days"
        explain = explainClassification.explainNotReadmitted(positiveCoefficients)

    else:
        resultStr = "As the probability of readmittance is the highest,
                     the patient will be readmitted within 30 days"
        result = "Readmitted within 30 days"
        explain = explainClassification.explainReadmitted(positiveCoefficients)
```

Fig. 6. Decision statements determining how to use *explainClassification* module to retrieve correct explanation statements

For initial implementation of explainability, simple statements such as "The probability of readmission increases as the patient's glucose serum levels were >200" were provided for each coefficient with the greatest impact on classification result. To further improve these explanations a degree of domain knowledge was also utilised, enabling the provision of explanations which considered the magnitude of change in the classification's result based on the change in data of the positive coefficients. The idea that one unit of change in one attribute will not have the same influence on the classification result as one unit of change in another attribute should be evident, as displayed by the coefficients bar chart, however this could not be integrated into the current explanations without developing the *explainClassification* module's decision statements. These statements were based on observational experimentation carried out by classifying a patient multiple times, each time slightly changing the value provided for one attribute at a time in order to determine its actual influence on the resulting probabilities. As an example, a patient who had neutral values for all positive coefficients provided a 26% probability of being readmitted. If the number of inpatient procedures is increased from 0 to 1, the probability of readmission increases to 32.7%, meaning a singular unit of change for this attribute is 6.7%. In comparison, if the number of inpatient procedures is reduced back to 0 and the number of emergency procedures is increased from 0 to 1, the probability of readmission is 28.8%, therefore a singular unit of change for this attribute causes a 2.8% increase in probability of readmission. The decision statements rely on boundary values determined by the unit changes for each positive coefficient attribute. When values for each positive coefficient fall into a boundary, a suitable statement is provided to the user. For instance, since we know that one unit of change in the number of inpatient procedures has a considerable effect on the classification result, we can determine that if the patient has 0 inpatient procedures, the classification result would not have been affected. However, if the patient has between 1 and 5 inpatient procedures, the readmission probability would have increased slightly, and inpatient procedures greater than 5 would have significantly affected the readmission probability. Similar statements were generated for each attribute with a positive coefficient. An example decision statement used to determine an explanation for a result of Readmitted using a positive coefficient attribute is shown in Fig. 7 below.

JQuery was then used to implement hotspots over each positive coefficient's bar on the bar chart. Upon hover of these hotspots, the statement related to the chosen attribute and the user-entered data for that attribute would display in and information box, providing the user with an explanation on that attribute's effect on the overall probabilities. Figure 8 demonstrates this functionality on a patient who is more likely to be readmitted.

The *explainClassification* module produced results with which the client was pleased, as during the final feedback stage they mentioned that it was easy for them to understand the classifier's result and how the data they entered affected it. Further in its success in providing understandable explanations, a less detailed yet more textual result was achieved when compared to explanations given by the SimMachines tool [18]. SimMachines classifies data and outputs a pie chart allowing the user to select each

```
if userEnteredCoef11 == "> 300" or userEnteredCoef11 == "> 200":

    response = "The patient's glucose serum test result was elevated
                at %s, therefore likelihood of readmission increases
                slightly." % userEnteredcoef11
    explain.append(response)

elif userEnteredCoef11 == "None" or userEnteredCoef11 == "Normal" :

    response = "The patient's glucose serum test result was %s,
                therefore this is unlikely to have affected the
                likelihood of readmission." % userEnteredCoef11
    explain.append(response)

else:
    response = " "
    explain.append(response)
```

Fig. 7. Decision statements within *explainClassification* used to determine correct explanation statements for a Readmitted result with the *max_glucose_serum* attribute

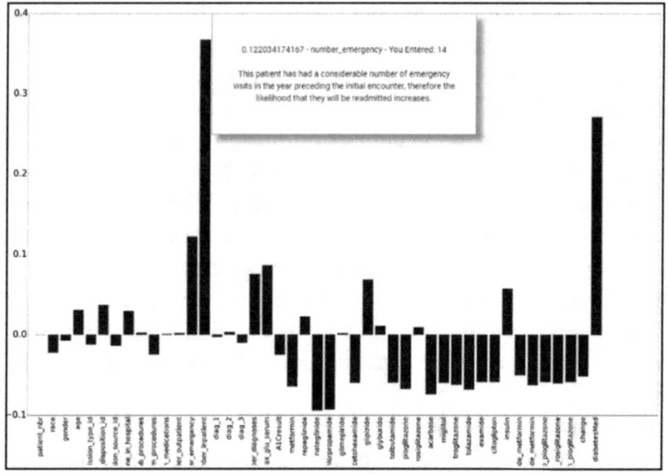

Fig. 8. Bar chart displaying logistic regression coefficients, providing explanations for classification result

segment of interest. This presents each piece of entered data which led to the chosen segment's result. For instance, the first split of the pie chart determines the overall binary result of the classification, the next provides a feature and its value that led to the binary result, and so on. Instead of focusing on each possible result as SimMachines does, the developed system only focuses on the final result and each of the important features. This allowed for the textual explanations behind the users' entered data and its effect on the overall result.

4 Evaluation of the Proposed Approach

4.1 Classifier Accuracy

According to Fig. 4 in Sect. 3.4, the classifier achieves an accuracy of 64.81%. It was identified that this inaccuracy could be due to dataset imbalance, as the confusion matrix was based on the 30% test split of the dataset, however as the emphasis of this research work was explainability of the machine learning algorithm and the implementation of the system was limited to a project timeframe, the accuracy of the classifier and imbalanced dataset were not prioritised, and a ~64% accuracy was deemed satisfactory.

The ROC Curve generated by a classification attempt, shown below in Fig. 9, demonstrates that the classifier is OK at separating readmitted patients from non-readmitted patients. A much more successful classifier would result in a curve following the left-hand border of the graph, indicating a higher true positive rate and more accurate results overall. Similarly, a graph with a larger area under the ROC curve would also indicate a more accurate classifier.

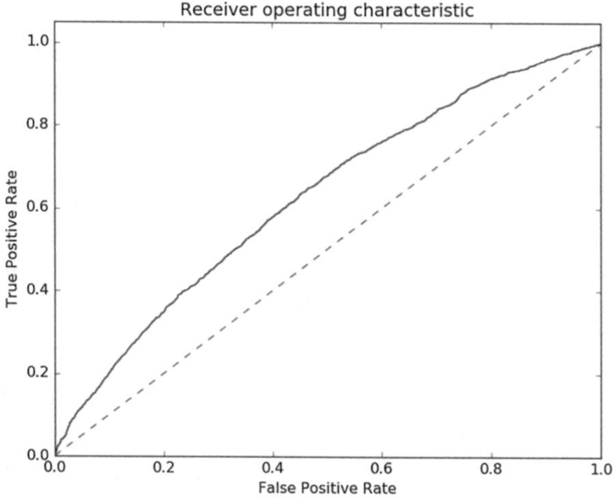

Fig. 9. ROC (Receiver Operating Characteristic) curve plotting the TPR (True Positive Rate) against the FPR (False Positive Rate) at differing thresholds

Again, the inaccuracy of the classifier is most likely due to the unbalanced nature of the dataset, however a satisfactory result in terms of the project's requirements was met.

4.2 Functional Correctness

System evaluation was performed as regards to functional correctness with functional test cases, which covered all aspects of the produced system, including login and registration, patient classification, previously classified patients and source data. From a total of 43 tests only 40 were executable due to reasons such as unimplemented or partly implemented features. Of the 40 executed tests the system achieved a success rate of 100% (Table 1).

Table 1. Record of test execution and success rates

Executed Tests	40/43 93%	Unexecuted Tests	3/43 7%
Passed Tests (of executed)	40/40 100%	Failed Tests (of executed)	0/40 0%

4.3 Explainability Overhead Experimentation

During implementation of explainability features into the logistic regression classifier, it was identified that a significant overhead exists pre-explainability upon classification using a local version of the system. An experiment into said overhead was undertaken, during which the following was attempted:

- Classification of the same patient was conducted 5 times pre-explainability, and another 5 times post-ante-hoc-explainability using a local version of the system and Internet Explorer.
- Internet Explorer's network console was utilised to identify the loading speeds of both the classifier and results page.

Pre-explainability, classification of a patient took an average of 4 s, resulting in an average total load time of 4.5 s. Post-explainability, the same classification actions resulted in an average time of 11.6 s, increasing the average total load time to 12.6 s. These results can be seen in Fig. 10.

Overall, the introduction of ante-hoc explainability resulted in a 176% increase of overhead. This finding would be crucial as, if the amount of data used for classification exceeded 70,000 rows, or more predictors are required to make a prediction – which are both likely in a real-world application – the total load time would increase further.

It is possible that chosen technologies could have influenced this increase in total load time. For the sake of machine learning systems in risky domains requiring explainability, this observed overhead is a consequence which must be understood and prepared for though careful optimization of both chosen technologies and code if timely results are relied on.

4.4 Client Feedback

Throughout the development of the system, the client (BT Adastral park, data science team) was consulted on a regular basis and expected to provide two types of feedback; written progress feedback throughout the lifetime of the project, and a final survey covering all system functions upon completion. Feedback throughout the project's

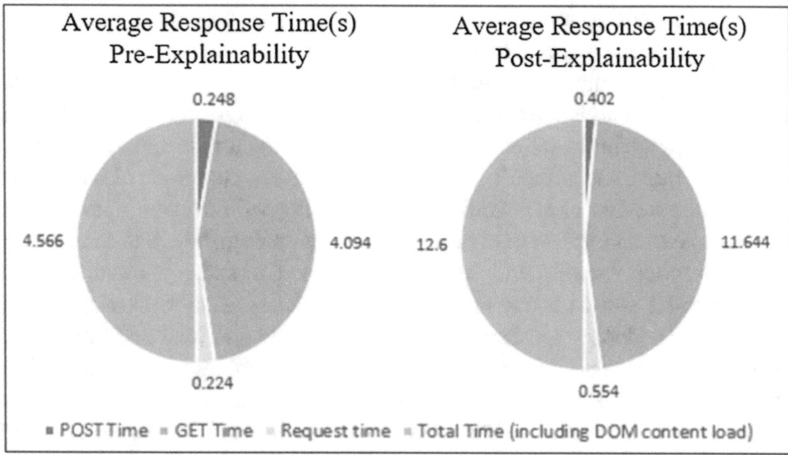

Fig. 10. Pre-explainability and post-explainability average timings

lifetime influenced both changes and completion of the system's features and played an integral role in the evaluation of the system.

Final feedback received from the client was very positive and took the form of an online survey created using MySurveyLab [19]. The client was asked to carry out several tasks which would invoke each of the system's features. Most importantly, they were able to understand certain classification and explainability aspects, such as; classifier performance and accuracy, feature importance, and explanations provided using the coefficients bar chart. Though the client had the ability to provide reasoning for their answers, they showed comfort in their understanding of explanations through only selecting "Yes" where asked whether they understood the above aspects. It can be assumed that, if the client was unsure about what was displayed to them, they would have selected the "Somewhat" option, and provided extra written feedback.

More work is required to take this research forward and accumulate feedback from the final users which is this case are the medical domain experts-doctors.

5 Conclusions and Future Work

In this paper, the design and development of web system used to perform classification and explanation of patient readmittance using machine learning has been explored and detailed. Technologies used in both design and development stages are discussed, with the aim of providing the reader with enough information to reproduce this system architecture in their own applications.

The resulting web system successfully provided a facility with which patients can be classified as either 'will be readmitted' or 'will not be readmitted' with satisfactory accuracy for the purposes of this research which focused explainability of the algorithm rather than its performance. More importantly, the system was able to provide the user with enough explanation with which an understanding of how changing user inputs – patient attributes – can affect the classification result.

This paper was only the beginning of the journey towards more explainable algorithms. More research is required in order to explore algorithmic properties that are amenable to explainability by extending the work to include different algorithms. As the problem is huge and difficult to address, we plan to start with classification-type algorithms and ante-hoc explainability. After working with several algorithms, trends and patterns to extract explainability are expected to emerge. We plan to use a bottom-up approach through working with several algorithms first and generalizing afterwards. Also, well-formed software engineering practices such as model-based design and domain-specific modelling are expected to be applicable and play an important role in this area. Identifying and appropriate describing the different domains such as the medical professional users-doctors and the machine-learning domain for developing explainable algorithms would be part of our research direction. Bringing explainability to the design level using the above methods will enable the final target of Explainability by design for the future trustworthy, accountable and explainable AI systems.

References

1. Dua, D., Karra Taniskidou, E.: UCI Machine Learning Repository. University of California, School of Information and Computer Science, Irvine, CA (2017). http://archive.ics.uci.edu/ml
2. Lichman, M.: UCI Machine Learning Repository. University of California, School of Information and Computer Science, Irvine, CA (2013). http://archive.ics.uci.edu/ml
3. Strack, B., DeShazo, J., Gennings, C., Olmo, J., Ventura, S., Cios, K., Clore, J.: Impact of HbA1c measurement on hospital readmission rates: analysis of 70,000 clinical database patient records. BioMed. Res. Int. **2014**, 1–11 (2014). http://dx.doi.org/10.1155/2014/781670
4. DARPA, Information Innovation Office. Broad Agency Announcement, Explainable Artificial Intelligence (XAI). DARPA, Arlington, VA (2016)
5. Launchburry, J.: A DARPA Perspective on Artificial Intelligence (2017)
6. Holzinger, A., Biemann, C., Pattichis, C., Kell, D.: What do we need to build explainable AI systems for the medical domain? (2017)
7. Ribeiro, M., Singh, S., Guestrin, C.: "Why should I trust you?": explaining the predictions of any classifier. In: Proceedings of the 2016 Conference of the North American Chapter of the Association for Computational Linguistics: Demonstrations (2016). http://aclweb.org/anthology/N16-3020
8. Oxborough, C., Birchall, A., Cameron, E., Townsend, A., Rao, A., Westermann, C.: Explainable AI: driving business value through greater understanding (2018). https://www.pwc.co.uk/audit-assurance/assets/explainable-ai.pdf. Last accessed 27 June 2018
9. Rudd, J., Priestley, J.A.: Comparison of decision tree with logistic regression model for prediction of worst non-financial payment status in commercial credit. Grey Literature from Ph.D. Candidates. 5 (2017). https://digitalcommons.kennesaw.edu/dataphdgreylit/5
10. Gunning, D.: Explainable Artificial Intelligence (XAI) (2016). https://www.darpa.mil/attachments/XAIIndustryDay_Final.pptx. Last accessed 5 July 2018
11. Kuhn, M., Johnson, K. Applied Predictive Modelling. Springer, New York (2013). https://doi.org/10.1007/978-1-4614-6849-3_12

12. Pedregosa, F., Varoquaux, G., Gramfort, A., Michel, V., Thirion, B., Grisel, O., Blondel, M., Prettenhofer, P., Weiss, R., Dubourg, V., Vanderplas, J., Passos, A., Cournapeau, D., Brucher, M., Perrot, M., Duchesnay, É.: Scikit-learn: machine learning in Python. J. Mach. Learn. Res. (2011). http://www.jmlr.org/papers/volume12/pedregosa11a/pedregosa11a.pdf. Last accessed 5 July 2018
13. Isaac, G., Meacham, S., Hamzeh, H., Stefanidis, A., Phalp, K.: An adaptive E-commerce application using web framework technology and machine learning. In: BCS Software Quality Management (SQM) Conference 2018, London, UK, 26–27 March 2018
14. Chapman, P., Clinton, J., Kerber, R., Khabaza, T., Reinartz, T., Shearer, C., Wirth, R.: CRISP-DM 1.0, p. 12. SPSS Inc., Chicago (2000) https://the-modeling-agency.com/crisp-dm.pdf. Last accessed 28 June 2018
15. Wainer, H.: How to display data badly. Am. Stat. **38**(2), 137 (1984). https://doi.org/10.1080/00031305.1984.10483186
16. Fawcett, T.: An introduction to ROC analysis. Pattern Recogn. Lett. **27**(8), 861–874 (2006). https://doi.org/10.1016/j.patrec.2005.10.010
17. Saabas, A.: Random forest interpretation with scikit-learn|Diving into data (2015). Blog. datadive.net
18. SimMachines.: Visualisation of TV data and Porsche purchase prediction [image] (2017). https://simmachines.com/focus-areas/media/demo-3/. Last accessed 2 July 2018
19. MySurveyLab.: MySurveyLab.com. 7 Points Ltd., Warsaw, Poland (2018)

A String Similarity Evaluation for Healthcare Ontologies Alignment to HL7 FHIR Resources

Athanasios Kiourtis[1(✉)], Argyro Mavrogiorgou[1], Sokratis Nifakos[2], and Dimosthenis Kyriazis[1]

[1] Department of Digital Systems, University of Piraeus, Piraeus, Greece
{kiourtis,margy,dimos}@unipi.gr

[2] Karolinska Institute, Solna, Sweden
sokratis.nifakos@ki.se

Abstract. Current healthcare services demand the transformation of health data into a mutual way, while respecting standards for making data exchange a reality, raising the needs of interoperability. Most of the developed techniques addressing this field are dealing only with specific one-to-one scenarios of data transformation. Among these solutions, the translation of healthcare data into ontologies is considered as an answer towards interoperability. However, during ontology transformations, different terms are produced for the same concept, resulting in clinical misinterpretations. In order to avoid that, ontology alignment techniques are used to match different ontologies based on specific string and semantic similarity metrics, where very little systematic analysis has been performed on which string similarity metrics behave better. To address this gap, in this paper we are investigating on finding the most efficient string similarity metric, based on an existing approach that can transform any healthcare dataset into HL7 FHIR, through the translation of the latter into ontologies, and their matching through syntactic and semantic similarities. The evaluation of this approach is being performed through the string similarity metrics of the Levenshtein distance, Cosine similarity, Jaro–Winkler distance and Jaccard similarity, resulting that the Levenshtein distance provides more reliable results when dealing with healthcare ontologies.

Keywords: Healthcare · Ontology alignment · String similarity · HL7 FHIR · Levenshtein distance · Cosine similarity · Jaro–Winkler distance · Jaccard similarity

1 Introduction

Moving into 2019, health analytics services such as predictive analytics, diagnostics, or personal healthcare records are able to offer a future with more powerful, independent and efficient health systems, with better and healthier life quality [1]. Nevertheless, what is mandatory for such a promise is that these systems are able to connect and interact with each other, by quickly and seamlessly exchanging data using open standards [2]. However, there still exist different challenges and difficulties that hospitals and healthcare systems need to address in freeing data from silos, affecting both patient care and medical research [3]. In order to share data with as many stakeholders

K. Arai et al. (Eds.): CompCom 2019, AISC 997, pp. 956–970, 2019.
https://doi.org/10.1007/978-3-030-22871-2_68

as possible, interoperability is the only sustainable way for letting systems intercommunicate and getting a patient's total image [4]. Interoperability is deeply being discussed by the global digital health community, mainly in the field of open standards, data, and sources. Henceforth, interoperability is being considered as crucial to overcome the today's fragmented and proprietary global health systems, providing the ability to various stakeholders to gain the most out of health data, such as extending, updating, and preserving it. According to [5], the Healthcare Industry will grow from $521.2 billion in 2017 to $674.5 billion by 2022, while the global healthcare market is expected to reach $3.07 billion by 2023 from $1.90 billion in 2018, in order to overcome healthcare interoperability issues. Such issues include exchanging health data among systems for supporting healthcare, population health management initiatives and to stay Health Insurance Portability and Accountability (HIPAA) compliant, aiming to support the use of data and to protect the privacy [6]. Based on [7], it costs annually almost $250 billion to analyse 30 billion of healthcare transactions, while according to [8] 86% of the healthcare errors are provoked by administrative reasons, where 30% of clinical tests have to be re-ordered since the results cannot be found. Moreover, the same report has showed that patient charts cannot be identified on 30% of visits, while about 80% of the most critical medical errors deal with bad communication during care transitions. Hence, by creating and implementing interoperability that contains the capabilities for capturing and integrating data from anywhere, it can be better understood how and why these medical errors are occurring, making health interoperability even more critical.

It is undisputable that healthcare and research require healthcare data to be shaped into a homogeneous format, respecting terminologies. By the time that the healthcare data is available electronically, each system has its own integration layer to communicate with another system, and in all cases, data has to be mapped or transformed and routed appropriately between the two systems. The latter is currently offered through the HL7 Fast Healthcare Interoperability (HL7 FHIR) standard [9] that despite its wide adoption, it still needs much time to become a global healthcare data exchange standard, as there exist systems that still produce data, which are not related to it. In this scenario, interoperability can be done by building medical domain ontologies for visualizing the different concepts, relationships, and axioms among the different healthcare datasets, and for representing medical terminology systems [10]. Creating new ontologies cannot be considered as a deterministic process, since different decisions must be considered, while the backgrounds and the overall goal are able to guide their final choices. The outcome is that two ontologies that belong to the same domain will not be equivalent. For that reason, apart from creating ontologies, ontology alignment must be implemented among the ontologies of the different healthcare datasets, for identifying the possible semantics and similarities, and for encoding the meaning of healthcare information among various overlapping demonstrations of the same domain.

Taking into consideration these challenges, in [11] a holistic approach has been illustrated for interoperability achievement through the transformation of health data into the HL7 FHIR structure. Shortly, the provided mechanism was building the healthcare ontologies that were primarily stored into a triplestore, in order to identify and compare their syntactic and semantic similarity with the HL7 FHIR Resources.

Consequently, based on the aggregation of the syntactic and semantic similarity results, the matching and translation to the HL7 FHIR was taking place. In this paper, we are going to fill the gap of the syntactic (i.e. string) similarity of the aforementioned mechanism, as multiple ontology alignment systems currently exist, with most of them using a string similarity metric. However, there has not been any systematic analysis on which metrics have better performance in the cases of ontology alignment. Hence, four of the most well-known string similarity metrics will be used under the same scenario, in order to result to the most efficient metric of syntactic similarity in the cases of healthcare ontologies alignment, and consequently to the most efficient HL7 FHIR transformation.

The rest of this paper is organized as follows. In Sect. 2, the related work is illustrated, while Sect. 3 depicts the overall approach with regards to the comparison of the different string similarity metrics. Section 4 includes the evaluation and the discussion of the derived results, whereas Sect. 5 states our conclusions and future steps.

2 Related Work

2.1 Ontology Alignment in Healthcare

Currently medical information systems have to communicate and exchange healthcare data in a secure and efficient way. Such thing can be achieved through healthcare related ontologies, which have been vastly used for the last few years, for different reasons [12], for representing medical terminology systems. An ontology can be considered as a common, shareable, and reusable view of a particular application domain that gives meaning to information structures that are exchanged among different information systems [13]. Therefore, the aim of the ontologies is to achieve information that can be shared and transferred between people and systems. Their aim is to gather domain knowledge, whilst their basic role is to capture the semantics in a generalized way, giving the means for agreement into a specific domain.

In the context of healthcare, among the most significant benefits that ontologies may provide has to do with the parallel integration of information and data [14]. However, the way that the healthcare domain is constructed results in creating different ontologies with multivariate or overlapping parts. Thus, these ontologies may result into semantic heterogeneity difficulties. A promising solution could be the development of methods for identifying matches among the multiple components of ontologies in healthcare towards enabling interoperability. Currently, several ontology alignment systems exist, where most of them are implementing string similarity metrics, for which there has been little analysis on the performance of these when applied to ontology alignment. In this area, the authors in [15] describe a proposal for normalizing HL7 messages and SNOMED-CT concept using ontology mapping, in an effort to achieve interoperability and seamless communication between healthcare entities. Moreover, the authors in [14] have built the required mapping rules to reconcile data from different ontological sources into a canonical format. What is more, the authors in [16] present a method that merges similarity measures without any ontology instances or user feedback regarding to the alignment of two provided ontologies. In this context, the authors in [17] have built an application including multiple constructed ontologies, which can be considered as unique but are related to each other, performing ontology alignment to allow inter-operation among them.

2.2 String Similarity Metrics

Currently medical information string similarity metrics are considered to have an major role in text related research, as they are widely used in the majority of Natural Language Processing (NLP) [18] tasks, including information retrieval, text classification, document clustering, topic detection, topic tracking, questions generation, question answering, essay scoring, short answer scoring, machine translation, text summarization and others. Identifying similarity between strings is an important part of string similarity that is then primarily used for identifying textual, paragraph and document similarities. In the case of string similarity, two strings can be considered as similar when they have similar or common character sequence. In order to calculate the string similarity between two different words, different string similarity metrics have been developed over the years in order to address different needs and applications. Among the most widely known string similarity metrics that are currently used are the Levenshtein distance, Sørensen–Dice coefficient, Cosine similarity, City block distance, Jaro–Winkler distance, Simple matching coefficient, Jaccard similarity, Tversky index, Overlap coefficient, Variational distance, Confusion probability, Tau metric, and Grammar-based distance, or TFIDF distance metric [19]. In more details, the authors in [20] demonstrate a hybrid plagiarism identification method by investigating the use of a diagonal line, which derives from the Levenshtein distance and the simplified Smith–Waterman algorithm, with a view to the application in the plagiarism detection. Moreover, the authors in [21] are using the City block distance to calculate the membership functions of fuzzy sets and identify the initial partition of a data set, in order to handle minute differences between two miRNA expression profiles. In the same context, the authors in [22] are using Cosine and Jaccard similarity for information retrieval so as to deal with the problem of document similarity out of a large amount of data. Furthermore, in [23] the authors are using Cosine distance of TF/IDF weighted document vectors for a fast and efficient way to align documents. In the same context, the authors in [24] propose the Jaro–Winkler distance and the Naive Bayes Classifier in order to identify pests and diseases of paddy, taking as an input the text that contains symptoms of the disease, and utilizing Jaro–Winkler distance to find symptoms from the user input. As it can be identified, it is clear that different string similarity metrics can be used in multiple domains and areas, based on the different needs and purposes of each domain. With that in mind, in this paper we are going to furtherly study and implement the string similarity metrics of (a) Levenshtein distance, (b) Cosine similarity, (c) Jaccard similarity, and (d) Jaro–Winkler similarity for identifying the syntactic similarity during the ontology alignment process.

2.2.1 Levenshtein Distance

The Levenshtein distance [25] is a string similarity metric for calculating the difference between two sequences or strings. The Levenshtein distance between any two strings is calculated as the minimum number of single-character edits (i.e. insertions, deletions or substitutions) required to change one string into the other. Levenshtein distance may also be referred to as edit distance, although it may also denote a larger family of distance metrics. Mathematically, the Levenshtein distance between two strings a, b (of length |a| and |b| respectively) is given by (1):

$$lev_{a,b}(i,j) = \begin{cases} \max(i,j) \\ min\begin{cases} lev_{a,b}(i-1,j)+1 \\ lev_{a,b}(i,j-1)+1 \\ lev_{a,b}(i-1,j-1)+1_{(a_i \neq b_j)} \end{cases} \end{cases} \tag{1}$$

$(a_i \neq b_j)$ is the indicator function equal to 0 when $a_i \neq b_j$ and equal to 1 otherwise, and $lev_{a,b}$ (i,j) is the distance between the first i characters of a and the first j characters of b. It should be noted that the first element in the minimum corresponds to deletion (from a to b), the second to insertion and the third to match or mismatch, depending on whether the respective symbols are the same.

2.2.2 Cosine Similarity

The Cosine similarity [26] is a string similarity metric for quantifying the similarity of sequences by treating them as vectors and calculating their cosine. This produces a value between 0 and 1, with 0 meaning no similarity, and 1 meaning that both sequences are exactly the same. In that case, the first step is to turn the strings that will be compared into vectors by considering the Term Frequency of unique letters in a given string. Shortly, vectors are geometric elements that have magnitude (length) and direction. Several mathematical operations can be performed on them but in the case of Cosine similarity, only the dot product and magnitude are going to be used.

The dot product of two vectors a and b can be measured as in (2):

$$\vec{a} * \vec{b} = a_1b_1 + a_2b_2 + \ldots + a_nb_n = \sum_{i=1}^{n} a_ib_i \tag{2}$$

The magnitude of these two vectors can be calculated as in (3):

$$\|\vec{a}\| = \sqrt{a_1^2 + a_2^2 + \ldots + a_n^2} \tag{3}$$

Consequently, for mathematically calculating the cosine similarity of these vectors, (4) is being used where the dot product of vectors a and b is divided by the product of magnitude a and magnitude b.

$$similarity\left(\vec{a}, \vec{b}\right) = cos\theta = \frac{\vec{a} * \vec{b}}{\|\vec{\alpha}\| * \left\|\vec{b}\right\|} \tag{4}$$

2.2.3 Jaccard Similarity

The Jaccard similarity [27] is a string similarity metric that compares members for two strings to see which members are shared and which are distinct. It is a measure of similarity for the two sets of strings, with a range from 0% to 100%. The higher the percentage, the more similar these two strings are. In the case of Jaccard similarity, (5) is being used for two strings a and b:

$$J(a, b) = \frac{|a \cap b|}{|a \cup b|} \tag{5}$$

$|a \cap b|$ provides the number of members that are shared between both sets of strings, and $|a \cup b|$ provides the total number of members in both sets of strings. The division of the number of shared members by the total number of members provides their Jaccard similarity.

2.2.4 Jaro–Winkler Similarity

The Jaro–Winkler distance [28] is a string similarity metric for quantifying the similarity between two strings. The higher the Jaro–Winkler distance for two strings is, the more identical the strings are. The Jaro–Winkler distance metric performs better for short strings, while in the case that its similarity score is about 0, it means that there is no similarity, while 1 means that there is an exact match. The Jaro–Winkler similarity is given by (6):

$$dw = dj + (lp(1 - dj)) \tag{6}$$

In that case, the Jaro distanced dj has to be firstly calculated, where given two strings s1 and s2, dj is given by (7):

$$dj = \begin{cases} 0 \\ \frac{1}{3}\left(\frac{m}{|s1|} + \frac{m}{|s2|} + \frac{m-t}{m}\right) \end{cases} \tag{7}$$

In that case, s1 and s2 are the total number of characters of each string accordingly, m is the number of matching characters, and t is half number of transpositions. What is more, l is the length of common prefix at the start of the string, up to a maximum of four characters, while p is a constant scaling factor for how much the score is adjusted upwards for having common prefixes. In general, p should not exceed 0.25, otherwise the distance can become larger than 1. The standard value for this constant in Winkler's work is p = 0.1.

2.3 Key Contributions of the Proposed Approach

Several researches have been proposed focusing on the healthcare interoperability field, dealing with specific one-to-one scenarios of data transformation. Among these solutions, the translation of healthcare data into ontologies is considered as a solution towards interoperability. However, during ontology transformations, different terms are produced for the same reason, resulting in clinical misinterpretations. Hence, in order to avoid that, ontology alignment techniques are used to match different ontologies together for getting their semantic similarity, which is however based on specific string and semantic similarity metrics, where very little systematic analysis has been performed to identify the string similarity metrics that behave better in the case of ontology alignment. To address that gap, in this paper we are investigating on finding the most efficient string similarity metric, for performing ontology alignment in healthcare ontologies. This study is based on an existing approach that can transform any healthcare datasets of any format and nature into HL7 FHIR through the translation of the latter into ontologies, and their matching through syntactic and semantic similarities.

3 Proposed Approach

In the context of ontology alignment of healthcare data, several techniques are presented that lead to better decision making, and interoperable exchange and use of data. However, these techniques are providing solutions that are adapted to deliver results under specific scenarios. To address this gap, in [11] a holistic approach is proposed that aims in building healthcare ontologies from different datasets, and then measuring the similarities and common links between these ontologies and the ontologies of the HL7 FHIR Resources. The first steps of the developed mechanism were dealing with the automated transformation of the healthcare dataset and the HL7 FHIR Resources into their ontological form (i.e. Ontology Building System). By the time that these ontologies hadbeen constructed, they were stored into a triplestore (i.e. Ontology Triplestore), containing the different concepts, relationships, and axioms, for each different ontology. Afterwards, two sub-mechanisms had occurred, providing the syntactic and the semantic similarity between two different ontologies (i.e. Syntactic Similarity Identifier and Semantic Similarity Identifier), according to their syntactic representations and their semantic knowledge. The final stage implemented an additional sub-mechanism (i.e. Overall Ontology Mapper) that aggregated and merged the results of the aforementioned sub-mechanisms, aiming to provide the overall alignment results, for the overall transformation of the healthcare dataset into HL7 FHIR. Henceforth based on this developed mechanism, in our case, we assume that healthcare ontologies have already been constructed for both the ingested healthcare dataset and the different HL7 FHIR Resources. Consequently, the four described string similarity metrics are going to be implemented, for measuring the syntactic similarity of the built ontologies through the Syntactic Similarity Identifier. Figure 1 illustrates the overall architecture of the mechanism as depicted in [11], where the components which are highlighted with grey color are going to be ignored since they have already been discussed in [11].

3.1 Syntactic Similarity Identifier

The objective of the Syntactic Similarity Identifier is to provide the means for identifying the syntactic similarity between two ontologies, and result into the probability that a specific ontology is the same – in terms of its syntactic interpretation – with another ontology. In this scenario, in order to calculate the syntactic similarity between two ontologies, the next three steps are followed (Fig. 2).

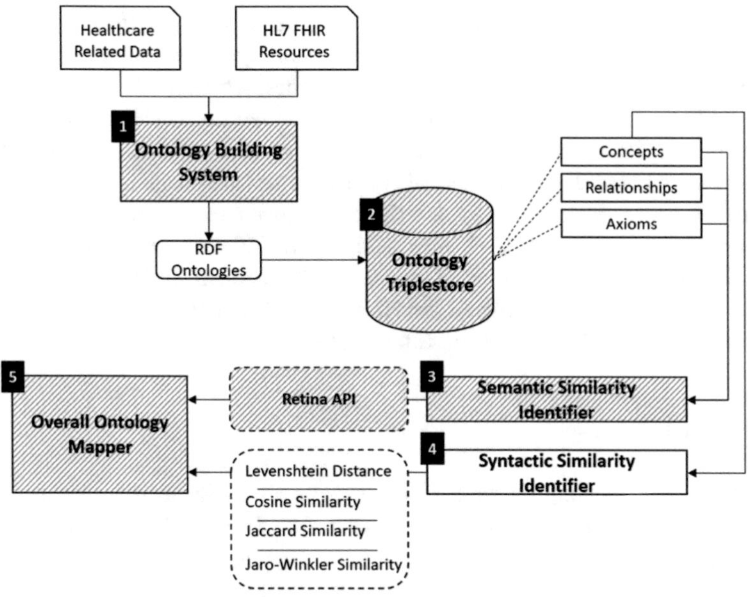

Fig. 1. Overall architecture

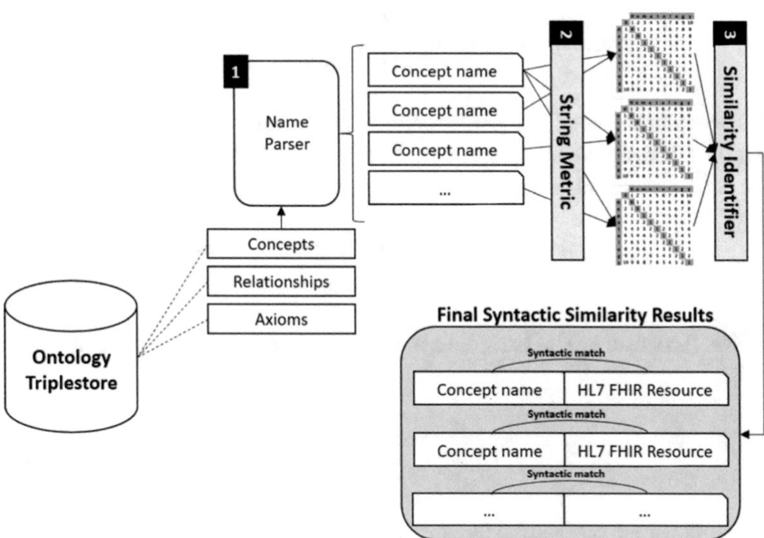

Fig. 2. Syntactic Similarity Identifier

Initially, the different ontologies that are stored into the Ontology Triplestore are being processed through the Name Parser that is responsible for getting the values (i.e. concept names) of the different ontological concepts. Sequentially, the concept names of the ontologies are compared one-by-one, in order to calculate their syntactic similarity. For that reason, the four different string similarity metrics are implemented, as a measure of similarity between the two concept names, repeating the same comparisons for each different case. By the time that all the different combinations of comparisons have occurred with the different string similarity metrics, different tables of the syntactic similarity between the ontological concept names are created. The purpose of the Syntactic Similarity Identifier is to identify the concept names of the healthcare dataset that have greater degree of resemblance with the HL7 FHIR Resources. Consequently, among the different degrees of similarities, this mechanism finally stores and syntactically maps the concept names and the HL7 FHIR Resources that syntactically match better, providing the Final Syntactic Similarity Results.

4 Evaluation

In our case, for evaluating the four different string similarity metrics, we are going to implement them under the same use case scenario where we will calculate their precision, recall, F-measure, and total errors, respecting the UMLS-based reference alignment [29]. Shortly, these comparison criterions are defined as follows:

(i) The results' precision (in percentage): The fraction of the data that is retrieved and is relevant to the query, visualizing the number of correct results divided by the number of all returned results, as in (8).

$$precision = \frac{|\{relevant\,data\} \cap \{retrieved\,data\}|}{|\{retrieved\,data\}|} \tag{8}$$

(ii) The results' recall (in percentage): The fraction of the relevant data that is successfully retrieved, visualizing the number of correct results divided by the number of results that should have been provided, as in (9).

$$recall = \frac{|\{relevant\,data\} \cap \{retrieved\,data\}|}{|\{relevant\,data\}|} \tag{9}$$

(iii) The F-measure: The harmonic mean (i.e. combination) of precision and recall, as in (10).

$$F = \frac{2 * precision * recall}{precision + recall} \tag{10}$$

(iv) Errors (in percentage): The percentage that displays the total amount of faulty data compared to the total amount of data, as in (11).

$$errors = \frac{|faulty\,data|}{|\{total\,data\}|} \tag{11}$$

After this calculation, a comparison between the results will take place in order to conclude to the string similarity metric that is more suitable for aligning ontologies and calculating their string similarity in the healthcare domain.

4.1 Use Case Description

The use case exploits a CSV formatted healthcare dataset, covering all the previous steps. The used dataset for evaluating the efficiency of the proposed approach (Fig. 3) is a sub-dataset of anonymized citizens' personal information, derived from Karolinska Institute [30]. In more detail, it consists of 5000 instances of the personal information of certain citizens about their: (i) personal identifier (*subject*), (ii) gender (*gender*), (iii) date of birth (*dateOfBirth*), (iv) date of death (*dateOfDeath*), and (v) cause of death (*causeOfDeath*).

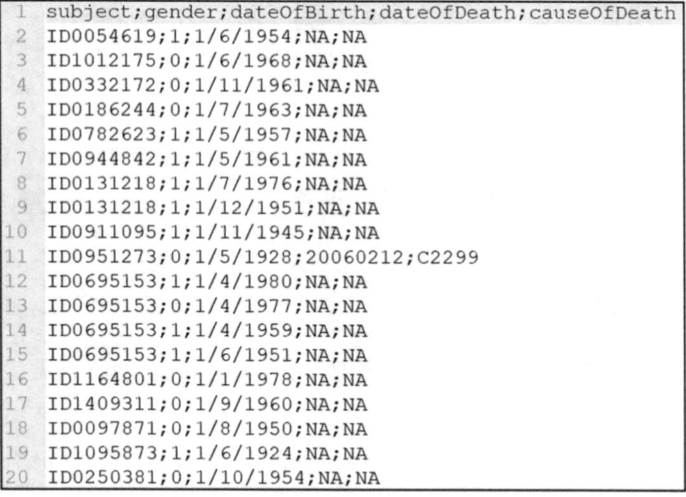

```
subject;gender;dateOfBirth;dateOfDeath;causeOfDeath
ID0054619;1;1/6/1954;NA;NA
ID1012175;0;1/6/1968;NA;NA
ID0332172;0;1/11/1961;NA;NA
ID0186244;0;1/7/1963;NA;NA
ID0782623;1;1/5/1957;NA;NA
ID0944842;1;1/5/1961;NA;NA
ID0131218;1;1/7/1976;NA;NA
ID0131218;1;1/12/1951;NA;NA
ID0911095;1;1/11/1945;NA;NA
ID0951273;0;1/5/1928;20060212;C2299
ID0695153;1;1/4/1980;NA;NA
ID0695153;0;1/4/1977;NA;NA
ID0695153;1;1/4/1959;NA;NA
ID0695153;1;1/6/1951;NA;NA
ID1164801;0;1/1/1978;NA;NA
ID1409311;0;1/9/1960;NA;NA
ID0097871;0;1/8/1950;NA;NA
ID1095873;1;1/6/1924;NA;NA
ID0250381;0;1/10/1954;NA;NA
```

Fig. 3. Use case dataset objective

4.2 Application of String Metrics

After performing the steps introduced into the Syntactic Similarity Identifier, the following tables are being created (Tables 1, 2, 3, 4). In these tables, there are depicted the HL7 FHIR Resources with higher syntactic similarity degrees, in comparison with the concept names of the use case dataset, through the four different string similarity metrics.

4.3 Discussion of Results

In order to evaluate these results with regards to the metrics of precision, recall, F-measure, and errors, the specific use case dataset was translated manually in

Table 1. Syntactic Similarity Identifier top results for Levenshtein distance

Use case dataset attribute	HL7 FHIR resources	Levenshtein distance
subject	*Patient.contact*	**0.237 (~24%)**
gender	*Patient.gender*	**0.743 (~74%)**
dateOfBirth	*Patient.birthDate*	**0.881 (~88%)**
dateOfDeath	*Patient.contact.relationship*	**0.250 (25%)**
causeOfDeath	*Patient.contact.gender*	**0.278 (~28%)**

Table 2. Syntactic Similarity Identifier top results for Cosine similarity

Use case dataset attribute	HL7 FHIR resources	Cosine similarity
subject	*Patient.contact*	**0.267 (~27%)**
gender	*Patient.gender*	**0.743 (~75%)**
dateOfBirth	*Patient.multipleBirth*	**0.591 (~59%)**
dateOfDeath	*Patient.animal.breed*	**0.250 (25%)**
causeOfDeath	*Patient.contact.address*	**0.128 (~13%)**

Table 3. Syntactic Similarity Identifier top results for Jaccard similarity

Use case dataset attribute	HL7 FHIR resources	Jacccard similarity
subject	*Patient.animal.species*	**0.293 (~29%)**
gender	*Patient.gender*	**0.620 (~62%)**
dateOfBirth	*Patient.multipleBirth*	**0.241 (~24%)**
dateOfDeath	*Patient.contact.address*	**0.221 (22%)**
causeOfDeath	*Patient.multipleBirth*	**0.218 (~22%)**

Table 4. Syntactic Similarity Identifier top results for Jaro–Winkler distance

Use case dataset attribute	HL7 FHIR resources	Jaro–Winkler distance
subject	*Patient.photo*	**0.462 (~46%)**
gender	*Patient.gender*	**0.543 (~54%)**
dateOfBirth	*Patient.birthDate*	**0.781 (~78%)**
dateOfDeath	*Patient.gender*	**0.186 (19%)**
causeOfDeath	*Patient.contact.gender*	**0.198 (~20%)**

HL7 FHIR, so as to compare the results of the developed mechanism, with the actual outcomes. This was the main reason that a small data sample was chosen for the mechanism's evaluation, in order to manually conclude more easily to the aforementioned results. For the results that resulted manually, it was assumed being of high accuracy, respecting the HL7 FHIR format guidelines, considered as a benchmark of high quality and precision. Table 5 illustrates the results of the manual transformation. It should be noted that in the case of the *causeOfDeath* attribute, the manual

transformation results did not provide any corresponding HL7 FHIR Resource, since in the current version of the HL7 FHIR (v3.0.1) there does not exist any resource describing this specific attribute yet.

Table 5. Manual derived results

Use case dataset attribute	Manual results	Similarity
subject	*Patient.identifier*	100%
gender	*Patient.gender*	100%
dateOfBirth	*Patient.birthDate*	100%
dateOfDeath	*Patient.deceased*	100%
causeOfDeath	NO MATCHING	NO MATCHING

Henceforth, based on the results of Table 5, we can easily calculate the different comparison criterions for each string similarity metric. Table 6 is providing us with these results.

Table 6. Comparison criterions of string similarity metrics

	Precision	Recall	F-measure	Errors
Manual Results	**100%**	**100%**	**100%**	**0%**
Levenshtein distance	48%	47%	48%	52%
Cosine similarity	39%	42%	41%	61%
Jaccard similarity	32%	38%	35%	68%
Jaro–Winkler distance	43%	49%	46%	67%

In Fig. 4, a bar chart of the aforementioned results are depicted, showing how the four different string similarity metrics behave in the total syntactic similarity identification, with regards to the manually provided results which did not contain any errors (0% of errors), while their precision, recall and F-measure had the value of 100%.

It is clear that there was not any string similarity metric that provided 100% accurate results. It is clear that through Table 6 and Fig. 3, among the different metrics, the Levenshtein Distance behave better since it provided less errors (52%) compared with the other string similarity metrics (61%, 68% and 67% accordingly). Moreover, the F-measure (a result of both the precision and recall) of the Levenshtein Distance had the greatest value (48%) in comparison with the other string similarity metrics (41%, 35% and 46% accordingly). Consequently, this lead us to the conclusion that for the Syntactic Similarity Identifier, the most convenient string similarity metric to use is the Levenshtein Distance measure, since it provides more efficient and reliable results.

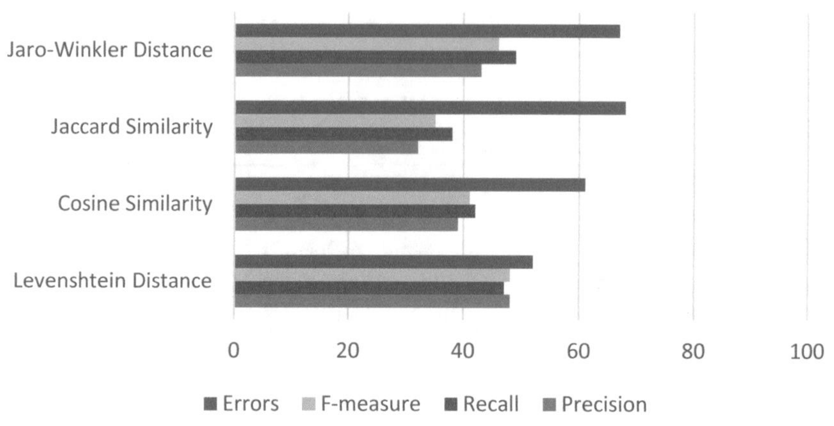

Fig. 4. Overall Ontology Mapper results

5 Conclusions

In the field of healthcare interoperability, multiple attempts have been provided for standardizing and covering most of the healthcare field since the existing standards use multiple terminologies for the same reason, which often results in clinical misunderstandings, knowledge mismanagement, misdiagnosis of a patient's illness, or even death. To resolve this problem ontologies have been developed, resulting in interoperability by knowledge sharing and reuse. A solution to this problem is the development of techniques for finding ontology matches among the different components of healthcare ontologies. However, when matching ontologies string similarity has to be identified, for which there has been little analysis to identify the metrics that perform well when applied to ontology alignment. For that reason, in order to address the gap of string similarity, in this paper a previous research was considered, about a healthcare ontology matching mechanism that has the ability of transforming healthcare data to HL7 FHIR format, by building healthcare ontologies from different datasets, and finding syntactic and semantic similarities among these ontologies and the ontologies of the HL7 FHIR Resources. Therefore, based upon this research, in order to identify the required syntactic similarity, four of the most well-known string similarity metrics were used under the same scenario, in order to result to the most efficient metric of syntactic similarity in the cases of healthcare ontologies alignment, and consequently to the most efficient HL7 FHIR transformation.

According to the evaluation results of the different string similarity metrics, the Levenshtein Distance behave better under the same circumstances, in terms of total errors, precision, recall and F-measure compared with the other metrics. Henceforth, even more reliable and accurate results were provided, a fact that was verified after the Levenshtein's Distance comparison with the manually provided ontology matching results.

In general, writing an automated ontology matching mechanism that is able to cover multiple domains and scenarios, is still a challenging research task, since different string similarity metrics behave in a different way in dissimilar scenarios. To this domain, we will continue working on evaluating the Syntactic Similarity Identifier mechanism with additional string similarity metrics, in order to identify the metric that is most suitable for the case of healthcare ontologies alignment. What is more, we want to perform a similar comparison and evaluation concerning the Semantic Similarity Identifier mechanism, in order to conclude to more reliable and efficient results. To this end, we will continue on the evaluation of the Syntactic Similarity Identifier with datasets of different sizes, standards and formats, respecting privacy topics, based on the developed mechanism in [31].

Acknowledgments. The CrowdHEALTH project has received funding from the European Union's Horizon 2020 research and innovation program under grant agreement No 727560.

References

1. Health care consumer engagement No "one-size-fits-all" approach. https://www2.deloitte.com/content/dam/Deloitte/us/Documents/life-sciences-health-care/us-dchs-consumer-engagement-healthcare.pdf. Accessed 8 Oct 2018
2. Data integration and interoperability: ISO/OGC. https://www.directionsmag.com/article/3396. Accessed 8 Oct 2018
3. So long to silos: why health systems must embrace interoperability now. https://www.webpt.com/blog/post/so-long-to-silos-why-health-systems-must-embrace-interoperability-now. Accessed 8 Oct 2018
4. Moving forward with interoperability: how patients will drive change. https://www.philips.com/a-w/about/news/archive/blogs/innovation-matters/moving-forward-with-interoperability-how-patients-will-drive-change.html. Accessed 8 Oct 2018
5. Healthy growth forecast for medical devices global market. https://www.bccresearch.com/pressroom/hlc/healthy-growth-forecast-for-medical-devices-global-market. Accessed 8 Oct 2018
6. Healthcare security: understanding HIPAA compliance and its role in patient data protection. https://digitalguardian.com/blog/healthcare-security-understanding-hipaa-compliance-and-its-role-patient-data-protection. Accessed 8 Oct 2018
7. 30 Healthcare statistics that keep hospital executives up at night. https://getreferralmd.com/2016/08/30-healthcare-statistics-keep-hospital-executives-night/. Accessed 8 Oct 2018
8. 3 Signs you should invest in healthcare automation. https://www.formstack.com/blog/2017/invest-in-healthcare-automation/. Accessed 8 Oct 2018
9. HL7 FHIR. https://www.hl7.org/fhir/. Accessed 8 Oct 2018
10. Gylys, B.A., Wedding, M.E.: Medical Terminology Systems: A Body Systems Approach. FA Davis, Philadelphia (2017)
11. Kiourtis, A., Mavrogiorgou, A., Kyriazis, D., Nifakos, S.: Aggregating the syntactic and semantic similarity of healthcare data towards their transformation to HL7 FHIR through ontology matching. Int. J. Med. Inform. (IJMI) (2018) (under review)
12. del Carmen Legaz-García, M., Martínez-Costa, C., Menárguez-Tortosa, M., Fernández-Breis, J.T.: A semantic web based framework for the interoperability and exploitation of clinical models and EHR data. Knowl.-Based Syst. **105**, 175–189 (2016)

13. Fernández-Breis, J.T., et al.: An ontological infrastructure for the semantic integration of clinical archetypes. In: Pacific Rim Knowledge Acquisition Workshop, pp. 156–167. Springer, Heidelberg (2006)
14. Sonsilphong, S., et al.: A semantic interoperability approach to health-care data: resolving data-level conflicts. Expert Syst. **33**(6), 531–547 (2016)
15. Oliveira, D., Pesquita, C.: Improving the interoperability of biomedical ontologies with compound alignments. J. Biomed. Semant. **9**(1), 1 (2018)
16. Nezhadi, A., Shadgar, B., Osareh, A.: Ontology alignment using machine learning techniques. Int. J. Comput. Sci. Inf. Technol. **3**(2), 139–150 (2011)
17. Mony, M., et al.: Semantic search based on ontology alignment for information retrieval. J. Comput. Appl. **107**(10) (2014)
18. Evolution of natural language processing. https://www.sas.com/en_us/insights/analytics/what-is-natural-language-processing-nlp.html. Accessed 8 Oct 2018
19. Cheatham, M., Hitzler, P.: String similarity metrics for ontology alignment. In: International Semantic Web Conference, pp. 294–309. Springer, Berlin, Heidelberg (2013)
20. Su, Z., Ahn, B.R., Eom, K.Y., Kang, M.K., Kim, J.P., Kim, M.K.: Plagiarism detection using the Levenshtein distance and Smith–Waterman algorithm. In: Innovative Computing Information and Control (ICICIC), International Conference on IEEE, p. 569 (2008)
21. Paul, S., Maji, P.: City block distance and rough-fuzzy clustering for identification of co-expressed micrornas. Mol. BioSyst. **10**(6), 1509–1523 (2014)
22. Jain, A., Jain, A., Chauhan, N., Singh, V., Thakur, N.: Information retrieval using cosine and Jaccard similarity measures in vector space model. Int. J. Comput. Appl. **164**(6), 28–30 (2017)
23. Buck, C., Koehn, P.: Quick and reliable document alignment via tf/idf-weighted cosine distance. In: Proceedings of the First Conference on Machine Translation: Volume 2, Shared Task Papers, vol. 2, pp. 672–678 (2016)
24. Sari, A.P., Saptono, R., Suryani, E.: The implementation of Jaro–Winkler distance and Naive Bayes classifier for identification system of pests and diseases on paddy. ITSMART: Jurnal Teknologi dan Informasi **7**(1), 1–7 (2018)
25. The Levenshtein Algorithm. https://www.cuelogic.com/blog/the-levenshtein-algorithm. Accessed 8 Oct 2018
26. Machine learning: cosine similarity for vector space models. http://blog.christianperone.com/2013/09/machine-learning-cosine-similarity-for-vector-space-models-part-iii/. Accessed 8 Oct 2018
27. Jaccard's Coefficient. https://people.revoledu.com/kardi/tutorial/Similarity/Jaccard.html. Accessed 8 Oct 2018
28. Jaro–Winkler distance. https://ipfs.io/ipfs/QmXoypizjW3WknFiJnKLwHCnL72vedxjQkDDP1mXWo6uco/wiki/Jaro%E2%80%93Winkler_distance.html. Accessed 8 Oct 2018
29. Ontology alignment evaluation initiative. http://oaei.ontologymatching.org/2018/. Accessed 8 Oct 2018
30. Karolinska Institutet. https://ki.se/start. Accessed 8 Oct 2018
31. Kiourtis, A., Mavrogiorgou, A., Kyriazis, D.: Towards a secure semantic knowledge of healthcare data through structural ontological transformations. In: Joint Conference on Knowledge-Based Software Engineering, pp. 178–188 (2018)

Detection of Distal Radius Fractures Trained by a Small Set of X-Ray Images and Faster R-CNN

Erez Yahalomi[1(✉)], Michael Chernofsky[2(✉)], and Michael Werman[1(✉)]

[1] School of Computer Science and Engineering, The Hebrew University of Jerusalem, Jerusalem, Israel
{erez.yahalomi,michael.werman}@mail.huji.ac.il
[2] Hadassah-Hebrew University Medical Center, Jerusalem, Israel
mac724@aol.com

Abstract. Distal radius fractures are the most common fractures of the upper extremity in humans. As such, they account for a significant portion of the injuries that present to emergency rooms and clinics throughout the world. We trained a Faster R-CNN, a machine vision neural network for object detection, to identify and locate distal radius fractures in anteroposterior X-ray images. We achieved an accuracy of 96% in identifying fractures and mean Average Precision, mAP of 0.866. This is significantly more accurate than the detection achieved by physicians and radiologists. These results were obtained by training the deep learning network with only 38 original images of anteroposterior hands X-ray images with fractures. This opens the possibility to detect rare diseases or rare symptoms of common diseases with this type of network, where there is only a small set of diagnosed X-ray images.

Keywords: Machine vision · Medical diagnostic · Deep learning network · Neural network · Object detection

1 Introduction

Distal radius fractures at the wrist are the most common fractures of the upper extremity in humans. As such, they account for a significant portion of the injuries that present to emergency rooms and clinics throughout the world. Every year tens of millions of people worldwide suffer hand trauma [5]. Diagnosis is made on the basis of X-ray imaging, a technology that remains prevalent despite being developed over one hundred years ago. In England alone about ten million visits per year to accident and emergency centers involve having an X-ray mostly to check for bone injury, [7]. A significant portion, up to 30%, of the hand wrist X-ray images are incorrectly diagnosed, [17], resulting in many people not receiving the medical treatment they need.

This research was supported by the Israel Science Foundation and by the Israel Ministry of Science and Technology.

K. Arai et al. (Eds.): CompCom 2019, AISC 997, pp. 971–981, 2019.
https://doi.org/10.1007/978-3-030-22871-2_69

Multiple X-rays of an injured wrist are typically performed, including anteroposterior (AP), lateral, and oblique images, in order to capture pictures of the bones from different points of view, and optimize fracture recognition.

Treatment of distal radius fractures depends first and foremost upon a diagnosis based on the X-rays. Once a diagnosis is made, the specific treatment depends upon many factors, including the radiographic nature of the fracture pattern. Treatment can include cast immobilization, closed reduction and casting, closed or open reduction and pin fixation, closed or open reduction and external fixation, as well as various techniques of open reduction and internal fixation. The fundamental principles of treatment are restoration of normal alignment and position of the fracture components, and maintenance of this condition until adequate healing has taken place.

Despite the exponential growth of technology, X-ray interpretation remains archaic. Throughout the world we rely on medical doctors to look at and accurately interpret these images. Most doctors who interpret X-rays are general practitioners, family doctors, or general orthopedic doctors, who may have very limited training in wrist X-ray interpretation. The images themselves are inherently difficult to interpret, and this is compounded in the case of distal radius fractures by the fact that the bones of the wrist obscure each other in X-ray images. Even radiologists, who are medical specialists whose job it is to interpret diagnostic images, make mistakes. Fractures are missed, proper diagnoses are not made, and patients suffer, with resulting economic and medicolegal consequences.

Machine learning can provide an effective way to automate and optimize diagnosis of distal radius fractures on the basis of X-rays images, minimizing the risk of misdiagnoses and the subsequent personal and economic damages.

In this paper, we demonstrate automatic computerized detection of anteroposterior (AP) distal radius fractures based on a Faster R-CNN neural network [19]. We trained a Faster R-CNN, a neural network for object detection, to identify and locate distal radius fractures in anteroposterior X-ray images achieving the excellent accuracy of 96% of identifying fractures and a mAP of 0.87.

Distal radius fractures are the most common hand fracture. The computerized fracture detection system we present has higher accuracy in detecting AP distal fractures than the average radiologist and physician.

Fracture detection by machine vision is a challenging task. In many cases the fracture's size is small and hard to detect. Moreover, the fractures have a wide range of different shapes. The classification of an image shape as a fracture also depends in its location. A shape diagnosed as a fracture in one location is diagnosed as a normal structure of the hand in a different location. To cope with these challenges, we trained a state of the art neural network for object detection, Faster R-CNN, for two tasks: a. classifying if there is a fracture in the distal radius. b. finding the fracture's location. The advantage of Faster R-CNN is that it can handle high-resolution images. We trained this network with images with a resolution of up to 1600×1600 pixels. The ability to process high-resolution images enables Faster R-CNN to successfully detect objects in X-ray images, which reveals less details and produces images with a lower signal to noise ratio than MRI or CT. In addition, this enables Faster R-CNN to detect small objects. With its unique network structure, described in Sect. 3, Faster R-CNN can be trained to a high accuracy in detecting objects with a small number of images. In this paper, we show that Faster R-CNN can be trained by a small set of X-ray images to produce high accuracy detection results. This makes Faster R-CNN an excellent tool to detect rare symptoms or rare diseases in X-ray, MRI and CT images where there are only a small number of available X-ray images of these rare medical cases.

2 Related Work

Vijakumar et al. discuss image pre-processing and enhancement techniques of medical images for removing different types of noise from the images such as Gaussian, salt and pepper etc. [24]. Edward et al. [3], introduced automated techniques and methods to verify the presence or absence of fractures. To make these images accurate they applied one or more steps of pre-processing to remove the noise. The existing scheme is modified by a better segmentation and edge detection algorithm to improve efficiency. Deshmukh et al. [16], conclude that Canny Edge detection can be used in detecting fractured bones from X-ray images. Wu et al. detected fractures in pelvic bones with traumatic pelvic injuries, by automated fracture detection from segmented tomography (CT) images [25], based on a hierarchical algorithm, adaptive windowing, boundary tracing, and the wavelet transform. Fracture detection was performed based on the results of prior pelvic bone segmentation via an active shape model (RASM), Rathode and Wahid used to train various classifiers [18]. Object detection is a major area in computer vision field. Traditional features are the histogram of oriented gradients, scale-invariant features, wavelet transforms etc, but the performance of traditional methods fall short of current work based on deep learning [9,14,19,23].

Convolutional neural networks have been applied to medical images imaged with different techniques in recent years. For example, computed tomography (CT) [11] and X-rays [1,4,21]. CNN models are successful in many medical imaging problems. Shin et al. [21] used CNNs to study specific detection problems, such as thoraco-abdominal lymph node (LN) detection and interstitial lung diesis (ILD)classification. Esteva et al. [4], used googlenet [23], for skin cancer classification, reaching a level of accuracy comparable to dermatologists. Dong et al. [2], use CNNs with 16,000 diagnosed X-ray images, to train a classification model to diagnose multi lung diseases such as bronchitis, emphysema and aortosclerosis. Sa et al. used a Faster R-CNN deep detection network to identify osseous landmark points in the lateral lumbar spine in X-ray images [20].

3 Background

Faster R-CNN is a state of the art detection network. It has three parts: 1. A convolutional deep neural network for classification and generating a feature map. 2. A regional proposal network, generating region proposals. 3. A regressor, finding by regression and additional convolutional layers, the precise location of each object and its classification. The neural network for classification we used is VGG 16. It contains 16 layers including 3×3 convolution layers, 2×2 pooling layers and fully connected layers with over 144 million parameters. The convolutional feature map can be used to generate rectangular object proposals. To generate region proposals a small network is slid over the convolutional feature map output. This feature is fed into two fully connected layers, a box regression layer, and a box classification layer. Figure 1 illustrates the parts in Faster-RCNN deep learning neural network [19].

Anchors, at each window location, up to 9 anchors with different aspect ratios and scales give region proposals. The RPN keeps anchors that either have the highest intersection over union (IOU) with the ground truth box or anchors that have IOU overlap of at least 70% with any positive ground truth. A single ground truth element may affect several anchors.

The loss function, Faster R-CNN is optimized for a multi-task loss function. The multi-task loss function [19], combines the losses of classification and bounding box regression.

$$L(\{p_i\},\{t_i\}) = \frac{1}{N_{cls}} \sum_i L_{cls}(p_i, p_i^*) + \lambda \frac{1}{N_{reg}} \sum_i p_i^* L_{reg}(t_i, t_i^*) \tag{1}$$

i - index of anchors in mini batch, p_i - predicted probabilities of anchor being an object, p_i^* - is 1 if anchor is positive and 0 if anchor is negative. t_i - is a vector representing the 4 parameterized coordinates of the predicted bounding box, t_i^* - the ground truth vector of the coordinates associated with positive anchors, L_{cls} - classification log loss over the two classes (object vs. no object).

Regression loss, the output of the regression determines a predicted bounding box and the regression loss indicates the offset from the true bounding box. The regression loss [3] is:

$$L_{reg}(t_i, t_i^*) = R(t_i - t_i^*) \tag{2}$$

Where R is the robust loss function (smooth L1) defined in [8].

Training RPN, RPN is trained by propagation and stochastic gradient descent (SGD). New layers are randomly initialized by zero-mean Gaussian distributions. All other layers are initialized using the pre-trained model of imagenet classification.

4 Methods

4.1 Image Pre-processing

The image set contained hand X-ray images, especially of the distal radius area. The X-ray images were taken at Hadassah Hospital, Jerusalem and analyzed by a hand orthopaedist expert at Hadassah Hospital. The ray images were taken

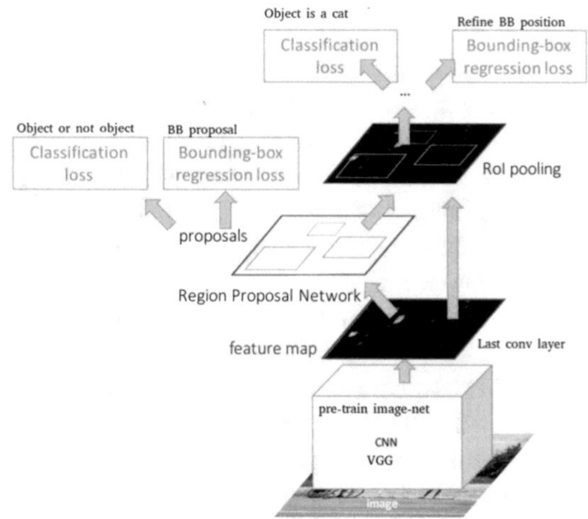

Fig. 1. Faster R-CNN scheme. Ren et al. [19].

and processed by GE Healthcare Revolution XR/d Digital Radiographic Imaging system, Agfa ADC Compact digitizer and GE Healthcare thunder platform DIACOM. The initial dataset pool before augmentation for training and test together contained 55 AP images with distal radius fractures and 40 AP images of hands with no fractures. In addition, there were 25 original images not showing hand bones, for the negative image set. These images were divided into training and test sets. About 80% for training and 20% for the test. The initial dataset images of the positive images were augmented to 4,476 images with labels and bounding boxes for each augmented image, using mirroring, sharpness, brightness and contrast augmentation. We didn't use shear, strain or spot noise augmentation since these could cause a normal hand image to be classified as a hand with a fracture. The images were annotated by VOTT software.

4.2 CNN Based Model, Research Approach

We used Faster R-CNN inside Microsoft's cognitive toolkit deep learning framework. The neural network for classification in the Faster R-CNN is VGG 16. For transfer learning the VGG 16 in the Faster R-CNN we used the image-net database. The X-ray images in the dataset came in couples: anteroposterior position image, AP and lateral position image, of the same hand. After some testing we found that the object detection neural network had better results when trained only on AP images instead of AP and lateral images together as the AP and lateral images are substantially different. Another reason is in some of the lateral images the fracture does not appear because the other bones hide the fracture.

Adjusting the hyperparameters such as, nms, roi, and bounding box scale and ratio in the region proposal, regression and evaluation stages significantly improved, by a factor of 2 the mAP score and in particular the network's precision of finding the location of the fractures.

To increase the classification accuracy of finding if fractures appear or not in the X-ray image; the X-ray images were tagged with two labels one for images with fractures and one for hand images with no fractures. This method was used since there is a large similarity between the hand X-ray images with fractures and without fractures. In this way, the deep neural network find and trains on the differences between the two kinds of images. For the negative images in the training, images of different kinds, not related to X-ray images were added.

To increase the detection accuracy, four types of image augmentation were created: sharpness, brightness, contrast, and mirror symmetry.

Other ways initially considered in the research are: Training a VGG only, network [22]. The disadvantage, it can only classify if there is a distal radius fracture or not in the image. It does not detect the fracture location in the image. Training an SSD 500 network for object detection [15]. Its disadvantage, the highest input image resolution allowed is 512×512 pixels resulting in lower mAP score and accuracy than Faster R-CNN, especially in detecting small objects like bones fractures.

5 Experiments

We ran the Faster R-CNN in Microsoft cognitive toolkit framework on NVIDIA GeForce GTX 1080 Ti graphic card. With 3584 gpu cores, 11.3 TFLOPS, 11 Gbps GDDR5X memory, 11 GB frame buffer and 484 GB/sec memory bandwidth. With CUDA version: 9.0.0 and CUDNN version: 7.0.4, on a computer with an Intel Core i7-4930K processor. Which has six cores each operating at a frequency of 3.4–3.9 GHz, twelve threads and bus speed 5 GT/s DMI2. The computer was configured to the Debian 9, operating system.

The possibility to train high-resolution images is an important advantage of Faster R-CNN when it used for X-ray images of fractures, as often the fracture is small compared to the total image. We tested image sets with different resolutions from 500×500 to 1600×1600 pixels. We got the best results when the image resolution is 1300×1300 pixels.

To increase the detection accuracy image augmentation is used, as the number of the augmented images for training increased the mAP increased as well, the results appear in Table 1.

Table 1. Object detection precision versus number of images augmentation

Mean AP	Loss	Number of augmented training images
0.657	0.061	552
0.825	0.026	4280
0.866	0.020	4476

Training the two label Faster R-CNN with a set of 4,476 augmented images, resulted in a training loss of 0.020. Testing this network with 1,312 augmentations of images, which were not included in training part of the data set. Received a mAP of 0.866. Using mean AP average accuracy as defined in The PASCAL Visual Object Classes challenge, VOC2007 [6]. The classification accuracy for a distal radius fracture is 96%. We ran the system for 45 epochs. It almost converged in the 22nd epoch.

Figure 2 shows hand X-ray images. On the left column are images before evaluating by deep neural network system for object detection. On the right columns are the same X-ray images after evaluation. The system detected the distal radius fracture and labeled as a fracture in the image. The system located the fracture locations and marked their locations in the image with a blue rectangle. The number in the images is the certainty of the fracture detection, as calculated by the system. The certainty ranges from 0 to 1 where 1 is full certainty.

In Fig. 2, the lower images are over-exposed. It can be seen that even in this case the deep network system was able to detect the fracture with very high certainty. On the other hand, a radiologist or physician would have asked to make another image, as by regular standards this image cannot be diagnosed.

Figure 3 shows detection of fractures by the object detection deep neural network in of X-ray images with enhanced contrast from augmented images in the test set.

6 Discussion

6.1 Conclusions

In this paper, the neural network, Faster R-CNN, was trained to detect distal radius fractures. The system was trained on 4,476 augmented anteroposterior hands X-ray images and pre-trained on the imagenet data set. Testing this deep neural network system on 1,312 anteroposterior hands X-ray images resulted in an mAP of 0.866 for object detection of distal radius fractures, With an accuracy of 96% in classifying if there is a distal radius fracture. The network accuracy is substantially higher than the average accuracy of fracture identification by qualified radiologists and physicians. This makes the Faster R-CNN an excellent tool to detect hand fractures in X-ray images. We obtained this accuracy using initial small data set of only 38 anteroposterior X-ray images, of hands diagnosed with fractures. As far as we know it is the smallest set of positively diagnosed images used to train a machine vision network successfully, for classification or objects detection on X-ray, CT or MRI medical images.

The ability, we showed in this paper, to train Faster R-CNN with a small number of X-ray images with distal radius fractures and get high accuracy detection may open the possibility for detecting rare diseases or rare symptoms of common diseases, by Faster R-CNN or related deep neural networks. Where the appearance of these diseases or symptoms is diagnosed today by X-ray, CT or MRI images. But since there is only a small number of diagnosed X-ray images available, of these medical cases, only experts can detect them. Rare diseases

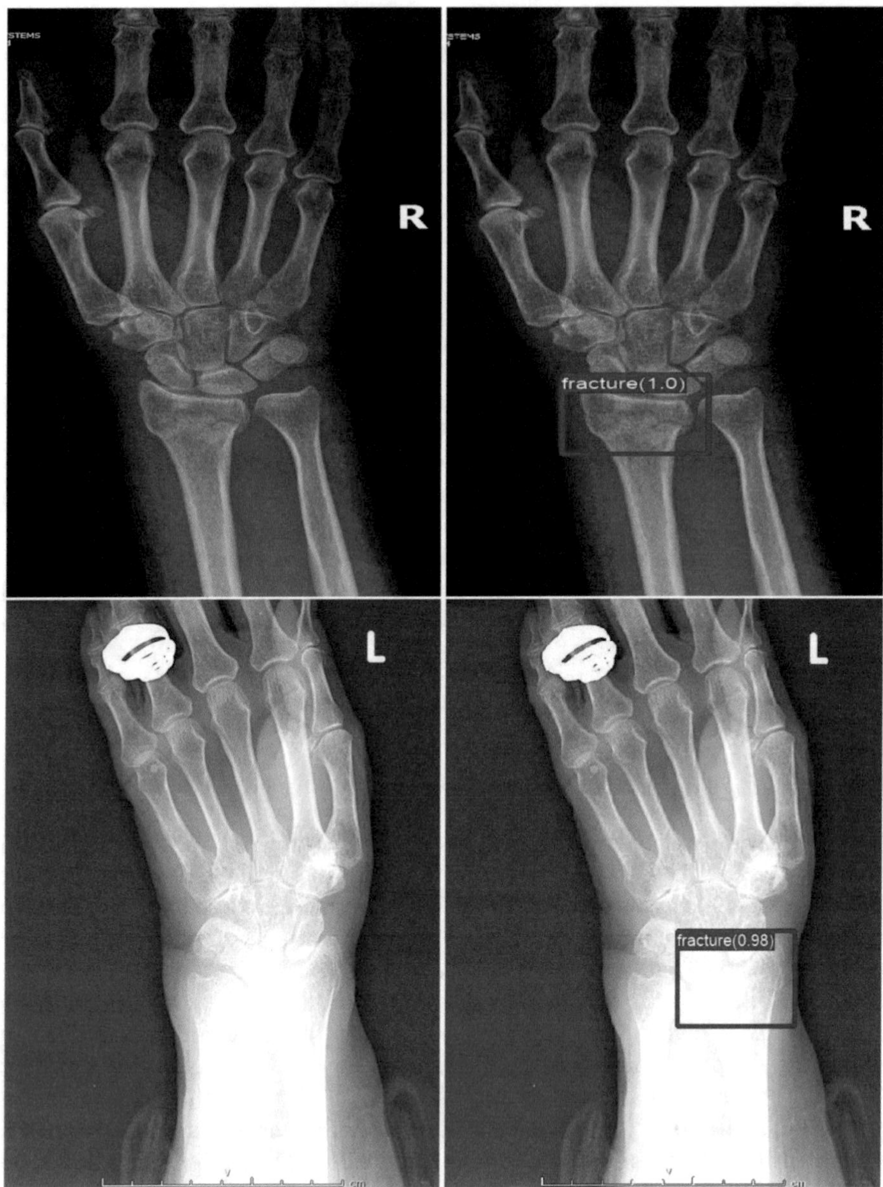

Fig. 2. X-ray images of hands bones. Before and after evaluation by the machine vision system. The distal radius fractures detected by the deep neural network are bounded by a blue rectangle.

conditions that affect less than 1 in 1800 individuals often go untreated. On average, it takes most rare disease patients eight years to receive an accurate diagnosis, [10]. By that time, they have typically seen 10 specialists and have

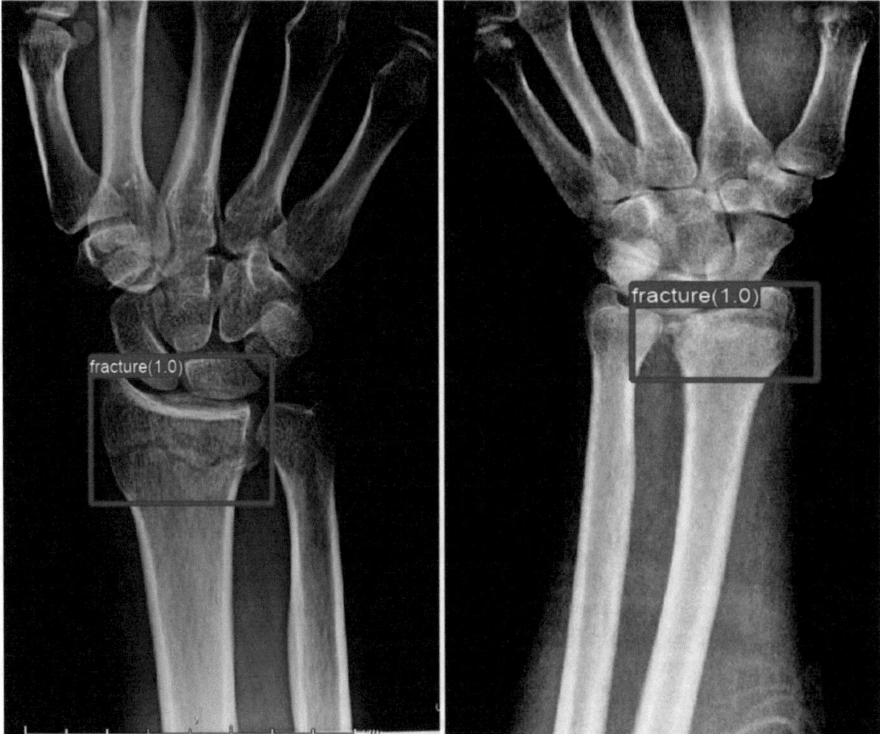

Fig. 3. Distal radius fracture detection in augmented X-ray images with enhanced contrast.

been misdiagnosed three times. The deep neural network object detection may make the diagnosis of some rare diseases available to every physician.

6.2 Limitations and Future Scope

The network was trained on a small set of images compared to the regular training for machine vision detection of X-ray images, which varied from about a thousand original images, [13], up to tens of thousands original images, [2]. For regular images, not X-ray images or medical images, Faster R-CNN already achieved good precision results for labels trained by a small set of images, [12]. In the future testing and training the Faster-RCNN network with thousands of images is suggested to verify the accuracy robustness, of the network and probably to increase the accuracy more. Although the network achieved top-level accuracy, much higher than regular physician or radiologist accuracy, every single percent increase in the accuracy means many more detections that are correct in a global pool of tens of millions of new X-ray bone fractures imaged every year.

Training the network for detection of rare diseases facing a challenge of finding positively diagnosed X-ray images. In the future, we suggest collecting positively diagnosed X-ray images of rare diseases from different hospitals and medical centers. The collection can be on a national scale or even international scale. Examples of rare diseases that are diagnosed by X-ray images, Pycnodysostosis, Erdheim-Chester disease, Gorham's disease, adult-onset Still's Disease (AOSD), Carcinoid tumors, Pulmonary hypertension and Chilaiditi's syndrome. Once enough X-ray images of rare diseases will be collected to train the objects detection neural network the system could be distributed to any medical facility, substantially aiding the detection of rare diseases and significantly reducing the time till people who have the rare disease get the proper treatment.

References

1. Bar, Y., Diamant, I., Wolf, L., Lieberman, S., Konen, E., Greenspan, H.: Chest pathology detection using deep learning with non-medical training. In: ISBI, pp. 294–297. Citeseer (2015)
2. Dong, Y., Pan, Y., Zhang, J., Xu, W.: Learning to read chest x-ray images from 16000+ examples using CNN. In: Proceedings of the Second IEEE/ACM International Conference on Connected Health: Applications, Systems and Engineering Technologies, pp. 51–57. IEEE Press (2017)
3. Edward, C.P., Hepzibah, H.: A robust approach for detection of the type of fracture from x-ray images. Int. J. Adv. Res. Comput. Commun. Eng. **4**(3), 479–482 (2015)
4. Esteva, A., Kuprel, B., Novoa, R.A., Ko, J., Swetter, S.M., Blau, H.M., Thrun, S.: Dermatologist-level classification of skin cancer with deep neural networks. Nature **542**(7639), 115 (2017)
5. Serviste et al.: Overlooked extremity fractures in the emergency department. Turk. J. Trauma Emerg. Surg. **19**(1), 25–28 (2013)
6. Everingham, M., Van Gool, L., Williams, C.K.I., Winn, J., Zisserman, A.: The Pascal visual object classes (VOC) challenge. Int. J. Comput. Vis. **88**(2), 303–338 (2010)
7. Feinmann: Why are so many broken bones being missed in a&e and thousands of patients being sent home in agony with just a paracetamol? Daily Mail (2015)
8. Girshick, R.: Fast R-CNN. In: Proceedings of the IEEE International Conference on Computer Vision, pp. 1440–1448 (2015)
9. He, K., Zhang, S., Ren, X., Sun, J.: Deep residual learning for image recognition. In: Proceedings of the IEEE Conference on Computer Vision and Pattern Recognition, pp. 770–778 (2016)
10. Hendriksz, C.: How do you explain a rare disease (2017). https://health.10ztalk.com/2018/03/15/how-do-you-explain-a-rare-disease-figure-1-medium
11. Shin, H., Roth, H.R., Gao, M., Lu, L., Xu, Z., Nogues, I., Yao, J., Mollura, D., Summers, R.M.: Deep convolutional neural networks for computer-aided detection: CNN architectures, dataset characteristics and transfer learning. IEEE Trans. Med. Imaging **35**(5), 1285 (2016)
12. Żak, K.: cntk-hotel-pictures-classificator (2017). https://github.com/karolzak/cntk-hotel-pictures-classificator

13. Kim, D.H., MacKinnon, T.: Artificial intelligence in fracture detection: transfer learning from deep convolutional neural networks. Clin. Radiol. **73**(5), 439–445 (2018)
14. Krizhevsky, A., Sutskever, I., Hinton, G.E.: Imagenet classification with deep convolutional neural networks. In: Proceedings of the 25th International Conference on Neural Information Processing Systems, NIPS 2012, vol. 1, pp. 1097–1105 (2012)
15. Liu, W., Anguelov, D., Erhan, D., Szegedy, C., Reed, S., Fu, C., Berg, A.C.: SSD: single shot multibox detector. In: European Conference on Computer Vision, pp. 21–37. Springer (2016)
16. Deshmukh, S., Zalte, S., Vaidya, S., Tangade, P.: Bone fracture detection using image processing in Matlab. Int. J. Advent Res. Comput. Electron. (IJARCE) 15–19 (2015)
17. Ootes, D., et al.: The epidemiology of upper extremity injuries presenting to the emergency department in the United States. Hand **7**(1), 18–22 (2011)
18. Rathode, D.H., Ali, W.: MRI brain image quantification using artificial neural networks a review report. ISOI J. Eng. Comput. Sci. **1**(1), 48–55 (2015)
19. Ren, S., He, K., Girshick, R., Sun, J.: Faster R-CNN: towards real-time object detection with region proposal networks. Adv. Neural Inf. Process. Syst. **28**, 91–99 (2015)
20. Sa, R., Owens, W., Wiegand, R., Studin, M., Capoferri, D., Barooha, K., Greaux, A., Rattray, R., Hutton, A., Cintineo, J., et al.: Intervertebral disc detection in x-ray images using faster R-CNN. In: 2017 39th Annual International Conference of the IEEE Engineering in Medicine and Biology Society (EMBC), pp. 564–567. IEEE (2017)
21. Shin, H., Roberts, K., Lu, L., Demner-Fushman, D., Yao, J., Summers, R.M.: Learning to read chest x-rays: recurrent neural cascade model for automated image annotation. In: Proceedings of the IEEE Conference on Computer Vision and Pattern Recognition, pp. 2497–2506 (2016)
22. Simonyan, K., Zisserman, A.: Very deep convolutional networks for large-scale image recognition. arXiv preprint arXiv:1409.1556 (2014)
23. Szegedy, C., Liu, W., Jia, Y., Sermanet, P., Reed, S., Anguelov, D., Erhan, D., Vanhoucke, V., Rabinovich, A.: Going deeper with convolutions. In: Proceedings of the IEEE Conference on Computer Vision and Pattern Recognition, pp. 1–9 (2015)
24. Vanathi, P., Vijaykumar, V., Kanagasabapathy, P.: Fast and efficient algorithm to remove Gaussian noise in digital images. IAENG Int. J. Comput. Sci. **37**(1), 300–302 (2010)
25. Wu, J., Davuluri, P., Ward, K.R., Cockrell, C., Hobson, R., Najarian, K.: Fracture detection in traumatic pelvic CT images. J. Biomed. Imaging **2012**, 1 (2012)

Cloud-Based Skin Lesion Diagnosis System Using Convolutional Neural Networks

E. Akar, O. Marques, W. A. Andrews, and B. Furht(✉)

Florida Atlantic University, 777 Glades Road, Boca Raton, FL 33431, USA
bfurht@fau.edu

Abstract. In this paper, we developed cloud-based skin lesion diagnosis system using convolutional neural networks, which consists of the following: (a) Deep learning based classifier that processes user submitted lesion images which runs on a server connected to the cloud based database. (b) Deep learning based classifier performs quality checks and filters user requests before the request is sent off to the diagnosis classifier. (c) A mobile application that runs on Android and iOS platforms to showcase the system. We designed and implemented the system's architecture.

Keywords: Skin lesion diagnosis · Early melanoma detection · Convolutional neural networks · Cloud-based system · Mobile application

1 Introduction

Development of AI may have significant impact in the future in important fields like medicine which may help to provide more efficient and affordable services to patients around the world. Intelligent diagnosis systems based on AI can aid doctors in detecting diseases. Recently, deep learning methods have been explored to provide solutions to image-based diagnosis tasks such as skin lesion classification. Effective diagnosis systems for lesion detection can improve diagnosis accuracy, which inherently reduces the number of biopsies made by dermatologists to determine the nature of the lesion, and can also aide with early detection of skin cancer like melanoma which can turn fatal if left undiagnosed and untreated. The diagnosis system may run on different types of personal devices like mobile phones and tablets to maximize its access to patients around the world.

On the research end of deep learning based diagnosis systems, there is progress being made, but a full, production-level diagnosis system is still lacking in the field of dermatology. As such, the focus of this thesis is the design and implementation of a skin lesion diagnosis system powered by convolutional neural networks and based on cloud architecture. Front-ends for this system can be made to run on multitude of devices including mobile phones, tablets, and personal computers. A mobile application has also been developed as a front-end to showcase the system. In this paper we developed cloud-based skin lesion diagnosis system using convolutional neural networks, which consist of: Background on skin cancer, the magnitude of the problem, and current deep learning based diagnosis solutions are described in Sect. 2. In Sect. 3,

K. Arai et al. (Eds.): CompCom 2019, AISC 997, pp. 982–1000, 2019.
https://doi.org/10.1007/978-3-030-22871-2_70

the design and implementation of the diagnosis system and its components are explained. In Sects. 4 and 5, the architecture, training, and results of the deep learning based diagnosis and preliminary classifiers are discussed. The design and implementation of the mobile application demo is presented in Sect. 6. Future work is discussed in Sect. 7.

2 Background

Skin cancer is a major medical problem. Every year in United States alone 5.4 million skin cancer cases are reported [1–3]. In United States $8.1 billion is spent annually for skin cancer treatment, $3.3 billion of which is the total cost of treatment for melanoma alone [4]. Melanoma is responsible for 75% of annual deaths due to skin cancer, claiming 10,000 lives annually although it only makes up 5% of skin cancer cases [1, 5]. Early detection for melanoma is vital for survival. If detected in its earliest stages, survival rate for melanoma is extremely high at 99% which fall down to only 14% in its latest stage.

As shown in Fig. 1, visual differences between different skin lesions, especially melanoma and benign lesions can be minute. There are several visual methods used by dermatologist to diagnose melanoma without biopsy, but the accuracy of these methods is poor, only around 60% [6]. Since these methods are not reliable, biopsies are generally required for diagnosis. As a result, computer aided visual diagnosis systems made available to dermatologists and patients may provide more accurate and convenient methods of diagnosis to improve the rate of early detection of skin cancer and reduce the number of diagnosis tests.

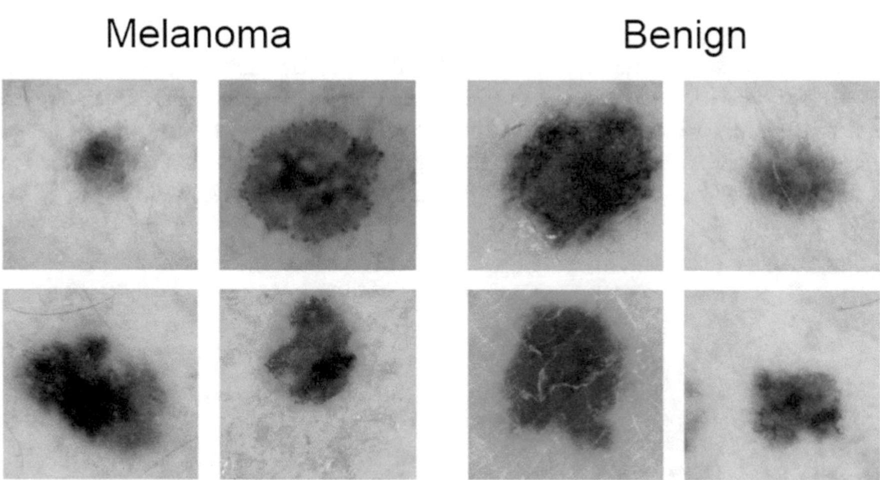

Fig. 1. Visual similarities between melanoma and benign lesions make it hard for dermatologists and machine learning algorithm to visually identify the lesions. Sample images taken from the "ISIC 2018: Skin Lesion Analysis Towards Melanoma Detection" grand challenge datasets [14, 15].

2.1 Deep Learning in Lesion Detection

Deep learning, specifically convolutional neural networks (CNNs) have rapidly become the technique of choice in computer vision. This is due to the contribution of Krizhevsky et al. winning the ImageNet competition in 2012 and other more recent successes of CNNs improving the state of the art in many domains. This is becoming increasingly the case for many visual problems in the field of medicine as well.

To tackle the problem of skin lesion diagnosis, several authors have used CNNs. Training the CNNs were done using only dermoscopic and photographic images as input along with image labels in a single CNN [7, 8]. In the case of Esteva et al., the trained network was tested against 21 board-certified dermatologists and achieved expert level performance; a feat, according to the authors, has not been achieved before [8].

One of the general downsides of training neural networks is the requirement for a large training set. Another challenge for this specific task is that images of lesions may vary in factors like lighting and zoom or even feature the same skin disease in different stages which may appear different. Esteva et al. were able to overcome these challenges by collecting almost 130,00 clinical images and 3370 dermoscopy images. They utilized GoogleNet Inception v3 CNN architecture that was pre-trained on 2014 ImageNet Large Scale Visual Recognition Challenge and trained the network with the collected lesion images using transfer learning [9–11].

Esteva et al. also point out that with billions of smartphones in use around the world that can be equipped with deep neural networks, low-cost access to diagnostic services can be provided universally.

3 The Diagnosis System Architecture

The system consists of the client device, cloud database, and server that is connected to both cloud database and the script that loads the CNN, as shown in Fig. 2. Any device that has internet connection could connect to the cloud database and upload lesion images to be processed. At the end of the diagnosis process, a skin lesion classification and the confidence score computed by the CNN is reported back to the client.

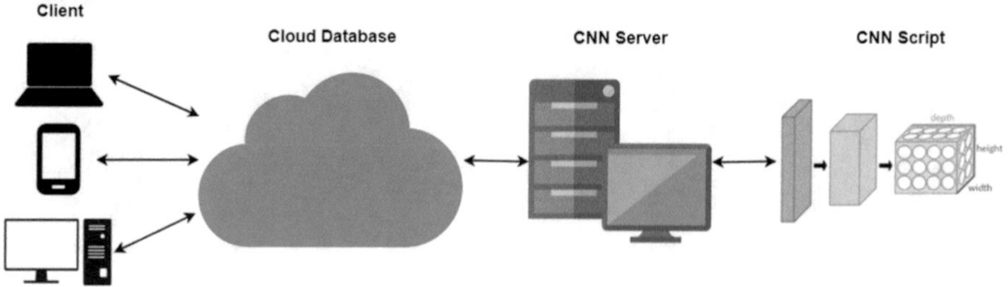

Fig. 2. Cloud-based diagnosis system architecture.

Once an image is uploaded from a device to the cloud database, the CNN server is notified which then downloads the image and informs the CNN script to process the image with the already trained CNN. Once a result is acquired from the CNN, such as melanoma or benign lesion, the result is sent back to the CNN server which updates the cloud database which then updates the client with the lesion results.

3.1 Client

Applications running on client devices are relatively simple as the only requirements are connection to the cloud database and client-side image resizing. The architecture of the system was designed to keep client memory and processor requirements minimal and be accessible to a large range of devices with varying specifications. The issue of slow internet upload and download speed is also mitigated by utilizing client-image image resizing which resizes lesion images to the exact size as required by the CNN. As a result, patients and physicians who may have access to different types of devices and internet connection can employ the system with ease.

The application running on devices with cameras can also utilize the camera and handle the task of taking images of lesions. This has been implemented in the mobile phone application written for this thesis. The details of the mobile app are discussed in the next paragraph.

3.2 Cloud Database

The task of the cloud database is to establish connection between client devices and the CNN server and store and serve data coming in from both directions. It must be able to connect to a large range of devices, store lesion images uploaded from clients into a filesystem and capture results for these images sent from the CNN server into a database. Google's Firebase has many services that offer these features so we elected to build our system with it, specifically with Cloud Firestore and Cloud Storage Firebase services [14]. Firebase provides client-side libraries for applications that run on Android, iOS, and JavaScript, allowing mobile, web, and server applications to integrate with this system. Firestore is a realtime database service that synchronizes data events across applications, and Cloud Storage handles the storage and serving of files which are lesions images in our case.

When a diagnosis for an image is requested, a new document with a unique id is created in Firestore, and the image is uploaded with this id as the filename. The document initially stores the id, timestamp, and the location of the uploaded image in Cloud Storage. In the next step, the CNN server gets notified of the document and handles the rest of the process.

3.3 CNN Server

The CNN server is written in JavaScript for Node.js, and acts as a task manager for the CNN script. Since Firestore synchronizes all connections, the CNN server which is already listening for changes in Firestore is notified immediately of requests in the form of newly created documents. The server then retrieves documents that have not yet been processed, and using the id of the documents, downloads the corresponding lesion images as shown in part 2 in the source code of the Node.js server shown in Fig. 3.

```
const admin = require('firebase-admin');
const keys = require('./key.json');
const spawn = require('child_process').spawn;
const fs = require('fs');
admin.initializeApp({
    credential: admin.credential.cert(keys),
    databaseURL: "https://melonoma-a896a.firebaseio.com",
    storageBucket: "gs://melonoma-a896a.appspot.com"
});
const db = admin.firestore();
function processUpload(id) {
    jobs[id] = true;
    py.stdin.write(JSON.stringify({id}) + '\n'); //send off to script to process
}

let jobs = {};

// 1) spawn the CNN script
const py = spawn("python", ["-u", "app.py"]);
py.stdout.setEncoding('utf-8');
py.on('exit', () => {
    console.log('PYTHON SCRIPT ERROR');
});

// 2) listen for new documents in Firestore and download images
db.collection('uploads').where('processed', '==', false).onSnapshot(docs => {
    docs.forEach(doc => {
        if (jobs[doc.id] !== undefined) return;
            admin.storage().bucket().file(doc.id).createReadStream()
            .pipe(fs.createWriteStream(`tmp/${doc.id}`).on('finish', () => {
              processUpload(doc.id);
        }));
    });
});

// 3) returned results from the CNN script
py.stdout.on('data', (data) => {
    let result = JSON.parse(data); // json result
    result.processed = true;
    console.log(result);
    fs.unlink(`tmp/${result.id}`, (error) => { /* handle error */ }); // delete file
    db.doc(`uploads/${result.id}`).update(result).then(() => { // update change
        delete jobs[result.id];
    });
});

process.stdin.resume();
```

Fig. 3. Source code of the Node.js.

Using the document IDs, the uploaded images stored in Cloud Storage are downloaded into a temporary location in the filesystem. The CNN script is a child process of the CNN server and is spawned at the initialization of the server in part 1 of the source code. The communication between the script and server is done using standard IO between parent and child processes. Notifying the CNN script to processes a request, handled by the process Upload function, is done by writing the id of the document to the standard input of the CNN script. In part 3, results from the script in the form of stringified JSON are written to the standard input of the server. Next, the JSON response is parsed, and with the extracted id, the corresponding document is updated with the results from the script. For clean-up, the downloaded image in the temporary location gets deleted.

As shown in Fig. 4, the updated document includes the label of the lesion, confidence of the CNN in the diagnosis, and a Boolean for recording the malignancy of the lesion. After this update, the client device is synchronized with the changes in Firestore, and receives the results of the diagnosis.

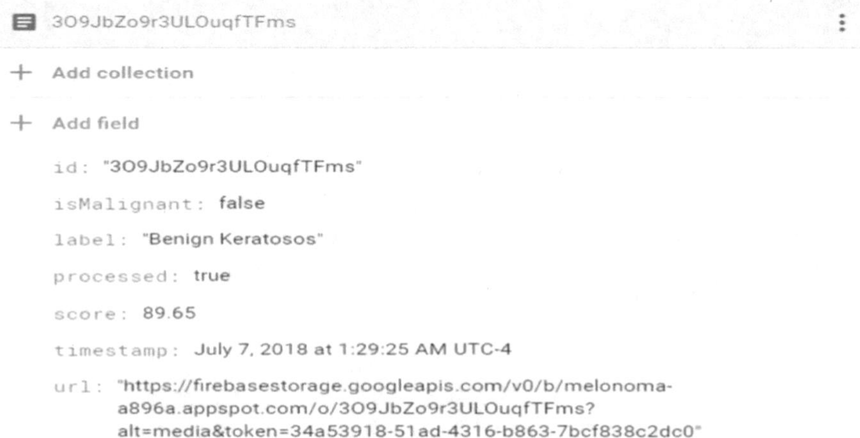

Fig. 4. A Firestore document updated by the CNN server with results from the CNN script.

3.4 CNN Script

The CNN script is a child process spawned by the CNN server that loads the trained CNN model to process the uploaded lesion images, and returns the output from the CNN. The script also features a preliminary CNN that was trained to detect images that do not contain lesions. Images that do not pass the preliminary CNN are rejected, and do not get processed by the diagnosis CNN. In that case, the CNN script returns an error message to the CNN server which then updates the Firestore document in the Cloud Database with the appropriate fields as shown in Fig. 5.

The preliminary CNN is tiny in number of parameters compared to the diagnosis CNN, and a quick check by this CNN increases the overall quality of the system and may even increase the overall efficiency of the system by preventing some images making their way to the much larger, slower diagnosis CNN. The flow of the script is given in Fig. 6.

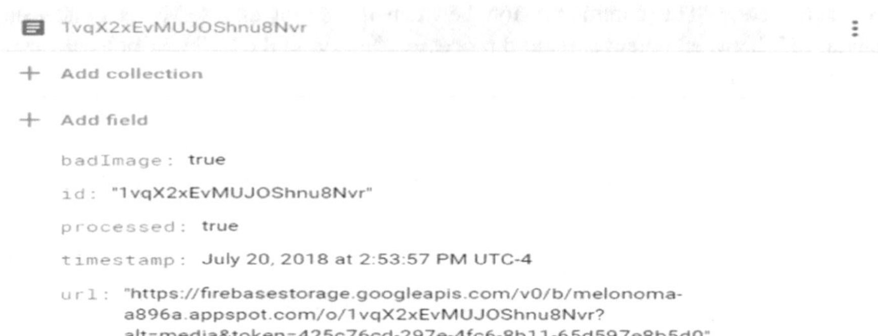

Fig. 5. A Firestore document that reports that an image request was declined by the preliminary CNN.

The script was written in Python, and for training and running the CNNs, the neural network library Keras was used [15]. The source code for the CNN script is given in Fig. 7.

In part 1 of the source code, the two models are loaded. The diagnosis model was trained on the ISIC 2018 dataset which contains images for several skin diseases including melanoma as can be observed in the array of labels in part 2 [12, 13]. The preliminary CNN model has two outputs for lesion and non-lesion images. The datasets used to the train the network were Caltech 101 for non-lesion and the ISIC dataset for lesion images [16].

The image IDs are passed to the standard input of the script in JSON form, and in part 3.1, the id is loaded from the JSON input, and the image corresponding to the id is decoded and preprocessed for both CNNs. Next, in part 3.2, the input image is processed by the preliminary CNN, and an output, either lesion or non-lesion, is acquired. In the following subsection, images labeled as lesion images by the preliminary CNN are processed by the diagnosis CNN, and the diagnosis results are printed out in JSON form back to the CNN server. Non-lesion images on the hand are reported back with an error field. The discussion on the architecture of the models and the training methods take place in the following section.

3.5 Alternative Architectures

There are several alternative architectures that were considered during the design stage of the system. Generally, there are two ways of incorporating deep neural networks in a mobile app: running the network locally on the native device or offloading the processing to servers. Running locally only became possible in recent years as training and running neural networks require huge amounts of computation power. In fact, one of the main reasons for the booming of deep learning is due to the recent growth in processing capabilities of processors, mainly graphics processing units (GPU) [17]. In addition to computation demands, the size of the model can put restraints on the device's memory. ResNet50 is the model that was used for the diagnosis CNN, and it's

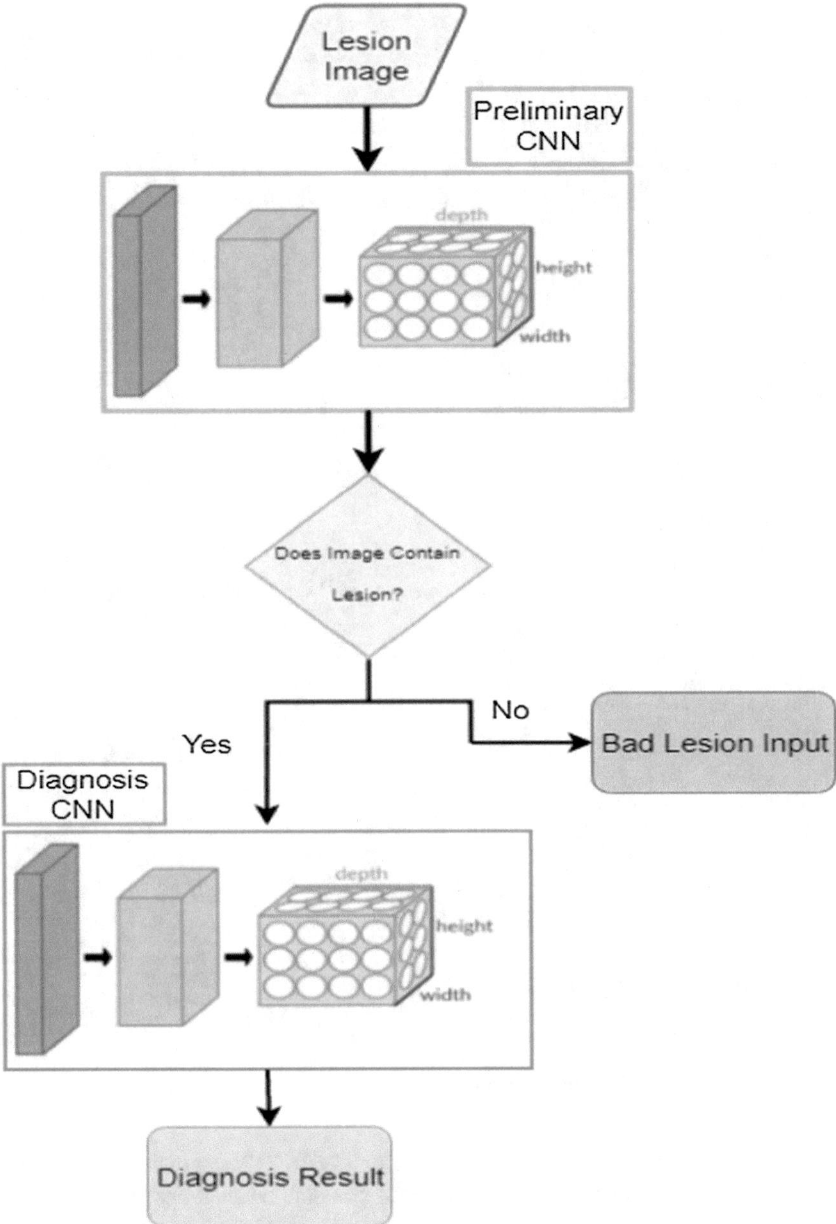

Fig. 6. The preliminary CNN prevents images that do not contain lesions from getting further processed by the diagnosis model.

```
import sys
import json

from keras.applications.resnet50 import preprocess_input
from keras.models import load_model
from keras.preprocessing import image
import numpy as np

# 1) Two trained models
model = load_model('models/model_resnet_best.hdf5')
isLesionModel = load_model('models/model_islesion.hdf5')

# 2) Labels CNN outputs
LABEL_STRS = ["AKIEC", "BCC", "BKL", "DF", "MEL", "NV", "VASC"]
LABEL_FULL = ["Actinic Keratosis", "Basal Cell Carcinoma",  "Benign Keratosos",
"Dermatofibroma", "Melanoma", "Nevus", "Vascular Lesion"]
ISMALIGNANT = [True, True, False, False, True, False, False]

# 3) Await IDs input from CNN server
for line in sys.stdin:

    # 3.1) Extract id from JSON string from server and load image with id
    id = json.loads(line)['id']
    PATH = "tmp/" + id
    img = np.array([preprocess_input(image.img_to_array(image.load_img(PATH)))])

    # 3.2) Run the image through the preliminary CNN to check if image contains lesion.
    pred = isLesionModel.predict([img])
    score = round(max(pred[0]) * 100.0,2) #score
    idx = pred.argmax(axis=-1)[0] #label index

    # 3.3) if image is good, run the diagnosis CNN, else report back with badImage label
    if idx == 0:
        result = {'badImage': True, 'id': id}
        print(json.dumps(result))
    else:
        pred = model.predict([img])
        score = round(max(pred[0]) * 100.0,2) #score
        idx = pred.argmax(axis=-1)[0] #label index
        result = {'score': score, 'isMalignant': ISMALIGNANT[idx], 'label': LABEL_FULL[idx],
                  'id': id}
        print(json.dumps(result))
```

Fig. 7. Source code for CNN script.

binary size is roughly 99 Megabytes (MBs) with 25,636,712 parameters, which makes it difficult to run on older mobile devices with constrained memory and computation capabilities [18, 19]. Another downside of running locally is that some application that is powered by neural networks may need frequent updates. Any updates to the model as a result of training as more and more medical data is gathered would need to be downloaded by the client device in the form 99 MBs patches in the least, putting users with weak and slow Internet connection out of reach.

For server architectures, a seemingly more simple, monolith architecture was also considered where client requests, database, and the CNN are handled by a single server instance. The cloud architecture, on the contrary, establishes connection between specialized programs written in different programming environments that solve different tasks and does not force the entire system to operate in one environment, as it is the case in a monolith server. It follows a software philosophy where a large system is composed of a number of components or programs that performs a simple task and pass on data to another component to complete another task. As a result, although monolith architecture has fewer independent components than the cloud architecture, the implementation may be more complex and inefficient.

An example of specialization is that many neural network libraries have been written in Python which makes Python a popular, specialized language of choice for implementing and training neural networks. Node.js is also another popular programming environment used for creating server which is featured in the cloud architecture. The cloud architecture takes advantage of segregating tasks to specialized components whereas a monolith system offers very little flexibility.

The cloud architecture can also grow more naturally and efficiently. To respond to high user demand, multiple CNN server nodes can be spawned in different geographical locations where each CNN server is also capable of spawning multiple CNN scripts to maximize throughput with little change to the source code. Also, since data storage and processing are separated in the form of the cloud database and CNN server, users attempting to access results from the cloud database are not affected by any amount of request traffic the CNN server may be under. The cloud architecture allows the client application to be light and simple which opens up the range of devices and environments the system can operate in, so this architecture was opted to be implemented.

4 Diagnosis Convolutional Neural Network

4.1 Dataset

The dataset containing lesion image samples used to train the preliminary and diagnosis CNNs come from the "ISIC 2018: Skin Lesion Analysis Towards Melanoma Detection" grand challenge datasets [12, 13], which contains a total of 10,015 samples of skin diseases shown in Table 1. The dataset is not balanced as there are significantly more nevus (benign, non-harmful) samples than any other samples, combined. Also, dermatofibroma and vascular lesion are severely underrepresented. An unbalanced dataset can make it difficult to train and assess the accuracy of a network so the dataset was rebalanced before training.

Table 1. Skin diseases and sample counts in the ISIC 2018 dataset (10,015 samples).

Disease	Actinic Keratosis	Basal Cell Carcinoma	Benign Keratosis	Dermatofibroma	Melanoma	Nevus	Vascular Lesion
Count	327	514	1099	115	1113	6705	142
Percentage of Total	3.3%	5.1%	11.1%	1.1%	11%	67%	1.4%

4.2 Training and Results

The dataset needs to be rebalanced by under-sampling overrepresented categories and split to create training and testing sets. For the split, 80% of the dataset was used for training and 20% for testing. Each split set was rebalanced by under-sampling so that only 20% of the set contained nevus samples. The balance of the training and training sets can be observed in Table 2.

Table 2. Nevus samples were under-sampled to balance the training and testing sets.

Train set							
Disease	Actinic Keratosis	Basal Cell Carcinoma	Benign Keratosis	Dermatofibroma	Melanoma	Nevus	Vascular Lesion
Count	265	408	893	93	874	662	112
Percentage of total	8%	12.3%	27%	2.8%	26.4%	20%	3.4%
Test set							
Disease	Actinic Keratosis	Basal Cell Carcinoma	Benign Keratosis	Dermatofibroma	Melanoma	Nevus	Vascular Lesion
Count	62	106	206	22	239	166	30
Percentage of total	7.5%	12.8%	24.8%	2.7%	28.8%	20%	3.6%

Because training from scratch isn't plausible, transfer learning on the ResNet50 architecture pre-trained for the 2014 ImageNet Large Scale Visual Recognition Challenge was used for training [11, 19]. The final layer of a ResNet network is a simple softmax dense layer which connects to a flattened average-pooled 2048 dimension layer connected to the final convolutional layer. The final softmax layer was replaced by two densely connected layers with 500 neurons in each and a new, 7 category softmax layer. During training, the entire networks besides the newly added layers were kept frozen. 25% dropout between every connection and batch normalization before every activation was added to the new layers for regularization [20, 21]. Stochastic gradient descent (SGD) with a learning rate of 0.01 was used to train the CNN. The code sample, shown in Fig. 8, is the implementation of the new dense layers in Keras.

The accuracy of the neural network after training on the test set is 77.4%. The rebalance of the dataset resulted in high accuracies for non-nevus categories, as shown in Figs. 9 and 10, and having more samples of melanoma than nevus also helped improved sensitivity of the network (lowered false negatives) when diagnosing melanoma vs nevus lesions which is especially vital for diagnosis systems. Although human level accuracy on this dataset has not been determined, it is safe to assume that as per [4] the accuracy of the diagnosis network, at least on melanoma detection, is on or very close to human level performance.

```
epochs = 350
drop = 0.25
dns = 500
batch_size = 16
rg = 1e-3

mc_top = ModelCheckpoint('weights/model_bottleneck_test.hdf5', monitor='val_acc', verbose=0,
save_best_only=True, save_weights_only=False, mode='auto', period=1)

model = Sequential()

model.add(Flatten(input_shape=train_data.shape[1:]))
model.add(Dropout(drop))

model.add(Dense(dns, kernel_initializer='he_normal', kernel_regularizer=l2(rg)))
model.add(BatchNormalization())
model.add(Activation('relu'))
model.add(Dropout(drop))

model.add(Dense(dns, kernel_initializer='he_normal', kernel_regularizer=l2(rg)))
model.add(BatchNormalization())
model.add(Activation('relu'))
model.add(Dropout(drop))

model.add(Dense(num_classes, kernel_initializer="he_normal", activation='softmax'))

opt = SGD(lr=0.01)
model.compile(optimizer=opt, loss='categorical_crossentropy', metrics=['accuracy'])
model.fit(train_data, train_labels,
          epochs=epochs,
          verbose=2,
          batch_size=batch_size,
          shuffle=True,
          validation_data=(test_data, test_labels),
          callbacks=[mc_top])
```

Fig. 8. The implementation of the new dense layers in Keras.

Further experiments like unfreezing final few layers to continue backpropagation over ResNet resulted in overfitting and lowered test accuracy. Despite a tiny amount of training data, several CNNs ranging from few layers to up to 18 layers were experimented with, and the best accuracy achieved was 66%. In the end, the network trained with transfer learning was chosen as the diagnosis CNN.

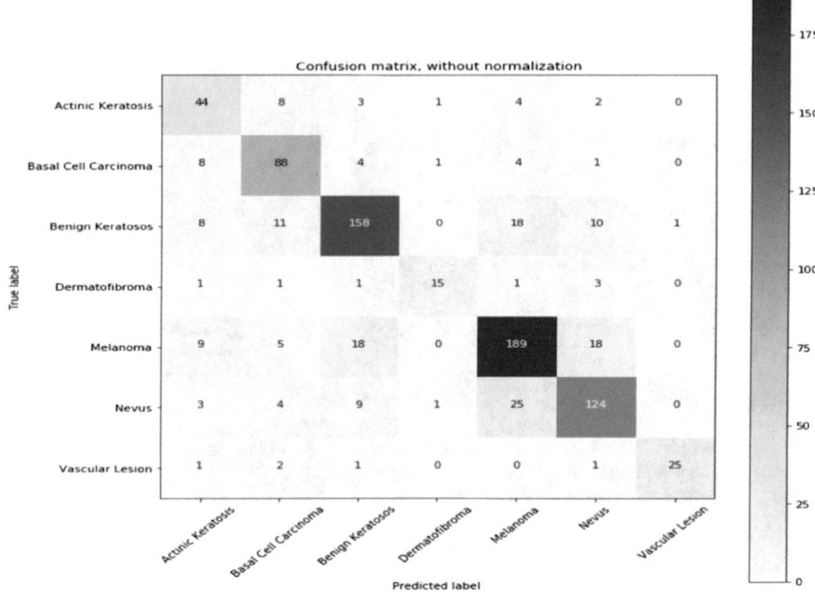

Fig. 9. Confusion matrix of the test set.

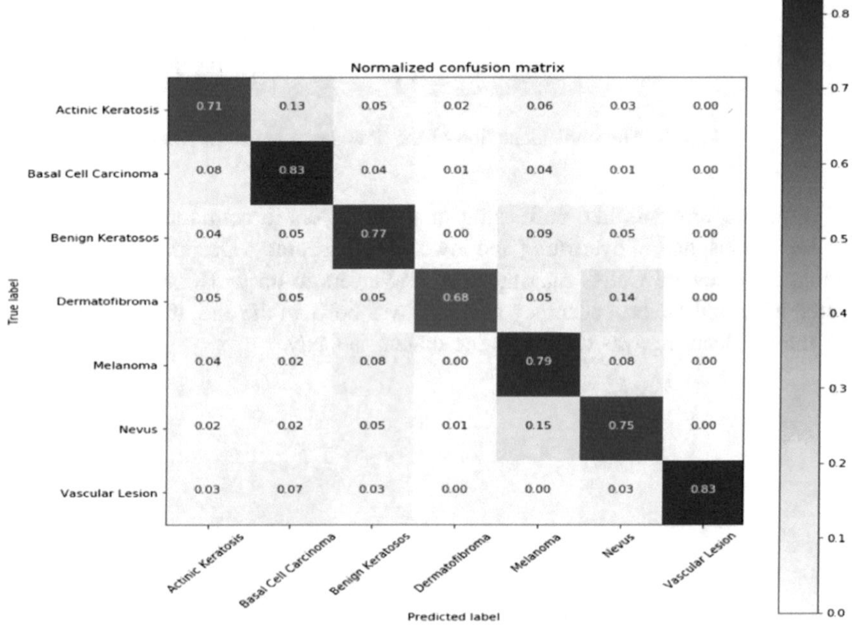

Fig. 10. Normalized confusion matrix of the test set.

5 Preliminary Convolutional Neural Network

The purpose of the preliminary CNN is to filter bad requests from the user. This way, the diagnosis system won't attempt to classify images that are not lesion images or do not meet the quality of the images in the training set and send back garbage results as a result. To train the CNN, any dataset containing categories of common office or household objects that a user might take a picture of to test the system, or any category that is exclusively not medical and skin lesion related to train against the lesion images could work. As counter to the ISIC 2018 dataset, the Caltech 101 dataset along with ISIC 2018 was used to train the CNN. Samples in the two categories in the training and testing sets were randomly chosen and contain roughly 8000 and 2000 sample images, respectively.

The architecture of the CNN is relatively small, containing roughly 73,000 parameters. After the input layer, there are 6 convolutional layers with kernel size of 3×3 and strides of 2×2. After the final convolution layer, an average pool is computed and the layer is flattened which is connected to the final double output softmax layer. Training was done with SGD and learning rate of 0.01. The architecture can be observed in detail in the Keras code sample shown in Fig. 11.

Since the images in both categories are very different from each other, the accuracy of the network is very high. After only three epochs, the accuracy on the test set was 99.4% so training was stopped early and the preliminary CNN was fitted with this network.

6 Mobile Phone Application

We developed an application for both Android and iOS platforms in order to complete the entire cloud-based skin lesion diagnosis system. The application connects to Firebase at startup to upload images and receive results from the diagnosis pipeline. The application is a hybrid mobile app built using Ionic SDK [22]. With Ionic, developers can build mobile and web applications using Web technologies like HTML, CSS, and Javascript with built-in native features that the device offers like camera and geolocation. Using the same source code, the application can be compiled to run on multiple platforms devices which include Android, iOS, and web browsers.

The application connects to the device's camera and displays the camera stream. The user is guided with a red circle to help place the skin lesion centered horizontally and vertically in the visual field of the camera (Fig. 12). This is an important feature because the lesion images in the dataset are centered on the lesion and replicating this property in the user requests may increase the accuracy of the diagnosis. Once an image is taken by the user, the users have the option to either retake the image or upload the image to Firebase. Before uploading, the image is cropped and resized to match the exact dimensions of the diagnosis and preliminary CNNs. Once uploaded, the user is taken to the History page, which displays results from the diagnosis CNN, as illustrated in Fig. 13.

```
ROW_AXIS = 1
COL_AXIS = 2
CHANNEL_AXIS = 3

def bn_rel (conv):
    norm = BatchNormalization(axis=CHANNEL_AXIS)(conv)
    return Activation("relu")(norm)

def conv2(prev, filters=64, kernel_size=(3,3), strides=(2,2), padding='same'):
        return Conv2D(filters=filters,
                      kernel_size=kernel_size,
                      strides=strides,
                      padding=padding,
                      kernel_initializer='he_normal',
                      kernel_regularizer=l2(1.e-4))(prev)
dim = 224
num_classes = 7

# input block
input = Input(shape=(dim,dim, 3))

filters = 16

conv = conv2(input,filters=filters)
actv = bn_rel(conv)
conv = conv2(actv, filters=filters)
actv = bn_rel(conv)

filters = 32

conv = conv2(actv, filters=filters)
actv = bn_rel(conv)
conv = conv2(actv, filters=filters)
actv = bn_rel(conv)

filters = 64

conv = conv2(actv, filters=filters)
actv = bn_rel(conv)
conv = conv2(actv, filters=filters)
actv = bn_rel(conv)

block_shape = K.int_shape(actv)
pool = AveragePooling2D(pool_size=(block_shape[ROW_AXIS], block_shape[COL_AXIS]),
strides=(1, 1))(actv)

flat = Flatten()(pool)

dnse = Dense(2, kernel_initializer="he_normal", activation="softmax")(flat)

model = Model(inputs=input, outputs=dnse)

opt = SGD(lr=0.01)
model.compile(optimizer=opt, loss='binary_crossentropy', metrics=['accuracy'])
model.summary()
```

Fig. 11. Keras code sample of the CN architecture.

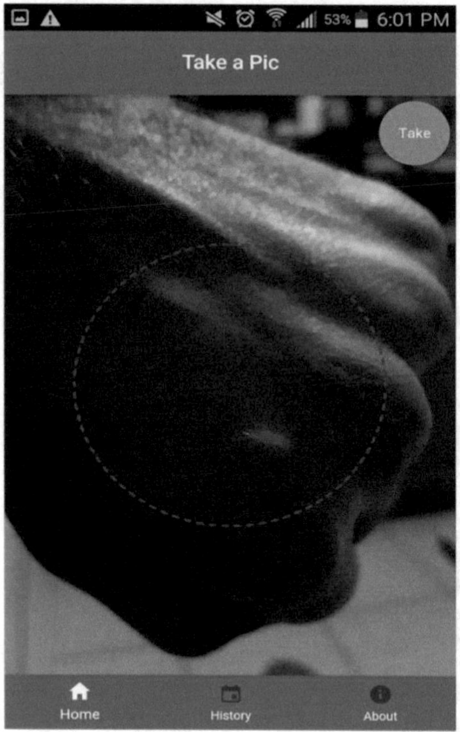

Fig. 12. The application utilizes the native camera on the mobile phone to takes pictures of lesions.

During the diagnosis process, the uploaded request is marked as 'processing'. Once a result is acquired, the app is updated with a diagnosis and confidence score of the diagnosis CNN. If an image does not pass the preliminary CNN, the user is notified with a warning text.

The diagnosis system being cloud-based, complexity of the application is minimal. The main requirements of the application are connection to the camera and Firebase and client-side image resizing which can be done with few lines of Javascript. The minimalist nature of the front-end application also allowed for a very quick design and implementation.

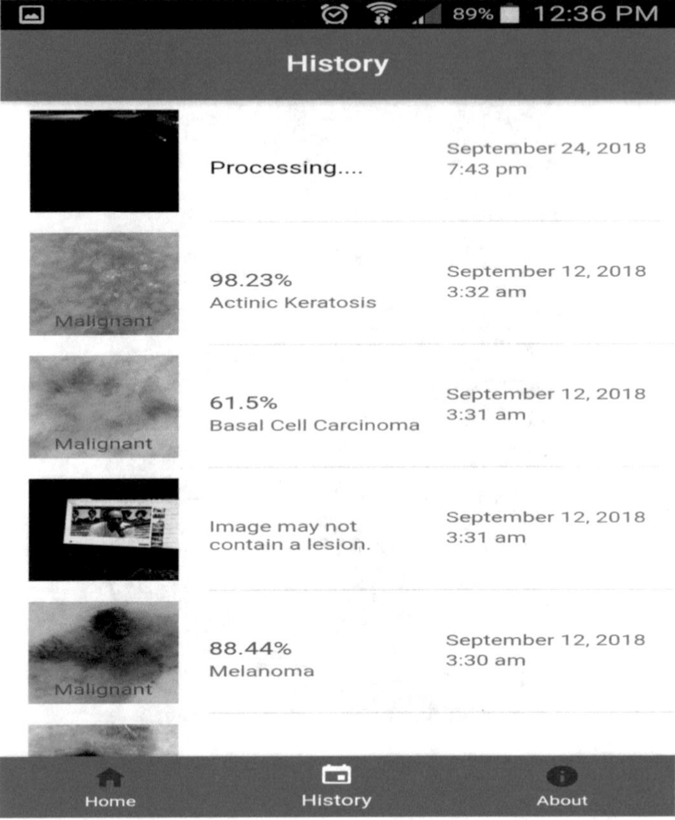

Fig. 13. History of results page displays diagnosis results.

7 Conclusion and Future Work

Skin cancer is a serious medical problem. Early diagnosis for type of cancer like melanoma is vital for survival. For this thesis, a skin lesion diagnosis system was built to help with diagnosis. The system is cloud-based and uses Firebase. The diagnosis process is offloaded to servers which allow developers to build front-ends for a large range of mobile devices that are lightweight and minimalistic in code complexity. The diagnosis is handled by a two-stage CNN pipeline where a preliminary CNN does quality check on user requests, and a diagnosis CNN, whose accuracy is near dermatologist level, outputs probabilities over seven different lesion categories corresponding to the categories in the ISIC 2018 dataset used for training. For training, transfer learning was applied to a ResNet50 network trained on the ImageNet competition dataset.

For future work, to improve accuracy of the diagnosis network, more data would need to be gathered. The diagnosis CNN was trained using transfer learning with roughly 3,000 images and that is not nearly enough to allow experimentation with different CNN architectures and training paradigms. One experiment, put under future work, is combining the diagnosis ResNet with a smaller network that is trained from scratch in multi-stream fashion where the outputs of both networks are connected to a combined final layer. This way, a unique version of ensemble of networks may result in higher accuracy and better generalization.

Acknowledgments. The authors gratefully acknowledge funding from NSF award No. 1464537, Industry/University Cooperative Research Center, Phase II under NSF 13-542. We are also thankful to Farris Foundation who also provided funds for this project.

References

1. Rogers, H.W., Weinstock, M.A., Feldman, S.R., Coldiron, B.M.: Incidence estimate of nonmelanoma skin cancer (keratinocyte carcinomas) in the US population, 2012. JAMA Dermatol. **151**(10), 1081–1086 (2015)
2. Cancer Facts and Figures 2018. American Cancer Society. https://www.cancer.org/content/dam/cancer-org/research/cancer-facts-and-statistics/annual-cancer-facts-and-figures/2018/cancer-facts-and-figures-2018.pdf. Accessed 3 May 2018
3. Stern, R.S.: Prevalence of a history of skin cancer in 2007: results of an incidence-based model. Arch. Dermatol. **146**(3), 279–282 (2010)
4. Guy, G.P., Machlin, S.R., Ekwueme, D.U., Yabroff, K.R.: Prevalence and costs of skin cancer treatment in the U.S., 2002–2006 and 2007–2011. Am. J. Prev. Med. **104**(4), e69–e74 (2014). https://doi.org/10.1016/j.amepre.2014.08.036
5. Siegel, R., Miller, K.D., Jemal, A.: Cancer statistics, 2016. CA Cancer J. Clin. **66**, 7–30 (2016)
6. Kittler, H., Pehamberger, H., Wolf, K., Binder, M.: Diagnostic of dermoscopy. Lancet Oncol. **3**, 159–165 (2002)
7. Yu, L., Chen, H., Dou, Q., Qin, J., Heng, P.A.: Automated melanoma recognition in dermoscopy images via very deep residual networks. IEEE Trans. Med. Imaging **36**(4), 994–1004 (2017). https://doi.org/10.1109/TMI.2016.2642839
8. Esteva, A., Kuprel, B., Novoa, R.A., Ko, J., Swetter, S.M., Blau, H.M., Thrun, S.: Dermatologist-level classification of skin cancer with deep neural networks. Nature **542** (7639), 115–118 (2017)
9. Szegedy, C., Vanhoucke, V., Ioffe, S., Shlens, J., Wojna, Z.: Rethinking the inception architecture for computer vision (2015). Preprint at https://arxiv.org/abs/1512.00567
10. Russakovsky, O., et al.: Imagenet large scale visual recognition challenge. Int. J. Comput. Vis. **115**, 211–252 (2015)
11. Pan, S.J., Yang, Q.: A survey on transfer learning. IEEE Trans. Knowl. Data Eng. **22**, 1345–1359 (2010)
12. Tschandl, P., Rosendahl, C., Kittler, H.: The HAM10000 dataset, a large collection of multi-source dermatoscopic images of common pigmented skin lesions. Sci. Data **5**, 180161 (2018). https://doi.org/10.1038/sdata.2018.161
13. Codella, N.C.F., Gutman, D., Celebi, M.E., Helba, B., Marchetti, M.A., Dusza, S.W., Kalloo, A., Liopyris, K., Mishra, N., Kittler, H., Halpern, A.: Skin lesion analysis toward Melanoma detection: a challenge. In: 2017 International Symposium on Biomedical Imaging (ISBI), Hosted by the International Skin Imaging Collaboration (ISIC) (2017). arXiv:1710.05006
14. Cloud Firestore.: (n.d.). https://firebase.google.com/docs/firestore/. Accessed 29 Aug 2018

15. Chollet, F., and others: Keras, GitHub repository (2018). https://github.com/keras-team/keras
16. Fei-Fei, L., Fergus, R., Perona, P.: Learning generative visual models from few training examples: an incremental Bayesian approach tested on 101 object categories. IEEE. CVPR 2004, Workshop on Generative-Model Based Vision (2004)
17. Deng, L.: A tutorial survey of architectures, algorithms, and applications for deep learning. APSIPA Trans. Sig. Inf. Process. **3**, e2 (2014)
18. "Documentation for individual models" (2018). https://keras.io/applications. Accessed 1 Aug 2018
19. He, K., et al.: Deep residual learning for image recognition. In: Proceedings of the IEEE Conference on Computer Vision and Pattern Recognition (2016)
20. Srivastava, N., et al.: Dropout: a simple way to prevent neural networks from overfitting. J. Mach. Learn. Res. **15**(1), 1929–1958 (2014)
21. Ioffe, S., Szegedy, C.: Batch normalization: Accelerating deep network training by reducing internal covariate shift. arXiv preprint (2015). arXiv:1502.03167
22. Drifty, Inc.: Ionic (2016). https://ionicframework.com

Color Correction for Stereoscopic Images Based on Gradient Preservation

Pengyu Liu[1], Yuzhen Niu[1,2](✉), Junhao Chen[1], and Yiqing Shi[1]

[1] College of Mathematics and Computer Science, Fuzhou University, Fuzhou, China
2521151088@qq.com, yuzhenniu@gmail.com, 593056806@qq.com, 459671456@qq.com

[2] Fujian Provincial Key Lab of the Network Computing and Intelligent Information Processing, Fujian, China

Abstract. Color correction can eliminate the color difference between similar images in image stitching and 3D video reconstruction. The result images generated by the local color correction algorithms usually show structure inconsistency problem with the input images. In order to solve this problem, we propose a structure consistent color correction algorithm for stereoscopic images based on gradient preservation. This method can not only eliminate color difference between reference and target images, but also optimize structure between the input target image and the result image. Firstly, the algorithm extracts the structure information of the target image and style information of the reference image using the SIFT algorithm and generates the structure image and the pixel matching image. Then an initial result image is generated by local pixel mapping. Finally the initial result image is iteratively optimized by the gradient preserving algorithm. The experimental results show that our algorithm can not only optimize the structure inconsistency, but also effectively process image pairs with large color difference.

Keywords: SIFT feature match · Color correction · Region mapping · Gradient preservation

1 Introduction

Color is one of the important attributes of images. Due to the differences of shooting time and shooting device, images capturing the same scene or similar scenes may show color difference. When stitching images or reconstructing stereo images from multiple images, color inconsistency between images usually affects the quality of the result.

Color correction, also known as color transfer, color mapping, or image style transfer, is designed to eliminate the color difference between two similar images, or apply the style of the reference image to the target image, so that the target image may show the style of the reference image. Color corrections can be divided into global color correction [1, 7, 9] and local color correction [2, 8] algorithms based on whether the color mapping is performed globally or locally. The global color correction algorithm is currently the most common color correction algorithm. Global color correction

K. Arai et al. (Eds.): CompCom 2019, AISC 997, pp. 1001–1011, 2019.
https://doi.org/10.1007/978-3-030-22871-2_71

algorithms are usually simple, efficient, and have no structure inconsistency problems between the target image and the result image. The n-dimensional probability density color correction algorithm proposed by Pitié *et al.* [3] uses the continuous iteration of the one-dimensional probability density function. But the grain noise artifacts are introduced to the result due to the stretching transformation in the image convergence mapping. Then Pitié *et al.* proposed automated color grading using color distribution transfer (ACDT) [4] and further optimized the result images using the gradient preserving algorithm. The optimization can not only eliminate the noise, but also preserve the gradient of the resulting images.

Wang *et al.* proposed a local color correction algorithm based on image segmentation and SIFT feature matching (SSFM) [2]. The method corrects the color in the image region by region. But it has the problem of structural misalignment as shown in Fig. 1(d). Zheng *et al.* proposed a local color correction algorithm (MO) [5]. MO first combines the region mapping with the result of global correction [1] to obtain an initial result image. However, structure inconsistency and the noise problem usually appear in the initial result image. Then Zheng *et al.* smoothed the initial result image using the similarity of color information in adjacent regions. MO solves the noise problem of the image, but cannot solve the structure problem of the result image, as shown in Fig. 1 (e). Furthermore, MO also loses image details and reduces the fidelity of the images.

Fig. 1. Video images example of color correction.

In summary, the global color correction algorithm uses the same mapping function and mapping parameters in all regions. Different color differences in different the regions in the image are neglected. When the image color information is rich, error mapping may occur. The local color correction algorithm can overcome this problem. However, the local mapping may lead to the inconsistent structure of the image, as shown in the yellow rectangular regions of Fig. 1(d) and (e).

In this paper, we propose a structure consistent color correction algorithm based on gradient preservation (SCGP) for video images and stereoscopic images. Our algorithm is a local color correction algorithm. We first use the SIFT algorithm to extract the structure information of the target image and style information of the reference image. Then we generate the structure image and the pixel matching image using the SIFT feature. The structure image and the match image are used for local mapping to generate an initial result image. To solve the problem of image's detail loss and structure inconsistency, we optimize the initial result image using the gradient preservation [4, 10]. The proposed SCGP algorithm can not only achieve color correction, but also solve structure misalignment problem, as shown in Fig. 1(f).

2 Color Correction Based on Gradient Preservation

In this section, we mainly introduce the framework and specific steps of the proposed SCGP algorithm. Figure 2 shows the framework of our algorithm, which is mainly divided into three steps: (1) Feature Extraction; (2) Result Initialization; (3) Structure Optimization.

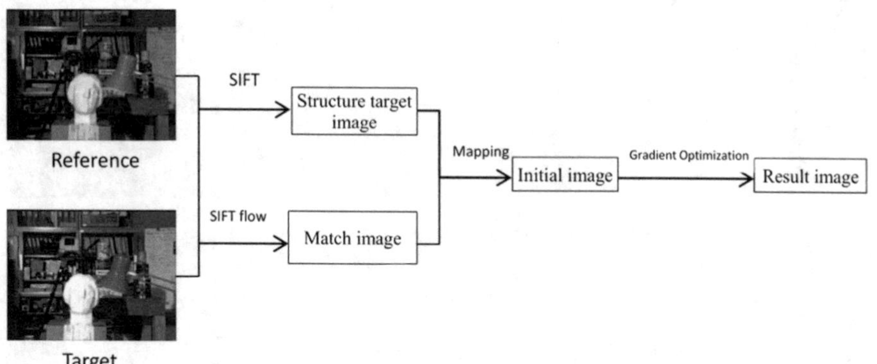

Fig. 2. Framework of the proposed SCGP algorithm.

2.1 Feature Extraction

Color correction result image needs to retain the structure feature of the target image and the color feature of the reference image. Therefore, we need to obtain the information of structure of the target image and the pixel information of the reference image. We obtain the structure image using SIFT feature shown in Fig. 3(h) and pixel matching image shown in Fig. 3(i).

Generate the structure image. The structure image generated not only has the identical structure with the target image, but also has a similar style with the reference image. The algorithm for generating structure images is inspired by the paper [2]. We also improve the segmentation and feature point addition methods to make the algorithm more efficient. The specific steps are as follows:

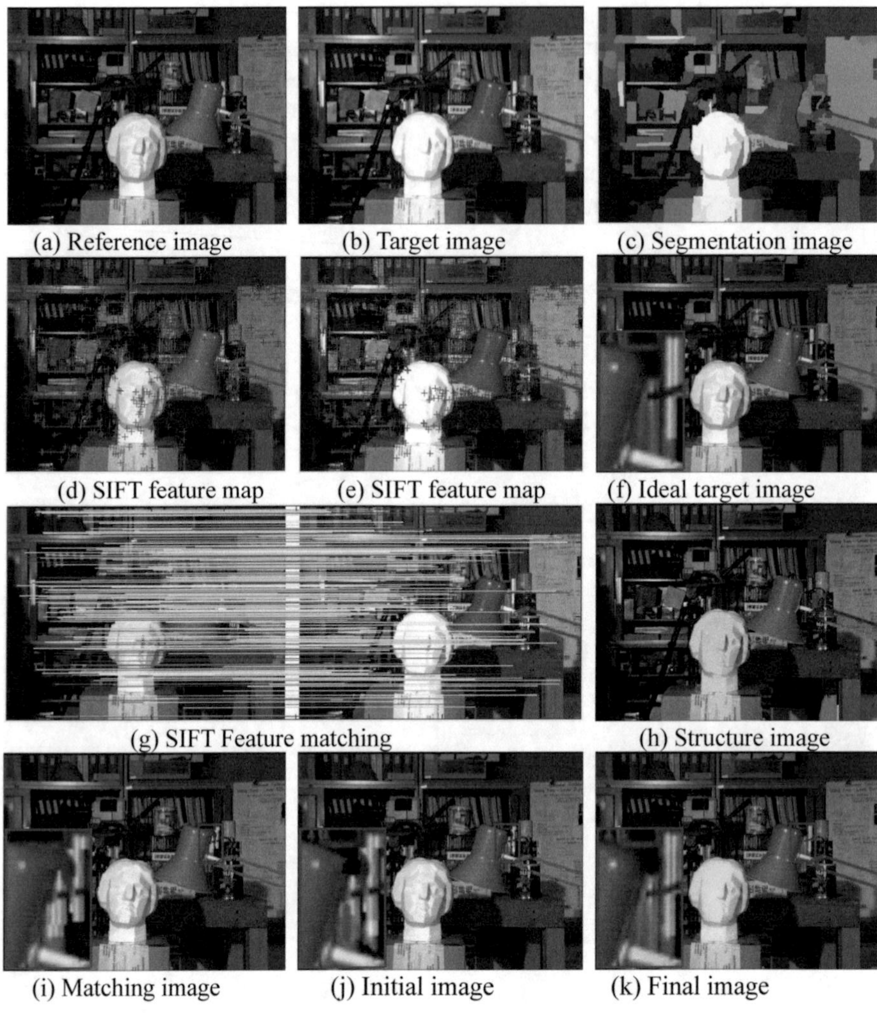

(a) Reference image (b) Target image (c) Segmentation image

(d) SIFT feature map (e) SIFT feature map (f) Ideal target image

(g) SIFT Feature matching (h) Structure image

(i) Matching image (j) Initial image (k) Final image

Fig. 3. Stereoscopic images example of the proposed SCGP algorithm.

(1) Firstly, we extract the SIFT feature points in the target image and the reference image separately. For example, in Fig. 3(d) and Fig. 3(e), the red dots in the feature maps represent the extracted SIFT feature points. Then the corresponding SIFT feature points of the two images are matched, as shown in Fig. 3(g).
(2) Secondly, we perform segmentation on the target image to allow color correction in each local region, as shown in Fig. 3(c).
(3) Thirdly, we perform an exhaustive search to find the region without feature points, and add feature points into each such region. The cross-division method is adopted to ensure that each segmentation region has at least 12 feature points. As shown in Fig. 4, we firstly add 12 feature points (red dots) to the region in the target image, and then add feature point of the corresponding region in the

reference image based on their relationship to the existing feature points (the black dots) closest to the region. This ensures that feature points of the reference image and the target image correspond to each other.

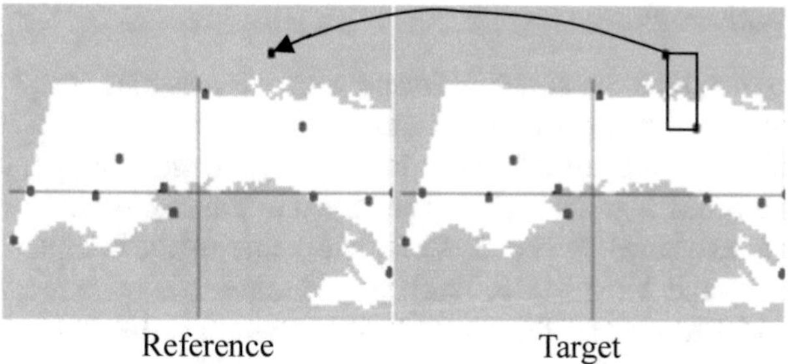

Fig. 4. Feature point addition diagram. Black dots represent existing feature points, and red dots represent the added feature points.

(4) Finally, calculating the mean values of the pixels in the neighborhood of 3×3 centered on the feature points, and then the difference of the pixels means between the reference image and the corresponding region in the target image is calculated. The mean of each color difference is added to the region corresponding to the target image, so as to get the structure image used to correct or fix the matching image's structure, as shown in Fig. 3(h). The mapping function can be expressed as follows:

$$I_f^j(x,y) = I_f^{j-1}(x,y) + \left(\sum_{\substack{(x_r,y_r) \in S_r^i \\ (x_t,y_t) \in S_t^i}} [M(x_r,y_r) - M(x_t,y_t)] \right) \times \frac{1}{k} \qquad (1)$$

where (x, y) is the spatial position in the image, $I_f(x, y)$ is the corrected pixel value at (x, y), j represents number of iterations, the initial value of $I_f^{j-1}(x, y)$ is $I_o(x, y)$ when $j = 1$, (x_r, y_r) is a SIFT feature point in the i-th segmentation region S_r^i of the reference image, (x_t, y_t) is a SIFT key point corresponding to (x_r, y_r) in the i-th segmentation region S_t^i of the target image, M represents the mean value of pixels in the neighborhood of 3×3, and K represents the number of feature points in the region S.

Generate matching image. Our final result image needs to preserve the style of the reference image, so we need the pixel information of the reference image. We use SIFT flow algorithm [6] to extract the feature of the reference image and the target image, and generate the matching image (that is, the pixel target image), as shown in Fig. 3(i). This image has both the color characteristics of the reference image and incomplete structure of the target image. However, the matching image has style information loss and the structure misalignment problems, as shown in the blue rectangular region of

Fig. 3(i). Finally we use the generated structure image and matching image to initialize the result image, so as to obtain the initial result image with complete structure and pixel information.

2.2 Result Initialization

In this step, we combine the structure image with the matching image using a confidence map to obtain the initial result image.

We calculate the structure difference image of the target image and the matching image using the SSIM algorithm [11]. The structure difference image can be called a confidence map, as shown in Fig. 5. The pixel value of the confidence map ranges between 0 (black) and 1 (white). A black pixel indicates a low structure consistency between the initial result and the target image. Therefore the structure of the black regions should be corrected. In contrast a white pixel shows a high structure consistency, so the original pixel value in the initial result should be retained. The result initialization formula is as follows:

Fig. 5. Confidence map

$$I_i(x, y) = \begin{cases} I_f(x, y)\, c(x, y) \le R \\ I_m(x, y)\, c(x, y) > R \end{cases}, \tag{2}$$

where I_i represents the mapped image, we call it the initial result image, as shown in Fig. 4. (10), $I_i(x, y)$ represents the pixel value corresponding to position (x, y), I_f represents the structure image, I_m represents the matching image, and $c(x, y)$ represents a confidence value of confidence map c, R is a threshold, we take 0.5.

2.3 Structure Optimization

After the above two steps, we have obtained an initial result image whose color is similar to the reference image, but the disadvantages of local color correction cannot be avoided. Moreover, because the confidence map is incomplete, the structure of the image may also have distortions that are recognizable by human eyes. In the section, we optimize the initial result image using the gradient preservation [4, 10] that is a reference optimization method.

The main idea of the gradient preservation method is to adjust the gradient of the initial result image to match the gradient of the target image [4]. Since the final result image is different from the target image in color, therefore the gradient matching should be a relaxation matching. The final condition for achieving gradient optimization is to find the optimal image J that minimizes the following integrals:

$$\min_{J^c} \iint_{\Omega} \phi \cdot \|\nabla J^c - \nabla I_t^c\|^2 + \varphi \cdot \|J^c - I_i^c\|^2 \, dxdy, \tag{3}$$

where the image is optimized in three channels in the RGB color space, c represents one of the three color channels. I_t^c represents the target image, Ω represents the entire image, and ∇ represents the gradient. The boundary condition is $\nabla J^c \mid_{\partial\Omega} = \nabla I_t^c \mid_{\partial\Omega}$, which is used to ensure that the gradients of J^c and I_t^c match at the boundary $\partial\Omega$. The first item $\|\nabla J^c - \nabla I_t^c\|^2$ is used to preserve the gradient of the image J^c, and the second item $\| J^c - I_i^c \|^2$ ensures that the color information of the final result image J^c is identical to the initial result image I_i^c. Weights ϕ and φ are used to balance the optimization of the gradient and color information. ϕ is used to adjust the gradient information at image boundary to make the edges in the image more realistic and the details of the image are preserved. φ is used to eliminate the distortion of the image structure. Weights ϕ and φ are defined as follows:

$$\phi(x,y) = \frac{30}{1 + 10 \| \nabla I_t^c \|}, \tag{4}$$

$$\varphi(x,y) = \begin{cases} 1 & \|\nabla I_t^c\| > 5 \\ \|\nabla I_t^c\|/5 & \|\nabla I_t^c\| \leq 5 \end{cases}, \tag{5}$$

where $\|\nabla I_t^c\| > 5$ means that in the region where the gradient is larger in the target image, we retain the pixels of the initial result, and $\|\nabla I_t^c\| \leq 5$ means that the region with a small gradient is slightly modified. And this makes the color information more natural and the image appears more realistic. The solution of Eq. (3) needs to satisfy the Euler-Lagrange equation [1] as follow:

$$\frac{\partial F}{\partial J^c} - \frac{d}{dx}\frac{\partial F}{\partial J_x^c} - \frac{d}{dx}\frac{\partial F}{\partial J_y^c} = 0, \tag{6}$$

where

$$F(J^c, \nabla J^c) = \phi \, ||\nabla J^c - \nabla I_t^c||^2 \, + \varphi \, ||J^c \, - I_i^c||^2. \tag{7}$$

3 Experiments

In this section, we show our experiment results. The video images and stereoscopic images required for the experiment are from the Middlebury website [19] and ICCD2015 dataset [12]. One of each pair of images is a reference image, as shown in Fig. 6(a) and the other is an original target image. The original target image can be called ideal target image, as shown in Fig. 6(c). Then we modify the hue, luminance, exposure, contrast and saturation of the ideal target image using Adobe Photoshop CS6. Finally, we get the target image with the same structure as the ideal target image, but large color difference with the reference image, as shown in Fig. 6(b). The color of the target images is modified up to 30%, as shown in Fig. 6(b) and Fig. 6(c). Therefore, all images we use in our experiment are color distorted images. Finally, we correct the color of the target image using the style of the reference image. The experimental results shown in Fig. 6(g) indicate that the proposed SCGP algorithm is effective. The final result images of the proposed SCGP algorithm can not only preserve the color of the reference image, but also optimize structure of the result image.

The proposed SCGP algorithm is a local color correction algorithm. We compare it with some state-of-the-art local color correction algorithms, including SSFM and MO. We also compare our method with the global color correction algorithm (ACDT), which is a top-ranking algorithm in the current color correction algorithm.

Figure 6 shows the experimental results of our method and several quality assessment methods, where (a) is the reference images, (b) is the target images, and (c) is the ideal target images. We mainly compare the experiment results of the four methods, including (d) SSFM [2], (e) MO [5], (f) ACDT [4] and (g) SCGP. Among the four methods, the SSFM algorithm (Fig. 6(d)) usually cannot correct the colors, especially in the regions marked using rectangles. And there are some structure problems in its result images.

Comparing MO (Fig. 6(e)) with SCGP (Fig. 6(g)), we find that the MO algorithm (Fig. 6(e)) usually has structure problems in the local regions, as shown in the blue rectangular regions of Fig. 6(e). The proposed SCGP algorithm can better maintain the structure, shown in the blue rectangular regions of Fig. 6(g).

Comparing the global color correction algorithms ACDT (Fig. 6(f)) and SCGP (Fig. 6(e)), we can see that the color in the results of the proposed SCGP algorithm is closer to the ideal target image (Fig. 6(c)), than that in the results of the ACDT algorithm, especially in the yellow rectangular regions.

As the experimental results show, the proposed SCGP algorithm can eliminate the color difference of the image well and does not cause structure problem. Compared to other algorithms, our experimental results are closer to the ideal target image. To further illustrate the effectiveness of our algorithm, we also perform an objective image quality assessment, as shown in Table 1.

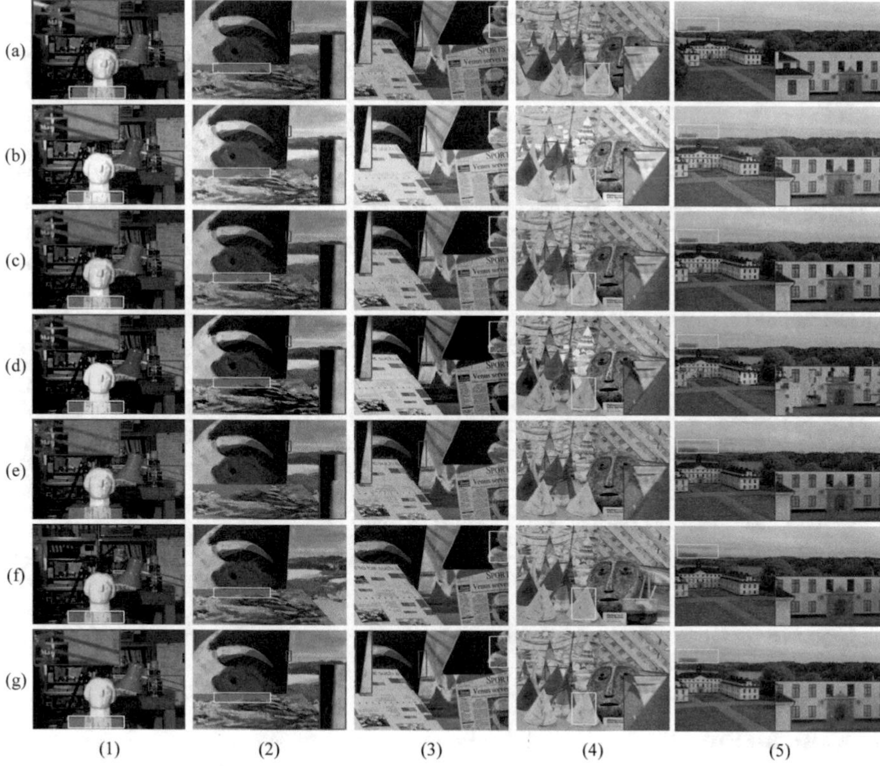

Fig. 6. Color correction results of the global and local algorithms: (a) Reference image, (b) Target image, (c) Ideal target image, (d) SSFM, (e) MO, (f) ACDT, (g) SCGP

We use quality assessment metrics such as structure similarity (SSIM) [11], mean square error (MSE), signal-to-noise ratio peak (PSNR), feature-based similarity index (FSIM) [13], visual significance induction index (VSI) [14], detection of distorted (MAD) [15], gradient similarity index (GSM) [16], color similarity index based on direction statistics (DSCSI) [17], improved color difference index (iCID) [18], UQI [20] and IFS [21]. We compare all experimental results with the ideal target image separately using that 11 quality assessment metrics. Among the 11 image quality assessment metrics, small CID, MSE, MAD values and large other values indicate good color correction results.

(1) ~ (5) in Table 1 correspond to (1) ~ (5) in Fig. 6 respectively. In Table 1, bold black numbers refer to the best performance value achieved by the four algorithms. As shown in Table 1, our algorithm is always almost ranked first among by 11 evaluation algorithms. Among these assessment metrics, GSM is a metric for detecting gradient similarity. Therefore GSM can better explain the effectiveness of our algorithm in structure preservation. In summary, the proposed SCGP algorithm is superior to other color correction algorithms in both subjective evaluation and objective data evaluation.

Table 1. Comparison results using 11 image quality assessments

		CID↓	SSIM↑	MSE↓	PSNR↑	VSI↑	FSIM↑	MAD↓	GSM↑	UQI↑	IFS↑	DSCSI↑
(1)	SSFM	0.4134	0.8222	188.73	25.372	0.9830	0.9600	2.6709	0.9935	0.8006	0.8959	0.6070
	MO	0.2607	0.9249	49.694	31.167	0.9849	0.9515	1.1895	0.9939	0.8415	0.9279	0.8309
	ACDT	0.1957	0.9631	30.449	33.295	0.9914	0.9742	1.3577	0.9965	0.9134	0.9380	0.7925
	SCGP	**0.1755**	**0.9647**	**25.850**	**34.006**	**0.9920**	**0.9751**	**1.1448**	**0.9966**	**0.9171**	**0.9775**	**0.8655**
(2)	SSFM	0.6016	0.8617	282.67	23.618	0.9771	0.9449	3.0474	0.9928	0.7461	0.9388	0.6267
	MO	0.3994	0.9181	32.254	33.044	0.9862	0.9569	**1.1861**	0.9953	0.7562	0.9638	0.8769
	ACDT	0.3534	0.9224	28.190	33.629	0.9868	0.9594	1.2643	0.9959	0.7964	0.9768	0.8754
	SCGP	**0.3428**	**0.9245**	**26.135**	**33.958**	**0.9875**	**0.9605**	1.1966	**0.9960**	**0.7990**	**0.9827**	**0.8994**
(3)	SSFM	0.7170	0.8028	433.16	21.764	0.9717	0.9614	2.9784	0.9933	0.7398	0.8679	0.3748
	MO	0.4613	0.8929	80.409	29.078	0.9805	0.9398	1.4506	0.9926	0.6773	0.9241	0.7686
	ACDT	0.3095	0.9544	40.159	32.093	0.9902	0.9742	1.2811	0.9967	0.8556	0.9716	0.7575
	SCGP	**0.2636**	**0.9567**	**26.319**	**33.928**	**0.9912**	**0.9752**	**1.2215**	**0.9969**	**0.8579**	**0.9756**	**0.8167**
(4)	SSFM	0.6785	0.9096	354.27	22.637	0.9714	0.9509	3.7380	0.9924	0.8844	0.9069	0.3308
	MO	0.3436	0.8877	86.608	28.755	0.9799	0.9351	**1.5831**	0.9933	0.8621	0.9345	0.7832
	ACDT	0.2714	0.9522	**44.981**	**31.600**	0.9880	0.9680	1.6639	0.9964	0.9326	0.8975	0.7190
	SCGP	**0.2492**	**0.9528**	49.935	31.147	**0.9891**	**0.9688**	1.6510	**0.9965**	**0.9351**	**0.9562**	**0.8154**
(5)	SSFM	0.3035	0.9668	103.17	27.995	0.9910	0.9825	**1.8537**	0.9965	**0.8557**	0.9619	0.7026
	MO	0.2491	0.9817	54.843	30.739	0.9955	0.9861	2.2183	0.9978	0.8503	0.9742	0.8793
	ACDT	**0.2490**	**0.9817**	54.844	30.740	0.9955	0.9861	2.2181	0.9979	0.8504	0.9742	0.8793
	SCGP	0.2930	0.9623	**21.845**	**34.737**	**0.9968**	**0.9867**	2.1777	**0.9981**	0.7803	**0.9800**	**0.9032**

4 Conclusion

In this paper, we propose a structure consistent color correction algorithm for video images and stereoscopic images based on gradient preservation. This algorithm mainly solves the structure inconsistency problems of images and the color problems of the regions. The algorithm combines gradient optimization, feature extraction and region mapping, and corrects the approximate region using local mapping. It is a local color correction algorithm. The experimental results validate that our algorithm is effective and avoids the structure problems of the local color correction algorithm. Moreover, compared with the global color correction algorithm, our algorithm can also solve the problem that the global algorithm cannot handle some local regions in color rich images.

Acknowledgments. The presented research work is supported by the National Natural Science Foundation of China under Grant 61672158, Grant 61671152, and Grant 61502105, in part by the Fujian Natural Science Funds for Distinguished Young Scholar under Grant 2015J06014, in part by the Technology Guidance Project of Fujian Province under Grant No. 2017H0015, and in part by the Fujian Collaborative Innovation Center for Big Data Application in government.

References

1. Reinhard, E., Ashikhmin, M., Gooch, B., Shirley, P.: Color transfer between images. IEEE Comput. Graphics **21**(5), 34–41 (2002)
2. Wang, Q., Yan, P., Yuan, Y., Li, X.: Robust color correction in stereo vision. In: IEEE International Conference on Image Processing, pp. 965–968 (2011)
3. Pitie, F., Kokaram, A.C., Dahyot, R.: N-dimensional probability density function transfer and its application to colour transfer. In: IEEE International Conference on Computer Vision, pp. 1434–1439 (2005)
4. Piti, F., Kokaram, A.C., Dahyot, R.: Automated colour grading using colour distribution transfer. Comput Vis. Image Underst. **107**(1C2), 123–137 (2007)
5. Zheng, X., Niu, Y., Chen, J., Chen, Y.: Color correction for stereo-scopic image based on matching and optimization. In International Conference on 3D Immersion, pp. 1–8 (2017)
6. Liu, C., Yuen, J., Torralba, A.: Sift flow: dense correspondence across scenes and its applications. IEEE Trans. Pattern Anal. **33**(5), 978–994 (2011)
7. Xiao, X., Ma, L.: Color transfer in correlated color space. In: International Conference on Virtual Reality Continuum, pp. 305–309 (2006)
8. Zhang, M., Georganas, N.D.: Fast color correction using principal regions mapping in different color spaces. Real-Time Imaging **10**(1), 23–30 (2004)
9. Fecker, U., Barkowsky, M., Kaup, A.: Histogram-based prefiltering for luminance and chrominance compensation of multiview video. IEEE Trans. on Circuits Syst. Video Technol. **18**(9), 1258–1267 (2008)
10. Xiao, X., Ma, L.: Gradient-Preserving Color Transfer. Comput. Graphics Forum **28**, 1879–1886 (2009). https://doi.org/10.1111/j.1467-8659.2009.01566.x
11. Wang, Z., Bovik, A.C., Sheikh, H.R., Simoncelli, E.P.: Image quality assessment: from error visibility to structural similarity. IEEE Trans. Image Process. **13**(4), 600–612 (2004)
12. Niu, Y., Zhang, H., Guo, W., Ji, R.: Image quality assessment for color correction based on color contrast similarity and color value difference. IEEE Trans. on Circuits Syst. Video Technol. **28**(4), 849–862 (2018)
13. Zhang, D.: Fsim: A feature similarity index for image quality assessment. IEEE Trans. Image Processing A Publication of the IEEE Signal Processing Society **20**(8), 2378 (2011)
14. Zhang, L., Shen, Y., Li, H.: Vsi: a visual saliency-induced index for perceptual image quality assessment. IEEE Trans. Image Process. **23**(10), 4270–4281 (2014)
15. Larson, E.C., Chandler, D.M.: Most apparent distortion: full-reference image quality assessment and the role of strategy. J. Electron. Imaging **19**(1), 011006 (2010)
16. Liu, A., Lin, W., Narwaria, M.: Image quality assessment based on gradient similarity. IEEE Trans. Image Process. **21**(4), 1500 (2012)
17. Lee, D., Plataniotis, K.N.: Towards a full-reference quality assessment for color images using directional statistics. IEEE Trans. Image Process. **24**(11), 3950–3965 (2015)
18. Preiss, J., Fernandes, F., Urban, P.: Color-image quality assessment: from prediction to optimization. IEEE Trans. Image Process. **23**(3), 1366–1378 (2014)
19. http://vision.middlebury.edu/stereo/data/
20. Wang, Z., Bovik, A.C.: A universal image quality index. IEEE Signal Process. Lett. **9**(3), 81–84 (2002)
21. Chang, H.-W., Zhang, Q.-W., Wu, Q.-G., Gan, Y.: Perceptual image quality assessment by independent feature detector. Neurocomputing **151**, 1142–1152, March (2015)

Exact NMF on Single Images via Reordering of Pixel Entries Using Patches

Richard M. Charles(✉) and James H. Curry

Department of Applied Mathematics, University of Colorado at Boulder, Box 526 UCB, Boulder, CO 80309-0526, USA
{Richard.Charles,James.H.Curry}@colorado.edu

Abstract. Non-negative Matrix Factorization (NMF) has been shown to be effective in providing low-rank, parts-based approximations to canonical datasets comprised of non-negative matrices. The approach involves the factorization of the non-negative matrix A into the product of two non-negative matrices W and H, where the columns of W serve as a set of dictionary vectors for approximating the matrix A. One drawback to this approach is the lack of an exact solution since the problem is not convex in both W and H simultaneously. Previous authors have shown that an exact solution can be achieved by using datasets with specified properties. In this paper we propose a factorial dataset for the use of NMF on patches of a single image. We show that when the multiplicative update is applied to a single image, we are successful in achieving a set of standard basis vectors for the image. We show that by reordering the patches of a specified dataset, the algorithm is successful in achieving exact approximations of single images while preserving the number of standard basis vectors. We use Mean Squared Error (MSE), Peak Signal-to-Noise Ratio (PSNR) and Mean Structured Similarity Index (MSSIM) as measures of the quality of the low rank approximations for a given rank k.

Keywords: Patches · Singular Value Decomposition (SVD) · Non-negative Matrix Factorization (NMF) · Peak Signal-to-Noise Ratio (PSNR) · Mean Structured Similarity Index (MSSIM)

1 Introduction

The creation of non-negative matrices formed from pixel intensities associated with natural images is often the first step when there is a need to analyze or process images. Depending on the application, these techniques are applied directly to the matrix of pixel intensities. Often times, applications involving the compression of datasets require low-rank approximations in the form of basis vectors associated with the datasets. Non-negative Matrix Factorization (NMF) provides one such approach and was first introduced by Paatero and Tapper in

K. Arai et al. (Eds.): CompCom 2019, AISC 997, pp. 1012–1026, 2019.
https://doi.org/10.1007/978-3-030-22871-2_72

[13] and subsequently made popular by Lee and Seung [11]. The problem that Non-negative Matrix Factorization (NMF) seeks to solve is the following:

$$A \approx WH \tag{1}$$

where $A \in \mathbb{R}^{n \times m}$, $A \geq 0$, $W \in \mathbb{R}^{n \times k}$, $H \in \mathbb{R}^{k \times m}$ and $W \geq 0$, and $H \geq 0$. The original algorithm proposed in [11] was applied to a matrix where each column vector represented a vectorized version of an image within the dataset. The authors were successful in achieving a parts-based, low-rank approximation yielding a set of sparse vectors in W.

Non-negative matrix factorization yields a rank-k approximation to a matrix, similar to the Singular Value Decomposition (SVD). In [4] we observe that the squared-error of the approximation produced by NMF will generally be larger than the squared-error of the approximation produced by the SVD. This is consistent with the Eckart-Young theorem, which asserts that the approximation to A obtained by truncating the SVD of A is the rank-k approximation with minimal squared-error [19]. Unlike the SVD, NMF has the additional features of creating a parts-based decomposition and preserving the sign structure of the original matrix. These features follow from the fact that the matrices consists of only non-negative entries, and that the NMF factorization produces a matrix product composed of non-negative weights and non-negative components of the image [9]. Nevertheless, we consider the NMF for the following three reasons: (1) it is capable of providing a low-rank approximation of an image, (2) it produces a *parts-based decomposition* of the whole image, and (3) it preserves the sign-structure of the original dataset.

In contrast, the use of *patches* of images in conjunction with Non-negative Matrix Factorization (NMF) as an alternative to achieving improved and computationally efficient, low-rank approximations was first proposed by the authors in [4]. The approach reorders the non-negative entries of a matrix based on patches. Thus, the NMF algorithm is applied to a matrix whose columns consists of vectorized versions of patches within a single image. Applications that have benefited from patch-based algorithms include denoising [3], feature extraction [15,16], and compression [1,14].

If we regard an intensity image as an $n \times m$ matrix, A, then a patch is equivalent to a submatrix of A. Typically, algorithms that are based on submatrices, are designed primarily to parallelize or optimize computations of matrix operations [19]. In contrast, patch-based algorithms are motivated by the fact that the set of patches provides a mechanism to emphasize and uncover structure in the local correlation of pixels and repeating patterns/textures that are characteristic of natural images.

In this paper, we present and analyze a factorial dataset to be used in conjunction with NMF on patches for factorizing a matrix and yielding an exact factorization with basis vectors akin to the standard Euclidean basis vectors. More precisely, we *factorize* a reordered version of the image derived from patches rather than factorize the original image itself. We are motivated by determining a set of Euclidean basis vectors of a non-negative matrix formed by pixel intensities of a single image. We believe that through this approach, numerical schemes may be developed which enable more efficient strategies for signal compression.

The outline of this paper is as follows: Sect. 2 will review the premise for using NMF in order to achieve a parts-based decomposition on a library of images. We use the swimmer dataset of images used in the paper by the authors of [6] in order to verify their results using our standard NMF algorithm. Section 3 describes the NMF Multiplicative Update algorithm used in [11]. Section 4 describes the theory behind the approach taken by [6] and its application to a library of images. In Sect. 5 we then propose a new technique for creating datasets from single images using patches and extend the theory proposed for the swimmer dataset to patches. In addition, Sect. 5 outlines the duality theory that is necessary for identifying the rules for exact patches. In Sect. 6 we construct three datasets for patches that achieves exact factorization into component parts. Section 7 discusses the results of the application of NMFP on the newly developed datasets.

2 Exact Parts-Based Decompositions of NMF

It has been shown by the authors in [6] that when the multiplicative form of NMF is applied to a dataset with certain properties, an exact decomposition into the component parts is achieved. The authors explored the details of these datasets and created a dataset that achieved an exact decomposition. This was a significant result since prior to their investigations, many thought that this was not achievable. Others [8] have since shown that an exact decomposition is achievable by means of including the *extreme vectors* as column vectors of the original matrix A. This was referred to as the Extreme Vector Property (EVP) and is attributed to matrices which include the vectors which define the cone of minimum volume for the dataset. Heuristic approaches to exact NMF by leveraging simulated annealing and greedy randomized adaptive search procedures have also been proposed in [17].

We explore the details of the method in [6] used to construct these datasets and recall that the original authors of NMF in [11] were able to achieve a parts-based representation for libraries of images with a reasonable amount of effort. The significant advantage in the use of NMF comes from the fact that intuitively, the sense of adding features of images is constructive and can provide a basis for including variation in datasets simply by adding these components. In contrast, factorization methods which are not parts-based and constructive do not reflect the additive manner in which the Human Visual System (HVS) interprets the world [2]. An example of such a system includes the SVD and Wavelet compression algorithms. Often when SVD is used for the purpose of compressing images, low-rank approximations to the datasets typically take on a global appearance involving matrices with many negative entries, thus not sign structure preserving.

The authors of [11] worked with images from the Center for Biological and Computational Learning (CBCL) facial dataset, where canonical images of faces were used for their trials. The dataset contained black and white images of faces which were 19×19 pixels in dimension. The authors vectorized each image into a column vector using lexicographic ordering. Each image of a face within the dataset had set locations for eyes, noses, mouths, etc. This was an important attribute in the original paper since subsequent papers [7] showed that if the images included faces that were not aligned in some manner, the parts-based representation and the sparsity of the matrix would be lost. In the section that follows, we outline the algorithm used in the original application of NMF.

3 NMF Multiplicative Update

In this section we discuss the NMF multiplicative update algorithm originally proposed by the authors in [11]. Using the technique of Lagrange multipliers to impose the positivity constraints, the algorithms developed by Lee and Seung using the Euclidean distance cost function requires the following updates:

$$H_{a\mu} \leftarrow H_{a\mu}\frac{(W^T A)_{a\mu}}{(W^T W H)_{a\mu}} \tag{2}$$

$$W_{ia} \leftarrow W_{ia}\frac{(AH^T)_{ia}}{(WHH^T)_{ia}} \tag{3}$$

However, the formulas for the updates of the H and W matrices using the divergence as a cost function, were as follows:

$$H_{a\mu} \leftarrow H_{a\mu}\frac{\sum_i W_{ia}A_{i\mu}/(WH)_{i\mu}}{\sum_k W_{ka}} \tag{4}$$

$$W_{ia} \leftarrow W_{ia}\frac{\sum_\mu H_{a\mu}A_{i\mu}/(WH)_{i\mu}}{\sum_\nu H_{a\nu}} \tag{5}$$

The authors compared multiplicative updates to that of additive updates such as found in gradient descent methods. For example,

$$H_{a\nu} \leftarrow H_{a\nu} + \eta_{a\nu}[(W^T A)_{a\mu} - (W^T W H)_{a\mu}] \tag{6}$$

by rescaling

$$\eta_{a\mu} = \frac{H_{a\mu}}{(W^T W H)_{a\mu}}, \tag{7}$$

we obtain the rule in Eq. 7.

For the divergence cost function, a simple substitution into the following additive update yields

$$H_{a\mu} \leftarrow H_{a\mu} + \eta_{a\mu}[\sum_i W_{ia}\frac{A_{i\mu}}{(WH)_{i\mu}} - \sum_i W_{ia}] \tag{8}$$

with

$$\eta_{a\mu} = \frac{H_{a\mu}}{\sum_i W_{ia}} \tag{9}$$

as a substitution that results in equation (2.13).

Although the method can be interpreted as having a specific additive update and associated step size, the authors in [11] expressed the distinct advantage of the multiplicative update approach. They claim that the multiplicative update reduces the need for projections into the non-negative orthant since it is a simple update to W and H and their entries are never negative. This works well provided that an entry does not go to zero as every successive iterate will remain zero in that entry. It should be noted that an NMF factorization can be achieved by other means: for example, multiplicative or additive gradient descent [10], 2nd order Newton [20], exponentiated gradient [5], and projected gradient [12]. These methods are beyond the scope of this paper. However, understanding the geometric interpretation of the factorization of a non-negative matrix will provide some insight into the approach used by the authors in [6]. We share this interpretation in the following section.

4 NMF Parts-Based Representation Generative Dataset

Donoho and Stodden [6] used a geometric interpretation of the NMF algorithm of Lee and Seung described above in order to develop the theory for the conditions for a given dataset to offer a parts-based decomposition using NMF. Specifically, the two questions that were posed were the following:

1. Under what assumptions is the notion of non-negative matrix factorization well defined, for example is the factorization in some sense *unique*?
2. Under what assumptions is the factorization *correct*, recovering the 'right answer'?

The authors gave examples of synthetic image articulation databases with specified conditions. They defined a series of black-and-white images with L parts, each having M articulations. Specifically, in order for NMF to factor a dataset of images exactly into its component parts, they defined a separable factorial articulation family, described below. We begin with a geometric interpretation of the factorization of a non-negative matrix which includes the concept of a simplicial cone.

Definition 1. *The* ***simplicial cone*** *generated by vectors* $\Phi = (\phi_j)_{j=1}^r$ *is*

$$\Gamma = \Gamma_\Phi = \{x : x = \sum_j \alpha_j \phi_j, \alpha_j \geq 0\} \tag{10}$$

For non-negative datasets such as images, the simplicial cones generated by the factorization all exist in the positive orthant. In addition, the factorization of A is not unique. As long as there exists a simplicial cone which contains the set of points created by the columns of W, it can serve as a solution to the factorization problem. This poses a significant challenge when seeking an exact solution. We now define what is meant by a simplicial hull.

Definition 2. *The **simplicial hull** generated by vectors $\Phi = (\phi_j)_{j=1}^{r}$ is simplicial cone of minimum volume containing the set of points created by the columns of W.*

The authors of [6] created a 'swimmer' dataset and showed that the following conditions to be necessary and sufficient for an exact solution to exist.

Definition 3. *A **Separable Factorial Articulation Family** is a collection X of points x obeying the following three conditions:*

R1. Generative Model. *Each image x in the dataset has a representation*

$$x = \sum_{q=1}^{L} \sum_{a=1}^{M} \alpha_{q,a} \psi_{q,a} \tag{11}$$

where the generators $\psi_{q,a} \in \Re^{L}$ obey the non-negativity constraint $\psi_{q,a} \geq 0$ along with the coefficients $\alpha_{q,a} \geq 0$.

R2. Separability. *For each q, a there exists a pixel $k_{q,a}$ such that*

$$\psi_{q',a'}(k_{q,a}) = 1_{\{a=a',q=q'\}} \tag{12}$$

i.e. Each part/articulation pair's presence or absence in the image is indicated by a certain pixel associated to that pair.

R3. Complete Factorial Sampling. *The dataset contains all M^L images in which the L parts appear in all combinations of M articulations.*

A few of the swimmer dataset images are shown below in Fig. 1. The figure illustrates the articulations that are possible with a limb of the stick figure. We can summarize these three characteristics of a dataset as follows:

1. each image represents a point in a high dimensional space with an associated vector and as such, linear combinations of these vectors will allow us to create any articulated image.
2. Each part's presence in an articulated image is uniquely defined by a single pixel. For example, if a swimmer body part is present, all other articulations of that same body part take on the value zero.
3. Also, every possible combination of body part articulations needs to be in the dataset.

Fig. 1. This figure illustrates 8 of the 32×32 images used in the swimmer dataset. The dataset is comprised of a stick figure with 4 limbs. Each limb has 4 possible articulations, therefore there are 256 possible combinations of limb articulations of the swimmer.

5 Separable Factorial Articulation Dataset on Images with Patches

One of the more interesting developments in NMF has been the development of algorithms on datasets which include vectors which define the simplicial hull of the dataset. Datasets satisfying this requirement are said to satisfy the Extreme Vector Property. This property was discussed in detail by Klingenberg, et al. in [8] and provides a methodology for getting an exact decomposition of a dataset. If a dataset has the extreme vector property, it contains the column vectors associated with the simplicial hull as part of the dataset. We've stated that the problem that is posed by the NMF algorithm is equivalent to finding the simplicial hull generated by the r column vectors of W when factoring a matrix A into two non-negative matrices W and H. This approach we will refer to as the Primal-Simplicial-Cone problem. Using constructs from convex duality theory, we can also state the problem as finding the simplicial hull generated by the r row vectors of W^*. The latter approach we will refer to as the Dual-Simplicial-Cone problem. Our work in this section will utilize the following definitions and lemmas in constructing a new theorem for generating factorial images using NMF on patches. We begin by defining the conical hull of a set of vectors.

Definition 4. *Given a point set (x^i), its* ***conical hull*** *is the simplicial hull generated by the vectors (x^i) themselves. We can denote this as $(x^i)^G$.*

Definition 5. *Given any square matrix A, whose dimension is greater than 1, we define the* ***component patch*** *of a matrix generated by pixel intensities to be those contiguous, non-overlapping sub-matrices of equal dimension $p \times p$, whose union results in the original matrix A.*

Lemma 1. *Given a $p \times p$ patch W^P of the matrix W, the p columns of patch W^p have a representation in $\mathbb{R}^p$. This representation defines the conical hull for the column space of the matrix patch W^P.*

Proof. Definition of the column space.

Lemma 2. *Given a $p \times p$ matrix patch W^P, the p rows of matrix W^{P^*} have a representation in $\mathbb{R}^p$. This representation defines the conical hull for the row space of the matrix patch.*

Proof. Definition of row space

Lemma 3. *Every square matrix in $\mathbb{R}^{n \times n}$, with dimension $n > 1$ can be partitioned into component patches, whose union forms the original matrix.*

Proof. Easily verified

We now use these lemmas in order to construct a theorem for the application of NMF on patches in order to achieve an exact factorization.

Definition 6. *A* ***Separable Factorial Articulation Family on Patches*** *is a collection Y of points P_i obeying the following three conditions:*

R1. Generative Model. *Each image patch P_i in the dataset has a representation*

$$P_i = \sum_{q=1}^{L_i} \sum_{a=1}^{M_i} \beta_{q,a} \phi_{q,a} \tag{13}$$

where the generators $\phi_{q,a} \in \Re^L$ obey the non-negativity constraint $\phi_{q,a} \geq 0$ along with the coefficients $\beta_{q,a} \geq 0$. Here L_i represents the parts (limbs) and M_i represents the articulations (moves) of the limbs of a single patch.

R2. Separability. *For each q, a there exists a pixel $k_{q,a}$ such that*

$$\phi_{q',a'}(k_{q,a}) = 1_{\{a=a',q=q'\}} \tag{14}$$

i.e. Each Limb/Mode pair's presence or absence in the image is indicated by a certain pixel associated to that pair.

R3. Complete Factorial Sampling. *The dataset contains all M^L patches within the image in which the L parts appear in all combinations of M articulations.*

Conjecture 1. Given an image whose component patches obey the rules [R1]-[R3], there is a unique simplicial hull for each patch with $p = M \cdot L$ generators which contains all of the points of the patch, and is contained in $L \cap V$, where V is a subspace such that $V \subset \mathbb{R}^L$.

The proof follows directly from the results in [6]. We need to show that we can combine patches from lower dimensional subspaces to create a higher dimensional subspace image and that simplicial cones from a lower dimensional subspace can be mapped to a higher dimensional subspace, while preserving volume and its extreme vectors. The latter part is much more challenging since there is no guarantee that this would occur. It is important to also note that the reordering of the entries of a matrix in order to apply a computational method is not rank-preserving. Thus, the ranks of matrices that have been reordered based on patches, are often full. In the following section we apply the NMF multiplicative update algorithm on our proposed dataset with the properties outlined above.

6 A Single Image Factorial Dataset

Donoho and Stodden [6] created an image database which resulted in basis vectors that were orthogonal and achieved an exact parts-based representation. Their dataset was designed to be a library of black and white images with a pixel dimension of 32×32. Each image represented a stick figure with four body parts P having four articulations. Inspired by this methodology, we propose the creation of a new class of single image factorial datasets utilizing NMF applied to patches of the image. The image has dimension 36×36 pixels and we describe the image as a representation of a turtle. The turtle is comprised of a 4 pixel square body and 4 limbs. Each limb has 3 possible articulations. A single patch mask is shown in Fig. 2 with its corresponding image in Fig. 3. A factorial of the articulations of this image was used to generate a single 36×36 image which is shown in Fig. 4. The images were constructed by mapping limb articulations to a number set provided by the mask. For example, the first patch in the image is set $\{0, 0, 0, 0\}$. Each entry in the set can have values ranging from 0 to 2. Each value represents each of the articulations of a limb where 0 represents the first position and '1' and '2' represent the subsequent positions for a clockwise orientation. The remaining patches are generated using the permutations of the four element dataset. The image constructed can provide recognizable structure which depends on the sequence of patches used for the construction. Figure 2 shows an example of the bit mask used to generate the patches within the single image, while Fig. 3 shows an example of one of the patches within the dataset that gets generated. It is important to note that the four pixels which represent the body of the turtle, remain constant in each image.

1	2	0	1
0			2
2			0
1	0	2	1

Fig. 2. This figure illustrates the bit mask used for each patch in order to create the factorial image. The image is comprised 36×36 pixels, with 4×4 patches. Each single image limb of the turtle has 3 articulations. The first entry of the mask begins in the upper right hand corner with subsequent entries in a clockwise rotation.

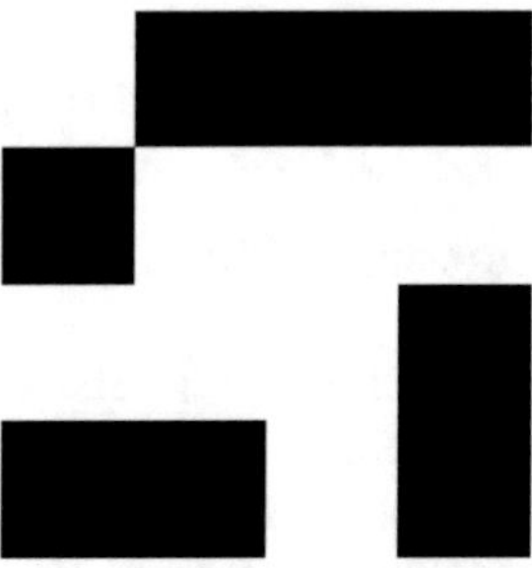

Fig. 3. This figure illustrates a patch in the single-image dataset. The image is comprised of a 4 pixel square body and 4 limbs, each with 3 articulations. A factorial of the articulations of this image was used to generate a single 36×36 image. This image shows the patch configuration for the mask {2, 2, 2, 1}.

Each of the patches that are generated as a result of the bit-allocation strategy is then combined to form a single image. In effect, the image becomes a quilt of patches of the turtle images. As a result of this strategy, the turtle dataset represents a factorial representation of the articulations of the turtle. In Fig. 4, we show an example of the type of image that can be generated using the approach. The image reflects the use of all permutations of parts, P and articulations A through the bitmask. It is important to note that this is one version of the image which can be formed using the Turtle dataset. In fact, any shuffling of the 81 distinct articulations will yield the exact decomposition. Clearly, these matrices can have varying ranks. Note that the rank of the Turtle dataset as shown in Fig. 4 is 6. Whereas, other examples of the types of images generated using the image dataset are illustrated in Fig. 5 have ranks of 36.

Fig. 4. This figure illustrates the single-image dataset created from factorial articulations of the Turtle dataset. The image is comprised of 36×36 pixels. Each 4×4 pixel patch contains a 2×2 pixel turtle body with 4 limbs, each with 3 articulations.

Fig. 5. This figure illustrates two single-image dataset variations created from factorial articulations of the Turtle dataset.

7 Results of the Single Image Factorial Dataset

The results of the application of NMFP on the single image factorial dataset indicate that we can develop a factorial dataset of image articulations on various scales that would yield the standard basis vectors for patches of the image. This scheme also requires some similarity within each of the patches. In this case, the four pixels comprising the body of the 'Turtle Dataset' serves the purpose. The resulting black and white image clearly illustrates this point. Figure 6 shows the standard basis vectors that are retrieved by the algorithm. As in the case of the swimmer dataset, portions of the image that are consistent in all articulations remain as a part of the basis set. We observe that each of the limbs and their articulations are represented. In effect, the 12 vectors serve as basis-vectors for the dataset. Using NMFP in conjunction with the Turtle Dataset yielded a set of vectors which defined all possible configurations of limbs and articulations possible. Also note that the four pixels representing the body of the Turtle is visible in each of the 12 vectors of the dictionary (Fig. 7).

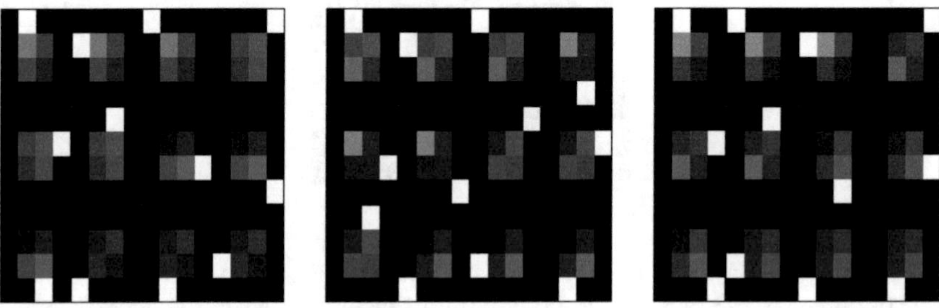

Fig. 6. This figure illustrates the standard basis created from factorial articulations of the turtle patch. The image is comprised of all 12 possible articulations of the individual limbs. Note that the Euclidean basis vectors are clearly present in each image along with a 4×4 body of the turtle.

7.1 MSSIM Results

We also use the Structural Similiarity Index (SSIM) [18] as an objective measure of how well the approximated image agrees with the original image. The SSIM is a well known alternative to providing an assessment of the preserved structures within an image as opposed to the MSE, which determines agreement of two images at the level of individual pixels. The SSIM separates the task of similarity measurement into three components: luminance, structure and contrast. These components are well defined and are combined in order to provide information about the structure of an image based on patches. The SSIM equation is as follows:

$$SSIM(x,y) = \frac{(2\mu_x\mu_y + C_1)(2\sigma_{xy} + C_2)}{(\mu_x^2 + \mu_y^2 + C_1)(\sigma_x^2 + \sigma_y^2 + C_2)} \tag{15}$$

Fig. 7. This figure illustrates the low rank approximations using NMF with patches on the datasets created from factorial articulations of the turtle patch.

In Eq. (15), μ_x represents the mean intensity for a path from image 'x', and σ_x represents the normalized unbiasd estimate of the intensities. In short, the estimate of the SSIM of two images satisfy the following conditions:

1. Symmetry: $S(x, y) = S(y, x)$;
2. Boundedness: $S(x, y) \leq 1$;
3. Unique maximum: $S(x, y) = 1$ if and only if $\mathbf{x} = \mathbf{y}$ ($x_i = y_i$, for all $i = 1, 2, ..., N$);

In Table 1 we show the mean SSIM values for the test images when the standard SVD and NMF algorithms are applied using $T_k(A)$ and $Q_{\hat{k},p}$. The mean SSIM (MSSIM) is calculated for the purpose of achieving an overall quality measure for an image. We define the MSSIM as follows:

$$MSSIM(\mathbf{X}, \mathbf{Y}) = \frac{1}{M} \sum_{j=1}^{M} SSIM(\mathbf{x}_j, \mathbf{y}_j), \tag{16}$$

where $\mathbf{X}$ and $\mathbf{Y}$ are the original and approximated images, respectively; $\mathbf{x}_j$ and $\mathbf{y}_j$ are the image contents at the j-th window; and M is the number of windows in the image. We define the local window to be a window where local statistics are measured. In this application, we use a two-dimensional Gaussian lowpass filter with a standard deviation of 1.5 pixels.

Table 1. Mean SSIM of relevant images is shown. Using a patch size of $p = 16$, images designated with 'p' indicate our proposed patch-based algorithm. Two identical images would have a structural similarity index of 1. Take special note of the mean SSIM scores for the Einstein and Clown standard and patch-based algorithms for k = 8.

Image	NMF			
	k = 4	k = 8	k = 12	k = 16
Turtle1	0.5873	0.6787	0.7892	0.5709
Turtle1-p	0.6682	0.7696	0.8583	0.6169
Turtle2	0.5576	0.6579	0.7671	0.5608
Turtle2-p	0.6421	0.7259	0.8138	0.5739
Turtle3	0.6859	0.7618	0.8647	0.6563
Turtle3-p	0.6917	0.7971	0.8889	0.6450

7.2 Resulting Conjecture

As a result of the performance of the algorithm, we can now provide the following conjecture:

Conjecture 2 (Separable Factorial Generative Set). Given an image A generated by pixel intensities, and a partitioning strategy for creating non-overlapping contiguous patches P, $\exists$ a Separable Factorial Generative Set with L parts and M articulations, such that application of the NMFP algorithm on the set P, yields an exact decomposition yielding a standard basis in $\mathbb{R}^{LM}$.

We were successful in generating a dictionary of elements which is not dissimilar to the standard Euclidean bases. This has significant implications for the use of NMF in solving equations in which the $Ax = b$ when A^{-1} is difficult to determine. In effect, one can use this strategy to transform the problem to the following:

$$WHx = b \tag{17}$$

By creating a generative set on patches, the problem becomes:

$$PIHx = b, \tag{18}$$

where P represents a permutation matrix and I, the identity matrix. This, finally yielding:

$$Hx = IP^{-1}b$$

If the matrix H is sparse, we now have a simpler solution for x, namely $x = (PH)^{-1}b$. In effect we find the inverse of the permuted columns of H.

8 Conclusion

We observed that the use of NMFP on a single image is effective in providing a sparse, parts-based representation of the components of the image. Results also indicate that when we construct a dataset satisfying the rules outlined in Definition 6 we are successful in achieving a set of dictionary elements which offer an exact decomposition into the components of the dataset. This idea of extracting information from single images seems promising in the area of pattern recognition. Further work must be performed in determining the ordering of the vectorized patches. We intend to investigate the use of Hilbert Space-filling Curves as a strategy for ordering the patches. Experiments revealed that the order of the patches has an impact on the convergence and the runtime. In addition, the reconstruction of the Turtle 3 image appears to have required a larger number of dictionary vectors than the anticipated 12. We also intend to explore why this occurred and will determine if there are properties of the Turtle 3 image which caused the failure of the algorithm to achieve the minimum number of dictionary vectors necessary for the exact approximation.

References

1. Aharon, M., Elad, M., Bruckstein, A.: K-SVD: an algorithm for designing overcomplete dictionaries for sparse representation. Trans. Sig. Process. **54**(11), 4311–4322 (2006)
2. Biederman, I.: Recognition-by-components: a theory of human image understanding. Psychol. Rev. **94**(2), 115 (1987)
3. Buades, A., Coll, B., Morel, J.M.: On image denoising methods. Technical report, Technical Note, CMLA (Centre de Mathematiques et de Leurs Applications (2004)
4. Charles, R.M., Taylor, K.M., Curry, J.H.: Nonnegative matrix factorization applied to reordered pixels of single images based on patches to achieve structured nonnegative dictionaries. arXiv preprint arXiv:1506.08110 (2015)
5. Cichocki, A., Amari, S., Zdunek, R., Kompass, R., Hori, G., He, Z.: Extended smart algorithms for non-negative matrix factorization. In: Artificial Intelligence and Soft Computing-ICAISC 2006, pp. 548–562. Springer (2006)
6. Donoho, D., Stodden, V.: When does non-negative matrix factorization give a correct decomposition into parts? In: Advances in Neural Information Processing Systems (2003)
7. Hoyer, P.O.: Non-negative matrix factorization with sparseness constraints. J. Mach. Learn. Res. **5**, 1457–1469 (2004)
8. Klingenberg, B., Curry, J., Dougherty, A.: Non-negative matrix factorization: Ill-posedness and a geometric algorithm. Pattern Recognit. **42**(5), 918–928 (2009)
9. Lee, D.D., Seung, H.S.: Algorithms for non-negative matrix factorization. In: NIPS, pp. 556–562. MIT Press (2000)
10. Lee, D.D., Seung, H.S.: Algorithms for non-negative matrix factorization. In: Advances in Neural Information Processing Systems, pp. 556–562 (2001)
11. Lee, D.D., Seung, H.S., et al.: Learning the parts of objects by non-negative matrix factorization. Nature **401**(6755), 788–791 (1999)
12. Lin, C.-J.: Projected gradient methods for nonnegative matrix factorization. Neural Comput. **19**(10), 2756–2779 (2007)

13. Paatero, P., Tapper, U.: Positive matrix factorization: a non-negative factor model with optimal utilization of error estimates of data values. Environmetrics **5**(2), 111–126 (1994)
14. Ranade, A., Mahabalarao, S.S., Kale, S.: A variation on SVD-based image compression. Image Vis. Comput. **25**(6), 771–777 (2007)
15. Shi, J., Malik, J.: Normalized cuts and image segmentation. IEEE Trans. Pattern Anal. Mach. Intell. **22**(8), 888–905 (2000)
16. Taylor, K.M., Meyer, F.G.: A random walk on image patches. SIAM J. Imaging Sci. **5**(2), 688–725 (2012)
17. Vandaele, A., Gillis, N., Glineur, F., Tuyttens, D.: Heuristics for exact nonnegative matrix factorization. J. Glob. Optim. **65**(2), 369–400 (2016)
18. Wang, Z., Bovik, A.C., Sheikh, H.R., Simoncelli, E.P.: Image quality assessment: from error visibility to structural similarity. IEEE Trans. Image Process. **13**(4), 600–612 (2004)
19. Watkins, D.S.: Fundamentals of Matrix Computations. Wiley, New York (1991)
20. Zdunek, R., Cichocki, A.: Non-negative matrix factorization with Quasi-Newton optimization. In: Artificial Intelligence and Soft Computing-ICAISC 2006, pp. 870–879 (2006)

Significant Target Detection of Traffic Signs Based on Walsh-Hadamard Transform

XiQuan Yang(✉) and Ying Sun(✉)

College of Information Sciences and Technology,
Northeast Normal University, Changchun, China
{Yangxq375, Suny303}@nenu.edu.cn

Abstract. This paper proposes a method based on Walsh-Hadamard transform to detect the significant target of traffic signs. This method uses the Walsh-Hadamard transform and normalized image binary spectrum to detect the traffic target with significant target. Experiments show that the algorithm is shorter than other algorithms and can quickly and effectively detect the significant area of traffic signs compared with other algorithms in the text.

Keywords: Significant detection · Binary spectrum · Walsh-Hadamard

1 Introduction

Road traffic signs use signals and graphical symbols to transmit signals, directions, warnings, bans, etc. to vehicles and pedestrians, and regulate, channel, and control the flow and flow of vehicles. It improves road traffic efficiency, predict road conditions and regulate traffic behavior. However, due to the restrict size of traffic sign and the complicated traffic environment, crowded vehicles on the street surrounded by various store advertising board, can ignore some important traffic signs, which undoubtedly bring danger to the driver and pedestrian, so significant target detection of traffic signs has a practical effect. The methods of visual saliency detection are mainly divided into two categories, one is the bottom-up data-driven visual attention mechanism, and the other is the top-down visual attention mechanism based on task-driven targets; most of the computable visual models are bottom-up. In the field of computer vision, many computable attention selection models have been generated for simulating human visual attention mechanisms. Models that commonly used are cognitive models, information theory models, graph theory models, and spectral domain models, all of which are directly or indirectly inspired by cognitive models. Among them, the ITTI model based on cognitive model in traditional model [1] uses multiple color, attribute and direction features to decompose by multiple feature channels and scale, then filter to obtain feature map, and finally perform fusion calculation on feature map to get significant figure. The typical model based on graph theory is GBVS [2], improved based on ITTI. By construct Markov chain of two-dimensional image using the characteristics of Markov random field, and the significant map is obtained by finding the equilibrium distribution. Based on the saliency model of spectrum analysis, SR [3]

K. Arai et al. (Eds.): CompCom 2019, AISC 997, pp. 1027–1035, 2019.
https://doi.org/10.1007/978-3-030-22871-2_73

model, which obtains a significant graph by performing inverse Fourier transform on the residual spectrum. The algorithm only needs to perform Fourier transform and inverse Fourier transform. Based on the local contrast AC [4] model, it uses the Lab color space to calculate the distance. The above traditional saliency detection has a good detection effect on simple background scene, but perform poorly for traffic sign in complicated background and restricted detection area. This paper adopts an algorithm based on Walsh-Hadamard transform [5] to detect the significant target of traffic signs. This algorithm is able to detect the significant target area with higher accuracy and simpler calculation: replace calculate multiples with add and subtraction. By doing so the amount of calculation is greatly reduced, and the speed of significant target detection is improved.

2 Walsh Transform

The Walsh function was proposed by American mathematician Walsh transform in 1923. He completed the incomplete Redmacher function and formed a complete set of orthogonal rectangular functions, known as Walsh function system [6]. Walsh functions can be divided into three categories, namely, Walsh function of Walsh order, Walsh function of paley order and Walsh function of Hadamard order. The difference between these three types of functions is that their respective functions appear differently. The nature of these three sequences is the same, and they can be converted to each other through a transformation matrix [7]. Walsh functions have three different function definitions, but they all can be made up of Radek functions. The Walsh function arranged by Hadamard can be defined as:

$$Wal_H(i,t) = \prod_{k=0}^{p-1} [R(k+1,t)]^{(i_k)} \tag{1}$$

R(k + 1, t) Is any Radmec function, (i_k) Is the kth digit of the reversed binary code. P is Positive integer, $(i_k) \in \{0,1\}$.

2.1 Discrete Walsh-Hadamard Transform

The one-dimensional discrete Walsh transform is defined as:

$$W(u) = \frac{1}{N}\sum_{x=0}^{N-1} f(x)Wal_H(u,x) \tag{2}$$

The one-dimensional discrete Walsh inverse transform is defined as:

$$f(x) = \sum_{u=0}^{N-1} W(u)wal_H(u,x) \tag{3}$$

It is known that the Walsh-Hadamard transform essentially adds and subtracts the sign of each value of the discrete sequence f(x) according to a certain law. Therefore, the DCT calculated by DFT and cosine operation are much simpler.

2.2 Two-Dimensional Discrete Walsh Transform

The definition of one-dimensional WHT is extended to two-dimensional WHT. The positive transform kernel and the inverse transform kernel of the two-dimensional WHT are:

$$W(u,v) = \frac{1}{MN}\sum_{x=0}^{M-1}\sum_{y=0}^{N-1} f(x,y) Wal_H(u,x) Wsl_H(v,y) \quad (4)$$

$$f(x,y) = \sum_{u=0}^{M-1}\sum_{v=0}^{N-1} W(u,v) Wal_H(u,x) Wsl_H(v,y) \quad (5)$$

After the two-dimensional WHT transformation (see Fig. 1(a)) is the original image Fig. 1(b) is the WHT result. As can be seen from the above examples, the two-dimensional WHT concentrate the energy and the more uniformly the numbers in the original data, the more concentrated the transformed data is on the corners of the matrix. Therefore, two-dimensional WHT can be used to compress image information.

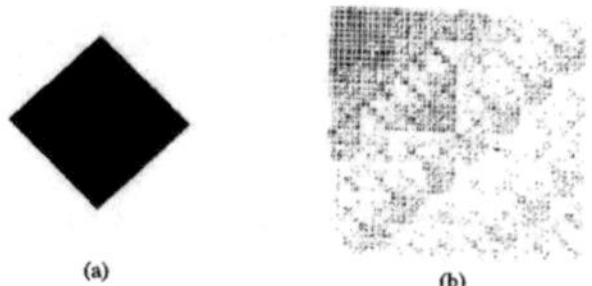

Fig. 1. Two-dimensional WHT transformation

3 Traffic Sign Significant Target Detection

3.1 Color Feature Extraction

Images are usually divided into salient regions and non-significant regions, by its underlying visual features such as color, texture, orientation, shape, etc. LUO et al. [8] demonstrated that the salient regions of the image are easily identified in a particular color channel, thereby using the color channel model to detect saliency. Ma et al. [9] proposed using RGB color channels to detect saliency. Zhang et al. [10] used LAB color channel to detect saliency. Cai et al. [11] proposed to calculate the significant maps of L, a, b and H, S, V color channels in two-color spaces of Lab and HSV, respectively, to calculate the salient regions of traffic signs.

The LAB color space is similar to human vision. It is a color mode developed by CIE (International Commission on Illumination). In nature, any color can be expressed in LAB space, and its color space is larger than RGB space. In addition, it is a device-independent color system and based on physiological characteristics. It describes the human visual sense in a digital way. It is suitable for the representation and calculation of all light source color or object color. At the same time, it covers the deficiency of RGB and CMYK mode, which must rely on the lack of color characteristics of the device. Therefore, this paper uses LAB color channel model to detect saliency by color feature extraction.

Calculate the saliency map corresponding to the three color spaces L, A, and B. Since the brightness and color of the Lab color mode feature are separate, the L channel has no color, and only the A channel and the B channel have colors, so only the two color channels of the images A and B are extracted.

Since RGB images cannot be directly converted to LAB color channels, they need to be converted to XYZ and then converted to LAB: RGB-XYZ-LAB.

So the conversion formula is divided into two parts:

$$\begin{bmatrix} X \\ Y \\ Z \end{bmatrix} = M \times \begin{bmatrix} R \\ G \\ B \end{bmatrix} \tag{6}$$

$$M = \begin{bmatrix} 0.412 & 0.357 & 0.180 \\ 0.212 & 0.715 & 0.072 \\ 0.019 & 0.119 & 0.950 \end{bmatrix} \tag{7}$$

In the first part, the M matrix is used to perform nonlinear tone editing on images to improve image contrast. After testing, the histogram obtained by this formula and the transformed image are very similar to the photo-sensing results after Photoshop conversion.

RGB to XYZ. Assume that RGB is a pixel of three channels, and the range of values is [0 255]. The conversion formula is as follows:

$$\begin{cases} L = 116f(Y/255) - 16 \\ A = 500[f(X/255) - f(Y/255)] \\ B = 200[f(Y/255) - f(Z/255)] \end{cases} \tag{8}$$

In the above two formulas, L, A, and B are the values of the three channels of the final LAB color space. X, Y, and Z are the values calculated by RGB to XYZ in (6).

3.2 Walsh-Hadamard Matrix

The saliency detection marks the area that best represents the image information. The compression of the image can compress the key information as much as possible. For non-critical information, it can be uncompressed or partially compressed [12], which can speed up the compression and reduce the compression space. Walsh-Hadamard transform is a typical non-sinusoidal function transform, which uses orthogonal right-angle function as the basis function. It has similar properties to the Fourier function. The more evenly distributed the image data, the more data the Hadamard transform is concentrated on the corners of the matrix. Therefore, the Walsh transform has the property of energy concentration and can be used to compress image information. The key information can be extracted by the compression algorithm and applied to the significance detection.

The Walsh-Hadamard transform essentially adds or subtracts the signs of the values of the discrete sequences according to a certain rules. It is simpler and faster than the discrete Fourier transform using complex operations and the discrete cosine transform using cosine operations.

The saliency of the image was detected by using the Walsh-Hadamard transform [13], Ahmed [14] et al. demonstrated that visual redundancy between pixels can be obtained by the Walsh-Hadamard transform of the image. In this paper, the Walsh-Hadamard transform is used to extract key information and perform forward transform and inverse transform. The saliency map extracted from the color features of the two channels A and B is subjected to a Walsh-Hadamard positive transformation, and the image matrix is normalized by a symbol function. The function of the symbolic function is to take the sign of each number and set the positive number to 1 and the negative number to −1. Its expression is:

$$B = sign(WHT(X)) \tag{9}$$

Where X represents the image matrix, WHT() represents the two-dimensional Walsh-Hadamard transform, sign represents the symbol function, and B is the image normalized by the Walsh-Hadamard transform, that is, the image is transformed and normalized. The resulting new image is called a binary spectrum [15]. The binary spectrum carries some significant information about the image. To recover the significant information in space, the Walsh-Hadamard inverse transform is performed on the binary spectrum carrying some significant information, which is defined as:

$$Y = abs(IWHT(X)) \tag{10}$$

Where IWHT() represents the two-dimensional Walsh-Hadamard inverse transform, abs() is an absolute value function, and Y is a significant graph after the saliency information is restored. The use of the Walsh-Hadamard transform can result in information loss. This article sets a threshold by experiment. When the brightness is greater than the threshold, the value remains, when the brightness is less than the threshold, its brightness part is set to 0. Since most traffic signs are red and blue, the A color channel indicates green to red, the value range is [127, −128], and the B color channel indicates yellow to blue. The value range is [127, −128], so the red part is higher, indicating that its brightness value is higher while the blue part is behind and its brightness value is lower. Figure 2 shows the change of the improved saliency map. The saliency diagram of the original method always lacks a road sign. The improved saliency map has significant information on the second road sign. Finally, the two channels of the salient map are fused based on the maximum value method to obtain the final saliency map.

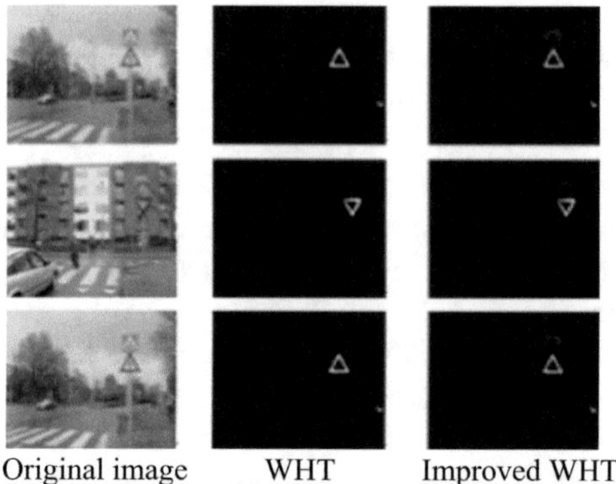

Fig. 2. Improved saliency map comparison

3.3 Gaussian Low-Pass Filtering

The Gaussian low-pass filtering is edge smoothing, and the Gaussian filtering of the significant image can remove the noise. The grayscale transformation of the edge region is increased, and the edge portion of the high-pass filter of the significant image is preserved, the edge is extracted, and the saliency map is sharpened.

4 Experimental Results

This article is tested on the i5-4200U dual-core CPU, clocked at 1.6 GHz, memory 4 GB, windows7 platform, code written on MATAB R2014. The test gallery comes from the dataset (http://http://www.datatang.com/), which is an image of 48 different road signs with different brightness and darkness. The image size is 256 pixels × 256 pixels. The ITTI, SR, AC, HC, and FT algorithms were selected as comparative experiments.

The experimental procedure is as follows:

In the pre-processing stage, adjust the image to the appropriate size for LAB color space conversion.

Perform two-dimensional Walsh-Hadamard transform normalization on the two-color channels of AB to generate two binary spectra carrying significant information.

Perform a two-dimensional Walsh-Hadamard inverse transform on the binary spectrum to recover the saliency information in the binary spectrum and obtain a significant picture of the A and B color channels.

The excess information is filtered according to the set threshold, and the filtered image is subjected to Gaussian filtering to generate a final saliency map.

In the comparative experiment, we compare the five algorithms that are more classic in recent years in the field of significant detection research. ITTI's pixel-based saliency detection model, SR and FT [16] are saliency algorithms based on spatial frequency domain analysis, HC [17] saliency algorithm based on color features, and AC is a saliency detection algorithm based on local contrast pairs. The experimental results show that the method is simple and effective, and the experimental results are shown in Fig. 3.

4.1 Evaluation Indicators

In this paper, the results are analyzed using indicators such as AUC value and running time. At the same time, the evaluation of these two evaluation indicators can make the experimental results more objective. AUC (Area Under Curve) is defined as the area under the ROC curve, and it is obvious that the value of this area will not be greater than 1. Since the ROC curve is generally above the line y = x, the AUC value ranges between 0.5 and 1. The AUC value is used as the evaluation criterion because many times the ROC curve does not clearly indicate which classifier works better, and as a numerical value, the classifier corresponding to the larger AUC is better. The AUC values of the respective comparison methods are shown in Table 1. The formula for calculating the AUC value is as follows:

$$AUC = \frac{\sum_{i \in positiveClass} rank_i - \frac{M(1+M)}{2}}{M \times N} \tag{11}$$

From Table 1, the AUC value of the algorithm in this paper is significantly higher than other traditional methods. In this paper, several significant detection results algorithms are compared to run-time comparison experiments. It can be seen from Table 2 that our method in this article runs for the shortest time.

Table 1. AUC values for various methods

	Our method	ITTI	AC	SR	HC	FT
AUC	0.9613	0.8896	0.8896	0.8621	0.96136	0.9015

Table 2. Running time comparison

	Our method	ITTI	AC	SR	HC	FT
Runtime/s	0.1713	0.4787	5.6327	0.3053	0.3151	0.2961

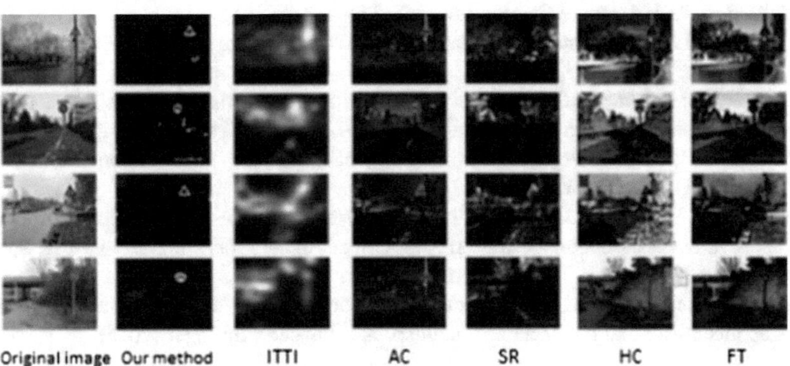

Fig. 3. Experimental result chart

5 Conclusion

This paper uses the Walsh-Hadamard transform method to detect the significant target of traffic signs. Since the Walsh-Hadamard transform has only addition and subtraction operations, it consumes little computational power and performs fast transformation, which saves computation time and improves the operation efficiency. However, there will be information loss in the process of image passing Walsh-Hadamard transform and inverse transform. In this paper, the threshold area is set to make the salient region have certain significance and improve its precision. The algorithm of this paper is especially suitable for application to some other real-time saliency detection, such as fire smoke detection.

References

1. Itti, L., Koch, C., Niebur, E.: A model of saliency-based visual attention for rapid scene analysis. In: IEEE TPAMI (1998)
2. Harel, J., Koch, C., Perona, P.: Graph-based visual saliency. In: NIPS, pp. 545–552 (2007)
3. Hou, X., Zhang, L.: Saliency detection: a spectral residual approach. In: IEEE Conference on Computer Vision and Pattern Recognition, pp. 1–8 (2007)
4. Achanta, R., Estrada, F., Wils, P., Süsstrunk, S.: Salient region detection and segmentation, pp. 4–5 (2008)
5. Yu, Y., Yang, J.: Visual saliency using binary spectrum of Walsh–Hadamard transform and its applications to ship detection in multispectral imagery, pp. 99–152. Springer, New York (2016)
6. Chaum, D.: Blind signatures for untraceable payments. In: Lecture Notes in Computer Science, pp. 199–203 (1982)
7. rui song, Y.: Simple generation of Walsh transform kernel matrix and its application. In: Communication and Computer, pp. 221–225 (2005)
8. Luo, W., Li, H., Liu, G., Ngan, KN.: Global salient information maximization for saliency detection. In: Signal Process. Image Commun., 238–248 (2012)
9. Ma, X., Xie, X., Lam, K-M., Zhong, Y.: Efficient saliency analysis based on wavelet transform and entropy theory. J. Vis. Commun. Image Represent. 201–207 (2015)
10. Zhang, H.: A significant region extraction algorithm based on color and texture. J. Huazhong Univ. Sci. Tech. (Natural Science Edition), 399–402 (2013)
11. Cai.: A method for detecting significant targets based on traffic signs. Photoelectric Eng. 84–87 (2013)
12. Kunz, H.O.: On the equivalence between one-dimensional discrete Walsh–Hadamard and multidimensional discrete Fourier transforms. IEEE Trans. Comput. 267–268 (1979)
13. Andrushia, A.D., Thangarjan, R.: Saliency-based image compression using Walsh–Hadamard Transform (WHT), pp. 21–41. Springer (2018)
14. Ahmed, N., Rao, K.R., Ahmed, N., Rao, K.R.: Walsh–Hadamard transform. In: Ahmed, N., Rao, K.R. (eds.) Orthogonal Transforms for Digital Signal Processing, pp. 99–152. Springer, New York (1975)
15. Ying, Y., Jian, Y.: Visual saliency using binary spectrum of Walsh–Hadamard transform and its applications to ship detection in multispectral imagery. Springer (2016)

16. Achanta, R., Hemami, S., Estrada, F., Susstrunk, S.: Frequency-tuned salient region detection. In Proc. IEEE International Conference on Computer Vision and Pattern Recognition, pp. 1597–1604 (2009)
17. Cheng, M.-M., Mitra, N.J., Huang, X., Torr, P.H.S., Hu, S.-M.: Global contrast based salient region detection. In: IEEE Transactions on Pattern Analysis and Machine Intelligence, pp. 569–582 (2015)

Fast Implementation of Face Detection Using LPB Classifier on GPGPUs

Mohammad Rafi Ikbal[1(✉)], Mahmoud Fayez[1], Mohammed M. Fouad[1], and Iyad Katib[2]

[1] Fujitsu Technology Solutions, Jeddah, Saudi Arabia
{mohammad.rafi,Mahmoud.Fayez, Mohammed.Fouad}@ts.fujitsu.com
[2] King Abdul Aziz University, Jeddah, Saudi Arabia
iakatib@kau.edu.sa

Abstract. Face detection is one of the classical computational challenges in computer vision. Detecting faces in an image has many applications in the field of security surveillance, marketing, biometric authentication, social media, photography and many more. Scanning window is most common technique used in face detection. In this method, an image is divided into multiple regions and each region is processed to detect presence of human face. To process an image of size 640 × 480 pixels, 178000 image regions must be processed and at-least 24 frames must be processed within a second, the number of regions to be processed grows significantly for images with higher dimensions and frame rates. With the advent of high resolution video capturing devices capable of recording at higher frame rates, real-time face detection is a challenge as it is computationally intensive process. In this paper, we present a framework which uses scanning-window technique, integral image and Local Binary Pattern (LBP) implemented on GPGPU to process images in short duration with high accuracy. In our experiments, we have achieved processing speed up to 287 frames per second on average with image dimensions of 640 × 480 pixels.

Keywords: GPGPU · LBP classifier · Face detection · Image processing · Image scanning

1 Introduction

Face detection has significant importance in many domains like Robotics, Computer vision, surveillance and many more. This is also critical step in many systems which process human faces like face recognition system, facial expression detection, etc. These systems depend on face detection system to identify facial region in given image or video sequence. Once facial region is identified, these sub-systems act upon these regions to implement the algorithms they are intended for. Speed and efficiency are prime factors of face detection applications. Face detection system should be able to detect facial regions of given input and should have enough time for the sub-systems like face recognition to process these region before next frame is generated by the input device. If the system fails to detect facial regions in real time before next frame is

K. Arai et al. (Eds.): CompCom 2019, AISC 997, pp. 1036–1047, 2019.
https://doi.org/10.1007/978-3-030-22871-2_74

generated, it will introduce lag and this lag will cumulatively carry forward throughout the input and will make the system inefficient. To detect facial regions in given frame, input image is divided into multiple sub-regions and then each region is processed to detect presence of face using pre-defined algorithms. Once all regions are processed, scale of the sub region or the image is changed and the processes are repeated. An image of dimensions 640 × 480 can produce approximately 178000 image regions. All these regions must be processes by the processor with in 41 ms considering that image capturing device is producing 24 frames per second. The number of frames to be processed is directly proportional to dimensions of the image and frame rate of video capturing device. The number of regions to be processed increases drastically if the video capturing device is able to capture higher resolution images at higher frame rates which is needed for increased accuracy.

Many image processing techniques and algorithms were developed in last few decades with varying complexities and methods. Few of these algorithms rely on complex floating point operations over multiple iterations on same region to detect presence of face while other algorithms use numerical operations along with other methods like boosting to detect facial regions in given image. Performance of each of these algorithms varies greatly on different systems they are intended to execute and how they were adopted on the system. Some algorithms may involve too many light weight tasks to be computed by a single processor but this will introduce too many task switching and will impact performance of the processor. We can use NVidia GPGPUs (General Purpose Graphics Processing Unit) for this purpose as these devices has capability to process thousands of threads in parallel. On the other hand, GPGPUs can process hundreds of threads simultaneously in parallel; the number of task switching involved in GPUs for this kind of problem is reduced greatly delivering higher performance and frame rates which can meet current demands of real-time applications. With these features of GPGPUs and advancements in CUDA (Compute Unified Device Architecture) programming model, this paper aims to implement face detection algorithm on GPGPUs.

Viola and Jones [1] pioneered the face detection algorithm which is based on integral image, cascade classifier and Ada-Boosting training algorithm. For feature detection, their method used Haar-like detector. In our paper, we have used Local Binary Pattern to detect faces instead of Haar-Like features.

This framework needs a pre-trained feature set like Local Binary Pattern (LBP) that could be acquired from open-source sources like OpenCV. This feature set was trained based on thousands of positive and negative images. Most notable modification of the algorithm was to implement a method to transfer, store and retrieve the feature set in most suitable way for GPGPU. This method minimizes the memory access time between global memory and shared memory of GPU. Other parts of the algorithm like scanning window and integral images were also modified in such a way that they can run efficiently on the GPGPU architecture. Though we have faced many challenges while implementing the application on GPU, we have added optimizations and workaround to address these challenges using best practices from NVIDIA.

The overview of our paper is as follows: Sect. 2 discusses the related work. Problem statement and implementation method is described in Sects. 3 and 4 respectively. Results of our work are briefly described in Sect. 5 and conclusion of our work is in Sect. 6.

2 Related Work

Viola and Jones pioneered face detection framework with three main features [1]. The first one was the use of integral image which improved the performance of face detection greatly by reducing calculation of sum of any region of image to 4 operations. Second feature was to use AdaBoost learning algorithm and the third feature was to combine classifiers in a cascade.

Marwa et al. [2] proposed optimized parallel implementation of face detection on GPU. Though their results look promising, the image samples are of dimensions 32×32, 64×64, 512×512 which are nonstandard image resolutions.

Vikram et al. [3] proposed a face detection system using GPU which works based on motion and skin color. Rainer et al. [4] proposed a system which uses rotated haar-like features to increase detectability of human face in given image. Soni et al. [5] work is based on viola-jones algorithm, their work is focused on increasing accuracy of the algorithm however, the detection rate of 2.89 s per frame is too slow for real time face detection systems. They have provided comparison of detection rate between Young people, children and old people which seems to be hypothetical since the feature evaluation process does not depend on age of the subject.

Oh et al. [6] have proposed heterogeneous system that uses CPU and GPU to process the given image. They have done a good work in implementing LBP on CPU and GPU. They have used multiple cores in CPU to pre-process the image and used GPU to detect faces. They have achieved 29fps for Full HD image.

Fayez et al. [7] proposed image scanning framework using GPGPU in which they have implemented viola-jones algorithm on GPGPU. They have used Haar-Like features to evaluate if the region has face. They have achieved 37fps for HD Images.

3 Problem Statement

Face detection is a classical computationally intensive problem in computer vision domain. Detecting faces in given image has many applications in diverse fields. With the advent of video capturing devices capable of generating high resolution images at high frame rates and steep increase in video and image database through social media, it is challenging to process these images in real-time and detect facial regions in input image. Though there are many algorithms developed in past few decades to solve this problem, performance of these algorithms varies greatly on different processors and hardware platforms. Performance of selected algorithm varies greatly on the implementation method on the hardware architecture it is being implemented on. Performance of the system also depends on how the image is being processed by the algorithm. Adopting selected algorithm on specific architecture is also quite challenging due to specifications and programming needs of specific architecture.

3.1 Image Scanning

Scanning an image to detect faces within has quite some complexities associated with it, like what region should be processed? What should be size of processing window? What if there is more than one face in the image and the faces are of different dimensions? A classifier can be applied to specific region in an image and can detect only one face if present in this region. This region may or may not have a face in it, but the system must process this region. The size and position of face is arbitrary. To solve this problem, sliding window technique is employed. Since the dimensions of the pre-trained classifier is fixed, there are two approaches to scan the image to detect faces with different size and position in image. The first approach is scaling the features from the classifier as shown in Fig. 1 and the second approach is to scale the image such that faces with-in the image is scaled and matches the dimension of the classifier at some scaling value as shown in Fig. 2. Usually scaling the image is preferred since many classifiers does not support scaling, however scaling down images will result in loss of data which classifiers rely on to evaluate face. In this paper we have employed scaling of features since LBP supports feature scaling. Secondly, scaling image will result in rebuilding of image and associated data like integral image with each frame.

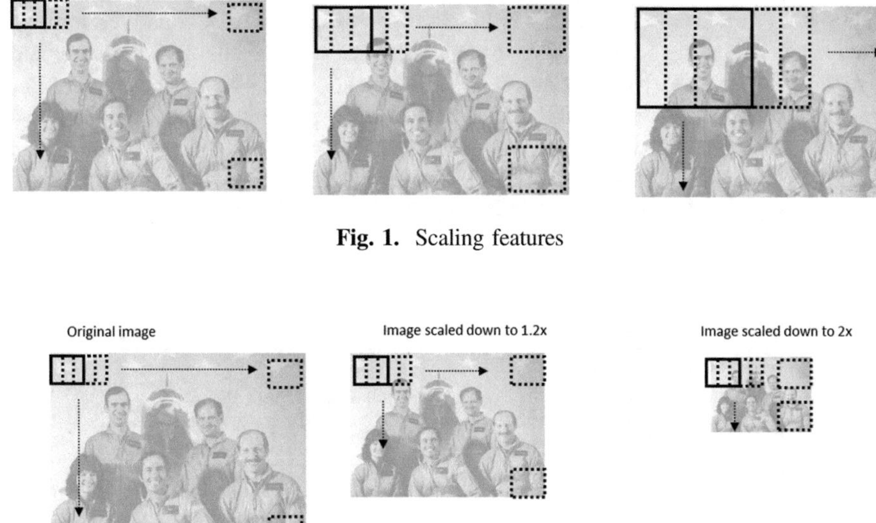

Fig. 1. Scaling features

Fig. 2. Scaling image

Given an image of dimensions *WxH*, where *W* is width of the image and *H* represents height of the image in pixels. We can use the algorithm presented in Fig. 3 to find the total number of scanning windows to be generated to process the given image. Calculating this number is crucial as we will scale features and launch CUDA threads based on this number.

```
Input : W (Image Width)
Input : H (Image Height)
Input : ScaleUpFactor
Output: CountOfWindows
Output: ListOfWidows
BEGIN
  FOR scale = InitialScale to MaxScale STEP ScaleUpFactor
   FOR x = 0 to W – WindowWidth STEP ScaleUpFactor
    FOR y = 0 to H – WindowHeight STEP ScaleUpFactor
           ListOfWidows.Add(x, y, scale)
           CountOfWindows++
     END
    END
  END
  return CountOfWIndows, ListOfWidows
END
```

Fig. 3. Calculate number of sliding windows

3.2 Integral Image

Using integral image to compute sum of pixels in given region of image was first proposed by Viola and Jones [1]. Evaluation of features in LBP involves multiple iteration of computing sum of pixels in region defined in the features. Computing sum of regions is most recurring operation in LBP operation, as a result these operations consumes most of the computation in face detection process. Using integral image reduces this to three operations per region irrespective of the dimensions of the region. For example, Fig. 4 shows the original image matrix, while Fig. 5 represents the integral image for the original image matrix that had been computed in a single pass over the image using Eq. (1).

6	4	8	8	4
3	8	10	7	5
1	3	8	10	8
6	4	6	10	4
6	2	7	4	0

Fig. 4. Original image matrix (I)

6	10	18	26	30
9	21	39	54	63
10	25	51	76	93
16	35	67	102	123
22	43	82	121	142

Fig. 5. Integral image (I')

$$I'(x,y) = I(x,y) + I'(x,y-1) + I'(x-1,y) - I'(x-1,y-1) \tag{1}$$

Where $I'(x, y)$ represents value of any pixel in integral image at location (x, y). I and I' are original image and Integral image respectively. The value of $I'(x, y)$ at any location contains the sum of pixels from $(0, 0)$ to (x, y) from the original image including value at (x, y) as shown in Eq. (2).

$$I'(x', y') = \sum_{0 \le x, 0 \le y} I(x, y) \tag{2}$$

We can use Eq. (3) to compute sum of values of pixels in the region (x_m, y_m) and (x_n, y_n) as shown in Fig. 6, where $m < n$. If we consider an example where difference between m and n is 3 as shown in Fig. 4, then we will end up performing addition of 9 pixels of this region. Another example where the difference between m and n is 10, then we end up performing addition of 100 pixels to compute the sum of this region.

$$sum = \sum_{(x_m \le x_n\, y_m \le y_n)} I(x, y) \tag{3}$$

We can compute the sum of values of pixels in the region with only three operations irrespective of the difference between m and n using Eq. (4). I' in this equation represents integral image.

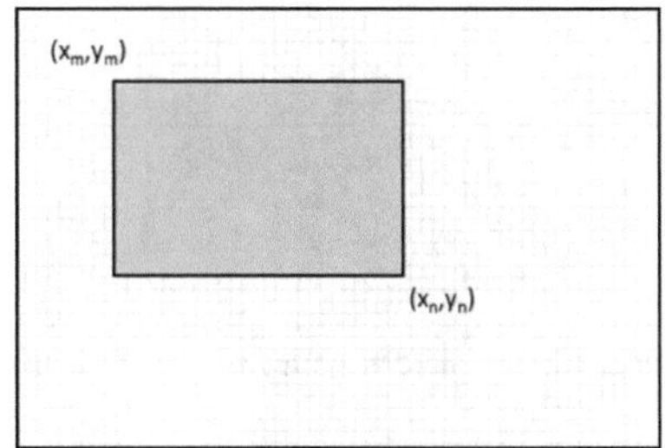

Fig. 6. Sum of region (co-ordinates)

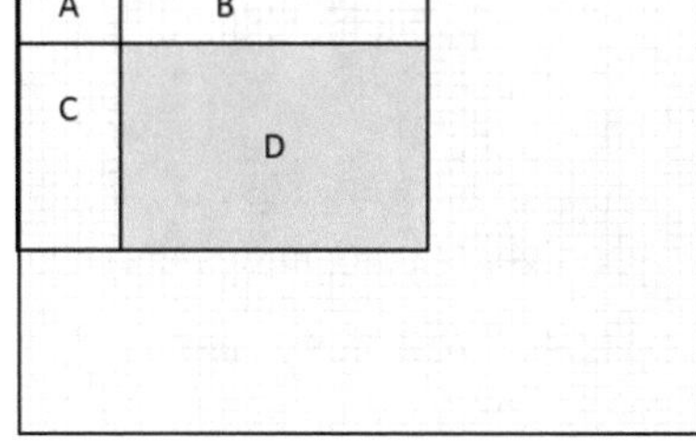

Fig. 7. Sum of region

$$sum = I'\left(x_n, y_n\right) + I'\left(x_m, y_m\right) - I'\left(x_{(m-n)}, y_m\right) - I'\left(x_m, y_{m-n)}\right) \tag{4}$$

If we represent this as sum of region as shown in Fig. 7, then the sum of pixels in region D can be computed using Eq. (5). As we can see, this equation can compute sum of any given region in only three operations irrespective of size of region. This feature will enable rapid computation of sum of pixels in given region of the image.

$$Sum = D + A - (B + C) \tag{5}$$

3.3 Local Binary Pattern (LBP)

LBP is an efficient, extremely fast and computationally simple texture operator introduced by Ojala et al. [8] that has been used predominantly in texture analysis. Initially LBP was implemented on 3×3 neighborhood as shown in Fig. 8 and later it was extended to neighborhood of any given dimensions as shown in Fig. 9. LBP works by comparing center pixel with neighbors and represent results in binary format, since the dimensions of the neighborhood is 3×3, this will result in 8-bit label which is used as texture descriptor. This texture descriptor has only $2^8 = 256$ possible values. In our experiments we are using a pre-trained classifier which we have obtained from OpenCV.

To compute LBP feature 3×3 (gray rectangle) shows in Fig. 8, we start by comparing value of cell *(x2, y2)* with value of center pixel which is *(x3, y3)* as in image Fig. 8. If *(x2, y2)* is greater than or equal to *(x3, y3)*, then we consider this value to be one else it is zero. Now we continue comparing next value of *(x3, y2)* with center pixel *(x3, y3)* in clock-wise direction. In case of computing LBP for region, we consider the sum of image region *(x2, y2)* and *(x3, y3)* and compare with sum of center region which is sum of *(x4, y4)* and *(x5, y5)* and if the value of latter is less than the former then we consider the result to be zero else the result will be one.

	x1	x2	x3	x4	x5
y1	6	4	8	8	4
y2	3	8	10	7	5
y3	1	3	8	10	8
y4	6	4	6	10	4
y5	6	2	7	4	0

Fig. 8. Local binary pattern

	x1	x2	x3	x4	x5	x6	x7	x8	x9	x10
y1	6	4	8	8	4	8	7	9	5	6
y2	3	8	10	7	5	8	2	4	7	9
y3	1	3	8	10	8	3	8	9	10	9
y4	6	4	6	10	4	5	0	6	1	8
y5	6	2	7	4	0	10	2	3	3	6
y6	0	4	3	3	10	6	9	2	5	5
y7	6	3	3	6	10	1	9	5	7	9
y8	9	7	1	2	6	0	1	3	10	0
y9	3	10	5	7	8	5	10	10	9	8
y10	6	2	1	6	4	9	3	7	5	3

Fig. 9. Local binary pattern using image region

Complexity of LBP can be calculated using Eq. (6). We can see from Table 1 that complexity of the problem increases with resolution and it is quite difficult to solve this problem on CPUs only especially if the algorithm is serial. Since GPGPUs have hundreds of cores which can process lot of threads in parallel, this problem fits on GPGPU architecture.

$$O(W, H, M, S) = M \times W \times H \times \log(S_{max}) \tag{6}$$

Where:

- W and H are width and height of the image
- M is number of features
- Smax is maximum scaling factor

Pre-trained LBP classifier contains stages, each stage consist of multiple features whose values are added together to be compared with the stage threshold. Feature has a rectangular region position and dimensions, 256-bit look up table represented as eight 32-bit signed integers and two leaf values. Each stage will have a floating-point stage threshold to evaluate if the stage is passed or not. To evaluate LBP, sum of each region surrounding rectangular region in LBP features is calculated using integral Eq. (1). The central region R4 from Fig. 9 is compared with surrounding regions clockwise to generate 8-bit LBP. Appropriate leaf value is added to stage sum based on this returned value. This stage sum is compared with stage threshold in LBP cascade to determine if the stage has passed or not. If one of the stages fails, remaining stages are not processed, else the region is returned as positive face as shown in Fig. 10.

Table 1. Complexity for different image resolutions

Width	Height	Features count	Smax	O(W, H, M, S)
320	240	139	1	1.07E + 07
640	480	139	1.30	5.56E + 07
1280	720	139	1.47	1.89E + 08
1920	1080	139	1.65	4.77E + 08

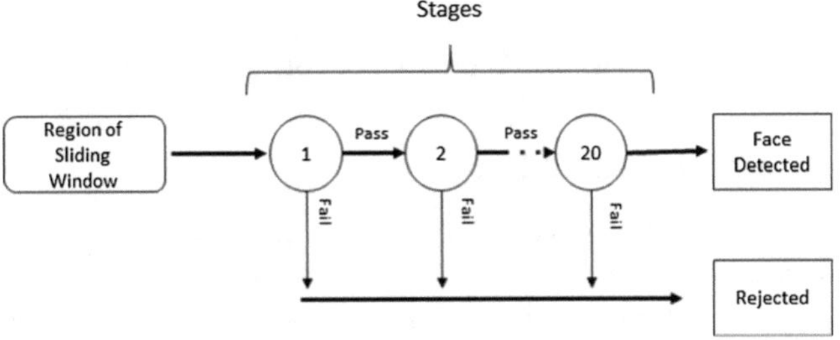

Fig. 10. Classifier stages

4 Implementation

Overview of our implementation is described in Fig. 11. Pre-trained feature set is converted from XML format to data structure suitable for CUDA architecture.

Given image is converted to integral image and stored in an array which is copied to GPGPU memory. All possible position and scales are computed based on dimension of input image as shown in Fig. 1. Integral image and feature-set is transferred to GPU memory and processed. Results are aggregated using results aggregator and faces are marked as shown in Fig. 11. Following sections describe these steps in detail.

4.1 Parsing Classifier

We have obtained pre-trained LBP classifier from OpenCV in XML format. XML file is converted to one-dimensional array since CUDA supports only basic data-types as shown in Fig. 11. This file has multiple stages and each stage has multiple weak classifiers and stage threshold. Each classifier has a reference to rectangle dimension, a lookup table in the form of signed integer and two leaf values. We have parsed all this information to one-dimensional array in such a way that each value can be referenced by stage and feature index.

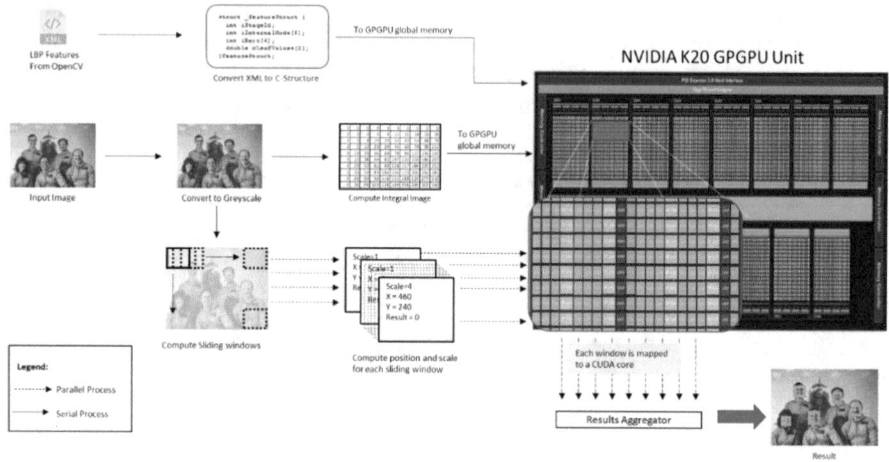

Fig. 11. Implementation overview

4.2 Parsing Image to Integral Image

Input image is converted to matrix and then converted to grey scale. This process will convert the pixel values to range of 0 and 255. We convert this image matrix to integral image using Eq. (1) as shown in Fig. 11. Integral image is converted to one-dimensional array and copied to GPU memory.

4.3 Computing Image Regions

We used NVIDIA k20 GPGPU for our experiments, which has 2496 CUDA cores.. To spawn one CUDA worker to process each region of the image, we compute number of possible regions of image for each scale and x, y positions. We use pseudo code in Fig. 3 to calculate the number of regions and also use the same pseudo code to create an array of tuples with x, y positions and scale parameters. LBP classifier has a dimension of 24×24 pixels, scaled window size can be computed by multiplying the starting dimension with scale. Each CUDA thread will have access to following parameters as outlined in Fig. 11:

- Integral image calculated from original image.
- Pre-trained LBP classifier in the form of an array.
- Scale and position of each scanning window region.

Once spawned, each CUDA thread acquires the position and scale of region it would process from the array by referencing to its thread id then the thread would start processing each feature in LBP feature set. Each feature needs to be scaled before processing by multiplying feature position and dimensions with scale value of the thread. If the stage passes, the thread would proceed processing another stage, else it would flag that the region with the specific scale does not contain a human face as shown in Fig. 11.

5 Results

We are using GCC compiler and Nvidia CUDA SDK version 7.5 to compile our proposed GPU implementation and OpenCV implantations for single CPU and GPU. The benchmarking approach we took was to execute each experiment 10 times with different images that has different number of faces. Each represented number in the

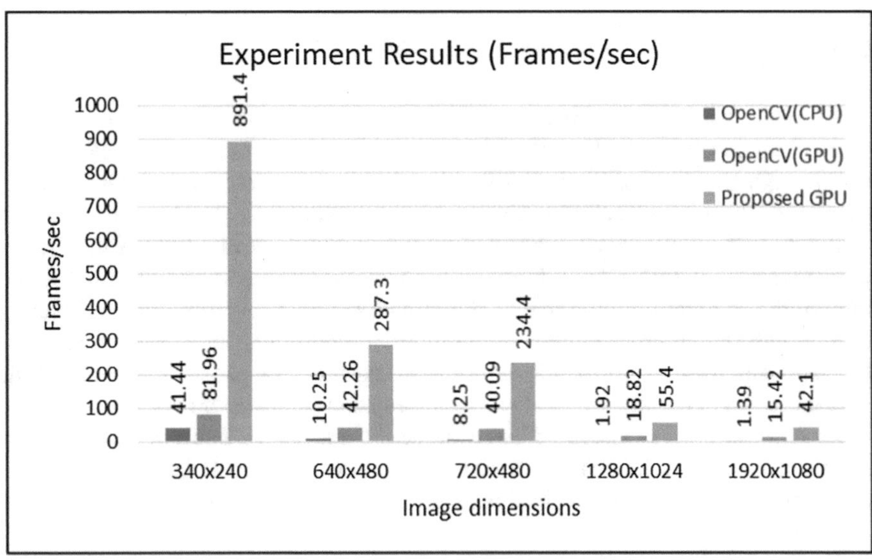

Fig. 12. Experiment results comparison

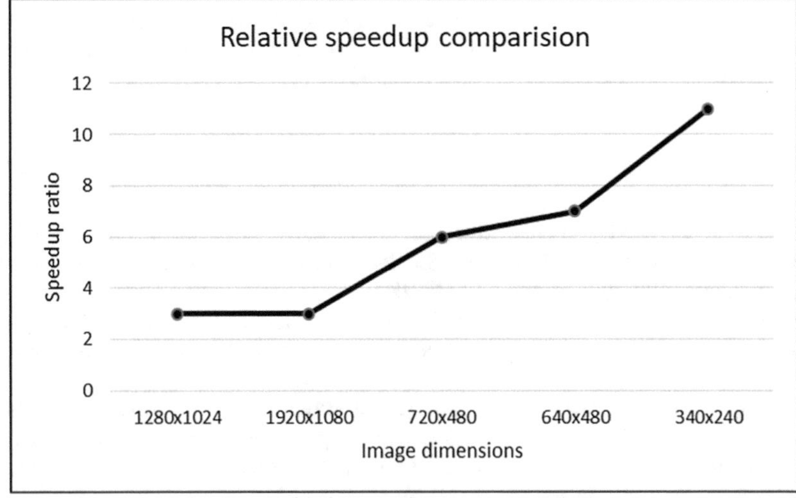

Fig. 13. Relative speedup comparison between OpenCV-GPU and proposed implementation

results graph shows the average of 10 experiments. We are using same configuration for all the experiments which includes window size, scaling up factor, step size, same GPU (NVidia Tesla K20), CPU (Intel Xeon E2950v2) and same LBP classifier dataset.

Figure 12 shows that our proposed LBP implementation out performs OpenCV CPU and GPU implementations by order of magnitude, at some image dimensions the relative speedup is more than 11x for OpenCV GPU implementation and 21x for OpenCV CPU implementation (Fig. 13).

6 Conclusion

In this work, we have presented a simple and efficient way to detect faces in given image using GPGPUs. We have used image scanning framework and LBP classifier to achieve high frame rates. LBP classifier does not use double-precision floating point operations which help greatly in achieving higher frame rates. Deep understanding of NVidia GPGPU architecture, better design and optimization may enable this framework to achieve even higher frame rates. However, current performance of our framework can process all the frames generated by current generation video capturing devices without any lag and leave ample time for intended application to process the detected facial regions. The results of our work excel on comparison with most of the work presented in this area so far. Our implementation details on procedure to generate image regions, process classifier based on scales and other details such that future implementations in this area can be compared. Other important facial features like mouth, nose, and eyes can be identified using our framework by using appropriate classifier.

We strongly believe that our work is a breakthrough in the field of face detection delivering performance and accuracy. We get an improvement of over 100% on comparison with performance of CPU

Acknowledgment. This work was supported in part by the High-Performance Computing Project in King Abdulaziz University and all experiments were conducted on Aziz supercomputer.

References

1. Viola, P., Jones, M.: Rapid object detection using a boosted cascade of simple features. In: Proceedings of the 2001 IEEE Computer Society Conference on Computer Vision and Pattern Recognition. CVPR 2001. p. I-511-I-518 (2004)
2. Chouchene, M., Sayadi, F.E., Bahri, H., Dubois, J., Miteran, J., Atri, M.: Optimized parallel implementation of face detection based on GPU component. Microprocess. Microsyst. **39**, 393–404 (2015)
3. Mutneja, V., Singh, S.: GPU accelerated face detection from low resolution surveillance videos using motion and skin color segmentation. Optik (Stuttg) **157**, 1155–1165 (2018)
4. Lienhart, R., Maydt, J.: An extended set of Haar-like features for rapid object detection. Proceedings. Int. Conf. Image Process. 1, I-900-I-903 (2002)

5. Soni, L.N., Datar, A., Datar, S.: Implementation of Viola-Jones algorithm based approach for human face detection. Int. J. Curr. Eng. Technol. 410677 (2017)
6. Oh, C., Yi, S., Yi, Y.: Real-time face detection in full HD images exploiting both embedded CPU and GPU. In: IEEE International Conference on Multimedia and Expo (ICME), pp. 1–6 (2015)
7. Fayez, M., Faheem, H.M., Katib, I., Aljohani, N.R.: Real-time image scanning framework using GPGPU – face detection case study. In: Proceedings of the International Conference on Image Processing, Computer Vision, and Pattern Recognition (IPCV), pp. 147–152 (2016)
8. Ojala, T., Pietikainen, M., Harwood, D.: A comparative study of texture measures with classification based on feature distributions. Pergamon Pattern Recognition. **29**, 51–59 (1996)

Optimized Grayscale Intervals Study of Leaf Image Segmentation

Jianlun Wang(✉), Shuangshuang Zhao, Rina Su, Hongxu Zheng, Can He, Chenglin Zhang, Wensheng Liu, Liangyu Jiang, and Yiyi Bu

College of Information and Electrical Engineering, China Agricultural University, Beijing, China
wangjianlun@cau.edu.cn, 1127381060@qq.com, 1601372955@qq.com, 496326832@qq.com, 348506582@qq.com, 787973615@qq.com, 550742142@qq.com, 2676117752@qq.com, 1273004647@qq.com

Abstract. In this paper, we propose a segmenting method for leaf images taken in the field. Instead of processing images in the whole grayscale range, we choose several subintervals of grayscale to operate on. The subintervals are chosen in accordance with the envelopes of image histograms. The subinterval is composed of random two local minimum points. In order to examine the effect of the method, four evaluation grades are settled as 'Perfect (P)', 'Well (W)', 'Average (A)' and 'Unsatisfying (U)'. 'Perfect' (P) represents the leaf edge segmented is complete; 'Well' (W) represents leaf edge segmented is almost complete; 'Average' (A) represents there is no complete leaf edge segmented and 'Unsatisfied' (U) indicates that the majority of the leaf edge is not segmented from the background. Operating the segmentation on different subintervals and clustering the segmenting results into these four grades, we could obtain the suitable grayscale subintervals; therefore the subintervals chosen before have been optimized. Experiments on four kinds of leaves including jujube, strawberry, begonia and jasmine leaves shows that optimized grayscale interval solutions are sufficient to segment target leaf from the background with high efficiency and accuracy.

Keywords: The envelope of image histogram · Optimized grayscale subinterval solutions · Leaf image segment

1 Introduction

Topology analysis and partial differentiation are widely used in image processing research in the computer vision and image processing fields. The previous research on partial differential image function analysis transfers the discrete digital image into a continuous mathematical model and regards the obtained solutions of the partial differential equation as the target results. The partial differential equation is applied in image de-noising [1], enhancement [2], segmentation [3], restoration [4] and other fields. Topological analysis also has many applications in image space in previous studies, for example, researchers use topological mapping to distinguish convex and concave points of image shapes [5] and use the topological mapping to acquire image

K. Arai et al. (Eds.): CompCom 2019, AISC 997, pp. 1048–1062, 2019.
https://doi.org/10.1007/978-3-030-22871-2_75

contours [6]. Both image analyzing and pattern recognition are based on segmentation. In order to analyze and measure leaves in the field, we have to segment the leaves out from the background. Segmentation of leaves is therefore an underlying step in dealing with images taken in fields. Requirements still need to meet in modern precise agriculture are digitalizing the plants, such as measuring the leaf area, height and other biomass of the plants. Such measurements will be precise only when the leaf is precisely segmented from the surroundings. Only in this way, the long term aim of monitoring plants growth conditions bypass remote online system could be achieved.

Vital though the segmentation is, segmenting leaves from an image taken in the field is still unsatisfying. The problem is the complicated background full of similar leaves. A plant in natural condition does not have single leaf, it has multiple similar leaves. Leaves of the same kind of plants share the high similarity in shape, color and even texture. Ordinary method such as only discriminating targets by shape, texture or color characteristics does not work in such situation. Besides, several overlapping leaves just makes it harder to segment one single leaf to analyze and measure.

Examples are auto machine picking by means of maximum class variance in segmenting cucumber images [7], or segmenting weeds from crops with the method of thresholding [8]. These segmentations are successful regardless of the background complexity, which is the overlapping problem of leaves.

In this paper, we propose an optimized segmenting method based on previous research [9]. Instead of processing images in the whole grayscale range, we choose several subintervals of grayscale to operate on. The subinterval is composed of random two local minimum points. In order to examine the effect of the method, four evaluation grades are settled as 'Perfect (P)', 'Well (W)', 'Average (A)' and 'Unsatisfying (U)'. Operating the segmentation on different subintervals and clustering the segmenting results into these four grades, we could obtain the suitable grayscale subintervals; therefore the subintervals chosen before have been optimized. Experiments on four kinds of leaves including jujube, begonia, strawberry and jasmine leaves shows that optimized grayscale intervals are sufficient to segment target leaf from the background with high effectiveness and accuracy.

The remainder of this paper is organized as follows: In the next section, we will outline the operators we use in the method and how the optimization work. The segmentation procedure will be summarized into five steps. Experimental results will be presented in Sect. 3 before we conclude in Sect. 4.

2 Methods and Algorithm

2.1 Grayscale Subinterval and Grayscale Histogram Envelope

In the field of image processing, segmentation is often operated on the whole grayscale range. Research [10] indicates that such segmentation applied on apples is insufficient. Therefore, the concepts of grayscale subinterval and the histogram envelope are introduced.

Let the grayscale function of image $F(x, y)$ be $g(x, y)$. $J(h)$ represents the original envelope of grayscale histogram where h represents specific grayscale value. Grayscale histogram is a chart illustrating the number of points related to specific grayscale value h. Since it's a histogram, there is no continuous curve showed on the chart. That's the reason why grayscale histogram envelope is seldom paid attention to. Once the envelope of a grayscale histogram is traced out, it's obvious that there are many local minimum points.

Table 1 shows a grayscale histogram envelope of an image. Local minimum points of the envelope are pointed out by the arrows [11], the dotted arrows indicate the local minimums, and the solid arrows indicate the global minimum.

The local minimums satisfy Euler-Passion equation, as follows:

$$F_y - \frac{d}{dh}F_{y'} + \frac{d^2}{dh^2}F_{y''} = 0 \tag{1}$$

When the first-order variation of (2.2) is 0, we obtain the boundary conditions of H_m, and then we obtain the minimum points set $\{W_l\}$ by the positive and negative characteristics of the second-order variation of (1). By annealing principle [12], we select the interval between any two points in set $\{W_l\}$ as the energy steady state subintervals corresponding to the optimized interval solutions of the image functional.

There are 62 local minimum points on the envelope shown in Table 1, therefore, the total number of subintervals would be $C_{62}^2 = 1891$.

Table 1. A sketch map of grayscale histogram envelope and grayscale subintervals

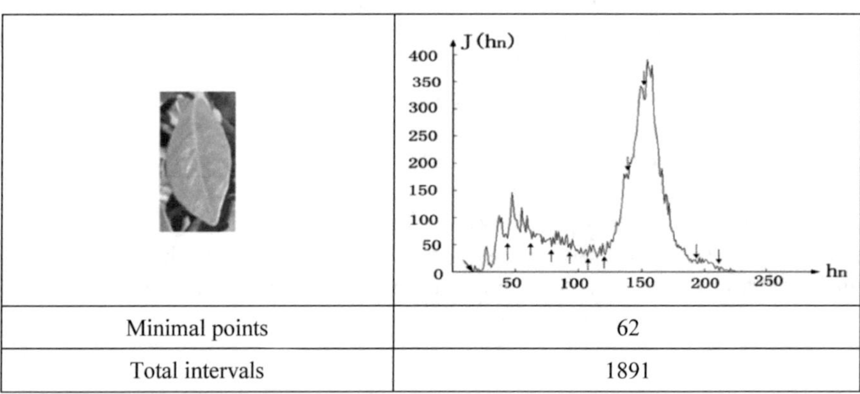

Minimal points	62
Total intervals	1891

2.2 Segmentation Operators

The Preprocess Operator

In the preprocess, we use operator T_p, which consists of image compression and wiener filtering. The comparison between images operated by T_p before and after is shown in Fig. 1.

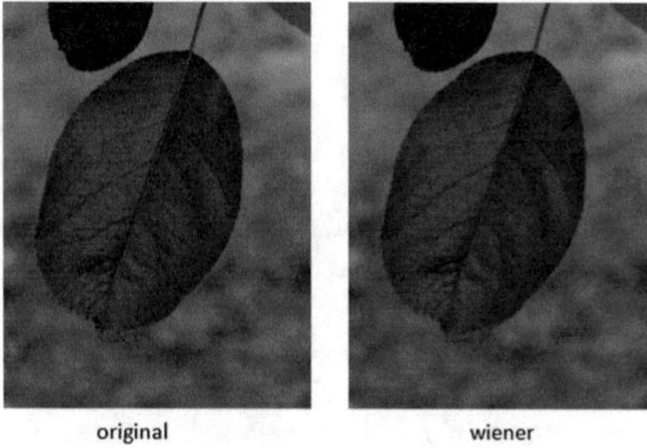

Fig. 1. The effect of preprocess operator T_p

Interval Operator Set T_i

Interval operator is an operator set represented by T_i. All the local minimum points on the grayscale histogram envelope are found though operator T_1. By the means of operator T_2, intervals between arbitrary two local minimum local points are settled. T_3 map the image into grayscale interval set by T_2. These intervals constitute grayscale subintervals which are preliminary intervals before the optimization. As shown in Table 1, through interval operator $T_1 \sim T_3$, the number of preliminary grayscale intervals is 1891.

Figure 2 shows the result of a strawberry leaf processed by T_p and T_i.

Fig. 2. The result of T_p and T_i applied on a strawberry leaf

Linear Segmentation Operator Set T_j

Just as interval operator set T_i, linear segmentation operator is also an operator set, which is represented by T_j. OTSU and CANNY are seperately denoted by T_4 and T_5. T_6 exerts logical operation on the image while T_7 imposes morphological process to the image.

Figure 3 shows a jujube leaf image which is mapped into grayscale interval [12,193] and its periodical and final results handled by linear segmentation operator $T_4 \sim T_7$.

Fig. 3. A mapped image of jujube leaf handled by operator T_j

2.3 Clustering and Optimization

In order to examine the effect of the segmentation method, four evaluation grades (PWAU) are settled as 'Perfect (P)', 'Well (W)', 'Average (A)' and 'Unsatisfying (U)'. The number of preliminary grayscale intervals is usually large. As shown in Table 1 through interval operator $T_1 \sim T_3$, the number of preliminary grayscale intervals obtained is 1891. Such scale of data is tremendous which would lead to the insufficiency of the algorithm, especially for real-time applications. In order to decrease the number of optional intervals among preliminary intervals, optimization is necessary. Clustering current operating results into the four evaluation grades PWAU, optimized grayscale intervals would be obtained. As a result, we no more need to segment images through the whole grayscale range, the optimized grayscale intervals are just sufficient and the efficiency of the algorithm would be raised.

2.4 Segmentation Procedure

In the optimized grayscale intervals segmentation procedure, we take the following steps:

Step 1 Preprocess the image with the operator T_p.
Step 2 Settle the preliminary grayscale intervals and map the image obtained in step 1 into these intervals by the means of operator set T_i.
Step 3 Impose linear segmentation on the image obtained in step 2 using linear segmentation operator set T_j.
Step 4 Cluster the images obtained in step 3 into the four evaluation grades PWAU and obtain the optimized grayscale intervals.
Further Step Once the optimized grayscale interval solutions are obtained, there is no need to segment images in the whole grayscale range, optimized grayscale intervals would be sufficient.

3 Results and Discussion

3.1 Segmentation Materials

The materials we use in this paper are four kinds of leaves, including strawberry leaf images, jujube leaf images, begonia leaf images and jasmine leaf images taken in the field. The cameras we use are a Nikon D200 and a Micro-vision VEM200sc online camera. The configuration of our computer is an Intel(R) Core(TM) i7 CPU, 960@3.20 GHz, with 6.00 GB of RAM.

The experimental materials used in the experiment are shown in Fig. 4.

Fig. 4. Experimental material images.

3.2 Image Segmentation

In order to examine the effect of the method, we classify the segmented edges of leaves into four grades by their visual appearances. The four evaluation grades are settled as 'Perfect (P)', 'Well (W)', 'Average (A)' and 'Unsatisfying (U)'.

The leaf edge appearances are classified as follows: 'Perfect' (P) represents leaf image segmentation edge results that are complete: 'Well' (W) represents leaf image segmentation edge results that are almost complete; 'Average' (A) represents leaf image segmentation edges that are not complete; and 'Unsatisfied' (U) indicates that the majority of the leaf edge is not segmented from the background.

In order to illustrate the effect of segmentation on preliminary subintervals, segmentation operated on both jasmine and strawberry leaves images are shown in Fig. 5 as an example. The preliminary experimental results of whole four kinds of leaves are shown in Table 3 and Table 4.

Figure 5 shows the segmentation operated on several different grayscale intervals of both jasmine and strawberry leaves. Table 2 shows the corresponding segmentation results classification.

Figure 5 and Table 2 enumerate four grayscale intervals to operate the segmentation method. Among all the four kinds of leaves, jujube, jasmine and begonia leaves, we choose 25 sample leaves to experiment. 9 sample leave images are taken by Nikon D200 and the other 16 sample leave images are taken by the Micro-vision VEM200sc online camera. Table 3 shows the segmentation appearance of the 16 sample leaves taken by the Micro-vision VEM200sc online camera. Table 4 shows the segmentation appearance of the 9 sample leaves taken by the Nikon D200.

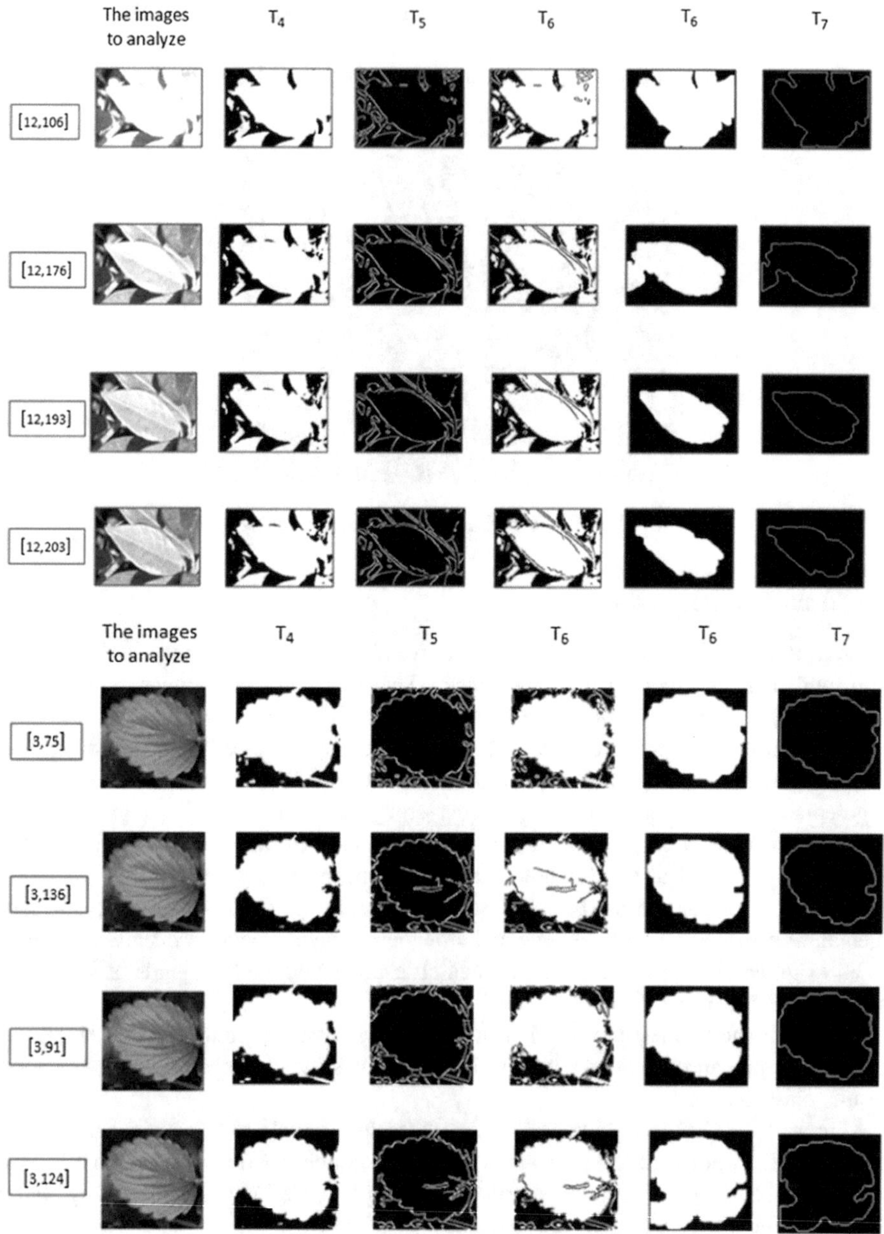

Fig. 5. Several grayscale intervals segmentation of jasmine and strawberry leaves

3.3 Optimization by Clustering

In the processing of obtaining the leaf shape from interval-mapping solutions, we analyzed the variation of the solution intervals by the four edge appearance grades and

Table 2. The corresponding classification of the results in Fig. 5

Jujube leaf mapping intervals		Segment results	Strawberry leaf mapping intervals		Segment results
h_u	h_v		h_u	h_v	
12	106	U	3	75	A
12	176	A	3	136	P
12	193	P	3	91	W
12	203	W	3	124	U

found that the probability density of these solutions has a clustering trend. The data shown in Fig. 6 were taken from different equipment (Micro-vision and Nikon). The abscissa and ordinate of Fig. 6 are the endpoint values of mapping intervals, Fig. 6 show the distribution of the interval points, and all the data in Fig. 6 are shown in Table 3 and Table 4.

Table 3. Classifications of 16 sample leaves images taken by micro-vision VEM200sc

Micro-vision		Interval number			
Segmentation appearance		Perfect	Well	Average	Unsatisfied
Interval images number	Leaf 1	3	7	0	75
	Leaf 2	2	5	10	67
	Leaf 3	2	18	4	51
	Leaf 4	108	391	74	1582
	Leaf 5	1	9	41	73
	Leaf 6	0	8	43	72
	Leaf 7	2	24	1	83
	Leaf 8	5	73	50	641
	Leaf 9	0	16	14	66
	Leaf 10	1	42	0	67
	Leaf 11	0	31	12	48
	Leaf 12	3	5	13	68
	Leaf 13	0	99	57	229
	Leaf 14	0	8	27	74
	Leaf 15	0	29	10	68
	Leaf 16	0	30	7	35
	Total	127	795	363	3299

To analyze the clustering trend, we calculated the probability density value of 10,930 mapping solution interval points from the 25 sample images above with a

Table 4. Classifications of 9 sample leaves images taken by Nikon D200

Nikon		Interval number			
Segmentation appearance		Perfect	Well	Average	Unsatisfied
Interval images number	Leaf 1	6	62	138	375
	Leaf 2	1	63	361	882
	Leaf 3	3	23	11	10
	Leaf 4	6	12	15	65
	Leaf 5	3	4	6	22
	Leaf 6	2	640	236	1283
	Leaf 7	12	32	59	869
	Leaf 8	77	255	14	703
	Leaf 9	6	25	16	49
	total	116	1116	856	4258

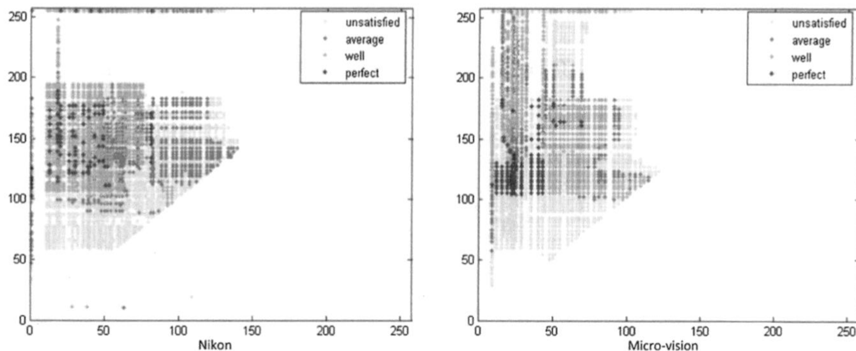

Fig. 6. The distribution of the solution interval points of 25 leaf images from two different cameras.

different clustering radius and found that the probability density clustering feature of the mapping interval points is more prominent within the radius of 10 integer point units. We defined the probability density as follows:

$$P_n = \sum_{i=1}^{R} \sum_{\substack{x \in R_i \\ y \in R_i}} \left(\frac{S_{R_i}}{N} \right) \tag{2}$$

$$P_n = \{ P_{Perfect}, \ P_{Well}, \ P_{Average}, \ P_{Unsatisfied} \} \tag{3}$$

P_n is the ratio of the sample point number to the overall integer unit point number within a radius of 10 integer point units around the common sample points. It represents the probability density of the clustering central sample point. S_{R_i} is the sample point number within the radius of R_i, and *N* is the overall unit points within the radius of R_i. R_i is the radius of the clustering, and R is 10 integer units in this paper. $P_{Perfect}$, P_{Well}, $P_{Average}$ and $P_{Unsatisfied}$ represents the probability density of the four segmentation appearance grades of sample points, for example, the probability density of the perfect segmentation appearance is as following:

$$P_{Perfect} = \sum_{i=1}^{R} \sum_{\substack{x \in R_i \\ y \in R_i}} \left(\frac{S_{PerfectinR_i}}{N} \right) \tag{4}$$

The three-dimensional distribution curved surfaces of the probability density of the solution interval points are shown in Fig. 7. The horizontal coordinates of Fig. 7 are the distributions of the solution interval points, as shown in Fig. 6, and they are divided into four segmentation appearance grades. The vertical coordinate is the probability density value of the clustering centroid sample point; the total represented the distributions of all solution interval points.

The clustering curved surfaces of the four kinds of interval points with different segmentation edge appearance and the clustering curved surfaces of the total segmented interval points are constructed to observe the clustering trend, which is shown in Fig. 7.

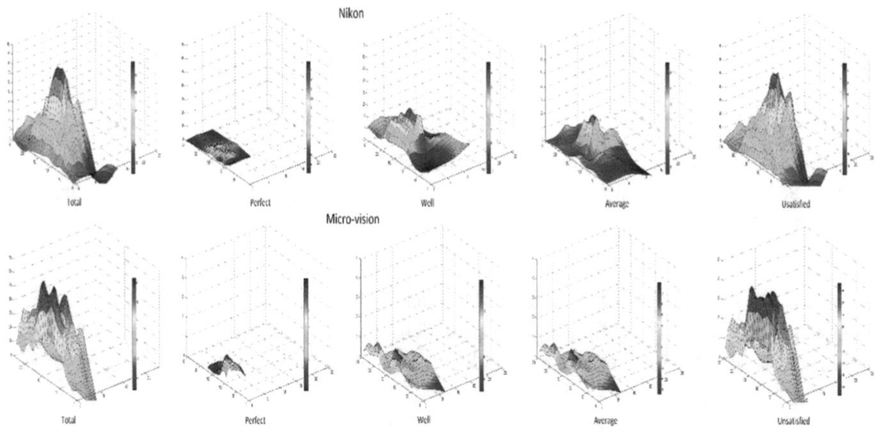

Fig. 7. Clustering curved surface of the probability density.

The probability density distributions of the P and W segmented interval points are shown in Fig. 8, and the probability density values have a clustering trend in the top 30%, 50% and 70% of the P and W appearance samples.

To judge the clustering effect, we obtained the polygonal edges *Polygon*(x, y) of the P and W segmented sample points in different top percentages of the probability density value by sorting the polar angles $\phi(i)$ of all vectors from the central point

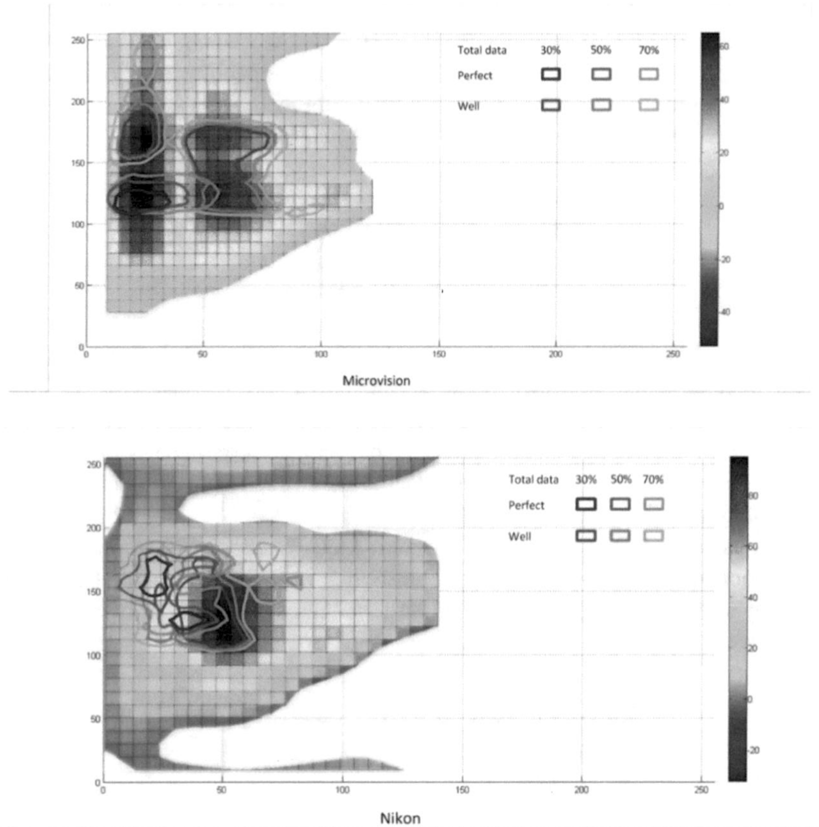

Fig. 8. The different top percentages of the probability density value of the P and W segmented samples are in blue and purple polygons, and the background is the distribution of the total segmented samples.

C(x,y) of the samples to arbitrary sample point (x_i, y_i). Then, we obtained the total sample points within the P and W segmented sample polygon regions by nonzero winding rule and by indicating the number of the total segmented points inside and on the polygonal edges [13], as shown in Table 5 and Fig. 9.

The ratios of the inner points of the P and W segmented to the inner points of the total segmented samples in the polygonal region increased, compared with the ratios of the inner points of P and W segmented to the total segmented sample points. The inner points are the points inside and on the P and W segmented samples polygonal edges. The sample numbers are shown in Table 6.

The data in Table 6 show the results of clustering. T/IT represents the increasing times from I/T to I/IT, as shown in Fig. 10.

The blue line is I/IT, which represent the ratio of P or W samples to the total samples inside the polygon edges; the red line is I/T, which represent the ratio of the inner P or W samples to the total samples. The ratio of I to IT is increased to

Table 5. Algorithm of calculating the total sample points inside and on the P and W polygonal edges

Algorithm 1. Calculating the total sample points inside and on the P and W polygonal edges
1: Calculate the probability density of common sample with $P_n = \sum_{i=1}^{R} \sum_{\substack{x \in R_i \\ y \in R_i}} \left(\frac{S_{R_i}}{N}\right)$
2: Fetch and sort the top percentage of the clustering samples with $Sort(P_n)$
3: Calculate the polar angles of the sorted clustering samples with $\phi(i) = \arctan((y(i) - C(y))/(x(i) - C(x)))$
4: Sort the polar angles of the fetched clustering samples $Sort(\phi(i))$
5: Draw the clustering polygon $Polygon(x, y)$ with the sorted sample points
6: Calculate the number of the total interval points inside and on the clustering polygonal edges by $In - polygong(total(x, y))$

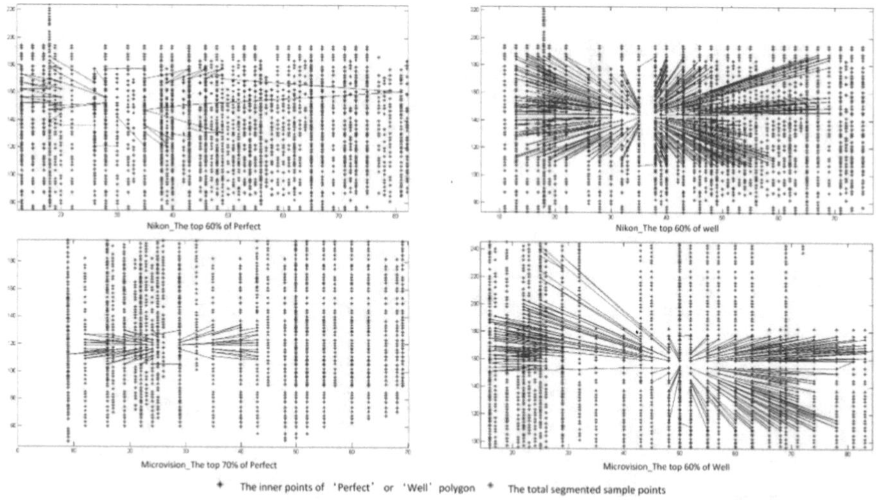

Fig. 9. The inner points of the P and W polygonsand the total segmented sample points.

significantly higher levels than the ratio of I to T, as shown in Fig. 10. Therefore, we could significantly decrease the traversal times of mapping to obtain a P or W segmented appearance leaf edge by selecting the clustering regions of the P or W polygons as traversal mapping and segmenting regions, for example, if we select the Microvision-perfect sample region in the top 71.65% of probability density value as the traversal mapping and segmenting area, we could segment only 321 times to obtain 91 perfect leaves edges instead of traversal segmenting a total of 4,584 times.

According to the analysis and the process above, we could simplify the solution process by selecting the mapped interval solutions in the regions of the P and W segmented appearance with higher clustering probability density values to obtain the segmented leaf edge, instead of traverse-segmenting all mapped interval solutions.

Table 6. The results of the clustering.

Micro-vision-perfect	%	8.66	20.47	30.71	39.37	53.54	62.20	71.65	79.53	93.70
	I	11	26	39	50	68	79	91	101	119
	IT	52	112	166	204	260	294	324	376	525
	T/IT	88.15385	40.9286	27.6145	22.4706	17.6308	15.5918	14.1482	12.1915	8.7314
Microvision-well	%	9.69	21.01	29.81	40.25	50.06	59.50	70.06	79.87	90.94
	I	77	167	237	320	398	473	557	635	723
	IT	208	524	690	908	1,150	1,358	1,587	1,774	2,115
	T/IT	22.0385	8.7481	6.6435	5.0485	3.9861	3.3756	2.8885	2.5840	2.1674
Nikon-perfect	%	9.48%	16.38	30.17	41.38	50.00	65.52	75.86	85.34	88.79
	I	11	19	35	48	58	76	88	99	103
	IT	119	270	481	653	758	905	1,081	1,303	1,333
	T/IT	53.3277	23.5037	13.1934	9.7182	8.3720	7.0122	5.8705	4.8703	4.7607
Nikon-well	%	9.68%	20.52	30.47	41.13	50.63	59.86	69.44	80.20	90.32
	I	108	229	340	459	565	668	775	895	1,008
	IT	331	635	946	1,201	1,451	1,732	1,993	2,316	2,689
	T/IT	19.1722	9.9937	6.7082	5.2839	4.37354	3.6640	3.18415	2.7401	2.3600

%: The top percentage of probability density value in P or W polygons.
I: The inner perfect sample points of the P or W polygons with different top percentages.
T: The total sample points: T-Nikon is 6,346, and T-Micro-vision is 4,584.
IT: The inner total sample points of the P or W polygons with different top percentages.

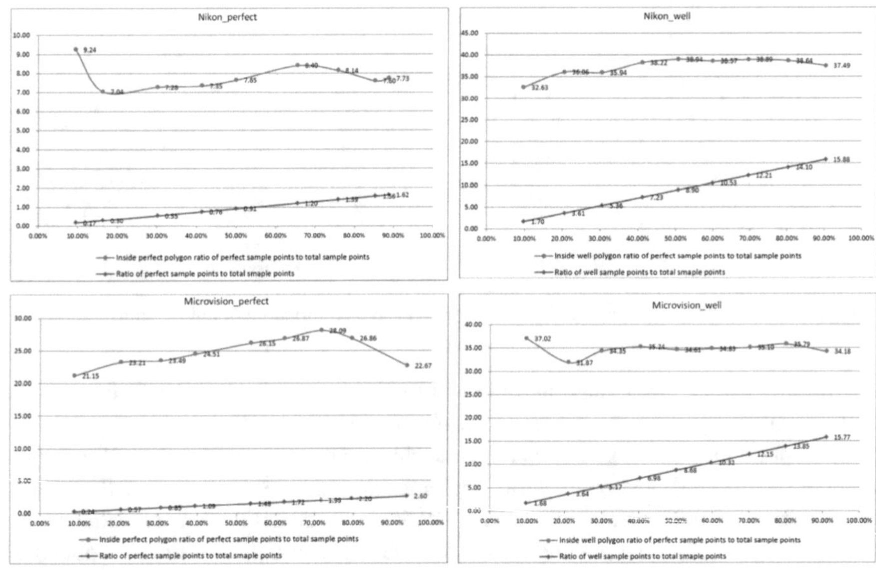

Fig. 10. The effect of probability density clustering of the P and W segmentation appearance.

4 Conclusion

In this paper, we propose an optimized segmenting method based on previous research [14]. In the previous study, we segment the jujube leaf image taken in the field in the whole grayscale range. In this paper, instead of processing images in the whole grayscale range, we choose several subintervals of grayscale to operate on.

Our optimized grayscale intervals segmentation method proposed in this paper, preprocess the original leaf images with preprocessing operator T_p. Grayscale subintervals are preliminarily set by the interval operator set T_i. Basically, the grayscale subintervals are chosen among the intervals between any arbitrary local minimum points of the histogram envelope. Original images are mapped into chosen grayscale intervals and subsequently linear segmented by the linear segmentation operator set T_j. The effect of operators T_p,T_i,T_j are shown in Sect. 2.

In order to examine the effect of the method, four evaluation grades are settled as 'Perfect (P)', 'Well (W)', 'Average (A)' and 'Unsatisfying (U)'. 25 sample leaf images are chosen to operate the segmentation on. Segmentation operated on the preliminary grayscale intervals are shown in Sect. 4, Table 3 and Table 4.

We optimize the method by the clustering analysis of mapping interval points. By analyzing the clustering of the probability density of interval points, the traversing segmentations of all mapped solutions could be replaced by the segmentations of the interval solutions within the clustering regions of the P and W appearance polygons shown in Table 6 and Fig. 10.

Clustering segmenting results into these four grades, optimized grayscale interval solutions are obtained. Experiments on three kinds of leaves including jujube, begonia and strawberry leaves shows that optimized grayscale intervals are sufficient to segment target leaf from the background with high effectiveness and accuracy.

Innovations of the optimized grayscale intervals segmentation method proposed in this paper are as follows:

(1) Instead of processing images in the whole grayscale range, we choose several subintervals of grayscale to operate on.
(2) The envelope of the grayscale histogram is seldom noticed in the image processing. In this paper, we choose the grayscale intervals in accordance to the grayscale histogram envelope.
(3) Through clustering analysis, we find a clustering trend in the segmentation. Experiments on three kinds of leaves including jujube, begonia and strawberry leaves show that optimized grayscale intervals are sufficient to segment target leaf from the background with high effectiveness and accuracy. Clear, smooth and accurate edges of the leaves could be segmented on these optimized grayscale interval solutions.

The method optimized grayscale intervals segmentation proposed in this paper provides a new thought to operate segmentation. Optimized grayscale interval solutions narrow the range on which segmentation operates, therefore, this method is much more efficient than the previous segmentation algorithm in previous research (Wang, 2013).

References

1. He, Y., Xu, Q., Xing, S.: The denoising method based on partial differential equation. J. Geomatics Sci. Technol. **4**(4), 284–286 (2007)
2. Wang, C., Ye, Z.: The image enhancement algorithm and false color mapping based on variation. J. Data Acquis. Process. **20**(1), 18–22 (2005)
3. Wang, W.: The study of image segmentation based on partial differential equation. Master Thesis of ChongQing University, 4 (2010)
4. Lu, Z.: The study of image restoration technique based on partial differential equation. Master Thesis of China University of Mining and Technology, 5 (2012)
5. Wu, C., Lu, G., S, Zhang: The concavo-convex discriminant algorithm of polygon vertices based on topological mapping. J. Comput.-Aided Des. Comput. Graphics **14**(9), 810–813 (2002)
6. Zhang, S., Tan, J., Peng, Q.: The view profile information automatically access algorithm that based on topological mapping. J. Image Graph. **6**(10), 1016–1020 (2001)
7. Jun, S.: Improved 2D maximum between-cluster variance algorithm and its application to cucumber target segmentation. Trans. CSAE **25**(10), 176–181 (2009)
8. Qian, D.: Research on influencing factors of image segmentation for crop and weed identification. Jiangsu University, Jiangsu (2006)
9. Wang, J., He, J., Han, Y., Ouyang, C., Li, D.: An adaptive thresholding algorithm of field leaf image. Comput. Electron. Agric. **96**, 23–39 (2013)
10. Bulanon, D.M., Kataoka, T., Ota, Y., et al.: A segmentation algorithm for the automatic recognition of Fuji apples at harvest. Biosys. Eng. **83**(4), 405–412 (2002)
11. Feijun, O.: Variation method and its application. Higher Education Press, Beijing (2013)
12. Li, H.: The image segmentation algorithm based on functional extremum. Master Thesis of Central South University (2009)
13. Hormann, K., Agathos, A.: The point in polygon problem for arbitrary polygons. Comput. Geom.-Theory and Application **20**(3), 131–144 (2001)
14. Wang, J., He, J., Han, Y., Ouyang, C., Li, D.: An adaptive thresholding algorithm of field leaf image. Comput. Electron. Agric. **96**, 23–39 (2013)

The Software System for Solving the Problem of Recognition and Classification

Askar Boranbayev[1(✉)], Seilkhan Boranbayev[2], Askar Nurbekov[2], and Roman Taberkhan[2]

[1] Nazarbayev University, Astana, Kazakhstan
aboranbayev@nu.edu.kz
[2] L.N. Gumilyov, Eurasian National University, Astana, Kazakhstan
sboranba@yandex.kz, nurbekoff@gmail.com, rn_82@bk.ru

Abstract. Data classification and data processing are relevant tasks in the field of automatic information processing. For successful functioning of the data classification and data processing software, the methods that provide high operation flow velocity and identification accuracy are necessary. The research showed that Viola-Jones and HOG (Histogram of Oriented Gradients) have good indexes of the effectiveness of identification and high working speed. Each of these methods are implemented by different approaches, and has their own advantages, disadvantages and limitations of applicability. The software system for solving the problem of recognition and classification has been developed based on these methods during work.

Keywords: Classification · Data · Algorithm · Method · Program · System

1 Introduction

This work was done as part of a research grant №AP05131784 of the Ministry of Education and Science of the Republic of Kazakhstan for 2018-2020.

Nowadays most of the areas of science, technology and manufacturing are oriented to develop systems in which information is transmitted by means of images. During this kind of information processing there is a variety of complex scientific, technical and technological issues. One of these issues is image handling and image recognition. Concern about image handling and image recognition is relevant because of increasing practical demands such as security systems, credit card verifications, forensic examinations, teleconferences etc. Despite that human can identifies people's faces, there is a question, how to teach this computer, also how to decode and store digital face images.

A recognition of a human face on an image is the main key concept in tasks such as emote recognitions and automated tracking after moving people in a camera view field. The task of an optimal search and a human face identification based on cybernetic vision can be considered as classical problem of perception and as new techniques. The issue of localization and face recognition was studied at an early stage of computer vision. For more than 30 years, many companies have been developing automatic systems for detecting and recognizing human faces: ASID, FaceID, Imagis, Epic Solutions, Spillman, Trueface system, SMMA (Shoot Me My Account) video capture

K. Arai et al. (Eds.): CompCom 2019, AISC 997, pp. 1063–1074, 2019.
https://doi.org/10.1007/978-3-030-22871-2_76

system, etc. Face recognition tasks can be implemented using several approaches based on the following methods: statistical methods, graph theory, neural networks. In the developed software system, HOG, Viola-Jones and convolutional neural networks were applied.

2 Methods for Solving the Recognition and Classification Problem

2.1 The HOG Method

In the HOG algorithm [1], the appearance and shape of the object in the image region are described by the distribution of the intensity gradients. The implementation of descriptors is done by dividing the image into small, connected cells. For each cell, histograms of the gradient directions are calculated. The combination of histograms is a descriptor. In order to improve the accuracy, the histograms are normalized in contrast and the intensity measure is calculated in large areas of the image, which are called blocks. The advantage of the HOG descriptor is that it operates locally and supports the invariance of geometric and photometric transformations, with the exception of the orientation of the object. The HOG descriptor is a good tool for finding people in images [2].

The algorithm implementation goes through several stages. The initial step calculates the gradient values. For this, a one-dimensional differentiating mask is used in the horizontal and/or vertical direction. In this method the color or luminance component is filtered using filtering nuclei.

Then histograms of cells are calculated, each pixel in which participates in weighted voting for the direction histogram channels based on the value of the gradients. Gradients are subject to local normalization, in order to take into account the brightness and contrast. For this, cells are grouped into larger connected blocks. The HOG descriptor is a vector of components of normalized cell histograms from all areas of the block. Rectangular R-HOGs and round C-HOGs are used - the two main block geometries. R-HOG blocks are usually square grids, characterized by three parameters: the number of cells per block, pixels per cell and the number of channels per cell histogram. The C-HOG blocks have 2 varieties: with a single central cell and divided into sectors. These blocks can be described by 4 parameters: the radius of the central ring, the expansion coefficient for the radii of the rings, the number of sectors and rings.

The final step in pattern recognition using HOG is the classification of descriptors. The system of training with the teacher is used (method of support vectors) [3, 4].

2.2 The Viola-Jones Method

Based on the principle of the scanning window The Viola-Jones method is one of the highly effective and popular methods for searching and classifying objects in images and video sequences in real time.

The research shows that this method works well and detects facial features even when observing an object at an angle of up to 30° [5], reaching a recognition accuracy value of over 85%. However, at an angle of inclination greater than 30°, the probability of face detection decreases. This disadvantage makes it difficult to use the algorithm in modern production systems, taking into account their growing needs [6].

The main concepts on which the Viola-Jones Method is based:

Use of Haar-like features. The desired object is searched applying Haar-like features.

Haar-like feature is a mapping of an image area to a set of acceptable values, that is $y : X = > D_y$ where X is the image area that computes the characteristic; D_y is the set of admissible characteristic values.

Use of an image in an integral representation. This concept is used to quickly calculate the required objects. The integral representation of the image is a matrix, the elements of which are calculated by the formula (1) and store the sum of the pixel brightness that are to the left and above this element:

$$R(x, y) = \sum_{i=0, j=0}^{i \le x, j \le y} r(i, j) \quad (1)$$

where R (x, y) is the image in the integral representation; r (i, j) is the original image.

The matrix is identical in size to the original image.

Applying of the cascading signs to quickly discard windows. This method discards windows in which no faces have been found. The advantage of the method is to increase the speed of detection, focusing the process of work on informative areas of the image. The cascade consists of several layers, which are classifiers trained with the help of the bootstrap procedure [7].

2.3 PCA (Principal Component Analysis)

In the PCA method faces are represented as a set (vector) of the main components of the images - "Eigenfaces". The image corresponding to each such vector has a face-like shape.

PCA has two useful properties when used in face recognition. First, it can be used to reduce the dimension of the feature vectors. The second useful feature is that the PCA eliminates all statistical covariance in the vectors of the transformed objects. This means that the covariance matrix for the vectors of the transformed (learning) features will always be diagonal [8].

To calculate the main components, it is necessary to calculate the eigenvectors and eigenvalues of the covariance matrix. The matrix is calculated from the image. The resulting sum of principal components will be the reconstruction of the image. The main components are calculated for each face image in the image. On average, 5 to 200 main components are used. In the recognition process, the main task is to compare the main components of an unknown image with components of known images, where the images of faces corresponding to one person are grouped into clusters in their own space. Further, the database searches and selects images having the smallest distance from the input image.

To implement the method, first of all, it is necessary to train Eigenfaces using a training sample containing images of faces that need to be recognized. The image of the trained model is fed and determined to which image from the training sample corresponds, or the image at the input does not match. The main purpose of the method is to represent the image as a sum of basic components (images):

$$y_i = \sum_{j=1}^{N} k_j l_j$$

Where Yi is the centered i-th image of the original sample, k_j - weights, l_j - eigenvectors or eigenfaces. Next, the training sample is projected into a new space. The basis of the new space is in such a way that the data in it is located optimally.

The principal component method was used to recognize faces. To compile a training set, the database was used - Olivetti Research Lab's (ORL) Face Database, which has 25 photographs of 50 different people.

Functions are called and processed:

- face_Nums - vector numbers of faces.
- variant_Nums - variant number (50 directories in each of 25 image files of the same face),

where the sample consists of 4 images.

- Read_Data_Vector. In this function, the data is sequentially read and the image is translated into a vector.

As a result, the output is a matrix whose columns are "deployed" into a vector image. Vector data are considered as points in a multidimensional space, in which the dimension is determined by the number of pixels. In this case, images of 80 × 112 size give a vector of 8960 elements or specify a point in the 8960-dimensional space.

Further, the images are normalized in the training sample, the elements common to all images are removed and only the unique information remains.

The following functions are called:

- Normalizeimg - returns the averaged sample.
- Averageimg - returns the mean vector. When this vector is collapsed in the image, you can see the "averaged face".

The next step is to calculate the eigenfaces, the weights for each image in the training sample, the covariance matrix, and identify the main components. The eigenvalues determine the variance over each of the axes of the principal components (each dimension corresponds to one dimension in space). A new selection of 29 elements for recognition is created. The first 4 elements are taken the same as in the training sample. The rest are different versions of images from the training sample. Data is processed by the Recognize procedure. In this procedure, the image is averaged, displayed in the space of the main components, and weights k are found. Then it is determined to which of the existing objects the vector k is closest to each other. To do this, use the dist function and find the minimum distances and indices of the object to which the image is located closest [9, 10].

3 The Software System for Solving the Recognition and Classification Problem

HOG and Viola-Jones methods are used to solve the recognition and classification problem in the software system.

Consider the application of the algorithm based on the HOG method for face recognition.

To detect faces in an image, it is made in black and white. color data is not needed to detect faces.

Then for each individual pixel in the image, its immediate surroundings.

It is necessary to find out how dark the current pixel is in comparison with the immediately adjacent pixels. Then the direction in which the image becomes darker is determined [11]. If you repeat this process, each pixel will be replaced with an arrow. The image is divided into small squares of 16 $\times$ 16 pixels in each. The square counts how many gradient arrows show in each direction (i.e. how many arrows point up, up-right, right, etc.). Then the considered square in the image is replaced by an arrow with the direction prevailing in this square. As a result, the original image becomes a representation that shows the basic structure of the face in a simple form.

Then an algorithm for estimating anthropometric points is used. There are 68 specific points (marks) on the face, the protruding part of the chin, the edges of each eye, the outer and inner edges of the eyebrows, and the like. Then the machine learning algorithm is set up to search for these 68 specific points on the face [11] (Fig. 1).

Fig. 1. Specific points on a human face, taken from [11]

Now that you know where the eyes and mouth are, you can rotate, resize and move the image so that your eyes and mouth are centered as best you can. Only basic image transformations are done, such as rotation and scaling, which preserve parallel lines. In this case, no matter how the face is rotated, the eyes and mouth can be centered so that they are approximately in the same position in the image. This will make the accuracy of the next step much higher.

The next step is, actually, the very face recognition. At this step, the main task is to train a deep convolutional neural network. The network creates 128 characteristics for each face. The learning process is valid when examining 3 faces at the same time:

1. The training image of a famous person's face is loaded
2. Another image of the same person's face is loaded
3. The image of the face of the new person is loaded.

Further, the algorithm studies the characteristics created for each of the three images, makes the neural network adjustment so that the characteristics created for images 1 and 2 are closer to each other, and for images 2 and 3 - further.

After repeating this step m times for n images of different people, the neural network is able to reliably create 128 characteristics for each person. Any 10–15 different images of the same person may well give the same characteristics.

The algorithm allows you to do in the database search for an image that has characteristics that are closest to the characteristics of the desired image [12–22].

Despite the existence of different algorithms, it is possible to distinguish the general scheme of the face recognition process as shown in Fig. 2.

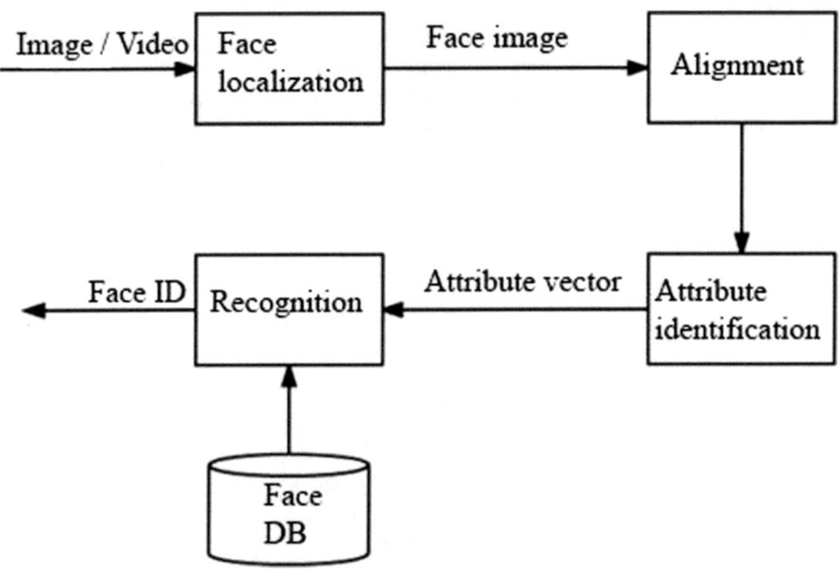

Fig. 2. The face recognition process schema

Before you begin the process of recognizing faces, you need to go through the following steps:

- localize faces on the image;
- align the face images (geometric and luminous);

- identify signs;
- directly the recognition of faces - a comparison of the found characteristics with the standards previously laid in the database.

To solve the problem of detecting faces on images and video streams in the software system, the HOG and Viola-Jones methods are used, for the recognition task, convolutional neural networks.

To carry out the work we used:

- Computer: Notebook HPEnvy 4, Intel Core i5-3317U CPU@1.70 GHz, 8 Gb RAM; Nvidia Geforce GT 740 M, 2 Gb graphics card.
- Operating system: Win10, 64-bit
- Language programming language Python 3.5; g ++ compiler; libraries for developing python-dev; library for scientific computing NumPy; library dlib, OpenFace [23], library for mathematics SciPy [24]; library BLAS - basic subroutines of linear algebra; Git - distributed version control system; Theano library for optimization and evaluation of mathematical expressions with multidimensional arrays; library cudnn for convolutional functions [25].

Figure 3 shows the interface of the software system.

Fig. 3. The graphical user interface of the software system (the image for recognition is taken from database [26]).

4 Analysis of the Obtained Results

In order to carry out a comparative analysis of the work of HOG and Viola-Jones methods, the program has a graph showing the processing time of the image. Having analyzed the methods on a large number of images, it was found that the HOG for accuracy of detection works better, is able to detect faces at various distortions, with a head rotation of up to 40°; however, according to the detection time, Viola-Jones works faster, as illustrated in Fig. 4.

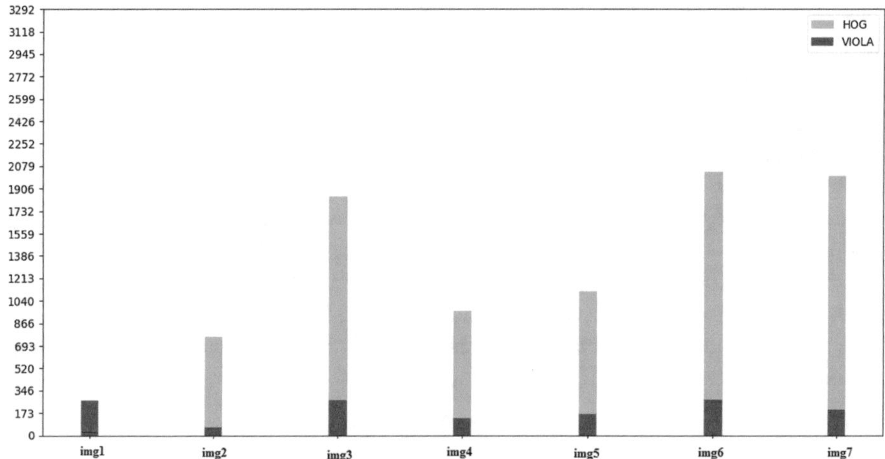

Fig. 4. Comparison of the detection algorithms running time

The Viola-Jones algorithm is trainable. For training it is necessary to have a base of positive and negative images. On positive images there are faces of people of different ages, with glasses, mustaches, etc., and on the negative - background. For successful execution of the algorithm, negative images should be larger. Since the training of classifiers is a long and complex process, the implementation of the selection stage takes place using a ready-made set of characteristics. The base is cascades from the OpenCV library: «haarcascade_frontalface_alt2.xml».

In Fig. 5, most people are found, despite the angles of the faces and blurred images. Not found only persons on which there are no characteristic points necessary for the identification of persons. On the left is the result of finding the Viola-Jones method, on the right is HOG.

To evaluate the algorithms, a sample of 50 color images totaling more than 500 faces was applied. The evaluation was carried out according to the following parameters:

CD (correct detection) - the number of persons detected by the algorithm in their presence on the images.

FPD (false positive detection) - the number of persons detected by the algorithm if they are absent in the image.

FND (false negative detection) - the number of persons present in the image and not detected by the algorithm.

The accuracy of detection is calculated [27]:

$$\Delta \text{ FPD} = \text{FPD} / (\text{CD} + \text{FND}$$
$$\Delta \text{ FND} = \text{FND} / (\text{CD} + \text{FND})$$

The final accuracy of detection of the algorithm can be calculated by the formula:

$$(1 - (\Delta\ FPD + \Delta\ FOD)) * 100\%$$

The experimental results show that the HOG method is more accurate than the Viola-Jones methods. The HOG method detects about 92% of the individuals represented on the test set of images, Viola-Jones - 85%. In addition, tests have shown that the HOG method significantly reduces the likelihood of false detection.

Fig. 5. The result of the executed algorithms

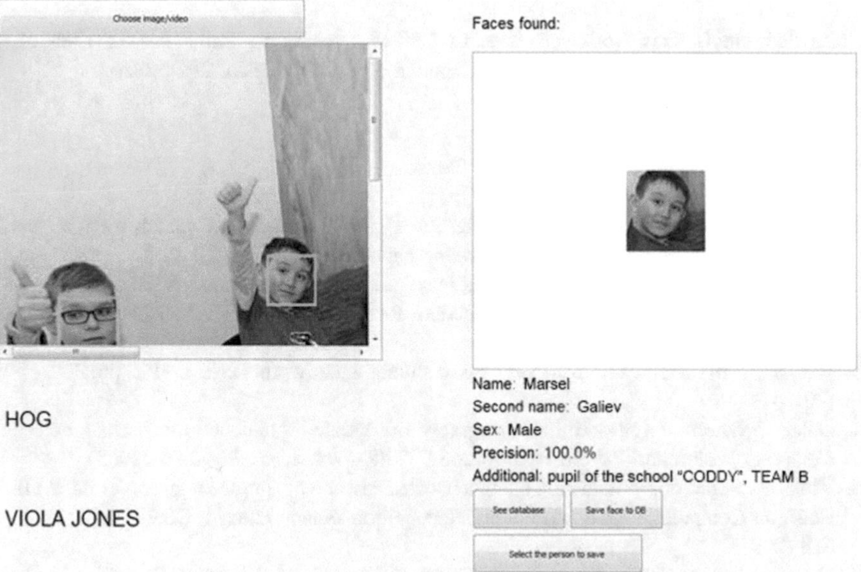

Fig. 6. The human recognition

To recognize faces in the image, a convolutional neural network was used. This method has certain advantages, such as the ability to recognize faces in real time from an input image or from a video stream; At the frontal position of the face and with a large scale, fully authentic recognition is performed; the software system gives positive results and when turning the face more than 20 degrees, as shown in Fig. 6, and under poor lighting conditions.

5 Conclusion

The Viola-Jones and HOG methods for face detection and convolutional neural networks for face recognition are used to implement the developed software system. Algorithm testing was conducted. The accuracy of image recognition by these algorithms is estimated. The software system allows you to select any of these algorithms to solve the detection and recognition problem.

The software system finds the area of the face, makes recognition of this face. Then there is a search in the database, and if the face is in the database, then there is taken available additional information (for example, name, age, etc.). A SQLite database was created, containing about 2000 images of people. Frames taken with different positions of the head and facial expressions were normalized. A set of data was created for training the neural network with variations in the form of distortions and color filters. In the program system, the existing base of famous personalities has been added. Based on the results of the research, the Viola-Jones and HOG methods produced a good detection rate, but the Viola-Jones method works faster. Accuracy of recognition of the convolutional neural network method varies from 70–100% depending on external conditions.

Acknowledgment. This work was done as part of a research grant №AP05131784 of the Ministry of Education and Science of the Republic of Kazakhstan for 2018-2020.

References

1. Extract HOG Features. [Electronic resource] (2017). http://www.mathworks.com/help/vision/ref/extracthogfeatures.html (browsing data: 25.01.2018)
2. Porikli, F.: Integral histogram: A fast way to extract histograms in cartesian spaces: Conference on Computer Vision and Pattern Recognition (CVPR), San Diego, CA, USA, pp. 829–836 (2005)
3. Wei, Y., Tao, L.: Efficient histogram-based sliding window. In: IEEE CVPR, pp. 3003–3010 (2010)
4. Dalal, Navneet, Triggs, Bill.: Histograms of oriented gradients for human detection: Computer Vision and Pattern Recognition (CVPR), vol. 1, pp. 886–893 (2005)
5. Viola, P.: Rapid object detection using a boosted cascade of simple features / Viola, P.: IEEE Conf. on Computer Vision and Pattern Recognition. Kauai, Hawaii, USA, vol. 1, pp. 511–518 (2001)
6. Viola, J.: Robust. Real-time object detection. Int. J. Comput. Vision **57**(2), 137–154 (2004)

7. Viola, P., Jones, M.J., Snow, D.: Detecting pedestrians using patterns of motion and appearance: The 9th ICCV, Nice, France, vol. 1, pp. 734–741 (2003)
8. «Method of main components», The digital library of the computer graphics and multimedia laboratory of the Computational Mathematics and Cybernetics faculty MSU, [Electronic resource]. http://library.graphicon.ru/catalog/217 (browsing date: 20.01.2018)
9. Shemi, P.M., Ali, M.A.: A principal component analysis method for recognition of human faces: eigenfaces approach. Int. J. Electron. Commun. Comput. Technol. (IJECCT) **2**(3) (2012)
10. Pahirka, A.I.: Application of the method of the main components for face recognition/Pahirka, A.I.: International Conference «Digital signal processing and its application» Moscow City, pp. 388–390 (2009)
11. Yang, M., Ahuja, N., Kriegman, D.: Face recognition using kernel eigenfaces. Image Process. IEEE Trans. **1**, 37– 40 (2000)
12. Boranbayev, S., Nurkas, A., Tulebayev, Y., Tashtai, B.: Method of Processing Big Data. Advances in Intelligent Systems and Computing, vol. 738, pp. 757–758 (2018)
13. Boranbayev, Askar, Boranbayev, Seilkhan, Nurusheva, Assel.: Analyzing Methods of Recognition, Classification and Development of a Software System // Proceedings of Intelligent Systems Conference (IntelliSys), 6–7 September 2018, London, UK, pp. 1055–1061 (2018)
14. Boranbayev, S., Altayev, S., Boranbayev, A.: Applying the method of diverse redundancy in cloud based systems for increasing reliability. Proceedings of the 12th International Conference on Information Technology: New Generations (ITNG 2015), April 13–15, Las Vegas, Nevada, USA, pp. 796–799 (2015)
15. Boranbayev, S., Boranbayev, A., Altayev, S., Nurbekov, A.: Mathematical model for optimal designing of reliable information systems. Proceedings of the 2014 IEEE 8th International Conference on Application of Information and Communication Technologies-AICT2014, Astana, Kazakhstan, October 15–17, pp. 123–127 (2014)
16. Boranbayev, A., Shuitenov, G., Boranbayev, S.: The method of data analysis from social networks using apache hadoop. Advances in Intelligent Systems and Computing, vol. 558, pp. 281–288 (2018)
17. Boranbayev, A., Boranbayev, S., Yersakhanov, K., Nurusheva, A., Taberkhan, R.: Methods of ensuring the reliability and fault tolerance of information systems. Advances in Intelligent Systems and Computing, vol. 738, pp. 729–730 (2018)
18. Boranbayev, A., Boranbayev, S., Nurusheva, A., Yersakhanov, K.: Development of a software system to ensure the reliability and fault tolerance in information systems. J. Eng. Appl. Sci. **13**(23), 10080–10085 (2018)
19. Boranbayev, S.N., Nurbekov, A.B.: Development of the methods and technologies for the information system designing and implementation. J. Theor. Appl. Inf. Technol. **82**(2), 212–220 (2015)
20. Boranbayev, S., Altayev, S., Boranbayev, A., Seitkulov, Y.: Application of diversity method for reliability of cloud computing. Proceedings of the 2014 IEEE 8th International Conference on Application of Information and Communication Technologies-AICT2014, Astana, Kazakhstan, October 15–17, pp. 244–248 (2014)
21. Boranbayev, A.S., Boranbayev, S.N.: Development and optimization of information systems for health insurance billing. ITNG2010 - 7th International Conference on Information Technology: New Generations, pp. 1282–1284 (2010)
22. Dalal, N., Triggs, B., Schimid, C.: Human detection using oriented histograms of flow and appearance: European Conference on Computer Vision (ECCV), pp. 428–441 (2006)
23. Howse, J.: OpenCV Computer vision with Python. Packt Publishing Ltd., UK, pp. 122 (2013)

24. O. B. F. B. P. L. R. P. G. D. J. T. D. W.-F. a. Y. B. J. Bergstra, «Theano: A CPU and GPU Math Expression Compiler»: Python for Scientific Computing Conference (SciPy), Austin, TX (2010)
25. Bradski, G., Kaebler, A.: Learning OpenCV. Published by O'Reilly Media, pp. 495–512 (2008)
26. The Database of Faces. [Electronic resource] http://www.cl.cam.ac.uk/research/dtg/attarchive/facedatabase.html (browsing date: 30.01.2018)
27. Omaima, N.A.: Review of face detection systems based artificial neural networks algorithms. Int. J. Multimedia & Its Appl. **6**, 448–455 (2014)

Vision Monitoring of Half Journal Bearings

Iman Abulwaheed(✉), Sangarappillai Sivaloganathan, and Khalifa Harib

Department of Mechanical Engineering, United Arab Emirates University, 15551, AlAin, United Arab Emirates
eng.iman@outlook.com,
{sangarappillai,k.harib}@uaeu.ac.ae

Abstract. This paper describes a Machine Vision system developed to continuously monitor the wear in a system made up of half journal bearings that can form part of a smart factory. The continuous monitoring facilitates the estimation of wear characteristics such as wear rate, and remaining life that is not possible in the traditional observation of Mean Time Between Failures. It was also found that the system is amenable to be part of the IoT network in a Smart Factory. A centrally loaded shaft supported by two half-journal bearings and driven at 28 rpm by an ordinary AC electric motor through a worm gear (reduction 50:1) is considered. A Logitech C920 camera was set to monitor a half-journal bearing. MATLAB Image Processing Toolbox was used to program the acquiring of images and the processing of the acquired images. Wear at specified time intervals were obtained and tabulated. The data in the Table were used to estimate and monitor wear characteristics. The system and the software developed can be used as part of a smart factory where journal bearings form part of it and the operation of the bearings can be made autonomous during their lifetime.

Keywords: Machine vision · Journal bearings · Image processing · Jackson structured programming

1 Introduction

A factory is an industrial site, consisting of buildings and large collection of machinery, where workers operate machines to manufacture goods. Smart factory is the integration of, advanced sensors, all recent IoT technological advances in computer networks, data integration, and analytics, to all manufacturing factories [1]. It represents a fully connected and flexible system that can use a constant stream of data from connected operations and production systems to learn and adapt to new demands [2]. But factories have several parts and units that wear and eventually break or become unsuitable for the intended application. Detection and monitoring of wear are rather important in tribological research as well as in industrial applications. Some typical examples are: measurement of dynamics of wear processes, engineering surface inspection, coating failure detection, tool wear monitoring and so on. Machine Vision is a non-contact and nondestructive way of measuring wear. Due to the dynamics and the complex nature of a wear process, measurement of wear is usually conducted offline [3–5] that means during measurement the wear process needs to be interrupted and the specimen needs

K. Arai et al. (Eds.): CompCom 2019, AISC 997, pp. 1075–1089, 2019.
https://doi.org/10.1007/978-3-030-22871-2_77

to be removed from the tester periodically to measure the evolution of wear as a function of time, number of cycles, or sliding distance.

A system to monitor the wear characteristics in a continuous manner is needed to (a) study the realizable benefits and (b) the possibility of incorporating it as part of a smart factory. In this research the wear of a half journal bearing was monitored using a computer vision system. The test rig consists of a motor, a gear box, a universal joint for correcting any misalignment between the gearbox and the shaft and, a pulley carrying a load of 7.5 kg fitted on the shaft using a ball bearing. The half journal bearings were monitored using a machine vision system (Logitech C920 camera). The data is acquired as an image signal and passed to the image processing system inside the computer. The digital image signal contains the pixel signal distribution and collection brightness, color and other information. The system will operate on the signal features and compare them with the data stored in the database to estimate and determine the wear. Machine vision increases production flexibility and degree of automation. It has several advantages such as relieving people from operation in hazardous positions, easy integration of information and easy implementation of intelligence in the system [6]. The schematic diagram of the test rig is shown in Fig. 1. Image Processing Toolbox of MATLAB software was used to process the images and get the wear. Image Processing has been playing an increasingly essential role in scientific areas [7, 8]. The reason for this is the ever-improving performance of computers that are now capable of quickly processing the large amount of data produced by images [9]. By converting the analogue image into the digital system and using digital image processing techniques, it is possible to extract various features from the image. Once the wear is measured data analytics can be

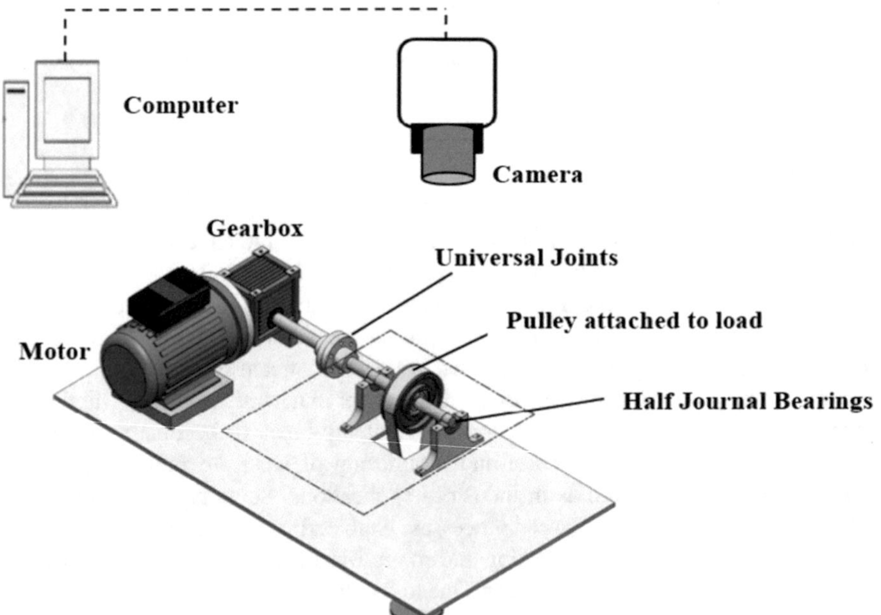

Fig. 1. System layout

applied on the data to compute items such as remaining life time of the journal bearing, the wear rate indicating the operation environment and the K factor. The values can be used to trigger remedial actions through wireless access to the Internet.

2 Literature Survey

The usage of machine vision in the determination of wear is fairly widespread in manufacturing industry. Kurada et al. [10] designed a machine vision system that can measure flank wear using a threshold to bring out the wear area. Kerr et al. [11] used a CCD camera to capture images from the tool nose by using edge operators, texture information, histogram analysis and Fourier transform. This information was used to extract the wear information from the tool. They noticed that texture information was found to be the most useful and accurate for measuring the wear. Selvaraj et al. [12] designed an image processing tool to determine the amount of wear accumulated on single point cutting tool after successive machining operations. The processing and analysis of the acquired image has been done using MATLAB software. Zhou Xing-lin and Peng Kai [13] studied the principle of subtraction to separate the image of LEDs from the original image efficiently, at the first step of image processing. Renata Klein et al. [14] used image-processing technique for bearing diagnostic. Gang Li et al. [15] discuss image processing for long distance precision inspection for bridge cracks. While these researchers concentrate on the use of machine vision measuring tool wear, no one has reported the monitoring of a half journal bearing. This paper shows the importance of using image-processing toolbox to monitor wear of half journal bearings in a continuous fashion.

3 Methodology for Measuring Wear

This section describes the methodology used for measuring the wear of bronze journal bearings using image-processing toolbox of MATLAB software. Diagrams and images are considered more communicative as compared to text. Following this the paper introduces how wear has been measured on half bearings using Jackson Structured Programming (JSP) diagrams [16]. JSP is basically a program design procedure that applies on systems with well-defined inputs and outputs. This design technique is language independent and can be used for any structured programming language. Table 1 shows the JSP symbols and description.

In this research the wear in the half journal bearing shown in Fig. 2 is monitored by a computer vision system. At specific time intervals a Logitech webcam camera takes an image or picture of the shown journal bearing assembly, which is stored in the computer memory. The vision software then analyzes the image of the region ABCD shown in Fig. 2 and estimates by how much the edge AD has moved from the first image. This is the wear. The wear is written to a file together with the lapsed time. The process continues with the image-acquiring and wear-calculating activities. The measured wear is then analyzed to monitor the wear rate, remaining life, etc. They can be used to trigger messages to the maintenance, supplies and other necessary parties.

Table 1. JSP symbols and description.

JSP Symbols	Description
	Sequence A component that has two or more parts occurring once each, and in order
o	**Selection** It is a composite task and consists of one or more parts, only one of them is executed
*	**Iteration** It is a composite task that repeats zero or more times

Fig. 2. The half journal bearing under investigation

The Methodology of measuring the wear consists of three parts:

1. Image acquisition
2. Creating the mask with first unworn image and
3. Obtaining the measurement of the wear and getting the wear versus time values table.

3.1 Image Acquisition

This is the process of taking pictures at a specified interval over a specified period of time, display the image on the computer screen, and write the picture in a file. For the process reported here the images were obtained for a period of three hours and the pictures were taken at 51 s interval so that with processing time one picture is taken every minute. To see significant amount of wear no lubricants were introduced between the journal and the bearing.

The software to execute this activity is written as a function in MATLAB whose structure is shown in Fig. 3. The function starts with setting up maximum time, pause time, resolution, file name and other similar variables under the title 'set parameters'. The camera, which is connected to form part of the image processing system, is then called to take pictures. In this research a Logitech C920 webcam camera is connected to the system. The camera in turn takes pictures in cycles. The picture taken is transferred to the computer, which displays the picture in the screen and writes it in the file present in the computer hardware during every cycle. The process is repeated until the specified maximum time is reached. Figure 3 shows the structure of the 'acquire image function'.

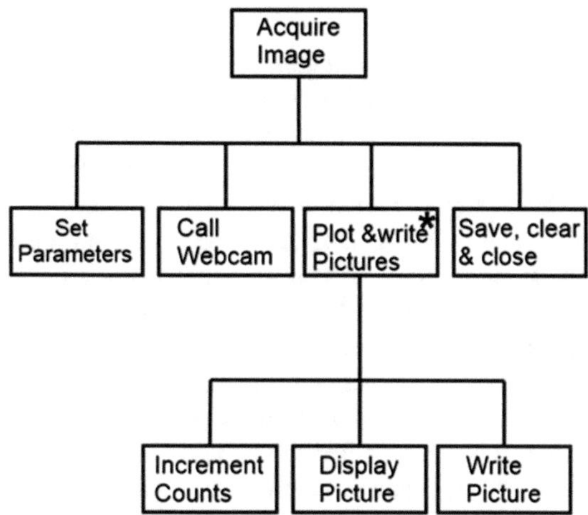

Fig. 3. Software for acquiring the image

3.2 Masking the Region of Interest

The picture covers a relatively large area in comparison to the bearings area of interest, which is marked by the reference rectangle ABCD in Fig. 2. This step is done only once on the first image in each set since the camera does not move and its position is constant for the same set. The process starts by drawing the rectangle ABCD (called polygon in Fig. 4) and setting this rectangle as the mask. Then the reference angle is set as one equal to the arc tan of y divided by the x values. The values of the drawn rectangle and the distance are set. These values are saved for use in the next program where the wear is measured.

3.3 Algorithm for Measuring Wear

Figure 5 shows the process of measuring the wear. The mask that has been created in the previous step is initialized first. Then the pictures are split into jpg and numbers. This is followed by the initialization of pixels and depth. The images are preprocessed so that wear can be obtained continuously. The required region is masked as shown in

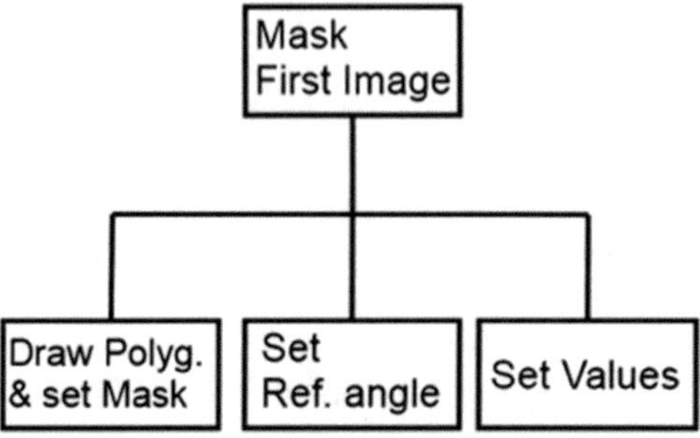

Fig. 4. Software for acquiring the image

Fig. 6(a) and cropped as in Fig. 6(b) and then the cropped image is rotated as in Fig. 6 (c) this results in adding more zeros, or in other words black edges around the image. The zeros are removed by taking row-wise and column-wise sum of the pixel values, and considering the first nonzero sum as the beginning and the end of each summed lines and the result of this step is shown in Fig. 6(d). The RGB image is converted to grayscale, after that a range filter is used to enhance only the important aspects like contour of wear. The filtered image is then binarized using a predefined threshold to capture much of the contour that indicates wear. Figure 6(e) shows the binarized image showing the boundaries of the bearing. Figure 7 shows a worn journal bearing and Fig. 8 shows the same steps for processing a worn journal bearing.

Then this binarized image is used to get the pixels' values versus rows graph. The highest peaks are detected by using differentiation command so that the boundaries of the region of interest are clearly shown. Figure 9 shows a typical 'pixel sum' vs 'Row' graph where the actual length is represented by the gap between the peaks. Every picture taken by the camera at specified time intervals will be having a similar graph from which the actual size of AB represented by this gap can be calculated during the

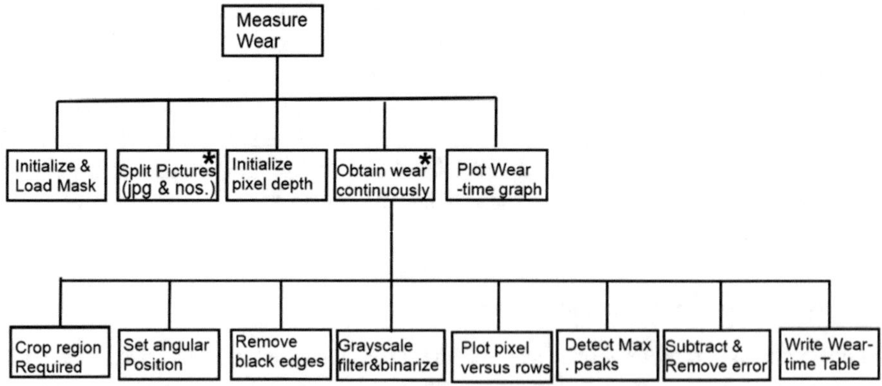

Fig. 5. Software for measuring wear

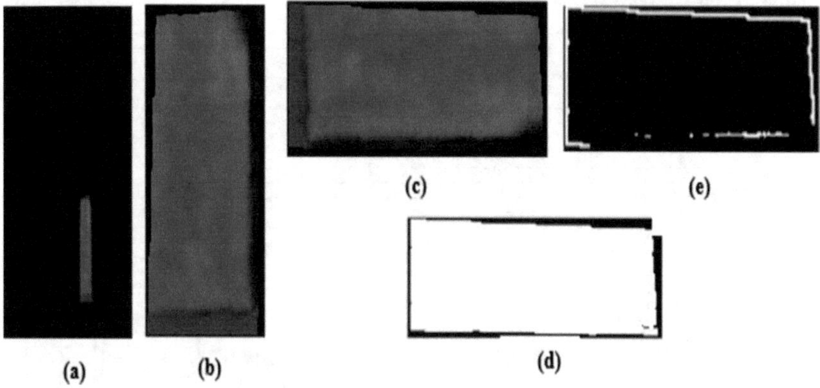

Fig. 6. Procedure of processing the unworn images

Fig. 7. The worn half journal bearing

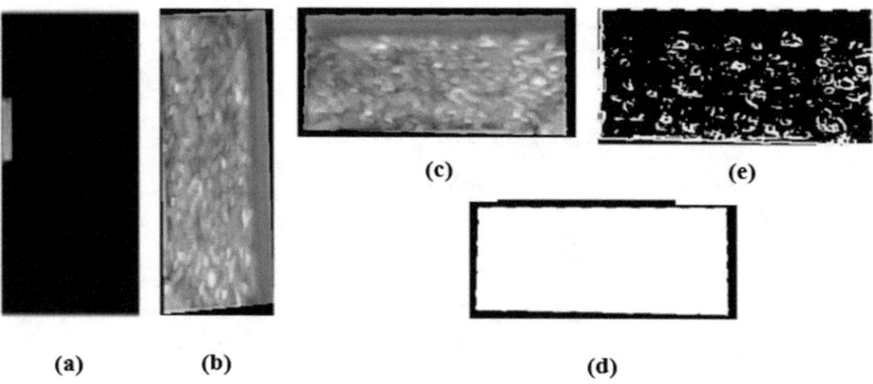

Fig. 8. Procedure of processing the worn images

processing of that picture. The difference between the reference size (from the first picture) and the current size is the cumulative wear.

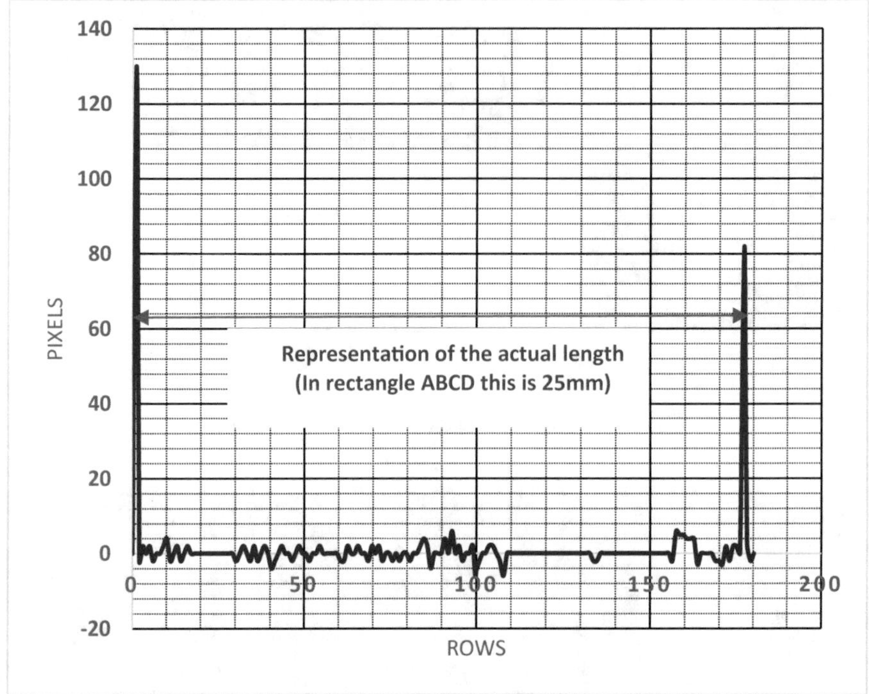

Fig. 9. Pixel sum Vs row number

The structure of the entire program to extract wear on a continuous manner is given in Fig. 10. It is a combination of the program structures discussed in Sects. 3.1, 3.2 and 3.3. The output of this is a table having wear and corresponding time as the columns. This table can be used for estimating values for various analytics such as wear rate and remaining life.

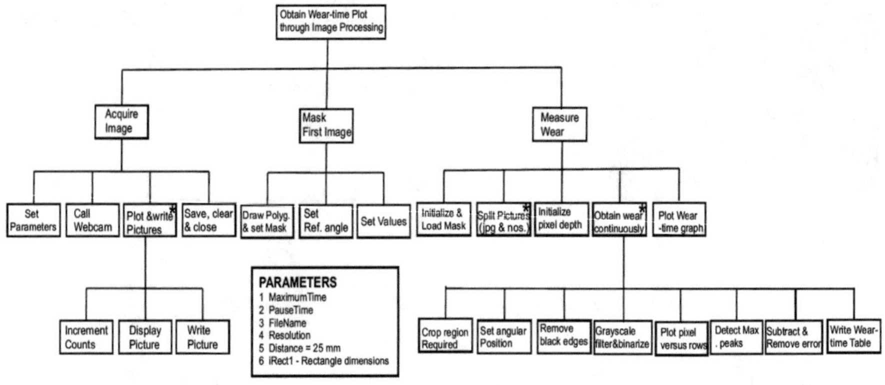

Fig. 10. Structure of the complete program

4 Results

4.1 Wear Versus Time Graph

Figure 11 shows the wear-time curve of a trial. It can be seen that the wear is increasing rapidly because no lubricant was introduced. Initially, the curve shows a rapid change in wear (0 to 15 min). The wear started to slow down after the 15 min. From 15 to 40 min the slope started to decline and from 40 to 160 min the wear was uniform. The wear reached a value of 2.25 mm at 164 min.

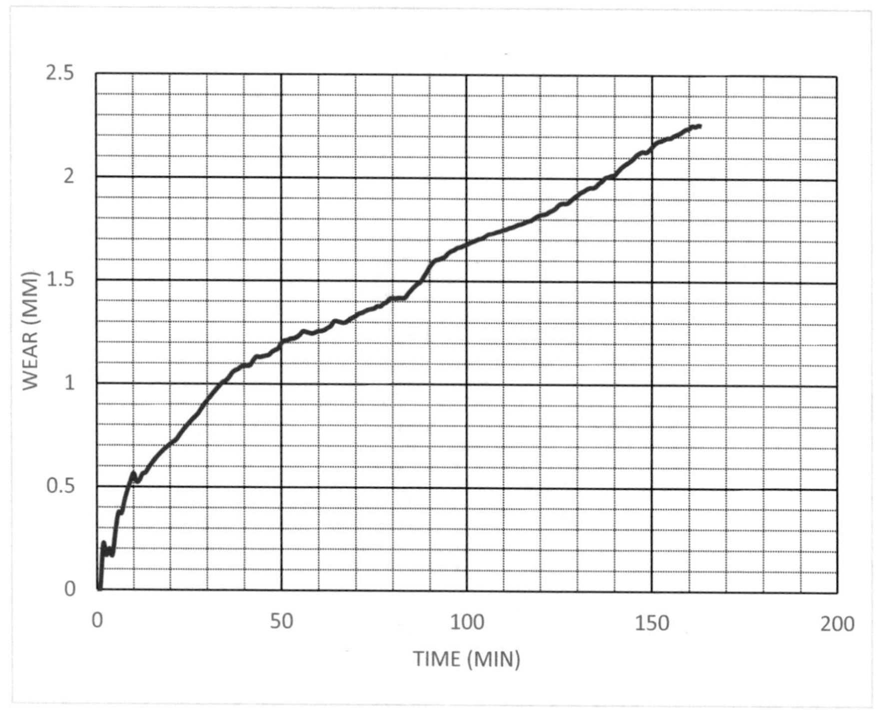

Fig. 11. Figure of wear versus time

Table 2 shows the time and wear extracted from the previous plot. Table 2 can be used for getting the analytics.

Table 2. Wear-time table

Time (min)	Wear (mm)
0	0
10.833	0.52381
15	0.61404
20	0.70667
25	0.80645
30	0.91892
35	1.0155
40	1.0884
45	1.1333
50	1.1967
55	1.2388
60	1.2557
65	1.3038
70	1.3294
75	1.3663
80	1.4158
85	1.4554
90	1.5677
95	1.6337
100	1.6766
105	1.7162
110	1.7492
115	1.7789
120	1.8218
130	1.9175
140	2.0165
150	2.1518
160	2.2409
163.33	2.2541

5 Analysis of the Results - Analytics

5.1 Analysis of the Wear

Budyanas and Nisbett in Shigley's Mechanical Engineering Design [17] explain the wear in the following way:

Consider the block of cross sectional area A, shown in Fig. 12(a) sliding through a distance S and reaches the position as shown in Fig. 9(b). In the process it undergoes a wear w as shown. Let the pressure on the wearing surface be P and the coefficient of friction be f_s.

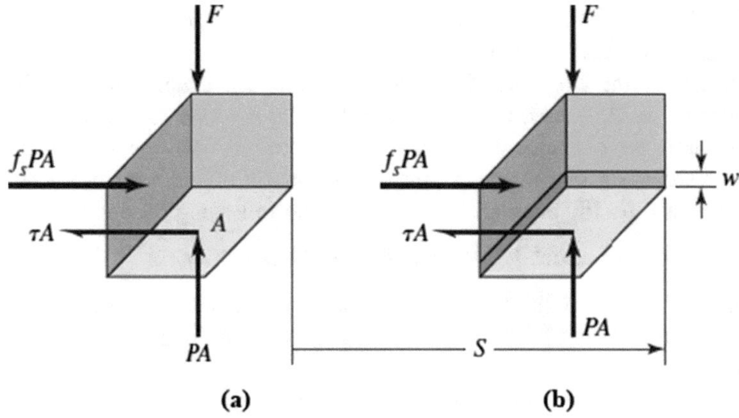

Fig. 12. Sliding block subjected to wear (adapted from [17])

The frictional force,

$$f_s P A \, Newtons \tag{1}$$

Work done in moving by a distance S is,

$$f_s P A S \tag{2}$$

But the work done is proportional to the volume of material removed. The material removed is,

$$\mathrm{wA\,mm^3} \tag{3}$$

Therefore,

$$f_s P A S \propto wA \tag{4}$$

This leads to

$$w = K_1 f_s P S \, mm \tag{5}$$

Also,

$$S = Vt \tag{6}$$

This leads to

$$w = KPVt \; mm \tag{7}$$

Where K is the combination of K_1 and f_s, called the 'Wear Factor' and is determined by experiments for different materials.

A journal bearing works satisfactorily until the wear reaches a limit at which point the sloppiness would increase vibrations and create damage to the rest of the plant units. Normally maintenance units replace the worn-out bearings. Continuous monitoring as described above would permit the estimation of the wear factor at frequent intervals for the specific bearing. This can give the following benefits:

i. Estimation of the Exact Value of K for the Given Bearing

Consider the bearing in the test rig. The motor has a speed of 1400 rpm and the gearbox has a reduction of 50. This leaves the speed of the shaft be 28 rpm. The shaft diameter is 25 mm.

Hence the peripheral velocity of rubbing equals,

$$(\pi \times 25) \times \left(\frac{28}{60}\right) \times 10^{-3} = 0.037\,\text{m/sec} \tag{8}$$

The load at the center is 7.5 kg. This can be considered as 75 N. The length of the bearing is 25 mm. Hence the,

$$\text{Average pressure} = \frac{75}{2 \times 25 \times 25} = 0.06\,\text{MPa} \tag{9}$$

Thus in general for this bearing the wear $w = K \times 0.06 \times 0.037 \times 10^3 \times t$

Now if the wear during the first 10 min is considered

$$0.52381 = \text{K} \times 0.06 \times 0.037 \times 10^3 \times 10.8333 \times 60 \tag{10}$$

Hence,

$$K = \frac{0.52381}{0.06 \times 0.037 \times 10^3 \times 10.833 \times 60} = 3.63 \times 10^{-4} \tag{11}$$

Now if the 30 min' interval from 20 min to 50 min is considered.

$$\text{Wear } w = 1.1967 - 0.70667 = K \times 0.06 \times 0.037 \times 10^3 \times 30 \times 60 \tag{12}$$

Hence,

$$K = \frac{0.49}{0.06 \times 0.037 \times 10^3 \times 30 \times 60} = 1.23 \times 10^{-4} \tag{13}$$

This tells that the K value under normal operations is much lower than the initial value. The initial K was high till the peaks and valleys in the two mating surfaces smooth themselves out. The variation of K values in 20 min' intervals is given in Table 3.

Table 3. Variation of K in 20 min interval

Time (min)	Wear (mm)	K
0	0	
20	0.70667	26.52×10^{-5}
40	1.0884	20.43×10^{-5}
60	1.2557	15.71×10^{-5}
80	1.4158	13.29×10^{-5}
100	1.6766	12.59×10^{-5}
120	1.8218	11.39×10^{-5}
140	2.0165	10.81×10^{-5}
160	2.2409	10.51×10^{-5}

Monitoring the K value can give an indication about the lubrication and environmental condition such as dust. Any increase in the value would indicate the need for checking the environment and lubrication instead of depending on Preventive maintenance.

ii. Estimation of Remaining Life

Estimation of remaining life based on specific measured values would be more reliable than those figures based on historical data. For the given setup let the permissible wear is 3 mm. If an average value of 7.2×10^{-5} is assumed for K the total life can be calculated as:

$$\begin{aligned} \textit{Total life} &= \frac{\textit{Permissible wear}}{K \times 0.06 \times 0.037 \times 10^3} = \frac{3}{7.2 \times 10^{-5} \times 0.06 \times 0.037 \times 10^3} \\ &= 312.8\,\text{Mins} \end{aligned} \tag{14}$$

As can be seen the estimated total life is 312.8 min. This value will change with the change in the K value. But it can be used as a more reliable estimate. Using this for example at the end of 140 min the remaining life can be estimated as:

$$\text{Remaining } \textit{life} = (312.8 - 140) = 172.8\,\text{mins} \tag{15}$$

iii. Application Basis for IoT

In a factory there can be several plant units that require routine replacement after certain time is elapsed. Coordinating them and replacing them during a single planned shutdown is a major task in conventional factory maintenance. With continuous monitoring and facility for the estimation of remaining lives of several units the plant units can communicate among themselves and plan an optimal time for a shutdown. Thus the vision monitoring is proving to be a powerful tool for making factories smart and journal bearings is a possible application area.

6 Conclusions

The objective of this research was to develop a system to monitor the wear characteristics in a continuous manner to (a) study the realizable benefits and (b) the possibility of incorporating it as part of a smart factory. In this research the wear of a half journal bearing was chosen to be monitored using a computer vision system. A test rig consisting of a motor, a gear box, a universal joint for correcting any misalignment between the gearbox and the shaft and, a pulley carrying a load of 7.5 kg fitted on the shaft using a ball bearing was designed and built for the investigation. The half journal bearings were monitored using a machine vision system (Logitech C920 camera). The data acquisition and processing was done in three stages. Jackson's structured program design method was employed for the program design. Image Processing Toolbox of the MATLAB software was used for coding. The rig was run for about three hours to monitor the wear and to collect data. The collected data was analyzed and the following conclusions were made:

i. Beneficial observation and measurement of the variation of the wear factor were possible with this system. This ability can make the system to operate more robustly if incorporated as part of the system.
ii. Realistic estimation of remaining life was possible with this method and this can remove a lot of uncertainties with which the conventional factories operate.
iii. The system can be incorporated as part of the IoT network in a new or retrofitted smart factory.

References

1. Jay, L.: Smart factory systems. Informatik Spektrum (2015)
2. Deloitte Development LLC: The smart factory Responsive, adaptive, connected manufacturing, A Deloitte series on Industry 4.0, digital manufacturing enterprises, and digital supply networks (2017)
3. Archard, J.F.: Contact and rubbing of flat surfaces. J. Appl. Phys. **24**(8), 981–988 (1953)
4. Glaeser, W.A.: Wear measurement techniques using surface replication. Wear **40**, 135–137 (1976)
5. Matsunaga, M., Ito, Y., Kobayashi, H.: Wear test of bucket teeth. Am. Soc. Mech. Eng. 336–342 (1979)
6. Honghui, R.: Rice seeds based on machine vision quality inspection machine. Agricultural Research (2009). (In Chinese)
7. Chelappa, R., et al.: The past, present, and future of image and multidimensional signal processing. IEEE Signal Process. Mag. **15**, 21–58 (1998)
8. Ballard, D.H., Brown, C.M.: Computer Vision. Prentice-Hall, Englewood Cliffs, NJ (1982)
9. Angrisani, L., Daponte, P., Liguori, C., Pietrosanto, A.: An image based measurement system for the characterization of automotive gaskets. Measurement **25**, 169–181 (1999)
10. Kurada, S., Bradley, C.: A review of machine vision sensors for tool condition monitoring. Comput. Ind. **34**(1), 55–72 (1997)
11. Kerr, D., Pengilley, J., Garwood, R.: Assessment and visualization of machine tool wear using computer vision. Int. J. Adv. Manuf. Technol. **28**(7–8), 781–791 (2006)

12. Selvaraj, T., Balasubramani, C., Hari Vignesh, S., Prabakaran, M.P.: Tool wear monitoring by image processing. Int. J. Eng. Res. & Technol. **2**(8), (2013)
13. Zhou, X.-l., Peng, K.: Image processing in vision 3D coordinate measurement system. In: IEEE (2009)
14. Klein, R., Masad, E., Rudyk, E., Winkler, I.: Bearing diagnostics using image processing methods. Mech. Syst. Signal Process. **45**, 105–113 (2014)
15. Li, G., He, S., Ju, Y., Du, K.: Long-distance precision inspection method for bridge cracks with image processing. Autom. Constr. **41**, 83–95 (2014)
16. Kang, R.S.: Introduction to Jackson Structured Programming (JSP), pp. 1–12 (2002)
17. Richard, G., Budynas, J.: Keith Nisbeth: Mechanical Engineering Design, 10th edn. McGraw-Hill Education, United States of America (2015)

Manual Tool and Semi-automated Graph Theory Method for Layer Segmentation in Optical Coherence Tomography

Dean Sayers[1], Maged Salim Habib[2], and Bashir AL-Diri[1](✉)

[1] University of Lincoln, Lincoln, UK
baldiri@lincoln.ac.uk
[2] Sunderland Eye Infirmary, Sunderland, UK

Abstract. Optical Coherence Tomography (OCT) is a major tool in the diagnosis of various diseases. Disease diagnosis is based on various features within the OCT images, including retinal layer positions and the distances between them and the build-up of fluid. All of these features require an expert marker in order to identify them so that the information can properly aid in the diagnosis for the patient. This process takes an incredible amount of time for the expert carry out as they need to manually trace the layers for every frame. This therefore indicates that there is a need for automation so that the expert can more easily and efficiently label the retinal layers. In this project two processes were developed. The first step is to use a semi-automated graph theory method to segment a specific layer given a rectangular region of interest, specified by the user. The output of the first process can then be corrected, where needed, using the manual tool. This method can segment layers with on average less than 1–2 pixels of error vs. two expert markers.

Keywords: Optical Coherence Tomography (OCT) · Image · Graph · Theory · Semi · Auto · Manual

1 Introduction

Optical Coherence Tomography (OCT) is a process which can capture a 3D image of the different layers of the eye including the disk, and macula. OCT is similar in concept to ultrasound, however differs as it makes use of light waves instead of sound waves [1]. The two main types of OCT are Time-Domain OCT and Spectral Domain OCT, each with advantages such as Time-Domain is easier to understand, and Spectral Domain is more sensitive and quicker [2].

OCT images are incredibly useful in the diagnosis of various diseases and pathologies. Various characteristics in the OCT images have been shown to directly correlate with disease, including retinal thickness, cystic spaces and posterior cortical vitreous [3]. These different aspects of OCT can therefore aid a clinician during the diagnosis of various diseases [3]. In total there are 11 different layers which can be isolated from an OCT image, these are in Fig. 1.

There are many different methods for the segmentation of layers with OCT images. The Graph Theory is one, and is the method chosen for use in this project. This method

K. Arai et al. (Eds.): CompCom 2019, AISC 997, pp. 1090–1109, 2019.
https://doi.org/10.1007/978-3-030-22871-2_78

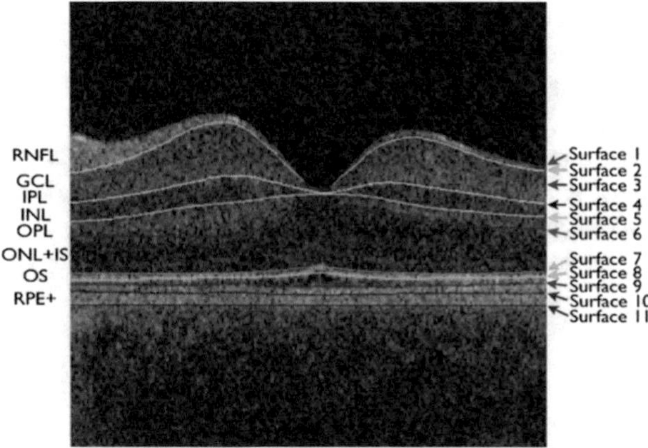

Fig. 1. Shows the labelled layers and regions of an OCT cross-sectional image [4].

has been previously shown to work fully-automatically with certain caveats. These make assumptions of the OCT beforehand. Graph Theory is good as it is very fast, which can come at a loss of accuracy which is why in this work we suggest a method for limiting the area the Graph Search can operate over, thus elevating some error. Recently Deep Learning has been used in this field for fully automatic segmentation, this has the advantage that once well-trained it should be able to handle a lot of the different diseases. However, the training can be slow and requires a lot of data. The goal of this work is to create tools which can speed up the process of labelling OCT images with both a semi-automated method and a manual tool. The manual tool may then be used for the easy creation of labelled data for deep learning methods.

This work introduces two algorithms: a semi-automated algorithm and a manual algorithm. The first stage is to use the semi-automated method for labelling OCT layers and then to correct the labelling using the manual method; which could also be used for the creation of a gold standard in order to evaluate future systems.

Section 2 presents the related work. Section 3 introduces our proposed methods. Section 4 reports the experiments and results. Finally, Sect. 5 concludes this work and discusses some considerations regarding directions for future work.

2 Related Work

The author in [5] explains the process of graph theory very well. Graph theory is the idea of representing an image as a graph of connected nodes rather than an array of pixels. This can either be 4 or 8 connected; including or not including diagonal connections. Along with considering all the pixels as nodes, two extra nodes are added to represent the foreground and background. These nodes are connected to all of the pixel nodes, as shown in Fig. 2. The aim of graph theory is to take this graph of nodes and to find the minimum path from a starting node to an end node. This is done using Dijkstra's algorithm. This algorithm will always return the shortest path across a graph of connected nodes. The aim is to have no direct path between the foreground and background nodes which were added. An important question for this method is related to how the edge weights connecting the nodes are calculated. There are three major ways this can be done; firstly, by using the intensity of the pixels alone. Secondly by creating a gradient image from the original image and then using that to calculate the edge weights. Finally, by creating a negative gradient image of the original image and then using that to calculate the edge weights.

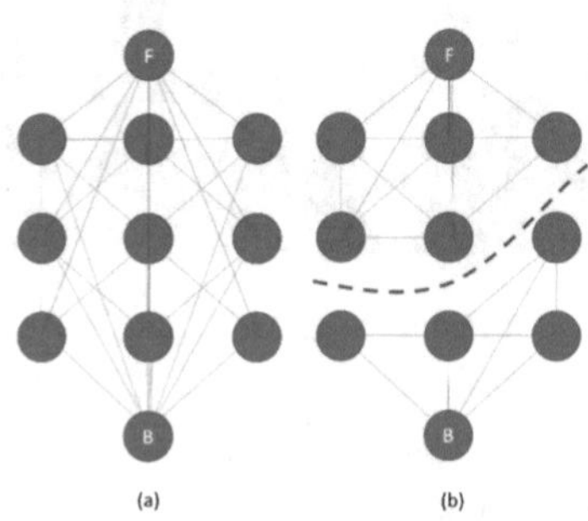

Fig. 2. A visual representation of the graph theory process of separating a connected graph (a) into foreground and background (b).

A quick and automated segmentation method using directional graph search to split the OCT into different layers given the presence of different pathologies was presented by [6]. OCT angiography uses the motion of red blood cells against static tissue as contrast. In healthy patients the vitreous is avascular, meaning that there is no blood flow above the Inner Limiting Membrane (ILM) boundary thus appearing black. Retinal Neovascularization (RNV) or the growth of new vessels above the ILM is a distinguishing trait of great diabetic retinopathy (PDR). The presence of RNV is linked to a higher risk of sight loss.

The method presented by [6] makes use of both the gradient and the inverse gradient of the OCT image, because the layer boundaries can behave in two types of ways. The first is the movement between a light intensity to a dark intensity. The second is the movement between a dark intensity to a light. Both gradients are used to build the graph which is used for segmentation. The graph search also uses a directed neighbour approach where each node only considers 5 neighbours around it.

The author in [6] explains that the use of a directional graph search in OCT data with significant tissue curvature can lead the graph search to fail. This can be resolved by flattening the OCT image based on the RPE layer.

The author in [7] presents a method for simultaneous segmentation and classification of image partitions using graph cuts. The core idea of graph cuts is presented explaining the theory of representing an image as a graph of nodes, along with a good explanation the author also demonstrates the idea visually with graphical representations of the method. The author also outlines the method for calculating the weights of each edge between the pixel nodes, this being:

$$W_{ij}^{I} = e^{\left(-\frac{\|w(i)-w(j)\|^2}{\sigma_W}\right)}$$

The numerator in this equation is simply the Euclidian distance between the two-pixel intensities. This is then divided by a tuning parameter; this tuning parameter weights the importance of each of the different aspects or features. The author concludes by demonstrating the method's ability to segment an underwater scene as well as a beach scene. Both tests show that the system is capable of the task initially outlined. The most important idea from this author, from the aspect of this work, is the formulae to obtain the weights of each of the inter pixel node edges. This is important as this work aims to compare different metrics for determining node weights when building the graph for each of the different layers. This is because it is highly likely that some metrics may perform better on some layers as opposed to others.

The author in [8] puts forward a method of using a structured learning algorithm to improve the results of a more traditional graph based retinal segmentation method. The method to do this was split into two main phases. Phase one was the training of structured forests for the edge for each layer and phase two was then to predict the edge maps for the upper boundaries of each layer. This feature can then be factored into any graph cut segmentation method along with other common features such as gradient or intensity.

The author in [8] also discusses the increasing interest for a system capable of layer segmentation. Firstly, automated width maps allow ophthalmologists to select a representative image for diagnosis. Secondly, there has been growing interest in multiple retinal layers for connecting the individual layer width or the relation between layer widths and various pathologies. An example of a pathology which results in a large deformation of the layers is diabetic macular edema (DME). In this case, support vector machines (SVM) have been used to strengthen the graph cut approach which has resulted in an improvement in performance with deformed layers. The results of this method were a mean distance error of 1.38 pixels. This was better than the state-of-the-art result which was 1.68 pixels. The dataset used to evaluate the performance of the system was the online Duke University dataset which contains 110 B-scans of 10 diabetic macular edema subjects with two experts annotating eight retinal layers [9]. Duke University's Retina OCT dataset was also used to test the proposed algorithm against a dynamic programming-based segmentation approach (AN). In this method, the image graph edges were constructed based on vertical gradients. Finally, [8] made use of two external MATLAB toolboxes to complete their system; Piotr's image and video toolbox and the edges toolbox.

A fully automated method based on path and sets (SETS) is presented by the author in [10] for the fluid segmentation in OCT volumes which contain age-related degeneration (EAMD). The author made use of two. The first was a local from UMN ophthalmology clinic containing 600 OCT scans collected from 24 EAMD subjects. The second was the OPTIMA cyst segmentation challenge which contains 4 subjects with 49 images per subject.

The author then outlines the difficulty of fluid segmentation in EAMD subjects, due to the following reasons. One reason is that the low contrast of retinal images along with the variation of size shape and location of the fluid regions makes them hard to obtain. Also, the similarity of foreground and background textures. Another key outlined feature is that RPE segmentation is challenging because sub-retinal and sub-RPE fluids are able to mislead the graph cut algorithm, whereas the segmentation of ILM is quite straightforward.

The method presented is split into three main steps:

The inner limiting membrane (ILM) and the retinal pigment epithelium (RPE) layers are both segmented using graph shortest path in the NS domain.

Seed points for the fluid and tissue are initialised for the graph cut by considering the fluid regions as object and tissue as background.

A cost function is used in kernel space and is minimised with max flow and min cut algorithms which results in a binary segmentation.

A study by [11] proposes an automated volumetric segmentation method to detect and quantify fluid in OCT images. This is an important topic as early detection and then monitoring is essential in preventing permanent visual impairment. The retinal thickness correlates with vascular leakage and can also decrease due to ischemic atrophy which occurs in some patients with DME. The method outlined by the author uses a fuzzy level-set method to both identify and quantify retinal fluid including intraretinal fluid (IRF) and subretinal fluid (SRF) on OCT structural images.

The first step was in pre-processing. The OCT images are first flattened using the ILM layer. This boundary is segmented using a directional graph search. The author lays out that DME makes automatic layer segmentation likely to fail even with robust methods. ILM ad BM can be automatically detected with a high degree of accuracy as opposed to IS and OS which can require manual intervention. Usually a Gaussian smoothing operator is used to calculate the boundary weight however the author instead opts to use a median operator as it can suppress the noise whilst maintaining the edge sharpness of the OCT images.

Finally, [11] lay out that there is a high amount of clinical relevance of IRF and SRF in OCT, as the resolution and stabilization of these is a main indicator of disease activity for the studies of DME and neovascular aged-related macular degeneration (AMD) and finally retinal vein occlusion.

The author in [12] presents a method for graph-search based segmentation of OCT images. The method is to represent the pixels in the image as nodes in a graph; each node is connected to the 8 surrounding pixels. The weights of these connections are obtained from a combination of various sources such as pixel intensity and gradient. This method differs from other graph search methods as rather than using Dijkstra's algorithm to find the shortest path, the method aims to minimise the AMAL metric. The method was evaluated on the publicly available dataset from Duke University.

$$AMAL = \frac{1}{N-1}(\sum_{i=1}^{N-1} wi)$$

The implementation of this metric over the more common shortest path metric (Dijkstra's) is significantly more challenging. This is because the segmentation algorithm needs to be able to segment different shapes, therefore the search area must be able to create cycles. The method greedily searches for the lowest AMAL, then at every iteration, the node with the lowest adjusted mean arc from the top of the queue was removed and made the top of the queue. A new AMAL is generated for each neighbour of the current node. Two additional nodes were added to the graph; these are used for automatic start and end node initialisation.

A study by [13] presents a technique which enables the general-purpose segmentation of N-dimensional images using graph theory. The presented method is an interactive segmentation method. This means that rather than a fully automated system which can determine the segmentation alone, interactive segmentation requires the addition of a user to assist the segmentation. The author explains how interactive segmentation is becoming increasingly popular as it circumnavigates problems which are present in fully automatic systems, namely that fully automatic segmentation seems to never provide perfect results for all cases.

The method explained by [13] works by requiring the user to select seed points, which are points which absolutely must be background and points which absolutely must be foreground. From these points, or hard constraints, the rest of the image is then segmented automatically by computing the best path across the image, using the max-flow algorithm, whilst also satisfying the hard constraints laid out be the user. The author shows multiple examples of the system working across multiple domains of interest, namely video sequences and medical images.

Interactive segmentation is highly relevant to this work. This is because as explained, fully automatic systems for segmentation seem to never be perfect for all cases. This is highly relevant to this project's domain of segmentation of OCT images as OCT images have several features which can make fully automatic segmentation problematic. Firstly, the similarity of layers means that they are often misclassified. Secondly, generally poor resolution of OCT images makes layer boundary harder to detect. Thirdly, presence of disease such as macular DME results in OCT images which wildly differ from normal OCT images thus resulting in poor segmentation performance.

The author in [14] set out the goal of creating a single system which can segment all 11 retinal layers from an OCT automatically. The author also lays out the importance for these types of systems, noting many important pieces of information. The first of these pieces of information is that macular edema is the swelling or thickening of the macula. This condition occurs frequently secondary to macular degeneration (AMD) which results from age. The problem with this swelling is that it affects the accumulation of fluid inside the intraretinal fluid (IRF) and underneath the neurosensory retina (subretinal fluid SRF). This then severely affects the layer structure of the retina which can lead to major visual impairment.

The method proposed by this author involves several steps. The first step is the extraction of image features from the raw PCT which are then used with manual labels to train a voxel classifier. This stage results in the creation of a probability map which is then used to perform the initial surface segmentation and then to obtain context features from the data. The surface segmentation used is a 3D graph cut method. In the second stage the features so far obtained are used to train a second classifier. The method outlined differs from other methods as it does not manually define image features. Instead, this method used the generation of convolutional kernels in an unsupervised method.

Two datasets were used through the development of this method. The first dataset was a large private dataset containing 100 OCT volumes. In addition to this first dataset, the method was also evaluated on a publicly available dataset from Duke University [9].

The author in [15] presents a method for the fully automatic segmentation of multiple retinal layers. This method uses graph theory with an iterative process of reducing the area in which the graph theory can operate over. Before this can be done this method shows the importance of first flattening the image by shifting the columns up or down, to improve the accuracy of the graph cut. This method includes two important systems which are of key importance in the improvement of the graph cut process to work on a wider range of OCT images.

1. The detection of Vessel locations, then set the nodes within the vessel locations to the lowest value for the graph cutting process. This is important as the vessels can swell beyond where the layer should be, thus reducing the accuracy of the graph cut.
2. The detection of the Fovea. This is an important step because the layers in OCT tend to converge when the fovea is present. This therefore means with Graph Theory methods, that the layers are offset when they should be touching the same point along the Fovea.

3 Proposed Methods

We are proposing two methods for labeling OCT layers; these methods could be used separately, or they could be used sequentially to save time and produce a more accurate output. The overview of this system is shown in Fig. 3. This system will be available for the public on www.xxx.lincoln.ac.uk.

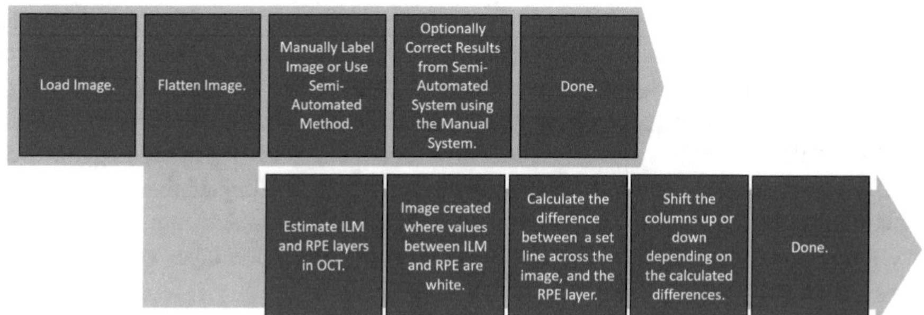

Fig. 3. Overview of semi-auto algorithm.

3.1 Flattening Algorithm

After the images have been loaded in, the second step is to flatten the OCT pseudo code shown in Fig. 4. This is because a big problem with using graph theory to determine layer boundaries in OCT images, is that OCT images can suffer from excessive curvature due to the eye being naturally curved. The more curvature present in the OCT images then the more likely the Graph Search is at making an error.

```
Attain bottom layer
    Apply [5, 40] Gaussian filter to blur the image
    Sharpen the image with a radius of 4 and an amount of 20
    Binarize the image, at a level of 0.9
    Remove any connected objects < 20pixels in area
    Apply the close operation onto the image with a [1, 12] kernel
    Remove any connected objects < 200pixels in area
    Apply the close operation onto the image with a [1, 20] kernel
    Remove any connected objects < 1000pixels in area
    For each column:
        Scan from the bottom row up, until the first white pixel is found
        Save the location

Attain top layer
    Binarize the image, at a level of 0.2
    Remove any connected objects < 50pixels in area
    Apply the open operation onto the image with a disk kernel of size 3
    For each column:
        Scan from the top row down, until the first white pixel is found
        Save the location

Smooth the top layer points using rloess at 0.06
Smooth the bottom layer points using rloess at 0.46

Create a image the same size as the input
Set all values to black
Set all values which lie between the top and bottom layer to white

Foreach column
    Calculate the difference between the bottom layer and a set points

Foreach column
    Move the column up or down depending on the current difference

Return flattened image
```

Fig. 4. Pseudo code for the flattening algorithm.

3.1.1 Flattening Algorithm Overview

The flattening algorithm takes in an RGB image; this is important as the algorithm flattens all of the colour dimensions at the same time, thus returning an RGB image. A problem with the OCT data which is available for this project is that it includes several graphical display elements which interfere with the segmentation of important features. The first step of this algorithm is to remove these graphical display elements from the top and bottom of the images.

Since the input to the algorithm is RGB, the second important step is to convert the image to being grayscale. This allows a much simpler and faster way to analyse the key features which are needed to be extracted from the image for flattening to be correctly applied. The output of this algorithm can be seen in Fig. 5.

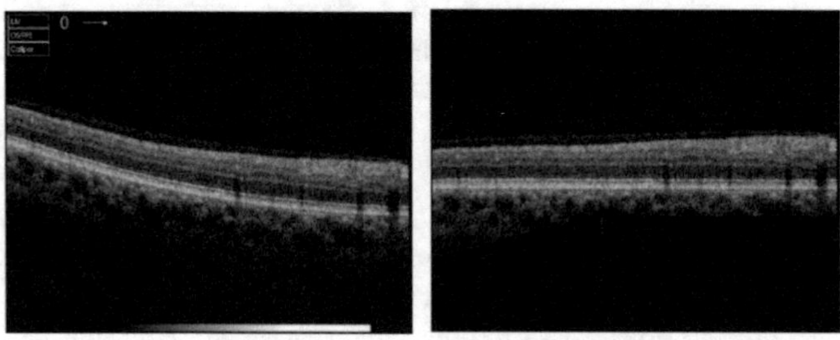

Fig. 5. OCT frame, before (left) and after (right) flattening has been applied.

3.1.2 Obtain Region of Interest

The first step of the flattening algorithm is the most fundamentally important part of the algorithm; the estimation of the ILM (top-most) and the RPE (bottom-most) layers. Of these layers, the RPE is the most significantly important layer due to the flattening of the image being applied based on the difference between the position of this layer and a fixed line across the image. The reason why both of these layers were calculated is so that a mask could be generated, as all points between them are the active region of interest for analysis. In the work presented here, the mask is not utilised as a mask to isolate the region but is utilised in the generation of the difference array which is described later. A key part of the segmentation of these two layers is that they are segmented differently from each other. This was due to a single algorithm not being able to work reliably across multiple different images.

The RPE layer is segmented by first applying a 5×40 gaussian filter to smooth the image. This is important to fill in any small holes and to connect the layers more horizontally in the case of shadows across the layers which could result in the inability to capture important information. The second step is to sharpen the image with a radius of 4 and an amount of 20 to ensure that the layers are very well defined and easy to detect. Thirdly, the image is converted to a binary image at a harsh threshold of 0.9, which is key as the bottom layer consists mainly of white pixels. The next step is to remove any small objects less than 20 pixels in area. This is of high importance as the next step closes the image, and excessive noise around the layer can distort its shape and therefore corrupt the flattening process. The next step involved applying the morphological close operation onto the image with a 1 by 12 rectangular kernel. This step ensures any small gaps between the bottom layer due to shadowing on the image are filled in. The next step involves the removal of any object which contains less than 200 pixels. This step removes any medium noise remaining in the image which would taint the accuracy of the system. The final step of the segmentation of the bottom layer is another close operation with a rectangular kernel of 1×20 to ensure that any large gaps are closed together. This is important to be applied after removing any large objects left in the image as these could be accidentally joined to the layer object thus ruining the segmentation attempt. The next step is to remove any object smaller than 1000 pixels. This ensures that there are no final pieces of noise which could affect the system. This is safe at this time as the bottom layer should be connected, thus the large size should not remove important information. The last part of this step is the identification of the location of the bottom of the object. This is done by scanning up the columns until the first white pixel is found and then storing that point. The assertation of these coordinates is important as it allows for further calculations to be much simpler.

The segmentation of the ILM layer has less noise above the ILM than there is below the RPE layer. The first step is to convert the image to binary with a threshold of 0.2, thus retaining much of the image. The second step is to remove all small objects which have less than 50 pixels. This removes much of the noise around the layer which could interfere with the result. The final step of the segmentation of the layer is to apply the morphological open operator with a disk kernel of size 3. This breaks any small gaps which may exist between the top layer and the noise, whilst also removing any of the small noise which may still exist. The last step is again to identify the coordinates of the top of the layer, similarly to the method carried out on the RPE layer, however travels down the columns to find the first white pixel rather than up.

The next step to obtain the ILM and RPE layers is to smooth the estimated points for the ILM and RPE layers. This step is highly important as it ensures that the curve properly represents the data and ignores anomalous results from areas where the segmentation may have been insufficient. The smoothing is applied using the MATLAB function smooth; this method uses the rloess method to smooth the points. For the ILM layer, a span of 6% of the total number of datapoints is used to smooth and for the RPE layer, a span of 46% of the total number of datapoints is used. 46% of the total number of datapoints is used as it is enough that the line represents the data but misses enough of the data out so that it ignores anomalous data thus giving a much truer representation in the data.

The final step to obtain the mask of the region of interest is to create a new image where pixels are white if they are between the smoothed ILM and RPE points. These points are rounded up so that they can be used as index into the image in order to easily set the points to white which are between them.

3.1.3 Calculate Difference Array

The second step of the flattening algorithm is to calculate a difference array. The difference array is an array which contains the difference between the bottom of the mask (the RPE layer boundary) and a fixed line across the image. This step is very important as the differences are used to move the columns up or down in the next step. The first step is to get the points along the bottom of the mask. This is the same method used previously to obtain the bottom where columns are scanned up until a white pixel is found and then the coordinates of that position are stored in an array. The next step is to calculate the difference between a set value of 320 and every row position found. This results in an array of how much each column needs to move in a direction to be aligned to a level across the image, in this case 320. This method was inspired by the method used in [15] however the implementation is unique.

3.1.4 Shifting the Columns

The final major step of the flattening algorithm is the moving of the columns up or down depending on the values held in the difference array previously calculated. In this step, each column of the image is processed in turn. For each column, the difference for that column is taken and tested to see if it falls into one of 3 categories. The first of these is if the current difference is 0, then the entire column should be copied as is into the output. The second is if it is less than 150 but greater than 0. In this case, the rows which are between '1 to 368' need to be copied to the rows in the output image which are '1+ current difference to 368+ current difference'. The third case is applied when the difference is less than 0. In this case, the rows of the output from '1 to 368' are populated with the rows in the image which are from '1-current difference to 368-current difference'. This method of moving the columns is done so that the image does not need to be padded and then cropped, thus saving a small amount of time to an already costly algorithm.

3.2 Semi-auto Method

The semi-autonomous method uses Graph Theory in order to attain the layers in the OCT. This step must be done after the previous flattening step has already been applied. The system could be applied to un-flattened images however the results may be less accurate.

The semi-automatic system works by iteratively attaining the layer positions of certain layers and then using those found layers to limit the region of interest of the next graph search. The order of the layers is shown in Appendix A. This ordering was important as some layers are easier to attain after others have already been located, such as the more interior layers 3, 4, and 5. These layers are retrieved much more efficiently after the layers around them have been found and then their areas ruled out of the next search.

This tool allows for a very rapid segmentation system meaning the user can quickly segment multiple layers. Furthermore, after the layers have been segmented the user can quite easily then use the manual system in order to make corrections if and where the semi-automatic system has made errors. This therefore means that overall the system can have a near perfect level of accuracy but also with a much lower amount of time needed in order to achieve.

3.3 Manual Method

One major aspect of the tool is the manual manipulation aspect. This aspect allows the user to place points onto the image and then move and manipulate those points. The points are stored so that they can be easily manipulated and retrieved. Firstly, a class is used to hold a list of points. This class allows for the addition of new points, removal of existing points, the finding of the closest point to a point passed and for the sorting of points. The retrieval of the closest point to the point passed into the function is important as it enables the system to be able to manipulate pre-existing points. The sorting of points is useful as it means new points added can be correctly rendered at the

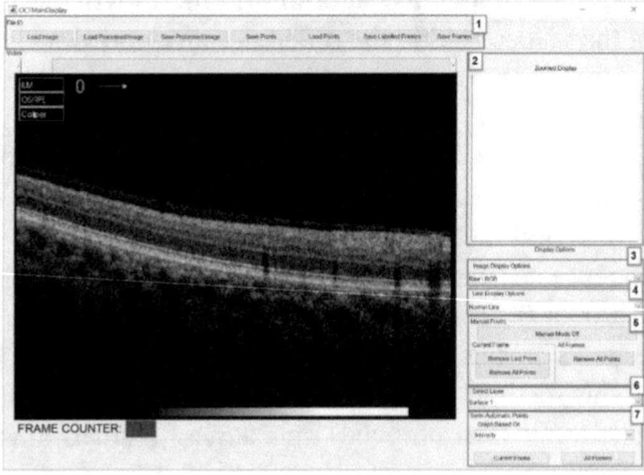

Fig. 6. Shows the tool which was created as part of this work. This tool combines most of the functions described in this chapter into one easy to use graphical interface.

correct location in the line. The instances of this class are stored in a large array. This array is 11, based on the number of layers in an OCT frame [4], by the number of frames in the OCT volume. This means that if layer 2 is being worked on, and the current frame being evaluated is 11, then the points which are needed can be quickly indexed and accessed using the class. The order of the layers is shown in Appendix B.

When manual mode is enabled via the pressing of a button shown on the graphical interface, the system goes into a while loop which is evaluated until the user taps the space key. While in this loop, the system asks the user to select a point in the image. If the user selects this point with a left mouse press, then a new point is added at the point pressed. However, if the user selects the point using the right mouse button, the system requests the frame points class, which corresponds to the current frame and layer being worked on, for the closest point that already exists to that point. The 'frame points' class then returns this as well as the distance between the two points. The distance is important because for ease of editing and manipulation it is important to only allow points to be edited if the distance is relatively small. This enables more accurate manipulation of the points. If the distance is small enough, the system asks the user to select another point in the image with the left mouse button. The closest point found previously is then updated to have the x and y value of the new point. This system allows for a simplified process of the manipulation of points thus making the system much faster and easier to use.

4 Experiments and Results

The dataset used in this work is the dataset from Duke University [9]. This dataset contains OCT frames with challenging layer boundaries due to various disease states causing extreme distortion. The file loaded contains various pieces of information; it contains a list of the OCT images and the manual labelling used in this project for all of the frames by two different manual markers.

Experiment One:

- Determine how long it takes on average to manually label an image. Note the accuracy of the manual system will not be measured, as the accuracy of this system entirely depends on the user's expertise.

Experiment Two:

- Determine how long it takes on average to use the semi-automatic algorithm to label an image. The accuracy of this method will be evaluated by testing it on 5 different OCT images.

Experiment Three:

- Determine how long it takes on average to use the semi-automatic algorithm and then the manual system to correct for mistakes. The accuracy of this method will be evaluated by testing it on 5 different OCT images.

The results were generated by finding the difference between the manual markers lines, and the lines gathered through the semi-automated algorithm. This gives an array of values, the mean, standard deviation and the root mean squared error are then generated from this.

Mean of the overall statistics across all 10 subjects.

	Mean	Median	STD
Layer 1	0.577836994	0.433096	0.435219062
Layer 2	0.853795637	0.702922	0.624440133
Layer 3	1.217442854	0.944345	0.951836906
Layer 4	1.132139027	0.851166	0.86686853
Layer 5	1.826576445	1.475508	1.397889431
Layer 6	0.632801116	0.526438	0.486162667
Layer 7	0.794172759	0.615309	0.604126209
Layer 8	0.678234218	0.521911	0.530812997
Average	0.964124881	0.758836875	0.737169492

Fig. 7. Manual Marker A vs Manual Marker B.

The timed experiments show that the automated methods, semi or semi with manual corrections, are faster than just manually labelling the image. This therefore shows how the system can increase the speed and efficiency of this process thus reducing the time and cost for the ophthalmologist.

C

TestA	RMSA	MEANA	STDA
"Layer 1 vs Manual A"	"1.2142"	"0.36425"	"1.1594"
"Layer 2 vs Manual A"	"2.1164"	"0.99094"	"1.8718"
"Layer 3 vs Manual A"	"2.2365"	"1.0605"	"1.971"
"Layer 4 vs Manual A"	"2.7678"	"1.5933"	"2.2654"
"Layer 5 vs Manual A"	"1.8758"	"0.11486"	"1.8743"
"Layer 6 vs Manual A"	"1.7566"	"1.4408"	"1.0058"
"Layer 7 vs Manual A"	"1.9597"	"1.649"	"1.0598"
"Layer 8 vs Manual A"	"1.5913"	"1.1135"	"1.138"

TestB	RMSB	MEANB	STDB
"Layer 1 vs Manual B"	"1.6933"	"1.4406"	"0.89073"
"Layer 2 vs Manual B"	"2.2986"	"1.3293"	"1.877"
"Layer 3 vs Manual B"	"2.6465"	"1.4638"	"2.2069"
"Layer 4 vs Manual B"	"3.2726"	"2.2011"	"2.4241"
"Layer 5 vs Manual B"	"3.0614"	"1.759"	"2.5081"
"Layer 6 vs Manual B"	"1.5219"	"1.2036"	"0.93224"
"Layer 7 vs Manual B"	"2.6466"	"2.3983"	"1.1201"
"Layer 8 vs Manual B"	"2.3545"	"1.8016"	"1.5174"

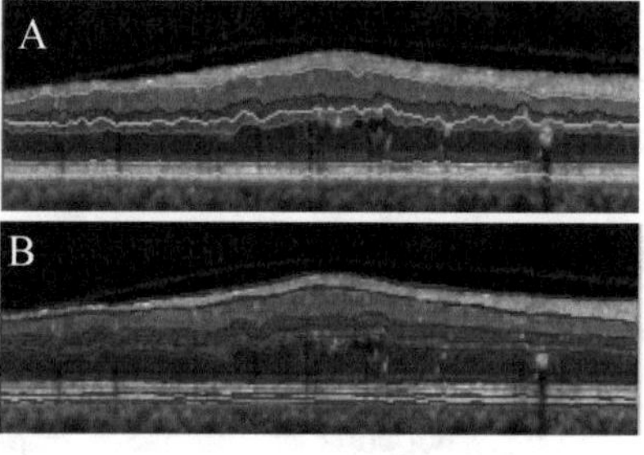

Fig. 8. Subject 3, Frame 3 From Duke dataset, layer segmentation results. "A" shows the Semi-Automated output, "B" shows both manual markings, and C shows the results.

C

TestA	RMSA	MEANA	STDA
"Layer 1 vs Manual A"	"1.9166"	"1.3504"	"1.3614"
"Layer 2 vs Manual A"	"2.6012"	"1.4646"	"2.1517"
"Layer 3 vs Manual A"	"2.3301"	"1.3399"	"1.9081"
"Layer 4 vs Manual A"	"3.8016"	"2.4376"	"2.92"
"Layer 5 vs Manual A"	"4.5557"	"2.7239"	"3.6552"
"Layer 6 vs Manual A"	"1.7676"	"1.2804"	"1.2197"
"Layer 7 vs Manual A"	"1.4859"	"0.3497"	"1.4456"
"Layer 8 vs Manual A"	"2.0612"	"1.6385"	"1.2517"

TestB	RMSB	MEANB	STDB
"Layer 1 vs Manual B"	"1.5638"	"1.0405"	"1.1686"
"Layer 2 vs Manual B"	"1.9558"	"0.2887"	"1.9362"
"Layer 3 vs Manual B"	"2.747"	"1.6229"	"2.2184"
"Layer 4 vs Manual B"	"2.9406"	"1.7165"	"2.3898"
"Layer 5 vs Manual B"	"3.8063"	"1.9607"	"3.2655"
"Layer 6 vs Manual B"	"2.6829"	"2.5268"	"0.90278"
"Layer 7 vs Manual B"	"1.6819"	"0.22955"	"1.6677"
"Layer 8 vs Manual B"	"1.919"	"1.4779"	"1.2253"

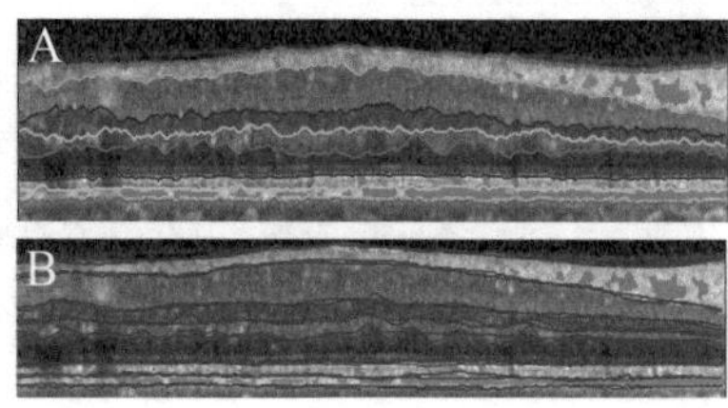

Fig. 9. Subject 2, Frame 3 From Duke dataset, layer segmentation results. "A" shows the Semi-Automated output, "B" shows both manual markings, and C shows the results.

C

TestA	RMSA	MEANA	STDA
"Layer 1 vs Manual A"	"1.725"	"0.37073"	"1.6864"
"Layer 2 vs Manual A"	"1.7889"	"0.49387"	"1.7211"
"Layer 3 vs Manual A"	"1.4075"	"0.56735"	"1.2894"
"Layer 4 vs Manual A"	"1.9708"	"1.1476"	"1.6038"
"Layer 5 vs Manual A"	"1.4716"	"0.94379"	"1.1302"
"Layer 6 vs Manual A"	"2.5849"	"2.363"	"1.0489"
"Layer 7 vs Manual A"	"3.207"	"2.7706"	"1.6167"
"Layer 8 vs Manual A"	"1.7974"	"1.4604"	"1.0488"

TestB	RMSB	MEANB	STDB
"Layer 1 vs Manual B"	"1.8342"	"1.3609"	"1.2309"
"Layer 2 vs Manual B"	"1.4282"	"0.063209"	"1.4282"
"Layer 3 vs Manual B"	"1.6963"	"0.67952"	"1.5558"
"Layer 4 vs Manual B"	"1.917"	"1.0513"	"1.6046"
"Layer 5 vs Manual B"	"1.4762"	"0.55902"	"1.3676"
"Layer 6 vs Manual B"	"2.8465"	"2.6797"	"0.96104"
"Layer 7 vs Manual B"	"3.155"	"2.404"	"2.0453"
"Layer 8 vs Manual B"	"1.6101"	"0.84925"	"1.3693"

Fig. 10. Subject 7, Frame 3 From Duke dataset, layer segmentation results. "A" shows the Semi-Automated output, "B" shows both manual markings, and C shows the results.

Figures 8, 9, and 10 demonstrate the system working at its peak performance, where it can get mean pixel difference of less than 3 pixels with most of the time getting values below 1 pixel in error. This therefore shows how accurate this system can be, as these error values are consistent with the difference between the two manual markers, shown in Fig. 7.

In comparison to the work carried out by the authors in [15], this system is unable to handle cases in which the fovea is present. The presence of this in an image makes layers tend to converge into each other in the centre of the image. This is a problem for this system as the limiting of the Graph Search rules out the layers already found so

C

TestA	RMSA	MEANA	STDA
"Layer 1 vs Manual A"	"3.2253"	"2.4129"	"2.1421"
"Layer 2 vs Manual A"	"2.3118"	"0.38096"	"2.2823"
"Layer 3 vs Manual A"	<missing>	<missing>	<missing>
"Layer 4 vs Manual A"	"4.5335"	"0.44096"	"4.5162"
"Layer 5 vs Manual A"	<missing>	<missing>	<missing>
"Layer 6 vs Manual A"	"2.5775"	"1.9629"	"1.672"
"Layer 7 vs Manual A"	"2.3377"	"1.5925"	"1.713"
"Layer 8 vs Manual A"	"1.8517"	"1.1407"	"1.46"
TestB	**RMSB**	**MEANB**	**STDB**
"Layer 1 vs Manual B"	"2.6934"	"2.0489"	"1.7499"
"Layer 2 vs Manual B"	"3.7309"	"0.42896"	"3.7096"
"Layer 3 vs Manual B"	<missing>	<missing>	<missing>
"Layer 4 vs Manual B"	"3.5787"	"0.32663"	"3.5674"
"Layer 5 vs Manual B"	<missing>	<missing>	<missing>
"Layer 6 vs Manual B"	"3.0751"	"2.7005"	"1.4722"
"Layer 7 vs Manual B"	"1.9547"	"1.3098"	"1.4523"
"Layer 8 vs Manual B"	"2.0227"	"0.79458"	"1.8619"

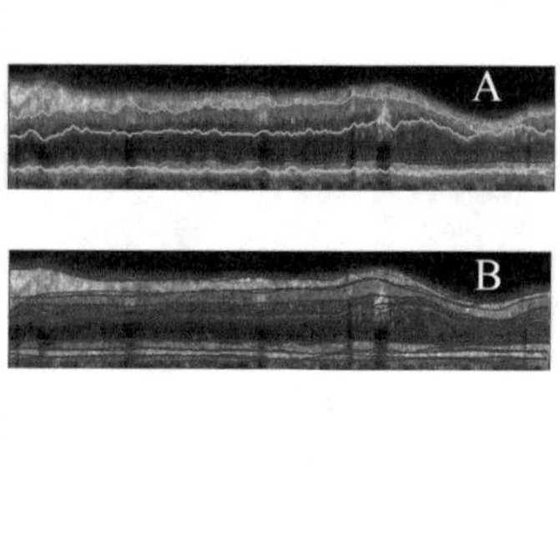

Fig. 11. Subject 5, Frame 11 From Duke dataset, layer segmentation results. "A" shows the Semi-Automated output, "B" shows both manual markings, and C shows the results.

C

TestA	RMSA	MEANA	STDA
"Layer 1 vs Manual A"	"1.8439"	"1.2858"	"1.3228"
"Layer 2 vs Manual A"	"7.9973"	"4.3533"	"6.7148"
"Layer 3 vs Manual A"	<missing>	<missing>	<missing>
"Layer 4 vs Manual A"	"5.6499"	"3.5401"	"4.4074"
"Layer 5 vs Manual A"	"8.2557"	"5.0063"	"6.5707"
"Layer 6 vs Manual A"	"2.5549"	"2.4052"	"0.8626"
"Layer 7 vs Manual A"	"3.0489"	"2.7482"	"1.3215"
"Layer 8 vs Manual A"	"1.1585"	"0.25754"	"1.1306"
TestB	**RMSB**	**MEANB**	**STDB**
"Layer 1 vs Manual B"	"1.9378"	"1.4116"	"1.3287"
"Layer 2 vs Manual B"	"9.0617"	"5.743"	"7.0159"
"Layer 3 vs Manual B"	<missing>	<missing>	<missing>
"Layer 4 vs Manual B"	"5.2443"	"3.0929"	"4.2391"
"Layer 5 vs Manual B"	"8.0751"	"5.3462"	"6.0605"
"Layer 6 vs Manual B"	"2.5592"	"2.4366"	"0.78329"
"Layer 7 vs Manual B"	"3.6437"	"3.3527"	"1.4282"
"Layer 8 vs Manual B"	"1.4362"	"0.49225"	"1.3504"

A

B

Fig. 12. Subject 8, Frame 5 From Duke dataset, layer segmentation results. "A" shows the Semi-Automated output, "B" shows both manual markings, and C shows the results.

therefore as can be seen in Figs. 11 and 12, the second layer is a few pixels below where it should be. The work carried out by the authors in [15] resolved this by locating the fovea and adjusting the layers to the edge of the fovea. Due to time constraints in this work, a similar system was unable to be developed.

When comparing the results produced by the system created in this work to others in the field, for example those produced by the authors in [15] as shown in Fig. 13, this system performed only slightly worse by approximately 1 mean pixel. However, the authors in [15] only used images from healthy subjects, whereas the system created in this work was tested using images of which a high proportion contained distortion due to disease. Therefore, a limitation in the comparison between this system and the one

	Retinal Layer Thickness Differences			
	Comparison of Automatic and Manual Segmentation			
	Column II - 29 B-scan		Column III - 100 B-scan	
Retinal Layer	Mean Difference	Standard	Mean Difference	Standard
NFL	0.99	0.76	0.88	0.68
GCLIPL	0.57	0.48	0.77	0.65
INL	1.07	0.87	0.98	0.74
OPL	1.64	1.08	1.48	1.05
ONS-IS	1.40	1.06	1.20	0.92
OS	0.92	0.80	0.88	0.73
RPE	0.96	0.69	0.99	0.87
Total Retina	0.92	0.83	0.94	0.82

Fig. 13. Snippet of results from [15]. Shows the retinal thickness of method from [15] against manual segmentation.

presented by the authors in [15] is the use of highly differing data sets. Post-processing the layers, for example by smoothing, could improve the accuracy of the system created in this work. Smoothing was applied by the authors in [15] to more accurately resemble the manual markings as due to the graph search the layers strictly follow the gradient change, whereas manual markings tend to be smoother [15].

A major difference between this work and the work presented in [15] is the inclusion of the manual tool for correcting the layers found from the semi-automated method. This is majorly important as it means corrections can be easily and quickly made, as shown in the timed experiments in Table 1. This is key as when dealing with eye conditions there may by any number of edge cases which were not accounted for and would be best resolved by a quick correction by an expert.

Table 1. Results from time measurements of each system.

	Time taken to complete action(s)
Manually label an image	~120 (~15 s a layer)
Semi-automatically label an image	~31
Semi-automatically label an image with manual corrections	~43

The method of evaluation could be improved by narrowing the type of the OCT images used for example: focusing only on images where the Macula is present in order to create a system which is much better at dealing with this case. Similarly, other categories could have been used such as presence of DME, etc. This is important as it would improve the systems capability in the focused cases of interest e.g. DME. Whereas in this work, the system was only evaluated generally on multiple types of images thus meaning it was not optimised for any edge case. The method of evaluation could also be improved by testing on a much larger set of OCT images. In this work the main issue was that there was not much time to test the system on a large set of OCT images, therefore a set of only 5 images were selected to try and cover as wide a field as possible. Therefore, the accuracy may be misleading due to the small number of images, and that they were all from one dataset. Also, other images should be included taken from different machines in order to test the generalisation of the system.

5 Conclusion and Further Work

This work presents both a Semi-Automated method for the segmentation of retinal layers in OCT, which has been shown to work well in several cases with only an average of a few pixels of error. A graphical tool which allows for the retinal layers to be manually identified was also presented as a means for correcting any errors produced in the first method. This is massively useful for the creation of expert datasets, as it allows a fast and efficient method for placing and updating the layers. The creation of larger expert datasets is important for the training of deep learning models which can require a large amount of training data to create a reliable model. Deep learning could also be used in order to increase the performance and reduce the correction time, by learning the rectangular areas so that the user does not need to manually select them each time. This would allow the system to run in a significantly lower amount of time.

Overall, the main limitation in this work was the time constraint which resulted in only being able to test a small number of OCT frames during the evaluation and the lack of implementation of post-processing and other algorithms, such as fovea location. A very limited number of OCT frames meant that although the results obtained by the system when used on these few frames were positive, an evaluation using a much larger set of OCT frames needs to be used in future work to further test the reliability of the system and to test its performance in a variety of situations. Another area of improvement would be the addition of layer post processing with either smoothing or another operation to remove small errors caused by shadowing or low resolution. This would remove anomalous data which may result in later layers being tarnished. In OCT frames containing the fovea, a useful addition may be to include a system similar to that shown in [15], where the fovea is identified, and the layers corrected accordingly. This would enable the system to work with less errors in images containing the fovea.

Appendix A

The Semi-Autonomous system operates over the following major steps, the output is shown in the image above:

1. Attain Layer 1 and 8 through graph cutting the whole frame based on a gradient image and negative gradient image respectively.
2. Attain Layer 7. Rule out areas above layer 1 and below layer 8. Ask the user to select a rectangular area (to further limit the graph cut). Graph cut the remaining area based on a gradient image.
3. Attain Layer 6. Rule out areas above layer 1 and below layer 7. Ask the user to select a rectangular area (to further limit the graph cut). Graph cut the remaining area based on a gradient image.
4. Attain Layer 2. Rule out areas above layer 1 and below layer 6. Ask the user to select a rectangular area (to further limit the graph cut). Graph cut the remaining area based on a gradient image.

5. Attain Layer 4. Rule out areas above layer 2 and below layer 6. Ask the user to select a rectangular area (to further limit the graph cut). Graph cut the remaining area based on a gradient image.
6. Attain Layer 3. Rule out areas above layer 2 and below layer 4. Ask the user to select a rectangular area (to further limit the graph cut). Graph cut the remaining area based on a gradient image.
7. Attain Layer 5. Rule out areas above layer 4 and below layer 6. Ask the user to select a rectangular area (to further limit the graph cut). Graph cut the remaining area based on a gradient image.
8. Remove all the points in the padded first and last column of the image from the found layers.
9. This process upon each iteration limits the frame, this has the following benefit, it enables less errors to be made as the graph search is being focused on a much more limited area of the image.

Appendix B

The manual tool consists of several key elements which are annotated in the image above. An image of this tool is shown in Fig. 6.

1. The file IO bar is where various saving and loading functions are carried out. The main two functions are load image and load processed image. Load image loads and flattens an OCT wmv volume into the system. This data can be saved to a.mat file using the save processed image function. This allows the data to be quickly re-loaded into the system for further inspection.
2. The zoomed display shows a zoomed version of the OCT image where a manual point has been placed or is being modified. This makes the manual manipulation aspect much more precise as it acts as a visual aid to the user.
3. This is a drop-down menu to allow the user to choose which of the versions of the OCT image they want to inspect. The options for this are; RGB, grayscale, flattened RGB and flattened grayscale. The flattened grayscale option is the mode that the system must be set to in order for the semi-automated system to work.
4. The line display drop down allows the user to modify how they want the lines to be drawn to the screen. There are 3 options for how the lines can be drawn to the screen. Normal is the mode which allows the nodes to be easily clicked and edited as square nodes are drawn for each point. Solid mode is the same as normal without the square nodes being drawn. Solid White is a significant feature as it was important that the system could be theoretically used to create training data for Deep Learning approaches easily.
5. This is the manual tool which allows points to be added or removed. This system is useful where a mistake made by the semi-automatic systems needs to be resolved. When clicking the manual mode off button, the system goes into modification mode allowing the user to add new points with left click and with right click select a point to modify then on the subsequent left click be modified to the new position.

6. The select layer drop down allows the user to select which layer they want to work on. The layer selected will be where all the points are stored, allowing the layers to be easily worked on separately.
7. This is the semi-automatic tool; the drop-down menu allows the user to select the graph building type that they wish to employ on the layer they are working on. The current frame button will run the semi-automatic procedure on the current frame. The all frames button is experimental. After one layer has been found for one frame it uses the layer to determine where the rectangle is placed on the following frame. This is repeated until that layer has been labelled across every frame.
8. The rest of the tool is just a simple video area with a track bar allowing the user to easily navigate through the frames, there is also a frame indicator so that the user is always aware of the frame that they are working on.

References

1. Huang, D., Swanson, E.A., Lin, C.P., Schuman, J.S., Stinson, W.G., Chang, W., Hee, M.R., Flotte, T., Gregory, K., Puliafito, C.A., Fujimoto, J.G.: Optical coherence tomography. Science **254**(5035), 1178–1181 (1991)
2. Sampson, D.D., Hillman, T.R.: Optical coherence tomography. In: Palumbo, G., Pratesi, R. (eds.) Lasers and Current Optical Techniques in Biology, ESP Comprehensive Series in Photosciences, vol. 4, pp. 481–571. Cambridge, UK (2004)
3. van Velthoven, M.E.J., Faber, D.J., Verbraak, F.D., van Leeuwen, T.G., de Smet, M.D.: Recent developments in optical coherence tomography for imaging the retina. Prog. Retin. Eye Res. **26**, 57–77 (2007)
4. Abramoff, M.D., Garvin, M.K., Sonka, M.: Retinal imaging and image analysis. IEEE Rev. Biomed. Eng. **3**, 169–208 (2010)
5. Radke, R.J.: Computer vision for visual effects. Cambridge University Press, New York, USA (2012)
6. Zhang, M., Wang, J., Pechauer, A.D., Hwang, T.S., Gao, S.S., Liu, L., Liu, L., Bailey, S.T., Wilson, D.J., Huang, D., Jia, Y.: Advanced image processing for optical coherence tomographic angiography of macular diseases. Biomed. Optics Exp. **6**(12), 4661–4675 (2015)
7. Eriksson, A.P., Barr, O. and Åström: Image segmentation using minimal graph cuts. Swedish Symp. Image Anal. 45–48 (2006)
8. Karri, S.P.K., Chakraborthi, D., Chatterjee, J.: Learning layer specific edges for segmenting retinal layers with large deformations. Biomed. Optics Exp. **7**(7), 2888–2901 (2016)
9. Chiu, S.J., Allingham, M.J., Mettu, P.S., Cousins, S.W., Izatt, J.A., Farsiu, S.: Kernel regression-based segmentation of optical coherence tomography images with diabetic macular edema. Biomed. Optics Exp. **6**(4), 1172–1194 (2015).http://www.osapublishing.org/boe/abstract.cfm?uri=boe-6-4-1172
10. Rashno, A., Nazari, B., Koozekanani, D.D., Drayna, P.M., Sadri, S., Rabbani, H., Parhi, K. K.: Fully-automated segmentation of fluid regions in exudative age-related macular degeneration subjects: Kernel graph cut in neutrosophic domain. PLoS One **12**(10), 1–26 (2017)

11. Wang, J., Zhang, M., Pechauer, A.D., Liu, L., Hwang, T.S., Wilson, D.J., Li, D., Jia, Y.: Automated volumetric segmentation of retinal fluid on optical coherence tomography. Biomed. Optics Exp. **7**(4), 1577–1589 (2016).http://www.osapublishing.org/boe/abstract.cfm?uri=boe-7-4-1577
12. Keller, B., Cunefare, D., Grewal, D.S., Mahmoud, T.H., Izatt, J.A., Farsiu, S.: Length-adaptive graph search for automatic segmentation of pathological features in optical coherence tomography images. J. Biomed. Optics **21**(7), (2016)
13. Boykov, Y.Y., Jolly, M.P.: Interactive graph cuts for optimal boundary & region segmentation of objects in N-D images. In: International Conference on Computer Vision, pp. 105–112 (2001)
14. Montuoro, A., Waldstein, S.M., Gerendas, B.S., Schmidt-Erfurth, U., Bogunović: Joint retinal layer and fluid segmentation in OCT scans of eyes with severe macular edema using unsupervised representation and auto-context. Biomed. Optics Exp. **8**(3), 1874–1888 (2017)
15. Chiu, S.J., Li, X.T., Nicholas, P., Toth, C.A., Izat, J.A. and Farsiu, S.: Automatic segmentation of seven retinal layers in SDOCT images congruent with expert manual segmentation. Optics Exp. **18**(18), 19 413–19 428 (2010). http://www.ncbi.nlm.nih.gov/pmc/articles/PMC3408910/

Author Index

A
AbouElNadar, Nour A., 887
Abubakar, Aliyu, 641
Abulwaheed, Iman, 1075
Aburukba, Raafat, 254
Ahmad, Jawad, 86
Akar, E., 982
Alakhras, Marwan, 149
AL-Diri, Bashir, 1090
Al-Graitti, A. J., 424
Alhoula, Wedad, 22
Ali, Najat, 929
Aljameel, Sumayh, 498
Aloul, Fadi, 254
Al-Ruzaiqi, Sara K., 785
Andressen, Andreas, 263
Andrews, W. A., 982
Aviles-Cruz, Carlos, 73

B
Bandala, Argel A., 725
Berthelson, P. R., 424
Bertocco, Sara, 179
Billones, Robert Kerwin C., 725
Bird, Jordan J., 593, 751
Bonakdari, Hossein, 607
Boranbayev, Askar, 1063
Boranbayev, Seilkhan, 1063
Braga, Juliao, 527
Brasher, Reuben, 648
Bu, Yiyi, 1048
Buckingham, Christopher D., 593, 751

C
Calimeri, Francesco, 900
Capp, Noah, 563
Carey, William, 275
Carter, A., 11
Charles, Richard M., 1012
Chatterjee, Niladri, 846
Chatterjee, Punyasha, 1
Chen, Junhao, 1001
Chernofsky, Michael, 971
Cordero-Sanchez, Salomon, 73
Cosatto, Eric, 688
Crockett, Keeley, 498
Curry, James H., 1012

D
Da Col, Giacomo, 410
Dadios, Elmer P., 725
Davari, Somayeh, 659
Dawson, Christian W., 785
De Vitis, Gabriele Antonio, 165
Debnath, Arpita, 1
Deshpande, S., 11
Durand-Dubief, Françoise, 900
Durbhakula, Murthy, 194, 243

E
Earle, Lesley, 263
Ekárt, Anikó, 593, 751
Endo, Patricia, 527
Engel, Georg, 776
Ennouamani, Soukaina, 625

K. Arai et al. (Eds.): CompCom 2019, AISC 997, pp. 1111–1114, 2019.
https://doi.org/10.1007/978-3-030-22871-2

Erdélyi, Katalin, 463
Exter, Emiel den, 275
Ezenkwu, Chinedu Pascal, 335

F
Fanah, Muna Al, 738
Faria, Diego R., 593, 751
Fayez, Mahmoud, 1036
Ferraz, Marta, 275
Ferreira, Edmundo, 275
Fielding, M., 11
Fillone, Alexis M., 725
Foglia, Pierfrancesco, 165
Fonseca Cacho, Jorge Ramón, 815
Fouad, Mohammed M., 1036
Furht, B., 982

G
Gan Lim, Laurence A., 725
Garcia, Carlos A., 376
Garcia, Marcelo V., 376
Ghadiri, Nasser, 659
Gharabaghi, Bahram, 607
Gholami, Azadeh, 607
Gikunda, Patrick Kinyua, 763
Golmohammadi, Meysam, 563
Goz, David, 179
Grabaskas, Nathaniel, 867
Grenouilleau, Jessica, 275

H
Habib, Maged Salim, 1090
Harib, Khalifa, 1075
Hartley, Joanna, 22
He, Can, 1048
Hegazy, Ola, 140
Hettiarachchi, I., 11
Hirata, Hayato, 319
Hu, Yue, 444
Hussein, Mousa, 149

I
Idris, Ahmad, 929
Ikbal, Mohammad Rafi, 1036
Isaac, Georgia, 939
Ivanov, A., 209

J
Ji, Philip, 688
Jiang, Liangyu, 1048
Jin, Caroline, 796
Jin, Jian, 126
John, Melford, 918
Jones, M. D., 424
Jones, R., 11
Joshi, Ajay, 918
Jouandeau, Nicolas, 763
Jungreis, David, 563

K
Kaleem, Mohammad, 498
Katib, Iyad, 1036
Kaur, Harpreet, 796
Khachaturyan, Mikhail Vladimirovich, 678
Khalid, G. A., 424
Khatun, Amena, 796
Kim, Iuliia, 59
Kiourtis, Athanasios, 956
Klicheva, Evgeniia Valeryevna, 678
Knizhnik, A., 209
Kor, Ah-Lian, 263
Korany, S., 11
Krueger, Thomas, 275
Kyriazis, Dimosthenis, 956

L
Laas-Billson, Liina, 299
Larijani, Hadi, 86
Latham, Annabel, 498
Latif, Muhammad, 573
Levchenko, V., 209
Liao, Yuanxiu, 517
Liu, Conggui, 474
Liu, Li, 126
Liu, Pengyu, 1001
Liu, Quanchao, 444
Liu, Wensheng, 1048
Liu, Yinhua, 474
Luo, Xudong, 517

M
Mahani, Zouhir, 625
Maia, Rodrigo Filev, 99
Mancas, Christian, 390
Mandal, Prasanta, 1
Marques, Margarida M., 113
Marques, O., 982
Marzullo, Aldo, 900
Mavrogiorgou, Argyro, 956
Meacham, Sofia, 939
Melendez, Roy, 804
Melissari, Giovanni, 900

Mendoza, Sonia, 232
Milione, Giovanni, 688
Molnár, Bálint, 463
Moseley, Ralph, 548
Mtetwa, Nhamoinesu, 86
Mu, Cun (Matthew), 485
Mullins, A., 11
Mullins, J., 11

N
Nahavandi, D., 11
Nahavandi, S., 11
Najdovski, Z., 11
Naranjo, Jose E., 376
Nauck, Detlef, 939
Neagu, Daniel, 929
Nifakos, Sokratis, 956
Niimi, Ayahiko, 860
Nitta, Katsumi, 319
Niu, Yuzhen, 1001
Norta, Alex, 299
Nurbekov, Askar, 1063

O
O'Shea, James, 498
Oita, Marilena, 834
Okada, Shogo, 319
Omar, Nizam, 527
Osman, Ahmed, 254
Osuna-Galan, Ismael, 73
Otremba, Frank, 44
Oussalah, Mourad, 149

P
Parve, Mart, 299
Pattinson, Colin, 263
Perez-Pimentel, Yolanda, 73
Pershin, I., 209
Picone, Joseph, 563
Pombo, Lúcia, 113
Potapkin, B., 209
Prabhu, R. K., 424
Prete, Cosimo Antonio, 165

Q
Qamar, Usman, 573
Qureshi, Ayyaz-ul-Haq, 86

R
Rado, Omesaad, 738, 929
Ramroach, Sterling, 918
Reyes, Matthew G., 705
Rodríguez, José, 232
Romero Navarrete, José A., 44
Rossar, Risto, 299
Roth, Nat, 648
Rovina, Hannes, 275

S
Saad, Amani A., 887
Sagahyroon, Assim, 254
Sánchez-Adame, Luis Martín, 232
Sani, Habiba Muhammad, 929
Sappey-Marinier, Dominique, 900
Sayers, Dean, 1090
Senbel, Samah, 359
Shi, Yiqing, 1001
Silva, Joao Nuno, 527
Simonette, Marcel, 99
Singh, Aashima, 688
Siu, Chapman, 697
Sivaloganathan, Sangarappillai, 1075
Spina, Edison, 99
Stamile, Claudio, 900
Starkey, Andrew, 335
Su, Rina, 1048
Sun, Feng, 126
Sun, Ying, 1027
Sybingco, Edwin, 725

T
Taberkhan, Roman, 1063
Taffoni, Giuliano, 179
Taghva, Kazem, 815
Taktek, Ebtesam, 738
Tanaka, Keisuke, 860
Teppan, Erich C., 410
Tornatore, Luca, 179

U
Ugail, Hassan, 641
Uppalapati, Sitara, 796

V
van der Hulst, Frank, 275
Viksnin, Ilya, 59

Villalba, Williams R., 376
Villegas-Cortez, Juan, 73
Virginas, Botond, 939
Viveros, Amilcar Meneses, 232

W
Wagle, Justin, 648
Wang, Jianlun, 1048
Watson, M., 11
Wei, L., 11
Werman, Michael, 971
Wu, Jingli, 517

Y
Yahalomi, Erez, 971
Yang, Guang, 485
Yang, XiQuan, 1027

Z
Zhang, Chenglin, 1048
Zhao, Shuangshuang, 1048
Zheng, Hongxu, 1048
Zheng, Yan (John), 485
Zuniga-Lopez, Arturo, 73

Printed in the United States
By Bookmasters